Texts in Applied Mathematics 49

Texts in Applied Mathematics

1. *Sirovich:* Introduction to Applied Mathematics.
2. *Wiggins:* Introduction to Applied Nonlinear Dynamical Systems and Chaos, 2nd ed.
3. *Hale/Koçak:* Dynamics and Bifurcations.
4. *Chorin/Marsden:* A Mathematical Introduction to Fluid Mechanics, 3rd ed.
5. *Hubbard/West:* Differential Equations: A Dynamical Systems Approach: Ordinary Differential Equations.
6. *Sontag:* Mathematical Control Theory: Deterministic Finite Dimensional Systems, 2nd ed.
7. *Perko:* Differential Equations and Dynamical Systems, 3rd ed.
8. *Seaborn:* Hypergeometric Functions and Their Applications.
9. *Pipkin:* A Course on Integral Equations.
10. *Hoppensteadt/Peskin:* Modeling and Simulation in Medicine and the Life Sciences, 2nd ed.
11. *Braun:* Differential Equations and Their Applications, 4th ed.
12. *Stoer/Bulirsch:* Introduction to Numerical Analysis, 3rd ed.
13. *Renardy/Rogers:* An Introduction to Partial Differential Equations, 2nd ed.
14. *Banks:* Growth and Diffusion Phenomena: Mathematical Frameworks and Applications.
15. *Brenner/Scott:* The Mathematical Theory of Finite Element Methods, 2nd ed.
16. *Van de Velde:* Concurrent Scientific Computing.
17. *Marsden/Ratiu:* Introduction to Mechanics and Symmetry, 2nd ed.
18. *Hubbard/West:* Differential Equations: A Dynamical Systems Approach: Higher-Dimensional Systems.
19. *Kaplan/Glass:* Understanding Nonlinear Dynamics.
20. *Holmes:* Introduction to Perturbation Methods.
21. *Curtain/Zwart:* An Introduction to Infinite-Dimensional Linear Systems Theory.
22. *Thomas:* Numerical Partial Differential Equations: Finite Difference Methods.
23. *Taylor:* Partial Differential Equations: Basic Theory.
24. *Merkin:* Introduction to the Theory of Stability of Motion.
25. *Naber:* Topology, Geometry, and Gauge Fields: Foundations.
26. *Polderman/Willems:* Introduction to Mathematical Systems Theory: A Behavioral Approach.
27. *Reddy:* Introductory Functional Analysis with Applications to Boundary-Value Problems and Finite Elements.
28. *Gustafson/Wilcox:* Analytical and Computational Methods of Advanced Engineering Mathematics.
29. *Tveito/Winther:* Introduction to Partial Differential Equations: A Computational Approach.
30. *Gasquet/Witomski:* Fourier Analysis and Applications: Filtering, Numerical Computation, Wavelets.

(continued after index)

Francesco Bullo   Andrew D. Lewis

Geometric Control of Mechanical Systems

Modeling, Analysis, and Design for Simple Mechanical Control Systems

With 102 Illustrations

Springer

Francesco Bullo
Department of Mechanical &
Environmental Engineering
University of California
at Santa Barbara
Santa Barbara, CA 93106-5070
USA
bullo@engineering.ucsb.edu

Andrew D. Lewis
Department of Mathematics
and Statistics
Queen's University
Kingston, ON K7L 3N6
Canada
andrew@mast.queensu.ca

Series Editors

J.E. Marsden
Control and Dynamical Systems, 107-81
California Institute of Technology
Pasadena, CA 91125
USA

L. Sirovich
Laboratory of Applied Mathematics
Department of Biomathematical Sciences
Mount Sinai School of Medicine
New York, NY 10029-6574
chico@camelot.mssm.edu

M. Golubitsky
Department of Mathematics
University of Houston
Houston, TX 77204-3476
USA

Mathematics Subject Classification (2000): 93XX

Library of Congress Cataloging-in-Publication Data
Bullo, Francesco.
Geometric control of mechanical systems : modeling, analysis, and design for simple mechanical control systems : monograph / Francesco Bullo and Andrew D. Lewis.
p. cm.
Includes bibliographical references and index.

1. Automatic control. 2. Geometry, Differential. I. Lewis, Andrew D. II. Title.
TJ213.B7835 2004
629.8–dc22 2004052220

ISBN 978-1-4419-1968-7

Printed in the United States of America. (MP)

9 8 7 6 5 4 3 2 1

springeronline.com

To our families

Series Preface

Mathematics is playing an ever more important role in the physical and biological sciences, provoking a blurring of boundaries between scientific disciplines and a resurgence of interest in the modern as well as the classical techniques of applied mathematics. This renewal of interest, both in research and teaching, has led to the establishment of the series *Texts in Applied Mathematics* (*TAM*).

The development of new courses is a natural consequence of a high level of excitement on the research frontier as newer techniques, such as numerical and symbolic computer systems, dynamical systems, and chaos, mix with and reinforce the traditional methods of applied mathematics. Thus, the purpose of this textbook series is to meet the current and future needs of these advances and encourage the teaching of new courses.

TAM will publish textbooks suitable for use in advanced undergraduate and beginning graduate courses, and will complement the *Applied Mathematical Sciences* (*AMS*) series, which will focus on advanced textbooks and research level monographs.

Pasadena, California	J.E. Marsden
Providence, Rhode Island	L. Sirovich
Houston, Texas	M. Golubitsky

Preface

In the course of the past decade, a great deal of progress has been made in the theory and application of ideas in nonlinear control theory to mechanical systems. The areas of application of control theory for mechanical systems are diverse and challenging, and this constitutes an important factor for the interest in these systems. Such areas of application include robotics and automation, autonomous vehicles in marine, aerospace, and other environments, flight control, problems in nuclear magnetic resonance, micro-electromechanical systems, and fluid mechanics. What is more, the areas of overlap between mechanics and control possess the sort of mathematical elegance that makes them appealing to study, independently of applications.

This book is an outgrowth of our research efforts in the area of mechanics and control, and is a development of separate courses offered by us at our respective institutions, at both the advanced undergraduate and graduate levels. The book reflects our point of view that differential geometric thinking is useful, both in nonlinear control theory and in mechanics. Indeed, the primary emphasis of the book is the exploration of areas of overlap between mechanics and control theory for which differential geometric tools are useful. This area of overlap can be essentially characterized by whether one adopts a Hamiltonian or a Lagrangian view of mechanics. In the former, symplectic or Poisson geometry dominates. On the Lagrangian side, it is less clear what is the most useful geometric framework. We sidestep this to some extent by restricting our attention to the special class of mechanical systems known in the literature as "simple."[1] Simple mechanical systems are characterized by the fact that their Lagrangian is kinetic energy minus potential energy. While it is certainly true that simple mechanical systems are not completely general, it is also true that a very large number of applications, indeed the majority

[1] This terminology appears to originate with Smale [1970]. As the reader might conclude after going through the material in our book, the word "simple" should not be taken to have its colloquial meaning. In some of the literature on mechanics, those systems that we call "simple" are called "natural."

considered in the research literature, fall into the “simple” category. Simple mechanical systems also offer the advantage of providing the useful geometric structure of a Riemannian metric, with its attendant Levi-Civita affine connection. It is this Riemannian geometric, or more generally affine differential geometric, point of view that dominates the presentation in this book.

Intended audience

This book is intended to serve as a reference book for graduate students and for researchers wishing to learn about the affine connection approach to control theory for mechanical systems. The book is also intended to be a textbook for graduate students and undergraduates with a suitable background. Let us say a few words about what a suitable background might be, and then a few words about which portion of the research audience we are aiming for.

Prerequisites. Students who use this book as a text should have a background in analysis, linear algebra, and differential equations. An analysis course, rigorously covering such topics as continuity, differentiability, and convergence, is essential. At Queen's University the book has been used as a text for students with a second-year course in analysis. Linear algebra beyond a first course is also essential. Students will be expected to have a knowledge of abstract vector spaces, linear maps, norms, and inner products. At Queen's a sufficient background is achieved with two courses in linear algebra. A basic course in differential equations may be sufficient, although an advanced course will be very useful. An undergraduate instructor teaching students not having seen this material can be sure to spend a significant amount of time in review using external sources. For graduate students in engineering, our experience is that they will either have had this material, or are capable of filling in the gaps as necessary. At the University of Illinois at Urbana-Champaign the course is offered as a first-year, second-semester graduate course. Students are required to take a prerequisite introductory graduate course on linear control theory. More will be said later in this preface on the use of the book as a text.

A note to researchers. We are certainly aware that a great deal of valuable research is done on control theory for mechanical systems without the aid of the mathematical methodologies advocated by this book. We acknowledge that certain examples of mechanical control systems, and certain general problems in the control theory of mechanical systems, can be treated effectively without the aid of differential geometry. Furthermore, we are sympathetic to the fact that a significant effort is required in order for the reader to become comfortable with the necessary mathematics. However, at some point the unity offered by a differential geometric treatment becomes advantageous and we feel that this is merely a necessary part of the subject, as we see it. It was our objective to write a book about a *class* of mechanical control systems, and not about specific examples. In our approach, examples provide motivation, and they provide a testing ground for general ideas. Nonetheless, we hope that the effectiveness of our ideas, as applied to specific examples, will provide some impetus to the reader unsure of whether they should invest the effort in learning the necessary background. Additionally, one of the features of much of the literature on geometric mechanics and on geometric nonlinear control theory, is that it is written in a precise, mathematical style. We have chosen not to deny this mathematical sophistication. Rather, we have adopted the approach of trying to prepare the uninitiated reader, by providing, under one cover, the necessary background as well as the motivation for what is a demanding mathematical formalism.

Objectives

We state our objectives roughly in descending degree of generality.

Broad aim. Broadly, our objectives were to write a book that could serve as a textbook for instructors with some knowledge of the subject area, and to write a book that would be a reference book, stating some of the important results in the field, and providing a guide to the literature for others. These objectives need not be at odds. Indeed, in terms of a graduate text, the objectives align quite nicely, since students can at the same time learn the basics of the subject, and get a glimpse at some advanced material. For example, at the University of Illinois at Urbana-Champaign this text has been used as an introduction to current research results on nonlinear control of mechanical systems. At the same time, the text has also been used for an advanced undergraduate course at Queen's University. In this capacity the book serves to illustrate to students the value of certain mathematical ideas in the physical sciences, and it serves as a spark to some students to pursue research in related fields.

Control theory for mechanics, and mechanics for control theory. Another objective of the book is to provide a background in (parts of) geometric mechanics for researchers familiar with geometric nonlinear control, and to provide a background in (parts of) geometric nonlinear control for readers familiar with geometric mechanics. This objective is served in two ways. For readers familiar with geometric nonlinear control theory, we describe in some detail the mathematical structure of the physical models we consider. We also clarify the connections between (some of) the tools of the nonlinear control theoretician (e.g., Lyapunov functions, Lie brackets) and (some of) the tools of the geometric mechanic (e.g., Riemannian metrics, nonholonomic constraints). For the reader with a background in geometric mechanics, we provide, in each chapter dealing with matters of control theory, a thorough overview of that area of control theory. It should be noted that the resulting coverage of control theory we provide is quite biased to serving our sometimes rather focused objectives. Therefore, a reader should not feel as if our treatment of nonlinear control is even close to comprehensive. We refer the reader to the texts and monographs [Agrachev and Sachkov 2004, Bloch 2003, Isidori 1995, 1999, Jurdjevic 1997, Khalil 2001, Nijmeijer and van der Schaft 1990,

Sastry 1999, Sontag 1998, van der Schaft 1999] for an overview of the subject from multiple points of view.

A unified treatment of the subject. At a more detailed level, the objective of the book is to present a unified treatment of modeling, analysis, and design for mechanical control systems, just as suggested by the title of the book. The unifying feature is our reliance on Riemannian and affine differential geometry to provide the structure in our models. Therefore, a significant portion of the book is devoted to drawing a clear line from physics to differential geometry. We are able to model a significantly large class of systems in our framework, and examples and exercises in each chapter amply illustrate this.

An outline of what is new in this book

It might be of interest to a researcher interested in mechanical control systems to know what material in this book is presented in a novel way or is not present in existing texts on geometry, mechanics, or control. One of the features of the book is that it is divided into three separate parts: Part I, dealing with modeling, Part II, dealing with analysis of mechanical systems and mechanical control systems, and Part III, dealing with design methodologies. In what follows we briefly overview each of these parts.

Part I. Modeling. As a part of modeling, we provide a self-contained review of the mathematical background for the book: linear algebra and differential geometry. In this review, we cover a couple of nonstandard topics that are useful for us later in the book. These topics include the structure of immersed submanifolds (useful for controllability), some background on the character of real-analytic maps, especially contrasted with the behavior of infinitely differentiable maps (useful for understanding certain of the results for which analyticity is required), and the structure of generalized subbundles (useful for controllability and for nonholonomic mechanics). Particularly noteworthy is our rather thorough treatment of distributions. The actual physical modeling is done in a manner that is, as far as we are aware, novel. We present a modeling methodology for mechanical systems that is extremely systematic, and which turns physical data into a differential geometric model. These models form the backbone of the material on analysis and design that follows.

Part II. Analysis. The analysis results we give are for stability, controllability, and perturbation theory. In all cases, we work as much as possible with our intrinsic geometric system models. For stability, we give a coordinate-free notion of exponential stability, and understand the global structure of stability for mechanical systems via Lyapunov analysis. The presentation we give for controllability differs from some standard treatments in that the development of accessibility has as its basis the classic results of Sussmann and Jurdjevic [1972]. We also provide some state-of-the-art local controllability tests, certain of which are not commonly found in texts, and certain of which are, at the time of press, very recent. The coordinate-free approach is carried over to the treatment of averaging and series expansions for mechanical systems. Of particular interest is the use of the intrinsic variation of constants formula for the averaging analysis.

Part III. Design. The geometric formulation of mechanical systems also has real value when it comes to control design. Of the four design chapters, three deal with stabilization, two deal with tracking, and one deals with motion planning. (For those counting, this means that two chapters deal with both stabilization and tracking.) The stabilization results are geometric in flavor, based as they are upon the intrinsic stability and averaging theory developed in the second part of the book. Similarly, the tracking results emphasize the geometry inherent in mechanical control systems. The power of the differential geometric formulation is particularly evident in our consideration of motion planning problems. Merely by understanding the geometry, one is led naturally to a series of seemingly solely mathematical questions. However, the answers to these questions provide simple, explicit motion planning algorithms for systems for which it is not *a priori* clear that such algorithms should exist.

What is not in this book

Although it may not seem commensurate with the length and density of this book, certain choices had to be made concerning what material to present. The need to make such choices was made more pressing in light of our intention to make the book as self-contained as possible. What follows is an incomplete list of topics which we might have included in a manuscript with no length restrictions.

Physics. In the text, we develop a mathematical modeling methodology which we claim is applicable to a large number of physical systems. However, we do not really justify this claim of applicability. We believe this is a serious omission. In our defense we make two points. The first is that Chapters 4 and 5 concerning mathematical modeling already take up a significant proportion of the book, and to develop the physics with an appropriate degree of rigor would have amounted to writing a separate book on mechanics. This leads to the second point of our defense, namely that the physical justification of the applicability of Lagrangian mechanics is taken up in many places, including in the books [Chetaev 1987, Goldstein 1980, Neĭmark and Fufaev 1972, Papastavridis 2002, Pars 1965, Rosenberg 1977, Whittaker 1904]. While none of these books makes the connection between physics and Lagrangian mechanics in the style of our book, perhaps the reader can obtain from them some comfort level with the use of Lagrangian mechanics for modeling physical systems.

Anything Hamiltonian. Many, including the authors, feel that the Hamiltonian setting has a significant role to play in control theory for mechanical systems. In this book, we have focused on a Lagrangian formulation for control theory of mechanical systems. We do this not because we think it superior in any way, but merely because this is where we have done most of our work. The Hamiltonian point of view is treated in Chapter 12 of [Nijmeijer and van der Schaft 1990], and in the recent book of Bloch [2003]. We refer the reader to Section 4.6.6 for additional references.

Complete treatment of symmetry. The subject of symmetry in mechanics is an important one, and is one that we barely scratch the surface of in Chapter 5, and in material related to that chapter. We feel that it is not necessary for us to provide a detailed treatment of symmetry because of the existence of very good books dealing with the subject, including those of Marsden and Ratiu [1999] and Bloch [2003].

Optimal control. For many researchers, particularly some who approach control theory from a certain mathematical point of view, "control theory" is synonymous with "optimal control theory." For such people, our omission of optimal control will be a serious one. The reasons for the omission from the book are exactly space related. However, the good news is that a discussion of optimal control will be included in the supplementary material on the website for the book (the URL is given below). We also refer the reader to the papers [Crouch and Silva Leite 1991, Noakes, Heinzinger, and Paden 1989, Silva Leite, Camarinha, and Crouch 2000] and to the book of Bloch [2003] for a treatment of some problems in optimal control, done using the sorts of tools used in our book.

Advanced control design methodologies. By focusing on control theory for mechanical systems, we have correspondingly focused our treatment of topics in nonlinear control theory. This means that we are not giving systematic treatment to enormous areas of nonlinear control theory. This is particularly true of stabilization theory, where much work has been done. Some standard references for nonlinear control are given in Section 1.4, and for references to additional topics in stabilization we refer the reader to Sections 10.1 and 10.5.

How to use this book as a text

As previously mentioned, various versions of this book have been used as a text, at both the undergraduate and graduate levels. In this section we quickly summarize the content of these courses, and provide some ideas to assist a lecturer who might wish to use this book as a text. We will also have on the webpage for the book (the URL is given below) an expanded version of this section, giving more details, and suggesting some natural variations of material that may make for a coherent course.

The book has been used as a text for an advanced undergraduate course, taught to fourth-year engineering students who are enrolled in a program that is mathematically enriched. There are occasionally mathematics undergraduates, and graduate students from both engineering and mathematics in the course. For this audience, in a twelve week course, one can cover a significant portion of Chapters 3 and 4, along with a collection of topics in control theory. The latter depend largely on the interest of the instructor. The most challenging part of teaching this course is the delivery of the background in differential geometry. This is done, quite effectively, by teaching the material from Chapters 3 and 4 concurrently. In this way, all mathematical concepts are given a fairly immediate physical basis.

A graduate course, taught to engineering students in their first-year, has also been taught from the book. Parts of this course can be sped up as compared to its undergraduate counterpart. It is also true that graduate students in the course will have had a course in linear systems theory, and this will enable the teaching of some deeper control theoretic material. This will be more so, the more familiar the instructor is with the material in the book.

A word of warning is also in order for instructors. The exercises in the book vary greatly in difficulty. In order to not turn students into enemies, instructors would be well-advised to carefully consider the problems they assign. While the occasional difficult problem is good for the constitution, a steady diet of these will probably not make for a good learning experience.

To conclude, let us also mention that we envision complementing and extending this material with internet supplements at the URL for the book:

`http://penelope.mast.queensu.ca/smcs/`

Acknowledgements

It is our great pleasure to thank the various collaborators that have contributed essentially to the material that forms part of this book. Particularly, we acknowledge the contributions of Jorge Cortés, Ron Hirschorn, Naomi Leonard, Kevin Lynch, Sonia Martínez, Richard Murray, David Tyner, and Miloš Žefran. Numerous topics covered in this book are the result of these fruitful collaborations.

We also owe a great debt of gratitude to Richard Murray, who put up with us as graduate students, contributed to our early work, and, most importantly, gave us our initiation to the beautiful subjects of geometric mechanics and mathematical control theory. While students at Caltech, we also had the very good fortune of interacting with Jerry Marsden. Jerry, both in person and through his written work, has provided us and many young academics a model to which to aspire. In the initial stages of development, both Jerry and Richard were also very encouraging, and this was of great value in providing an impetus to continue the work on the book.

We have each used some parts of this book as texts for courses, and we would like to extend our thanks to the students for putting up with the roughness of these early versions of the manuscript, and for finding some of the many small errors that appear in a written work of this size. We also distributed early versions of the manuscript to various graduate students and colleagues, and we received from them a good deal of feedback. In this regard, we particularly wish to thank Ajit Bhand, John Chapman, Peng Cheng, Jorge Cortés,

Elsa Hansen, Sonia Martínez, Sina Ober-Bloebaum, Ashoka Polpitiya, Witold Respondek, Jinglai Shen, Stefano Stramigioli, David Tyner, and Miloš Žefran. While many people found errors in the text, we, of course, are responsible for those that remain. On the webpage maintained by the authors can be found errata.

The anonymous reviewers for the book were helpful in providing advice on structure and content. They, along with Achi Dosanjh at Springer, also provided encouragement for the project in the crucial early stages. We would like to thank Manuel de León for his hospitality at the Consejo Superior de Investigaciones Científicas in Madrid during part of the summer of 2003. We are grateful for support from the National Science Foundation in the United States, the Engineering and Physical Sciences Research Council in the United Kingdom, and the Natural Sciences and Engineering Research Council in Canada.

Finally, we acknowledge the use of open source software in the preparation of this book.

Francesco's acknowledgements: I would like to take this opportunity to truly thank my wife, Lily, for her patience and support during all the long hours this project required; her loving presence made all the difference. I sincerely thank my parents, Carla and Aurelio, and my siblings, Valentina and Federico, for all their love and encouragement. Although we were geographically separated, they have always been there for me. Finally, I extend a world of thanks to my second set of parents, Martha and Juan, and to my brother-in-law, Carlos, for sharing so many memorable moments during the writing of this book.

Andrew's acknowledgements: I extend my gratitude to Laurie for putting up with the long hours required to write this book. I hope it proves to be worthwhile. While this book was in preparation, I experienced two singular events. In March 2001, Laurie and I gave birth to our son Gabriel, who has become my very best friend. His presence has provided me with a much needed distraction of pure happiness. In February 2003 my mother Elizabeth passed away. It is not possible to express the positive influence she has had, and will continue to have, on my life. She will be greatly missed, particularly by my sister Cathryn and myself, and it is to her that I primarily dedicate my share in this book.

FB
Urbana, IL
United States

ADL
Kingston, ON
Canada

Contents

Part III A sampling of design methodologies

Part I

Modeling of mechanical systems

Mechanics is the paradise of the mathematical sciences,
because by means of it one comes to the fruits of mathematics.

Leonardo da Vinci

1

Introductory examples and problems

To motivate the kinds of problems we consider in this book, we use this introductory chapter to present some simple examples for which the problems are fairly easily understood. The short presentation in this chapter is more informal than will be encountered in the remainder of the book. We draw the examples in this chapter from three loosely defined collections of physical systems: aerospace and underwater vehicles, robotic manipulators and multi-body systems, and constrained systems. These examples will give us an opportunity to discuss a variety of topics, and motivate the introduction of appropriate mathematical tools.

For each example, we will introduce the notion of degrees-of-freedom, configuration, velocity, state variables, forces, and constraints. We also pose some natural control theoretic questions that arise naturally for the example systems. This allows us to introduce some of the sorts of questions we address in the book, although we delay answering these questions for the actual text of the book.

Many readers might be familiar with "vector mechanics." This sort of mechanics is useful for modeling, say, a point mass moving in the plane $\mathbb{R}^2$ or in the three-dimensional space $\mathbb{R}^3$. We refer to these vector spaces as Euclidean spaces. However, only in exceptional circumstances can the configuration of a Lagrangian system be described by a vector in a vector space. In the natural mathematical setting, the system's configuration space is described loosely as a curved space, or more accurately as a differentiable manifold. In Chapter 4 we shall be precise about what we mean by this, and about how the mathematical objects of differential geometry represent the physical objects of mechanics. In this chapter we will be a little vague and descriptive about what we might mean by this correspondence. We hope, however, that the illustrative character of the examples we give exhibits the value of taking an approach that will unify all of these examples in one framework.

We close the chapter with a broad overview of the research literature on topics related to those we cover. A reader who is new to the subject can look here to get started in reading the literature. In the text of the book are contained many more references.

1.1 Rigid body systems

The first example we consider is from the class of systems modeled by rigid bodies. In applications, such systems include many aerospace and marine vehicle systems. The system we consider here is about the simplest example in this class, and can be thought of as a model for a simplified hovercraft, as depicted in Figure 1.1. Let us go through the elements of the model. The planar

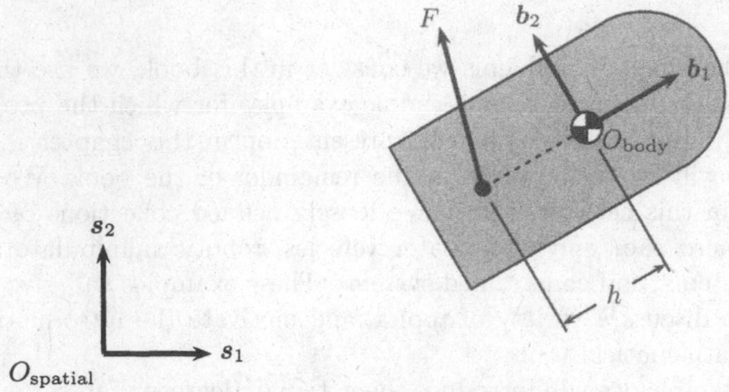

Figure 1.1. Planar rigid body

system has obviously three degrees-of-freedom, since it can translate in the plane, and rotate about its center of mass. The configuration is given by the following variables: θ describes the relative orientation of the body reference frame $\Sigma_{\text{body}} = (O_{\text{body}}, \{\boldsymbol{b}_1, \boldsymbol{b}_2\})$ with respect to the inertial reference frame $\Sigma_{\text{spatial}} = (O_{\text{spatial}}, \{\boldsymbol{s}_1, \boldsymbol{s}_2\})$. The vector (x, y) denotes the position of the center of mass measured with respect to the inertial reference frame Σ_{spatial}. We shall write $q = (\theta, x, y)$, but sometimes it will also be convenient to represent q as the matrix

$$\begin{bmatrix} \cos\theta & -\sin\theta & x \\ \sin\theta & \cos\theta & y \\ 0 & 0 & 1 \end{bmatrix}.$$

This representation emphasizes the matrix group structure that the configuration space enjoys; we refer the reader to the discussion in Chapter 5 where this structure is examined systematically under the name of "homogeneous representation."

The velocity of the system can be written either with respect to the inertial coordinate system Σ_{spatial}, or with respect to the body-fixed coordinate system

Σ_{body}. If we denote the components of the spatial velocity as $\dot{q} = (\dot{\theta}, \dot{x}, \dot{y})$ and the components of the body velocity as (ω, v_x, v_y), we have

$$\begin{bmatrix} \dot{\theta} \\ \dot{x} \\ \dot{y} \end{bmatrix} = \begin{bmatrix} 1 & 0 & 0 \\ 0 & \cos\theta & -\sin\theta \\ 0 & \sin\theta & \cos\theta \end{bmatrix} \begin{bmatrix} \omega \\ v_x \\ v_y \end{bmatrix}.$$

The state of the system is given by q along with its spatial velocity $\dot{q}$. However, it will sometimes be convenient to work with body rather than spatial velocities, although the reasons for this choice are not obvious, at this time.

Having determined the system's state, it is possible now to present the system's total energy as the sum of kinetic and potential energy. The kinetic energy is equal to

$$\text{KE} = \frac{1}{2}\dot{q}^T \mathbb{G}(q)\dot{q}, \quad \text{where} \quad \mathbb{G}(q) = \begin{bmatrix} J & 0 & 0 \\ 0 & m & 0 \\ 0 & 0 & m \end{bmatrix},$$

and where m is the mass of the body and J is its moment of inertia about the center of mass. If we assume that the body moves in a plane perpendicular to the direction of the gravitational forces, the potential energy is zero. As we shall see in Section 4.3, the kinetic and potential energies together allow us to write the equations of motion for the system, at least in the absence of external forces.

Since our interest is primarily with control systems, we will certainly have external forces, and these will be forces that the user can specify. For the planar body, the force we consider is applied to a point on the body that is a distance $h > 0$ from the center of mass, along the body $\boldsymbol{b}_1$-axis, as shown in Figure 1.1. Physically, this force might be thought of as being supplied by a variable-direction thruster on the body. By resolving the force into components in the body $\boldsymbol{b}_1$ and $\boldsymbol{b}_2$ directions, we consider this as a two-input system. In this case, one can readily ascertain, either "by hand," or by applying the methods of Section 4.3, that the equations of motion are

$$\begin{aligned} J\ddot{\theta} &= -hu_2, \\ m\ddot{x} &= u_1 \cos\theta - u_2 \sin\theta, \\ m\ddot{y} &= u_1 \sin\theta + u_2 \cos\theta, \end{aligned}$$

where u_a is the component of F in the body $\boldsymbol{b}_a$-direction, $a \in \{1, 2\}$. These equations provide a model for planar vehicles, for example, a hovercraft that glides on the surface of a body of water with negligible friction.

For these innocuous looking equations, one can ask the following control-theoretic questions.

1. Is it possible to steer from a given initial state to any desired final state?
2. Is it possible to steer the system from rest at an initial configuration q_1 to a final configuration q_2, also at rest?

3. If the answer to either of the first two questions is, "Yes," how does one accomplish the stated objective?

Questions 1 and 2 are referred to in the control theory literature as controllability questions. These and related issues shall be addressed in Chapter 7, and also in Chapter 8. Question 3 is a design question, dealing with what is called motion planning in the literature. We shall provide partial answers to this question in Chapter 13.

One may also wish to consider these questions in the event that the direction of the force is fixed to the body, i.e., the ratio between u_1 and u_2 is specified. Physically, this may happen if one is no longer able to vary the direction of the thruster. We note that in the original case there are two inputs, while in the latter case, there is only one. In either case, there are fewer inputs than degrees-of-freedom. As we shall see, that this has the effect of making the problem significantly more difficult than a system that has as many actuators as degrees-of-freedom. A system is "fully actuated" if each degree-of-freedom is actuated, and is otherwise "underactuated." (This is slightly imprecise, and we shall be more precise about this in Section 4.6.)

1.2 Manipulators and multi-body systems

The next example is drawn from robotic manipulation. These problems generally include robotic arms composed of rigid links and joints. Joints can be revolute, prismatic, or spherical. We present a planar chain for simplicity; a treatment of manipulators in three-dimensional Euclidean space is provided via the product of exponentials formula [Murray, Li, and Sastry 1994]. The system we consider is a planar two-link manipulator; see Figure 1.2. For $i \in \{1, 2\}$,

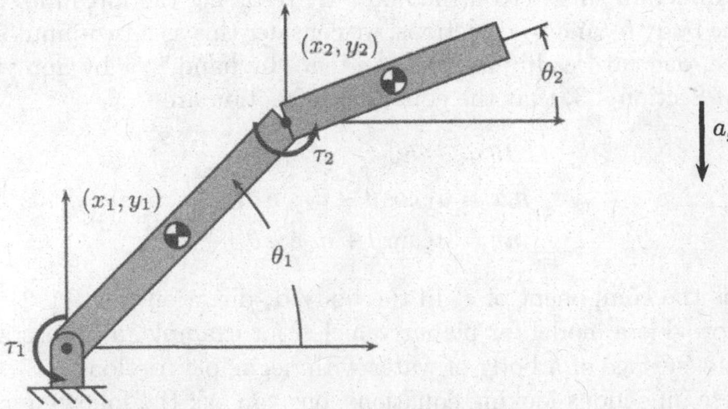

Figure 1.2. A two-link planar manipulator

we let θ_i denote the orientation of the ith link measured counterclockwise from the positive horizontal axis. If we suppose the bottom of the first link to be stationary in an inertial reference frame, the configuration of the system is specified by $q = (\theta_1, \theta_2)$. Thus this system has two degrees-of-freedom. The kinetic and potential energy of the ith link are

$$\mathrm{KE}_i = \tfrac{1}{2}J_i\dot{\theta}_i^2 + \tfrac{1}{2}m_i(\dot{x}_i^2 + \dot{y}_i^2), \quad \mathrm{PE}_i = m_i a_g\, y_i, \qquad i \in \{1, 2\}, \tag{1.1}$$

where (x_i, y_i) is the position of the center of mass, J_i is the moment of inertia of the ith link about its center of mass, and m_i is the mass of the ith link. Also, a_g denotes the acceleration due to gravity. While these equalities for kinetic and potential energy are easy to write down, they involve quantities such as (x_1, y_1, x_2, y_2) that are related explicitly to the actual configuration $q = (\theta_1, \theta_2)$. It is important to emphasize that, for any mechanical system, given the configuration q, one must be able to uniquely compute the position and orientation of each component of the system as a function of q. For the two-link manipulator, we must be able to write (x_1, y_1, x_2, y_2) as a function of (θ_1, θ_2). Indeed, we clearly have

$$\begin{aligned} x_1 &= \tfrac{1}{2}\ell_1 \cos\theta_1, & x_2 &= \ell_1\cos\theta_1 + \tfrac{1}{2}\ell_2\cos\theta_2, \\ y_1 &= \tfrac{1}{2}\ell_1 \sin\theta_1, & y_2 &= \ell_1\sin\theta_1 + \tfrac{1}{2}\ell_2\sin\theta_2, \end{aligned} \tag{1.2}$$

where ℓ_1 and ℓ_2 are the lengths of the links. Furthermore, it is possible to differentiate these relationships with respect to time and obtain an expression for $(\dot{x}_1, \dot{y}_1, \dot{x}_2, \dot{y}_2)$ as a function of the state $(\theta_1, \theta_2, \dot{\theta}_1, \dot{\theta}_2)$.

In summary, the only necessary variables to describe the system's kinetic and potential energy are its configuration variables. After some simplification, one can show that

$$\mathrm{KE} = \tfrac{1}{2}\dot{q}^T\mathbb{G}(q)\dot{q}, \tag{1.3}$$

where

$$\mathbb{G}(q) = \begin{bmatrix} J_1 + \frac{1}{4}(m_1 + 4m_2)\ell_1^2 & \frac{1}{2}m_2\ell_1\ell_2\cos(\theta_1 - \theta_2) \\ \frac{1}{2}m_2\ell_1\ell_2\cos(\theta_1 - \theta_2) & J_2 + \frac{1}{4}m_2\ell_2^2 \end{bmatrix}.$$

The explicit expression for the total gravitational potential energy can also be also easily written, although we do not give it here. As with the planar rigid body of Section 1.1, the total kinetic and potential energies suffice to ascertain the unforced equations of motion. We do not provide these equations here, since they are a bit cumbersome. The reader is asked to derive the forced version of these equations using Newtonian mechanics in Exercise E1.3.

Again, we are interested in applying forces to the system. For the system of Figure 1.2, natural input forces to the system are torques applied at the base of the first link, and/or at the joint between the two links (in our general setup, forces and torques are both called "forces"). This allows a possibility of at least three natural input configurations for the system. For any of the input configurations, we may ask the questions posed at the end of Section 1.1. Furthermore, we will also be interested in the following analysis and design questions.

1. Does the system, subject to no control forces, have any equilibrium configurations?
2. When is an equilibrium configuration stable? More precisely, for all initial conditions near the equilibrium configuration, when can it be guaranteed that the trajectories of the mechanical system remain near the equilibrium configuration?
3. Which configurations can be turned into stable equilibrium configurations by means of control forces? Can this be done in such a manner that the mechanical structure of the system is preserved?
4. Is it possible to specify the controls as functions of q and $\dot{q}$ in order to ensure that the resulting set of second-order differential equations will asymptotically approach the equilibrium configuration as $t \to +\infty$?
5. Given a reference trajectory, is it possible to specify the controls as functions of q and $\dot{q}$, as well as of the reference trajectory, in order to ensure that the resulting set of second-order differential equations will asymptotically follow the reference trajectory as $t \to +\infty$?

Questions 1 and 2 fall under the umbrella of stability theory, and are dealt with in Chapter 6. Questions 3, 4, and 5 are design questions. The first two of these problems are dealt with in Chapters 10, 11, and 12. Question 5 is considered in Chapters 11 and 12.

1.3 Constrained mechanical systems

The third class of systems from which we pull an example is those systems subject to velocity constraints. In particular, we focus on systems with rolling or skating constraints, that is, constraints on the instantaneous velocity of the system. In many physical applications, such systems arise in the case of vehicles with wheels. In this case, the constraint imposed is that the wheel roll without sliding on a horizontal plane.

The simple system we consider here is a disk of radius ρ shown in Figure 1.3. One may wish to think of this as a simple model for a unicycle. We assume that the rider is able to perfectly maintain their balance, so that the disk always remains exactly upright. The coordinates $q = (x, y, \theta, \phi)$ as shown in Figure 1.3 describe the configuration of the system ((x, y) being taken relative to the coordinate frame $\{s_1, s_2\}$ in the plane). Since these four coordinates uniquely characterize the position of the disk, the system has four degrees-of-freedom. Note, however, that not all velocities of the system are admissible. Indeed, the constraint that the disk roll without slipping means exactly that

$$\dot{x} - \rho \cos\theta \dot{\phi} = 0, \quad \dot{y} - \rho \sin\theta \dot{\phi} = 0.$$

This can be determined from Figure 1.4. One may also express this by saying that every admissible velocity is a linear combination of velocities of the form

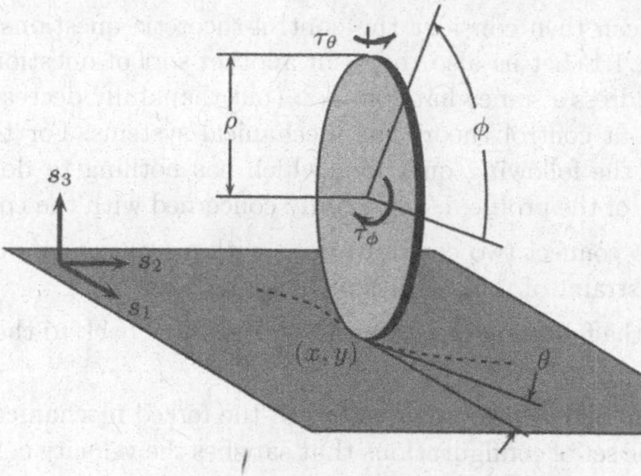

Figure 1.3. The rolling disk

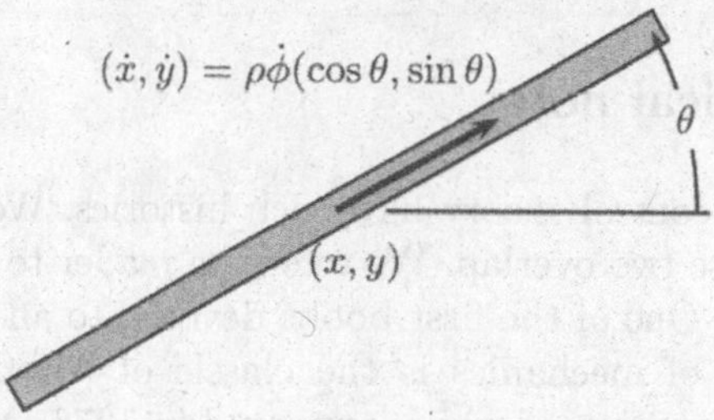

Figure 1.4. Depiction of constraints for rolling disk (the view is from above)

$$(0, 0, 1, 0), \quad (\rho\cos\theta, \rho\sin\theta, 0, 1).$$

This means that at each configuration (x, y, θ, ϕ), there are only two of the possible four directions available for the system to move. Note, however, that this does *not* mean that the configurations of the system are restricted. In Section 4.5 we will see how to systematically derive the equations of motion for systems with velocity constraints.

For the rolling disk, the kinetic energy is easy to compute, and is given by

$$\mathrm{KE} = \frac{1}{2}\dot{q}^T \mathbb{G}(q)\dot{q}, \quad \text{where} \quad \mathbb{G} = \begin{bmatrix} m & 0 & 0 & 0 \\ 0 & m & 0 & 0 \\ 0 & 0 & J_{\text{spin}} & 0 \\ 0 & 0 & 0 & J_{\text{roll}} \end{bmatrix}.$$

Here m is the mass of the disk, J_{roll} is the moment of inertia of the disk about its center, and J_{spin} is the moment of inertia of the disk about the vertical axis. Note that this function is defined without regard for the velocity constraints.

As forces applied to the rolling disk, one may consider combinations of two torques, one that "rolls" the disk, and the other that "spins" the disk. With these inputs, one can then consider the control-theoretic questions posed at the end of Section 1.1. Let us also throw in another sort of question, so that we can early on address a somewhat common (but thankfully decreasingly so) misconception about control theory for mechanical systems. For the rolling disk, one can ask the following question, which has nothing to do with the mechanical nature of the problem, and is only concerned with the constraints.

1. Is it possible to connect two configurations with a sequence of curves that satisfy the constraint of rolling without slipping?

Related to this is the following question, which brings us back to the mechanical realm.

2. Is it possible to follow, with a trajectory of the forced mechanical system, any curve in the set of configurations that satisfies the velocity constraints?

This latter question is dealt with systematically in our framework in Chapter 8, using the language of kinematic reductions.

1.4 Bibliographical notes

Both mechanics and control theory have rich histories. We intend to restrict ourselves to how these two overlap. We refer the reader to [Dugas 1957] for a history of mechanics. One of the first books devoted to an uncompromisingly geometric treatment of mechanics is the classic of Abraham and Marsden [1967], a substantial revision of which appeared in 1978. Another classic text in geometric methods in mechanics is that of Arnol'd [1978]. The history of feedback control is surveyed in [Mayr 1970]. Mathematical control theory is a younger subject, and a good source with which to judge its progress is the collection of seminal papers in [Başar 2001]. Recent differential geometric treatments of control theory may be found in the books [Agrachev and Sachkov 2004, Bloch 2003, Isidori 1995, 1999, Jurdjevic 1997, Nijmeijer and van der Schaft 1990, Sastry 1999]. In particular, the book of Agrachev and Sachkov [2004] provides a rather thorough discussion of many core ideas in mathematical geometric control theory.

An early paper that explicitly identifies the differential geometric bond between mechanics and control theory is that of Brockett [1977]. In this paper, Brockett looks at the forced spherical pendulum and a few single degree-of-freedom systems as motivational examples. Control problems in both the Lagrangian and Hamiltonian framework are identified. On the Hamiltonian side, some connections are drawn between Hamiltonian and gradient systems, something followed up on by Crouch [1981]. Brockett also considers passivity methods for stabilization, these having been introduced by Willems [1972]. The importance of the controllability problem, made more appealing for mechanical systems by their differential geometric structure, was also commented upon by Brockett.

During the course of the next fifteen years, there was only essentially isolated activity in terms of the development of a general theory of mechanics and control theory. An important contribution is that by Takegaki and Arimoto [1981] on the stabilization via so-called proportional-derivative control of certain robotic manipulators. These results led to a successful line of research, and we refer the reader to Chapters 10, 11, and 12 for details concerning this literature. One paper of a general nature was that of Bonnard [1984], which touches on two important subjects, one being controllability of mechanical systems, and the other being mechanical systems on Lie groups (considered in detail by us in Chapter 5). The work of Crouch [1981] also addresses the relationships between ideas in control theory and differential geometry, with many of the geometric ideas being those that arise in mechanics. Another early paper where geometric techniques are prominent is the that of [Crouch 1984], where stabilization is addressed. Constituting one of the few fully developed forays into the theory of control theory and mechanics during this period is the sequence of papers, [van der Schaft 1981/82, 1982, 1983, 1985, 1986], that developed a fairly complete picture of Hamiltonian control theory. A treatment of Hamiltonian control systems also appears as part of the work of Willems [1979] on physical systems modeling.

Around 1990 there began to appear some concentrated interest in developing the theory of mechanical control systems. This activity came from two directions.

On the one hand, in the geometric mechanics community there arose an interest in understanding the role of external forces and constraints in geometric mechanics, since these had largely been ignored in the geometrization of mechanics (this was pointed out by Brockett [1977]). Papers dealing with the inclusion of constraints in the modern geometric framework include [Bloch, Krishnaprasad, Marsden, and Murray 1996, Koiller 1992] in the Lagrangian setting, and [Bates and Śniatycki 1993, van der Schaft and Maschke 1994] in the Hamiltonian setting. When one considers external forces for a mechanical system, questions from control theory arise quite naturally. An early representative of this development is the work of van der Schaft on Hamiltonian control systems mentioned above, and the work of Bloch and Crouch [1992] (developed further in [Bloch and Crouch 1995]) on control for mechanical systems with nonholonomic constraints. This latter work is, as far as we know, the first place where the affine connection features prominently in the development of control theory for mechanical systems. Work on controllability of mechanical systems was done by Bloch, Reyhanoglu, and McClamroch [1992], a geometric mechanics based treatment of stabilization appears in the paper of Bloch, Krishnaprasad, Marsden, and Sánchez de Alvarez [1992], and an approach to stabilization of mechanical systems using vibrational methods is undertaken by Baillieul [1993]. From this point on, there has been a fairly steady growth in the development of the geometric features of control theory for mechanical systems, including, for example, the plenary presentations of Murray [1995] and Leonard [1998], a new series of IFAC Workshops on Lagrangian and Hamiltonian Methods for Nonlinear Control, and the PhD theses of the authors [Bullo 1998, Lewis 1995]. Indeed, since about 1995, there has been a significant growth in the research literature on the subject of geometric control theory for mechanical systems. We shall not comment at this point on specific research papers; this shall be done in the text at relevant junctures. However, we do point out the recent appearance of a few books in the area that represent at least partial culminations of efforts by various groups. An early such representative is Chapter 12 of Nijmeijer and van der Schaft [1990], which gives a rather complete picture of Hamiltonian control theory from the point of view of Poisson geometry. A recent book with a treatment similar in spirit with ours is the book by Bloch [2003]. This book represents the work of Bloch with various coworkers, principally Baillieul, Crouch, Krishnaprasad, Marsden, Murray, and Zenkov. As such, it covers a variety of perspectives in the control theory of mechanical systems, particularly those with nonholonomic constraints. Systems are treated from both the Hamiltonian and Lagrangian points of view, and special emphasis is given to systems with symmetry. The aims of our book are less sweeping, generally focusing on the geometric perspective offered by the simple mechanical structure.

A second point of view from which arose the increased interest in control theory for mechanical systems is from the applications side, with a focus on stabilization and passivity techniques. Representative of this, the books [Arimoto 1996, Ortega, Loria, Nicklasson, and Sira-Ramirez 1998] provide a thorough account of stabilization for Lagrangian systems using passivity methods. The methodology in these books differs from ours in that there is less reliance on differential geometry and more focus on electrical and electromechanical systems. This makes the books amenable to researchers looking to quickly apply tools to problems.

The above cited works form an extremely incomplete overview of the existing and ongoing research in the area of mechanical control systems. In the body of the text we have attempted to provide references to specific papers that are related to topics in the book. As such, the bibliography at the end of the book is incomplete. We hope that the interested reader can use it as a starting point, albeit a biased one, for entering the research literature.

Exercises

E1.1 Use Newton's and Euler's laws to derive the equations of motion for the planar body subject to the variable-direction force in Section 1.1.

E1.2 Suppose that the force applied to the planar body of Section 1.1 is restricted so as to always point in the direction of the body $\boldsymbol{b}_1$-axis, making the system a single-input system.

(a) Try to deduce the character of the motion of the body subjected to such a force. Test your intuition with numerical simulations using various inputs $u_1(t)$, including $u_1(t) = 1$ and $u_1(t) = \cos(t)$, for $t \in [0, 2\pi]$. (Assume $J = m = 1$.)

(b) Next, do the same when the force applied to the planar body is restricted so as to always point in the direction of the body $\boldsymbol{b}_2$-axis. Can you detect any qualitative difference between the two cases?

E1.3 Use Newton's and Euler's laws to derive the equations of motion for the two-link manipulator subjected to two torques in Section 1.2.

E1.4 For the two-link manipulator of Section 1.2, consider the question, "Can one steer the system from a given configuration at rest to another configuration, also at rest?"

(a) If the system is fully actuated, why is this question trivial?

(b) What do you think is the answer to the question when the system is underactuated?

E1.5 Use Newton's and Euler's laws to derive the equations of motion for the rolling disk subjected to two torques in Section 1.3.

E1.6 Consider a spherical pendulum; that is a point mass in three-dimensional space constrained to move on a spherical surface. Let (x, y, z) denote Cartesian coordinates relative to the orthonormal frame $\{\boldsymbol{s}_1, \boldsymbol{s}_2, \boldsymbol{s}_3\}$ in the figure. The system has two degrees-of-freedom with coordinates (θ, ϕ) as in Figure E1.1 (the distance from the origin to the point mass is equal to 1). Answer the following questions.

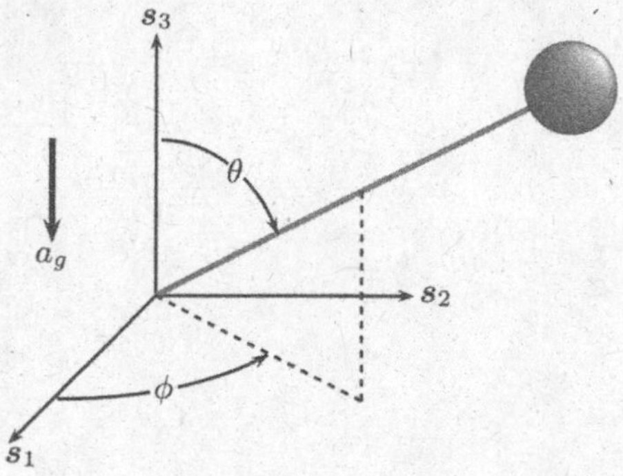

Figure E1.1. Coordinates for spherical pendulum

(a) Write the (x, y, z) position of the point mass in terms of the coordinates (θ, ϕ).

(b) Compute the kinetic energy in terms of the coordinates (θ, ϕ) and their time derivatives.

(c) Compute the potential energy assuming a gravitational field along the $\boldsymbol{s}_3$-axis.

(d) Show that the system satisfies the constraint $\dot{x}x + \dot{y}y + \dot{z}z = 0$. Is this a constraint only on velocities, or does it also constrain the evolution of (x, y, z)?

E1.7 For a realistic model of a bicycle, answer the following questions.

(a) Determine the configuration space.
(b) Write down a set of coordinates.
(c) Determine the velocity constraints that result from the wheels rolling without slipping.

E1.8 For which (if any) of the examples presented in Sections 1.1, 1.2, and 1.3 is it true that the coordinate "vector" q describing the configuration of the system can be legitimately thought of as being an element of Euclidean space? For those systems for which this is not true, indicate why.

2

Linear and multilinear algebra

The study of the geometry of Lagrangian mechanics requires that one be familiar with basic concepts in abstract linear and multilinear algebra. The reader is expected to have encountered at least some of these concepts before, so this chapter serves primarily as a refresher. We also use our discussion of linear algebra as a means of introducing the summation convention in a systematic manner. Since this gets used in computations, the reader may wish to take the opportunity to become familiar with it.

We suppose the reader to have had some exposure to basic concepts in linear algebra, and, for this reason, the pace in this chapter will be rather hurried. For the standard material, we shall simply give the definitions, state some results that we will use in the text, and occasionally give a few elementary examples. For readers looking to improve their facility with basic linear algebra, there is an enormous number of good sources. One that is amenable in depth to our needs, as well as being a fine work of exposition, is the text of Halmos [1996]. Linear algebra and its applications are the subject of Strang [1980]. The basics of tensors on vector spaces and on manifolds are well covered by Abraham, Marsden, and Ratiu [1988, Chapters 2 and 5]. A more in-depth analysis of tensors, especially on manifolds and in infinite dimensions, is the account of Nelson [1967]. We remark that in infinite dimensions, even in Banach spaces, there are some technical issues of which one must be careful.

2.1 Basic concepts and notation

Before we begin, since this is our first chapter with technical content, we take the opportunity to give elementary notation. We shall suppose the reader to be familiar with much of this, so our presentation will be quick and informal. Some of this introductory material can be found in [Halmos 1974b]; see also [Abraham, Marsden, and Ratiu 1988].

2.1.1 Sets and set notation

If x is a point in a set S, then we write $x \in S$. When it is syntactically convenient, we may also write $S \ni x$ rather than $x \in S$. For a set S, $\mathbf{2}^S$ denotes the set of subsets of S. The empty set is denoted by $\emptyset$, and we note that $\emptyset \in \mathbf{2}^S$ for any set S. If A is a subset of S, then we write $A \subset S$. We adopt the convention that, if $A \subset S$, then it is possible that $A = S$. If we wish for A to be a proper subset of S, then we write $A \subsetneq S$. For $A \subset S$, $S \setminus A$ is the ***complement*** of A in S, meaning the set of points in S that are not in A. We shall often define a subset of a set by specifying conditions that must be satisfied by elements of the subset. In this case, the notation

$$\{x \in S \mid \text{ condition(s) on } x\}$$

is used.

If S and T are sets, then $S \times T$ denotes the ***Cartesian product*** of the two sets, which consists of ordered pairs (x, y) with $x \in S$ and $y \in T$. If S is a set, then we will sometimes denote the n-fold Cartesian product of S with itself as S^n:

$$S^n = \underbrace{S \times \cdots \times S}_{n\text{-times}}.$$

If A and B are sets, then $A \cup B$ denotes the ***union*** of A and B, i.e., the new set formed by points that lie in either A or B, and $A \cap B$ denotes the ***intersection*** of A and B, i.e., the set of points lying in both A and B. If $S = A \cup B$ and if $A \cap B = \emptyset$, then we say that S is the ***disjoint union*** of A and B. One can also use disjoint union in a different context. Suppose that J is an arbitrary set, and $\{S_j \mid j \in J\}$ is a collection of sets, indexed by the set J. Thus we have a set assigned to each element of J. The ***disjoint union*** of all these sets is the set

$$\mathring{\bigcup_{j \in J}} S_j \triangleq \bigcup_{j \in J} (\{j\} \times S_j).$$

The idea is that one combines all the sets S_j, $j \in J$, but one also retains the notion of membership in a particular one of these sets. Put otherwise, a point in the disjoint union specifies two things: (1) the index j indicating which of the sets the point lies in, and (2) an element in S_j.

2.1.2 Number systems and their properties

The set of integers is denoted by $\mathbb{Z}$, and the set of natural numbers, i.e., $\{1, 2, \dots\}$, is denoted by $\mathbb{N}$. The set $\mathbb{N} \cup \{0\}$ will be denoted by $\mathbb{Z}_+$. The rational numbers are denoted by $\mathbb{Q}$, and consist of fractions of integers. The set of real numbers is denoted by $\mathbb{R}$. The strictly positive real numbers are denoted by $\mathbb{R}_+$, and the nonnegative real numbers are denoted by $\bar{\mathbb{R}}_+$. A subset $I \subset \mathbb{R}$ is an ***interval*** if it has one of the following forms:

1. $]-\infty, a[= \{x \in \mathbb{R} \mid x < a\}$;
2. $]-\infty, a] = \{x \in \mathbb{R} \mid x \leq a\}$;
3. $]a, b[= \{x \in \mathbb{R} \mid a < x < b\}$;
4. $]a, b] = \{x \in \mathbb{R} \mid a < x \leq b\}$;
5. $]a, +\infty[= \{x \in \mathbb{R} \mid x > a\}$;
6. $[a, b[= \{x \in \mathbb{R} \mid a \leq x < b\}$;
7. $[a, b] = \{x \in \mathbb{R} \mid a \leq x \leq b\}$;
8. $[a, +\infty[= \{x \in \mathbb{R} \mid x \geq a\}$;
9. $\mathbb{R}$.

The set of intervals is denoted by $\mathscr{I}$, and we shall reserve the letter I for a typical interval.

If $S \subset \mathbb{R}$, then an ***upper bound*** for S is an element $b \in \mathbb{R} \cup \{+\infty\}$ for which $x < b$ for each $x \in S$. Then $\sup S \in \mathbb{R} \cup \{+\infty\}$ denotes the least upper bound for S. In like manner, a ***lower bound*** for S is an element $a \in \{-\infty\} \cup \mathbb{R}$ for which $a < x$ for each $x \in S$, and $\inf S \in \{-\infty\} \cup \mathbb{R}$ denotes the greatest lower bound for S. We call $\sup S$ the ***supremum*** of S and $\inf S$ the ***infimum*** of S. It may be the case that $\sup S = +\infty$ and/or $\inf S = -\infty$, but it is always the case that $\sup S$ and $\inf S$ exist; see [Halmos 1974b]. In cases where it is certain that $\sup S \in S$ (resp. $\inf S \in S$)—for example if S is finite—we may write $\max S$ for $\sup S$ (resp. $\min S$ for $\inf S$).

The set of complex numbers is denoted by $\mathbb{C}$. We abbreviate $\sqrt{-1}$ as i. For $z \in \mathbb{C}$, $\mathrm{Re}(z)$ denotes the real part of z, and $\mathrm{Im}(z)$ denotes the imaginary part of z. It is also useful to split up the complex plane, and we do this as follows:

$$\mathbb{C}_+ = \{z \in \mathbb{C} \mid \mathrm{Re}(z) > 0\}, \qquad \mathbb{C}_- = \{z \in \mathbb{C} \mid \mathrm{Re}(z) < 0\},$$
$$\overline{\mathbb{C}}_+ = \{z \in \mathbb{C} \mid \mathrm{Re}(z) \geq 0\}, \qquad \overline{\mathbb{C}}_- = \{z \in \mathbb{C} \mid \mathrm{Re}(z) \leq 0\},$$
$$\mathrm{i}\mathbb{R} = \{z \in \mathbb{C} \mid \mathrm{Re}(z) = 0\}.$$

2.1.3 Maps

For sets S and T, a ***map*** f from S to T assigns to each element $x \in S$ one and only one element $f(x) \in T$. We write a map as $f\colon S \to T$, calling S the ***domain*** and T the ***codomain*** of f. If we wish to define a map, along with its domain and codomain, all at once, then we will sometimes employ the notation

$$\begin{aligned} f\colon S &\to T \\ x &\mapsto \text{what } f \text{ maps } x \text{ to.} \end{aligned}$$

Thus S is the domain of f, T is the codomain of f, and the expression on the right on the last line defines the map, whatever that definition might be.

We are careful to distinguish notationally between a map f and its value at a point, denoted by $f(x)$. We define the ***image*** of f by $\mathrm{image}(f) = \{f(x) \mid x \in S\}$. The map $\mathrm{id}_S\colon S \to S$ defined by $\mathrm{id}_S(x) = x$, for all $x \in S$, is the ***identity map***. If $f\colon S \to T$ and $g\colon T \to U$ are maps, then $g \circ f\colon S \to U$ is the ***composition*** of f and g, defined by $g \circ f(x) = g(f(x))$. If $f\colon S \to T$ is a map, and if $A \subset S$, then we denote by $f|A\colon A \to T$ the ***restriction*** of f, which is defined by $(f|A)(x) = f(x)$, $x \in A$.

A map $f: S \to T$ is ***injective*** (or is an ***injection***) if the equality $f(x_1) = f(x_2)$ for $x_1, x_2 \in S$ implies that $x_1 = x_2$. A map $f: S \to T$ is ***surjective*** (or is a ***surjection***) if, for each $y \in T$, there exists at least one $x \in S$ such that $f(x) = y$. It is common to see an injective map called ***one-to-one*** or ***1–1***, and a surjective map called ***onto***. A map that is both injective and surjective is ***bijective*** (or is a ***bijection***). Sets S and T for which there exists a bijection $f: S \to T$ are sometimes said to be in ***1–1 correspondence***. One verifies the following equivalent characterizations of injective, surjective, and bijective maps:

1. a map $f: S \to T$ is injective if and only if there exists a map $f_L: T \to S$, called a ***left inverse*** of f, such that $f_L \circ f = \mathrm{id}_S$;
2. a map $f: S \to T$ is surjective if and only if there exists a map $f_R: T \to S$, called a ***right inverse*** of f, such that $f \circ f_R = \mathrm{id}_T$;
3. a map $f: S \to T$ is bijective if and only if there exists a unique map $f^{-1}: T \to S$, called the ***inverse*** of f, having the property that $f \circ f^{-1} = \mathrm{id}_T$ and $f^{-1} \circ f = \mathrm{id}_S$.

If $A \subset S$, then $i_A: A \to S$ denotes the ***inclusion map***, which assigns to a point $x \in A$ the same point, but thought of as being in S. For a map $f: S \to T$ and for $B \subset T$, we write

$$f^{-1}(B) = \{x \in S \mid f(x) \in B\},$$

and call this the ***preimage*** of B under f. In the event that $B = \{y\}$ is a singleton, we write $f^{-1}(\{y\}) = f^{-1}(y)$. If $T = \mathbb{R}$ and $y \in \mathbb{R}$, then it is common to call $f^{-1}(y)$ a ***level set*** of f. If $S_1, \dots, S_k$ are sets, then, for $i \in \{1, \dots, k\}$, the map $\mathrm{pr}_i: S_1 \times \cdots \times S_k \to S_i$ is the ***projection onto the ith factor***, which assigns $x_i \in S_i$ to $(x_1, \dots, x_i, \dots, x_k) \in S_1 \times \cdots \times S_i \times \cdots \times S_k$.

If one has a collection of maps that are related, in some way, then a useful way of displaying these relationships is with a ***commutative diagram***. This is best illustrated with examples; consider the two diagrams

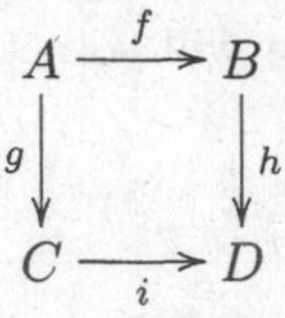

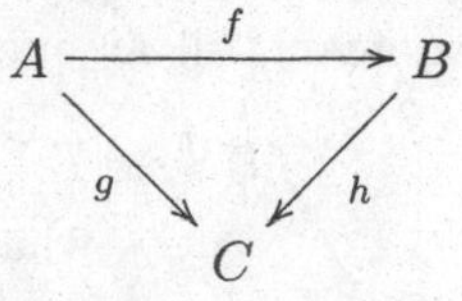

The diagram on the left commutes if $h \circ f = i \circ g$, and the diagram on the right commutes if $h \circ f = g$. More generally, and somewhat imprecisely, a diagram ***commutes*** if all possible maps between two sets, obtained by composing maps on the arrows in the diagram, are equal. Note that, in a commutative diagram, unless it is otherwise stated, the maps are not invertible, so one can follow the arrows only in the forward direction.

2.1.4 Relations

If S is a set, then a ***relation*** in S is a subset $R \subset S \times S$, and two points $x_1, x_2 \in S$ are ***R-related*** if $(x_1, x_2) \in R$. A relation R is an ***equivalence relation*** if

1. for each $x \in S$, $(x, x) \in R$ (***reflexivity***),
2. if $(x_1, x_2) \in R$, then $(x_2, x_1) \in R$ (***symmetry***), and
3. if $(x_1, x_2) \in R$ and $(x_2, x_3) \in R$, then $(x_1, x_3) \in R$ (***transitivity***).

For an equivalence relation R, two members $x_1, x_2 \in S$ are ***equivalent*** if $(x_1, x_2) \in R$. We often write $x_1 \sim x_2$ in this case. Indeed, we will on occasion, with a slight abuse of notation, define an equivalence relation in S by indicating when it holds that $x_1 \sim x_2$ for $x_1, x_2 \in S$. For $x_0 \in S$, $[x_0] = \{x \in S \mid x \sim x_0\}$ typically denotes the ***equivalence class*** of x_0, i.e., all those points in S equivalent to x. The set of all equivalence classes is written $S/\sim$. Thus $(S/\sim) \subset \mathbf{2}^S$. The map assigning to $x \in S$ its equivalence class $[x] \in S/\sim$ is called the ***canonical projection*** for the equivalence relation.

Let us give an example of an equivalence relation to illustrate the notation and concepts.

Example 2.1. We take $S = \mathbb{R}^2$ and define a relation in S by

$$R = \{((x, y), (x + t, y + t)) \mid t \in \mathbb{R}\}.$$

Thus $(x_1, y_1) \sim (x_2, y_2)$ if and only if $x_1 - x_2 = y_1 - y_2$. The equivalence class of $(x, y) \in \mathbb{R}^2$ is the line through (x, y) having slope 1:

$$[(x, y)] = \{(x + t, y + t) \mid t \in \mathbb{R}\}.$$

The set of equivalence classes, $\mathbb{R}^2/\sim$, is therefore the set of lines in $\mathbb{R}^2$ with slope 1. To make this more concrete, we note that the equivalence class $[(x, y)]$ is uniquely determined by the point $(x - y, 0) \in [(x, y)]$, i.e., by the point in $[(x, y)]$ on the x-axis. Thus there is a natural bijection between $\mathbb{R}^2/\sim$ and $\mathbb{R}$. •

2.1.5 Sequences and permutations

A set S is ***finite*** if there exist $n \in \mathbb{N}$ and a bijection $f: S \to \{1, \dots, n\}$. If S is not finite, it is ***infinite***. A set S is ***countable*** if there exists a bijection $f: S \to \mathbb{N}$, and is ***uncountable*** if it is infinite and not countable. A ***sequence*** in a set S is a map $f: \mathbb{N} \to S$, and we typically denote a sequence by writing all of its elements as $\{f(n)\}_{n \in \mathbb{N}}$, or more typically as $\{x_n\}_{n \in \mathbb{N}}$, where $x_n = f(n)$. We shall on occasion want collections of sets more general than sequences, and in such cases we may write $\{x_a\}_{a \in A}$, where A is an arbitrary ***index set*** (e.g., see the discussion on disjoint union). If $\{x_n\}_{n \in \mathbb{N}}$ is a sequence, then a ***subsequence*** is a subset $\{x_{n_k}\}_{k \in \mathbb{N}}$ where $\{n_k\}_{k \in \mathbb{N}}$ is a sequence in $\mathbb{N}$ satisfying

$$n_1 < n_2 < \cdots < n_k < n_{k+1} < \cdots.$$

Let us recall some facts about permutations. For $m \in \mathbb{N}$, let S_m be the set of bijections of $\{1, \dots, m\}$ with itself, i.e., the set of permutations of $\{1, \dots, m\}$. We call S_m the ***permutation group*** of order m. If $\sigma \in S_m$, then we represent it as

$$\begin{pmatrix} 1 & 2 & \cdots & m \\ \sigma(1) & \sigma(2) & \cdots & \sigma(m) \end{pmatrix}.$$

Thus, for $j \in \{1, \dots, m\}$, below the jth element in the $2 \times m$ matrix is $\sigma(j)$. A ***transposition*** is a permutation of the form

$$\begin{pmatrix} 1 & \cdots & i & \cdots & j & \cdots & m \\ 1 & \cdots & j & \cdots & i & \cdots & m \end{pmatrix},$$

i.e., a swapping of two elements of $\{1, \dots, m\}$. A permutation $\sigma \in S_m$ is ***even*** if it is the composition of an even number of transpositions, and is ***odd*** if it is the composition of an odd number of transpositions. This definition may be shown to not depend on the choice of transpositions into which a permutation is decomposed. We define $\operatorname{sgn}: S_m \to \{-1, 1\}$ by

$$\operatorname{sgn}(\sigma) = \begin{cases} 1, & \sigma \text{ is even}, \\ -1, & \sigma \text{ is odd}. \end{cases}$$

The number $\operatorname{sgn}(\sigma)$ is the ***sign*** of σ.

2.1.6 Zorn's Lemma

In this section we present Zorn's Lemma, which can probably be safely omitted on a first reading. We make only one nontrivial application of Zorn's Lemma, that being in the proof of Lemma 3.94. However, we make many (seemingly) trivial uses of Zorn's Lemma, some of which we point out. We encourage the interested reader to find as many of these occurrences as they can. Some of them are subtle, as may be ascertained from the fact that Zorn's Lemma may be shown to be equivalent to the seemingly obvious Axiom of Choice.[1] There is a fascinating mathematical tale behind these matters, and we refer to [Moore 1982] for an account of this. A proof of Zorn's Lemma may be found in [Halmos 1974b].

To state Zorn's Lemma, we need to state some definitions from set theory. In the following definition, it is convenient to use the notation for a relation that defines what "$x_1 \sim x_2$" means, rather than defining a subset R of $S \times S$.

[1] The Axiom of Choice says that, given any set of mutually exclusive nonempty sets, there exists at least one set that contains exactly one element in common with each of the nonempty sets.

Definition 2.2. Let S be a set.

(i) A ***partial order*** in a set S is a relation $\preceq$ with the properties that
  (a) $x \preceq x$,
  (b) $x \preceq y$ and $y \preceq z$ implies $x \preceq z$, and
  (c) $x \preceq y \preceq x$ implies $x = x$.

A ***partially ordered set*** is a pair $(S, \preceq)$, where $\preceq$ is a partial order in S.

(ii) A ***total order*** on S is a partial order for which either $x \preceq y$ or $y \preceq x$ for all distinct $x, y \in S$. A set S equipped with a total order is called a ***chain***.

(iii) An ***upper bound*** for a chain S is an element $x \in S$ for which $y \preceq x$ for all $x \in S$. •

With these definitions, we state Zorn's Lemma.

Theorem 2.3 (Zorn's Lemma). *A partially ordered set $(S, \preceq)$ in which every chain has an upper bound contains at least one maximal element.*

2.2 Vector spaces

This section introduces the notions of vector space, linear map, and dual space, with the objective of presenting the following key concepts. First, it is convenient to consider vector spaces in more abstract terms than to simply think of $\mathbb{R}^n$. In other words, a vector space is a set with two operations satisfying certain properties, not a set of n-tuples. Second, vectors, linear maps, and other objects defined on vector spaces can be written in components, once a basis is available, and the summation convention is a convenient procedure to do so. Third, naturally associated to each vector space is the vector space of linear functions on the vector space; this vector space is called the dual space and is analyzed in Section 2.2.5. We refer the reader to [Halmos 1996] and [Strang 1980] for more detailed expositions.

Vector spaces over the field of real numbers arise in many ways in the book. On occasion, we will also find it useful to have on hand some properties of complex vector spaces. We define $\mathbb{F} \in \{\mathbb{R}, \mathbb{C}\}$; thus $\mathbb{F}$ will be used whenever we intend to say that either $\mathbb{R}$ or $\mathbb{C}$ is allowed. If $a \in \mathbb{C}$, then $\bar{a}$ is the complex conjugate, and if $a \in \mathbb{R}$, then $\bar{a} = a$. In like manner, for $a \in \mathbb{F}$, $|a|$ is the absolute value when $\mathbb{F} = \mathbb{R}$, and the complex modulus when $\mathbb{F} = \mathbb{C}$.

2.2.1 Basic definitions and concepts

We begin with a definition.

Definition 2.4 (Vector space). An $\mathbb{F}$-***vector space*** (or simply a ***vector space*** if $\mathbb{F}$ is understood, or if it is immaterial whether $\mathbb{F} = \mathbb{R}$ or $\mathbb{F} = \mathbb{C}$) is a set V equipped with two operations: (1) vector addition, denoted by $v_1 + v_2 \in \mathsf{V}$ for $v_1, v_2 \in \mathsf{V}$, and (2) scalar multiplication, denoted by $a\,v \in \mathsf{V}$ for $a \in \mathbb{F}$ and $v \in \mathsf{V}$. Vector addition must satisfy the rules

(i) $v_1 + v_2 = v_2 + v_1$, $v_1, v_2 \in \mathsf{V}$ (***commutativity***),
(ii) $v_1 + (v_2 + v_3) = (v_1 + v_2) + v_3$, $v_1, v_2, v_3 \in \mathsf{V}$ (***associativity***),
(iii) there exists a unique vector $0 \in \mathsf{V}$ with the property that $v + 0 = v$ for every $v \in \mathsf{V}$ (***zero vector***), and
(iv) for every $v \in \mathsf{V}$, there exists a unique vector $-v \in \mathsf{V}$ such that $v+(-v) = 0$ (***negative vector***),

and scalar multiplication must satisfy the rules

(v) $a_1(a_2\, v) = (a_1 a_2)v$, $a_1, a_2 \in \mathbb{F}$, $v \in \mathsf{V}$ (***associativity***),
(vi) $1\, v = v$, $v \in \mathsf{V}$,
(vii) $a(v_1 + v_2) = a\, v_1 + a\, v_2$, $a \in \mathbb{F}$, $v_1, v_2 \in \mathsf{V}$ (***distributivity***), and
(viii) $(a_1 + a_2)v = a_1\, v + a_2\, v$, $a_1, a_2 \in \mathbb{F}$, $v \in \mathsf{V}$ (***distributivity*** again). •

Example 2.5. The prototypical $\mathbb{F}$-vector space is $\mathbb{F}^n$, $n \in \mathbb{Z}_+$, the collection of ordered n-tuples, $(x^1, \dots, x^n)$, of elements of $\mathbb{F}$. On this space vector, addition and scalar multiplication are defined by

$$(x^1, \dots, x^n) + (y^1, \dots, y^n) = (x^1 + y^1, \dots, x^n + y^n),$$
$$a(x^1, \dots, x^n) = (ax^1, \dots, ax^n).$$

It is an easy exercise to verify that these operations satisfy the conditions of Definition 2.4. We shall normally denote a typical vector in $\mathbb{F}^n$ by a boldface letter, e.g., $\boldsymbol{x} = (x^1, \dots, x^n)$. If $n = 0$, then we adopt the convention that $\mathbb{F}^n$ is the trivial vector space consisting of only the zero vector. •

The set $\mathbb{R}^n$ will be referred to frequently as the ***n-dimensional Euclidean space***. While it has a vector space structure, it also has other structure that will be of interest to us. In particular, it is possible to talk about differentiable functions on Euclidean space, and this makes possible the developments of Chapter 3 that are integral to our approach.

Let us list some concepts that can be immediately defined once one embraces the notion of a vector space.

Definition 2.6. Let V be an $\mathbb{F}$-vector space.

(i) For subsets $S_1, S_2 \subset \mathsf{V}$, the ***sum*** of S_1 and S_2 is the set

$$S_1 + S_2 = \{v_1 + v_2 \mid v_1 \in S_1,\ v_2 \in S_2\}.$$

(ii) A subset $\mathsf{U} \subset \mathsf{V}$ of a vector space is a ***subspace*** if $u_1 + u_2 \in \mathsf{U}$ for all $u_1, u_2 \in \mathsf{U}$ and if $a\, u \in \mathsf{U}$ for every $a \in \mathbb{F}$ and $u \in \mathsf{U}$.
(iii) If V_1 and V_2 are vector spaces, the ***direct sum*** of V_1 and V_2 is the vector space $\mathsf{V}_1 \oplus \mathsf{V}_2$ whose set is $\mathsf{V}_1 \times \mathsf{V}_2$ (the Cartesian product), and with vector addition defined by $(u_1, u_2) + (v_1, v_2) = (u_1 + v_1, u_2 + v_2)$ and scalar multiplication defined by $a(v_1, v_2) = (av_1, av_2)$.
(iv) If U_1 and U_2 are subspaces of V, then we shall also write $\mathsf{V} = \mathsf{U}_1 \oplus \mathsf{U}_2$ if $\mathsf{U}_1 \cap \mathsf{U}_2 = \{0\}$ and if every vector $v \in \mathsf{V}$ can be written as $v = u_1 + u_2$

for some $u_1 \in \mathsf{U}_1$ and $u_2 \in \mathsf{U}_2$.[2] It is often convenient to denote a typical vector in a direct sum $\mathsf{V}_1 \oplus \mathsf{V}_2$ by $v_1 \oplus v_2$.

(v) If $\mathsf{V}_1, \dots, \mathsf{V}_k$ are $\mathbb{F}$-vector spaces, then we may write

$$\bigoplus_{j=1}^{k} \mathsf{V}_j \triangleq \mathsf{V}_1 \oplus \cdots \oplus \mathsf{V}_k.$$

•

The next definition collects some language surrounding the notions of linear independence and bases.

Definition 2.7 (Linear independence and bases). Let V be an $\mathbb{F}$-vector space.

(i) A set $S \subset \mathsf{V}$ of vectors is ***linearly independent*** if, for every finite subset $\{v_1, \dots, v_k\} \subset S$, the equality $c^1 v_1 + \cdots + c^k v_k = 0$ implies that $c^1 = \cdots = c^k = 0$.

(ii) A set of vectors $S \subset \mathsf{V}$ ***generates*** a vector space V if every vector $v \in \mathsf{V}$ can be written as $v = c^1 v_1 + \cdots + c^k v_k$ for some choice of constants $c^1, \dots, c^k \in \mathbb{F}$, and for some $v_1, \dots, v_k \in S$. In this case we write $\mathsf{V} = \operatorname{span}_{\mathbb{F}}\{S\}$.

(iii) A ***basis*** for a vector space V is a collection of vectors that is linearly independent and that generates V.

(iv) A vector space is ***finite-dimensional*** if it possesses a basis with a finite number of elements, and the number of basis elements is the ***dimension*** of V, denoted by $\dim(V)$ (one can prove that this is independent of basis).

(v) If U is a subspace of a finite-dimensional vector space V, then the ***codimension*** of U is $\dim(\mathsf{V}) - \dim(\mathsf{U})$. •

The following result indicates why the notion of a basis is so useful.

Proposition 2.8 (Components of a vector). *If $\{e_1, \dots, e_n\}$ is a basis for an $\mathbb{F}$-vector space V, then, for any $v \in \mathsf{V}$, there exist unique constants $v^1, \dots, v^n \in \mathbb{F}$ such that $v = v^1 e_1 + \cdots + v^n e_n$. These constants are called the* ***components*** *of v relative to the basis.*

Here we begin to adopt the convention that components of vectors are indexed with *super*scripts, while lists of vectors are indexed with *sub*scripts. Let us use this chance to introduce the summation convention, first employed by Einstein [1916], that we shall use.

The summation convention. Whenever an expression contains a repeated index, one as a subscript and the other as a superscript, summation is implied over this index. Thus, for example, we have

[2] Some authors use "external direct sum" for the direct sum in (iii) and "internal direct sum" for the direct sum in (iv).

$$v^i e_i = \sum_{i=1}^{n} v^i e_i,$$

since summation over i is implied. •

Example 2.9. The ***standard basis*** for $\mathbb{F}^n$ is given by

$$e_1 = (1, 0, \ldots, 0),\; e_2 = (0, 1, \ldots, 0),\; \ldots,\; e_n = (0, 0, \ldots, 1).$$

The choice of a basis $\{e_1, \ldots, e_n\}$ for a general n-dimensional vector space V makes V "look like" $\mathbb{F}^n$, in that every vector in V is uniquely represented by its components $(v^1, \ldots, v^n) \in \mathbb{F}^n$ relative to this basis. •

We will have occasion to use the notion of a quotient space.

Definition 2.10 (Quotient space). Let $\mathsf{U} \subset \mathsf{V}$ be a subspace and consider the equivalence relation $\sim_{\mathsf{U}}$ in V given by $v_1 \sim_{\mathsf{U}} v_2$ if $v_2 - v_1 \in \mathsf{U}$. Denote by V/U the set of equivalence classes under this equivalence relation, and call V/U the ***quotient space*** of U in V. The equivalence class containing $v \in \mathsf{V}$ will be denoted by $v + \mathsf{U}$. •

The following result gives some useful properties of the quotient space.

Proposition 2.11. *For a finite-dimensional $\mathbb{F}$-vector space V with $\mathsf{U} \subset \mathsf{V}$ a subspace, the quotient space V/U has the following properties:*

(i) the operations

$$(v_1 + \mathsf{U}) + (v_2 + \mathsf{U}) = (v_1 + v_2) + \mathsf{U}, \qquad v_1, v_1 \in \mathsf{V},$$
$$a(v + \mathsf{U}) = a\,v + \mathsf{U}, \qquad a \in \mathbb{F},\; v \in \mathsf{V},$$

make V/U an $\mathbb{F}$-vector space;

(ii) $\dim(\mathsf{V}/\mathsf{U}) = \dim(\mathsf{V}) - \dim(\mathsf{U})$;

(iii) if U' is any complement to U in V (i.e., U' has the property that $\mathsf{V} = \mathsf{U} \oplus \mathsf{U}'$), then U' is naturally isomorphic (a notion we define formally below) to V/U.

Proof. This is left to the reader (Exercise E2.1). ■

2.2.2 Linear maps

We once again begin with the definition.

Definition 2.12 (Linear map). A map $A\colon \mathsf{U} \to \mathsf{V}$ between $\mathbb{F}$-vector spaces U and V is ***linear*** if $A(au) = aA(u)$ and if $A(u_1 + u_2) = A(u_1) + A(u_2)$ for each $a \in \mathbb{F}$ and $u, u_1, u_2 \in \mathsf{U}$. If $\mathsf{U} = \mathsf{V}$, then A is sometimes called a ***linear transformation***. •

Remarks 2.13. 1. Sometimes, when there are multiple algebraic structures present in a problem, it is convenient to call a linear map between $\mathbb{F}$-vector spaces ***$\mathbb{F}$-linear***. However, we shall only occasionally make use of this terminology.

2. On occasion we shall make use of the notion of an ***affine map*** between vector spaces U and V, by which we mean a map of the form $u \mapsto A(u)+b$, where $A\colon \mathsf{U} \to \mathsf{V}$ is linear and $b \in \mathsf{V}$. •

The set of linear maps from a vector space U to a vector space V is itself a vector space, which we denote by $L(\mathsf{U};\mathsf{V})$. Vector addition in $L(\mathsf{U};\mathsf{V})$ is given by

$$(A+B)(u) = A(u) + B(u),$$

and scalar multiplication is defined by

$$(aA)(u) = a(A(u)).$$

Note that what is being defined in these two equations is $A+B \in L(\mathsf{U};\mathsf{V})$ in the first case, and $aA \in L(\mathsf{U};\mathsf{V})$ in the second case. One verifies that $\dim(L(\mathsf{U};\mathsf{V})) = \dim(\mathsf{U})\dim(\mathsf{V})$, provided that U and V are finite-dimensional.

Example 2.14. Let us consider linear maps from $\mathbb{F}^n$ to $\mathbb{F}^m$. We first note that if $A^a_i \in \mathbb{F}$, $a \in \{1,\dots,m\}$, $i \in \{1,\dots,n\}$, then the map $\boldsymbol{A}\colon \mathbb{F}^n \to \mathbb{F}^m$ defined by

$$\boldsymbol{A}(\boldsymbol{x}) = \Big(\sum_{i=1}^n A^1_i x^i, \dots, \sum_{i=1}^n A^m_i x^i\Big),$$

is readily verified to be linear. What is more, one can also verify (see Exercise E2.2) that *every* linear map from $\mathbb{F}^n$ to $\mathbb{F}^m$ is of this form. •

The image of a linear map is simply its image as defined in Section 2.1. Let us define some useful related concepts.

Definition 2.15. Let U and V be $\mathbb{F}$-vector spaces and let $A \in L(\mathsf{U};\mathsf{V})$.

(i) The ***kernel*** of A, denoted by $\ker(A)$, is the subset of U defined by $\{u \in \mathsf{U} \mid A(u) = 0\}$.

(ii) The ***rank*** of A is defined to be $\operatorname{rank}(A) = \dim(\operatorname{image}(A))$. (We shall see below that $\operatorname{image}(A)$ is a subspace, so its dimension is well-defined.)

(iii) If U and V are finite-dimensional, then A has ***maximal rank*** if $\operatorname{rank}(A) = \min\{\dim(\mathsf{U}), \dim(\mathsf{V})\}$.

(iv) If A is invertible, it is called an ***isomorphism***, and U and V are said to be ***isomorphic***. •

In some texts on linear algebra, what we call the kernel of a linear map is called the ***null space***, and what we call the image of a linear map is called the ***range***.

Often, the context might imply the existence of a *natural* isomorphism between vector spaces U and V (e.g., the existence of a basis $\{e_1,\dots,e_n\}$ for an $\mathbb{F}$-vector space V implies a natural isomorphism from V to $\mathbb{R}^n$), and in such cases we may write $\mathsf{U} \simeq \mathsf{V}$.

We record some useful properties of the kernel and image of a linear map [Halmos 1996].

Proposition 2.16. *Let* U *and* V *be* $\mathbb{F}$*-vector spaces and let* $A \in L(\mathsf{U};\mathsf{V})$*. The following statements hold:*

(i) $\ker(A)$ *is a subspace of* U*, and* $\operatorname{image}(A)$ *is a subspace of* V*;*
(ii) if U *is finite-dimensional, then* $\dim(\ker(A)) + \operatorname{rank}(A) = \dim(\mathsf{U})$ *(this is the* ***rank–nullity formula****).*

If $\mathsf{U} = \mathsf{U}_1 \oplus \cdots \oplus \mathsf{U}_k$ and $\mathsf{V} = \mathsf{V}_1 \oplus \cdots \oplus \mathsf{V}_l$, then $A \in L(\mathsf{U};\mathsf{V})$ may be represented in block form by

$$\begin{bmatrix} A_{11} & \cdots & A_{1k} \\ \vdots & \ddots & \vdots \\ A_{l1} & \cdots & A_{lk} \end{bmatrix},$$

where $A_{rs} \in L(\mathsf{U}_s;\mathsf{V}_r)$, $r \in \{1,\dots,l\}$, $s \in \{1,\dots,k\}$. At times it will be convenient to simply write a linear map in $L(\mathsf{U};\mathsf{V})$ in this form without comment.

2.2.3 Linear maps and matrices

In Example 2.14 we examined the structure of linear maps between the vector spaces $\mathbb{F}^n$ and $\mathbb{F}^m$. It is natural to think of such linear maps as matrices. Let us formally discuss matrices and some of their properties. An ***m × n matrix with entries in*** $\mathbb{F}$ is a map from $\{1,\dots,m\} \times \{1,\dots,n\}$ to $\mathbb{F}$. The image of $(a,i) \in \{1,\dots,m\} \times \{1,\dots,n\}$ is called the ***(a, i)th component*** of the matrix. A typical matrix is denoted by $\boldsymbol{A}$, i.e., we use a bold font to denote a matrix. It is common to represent a matrix as an array of numbers in the form

$$\boldsymbol{A} = \begin{bmatrix} A_{11} & \cdots & A_{1n} \\ \vdots & \ddots & \vdots \\ A_{m1} & \cdots & A_{mn} \end{bmatrix}. \tag{2.1}$$

The reader should not be overly concerned with the location of the indices in the preceding equation. There will be times when it will be natural to think of the indices of the components of a matrix as being both up, both down, or one up and one down, depending on what sort of object is being represented by the matrix. This use of matrices is multiple contexts is a potential point of confusion for newcomers, so we recommend paying some attention to understanding this when it comes up. The set of $m \times n$ matrices with entries in $\mathbb{F}$ we denote by $\mathbb{F}^{m \times n}$, and we note that these matrices can be naturally regarded as linear maps from $\mathbb{F}^n$ to $\mathbb{F}^m$ (cf. Example 2.14). Certain matrices will be of particular interest. An $n \times n$ matrix $\boldsymbol{A}$ is ***diagonal*** if $A_{ij} = 0$ for $i \neq j$. The $n \times n$ identity matrix (i.e., the diagonal matrix with all 1's on the diagonal) is denoted by $\boldsymbol{I}_n$. The $m \times n$ matrix whose entries are all zero is denoted by $\mathbf{0}_{m\times n}$. There will be times when the size of a matrix of zeros will be evident from context, and we may simply write $\mathbf{0}$ in such cases, to avoid clutter.

Remark 2.17 (Elements of $\mathbb{F}^n$ as column vectors). On occasion, it will be convenient to adopt the standard convention of regarding vectors in $\mathbb{F}^n$ as being $n \times 1$ matrices, i.e., as column vectors. Like writing a matrix as (2.1), this is an abuse, although a convenient one. •

For $\boldsymbol{A} \in \mathbb{F}^{n\times n}$, we denote by

$$\operatorname{tr} \boldsymbol{A} = \sum_{j=1}^{n} A_{jj}$$

the ***trace*** of $\boldsymbol{A}$. The ***determinant*** of $\boldsymbol{A} \in \mathbb{F}^{n\times n}$ is defined by

$$\det \boldsymbol{A} = \sum_{\sigma \in S_n} \operatorname{sgn}(\sigma) A_{\sigma(1)1} \cdots A_{\sigma(n)n}.$$

It may be shown that $\det \boldsymbol{A} \neq 0$ if and only if the linear transformation of $\mathbb{F}^n$ defined by $\boldsymbol{A}$ is invertible, in which case we say the matrix $\boldsymbol{A}$ is ***invertible***.

Let us now discuss a useful matrix operation.

Definition 2.18 (Transpose and symmetry). If $\boldsymbol{A} \in \mathbb{F}^{m\times n}$, we denote the ***transpose*** of $\boldsymbol{A}$ by $\boldsymbol{A}^T \in \mathbb{F}^{n\times m}$. Thus,

$$\boldsymbol{A} = \begin{bmatrix} A_{11} & \cdots & A_{1n} \\ \vdots & \ddots & \vdots \\ A_{m1} & \cdots & A_{mn} \end{bmatrix} \implies \boldsymbol{A}^T = \begin{bmatrix} A_{11} & \cdots & A_{m1} \\ \vdots & \ddots & \vdots \\ A_{1n} & \cdots & A_{mn} \end{bmatrix}.$$

If $\mathbb{F} = \mathbb{R}$ and $\boldsymbol{A} = \boldsymbol{A}^T$ (resp. $\boldsymbol{A} = -\boldsymbol{A}^T$), then $\boldsymbol{A}$ is ***symmetric*** (resp. ***skew-symmetric***). •

Next we observe that, by choosing bases, *every* linear map $A \in L(\mathsf{U}; \mathsf{V})$ between finite-dimensional vector spaces can be represented as in Example 2.14. If $\{f_1, \dots, f_n\}$ is a basis for U and $\{e_1, \dots, e_m\}$ is a basis for V, for each $i \in \{1, \dots, n\}$ we may write

$$A(f_i) = A_i^1 e_1 + \cdots + A_i^m e_m,$$

for some unique choice of constants $A_i^1, \dots, A_i^m \in \mathbb{F}$. By letting i run from 1 to n, we thus define nm constants $A_i^a \in \mathbb{F}$, $i \in \{1, \dots, n\}$, $a \in \{1, \dots, m\}$. If we think of $A_i^a \in \mathbb{F}$ as being the (a, i)th component of a matrix, then this defines the ***matrix representation*** or ***matrix representative*** of A relative to the two bases. In identifying the rows and columns of the matrix representative, we note that the superscript is the row index, and the subscript is the column index. We write this matrix as $[A]$, it being understood in any context what bases are being used. If $u \in \mathsf{U}$ is written as $u = u^1 f_1 + \cdots + u^n f_n$, one readily ascertains that

$$A(u) = \sum_{a=1}^{m} \sum_{i=1}^{n} A_i^a u^i e_a.$$

Thus the components of $A(u)$ are written using the summation convention as $A_i^1 u^i, \dots, A_i^m u^i$. We see that the components of $A(u)$ are obtained from those of u by the usual matrix-vector multiplication, using the matrix representation of A, and thinking of the components of u as forming a column vector.

Let us say a few more things about the summation convention we use.

More on the summation convention. 1. In the usual notion of matrix-vector multiplication, the "up" index for A is the row index, and the "down" index is the column index. Note that we can also compactly write

$$\sum_{a=1}^{m} \sum_{i=1}^{n} A_i^a u^i e_a = A_i^a u^i e_a.$$

2. For $i \in \{1, \dots, n\}$, $a \in \{1, \dots, m\}$, let A_i^a and B_a^i be the components of $A \in L(\mathsf{U}; \mathsf{V})$ and $B \in L(\mathsf{V}; \mathsf{U})$, respectively. If A is the inverse of B, then

$$A_i^a B_a^j = \delta_i^j, \quad \text{and} \quad A_i^a B_b^i = \delta_b^a,$$

where δ_i^j, $i, j \in \mathbb{Z}_+$, denotes the ***Kronecker delta***, defined by

$$\delta_i^j = \begin{cases} 1, & i = j, \\ 0, & \text{otherwise.} \end{cases}$$

We will at times also find use for the symbol δ_{ij} that has the same meaning as δ_j^i, i.e., it is 1 when i and j are equal, and 0 otherwise. •

Since a matrix representation depends on a choice of basis, it is sometimes interesting to understand how the matrix representation changes when one changes basis. Let us consider the simplest case when $A \in L(\mathsf{V}; \mathsf{V})$, i.e., A is a linear map from a vector space to itself. Let $\mathscr{B} = \{e_1, \dots, e_n\}$ and $\tilde{\mathscr{B}} = \{\tilde{e}_1, \dots, \tilde{e}_n\}$ be bases for V, and let $[A]_{\mathscr{B}}$ and $[A]_{\tilde{\mathscr{B}}}$ be the corresponding matrix representatives for A. The bases $\mathscr{B}$ and $\tilde{\mathscr{B}}$ can be related by

$$e_i = P_i^j \tilde{e}_j, \qquad i \in \{1, \dots, n\},$$

so defining an invertible $n \times n$ matrix $\boldsymbol{P}$ whose (i, j)th component is P_j^i. One then readily checks that the matrix representatives are related by

$$[A]_{\tilde{\mathscr{B}}} = \boldsymbol{P}[A]_{\mathscr{B}} \boldsymbol{P}^{-1}, \tag{2.2}$$

which is called the ***change of basis formula*** for the matrix representative for A. The transformation of $\mathbb{R}^{n \times n}$ given by $\boldsymbol{A} \mapsto \boldsymbol{PAP}^{-1}$ is called a ***similarity transformation***.

Finally, we comment that the notions of trace and determinant for matrices can be extended to linear maps $A \in L(\mathsf{V};\mathsf{V})$. Indeed, one can define the determinant (resp. trace) of A as the determinant (resp. trace) of its matrix representation in a basis. In basic texts on linear algebra [e.g., Halmos 1996] it is shown that these definitions are independent of basis.

2.2.4 Invariant subspaces, eigenvalues, and eigenvectors

We will on occasion encounter linear maps that have an interesting property relative to some subspace.

Definition 2.19 (A-invariant subspace). For V an $\mathbb{F}$-vector space and for $A \in L(\mathsf{V};\mathsf{V})$, a subspace $\mathsf{U} \subset \mathsf{V}$ is ***A-invariant*** if $A(u) \in \mathsf{U}$ for all $u \in \mathsf{U}$. •

The next result follows immediately from the definition of the matrix representation for a linear map.

Proposition 2.20. *Let V be an $\mathbb{F}$-vector space, let $A \in L(\mathsf{V};\mathsf{V})$, and let $\mathsf{U} \subset \mathsf{V}$ be an A-invariant subspace. If $\{e_1,\dots,e_n\}$ is a basis for V with the property that $\{e_1,\dots,e_k\}$ is a basis for U, then*

$$[A] = \begin{bmatrix} \boldsymbol{A}_{11} & \boldsymbol{A}_{12} \\ \boldsymbol{0}_{n-k,k} & \boldsymbol{A}_{22} \end{bmatrix},$$

for $\boldsymbol{A}_{11} \in \mathbb{F}^{k\times k}$, $\boldsymbol{A}_{12} \in \mathbb{F}^{k\times(n-k)}$, and $\boldsymbol{A}_{22} \in \mathbb{F}^{(n-k)\times(n-k)}$.

We shall also require the following construction involving the notion of an invariant subspace. For an $\mathbb{F}$-vector space V, an arbitrary subset $\mathscr{L} \subset L(\mathsf{V};\mathsf{V})$, and a subspace $\mathsf{U} \subset \mathsf{V}$, we denote by $\langle \mathscr{L}, \mathsf{U}\rangle$ the smallest subspace of V containing U that is as well an invariant subspace for each of the linear maps from $\mathscr{L}$. One readily verifies that $\langle \mathscr{L}, \mathsf{U}\rangle$ is generated by vectors of the form

$$L_1 \circ \cdots \circ L_k(u), \qquad L_1,\dots,L_k \in \mathscr{L} \cup \{\mathrm{id}_{\mathsf{V}}\},\ u \in \mathsf{U},\ k \in \mathbb{N}.$$

At times it will be useful to extend a $\mathbb{R}$-vector space to a $\mathbb{C}$-vector space.

Definition 2.21 (Complexification). If V is a $\mathbb{R}$-vector space, its ***complexification*** is the $\mathbb{C}$-vector space $\mathsf{V}_{\mathbb{C}}$ defined by $\mathsf{V}_{\mathbb{C}} = \mathsf{V} \times \mathsf{V}$, with vector addition and scalar multiplication defined by

$$(u_1, v_1) + (u_2, v_2) = (u_1 + u_2, v_1 + v_2),$$
$$(\alpha + \mathrm{i}\beta)(u, v) = (\alpha u - \beta v, \beta u + \alpha v). \quad \bullet$$

One may verify that these operations satisfy the axioms of a $\mathbb{C}$-vector space. For convenience, if V is a $\mathbb{C}$-vector space, then we write $\mathsf{V}_{\mathbb{C}} = \mathsf{V}$. A $\mathbb{R}$-linear map $A \in L(\mathsf{U};\mathsf{V})$ extends to a $\mathbb{C}$-linear map $A_{\mathbb{C}} \in L(\mathsf{U}_{\mathbb{C}};\mathsf{V}_{\mathbb{C}})$ by $A_{\mathbb{C}}(u,v) = (A(u), A(v))$. If U and V are $\mathbb{C}$-vector spaces, for convenience, we take $A_{\mathbb{C}} = A$.

Of special interest are linear maps from an $\mathbb{F}$-vector space V to itself.

Definition 2.22 (Eigenvalues and eigenvectors). Let V be an $\mathbb{F}$-vector space and let $A \in L(\mathsf{V};\mathsf{V})$. An ***eigenvalue*** for A is an element $\lambda \in \mathbb{C}$ with the property that $A_{\mathbb{C}}(v) = \lambda v$ for some nonzero vector $v \in \mathsf{V}_{\mathbb{C}}$, called an ***eigenvector*** for λ. The set of eigenvalues of A is denoted by $\mathrm{spec}(A)$. •

Remark 2.23. We shall sometimes speak of the eigenvalues and eigenvectors of an $n \times n$ matrix. In so doing, we are thinking of such a matrix as being an element of $L(\mathbb{R}^n;\mathbb{R}^n)$. In elementary courses on linear algebra, one talks about eigenvalues and eigenvectors *only* for matrices. It is our desire to talk about these concepts for general vector spaces that necessitates the idea of complexification. •

In the subsequent discussion, we fix a linear map $A \in L(\mathsf{V};\mathsf{V})$, and assume that $\dim(\mathsf{V}) < +\infty$. The eigenvalues of A are the roots of the ***characteristic polynomial*** $P_A(\lambda) \triangleq \det(\lambda \mathrm{id}_{\mathsf{V}_{\mathbb{C}}} - A_{\mathbb{C}})$, which is a monic polynomial[3] having degree equal to the dimension of V. If $\det(\lambda \mathrm{id}_{\mathsf{V}_{\mathbb{C}}} - A_{\mathbb{C}}) = (\lambda - \lambda_0)^k P(\lambda)$ for a polynomial $P(\lambda)$ having the property that $P(\lambda_0) \neq 0$, then the eigenvalue λ_0 has ***algebraic multiplicity*** k. The eigenvectors for an eigenvalue λ_0 are nonzero vectors in the subspace $\mathsf{U}_{\lambda_0} = \ker(\lambda_0 \mathrm{id}_{\mathsf{V}_{\mathbb{C}}} - A_{\mathbb{C}})$. The ***geometric multiplicity*** of an eigenvalue λ_0 is $\dim(\mathsf{U}_{\lambda_0})$. We let $m_a(\lambda_0)$ denote the algebraic multiplicity and $m_g(\lambda_0)$ denote the geometric multiplicity of λ_0. It is always the case that $m_a(\lambda_0) \geq m_g(\lambda_0)$, and both equality and strict inequality can occur. A useful result concerning a linear map and its characteristic polynomial is the following.

Theorem 2.24 (Cayley–Hamilton Theorem). *Let V be a finite-dimensional $\mathbb{F}$-vector space and let $A \in L(\mathsf{V};\mathsf{V})$. If P_A is the characteristic polynomial of A, then $P_A(A) = 0$. That is to say, A satisfies its own characteristic polynomial.*

Finally, we comment that there is an important normal form associated with the eigenvalues and eigenvectors of a linear map. This normal form is called the ***Jordan normal form*** and we shall on occasion refer to it in exercises. Readers with a good course in linear algebra behind them will have encountered this. Others may refer to [Halmos 1996] or [Horn and Johnson 1990].

2.2.5 Dual spaces

The notion of a dual space to a vector space V is extremely important in mechanics. It can also be a potential point of confusion, since there is often a reflex to identify the dual of a vector space with the vector space itself. It will be important for us to understand the distinction between a vector space and its dual. Indeed, the reader should be forewarned that certain physical concepts (e.g., velocity) are naturally regarded as living in a certain vector

[3] A polynomial is ***monic*** when the coefficient of the highest-degree term is 1.

space, while others (e.g., force) are naturally regarded as living in the dual of that vector space. The easiest thing to do is to accept objects living in duals of vector spaces as "brute facts" arising from the mathematical development we utilize.

Definition 2.25 (Dual space). If V is a finite-dimensional $\mathbb{F}$-vector space, the ***dual space*** to V is the set $\mathsf{V}^* = L(\mathsf{V};\mathbb{F})$ of linear maps from V to $\mathbb{F}$. •

If $\alpha \in \mathsf{V}^*$, we shall alternately write $\alpha(v)$, $\alpha \cdot v$, or $\langle \alpha; v\rangle$ to denote the image in $\mathbb{F}$ of $v \in \mathsf{V}$ under α. If $S \subset \mathsf{V}$, we denote by $\operatorname{ann}(S)$ the ***annihilator*** of S, which is defined by $\operatorname{ann}(S) = \{\alpha \in \mathsf{V}^* \mid \alpha(v) = 0,\ v \in S\}$. Note that, for any nonempty set S, $\operatorname{ann}(S)$ is a subspace. By symmetry (at least for finite-dimensional vector spaces), for $T \subset \mathsf{V}^*$ we define the ***coannihilator*** of T to be the subspace of V defined by $\operatorname{coann}(T) = \{v \in \mathsf{V} \mid \alpha(v) = 0,\ \alpha \in T\}$.

Note that, since $\dim(\mathbb{F}) = 1$, V^* is a vector space having dimension equal to that of V. We shall typically call elements of V^* ***covectors***, although the term ***one-form*** is also common.

Let us see how to represent elements in V^* using a basis for V. Given a basis $\{e_1, \dots, e_n\}$ for V, we define n elements of V^*, denoted by $e^1, \dots, e^n$, by $e^i(e_j) = \delta^i_j$, $i, j \in \{1, \dots, n\}$.

The following result is useful, albeit simple.

Proposition 2.26 (Dual basis). *If* $\{e_1, \dots, e_n\}$ *is a basis for* V, *then* $\{e^1, \dots, e^n\}$ *is a basis for* V^*, *called the* ***dual basis***.

Proof. First let us show that the dual vectors $\{e^1, \dots, e^n\}$ are linearly independent. Let $c_1, \dots, c_n \in \mathbb{F}$ have the property that $c_i e^i = 0$. For each $j \in \{1, \dots, n\}$, we must therefore have $c_i e^i(e_j) = c_i \delta^i_j = c_j = 0$. This gives linear independence. Now let us show that each dual vector $\alpha \in \mathsf{V}^*$ can be expressed as a linear combination of $\{e^1, \dots, e^n\}$. For $\alpha \in \mathsf{V}^*$, define $\alpha_1, \dots, \alpha_n \in \mathbb{F}$ by $\alpha_i = \alpha(e_i)$, $i \in \{1, \dots, n\}$. We claim that $\alpha = \alpha_i e^i$. To prove this, it suffices to check that the two covectors α and $\alpha_i e^i$ agree when applied to any of the basis vectors $\{e_1, \dots, e_n\}$. However, this is obvious, since, for $j \in \{1, \dots, n\}$, we have $\alpha(e_j) = \alpha_j$ and $\alpha_i e^i(e_j) = \alpha_i \delta^i_j = \alpha_j$. ■

If $\{e_1, \dots, e_n\}$ is a basis for V with dual basis $\{e^1, \dots, e^n\}$, then we may write $\alpha \in \mathsf{V}^*$ as $\alpha = \alpha_i e^i$ for some uniquely determined $\alpha_1, \dots, \alpha_n \in \mathbb{F}$, called the ***components*** of α in the given basis. If $v \in \mathsf{V}$ is expressed as $v = v^i e_i$, then we have

$$\alpha(v) = \alpha_i e^i(v^j e_j) = \alpha_i v^j e^i(e_j) = \alpha_i v^j \delta^i_j = \alpha_i v^i.$$

Note that this makes the operation of feeding a vector to a covector look very much like the "dot product," but it is in the best interests of the reader to refrain from thinking this way. One cannot take the dot product of objects in different spaces, and this is the case with $\alpha(v)$, since $\alpha \in \mathsf{V}^*$ and $v \in \mathsf{V}$. The proper generalization of the dot product, called an inner product, is given in Section 2.3.

Remark 2.27 (Elements of $(\mathbb{F}^n)^*$ as row vectors). Let us follow up on Remark 2.17. Elements of $(\mathbb{F}^n)^*$ can be thought of as being $1 \times n$ matrices, i.e., as row vectors. Thus, given $\boldsymbol{\alpha} \in (\mathbb{F}^n)^*$, we can write $\boldsymbol{\alpha} = \begin{bmatrix}\alpha_1 & \cdots & \alpha_n\end{bmatrix}$. Moreover, given $\boldsymbol{v} = (v^1, \dots, v^n) \in \mathbb{F}^n$, and thinking of this vector as a column vector, we have

$$\boldsymbol{\alpha}(\boldsymbol{v}) = \begin{bmatrix}\alpha_1 & \cdots & \alpha_n\end{bmatrix} \begin{bmatrix} v^1 \\ \vdots \\ v^n \end{bmatrix}.$$

This representation serves two purposes. First, it makes the notion of a dual space concrete in the case of $\mathbb{F}^n$. Second, it emphasizes the fact that $(\mathbb{F}^n)^*$ is a different object than $\mathbb{F}^n$. The latter is naturally thought of as being the collection of column vectors, while the former is naturally thought of as being the collection of row vectors. •

Given a linear map between finite-dimensional vector spaces, there is a naturally induced map between the dual spaces.

Definition 2.28 (Dual of a linear map). If $A \colon \mathsf{U} \to \mathsf{V}$ is a linear map between finite-dimensional $\mathbb{F}$-vector spaces U and V, then the linear map $A^* \colon \mathsf{V}^* \to \mathsf{U}^*$ defined by

$$\langle A^*(\alpha); u\rangle = \langle \alpha; A(u)\rangle, \qquad \alpha \in V^*,\ u \in \mathsf{U},$$

is the ***dual*** of A. •

Of course, one should verify that this definition makes sense (i.e., is independent of u) and that A^* is linear. This, however, is easily done. The following result records the matrix representative of the dual of a linear map.

Proposition 2.29 (Matrix representation of the dual). *Let U and V be finite-dimensional $\mathbb{F}$-vector spaces, let $A \in L(\mathsf{U}; \mathsf{V})$, and let $\{f_1, \dots, f_n\}$ and $\{e_1, \dots, e_m\}$ be bases for U and V, respectively, with $\{f^1, \dots, f^n\}$ and $\{e^1, \dots, e^m\}$ the dual bases. Then the matrix representation for A^* is given by $[A^*] = [A]^T$. In particular, if we write $\alpha = \alpha_a e^a$, we have $A^*(\alpha) = A^a_i \alpha_a f^i$.*

Remark 2.30. In assigning the rows and columns of the matrix representative of A^*, the subscripts are row indices and the superscripts are column indices. This is the opposite of the convention used for the matrix representative of A. This reflects the general fact that, when dealing with duals of vector spaces, the location of indices is swapped from when one is dealing with the vector spaces themselves. •

This result leads to the following extension of our summation convention.

More on the summation convention. As previously observed, when we write a collection of elements of a vector space, we use subscripts to enumerate them, e.g., $v_1, \dots, v_k$. For collections of elements of the dual space, we use *super*scripts to enumerate them, e.g., $\alpha^1, \dots, \alpha^k$. By contrast, we saw that the components of a vector with respect to a basis are written with indices as superscripts. In keeping with these conventions, the components of a covector with respect to a basis for the dual space are written with indices as subscripts. •

2.3 Inner products and bilinear maps

Readers may have encountered the notion of an inner product, perhaps as the dot product in $\mathbb{R}^n$, or perhaps in the context of Hilbert spaces arising in Fourier analysis. The inner product will arise in its most important role for us as a model for the inertial properties of a mechanical system. We shall also require inner products and norms in certain technical developments, so we present here the basic underlying ideas. We also present a generalization of an inner product that will arise in numerous instances: a symmetric bilinear map. The discussion of symmetric bilinear maps provides a nice segue into the discussion of tensors in Section 2.4.

One key idea arising from this section is that inner products, orthogonality, and bilinear maps on vector spaces are defined in general terms, and should not only be understood as the dot product between elements of $\mathbb{R}^n$. A second important point is the basic distinction between linear maps between vector spaces and bilinear maps on a vector space. Despite their similar appearances (their representations in components are similar), these are different objects. A third key concept in this section is that any nondegenerate bilinear map induces two associated linear maps that naturally transform vectors into covectors, and vice versa. These maps can be understood easily when written in components.

2.3.1 Inner products and norms

The notion of an inner product is fundamental to the approach we take in this book. While we shall see that an inner product naturally arises for us in the differential geometric setting, in this section we consider the linear situation.

Definition 2.31 (Inner product). An ***inner product*** on an $\mathbb{F}$-vector space V assigns to each pair of vectors $v_1, v_2 \in \mathsf{V}$ a number $\langle\langle v_1, v_2\rangle\rangle \in \mathbb{F}$, and this assignment satisfies

(i) $\langle\langle v_1, v_2\rangle\rangle = \overline{\langle\langle v_2, v_1\rangle\rangle}$, $v_1, v_2 \in \mathsf{V}$ (***symmetry***),

(ii) $\langle\langle c_1 v_1 + c_2 v_2, v_3\rangle\rangle = c_1 \langle\langle v_1, v_3\rangle\rangle + c_2 \langle\langle v_2, v_3\rangle\rangle$, $v_1, v_2, v_3 \in \mathsf{V}$, $c_1, c_2 \in \mathbb{F}$ (***bilinearity***), and

(iii) $\langle\langle v, v\rangle\rangle \geq 0$, $v \in \mathsf{V}$, and $\langle\langle v, v\rangle\rangle = 0$ if and only if $v = 0$ (***positive-definiteness***). •

A structure more general than an inner product is that of a norm.

Definition 2.32 (Norm). A ***norm*** on a vector space V assigns to each vector $v \in \mathsf{V}$ a number $\|v\| \in \mathbb{R}$, and this assignment satisfies

(i) $\|v\| \geq 0$, $v \in \mathsf{V}$, and $\|v\| = 0$ if and only if $v = 0$ (***positivity***),

(ii) $\|\lambda v\| = |\lambda|\,\|v\|$, $\lambda \in \mathbb{F}$, $v \in \mathsf{V}$ (***homogeneity***), and

(iii) $\|v_1 + v_2\| \leq \|v_1\| + \|v_2\|$, $v_1, v_2 \in \mathsf{V}$ (***triangle inequality***). •

An ***inner product space*** is a pair $(\mathsf{V}, \langle\!\langle\cdot,\cdot\rangle\!\rangle)$, where $\langle\!\langle\cdot,\cdot\rangle\!\rangle$ is an inner product on V. A ***normed vector space*** is a pair $(\mathsf{V}, \|\cdot\|)$, where $\|\cdot\|$ is a norm on V. An inner product space is an example of a normed vector space with the norm given by $\|v\| = \sqrt{\langle\!\langle v, v\rangle\!\rangle}$. To show that this proposed norm satisfies the triangle inequality is not perfectly trivial, but requires the Cauchy–Schwarz Inequality (given as Theorem 2.36 below).

Examples 2.33. 1. Let $\{\boldsymbol{e}_1, \dots, \boldsymbol{e}_n\}$ be the standard basis for $\mathbb{F}^n$. The ***standard inner product*** on $\mathbb{F}^n$ is defined by

$$\langle\!\langle \boldsymbol{x}, \boldsymbol{y}\rangle\!\rangle_{\mathbb{F}^n} = \sum_{i=1}^{n} x^i \bar{y}^i,$$

where $\boldsymbol{x} = (x^1, \dots, y^n)$ and $\boldsymbol{y} = (y^1, \dots, y^n)$. Note that this inner product is defined uniquely (by linearity) by the property that $\langle\!\langle \boldsymbol{e}_i, \boldsymbol{e}_j\rangle\!\rangle = \delta_{ij}$. The corresponding norm on $\mathbb{F}^n$ is called the ***standard norm***, and is denoted by $\|\cdot\|_{\mathbb{F}^n}$.

2. Let us consider a generalization of the standard inner product on $\mathbb{R}^n$. Let $\boldsymbol{W} \in \mathbb{R}^{n\times n}$ be a symmetric matrix. One may verify (see Proposition 2.39 below) that $\boldsymbol{W}$ has real eigenvalues. Let us further suppose that $\boldsymbol{W}$ has *positive* eigenvalues. Then we define an inner product on $\mathbb{R}^n$ by $\langle\!\langle \boldsymbol{x}, \boldsymbol{y}\rangle\!\rangle_{\boldsymbol{W}} = \boldsymbol{x}^T\boldsymbol{W}\boldsymbol{y}$. This is indeed an inner product. What is more, *every* inner product on $\mathbb{R}^n$ has this form for some suitable $\boldsymbol{W}$, as the reader may verify in Exercise E2.15. The norm associated with the inner product $\langle\!\langle\cdot,\cdot\rangle\!\rangle_{\boldsymbol{W}}$ is denoted by $\|\cdot\|_{\boldsymbol{W}}$. •

An important concept associated with an inner product space is that of orthogonality.

Definition 2.34 (Orthogonality). Let V be an $\mathbb{F}$-vector space with inner product $\langle\!\langle\cdot,\cdot\rangle\!\rangle$.

(i) Two vectors v and u are ***orthogonal*** if $\langle\!\langle v, u\rangle\!\rangle = 0$.

(ii) A collection of nonzero vectors $\{v_1, \dots, v_m\}$ is ***orthogonal*** if $\langle\!\langle v_i, v_j\rangle\!\rangle = 0$ for $i \neq j$. If additionally $\|v_i\| = 1$, $i \in \{1, \dots, m\}$, then the collection of vectors is orthonormal.

(iii) An ***orthogonal basis*** (resp. ***orthonormal basis***) for V is a basis $\{e_1, \dots, e_n\}$ that is orthogonal (resp. orthonormal).

(iv) An ***orthogonal family*** in V is a countable collection $\{e_j\}_{j\in\mathbb{N}}$ of nonzero vectors for which $\langle\!\langle e_i, e_j\rangle\!\rangle = 0$ whenever $i \neq j$. If additionally $\langle\!\langle e_i, e_i\rangle\!\rangle = 1$ for $i \in \mathbb{N}$, then the family is ***orthonormal***.

(v) Given a subset S of V, we define its ***orthogonal complement*** $S^\perp$ as the subspace $\{ v \in \mathsf{V} \mid \langle\langle v, w \rangle\rangle = 0 \text{ for all } w \in S \}$. •

The situation in which one has a subspace of an inner product space is particularly interesting.

Proposition 2.35. *Let* V *be an* $\mathbb{F}$*-vector space with inner product* $\langle\langle \cdot, \cdot \rangle\rangle$*, and let* $\mathsf{U} \subset \mathsf{V}$ *be a subspace with* $\mathsf{U}^\perp$ *its orthogonal complement. Then* $\mathsf{V} = \mathsf{U} \oplus \mathsf{U}^\perp$*.*

Let V, $\langle\langle \cdot, \cdot \rangle\rangle$, and U be as in the proposition. Then, for any $v \in \mathsf{V}$, we can write $v = v_1 \oplus v_2$ with $v_1 \in \mathsf{U}$ and $v_2 \in \mathsf{U}^\perp$. The element of $L(\mathsf{V}; \mathsf{V})$ given by $v = v_1 \oplus v_2 \mapsto v_1 \oplus 0$ is the ***orthogonal projection*** onto U.

In some of our analysis it will be useful to have the following result, for whose proof we refer to [Abraham, Marsden, and Ratiu 1988].

Theorem 2.36 (Cauchy–Schwarz Inequality). *In an inner product space* $(\mathsf{V}, \langle\langle \cdot, \cdot \rangle\rangle)$ *we have* $|\langle\langle v_1, v_2 \rangle\rangle| \leq \|v_1\| \, \|v_2\|$*. Furthermore, equality holds if and only if* v_1 *and* v_2 *are collinear (i.e., if and only if* $\dim(\operatorname{span}_{\mathbb{F}} \{v_1, v_2\}) \leq 1$*).*

2.3.2 Linear maps on inner product spaces

On an inner product space $(\mathsf{V}, \langle\langle \cdot, \cdot \rangle\rangle)$ one can define special types of linear maps.

Definition 2.37 (Symmetric and skew-symmetric linear maps). Let V be a $\mathbb{R}$-vector space with inner product $\langle\langle \cdot, \cdot \rangle\rangle$, and let $A \in L(\mathsf{V}; \mathsf{V})$.

(i) A is ***symmetric*** if $\langle\langle A(v_1), v_2 \rangle\rangle = \langle\langle v_1, A(v_2) \rangle\rangle$ for all $v_1, v_2 \in \mathsf{V}$.

(ii) A is ***skew-symmetric*** if $\langle\langle A(v_1), v_2 \rangle\rangle = - \langle\langle v_1, A(v_2) \rangle\rangle$ for all $v_1, v_2 \in \mathsf{V}$. •

Remarks 2.38. 1. Without the presence of an inner product, note that these notions *do not* make sense.

2. The notion of symmetric and skew-symmetric linear maps are generalizations of the notions of symmetric and skew-symmetric matrices. Indeed, if ones makes the identification between elements of $\mathbb{R}^{n \times n}$ and elements of $L(\mathbb{R}^n; \mathbb{R}^n)$, as indicated in Example 2.14, then one can easily show that $\boldsymbol{A} \in \mathbb{R}^{n \times n}$ is symmetric (resp. skew-symmetric) if and only if $\boldsymbol{A}$ is a symmetric (resp. skew-symmetric) linear map on the $\mathbb{R}$-inner product space $(\mathbb{R}^n, \langle\langle \cdot, \cdot \rangle\rangle_{\mathbb{R}^n})$.

3. It is possible to extend the notion of a symmetric linear map to a $\mathbb{C}$-vector space. In this case, a $\mathbb{C}$-linear map $A \in L(\mathsf{V}; \mathsf{V})$ is ***Hermitian*** if $\langle\langle A(v_1), v_2 \rangle\rangle = \langle\langle v_1, A(v_2) \rangle\rangle$. We shall not have much occasion to use Hermitian linear maps, although they do come up in the proof of Theorem 6.42. •

The eigenvalues of symmetric linear maps have useful properties, as recorded by the following results.

Proposition 2.39 (Eigenvalues of symmetric linear maps). *Let* $(\mathsf{V}, \langle\!\langle \cdot, \cdot \rangle\!\rangle)$ *be a* $\mathbb{R}$*-inner product space and let* $A \in L(\mathsf{V}; \mathsf{V})$ *be symmetric. The following statements hold:*

(i) $\operatorname{spec}(A) \subset \mathbb{R}$*;*

(ii) if $\lambda_1, \lambda_2 \in \operatorname{spec}(A)$ *are distinct, and if* v_i *is an eigenvector for* λ_i*,* $i \in \{1, 2\}$*, then* $\langle\!\langle v_1, v_2 \rangle\!\rangle = 0$*;*

(iii) more generally, if V *is finite-dimensional, then there exists an orthonormal basis for* V *consisting of eigenvectors for* A*.*

If $A \in L(\mathsf{V}; \mathsf{V})$ is symmetric, we denote by $\lambda_{\min}(A)$ (resp. $\lambda_{\max}(A)$) the smallest (resp. largest) eigenvalue of A.

2.3.3 Bilinear maps

An inner product is an example of a more general algebraic object that we now introduce. In this section we shall restrict our discussion to $\mathbb{F} = \mathbb{R}$, unless otherwise stated.

Definition 2.40 (Symmetric and skew-symmetric bilinear maps). If V is a $\mathbb{R}$-vector space, a ***bilinear map*** on V is a map $B \colon \mathsf{V} \times \mathsf{V} \to \mathbb{R}$ with the property that

$$\begin{aligned} B(c_1 v_1 + c_2 v_2, c_3 v_3 + c_4 v_4) \\ = c_1 c_3 B(v_1, v_3) + c_2 c_3 B(v_2, v_3) + c_1 c_4 B(v_1, v_4) + c_2 c_4 B(v_2, v_4), \end{aligned}$$

for all $v_1, v_2, v_3, v_4 \in \mathsf{V}$, and $c_1, c_2, c_3, c_4 \in \mathbb{R}$. If

(i) $B(v_1, v_2) = B(v_2, v_1)$ for all $v_1, v_2 \in \mathsf{V}$, then B is ***symmetric***, and if

(ii) $B(v_1, v_2) = -B(v_2, v_1)$ for all $v_1, v_2 \in \mathsf{V}$, then B is ***skew-symmetric***.

•

Remark 2.41. As with inner products, the notion of a symmetric bilinear map makes sense for $\mathbb{C}$-vector spaces. In the case that V is a $\mathbb{C}$-vector space, a map $B \colon \mathsf{V} \times \mathsf{V} \to \mathbb{C}$ is ***Hermitian*** if it has the properties of symmetry and bilinearity from Definition 2.31 for $\mathbb{C}$-inner products. This idea will come up only in the proof of Theorem 6.42. •

Note that a symmetric bilinear map has exactly the properties of an inner product, except that we no longer require positive-definiteness. The set of symmetric bilinear maps on V is denoted by $\Sigma_2(\mathsf{V})$. Note that $\Sigma_2(\mathsf{V})$ is a $\mathbb{R}$-vector space with the operations of vector addition and scalar multiplication given by

$$(B_1 + B_2)(v_1, v_2) = B_1(v_1, v_2) + B_2(v_1, v_2), \quad (aB)(v_1, v_2) = a(B(v_1, v_2)).$$

Just as a basis $\{e_1, \dots, e_n\}$ for V induces a basis for V^*, it also induces a basis for $\Sigma_2(\mathsf{V})$. The following result is straightforward to prove (Exercise E2.9).

Proposition 2.42 (Basis for $\Sigma_2(\mathsf{V})$). *Let* V *be a* $\mathbb{R}$*-vector space with* $\{e_1,\dots,e_n\}$ *a basis for* V*. For* $i,j\in\{1,\dots,n\}$*,* $i\le j$*, define* $s^{ij}\in\Sigma_2(\mathsf{V})$ *by*

$$s^{ij}(u,v)=\begin{cases}u^iv^j, & i=j,\\ u^iv^j+v^iu^j, & i\neq j.\end{cases}$$

Then $\{\,s^{ij}\mid i,j\in\{1,\dots,n\},\ i\le j\}$ *is a basis for* $\Sigma_2(\mathsf{V})$*.*

Note that, for $B\colon\mathsf{V}\times\mathsf{V}\to\mathbb{R}$ a bilinear map, for $u,v\in\mathsf{V}$, and for $\{e_1,\dots,e_n\}$ a basis for V, we have

$$B(u,v)=B(u^ie_i,v^je_j)=B(e_i,e_j)u^iv^j.$$

Motivated by this, let us call $B_{ij}=B(e_i,e_j)$, $i,j\in\{1,\dots,n\}$, the ***components*** of B in the given basis. Note that there are n^2 components, and these are naturally arranged in a matrix, which we denote by $[B]$, called the ***matrix representation*** of B, whose (i,j)th component is B_{ij}. Clearly, B is symmetric (resp. skew-symmetric) if and only if $[B]$ is symmetric (resp. skew-symmetric). It is important to distinguish between linear maps on V and bilinear maps on V. Although each type of map is represented in a basis by an $n\times n$ matrix, these matrices represent different objects. That these objects are not the same is most directly seen by the manner in which the matrix components are represented; for linear maps there is an up index and a down index, whereas for bilinear maps, both indices are down.

To write a symmetric bilinear map $B\in\Sigma_2(\mathsf{V})$ in terms of the basis of Proposition 2.42, one writes

$$B=\sum_{1\le i\le j\le n}B_{ij}s^{ij}\quad\text{or simply}\quad B=B_{ij}s^{ij},\ i\le j. \tag{2.3}$$

Note the slight abuse in the summation convention here resulting from the way in which we defined our basis for $\Sigma_2(\mathsf{V})$. Let us summarize some features of the summation convention as they relate to symmetric bilinear maps.

More on the summation convention. The indices for the matrix of a symmetric bilinear map are both subscripts. This should help distinguish bilinear maps from linear maps, since in the latter there is one index up and one index down. If B is a bilinear map with components B_{ij}, $i,j\in\{1,\dots,n\}$, and if u and v are vectors with components u^i,v^i, $i\in\{1,\dots,n\}$, then

$$B(u,v)=B_{ij}u^iv^j.$$

Note that, in contrast with (2.3), the summation over i and j is unrestricted in the previous expression. •

Example 2.43. Let $\langle\!\langle\cdot,\cdot\rangle\!\rangle_{\mathbb{R}^n}$ be the standard inner product on $\mathbb{R}^n$. Its matrix representation with respect to the standard basis $\{e_1,\dots,e_n\}$ is the $n\times n$ identity matrix $\boldsymbol{I}_n$. •

For symmetric bilinear maps, there are concepts of definiteness that will be useful for us. To make these definitions, it is convenient to associate with $B \in \Sigma_2(\mathsf{V})$ the function $Q_B \colon \mathsf{V} \to \mathbb{R}$ defined by $Q_B(v) = B(v, v)$, sometimes called the ***quadratic form*** associated with B. Furthermore, a symmetric bilinear map is uniquely determined by its quadratic form (see Exercise E2.12), so we will interchangeably use the expressions "symmetric bilinear map" and "quadratic form" for the same thing.

Definition 2.44 (Definiteness of symmetric bilinear maps). With the preceding notation, $B \in \Sigma_2(\mathsf{V})$ is

(i) ***positive-semidefinite*** if $\mathrm{image}(Q_B) \subset \bar{\mathbb{R}}_+$, is
(ii) ***positive-definite*** if it is positive-semidefinite and $Q_B^{-1}(0) = \{0\}$, is
(iii) ***negative-semidefinite*** if $-B$ is positive-semidefinite, is
(iv) ***negative-definite*** if $-B$ is positive-definite, is
(v) ***semidefinite*** if it is either positive- or negative-semidefinite, is
(vi) ***definite*** if it is either positive- or negative-definite, and is
(vii) ***indefinite*** if it is neither positive- nor negative-semidefinite. •

Remarks 2.45. 1. Note that if V is the zero vector space, i.e., $\mathsf{V} = \{0\}$, then there is only one element of $\Sigma_2(\mathsf{V})$, and that is the zero map. By definition this is positive-definite. This degenerate case will actually come up in the treatment of controllability using vector-valued quadratic forms in Chapter 8, as well as in our treatment of stabilization in Chapter 10.

2. The above definitions can also be applied to a linear map $A \in L(\mathsf{V}; \mathsf{V})$ that is symmetric with respect to an inner product $\langle\!\langle \cdot, \cdot \rangle\!\rangle$ on V. To do this, one considers on V the symmetric bilinear map given by $B_A(v_1, v_2) = \langle\!\langle A(v_1), v_2 \rangle\!\rangle$, and then Definition 2.44 applies to B_A. In particular, if we take $\mathsf{V} = \mathbb{R}^n$ and $\langle\!\langle \cdot, \cdot \rangle\!\rangle = \langle\!\langle \cdot, \cdot \rangle\!\rangle_{\mathbb{R}^n}$ then we see that it is possible to talk about symmetric matrices having the various definiteness properties ascribed to symmetric bilinear maps. This is somewhat related to Exercise E2.22. •

It is in fact easy in principle to determine whether a given $B \in \Sigma_2(\mathsf{V})$ satisfies any of the above conditions. The following theorem indicates how to do this.

Theorem 2.46 (Normal form for symmetric bilinear map). *If B is a symmetric bilinear map on a finite-dimensional $\mathbb{R}$-vector space V, then there exists a basis $\{e_1, \dots, e_n\}$ for V such that the matrix representative for B in this basis is given by*

$$[B] = \begin{bmatrix} \boldsymbol{I}_k & \boldsymbol{0} & \boldsymbol{0} \\ \boldsymbol{0} & -\boldsymbol{I}_l & \boldsymbol{0} \\ \boldsymbol{0} & \boldsymbol{0} & \boldsymbol{0} \end{bmatrix}.$$

The number of nonzero elements on the diagonal of this matrix (i.e., $k+l$) is the ***rank*** *of B, denoted by* $\mathrm{rank}(B)$. *The number of -1's along the diagonal (i.e., l) is the* ***index*** *of B, denoted by* $\mathrm{ind}(B)$. *The number of $+1$'s along* *the diagonal minus the number of* -1*'s along the diagonal (i.e.,* $k-l$*) is the* ***signature*** *of* B*, denoted by* $\operatorname{sig}(B)$*. The map* B *is* ***nondegenerate*** *if* $\operatorname{rank}(B) = \dim(\mathsf{V})$.

This theorem is proved by the Gram–Schmidt Procedure that may be familiar to students having had a good course in linear algebra [Halmos 1996]. In Exercise E2.20 the reader may work out the fairly obvious correspondences between rank, index, and signature, and the various forms of definiteness. In Exercise E2.22 we also provide an alternative method for determining the definiteness of a symmetric bilinear map.

2.3.4 Linear maps associated with bilinear maps

In this section we introduce notation for some operations that are natural in coordinates, but which can provide confusion when thought of abstractly. Part of the difficulty may be that the notation fits together in a slick way, and the slickness may cause confusion.

For $B\colon \mathsf{V}\times\mathsf{V}\to\mathbb{R}$ a bilinear map, we define the ***flat map*** $B^\flat\colon \mathsf{V}\to\mathsf{V}^*$ by asking that, for $v\in\mathsf{V}$, $B^\flat(v)$ satisfy $\langle B^\flat(v);u\rangle = B(u,v)$ for all $u\in\mathsf{V}$. If $\{e_1,\dots,e_n\}$ is a basis for V with B_{ij}, $i,j\in\{1,\dots,n\}$, the components of B in this basis, then the map $B^\flat\colon\mathsf{V}\to\mathsf{V}^*$ is given simply by $B^\flat(v) = B_{ij}v^j e^i$. In other words, relative to the basis $\{e_1,\dots,e_n\}$ for V and the dual basis $\{e^1,\dots,e^n\}$ for V^*, the components of the matrix representative for the linear map $B^\flat\colon\mathsf{V}\to\mathsf{V}^*$ are exactly the components of B.

If $B^\flat$ is invertible, then we denote the inverse, called the ***sharp map***, by $B^\sharp\colon\mathsf{V}^*\to\mathsf{V}$. Of course, given a basis $\{e_1,\dots,e_n\}$ for V inducing the dual basis $\{e^1,\dots,e^n\}$ for V^*, the matrix $[B^\sharp]$ of the linear map $B^\sharp$ is exactly the inverse of the matrix $[B^\flat]$ for $B^\flat$. It is convenient to denote the components of the matrix $[B^\sharp]$ by B^{ij}, $i,j\in\{1,\dots,n\}$, if B_{ij}, $i,j=\{1,\dots,n\}$, are the components for B. Thus $B^{ij}B_{jk}=\delta^i_k$, $i,k\in\{1,\dots,n\}$. In this case we have $B^\sharp(\alpha)=B^{ij}\alpha_j e_i$.

Remarks 2.47. 1. Some authors use a different convention for defining $B^\flat$, using $\langle B^\flat(v);u\rangle = B(v,u)$ for all $u\in\mathsf{V}$ to define $B^\flat(v)$. In the text, this will not come up since we will only use $B^\flat$ in the case that B is symmetric. However, there are instances (e.g., Hamiltonian mechanics) where it will make a difference.

2. In the case that the vector space under consideration is itself a dual space, say $\mathsf{V}=\mathsf{U}^*$, and if $B\colon\mathsf{V}\times\mathsf{V}\to\mathbb{R}$ is a bilinear map, then $B^\flat$ is a linear map between $\mathsf{V}=\mathsf{U}^*$ and $\mathsf{V}^*=\mathsf{U}^{**}$. Provided that U is finite-dimensional, it is easy to show that $\mathsf{U}^{**}\simeq\mathsf{U}$ (see Exercise E2.4). For this reason, we will actually write the map induced by B from U^* to U as $B^\sharp$. This will come up in Chapters 10 and 12. •

More on the summation convention. The maps $B^\flat\colon\mathsf{V}\to\mathsf{V}^*$ and $B^\sharp\colon\mathsf{V}^*\to\mathsf{V}$ have the following devices for easily remembering their form in coordinates. The map $B^\flat$ in words is "B-flat," and in coordinates it takes a vector (with indices up) and returns a covector (with indices down). Conversely, the map $B^\sharp$ in words is "B-sharp," and in coordinates it takes a covector (with indices down) and returns a vector (with indices up). Thus the notation for the two maps is intended to reflect what they do to indices. In order for the expression $B^\flat(v) = B_{ij}v^j c^i$ to represent usual matrix-vector multiplication between the matrix representative of B and the components of v, one should interpret the first index in B_{ij} as being the row index, and the second as being the column index. The same is true when interpreting $B^\sharp(\alpha) = B^{ij}\alpha_j e_i$ as matrix-vector multiplication. •

We conclude this section with some remarks on inner products. Since an inner product is an element of the vector space $\Sigma_2(\mathsf{V})$, it too has components. For this reason, it is often convenient to write an inner product not as $\langle\langle\cdot,\cdot\rangle\rangle$, but with a character; we choose $\mathbb{G}$ to represent a generic inner product. With this notation, we let $\mathbb{G}_{\mathbb{R}^n}$ denote the standard inner product on $\mathbb{R}^n$. We can now talk about the components of an inner product $\mathbb{G}$, and talk about its matrix representation $[\mathbb{G}]$. The discussion of the preceding paragraph can be specialized to the case of an inner product, yielding the maps $\mathbb{G}^\flat\colon \mathsf{V} \to \mathsf{V}^*$ and $\mathbb{G}^\sharp\colon \mathsf{V}^* \to \mathsf{V}$, called the ***associated isomorphisms*** for the inner product $\mathbb{G}$. Note that $\mathbb{G}^\flat$ is indeed an isomorphism by virtue of the positive-definiteness of an inner product. The isomorphism $\mathbb{G}^\flat$ allows us to define an inner product on V^* that we denote by $\mathbb{G}^{-1}$. We define this inner product by

$$\mathbb{G}^{-1}(\alpha^1,\alpha^2) = \mathbb{G}(\mathbb{G}^\sharp(\alpha^1),\mathbb{G}^\sharp(\alpha^2)). \tag{2.4}$$

If $\{e_1,\dots,e_n\}$ is a basis for V, then the components of the inner product $\mathbb{G}^{-1}$ relative to the dual basis $\{e^1,\dots,e^n\}$ are $\mathbb{G}^{ij}$, $i,j\in\{1,\dots,n\}$, i.e., the components of the inverse of the matrix of components of $\mathbb{G}$.

2.4 Tensors

Without actually saying so, this has, thus far, been a chapter on tensors. That is, many of the objects in the preceding sections on linear algebra are part of a more general class of objects called tensors. Now we will talk about tensors in a general way. Readers having the benefit of a good algebra course may have seen tensors presented in a different way than is done here. We do not comment on this except to say that the two approaches are equivalent in finite dimensions, and that Nelson [1967] and Abraham, Marsden, and Ratiu [1988, Section 5.1] have more to say about this. Throughout this section, we restrict consideration to $\mathbb{R}$-vector spaces.

Roughly speaking, tensors are multilinear maps whose domain is multiple copies of a vector space and of its dual space, and whose codomain is $\mathbb{R}$. While the essential idea of multilinearity is straightforward, some careful bookkeeping is required to perform computations on tensors, such as the tensor product, and the operations of pull-back and push-forward. The latter two operations arise when, given a linear map between two vector spaces, one wants to "map" the tensor from one vector space to the other. These operations will be natural and important when dealing with more complex objects in later chapters.

2.4.1 Basic definitions

Let us begin by defining multilinear maps. Let $\mathsf{U}_1, \dots, \mathsf{U}_k$ and V be $\mathbb{R}$-vector spaces. A map $A\colon \mathsf{U}_1 \times \cdots \times \mathsf{U}_k \to \mathsf{V}$ is called ***multilinear*** if it is linear in each of the k arguments separately. Precisely, A is multilinear if, for each $i \in \{1, \dots, k\}$, and for each $u_j \in \mathsf{U}_j$, $j \in \{1, \dots, i-1, i+1, \dots, k\}$, the map

$$\mathsf{U}_i \ni u_i \mapsto A(u_1, \dots, u_{i-1}, u_i, u_{i+1}, \dots, u_k) \in \mathsf{V}$$

is linear. We let $L(\mathsf{U}_1, \dots, \mathsf{U}_k; \mathsf{V})$ denote the set of multilinear maps from $\mathsf{U}_1 \times \cdots \times \mathsf{U}_k$ to V. If $\mathsf{U}_1 = \cdots = \mathsf{U}_k = \mathsf{U}$, we abbreviate the set of multilinear maps by $L^k(\mathsf{U}; \mathsf{V})$. As with linear maps, sometimes it is useful to refer to a multilinear map between $\mathbb{R}$-vector spaces as being $\mathbb{R}$-***multilinear***, to distinguish between possible multiple algebraic structures in the same setup.

Definition 2.48 (Tensors on vector spaces). Let V be a $\mathbb{R}$-vector space and let $r, s \in \mathbb{Z}_+$. We define the ***tensors of type*** (r, s) to be the set

$$T^r_s(\mathsf{V}) = L(\underbrace{\mathsf{V}^* \times \cdots \times \mathsf{V}^*}_{r \text{ copies}} \times \underbrace{\mathsf{V} \times \cdots \times \mathsf{V}}_{s \text{ copies}}; \mathbb{R}). \qquad \bullet$$

Therefore, a tensor of type (r, s), or simply an (r, s)-tensor, is a $\mathbb{R}$-valued multilinear map taking as inputs r covectors and s vectors. For example, symmetric bilinear maps, and in particular inner products, are $(0, 2)$-tensors, since they bilinearly map elements in $\mathsf{V} \times \mathsf{V}$ into $\mathbb{R}$. In the case in which $r = s = 0$, we adopt the convention that $T^0_0(\mathsf{V}) = \mathbb{R}$.

A basic operation one can perform with tensors is to take the product of $t_1 \in T^{r_1}_{s_1}(\mathsf{V})$ and $t_2 \in T^{r_2}_{s_2}(\mathsf{V})$ to obtain a tensor $t_1 \otimes t_2 \in T^{r_1+r_2}_{s_1+s_2}(\mathsf{V})$ as follows:

$$\begin{aligned}(t_1 \otimes t_2)(\alpha^1, \dots, \alpha^{r_1}, \beta^1, \dots, \beta^{r_2}, v_1, \dots, v_{s_1}, u_1, \dots, u_{s_2}) \\ = t_1(\alpha^1, \dots, \alpha^{r_1}, v_1, \dots, v_{s_1}) t_2(\beta^1, \dots, \beta^{r_2}, u_1, \dots, u_{s_2}).\end{aligned}$$

This is the ***tensor product*** of t_1 and t_2. Note that generally $t_1 \otimes t_2 \neq t_2 \otimes t_1$. That is to say, the tensor product is *not* commutative. The tensor product is, however, associative, meaning that $(t_1 \otimes t_2) \otimes t_3 = t_1 \otimes (t_2 \otimes t_3)$. It is also linear with respect to the operations of vector addition and scalar multiplication we defined below (see (2.5)). Associativity and linearity are often useful in computations involving the tensor product.

What we called in Section 2.3 a symmetric bilinear map on V is nothing more than a $(0, 2)$ tensor B that satisfies $B(v_1, v_2) = B(v_2, v_1)$ for $v_1, v_2 \in$

V. We call such $(0,2)$-tensors ***symmetric***. In like manner, a $(0,2)$-tensor B is ***skew-symmetric*** if $B(v_1, v_2) = -B(v_2, v_1)$ for $v_1, v_2 \in \mathsf{V}$. We denote by $\Sigma_2(\mathsf{V})$ the symmetric $(0,2)$-tensors, just as when we discussed symmetric bilinear maps above. Of course, similar definitions can be made for tensors of type $(2,0)$, and the resulting symmetric $(2,0)$-tensors are denoted by $\Sigma_2(\mathsf{V}^*)$.

These notions of symmetry and skew-symmetry may be carried over to more general tensors, namely, to $(r,0)$-tensors and to $(0,s)$-tensors for any $r, s \in \mathbb{N}$. However, we shall not need this degree of generality.

2.4.2 Representations of tensors in bases

Let us now see how one may represent tensors in bases, just as we have done for linear maps, covectors, and symmetric bilinear maps. First one should note that $T^r_s(\mathsf{V})$ is indeed a $\mathbb{R}$-vector space with vector addition and scalar multiplication defined by

$$\begin{aligned}(t_1 + t_2)(\alpha^1, \dots, \alpha^r, v_1, \dots, v_s)& \\ = t_1(\alpha^1, \dots, \alpha^r, v_1, \dots, v_s) &+ t_2(\alpha^1, \dots, \alpha^r, v_1, \dots, v_s), \\ (at)(\alpha^1, \dots, \alpha^r, v_1, \dots, v_s) &= a(t(\alpha^1, \dots, \alpha^r, v_1, \dots, v_s)).\end{aligned} \tag{2.5}$$

Now let $\{e_1, \dots, e_n\}$ be a basis for V and let $\{e^1, \dots, e^n\}$ be the dual basis for V^*. The reader is invited to prove the following result, which gives an induced basis for $T^r_s(\mathsf{V})$ (Exercise E2.25).

Proposition 2.49 (Basis for $T^r_s(\mathsf{V})$). *If V is a finite-dimensional $\mathbb{R}$-vector space with basis $\{e_1, \dots, e_n\}$ and dual basis $\{e^1, \dots, e^n\}$, then the set*

$$\{e_{i_1} \otimes \cdots \otimes e_{i_r} \otimes e^{j_1} \otimes \cdots \otimes e^{j_s} \mid i_1, \dots, i_r, j_1, \dots, j_s \in \{1, \dots, n\}\}$$

is a basis for $T^r_s(\mathsf{V})$.

The ***components*** of $t \in T^r_s(\mathsf{V})$ relative to $\{e_1, \dots, e_n\}$ are

$$t^{j_1 \cdots j_r}_{i_1 \cdots i_s} = t(e^{j_1}, \dots, e^{j_r}, e_{i_1}, \dots, e_{i_s}), \qquad i_1, \dots, i_s, j_1, \dots, j_r \in \{1, \dots, n\}.$$

Thus an (r,s)-tensor has n^{r+s} components. Indeed, one easily verifies that we in fact have

$$t = t^{j_1 \cdots j_r}_{i_1 \cdots i_s} e_{j_1} \otimes \cdots \otimes e_{j_r} \otimes e^{i_1} \otimes \cdots \otimes e^{i_s}.$$

Note here that the summation convention is in full force, taking the place of $r+s$ summation signs. Also note that we can write

$$t(\alpha^1, \dots, \alpha^r, v_1, \dots, v_s) = t^{j_1 \cdots j_r}_{i_1 \cdots i_s} \alpha^1_{j_1} \cdots \alpha^r_{j_r} v^{i_1}_1 \cdots v^{i_s}_s,$$

where again the summation convention is implied with respect to the indices $j_1, \dots, j_r$ and $i_1, \dots, i_s$.

2.4.3 Behavior of tensors under linear maps

Given two $\mathbb{R}$-vector spaces U and V and a linear map $A \in L(\mathsf{U};\mathsf{V})$, there is an induced map between certain of the tensors on U and V. This will be particularly useful to us in the geometric setting of Chapter 3, but it is convenient to initiate the presentation in the context of linear algebra.

Definition 2.50 (Push-forward and pull-back on vector spaces). Let U and V be finite-dimensional $\mathbb{R}$-vector spaces and let $A \in L(\mathsf{U};\mathsf{V})$.

(i) If $t \in T^r_0(\mathsf{U})$, then the ***push-forward*** of t by A is $A_*t \in T^r_0(\mathsf{V})$, defined by

$$A_*t(\beta^1,\dots,\beta^r) = t(A^*(\beta^1),\dots,A^*(\beta^r)).$$

(ii) If $t \in T^0_s(\mathsf{V})$, then the ***pull-back*** of t by A is $A^*t \in T^0_s(\mathsf{U})$, defined by

$$A^*t(u_1,\dots,u_s) = t(A(u_1),\dots,A(u_s)).$$

(iii) If A is an isomorphism and if $t \in T^r_s(\mathsf{U})$, then the ***push-forward*** of t by A is $A_*t \in T^r_s(\mathsf{V})$, defined by

$$A_*t(\beta^1,\dots,\beta^r,v_1,\dots,v_s) \\ = t(A^*(\beta^1),\dots,A^*(\beta^r),A^{-1}(v_1),\dots,A^{-1}(v_s)).$$

(iv) If A is an isomorphism and if $t \in T^r_s(\mathsf{V})$, then the ***pull-back*** of t by A is $A^*t \in T^r_s(\mathsf{U})$, defined by

$$A^*t(\alpha^1,\dots,\alpha^r,u_1,\dots,u_s) \\ = t((A^{-1})^*(\alpha^1),\dots,(A^{-1})^*(\alpha^r),A(u_1),\dots,A(u_s)).$$ •

Remark 2.51. Note the convenient confluence of notation for A^*t when $A \in L(\mathsf{U};\mathsf{V})$ and $t \in T^0_1(\mathsf{V})$. One can think of A^*t as the pull-back of t by A, or as the dual of A applied to t. •

It is important to note in the above definitions that, unless the linear map A is an isomorphism, the notions of push-forward and pull-back are genuinely restricted to tensors of type $(r,0)$ and $(0,s)$, respectively. This will arise also in the geometric notions of push-forward and pull-back, but with additional restrictions being present in that context.

For completeness, let us provide coordinate formulae for push-forward and pull-back. Let U, V, and A be as above, and let $\{f_1,\dots,f_n\}$ and $\{e_1,\dots,e_m\}$ be bases for U and V, respectively. Let A^a_i, $a \in \{1,\dots,m\}$, $i \in \{1,\dots,n\}$, be the components of A with respect to these bases. If A is invertible, let B^i_a, $a,i \in \{1,\dots,n\}$, be the components of $A^{-1} \in L(\mathsf{V};\mathsf{U})$ with respect to the given bases.

1. If $t \in T^r_0(\mathsf{U})$ has components $t^{i_1\cdots i_r}$, then the components of A_*t are

$$(A_*t)^{a_1\cdots a_r} = A^{a_1}_{i_1}\cdots A^{a_r}_{i_r}t^{i_1\cdots i_r}.$$

2. If $t \in T^0_s(\mathsf{V})$ has components $t_{a_1 \cdots a_s}$, then the components of A^*t are

$$(A^*t)_{i_1 \cdots i_s} = A^{a_1}_{i_1} \cdots A^{a_s}_{i_s} t_{a_1 \cdots a_s}.$$

3. If A is an isomorphism and if $t \in T^r_s(\mathsf{U})$ has components $t^{i_1 \cdots i_r}_{j_1 \cdots j_s}$, then the components of A_*t are

$$(A_*t)^{a_1 \cdots a_r}_{b_1 \cdots b_s} = A^{a_1}_{i_1} \cdots A^{a_r}_{i_r} B^{j_1}_{b_1} \cdots B^{j_s}_{b_s} t^{i_1 \cdots i_r}_{j_1 \cdots j_s}.$$

4. If A is an isomorphism and if $t \in T^r_s(\mathsf{V})$ has components $t^{a_1 \cdots a_r}_{b_1 \cdots b_s}$, then the components of A^*t are

$$(A^*t)^{i_1 \cdots i_r}_{j_1 \cdots j_s} = B^{i_1}_{a_1} \cdots B^{i_r}_{a_r} A^{b_1}_{j_1} \cdots A^{b_s}_{j_s} t^{a_1 \cdots a_r}_{b_1 \cdots b_s}.$$

Push-forwards and pull-backs have certain useful properties with respect to composition of linear maps [Abraham, Marsden, and Ratiu 1988].

Proposition 2.52 (Push-forward, pull-back, and composition). *If* U, V, *and* W *are* $\mathbb{R}$*-vector spaces, if* $t_1 \in T^r_s(\mathsf{U})$ *and* $t_2 \in T^r_s(\mathsf{W})$, *and if* $A \in L(\mathsf{U};\mathsf{V})$ *and* $B \in L(\mathsf{V};\mathsf{W})$, *then the formulae*

$$(B \circ A)_* t_1 = B_*(A_* t_1) \quad \textit{and} \quad (B \circ A)^* t_2 = A^*(B^* t_2)$$

hold, whenever they are well-defined.

2.5 Convexity

While convexity, the topic of this section, is not quite in the domain of linear algebra, now is as good a time as any to present these ideas. We barely graze the surface of the subjects considered in this section, since we shall need only the most basic of concepts. We refer interested readers to the classic text of Rockafellar [1970] on convex analysis, and to the recent text of Boyd and Vandenberghe [2004] on convex optimization.

Definition 2.53 (Convex set). Let V be a $\mathbb{R}$-vector space.

(i) A subset $A \subset \mathsf{V}$ is ***convex*** if $v_1, v_2 \in A$ implies that

$$\{(1-t)v_1 + tv_2 \mid t \in [0,1]\} \subset A.$$

(ii) If $A \subset \mathsf{V}$ is a general subset, a ***convex combination*** of vectors $v_1, \dots, v_k \in A$ is a linear combination of the form

$$\sum_{j=1}^k \lambda_j v_j, \qquad \lambda_1, \dots, \lambda_k \geq 0, \ \sum_{j=1}^k \lambda_j = 1, \ k \in \mathbb{N}.$$

(iii) The ***convex hull*** of a general subset $A \subset \mathsf{V}$, denoted by $\operatorname{conv}(A)$, is the smallest convex set containing A. •

Thus a set is convex when the line segment connecting any two points in the set lies within the set. It may be verified that a set is convex if and only if it contains all convex combinations of its points. Our definition of convex hull makes sense since the intersection of convex sets is convex. This allows one to assert the existence of the convex hull. One may also show that $\mathrm{conv}(A)$ consists of the union of all convex combinations of elements of A.

The notion of an affine space is also interesting for us.

Definition 2.54 (Affine subspace). Let V be a $\mathbb{R}$-vector space.

(i) A subset A of V is an ***affine subspace*** of V if there exists $v \in \mathsf{V}$ and a subspace U of V for such that $\mathsf{A} = \{v + u \mid u \in \mathsf{U}\}$.

(ii) If $A \subset \mathsf{V}$ is a general subset, an ***affine combination*** of vectors $v_1, \dots, v_k \in A$ is a linear combination of the form

$$\sum_{j=1}^{k} \lambda_j v_j, \qquad \lambda_1, \dots, \lambda_k \in \mathbb{R}, \ \sum_{j=1}^{k} \lambda_j = 1, \ k \in \mathbb{N}.$$

(iii) The ***affine hull*** of a general subset $A \subset \mathsf{V}$, denoted by $\mathrm{aff}(A)$, is the smallest affine subspace of V containing A. •

Thus an affine subspace is a "shifted subspace," possibly shifted by the zero vector. Analogously to the convex hull, the definition we give for the affine hull makes sense, since the intersection of affine subspaces is again an affine subspace. This allows one to conclude the existence of the affine hull. One may also show that $\mathrm{aff}(A)$ consists of the union of all affine combinations of elements of A. We refer to Exercise E2.29 for an alternative characterization of an affine subspace.

Exercises

E2.1 Let V be an $\mathbb{F}$-vector space with $\mathsf{U} \subset \mathsf{V}$ a subspace and $\mathsf{U}' \subset \mathsf{V}$ a complement to U in V (i.e., $\mathsf{V} = \mathsf{U} \oplus \mathsf{U}'$). Show that there is a natural isomorphism between U' and V/U.

E2.2 Show that every linear map from $\mathbb{F}^n$ to $\mathbb{F}^m$ has the form exhibited in Example 2.14.

E2.3 Let $\mathcal{H}_2^n$ denote the set of functions $\mathbb{R}^n \to \mathbb{R}$ that are polynomial functions, homogeneous of degree 2, in the variables $(x^1, \dots, x^n)$.

(a) Show that $\mathcal{H}_2^n$ is a $\mathbb{R}$-vector space.

(b) Provide a basis for $\mathcal{H}_2^n$.

(c) Compute the dimension of $\mathcal{H}_2^n$ as a function of n.

E2.4 Let V be a normed $\mathbb{R}$-vector space, and define a linear map ι_{V} from V to V^{**} by $\iota_{\mathsf{V}}(v)(\alpha) = \alpha(v)$. Show that, if V is finite-dimensional, then ι_{V} is an isomorphism.

E2.5 Let V be an $\mathbb{F}$-vector space with $\mathsf{U} \subset \mathsf{V}$ a subspace. Show that $(\mathsf{V}/\mathsf{U})^* \simeq \mathrm{ann}(\mathsf{U})$.

E2.6 Show that a norm $\|\cdot\|$ on a vector space V is the natural norm associated to an inner product if and only if $\|\cdot\|$ satisfies the ***parallelogram law***:

$$\|u+v\|^2 + \|u-v\|^2 = 2(\|u\|^2 + \|v\|^2).$$

Hint: *"Only if" is easy, but "if" is not; see [Yosida 1980].*

E2.7 Consider $\mathbb{R}^3$ with the standard inner product and let $\mathsf{U} \subset \mathbb{R}^3$ be a two-dimensional subspace.
(a) Define explicitly the orthogonal projection P_{U} onto U.
(b) Define explicitly the reflection map R_{U} that reflects vectors about U.

E2.8 Let $\boldsymbol{A}, \boldsymbol{B} \in \mathbb{R}^{n\times n}$ and define $\langle\langle \boldsymbol{A}, \boldsymbol{B}\rangle\rangle = \operatorname{tr}(\boldsymbol{A}\boldsymbol{B}^T)$.
(a) Show that $\langle\langle\cdot,\cdot\rangle\rangle$ is an inner product on $\mathbb{R}^{n\times n}$.
(b) Show that the subspaces of symmetric and skew-symmetric matrices are orthogonal with respect to this inner product.
(c) Show that the orthogonal projection $\operatorname{sym}\colon \mathbb{R}^{n\times n} \to \mathbb{R}^{n\times n}$ onto the set of symmetric matrices is given by $\operatorname{sym}(\boldsymbol{A}) = \frac{1}{2}(\boldsymbol{A}+\boldsymbol{A}^T)$.
(d) Show that the orthogonal projection $\operatorname{skew}\colon \mathbb{R}^{n\times n} \to \mathbb{R}^{n\times n}$ onto the set of skew-symmetric matrices is given by $\operatorname{skew}(\boldsymbol{A}) = \frac{1}{2}(\boldsymbol{A}-\boldsymbol{A}^T)$.

E2.9 Prove Proposition 2.42.

E2.10 If $B \in \Sigma_2(\mathsf{V})$, show that, thinking of B as a linear map from V to V^*, we have $B = B^*$.

E2.11 Let $\mathbb{G}$ be an inner product on a vector space V, let $\{e_1,\dots,e_n\}$ be a basis for V, and let $\{e^1,\dots,e^n\}$ be the corresponding dual basis.
(a) Show that the matrix representations of $\mathbb{G}$ and $\mathbb{G}^\flat$ agree.
(b) Show that the matrix representations of $\mathbb{G}^{-1}$ and $\mathbb{G}^\sharp$ agree.

E2.12 Let V be a $\mathbb{R}$-vector space and let $B \in \Sigma_2(\mathsf{V})$.
(a) Prove the ***polarization identity***:

$$4B(v_1,v_2) = B(v_1+v_2, v_1+v_2) - B(v_1-v_2, v_1-v_2).$$

A ***quadratic function*** on V is a function $Q\colon \mathsf{V} \to \mathbb{R}$ satisfying $Q(\lambda v) = \lambda^2 Q(v)$ for each $\lambda \in \mathbb{R}$ and $v \in \mathsf{V}$.
(b) Show how the polarization identity provides a means of defining a symmetric bilinear map given a quadratic function on a vector space.

E2.13 Describe the Gram–Schmidt Procedure for obtaining an orthonormal set of vectors starting from a set of linearly independent vectors, and use this procedure to prove Theorem 2.46.

E2.14 Let $[a,b]$ be an interval and let $L_2([a,b])$ be the set of measurable[4] functions $f\colon [a,b] \to \mathbb{R}$ whose square is integrable over $[a,b]$. We refer to such functions as ***square-integrable***. Show that the map $(f,g) \mapsto \int_a^b f(t)g(t)\,dt$ is an inner product. This map is referred to as the $L_2([a,b])$-inner product; it is convenient to denote it by $\langle\langle f,g\rangle\rangle_{L_2([a,b])}$ and its norm by $\|f\|_{L_2([a,b])}$.

E2.15 A matrix $\boldsymbol{W} \in \mathbb{F}^{n\times n}$ is ***Hermitian*** if $\boldsymbol{W} = \bar{\boldsymbol{W}}^T$.

---

[4] Readers not familiar with Lebesgue integration can omit "measurable" without conceptual loss.

(a) Show that a Hermitian matrix $\boldsymbol{W}$ has real eigenvalues.
Suppose now that all eigenvalues of the Hermitian matrix $\boldsymbol{W}$ are positive.
(b) Show that $\langle\langle \boldsymbol{x}, \boldsymbol{y} \rangle\rangle_{\boldsymbol{W}} = \boldsymbol{x}^T \boldsymbol{W} \bar{\boldsymbol{y}}$ defines an inner product on $\mathbb{F}^n$.
(c) Show that every inner product on $\mathbb{F}^n$ is obtained as in (b) for some Hermitian matrix $\boldsymbol{W}$ with positive eigenvalues.

E2.16 Given a symmetric linear map A on a finite-dimensional $\mathbb{R}$-inner product space $(\mathsf{V}, \langle\langle \cdot, \cdot \rangle\rangle)$, let $\{v_1, \dots, v_n\}$ be an orthonormal basis of eigenvectors of A with $\{\lambda_1, \dots, v_n\}$ the corresponding eigenvalues.
(a) Show that

$$A = \sum_{j=1}^{n} \lambda_j v_j \otimes v_j^* \tag{E2.1}$$

(by $v_j^* \in V^*$, $j \in \{1, \dots, n\}$, we mean the dual of the linear map from $\mathbb{R}$ to V defined by $a \mapsto a v_j$).
(b) Give a matrix version of (E2.1) in the case that $\mathsf{V} = \mathbb{R}^n$, $\langle\langle \cdot, \cdot \rangle\rangle$ is the standard inner product on $\mathbb{R}^n$, and (by implication) A is defined by a symmetric matrix.

E2.17 Let $(\mathsf{V}, \langle\langle \cdot, \cdot \rangle\rangle)$ be a $\mathbb{R}$-inner product space with $A \in L(\mathsf{V}; \mathsf{V})$ a symmetric linear map. Show that $\ker(A) = \mathrm{image}(A)^\perp$.

E2.18 Given two positive-definite symmetric bilinear maps $\mathbb{G}_1$ and $\mathbb{G}_2$ on $\mathbb{R}^n$, show that there exist two constants $0 < \lambda_1 \leq \lambda_2$ such that, for all $\boldsymbol{x} \in \mathbb{R}^n$, we have

$$\lambda_1 \mathbb{G}_1(\boldsymbol{x}, \boldsymbol{x}) \leq \mathbb{G}_2(\boldsymbol{x}, \boldsymbol{x}) \leq \lambda_2 \mathbb{G}_1(\boldsymbol{x}, \boldsymbol{x}).$$

E2.19 Recall the list of seven definiteness properties for symmetric bilinear maps from Definition 2.44. Show that this list provides a complete classification of $\Sigma_2(\mathsf{V})$ if V is finite-dimensional.

E2.20 Let V be a finite-dimensional $\mathbb{R}$-vector space and let $B \in \Sigma_2(\mathsf{V})$.
(a) Show that B is positive-semidefinite if and only if $\mathrm{sig}(B) + \mathrm{ind}(B) = \mathrm{rank}(B)$.
(b) Show that B is positive-definite if and only if $\mathrm{sig}(B) + \mathrm{ind}(B) = n$.
(c) Show that B is negative-semidefinite if and only if $\mathrm{ind}(B) = \mathrm{rank}(B)$.
(d) Show that B is negative-definite if and only if $\mathrm{ind}(B) = n$.

E2.21 Let V be an n-dimensional $\mathbb{R}$-vector space and let $B \in \Sigma_2(\mathsf{V})$.
(a) Show that $\dim(\mathrm{image}(B^\flat)) = \mathrm{rank}(B)$.
(b) Show that $\dim(\ker(B^\flat)) = n - \mathrm{rank}(B)$.

E2.22 Let V be a finite-dimensional $\mathbb{R}$-vector space, let $\{e_1, \dots, e_n\}$ be a basis for V, and let $B \in \Sigma_2(\mathsf{V})$. Let $\{\lambda_1, \dots, \lambda_n\}$ be the eigenvalues of the matrix $[B]$.
(a) Show that $\mathrm{ind}(B)$ is equal to the number of negative eigenvalues of $[B]$.
(b) Show that $\mathrm{sig}(B) + \mathrm{ind}(B)$ is equal to the number of positive eigenvalues of $[B]$.
(c) Show that $n - \mathrm{rank}(B)$ is equal to the number of zero eigenvalues of $[B]$.
(d) Are the numerical values (apart from their sign) of the eigenvalues of B meaningful?

E2.23 For V a finite-dimensional $\mathbb{R}$-vector space, show that the following pairs of vector spaces are naturally isomorphic by providing explicit isomorphisms:
(a) $T_0^1(\mathsf{V})$ and V;

(b) $T_1^0(\mathsf{V})$ and V^*;
(c) $T_1^1(\mathsf{V})$ and $L(\mathsf{V};\mathsf{V})$;
(d) $T_2^0(\mathsf{V})$ and $L(\mathsf{V};\mathsf{V}^*)$;
(e) $T_0^2(\mathsf{V})$ and $L(\mathsf{V}^*;\mathsf{V})$.

E2.24 If the components of $t \in T_2^0(\mathbb{R}^2)$ in the standard basis are

$$\begin{bmatrix} 1 & 2 \\ 3 & 4 \end{bmatrix},$$

what is the representation of t as a sum of tensor products?

E2.25 Prove Proposition 2.49.

E2.26 Let $A \in L(\mathbb{R}^2;\mathbb{R}^2)$ have matrix representation

$$[A] = \begin{bmatrix} 1 & 2 \\ 0 & 2 \end{bmatrix}$$

in the standard basis. Next, let

$$\boldsymbol{T} = \begin{bmatrix} 3 & 0 \\ 1 & 2 \end{bmatrix}.$$

Answer the following questions.
(a) If $\boldsymbol{T}$ is the matrix representation of a $(1,1)$-tensor t, what are the push-forward and pull-back of t by A?
(b) If $\boldsymbol{T}$ is the matrix representation of a $(2,0)$-tensor t, what are the push-forward and pull-back of t by A?
(c) If $\boldsymbol{T}$ is the matrix representation of a $(0,2)$-tensor t, what are the push-forward and pull-back of t by A?

E2.27 For each of the following subsets of $\mathbb{R}^2$, determine its convex hull, its affine hull, and the subspace generated by the set:
(a) $\{(1,0),(0,1)\}$;
(b) $\operatorname{span}_{\mathbb{R}}\{(1,0)\}$;
(c) $\{(1,1)+a(1,-1) \mid a \in \mathbb{R}\}$;
(d) $\{(0,1),(-\frac{\sqrt{3}}{2},-\frac{1}{2}),(\frac{\sqrt{3}}{2},-\frac{1}{2})\}$.

E2.28 Let S be a subset of a $\mathbb{R}$-vector space V and let $\operatorname{aff}(S)$ be the affine hull of S. Since $\operatorname{aff}(S)$ is an affine subspace, there exists a subspace $\mathsf{U}_S \subset \mathsf{V}$ such that $\operatorname{aff}(S) = \{v+u \mid u \in \mathsf{U}_S\}$ for some $v \in V$. Show that U_S is generated by the vectors of the form

$$\sum_{j=1}^{k} \lambda_j v_j, \qquad \lambda_1,\dots,\lambda_k \in \mathbb{R}, \; \sum_{j=1}^{k} \lambda_j = 0, \; v_1,\dots,v_k \in S, \; k \in \mathbb{N}.$$

E2.29 Let V be a $\mathbb{R}$-vector space. Show that $\mathsf{A} \subset \mathsf{V}$ is an affine subspace if and only if there exists a subspace U of V and a vector $[v] \in \mathsf{V}/\mathsf{U}$ for which $\mathsf{A} = \pi_{\mathsf{U}}^{-1}([v])$, where $\pi_{\mathsf{U}}\colon \mathsf{V} \to \mathsf{V}/\mathsf{U}$ denotes the canonical projection.

3

Differential geometry

As can be seen from the title of this book, our treatment of mechanical control systems is geometric, more precisely, differential geometric. In this chapter we review the essential differential geometric tools used in the book. Since a thorough treatment of all parts of the subject necessary for our objectives would be even more lengthy than what we currently have, we essentially present a list of definitions and facts that follow from these definitions. The treatment we give, while comparatively brief, is not as fast-paced as was the case in Chapter 2. In this chapter we spend more time on examples that illustrate the main ideas. Nevertheless, a reader who knows no differential geometry can expect to invest some time in learning the subject, and might find the review in this chapter too hasty. In this case, additional references (see below) will be helpful in providing context and further examples. Readers having a passing acquaintance with differential geometry may wish to quickly scan this chapter to see if there are any major topics that are unfamiliar to them.

We should say a few words about why we think the differential geometric point of view is useful, despite the effort required to properly take advantage of it. The basic object of differential geometry is a manifold, which is, roughly speaking, a set that locally "looks like" an open subset of Euclidean space. By making precise the notion of "looks like," one is able to make sensible on manifolds many of the familiar constructions from analysis in Euclidean space. There is, however, one essential difference, and that is the desirability of having everything one uses be "coordinate-invariant," or "intrinsic." For a given manifold, there are many ways of making it locally "look like" an open subset of Euclidean space, and one does not want to rely on concepts that are dependent on a specific choice. This is important not only for aesthetic reasons, but for physical reasons. To wit, a physical problem typically comes equipped with some physical data, and normally, if the modeling is honestly carried out, *not* included in this physical data is a specific set of coordinates. A certain user may have a bias towards certain coordinates, but it is inappropriate to transfer this bias to the actual physics of the problem. Note that this does *not* mean that one should not use coordinates! Indeed, in most of the examples we give, a specific set of coordinates is fixed once and for all. However, we are careful that we only use concepts that are coordinate-invariant, albeit represented in the coordinates that we have chosen. Simply put, *one should deal with concepts that are coordinate-invariant*. If one chooses to properly use differential geometry, then coordinate-invariance is a consequence of this choice. This is one of the principal motivations for our reliance on a differential geometric treatment. Much of our modeling in Chapters 4 and 5 consists of translating physical quantities to differential geometric quantities, whereupon they then fit into a precise mathematical framework.

Readers who have no background on differential geometry might find [Boothby 1986] a well-paced introduction. An excellent thorough account of basic differential geometric ideas, presented in a spirit similar to ours here, is the text of Abraham, Marsden, and Ratiu [1988]. Much of our notation follows this book. The one topic needed by us, and not covered in [Abraham, Marsden, and Ratiu 1988], is affine differential geometry. However, it is possible to give a short, self-contained treatment, and we do this in Section 3.8. A good basic reference for Riemannian geometry, itself a subset of affine differential geometry, is the text of Do Carmo [1992]. A thorough account of affine differential geometry, certainly more than ample for our needs, is the classic two volumes of Kobayashi and Nomizu [1963]. The reader familiar with these volumes, particularly the first, can happily proceed directly to the control theoretic material in the book. Other references for various aspects of differential geometry include [Gallot, Hulin, and Lafontaine 1987, Helgason 1978, Klingenberg 1982, Lang 1995, Warner 1971], but there are many more such references. We encourage the reader to explore these, and find the ones that are most suitable for them.

3.1 The prelude to differential geometry

Before we can proceed with differential geometry proper, we need to discuss some preliminary ideas concerning topology and elementary analysis in Euclidean space. Our discussion here is a condensed version of what may be found in the preliminary chapters of [Abraham, Marsden, and Ratiu 1988], and in the undergraduate texts [Marsden and Hoffman 1993, Mendelson 1990].

In our brief about topology, we review definitions and properties of open sets, sequences, and continuous maps. In the calculus section we review operator norms, differentiability, and Taylor series. One important notion is the classification of functions based on their differentiability properties. We will be fairly careful in the chapter to keep track of the degree of differentiability of the objects we consider. Readers may find some of the comments in Section 3.9.2 helpful in understanding why we do this.

3.1.1 Topology

The reader is probably familiar with the notion of open and closed sets in $\mathbb{R}^n$, and with what is meant by convergence, continuity, and the like. In this section we review an abstract setting designed to generalize these notions to spaces more general than $\mathbb{R}^n$, since most of the time we will be working in such settings.

The basic object in the discussion is the following.

Definition 3.1 (Topological space). A ***topological space*** is a pair $(S, \mathscr{O})$, where S is a set and $\mathscr{O} \subset \mathbf{2}^S$ is a collection of subsets, called ***open sets***, satisfying

(i) $\emptyset \in \mathscr{O}$ and $S \in \mathscr{O}$,
(ii) if A is an arbitrary index set and $\{\mathcal{O}_a\}_{a \in A} \subset \mathscr{O}$ is an arbitrary collection of open sets, then $\cup_{a \in A} \mathcal{O}_a \in \mathscr{O}$,
(iii) if $\mathcal{O}_1, \mathcal{O}_2 \in \mathscr{O}$, then $\mathcal{O}_1 \cap \mathcal{O}_2 \in \mathscr{O}$.

Given a set S, a collection of subsets $\mathscr{O}$, for which $(S, \mathscr{O})$ is a topological space, is a ***topology*** on S. •

Note that we get to *declare* what we mean by open sets in a topological space, provided that they obey the stated rules. This allows some odd notions of open sets. We do not spend much time talking about these, since our open sets will generally behave in a manner with which one is accustomed. However, just for fun, here are two topologies possessed by every set.

Examples 3.2. Let S be an arbitrary set.

1. The ***trivial topology*** on S is given by $\mathscr{O}_{\text{trivial}} = \{\emptyset, S\}$.
2. The ***discrete topology*** on S is given by $\mathscr{O}_{\text{discrete}} = \mathbf{2}^S$. •

From the bare notion of a topological space follows a host of other notions. Let us simply list these, referring to the references for details. Many of these will be used only infrequently in the text, so we recommend that readers unfamiliar with these concepts refer to them when needed, at least on a first read.

Definition 3.3. Let S be a set with a topology $\mathscr{O}$.

(i) A subset $C \subset S$ is ***closed*** if $S \setminus C$ is open.
(ii) For $x \in S$, a ***neighborhood*** of x is an open set $\mathcal{U} \in \mathscr{O}$ for which $x \in \mathcal{U}$.
(iii) If $A \subset S$, then the ***interior*** of A is the subset of A defined by

$$\operatorname{int}(A) = \bigcup \{\mathcal{O} \in \mathscr{O} \mid \mathcal{O} \subset A\}.$$

(iv) If $A \subset S$, then a point $x \in S$ is a ***limit point*** of A if, for any neighborhood $\mathcal{U}$ of x, the set $\mathcal{U} \cap A$ is nonempty.
(v) If $A \subset S$, then the ***closure*** of A is the subset of S defined by

$$\operatorname{cl}(A) = \bigcap \{C \mid C \text{ is closed and } A \subset C\}.$$

(vi) If $A \subset S$, then the ***boundary*** of A is the subset of S defined by $\mathrm{bd}(A) = \mathrm{cl}(A) \cap \mathrm{cl}(S \setminus A)$.

(vii) A subset $\mathscr{B} \subset \mathscr{O}$ is a ***basis*** for $\mathscr{O}$ if, for every $\mathcal{O} \in \mathscr{O}$, there exist an index set A and a collection of sets $\{B_a\}_{a\in A} \subset \mathscr{B}$ such that $\mathcal{O} = \cup_{a\in A} B_a$. We say in this case that $\mathscr{B}$ ***generates*** $\mathscr{O}$.

(viii) A ***cover*** of $(S, \mathscr{O})$ is a subset $\{\mathcal{O}_a\}_{a\in A} \subset \mathscr{O}$ with the property that $S = \cup_{a\in A}\mathcal{O}_a$.

(ix) A cover $\{\tilde{\mathcal{O}}_{\tilde{a}}\}_{\tilde{a}\in\tilde{A}}$ is a ***refinement*** of a cover $\{\mathcal{O}_a\}_{a\in A}$ if, for every $a \in A$, there exists $\tilde{a} \in \tilde{A}$ such that $\tilde{\mathcal{O}}_{\tilde{a}} \subset \mathcal{O}_a$.

(x) A subset $\{\mathcal{O}_a\}_{a\in A} \subset \mathscr{O}$ is ***locally finite*** if, for each $x \in S$, there is a neighborhood $\mathcal{U}$ such that the set $\{a \in A \mid \mathcal{U} \cap \mathcal{O}_a \neq \emptyset\}$ is finite.

(xi) If $A \subset S$, then one defines a topology on A by $\{A \cap \mathcal{O} \mid \mathcal{O} \in \mathscr{O}\}$. This is called the ***subspace topology***.

(xii) If $B \subset A \subset S$, then $\mathrm{int}_A(B)$ denotes the interior of B in the subspace topology on A. •

Let us make concrete the preceding abstraction with something familiar.

Example 3.4. Let $(\mathsf{V}, \|\cdot\|)$ be a normed vector space. The ***open ball*** of radius r centered at $v_0 \in \mathsf{V}$ is given by

$$B_r(v_0) = \{v \in \mathsf{V} \mid \|v - v_0\| < r\}.$$

The set of all open balls generates a topology on V called the ***norm topology***. In the case in which $(\mathsf{V}, \|\cdot\|) = (\mathbb{R}^n, \|\cdot\|_{\mathbb{R}^n})$, the resulting topology is the ***standard topology*** for $\mathbb{R}^n$. One can verify that the open sets in this topology are the usual ones. That is, a set $\mathcal{U} \subset \mathbb{R}^n$ is open in the standard topology if and only if, for each $\boldsymbol{x} \in \mathcal{U}$, there exists $\varepsilon > 0$ such that $B_\varepsilon(\boldsymbol{x}) \subset \mathcal{U}$.

Let us also define the ***closed ball*** of radius r centered at $v_0 \in \mathsf{V}$ to be

$$\bar{B}_r(v_0) = \{v \in \mathsf{V} \mid \|v - v_0\| \leq r\}.$$

In the case in which $(\mathsf{V}, \|\cdot\|) = (\mathbb{R}^n, \|\cdot\|_{\mathbb{R}^n})$, one may verify that this set is indeed closed in the standard topology. A point $\boldsymbol{x} \in \mathbb{R}^n$ is a limit point of a set $A \subset \mathbb{R}^n$ if and only if, for every $\varepsilon > 0$, the set $B_\varepsilon(\boldsymbol{x}) \cap A$ is nonempty.

Now let us consider $\mathsf{V} = \mathbb{R}^3$ and let $\mathsf{W} = \mathrm{span}_{\mathbb{R}}\{(1,0,0),(0,1,0)\}$. Since the open balls generate the standard topology on $\mathbb{R}^3$, the sets

$$\{B_r(\boldsymbol{x}) \cap \mathsf{W} \mid r > 0,\ \boldsymbol{x} \in \mathsf{W}\}$$

generate the subspace topology on W. However, thinking of W as being essentially $\mathbb{R}^2$, this is also the standard topology for W. •

By itself, a topological space is a rather unstructured object. One normally deals with topological spaces having certain pleasant attributes. Here we merely give a list of some such attributes, offering little in the way of interpretations of what they mean. These definitions can be bypassed initially, and then referred to when they subsequently come up in the text. To fully understand these concepts, the reader should be prepared to allocate some time to read the supplementary texts mentioned at the beginning of the section.

Definition 3.5. Let $(S, \mathscr{O})$ be a topological space.

(i) $(S, \mathscr{O})$ is ***first-countable*** if, for every $x \in S$, there exists a sequence of neighborhoods $\{\mathcal{U}_j\}_{j\in\mathbb{N}}$ of x such that, for any neighborhood $\mathcal{U}$ of x, there exists $j \in \mathbb{N}$ with $\mathcal{U}_j \subset \mathcal{U}$.

(ii) $(S, \mathscr{O})$ is ***second-countable*** if it has a countable basis.

(iii) $(S, \mathscr{O})$ is ***compact*** if, for every collection $\{\mathcal{O}_a\}_{a\in A}$ of open sets for which $S = \cup_{a\in A}\mathcal{O}_a$, there exists a finite subset $\{a_1, \dots, a_k\} \subset A$ such that $S = \cup_{j=1}^k \mathcal{O}_{a_j}$.

(iv) A subset $A \subset S$ is ***compact*** if A is compact with the subspace topology.

(v) An open set $\mathcal{U} \subset S$ is ***relatively compact*** if its closure is compact.

(vi) $(S, \mathscr{O})$ is ***Hausdorff*** if, for each distinct $x_1, x_2 \in S$, there exist neighborhoods $\mathcal{U}_1$ of x_1 and $\mathcal{U}_2$ of x_2 such that $\mathcal{U}_1 \cap \mathcal{U}_2 = \emptyset$.

(vii) $(S, \mathscr{O})$ is ***connected*** if, when $A \in \mathbf{2}^S$ has the property that A is both open and closed, then $A \in \{\emptyset, S\}$.

(viii) If $(S, \mathscr{O})$ is not connected, then it is ***disconnected***, and one can show that it is a disjoint union of connected sets, each of which is called a ***connected component***.

(ix) $(S, \mathscr{O})$ is ***paracompact*** if it is Hausdorff and if every cover possesses a locally finite refinement.

(x) A subset $A \subset S$ of a topological space is ***dense*** if $\mathrm{cl}(A) = S$. •

Again, these notions are best illustrated in the context of a familiar example.

Example 3.6 (Example 3.4 cont'd). Consider $\mathbb{R}^n$ with its standard topology. It is possible to show that the open balls $B_r(\boldsymbol{x})$ with $r \in \mathbb{Q}$ and with $\boldsymbol{x}$ having rational coordinates form a countable basis for the standard topology. Thus $\mathbb{R}^n$ is second-countable, using the fact that a countable union of countable sets is again countable. If $\boldsymbol{x}_1, \boldsymbol{x}_2 \in \mathbb{R}^n$ are distinct and if $\varepsilon < \frac{1}{2}\|\boldsymbol{x}_1 - \boldsymbol{x}_2\|_{\mathbb{R}^n}$, then one can readily see that $B_\varepsilon(\boldsymbol{x}_1) \cap B_\varepsilon(\boldsymbol{x}_2) = \emptyset$. Thus the standard topology on $\mathbb{R}^n$ is Hausdorff. It is also fairly easy to characterize the compact subsets of $\mathbb{R}^n$ with the standard topology. Let us say that a subset $A \subset \mathbb{R}^n$ is ***bounded*** if there exists $R > 0$ such that $A \subset B_R(\mathbf{0})$. The Heine–Borel Theorem then says that a subset $A \subset \mathbb{R}^n$ is compact if and only if A is closed and bounded [Abraham, Marsden, and Ratiu 1988]. This characterization of compactness does not hold in general, even for normed vector spaces. In general, one must revert to the definition of compactness. •

With the notion of a topological space, one can define the concepts of convergence and continuity. First we consider convergence.

Definition 3.7. Let $(S, \mathscr{O})$ be a topological space.

(i) A sequence $\{x_j\}_{j\in\mathbb{N}}$ ***converges*** to $x \in S$ in the topology $\mathscr{O}$ if, for each neighborhood $\mathcal{U}$ of x, there exists $N \in \mathbb{N}$ such that $x_j \in \mathcal{U}$ for each $j \geq N$. If $\{x_j\}_{j\in\mathbb{N}}$ converges to x, we may write $\lim_{j\to+\infty} x_j = x$.

(ii) A sequence $\{x_j\}_{j\in\mathbb{N}}$ is ***convergent*** if it converges to some point in S. •

Sometimes it is convenient to discuss limits in situations more general that we just gave for sequences. Let us provide one commonly encountered such situation. Suppose that $(S, \mathscr{O}_S)$ and $(T, \mathscr{O}_T)$ are topological spaces, and let $f\colon S \to T$. For $x_0 \in S$, we wish to make sense of the expression $\lim_{x\to x_0} f(x)$. We do this by requiring that, for every sequence $\{x_j\}_{j\in\mathbb{N}}$ converging to x_0, the sequence $\{f(x_j)\}_{j\in\mathbb{N}}$ converges, and the limit is independent of the sequence $\{x_j\}_{j\in\mathbb{N}}$. The value of this limit is, by definition, $\lim_{x\to x_0} f(x)$.

Next we talk about the important concept of continuity.

Definition 3.8. Let $(S, \mathscr{O}_S)$ and $(T, \mathscr{O}_T)$ be topological spaces, and let $f\colon S \to T$ be a map.

(i) The map f is ***continuous at*** $\boldsymbol{x_0}$ for $x_0 \in S$ if, for each neighborhood $\mathcal{V}$ of $f(x_0)$, there exists a neighborhood $\mathcal{U}$ of x_0 such that $f(\mathcal{U}) \subset \mathcal{V}$.

(ii) If f is continuous at each $x \in S$, then it is ***continuous***. One may verify that f is continuous if and only if, for each $\mathcal{O} \in \mathscr{O}_T$, $f^{-1}(\mathcal{O}) \in \mathscr{O}_S$.

(iii) If f is a bijection, is continuous, and if its inverse is also continuous, then it is a ***homeomorphism***. •

Concerning maps, let us give a few other properties that will come up in the book.

Definition 3.9. Let $(S, \mathscr{O}_S)$ and $(T, \mathscr{O}_T)$ be topological spaces, and let $f\colon S \to T$ be a map.

(i) The map f is ***proper*** if $f^{-1}(K)$ is compact for every compact subset $K \subset T$.

(ii) The map f is ***locally constant*** if, for every connected subset $A \subset S$, $f|A$ is constant. •

These notions have the usual interpretation in the familiar context.

Example 3.10 (Example 3.4 cont'd). A sequence $\{\boldsymbol{x}_j\}_{j\in\mathbb{N}}$ in $\mathbb{R}^n$ converges to $\boldsymbol{x} \in \mathbb{R}^n$ if and only if, for each $\varepsilon > 0$, there exists $N \in \mathbb{N}$ such that $\|\boldsymbol{x} - \boldsymbol{x}_j\|_{\mathbb{R}^n} < \varepsilon$ for all $j \geq N$. This is the usual notion of "the tail of the sequence getting arbitrarily close to $\boldsymbol{x}$." A map $f\colon \mathbb{R}^n \to \mathbb{R}^m$ is continuous at $\boldsymbol{x}_0$ if and only if, for each $\varepsilon > 0$, there exists $\delta > 0$ such that $\|\boldsymbol{x} - \boldsymbol{x}_0\|_{\mathbb{R}^n} < \delta$ implies that $\|f(\boldsymbol{x}) - f(\boldsymbol{x}_0)\|_{\mathbb{R}^n} < \varepsilon$. This is not quite an obvious consequence of the basic definition of continuity, and we encourage the reader to make this connection.

The reader may give a characterization of proper functions on $\mathbb{R}^n$ in Exercise E3.5. •

Remark 3.11 (Convergence of sums). We shall often require the notion of convergence, not of a sequence, but of a sum $\sum_{j=1}^{+\infty} \boldsymbol{x}_j$, where $\boldsymbol{x}_j \in \mathbb{R}^n$, $j \in \mathbb{N}$. Corresponding to a sum of this form, one defines the sequence of ***partial sums***, denoted by $\{\boldsymbol{s}_k\}_{k\in\mathbb{N}}$, to be $\boldsymbol{s}_k = \sum_{j=1}^{k} \boldsymbol{x}_j$. Convergence of the sum is then, by definition, convergence of the sequence of partial sums. For sums, one also has an additional notion of convergence. A sum $\sum_{j=1}^{+\infty} \boldsymbol{x}_j$ ***converges***

absolutely if the sum $\sum_{j=1}^{+\infty}\|\boldsymbol{x}_j\|_{\mathbb{R}^n}$ converges. One can show fairly easily that a sum that is absolutely convergent is also convergent [Marsden and Hoffman 1993]. The converse is generally false, as there are alternating sums of real numbers that converge, but not absolutely. •

Remarks 3.12. Some of the various notions stated above are related in useful ways. Rather than state these relationships formally, let us merely list the ones that will be helpful to us.

1. If $(S, \mathscr{O})$ is a compact topological space with $f\colon S \to \mathbb{R}$ a continuous function (with respect to the standard topology on $\mathbb{R}$, of course), then f is bounded. That is, there exists $M > 0$ such that, for each $x \in S$, $|f(x)| < M$.
2. For $(S, \mathscr{O})$ a compact topological space and $f\colon S \to \mathbb{R}$ a continuous function, let $f_{\max} = \sup(\text{image}(f))$. Then there exists a point $x_0 \in S$ for which $f(x_0) = f_{\max}$.
3. If $(S, \mathscr{O}_S)$ and $(T, \mathscr{O}_T)$ are topological spaces with $(S, \mathscr{O}_S)$ first-countable, then a map $f\colon S \to T$ is continuous if and only if, for every convergent sequence $\{x_j\}_{j\in\mathbb{N}}$, the sequence $\{f(x_j)\}_{j\in\mathbb{N}}$ is also convergent.
4. If $(S, \mathscr{O})$ is compact, first-countable, and Hausdorff, and if $\{x_j\}_{j\in\mathbb{N}}$ is a sequence in S, then there exists a convergent subsequence $\{x_{j_k}\}_{k\in\mathbb{N}}$. This is the ***Bolzano–Weierstrass Theorem*** (see [Abraham, Marsden, and Ratiu 1988, Marsden and Hoffman 1993]). •

Remark 3.13. Continuity also allows a refined notion of connectedness. A topological space $(S, \mathscr{O})$ is ***path connected*** if, for $x_1, x_2 \in S$, there exists a continuous map $\gamma\colon [0,1] \to S$ for which $\gamma(0) = x_1$ and $\gamma(1) = x_2$. The space $(S, \mathscr{O})$ is ***locally path connected*** if, for each $x \in S$ and for each neighborhood $\mathcal{U}$ of x, there exists a neighborhood $\mathcal{V}$ of x such that, if $x_1, x_2 \in \mathcal{V}$, then there exists a continuous map $\gamma\colon [0,1] \to \mathcal{U}$ for which $\gamma(0) = x_1$ and $\gamma(1) = x_2$. It is true that, if $(S, \mathscr{O})$ is connected and locally path connected, then it is path connected [Abraham, Marsden, and Ratiu 1988]. •

The final topological notion we discuss concerns the global topology of a space. Let $(S, \mathscr{O})$ be a topological space. A ***loop*** in S is a continuous map $\gamma\colon [0,1] \to S$ having the property that $\gamma(0) = \gamma(1)$. If $x = \gamma(0) = \gamma(1)$, then x is the ***base point*** for the loop γ. The ***trivial loop at*** $\boldsymbol{x}$ is the loop $\gamma_x\colon t \mapsto x$. Two loops γ_1 and γ_2 with the same base point x are ***homotopic*** if there exists a continuous map $H\colon [0,1] \times [0,1] \to S$ with the following properties:

1. $H(t, 0) = \gamma_1(t)$ for all $t \in [0,1]$;
2. $H(t, 1) = \gamma_2(t)$ for all $t \in [0,1]$;
3. for each $s \in [0,1]$, the map $t \mapsto H(t,s)$ is a loop with base point x.

The topological space $(S, \mathscr{O})$ is then ***simply connected*** if every loop γ is homotopic to the trivial loop γ_x, where x is the base point for γ. Intuitively, a topological space is simply connected if it has "no holes," the idea being that a loop around a hole will not be homotopic to a trivial loop.

3.1.2 Calculus in $\mathbb{R}^n$

We next use some of the ideas of the preceding section as a basis for discussions of calculus in Euclidean space. Throughout this section we consider the standard topology on $\mathbb{R}^n$ induced by the standard norm $\|\cdot\|_{\mathbb{R}^n}$.

First we consider the properties of linear maps between Euclidean spaces; see [Abraham, Marsden, and Ratiu 1988].

Proposition 3.14 (Continuity and boundedness of linear maps). *For $A \in L(\mathbb{R}^n; \mathbb{R}^m)$, the following statements hold and are equivalent:*

(i) A is continuous with respect to the standard topologies on $\mathbb{R}^n$ and $\mathbb{R}^m$;
(ii) A is continuous at $\mathbf{0} \in \mathbb{R}^m$;
(iii) there exists $M > 0$ such that, for each $\boldsymbol{x} \in \mathbb{R}^n$, $\|A(\boldsymbol{x})\|_{\mathbb{R}^m} \leq M\|\boldsymbol{x}\|_{\mathbb{R}^n}$; that is to say, A is ***bounded****.*

Now let us look at some structure of the set of all linear maps.

Definition 3.15 (Operator norm). For $A \in L(\mathbb{R}^n; \mathbb{R}^m)$, the expression

$$\|A\|_{\mathbb{R}^n,\mathbb{R}^m} = \sup\{\, \|A(\boldsymbol{x})\|_{\mathbb{R}^m} \mid \|\boldsymbol{x}\|_{\mathbb{R}^n} = 1\},$$

is the ***operator norm*** of A. •

One can verify that the operator norm does indeed define a norm on the vector space $L(\mathbb{R}^n; \mathbb{R}^m)$ (Exercise E3.7). This then defines a topology on $L(\mathbb{R}^n; \mathbb{R}^m)$ generated by open balls using the operator norm. Thus $L(\mathbb{R}^n; \mathbb{R}^m)$ becomes a candidate for being the domain or codomain of a continuous map. More generally, one can verify that, if $A \in L^k(\mathbb{R}^n; \mathbb{R}^m)$, then

$$\|A\|_{(\mathbb{R}^n)^k,\mathbb{R}^m} = \sup\{\, \|A(\boldsymbol{x}_1, \dots, \boldsymbol{x}_k)\|_{\mathbb{R}^m} \mid \|\boldsymbol{x}_1\|_{\mathbb{R}^n} = 1, \dots, \|\boldsymbol{x}_k\|_{\mathbb{R}^n} = 1\} \tag{3.1}$$

defines a norm on $L^k(\mathbb{R}^n; \mathbb{R}^m)$.

With these properties of linear and multilinear maps it is possible to discuss differentiation in a systematic manner. Let $\mathcal{U} \subset \mathbb{R}^n$ be an open set and let $f\colon \mathcal{U} \to \mathbb{R}^m$ be a map. We wish to determine what it means for f to be differentiable at $\boldsymbol{x}_0 \in \mathcal{U}$. The following lemma gets us started.

Lemma 3.16. *Let $\mathcal{U} \subset \mathbb{R}^n$ be an open set and let $f\colon \mathcal{U} \to \mathbb{R}^m$ be a map. For $\boldsymbol{x}_0 \in \mathcal{U}$, there exists at most one $A \in L(\mathbb{R}^n; \mathbb{R}^m)$ such that*

$$\lim_{\boldsymbol{x} \to \boldsymbol{x}_0} \frac{\|f(\boldsymbol{x}) - (f(\boldsymbol{x}_0) + A(\boldsymbol{x} - \boldsymbol{x}_0))\|_{\mathbb{R}^m}}{\|\boldsymbol{x} - \boldsymbol{x}_0\|_{\mathbb{R}^n}} = 0. \tag{3.2}$$

Proof. Suppose there are two such maps A_1 and A_2. For any $\boldsymbol{x} \in \mathcal{U}$, we may write $\boldsymbol{x} = \boldsymbol{x}_0 + a\boldsymbol{v}$ for some $a > 0$ and $\boldsymbol{v} \in \mathbb{R}^n$ such that $\|\boldsymbol{v}\|_{\mathbb{R}^n} = 1$. We compute

$$\|A_1(\boldsymbol{v}) - A_2(\boldsymbol{v})\|_{\mathbb{R}^m} = \frac{\|A_1(\boldsymbol{x}-\boldsymbol{x}_0) - A_2(\boldsymbol{x}-\boldsymbol{x}_0)\|_{\mathbb{R}^m}}{\|\boldsymbol{x}-\boldsymbol{x}_0\|_{\mathbb{R}^n}}$$
$$= \frac{\|-f(\boldsymbol{x}) + f(\boldsymbol{x}_0) + A_1(\boldsymbol{x}-\boldsymbol{x}_0) + f(\boldsymbol{x}) - f(\boldsymbol{x}_0) - A_2(\boldsymbol{x}-\boldsymbol{x}_0)\|_{\mathbb{R}^m}}{\|\boldsymbol{x}-\boldsymbol{x}_0\|_{\mathbb{R}^n}}$$
$$\leq \frac{\|f(\boldsymbol{x}) - f(\boldsymbol{x}_0) - A_1(\boldsymbol{x}-\boldsymbol{x}_0)\|_{\mathbb{R}^m}}{\|\boldsymbol{x}-\boldsymbol{x}_0\|_{\mathbb{R}^n}} + \frac{\|f(\boldsymbol{x}) - f(\boldsymbol{x}_0) - A_2(\boldsymbol{x}-\boldsymbol{x}_0)\|_{\mathbb{R}^m}}{\|\boldsymbol{x}-\boldsymbol{x}_0\|_{\mathbb{R}^n}}.$$

Since A_1 and A_2 both satisfy (3.2), as we let $\boldsymbol{x} \to \boldsymbol{x}_0$ the right-hand side goes to zero showing that $\|A_1(\boldsymbol{v}) - A_2(\boldsymbol{v})\|_{\mathbb{R}^m} = \|(A_1 - A_2)(\boldsymbol{v})\|_{\mathbb{R}^m} = 0$ for every $\boldsymbol{v}$ with $\|\boldsymbol{v}\|_{\mathbb{R}^n} = 1$. Thus $A_1 - A_2$ is the trivial map sending any vector to zero, or equivalently $A_1 = A_2$. ■

If a map A exists satisfying (3.2), then the idea is that the map $\boldsymbol{x} \mapsto f(\boldsymbol{x}_0) + A(\boldsymbol{x}-\boldsymbol{x}_0)$ provides a linear approximation for f at $\boldsymbol{x}_0$. It is convenient to introduce notation describing approximations of various forms. When we write "$o(\|\boldsymbol{x}\|^k_{\mathbb{R}^n})$," we mean a continuous map from a neighborhood of $\boldsymbol{0} \in \mathbb{R}^n$ to $\mathbb{R}^m$ that satisfies

$$\lim_{\boldsymbol{x}\to\boldsymbol{0}} \frac{o(\|\boldsymbol{x}\|^k_{\mathbb{R}^n})}{\|\boldsymbol{x}\|^k_{\mathbb{R}^n}} = 0.$$

With this notation, Lemma 3.16 says that there exists at most one $A \in L(\mathbb{R}^n; \mathbb{R}^m)$ such that

$$f(\boldsymbol{x}) = f(\boldsymbol{x}_0) + A(\boldsymbol{x}-\boldsymbol{x}_0) + o(\|\boldsymbol{x}-\boldsymbol{x}_0\|_{\mathbb{R}^n}).$$

We shall also use the notation $O(\|\boldsymbol{x}\|^k_{\mathbb{R}^n})$, which stands for a map defined in the neighborhood of $\boldsymbol{0} \in \mathbb{R}^n$, taking values in $\mathbb{R}^m$, and satisfying

$$\|O(\|\boldsymbol{x}\|^k_{\mathbb{R}^n})\|_{\mathbb{R}^m} \leq M\|\boldsymbol{x}\|^k_{\mathbb{R}^n}$$

for some $M > 0$. The symbols "o" and "O" are called the ***Landau symbols***, and one normally pronounces these as "little oh" and "big oh," respectively.

Let us define the derivative and differentiability.

Definition 3.17 (Differentiable maps). Let $\mathcal{U} \subset \mathbb{R}^n$ be an open set and let $f\colon \mathcal{U} \to \mathbb{R}^m$ be a map.

(i) If there exists $A \in L(\mathbb{R}^n; \mathbb{R}^m)$ such that (3.2) holds, then we call A the ***derivative*** of f at $\boldsymbol{x}_0$, and we say that f is ***differentiable*** at $\boldsymbol{x}_0$.

(ii) If f is differentiable at $\boldsymbol{x}_0$, then we denote its derivative by $\boldsymbol{D}f(\boldsymbol{x}_0)$.

(iii) If f is differentiable at each $\boldsymbol{x} \in \mathcal{U}$, then f is ***differentiable***.

(iv) If $\boldsymbol{D}f\colon \mathcal{U} \to L(\mathbb{R}^n; \mathbb{R}^m)$ is continuous, then f is ***continuously differentiable***, and $\boldsymbol{D}f(\boldsymbol{x})$ is the ***Jacobian*** of f at $\boldsymbol{x} \in \mathcal{U}$. •

Note that, in the last part of the definition, we use the topology on $L(\mathbb{R}^n; \mathbb{R}^m)$ defined by the operator norm. In the event that we wish to evaluate $\boldsymbol{D}f(\boldsymbol{x})$ at $\boldsymbol{v} \in \mathbb{R}^n$ we shall write $\boldsymbol{D}f(\boldsymbol{x}) \cdot \boldsymbol{v} \in \mathbb{R}^m$. This avoids the awkward notation $\boldsymbol{D}f(\boldsymbol{x})(\boldsymbol{v})$. The same convention will apply to the higher-order derivatives we are about to define.

Since the derivative is itself a map between normed vector spaces, we can bootstrap the process to define higher-order derivatives. Let us see what the resulting constructions give us. A continuously differentiable map $f\colon \mathcal{U} \subset \mathbb{R}^n \to \mathbb{R}^m$ is ***twice differentiable*** at $\boldsymbol{x}_0 \in \mathcal{U}$ if $\boldsymbol{D}f$ is differentiable at $\boldsymbol{x}_0$. The second derivative is then $\boldsymbol{D}(\boldsymbol{D}f)(\boldsymbol{x}_0) \in L(\mathbb{R}^n; L(\mathbb{R}^n;\mathbb{R}^m))$. This can be understood in a more convenient manner if we recognize that $L(\mathbb{R}^n; L(\mathbb{R}^n;\mathbb{R}^m))$ is isomorphic to $L^2(\mathbb{R}^n;\mathbb{R}^m)$ via the isomorphism that assigns to $L \in L(\mathbb{R}^n; L(\mathbb{R}^n;\mathbb{R}^m))$ the multilinear map $(\boldsymbol{v}_1, \boldsymbol{v}_2) \mapsto L(\boldsymbol{v}_1)\cdot\boldsymbol{v}_2$ in $L^2(\mathbb{R}^n;\mathbb{R}^m)$. If the second derivative is continuous, then f is ***twice continuously differentiable***. We write the second derivative as $\boldsymbol{D}^2 f\colon \mathcal{U} \to L^2(\mathbb{R}^n;\mathbb{R}^m)$. One can additionally verify that, if f is twice continuously differentiable, then $\boldsymbol{D}^2 f(\boldsymbol{x})$ is symmetric for each $\boldsymbol{x} \in \mathcal{U}$. By repeating this process, we inductively define the rth derivative $\boldsymbol{D}^r f\colon \mathcal{U} \to L^r(\mathbb{R}^n;\mathbb{R}^m)$, if this derivative exists. Let us define some language associated with these higher-order derivatives.

Definition 3.18 (High-order differentiable maps). Let $\mathcal{U}$ be an open subset of $\mathbb{R}^n$ and let $f\colon \mathcal{U} \to \mathbb{R}^m$.

(i) If $\boldsymbol{D}^r f$ exists and is continuous, then f is ***r-times continuously differentiable***, or of ***class*** $\boldsymbol{C^r}$ (continuous maps are of ***class*** $\boldsymbol{C^0}$).
(ii) If f is of class C^r for each $r \in \mathbb{N}$, then it is ***infinitely differentiable***, or of ***class*** $\boldsymbol{C^\infty}$. We will also say that f is ***smooth*** if it is of class C^∞.
(iii) A bijection of open sets, $f\colon \mathcal{U} \subset \mathbb{R}^n \to \mathcal{V} \subset \mathbb{R}^m$, which is of class C^r, and for which f^{-1} is also of class C^r, is a $\boldsymbol{C^r}$***-diffeomorphism***. •

It will also be useful to consider maps that are "smoother" than class C^∞, although it is not immediately apparent that such objects should exist. To describe such functions, one uses the idea of a Taylor series, which, as we see in the following definition, is a formal power series, by which we mean that we do not care *a priori* about whether the series converges.

Definition 3.19 (Taylor series and real-analyticity). Let $\mathcal{U} \subset \mathbb{R}^n$ be an open set and let $f\colon \mathcal{U} \to \mathbb{R}^m$ be of class C^∞.

(i) The ***Taylor series*** for f about $\boldsymbol{x}_0$ is the formal power series

$$\sum_{r=0}^{+\infty} \frac{1}{r!}\boldsymbol{D}^r f(\boldsymbol{x}_0)\cdot(\underbrace{\boldsymbol{v},\dots,\boldsymbol{v}}_{r \text{ copies}}), \qquad \boldsymbol{v} \in \mathbb{R}^n. \tag{3.3}$$

(ii) The function f is ***real-analytic***, or simply ***analytic***, if, for every $\boldsymbol{x}_0 \in \mathcal{U}$, there exists $\rho > 0$ such that the expression (3.3) converges to $f(\boldsymbol{x}_0 + \boldsymbol{v})$, provided that $\|\boldsymbol{v}\|_{\mathbb{R}^n} < \rho$. A real-analytic map is said to be of ***class*** $\boldsymbol{C^\omega}$. •

Remarks 3.20. 1. It can be shown that the convergence to f of the Taylor series for f about $\boldsymbol{x}_0$ is uniform in a sufficiently small neighborhood of $\boldsymbol{x}_0$. That is to say, if f is real-analytic, then there exists $\rho > 0$ so that, for each $\varepsilon > 0$, there exists $N \in \mathbb{N}$ for which

$$\left\| \sum_{r=0}^{k} \boldsymbol{D}^r f(\boldsymbol{x}_0)(\boldsymbol{v}, \ldots, \boldsymbol{v}) - f(\boldsymbol{x}_0 + \boldsymbol{v}) \right\|_{\mathbb{R}^n} < \varepsilon$$

for each $k \geq N$ and provided that $\|\boldsymbol{v}\|_{\mathbb{R}^n} < \rho$. We shall give the general definition of uniform convergence in Section 3.1.3.

2. Note that f is real-analytic if it is equal to its own Taylor series (this, therefore, adapts the notion of analyticity from complex variable theory). What is perhaps not obvious is that analytic is actually different from C^∞. The reader is asked to consider a specific occurrence of this in Exercise E3.8. In Section 3.9.2 we consider in some detail the differences between C^ω and C^∞. •

For C^∞-functions, even though their Taylor series may not be equal to the function, one can place some restrictions on the remainder of any order.

Theorem 3.21 (Taylor's Theorem). *If $\mathcal{U} \subset \mathbb{R}^n$ and if $f : \mathcal{U} \to \mathbb{R}^m$ is of class C^∞, then, for any $r \in \mathbb{N}$ and for any $\boldsymbol{x}_0 \in \mathcal{U}$,*

$$f(\boldsymbol{x}_0 + \boldsymbol{v}) = \sum_{k=0}^{r} \frac{1}{k!} \boldsymbol{D}^k f(\boldsymbol{x}_0) \cdot (\boldsymbol{v}, \ldots, \boldsymbol{v}) + o(\|\boldsymbol{v}\|_{\mathbb{R}^n}^{r+1}).$$

One wishes to often differentiate compositions of maps.

Theorem 3.22 (Chain Rule). *Let $\mathcal{U} \subset \mathbb{R}^n$ and $\mathcal{V} \subset \mathbb{R}^m$ be open sets and let $f : \mathcal{U} \to \mathcal{V}$ and $g : \mathcal{V} \to \mathbb{R}^l$ be maps of class C^r. Then the composition $g \circ f : \mathcal{U} \to \mathbb{R}^l$ is of class C^r, and its derivative is given by*

$$\boldsymbol{D}(g \circ f)(\boldsymbol{x}) = \boldsymbol{D}g(f(\boldsymbol{x})) \circ \boldsymbol{D}f(\boldsymbol{x}) \in L(\mathbb{R}^n; \mathbb{R}^l).$$

There are also expressions for higher-order derivatives of the composition, but these are complicated and we will not need them; see [Abraham, Marsden, and Ratiu 1988].

We have gone rather quickly in this section through some concepts that are nontrivial, but are still probably familiar in principle to most readers. The difficulty lies in making the concepts like high-order differentiability precise. However, this clarity is essential, since we are shortly going to use these ideas to talk about objects more abstract, and more useful, than open subsets of Euclidean space.

3.1.3 Convergence of sequences of maps

In the presentation of the series expansion results in Chapter 9, we shall encounter sequences of functions defined on an interval $I \subset \mathbb{R}$, and taking values in $\mathbb{R}^n$. We shall also be interested in convergence for sequences of functions defined on subsets of Euclidean space. In this section we shall review some ideas of convergence associated to such situations. We refer to [Marsden and Hoffman 1993] for additional discussion.

To keep the presentation sufficiently general, we let $(\mathsf{V}, \|\cdot\|)$ be a finite-dimensional normed vector space. We then let $\{f_j\}_{j\in\mathbb{N}}$ be a sequence of maps from a subset $A \subset \mathbb{R}^n$ into V. Often the set A will have some nice properties (e.g., open, compact, etc.), as will the functions $\{f_j\}_{j\in\mathbb{N}}$ (e.g., continuous, bounded, etc.), but for the purposes of the definitions this does not matter. For the sequence of maps $\{f_j\}_{j\in\mathbb{N}}$, we have the following notions of convergence.

Definition 3.23. Let A and $\{f_j\}_{j\in\mathbb{N}}$ be as above, and let f be a map from A to V. The sequence $\{f_j\}_{j\in\mathbb{N}}$

(i) ***converges*** to f if, for each $\boldsymbol{x} \in A$, the sequence $\{f_j(\boldsymbol{x})\}_{j\in\mathbb{N}}$ converges to $f(\boldsymbol{x})$,

(ii) ***converges uniformly*** to f if, for each $\varepsilon > 0$, there exists $N \in \mathbb{N}$ such that $\|f_j(\boldsymbol{x}) - f(\boldsymbol{x})\| < \varepsilon$ for each $j \geq N$ and for each $\boldsymbol{x} \in A$, and

(iii) ***converges absolutely*** if, for each $\boldsymbol{x} \in A$, the sequence $\{f_j(\boldsymbol{x})\}_{j\in\mathbb{N}}$ converges absolutely. •

These notions of convergence are related as follows:

$$\text{uniform convergence} \implies \text{convergence} \impliedby \text{absolute convergence}.$$

3.2 Manifolds, maps, and submanifolds

In this section we introduce the fundamental object in differential geometry, a differentiable manifold, and consider some naturally associated constructions. The key points in this section are as follows. As mentioned in the introduction of this chapter, a manifold is a set that locally "looks like" an open subset of Euclidean space. To define a manifold, one needs the notion of (1) a local chart, i.e., a map from a subset of the manifold to an open subset of Euclidean space, and (2) an atlas, i.e., a collection of charts that covers the entire manifold, while satisfying certain overlap conditions. Given these constructions, one can then talk about curves on manifolds, maps between manifolds, smoothness of maps, and manifolds that are diffeomorphic, i.e., roughly speaking, identical. The essential idea is the following: given an object on a manifold, it is always possible (and often easy) to use different choices of overlapping charts to represent the object in different ways. The manifold structure allows one to carefully focus on the properties of the object, and not of its representation in local charts.

3.2.1 Charts, atlases, and differentiable structures

We begin by making precise the idea of a set that locally "looks like" an open subset of Euclidean space.

Definition 3.24 (Charts, atlases, and differentiable structures). Let S be a set. A ***chart*** for S is a pair $(\mathcal{U}, \phi)$ with

(i) $\mathcal{U}$ a subset of S, and
(ii) $\phi\colon \mathcal{U} \to \mathbb{R}^n$ an injection for which $\phi(\mathcal{U})$ is an open subset of $\mathbb{R}^n$.

For $r \in \mathbb{N} \cup \{\infty\} \cup \{\omega\}$, a ***$C^r$-atlas*** for S is a collection $\mathscr{A} = \{(\mathcal{U}_a, \phi_a)\}_{a \in A}$ of charts with the properties that $S = \cup_{a \in A} \mathcal{U}_a$, and that, whenever $\mathcal{U}_a \cap \mathcal{U}_b \neq \emptyset$, we have

(iii) $\phi_a(\mathcal{U}_a \cap \mathcal{U}_b)$ and $\phi_b(\mathcal{U}_a \cap \mathcal{U}_b)$ are open subsets of $\mathbb{R}^n$, and
(iv) the ***overlap map*** $\phi_{ab} \triangleq \phi_b \circ \phi_a^{-1} | \phi_a(\mathcal{U}_a \cap \mathcal{U}_b)$ is a C^r-diffeomorphism from $\phi_a(\mathcal{U}_a \cap \mathcal{U}_b)$ to $\phi_b(\mathcal{U}_a \cap \mathcal{U}_b)$.

Two C^r-atlases $\mathscr{A}_1$ and $\mathscr{A}_2$ are ***equivalent*** if $\mathscr{A}_1 \cup \mathscr{A}_2$ is also a C^r-atlas. A ***C^r-differentiable structure*** on S is an equivalence class of atlases under this equivalence relation. A ***C^r-differentiable manifold***, or simply a ***C^r-manifold***, M is a set S with a C^r-differentiable structure. An ***admissible chart*** for a manifold M is a pair $(\mathcal{U}, \phi)$ that is a chart for some atlas defining the differentiable structure. If all charts take values in $\mathbb{R}^n$ for some fixed n, then n is the ***dimension*** of M, denoted by $\dim(\mathsf{M})$. The ***manifold topology*** on a set S with a differentiable structure is the topology generated by the domains of the admissible charts. •

Unless we explicitly state otherwise, we will generally deal with paracompact, second-countable differentiable manifolds. Manifolds that are not paracompact are generally regarded as being pathological. Manifolds that are not second-countable are also often somewhat unusual. Chevalley [1946, Section III.IX] has a nice discussion of some of the useful ideas that involve second-countability. We comment that non-Hausdorff (and, therefore, non-paracompact) manifolds can arise in applications, although such applications are not important in this book. We also sometimes assume without comment that manifolds have a well-defined dimension. That is to say, all charts take values in $\mathbb{R}^n$ for some fixed n.

The condition (iv) is the ***overlap condition***, and is represented in Figure 3.1.

If $(\mathcal{U}, \phi)$ is a chart for a manifold M, then, for $x \in \mathcal{U}$, we can write $\phi(x) = (x^1(x), \dots, x^n(x))$, thus defining functions $x^i\colon \mathcal{U} \to \mathbb{R}$, $i \in \{1, \dots, n\}$. Motivated by this, we will often use language like "let $(x^1, \dots, x^n)$ be coordinates in a chart $(\mathcal{U}, \phi)$," by which we mean that $(x^1, \dots, x^n) \in \mathbb{R}^n$ will denote a typical point in $\phi(\mathcal{U})$, this point being uniquely defined by its preimage in $\mathcal{U}$. Note that we use superscripts for the coordinate indices. This convention will prove helpful when we deal with tensors on manifolds.

Let us utilize a simple, familiar example that will serve to illustrate certain concepts as we go along.

Example 3.25 (Polar coordinates). We take $\mathsf{M} = \mathbb{R}^2$. Our first coordinate chart is the "obvious" one. We define a chart $(\mathcal{U}_1, \phi_1)$ by $\mathcal{U}_1 = \mathsf{M}$ and $\phi_1(x, y) = (x, y)$. Next we define $(\mathcal{U}_2, \phi_2)$ by

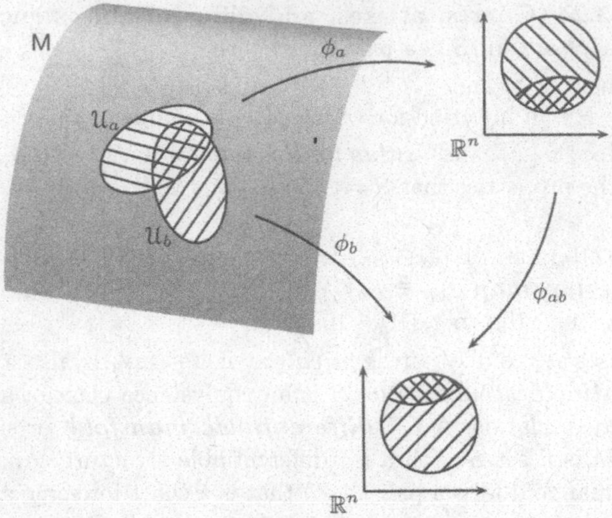

Figure 3.1. An interpretation of the overlap condition

$$\mathcal{U}_2 = \mathbb{R}^2 \setminus \{(x,0) \mid x \leq 0\}, \qquad \phi_2(x,y) = (\sqrt{x^2+y^2}, \operatorname{atan}(x,y)).$$

Here, atan: $\mathbb{R}^2 \setminus \{(0,0)\} \to]-\pi,\pi]$ is the "smart" arctangent function that is able to discern the appropriate quadrant. This second coordinate chart is illustrated in Figure 3.2. These coordinates are, of course, the usual polar

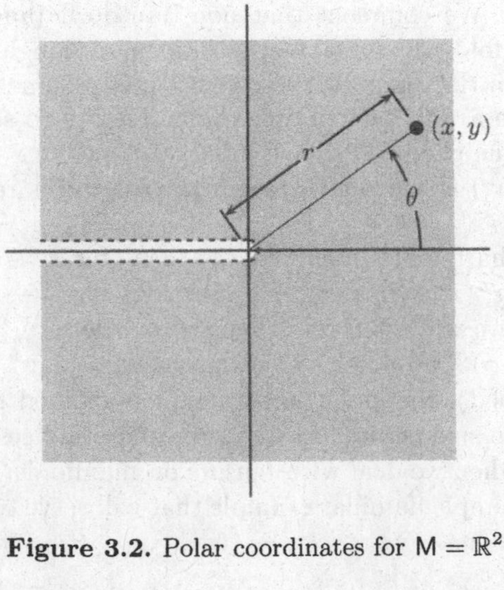

Figure 3.2. Polar coordinates for $\mathsf{M} = \mathbb{R}^2$

coordinates with which the reader is doubtlessly familiar. We are simply being more careful with their description.

Let us now see whether these two coordinate charts satisfy the overlap condition of Definition 3.24. We have

$$\phi_1(\mathcal{U}_1 \cap \mathcal{U}_2) = \mathbb{R}^2 \setminus \{(x,0) \mid x \leq 0\},$$
$$\phi_2(\mathcal{U}_1 \cap \mathcal{U}_2) = \{(r,\theta) \mid r > 0,\ \theta \in]-\pi,\pi[\}.$$

Note that we use coordinates (x, y) for $\phi_1(\mathcal{U}_1 \cap \mathcal{U}_2)$ and coordinates (r,θ) for $\phi_2(\mathcal{U}_1 \cap \mathcal{U}_2)$. The map ϕ_{12} is then computed to be

$$\phi_{12}(x,y) = (\sqrt{x^2+y^2}, \operatorname{atan}(x,y)).$$

This map is certainly a bijection. Let us also check that ϕ_{12} satisfies condition (iv) of Definition 3.24. Its Jacobian is

$$\boldsymbol{D}\phi_{12}(x,y) = \begin{bmatrix} \frac{x}{\sqrt{x^2+y^2}} & \frac{y}{\sqrt{x^2+y^2}} \\ -\frac{y}{x^2+y^2} & \frac{x}{x^2+y^2} \end{bmatrix}.$$

This matrix has continuous entries, so ϕ_{12} is differentiable. The inverse of ϕ_{12} is readily determined to be

$$\phi_{12}^{-1}(r,\theta) = (r\cos\theta, r\sin\theta),$$

so that

$$\boldsymbol{D}\phi_{12}^{-1}(r,\theta) = \begin{bmatrix} \cos\theta & -r\sin\theta \\ \sin\theta & r\cos\theta \end{bmatrix}.$$

Since this matrix has continuous entries, ϕ_{12}^{-1} is differentiable, and so ϕ_{12} is a C^1-diffeomorphism. In fact, this overlap map is analytic, so we have provided a C^ω-differentiable structure for $\mathbb{R}^2$. •

Let us consider some additional examples of manifolds, some of which will arise frequently in the book.

Examples 3.26. 1. We let S be an arbitrary set, and we define an atlas on S by $\{\mathcal{U}_x, \phi_x\}_{x\in S}$, where $\mathcal{U}_x = \{x\}$ and $\phi_x(x) = \mathbf{0}$. This does indeed define an atlas, and the resulting manifold is zero-dimensional. The manifold topology defined by the differentiable structure is the discrete topology, and so will not be second-countable, unless S is itself countable.

This is a rather degenerate example that illustrates the generality allowed by Definition 3.24. This generality is useful, but it also necessitates that care be used in dealing with statements concerning general manifolds.

2. If $\mathcal{U} \subset \mathbb{R}^n$ is open, then this set is a manifold with an atlas defined by $\mathscr{A} = \{(\mathcal{U}, \mathrm{id}_{\mathbb{R}^n}|\mathcal{U})\}$. Thus, in this case, we have a single chart that covers all of the manifold. This will generally not be the case, however. From this example we also see that an open subset of $\mathbb{R}^n$ comes equipped with a natural differentiable structure, and we will always suppose that this is the differentiable structure we use in this case.

3. If M and N are C^r-manifolds of dimension n and m, respectively, then we define their ***product*** by defining a C^r-atlas on the set $\mathsf{M} \times \mathsf{N}$. Let $\mathscr{A}_{\mathsf{M}} = \{(\mathcal{U}_a, \phi_a)\}_{a\in A}$ and $\mathscr{A}_{\mathsf{N}} = \{(\mathcal{V}_b, \psi_b)\}_{b\in B}$ be atlases for M and N, respectively. We define an atlas $\mathscr{A}_{\mathsf{M}} \times \mathscr{A}_{\mathsf{N}} = \{(\mathcal{U}_a \times \mathcal{V}_b, \phi_a \times \psi_b)\}_{(a,b)\in A\times B}$ for $\mathsf{M} \times \mathsf{N}$ by letting $\mathcal{U}_a \times \mathcal{V}_b$ be the usual Cartesian product, and by defining
$$\begin{aligned} \phi_a \times \psi_b \colon \mathcal{U}_a \times \mathcal{V}_b &\to \mathbb{R}^n \times \mathbb{R}^m \\ (x, y) &\mapsto (\phi_a(x), \psi_b(y)). \end{aligned}$$

4. The ***n-dimensional sphere*** is the subset $\mathbb{S}^n$ of $\mathbb{R}^{n+1}$ defined by
$$\mathbb{S}^n = \left\{ \boldsymbol{x} \in \mathbb{R}^{n+1} \;\middle|\; \|\boldsymbol{x}\|_{\mathbb{R}^{n+1}} = 1 \right\}.$$
Let us show how this is a manifold by providing an atlas. On $\mathbb{S}^n$ consider the two charts defined as follows. Let $\mathcal{U}_1 = \mathbb{S}^n \setminus \{(0,\dots,0,1)\}$ and $\mathcal{U}_2 = \mathbb{S}^n \setminus \{(0,\dots,0,-1)\}$. One can regard $\mathcal{U}_1$ as the sphere minus the "north pole," and $\mathcal{U}_2$ as the sphere minus the "south pole." Define
$$\begin{aligned} \phi_1 \colon \mathcal{U}_1 &\to \mathbb{R}^n \\ (x^1, \dots, x^{n+1}) &\mapsto \left(\frac{x^1}{1 - x^{n+1}}, \dots, \frac{x^n}{1 - x^{n+1}} \right) \end{aligned}$$
and
$$\begin{aligned} \phi_2 \colon \mathcal{U}_2 &\to \mathbb{R}^n \\ (x^1, \dots, x^{n+1}) &\mapsto \left(\frac{x^1}{1 + x^{n+1}}, \dots, \frac{x^n}{1 + x^{n+1}} \right). \end{aligned}$$
(See Figure 3.3 for $n = 1$.) It may be verified that $\phi_i(\mathcal{U}_i) = \mathbb{R}^n \setminus \{\mathbf{0}\}$,

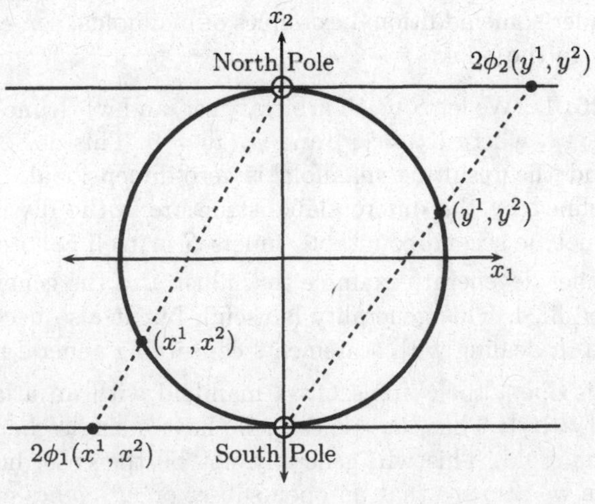

Figure 3.3. Stereographic coordinates for $\mathbb{S}^1$

$i \in \{1,2\}$, and that the overlap map is given by $(\phi_2 \circ \phi_1^{-1})(\boldsymbol{y}) = \frac{\boldsymbol{y}}{\|\boldsymbol{y}\|^2_{\mathbb{R}^n}}$ for $\boldsymbol{y} \in \mathbb{R}^n \setminus \{\boldsymbol{0}\}$. This map is easily seen to be of class C^ω and, since $\mathbb{S}^n = \mathcal{U}_1 \cup \mathcal{U}_2$, this makes $\mathbb{S}^n$ a C^ω-differentiable manifold.

We note that there is a possible notational confusion here for the newcomer. The set $\mathbb{S}^n$ is *not* the n-fold Cartesian product of $\mathbb{S}^1$ with itself. In fact, let us turn our attention to this Cartesian product right now.

5. In $\mathbb{R}^n$ consider the equivalence relation $\sim$ defined by asserting that $\boldsymbol{x}_1 \sim \boldsymbol{x}_2$ if and only if $\boldsymbol{x}_1 - \boldsymbol{x}_2 \in \mathbb{Z}^n$. We define $\mathbb{T}^n = \mathbb{R}^n/\sim$, which is called the ***n-dimensional torus***, or simply the ***n-torus***. Often one writes $\mathbb{T}^n = \mathbb{R}^n/\mathbb{Z}^n$. An equivalent way of thinking about $\mathbb{T}^n$ is as follows. Let $S = [0,1]^n$ be the unit cube in $\mathbb{R}^n$. For $i \in \{1,\dots,n\}$, write

$$F_{0,i} \triangleq \{(x^1,\dots,x^{i-1},0,x^{i+1},\dots,x^n) |\ x^1,\dots,x^{i-1},x^{i+1},\dots,x^n \in [0,1]\},$$
$$F_{1,i} \triangleq \{(x^1,\dots,x^{i-1},1,x^{i+1},\dots,x^n) |\ x^1,\dots,x^{i-1},x^{i+1},\dots,x^n \in [0,1]\}.$$

Thus $(\cup_{i=1}^n F_{0,i}) \cup (\cup_{i=1}^n F_{1,i})$ is the boundary of S. In S consider the equivalence relation

$$R = \cup_{i=1}^n \{(\boldsymbol{x}, \boldsymbol{x} + \boldsymbol{e}_i) \mid \ \boldsymbol{x} \in F_{0,i}\} \cup (\operatorname{int}(S) \times \operatorname{int}(S)).$$

This is visualized fairly easily for the 2-torus, as shown in Figure 3.4. We

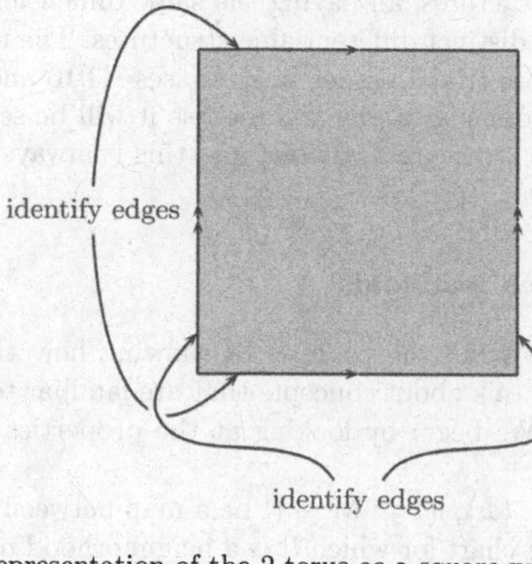

Figure 3.4. A representation of the 2-torus as a square with edges identified

see that, under the equivalence relation R, the left and right edges, and the top and bottom edges of S are identified. This representation of Figure 3.4 will be more convenient for us at various points. For example, note that the projection from $\mathbb{R}^n$ to $\mathbb{R}^n/\mathbb{Z}^n$ when restricted to $\operatorname{int}(S)$ is a bijection, and so establishes a natural chart for $\mathbb{T}^n$. The coordinates in this chart are the ***natural coordinates*** for $\mathbb{T}^n$ and are denoted by $(x^1,\dots,x^n)$.

Figure 3.5. A representation of the 2-torus as the surface of a "donut"

Another way of visualizing the 2-torus, one often used in blackboard discussions, is as the surface of a donut, depicted in Figure 3.5. The reader can explore various equivalent representations for $\mathbb{T}^1$ in Exercise E3.14. •

Remark 3.27. We shall from now on write a generic manifold as M, thinking of this as being the underlying set with the differentiable structure being understood. Note that this does *not* mean that given a set S there is a natural differentiable structure on this set. Indeed, even on such sets as certain spheres [Milnor 1956, 1959] or $\mathbb{R}^4$ [Donaldson 1983], there exist multiple, distinct differentiable structures, all having the same dimension. Indeed, $\mathbb{R}^4$ has an infinite number of distinct differentiable structures. The reader can explore some related, but more trivial, issues in Exercises E3.10 and E3.11. This being said, in our development of physical models it will be seen that there is a natural differentiable structure available, and this is always the one that will be used. •

3.2.2 Maps between manifolds

Let us now begin to fulfill the promise of showing how the definition of a manifold allows us to talk about concepts that are familiar to us from analysis in Euclidean space. We begin by looking at the properties of maps between manifolds.

For $r \in \mathbb{N} \cup \{\infty\} \cup \{\omega\}$, let $f\colon \mathsf{M} \to \mathsf{N}$ be a map between C^r-manifolds, let $x \in \mathsf{M}$, let $(\mathcal{U}, \phi)$ be a chart for which $\mathcal{U}$ is a neighborhood of x, and let $(\mathcal{V}, \psi)$ be a chart for which $\mathcal{V}$ is a neighborhood of $f(x)$, assuming that $f(\mathcal{U}) \subset \mathcal{V}$. If f is continuous, we can always shrink $\mathcal{U}$ so that this last condition is satisfied.

Definition 3.28 (Local representative of a map). With $\mathcal{U}$, $\mathcal{V}$, and $f\colon \mathcal{U} \to \mathcal{V}$ as above, the ***local representative*** of f with respect to these two charts is the map $f_{\phi\psi}\colon \phi(\mathcal{U}) \to \psi(\mathcal{V})$ given by

$$f_{\phi\psi}(\boldsymbol{x}) = \psi \circ f \circ \phi^{-1}(\boldsymbol{x}).$$

•

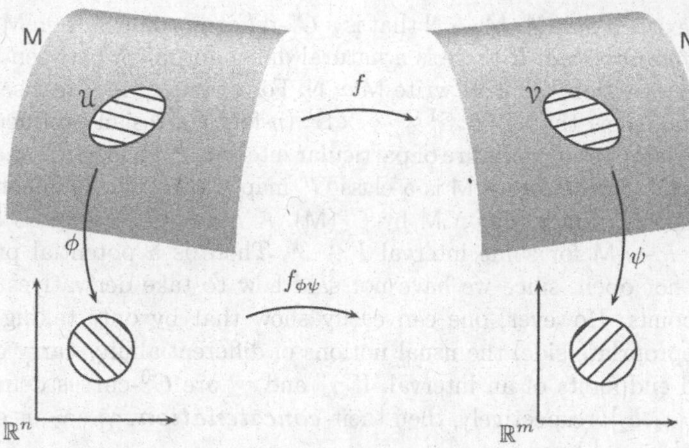

Figure 3.6. The local representative of a map

Thus the local representative is a map between open sets in Euclidean space (see Figure 3.6). Thanks to our work in Section 3.1.2, we know what it means for such a map to be of class C^r.

Definition 3.29 (C^r-map between manifolds). Let $f\colon \mathsf{M} \to \mathsf{N}$ be a continuous map between manifolds.

(i) We say that $f\colon \mathsf{M} \to \mathsf{N}$ is of ***class C^r***, $r \in \mathbb{N} \cup \{\infty\} \cup \{\omega\}$, if, for every point $x \in \mathsf{M}$, there exist coordinate charts $(\mathcal{U}, \phi)$ for M and $(\mathcal{V}, \psi)$ for N, as in Definition 3.28, and for which the local representative $f_{\phi\psi}$ is of class C^r.
(ii) If f is continuous, it is of ***class C^0***.
(iii) If f is of class C^∞, then we shall say that f is ***smooth***.
(iv) The set of class C^r maps from M to N is denoted by $C^r(\mathsf{M};\mathsf{N})$.
(v) If f is a bijection of class C^r, and if f^{-1} is also of class C^r, then f is a ***C^r-diffeomorphism***. •

One can verify that, since the overlap maps are C^r-diffeomorphisms, a map is of class C^r if and only if, for every pair of atlases $\mathscr{A}_{\mathsf{M}} = \{(\mathcal{U}_a, \phi_a)\}_{a\in A}$ for M and $\mathscr{A}_{\mathsf{N}} = \{(\mathcal{V}_b, \psi_b)\}_{b\in B}$ for N, the local representatives f_{ϕ_a,ψ_b} are of class C^r, whenever they are well-defined. That is to say, the notion of a map being of class C^r does not depend on how one chooses to locally represent it. If $(\mathcal{U}, \phi)$ is a chart for M with coordinates $(x^1, \dots, x^n)$, and if $(\mathcal{V}, \psi)$ is a chart for N with coordinates $(y^1, \dots, y^m)$, and for which $f(\mathcal{U}) \subset \mathcal{V}$, then the local representative for $f \in C^r(\mathsf{M};\mathsf{N})$ can be written as

$$f_{\phi\psi}(x^1, \dots, x^n) = (f^1(x^1, \dots, x^n), \dots, f^m(x^1, \dots, x^n)),$$

so defining functions $f^1, \dots, f^m$ called the ***components*** of the local representative.

If there exists a map $f\colon \mathsf{M} \to \mathsf{N}$ that is a C^r-diffeomorphism, then M and N are ***C^r-diffeomorphic***. If there is a natural diffeomorphism between M and N in a given situation, we may write $\mathsf{M} \simeq \mathsf{N}$. For example, in Exercise E3.14 the reader can show that $\mathbb{T}^n \simeq \mathbb{S}^1 \times \cdots \times \mathbb{S}^1$ (n-fold Cartesian product).

Two special types of maps are of particular interest. For a class C^r manifold M, a ***class C^r function*** on M is a class C^r map $f\colon \mathsf{M} \to \mathbb{R}$. We denote the set of all class C^r functions on M by $C^r(\mathsf{M})$. A ***class C^r curve*** is a class C^r map $\gamma\colon I \to \mathsf{M}$ for some interval $I \in \mathscr{I}$. There is a potential problem here if I is not open, since we have not said how to take derivatives at the closed endpoints. However, one can easily show that by only taking limits from the appropriate side, the usual notions of differentiability carry over to allow closed endpoints of an interval. If γ_1 and γ_2 are C^0-curves defined on $[a_1, b_1]$ and $[a_2, b_2]$, respectively, then their ***concatenation***, $\gamma_1 * \gamma_2$, is defined on $[a_1, b_1 + b_2 - a_2]$ by

$$\gamma_1 * \gamma_2(t) = \begin{cases} \gamma_1(t), & t \in [a_1, b_1], \\ \gamma_2(t + a_2 - b_1), & t \in]b_1, b_1 + b_2 - a_2]. \end{cases}$$

If $\gamma_1(b_1) = \gamma_2(a_2)$, then $\gamma_1 * \gamma_2$ is also a C^0-curve. However, in this case it will not generally be true that the concatenation of C^r-curves will be C^r when $r \geq 1$.

There will be occasions when we need to talk about curves that have the rather general smoothness property of being ***locally absolutely continuous***. The precise definition for this is given in Section A.2. An intuitive definition is that a locally absolutely continuous curve is continuous, and differentiable "almost everywhere." Curves of this sort are actually important in this book, since they are what arise when one allows a broad class of control forces. However, we are not yet prepared to talk about this. The detailed discussion is postponed to Section 3.5.2, and to Section A.2.

3.2.3 Submanifolds

When one studies vector spaces, the notion of a subspace comes up when one has a subset of the vector space to which the vector space structure restricts. For manifolds, one has a similar notion of subsets that acquire a differentiable structure from the manifold of which they are subset.

Definition 3.30 (Submanifold). Let $r \in \mathbb{N} \cup \{\infty\} \cup \{\omega\}$. A subset S of a C^r-manifold M is a ***C^r-submanifold*** if, for each point $x \in \mathsf{S}$, there is an admissible chart $(\mathcal{U}, \phi)$ for M with $x \in \mathcal{U}$, and such that

(i) ϕ takes its values in a product $\mathbb{R}^k \times \mathbb{R}^{n-k}$, and
(ii) $\phi(\mathcal{U} \cap \mathsf{S}) = \phi(\mathcal{U}) \cap (\mathbb{R}^k \times \{0\})$.

A chart with these properties is a ***submanifold chart*** for S. •

Remark 3.31. Note that, if $(\mathcal{U}, \phi)$ is a submanifold chart for S, then $(\mathcal{U} \cap \mathsf{S}, \phi|(\mathcal{U} \cap \mathsf{S}))$ is a chart for S. One can easily verify that the overlap condition for two such charts, if they intersect, is satisfied. Therefore, the collection of submanifold charts forms an atlas for S. Therefore, submanifolds are manifolds by our usual definition. If a submanifold S has a well-defined dimension as a manifold, (i.e., the integer k in the definition is independent of $x \in \mathsf{S}$), then we denote this dimension by $\dim(\mathsf{S})$, which is the ***dimension*** of S. If it further holds that the manifold M of which S is a submanifold also has a well defined dimension, then $\dim(\mathsf{M}) - \dim(\mathsf{S})$ is the ***codimension*** of S. •

Figure 3.7 gives the intuition for submanifold charts. We can also be con-

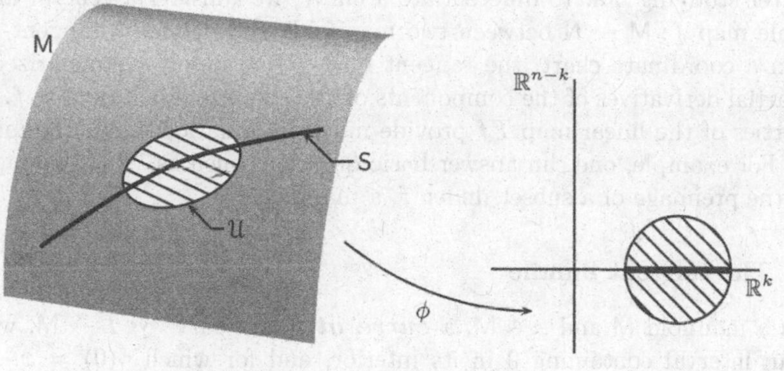

Figure 3.7. A submanifold chart

crete about this in a particular example.

Example 3.32. We let our manifold be $\mathsf{M} = \mathbb{R}^2$ with its natural differentiable structure, and consider the one-dimensional sphere:

$$\mathbb{S}^1 = \{ (x, y) \in \mathsf{M} \mid x^2 + y^2 = 1 \} .$$

A simple adaptation of the polar coordinate chart of Example 3.25 suffices to exhibit a submanifold chart for $\mathbb{S}^1$. Thus we consider the chart $(\mathcal{U}, \phi)$ for M with

$$\mathcal{U} = \mathbb{R}^2 \setminus \{ (x, 0) \mid x \leq 0 \} , \qquad \phi(x, y) = (\sqrt{x^2 + y^2} - 1, \operatorname{atan}(x, y)).$$

Note that

$$\phi(\mathcal{U} \cap \mathbb{S}^1) = \{ (r, \theta) \in \phi(\mathcal{U}) \mid r = 0 \} ,$$

just as required by the definition of a submanifold chart. •

In Definition 3.43 and in the discussion following it, we shall encounter an alternative characterization of a submanifold, along with a useful generalization.

3.3 Tangent bundles and more about maps

This section introduces the valuable notion of tangent bundle. Roughly speaking, the tangent bundle is the set of all possible velocities at all points of a manifold. Remarkably, the tangent bundle is a manifold itself. Looking ahead to the context of Lagrangian mechanics, if the set of positions of a system is a manifold, then the tangent bundle of this manifold should be regarded as the set of positions and velocities. Therefore, the state manifold of the system is the tangent bundle. For readers not familiar with differential geometric ideas, perhaps the most surprising outcome of the introduction of the tangent bundle is that velocities do not have a life of their own, and one can only talk about velocity *at a position*.

After studying how to differentiate a curve, we consider a general differentiable map $f\colon \mathsf{M} \to \mathsf{N}$ between two manifolds and define its tangent map Tf. In a coordinate chart, the tangent map Tf is simply characterized by the partial derivatives of the components of the local representative of f. The properties of the linear map Tf provide much information about the map f itself. For example, one can answer basic questions such as, "Is the image of f, or the preimage of a subset under f, a manifold?"

3.3.1 The tangent bundle

Given a manifold M and $x \in \mathsf{M}$, a ***curve at x*** is a curve $\gamma\colon I \to \mathsf{M}$, where I is an interval containing 0 in its interior, and for which $\gamma(0) = x$. Two curves γ_1 and γ_2 at x are ***equivalent*** at x if, in a coordinate chart $(\mathcal{U}, \phi)$ with $x \in \mathcal{U}$, the local representatives of γ_1 and γ_2 have the same derivative at 0 (see Figure 3.8). Since the overlap maps are diffeomorphisms, this notion of

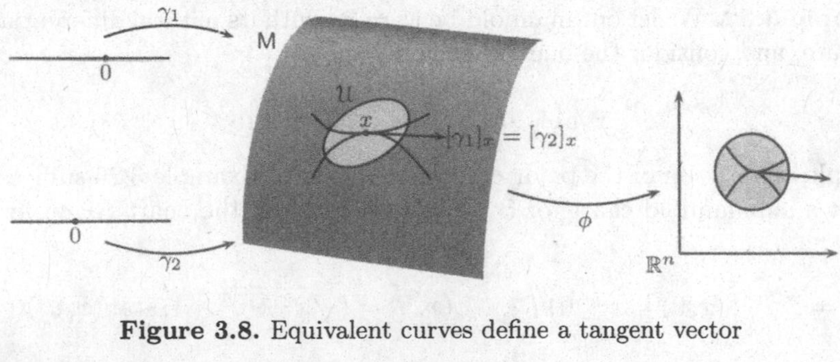

Figure 3.8. Equivalent curves define a tangent vector

equivalence is independent of coordinate chart. Let us write $\gamma_1 \sim_x \gamma_2$ if γ_1 and γ_2 are equivalent at x. This notion of equivalence is readily verified to be an equivalence relation. The equivalence class of γ will be denoted by $[\gamma]_x$.

Definition 3.33 (Tangent bundle). A ***tangent vector*** at x is an equivalence class of curves under the equivalence relation $\sim_x$. The collection of all tangent vectors at x is the ***tangent space*** at x, and is denoted by $\mathsf{T}_x\mathsf{M}$. The disjoint union (see Section 2.1)

$$\mathsf{TM} = \mathring{\bigcup_{x\in\mathsf{M}}} \mathsf{T}_x\mathsf{M}$$

of all tangent spaces is called the ***tangent bundle***. The ***tangent bundle projection*** is the map $\pi_{\mathsf{TM}}\colon \mathsf{TM} \to \mathsf{M}$ defined by $\pi_{\mathsf{TM}}(v) = x$ when $v \in \mathsf{T}_x\mathsf{M}$. •

The abstractness of this definition is necessitated by our rather featureless definition of a manifold. If a manifold is thought of as being a submanifold of Euclidean space, then one can be more concrete about what one means by a tangent space. For example, the tangent bundle of a sphere is the collection of planes tangent to the sphere. However, since we have defined a manifold in great generality, the tangent space must be constructed "out of thin air," and the method in which we have done this provides a fairly intuitive means of understanding what we mean by a tangent space. We ask the reader to verify in Exercise E3.16 that $\mathsf{T}_x\mathsf{M}$ is naturally a $\mathbb{R}$-vector space.

The tangent bundle of a manifold is itself a manifold. To show this, we will show how to construct an atlas for TM given an atlas $\mathscr{A} = \{(\mathcal{U}_a, \phi_a)\}_{a\in A}$ for M. To do this, we show how to associate a chart for TM to a chart $(\mathcal{U}, \phi)$ for M. Let us denote this chart for TM by $(\mathsf{T}\mathcal{U}, T\phi)$. First we take $\mathsf{T}\mathcal{U} = \mathring{\cup}_{x\in\mathcal{U}}\mathsf{T}_x\mathsf{M}$. Next, we define $T\phi\colon \mathsf{T}\mathcal{U} \to \mathbb{R}^n \times \mathbb{R}^n$ as follows. Let $x \in \mathcal{U}$ and let $[\gamma]_x \in \mathsf{T}_x\mathsf{M}$. Then $\phi\circ\gamma$ is a curve at $\phi(x) \in \mathbb{R}^n$, possibly after restricting the domain of γ. Now, one can easily show that there exists a unique $\boldsymbol{v}_\gamma \in \mathbb{R}^n$ such that $\phi\circ\gamma$ is equivalent at $\phi(x)$ to the curve $t \mapsto \phi(x) + t\boldsymbol{v}_\gamma$. We now define $T\phi([\gamma]_x) = (\phi(x), \boldsymbol{v}_\gamma) \in \mathbb{R}^n \times \mathbb{R}^n$. This does indeed define a chart $(\mathsf{T}\mathcal{U}, T\phi)$ for TM, which we call a ***natural chart***. One must then show that if $(\mathcal{U}_1, \phi_1)$ and $(\mathcal{U}_2, \phi_2)$ are charts for which $\mathcal{U}_1 \cap \mathcal{U}_2 \neq \emptyset$, then the resulting charts $(\mathsf{T}\mathcal{U}_1, T\phi_1)$ and $(\mathsf{T}\mathcal{U}_2, T\phi_2)$ satisfy the overlap condition. For convenience, and without loss of generality, let us suppose that $\mathcal{U}_1 = \mathcal{U}_2$. One can compute the overlap map explicitly as

$$T\phi_2 \circ T\phi_1^{-1}(\boldsymbol{x}, \boldsymbol{v}) = (\phi_{12}(\boldsymbol{x}), \boldsymbol{D}\phi_{12}(\boldsymbol{x}) \cdot \boldsymbol{v}), \qquad \boldsymbol{x} \in \phi_1(\mathcal{U}_1),\ \boldsymbol{v} \in \mathbb{R}^n. \tag{3.4}$$

From this, we readily see that the overlap map is indeed a diffeomorphism. If we are counting degrees of differentiability, then M being C^r gives the overlap maps for TM as being of class C^{r-1}. Therefore, TM is a C^{r-1}-manifold. Finally, we see that $\mathsf{T}\mathscr{A} = \{(\mathsf{T}\mathcal{U}_a, T\phi_a)\}_{a\in A}$ is indeed an atlas for TM, since $\mathsf{TM} = \cup_{a\in A}\mathsf{T}\mathcal{U}_a$.

Charts on the manifold M thus give rise in a natural way to charts for the tangent bundle in exactly the manner we have discussed above. If $(x^1, \dots, x^n)$ are coordinates for M, we will denote coordinates for TM in the associated natural chart by $((x^1, \dots, x^n), (v^1, \dots, v^n))$. In this case, if we have coordinates $(\tilde{x}^1, \dots, \tilde{x}^n)$ in an overlapping chart, then, by (3.4), the tangent vector coordinates $(\tilde{v}^1, \dots, \tilde{v}^n)$ in this new chart are related to the original tangent vector coordinates by the formula

$$\tilde{v}^i = \frac{\partial \tilde{x}^i}{\partial x^j} v^j, \qquad i \in \{1, \dots, n\}. \tag{3.5}$$

Note that we are using the summation convention of Chapter 2 with one little sleight of hand: the *super*script in the denominator of $\frac{\partial}{\partial x^j}$ is to be thought of as a *sub*script as far as the summation convention is concerned.

Although the above discussion may appear to contain a great deal of information, in practice the ideas come up quite naturally. Let us demonstrate this with our polar coordinate example.

Example 3.34 (Polar coordinates cont'd). Corresponding to the Cartesian coordinates (x, y), we have the tangent bundle coordinates $((x, y), (v_x, v_y))$, and corresponding to the polar coordinates (r, θ), we have the tangent bundle coordinates $((r, \theta), (v_r, v_\theta))$. These coordinates are related by

$$\begin{aligned}
v_r &= \frac{\partial r}{\partial x} v_x + \frac{\partial r}{\partial y} v_y = \frac{x}{\sqrt{x^2 + y^2}} v_x + \frac{y}{\sqrt{x^2 + y^2}} v_y, \\
v_\theta &= \frac{\partial \theta}{\partial x} v_x + \frac{\partial \theta}{\partial y} v_y = -\frac{y}{x^2 + y^2} v_x + \frac{x}{x^2 + y^2} v_y.
\end{aligned}$$

Note that this is merely how velocities would transform via the Chain Rule, so this agrees with what one would naïvely use, except that we use v's for velocities rather than using the "dot" notation. •

Let us look at some examples of tangent bundles.

Examples 3.35. 1. If $\mathsf{S} \subset \mathsf{M}$ is a submanifold, then we may naturally regard TS as a submanifold of TM. Indeed, for $x \in \mathsf{S}$, $\mathsf{T}_x\mathsf{S}$ consists of those equivalence classes of curves $[\gamma]_x$ at x for which γ takes values in S. Such equivalence classes also lie in $\mathsf{T}_x\mathsf{M}$, so TS is certainly a *subset* of TM. A simple computation using submanifold charts for S shows that TS is actually a submanifold.

2. If $\mathcal{U} \subset \mathbb{R}^n$ is an open subset, then our discussion concerning the differentiable structure of the tangent bundle directly gives a natural diffeomorphism of $\mathsf{T}\mathcal{U}$ with $\mathcal{U} \times \mathbb{R}^n$. Thus the tangent bundle in this case may be naturally regarded as a product. As we shall see, it is not generally the case that a tangent bundle is a product manifold, and even when it is, it is not always advisable to explicitly recognize the product structure. We shall have more to say about this in Remark 3.36 below.
3. Consider the submanifold $\mathbb{S}^1$ of $\mathbb{R}^2$. Using the natural identification of $\mathsf{T}\mathbb{R}^2$ with $\mathbb{R}^2 \times \mathbb{R}^2$, as a submanifold of $\mathsf{T}\mathbb{R}^2$ the tangent bundle of $\mathbb{S}^1$ is given by

$$\mathsf{T}\mathbb{S}^1 = \{((x,y),(u,v)) \in \mathbb{R}^2 \times \mathbb{R}^2 \mid x^2+y^2=1,\ ux+vy=0\}.$$

Thus, for $(x,y) \in \mathbb{S}^1$, a tangent vector $(u,v) \in \mathsf{T}_{(x,y)}\mathbb{R}^2$ is in $\mathsf{T}_{(x,y)}\mathbb{S}^1$ if and only if it is orthogonal to (x,y). This should make sense, intuitively. Let us now illustrate that $\mathsf{T}\mathbb{S}^1$ is diffeomorphic to $\mathbb{S}^1 \times \mathbb{R}$. First note that, for $((x,y),(u,v)) \in \mathsf{T}\mathbb{S}^1$, the vector in $\mathbb{R}^2$ given by

$$\begin{bmatrix} \frac{x}{\sqrt{x^2+y^2}} & \frac{y}{\sqrt{x^2+y^2}} \\ -\frac{y}{\sqrt{x^2+y^2}} & \frac{x}{\sqrt{x^2+y^2}} \end{bmatrix} \begin{bmatrix} u \\ v \end{bmatrix} \tag{3.6}$$

is collinear with $(0,1)$. Geometrically, this vector is obtained by rotating the vector $(u,v) \in \mathbb{R}^2$ by the angle $\theta = \operatorname{atan}(x,y)$. Thus let us define $a(x,y,u,v) \in \mathbb{R}$ so that the vector in (3.6) is exactly $(0, a(x,y,u,v))$. One can then see that the map

$$\mathsf{T}\mathbb{S}^1 \ni ((x,y),(u,v)) \mapsto ((x,y), a(x,y,u,v)) \in \mathbb{S}^1 \times \mathbb{R}$$

is a diffeomorphism. This shows that $\mathbb{S}^1$ is ***parallelizable***, meaning that its tangent bundle is diffeomorphic to a product of it with a typical tangent space. We shall have more to say about this in Remark 3.53. •

Remark 3.36 (Notation for tangent vectors). It is not uncommon to see authors write a typical point in the tangent bundle of a manifold M as (x,v), with the intended meaning that $v \in \mathsf{T}_x\mathsf{M}$. This is deceptive notation, since it implies that the tangent bundle is a product manifold. We shall, therefore, write a typical element of the tangent bundle of M as v_x. In this way, we still encode in our notation the point in M in whose tangent space v_x resides. Only when we are dealing with open subsets of Euclidean space (as, for example, when doing local coordinate calculations) will we write tangent vectors as pairs. •

3.3.2 More about maps

We already have on hand the notion of what it means for a map $f\colon \mathsf{M} \to \mathsf{N}$ between manifolds to be of class C^r, for $r \in \mathbb{N} \cup \{\infty\} \cup \{\omega\}$. Let us give some additional properties of such maps.

First we define the derivative of a map $f\colon \mathsf{M} \to \mathsf{N}$. Let $x \in M$ and let $[\gamma]_x \in \mathsf{T}_x\mathsf{M}$. Then $f \circ \gamma$ is a curve at $f(x)$, and we define $Tf([\gamma]_x) = [f \circ \gamma]_{f(x)}$. This then defines a map $Tf\colon \mathsf{TM} \to \mathsf{TN}$, called the ***tangent map*** of f. One verifies that, if $f \in C^r(\mathsf{M};\mathsf{N})$, then $Tf \in C^{r-1}(\mathsf{TM};\mathsf{TN})$. The restriction of Tf to $\mathsf{T}_x\mathsf{M}$ is denoted by T_xf. Note that $T_xf \in L(\mathsf{T}_x\mathsf{M}; \mathsf{T}_{f(x)}\mathsf{N})$.

Remark 3.37 (Notation for the tangent map). One sees a variety of notation for the map that we denote by Tf. Common examples include f_*, $\mathrm{d}f$, and $\mathrm{D}f$. Of course, the choices made here are largely a matter of taste.

However, we would be remiss if we did not point out the obvious notational logic of $Tf\colon \mathsf{TM} \to \mathsf{TN}$ being the derivative of $f\colon \mathsf{M} \to \mathsf{N}$. The notation for the restriction to the tangent space, $T_x f\colon \mathsf{T}_x\mathsf{M} \to \mathsf{T}_{f(x)}\mathsf{N}$, is similarly natural. We will certainly have to ask the reader to steer clear of using f_* (which is quite popular), since we have already used this for push-forward in Section 2.4.3. We will also be extending the operation of push-forward to manifolds, so the use of f_* to denote the derivative of f will be genuinely confusing. •

Let us provide the representation of Tf in coordinates. Let $(x^1,\dots,x^n)$ be coordinates in a chart $(\mathcal{U},\phi)$ defined in a neighborhood of $x \in \mathsf{M}$, and let $(y^1,\dots,y^n)$ be coordinates in a chart $(\mathcal{V},\psi)$ defined in a neighborhood of $f(x) \in \mathsf{N}$. Also assume that $f(\mathcal{U}) \subset \mathcal{V}$. The local representative $f_{\phi\psi}$ is then a map from the open subset $\phi(\mathcal{U}) \subset \mathbb{R}^n$ to the open subset $\psi(\mathcal{V}) \subset \mathbb{R}^m$. The local representative of the tangent map is then

$$\phi(\mathcal{U}) \times \mathbb{R}^n \ni (\boldsymbol{x},\boldsymbol{v}) \mapsto (f_{\phi\psi}(\boldsymbol{x}), \boldsymbol{D}f_{\phi\psi}(\boldsymbol{x})\cdot\boldsymbol{v}) \in \psi(\mathcal{V}) \times \mathbb{R}^m,$$

where we assume, of course, that we are using the natural charts for TM and TN. In component notation, if $f_{\phi\psi}$ looks like

$$(x^1,\dots,x^n) \mapsto (f^1(\boldsymbol{x}),\dots,f^m(\boldsymbol{x})),$$

then the local representative of Tf looks like

$$\begin{aligned}&((x^1,\dots,x^n),(v^1,\dots,v^n))\\ &\qquad\mapsto \Big((f^1(\boldsymbol{x}),\dots,f^m(\boldsymbol{x})), \Big(\frac{\partial f^1(\boldsymbol{x})}{\partial x^j}v^j,\dots,\frac{\partial f^m(\boldsymbol{x})}{\partial x^j}v^j\Big)\Big).\end{aligned}$$

If the map under consideration is a curve $\gamma\colon I \to \mathsf{M}$, then we will use some special notation associated with the derivative. First note that, since $\mathsf{T}\mathbb{R} \simeq \mathbb{R}\times\mathbb{R}$, it makes sense to write $\gamma'(t) = T\gamma(t,1) \in \mathsf{T}_{\gamma(t)}\mathsf{M}$. We call $\gamma'\colon I \to \mathsf{TM}$ the ***tangent vector field*** of the curve γ. We will also refer to γ' as the ***velocity curve*** of γ. If the curve γ is represented in coordinates $(x^1,\dots,x^n)$ by $t \mapsto (x^1(t),\dots,x^n(t))$, then the local representative of γ' is

$$t \mapsto ((x^1(t),\dots,x^n(t),(\dot{x}^1(t),\dots,\dot{x}^n(t))),$$

where "$\dot{\ }$" means derivative with respect to t. Thus γ' is merely the usual velocity along the curve γ when written in coordinates.

Now that we have on hand the notion of the derivative of a map between manifolds, let us provide some properties of maps using this derivative.

Definition 3.38 (Submersion and immersion). Let $f \in C^r(\mathsf{M};\mathsf{N})$ for $r \in \mathbb{N}\cup\{\infty\}\cup\{\omega\}$.

(i) A ***regular value*** of f is a point $y \in \mathsf{N}$ with the property that, for every $x \in f^{-1}(y)$, $T_x f$ is surjective.

(ii) For a subset $A \subset \mathsf{M}$, we say that f is a ***submersion on A*** if, for each $x \in A$, $T_x f$ is surjective. If f is a submersion on M, then it is simply a ***submersion***.

(iii) If $T_x f$ is surjective, then x is a ***regular point*** for f. Otherwise, x is a ***critical point*** and $f(x)$ is a ***critical value***.

(iv) If $T_x f$ is injective, we say that f is an ***immersion at x***. If f is an immersion at every $x \in \mathsf{M}$, we say that it is an ***immersion***. •

Examples 3.39. 1. Consider the C^ω-map $f\colon \mathbb{R}^{n+1} \to \mathbb{R}$ defined by $f(\boldsymbol{x}) = \|\boldsymbol{x}\|^2_{\mathbb{R}^{n+1}}$. We claim that f is a submersion on $\mathbb{S}^n$. Indeed, we readily compute

$$Tf(\boldsymbol{x}, \boldsymbol{v}) = (\|\boldsymbol{x}\|^2_{\mathbb{R}^{n+1}}, 2\,\langle\!\langle \boldsymbol{x}, \boldsymbol{v}\rangle\!\rangle_{\mathbb{R}^{n+1}}).$$

In order for $T_{\boldsymbol{x}} f$ to be surjective, the map $\boldsymbol{v} \mapsto \langle\!\langle \boldsymbol{x}, \boldsymbol{v}\rangle\!\rangle$ must not be the zero linear map. This is true exactly when $\boldsymbol{x} \neq 0$. In particular, it is true on $\mathbb{S}^n$.

2. Define a map $\tilde{f}_a\colon \mathbb{R} \to \mathbb{R}^2$ by $\tilde{f}_a(x) = (x, ax)$ for some $a > 0$. Now define $f_a\colon \mathbb{R} \to \mathbb{T}^2$ by $f_a = \pi \circ \tilde{f}_a$, where $\pi\colon \mathbb{R}^2 \to \mathbb{T}^2 = \mathbb{R}^2/\mathbb{Z}^2$ is the canonical projection. Clearly $T_x f_a$ is injective for each $x \in \mathbb{R}$, and so f_a is an immersion. •

The basic idea behind the notions of immersion and submersion is that, around points where f has either of these properties, its behavior is "simple." In fact, we have the following result whose proof may be found in [Abraham, Marsden, and Ratiu 1988].

Proposition 3.40 (Local representatives for submersions and immersions). *Let $f \in C^r(\mathsf{M}; \mathsf{N})$ for $r \in \mathbb{N} \cup \{\infty\} \cup \{\omega\}$.*

(i) A point $x \in \mathsf{M}$ is a regular point of f if and only if there exist a chart $(\mathcal{U}, \phi)$ around x, and a chart $(\mathcal{V}, \psi)$ around $f(x)$, such that

(a) ϕ takes its values in $\mathbb{R}^{n-m} \times \mathbb{R}^m$ and ψ takes its values in $\mathbb{R}^m$, and

(b) the local representative $f_{\phi\psi}$ is given by $(\boldsymbol{x}_1, \boldsymbol{x}_2) \mapsto \boldsymbol{x}_2$.

(ii) f is an immersion at $x \in \mathsf{M}$ if and only if there exist a chart $(\mathcal{U}, \phi)$ around x, and a chart $(\mathcal{V}, \psi)$ around $f(x)$, such that

(a) ϕ takes its values in $\mathbb{R}^n$ and ψ takes its values in $\mathbb{R}^n \times \mathbb{R}^{m-n}$, and

(b) the local representative $f_{\phi\psi}$ is given by $\boldsymbol{x} \mapsto (\boldsymbol{x}, \boldsymbol{0})$.

The picture one should have in mind locally for submersions and immersions is depicted in Figure 3.9. The idea is that a submersion locally looks like a projection onto a subspace, and an immersion locally looks like the inclusion of a subspace. At points where the derivative does not have maximal rank, then the local behavior of the map is more difficult to classify and is the topic of singularity theory (for example, see [Golubitsky and Guillemin 1973]). The local behavior of a $\mathbb{R}$-valued function about a critical point is discussed in the Section 6.1.4 via Morse theory.

An important and useful result is the following.

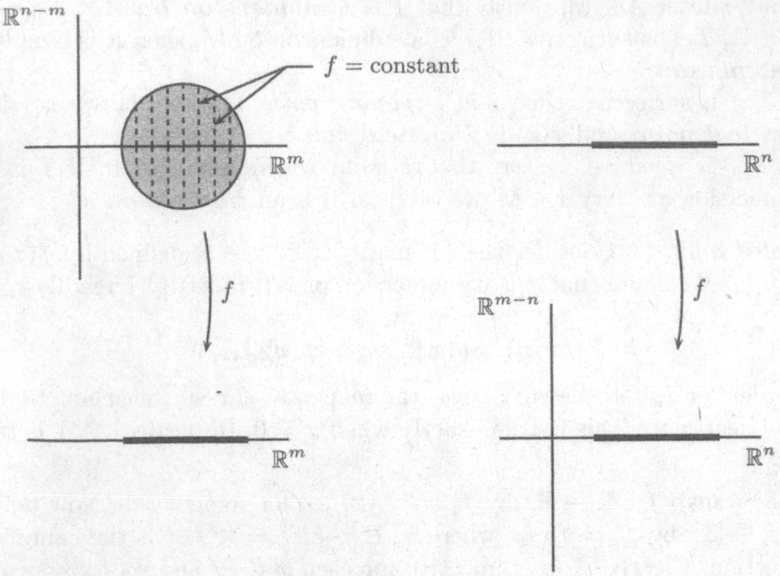

Figure 3.9. Local behavior of submersions (on the left) and immersions (on the right)

Theorem 3.41 (Inverse Function Theorem). *Let $r \in \mathbb{N} \cup \{\infty\} \cup \{\omega\}$ and let $f \in C^r(\mathsf{M}; \mathsf{N})$. If $T_{x_0}f$ is an isomorphism, then there exists a neighborhood $\mathcal{U}$ of x_0 for which $f|\mathcal{U}\colon \mathcal{U} \to f(\mathcal{U})$ is a C^r-diffeomorphism. Furthermore, if $g\colon f(\mathcal{U}) \to \mathcal{U}$ is the inverse of $f|\mathcal{U}$, then $T_{f(x_0)}g = (T_{x_0}f)^{-1}$.*

The next result follows from the Implicit Function Theorem (which itself follows from the Inverse Function Theorem; see [Abraham, Marsden, and Ratiu 1988]) and provides an easy way of ascertaining when certain subsets of manifolds are submanifolds.

Proposition 3.42 (Submersion Theorem). *If, for $r \in \mathbb{N} \cup \{\infty\} \cup \{\omega\}$, a map $f \in C^r(\mathsf{M}; \mathsf{N})$ is a submersion on $S \subset \mathsf{N}$, then $f^{-1}(y)$ is a closed C^r-submanifold of M for each $y \in S$. Furthermore, for each $y \in S$, $\mathsf{T}_x(f^{-1}(y)) = \ker(T_x f)$ for each $x \in f^{-1}(y)$.*

Since the preimage of a point under a submersion is a submanifold, it seems natural to ask if the image of an immersion is a submanifold. Proposition 3.40 indicates that this is locally the case. However, it is not in general globally true. The following definitions make some useful distinctions.

Definition 3.43 (Embedded and immersed submanifolds). Let N and M be C^r-manifolds, $r \in \mathbb{N} \cup \{\infty\} \cup \{\omega\}$.

(i) A C^r-injective immersion $f\colon \mathsf{N} \to \mathsf{M}$ is a ***C^r-embedding*** if $f(\mathsf{N})$ is a submanifold.

(ii) A subset $\mathsf{S} \subset \mathsf{M}$ is a ***C^r-immersed submanifold*** if there exists a manifold N and a C^r-injective immersion $f\colon \mathsf{N} \to \mathsf{M}$ for which $\mathsf{S} = \text{image}(\mathsf{N})$. •

One may show that an injective immersion f is an embedding if and only if f is a homeomorphism onto its image, where we equip $f(\mathsf{M})$ with the subspace topology acquired from N. Thus a submanifold $\mathsf{S} \subset \mathsf{M}$ has the property that the inclusion map $i_{\mathsf{S}}\colon \mathsf{S} \to \mathsf{M}$ is an embedding. What we call a submanifold is also sometimes explicitly called an ***embedded submanifold***, to distinguish it from an immersed submanifold. Some discussion of the differences between immersed and embedded submanifolds is given in Section 3.9.1. Particularly, there we indicate how one can put a differentiable structure on an immersed submanifold, using the differentiable structure of the larger manifold.

Examples 3.44. 1. As an example of a map that illustrates the subtleties concerning immersions and embeddings, recall the map $f_a\colon \mathbb{R} \to \mathbb{T}^2$ defined in Example 3.39–2. First we note that $T_x f_a$ is injective for all $x \in \mathbb{R}$, so that f_a is an immersion, no matter the value of a. A little thought indicates that, if a is rational, then f_a is not injective and so f_a is not an *injective* immersion, and so not an embedding. In this case it is true, however, that image(f_a) is an embedded submanifold of $\mathbb{T}^2$. For a irrational, f_a *is* injective (the image of $\mathbb{R}$ under f_a in this case is a curve winding around the torus and coming arbitrarily close to, or passing through, every point). Thus image(f_a) is an immersed submanifold. However, f_a is not a homeomorphism onto its image, since, in the subspace topology on $f_a(\mathbb{R}) \subset \mathbb{T}^2$, every neighborhood of a point $x \in \mathbb{R}$ contains points arbitrarily far from x. This is clearly not the normal topology on $\mathbb{R}$, thus showing that image(f_a) is not an embedded submanifold.

2. As we saw in Example 3.39–1, the map $f(\boldsymbol{x}) = \|\boldsymbol{x}\|^2_{\mathbb{R}^{n+1}}$ is a submersion on $\mathbb{S}^n$. Therefore, 1 is a regular value for f, and so $\mathbb{S}^n$ is a submanifold of $\mathbb{R}^{n+1}$ by Proposition 3.42. •

3.4 Vector bundles

The tangent bundle is an example of a particularly useful concept in differential geometry: a bundle. In this section we define vector bundles, vector subbundles, and generalized vector subbundles. Intuitively, a bundle consists of a "total space" (TM in the case of the tangent bundle), a "base space" (M in the case of the tangent bundle), and a projection (π_{TM} in the case of the tangent bundle). The "fiber" of a point in the base space (the tangent space in the case of the tangent bundle) is the preimage of the point under the projection map. In a vector bundle, the fiber for each point b of the base space is a vector space. We refer the reader to [Abraham, Marsden, and Ratiu 1988] for a discussion of vector bundles. A discussion of generalized subbundles of tangent bundles can be found in [Elkin 1999], and in Section 3.9.3 below.

Two important topics in any discussion vector bundles are the notions of a section and of a subbundle. A section is a map from the base space to the total space with the following property: if ξ is a section and b is a point in the base space, then $\xi(b)$ belongs to the fiber of b. In the case of the tangent bundle, a section associates to a point, a tangent vector at that point; we will study such maps in the next section and call them vector fields. A subbundle of a vector bundle is a pointwise assignment of subspaces. As such, subbundles of vector bundles are the analogue of subspaces of vector spaces.

With all the required definitions in place, we introduce the notion of tensor bundles, and define many important examples of tensors in Example 3.55, including vector fields, covector fields and tensor fields. These objects will be studied in more detail in the later sections.

3.4.1 Vector bundles

We start by looking at a specific type of bundle, one that attaches a vector space to each point of a base manifold. Just as the local model for a manifold is an open subset of Euclidean space, we must provide a local model for a vector bundle.

Definition 3.45 (Local vector bundle). Let $\mathcal{U} \subset \mathbb{R}^n$ and $\mathcal{V} \subset \mathbb{R}^m$ be open sets and let $r \in \mathbb{N} \cup \{\infty\} \cup \{\omega\}$.

(i) A ***local vector bundle*** is a product $\mathcal{U} \times \mathbb{R}^k$.

(ii) If $\mathcal{U} \times \mathbb{R}^k$ and $\mathcal{V} \times \mathbb{R}^l$ are local vector bundles, then a map $g\colon \mathcal{U} \times \mathbb{R}^k \to \mathcal{V} \times \mathbb{R}^l$ is a ***C^r-local vector bundle map*** if it has the form $g(\boldsymbol{x}, \boldsymbol{v}) = (g_1(\boldsymbol{x}), g_2(\boldsymbol{x}) \cdot \boldsymbol{v})$, where $g_1\colon \mathcal{U} \to \mathcal{V}$ and $g_2\colon \mathcal{U} \to L(\mathbb{R}^k; \mathbb{R}^l)$ are of class C^r.

(iii) If, in part (ii), g_1 is a C^r-diffeomorphism and $g_2(\boldsymbol{x})$ is an isomorphism for each $\boldsymbol{x} \in \mathcal{U}$, then we say that g is a ***C^r-local vector bundle isomorphism***. •

A vector bundle is constructed, just as was a manifold, by patching together local objects.

Definition 3.46 (Vector bundle). A ***C^r-vector bundle***, $r \in \mathbb{N} \cup \{\infty\} \cup \{\omega\}$, is a set S that has an atlas $\mathscr{A} = \{\mathcal{U}_a, \phi_a\}_{a \in A}$ where image(ϕ_a) is a local vector bundle, $a \in A$, and for which the overlap maps are C^r-local vector bundle isomorphisms. Such an atlas is a ***C^r-vector bundle atlas***. Two C^r-vector bundle atlases, $\mathscr{A}_1$ and $\mathscr{A}_2$, are ***equivalent*** if $\mathscr{A}_1 \cup \mathscr{A}_2$ is a C^r-vector bundle atlas. A ***C^r-vector bundle structure*** is an equivalence class of such atlases. A chart in one of these atlases is called an ***admissible vector bundle chart***. A typical vector bundle will be denoted by V. •

The ***base space*** B of a vector bundle V is given by all points $v \in \mathsf{V}$ having the property that there exists an admissible vector bundle chart $(\mathcal{V}, \psi)$ such that $\psi(v) = (\boldsymbol{x}, \boldsymbol{0}) \in \mathcal{U} \times \mathbb{R}^k$. This definition may be shown to make sense, since the overlap maps are local vector bundle isomorphisms that map the zero vector in one local vector bundle to the zero vector in another. To any point $v \in \mathsf{V}$ we associate a point $b \in \mathsf{B}$ as follows. Let $(\mathcal{V}, \psi)$ be a local vector bundle chart for V around v. Thus $\psi(v) = (\boldsymbol{x}, \boldsymbol{v}) \in \mathcal{U} \times \mathbb{R}^k$. Define $b = \psi^{-1}(\boldsymbol{x}, \boldsymbol{0})$. Once again, since the overlap maps are local vector bundle isomorphisms, this definition makes sense. We denote the resulting map from V to B by π and we call this the ***vector bundle projection***. Sometimes we will write a vector bundle as $\pi \colon \mathsf{V} \to \mathsf{B}$.

One may verify the following properties of vector bundles.

Proposition 3.47. *Let $\pi \colon \mathsf{V} \to \mathsf{B}$ be a C^r-vector bundle, $r \in \mathbb{N} \cup \{\infty\} \cup \{\omega\}$. Then*

(i) B is a C^r-submanifold of V, and
(ii) π is a surjective submersion of class C^r.

When we wish to think of the base space B as a submanifold of V, we shall call it the ***zero section*** and denote it by $Z(\mathsf{V})$. For $b \in \mathsf{B}$, the set $\pi^{-1}(b)$ is the ***fiber*** over b, and is often written V_b. One may verify that the operations of vector addition and scalar multiplication defined on V_b in a fixed vector bundle chart are actually independent of the choice made for this chart. Thus V_b is indeed a vector space. We will sometimes denote the zero vector in V_b as 0_b.

Examples 3.48. 1. Let M be a manifold. We define a vector bundle structure on $\mathsf{M} \times \mathbb{R}^n$ as follows. We let $\mathscr{A} = \{(\mathcal{U}_a, \phi_a)\}_{a \in A}$ be an atlas for M and define an associated vector bundle atlas by $\mathscr{A}_{\text{trivial}} = \{(\mathcal{U}_a \times \mathbb{R}^n, \psi_{a,\text{trivial}})\}_{a \in A}$, where $\psi_{a,\text{trivial}}(x, \boldsymbol{v}) = (\phi_a(x), \boldsymbol{v})$. A vector bundle that is a product manifold, and with the vector bundle structure as defined above, is called ***trivial***.

2. If $\pi \colon \mathsf{V} \to \mathsf{B}$ is a vector bundle, and if $\mathsf{S} \subset \mathsf{B}$ is a submanifold, then $\pi|\pi^{-1}(\mathsf{S}) \colon \pi^{-1}(\mathsf{S}) \to \mathsf{S}$ may be shown to inherit a natural vector bundle structure from V using submanifold charts. This is called the ***restricted bundle***, and is denoted by $\mathsf{V}|\mathsf{S}$.
3. The tangent bundle of a C^{r+1} manifold M is a C^r-vector bundle with an atlas of local vector bundle charts being given by the natural atlas for TM.
4. Let us construct a vector bundle that is not trivial. We do this by defining an equivalence relation R in the set $[0,1] \times \mathbb{R}$ by

$$R = \{\, ((0,y),(1,-y)) \mid\ y \in \mathbb{R}\} \cup \{\, ((x,y),(x,y)) \mid\ (x,y) \in]0,1[\times \mathbb{R}\}.$$

This construction resembles that for the 2-torus in Example 3.26–5. The ***Möbius vector bundle*** is then $([0,1] \times \mathbb{R})/\sim$, where $\sim$ means equivalence as defined by R. The equivalence classes of points in $[0,1] \times \{0\}$ form the zero section. The base space is, therefore, diffeomorphic to $\mathbb{S}^1$, and the fibers are isomorphic to $\mathbb{R}$. (Strictly speaking, we should define a vector bundle atlas, but we are just trying to informally understand this example.) Intuitively, since the edges of $[0,1] \times \mathbb{R}$ are identified with opposite orientation, the vector bundle has a "twist," much like the well-known Möbius strip. Because of this twist, the resulting vector bundle is not "equivalent to" $\mathbb{S}^1 \times \mathbb{R}$. (We shall make precise below what we mean by "equivalent to." It means "isomorphic as vector bundles.") We illustrate the Möbius vector bundle in Figure 3.10, where the lines on the surface

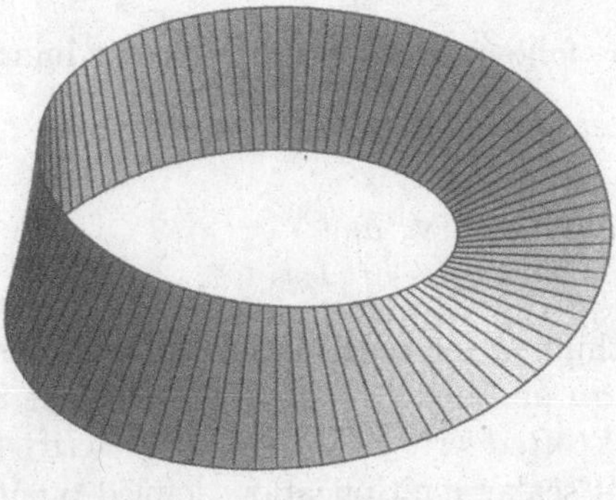

Figure 3.10. The Möbius vector bundle

represent the fibers of the bundle. •

An important construction is the choice of a single vector from the fiber V_b for each point $b \in \mathsf{B}$.

Definition 3.49 (Section of a vector bundle). For $r \in \mathbb{N} \cup \{\infty\} \cup \{\omega\}$, a ***$C^r$-section*** of a vector bundle $\pi \colon \mathsf{V} \to \mathsf{B}$ is a class C^r map $\xi \colon \mathsf{B} \to \mathsf{V}$ for which $\pi \circ \xi = \mathrm{id}_{\mathsf{B}}$. The set of C^r-sections of V is denoted by $\Gamma^r(\mathsf{V})$. •

The last property in the definition ensures that $\xi(b) \in \mathsf{V}_b$, as desired. The set of sections is a $\mathbb{R}$-vector space with vector addition and scalar multiplication defined by

$$(\xi_1 + \xi_1)(b) = \xi_1(b) + \xi_2(b), \quad (a\xi)(b) = a(\xi(b)).$$

This vector space will generally be infinite-dimensional. The set of sections also admits multiplication by scalars that are functions on B as follows. For $f \in C^r(\mathsf{B})$ and $\xi \in \Gamma^r(\mathsf{V})$, define $f\xi \in \Gamma^r(\mathsf{V})$ by $(f\xi)(b) = f(b)\xi(b)$. This gives the set of sections the structure of a module over the ring $C^r(\mathsf{B})$, for those who know a little algebra (we refer the reader to Section 3.9.4 for the definitions of a ring and a module). We will sometimes wish to consider the span in this case, and this is defined as follows. Let $\mathscr{V} \subset \Gamma^r(\mathsf{V})$. We denote by $\mathrm{span}_{C^r(\mathsf{B})}\{\mathscr{V}\}$ the smallest module over $C^r(\mathsf{B})$ containing $\mathscr{V}$. Explicitly, $\mathrm{span}_{C^r(\mathsf{B})}\{\mathscr{V}\}$ consists of sections of V of the form

$$f^1\xi_1 + \cdots + f^k\xi_k, \qquad k \in \mathbb{N},\ f^1, \dots, f^k \in C^r(\mathsf{B}),\ \xi_1, \dots, \xi_k \in \mathscr{V}.$$

Issues related to this, in the case where V is a tangent bundle, are the subject of Section 3.9.4.

We sometimes also wish to talk about sections along curves. For $\gamma\colon I \to \mathsf{B}$ a C^r-curve, a ***C^r-section along γ*** is a C^r-map $\xi\colon I \to \mathsf{V}$ having the property that $\xi(t) \in \mathsf{V}_{\gamma(t)}$ for each $t \in I$.

Just as submanifolds are distinguished subsets of manifolds, so too do vector bundles possess distinguished subsets.

Definition 3.50 (Subbundle of a vector bundle). Let $\pi\colon \mathsf{V} \to \mathsf{B}$ be a C^r-vector bundle, $r \in \mathbb{N} \cup \{\infty\} \cup \{\omega\}$. A subset $\mathsf{W} \subset \mathsf{V}$ is a ***C^r-vector subbundle***, or simply a ***C^r-subbundle***, if, for every $b \in \mathsf{B}$, there exists a chart $(\mathcal{U}, \phi)$ for B, and an admissible vector bundle chart $(\pi^{-1}(\mathcal{U}), \psi)$ for V, such that

(i) $b \in \mathcal{U}$,
(ii) ψ takes its values in $\phi(\mathcal{U}) \times \mathbb{R}^l \times \mathbb{R}^{k-l}$,
(iii) the local representative of π is

$$\phi(\mathcal{U}) \times \mathbb{R}^l \times \mathbb{R}^{k-l} \ni (\boldsymbol{x}, \boldsymbol{u}, \boldsymbol{v}) \mapsto \boldsymbol{x} \in \phi(\mathcal{U}),$$

and
(iv) $\psi(\pi^{-1}(\mathcal{U}) \cap \mathsf{W}) = \phi(\mathcal{U}) \times \mathbb{R}^l \times \{\boldsymbol{0}\}$. •

Note that a vector subbundle of a vector bundle is a submanifold. The essential idea of a vector subbundle is that one can "smoothly" select a subspace $\mathsf{W}_b = \mathsf{V}_b \cap \mathsf{W}$ from each fiber V_b. The dimension of the subspace W_b is called the ***rank*** of the subbundle at b, denoted by $\operatorname{rank}(\mathsf{W}_b)$. The notion of subbundle we give has the property that around each point $b \in \mathsf{B}$, the rank is constant. In particular, if B is connected, then the rank is constant on the entire base space. This is actually a severe restriction, both in terms of the theory, and in practice. In practice, we will often encounter objects, similar to subbundles, that come from the physics of the problem, but that do not have locally constant rank. Furthermore, we will be computing derived subbundles, and sequences of derived subbundles, any of which might have locally variable rank. For this reason, we will often employ a more general notion of a subbundle as follows.

Definition 3.51 (Generalized subbundle of a vector bundle). For $r \in \mathbb{N} \cup \{\infty\} \cup \{\omega\}$, a ***$C^r$-generalized subbundle*** of a vector bundle $\pi\colon \mathsf{V} \to \mathsf{B}$ is a subset $\mathsf{W} \subset \mathsf{V}$ with the properties

(i) for each $b_0 \in \mathsf{B}$, $\mathsf{V}_{b_0} \cap \mathsf{W}$ is a subspace, and
(ii) for each $b_0 \in \mathsf{B}$, there exist a neighborhood $\mathcal{U} \subset \mathsf{B}$ of b_0 and a family $\{\xi_a\}_{a \in A}$ of C^r-sections of $\mathsf{V}|\mathcal{U}$, called ***local generators*** of W, such that $\mathsf{V}_b \cap \mathsf{W} = \operatorname{span}_{\mathbb{R}} \{\xi_a(b) \mid a \in A\}$ for each $b \in \mathcal{U}$.

A C^r-generalized subbundle W is ***locally finitely generated*** if, for each $b_0 \in \mathsf{B}$, there exist a neighborhood $\mathcal{U} \subset \mathsf{B}$ of b_0 and a finite collection $\{\xi_1, \dots, \xi_k\}$ of C^r-sections of V such that $\mathsf{V}_b \cap \mathsf{W} = \operatorname{span}_{\mathbb{R}} \{\xi_1(b), \dots, \xi_k(b)\}$ for each $b \in \mathcal{U}$. If we can take $\mathcal{U} = \mathsf{B}$ in the preceding definition, then W is ***globally finitely generated***. •

Matters concerning the nature and existence of local and global generators are discussed further in Sections 3.9.3 and 3.9.4 below.

Given two C^r-generalized subbundles W_1 and W_2 of $\pi\colon \mathsf{V} \to \mathsf{B}$, we may define their ***sum*** to be the C^r-generalized subbundle $\mathsf{W}_1 + \mathsf{W}_2$ whose fiber at $b \in \mathsf{B}$ is

$$(\mathsf{W}_1 + \mathsf{W}_2)_b = \{ w_1 + w_2 \mid w_1 \in \mathsf{W}_{1,b},\ w_2 \in \mathsf{W}_{2,b} \}.$$

It is easy to see that $\mathsf{W}_1 + \mathsf{W}_2$ is indeed a C^r-generalized subbundle. However, it may be the case that even if b is a regular point for both W_1 and W_2, it may not be a regular point for $\mathsf{W}_1 + \mathsf{W}_2$. If $\mathsf{W}_1 \cap \mathsf{W}_2 = Z(\mathsf{V})$, then we write the sum as $\mathsf{W}_1 \oplus \mathsf{W}_2$, which is the ***direct sum*** of W_1 and W_2. The direct sum of vector bundles is sometimes called the ***Whitney sum***.

Natural maps between vector bundles are defined as follows.

Definition 3.52 (Vector bundle map). If $\pi\colon \mathsf{V} \to \mathsf{B}$ and $\sigma\colon \mathsf{W} \to \mathsf{A}$ are C^r-vector bundles, $r \in \mathbb{N} \cup \{\infty\} \cup \{\omega\}$, a C^r-map $f\colon \mathsf{V} \to \mathsf{W}$ is a ***C^r-vector bundle map*** if

(i) there exists a C^r-map $f_0\colon \mathsf{B} \to \mathsf{A}$ so that the diagram

$$\begin{array}{ccc} \mathsf{V} & \xrightarrow{\ f\ } & \mathsf{W} \\ {\scriptstyle \pi}\big\downarrow & & \big\downarrow{\scriptstyle \sigma} \\ \mathsf{B} & \xrightarrow[\ f_0\ ]{} & \mathsf{A} \end{array}$$

commutes, and

(ii) $f|\mathsf{V}_b \in L(\mathsf{V}_b; \mathsf{W}_{f_0(b)})$.

If f is a C^r-diffeomorphism, we call it a ***C^r-vector bundle isomorphism***. •

If V and W are vector bundles between which there is a natural isomorphism, then we will often write $\mathsf{V} \simeq \mathsf{W}$. For example, we can write $\mathsf{T}\mathbb{S}^1 \simeq \mathbb{S}^1 \times \mathbb{R}$, following Example 3.35–3.

Remark 3.53. A vector bundle that is isomorphic to a trivial vector bundle is called ***trivializable***. The vector bundle of Example 3.48–4 is not trivializable. An interesting question is, "Are tangent bundles trivializable?" The answer is, "Not generally," and a manifold whose tangent bundle is trivializable is called ***parallelizable***. It is well-known that $\mathbb{S}^{2n}$, $n \in \mathbb{N}$, is not parallelizable. The theorem stating this fact is called the "Hairy Ball Theorem," the name arising because one cannot "comb" smoothly the tangent vectors on spheres of even dimension. A fairly elementary proof can be found in the paper of Milnor [1978]. This discussion illustrates why one should refrain from thinking of the tangent bundle as a product manifold, except in the exceptional case of open subsets of Euclidean space. •

Examples 3.54. 1. If $f\colon \mathsf{V} \to \mathsf{W}$ is a vector bundle map of vector bundles $\pi\colon \mathsf{V} \to \mathsf{B}$ and $\sigma\colon \mathsf{W} \to \mathsf{A}$, and if the dimension of image$(f|\mathsf{V}_b)$ is independent of $b \in B$, then image(f) is a vector subbundle of W. If dim(image$(f|\mathsf{V}_b)$) is not independent of b, then image(f) is a generalized subbundle (see Exercise E3.18).

2. Similarly, if dim(ker$(f|\mathsf{V}_b)$) is independent of $b \in B$, then $\ker(f) = \{v \mid f(v) \in Z(\mathsf{W})\}$ is a subbundle of V. It is *not* generally true that if $\ker(f)$ does not have locally constant rank, then it is a generalized subbundle (again, see Exercise E3.18). •

3.4.2 Tensor bundles

Given a vector bundle $\pi\colon \mathsf{V} \to \mathsf{B}$, one can naturally associate to it various other vector bundles over B, constructed from the vector bundle V in a "natural" way. For these new vector bundles, there are unified ways of constructing natural vector bundle atlases. We do not wish to undertake the development necessary to make this precise in a general way, but merely remark that, for the examples we consider in this section, it is easy to see how one can use the underlying vector bundle structure on $\pi\colon \mathsf{V} \to \mathsf{B}$ to induce a vector bundle structure on the new vector bundle. For tangent bundles, this is essentially implicit in the discussion about local representations of tensors at the beginning of Section 3.6.2. For details, we refer to [Abraham, Marsden, and Ratiu 1988, Lang 1995].

Having relieved ourselves of the burden of having to fully justify that the objects we are about to construct are indeed vector bundles, let us merely list the examples that will be most useful to us. We give only the briefest hint as to the character of the vector bundle structure. For the following examples, we let $k \in \mathbb{N} \cup \{\infty\} \cup \{\omega\}$.

Examples 3.55. 1. Let $\pi\colon \mathsf{V} \to \mathsf{B}$ and $\sigma\colon \mathsf{W} \to \mathsf{B}$ be C^k-vector bundles over B. One can then define a C^k-vector bundle $L(\mathsf{V};\mathsf{W})$ over B by asking that the fiber over $b \in \mathsf{B}$ consist of linear maps from V_b to W_b. The vector bundle structure is induced by the change of basis formula derived from the overlap conditions for the vector bundles V and W.

   If $\mathsf{W} = \mathsf{B} \times \mathbb{R}$, then the bundle $L(\mathsf{V};\mathsf{W})$ is the ***dual vector bundle*** of V, which we denote V^*. In the case in which $\mathsf{B} = \mathsf{M}$ and $\mathsf{V} = \mathsf{TM}$ is the tangent bundle, the resulting vector bundle over M is the ***cotangent bundle***, denoted by $\mathsf{T}^*\mathsf{M}$. The projection from $\mathsf{T}^*\mathsf{M}$ to M we denote by $\pi_{\mathsf{T}^*\mathsf{M}}\colon \mathsf{T}^*\mathsf{M} \to \mathsf{M}$.

2. For $r, s \in \mathbb{Z}_+$, we let $T^r_s(\mathsf{V})$ be the vector bundle over B whose fiber over $b \in \mathsf{B}$ is the set of (r,s)-tensors on V_b. The vector bundle structure of $T^r_s(\mathsf{V})$ is derived from the vector bundle structure for V using the push-forward by the linear part of the overlap maps. Sections of $T^r_s(\mathsf{V})$ that are of class C^k will simply be said to be of ***class C^k***; we will not have special notation to indicate the degree of differentiability.

In the case that $\mathsf{B} = \mathsf{M}$ and $\mathsf{V} = \mathsf{TM}$ is the tangent bundle, the C^k-sections of $T^r_s(\mathsf{TM})$ are called ***class C^k (r,s)-tensor fields*** on M. One defines the ***tensor product*** of tensor fields $t_1 \in \Gamma^\infty(T^{r_1}_{s_1}(\mathsf{TM}))$ and $t_2 \in \Gamma^\infty(T^{r_2}_{s_2}(\mathsf{TM}))$ by $(t_1 \otimes t_2)(x) = t_1(x) \otimes t_2(x)$.

3. A ***C^k-vector field*** on M is a class C^k $(1,0)$-tensor field on M.
4. A ***C^k-covector field*** on M is a class C^k $(0,1)$-tensor field on M.
5. A ***C^k-Riemannian metric*** on M is a class C^k $(0,2)$-tensor field $\mathbb{G}$ on M having the property that $\mathbb{G}(x)$ is an inner product on $\mathsf{T}_x\mathsf{M}$. A C^k-manifold M with a C^k-Riemannian metric $\mathbb{G}$ is called a ***C^k-Riemannian manifold***, and we will often denote a Riemannian manifold as the pair $(\mathsf{M}, \mathbb{G})$. As we shall see in Chapter 4, Riemannian metrics will be an important element in the physical data for the problems from mechanics that we consider in this book. Following our notation of Section 2.3.4, for each $x \in \mathsf{M}$, we have isomorphisms $\mathbb{G}(x)^\flat \colon \mathsf{T}_x\mathsf{M} \to \mathsf{T}^*_x\mathsf{M}$ and $\mathbb{G}(x)^\sharp \colon \mathsf{T}^*_x\mathsf{M} \to \mathsf{T}_x\mathsf{M}$. By taking these maps at each point in M, this defines two vector bundle isomorphisms $\mathbb{G}^\flat \colon \mathsf{TM} \to \mathsf{T}^*\mathsf{M}$ and $\mathbb{G}^\sharp \colon \mathsf{T}^*\mathsf{M} \to \mathsf{TM}$ called the ***associated isomorphisms*** for the Riemannian metric $\mathbb{G}$. •

Given $t \in \Gamma^k(T^r_s(\mathsf{TM}))$, $\alpha^1, \dots, \alpha^r \in \Gamma^k(\mathsf{T}^*\mathsf{M})$, and $X_1, \dots, X_s \in \Gamma^k(\mathsf{TM})$, we define a function on M by

$$x \mapsto t(x) \cdot (\alpha^1(x), \dots, \alpha^r(x), X_1(x), \dots, X_s(x)).$$

In this way, we can think of t as being a map

$$t \colon \Gamma^k(\mathsf{T}^*\mathsf{M}) \times \cdots \times \Gamma^k(\mathsf{T}^*\mathsf{M}) \times \Gamma^k(\mathsf{TM}) \times \cdots \times \Gamma^k(\mathsf{TM}) \to C^\infty(\mathsf{M}). \quad (3.7)$$

Note that the map t in (3.7) is linear with respect to multiplication by elements in $C^k(\mathsf{M})$. In this case, we say that the map is ***$C^\infty(\mathsf{M})$-linear***. It is not uncommon to say that a map that is $C^\infty(\mathsf{M})$-linear is ***tensorial*** in its arguments. It will be convenient, at times, to use this language. With this language, there is a 1-1 correspondence between tensors and maps with the domain and codomain in (3.7) that are tensorial.

3.5 Vector fields

One of our more commonly utilized geometric objects will be the vector field. As we saw in Example 3.55–3, a ***C^r-vector field*** on M is an element of $\Gamma^r(\mathsf{TM})$, i.e., a C^r-section of the tangent bundle of M. In other words, a vector field assigns a tangent vector to each point on the manifold in an appropriately smooth way. Associated with vector fields is an extraordinarily rich and diverse structure. We shall be thinking of vector fields in a multitude of contexts, and using them in many seemingly distinct ways. In particular, we shall extensively study them as differential operators in Section 3.5.1 and show how to use them to differentiate functions and other vector fields. In Section 3.5.2 we shall see how there is a 1–1 correspondence between ordinary differential equations and vector fields. Also of key importance is the operation of Lie bracket between vector fields: it is defined as an appropriate composition of differential operators, and it frequently arises in the composition of flows of differential equations.

After reading this section the reader should be comfortable with the many characters that a vector field may assume. For proofs and additional examples we refer, as usual, to [Abraham, Marsden, and Ratiu 1988].

3.5.1 Vector fields as differential operators

Let us first consider differential equations as operators, first on functions, then on other vector fields, and finally on general tensor fields. First, let us look a little at the character of C^r-functions for $r \in \mathbb{N} \cup \{\infty\} \cup \{\omega\}$. For $f \in C^r(\mathsf{M})$, we have $Tf\colon \mathsf{TM} \to \mathsf{T}\mathbb{R}$. Since $\mathsf{T}\mathbb{R} \simeq \mathbb{R} \times \mathbb{R}$, we may write

$$Tf(v_x) = (f(x), \alpha_f(x) \cdot v_x)$$

for some $\alpha_f(x) \in L(\mathsf{T}_x\mathsf{M}; \mathbb{R}) = \mathsf{T}_x^*\mathsf{M}$. This cotangent vector is important, so we give it a name.

Definition 3.56 (Differential of a function). With the above notation, the element $\alpha_f(x) \in \mathsf{T}_x^*\mathsf{M}$ is the ***differential*** of f at x, and is denoted by $\mathrm{d}f(x)$. The C^{r-1} section $x \mapsto \mathrm{d}f(x)$ of $\mathsf{T}^*\mathsf{M}$ is denoted by $\mathrm{d}f$. •

Using the differential, one may contract $\mathrm{d}f$ with X to get a useful object.

Definition 3.57 (Lie derivative of a function). If $X \in \Gamma^r(\mathsf{TM})$ and $f \in C^r(\mathsf{M})$, $r \in \mathbb{N} \cup \{\infty\} \cup \{\omega\}$, the function $\mathscr{L}_X f \in C^{r-1}(\mathsf{M})$ defined by $x \mapsto \mathrm{d}f(x) \cdot X(x)$ is the ***Lie derivative*** of f with respect to X. •

One may easily verify the following result by applying the definition of $\mathrm{d}f$.

Proposition 3.58 (Properties of the Lie derivative of a function). *Let $X \in \Gamma^r(\mathsf{TM})$ and $f \in C^r(\mathsf{M})$, $r \in \mathbb{N} \cup \{\infty\} \cup \{\omega\}$. Then*

(i) the map $f \mapsto \mathscr{L}_X f$ is $\mathbb{R}$-linear with respect to the operations of vector addition and scalar multiplication on $C^{r-1}(\mathsf{M})$,[1] *and*

(ii) $\mathscr{L}_X(fg) = (\mathscr{L}_X f)g + (\mathscr{L}_X g)f$.

One may less readily verify that, in the case in which $r = \infty$, there is a 1–1 correspondence between $\Gamma^\infty(\mathsf{TM})$ and operators on $C^\infty(\mathsf{M})$ having the

[1] The operations of vector addition and scalar multiplication on $C^{r-1}(\mathsf{M})$ are defined by

$$(f_1 + f_2)(x) = f_1(x) + f_2(x), \quad (af)(x) = a(f(x)),$$

where $a \in \mathbb{R}$ and $f, f_1, f_2 \in C^{r-1}(\mathsf{M})$.

properties given in Proposition 3.58. Indeed, some authors *define* vector fields as differential operators on functions.

For us, this differential operator property of vector fields allows a convenient manner of establishing local representations for vector fields. Indeed, let $(\mathcal{U}, \phi)$ be a chart with coordinates $(x^1, \dots, x^n)$. The coordinates may be regarded as functions on $\mathcal{U}$, and we define n distinguished vector fields, denoted by $\frac{\partial}{\partial x^1}, \dots, \frac{\partial}{\partial x^n}$, on $\mathcal{U}$ by defining the Lie derivative of the coordinate functions with respect to these vector fields. Precisely, we define $\frac{\partial}{\partial x^i} \in \Gamma^\infty(\mathsf{T}\mathcal{U})$ by

$$\mathscr{L}_{\frac{\partial}{\partial x^i}} x^j = \delta_i^j, \qquad i, j \in \{1, \dots, n\}.$$

It is not completely obvious that this uniquely defines the vector fields $\frac{\partial}{\partial x^1}, \dots, \frac{\partial}{\partial x^n}$, but this is indeed the case. Furthermore, at each point in $\mathcal{U}$, these vector fields are linearly independent. This means that, if $X \in \Gamma^\infty(\mathsf{T}\mathcal{U})$, then we may write $X = X^i \frac{\partial}{\partial x^i}$ for functions $X^i \in C^\infty(\mathcal{U})$, $i \in \{1, \dots, n\}$, called the ***components*** of X in the chart $(\mathcal{U}, \phi)$. Note again that we use the summation convention, with superscripts in the denominator being regarded as subscripts. The partial derivative notation for the basis vector fields $\frac{\partial}{\partial x^1}, \dots, \frac{\partial}{\partial x^n}$ may seem strange initially. However, the means we have used to make these definitions hopefully provides some motivation for the notation. To strengthen the case for this notation, we comment that one can easily show that the local representative of the function $\mathscr{L}_X f$ in a chart $(\mathcal{U}, \phi)$ is

$$\mathscr{L}_X f(x) = \frac{\partial (f \circ \phi^{-1})}{\partial x^i} (\phi(x)) X^i(x).$$

The local representative of the vector field $\frac{\partial}{\partial x^i}$ is $\boldsymbol{x} \mapsto (\boldsymbol{x}, \boldsymbol{e}_i)$, $i \in \{1, \dots, n\}$, where $\{\boldsymbol{e}_1, \dots, \boldsymbol{e}_n\}$ is the standard basis for $\mathbb{R}^n$.

It is useful to know how the basis vector fields $\frac{\partial}{\partial x^1}, \dots, \frac{\partial}{\partial x^n}$, and how the components $X^1, \dots, X^n$ of a vector field, transform under changes of coordinate. The following lemma records this, and may be proved easily by applying the definitions.

Lemma 3.59. *Let $(\mathcal{U}, \phi)$ and $(\mathcal{V}, \psi)$ be overlapping charts on a C^∞-manifold M with coordinates denoted by $(x^1, \dots, x^n)$ and $(\tilde{x}^1, \dots, \tilde{x}^n)$, respectively. Also, let X be a vector field with components $(X^1, \dots, X^n)$ and $(\tilde{X}^1, \dots, \tilde{X}^n)$, respectively, in the charts $(\mathcal{U}, \phi)$ and $(\mathcal{V}, \psi)$. Then*

(i) $\dfrac{\partial}{\partial \tilde{x}^i} = \dfrac{\partial x^j}{\partial \tilde{x}^i} \dfrac{\partial}{\partial x^j}$, *$i \in \{1, \dots, n\}$, and*

(ii) $\tilde{X}^i = \dfrac{\partial \tilde{x}^i}{\partial x^j} X^j$, *$i \in \{1, \dots, n\}$.*

Note that the basis vectors transform just like the differential operators, as suggested by the notation. Also, components of vector fields transform like the fiber coordinates for TM, or like velocities if one wishes to think in this way.

Let us see how this works out in our polar coordinate example.

Example 3.60 (Polar coordinates cont'd). Recall that we had defined two coordinate charts on $\mathsf{M} = \mathbb{R}^2$; $(\mathcal{U}_1, \phi_1)$ were the standard Cartesian coordinates, and $(\mathcal{U}_2, \phi_2)$ were polar coordinates. We shall define vector fields in each coordinate chart, and show that they represent the same vector field on M.

Suppose that we first consider the vector field whose representation in Cartesian coordinates is $X_{\phi_1} = -y\frac{\partial}{\partial x} + x\frac{\partial}{\partial y}$. We wish to show that this is the same as the vector field represented in polar coordinates by $X_{\phi_2} = \frac{\partial}{\partial \theta}$. By Lemma 3.59 we have

$$\begin{aligned}\frac{\partial}{\partial r} &= \frac{\partial x}{\partial r}\frac{\partial}{\partial x} + \frac{\partial y}{\partial r}\frac{\partial}{\partial y} = \cos\theta\frac{\partial}{\partial x} + \sin\theta\frac{\partial}{\partial y}\\ &= \frac{x}{\sqrt{x^2+y^2}}\frac{\partial}{\partial x} + \frac{y}{\sqrt{x^2+y^2}}\frac{\partial}{\partial y},\end{aligned}$$

and

$$\begin{aligned}\frac{\partial}{\partial \theta} &= \frac{\partial x}{\partial \theta}\frac{\partial}{\partial x} + \frac{\partial y}{\partial \theta}\frac{\partial}{\partial y} = -r\sin\theta\frac{\partial}{\partial x} + r\cos\theta\frac{\partial}{\partial y}\\ &= -y\frac{\partial}{\partial x} + x\frac{\partial}{\partial y}.\end{aligned}$$

Therefore, we directly see that

$$X_{\phi_2} = \frac{\partial}{\partial \theta} = -y\frac{\partial}{\partial x} + x\frac{\partial}{\partial y} = X_{\phi_1},$$

which means that the two representations are of the same vector field on M.

One can also arrive at the same result by using components rather than basis vector fields. Indeed, the components of X in the coordinates (x, y) are $(-y, x)$ and the components in the coordinates (r, θ) are $(0, 1)$. Following Lemma 3.59 we have

$$\frac{\partial r}{\partial x}(-y) + \frac{\partial r}{\partial y}(x) = \frac{x}{\sqrt{x^2+y^2}}(-y) + \frac{y}{\sqrt{x^2+y^2}}(x) = 0,$$

and

$$\frac{\partial \theta}{\partial x}(-y) + \frac{\partial \theta}{\partial y}(x) = -\frac{y}{x^2+y^2}(-y) + \frac{x}{x^2+y^2}(x) = 1.$$

Again we see that the two coordinate representations agree. •

The n-torus provides another concrete illustration of the notion of a vector field.

Example 3.61. We let $(x^1, \dots, x^n)$ be the natural coordinates for $\mathbb{T}^n$ introduced in Example 3.26–5. Corresponding to these coordinates are the basis vector fields $\frac{\partial}{\partial x^i}$, $i \in \{1, \dots, n\}$, defined on the domain of the chart. However, one can easily see that each of these vector fields extends uniquely to a vector field on all of $\mathbb{T}^n$. For example, when $n = 1$ then $\mathbb{T}^1 \simeq \mathbb{S}^1$, and the resulting vector field is $(x, y) \mapsto ((x, y), (-y, x))$, i.e., $\frac{\partial}{\partial \theta}$. •

Now let us introduce another important operation involving vector fields. Given $X, Y \in \Gamma^\infty(\mathsf{TM})$, one can verify that the map

$$C^\infty(\mathsf{M}) \ni f \mapsto \mathscr{L}_X \mathscr{L}_Y f - \mathscr{L}_Y \mathscr{L}_X f \in C^\infty(\mathsf{M})$$

satisfies the two differential operator properties of the Lie derivative with respect to a vector field given in Proposition 3.58. This essentially amounts to showing that the expression on the right depends only on the first derivatives of the function f, since the terms involving second derivatives cancel. This allows us to define, associated with vector fields X and Y, a new vector field, the importance of which is, at this point, not clear. However, we shall see in our subsequent developments in both mechanics and control theory that it is fundamental.

Definition 3.62 (Lie bracket). For $X, Y \in \Gamma^\infty(\mathsf{TM})$, the vector field $[X, Y]$ defined by

$$\mathscr{L}_{[X,Y]} f = \mathscr{L}_X \mathscr{L}_Y f - \mathscr{L}_Y \mathscr{L}_X f, \qquad f \in C^\infty(\mathsf{M}),$$

is the ***Lie bracket*** of X and Y, or the ***Lie derivative*** of Y with respect to X. •

Some authors refer to the Lie bracket as the ***Jacobi bracket***, or as the ***Jacobi–Lie bracket***. We will sometimes use the notation $\mathrm{ad}_X Y = [X, Y]$, which is useful in control theoretic applications, or the notation $\mathscr{L}_X Y = [X, Y]$. A slightly messy coordinate calculation (the reader is asked to produce this in Exercise E3.20) shows that in coordinates $(x^1, \dots, x^n)$, the components of $[X, Y]$ are

$$[X, Y]^i = \frac{\partial Y^i}{\partial x^j} X^j - \frac{\partial X^i}{\partial x^j} Y^j, \qquad i \in \{1, \dots, n\}. \tag{3.8}$$

The abstract Definition 3.62 only applies when $r = \infty$, since the correspondence between differential operators and vector fields only holds when $r = \infty$. Note that the coordinate formula (3.8) allows us the define the Lie bracket between C^r-vector fields when $r < \infty$, with the resulting vector field being of class C^{r-1}.

From its definition, it is not clear what is the "meaning" of the Lie bracket. In actuality, it has many interpretations, and the most useful one for us will be given in Proposition 3.69. For now, let us record some properties of the Lie bracket that essentially follow directly from its coordinate expression (and can also easily be deduced from its definition).

Proposition 3.63 (Properties of the Lie bracket). *For all $f, g \in C^\infty(\mathsf{M})$ and for all $X, Y, Z \in \Gamma^\infty(\mathsf{TM})$, we have*

(i) $[X, Y] = -[Y, X]$ *(**skew-symmetry**),*
(ii) $[X + Y, Z] = [X, Z] + [Y, Z]$ *(**linearity**),*
(iii) $[fX, gY] = fg[X, Y] + f(\mathscr{L}_X g)Y - g(\mathscr{L}_Y f)X$, *and*
(iv) $[X, [Y, Z]] + [Y, [Z, X]] + [Z, [X, Y]] = 0$ *(**Jacobi identity**).*

3.5.2 Vector fields and ordinary differential equations

Associated with a vector field is a notion with which most readers are probably familiar, the notion of an ordinary differential equation. The relationship goes as follows.

Definition 3.64 (Integral curve of a vector field). Let $r \in \mathbb{N} \cup \{\infty\} \cup \{\omega\}$. A differentiable curve $\gamma\colon I \to \mathsf{M}$ at $x \in M$ is an ***integral curve at x*** for a C^r-vector field X if $\gamma'(t) = X(\gamma(t))$ and $\gamma(0) = x$ (it is implicit that $0 \in I$). •

The idea is simply that the tangent vector to the curve γ is equal to the tangent vector specified by the vector field at each point along the curve (see Figure 3.11). The notion of an integral curve is quite concrete in coordi-

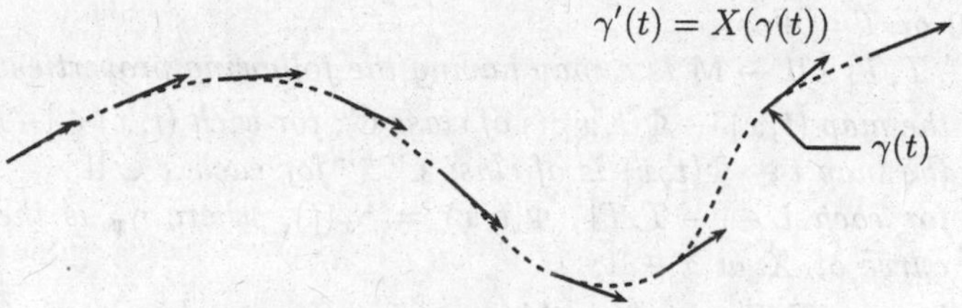

Figure 3.11. An integral curve

nates. The next result follows from a direct coordinate computation and an application of the definition of an integral curve.

Proposition 3.65 (Relationship between vector fields and ordinary differential equations). *Suppose that a curve $\gamma\colon I \to \mathsf{M}$ is represented in a chart $(\mathcal{U}, \phi)$ by $t \mapsto (x^1(t), \ldots, x^n(t))$, and that $X^1, \ldots, X^n$ are the components of X in the same set of coordinates. Then γ is an integral curve for X at $x \in \mathsf{M}$ if and only if $t \mapsto (x^1(t), \ldots, x^n(t))$ satisfies the initial value problem*

$$\begin{aligned}
\dot{x}^1(t) &= X^1(x^1(t), \ldots, x^n(t)), \\
&\vdots \\
\dot{x}^n(t) &= X^n(x^1(t), \ldots, x^n(t)), \\
\boldsymbol{x}(0) &= \phi(x).
\end{aligned}$$

Thus, in a coordinate chart, to determine integral curves one obtains the solution of an ordinary differential equation. While it is generally not possible to do this explicitly, one can at least resort to the theory of existence and uniqueness of solutions of ordinary differential equations to assert the local existence and uniqueness of integral curves for vector fields. Nice introductions to the theory of ordinary differential equations are the books of Arnol'd [1992] and Hirsch and Smale [1974]. More advanced treatments include [Coddington and Levinson 1955, Hartman 1964], and control theoretic treatments may be found in the texts [Khalil 2001, Sontag 1998]. In Section A.2 we present some results for ordinary differential equations in a somewhat more general setting, and we refer the reader there for some discussion, and for references. The local characterization offered by the relationship between integral curves and ordinary differential equations can be extended from a given chart to the manifold using the following result (see [Abraham, Marsden, and Ratiu 1988] for a proof).

Theorem 3.66 (Flow Box Theorem). *Let* $X \in \Gamma^r(\mathsf{TM})$, $r \in \mathbb{N} \cup \{\infty\} \cup \{\omega\}$, *and let* $x_0 \in \mathsf{M}$. *Then there exists a triple* $(\mathcal{U}, T, \Phi)$, *called a* ***flow box*** *of* X *at* x_0, *with the following properties:*

(i) $\mathcal{U}$ *is an open subset of* M *containing* x_0;
(ii) $T > 0$ *or* $T = +\infty$;
(iii) $\Phi\colon\,]-T,T[\, \times \mathcal{U} \to \mathsf{M}$ *is a map having the following properties:*
  (a) the map $(t,x) \mapsto \Phi(t,x)$ *is of class* C^r *for each* $(t,x) \in\,]-T,T[\, \times \mathcal{U}$;
  (b) the map $t \mapsto \Phi(t,x)$ *is of class* C^{r+1} *for each* $x \in \mathcal{U}$;
  (c) for each $t \in\,]-T,T[$, $\Phi(t,x) = \gamma_x(t)$, *where* γ_x *is the integral curve of* X *at* $x \in \mathcal{U}$;
(iv) for all $t \in\,]-T,T[$, $\Phi_t\colon \mathcal{U} \to \mathsf{M}$ *is a* C^r*-diffeomorphism onto its image, where* $\Phi_t(x) = \Phi(t,x)$.

Furthermore, if $(\tilde{\mathcal{U}}, \tilde{T}, \tilde{\Phi})$ *is another such triple, then* Φ *and* $\tilde{\Phi}$ *agree when restricted to* $(\,]-T,T[\, \cap\,]-\tilde{T},\tilde{T}[\,) \times (\mathcal{U} \cap \tilde{\mathcal{U}})$.

A consequence of the Flow Box Theorem is that integral curves for a C^r-vector field are C^{r+1}-curves. Note that the integral curves for a vector field may not extend to be defined on all of $\mathbb{R}$. For this reason, we define $I(X,x) \in \mathscr{I}$ as the largest interval on which an integral curve for X at x can be defined. The integral curve at x defined on $I(X,x)$ is called the ***maximal integral curve*** of X at x. We also define

$$\sigma_+(X,x) = \sup(I(X,x)), \quad \sigma_-(X,x) = \inf(I(X,x)).$$

Thus $\sigma_+(X,x)$ is the maximal time for which an integral curve at x can be extended forward in time, and $\sigma_-(X,x)$ in the maximal time for which an integral curve at x can be extended backward in time. Often $\sigma_+(X,x) = -\sigma_-(X,x) = +\infty$ for each $x \in \mathsf{M}$, in which case X is ***complete***. If only $\sigma_+(X,x) = +\infty$ for each $x \in \mathsf{M}$, then X is said to be ***positively complete***. Even in simple examples it can be the case that either or both of $\sigma_+(X,x)$ or $\sigma_-(X,x)$ are finite (see Exercise E3.25). The ***domain*** of the vector field is the subset $\mathrm{dom}(X) \subset \mathbb{R} \times \mathsf{M}$ given by $\mathrm{dom}(X) = \{(t,x) \mid t \in I(X,x)\}$. The ***flow*** of X is the map from $\mathrm{dom}(X)$ to M that assigns to $(t,x) \in \mathrm{dom}(X)$ the point $\gamma(t) \in \mathsf{M}$, where γ is the maximal integral curve of X at x. We shall write $\Phi_t^X(x)$ for the image of (t,x) under the flow.

Example 3.67 (Polar coordinates cont'd). We again have the two charts $(\mathcal{U}_1, \phi_1)$ and $(\mathcal{U}_2, \phi_2)$ that have their previous meanings. We also have the vector field X, that in the chart $(\mathcal{U}_1, \phi_1)$ is given by $X = -y\frac{\partial}{\partial x} + x\frac{\partial}{\partial y}$, and in the chart $(\mathcal{U}_2, \phi_2)$ is given by $X = \frac{\partial}{\partial \theta}$. The ordinary differential equation in the chart $(\mathcal{U}_1, \phi_1)$ is then

$$\dot{x} = -y, \quad \dot{y} = x,$$

which is linear, and for the initial conditions $(x(0), y(0))$ has the solution

$$(x(t), y(t)) = (x(0)\cos t - y(0)\sin t, x(0)\sin t + y(0)\cos t).$$

Note that solutions exist for all time so X is complete. The flow is then defined by

$$\Phi_t^X(x, y) = (x\cos t - y\sin t, x\sin t + y\cos t).$$

Let us look at the same ideas in the other coordinate chart. The ordinary differential equations in this coordinate chart are

$$\dot{r} = 0, \quad \dot{\theta} = 1.$$

Note that these equations are no longer linear, but are trivially solved for the initial condition $(r(0), \theta(0))$ as

$$(r(t), \theta(t)) = (r(0), \theta(0) + t).$$

From these computations, one might be tempted to deduce that X is complete. However, one should be careful, because this coordinate chart does not cover all of M. To verify that all integral curves may be extended for all time, one actually needs to find another chart to extend the integral curves (this motivates why one needs the Flow Box Theorem). We do not need to do this here, since we already know from our use of the coordinate chart $(\mathcal{U}_1, \phi_1)$ that we may extend all integral curves for infinite time. •

The following result is also a useful one when dealing with vector fields. We again refer to [Abraham, Marsden, and Ratiu 1988] for a proof.

Theorem 3.68 (Straightening Out Theorem). *If* $X \in \Gamma^r(\mathsf{TM})$, $r \in \mathbb{N} \cup \{\infty\} \cup \{\omega\}$, *and if* $X(x_0) \neq 0_{x_0}$, *then there exists a chart* $(\mathcal{U}, \phi)$ *for which* $x_0 \in \mathcal{U}$ *and for which the components of* X *in the chart are* $(1, 0, \ldots, 0)$.

Let us give a flow interpretation of the Lie bracket. The result can be fairly easily, albeit tediously, proved using Taylor series expansions for the flow [Nijmeijer and van der Schaft 1990]. A coordinate-free proof, using the definition of the Lie bracket, can be found in [Abraham, Marsden, and Ratiu 1988]. Nelson [1967] makes a connection between the formula we give here, and the problem of steering an automobile.

Proposition 3.69 (Flow interpretation of the Lie bracket). *Let $X, Y \in \Gamma^\infty(\mathsf{TM})$ and let $x \in \mathsf{M}$. Define a curve γ at x by*

$$\gamma(t) = \Phi^{-Y}_{\sqrt{t}} \circ \Phi^{-X}_{\sqrt{t}} \circ \Phi^{Y}_{\sqrt{t}} \circ \Phi^{X}_{\sqrt{t}}(x).$$

Then γ is differentiable and $\gamma'(0) = [X, Y](x)$. Moreover, if X and Y are complete, then $[X, Y](x) = 0$ for all $x \in \mathsf{M}$ if and only if $\Phi^X_t \circ \Phi^Y_s = \Phi^Y_s \circ \Phi^X_t$ for all $s, t \in \mathbb{R}$.

In particular, if we flow along X, then Y, then $-X$, then $-Y$, all for the same time, and we return to the same point, then $[X, Y](x) = 0$. The general situation is depicted in Figure 3.12. If $[X, Y](x) = 0$ for all $x \in \mathsf{M}$, then the

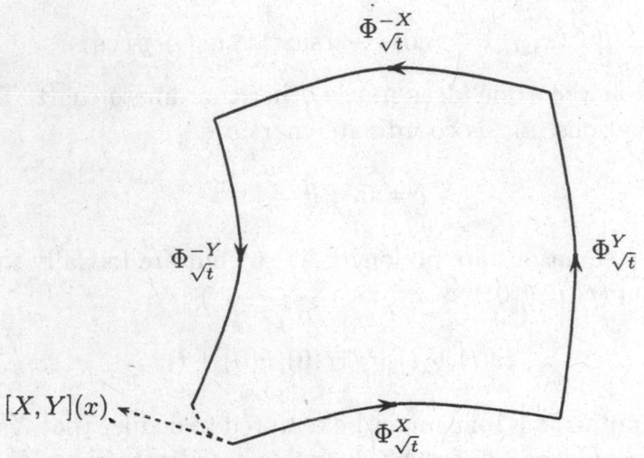

Figure 3.12. The flow interpretation of the Lie bracket

vector fields are said to ***commute***. A family $\mathscr{X}$ of vector fields ***commutes***, or is ***commutative***, if each pair of vector fields in the family commutes. This sort of interpretation of the Lie bracket lies at the heart of certain aspects of nonlinear control theory, particularly controllability. For this reason alone, the Lie bracket will be important for us. For now, let us give an example that illustrates Proposition 3.69.

Example 3.70. On $\mathsf{M} = \mathbb{R}^3$ consider the vector fields $X = \frac{\partial}{\partial y}$ and $Y = \frac{\partial}{\partial x} + y\frac{\partial}{\partial z}$, where (x, y, z) denote the standard coordinates for $\mathbb{R}^3$. We compute $[X, Y] = \frac{\partial}{\partial z}$. This means that if we flow along X, then along Y, then along $-X$, and then along $-Y$, all for the same amount of time, then we should end up moving, to lowest order, in the z-direction. Indeed, one can explicitly compute (by simply solving the associated ordinary differential equations)

$$\Phi^{-Y}_{\sqrt{t}} \circ \Phi^{-X}_{\sqrt{t}} \circ \Phi^{Y}_{\sqrt{t}} \circ \Phi^{X}_{\sqrt{t}}(0, 0, 0) = (0, 0, t),$$

and so in this case the resulting motion is *exactly* in the direction of the Lie bracket. This is not the usual situation, however. We encourage the reader to sketch the specific version of Figure 3.12 for this example to gain some insight into the Lie bracket. •

Let us also consider a generalization of vector fields for which time-dependence is allowed. This will be essential in the applications we consider, and we refer to Section A.2 for a more complete discussion of what we say here.

Definition 3.71 (Time-dependent vector field). For $r \in \mathbb{N} \cup \{\infty\} \cup \{\omega\}$, a ***$C^r$-time-dependent vector field*** is a locally integrally class C^r section of TM (see Section A.2). We denote by X_t, $t \in \mathbb{R}$, the C^r-vector field on M defined by $X_t(x) = X(t, x)$. •

As the reader can see by glimpsing at Section A.2, the time dependence specified by "locally integrally class C^r" is quite general, and the manner in which the differentiability in x and the dependence on t interact is a little complicated. However, the notion of "locally integrally class C^r," like it or not, turns out to be just what is needed in the text. It is the details surrounding this sort of generality that are dealt with in Section A.2. Nonetheless, the reader will not lose much conceptually by thinking of the time dependence as being something stronger like piecewise differentiable.

Just as for time-independent vector fields, we have the notion of integral curves. However, we must now account for the time-dependence.

Definition 3.72 (Integral curve for a time-dependent vector field). For $x \in \mathsf{M}$ and $t_0 \in \mathbb{R}$, a curve $\gamma\colon I \to \mathsf{M}$ is an ***integral curve at*** (t_0, x) for a C^r-time-dependent vector field X, $r \in \mathbb{N} \cup \{\infty\} \cup \{\omega\}$, if $\gamma'(t) = X(t, \gamma(t))$, and if $\gamma(t_0) = x$ (it is implicit that $t_0 \in I$). •

Let $(t_0, x) \in \mathbb{R} \times \mathsf{M}$ and let $(\mathcal{U}, \phi)$ be a chart for which $x \in \mathcal{U}$. If $t \mapsto (x^1(t), \dots, x^n(t))$ is the local representative of the integral curve γ at (t_0, x) for X, then this local representative must be a solution to the initial value problem

$$\begin{aligned}
\dot{x}^1(t) &= X^1(t, x^1(t), \dots, x^n(t)), \\
&\vdots \\
\dot{x}^n(t) &= X^n(t, x^1(t), \dots, x^n(t)), \\
\boldsymbol{x}(t_0) &= \phi(x).
\end{aligned}$$

As in the time-independent case, one can prove local existence and uniqueness of integral curves, as is explained in Section A.2. The resulting curves will not generally be differentiable, but only locally absolutely continuous. We shall in the sequel take these hypotheses and conclusions for granted, and we denote by $\Gamma^r(\mathbb{R}; \mathsf{TM})$ the set of time-dependent vector fields that are, in the parlance of Section A.2, locally integrally bounded of class C^r, where $r \in \mathbb{N} \cup \{\infty\} \cup \{\omega\}$. If we wish to only allow the time parameter to lie in an interval $I \subset \mathbb{R}$, we shall write $\Gamma^r(I; \mathsf{TM})$. (We may also talk more generally about the set $\Gamma^r(I; \mathcal{D})$ of locally integrally bounded vector fields taking values in subbundles $\mathcal{D}$ of TM. These ideas are discussed in some generality in Section A.2.1.) We also note that there is a Flow Box Theorem in the time-dependent case, and it is stated as Theorem A.11. We can now proceed as we did in the time-independent case to define the notation to give the flow. Thus $I(X, t_0, x) \in \mathscr{I}$ is the largest interval on which one may define an integral curve at (t_0, x) for X, and we can then define

$$\sigma_+(X, t_0, x) = \sup(I(X, t_0, x)), \quad \sigma_-(X, t_0, x) = \inf(I(X, t_0, x)).$$

The ***maximal integral curve*** for X at (t_0, x) is then the unique integral curve $\gamma \colon I(X, t_0, x) \to \mathsf{M}$ for X at (t_0, x). The ***domain*** of a time-dependent vector field X is the subset $\mathrm{dom}(X) \subset \mathbb{R} \times \mathbb{R} \times \mathsf{M}$ given by $\mathrm{dom}(X) = \{ (t_0, t, x) \mid t \in I(X, t_0, X)\}$. Finally, the ***flow*** of a time-dependent vector field X assigns to each point $(t, t_0, x) \in \mathrm{dom}(X)$ the point $\gamma(t) \in \mathsf{M}$, where γ is the maximal integral curve for X at (t_0, x). We denote by $\Phi^X_{t_0,t}(x)$ the point $\gamma(t) \in \mathsf{M}$, where γ is the maximal integral curve of X at (t_0, x).

3.5.3 Lifts of vector fields to the tangent bundle

One of the features of the approach we take to control theory for mechanical systems is that the systems we consider have a state manifold that is a tangent bundle, but for which it is possible to state many interesting control theoretic results on the base manifold. Part of doing this includes being able to "lift" objects from the base manifold to its tangent bundle. In this section we look at two ways of lifting a vector field. This idea of lifting is one that is of independent interest in differential geometry, and forms a significant portion of the book of Yano and Ishihara [1973].

We first define the vertical lift.

Definition 3.73 (Vertical lift). For $v_x \in \mathsf{TM}$, define a map $\mathrm{vlft}_{v_x} \colon \mathsf{T}_x\mathsf{M} \to \mathsf{T}_{v_x}\mathsf{TM}$ by

$$\mathrm{vlft}_{v_x}(X_x) = \left.\frac{\mathrm{d}}{\mathrm{d}t}\right|_{t=0} (v_x + tX_x).$$

The ***vertical lift*** of a vector field X on M is the vector field $\mathrm{vlft}(X)$ on TM defined by $\mathrm{vlft}(X)(v_x) = \mathrm{vlft}_{v_x}(X(x))$. •

The vertical lift acquires its name from the fact that it is "vertical" in the sense that it is annihilated by $\mathsf{T}\pi_{\mathsf{TM}}$, cf. Section 5.5. This may be verified directly from the definition of the vertical lift, or from the coordinate formula for $\mathrm{vlft}(X)$, which is

$$\mathrm{vlft}(X) = X^i \frac{\partial}{\partial v^i},$$

where $(X^1, \dots, X^n)$ are the components of X in the chosen coordinate chart.

Let us present another method for lifting a vector field.

Definition 3.74 (Tangent lift). The ***tangent lift*** of $X \in \Gamma^r(\mathsf{TM})$, $r \in \mathbb{N} \cup \{\infty\} \cup \{\omega\}$, is the C^{r-1}-vector field X^T on TM defined by

$$X^T(v_x) = \frac{\mathrm{d}}{\mathrm{d}t}\Big|_{t=0} (T_x \Phi_t^X(v_x)),$$ •

What we call the tangent lift is also frequently called the "complete lift" by some authors. By writing the flow in coordinates as a map from $\mathbb{R} \times \mathcal{U}$, where $\mathcal{U}$ is an open subset of $\mathbb{R}^n$, one can verify that the coordinate formula for X^T is

$$X^T = X^i \frac{\partial}{\partial x^i} + \frac{\partial X^i}{\partial x^j} v^j \frac{\partial}{\partial v^i}. \tag{3.9}$$

Note that, while $\mathrm{vlft}(X)(v_x)$ depends only on the value of X at x, $X^T(v_x)$ depends also on the derivative of X at x.

The following result gives a useful characterization of the tangent lift at an equilibrium point for X, i.e., a point x_0, where $X(x_0) = 0_{x_0}$. The result is most directly proved using the coordinate formula (3.9) for the tangent lift.

Proposition 3.75 (Tangent lift at an equilibrium point). *If $x_0 \in \mathsf{M}$ satisfies $X(x_0) = 0_{x_0}$, then $T_{v_{x_0}} \pi_{\mathsf{TM}}(X^T(v_{x_0})) = 0_{x_0}$ for each $v_{x_0} \in \mathsf{T}_{x_0}\mathsf{M}$. Furthermore, there exists a unique $A_X(x_0) \in L(\mathsf{T}_{x_0}\mathsf{M}; \mathsf{T}_{x_0}\mathsf{M})$ for which $X^T(v_{x_0}) = \mathrm{vlft}_{v_{x_0}}(A_X(x_0) \cdot v_{x_0})$. We call $A_X(x_0)$ is the* ***linearization*** *of X at x_0.*

Remark 3.76. The term "linearization" suggests that $A_X(x_0)$ represents the "first-order" behavior of X near x_0. This is indeed the case, although one should be careful when making this precise. It is also worth mentioning that this interpretation of linearization carries over to a general characterization of X^T: the flow of X^T is to be thought of as the linearization of the flow of X. Somewhat more precisely, the integral curve of X^T through v_{x_0} provides information on what happens to integral curves of X when one perturbs the initial x_0 in the direction v_{x_0}. This is discussed by Crampin and Pirani [1986, Section 3.5]. Also, this way of thinking about the tangent lift is considered as part of the supplementary material that will appear on the webpage for the book. •

3.6 Tensor fields

After completing our study of vector fields in the preceding section, here we investigate more general tensor fields. The key points in this section are as follows. Building on the definition of tensor field in Example 3.55, we will study how general tensor fields transform under changes of coordinate. Certain tensor fields, for example, covector fields, and Riemannian metrics, will arise often in our developments, so these get special attention. Finally, the operations of pull-back and push-forward of a a tensor by a diffeomorphism explain how tensor fields transform under maps between manifolds.

3.6.1 Covector fields

As we saw in Example 3.55–4, a ***C^r-covector field*** is an element of $\Gamma^r(\mathsf{T}^*\mathsf{M})$. In this section we concentrate on local representations of covector fields, introducing notation that will be used frequently in the remainder of the book. Just as the tangent bundle comes equipped with a natural atlas, so too does the cotangent bundle. We shall describe how the overlap conditions for admissible charts in this atlas behave. As should be expected, the overlap conditions for $\mathsf{T}^*\mathsf{M}$ will have a relationship to the overlap conditions for TM that resembles the relationship between the change of basis formulae for a vector space and its dual. With that said, let us simply jump to the punchline. Suppose that $(\mathcal{U}, \phi)$ and $(\mathcal{V}, \psi)$ are overlapping charts for M with coordinates given by $(x^1, \dots, x^n)$ and $(\tilde{x}^1, \dots, \tilde{x}^n)$, respectively. If the induced local coordinates for $\mathsf{T}^*\mathsf{M}$ are denoted by $((x^1, \dots, x^n), (p_1, \dots, p_n))$ and $((\tilde{x}^1, \dots, \tilde{x}^n), (\tilde{p}_1, \dots, \tilde{p}_n))$, respectively, then one can verify that

$$\tilde{p}_i = \frac{\partial x^j}{\partial \tilde{x}^i} p_j, \qquad i \in \{1, \dots, n\}.$$

Note that the fiber coordinates for $\mathsf{T}^*\mathsf{M}$ are written with subscript indices. This is consistent with their being components of vectors in $\mathsf{T}^*_x\mathsf{M}$.

This works out as we might expect in our polar coordinate example.

Example 3.77 (Polar coordinates cont'd). We again have $\mathsf{M} = \mathbb{R}^2$, with $(\mathcal{U}_1, \phi_1)$ the Cartesian coordinate chart and $(\mathcal{U}_2, \phi_2)$ the polar coordinate chart. Thus natural Cartesian coordinates for $\mathsf{T}^*\mathsf{M}$ are denoted by $((x, y), (p_x, p_y))$ and natural polar coordinates are denoted by $((r, \theta), (p_r, p_\theta))$. These coordinates are related by

$$\begin{aligned} p_r &= \frac{\partial x}{\partial r} p_x + \frac{\partial y}{\partial r} p_y = \frac{x}{\sqrt{x^2 + y^2}} p_x + \frac{y}{\sqrt{x^2 + y^2}} p_y, \\ p_\theta &= \frac{\partial x}{\partial \theta} p_x + \frac{\partial y}{\partial \theta} p_y = -y p_x + x p_y. \end{aligned}$$ •

Let $(\mathcal{U}, \phi)$ be a coordinate chart for M with coordinates $(x^1, \dots, x^n)$. We saw how the family of vector fields $\{\frac{\partial}{\partial x^1}, \dots, \frac{\partial}{\partial x^n}\}$, when evaluated at $x \in \mathcal{U}$, is a basis for the tangent space $\mathsf{T}_x\mathsf{M}$. We wish to do the same for cotangent spaces. To do this, we note that, by regarding the coordinates $x^1, \dots, x^n$ as functions on $\mathcal{U}$, the identity

$$\mathscr{L}_{\frac{\partial}{\partial x^i}} x^j = \left\langle \mathrm{d}x^j; \frac{\partial}{\partial x^i} \right\rangle$$

holds when evaluated at each point $x \in \mathcal{U}$. Thus the cotangent vectors $\{\mathrm{d}x^1(x), \dots, \mathrm{d}x^n(x)\}$ are exactly the dual basis to $\{\frac{\partial}{\partial x^1}(x), \dots, \frac{\partial}{\partial x^n}(x)\}$. The local representative of $\mathrm{d}x^i$ is $\boldsymbol{x} \mapsto (\boldsymbol{x}, \boldsymbol{e}^i)$, $i \in \{1, \dots, n\}$, where $\{\boldsymbol{e}^1, \dots, \boldsymbol{e}^n\}$ is the standard basis for $(\mathbb{R}^n)^*$. In particular, if $\alpha \in \Gamma^\infty(\mathsf{T}^*\mathcal{U})$, then we may write $\alpha = \alpha_i \mathrm{d}x^i$ for functions $\alpha_i \colon \mathcal{U} \to \mathbb{R}$ called the ***components*** of α in the given chart. Just as we did for vector fields and their components, we can say how the bases for covectors and components of covectors change when we change coordinates.

Lemma 3.78. *Let $(\mathcal{U}, \phi)$ and $(\mathcal{V}, \psi)$ be overlapping charts for a manifold M with coordinates denoted by $(x^1, \dots, x^n)$ and $(\tilde{x}^1, \dots, \tilde{x}^n)$, respectively, in the charts $(\mathcal{U}, \phi)$ and $(\mathcal{V}, \psi)$. Also let α be a covector field with components $(\alpha_1, \dots, \alpha_n)$ and $(\tilde{\alpha}_1, \dots, \tilde{\alpha}_n)$, respectively. Then*

(i) $\mathrm{d}\tilde{x}^i = \dfrac{\partial \tilde{x}^i}{\partial x^j} \mathrm{d}x^j$, $i \in \{1, \dots, n\}$, *and*

(ii) $\tilde{\alpha}_i = \dfrac{\partial x^j}{\partial \tilde{x}^i} \alpha_j$, $i \in \{1, \dots, n\}$.

Again, the notation works, since the change of coordinate formulae for the basis covector fields look just like the same one would use for "infinitesimals," except now the infinitesimals have precise meaning as covector fields in a coordinate chart.

Remark 3.79. Often, in applications of differential geometry, one is faced with an "object" and wants to know what it is. Indeed, geometric mechanics is exactly the act of answering these sort of questions, and understanding the implications of the answer, when the application is "classical mechanics." One might then come across a vector quantity having as many components as dimensions of the manifold, and would then want to know, "Is the object a vector or a covector field (or possibly neither)?" The most direct answer to this question is to make a change of coordinates, and see how the physical object changes when this change of coordinates is imposed.

For example, velocity is represented in coordinates by $(\dot{x}^1, \dots, \dot{x}^n)$. Upon changing coordinates, the Chain Rule tells us that $\dot{\tilde{x}}^i = \frac{\partial \tilde{x}^i}{\partial x^j} \dot{x}^j$, $i \in \{1, \dots, n\}$. Thus velocity changes coordinates like a tangent vector, and so is to be thought of as living in the tangent space.

As another example, let us consider a function f and its "gradient" consisting in coordinates of the components $\left(\frac{\partial f}{\partial x^1}, \dots, \frac{\partial f}{\partial x^n}\right)$. If one believes in the location of the index (in this case "down") as telling us something about the nature of the object, then we would be forced to say that these were the components of a covector field. Let us check thus by looking at the change of coordinate formula for these components. Again the Chain Rule gives $\frac{\partial f}{\partial \tilde{x}^i} = \frac{\partial x^j}{\partial \tilde{x}^i} \frac{\partial f}{\partial x^j}$, $i \in \{1, \dots, n\}$. Thus the components do indeed transform like those of a covector field. This particular covector field is, of course, the differential of f and has previously been denoted by $\mathrm{d}f$. Note that we do *not* call this the gradient. The ***gradient*** of f is only defined in the presence of a Riemannian metric $\mathbb{G}$, and is then given by $\operatorname{grad} f(x) = \mathbb{G}(x)^\sharp(\mathrm{d}f(x))$. Thus the gradient defined in this way is a vector field.

One should, however, be aware of identifying an object as being a certain geometric object, merely on the basis of the location of its indices. A perfect example of a case when this is not valid occurs with the Christoffel symbols of an affine connection (see Section 3.8.3). •

Example 3.80 (Polar coordinates cont'd). As usual in our running example, we have $\mathsf{M} = \mathbb{R}^2$ with two coordinate charts $(\mathcal{U}_1, \phi_1)$ and $(\mathcal{U}_2, \phi_2)$. First let us see how the basis covector fields in each coordinate chart are related. We have

$$\mathrm{d}x = \frac{\partial x}{\partial r}\mathrm{d}r + \frac{\partial x}{\partial \theta}\mathrm{d}\theta = \cos\theta \mathrm{d}r - r\sin\theta\mathrm{d}\theta,$$

and

$$\mathrm{d}y = \frac{\partial y}{\partial r}\mathrm{d}r + \frac{\partial y}{\partial \theta}\mathrm{d}\theta = \sin\theta \mathrm{d}r + r\cos\theta\mathrm{d}\theta.$$

Now let us consider a covector field that is the differential of a function. In the coordinate chart $(\mathcal{U}_1, \phi_1)$, we consider the function whose local representative is $(x, y) \mapsto \frac{1}{2}(x^2 + y^2)$. In the chart $(\mathcal{U}_1, \phi_1)$ we have

$$\mathrm{d}f = \frac{\partial f}{\partial x}\mathrm{d}x + \frac{\partial f}{\partial y}\mathrm{d}y = x\mathrm{d}x + y\mathrm{d}y.$$

In polar coordinates, the local representative of f is $(r, \theta) \mapsto \frac{1}{2}r^2$, so that $\mathrm{d}f = r\mathrm{d}r$. To compute how the components are related we have

$$\frac{\partial x}{\partial r}(x) + \frac{\partial y}{\partial r}(y) = \cos\theta(r\cos\theta) + \sin\theta(r\sin\theta) = r$$

and

$$\frac{\partial x}{\partial \theta}(x) + \frac{\partial y}{\partial \theta}(y) = -r\sin\theta(\cos\theta) + r\cos\theta(\sin\theta) = 0,$$

and these are exactly $\frac{\partial f}{\partial r}$ and $\frac{\partial f}{\partial \theta}$, respectively, is as they should be. •

3.6.2 General tensor fields

Having discussed the local representation of vector and covector fields, in this section we give the local representations of tensor fields, as well as discuss some additional properties of tensor fields on manifolds, following Section 2.4.

First let us indicate how one obtains local representations for tensor fields. Let $t \in \Gamma^\infty(T^r_s(\mathsf{TM}))$ and let $(\mathcal{U}, \phi)$ be a chart for M with coordinates $(x^1, \dots, x^n)$. The ***components*** of t in the chart $(\mathcal{U}, \phi)$ are the n^{r+s} functions $t^{i_1\cdots i_r}_{j_1\cdots j_s}$ defined on $\mathcal{U}$ by

$$t^{i_1\cdots i_r}_{j_1\cdots j_s}(x) = t\Big(\mathrm{d}x^{i_1}(x), \dots, \mathrm{d}x^{i_r}(x), \frac{\partial}{\partial x^{j_1}}(x), \dots, \frac{\partial}{\partial x^{j_s}}(x)\Big),$$

for $i_1, \dots, i_r, j_1, \dots, j_s \in \{1, \dots, n\}$. Mimicking the notation for tensors on vector spaces as given in Section 2.4, we write a tensor in coordinates as

$$t = t^{i_1 \cdots i_r}_{j_1 \cdots j_s} \frac{\partial}{\partial x^{i_1}} \otimes \cdots \otimes \frac{\partial}{\partial x^{i_r}} \otimes \mathrm{d}x^{j_1} \otimes \cdots \otimes \mathrm{d}x^{j_s}.$$

Thus, again mimicking the vector space case, we regard the tensors

$$\left\{ \frac{\partial}{\partial x^{i_1}} \otimes \cdots \otimes \frac{\partial}{\partial x^{i_r}} \otimes \mathrm{d}x^{j_1} \otimes \cdots \otimes \mathrm{d}x^{j_s} \;\middle|\; i_1, \dots, i_r, j_1, \dots, j_s \in \{1, \dots, n\} \right\},$$

as a local basis for (r, s)-tensor fields.

We saw in Section 2.4 how tensors on vector spaces transform in the presence of linear maps. Those constructions are easily adapted to the case in which we have a map $f \in C^\infty(\mathsf{M}; \mathsf{N})$.

Definition 3.81 (Push-forward and pull-back on manifolds). Let $f \in C^{k+1}(\mathsf{M}; \mathsf{N})$, $k \in \mathbb{N} \cup \{\infty\} \cup \{\omega\}$.

(i) If f is a diffeomorphism and $t \in \Gamma^k(T^r_s(\mathsf{TM}))$, then the ***push-forward*** of t by f is $f_* t \in \Gamma^k(T^r_s(\mathsf{TN}))$ defined by

$$f_* t(y) = (T_{f^{-1}(y)} f)_* t(f^{-1}(y)).$$

(ii) If f is a diffeomorphism and $t \in \Gamma^k(T^r_s(\mathsf{TN}))$, then the ***pull-back*** of t by f is $f^* t \in \Gamma^k(T^r_s(\mathsf{TM}))$ defined by

$$f^* t(x) = (T_x f)^* t(f(x)).$$

(iii) For any f of class C^{k+1}, if $t \in \Gamma^k(T^0_s(\mathsf{TN}))$, then the ***pull-back*** of t by f is $f^* t \in \Gamma^k(T^0_s(\mathsf{TM}))$ defined by

$$f^* t(x) = (T_x f)^* t(f(x)).$$ •

Note that, unlike in the linear case, we cannot generally push-forward a $(s, 0)$-tensor field on M to a $(s, 0)$-tensor field on N. One reason is that, to define a tensor field on N, we need to be able to define it at *every* point in N. But some points in N may not lie in the image of f so there is no way to define the push-forward at these points.

Let us provide coordinate expressions for the push-forward and pull-back. We take coordinates $(x^1, \dots, x^n)$ for M and $(y^1, \dots, y^m)$ for N, and we write the local representative of $f \in C^\infty(\mathsf{M}; \mathsf{N})$ in these coordinates as

$$f_{\phi\psi}(x^1, \dots, x^n) = (f^1(x^1, \dots, x^n), \dots, f^m(x^1, \dots, x^n)),$$

and the local representative of $f^{-1} \in C^\infty(\mathsf{N}; \mathsf{M})$ as

$$f^{-1}_{\psi\phi}(y^1, \dots, y^m) = (g^1(y^1, \dots, y^m), \dots, g^n(y^1, \dots, y^m)).$$

We shall assume that f is a diffeomorphism only when required.

Proposition 3.82 (Coordinate formulae for push-forward and pull-back). *If $f \in C^{k+1}(\mathsf{M};\mathsf{N})$, if $t \in \Gamma^k(T^r_s(\mathsf{TM}))$, and if $\tau \in \Gamma^k(T^r_s(\mathsf{TN}))$, $k \in \mathbb{N} \cup \{\infty\} \cup \{\omega\}$, then, with the above notation,*

(i) $(f_* t)^{i_1 \cdots i_r}_{j_1 \cdots j_s} = \Big(\frac{\partial f^{i_1}}{\partial x^{k_1}} \circ f^{-1}\Big) \cdots \Big(\frac{\partial f^{i_r}}{\partial x^{k_r}} \circ f^{-1}\Big) \frac{\partial g^{l_1}}{\partial y^{j_1}} \cdots \frac{\partial g^{l_s}}{\partial y^{j_s}} t^{k_1 \cdots k_r}_{l_1 \cdots l_s} \circ f^{-1}$

and

(ii) $(f^* \tau)^{i_1 \cdots i_r}_{j_1 \cdots j_s} = \Big(\frac{\partial g^{i_1}}{\partial y^{k_1}} \circ f\Big) \cdots \Big(\frac{\partial g^{i_r}}{\partial y^{k_r}} \circ f\Big) \frac{\partial f^{l_1}}{\partial x^{j_1}} \cdots \frac{\partial f^{l_s}}{\partial x^{j_s}} \tau^{k_1 \cdots k_r}_{l_1 \cdots l_s} \circ f.$

As expected, push-forward and pull-back obey the same composition properties as in the linear case. Thus if $f\colon \mathsf{M} \to \mathsf{N}$ and $g\colon \mathsf{N} \to \mathsf{P}$ are C^{k+1}-diffeomorphisms, and if $t_1 \in \Gamma^k(T^r_s(\mathsf{TM}))$ and $t_2 \in \Gamma^k(T^r_s(\mathsf{TP}))$, then

$$(g \circ f)_* t_1 = g_*(f_* t_1), \quad (g \circ f)^* t_2 = f^*(g^* t_2).$$

If $t_2 \in \Gamma^k(T^0_s(\mathsf{TP}))$, then the second formula is true, even if f and g are not diffeomorphisms. Also, push-forward and pull-back commute with the tensor product in all cases for which they can be defined. For proofs of these assertions, we refer to [Abraham, Marsden, and Ratiu 1988].

Next, let us show how the concept of Lie differentiation can be extended from functions and vector fields to tensor fields on M. First, let us show how to Lie differentiate a covector field $\alpha \in \Gamma^{k+1}(\mathsf{T}^*\mathsf{M})$ with respect to a vector field $X \in \Gamma^{k+1}(\mathsf{TQ})$, $k \in \mathbb{N} \cup \{\infty\} \cup \{\omega\}$. We do this by asking that the Lie derivative behave in a natural way relative to the relationship between vector fields and covector fields. That is, we ask that the Lie derivative $\mathscr{L}_X \alpha \in \Gamma^k(\mathsf{T}^*\mathsf{M})$ satisfy

$$\mathscr{L}_X \langle \alpha; Y \rangle = \langle \mathscr{L}_X \alpha; Y \rangle + \langle \alpha; \mathscr{L}_X Y \rangle, \qquad Y \in \Gamma^k(\mathsf{TM}).$$

One can simply rearrange this to give an implicit definition of $\mathscr{L}_X \alpha$ as

$$\langle \mathscr{L}_X \alpha; Y \rangle = \mathscr{L}_X \langle \alpha; Y \rangle - \langle \alpha; \mathscr{L}_X Y \rangle, \qquad Y \in \Gamma^k(\mathsf{TM}).$$

Note that we know how to compute everything on the right-hand side, so this definition of $\mathscr{L}_X \alpha$ is possibly sensible. To show that it is actually sensible, since the expression $\langle \mathscr{L}_X \alpha; Y \rangle$ is $C^k(M)$-linear in Y, the right-hand side should also be verified as being $C^k(M)$-linear in Y. Thus we should check that, for every $Y \in \Gamma^k(\mathsf{TM})$ and for every $f \in C^k(\mathsf{M})$, we have

$$\mathscr{L}_X \langle \alpha; fY \rangle - \langle \alpha; \mathscr{L}_X fY \rangle = f\big(\mathscr{L}_X \langle \alpha; Y \rangle - \langle \alpha; \mathscr{L}_X Y \rangle\big).$$

This can be directly checked using the properties of the Lie derivative on functions and vector fields. We recall that the extension of these properties to vector fields that are not C^∞ follows from the coordinate formula (3.8) for the Lie bracket. Now that we have defined the Lie derivative on functions, vector fields, and covector fields, we can easily extend it to arbitrary tensors as follows.

Definition 3.83 (Lie derivative of tensor fields). Let $k \in \mathbb{N} \cup \{\infty\} \cup \{\omega\}$. For $X \in \Gamma^{k+1}(\mathsf{TM})$ and $t \in \Gamma^{k+1}(T^r_s(\mathsf{TM}))$, the ***Lie derivative*** of t with respect to X is the class C^k-tensor field of type (r, s) defined by

$$\begin{aligned}(\mathscr{L}_X t)(\alpha^1, \dots, \alpha^r, X_1, \dots, X_s) &= \mathscr{L}_X(t(\alpha^1, \dots, \alpha^r, X_1, \dots, X_s)) \\ &\quad - \sum_{i=1}^{r} t(\alpha^1, \dots, \mathscr{L}_X \alpha^i, \dots, \alpha^r, X_1, \dots, X_s) \\ &\quad - \sum_{j=1}^{s} t(\alpha^1, \dots, \alpha^r, X_1, \dots, \mathscr{L}_X X_j, \dots, X_s), \\ &\qquad \alpha^1, \dots, \alpha^r \in \Gamma^k(\mathsf{T}^*\mathsf{M}),\ X_1, \dots, X_s \in \Gamma^k(\mathsf{TM}). \quad \bullet\end{aligned}$$

Note that this is the same idea we used in defining the Lie derivative of a covector field, in that we require the Lie derivative to behave naturally with respect to the action of tensors on vector fields and covector fields. A messy computation can be employed to verify the following coordinate formula for the Lie derivative of a general tensor field.

Proposition 3.84 (Components of the Lie derivative of a tensor field). *Let $k \in \mathbb{N} \cup \{\infty\} \cup \{\omega\}$. Let $t \in \Gamma^{k+1}(T^r_s(\mathsf{TM}))$ and $X \in \Gamma^{k+1}(\mathsf{TM})$, and let $(\mathcal{U}, \phi)$ be a chart with coordinates $(x^1, \dots, x^n)$. The components of $\mathscr{L}_X t$ are*

$$(\mathscr{L}_X t)^{i_1 \cdots i_r}_{j_1 \cdots j_s} = X^k \frac{\partial t^{i_1 \cdots i_r}_{j_1 \cdots j_s}}{\partial x^k} - \sum_{\rho=1}^{r} \frac{\partial X^{i_\rho}}{\partial x^k} t^{i_1 \cdots k \cdots i_r}_{j_1 \cdots j_s} + \sum_{\sigma=1}^{s} \frac{\partial X^k}{\partial x^{j_\sigma}} t^{i_1 \cdots i_r}_{j_1 \cdots k \cdots j_s},$$

for $i_1, \dots, i_r, j_1, \dots, j_s \in \{1, \dots, n\}$.

For example, the coordinate formula for the Lie differentiation of a covector field $\alpha = \alpha_i \mathrm{d}x^i$ with respect to a vector field $X = X^i \frac{\partial}{\partial x^i}$ is

$$\mathscr{L}_X \alpha = X^k \frac{\partial \alpha_i}{\partial x^k} \mathrm{d}x^i + \alpha_k \frac{\partial X^k}{\partial x^i} \mathrm{d}x^i.$$

The following gives an interpretation of the Lie derivative in terms of the pull-back by the flow. For the proof, we refer to [Abraham, Marsden, and Ratiu 1988].

Proposition 3.85 (Flow interpretation of the Lie derivative of a tensor field). *Let $k \in \mathbb{N} \cup \{\infty\} \cup \{\omega\}$. Let $X \in \Gamma^{k+1}(\mathsf{TM})$ and $\tau \in \Gamma^{k+1}(T^r_s(\mathsf{TM}))$. Then*

$$\frac{\mathrm{d}}{\mathrm{d}t} (\Phi^X_t)^* \tau = (\Phi^X_t)^* \mathscr{L}_X \tau.$$

In particular, $\mathscr{L}_X \tau = 0$ if and only if $\frac{\mathrm{d}}{\mathrm{d}t} (\Phi^X_t)^* \tau(x) = 0$ for each $x \in \mathsf{M}$.

A particular consequence of this result is that the time derivative of a function, evaluated along an integral curve, is the Lie derivative of the function, i.e.,

$$\frac{\mathrm{d}}{\mathrm{d}t} f(\Phi_t^X(x)) = \mathscr{L}_X f(\Phi_t^X(x)).$$

The following result is sometimes useful. We again refer to [Abraham, Marsden, and Ratiu 1988] for a proof.

Proposition 3.86 (Lie differentiation commutes with push-forward and pull-back). *Let $k \in \mathbb{N} \cup \{\infty\} \cup \{\omega\}$. If $X \in \Gamma^{k+1}(\mathsf{TM})$ and $Y \in \Gamma^{k+1}(\mathsf{TN})$, $t_1 \in \Gamma^{k+1}(T^r_s(\mathsf{TM}))$ and $t_2 \in \Gamma^{k+1}(T^r_s(\mathsf{TN}))$, and if $f \in C^{k+1}(\mathsf{M};\mathsf{N})$, then the formulae*

$$f_*(\mathscr{L}_X t_1) = \mathscr{L}_{f_*X} f_* t_1 \quad \textit{and} \quad f^*(\mathscr{L}_Y t_2) = \mathscr{L}_{f^*Y} f^* t_2$$

hold, whenever they are well-defined.

In practice, "naïve" computation of the transformation of tensors under maps are correct. To see what this means, let $f\colon \mathsf{M} \to \mathsf{N}$ be a diffeomorphism, and let $(\mathcal{U}, \phi)$ be a chart around $x \in \mathsf{M}$ and $(\mathcal{V}, \psi)$ be a chart around $f(x)$. To simplify matters suppose that $\mathcal{V} = f(\mathcal{U})$. We denote by $(x^1, \dots, x^n)$ coordinates in the chart $(\mathcal{U}, \phi)$ and $(y^1, \dots, y^n)$ coordinates in the chart $(\mathcal{V}, \psi)$. Rather than compute how the components of a tensor transform, it is often easier to do the equivalent computation, transform the basis vector fields $\{\frac{\partial}{\partial x^1}, \dots, \frac{\partial}{\partial x^n}\}$ and $\{\frac{\partial}{\partial y^1}, \dots, \frac{\partial}{\partial y^n}\}$, and the basis covector fields $\{\mathrm{d}x^1, \dots, \mathrm{d}x^n\}$ and $\{\mathrm{d}y^1, \dots, \mathrm{d}y^n\}$. These transform just like you would expect them to. That is,

$$\begin{aligned}
\frac{\partial}{\partial x^i}(x) &= \frac{\partial y^j}{\partial x^i}(x) \frac{\partial}{\partial y^j}(f(x)), && i \in \{1, \dots, n\}, \\
\frac{\partial}{\partial y^i}(y) &= \frac{\partial x^j}{\partial y^i} \circ f^{-1}(y) \frac{\partial}{\partial x^j}(f^{-1}(y)), && i \in \{1, \dots, n\},
\end{aligned}$$

and

$$\begin{aligned}
\mathrm{d}x^i(x) &= \frac{\partial x^i}{\partial y^j} \circ f(x) \mathrm{d}y^j(f(x)), && i \in \{1, \dots, n\}, \\
\mathrm{d}y^i(y) &= \frac{\partial y^i}{\partial x^j}(y) \mathrm{d}x^j(f^{-1}(y)), && i \in \{1, \dots, n\}.
\end{aligned}$$

Note that only the computation of $\mathrm{d}x^i$ does not require the inversion of f. This is why we can always pull-back $(0, s)$-tensors, but all other push-forwards and pull-backs require diffeomorphisms. With these formulae for basis elements, one can then use linearity of the tensor product under multiplication by functions to compute how an arbitrary tensor transforms. The preceding computations specialize to the change of coordinate formulae we have already encountered, in the case that f is the overlap map associated with two charts.

The following example shows that this is all not so bad in practice. In the example, we illustrate how one often deals with Riemannian metrics in practice.

Example 3.87 (Polar coordinates cont'd). We again take $\mathsf{M} = \mathbb{R}^2$ with its two charts $(\mathcal{U}_1, \phi_1)$ and $(\mathcal{U}_2, \phi_2)$. Let us define a Riemannian metric on M by defining it in the chart $(\mathcal{U}_1, \phi_1)$. We define

$$\mathbb{G} = \mathrm{d}x \otimes \mathrm{d}x + \mathrm{d}y \otimes \mathrm{d}y.$$

Let us first see that this Riemannian metric is a familiar object. Let X and Y be general vector fields given by

$$X = X_1 \frac{\partial}{\partial x} + X_2 \frac{\partial}{\partial y}, \qquad Y = Y_1 \frac{\partial}{\partial x} + Y_2 \frac{\partial}{\partial y}.$$

We then directly compute

$$\mathbb{G}(X, Y) = X_1 Y_1 + X_2 Y_2,$$

using the rules

$$\mathrm{d}x\Big(\frac{\partial}{\partial x}\Big) = 1, \quad \mathrm{d}x\Big(\frac{\partial}{\partial y}\Big) = 0, \quad \mathrm{d}y\Big(\frac{\partial}{\partial x}\Big) = 0, \quad \mathrm{d}y\Big(\frac{\partial}{\partial y}\Big) = 1.$$

Note that $\mathbb{G}$ is just the usual inner product on $\mathbb{R}^2$, where we use the natural identification $\mathsf{T}_{(x,y)}\mathbb{R}^2 \simeq \mathbb{R}^2$.

Now let us see how this Riemannian metric looks in polar coordinates. We shall do this in two ways. First let us use the change of basis formula. In practice, one does this as follows:

$$\begin{aligned}
\mathrm{d}x \otimes \mathrm{d}x + \mathrm{d}y \otimes \mathrm{d}y &= \mathrm{d}(r\cos\theta) \otimes \mathrm{d}(r\cos\theta) + \mathrm{d}(r\sin\theta) \otimes \mathrm{d}(r\sin\theta) \\
&= \cos^2\theta \mathrm{d}r \otimes \mathrm{d}r - r\sin\theta\cos\theta \mathrm{d}\theta \otimes \mathrm{d}r - r\sin\theta\cos\theta \mathrm{d}r \otimes \mathrm{d}\theta \\
&\quad + r^2\sin^2\theta \mathrm{d}\theta \otimes \mathrm{d}\theta + \sin^2\theta \mathrm{d}r \otimes \mathrm{d}r + r\sin\theta\cos\theta \mathrm{d}r \otimes \mathrm{d}\theta \\
&\quad + r\sin\theta\cos\theta \mathrm{d}\theta \otimes \mathrm{d}r + r^2\cos^2\theta \mathrm{d}\theta \otimes \mathrm{d}\theta \\
&= \mathrm{d}r \otimes \mathrm{d}r + r^2 \mathrm{d}\theta \otimes \mathrm{d}\theta.
\end{aligned}$$

One can also directly use Proposition 3.82, applied to the overlap map, to compute, for example,

$$\begin{aligned}
\mathbb{G}_{rr} &= \frac{\partial x}{\partial r}\frac{\partial x}{\partial r}\mathbb{G}_{xx} + \frac{\partial x}{\partial r}\frac{\partial y}{\partial r}\mathbb{G}_{xy} + \frac{\partial y}{\partial r}\frac{\partial x}{\partial r}\mathbb{G}_{yx} + \frac{\partial y}{\partial r}\frac{\partial y}{\partial r}\mathbb{G}_{yy} \\
&= \cos^2\theta(1) + \cos\theta\sin\theta(0) + \sin\theta\cos\theta(0) + \sin^2\theta(1) = 1.
\end{aligned}$$

Similar computations give $\mathbb{G}_{r\theta} = \mathbb{G}_{\theta r} = 0$ and $\mathbb{G}_{\theta\theta} = r^2$. In either case we have

$$\mathbb{G} = \mathrm{d}r \otimes \mathrm{d}r + r^2 \mathrm{d}\theta \otimes \mathrm{d}\theta. \qquad \bullet$$

As a final construction concerning tensors on manifolds, we indicate how, given a Riemannian metric $\mathbb{G}$ on M, there is an induced norm on the fibers of $T^r_s(\mathsf{TM})$ for any $r, s \in \mathbb{Z}_+$. The definitions follow along the lines of (3.1). First we note that one can define an inner product $\mathbb{G}^{-1}$ on the fibers of $\mathsf{T}^*\mathsf{M}$. This is defined as in (2.4):

$$\mathbb{G}^{-1}(\alpha_x, \beta_x) = \mathbb{G}(\mathbb{G}^\sharp(\alpha_x), \mathbb{G}^\sharp(\beta_x)).$$

We therefore have, for each $x \in \mathsf{M}$, a norm on $\mathsf{T}_x\mathsf{M}$ and $\mathsf{T}^*_x\mathsf{M}$. Let us denote both of these norms by $\|\cdot\|_{\mathbb{G}}$, it being clear from context whether we are considering TM or $\mathsf{T}^*\mathsf{M}$. Now we define the norm of $t \in T^r_s(\mathsf{T}_x\mathsf{M})$ by

$$\begin{aligned}\|t\|_{\mathbb{G}} = \sup\big\{|t(\alpha^1, \dots, \alpha^r, v_1, \dots, v_s)| \ \big|\ & \alpha^j \in \mathsf{T}^*_x\mathsf{M},\ \|\alpha^j\|_{\mathbb{G}} = 1,\ j \in \{1, \dots, r\}, \\ & v_k \in \mathsf{T}_x\mathsf{M},\ \|v_k\|_{\mathbb{G}} = 1,\ k \in \{1, \dots, s\}\big\}. \qquad (3.10)\end{aligned}$$

3.7 Distributions and codistributions

In this section we look at some special structures, called distributions, associated with subbundles and generalized subbundles of TM and $\mathsf{T}^*\mathsf{M}$. Roughly speaking, a distribution is a pointwise assignments of a subspace of the tangent space. Insight into the notion of a distribution comes from mechanics. In a mechanical system subject to velocity constraints, a distribution specifies at each point in the configuration space what velocities are allowed. Distributions also play a fundamental role in the theory of controllability, as we shall see in Chapter 7. Some readers may have encountered the word "distribution" in functional analysis (for example, the Dirac δ-function is a distribution in the function analytic sense). This notion of distribution is unrelated to the notion of distribution we discuss here.

3.7.1 Definitions and basic properties

We first give the formal definitions of distributions and codistributions.

Definition 3.88 (Distributions and codistributions). Let M be a C^r-manifold.

(i) A ***distribution*** $\mathcal{D}$ (resp. a ***codistribution*** Λ) is an assignment, to each point $x \in \mathsf{M}$, of a subspace $\mathcal{D}_x$ of $\mathsf{T}_x\mathsf{M}$ (resp. a subspace Λ_x of $\mathsf{T}^*_x\mathsf{M}$).
(ii) A distribution (resp. codistribution) is $\boldsymbol{C^r}$, $r \in \mathbb{N} \cup \{\infty\} \cup \{\omega\}$, if it is a C^r-generalized subbundle of TM (resp. $\mathsf{T}^*\mathsf{M}$).
(iii) A C^r-distribution (resp. C^r-codistribution) is ***regular*** if it is a C^r-subbundle of TM (resp. $\mathsf{T}^*\mathsf{M}$). •

Distributions and codistributions that are C^r are examples of C^r-generalized subbundles. The following is a discussion of some facts that arise from this. Some of these facts rely on the treatment of generalized subbundles in Section 3.9.3. We state these facts here, for convenient reference, since they will come up at various times in our consideration of distributions and codistributions.

Remarks 3.89. 1. Recall that, if $\mathcal{D}$ is a C^∞-distribution, then this implies the existence of C^∞-vector fields $\{X_a\}_{a\in A}$ around every point $x_0 \in \mathsf{M}$ such that, in a neighborhood $\mathcal{U}$ of x_0, we have

$$\mathcal{D}_x = \operatorname{span}_{\mathbb{R}} \{X_a(x) |\ a \in A\}, \qquad x \in \mathcal{U}.$$

Let us call these vector fields ***local generators*** of $\mathcal{D}$ about x_0.

2. Regularity of a distribution or codistribution means exactly that the rank is locally constant, i.e., constant on each connected component.
3. Let us make some important comments concerning C^r-distributions and codistributions. We refer the reader to Section 3.9.3 for a more complete discussion of these issues in the context of C^r-generalized subbundles of vector bundles. In what follows we let $r \in \mathbb{N} \cup \{\infty\} \cup \{\omega\}$ unless otherwise stated.
   (a) Under mild topological conditions, if a C^r-distribution or codistribution possesses no singular points, then one can find a finite set of global generators (cf. Theorem 3.114).
   (b) The set of regular points of a C^r-distribution or codistribution is open and dense. The reader can explore the differences in the character of the set of singular points between the cases $r = \infty$ and $r = \omega$ in Exercise E3.31.
   (c) If $r = \omega$, then a C^r-distribution or codistribution is locally finitely generated.
4. As we mentioned after Definition 3.49, the set of C^r-sections of a vector bundle possesses the structure of a module over the ring of C^r-functions on the base manifold. This leads to an alternative, but not generally equivalent, definition of a C^r-distribution as a submodule of $\Gamma^r(\mathsf{TQ})$. We discuss this in some detail in Section 3.9.4, and a related exercise is Exercise E3.42.
5. Following our usual assumptions, unless we state otherwise, we shall assume that distributions and codistributions are C^∞. An important case where distributions may not be C^∞ (or even C^0) occurs with nonholonomic constraints that do not have locally constant rank (e.g., the snakeboard of Section 13.4). •

3.7.2 Integrable distributions

In this section we consider two properties of distributions, involutivity and integrability, and show that, under some assumptions, they are equivalent.

We mention that the subject of this section is the first step in a larger theory, that of overdetermined nonlinear partial differential equations [Bryant, Chern, Gardner, Goldschmidt, and Griffiths 1991].

Let $r \in \mathbb{N} \cup \{\infty\} \cup \{\omega\}$. A C^r-distribution $\mathcal{D}$ is ***involutive*** if, for every $x_0 \in \mathsf{M}$, and for any pair of vector fields $X, Y \in \Gamma^r(\mathsf{TM})$ taking values in $\mathcal{D}$, it holds that the vector field $[X, Y]$ also takes values in $\mathcal{D}$. If $\mathcal{D}$ is smooth, then it is easy to check involutivity in a neighborhood of a regular point x_0 of $\mathcal{D}$. Since x_0 is a regular point, we can choose a finite set of C^r-local generators $\{X_1, \dots, X_l\}$ around x_0. If and only if it holds that

$$[X_i, X_j](x) = c_{ij}^k(x) X_k(x), \qquad i, j \in \{1, \dots, l\},$$

for some C^{r-1} (by Theorem 3.114) functions c_{ij}^k, $i, j, k \in \{1, \dots, l\}$, defined around x_0, then $\mathcal{D}$ is involutive in a neighborhood of x_0. Note that we do not require that $l = \operatorname{rank}(\mathcal{D})$ in the preceding discussion, although this special case is included. As we shall see in Remark 3.91–1 below, for analytic involutive distributions, our preceding discussion holds, even at singular points. The matter of checking involutivity of C^r-distributions, $r \neq \omega$, at singular points is not straightforward, and we refer to Elkin [1999] for a discussion of some aspects of this.

A notion related, but not obviously so, to involutivity is integrability. Let $\mathcal{D}$ be a C^r-distribution on a C^{r+1}-manifold M and let $x_0 \in \mathsf{M}$. A ***local integral manifold*** through x_0 for $\mathcal{D}$ is an immersed C^{r+1}-submanifold S of a neighborhood $\mathcal{U}$ of x_0 with the property that, for each $x \in \mathsf{S}$, $\mathsf{T}_x\mathsf{S} \subset \mathcal{D}_x$. A local integral manifold S is ***maximal*** if $\mathsf{T}_x\mathsf{S} = \mathcal{D}_x$ for each $x \in \mathsf{S}$. A maximal local integral manifold for $\mathcal{D}$ containing $x \in \mathsf{M}$ is the ***maximal integral manifold*** for $\mathcal{D}$ through x if it contains any maximal local integral manifold through x. The distribution $\mathcal{D}$ is ***integrable*** if there exists a maximal local integral manifold through each $x \in \mathsf{M}$.

The following important theorem is due to Frobenius [1877] for regular distributions in the C^∞-case, to Hermann [1962] for the singular C^∞-case (not stated here), and to Nagano [1966] in the analytic case. This is the first result we encounter where analyticity plays a key role.

Theorem 3.90 (Frobenius's Theorem). *The following statements hold:*

(i) a regular C^∞-distribution is integrable if and only if it is involutive;
(ii) a C^ω-distribution is integrable if and only if it is involutive.

Remarks 3.91. 1. The result stated by Nagano [1966] asserts more than we do in part (ii) in the preceding theorem. Indeed, Nagano shows that if $\mathcal{D}$ is an involutive analytic distribution on M, then M is a disjoint union of the maximal integral manifolds for $\mathcal{D}$. This statement follows from Theorem 7.5 that arises in our discussion of controllability. This stronger result allows us to assert that involutivity of analytic distributions can always be checked by determining that $[X_i, X_j](x) \in \mathcal{D}_x$, $i, j \in \{1, \dots, l\}$, where $\{X_1, \dots, X_l\}$ is a set of local generators for $\mathcal{D}$ around x. This can be proved by considering the following two facts: (1) analytic distributions are locally finitely generated (see Theorem 3.112) and (2) if two vector fields are tangent to a submanifold S, then their Lie bracket is also tangent to S.

2. The reader should exercise some care when reading about Frobenius's Theorem in the literature, especially concerning the relationship between the rank assumptions and the analyticity of the distributions. For example, part (ii) does not hold for C^∞-distributions (see Exercise E3.32). A version of Frobenius's Theorem for C^∞-distributions, not necessarily regular, is given in the internet supplement to the book of [Marsden and Ratiu 1999]. •

Let us illustrate this theorem with a simple example.

Example 3.92. We take $\mathsf{M} = \mathbb{R}^2$ and we define vector fields X and Y on M by

$$X = x\frac{\partial}{\partial x}, \quad Y = y\frac{\partial}{\partial y}.$$

We then take $\mathcal{D}_{(x,y)} = \operatorname{span}_{\mathbb{R}}\{X(x,y), Y(x,y)\}$ for each $(x,y) \in \mathsf{M}$. One may readily compute that $[X,Y] = 0$, so $\mathcal{D}$ is involutive, and hence integrable, by virtue of its being analytic. There are nine connected maximal integral manifolds:

1. $\{(0,0)\}$;
2. $\{(x,0) \mid x > 0\}$;
3. $\{(x,0) \mid x < 0\}$;
4. $\{(0,y) \mid y > 0\}$;
5. $\{(0,y) \mid y < 0\}$;
6. $\{(x,y) \mid x > 0,\ y > 0\}$;
7. $\{(x,y) \mid x > 0,\ y < 0\}$;
8. $\{(x,y) \mid x < 0,\ y > 0\}$;
9. $\{(x,y) \mid x < 0,\ y < 0\}$.

Also see Figure 3.13. Note that we have integral manifolds of dimension zero, one, and two, and that the union of the integral manifolds of dimension two is open and dense. This latter attribute is a consequence of the analyticity of the distribution. •

In the case of a regular integrable distribution, there is more one can say in terms of the existence of local coordinates adapted to the distribution.

Proposition 3.93. *If $\mathcal{D}$ is an integrable C^∞-distribution on M and if x_0 is a regular point for $\mathcal{D}$, then there exists a coordinate chart $(\mathcal{U}, \phi)$ with $x_0 \in \mathcal{U}$, and with the property that, for each $x \in \mathcal{U}$, we have*

$$\mathcal{D}_x = \operatorname{span}_{\mathbb{R}}\{\tfrac{\partial}{\partial x^1}(x), \dots, \tfrac{\partial}{\partial x^k}(x)\}.$$

Conversely, if such a coordinate chart exists, then $\mathcal{D}|\mathcal{U}$ is integrable.

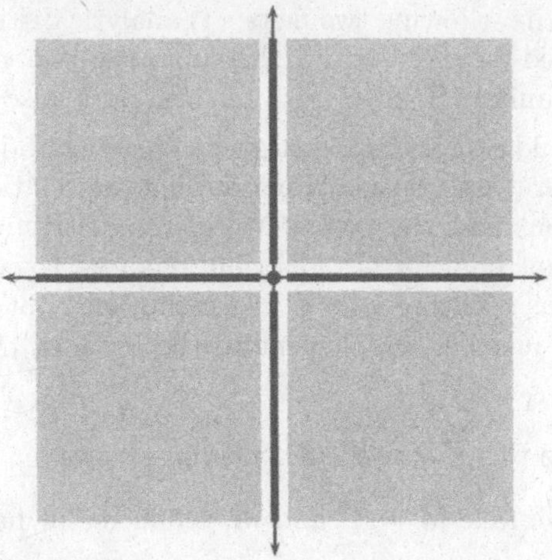

Figure 3.13. Integral manifolds for an integrable distribution

3.7.3 The Orbit Theorem for distributions

We now state a theorem due to Chow [1939] that might not be considered a fundamental part of differential geometry. However, since it has a close relationship to the material in the preceding section, we take the opportunity to present it here. This also allows us to take Chow's Theorem as background when we talk about nonholonomic constraints in Section 4.5, and controllability in Chapter 7.

We let $\mathcal{D}$ be a C^∞-distribution. Although $\mathcal{D}$ may not be a subbundle, let us denote by $\Gamma^\infty(\mathcal{D})$ those vector fields X for which $X(x) \in \mathcal{D}_x$. We denote by $\mathrm{Diff}(\mathcal{D})$ the set of diffeomorphisms of M generated by diffeomorphisms of the form

$$\Phi^{X_1}_{t_1} \circ \cdots \circ \Phi^{X_k}_{t_k}, \qquad t_1, \ldots, t_k \in \mathbb{R},\ X_1, \ldots, X_k \in \Gamma^\infty(\mathcal{D}),\ k \in \mathbb{N}.$$

Thus, a generator of this form, applied to x, sends x to the point obtained by flowing along X_k for time t_k, then along X_{k-1} for time t_{k-1}, and so on, down to flowing along X_1 for time t_1. The ***$\mathcal{D}$-orbit*** through x_0 is the set

$$\mathcal{O}(x_0, \mathcal{D}) = \{\varphi(x_0) \mid \varphi \in \mathrm{Diff}(\mathcal{D})\}.$$

Note that we do allow flow backwards in time along vector fields in $\Gamma^\infty(\mathcal{D})$. Thus one should simply think of the $\mathcal{D}$-orbit through x_0 as being those points that can be reached from x_0 by finite concatenations of C^∞-curves $\gamma_1, \ldots, \gamma_k$, defined intervals $[0, t_1], \ldots, [0, t_k]$, satisfying $\gamma_i'(t) \in \mathcal{D}_{\gamma_i(t)}$, $t \in [0, t_i]$, $i \in \{1, \ldots, k\}$, and for which the concatenated curve is continuous.

We wish to provide a description for $\mathcal{O}(x_0, \mathcal{D})$. To do this, we make a construction involving the distribution $\mathcal{D}$. The construction here is an abstract one, involving the use of Zorn's Lemma (see Section 2.1.6).

Lemma 3.94 (Existence of the involutive closure). *Let $r \in \{\infty\} \cup \{\omega\}$. If $\mathcal{D}$ is a distribution, then there exists a minimal involutive C^r-distribution containing $\mathcal{D}$. (Here, minimal means that it is contained in every other involutive distribution containing $\mathcal{D}$.)*

Proof. Let $\mathscr{I}_{\mathcal{D}}$ denote the collection of subsets of $\Gamma^r(\mathsf{TM})$ with the following properties:

1. if $\mathscr{V} \in \mathscr{I}_{\mathcal{D}}$ and if $X, Y \in \mathscr{V}$, then $[X, Y] \in \mathscr{V}$;
2. if $\mathscr{V} \in \mathscr{I}_{\mathcal{D}}$, then $\mathscr{V}$ contains the set of $\mathcal{D}$-valued C^r-vector fields.

Note that, if $\mathcal{D}'$ is an involutive C^r-distribution containing $\mathcal{D}$, then the set of $\mathcal{D}'$-valued C^r-vector fields is an element of $\mathscr{I}_{\mathcal{D}}$. Also note that $\mathscr{I}_{\mathcal{D}}$ is nonempty, since it contains $\Gamma^r(\mathsf{TM})$. The set $\mathscr{I}_{\mathcal{D}}$ is partially ordered as follows. If $\mathscr{V}_1, \mathscr{V}_2 \in \mathscr{I}_{\mathcal{D}}$, then $\mathscr{V}_1 \preceq \mathscr{V}_2$ if, for each $x \in \mathsf{M}$,

$$\operatorname{span}_{\mathbb{R}}\{X_2(x) \mid X_2 \in \mathscr{V}_2\} \subset \operatorname{span}_{\mathbb{R}}\{X_1(x) \mid X_1 \in \mathscr{V}_1\}.$$

Also, if $\mathscr{V}_1, \mathscr{V}_2 \in \mathscr{I}_{\mathcal{D}}$, then it is easy to check that $\mathscr{V}_1 \cap \mathscr{V}_2 \in \mathscr{I}_{\mathcal{D}}$. Next, we let $\{\mathscr{V}_a\}_{a \in A}$ be a chain in $\mathscr{I}_{\mathcal{D}}$, and take $\mathscr{V}_\infty = \cap_{a \in A} \mathscr{V}_a$. Since intersections of elements of $\mathscr{I}_{\mathcal{D}}$ are again elements of $\mathscr{I}_{\mathcal{D}}$, it follows that $\mathscr{V}_\infty$ is an upper bound for $\{\mathscr{V}_a\}_{a \in A}$. Thus Zorn's Lemma allows us to conclude that $\mathscr{I}_{\mathcal{D}}$ has a maximal (with respect to our partial order) element which we denote by $\mathscr{V}_{\mathcal{D}}$. The distribution whose fiber at $x \in \mathsf{M}$ is given by $\{X(x) \mid X \in \mathscr{V}_{\mathcal{D}}\}$ is then C^r and involutive, and furthermore is contained in every involutive distribution containing $\mathcal{D}$. ■

The minimal involutive distribution containing $\mathcal{D}$, whose existence is guaranteed by the lemma, is called the ***involutive closure*** of $\mathcal{D}$, and is denoted by $\mathrm{Lie}^{(\infty)}(\mathcal{D})$. The computation of $\mathrm{Lie}^{(\infty)}(\mathcal{D})$ is, in general, nontrivial, especially for C^∞-distributions. Often in practice, however, the computation of $\mathrm{Lie}^{(\infty)}(\mathcal{D})$ is not problematic. Let us indicate a computation that works in many cases.

We let $\mathcal{D}$ be a C^r-distribution, $r \in \{\infty\} \cup \{\omega\}$. We denote $\mathrm{Lie}^{(0)}(\mathcal{D}) = \mathcal{D}$, and inductively define C^∞-distributions $\mathrm{Lie}^{(l)}(\mathcal{D})$ on M by

$$\begin{aligned}\mathrm{Lie}^{(l)}(\mathcal{D})_x = \mathrm{Lie}^{(l-1)}(\mathcal{D})_x + \operatorname{span}_{\mathbb{R}}\{[X, Y](x) \mid\ & X \in \Gamma^\infty(\mathrm{Lie}^{(l_1)}(\mathcal{D})),\\ & Y \in \Gamma^\infty(\mathrm{Lie}^{(l_2)}(\mathcal{D})),\ l_1 + l_2 = l - 1\},\end{aligned}$$

for $l \in \mathbb{N}$. For fixed $x \in \mathsf{M}$, there obviously exists a smallest integer $N(x)$ for which $\mathrm{Lie}^{(l)}(\mathcal{D})_x = \mathrm{Lie}^{(N(x))}(\mathcal{D})_x$ for all $l \geq N(x)$. Doing this for each $x \in \mathsf{M}$ defines a distribution. The question now is, "What is the relationship between $\mathrm{Lie}^{(N(x))}(\mathcal{D})_x$ and $\mathrm{Lie}^{(\infty)}(\mathcal{D})_x$?" If $r = \infty$ and if x_0 is a regular point for each of the distributions $\mathrm{Lie}^{(l)}(\mathcal{D})$, $l \in \mathbb{Z}_+$, then there exists a neighborhood $\mathcal{U}$ of x_0 for which $\mathrm{Lie}^{(N(x_0))}(\mathcal{D})_x = \mathrm{Lie}^{(\infty)}(\mathcal{D})_x$ for each $x \in \mathcal{U}$. If $r = \omega$, then $\mathrm{Lie}^{(N(x))}(\mathcal{D})_x = \mathrm{Lie}^{(\infty)}(\mathcal{D})_x$ for each $x \in \mathsf{M}$. In cases other than these two, there appears to be little one can conclude, except that $\mathrm{Lie}^{(N(x))}(\mathcal{D})_x \subset$

$\text{Lie}^{(\infty)}(\mathcal{D})_x$ for each $x \in \mathsf{M}$. We shall have more to say about these matters in Section 3.9.4.

We now are able to describe the orbits of a distribution. The following result is part of the paper of Sussmann and Jurdjevic [1972], and provides a more general and detailed statement than the original theorem of Chow [1939]. The theorem was generalized to the C^∞-case by Sussmann [1973] and Stefan [1974].

Theorem 3.95 (Orbit Theorem for distributions). *If $\mathcal{D}$ is an analytic distribution on M and $x_0 \in \mathsf{M}$, then the following statements hold:*

(i) $\mathcal{O}(x_0, \mathcal{D})$ is an analytic immersed submanifold;

(ii) for each $x \in \mathcal{O}(x_0, \mathcal{D})$, $\mathsf{T}_x(\mathcal{O}(x, \mathcal{D})) = \text{Lie}^{(\infty)}(\mathcal{D})_x$;

(iii) M is the disjoint union of all orbits of $\mathcal{D}$.

Remarks 3.96. 1. The name ***Chow's Theorem*** is often given to the following corollary of the theorem. If M is connected and if, for a C^∞-distribution $\mathcal{D}$ on M, $\text{Lie}^{(\infty)}(\mathcal{D}) = \mathsf{TM}$, then $\mathcal{O}(x_0, \mathcal{D}) = \mathsf{M}$. The result is often jointly attributed to Chow [1939] and Rashevsky [1938], and has its roots in work of Carathéodory [1909] on the second law of thermodynamics.

2. The theorem proved by Sussmann and Jurdjevic [1972] is, in actuality, more general than the one we state here. For example, in the theorem of Sussmann and Jurdjevic, one can replace $\Gamma^\omega(\mathcal{D})$ with an arbitrary family $\mathscr{V}$ of analytic vector fields. This generalization will be reviewed in Section 7.1.3, along with other interesting aspects of this problem.

3. The C^∞-version of the Orbit Theorem is a little more complicated than the analytic version. In particular, the tangent spaces to the orbits include the subspace $\text{Lie}^{(\infty)}(\mathcal{D})_x$, but the inclusion may be strict. See Exercise E3.33, and the papers by Sussmann [1973] and Stefan [1974]. •

3.7.4 Codistributions

In this section we explore some relationships between distributions and codistributions. Given a distribution $\mathcal{D}$ on M, we define a codistribution $\text{ann}(\mathcal{D})$ on M, called the ***annihilator*** of $\mathcal{D}$, by $(\text{ann}(\mathcal{D}))_x = \text{ann}(\mathcal{D}_x)$. In like manner, if Λ is a codistribution on M, then we define a distribution $\text{coann}(\Lambda)$ on M, called the ***coannihilator*** of Λ, by $(\text{coann}(\Lambda))_x = \text{coann}(\Lambda_x)$. Unfortunately, it is not generally true that $\text{ann}(\mathcal{D})$ is a C^r-codistribution if $\mathcal{D}$ is a C^r-distribution, nor is it generally true that $\text{coann}(\Lambda)$ is a C^r-distribution if Λ is a C^r-codistribution. Indeed, one can show that if a C^r-distribution $\mathcal{D}$ has singular points, then $\text{ann}(\mathcal{D})$ is not even C^0, with a similar statement holding for codistributions. This is illustrated by a simple example, and we leave the general case to the reader (the proof of which involves the characterization of regular points for a distribution in Theorem 3.111).

Example 3.97 (Example 3.92 cont'd). On $\mathsf{M} = \mathbb{R}^2$ consider the C^ω-distribution $\mathcal{D}$ on M generated by the vector fields $X = x\frac{\partial}{\partial x}$ and $Y = y\frac{\partial}{\partial y}$. We note that

$$\operatorname{ann}(\mathcal{D})_{(x,y)} = \begin{cases} \mathsf{T}^*_{(x,y)}\mathsf{M}, & x = y = 0, \\ \operatorname{span}_{\mathbb{R}}\{dx\}, & x = 0,\ y \neq 0, \\ \operatorname{span}_{\mathbb{R}}\{dy\}, & x \neq 0,\ y = 0, \\ \{0\}, & \text{otherwise}. \end{cases}$$

To see that this is not even a C^0-codistribution, we note that it is not possible to find continuous covector fields α^1 and α^2, defined in a neighborhood of $(0,0)$, and such that $\operatorname{ann}(\mathcal{D})_{(0,0)} = \operatorname{span}_{\mathbb{R}}\{\alpha^1(0,0), \alpha^2(0,0)\}$. •

The following result indicates that, in the regular case, the situation is better.

Proposition 3.98. *Let $r \in \mathbb{N} \cup \{\infty\} \cup \{\omega\}$. If $\mathcal{D}$ is a C^r-distribution on M, and if $x_0 \in \mathsf{M}$ is a regular point for $\mathcal{D}$, then $\operatorname{ann}(\mathcal{D})$ is C^r in a neighborhood of x_0, and x_0 is a regular point for $\operatorname{ann}(\mathcal{D})$.*

Proof. Around a regular point x_0 for $\mathcal{D}$, we may choose C^r-vector fields $\{X_1, \dots, X_n\}$ having the properties that $\{X_1(x), \dots, X_n(x)\}$ forms a basis for $\mathsf{T}_x\mathsf{M}$ and that $\{X_1(x), \dots, X_k(x)\}$ form a basis for $\mathcal{D}_x$ for each x in a neighborhood $\mathcal{U}$ of x_0. Let $\{\alpha^1, \dots, \alpha^n\}$ be the C^r-covector fields dual to $\{X_1, \dots, X_n\}$. One then checks that $\{\alpha^{k+1}(x), \dots, \alpha^n(x)\}$ forms a basis for $\operatorname{ann}(\mathcal{D})_x$ for each $x \in \mathcal{U}$, thus showing that x_0 is indeed a regular point for $\operatorname{ann}(\mathcal{D})$, and that $\operatorname{ann}(\mathcal{D})$ is C^r in a neighborhood of x_0. ■

Because of the lack of a perfect correspondence between distributions and codistributions, except in the regular case, the theory of integrable codistributions does not just follow from that of distributions. We shall say that a regular codistribution Λ is ***integrable*** when the distribution $\operatorname{coann}(\Lambda)$ is integrable. Note that it is not necessarily clear how to define integrable codistributions when they are not regular; indeed, there are multiple ways to make such a definition. The matter of integrable codistributions is discussed in the papers [Freeman 1984, Malgrange 1976, 1977], for example.

3.8 Affine differential geometry

Lying at the very heart of much of what we do in this book is the notion of an affine connection. This object is part of the classic subjects of Riemannian geometry and affine differential geometry. Historically, important parts of these subjects were developed by, for example, Gauss [1828], Riemann [1854], Christoffel [1869], and Levi-Civita and Ricci [1901] (see also [Levi-Civita 1925]). We refer the reader to [Do Carmo 1992] for a modern introduction, and to [Gallot, Hulin, and Lafontaine 1987, Kobayashi and Nomizu 1963, Lang 1995] as additional references. As we shall see in Section 4.6, the affine connection is one of the basic ingredients in our model of mechanical control systems. In this sense, this section contains material directly relevant to our Lagrangian, as opposed to a Hamiltonian, view of mechanics.

Here we shall give only a cursory, although self-contained, introduction to the basic definitions of affine connection and covariant derivatives. Examples of affine connections will be plentiful later in the book. One important result, with ramifications in mechanics, is the beautiful 1–1 correspondence between Riemannian metrics and affine connections with certain properties (see Theorem 3.104). Along with the standard notions in affine differential geometry, we introduce the important operation of symmetric product and the notion of geodesic invariance in Section 3.8.4. Additional topics are presented in Section 3.9.6.

3.8.1 Definitions and general concepts

We begin with the basic definition.

Definition 3.99 (Affine connection). For $r \in \mathbb{N} \cup \{\infty\} \cup \{\omega\}$, a ***$C^r$-affine connection*** on M assigns to the pair $(X, Y) \in \Gamma^r(\mathsf{TM}) \times \Gamma^{r+1}(\mathsf{TM})$ a vector field $\nabla_X Y \in \Gamma^r(\mathsf{TM})$, and the assignment satisfies

(i) the map $(X, Y) \mapsto \nabla_X Y$ is $\mathbb{R}$-bilinear,

(ii) $\nabla_{fX} Y = f \nabla_X Y$ for each $X \in \Gamma^r(\mathsf{TM})$, $Y \in \Gamma^{r+1}(\mathsf{TM})$, and $f \in C^r(\mathsf{M})$, and

(iii) $\nabla_X fY = f \nabla_X Y + (\mathscr{L}_X f) Y$ for each $X \in \Gamma^r(\mathsf{TM})$, $Y \in \Gamma^{r+1}(\mathsf{TM})$, and $f \in C^{r+1}(\mathsf{M})$.

The vector field $\nabla_X Y$ is called the ***covariant derivative*** of Y with respect to X. •

The property (ii) is perhaps the most important, in that it distinguishes the covariant derivative from the Lie derivative. One should observe that, unlike the Lie derivative, the covariant derivative represents *extra* structure, and is not intrinsic to the manifold itself. As we shall see, in the situations in which we encounter an affine connection, it is specified by the physics of the problem.

Let us turn to providing some constructions that can be made with an affine connection. Since $\nabla_X Y$ is tensorial in X (that is, it depends only on the values of the vector field X, and not on its derivatives), if $v_x \in \mathsf{T}_x\mathsf{M}$ and if $Y \in \Gamma^{r+1}(\mathsf{TM})$, then we may define $\nabla_{v_x} Y(x) = \nabla_X Y(x) \in \mathsf{T}_x\mathsf{M}$, where X is any C^r-vector field having the property that $X(x) = v_x$. An important application of this observation is the following construction.

Given a curve $\gamma \colon I \to \mathsf{M}$, recall that a vector field along γ is a map that assigns to every $t \in I$ an element of $\mathsf{T}_{\gamma(t)}\mathsf{M}$. If γ is of class C^1 and if $Y \in \Gamma^1(\mathsf{TM})$, then we may define a C^0-vector field along γ by

$$I \ni t \mapsto \nabla_{\gamma'(t)} Y(\gamma(t)) \in \mathsf{T}_{\gamma(t)}\mathsf{M}.$$

It is also possible to verify [see Do Carmo 1992] that this construction makes sense, even if Y is not a vector field defined on all of M, but only defined along γ.

Lemma 3.100 (Covariant derivative of a vector field along a curve). *Let ∇ be a C^r-affine connection, $r \in \mathbb{N} \cup \{\infty\} \cup \{\omega\}$, and let $\gamma\colon I \to \mathsf{M}$ be a C^k-curve, $k \leq r+1$. Then there exists a unique map that assigns to each C^k-vector field ξ along γ a C^{k-1}-vector field along γ, denoted by $t \mapsto \nabla_{\gamma'(t)}\xi(t)$, with the following properties:*

(i) $\nabla_{\gamma'(t)}(\xi(t) + \eta(t)) = \nabla_{\gamma'(t)}\xi(t) + \nabla_{\gamma'(t)}\eta(t)$, for any C^k-vector fields ξ and η along γ;

(ii) $\nabla_{\gamma'(t)}(f(t)\xi(t)) = \dot{f}(t)\xi(t) + f(t)\nabla_{\gamma'(t)}\xi(t)$, for any C^k-vector field ξ along γ and any C^k-function $f\colon I \to \mathbb{R}$;

(iii) $\nabla_{\gamma'(t)}\xi(t) = \nabla_{\gamma'(t)}Y(\gamma(t))$, for any vector field Y satisfying $Y \circ \gamma = \xi$.

*The vector field $t \mapsto \nabla_{\gamma'(t)}\xi(t)$ along γ is the **covariant derivative** of ξ along γ.*

The reader is invited to prove the lemma in Exercise E3.36, along with some related formulae.

Remark 3.101 (Notation for covariant derivative along a curve). It is common to see different notation for the covariant derivative along a curve than that used for the covariant derivative between vector fields. A popular choice for this notation is to denote by $\frac{D\xi}{dt}$ the covariant derivative of ξ along a curve γ. •

The preceding discussion leads naturally to the following important definition.

Definition 3.102 (Geodesic). Let $r \in \mathbb{N} \cup \{\infty\} \cup \{\omega\}$. A ***geodesic*** of a C^r-affine connection ∇ on M is a curve $\gamma\colon I \to \mathsf{M}$ satisfying $\nabla_{\gamma'(t)}\gamma'(t) = 0$. •

Remark 3.103 (Geodesic spray). When we write the coordinate expression for $\nabla_{\gamma'(t)}\gamma'(t)$ below, we shall see that, if ∇ is a C^r-affine connection, $r \in \mathbb{N} \cup \{\infty\} \cup \{\omega\}$, then the geodesics satisfy a second-order differential equation that depends on points in Q in a C^r-manner. From this, one deduces that geodesics of a C^r-affine connection are C^{r+2}-curves. The second-order nature of the equations for geodesics allows one to infer the existence of a vector field $S \in \Gamma^r(\mathsf{TTM})$ having the property that the integral curves of S, when projected to M by π_{TM}, are geodesics for ∇. We refer the reader to [Abraham and Marsden 1978] for a discussion of second-order vector fields. The vector field S is the ***geodesic spray*** for ∇. An affine connection is ***complete*** if the vector field S is complete. •

For many of the control problems we consider, the unforced dynamics will be geodesic for some affine connection arising from the physics of the problem. This shall be made explicit in Sections 4.6 and 4.5.

As a final ingredient in our general discussion of affine connections, we indicate how one can define covariant differentiation for tensors of arbitrary degree. The construction here goes very much like that for Lie differentiation as given in Definition 3.83. We let ∇ be a C^k-affine connection and let X be a C^k-vector field. We take the covariant derivative of $f \in C^{k+1}(\mathsf{M})$ with respect to X to be $\nabla_X f = \mathscr{L}_X f$. With this as motivation, we implicitly define the covariant derivative of $\alpha \in \Gamma^{k+1}(\mathsf{T}^*\mathsf{M})$ with respect to X by

$$\langle \nabla_X \alpha; Y \rangle = \mathscr{L}_X \alpha(Y) - \langle \alpha; \nabla_X Y \rangle, \qquad Y \in \Gamma^k(\mathsf{TM}).$$

Then the ***covariant derivative*** of a class C^{k+1}-tensor field t of type (r, s) is defined by

$$\begin{aligned}(\nabla_X t)(\alpha^1, \dots, \alpha^r, X_1, \dots, X_s) &= \mathscr{L}_X(t(\alpha^1, \dots, \alpha^r, X_1, \dots, X_s)) \\ &\quad - \sum_{i=1}^{r} t(\alpha^1, \dots, \nabla_X \alpha^i, \dots, \alpha^r, X_1, \dots, X_s) \\ &\quad - \sum_{j=1}^{s} t(\alpha^1, \dots, \alpha^r, X_1, \dots, \nabla_X X_j, \dots, X_s), \\ &\qquad \alpha^1, \dots, \alpha^r \in \Gamma^k(\mathsf{T}^*\mathsf{M}),\ X_1, \dots, X_s \in \Gamma^k(\mathsf{TM}). \qquad (3.11)\end{aligned}$$

Thus $\nabla_X t$ is of class C^k. Because $\nabla_X t$ is tensorial in X itself, it make sense, and it is sometimes convenient, to consider the $(r, s+1)$-tensor field ∇t of class C^k defined by

$$\nabla t(\alpha^1, \dots, \alpha^r, X_1, \dots, X_s; X) = (\nabla_X t)(\alpha^1, \dots, \alpha^r, X_1, \dots, X_s).$$

Note that we are using somewhat nonstandard notation here, and in doing so, we follow the convention of at least some authors. At times we will also write

$$\nabla t(\alpha^1, \dots, \alpha^r, X_1, \dots, X_s) \in \Gamma^k(\mathsf{T}^*\mathsf{M}).$$

The resulting map ∇t from $(\Gamma^k(\mathsf{T}^*\mathsf{M}))^r \times (\Gamma^k(\mathsf{TM}))^s$ to $\Gamma^k(\mathsf{T}^*\mathsf{M})$ is called the ***covariant differential*** of t.

3.8.2 The Levi-Civita affine connection

Of particular interest for us will be a certain affine connection that is determined uniquely from a Riemannian metric $\mathbb{G}$ on M. Before we introduce this affine connection, let us consider a slightly more general type of affine connection. On a Riemannian manifold $(\mathsf{M}, \mathbb{G})$, an affine connection ∇ is a ***metric connection*** if $\nabla \mathbb{G} = 0$ (see Exercise E3.39 for an alternative characterization). The following theorem tells us that there is a unique metric connection that satisfies an additional condition.

Theorem 3.104 (Characterization of Levi-Civita connection). *For $r \in \mathbb{N} \cup \{\infty\} \cup \{\omega\}$, if $(\mathsf{M}, \mathbb{G})$ is a C^{r+1}-Riemannian manifold, then there exists a unique C^r-affine connection $\overset{\mathrm{G}}{\nabla}$ on M such that*

(i) $\overset{\mathrm{G}}{\nabla}$ is a metric connection, and

(ii) $\overset{\mathrm{G}}{\nabla}_X Y - \overset{\mathrm{G}}{\nabla}_Y X = [X, Y]$ for all $X, Y \in \Gamma^{r+1}(\mathsf{TM})$.

The affine connection $\overset{\mathrm{G}}{\nabla}$ is called the ***Levi-Civita affine connection*** *associated with $(\mathsf{M}, \mathbb{G})$.*

Proof. First we establish existence. We define $\overset{\mathrm{G}}{\nabla}$ by asking that, for vector fields X, Y, and Z on M, we have

$$\mathbb{G}(\overset{\mathrm{G}}{\nabla}_X Y, Z) = \tfrac{1}{2}\big(\mathscr{L}_X(\mathbb{G}(Y,Z)) + \mathscr{L}_Y(\mathbb{G}(Z,X)) - \mathscr{L}_Z(\mathbb{G}(X,Y)) \\ + \mathbb{G}([X,Y],Z) - \mathbb{G}([X,Z],Y) - \mathbb{G}([Y,Z],X)\big). \quad (3.12)$$

To ensure that this definition make sense one must verify that it is linear with respect to multiplication by functions in the second argument. Thus, for $f \in C^{r+1}(\mathsf{M})$, we must verify that $\mathbb{G}(\overset{\mathrm{G}}{\nabla}_X Y, fZ) = f\mathbb{G}(\overset{\mathrm{G}}{\nabla}_X Y, Z)$, using the relation (3.12). This is straightforwardly done using the derivation properties of the Lie bracket. These hold, even for vector fields that are not of class C^∞, as a result of the coordinate formula (3.8). We must also ensure that (3.12) does define an affine connection. This consists of checking that $\overset{\mathrm{G}}{\nabla}$ satisfies the three properties of Definition 3.99. The property (i) is obvious. To check (ii) let $f \in C^{r+1}(\mathsf{M})$ and then verify that $\mathbb{G}(\overset{\mathrm{G}}{\nabla}_{fX} Y, Z) = \mathbb{G}(f\overset{\mathrm{G}}{\nabla}_X Y, Z)$, which one does easily by using the derivation properties of the Lie derivative. In a similar manner one may show that $\mathbb{G}(\overset{\mathrm{G}}{\nabla}_X fY, Z) = \mathbb{G}(f\overset{\mathrm{G}}{\nabla}_X Y + (\mathscr{L}_X f)Y, Z)$. One also needs to verify that $\overset{\mathrm{G}}{\nabla}$ so defined satisfies (ii). It is straightforward to see that indeed

$$\mathbb{G}(\overset{\mathrm{G}}{\nabla}_X Y - \overset{\mathrm{G}}{\nabla}_Y X, Z) = \mathbb{G}([X,Y], Z),$$

which is exactly condition (ii). Finally, one needs to show that $\overset{\mathrm{G}}{\nabla}$ is a metric affine connection. To verify this, one can use (3.12) to show that

$$\mathscr{L}_X(\mathbb{G}(Y,Z)) = \mathbb{G}(\overset{\mathrm{G}}{\nabla}_X Y, Z) + \mathbb{G}(Y, \overset{\mathrm{G}}{\nabla}_X Z).$$

From the derivation properties of $\overset{\mathrm{G}}{\nabla}$, one then sees that this implies that $\overset{\mathrm{G}}{\nabla}_X \mathbb{G}(Y,Z) = 0$ for all vector fields X, Y, and Z. That is, $\overset{\mathrm{G}}{\nabla}$ is a metric connection. This establishes the existence of $\overset{\mathrm{G}}{\nabla}$.

Now we show that conditions (i) and (ii) uniquely define $\overset{\mathrm{G}}{\nabla}$. Suppose that ∇ is an affine connection satisfying these conditions. Then, for C^{r+1}-vector fields X, Y, and Z, one has

$$\mathscr{L}_X(\mathbb{G}(Y,Z)) = \mathbb{G}(\nabla_X Y, Z) + \mathbb{G}(Y, \nabla_X Z),$$
$$\mathscr{L}_Y(\mathbb{G}(Z,X)) = \mathbb{G}(\nabla_Y Z, X) + \mathbb{G}(Z, \nabla_Y X),$$
$$\mathscr{L}_Z(\mathbb{G}(X,Y)) = \mathbb{G}(\nabla_Z X, Y) + \mathbb{G}(X, \nabla_Z Y).$$

Now add the first two equations, and subtract from this the third to get

$$\begin{aligned}\mathscr{L}_X(\mathbb{G}(Y,Z)) &+ \mathscr{L}_Y(\mathbb{G}(Z,X)) - \mathscr{L}_Z(\mathbb{G}(X,Y)) \\ &= \mathbb{G}(\nabla_X Y, Z) + \mathbb{G}(\nabla_Y X, Z) + \mathbb{G}([X,Z],Y) + \mathbb{G}([Y,Z],X).\end{aligned}$$

In arriving at this formula we have used the relation $\nabla_X Y - \nabla_Y X = [X,Y]$, since ∇ satisfies (ii). Now we use the same relation to note that $\nabla_Y X = \nabla_X Y - [X,Y]$ to obtain

$$\begin{aligned}2\mathbb{G}(\nabla_X Y, Z) = \mathscr{L}_X(\mathbb{G}(Y,Z)) &+ \mathscr{L}_Y(\mathbb{G}(Z,X)) - \mathscr{L}_Z(\mathbb{G}(X,Y)) \\ &+ \mathbb{G}([X,Y],Z) - \mathbb{G}([X,Z],Y) - \mathbb{G}([Y,Z],X).\end{aligned}$$

Thus ∇ satisfies (3.12).

That $\overset{\mathbb{G}}{\nabla}$ is of class C^r if $\mathbb{G}$ is of class C^{r+1} follows most easily from our coordinate expression below for the Christoffel symbols for $\overset{\mathbb{G}}{\nabla}$. (This necessarily has to be postponed until our discussion of Christoffel symbols in Section 3.8.3.) ■

We shall write the geodesic spray for the Levi-Civita connection on a Riemannian manifold $(\mathsf{M}, \mathbb{G})$ as $\overset{\mathbb{G}}{S}$. A Riemannian manifold $(\mathsf{M}, \mathbb{G})$ is ***complete*** if its Levi-Civita affine connection is complete.

3.8.3 Coordinate formulae

Although an affine connection is not a tensor, there is still a convenient manner for representing it in local coordinates. Thus let ∇ be a C^r-affine connection on M and let $(x^1, \dots, x^n)$ be coordinates in a chart $(\mathcal{U}, \phi)$. For any $i, j \in \{1, \dots, n\}$, $\nabla_{\frac{\partial}{\partial x^i}} \frac{\partial}{\partial x^j}$ is a C^r-vector field on $\mathcal{U}$. Thus it is a $C^r(\mathcal{U})$-linear combination of the vector fields $\{\frac{\partial}{\partial x^1}, \dots, \frac{\partial}{\partial x^n}\}$. We can, therefore, write

$$\nabla_{\frac{\partial}{\partial x^i}} \frac{\partial}{\partial x^j} = \Gamma^k_{ij} \frac{\partial}{\partial x^k}$$

for n^3 uniquely defined C^r-functions $\Gamma^k_{ij} \colon \mathcal{U} \to \mathbb{R}$, $i, j, k \in \{1, \dots, n\}$. These functions are called the ***Christoffel symbols*** for ∇ in the chart $(\mathcal{U}, \phi)$. These are *not* the components of a $(1,2)$-tensor field on M (see Exercise E3.34).

Using the Christoffel symbols, one may represent all of the operations and objects associated with an affine connection. We merely present the end result of the necessary computations, leaving the details for Exercise E3.35. In the following, we consider coordinates $(x^1, \dots, x^n)$ in a chart $(\mathcal{U}, \phi)$, and all objects are restricted to $\mathcal{U}$. The Christoffel symbols for ∇ in these coordinates are written as Γ^k_{ij}, $i, j, k \in \{1, \dots, n\}$.

1. The coordinate formula for the covariant derivative of Y with respect to X is

$$\nabla_X Y = \Big(\frac{\partial Y^k}{\partial x^i} X^i + \Gamma^k_{ij} X^i Y^j\Big)\frac{\partial}{\partial x^k}.$$

2. The covariant derivative of a vector field Y along a curve γ is given by

$$\nabla_{\gamma'(t)} Y(t) = \Big(\dot{Y}^k(t) + \Gamma^k_{ij}(\gamma(t))\dot{x}^i(t) Y^j(t)\Big)\frac{\partial}{\partial x^k}.$$

Here $t \mapsto (x^1(t), \dots, x^n(t))$ is the local representative for γ.

3. The differential equations determining a geodesic $\gamma\colon I \to \mathsf{M}$ are then readily seen to be

$$\ddot{x}^k(t) + \Gamma^k_{ij}(\gamma(t))\dot{x}^i(t)\dot{x}^j(t) = 0, \qquad k \in \{1, \dots, n\}.$$

Here again, $t \mapsto (x^1(t), \dots, x^n(t))$ is the local representative for γ.

4. The geodesic spray has the representation

$$S = v^i \frac{\partial}{\partial x^i} - \Gamma^i_{jk} v^j v^k \frac{\partial}{\partial v^i}$$

in natural coordinates.

5. If ∇ is of class C^l and if $t \in \Gamma^{l+1}(T^r_s(\mathsf{TM}))$ has components $t^{i_1\cdots i_r}_{j_1\cdots j_s}$, then $\nabla_X t$ has components

$$(\nabla_X t)^{i_1\cdots i_r}_{j_1\cdots j_s} = \frac{\partial t^{i_1\cdots i_r}_{j_1\cdots j_s}}{\partial x^k} X^k + \sum_{\rho=1}^{r} \Gamma^{i_\rho}_{ki} t^{i_1\cdots i \cdots i_r}_{j_1\cdots j_s} X^k - \sum_{\sigma=1}^{s} \Gamma^j_{kj_\sigma} t^{i_1\cdots i_r}_{j_1\cdots j\cdots j_s} X^k.$$

From this expression, one can readily derive the expression for the components of the covariant differential as well.

6. The Christoffel symbols for $\overset{\mathbb{G}}{\nabla}$ are given by

$$\overset{\mathbb{G}}{\Gamma}{}^k_{ij} = \frac{1}{2}\mathbb{G}^{kl}\Big(\frac{\partial \mathbb{G}_{il}}{\partial x^j} + \frac{\partial \mathbb{G}_{jl}}{\partial x^i} - \frac{\partial \mathbb{G}_{ij}}{\partial x^l}\Big).$$

Let us look at our running polar coordinate example to illustrate the coordinate formula for the Christoffel symbols for the Levi-Civita affine connection.

Example 3.105 (Polar coordinates cont'd). On $\mathsf{M} = \mathbb{R}^2$ we have our two charts $(\mathcal{U}_1, \phi_1)$ and $(\mathcal{U}_1, \phi_2)$, representing Cartesian and polar coordinates, respectively. We consider the Riemannian metric $\mathbb{G}$ on M of Example 3.87. Thus, in Cartesian coordinates we have

$$\mathbb{G} = \mathrm{d}x \otimes \mathrm{d}x + \mathrm{d}y \otimes \mathrm{d}y,$$

and in polar coordinates

$$\mathbb{G} = \mathrm{d}r \otimes \mathrm{d}r + r^2 \mathrm{d}\theta \otimes \mathrm{d}\theta.$$

Since the components of $\mathbb{G}$ in Cartesian coordinates are constant, the Christoffel symbols in these coordinates are all zero. A direct computation gives the nonzero Christoffel symbols in polar coordinates as

$$\overset{\mathbb{G}}{\Gamma}{}^r_{\theta\theta} = -r, \qquad \overset{\mathbb{G}}{\Gamma}{}^\theta_{r\theta} = \overset{\mathbb{G}}{\Gamma}{}^\theta_{\theta r} = \frac{1}{r}.$$

Note that we adopt the convention of indexing the Christoffel symbols, in this case, by the names, rather than the numbers, of the coordinates. We will adopt this convention freely throughout the book, whenever this is convenient. Thus, for example, the geodesic equations in Cartesian coordinates are

$$\ddot{x} = 0, \quad \ddot{y} = 0,$$

and the geodesic equations in polar coordinates are

$$\ddot{r} - r\dot{\theta}^2 = 0, \quad \ddot{\theta} + \frac{2}{r}\dot{r}\dot{\theta} = 0.$$

We shall explore such issues in more detail in Section 4.3.4. •

3.8.4 The symmetric product

Next we consider the symmetric product that will be fundamental in the controllability results of Chapters 7 and 8, and the averaging results of Chapter 9.

Let us begin with the formal definition.

Definition 3.106 (Symmetric product). If ∇ is an affine connection on M, the ***symmetric product*** corresponding to ∇ is the operation that assigns to vector fields X and Y on M the vector field $\langle X : Y \rangle = \nabla_X Y + \nabla_Y X$. •

The symmetric product first appeared in the paper of Crouch [1981] in the description of gradient control systems (see also [Cortés and van der Schaft 2003]). The key idea in understanding the geometric significance of the symmetric product is the following.

Definition 3.107 (Geodesic invariance). A C^∞-distribution $\mathcal{D}$ on M is ***geodesically invariant*** with respect to a C^∞-affine connection ∇ if every geodesic $\gamma\colon I \to \mathsf{M}$, having the property that $\gamma'(t_0) \in \mathcal{D}_{\gamma(t_0)}$ for *some* $t_0 \in I$, satisfies $\gamma'(t) \in \mathcal{D}_{\gamma(t)}$ for *all* $t \in I$. •

We now have the following result of Lewis [1998].

Theorem 3.108 (Characterization of geodesic invariance). *A regular C^∞-distribution $\mathcal{D}$ is geodesically invariant with respect to a C^∞-affine connection ∇ if and only if $\langle X : Y \rangle \in \Gamma^\infty(\mathcal{D})$ for all vector fields X, Y taking values in $\mathcal{D}$.*

Proof. First, we gather three facts for the purpose of reference later in the proof. Recall that S is the geodesic spray for ∇.

1. Regard $\mathcal{D}$ as a submanifold of TQ. Then $\mathcal{D}$ is geodesically invariant if and only if S is tangent to $\mathcal{D} \subset \mathsf{TQ}$. This follows directly from the definitions of the concepts involved.
2. Let X and Y be vector fields on Q. Then $[\text{vlft}(X), [S, \text{vlft}(Y)]] = \text{vlft}(\langle X : Y \rangle)$. This can be proved directly in natural coordinates for TQ (see part (ii a) of Lemma B.3).
3. A vector field X on Q is a section of $\mathcal{D}$ if and only if $\text{vlft}(X)$ is tangent to $\mathcal{D}$ thought of as a submanifold of TQ. Again, this follows from the definitions.

Suppose that $\mathcal{D}$ is geodesically invariant. Thus S is tangent to $\mathcal{D} \subset \mathsf{TQ}$ by fact 1. Now let X and Y be sections of $\mathcal{D}$. By 3, $\text{vlft}(X)$ and $\text{vlft}(Y)$ are tangent to $\mathcal{D}$. From the characterization of the Lie bracket in Proposition 3.69, any iterated Lie bracket of $\text{vlft}(X)$, $\text{vlft}(Y)$, and S will also be tangent to $\mathcal{D}$. In particular, by fact 2, $[\text{vlft}(X), [S, \text{vlft}(Y)]] = \text{vlft}(\langle X : Y \rangle)$ is tangent to $\mathcal{D}$. Therefore, by fact 3, $\langle X : Y \rangle$ is a section of $\mathcal{D}$.

Now suppose that $\langle X : Y \rangle \in \Gamma^\infty(\mathcal{D})$ for sections X and Y of $\mathcal{D}$. Thus $[\text{vlft}(X), [S, \text{vlft}(Y)]]$ is tangent to $\mathcal{D} \subset \mathsf{TQ}$ for each pair of sections X and Y of $\mathcal{D}$ by facts 2 and 3. Let $\mathcal{U}$ be an open subset of Q. By shrinking $\mathcal{U}$, we can ensure the existence of linearly independent vector fields $\mathscr{X} = \{X_1, \dots, X_n\}$ on $\mathcal{U}$ having the property that these vector fields form a basis for each tangent space, and that $\{X_1, \dots, X_r\}$ are generators for $\mathcal{D}|\mathcal{U}$. Let us define generalized Christoffel symbols, $\overset{\mathscr{X}}{\Gamma}{}^i_{jk}$, $i, j, k \in \{1, \dots, n\}$, for ∇ on $\mathcal{U}$ by

$$\nabla_{X_i} X_j = \overset{\mathscr{X}}{\Gamma}{}^k_{ij} X_k, \qquad i, j \in \{1, \dots, n\}$$

(these will be discussed in more detail in Section 4.5.6). Since $\mathcal{D}$ is invariant under the symmetric product, we have

$$\overset{\mathscr{X}}{\Gamma}{}^\alpha_{ab} + \overset{\mathscr{X}}{\Gamma}{}^\alpha_{ba} = 0, \qquad \alpha \in \{r+1, \dots, n\},\ a, b \in \{1, \dots, r\}. \tag{3.13}$$

Now let $\gamma\colon I \to \mathcal{U}$ be a geodesic for ∇ and define $v^i\colon I \to \mathbb{R}$, $i \in \{1, \dots, n\}$, by $\gamma'(t) = v^i(t) X_i(\gamma(t))$. We then have

$$\nabla_{\gamma'(t)} \gamma'(t) = \Big(\dot{v}^i + \overset{\mathscr{X}}{\Gamma}{}^i_{jk}(\gamma(t)) v^j(t) v^k(t)\Big) X_i(\gamma(t)) = 0,$$

using Exercise E3.37. Now, (3.13) ensures that, if $v^\alpha(t_0) = 0$, $\alpha \in \{r+1, \dots, n\}$, for some $t_0 \in I$, then $v^\alpha(t) = 0$, $\alpha \in \{r+1, \dots, n\}$, for all $t \in I$. This means that $\mathcal{D}$ is geodesically invariant. ■

3.9 Advanced topics in differential geometry

In this section we discuss some advanced topics in differential geometry. These topics will, on occasion, play an important role in what we do. However, for a first-time reader, these can be distractions, and for, this reason, we choose to collect these topics in one section. This material can be safely skipped on a first reading.

We begin with a discussion of immersed submanifolds. These play a role in various places, most prominently in our study of controllability. Next, we point out some of the important distinctions between C^∞ and analytic geometry. A good introduction to real-analytic geometry is the book of Krantz and Parks [2002]. Next we collect some facts about generalized subbundles. This material appears not to have been published in book form, as least as far as we have been able to ascertain. Certain of the ideas, in the context of generalized subbundles of tangent and cotangent bundles, can be found in the monograph of Elkin [1999]. We follow this discussion of generalized subbundles with an alternative characterization of generalized subbundles of tangent bundles, i.e., distributions. Again, this material appears not to have been published in texts. The next section is perhaps the most standard, and concerns fiber bundles, which are generalizations of vector bundles, where fibers have the structure of a manifold, rather than a vector bundle. Readers wishing to learn more about fiber bundles are referred to the books [Husemoller 1994, Milnor and Stasheff 1974, Steenrod 1951]. The final section is perhaps the most optional, covering some topics in affine differential geometry not covered in Section 3.8.

3.9.1 The differentiable structure of an immersed submanifold

It is quite easy to see that the collection of submanifold charts for an embedded submanifold S of a manifold M can be used to define an atlas for S. In this way, the embedded submanifold S inherits a differentiable structure from M. Moreover, the manifold topology on an embedded submanifold, with the differentiable structure just described, agrees with the subspace topology of $\mathsf{S} \subset \mathsf{M}$. For immersed submanifolds, the picture is a little more complicated. However, it will come up at various points in the text, so let us consider it here.

Let $\mathsf{S} \subset \mathsf{M}$ be a C^r-immersed submanifold, and let $x \in \mathsf{S}$. By Proposition 3.40, there exists a chart $(\mathcal{U}, \phi)$ around x with the following properties:

1. ϕ takes values in $\mathbb{R}^{n-m} \times \mathbb{R}^m$;
2. $\phi(\mathcal{U} \cap \mathsf{S})$ consists of possibly infinitely many disconnected components (in the standard topology on $\mathbb{R}^{n-m} \times \mathbb{R}^m$), exactly one of which contains $\phi(x)$;
3. the connected component of $\phi(\mathcal{U} \cap \mathsf{S})$ containing $\phi(x)$ is an open subset of $\{ (\boldsymbol{x}, \boldsymbol{0}) \mid \boldsymbol{x} \in \mathbb{R}^{n-m}\}$.

We then define a chart for S around x as follows. We define

$$\mathcal{U}_x = \phi^{-1}(\{ (\boldsymbol{x}, \boldsymbol{0}) \mid \boldsymbol{x} \in \mathbb{R}^{n-m}\} \cap \phi(\mathcal{U})),$$

and we define $\phi_x(y) = \mathrm{pr}_1 \circ \phi(y)$ for each $y \in \mathcal{U}_x$, where $\mathrm{pr}_1 : \mathbb{R}^{n-m} \times \mathbb{R}^m \to \mathbb{R}^{n-m}$ is projection onto the first factor. One can readily verify that $(\mathcal{U}_x, \phi_x)$

is a chart. Furthermore, two such charts whose domains intersect will satisfy the overlap condition, by virtue of the fact that the corresponding charts for M satisfy the overlap condition. It is clear that $\mathscr{A}_{\mathsf{S}} = \{\mathcal{U}_x, \phi_x\}_{x\in\mathsf{S}}$ defines an atlas for S. Thus we see that an immersed submanifold obtains a differentiable structure from M. Since an immersed submanifold is defined to be the image of a manifold N under an injective immersion $f\colon \mathsf{N} \to \mathsf{M}$, it also follows that S inherits a differentiable structure from N via f. One can show that these two differentiable structures for S agree.

Now let us say a few words about the topology on an immersed submanifold $\mathsf{S} \subset \mathsf{M}$. Such an immersed submanifold possesses two natural topologies: (1) the manifold topology for the differentiable structure defined above and (2) the subspace topology inherited from M. These topologies agree if and only if S is an embedded submanifold. Moreover, the manifold topology is stronger than the subspace topology. By this we mean that all subsets that are open in the subspace topology are also open in the manifold topology, but there may be subsets that are open in the manifold topology that are not open in the subspace topology. The reader might benefit by trying to understand this in the case of Example 3.44–1 when a is irrational.

3.9.2 Comments on smoothness, in particular analyticity

To the reader unused to constant references to smoothness, our insistence in this book on referring to objects as being "class C^r," "smooth," or "analytic" might seem unnecessarily complicated. Except for the few occasions where analytic data is required (in Chapters 7, 8, and 9), we could get away with a blanket assumption of all data being C^∞. However, we *do* need analyticity on occasion, and this requires us to be careful in stating hypotheses involving degrees of differentiability. We refer to [Sussmann 1990b] for a discussion of the relevance of analyticity in control theory.

One of the places where C^∞-geometry and analytic geometry differ the most is in the character of the level sets of functions. For analytic functions, we have the following statement, more general versions of which were first proved by Hironaka [1973, 1975] (see also [Bierstone and Milman 1988, Sussmann 1990a]).

Theorem 3.109 (Zeros of analytic functions). *Let* M *be an analytic manifold with* $f\colon \mathsf{M} \to \mathbb{R}$ *an analytic function. Let* $Z_f = f^{-1}(0)$. *Then there exists a finite or countable collection* $\{\mathsf{S}_a\}_{a\in A}$ *of disjoint* C^ω*-submanifolds of* M *with the following properties:*

(i) $Z_f = \cup_{a\in A}\mathsf{S}_a$;

(ii) for each $x \in Z_f$, *there exists a neighborhood* $\mathcal{U}$ *of* x *for which the set*

$$\{a \in A \mid \mathcal{U} \cap \mathsf{S}_a \neq \emptyset\}$$

is finite.

Thus the zeros (or more generally, the level sets) of analytic functions are "nice" unions of analytic submanifolds. For C^∞-functions, the situation is not so nice. For example, it is possible to show that if $A \subset \mathbb{R}^n$ is a closed set, then there exists a C^∞-function $f\colon \mathbb{R}^n \to \mathbb{R}$ for which $f^{-1}(0) = A$. For $n = 1$, the reader can explore this in Exercise E3.15. For $n > 1$, one can make use of the Whitney Decomposition of an open subset of $\mathbb{R}^n$ [e.g., Federer 1969]. Many of the distinctions that arise between infinitely differentiable objects and analytic objects arise exactly because Theorem 3.109 does not hold for C^∞-functions. We refer the reader to [Krantz and Parks 2002] for additional information.

One of the other distinctions between C^∞ and analytic differential geometry, related to our preceding discussions of analytic functions, concerns partitions of unity. Roughly speaking, a ***partition of unity*** on a C^∞-manifold M is a decomposition of the function $x \mapsto 1$ as a sum of C^∞-functions, each having compact support,[2] and having the property that every point in M is contained in the support of at most a finite number of these functions. One may show [Abraham, Marsden, and Ratiu 1988, Section 5.5] that paracompact C^∞-manifolds possess partitions of unity. Partitions of unity are useful for constructing global data using local data. For example, one can use a partition of unity to show the existence of a Riemannian metric on any paracompact manifold. However, analytic partitions of unity do not exist. Indeed, any analytic function that vanishes on an open set must vanish on the connected component of that set. This creates certain difficulties in analytic differential geometry. For example, it is not easy to establish whether analytic manifolds possess analytic Riemannian metrics. This was proved in a difficult theorem by Grauert [1958], who showed that analytic manifolds can be analytically embedded in Euclidean space, and thus acquire an analytic Riemannian metric by restriction. The similar embedding theorem by Whitney [1936] in the C^∞ case is rather easier. A related question is whether a Riemannian manifold can be embedded in Euclidean space in such a way that the metric is preserved. In the C^∞ case, this is already hard, and was proved by Nash [1956]. In the analytic case, the complete resolution of this problem is outstanding, although there is partial progress (e.g., in the compact case [Greene and Jacobowitz 1971]).

Physically, there is not much reason for us to *not* make a blanket assumption that all data be analytic. However, since the nature of infinitely differentiable data and analytic data can sometimes be quite different (see Exercises E3.15, E3.30, E3.31, E3.33, and E6.17 for examples), it is mathematically unnatural to assume analyticity unless it is actually needed. We therefore offer the following rules for the reader to employ when parsing statements concerning degrees of differentiability.

[2] Recall that the ***support*** of a continuous function is the closure of the set of points at which the function is nonzero.

The role of smoothness hypotheses in this book. 1. In this chapter, for consistency of presentation, we have stated results and definitions using objects that are C^r, where $r \in \mathbb{N} \cup \{\infty\} \cup \{\omega\}$. Certain definitions even make sense when $r = 0$. On occasion we will write $r - 1$ where $r \in \mathbb{N} \cup \{\infty\} \cup \{\omega\}$. This has the usual meaning when $r \in \mathbb{N}$, and we define $r - 1 = \infty$ (resp. $r - 1 = \omega$) when $r = \infty$ (resp. $r = \omega$).

2. For all chapters, unless otherwise stated, all data can be assumed to be of class C^∞. We have tried to be careful in our statements to explicitly indicate the required degree of differentiability, however, we may have lapsed on occasion. Such a lapse can be taken to mean C^∞ data. When analyticity is needed (e.g., for some results in Chapters 7, 8, and 9), this hypothesis will be essential, and it will be clearly stated.
3. One place where the "C^∞ unless otherwise stated" assumption cannot be applied without due attention is with integral curves of time-dependent vector fields. As mentioned briefly at the end of Section 3.2.2, and explained in more detail in Sections 3.5.2 and A.2.1, we will have many occasions to consider curves that are not even class C^1. The reader should be aware that when general control inputs are present, the trajectories of the resulting system will generally only be locally absolutely continuous. •

3.9.3 Properties of generalized subbundles

Generalized subbundles can exhibit quite complicated behavior as compared to vector subbundles. Even for generalized subbundles of locally constant rank, i.e., for subbundles, there are interesting questions one can ask concerning the character of generators. In this section we provide a few useful results, and point the interested reader to appropriate places in the literature.

First let us give definitions that characterize some of the interesting features of generalized subbundles.

Definition 3.110. Let W be a C^r-generalized subbundle of a vector bundle $\pi\colon \mathsf{V} \to \mathsf{B}$.

(i) A point $b \in \mathsf{B}$ is a ***regular point*** for W if there exists a neighborhood $\mathcal{U} \subset \mathsf{B}$ of b such that the rank of the fiber is constant on $\mathcal{U}$.
(ii) If b is not a regular point, then it is a ***singular point*** for W.
(iii) The generalized subbundle W is ***regular*** if all points in B are regular points. •

Note that a regular C^r-generalized subbundle is a C^r-subbundle, in the language of Definition 3.50.

The following characterization of regular points is sometimes useful, and its proof is quite straightforward, cf. [Elkin 1999, Proposition 1.11].

Theorem 3.111. *The regular points for a C^r-generalized subbundle W of a vector bundle $\pi\colon \mathsf{V} \to \mathsf{B}$, $r \in \mathbb{N} \cup \{\infty\} \cup \{\omega\}$, form an open dense subset of B.*

Analytic generalized subbundles are particularly structured. Indeed, if the base space B is connected, then the set of points for which the rank of a generalized subbundle is *maximal* is open and dense.[3] Also, the set of regular points of an analytic generalized subbundle have characterizations that are far stronger than "open and dense." These characterizations are a consequence of the next important theorem.

Theorem 3.112. *An analytic generalized subbundle* W *of a vector bundle* $\pi\colon \mathsf{V} \to \mathsf{B}$ *is locally finitely generated.*

Since it appears difficult to find a complete proof of this rather important fact in one place, let us outline the essential ideas, so that a reader interested in looking this up will know where to start.

Remark 3.113. We do not intend that the following discussion be complete, self-contained, or perfectly precise. We refer the reader to Definitions 3.115 and 3.116 for the definitions of a ring and a module.

1. The property of analytic generalized subbundles on a base space B being locally finitely generated follows from the fact that the ring of analytic functions, defined in a sufficiently small neighborhood of a point $b \in \mathsf{B}$, is Noetherian. This means, by definition, that every ideal in this ring is finitely generated. It turns out that every submodule of a module defined over a Noetherian ring is finitely generated. Thus, the desired property of analytic generalized subbundles will follow if one can show the asserted Noetherian ring property. Facts about general Noetherian rings can be found in a good algebra text [e.g., Hungerford 1980].
2. The first step in the proof is, therefore, to show that the ring of analytic functions defined in a sufficiently small neighborhood of a point is Noetherian. One notes that analytic functions defined on a neighborhood of $b \in \mathsf{B}$ are in 1–1 correspondence with Taylor series which converge in a neighborhood of b to the function defining them.[4] Here one uses a coordinate chart around b to produce Taylor series, and it can be verified that convergence of a Taylor series is independent of coordinates, provided charts with sufficiently small domain are used. We refer to [Krantz and Parks 2002] for this sort of discussion.
3. Thus the property of being locally finitely generated will follow if one can show that a certain ring of power series, those corresponding to the Taylor series as above, is a Noetherian ring. This rests on the Weierstrass

[3] If B is not connected, then on each connected component of B, the set of points for which the rank of an analytic generalized subbundle is maximal is open and dense.

[4] It is a theorem of Borel that *every* power series is the Taylor series for some C^∞-map. Thus, it is important that, when identifying those power series associated to analytic maps, one indicates that the Taylor series converges to the function defining it.

Preparation Theorem. A proof of this in the real case is given by Krantz and Parks [2002]. The deduction of the Noetherian property of the suitable power series ring, given the Weierstrass Preparation Theorem (and also the Weierstrass Preparation Theorem in the complex case), can be found in the complex setting in [Hörmander 1973]. •

The property of being locally finitely generated leads to some important distinctions in the character of the open and dense set of regular points in C^∞ as opposed to analytic generalized subbundles. In particular, the set of singular points of a C^ω-generalized subbundle will be quite "small," in a way that is not possible for C^∞-generalized subbundles. The reader may explore these issues in Exercises E3.30 and E3.31.

Generally speaking, regular generalized subbundles (i.e., subbundles) are far better behaved than generalized subbundles with singular points. One of the ways in which this nicer behavior is manifested is in terms of the character of generators. For a C^r-generalized subbundle W of a vector bundle $\pi\colon \mathsf{V} \to \mathsf{B}$, a set $\{\xi_a\}_{a\in A}$ of C^r-sections of V are ***global generators*** for W if $\mathsf{V}_b \cap \mathsf{W} = \operatorname{span}_{\mathbb{R}} \{\xi_a(b) |\ a \in A\}$ for each $b \in \mathsf{B}$. Let us discuss the matter of the existence of a finite set of global generators. First of all, if W possesses singular points, it is generally not the case that one can find a finite set of global generators, no matter the smoothness of W. In the regular case, the situation is better. The following result is essentially due to Swan [1962].[5]

Theorem 3.114. *Let W be a C^r-generalized subbundle of a vector bundle $\pi\colon \mathsf{V} \to \mathsf{B}$, $r \in \mathbb{N} \cup \{\infty\} \cup \{\omega\}$. If B is paracompact, if all points in B are regular points for W, and if the dimension of the fibers of W is bounded, then there exists a finite set of global generators for W.*

Swan shows something more than the existence of global generators. Indeed, Swan shows that the set of sections of a regular generalized subbundle W is a "finitely generated projective module" over $C^r(\mathsf{B})$ in this case. The reader can explore aspects of this in Exercise E3.42 in the case of subbundles of the tangent bundle.

3.9.4 An alternative notion of distribution

We next consider some issues associated with those in the previous section, but now specialized to distributions. First, we provide the formal definitions of rings and modules, since these are fundamental in this section.

Definition 3.115 (Ring). A ***ring*** is a set R with two binary operations, $(r_1, r_2) \mapsto r_1 + r_2$ and $(r_1, r_2) \mapsto r_1 \star r_2$, called ***addition*** and ***multiplication***, respectively, and which together satisfy the following rules:

[5] Swan considers only the case where $r = 0$ and B is compact. At the end of the paper, it is indicated where the modifications need to be made to allow B to be paracompact. It is readily seen that Swan's constructions carry over immediately to the case where $r \in \mathbb{N} \cup \{\infty\} \cup \{\omega\}$.

(i) $(r_1+r_2)+r_3 = r_1+(r_2+r_3)$, $r_1, r_2, r_3 \in \mathsf{R}$ (***associativity*** of addition);
(ii) $r_1 + r_2 = r_2 + r_1$, $r_1, r_2 \in \mathsf{R}$ (***commutativity*** of addition);
(iii) there exists $0 \in \mathsf{R}$ such that $r + 0 = r$, $r \in \mathsf{R}$ (***additive identity***);
(iv) for $r \in \mathsf{R}$, there exists $-r \in \mathsf{R}$ such that $r + (-r) = 0$ (***additive inverse***);
(v) $(r_1 \star r_2) \star r_3 = r_1 \star (r_2 \star r_3)$, $r_1, r_2, r_3 \in \mathsf{R}$ (***associativity*** of multiplication);
(vi) $r_1 \star (r_2 + r_3) = (r_1 \star r_2) + (r_1 \star r_3)$, $r_1, r_2, r_3 \in \mathsf{R}$ (***left distributivity***);
(vii) $(r_1 + r_2) \star r_3 = (r_1 \star r_3) + (r_2 \star r_3)$, $r_1, r_2, r_3 \in \mathsf{R}$ (***right distributivity***).

If there exists $1 \in \mathsf{R}$ such that $1 \star r = r \star 1$ for all $r \in \mathsf{R}$, then the ring is a ***unit ring***. If $r_1 \star r_2 = r_2 \star r_1$ for all $r_1, r_2 \in \mathsf{R}$, then the ring is ***commutative***. •

The ring of interest to us in this section will be the ring $C^r(\mathsf{M})$, $r \in \mathbb{N} \cup \{\infty\} \cup \{\omega\}$, where M is a C^r-manifold. To be clear, let us provide the operations of addition and multiplication in this case. For $f_1, f_2 \in C^r(\mathsf{M})$ we define $f_1 + f_2 \in C^r(\mathsf{M})$ by $(f_1 + f_2)(x) = f_1(x) + f_2(x)$, and we define $f_1 \star f_2 \in C^r(\mathsf{M})$ by $(f_1 \star f_2)(x) = f_1(x) f_2(x)$. We shall adopt the convention of omitting the "$\star$" to denote multiplication, in this case. Note that this ring is a commutative unit ring.

We hope the reader will allow us a momentary use of poor notation. The next definition is the only place in the book where we will use M and N to denote modules. At all other times, M and N denote manifolds.

Definition 3.116 (Module and submodule). Let R be a commutative unit ring. An **R-*module*** is a set M equipped with two operations: (1) M-addition, denoted by $x_1 + x_2 \in \mathsf{M}$ for $x_1, x_2 \in \mathsf{M}$, and (2) R-multiplication, denoted by $r\,x \in \mathsf{M}$ for $r \in \mathsf{R}$ and $x \in \mathsf{M}$. M-addition must satisfy the rules

(i) $x_1 + x_2 = x_2 + x_1$, $x_1, x_2 \in \mathsf{M}$ (***commutativity***),
(ii) $x_1 + (x_2 + x_3) = (x_1 + x_2) + x_3$, $x_1, x_2, x_3 \in \mathsf{M}$ (***associativity***),
(iii) there exists a unique element $0 \in \mathsf{M}$ with the property that $x + 0 = x$ for every $x \in \mathsf{M}$ (***zero element***), and
(iv) for every $x \in \mathsf{M}$, there exists a unique element $-x \in \mathsf{M}$ such that $x + (-x) = 0$ (***negative element***),

and R-multiplication must satisfy the rules

(v) $r_1(r_2\,x) = (r_1 \star r_2)x$, $r_1, r_2 \in \mathsf{R}$, $x \in \mathsf{M}$ (***associativity***),
(vi) $1\,x = x$, $x \in \mathsf{M}$,
(vii) $r(x_1 + x_2) = r\,x_1 + r\,x_2$, $r \in \mathsf{R}$, $x_1, x_2 \in \mathsf{M}$ (***distributivity***), and
(viii) $(r_1 + r_2)x = r_1\,x + r_2\,x$, $r_1, r_2 \in \mathsf{R}$, $x \in \mathsf{M}$ (***distributivity*** again).

A subset N of a module M is a ***submodule*** if $x_1 + x_2 \in \mathsf{N}$ for all $x_1, x_2 \in \mathsf{N}$, and if $r\,x \in \mathsf{N}$ for every $r \in \mathsf{R}$ and for every $x \in \mathsf{N}$. •

It is possible to define modules over noncommutative rings, but we will not use these. Obviously, R-modules are generalizations of $\mathbb{F}$-vector spaces. We shall therefore use some of the language of Chapter 2 without formally defining it in the context of modules. The module of interest to us in this section is the $C^r(\mathsf{M})$-module of C^r-vector fields, $\Gamma^r(\mathsf{TM})$. Let us indicate what the module structure is. For $f \in C^r(\mathsf{M})$ and $X \in \Gamma^r(\mathsf{TM})$, we define $fX \in \Gamma^r(\mathsf{TM})$ by

$(fX)(x) = f(x)X(x)$. For $X_1, X_2 \in \Gamma^r(\mathsf{TM})$, we define $X_1 + X_2 \in \Gamma^r(\mathsf{TM})$ by $(X_1 + X_2)(x) = X_1(x) + X_2(x)$.

The next definition provides an alternative manner of thinking about distributions, one that involves the $C^r(\mathsf{M})$-module structure of $\Gamma^r(\mathsf{TM})$.

Definition 3.117 (Algebraic distribution). Let $r \in \mathbb{N} \cup \{\infty\} \cup \{\omega\}$ and let M be a C^{r+1}-manifold. A ***C^r-algebraic distribution*** on M is a submodule of $\Gamma^r(\mathsf{TM})$. If $\mathscr{X} \subset \Gamma^r(\mathsf{TM})$, then the C^r-algebraic distribution ***generated*** by $\mathscr{X}$ is the submodule $\operatorname{span}_{C^r(\mathsf{M})}\{\mathscr{X}\}$ consisting of all finite $C^r(\mathsf{M})$-linear combinations of vector fields from $\mathscr{X}$. •

Let us list some points concerning this alternative notion of distribution, especially as it compares to our definition, given in Definition 3.88.

Remarks 3.118. 1. Associated to every C^r-distribution $\mathcal{D}$ on M, there is a natural C^r-algebraic distribution, which we denote by $\Gamma^r(\mathcal{D})$, consisting of all C^r-vector fields on M taking values in $\mathcal{D}$. It is easy to check that $\Gamma^r(\mathcal{D})$ is a submodule of $\Gamma^r(\mathsf{TM})$.

2. Associated to every C^r-algebraic distribution $\mathscr{V} \subset \Gamma^r(\mathsf{TM})$, there is a natural C^r-distribution, which we denote by $\mathcal{D}(\mathscr{V})$, defined by

$$\mathcal{D}(\mathscr{V})_x = \operatorname{span}_{\mathbb{R}}\{X(x) \mid X \in \mathscr{V}\}.$$

That this is a C^r-distribution follows since $\mathscr{V}$ serves as a set of global generators.

3. The preceding two constructions relating distributions and algebraic distributions may lead one to think the two notions are exactly equivalent. They are generally not. There are two questions one can ask that help to understand the differences.

   (a) *Given a C^r-distribution $\mathcal{D}$, it is true that $\mathcal{D}(\Gamma^r(\mathcal{D})) = \mathcal{D}$?* The answer to this question is, "Yes, if M is paracompact." Thus, if one defines a distribution, then associates to this distribution the submodule of sections of this distribution, and then produces the distribution corresponding to this submodule, one ends up with the original distribution. In the C^∞-case, this is fairly easily proved using partitions of unity and local generators. In the analytic case, it is not so clear, since there might arise the problem of extending local analytic generators to global analytic generators. However, it is possible to resolve this matter by using the embedding theorem for real-analytic manifolds by Grauert [1958].

   (b) *Given a C^r-algebraic distribution $\mathscr{V}$, it is true that $\Gamma^r(\mathcal{D}(\mathscr{V})) = \mathscr{V}$?* The answer to this question is, "No." The inclusion $\mathscr{V} \subset \Gamma^r(\mathcal{D}(\mathscr{V}))$ holds, but the opposite inclusion does not generally hold, as may be seen with a trivial counterexample in the case in which $r = \omega$. We take $\mathsf{M} = \mathbb{R}$ and we let $\mathscr{V}$ be the submodule of $\Gamma^\omega(\mathsf{TM})$ generated by the vector field $X = x^2 \frac{\partial}{\partial x}$. Thus $\mathscr{V}$ consists of all vector fields on $\mathbb{R}$

of the form fX, where $f \in C^\omega(\mathsf{M})$. The distribution corresponding to the algebraic distribution $\mathscr{V}$ is

$$\mathcal{D}(\mathscr{V})_x = \begin{cases} 0, & x = 0, \\ \mathsf{T}_x\mathsf{M}, & x \neq 0. \end{cases}$$

Note that the vector field $x\frac{\partial}{\partial x}$ is in $\Gamma^\omega(\mathcal{D}(\mathscr{V}))$, but is not in $\mathscr{V}$. The reader can look at another example illustrating this phenomenon in Exercise E3.43. •

The difficulties arising in the simple counterexample above arise because of the choice of generator for $\mathscr{V}$. Had we instead chosen the generator $X = x\frac{\partial}{\partial x}$, then the equality $\mathscr{V} = \Gamma^r(\mathcal{D}(\mathscr{V}))$ would indeed hold. This suggests a distinguished sort of algebraic distribution, one that we record in the following definition, along with a similar notion for local generators of a distribution.

Definition 3.119 (Nondegenerate algebraic distribution). A C^r-algebraic distribution $\mathscr{V}$ is ***nondegenerate*** if $\mathscr{V} = \Gamma^r(\mathcal{D}(\mathscr{V}))$. Also, if $\mathcal{D}$ is a C^r-distribution, and if $\{X_a\}_{a\in A}$ is a collection of local generators defined on a neighborhood $\mathcal{U} \subset \mathsf{M}$, these generators are ***nondegenerate*** if $\text{span}_{C^r(\mathcal{U})}\{X_a \mid a \in A\} = \Gamma^r(\mathcal{D}|\mathcal{U})$, i.e., if $\{X_a\}_{a\in A}$ generates the submodule of C^r-sections of $\mathcal{D}|\mathcal{U}$. •

Let us record some facts about nondegenerate algebraic distributions and nondegenerate local generators for a distribution.

Remarks 3.120. 1. Clearly, by definition, the notions of nondegenerate algebraic distributions, and distributions in the sense we define them, are equivalent.

2. Next let us address the matter of the existence of nondegenerate local generators.
   (a) If $\mathcal{D}$ is a C^ω-distribution on M, then, around every point $x \in \mathsf{M}$, there exists a neighborhood $\mathcal{U}$ of x and a finite collection of nondegenerate local generators for $\mathcal{D}$ defined on $\mathcal{U}$. This follows from the discussion following the statement of Theorem 3.112 above, from which we see that $\Gamma^\omega(\mathcal{D}|\mathcal{U})$ is locally finitely generated, by virtue of its being a module over the Noetherian ring $C^\omega(\mathcal{U})$, provided that $\mathcal{U}$ is sufficiently small.
   (b) If M is paracompact, then, for $r \in \mathbb{N} \cup \{\infty\} \cup \{\omega\}$, a regular C^r-distribution $\mathcal{D}$ of bounded rank on M possesses a finite set of nondegenerate global generators. This follows from Theorem 3.114. In fact, more is true in this case. It turns out that *any* finite set of global generators will be nondegenerate. This follows essentially from the argument that the reader can reproduce in Exercise E3.42.
   (c) The authors are not aware of anything that can be said about the existence of nondegenerate generators, local or global, in cases other than the two given above.

3. One of the useful features of nondegenerate generators is that two sets of them will have a nice relationship to one another, a relationship that generally will not hold for sets of generators that are not nondegenerate. The relationship, in the case of finite sets of generators, is as follows. Suppose that $\{X_1, \dots, X_k\}$ and $\{\tilde{X}_1, \dots, \tilde{X}_{\tilde{k}}\}$ are two sets of nondegenerate local generators for a C^r-distribution $\mathcal{D}$, both defined on $\mathcal{U}$. The definition of nondegeneracy ensures that there exists $\Lambda_a^i \in C^r(\mathcal{U})$, $i \in \{1, \dots, k\}$, $a \in \{1, \dots, \tilde{k}\}$, and $\tilde{\Lambda}_i^a \in C^r(\mathcal{U})$, $a \in \{1, \dots, \tilde{k}\}$, $i \in \{1, \dots, k\}$, satisfying

$$\tilde{X}_a(x) = \Lambda_a^i(x) X_i(x), \qquad a \in \{1, \dots, \tilde{k}\},\ x \in \mathcal{U},$$
$$X_i(x) = \tilde{\Lambda}_i^a(x) \tilde{X}_a(x), \qquad i \in \{1, \dots, k\},\ x \in \mathcal{U}.$$

This relationship will be useful below. •

Now we use our above constructions to shed some light on the computation of the involutive closure of a C^r-distribution $\mathcal{D}$, $r \in \{\infty\} \cup \{\omega\}$. Following Lemma 3.94 we had defined a sequence of distributions. Let us reproduce this discussion, but now allowing the use of local generators, so that the notion of nondegenerate generators can be used to advantage. If $\mathcal{U}$ is an open subset of M and if $\mathscr{X} = \{X_a\}_{a \in A}$ is a collection of generators for $\mathcal{D}$ defined on $\mathcal{U}$, then we define a sequence of C^r-distributions $\mathrm{Lie}^{(l)}(\mathcal{D}, \mathcal{U}, \mathscr{X})$, $l \in \mathbb{Z}_+$, on $\mathcal{U}$, by $\mathrm{Lie}^{(0)}(\mathcal{D}, \mathcal{U}, \mathscr{X}) = \mathcal{D}|\mathcal{U}$, and

$$\begin{aligned} \mathrm{Lie}^{(l)}(\mathcal{D}, \mathcal{U}, \mathscr{X})_x = \mathrm{Lie}^{(l-1)}(\mathcal{D}, \mathcal{U}, \mathscr{X})_x + \mathrm{span}_{\mathbb{R}}\{[X, Y](x) |\ \\ X \in \Gamma^\infty(\mathrm{Lie}^{(l_1)}(\mathcal{D}, \mathcal{U}, \mathscr{X})),\ Y \in \Gamma^\infty(\mathrm{Lie}^{(l_2)}(\mathcal{D}, \mathcal{U}, \mathscr{X})),\ l_1 + l_2 = l - 1\}, \end{aligned} \tag{3.14}$$

for $l \in \mathbb{N}$ and $x \in \mathcal{U}$. For fixed $x \in \mathcal{U}$, there clearly exists a smallest integer $N(x, \mathscr{X})$ with the property that $\mathrm{Lie}^{(l)}(\mathcal{D}, \mathcal{U}, \mathscr{X}) = \mathrm{Lie}^{(N(x,\mathscr{X}))}(\mathcal{D}, \mathcal{U}, \mathscr{X})$ for $l \geq N(x, \mathscr{X})$. As we asserted after Lemma 3.94, if a point $x_0 \in \mathcal{U}$ is a regular point for all of these distributions, then one can readily ascertain that $\mathrm{Lie}^{(\infty)}(\mathcal{D}, \mathcal{U}, \mathscr{X})_{x_0} = \mathrm{Lie}^{(N(x,\mathscr{X}))}(\mathcal{D}, \mathcal{U}, \mathscr{X})_{x_0}$. Also, from Remark 3.120–3, one can readily determine that this sequence of distributions $\mathrm{Lie}^{(l)}(\mathcal{D}, \mathcal{U}, \mathscr{X})$, $l \in \mathbb{Z}_+$, on $\mathcal{U}$ is independent of choice of local generators, provided we restrict consideration to finite sets of nondegenerate local generators. For analytic distributions this is not a restriction. For C^∞-distributions, it is a significant restriction.

Let us record the result in the analytic case.

Proposition 3.121. *If $\mathcal{D}$ is an analytic distribution on M, then there is a well-defined sequence of analytic distributions,* $\mathrm{Lie}^{(l)}(\mathcal{D})$, $l \in \mathbb{Z}_+$, *defined according to* (3.14), *where, in that construction, finite sets of nondegenerate local generators are used.*

Now let us explore the relationship between the sequence of distributions $\mathrm{Lie}^{(l)}(\mathcal{D})$, $l \in \mathbb{Z}_+$, and the involutive closure $\mathrm{Lie}^{(\infty)}(\mathcal{D})$. For arbitrary C^∞-distributions, not much can be said. For analytic distributions, however, the association goes as well as one might hope.

Theorem 3.122. *Let $\mathcal{D}$ be an analytic distribution on* M*, let* $\mathcal{U}$ *be an open subset of* M*, and let* $\mathscr{X}$ *be a set of local generators defined on* $\mathcal{U}$*. Then, for each* $x \in \mathcal{U}$*,* $\cup_{l \in \mathbb{Z}_+} \mathrm{Lie}^{(l)}(\mathcal{D}, \mathcal{U}, \mathscr{X})_x = \mathrm{Lie}^{(\infty)}(\mathcal{D})_x$.

Thus, for analytic distributions, one can define the involutive closure by computing sequences of Lie brackets of local generators, and one can be guaranteed that, for a given point x, the construction will require only a finite number of Lie brackets.

3.9.5 Fiber bundles

In this section we briefly discuss a generalization of the concept of a vector bundle, where we now allow the fiber to be, not a vector space, but a manifold.

Definition 3.123 (Locally trivial fiber bundle). For $r \in \mathbb{N} \cup \{\infty\} \cup \{\omega\}$, a ***$C^r$-locally trivial fiber bundle*** is a 4-tuple $(\pi, \mathsf{M}, \mathsf{B}, \mathsf{F})$, where

(i) M, B, and F are C^r-manifolds,
(ii) $\pi\colon \mathsf{M} \to \mathsf{B}$ is a surjective submersion of class C^r, and
(iii) there exists an atlas $\mathscr{A} = \{(\mathcal{U}_a, \phi_a)\}_{a \in A}$ for B such that, for each $a \in A$, there is a C^r-diffeomorphism $\psi_a\colon \pi^{-1}(\mathcal{U}_a) \to \mathcal{U}_a \times \mathsf{F}$ with the property that $\mathrm{pr}_1 \circ \psi_a = \pi$.

We call M the ***total space***, B the ***base space***, and F the ***typical fiber***. The ***fiber*** over $b \in \mathsf{B}$ is $\pi^{-1}(b)$. The map ψ_a in the definition is called a ***local trivialization***. •

We shall often use the words "fiber bundle" to refer to a locally trivial fiber bundle, with the understanding that local triviality is implicit. A fiber bundle might be thought of as in Figure 3.14. Note that while each fiber $\pi^{-1}(b)$ is indeed diffeomorphic to F, there is not generally a natural diffeomorphism.

Examples 3.124. 1. The product manifold $\mathsf{M} = \mathsf{B} \times \mathsf{F}$ with $\pi = \mathrm{pr}_1$ is a locally trivial fiber bundle. It is obviously special and is called a ***trivial bundle***. More generally, a locally trivial fiber bundle $(\pi, \mathsf{M}, \mathsf{B}, \mathsf{F})$ is called ***trivializable*** if there exists a diffeomorphism $f\colon \mathsf{M} \to \mathsf{B} \times \mathsf{F}$ such that the following diagram commutes:

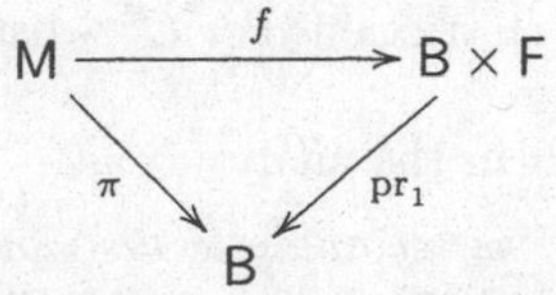

2. Any vector bundle is obviously a locally trivial fiber bundle. Example 3.48–4 shows that not all locally trivial fiber bundles are trivial.
3. If S is a submanifold of B, then the set $\pi^{-1}(\mathsf{S})$ is the total space of a locally trivial fiber bundle with base space S. This is the ***restricted bundle***, and is denoted $\mathsf{M}|\mathsf{S}$. •

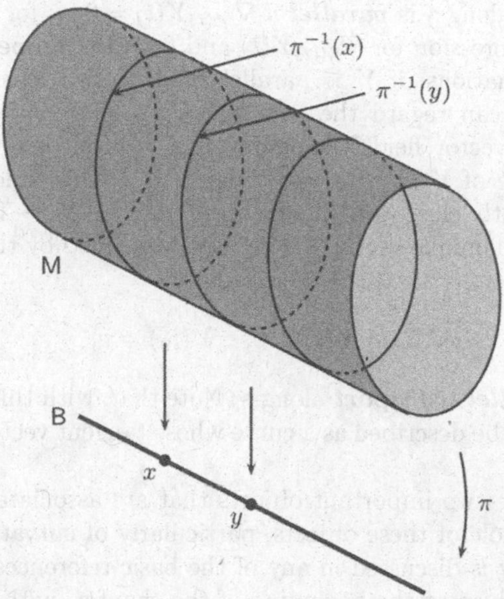

Figure 3.14. A representation of a fiber bundle

Definition 3.125 (Fiber bundle map). For $r \in \mathbb{N} \cup \{\infty\} \cup \{\omega\}$, let $(\pi, \mathsf{M}, \mathsf{B}, \mathsf{F})$ and $(\sigma, \mathsf{N}, \mathsf{A}, \mathsf{G})$ be C^r-locally trivial fiber bundles. A map $f\colon \mathsf{M} \to \mathsf{N}$ is called a ***C^r-fiber bundle map***, or simply a ***C^r-bundle map***, if there exists a map $f_0\colon \mathsf{B} \to \mathsf{C}$ of class C^r such that the following diagram commutes:

$$\begin{array}{ccc} \mathsf{M} & \xrightarrow{\ f\ } & \mathsf{N} \\ {\scriptstyle\pi}\downarrow & & \downarrow{\scriptstyle\sigma} \\ \mathsf{B} & \xrightarrow[\ f_0\ ]{} & \mathsf{A} \end{array}$$

We shall often say that f is a fiber bundle map ***over*** f_0. A fiber bundle map that is a diffeomorphism is called a ***fiber bundle isomorphism***. •

Note that a fiber bundle map maps $\pi^{-1}(b)$ to $\sigma^{-1}(f_0(b))$.

3.9.6 Additional topics in affine differential geometry

This final advanced topics section is the only one that can really be considered optional, in that we shall never make essential use of the ideas we discuss here. However, the topics in this section are of fundamental importance in affine differential geometry, so we include them for the sake of completeness.

Let ∇ be a C^r-affine connection on M, $r \in \mathbb{N} \cup \{\infty\} \cup \{\omega\}$. Associated with ∇ is the notion of parallel transport. If $\gamma\colon I \to \mathsf{M}$ is a C^{r+1}-curve, a vector field $Y: I \to \mathsf{TM}$ along γ is ***parallel*** if $\nabla_{\gamma'(t)}Y(t) = 0_{\gamma(t)}$ for each $t \in I$. From the coordinate expression for $\nabla_{\gamma'(t)}Y(t)$ and from the properties of solutions to differential equations, if Y is parallel then it is of class C^{r+1}. Turning this around, one can regard the equation $\nabla_{\gamma'(t)}Y(t) = 0$ as a differential equation for the vector field Y along γ. If γ is of class C^{r+1}, then, if one specifies the value of Y at $\gamma(t_0)$ for some $t_0 \in I$, this uniquely determines $Y(t)$ for t sufficiently close to t_0. The map $\tau^{\gamma}_{t_0,t}: \mathsf{T}_{\gamma(t_0)}\mathsf{M} \to \mathsf{T}_{\gamma(t)}\mathsf{M}$ that sends $v \in \mathsf{T}_{\gamma(t_0)}\mathsf{M}$ to the unique vector $Y(t) \in \mathsf{T}_{\gamma(t)}\mathsf{M}$ defined by the solution to the initial value problem

$$\nabla_{\gamma'(t)}Y(t) = 0, \quad Y(t_0) = v,$$

is called the ***parallel transport*** along γ. Note that, with this notion of parallel, a geodesic can be described as a curve whose tangent vector field is parallel along itself.

Next we define two important objects that are associated to every affine connection. The role of these objects, particularly of curvature, in affine differential geometry is discussed in any of the basic references for affine differential geometry given at the beginning of the chapter, with the two volumes of Kobayashi and Nomizu [1963] being perhaps the most comprehensive treatment.

Definition 3.126. Let ∇ be a C^r-affine connection on M, $r \in \mathbb{N} \cup \{\infty\} \cup \{\omega\}$.

(i) The ***torsion tensor*** associated to ∇ is the class C^r tensor field T of type $(1,2)$ on M defined by

$$T(X,Y) = \nabla_X Y - \nabla_Y X - [X,Y], \qquad X, Y \in \Gamma^\infty(\mathsf{TM}).$$

(ii) The ***curvature tensor*** associated to ∇ is the class C^{r-1} tensor field R of type $(1,3)$ on M defined by

$$R(X,Y)Z = \nabla_X\nabla_Y Z - \nabla_Y\nabla_X Z - \nabla_{[X,Y]}Z, \qquad X, Y, Z \in \Gamma^\infty(\mathsf{TM}). \quad \bullet$$

Note that the torsion tensor is defined as a $(1,2)$-tensor field by indicating how it takes two vector fields and returns another vector field. This is consistent with Exercise E3.19. Also note that is it not obvious from the definition of T that it is indeed a tensor, i.e., that $T(X,Y)$ does not depend on the derivatives of the vector fields X and Y. However, it is an easy matter to check that this is indeed the case (see Exercise E3.44). Similarly to our definition of the torsion tensor, the curvature tensor takes three vector fields and returns a vector field to define a $(1,3)$-tensor field. As with the definition of the torsion, one needs to check that the definition depends on the values of the vector field arguments, and not on their derivatives (see Exercise E3.44, again). We should also mention the nonstandard notation we use in writing down the curvature tensor: $R(X,Y)Z$ rather than $R(X,Y,Z)$. For fixed $X, Y \in \Gamma^\infty(\mathsf{TM})$, note that $Z \mapsto R(X,Y)Z$ defines a $(1,1)$-tensor field, i.e., a vector bundle map from TM to TM over id_{M}. In affine differential geometry the collection of these vector bundle maps is useful in studying the "holonomy" of an affine connection [see Kobayashi and Nomizu 1963, volume 1], and it is for this reason that the curvature tensor is often written as we write it.

Remark 3.127. For those familiar with the theory of curves in three-dimensional Euclidean spaces, the terms torsion and curvature will be familiar. Also, curvature appears in the study of two-dimensional surfaces in three-dimensional Euclidean spaces. One may then wonder if there is a relationship between these classic notions of torsion and curvature and the more abstract notions we utilize. If the affine connection is Levi-Civita (see Section 3.8.2), then there are relationships between the curvature tensor and curvature of surfaces, and this is made precise through the notion of "sectional curvature" [see Kobayashi and Nomizu 1963, volume 1]. For torsion, the relationship is not precise. The vanishing of torsion, from one point of view, at least, has to do with a certain Ehresmann connection on the bundle $\pi_{\mathsf{TM}}\colon \mathsf{TM} \to \mathsf{M}$. •

Let us produce the coordinate expressions for the components of the torsion and curvature tensors.

1. The components of the torsion tensor are

$$T^k_{ij} = \Gamma^k_{ij} - \Gamma^k_{ji}, \qquad i,j,k \in \{1,\dots,n\}.$$

2. The components of the curvature tensor are

$$R^i_{jkl} = \frac{\partial \Gamma^i_{lj}}{\partial x^k} - \frac{\partial \Gamma^i_{kj}}{\partial x^l} + \Gamma^i_{km}\Gamma^m_{lj} - \Gamma^i_{km}\Gamma^m_{kj}, \qquad i,j,k,l \in \{1,\dots,n\}.$$

Note that these expressions affirm our assertions in Definition 3.126 concerning the smoothness of torsion and curvature.

Exercises

E3.1 Let $(S, \mathscr{O}_{\mathrm{discrete}})$ and $(T, \mathscr{O}_T)$ be topological spaces, S having the discrete topology.
(a) Show that every map $f\colon S \to T$ is continuous.
(b) What properties should $(T, \mathscr{O}_T)$ have if f is to be a homeomorphism?

E3.2 Let S be a set.
(a) Show that, if S has the trivial topology, then every sequence in S converges.
(b) If S has the discrete topology, describe the sequences in S that converge.

E3.3 Let $[0, 2\pi[\subset \mathbb{R}$ and $\mathbb{S}^1 \subset \mathbb{R}^2$ be endowed with the subspace topologies, and consider the map $f\colon [0,2\pi[\to \mathbb{S}^1$ defined by $f(\theta) = (\cos\theta, \sin\theta)$.
(a) Is f a bijection?
(b) Is f continuous?

(c) Is f a homeomorphism?

E3.4 In this exercise we consider the standard topologies on $\mathbb{R}^2$ and $\mathbb{R}$.
(a) Give an example of a continuous map $f\colon \mathbb{R}^2 \to \mathbb{R}$ that is proper.
(b) Give an example of a continuous map $f\colon \mathbb{R}^2 \to \mathbb{R}$ that is not proper.

E3.5 Show that a continuous function $f\colon \mathbb{R}^n \to \mathbb{R}_+$ is proper if and only if $\lim_{\|\boldsymbol{x}\|_{\mathbb{R}^n}\to\infty} f(\boldsymbol{x}) = +\infty$.

E3.6 Indicate which of the following functions $f\colon \mathbb{R} \to \mathbb{R}$ is proper:

(a) $f(x) = x^2$;
(b) $f(x) = x^3$;
(c) $f(x) = \tan^{-1}(x)$;
(d) $f(x) = x^2 - x^4$;
(e) $f(x) = x^4$.

E3.7 Show that the operator norm of Definition 3.15 is a norm on $L(\mathbb{R}^n; \mathbb{R}^m)$.

E3.8 Consider the function $f\colon \mathbb{R} \to \mathbb{R}$ defined by

$$f(x) = \begin{cases} \exp(-\frac{1}{1-x^2}), & -1 < x < 1, \\ 0, & \text{otherwise.} \end{cases}$$

Answer the following questions.
(a) Draw the graph of f.
(b) Show that f is of class C^∞.
(c) Compute the Taylor series of f at the points -1 and 1.
(d) Is f real-analytic?
(e) Does the set of zeros of f satisfy the conclusions of Theorem 3.109?

E3.9 Let $\mathcal{U}, \mathcal{V} \subset \mathbb{R}^n$ be open sets, let $\phi\colon \mathcal{V} \to \mathcal{U}$ be a diffeomorphism, and let $f\colon \mathcal{U} \to \mathbb{R}$ be of class C^2. Let coordinates in $\mathcal{U}$ and $\mathcal{V}$ be denoted by $(x^1, \dots, x^n)$ and $(y^1, \dots, y^n)$, respectively.
(a) Show that the matrix representation of $\boldsymbol{D}^2 f(\boldsymbol{x})$, relative to the standard basis for $\mathbb{R}^n$, is

$$[\boldsymbol{D}^2 f(\boldsymbol{x})]_{ij} = \frac{\partial^2 f}{\partial x^i \partial x^j}(\boldsymbol{x}), \qquad i, j \in \{1, \dots, n\}.$$

Of course, if we define $\tilde{f} = f \circ \phi\colon \mathcal{V} \to \mathbb{R}$, then the matrix representation of $\boldsymbol{D}^2 \tilde{f}(\boldsymbol{y})$ in the standard basis is also

$$[\boldsymbol{D}^2 \tilde{f}(\boldsymbol{y})]_{ij} = \frac{\partial^2 \tilde{f}}{\partial y^i \partial y^j}(\boldsymbol{y}), \qquad i, j \in \{1, \dots, n\}.$$

(b) What is the relationship between $[\boldsymbol{D}^2 \tilde{f}(\boldsymbol{y})]$ and $[\boldsymbol{D}^2 f(\boldsymbol{x})]$ that is induced by the Chain Rule?
(c) Show that, if $\boldsymbol{D}\tilde{f}(\boldsymbol{y}_0) = \boldsymbol{0}$, then $\boldsymbol{D}f(\phi(\boldsymbol{y}_0)) = \boldsymbol{0}$.
(d) Show that, if $\boldsymbol{D}\tilde{f}(\boldsymbol{y}_0) = \boldsymbol{0}$, then $\boldsymbol{D}\phi(\boldsymbol{y}_0)_* \boldsymbol{D}^2 \tilde{f}(\boldsymbol{y}_0) = \boldsymbol{D}^2 f(\phi(\boldsymbol{y}_0))$.

E3.10 On the set $\mathbb{R}$ consider the two atlases

$$\mathscr{A}_1 = \{(\mathcal{U}_1 = \mathbb{R}, \phi_1 = \mathrm{id}_{\mathbb{R}})\},$$
$$\mathscr{A}_2 = \{(\mathcal{U}_2 = \mathbb{R}, \phi_2 = \{x \mapsto x^3\})\}.$$

Denote by M_1 and M_2 the C^∞-manifolds whose differentiable structures are generated by $\mathscr{A}_1$ and $\mathscr{A}_2$, respectively.

(a) Show that these atlases are not equivalent.
(b) Show that there exists a map $f\colon \mathbb{R} \to \mathbb{R}$ that is a C^∞-diffeomorphism from M_1 to M_2.

E3.11 Consider the set $\mathbb{R}^2$. For $y_0 \in \mathbb{R}$, define a subset of $\mathbb{R}^2$ by

$$\mathcal{U}_{y_0} = \{(x, y) \in \mathbb{R}^2 \mid y = y_0\}$$

and a map

$$\begin{aligned}\phi_{y_0} \colon U_{y_0} &\to \mathbb{R}\\ (x, y_0) &\mapsto x.\end{aligned}$$

Answer the following questions.
(a) Verify that $\mathscr{A} = \{(\mathcal{U}_y, \phi_y)\}_{y\in\mathbb{R}}$ is an atlas for $\mathbb{R}^2$.
(b) Does the resulting differentiable structure make $\mathbb{R}^2$ a one-dimensional manifold?

The remainder of the questions have to do with the topology on $\mathsf{M} = \mathbb{R}^2$. We take as the topology the manifold topology on the manifold we just defined.

(c) Is M first-countable?
(d) Is M second-countable?
(e) Is M Hausdorff?
(f) Is M connected?
(g) Is M compact?
(h) Is M paracompact?

E3.12 Using Remark 3.13, show that a manifold is connected if and only if it is path connected.

E3.13 Consider the set $\mathsf{GL}(n; \mathbb{R})$ of invertible $n \times n$ matrices with real entries (this is the ***general linear group***). Show that this is an open submanifold of $\mathbb{R}^{n\times n}$.

E3.14 Show that the following sets possess natural differentiable structures that render them diffeomorphic to one another:

1. $\mathbb{T}^1 = \mathbb{R}/\sim$, where $\sim$ is the equivalence relation $x \sim y$ if $x - y \in \mathbb{Z}$;
2. $\mathbb{S}^1 = \{(x, y) \in \mathbb{R}^2 \mid x^2 + y^2 = 1\}$;
3. $\mathsf{SO}(2) = \{\boldsymbol{R} \in \mathsf{GL}(2; \mathbb{R}) \mid \boldsymbol{R}^T\boldsymbol{R} = \boldsymbol{I}_2,\ \det \boldsymbol{R} = 1\}$ (see Exercise E3.13 for the definition of $\mathsf{GL}(2; \mathbb{R})$);
4. $\mathbb{S}^1_{\mathbb{C}} = \{z \in \mathbb{C} \mid |z| = 1\}$.

E3.15 Let $A \subset \mathbb{R}$ be closed.
(a) Show that $\mathbb{R} \setminus A$ is a finite or countable union of disjoint open intervals.
(b) Construct a function $f\colon \mathbb{R} \to \mathbb{R}$ so that $f^{-1}(0) = A$.
Hint: *Note that the function of Exercise E3.8 is nonzero exactly on an open interval.*

E3.16 Let M be a manifold with TM its tangent bundle.
(a) Using the definition of the tangent space, define natural operations of vector addition and scalar multiplication on $\mathsf{T}_x\mathsf{M}$, and verify that these operations make $\mathsf{T}_x\mathsf{M}$ is $\mathbb{R}$-vector space for each $x \in \mathsf{M}$.
(b) Is $\mathsf{T}_x\mathsf{M}$ a normed vector space? If so, what is the norm?

In the next exercise, you will show that the set of $n \times n$ orthogonal matrices is a submanifold of $\mathbb{R}^{n\times n}$. Let $\mathsf{O}(n) = \{\boldsymbol{R} \in \mathbb{R}^{n\times n} \mid \boldsymbol{R}\boldsymbol{R}^T = \boldsymbol{I}_n\}$ be the ***orthogonal group*** and let $\mathsf{SO}(n) = \{\boldsymbol{R} \in \mathsf{O}(n) \mid \det \boldsymbol{R} = 1\}$ be the ***special orthogonal group***. Thus $\mathsf{O}(n)$ consists of all orthogonal matrices, and $\mathsf{SO}(n)$ consists of all orthogonal matrices that preserve orientation. The use of the word "group" in the definitions will be explained in Chapter 5.

E3.17 Let $S(\mathbb{R}^{n\times n})$ denote the symmetric $n \times n$ matrices. Consider the map $f\colon \mathbb{R}^{n\times n} \to S(\mathbb{R}^{n\times n})$ defined by $f(\boldsymbol{A}) = \boldsymbol{A}\boldsymbol{A}^T$.
(a) Show that f is a submersion on $f^{-1}(\boldsymbol{I}_n)$.
(b) Conclude that $\mathsf{O}(n)$ is a closed submanifold of $\mathbb{R}^{n\times n}$. What is $\dim(\mathsf{O}(n))$?
(c) Show that $\mathsf{O}(n)$ is the union of two compact connected sets, and that one of these sets is $\mathsf{SO}(n)$.

E3.18 Let $f\colon \mathsf{V} \to \mathsf{W}$ be a vector bundle map.
(a) Show that image(f) is a generalized subbundle of W.
(b) Give an example of a V, W, and f for which $\ker(f)$ is not a generalized subbundle of V.

E3.19 For a manifold M and $s \in \mathbb{N}$, let $L^s(\mathsf{TM};\mathsf{TM})$ denote the vector bundle whose fiber over $x \in \mathsf{M}$ consists of the s-multilinear maps from $(\mathsf{T}_x\mathsf{M})^s$ to $\mathsf{T}_x\mathsf{M}$. Show that this vector bundle, with its natural vector bundle structure induced by that of TM, is isomorphic to $T^1_s(\mathsf{TM})$.

E3.20 Verify the coordinate expression for the components of the Lie bracket using the definition of the Lie bracket.

E3.21 (Exercise E2.3 cont'd) Let $\boldsymbol{A} \in \mathbb{R}^{2\times 2}$, and consider the linear vector field $X_{\boldsymbol{A}}$ defined by $X_{\boldsymbol{A}}(\boldsymbol{x}) = \boldsymbol{A}\boldsymbol{x}$ on $\mathbb{R}^2$.
(a) Show that $h \mapsto \mathscr{L}_{X_{\boldsymbol{A}}} h$ is a linear map on $\mathcal{H}_2$; that is, show that it belongs to $L(\mathcal{H}_2, \mathcal{H}_2)$.
(b) Compute its matrix representation with respect to the basis you designed in Exercise E2.3.

E3.22 Given three vector fields X, Y, Z and two scalar functions f, g, verify, via direct computation in a coordinate chart, the following equalities:

$$\begin{aligned}
&\mathscr{L}_{[X,Y]} f = \mathscr{L}_X \mathscr{L}_Y f - \mathscr{L}_Y \mathscr{L}_X f;\\
&[X,[Y,Z]] + [Y,[Z,X]] + [Z,[X,Y]] = 0;\\
&[fX, gY] = fg[X,Y] + f(\mathscr{L}_X g)Y - g(\mathscr{L}_Y f)X.
\end{aligned}$$

E3.23 Let $\boldsymbol{b}_1, \boldsymbol{b}_2 \in \mathbb{R}^n$ and let $\boldsymbol{A} \in \mathbb{R}^{n\times n}$. On $\mathsf{M} = \mathbb{R}^n$ define the three vector fields $X_0(\boldsymbol{x}) = \boldsymbol{A}\boldsymbol{x}$ and $X_i(\boldsymbol{x}) = \boldsymbol{b}_i$, $i \in \{1,2\}$.
(a) Show that $[X_1, X_2] = 0$.
(b) Show that $[X_0,[X_0,\cdots[X_0,X_i]]] = (-1)^k \boldsymbol{A}^k \boldsymbol{b}_i$ for $i \in \{1,2\}$ and $k \in \mathbb{Z}_+$, and where the expression on the left hand side of the equation contains k Lie brackets by X_0.
(c) Show that $[X_i,[X_0,X_j]] = 0$ for $i,j \in \{1,2\}$.

E3.24 Consider the three vector fields X_1, X_2, and X_3 on $\mathsf{M} = \mathbb{R}^3$ defined by

$$X_k = x^j \frac{\partial}{\partial x^i} - x^i \frac{\partial}{\partial x^j},$$

where $(i,j,k) \in \{(1,2,3),(2,3,1),(3,1,2)\}$. Show that $[X_1, X_2] = X_3$, $[X_2, X_3] = X_1$ and $[X_3, X_1] = X_2$.

E3.25 For the vector field X on $\mathsf{M} = \mathbb{R}$ given by $X = x^2 \frac{\partial}{\partial x}$, compute $\sigma_+(X, x)$ and $\sigma_-(X, x)$ for each $x \in \mathsf{M}$.

E3.26 Let $\mathcal{U} \subset \mathsf{M}$ be an open subset and suppose that the vector fields $\{X_1, \dots, X_n\}$ are vector fields on $\mathcal{U}$ having the property that $\mathsf{T}_x\mathsf{M} = \operatorname{span}_{\mathbb{R}}\{X_1(x), \dots, X_n(x)\}$ for each $x \in \mathcal{U}$. Show that

$$\mathsf{T}_{v_x}\mathsf{TM} = \operatorname{span}_{\mathbb{R}}\{\operatorname{vlft}(X_1)(v_x), \dots, \operatorname{vlft}(X_n)(v_x), X_1^T(v_x), \dots, X_n^T(v_x)\}$$

for every $v_x \in \mathsf{T}\mathcal{U}$.

E3.27 Let $X \in \Gamma^\infty(\mathsf{M})$ and suppose that $X(x_0) = 0_{x_0}$. Let $A_X(x_0)$ be the linearization of X at x_0 as defined in Proposition 3.75. Show that, for $v \in \mathsf{T}_{x_0}\mathsf{M}$, it holds that $A_X(x_0) \cdot v = [V, X](x_0)$, where $V \in \Gamma^\infty(\mathsf{TM})$ is any vector field with the property that $V(x_0) = v$.

E3.28 Let $X \in \Gamma^\infty(\mathsf{TM})$ and let $f\colon \mathsf{M} \to \mathsf{M}$ be a diffeomorphism. Show that $f_*X = X$ if and only if $f \circ \gamma$ is an integral curve for X for all integral curves γ of X.

E3.29 Provide a C^∞-distribution on $\mathbb{R}^3$ that is not a regular distribution, and that has rank 2 on an open dense set.

E3.30 Answer the following questions.

(a) Let $\mathsf{M} = \mathbb{R}^2$. Define a C^∞-generalized subbundle $\mathsf{V} \subset \mathsf{TM}$ having the following properties:

1. $\operatorname{rank}(\mathsf{V}_x) = 2$ for points x in the interior of the unit ball centered at $(0, 0)$ in $\mathbb{R}^2$;
2. $\operatorname{rank}(\mathsf{V}_x) = 0$ for points x in the complement of the closed unit ball centered at $(0, 0)$ in $\mathbb{R}^2$.

Hint: *Make use of the function defined in Exercise E3.8.*

(b) What is the set of regular points for V?

(c) What is the set of singular points for V?

(d) Is the set of regular points open and dense, consistent with V being C^∞?

(e) Is V analytic? Can this be predicted simply by looking at the function $x \mapsto \operatorname{rank}(\mathsf{V}_x)$?

In the next exercise you will investigate the character of the regular points of a distribution. The exercise requires some familiarity with the Lebesgue measure; this outlined in Appendix A, and discussed in detail in [Halmos 1974a].

E3.31 Let $\mathcal{U}$ be an open subset of $\mathbb{R}^n$ and let $\mathcal{D}$ be an analytic distribution on $\mathcal{U}$. Let $R_{\mathcal{D}}$ denote the set of regular points of $\mathcal{D}$ and let $S_{\mathcal{D}}$ denote the set of singular points of $\mathcal{D}$.

(a) Let $\boldsymbol{x} \in \mathcal{U}$. Using the fact that analytic distributions are locally finitely generated, show that there is a neighborhood $\mathcal{U}_{\boldsymbol{x}}$ of $\boldsymbol{x}$ so that $S_{\mathcal{D}} \cap \mathcal{U}_{\boldsymbol{x}}$ is the intersection of the zeros of a finite set of analytic functions defined on $\mathcal{U}_{\boldsymbol{x}}$.

(b) Using Theorem 3.109, conclude that the Lebesgue measure of $S_{\mathcal{D}}$ is zero.

Now we consider a specific C^∞-distribution on $\mathcal{U} =]0,1[$. We first construct a closed subset $A \subset \mathcal{U}$ as follows. Let $B_0 = [0, 1]$ and define B_1 by deleting from B_0 an open interval of length $\frac{\epsilon}{2}$ centered at the midpoint of B_0. Note that B_1

consists of two disjoint closed intervals whose total Lebesgue measure $1 - \frac{\varepsilon}{2}$. Next define B_2 by deleting from B_1 two open intervals, each of length $\frac{\varepsilon}{8}$, centered at the midpoints of each of the intervals comprising B_1. Note that B_2 consists of four disjoint closed intervals whose total Lebesgue measure is $1 - \frac{\varepsilon}{4}$. Proceed in this way, defining a sequence of sets $\{B_n\}_{n\in\mathbb{N}}$, where B_n consists of 2^n disjoint closed intervals whose total Lebesgue measure is $1 - \sum_{j=1}^{n} \frac{\varepsilon}{2^j}$. Take $A = \mathcal{U} \cap (\cap_{n=1}^{+\infty} B_n)$.

(c) Show that $\mathcal{U} \setminus A$ is open and dense in $\mathcal{U}$.

(d) Show that the Lebesgue measure of $\mathcal{U} \setminus A$ is ε.

From Exercise E3.15(b), let $f\colon \mathcal{U} \to \mathbb{R}$ be a function for which $f^{-1}(0) = A$. Let $\mathcal{D}$ be the C^∞-distribution on $\mathcal{U}$ generated by the vector field $X(x) = f(x)\frac{\partial}{\partial x}$.

(e) Is $R_{\mathcal{D}}$ open and dense?

(f) What is the Lebesgue measure of $R_{\mathcal{D}}$?

E3.32 On $\mathsf{M} = \mathbb{R}^2$ consider the C^∞-distribution $\mathcal{D}$ generated by the vector fields

$$X = \frac{\partial}{\partial x}, \quad Y = f(x)\frac{\partial}{\partial y},$$

where $f\colon \mathbb{R} \to \mathbb{R}$ is the C^∞-function

$$f(x) = \begin{cases} \mathrm{e}^{-\frac{1}{x^2}}, & x > 0, \\ 0, & x \leq 0. \end{cases}$$

Show that

(a) $\mathcal{D}$ is involutive, and that

(b) $\mathcal{D}$ is not integrable.

E3.33 On $\mathsf{M} = \mathbb{R}^2$ consider the C^∞-distribution $\mathcal{D}$ generated by the vector fields

$$X = \frac{\partial}{\partial x}, \quad Y = f(x)\frac{\partial}{\partial y},$$

where $f\colon \mathbb{R} \to \mathbb{R}$ is the C^∞-function of Exercise E3.8.

(a) Show that $\mathcal{O}((x,y),\mathcal{D}) = \mathsf{M}$ for each $(x,y) \in \mathsf{M}$.

(b) Show that $\mathrm{Lie}^{(\infty)}(\mathcal{D})_{(x,y)} \subset \mathsf{T}_{(x,y)}\mathcal{O}((x,y),\mathcal{D})$ for each $(x,y) \in \mathsf{M}$.

(c) Find the set of points (x,y) for which the inclusion of part (b) is equality.

E3.34 Let ∇ be an affine connection on M and let $(\mathcal{U}_1, \phi_1)$ and $(\mathcal{U}_2, \phi_2)$ be overlapping coordinate charts with coordinates $(x^1, \dots, x^n)$ and $(\tilde{x}^1, \dots, \tilde{x}^n)$. If Γ^i_{jk}, $i,j,k \in \{1,\dots,n\}$, are the Christoffel symbols for ∇ in the coordinate chart $(\mathcal{U}_1, \phi_1)$, and $\tilde{\Gamma}^i_{jk}$, $i,j,k \in \{1,\dots,n\}$, are the Christoffel symbols for ∇ in the coordinate chart $(\mathcal{U}_2, \phi_2)$, show that

$$\tilde{\Gamma}^i_{jk} = \frac{\partial \tilde{x}^i}{\partial x^l}\frac{\partial x^r}{\partial \tilde{x}^j}\frac{\partial x^s}{\partial \tilde{x}^k}\Gamma^l_{rs} + \frac{\partial \tilde{x}^i}{\partial x^l}\frac{\partial^2 x^l}{\partial \tilde{x}^j \partial \tilde{x}^k}.$$

E3.35 For each of the following operations and objects, verify the coordinate formulae of Section 3.8.3:

(a) the covariant derivative of a vector field, $\nabla_X Y$;

(b) the covariant derivative of a vector field along a curve, $\nabla_{\gamma'(t)} Y(t)$;

(c) the geodesic equations, $\nabla_{\gamma'(t)}\gamma'(t) = 0$;

(d) the covariant derivative of a general tensor, $\nabla_X t$.
(e) The Christoffel symbols for the Levi-Civita affine connection;
(f) the torsion tensor, T (optional);
(g) the curvature tensor, R (optional).

E3.36 Prove Lemma 3.100.

E3.37 Let ∇ be an affine connection on a manifold M, and suppose that, on an open subset $\mathcal{U} \subset \mathsf{M}$, the vector fields $\{X_1, \dots, X_n\}$ form a basis for each tangent space. Let $\gamma\colon I \to \mathsf{M}$ be a C^1-curve with ξ a differentiable vector field along γ. Define $u^i, v^i\colon I \to \mathbb{R}$, $i \in \{1, \dots, n\}$, by

$$\gamma'(t) = u^i(t) X_i(\gamma(t)), \quad \xi(t) = v^i(t) X_i(\gamma(t)).$$

Show that

$$\nabla_{\gamma'(t)}\xi(t) = \dot{v}^i(t) X_i(\gamma(t)) + u^i(t) v^j(t) \nabla_{X_i} X_j(\gamma(t)).$$

Hint: *Think of γ as an integral curve of a vector field X and define a vector field Y with the property that $Y(\gamma(t)) = \xi(t)$. Then use the properties of an affine connection.*

E3.38 Let $(\mathsf{M}, \mathbb{G})$ be a Riemannian manifold. Let $\gamma\colon I \to \mathsf{M}$ be a C^1-curve with ξ and η differentiable vector fields along γ. Show that

$$\frac{\mathrm{d}}{\mathrm{d}t}\mathbb{G}(\xi(t), \eta(t)) = \mathbb{G}(\overset{\mathbb{G}}{\nabla}_{\gamma'(t)}\xi(t), \eta(t)) + \mathbb{G}(\xi(t), \overset{\mathbb{G}}{\nabla}_{\gamma'(t)}\eta(t)).$$

E3.39 Let ∇ be an affine connection on a Riemannian manifold $(\mathsf{M}, \mathbb{G})$. Show that $\nabla\mathbb{G} = 0$ if and only if

$$\mathscr{L}_Z(\mathbb{G}(X, Y)) = \mathbb{G}(\nabla_Z X, Y) + \mathbb{G}(X, \nabla_Z Y)$$

for $X, Y, Z \in \Gamma^\infty(\mathsf{TM})$.

E3.40 Let X be a C^∞-vector field on a Riemannian manifold $(\mathsf{M}, \mathbb{G})$. Show that $\overset{\mathbb{G}}{\nabla}_X X = \mathbb{G}^\sharp(\mathscr{L}_X \mathbb{G}^\flat(X)) - \frac{1}{2}\operatorname{grad}\|X\|_{\mathbb{G}}^2$.

E3.41 Show that a regular distribution $\mathcal{D}$ is geodesically invariant with respect to an affine connection ∇ if and only if $\nabla_X X \in \Gamma^\infty(\mathcal{D})$ for every $X \in \Gamma^\infty(\mathcal{D})$.

In the next exercise, you will be led through a case where the two notions of distribution, the one we use and the notion of an algebraic distribution introduced in Section 3.9.4, are equivalent, namely the case when the distribution has locally constant rank. The construction here follows parts of the paper by Swan [1962], where it is shown that the set of sections of a vector bundle is a "finitely generated projective module."

E3.42 Let M be a connected, Hausdorff, C^r-manifold, and let $\mathscr{X} = \{X_1, \dots, X_k\}$ be a family of C^r-vector fields on M, $r \in \mathbb{N} \cup \{\infty\} \cup \{\omega\}$, and suppose that the distribution $\mathcal{D}$ generated by these vector fields has constant rank.

Consider the trivial vector bundle $\mathbb{R}^k_{\mathsf{M}} = \mathsf{M} \times \mathbb{R}^k$, and let $E_1, \dots, E_k$ be the sections of this vector bundle defined by $E_a(x) = (x, \boldsymbol{e}_a)$.

(a) Show that every C^r-section ξ of $\mathbb{R}^k_{\mathsf{M}}$ can be written as

$$\xi(x) = \xi^1(x)E_1(x) + \cdots + \xi^k(x)E_k(x)$$

for some uniquely defined C^r-functions $\xi^1, \ldots, \xi^k$ on M. (Algebraically, this means that $\Gamma^r(\mathbb{R}^k_{\mathsf{M}})$ is a "free" $C^r(\mathsf{M})$-module.)

Define a map $\tilde{\pi}$ from $\Gamma^r(\mathbb{R}^k_{\mathsf{M}})$ to $\Gamma^r(\mathsf{TM})$ by

$$\xi^1 E_1 + \cdots + \xi^k E_k \mapsto \xi^1 X_1 + \cdots + \xi^k X_k.$$

Answer the following questions.

(b) Show that $\tilde{\pi}(E_a) = X_a$, $a \in \{1, \ldots, k\}$.

(c) Show that $\tilde{\pi}$ is surjective onto $\Gamma^r(\mathcal{D})$.

(d) Show that there exists a unique surjective C^r-vector bundle map $\pi\colon \mathbb{R}^k_{\mathsf{M}} \to \mathcal{D}$ for which $\pi(\xi(x)) = \tilde{\pi}(\xi)(x)$.
Hint: *See Theorem 1 in [Swan 1962].*

(e) Conclude that, for every C^r-vector field X taking values in the distribution $\mathcal{D}$, there are C^r-functions $f^1, \ldots, f^k$ such that

$$X = f^1 X_1 + \cdots + f^k X_k.$$

E3.43 Let $\mathsf{M} = \mathbb{R}^2$, with $X_1 = x\frac{\partial}{\partial x}$ and $X_2 = \frac{\partial}{\partial x} + x\frac{\partial}{\partial y}$.

(a) Show that the vector field $X = \frac{\partial}{\partial x}$ has the property that $X(x, y) \in \operatorname{span}_{\mathbb{R}}\{X_1(x, y), X_2(x, y)\}$ for each $(x, y) \in \mathsf{M}$.

(b) Show that there are no C^0-functions f^1 and f^2 for which $X = f^1 X_1 + f^2 X_2$.

(c) Can you think of generators $\{\tilde{X}_1, \tilde{X}_2\}$ for the distribution generated by $\{X_1, X_2\}$ which have the property that, if X is any vector field taking values in this distribution, then there exists C^ω-functions f^1 and f^2 with the property that $X = f^1 \tilde{X}_1 + f^2 \tilde{X}_2$.

E3.44 For $X, Y \in \Gamma^\infty(\mathsf{TM})$, show that the expressions

(a) $\nabla_X Y - \nabla_Y X - [X, Y]$ and

(b) $\nabla_X \nabla_Y Z - \nabla_Y \nabla_X Z - \nabla_{[X,Y]} Z$

are $C^\infty(\mathsf{M})$-linear in X and Y in the first case, and in X, Y, and Z in the second case.

4

Simple mechanical control systems

The word "simple" as used in the title of this chapter has a technical meaning that we shall provide during the course of the chapter. In this context, "simple" does not mean "easy."

The preceding two chapters were preparatory material, and were presented fairly hastily, with the reader being directed to references for further details. In this chapter, we begin the presentation of the material that can be considered the core of the book. The aim in this chapter is to provide the methodology for going from a physical problem to a mathematical model. What is more, the mathematical model we consider uses the tools described in Chapter 3, particularly the affine connection formalism of Section 3.8. We are extremely systematic in our presentation, far more so than is the norm. This has the disadvantage of making the presentation lengthy. To this end, we should mention that it is not normal, when modeling a mechanical system, to go systematically through the steps we outline in this chapter. Indeed, someone experienced with the process can perform many of the steps we describe "just by looking at the system." The systematic presentation we provide has two advantages to counterbalance its length.

1. For complicated systems, one cannot always describe the mathematical model "just by looking at the system." In these cases, even for experienced researchers, it is useful to have a procedure to fall back on, and to prevent the making of careless modeling errors.
2. By making all constructions perfectly explicit, we draw very clear lines between physical concepts and mathematical objects (more precisely, differential geometric objects). Indeed, it is our (admittedly somewhat Platonic) belief that the mathematical models we arrive at are, along with equivalent models, e.g., Hamiltonian models, the unique correct mathematical models. That is to say, other ways of thinking about these models are approximations to what we present. Some approximations are still useful, and perhaps easier to understand, but are still approximations, and so subject to limitations as such.

After stating our point of view, let us outline the content of the chapter. In the course of the following discussion, the reader can get an idea of what parts of the preceding two chapters are most important for modeling mechanical systems.

The steps we describe for turning a mechanical control system into a mathematical model are roughly these.

1. Determine the configuration manifold of the system. This will be a differentiable manifold, usually analytic, whose points are in 1–1 correspondence with the configurations of the system. This is typically a simple matter, particularly after a little practice.
2. If necessary, choose convenient coordinates. Most often this *will* be necessary, but sometimes it will not be. For example, one can be a little more clever regarding the matter of choosing coordinates when the configuration manifold is a Lie group, but this is postponed until Chapter 5.
3. Construct the coordinate representative of the forward kinematic map. Using this, compute the coordinate representatives of body and spatial velocities.
4. Determine the kinetic energy of the system as a function on the tangent bundle of the configuration manifold, and then derive from this the Riemannian metric that defines this kinetic energy function.
5. If the system has external forces over which the user has no control, determine these as cotangent bundle-valued maps on the tangent bundle. Such forces may depend on time, and include potential forces, dissipative forces, uncontrolled excitation forces, etc.
6. If the system is subject to linear velocity constraints, determine these as a distribution on the configuration manifold.
7. Model the control forces on the system as covector fields on the configuration manifold.

Collecting all of this data together yields what we call a "simple mechanical control system." As we shall see, the class of systems that can be modeled in this way is large, and includes systems from many application areas, such as those outlined in Chapter 1. Our techniques do not allow modeling of infinite-dimensional systems. Even in the finite-dimensional case, our methods are not wholly inclusive, since they do not include, for example, certain models arising in flight dynamics. In aircraft models, it can happen that the control forces exerted by aerodynamics surfaces depend on velocity in a manner not allowed by our models [Etkin 1995]. Also, just because a system *can* be modeled as we outline above, this does not mean that the tools in this book can be used to "solve" the problem. What our methods offer are techniques that can be used to model, analyze, and/or do control design for certain systems. As we go along, we shall try to point out some of the more obvious limitations of the theory we present, although many readers can find others that we do not mention.

The material in this chapter can essentially be considered "geometric mechanics." There exists an enormous existing literature on this subject, so let us survey this here. A very good introductory text on geometric mechanics is that of Arnol'd [1978]. More advanced texts include [Abraham and Marsden 1978, Arnol'd 1988, Bloch 2003, Godbillon 1969, Guillemin and Sternberg 1990, Libermann and Marle 1987, Marsden 1992, Marsden and Ratiu 1999, Souriau 1970]. Much of the existing literature on geometric mechanics, particularly before, say, 1990, had a Hamiltonian emphasis. One of the reasons for this emphasis was the striking success of the Hamiltonian point of view for addressing problems in dynamics. This success goes back at least to Poincaré [1892-1899] and Birkhoff [1927], and includes important contributions by Kolmogorov, Arnol'd, and Moser (see the exposition in [Arnol'd and Avez 1989]) and Nekhoroshev [1977] to the dynamics of systems that are close to integrable. More recently there has been a renewed interest in the Lagrangian approach, in part initiated by the recognition of the importance played by external forces and constraints in mechanics. While it is certainly possible to take a Hamiltonian approach to these matters [see, for example, Bates and Śniatycki 1993, van der Schaft and Maschke 1994, Weber 1986], they arise most naturally in the Lagrangian setting.

In closing, we should comment that one should not equate "geometric mechanics" with "classical mechanics" or "analytical mechanics." That is to say, if something is known about the latter, it should not be supposed that something in known about the former, and it is the former that we shall be using in this book. Moreover, someone can understand geometric mechanics as a mathematical discipline without any knowledge of "classical mechanics" or "analytical mechanics." However, some familiarity with physics goes a long way towards providing some context and intuition to the mathematical subject of geometric mechanics. Readers feeling as if they might benefit from a less geometric treatment of mechanics are referred to the texts [Chetaev 1987, Goldstein 1980, Neĭmark and Fufaev 1972, Papastavridis 2002, Pars 1965, Rosenberg 1977, Whittaker 1904], among others.

4.1 The configuration manifold

In this section we will see how one might assign to a given mechanical system a differentiable manifold that serves to represent the configurations of the system. We shall use the three examples of Chapter 1 to illustrate the process, now doing things in a more systematic manner. These examples will be used not just in this section, but in the remaining sections in this chapter to illustrate the various mechanical concepts we introduce.

4.1.1 Interconnected mechanical systems

We begin with a description of the sorts of physical systems that we have in mind in this chapter. It is this sort of system that we will be interested in modeling, although by the end of the chapter the models we will introduce will be rather abstract, and not necessarily restricted to the physical systems we consider here.

First we give rather imprecise definitions of what we mean by a particle and a rigid body. These definitions will suffice for the moment, but more precise versions are given in Section 4.2.1. A ***particle*** is an object having mass and position, but no volume. A ***rigid body*** is a (possibly uncountable) collection of particles whose relative position to one another is fixed. We shall not be concerned at present with the inertial properties of particles and rigid bodies. Instead we will be concerned with kinematic descriptions. This subject will be taken up in more detail in Section 5.1, so here we will say only what is necessary for our present purposes. To describe the position of a particle or rigid body, one needs a reference from which to measure distances and angles. To this end, we fix in space a reference frame $\Sigma_{\text{spatial}} = (O_{\text{spatial}}, \{\boldsymbol{s}_1, \boldsymbol{s}_2, \boldsymbol{s}_3\})$. This reference frame allows us to treat points in space as vectors in $\mathbb{R}^3$, measured from the origin O_{spatial}, and with components taken relative to the basis $\{\boldsymbol{s}_1, \boldsymbol{s}_2, \boldsymbol{s}_3\}$. We additionally require that the frame O_{spatial} be "inertial." It is not really possible to define what is meant by an inertial reference frame except to say that it is one in which the Newton–Euler equations hold.[1] We refer to Chapter 1 of [Arnol'd 1978] for a discussion of matters related to space, time, and Galilean relativity.

To uniquely specify the position of a particle, one needs only to specify a point in space. To do this, one specifies a vector $\boldsymbol{r}$ from O_{spatial} to the position of the particle. This vector is written relative to the basis $\{\boldsymbol{s}_1, \boldsymbol{s}_2, \boldsymbol{s}_3\}$. To uniquely specify the position of a rigid body, one generally needs to specify its orientation as well as its position of a single point in the body. Thus one chooses a frame $\Sigma_{\text{body}} = (O_{\text{body}}, \{\boldsymbol{b}_1, \boldsymbol{b}_2, \boldsymbol{b}_3\})$ that moves with the body. In Exercise E3.17 the manifolds of matrices

$$\mathsf{O}(n) = \{\boldsymbol{R} \in \mathbb{R}^{n\times n} \mid \boldsymbol{R}\boldsymbol{R}^T = \boldsymbol{I}_n\}, \quad \mathsf{SO}(n) = \{\boldsymbol{R} \in \mathsf{O}(n) \mid \det \boldsymbol{R} = +1\}$$

were defined. These are, respectively, the sets of orthogonal and special orthogonal matrices. To specify the position of the body, one specifies the vector $\boldsymbol{r} = O_{\text{body}} - O_{\text{spatial}}$, with components relative to the basis $\{\boldsymbol{s}_1, \boldsymbol{s}_2, \boldsymbol{s}_3\}$, and the matrix $\boldsymbol{R} \in \mathsf{O}(3)$ whose ith column is comprised of the components of $\boldsymbol{b}_i$ in the basis $\{\boldsymbol{s}_1, \boldsymbol{s}_2, \boldsymbol{s}_3\}$. If the bases $\{\boldsymbol{s}_1, \boldsymbol{s}_2, \boldsymbol{s}_3\}$ and $\{\boldsymbol{b}_1, \boldsymbol{b}_2, \boldsymbol{b}_3\}$ share the same orientation (typically right-handed), then $\boldsymbol{R} \in \mathsf{SO}(3)$. We will suppose this to be the case in the sequel. The typical situation is depicted in Figure 4.1.

A ***free mechanical system*** is a collection $\{P_\alpha\}_{\alpha\in\{1,\dots,N_p\}} \cup \{B_\beta\}_{\beta\in\{1,\dots,N_b\}}$ of N_p particles and N_b rigid bodies. The possible positions of all particles and bodies are then described by the set

$$\mathsf{Q}_{\text{free}} = \underbrace{(\mathsf{SO}(3)\times\mathbb{R}^3)\times\cdots\times(\mathsf{SO}(3)\times\mathbb{R}^3)}_{N_b \text{ copies}} \times \underbrace{\mathbb{R}^3\times\cdots\times\mathbb{R}^3}_{N_p \text{ copies}}.$$

[1] One could replace "hold" with "hold approximately" if one is committed to relativistic mechanics.

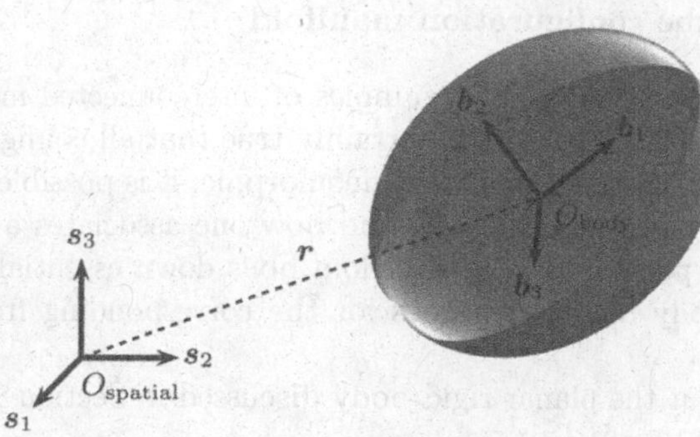

Figure 4.1. The specification of position and orientation of a body by spatial and body reference frames

Note that this specification of Q_{free} depends on a choice of spatial frame Σ_{spatial} and on a choice of body frame $\Sigma_{\text{body},a}$, $a \in \{1, \dots, k\}$, for each body. Different choices will give the same Q_{free}, but the correspondence between points in Q_{free} and configurations of the bodies and particles will be different.

Generally our interest will be in systems that are not free. To this end, we make the following definition.

Definition 4.1 (Interconnected mechanical system). An ***interconnected mechanical system*** is a collection $\{P_\alpha\}_{\alpha\in\{1,\dots,N_p\}} \cup \{B_\beta\}_{\beta\in\{1,\dots,N_b\}}$ of N_p particles and N_b rigid bodies restricted to move on a C^∞-immersed submanifold Q of Q_{free}. The manifold Q is the ***configuration manifold*** for the system. If $\dim(\mathsf{Q}) = n$, then the system is said to have ***n degrees-of-freedom***. •

Remarks 4.2. 1. We refer the reader to Exercise E4.2 for an example of a system where Q is an immersed, but not embedded, submanifold of Q_{free}. Generally, we shall think of configuration manifolds as being embedded for simplicity. However, there is no difficulty considering immersed submanifolds, for these come equipped naturally with a differentiable structure (see Section 3.9.1).

2. It can happen that the set Q that is in 1–1 correspondence with configurations of a mechanical system, is not a differentiable manifold. For example, if there are inequality constraints on the configurations, then the configuration manifold may have boundaries, and these boundaries may further not be smooth, i.e., it may have "corners." For systems like this, one can talk formally about manifolds with boundary [Abraham, Marsden, and Ratiu 1988], or one can take the configuration manifold as being a stratified set [Goodwine and Burdick 2001]. In such cases, the theory will change somewhat, but some of the same basic ideas still apply. That said, in this book we will concentrate on systems whose configuration manifold is a C^∞-manifold without boundary. •

4.1.2 Finding the configuration manifold

In this section we consider a few examples of interconnected mechanical systems. We shall see that, while it is certainly true that all configuration manifolds for a given physical system are diffeomorphic, it is possible to have some freedom both in how one defines Q, and how one associates a configuration in the body to a point in Q. This freedom boils down essentially to a choice of spatial and body reference frames for the corresponding free mechanical system.

We first look at the planar rigid body discussed in Section 1.1.

Example 4.3 (Planar rigid body). We recall that the physical system is a rigid body moving in the plane as depicted in Figure 4.2. This system is then comprised of a single body, and we denote by P the two-dimensional plane in which the body moves. We choose a spatial reference frame $(O_{\text{spatial}}, \{\boldsymbol{s}_1, \boldsymbol{s}_2, \boldsymbol{s}_3\})$ and a body reference frame $(O_{\text{body}}, \{\boldsymbol{b}_1, \boldsymbol{b}_2, \boldsymbol{b}_3\})$ with the following properties:

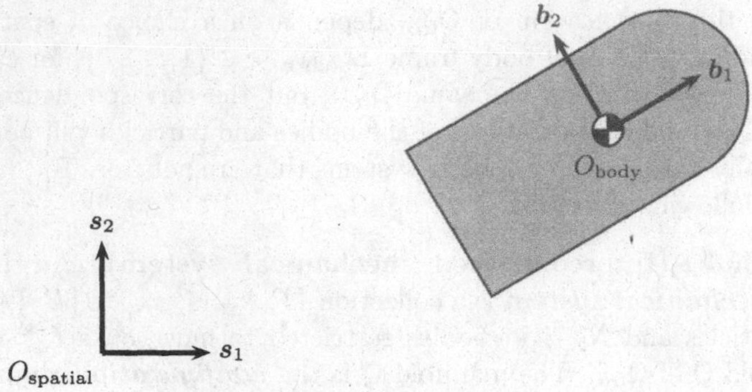

Figure 4.2. A planar rigid body

1. $O_{\text{spatial}}, O_{\text{body}} \in P$;
2. the vectors $\boldsymbol{s}_3$ and $\boldsymbol{b}_3$ are $\mathbb{G}_{\mathbb{R}^3}$-orthogonal to P.

Certainly other choices of spatial and body reference frames are possible, but frames having the above two properties are most convenient. We have $\mathsf{Q}_{\text{free}} = \mathsf{SO}(3) \times \mathbb{R}^3$. The body is restricted to move in the plane P, which means that the vector $\boldsymbol{r} = O_{\text{body}} - O_{\text{spatial}}$ has the form $(x, y, 0)$. For an admissible motion of the body, we will have $\boldsymbol{b}_3 = \boldsymbol{s}_3$. Therefore, the matrix $\boldsymbol{R} \in \mathsf{SO}(3)$ will have the form

$$\boldsymbol{R} = \begin{bmatrix} \cos\theta & -\sin\theta & 0 \\ \sin\theta & \cos\theta & 0 \\ 0 & 0 & 1 \end{bmatrix}, \tag{4.1}$$

where $\theta \in \mathbb{R}$. We can think of this matrix as being an element of the subset of $\mathsf{SO}(3)$ consisting of those matrices that fix $\boldsymbol{s}_3$. This is a submanifold of $\mathsf{SO}(3)$ that is clearly diffeomorphic to $\mathsf{SO}(2)$. Thus the map

$$\mathsf{SO}(2) \times \mathbb{R}^2 \ni \left(\begin{bmatrix} \cos\theta & -\sin\theta \\ \sin\theta & \cos\theta \end{bmatrix}, (x, y) \right) \mapsto (\boldsymbol{R}, (x, y, 0)) \in \mathsf{SO}(3) \times \mathbb{R}^3,$$

with $\boldsymbol{R}$ given by (4.1), gives an embedding from $\mathsf{SO}(2) \times \mathbb{R}^2$ into the submanifold Q of admissible configurations for the system. Thus we simply choose to define $\mathsf{Q} = \mathsf{SO}(2) \times \mathbb{R}^2$. In fact, in this chapter we shall use Exercise E3.14 to take $\mathsf{Q} = \mathbb{S}^1 \times \mathbb{R}^2$.

One can see that in our description of Q there was some freedom. This freedom is essentially described by possible variations in the choice of the frames $(O_{\text{spatial}}, \{\boldsymbol{s}_1, \boldsymbol{s}_2, \boldsymbol{s}_3\})$ and $(O_{\text{body}}, \{\boldsymbol{b}_1, \boldsymbol{b}_2, \boldsymbol{b}_3\})$. In the sequel, we shall further fix our choice of the body frame by asking that O_{body} lie at the center of mass of the body (readers wishing to see a precise definition of center of mass are referred to Section 4.2.1), and that the basis $\{\boldsymbol{b}_1, \boldsymbol{b}_2, \boldsymbol{b}_3\}$ be chosen to simplify our representation of the forces. •

Now we consider the example of two interconnected links first considered in Section 1.2.

Example 4.4 (Two-link manipulator). The example is the two interconnected links shown in Figure 4.3. We assume that the base point of the first

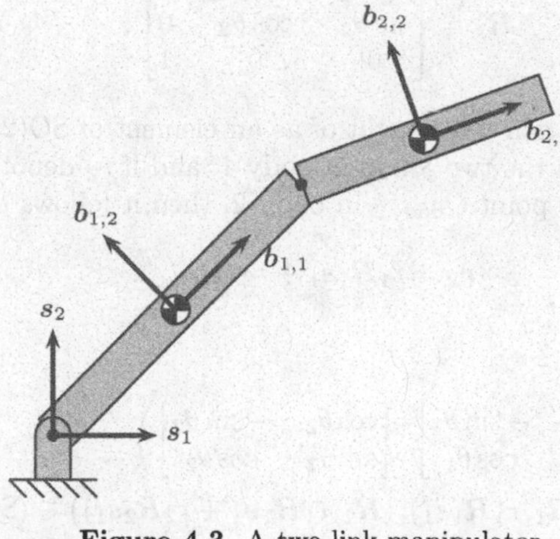

Figure 4.3. A two-link manipulator

link (let us say that this is the leftmost link in Figure 4.3) is fixed spatially, and that the two links are pinned together. Let us choose spatial and body reference frames. We shall call the body whose one end is fixed spatially "body 1," and the other body "body 2." We choose the spatial reference frame $(O_{\text{spatial}}, \{\boldsymbol{s}_1, \boldsymbol{s}_2, \boldsymbol{s}_3\})$ so that O_{spatial} is at the point where body 1 is pinned, and so that $\boldsymbol{s}_3$ points in the direction orthogonal to the motion allowed by the joint. We choose the body reference frames $(O_{\text{body},a}, \{\boldsymbol{b}_{a,1}, \boldsymbol{b}_{a,2}, \boldsymbol{b}_{a,3}\})$, $a \in \{1,2\}$, so that $O_{\text{body},a}$ is positioned at the center of mass of body a, and so that the vectors $\boldsymbol{b}_{a,3}$, $a \in \{1,2\}$, are orthogonal to the plane of motion allowed by the joints. We further choose $\boldsymbol{b}_{1,1}$ to point along the line connecting the joint to the center of mass, and similarly for $\boldsymbol{b}_{2,1}$. We also assume that the center of mass of body 1 lies on the line between the two joints. These choices are depicted in Figure 4.3.

We obviously have $\mathsf{Q}_{\text{free}} = (\mathsf{SO}(3) \times \mathbb{R}^3) \times (\mathsf{SO}(3) \times \mathbb{R}^3)$. For body 1, the matrix $\boldsymbol{R}_1$ describing the orientation of the body will have the form

$$\boldsymbol{R}_1 = \begin{bmatrix} \cos\theta_1 & -\sin\theta_1 & 0 \\ \sin\theta_1 & \cos\theta_1 & 0 \\ 0 & 0 & 1 \end{bmatrix}, \tag{4.2}$$

for $\theta_1 \in \mathbb{R}$. Thus $\boldsymbol{R}_1$ can be thought of as being an element of $\mathsf{SO}(2)$. The vector $\boldsymbol{r}_1 = O_{\text{body},1} - O_{\text{spatial}}$ will be constant of length, say r_1, and will have the form

$$\boldsymbol{r}_1 = r_1 \boldsymbol{R}_1 \boldsymbol{s}_1,$$

by virtue of the manner in which we have chosen our reference frames. Therefore, $\boldsymbol{r}_1$ is exactly prescribed by $\boldsymbol{R}_1$, and so the position and orientation of body 1 is exactly determined by $\boldsymbol{R}_1 \in \mathsf{SO}(2)$. For body 2, we have

$$\boldsymbol{R}_2 = \begin{bmatrix} \cos\theta_2 & -\sin\theta_2 & 0 \\ \sin\theta_2 & \cos\theta_2 & 0 \\ 0 & 0 & 1 \end{bmatrix}, \tag{4.3}$$

for $\theta_2 \in \mathbb{R}$. Thus $\boldsymbol{R}_2$ can be thought of as an element of $\mathsf{SO}(2)$. If ℓ_1 denotes the distance between the two joints in body 1, and if r_2 denotes the distance from the joint to the point $O_{\text{body},2}$ in body 2, then it follows that

$$\boldsymbol{r}_2 = \ell_1 \boldsymbol{R}_1 \boldsymbol{s}_1 + r_2 \boldsymbol{R}_2 \boldsymbol{s}_1.$$

Therefore, the map

$$\mathsf{SO}(2)^2 \ni \left(\begin{bmatrix} \cos\theta_1 & -\sin\theta_1 \\ \sin\theta_1 & \cos\theta_1 \end{bmatrix}, \begin{bmatrix} \cos\theta_2 & -\sin\theta_2 \\ \sin\theta_2 & \cos\theta_2 \end{bmatrix} \right)$$
$$\mapsto \big((\boldsymbol{R}_1, r_1 \boldsymbol{R}_1 \boldsymbol{s}_1), (\boldsymbol{R}_2, \ell_1 \boldsymbol{R}_1 \boldsymbol{s}_1 + r_2 \boldsymbol{R}_2 \boldsymbol{s}_1)\big) \in (\mathsf{SO}(3) \times \mathbb{R}^3)^2,$$

with $\boldsymbol{R}_1, \boldsymbol{R}_2 \in \mathsf{SO}(3)$ defined as in equations (4.2) and (4.3), is an embedding of $\mathsf{SO}(2)^2$ into the submanifold $\mathsf{Q} \subset \mathsf{Q}_{\text{free}}$ of admissible configurations of the system. Thus we say that $\mathsf{Q} = \mathsf{SO}(2)^2$ is the configuration manifold for the system. Since $\mathsf{SO}(2)^2 \simeq \mathbb{S}^1 \times \mathbb{S}^1$ (Exercise E3.14), we shall write $\mathsf{Q} = \mathbb{S}^1 \times \mathbb{S}^1$ below.

As with the planar body example, there is some freedom here, revolving around the choices of the spatial and body reference frames. The choices made above lead to certain simplifications. However, other natural choices are possible. •

As our final example in this section, let us consider the upright rolling disk considered in Section 1.3.

Example 4.5 (Rolling disk). The system we consider is the disk rolling on a plane P, constrained to remain in the upright position as shown in Figure 4.4. Again, the place to start is by choosing spatial and body frames.

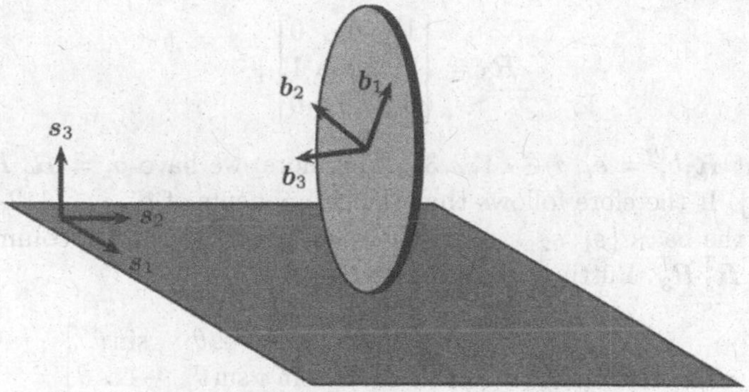

Figure 4.4. The upright rolling disk

We choose the frames $(O_{\text{spatial}}, \{\boldsymbol{s}_1, \boldsymbol{s}_2, \boldsymbol{s}_3\})$ and $(O_{\text{body}}, \{\boldsymbol{b}_1, \boldsymbol{b}_2, \boldsymbol{b}_3\})$ to have the following properties:

1. $O_{\text{spatial}} \in P$;
2. O_{body} lies at the center of rotation of the disk;
3. $\boldsymbol{s}_3$ is orthogonal to P;
4. $\boldsymbol{b}_3$ is parallel to P, and orthogonal to the direction of motion allowed by the rolling motion of the disk on the plane.

Again, these choices are not unique, but they are convenient. With these choices we have $\boldsymbol{r} = O_{\text{body}} - O_{\text{spatial}} = (x, y, \rho)$, where ρ is the radius of the disk and where $(x, y) \in \mathbb{R}^2$ is arbitrary. In this example it is not perfectly straightforward to write the matrix $\boldsymbol{R}$ describing the relative orientation of the two frames. After one gets some experience with this sort of construction, it is possible to simply write down the appropriate rotation matrix by looking at the system. However, let us do this here in a systematic manner.

The objective is to relate the basis $\{\boldsymbol{b}_1, \boldsymbol{b}_2, \boldsymbol{b}_3\}$ to the basis $\{\boldsymbol{s}_1, \boldsymbol{s}_2, \boldsymbol{s}_3\}$. Denote by

$$\boldsymbol{R}_1 = \begin{bmatrix} \cos\theta & \sin\theta & 0 \\ -\sin\theta & \cos\theta & 0 \\ 0 & 0 & 1 \end{bmatrix}$$

the matrix with the property that $\boldsymbol{R}_1\boldsymbol{b}_3 = -\boldsymbol{s}_2$. This always makes sense, since $\boldsymbol{b}_3$ lies in the subspace spanned by $\boldsymbol{s}_1$ and $\boldsymbol{s}_2$. Now define $\boldsymbol{b}'_i = \boldsymbol{R}_1\boldsymbol{b}_i$, $i \in \{1,2,3\}$, and note that $\boldsymbol{b}'_1$ and $\boldsymbol{b}'_2$ both lie in the subspace spanned by $\boldsymbol{s}_1$ and $\boldsymbol{s}_3$. What is more, one easily sees that the vectors $\{\boldsymbol{b}'_1, \boldsymbol{b}'_2\}$ have the same orientation as the vectors $\{\boldsymbol{s}_1, \boldsymbol{s}_3\}$. Thus there exists a matrix

$$\boldsymbol{R}_2 = \begin{bmatrix} \cos\phi & 0 & -\sin\phi \\ 0 & 1 & 0 \\ \sin\phi & 0 & \cos\phi \end{bmatrix}$$

such that $\boldsymbol{R}_2\boldsymbol{b}'_1 = \boldsymbol{s}_1$ and $\boldsymbol{R}_2\boldsymbol{b}'_2 = \boldsymbol{s}_3$. Define $\boldsymbol{b}''_i = \boldsymbol{R}_2\boldsymbol{b}'_i$, $i \in \{1,2,3\}$. Finally, taking

$$\boldsymbol{R}_3 = \begin{bmatrix} 1 & 0 & 0 \\ 0 & 0 & 1 \\ 0 & -1 & 0 \end{bmatrix},$$

we see that $\boldsymbol{R}_3\boldsymbol{b}''_i = \boldsymbol{s}_i$, $i \in \{1,2,3\}$. Therefore, we have $\boldsymbol{b}_i = \boldsymbol{R}_1^T\boldsymbol{R}_2^T\boldsymbol{R}_3^T\boldsymbol{s}_i$, $i \in \{1,2,3\}$. It therefore follows that the components of $\boldsymbol{b}_i$, $i \in \{1,2,3\}$, with respect to the basis $\{\boldsymbol{s}_1, \boldsymbol{s}_2, \boldsymbol{s}_3\}$ are the components of the ith column of the matrix $\boldsymbol{R}_1^T\boldsymbol{R}_2^T\boldsymbol{R}_3^T$. Putting this together we obtain

$$\boldsymbol{R} = \boldsymbol{R}_1^T\boldsymbol{R}_2^T\boldsymbol{R}_3^T = \begin{bmatrix} \cos\phi\cos\theta & \sin\phi\cos\theta & \sin\theta \\ \cos\phi\sin\theta & \sin\phi\sin\theta & -\cos\theta \\ -\sin\phi & \cos\phi & 0 \end{bmatrix}. \tag{4.4}$$

In Figure 4.5 we show the sequence of operations used in going from

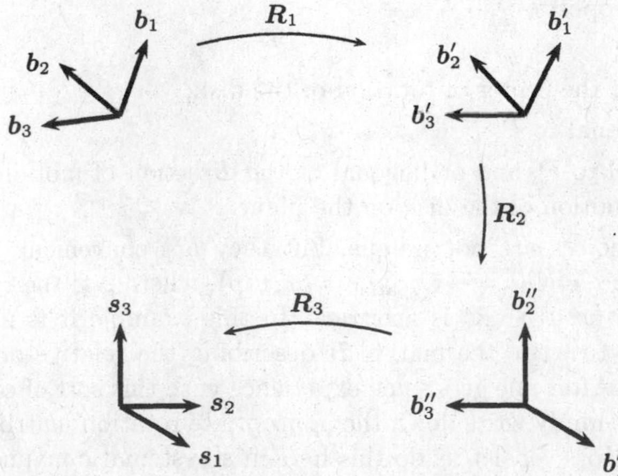

Figure 4.5. The operations used to construct $\boldsymbol{R}$ for the rolling disk

$\{\boldsymbol{b}_1, \boldsymbol{b}_2, \boldsymbol{b}_3\}$ to $\{\boldsymbol{s}_1, \boldsymbol{s}_2, \boldsymbol{s}_3\}$.

What our computations show is that the map

$$\mathbb{R}^2 \times \mathsf{SO}(2)^2 \ni \left((x,y), \begin{bmatrix} \cos\theta & -\sin\theta \\ \sin\theta & \cos\theta \end{bmatrix}, \begin{bmatrix} \cos\phi & -\sin\phi \\ \sin\phi & \cos\phi \end{bmatrix} \right) \mapsto ((x,y,\rho), \boldsymbol{R}) \in \mathbb{R}^3 \times \mathsf{SO}(3),$$

with $\boldsymbol{R}$ as defined in (4.4), is an embedding from $\mathbb{R}^2 \times \mathsf{SO}(2)^2$ to the submanifold Q of admissible configurations. We thus have $\mathsf{Q} \simeq \mathbb{R}^2 \times \mathsf{SO}(2)^2$, and in the sequel we shall take $\mathsf{Q} = \mathbb{R}^2 \times \mathbb{S}^1 \times \mathbb{S}^1$. •

Let us make some comments on the procedure we have discussed in the preceding three examples.

Remarks 4.6. 1. Note that the manifolds chosen for configuration manifolds in the above examples, $\mathbb{S}^1 \times \mathbb{R}^2$, $\mathbb{S}^1 \times \mathbb{S}^1$, and $\mathbb{R}^2 \times \mathbb{S}^1 \times \mathbb{S}^1$, were chosen abstractly. That is to say, we chose not to use the configuration manifold as a submanifold of $\mathsf{Q}_{\mathrm{free}}$, but rather to work with a manifold whose image, under an appropriate embedding, is the set of admissible configurations. It is often easier to work with this abstract manifold in practice. One should keep in mind, however, that the procedure we have described really gives an embedding (or more generally, an immersion) of this abstract manifold into Q.

2. In the above examples, we have made more work out of determining the configuration manifold than is the norm. From the examples we have given, we hope that the reader will learn some typical tricks for determining configuration manifolds. However, after some practice, it is not uncommon to simply "know" what the configuration manifold must be, and begin working with that immediately.

3. The preceding comment notwithstanding, one should initially exercise care in determining Q. Let us consider an example that, when encountered for the first time, often causes confusion. We consider a rigid body in space (i.e., not the plane) that is fixed at a point. Thus the body can rotate freely about a fixed point. We place at this fixed point two frames $\Sigma_{\mathsf{spatial}} = (O_{\mathrm{spatial}}, \{\boldsymbol{s}_1, \boldsymbol{s}_2, \boldsymbol{s}_3\})$ and $\Sigma_{\mathsf{body}} = (O_{\mathrm{body}}, \{\boldsymbol{b}_1, \boldsymbol{b}_2, \boldsymbol{b}_3\})$, the former being fixed in space and the latter fixed to the body. The procedure above then yields $\mathsf{Q} = \mathsf{SO}(3)$. Newcomers seldom have difficulty accepting that this is what Q *is*. What can be confusing is recognizing what Q is *not*. For example, the guess that Q could be $\mathbb{T}^3$ is common, the idea being that there are rotations about three axes that must be accounted for. Another popular guess is $\mathbb{S}^2 \times \mathbb{S}^1$, the idea here being that positions of the body can be assigned to the position of a unit vector (i.e., a point in $\mathbb{S}^2$) and a rotation about that vector (i.e., a point in $\mathbb{S}^1$). In each case, the problems arise for large changes of orientation of the body. That is to say, while the incorrect intuitive descriptions are reasonable for small changes of orientation near a reference configuration, they will break down for large orientation

changes. More specifically, the specifications of configurations in $\mathbb{S}^2 \times \mathbb{S}^1$ or $\mathbb{T}^3$ that seem reasonable for small orientation changes develop singularities for large orientation changes. The most convincing argument against the incorrect choices of $\mathbb{T}^3$ or $\mathbb{S}^2 \times \mathbb{S}^1$, however, is differential geometric (or topological, more precisely). The fact is that neither $\mathbb{T}^3$ nor $\mathbb{S}^2 \times \mathbb{S}^1$ is diffeomorphic to $\mathsf{SO}(3)$, although neither of these assertions is entirely obvious. Thus, if one accepts that our prescription for taking $\mathsf{Q} = \mathsf{SO}(3)$ is correct, this necessarily eliminates both $\mathbb{T}^3$ and $\mathbb{S}^2 \times \mathbb{S}^1$ as valid choices for the configuration manifold.

This points out that while it is often easy to determine Q, it is just as easy to make an incorrect choice. The only recourse is to have a convincing argument for establishing the 1–1 correspondence between points in Q and configurations of the system. •

4.1.3 Choosing coordinates

Now that we have made a choice for Q, and we understand precisely the manner in which points in Q correspond to configurations of the physical system, we indicate how one might choose coordinates. It is often convenient to do this, although in some cases, the rush to coordinates is best resisted. Note that our definition of the configuration manifold Q makes it an immersed submanifold of $\mathsf{Q}_{\mathrm{free}}$, and therefore gives Q a natural differentiable structure arising from the natural differentiable structure on $\mathsf{Q}_{\mathrm{free}}$. In other words, the issues raised in Remark 3.27 concerning uniqueness of differentiable structures do not arise in the present context.

In practice, one often does not choose an atlas, but merely a single coordinate chart that, while not covering all of Q, covers "most" of it. It is then left unstated that there is at least one other chart that, when added to the chosen one, will provide an atlas compatible with the natural differentiable structure. In this section we shall go through our examples and find this one coordinate chart, doing so in a way that is physically "reasonable."

Example 4.7 (Planar rigid body cont'd). For this example, we had decided to take $\mathsf{Q} = \mathbb{S}^1 \times \mathbb{R}^2$. Let us first define our chart, and then see what it means in terms of the physics of the problem. We take $(\mathcal{U}, \phi)$ to be

$$\mathcal{U} = (\mathbb{S}^1 \setminus \{(-1,0)\}) \times \mathbb{R}^2,$$
$$\phi((\xi,\eta),(x,y)) = (\mathrm{atan}(\xi,\eta), x, y) = (\theta, x, y) \in]-\pi,\pi[\times \mathbb{R}^2 \subset \mathbb{R}^3.$$

Note that this chart covers all of Q except for, essentially, a plane. One could define another chart to cover this omitted region, but in practice, this complication is not needed.

The coordinates (θ, x, y) are easy to understand physically. As we see in Figure 4.6, (x, y) is the position of the center of mass relative to the origin of $\Sigma_{\mathrm{spatial}}$, and θ is the angle the body $\boldsymbol{b}_1$-axis forms with the spatial $\boldsymbol{s}_1$-axis.

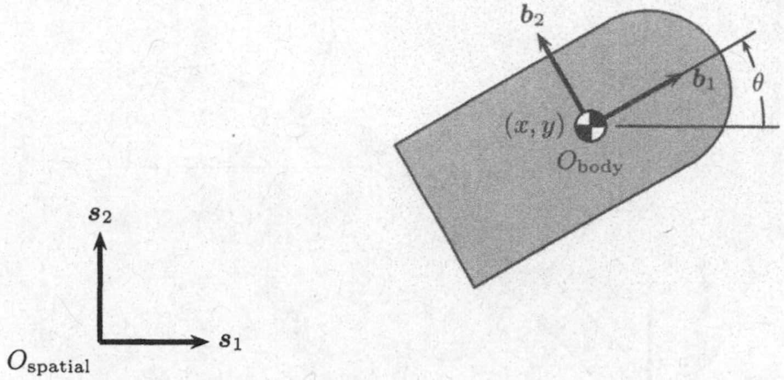

Figure 4.6. Coordinates for the planar rigid body

If one were handed this example, these would very likely be the coordinates that would naturally be selected. •

Now we look at coordinates for the two-link manipulator.

Example 4.8 (Two-link manipulator cont'd). For this example, we had $\mathsf{Q} = \mathbb{S}^1 \times \mathbb{S}^1$. We choose a coordinate chart $(\mathcal{U}, \phi)$ as follows:

$$\begin{aligned}\mathcal{U} &= \mathbb{S}^1 \times \mathbb{S}^1 \setminus \big(\{((x_1, y_1), (x_2, y_2)) \in \mathbb{S}^1 \times \mathbb{S}^1 \mid x_1 = -1\} \\ &\qquad \cup \{((x_1, y_1), (x_2, y_2)) \in \mathbb{S}^1 \times \mathbb{S}^1 \mid x_2 = -1\}\big), \\ \phi((x_1, y_1), (x_2, y_2)) &= (\operatorname{atan}(x_1, y_1), \operatorname{atan}(x_2, y_2)) \\ &= (\theta_1, \theta_2) \in]-\pi, \pi[\times]-\pi, \pi[\subset \mathbb{R}^2.\end{aligned}$$

The set of points in Q not covered by this chart consists of the union of two circles, as depicted in Figure 4.7, where we think of $\mathbb{S}^1 \times \mathbb{S}^1 \simeq \mathbb{T}^2$. It is easy

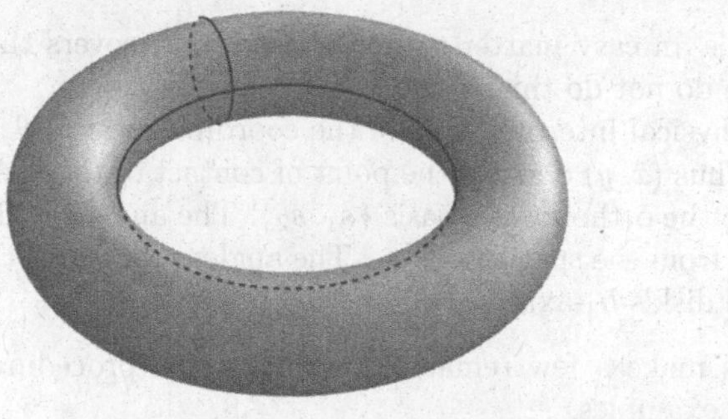

Figure 4.7. The set not covered by the chart $(\mathcal{U}, \phi)$ for the torus

to construct another chart to cover these points, although we do not do so.

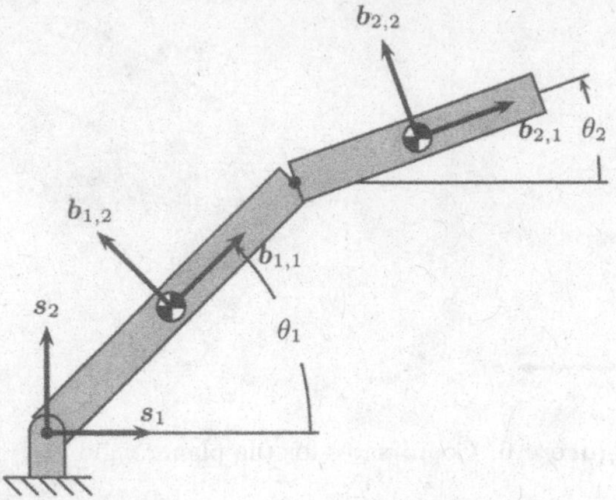

Figure 4.8. Coordinates for the two-link manipulator

Again, the meaning of the coordinates (θ_1, θ_2) is easy to understand in terms of the physics of the problem. As we see in Figure 4.8, the coordinates are simply the angles the link axes $\boldsymbol{b}_{1,1}$ and $\boldsymbol{b}_{2,1}$ form with spatial axis $\boldsymbol{s}_1$. •

Finally, we consider the coordinates for the rolling disk.

Example 4.9 (Rolling disk cont'd). The rolling disk has the configuration manifold $\mathsf{Q} = \mathbb{R}^2 \times \mathbb{S}^1 \times \mathbb{S}^1$, and we take the coordinate chart $(\mathcal{V}, \psi)$ defined by

$$\begin{aligned}\mathcal{V} = \mathbb{R}^2 \times \mathbb{S}^1 \times \mathbb{S}^1 \setminus \big(\mathbb{R}^2 \times \big(\{\, ((x_1, y_1), (x_2, y_2)) \in \mathbb{S}^1 \times \mathbb{S}^1 \mid x_1 = -1\}\\ \cup \{\, ((x_1, y_1), (x_2, y_2)) \in \mathbb{S}^1 \times \mathbb{S}^1 \mid x_2 = -1\}\big)\big),\\ \psi((x, y), ((x_1, y_1), (x_2, y_2)) = ((x, y), (\mathrm{atan}(x_1, y_1), \mathrm{atan}(x_2, y_2)))\\ = (x, y, \theta, \phi) \in \mathbb{R}^2\,]-\pi, \pi[\, \times\,]-\pi, \pi[\, \subset \mathbb{R}^4.\end{aligned}$$

Again, it is an easy matter to find a chart that covers the uncovered bits of Q, but we do not do this.

The physical interpretation of the coordinates (x, y, θ, ϕ) is shown in Figure 4.9. Thus (x, y) denotes the point of contact of the disk with the plane P, relative to the orthonormal basis $\{\boldsymbol{s}_1, \boldsymbol{s}_2\}$. The angle θ is the "heading angle," measured from the spatial axis $\boldsymbol{s}_1$. The angle ϕ measures the angle of rotation about the disk's $\boldsymbol{b}_3$-axis. •

Let us make a few remarks concerning the procedure illustrated in the preceding examples.

Remarks 4.10. 1. In our presentation of these examples, we have been careful to separate the procedure of defining Q and assigning to points in Q configurations of the system, and the procedure of defining coordinates. In practice the two procedures are typically not so systematically separated.

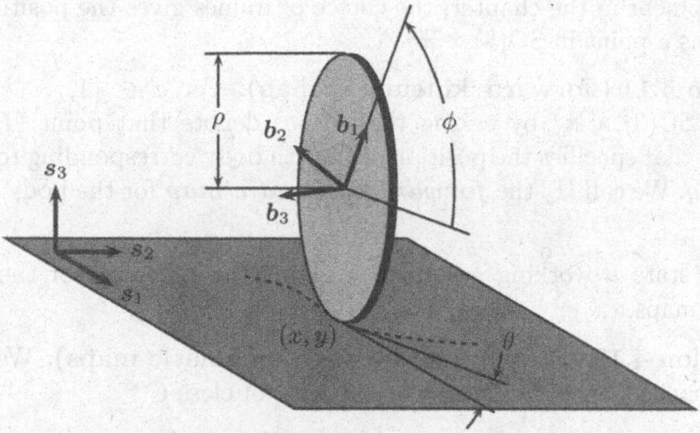

Figure 4.9. Coordinates for the rolling disk

Indeed, with practice, one often together defines coordinates, decides what Q is, and determines how points in Q correspond to configurations of the system.

2. In non-geometric treatments of "classical mechanics" (see the references given at the beginning of the chapter), one typically starts (and very often stops) with a coordinate chart, giving so-called "generalized coordinates," typically denoted by $(q^1, \ldots, q^n)$. There is nothing intrinsically wrong with such an approach, but it can deceive the uninitiated into thinking that the configuration manifold is an open subset of Euclidean space, or even more naïvely, Euclidean space itself. Again, this is fine provided that one is careful to deal with objects that are coordinate-invariant. We prefer the geometric approach, where the explicit realization of the configuration manifold as a differentiable manifold *forces* one to consider objects that are coordinate-invariant. •

4.1.4 The forward kinematic map

In this section, we introduce a simple, but key, ingredient in our development, the forward kinematic map. From the forward kinematic map, almost all subsequent constructions follow in a more or less direct manner. Much is said about forward kinematic maps in the robotics literature [e.g., Craig 1989, Murray, Li, and Sastry 1994, Selig 1996, Spong and Vidyasagar 1989].

We consider an interconnected mechanical system. To simplify the exposition, we suppose that the system is comprised of k rigid bodies. As we shall see in Section 4.2, particles can be essentially modeled as rigid bodies, so we lose no generality by not allowing particles in the formulation. We choose a spatial frame $\Sigma_{\text{spatial}} = (O_{\text{spatial}}, \{\boldsymbol{s}_1, \boldsymbol{s}_2, \boldsymbol{s}_2\})$ and body frames $\Sigma_{\text{body},a} = (O_{\text{body},a}, \{\boldsymbol{b}_{a,1}, \boldsymbol{b}_{a,2}, \boldsymbol{b}_{a,3}\})$, $a \in \{1, \ldots, k\}$. According to the preceding development in the chapter, the choice of frames gives the position of the ath body as a point in $\mathsf{SO}(3) \times \mathbb{R}^3$.

Definition 4.11 (Forward kinematic map). For $a \in \{1, \dots, k\}$, define $\Pi_a \colon \mathsf{Q} \to \mathsf{SO}(3) \times \mathbb{R}^3$ by asking that $\Pi_a(q)$ denote that point $(\boldsymbol{R}_a, \boldsymbol{r}_a) \in \mathsf{SO}(3) \times \mathbb{R}^3$ that specifies the position of the ath body corresponding to the configuration q. We call Π_a the ***forward kinematic map*** for the body $(\mathcal{B}_a, \mu_a)$. •

Let us state a working assumption about the character of the forward kinematic maps.

Assumption 4.12 (Property of forward kinematic maps). We suppose that the maps Π_a, $a \in \{1, \dots, k\}$, are at least of class C^∞. •

The determination of the forward kinematic map in examples is straightforward, particularly if one has been systematic about the assignment of configurations of the system to points in the configuration manifold. This is easily illustrated in our running examples. We begin with the planar rigid body.

Example 4.13 (Planar rigid body cont'd). The configuration manifold is $\mathsf{Q} = \mathbb{S}^1 \times \mathbb{R}^2$, and we use coordinates (θ, x, y), as defined in Example 4.7. The position of the origin of the body frame is given in terms of the coordinates (θ, x, y) by $\boldsymbol{r} = (x, y, 0)$ and the orientation of the body is given by

$$\boldsymbol{R} = \begin{bmatrix} \cos\theta & -\sin\theta & 0 \\ \sin\theta & \cos\theta & 0 \\ 0 & 0 & 1 \end{bmatrix}.$$

This means that the local representative of the forward kinematic map $\Pi_1 \colon \mathsf{Q} \to \mathsf{SO}(3) \times \mathbb{R}^3$ is $(\theta, x, y) \mapsto (\boldsymbol{R}, \boldsymbol{r})$, with $\boldsymbol{R}$ and $\boldsymbol{r}$ as given. •

Next we look at the two-link manipulator.

Example 4.14 (Two-link manipulator cont'd). Here $\mathsf{Q} = \mathbb{S}^1 \times \mathbb{S}^1$, and we use the coordinates (θ_1, θ_2) defined in Example 4.8. The positions of the origins of the body frames for the two links are given in terms of the coordinates (θ_1, θ_2) by

$$\begin{aligned} \boldsymbol{r}_1 &= (r_1 \cos\theta_1, r_1 \sin\theta_1, 0), \\ \boldsymbol{r}_2 &= (\ell_1 \cos\theta_1 + r_2 \cos\theta_2, \ell_1 \sin\theta_1 + r_2 \sin\theta_2, 0). \end{aligned}$$

The orientation of the links is given in terms of the coordinates (θ_1, θ_2) by

$$\boldsymbol{R}_a = \begin{bmatrix} \cos\theta_a & -\sin\theta_a & 0 \\ \sin\theta_a & \cos\theta_a & 0 \\ 0 & 0 & 1 \end{bmatrix}, \qquad a \in \{1, 2\}.$$

Thus the local representatives of the forward kinematic maps $\Pi_a \colon \mathsf{Q} \to \mathsf{SO}(3) \times \mathbb{R}^3$, $a \in \{1, 2\}$, are $(\theta_1, \theta_2) \mapsto (\boldsymbol{R}_a, \boldsymbol{r}_a)$, with the expressions as given for $\boldsymbol{R}_1$, $\boldsymbol{R}_2$, $\boldsymbol{r}_1$, and $\boldsymbol{r}_2$. •

The final example we consider is the rolling disk.

Example 4.15 (Rolling disk cont'd). For this example, we have $\mathsf{Q} = \mathbb{R}^2 \times \mathbb{S}^1 \times \mathbb{S}^1$, and we use coordinates (x, y, θ, ϕ) as described in Example 4.9. In Example 4.5 we determined that the position and orientation of the body in the coordinates (x, y, θ, ϕ) are given by

$$\boldsymbol{r} = (x, y, \rho), \quad \boldsymbol{R} = \begin{bmatrix} \cos\phi\cos\theta & \sin\phi\cos\theta & \sin\theta \\ \cos\phi\sin\theta & \sin\phi\sin\theta & -\cos\theta \\ -\sin\phi & \cos\phi & 0 \end{bmatrix},$$

where ρ is the radius of the disk. This gives the local representative of the forward kinematic map $\Pi_1 \colon \mathsf{Q} \to \mathsf{SO}(3) \times \mathbb{R}^3$ to be $(x, y, \theta, \phi) \mapsto (\boldsymbol{R}, \boldsymbol{r})$, with the expressions as given above. •

4.1.5 The tangent bundle of the configuration manifold

In this section, as in the preceding section, we assume an interconnected mechanical system comprised only of rigid bodies. Since the configuration manifold Q of an interconnected mechanical system is an immersed submanifold of the free configuration space Q_{free}, it follows that TQ is also an immersed submanifold of $\mathsf{TQ}_{\text{free}}$. Furthermore, TQ is an embedded submanifold if and only if Q is an embedded submanifold. In this section we provide some structure for TQ that is inherited from the structure of Q_{free}. The presentation in this section will be lacking a little in context. Some of this can be had be referring to Section 5.1. In this section, we are primarily interested in simply being able to do computations. To organize these computations, some notation is required. However, as we shall see in the examples at the end of the section, the computations are straightforward. Furthermore, they will form the backbone of many of our subsequent constructions concerning kinetic energy, forces, and constraints.

We first construct two vector bundle isomorphisms from $\mathsf{TSO}(3)$ to the trivial vector bundle $\mathsf{SO}(3) \times \mathbb{R}^3$. Note that since $\mathsf{SO}(3)$ is a submanifold of $\mathbb{R}^{3\times 3}$, we can write a tangent vector at $\boldsymbol{R} \in \mathsf{SO}(3)$ as a 3×3 matrix (cf. Remark 5.34). Let us denote a typical such element of $\mathsf{T}_{\boldsymbol{R}}\mathsf{SO}(3)$ by $\boldsymbol{A}_{\boldsymbol{R}}$.

Definition 4.16. Let $\boldsymbol{R} \in \mathsf{SO}(3)$ and let $\boldsymbol{A}_{\boldsymbol{R}} \in \mathsf{T}_{\boldsymbol{R}}\mathsf{SO}(3)$.

(i) The ***body angular velocity*** corresponding to $\boldsymbol{A}_{\boldsymbol{R}}$ is $\boldsymbol{R}^T\boldsymbol{A}_{\boldsymbol{R}}$.
(ii) The ***spatial angular velocity*** corresponding to $\boldsymbol{A}_{\boldsymbol{R}}$ is $\boldsymbol{A}_{\boldsymbol{R}}\boldsymbol{R}^T$. •

The body and spatial angular velocities are each 3×3 matrices. However, they are rather particular sorts of 3×3 matrices. Let us introduce some notation related to this. For $n \in \mathbb{N}$, we denote by $\mathfrak{so}(n)$ the set of $n \times n$ skew-symmetric matrices.

Lemma 4.17 (Skew-symmetry of angular velocities). *The body and spatial angular velocities lie in* $\mathfrak{so}(3)$.

Proof. Let $\boldsymbol{R}_0 \in \mathsf{SO}(3)$ and let $\boldsymbol{A}_{\boldsymbol{R}_0} \in \mathsf{T}_{\boldsymbol{R}_0}\mathsf{SO}(3)$. Let $t \mapsto \boldsymbol{R}(t)$ be a curve in $\mathsf{SO}(3)$ at $\boldsymbol{R}_0$ for which $[\boldsymbol{R}]_{\boldsymbol{R}_0} = \boldsymbol{A}_{\boldsymbol{R}_0}$. Differentiating the equality $\boldsymbol{R}^T(t)\boldsymbol{R}(t) = \boldsymbol{I}_3$ with respect to t gives

$$\dot{\boldsymbol{R}}^T(t)\boldsymbol{R}(t) + \boldsymbol{R}^T(t)\dot{\boldsymbol{R}}(t) = \boldsymbol{0}.$$

Therefore, evaluating at $t = 0$,

$$(\boldsymbol{R}_0^T\boldsymbol{A}_{\boldsymbol{R}})^T = -\boldsymbol{R}_0^T\boldsymbol{A}_{\boldsymbol{R}}.$$

This gives skew-symmetry the body angular velocity. Skew-symmetry of the spatial angular velocity follows similarly by differentiating $\boldsymbol{R}(t)\boldsymbol{R}^T(t) = \boldsymbol{I}_3$ with respect to t. ■

We adopt the convention of denoting a typical body (resp. spatial) angular velocity at $\boldsymbol{R}$ by $\widehat{\boldsymbol{\Omega}}$ (resp. $\widehat{\boldsymbol{\omega}}$). Associated to the matrices $\widehat{\boldsymbol{\Omega}}, \widehat{\boldsymbol{\omega}} \in \mathfrak{so}(3)$ are the vectors $\boldsymbol{\Omega}, \boldsymbol{\omega} \in \mathbb{R}^3$ defined by

$$\widehat{\boldsymbol{\Omega}} = \begin{bmatrix} 0 & -\Omega^3 & \Omega^2 \\ \Omega^3 & 0 & -\Omega^1 \\ -\Omega^2 & \Omega^1 & 0 \end{bmatrix}, \quad \widehat{\boldsymbol{\omega}} = \begin{bmatrix} 0 & -\omega^3 & \omega^2 \\ \omega^3 & 0 & -\omega^1 \\ -\omega^2 & \omega^1 & 0 \end{bmatrix}. \tag{4.5}$$

Thus we think of $\widehat{\cdot}$ as being an isomorphism from $\mathbb{R}^3$ to $\mathfrak{so}(3)$. One readily verifies that this correspondence gives $\widehat{\boldsymbol{\Omega}}\boldsymbol{v} = \boldsymbol{\Omega} \times \boldsymbol{v}$, for $\boldsymbol{v} \in \mathbb{R}^3$, and where $\times$ is the usual vector product in $\mathbb{R}^3$. These constructions give two vector bundle isomorphisms $(\boldsymbol{R}, \boldsymbol{A}_{\boldsymbol{R}}) \mapsto (\boldsymbol{R}, (\boldsymbol{R}^T\boldsymbol{A}_{\boldsymbol{R}})^\vee)$ and $(\boldsymbol{R}, \boldsymbol{A}_{\boldsymbol{R}}) \mapsto (\boldsymbol{R}, (\boldsymbol{A}_{\boldsymbol{R}}\boldsymbol{R}^T)^\vee)$ from $\mathsf{TSO}(3)$ to $\mathsf{SO}(3) \times \mathbb{R}^3$, where $\cdot^\vee : \mathfrak{so}(3) \to \mathbb{R}^3$ is the inverse of $\widehat{\cdot}$. (Note that this shows, in particular, that $\mathsf{SO}(3)$ is trivializable.)

Now we apply the preceding constructions to an interconnected mechanical system with configuration manifold Q. As with our discussion of the forward kinematic map, we assume that the system is comprised of k rigid bodies, so that $\mathsf{Q}_{\mathrm{free}} = (\mathsf{SO}(3) \times \mathbb{R}^3)^k$. Let $(\mathbb{R}^3 \oplus \mathbb{R}^3)^k_{\mathsf{Q}_{\mathrm{free}}}$ be the trivial vector bundle

$$\underbrace{((\mathsf{SO}(3) \times \mathbb{R}^3) \times \cdots \times (\mathsf{SO}(3) \times \mathbb{R}^3))}_{\mathsf{Q}_{\mathrm{free}}} \times \underbrace{((\mathbb{R}^3 \oplus \mathbb{R}^3) \oplus \cdots \oplus (\mathbb{R}^3 \oplus \mathbb{R}^3))}_{k \text{ copies}},$$

and denote by $(\mathbb{R}^3 \oplus \mathbb{R}^3)^k_{\mathsf{Q}}$ the restriction of this vector bundle to $\mathsf{Q} \subset \mathsf{Q}_{\mathrm{free}}$. We then have two vector bundle isomorphisms $\iota_{\mathrm{body}}, \iota_{\mathrm{spatial}} : \mathsf{TQ}_{\mathrm{free}} \to (\mathbb{R}^3 \oplus \mathbb{R}^3)^k_{\mathsf{Q}_{\mathrm{free}}}$ defined by

$$\begin{aligned} &\iota_{\mathrm{body}}\big(((\boldsymbol{R}_1, \boldsymbol{r}_1), \dots, (\boldsymbol{R}_k, \boldsymbol{r}_k)), ((\boldsymbol{A}_{\boldsymbol{R}_1}, \boldsymbol{v}_1), \dots, (\boldsymbol{A}_{\boldsymbol{R}_k}, \boldsymbol{v}_k))\big) \\ &\quad = (((\boldsymbol{R}_1, \boldsymbol{r}_1), \dots, (\boldsymbol{R}_k, \boldsymbol{r}_k)), (((\boldsymbol{R}_1^T\boldsymbol{A}_{\boldsymbol{R}_1})^\vee, \boldsymbol{v}_1), \dots, ((\boldsymbol{R}_k^T\boldsymbol{A}_{\boldsymbol{R}_k})^\vee, \boldsymbol{v}_k))), \end{aligned}$$

and

$$\begin{aligned} &\iota_{\mathrm{spatial}}\big(((\boldsymbol{R}_1, \boldsymbol{r}_1), \dots, (\boldsymbol{R}_k, \boldsymbol{r}_k)), ((\boldsymbol{A}_{\boldsymbol{R}_1}, \boldsymbol{v}_1), \dots, (\boldsymbol{A}_{\boldsymbol{R}_k}, \boldsymbol{v}_k))\big) \\ &\quad = (((\boldsymbol{R}_1, \boldsymbol{r}_1), \dots, (\boldsymbol{R}_k, \boldsymbol{r}_k)), (((\boldsymbol{A}_{\boldsymbol{R}_1}\boldsymbol{R}_1^T)^\vee, \boldsymbol{v}_1), \dots, ((\boldsymbol{A}_{\boldsymbol{R}_k}\boldsymbol{R}_k^T)^\vee, \boldsymbol{v}_k))). \end{aligned}$$

Now let us see how the preceding constructions fit together with the forward kinematic map. To do this, we introduce some notation. Maps $\rho_{\text{body}}\colon \mathsf{T}(\mathsf{SO}(3)\times\mathbb{R}^3)\to\mathbb{R}^3$ and $\lambda_{\text{body}}\colon \mathsf{T}(\mathsf{SO}(3)\times\mathbb{R}^3)\to\mathbb{R}^3$ are defined by

$$\rho_{\text{body}}((\boldsymbol{R},\boldsymbol{r}),(\boldsymbol{A_R},\boldsymbol{v}))=\boldsymbol{v},\quad \lambda_{\text{body}}((\boldsymbol{R},\boldsymbol{r}),(\boldsymbol{A_R},\boldsymbol{v}))=(\boldsymbol{R}^T\boldsymbol{A_R})^{\vee}.$$

Similarly, we define $\rho_{\text{spatial}}\colon \mathsf{T}(\mathsf{SO}(3)\times\mathbb{R}^3)\to\mathbb{R}^3$ and $\lambda_{\text{spatial}}\colon \mathsf{T}(\mathsf{SO}(3)\times\mathbb{R}^3)\to\mathbb{R}^3$ by

$$\rho_{\text{spatial}}((\boldsymbol{R},\boldsymbol{r}),(\boldsymbol{A_R},\boldsymbol{v}))=\boldsymbol{v},\quad \lambda_{\text{spatial}}((\boldsymbol{R},\boldsymbol{r}),(\boldsymbol{A_R},\boldsymbol{v}))=(\boldsymbol{A_R}\boldsymbol{R}^T)^{\vee}.$$

We note that $\lambda_{\text{body}}\circ T\Pi_a(v_q)=\lambda_{\text{spatial}}\circ \mathsf{T}\Pi_a(v_q)$ is the linear velocity of the origin of the ath body frame when the configuration passes through $q\in\mathsf{Q}$ with velocity v_q. The quantity $\rho_{\text{body}}\circ T\Pi_a(v_q)$ (resp. $\rho_{\text{spatial}}\circ T\Pi_a(v_q)$) is the body (resp. spatial) angular velocity of body a when the configuration passes through $q\in\mathsf{Q}$ with velocity v_q. The following result summarizes the relationships between the preceding notation, and follows directly from the definitions.

Lemma 4.18. *Let* Q *be the configuration manifold of an interconnected mechanical system comprised of* k *rigid bodies. Then, for* $v_q\in\mathsf{TQ}$*, the following statements hold:*

(i) $\iota_{\text{body}}(v_q)=\bigoplus_{a=1}^{k}(\rho_{\text{body}}\circ T\Pi_a(v_q)\oplus\lambda_{\text{body}}\circ T\Pi_a(v_q))$;

(ii) $\iota_{\text{spatial}}(v_q)=\bigoplus_{a=1}^{k}(\rho_{\text{spatial}}\circ T\Pi_a(v_q)\oplus\lambda_{\text{spatial}}\circ T\Pi_a(v_q))$.

There is a great deal of notation in the preceding discussion. Let us see how to implement it in examples.

Example 4.19 (Planar rigid body cont'd). Here we have $\mathsf{Q}=\mathbb{S}^1\times\mathbb{R}^2$ and coordinates (θ,x,y), as described in Example 4.7. We denote a typical point in TQ in coordinates by $((\theta,x,y),(\dot\theta,\dot x,\dot y))$. While the "$\dot{}$" notation for velocities is a little imprecise, it is convenient for doing the computations, and it sets the stage for how we do certain of the calculations in the sections to follow.

Given the coordinate representation of the forward kinematic map in Example 4.13, we compute the body angular velocity to be

$$\widehat{\boldsymbol{\Omega}}=\boldsymbol{R}^T\dot{\boldsymbol{R}}=\dot\theta\begin{bmatrix}0&-1&0\\1&0&0\\0&0&0\end{bmatrix},$$

so that $\boldsymbol{\Omega}=(0,0,\dot\theta)$. The spatial angular velocity is

$$\widehat{\omega} = \dot{R}R^T = \dot{\theta}\begin{bmatrix} 0 & -1 & 0 \\ 1 & 0 & 0 \\ 0 & 0 & 0 \end{bmatrix},$$

and so $\omega = (0, 0, \dot{\theta})$. Let us offer a physical interpretation of these quantities in this example. The form of the spatial angular velocity reflects the fact that the body rotates about the s_3-axis. As we shall see in Section 5.1, the spatial angular velocity ω should be interpreted as an instantaneous rotation of the body in space about the axis ω with speed $\|\omega\|_{\mathbb{R}^3}$. The body angular velocity is the representation of the spatial angular velocity in the body reference frame. In this example, since the body b_3-axis agrees with the spatial s_3-axis, the body angular velocity agrees with the spatial angular velocity.

We then compute body translational velocity to be

$$\lambda_{\text{body}} \circ T\Pi_1((\theta, x, y), (\dot{\theta}, \dot{x}, \dot{y})) = (\dot{x}, \dot{y}, 0),$$

and the body angular velocity to be

$$\rho_{\text{body}} \circ T\Pi_1((\theta, x, y), (\dot{\theta}, \dot{x}, \dot{y})) = (0, 0, \dot{\theta}).$$

Similarly, for the spatial velocities we have

$$\lambda_{\text{spatial}} \circ T\Pi_1((\theta, x, y), (\dot{\theta}, \dot{x}, \dot{y})) = (\dot{x}, \dot{y}, 0),$$

and

$$\rho_{\text{spatial}} \circ T\Pi_1((\theta, x, y), (\dot{\theta}, \dot{x}, \dot{y})) = (0, 0, \dot{\theta}).$$

The maps $\iota_{\text{body}}|\mathsf{TQ}$ and $\iota_{\text{spatial}}|\mathsf{TQ}$ are simply notation now, since the computations are already done. Indeed, we have

$$\iota_{\text{body}}((\theta, x, y), (\dot{\theta}, \dot{x}, \dot{y})) = \left(\left(\begin{bmatrix} \cos\theta & -\sin\theta & 0 \\ \sin\theta & \cos\theta & 0 \\ 0 & 0 & 1 \end{bmatrix}, (x, y, 0)\right), ((0, 0, \dot{\theta}), (\dot{x}, \dot{y}, 0))\right).$$

In this case, the expression for ι_{spatial} is the same. •

Next we look at the two-link manipulator.

Example 4.20 (Two-link manipulator cont'd). We have $\mathsf{Q} = \mathbb{S}^1 \times \mathbb{S}^1$, and we use the coordinates (θ_1, θ_2) defined in Example 4.8. The corresponding coordinates for TQ are denoted by $((\theta_1, \theta_2), (\dot{\theta}_1, \dot{\theta}_2))$. Using the coordinate formula for the forward kinematic map given in Example 4.14, it is a simple matter to compute the body and spatial angular velocities to be

$$\widehat{\Omega}_1 = R_1^T \dot{R}_1 = \dot{\theta}_1 \begin{bmatrix} 0 & -1 & 0 \\ 1 & 0 & 0 \\ 0 & 0 & 0 \end{bmatrix}, \quad \widehat{\Omega}_2 = R_2^T \dot{R}_2 = \dot{\theta}_2 \begin{bmatrix} 0 & -1 & 0 \\ 1 & 0 & 0 \\ 0 & 0 & 0 \end{bmatrix},$$

and

$$\widehat{\boldsymbol{\omega}}_1 = \dot{\boldsymbol{R}}_1\boldsymbol{R}_1^T = \dot{\theta}_1 \begin{bmatrix} 0 & -1 & 0 \\ 1 & 0 & 0 \\ 0 & 0 & 0 \end{bmatrix}, \quad \widehat{\boldsymbol{\omega}}_2 = \dot{\boldsymbol{R}}_2\boldsymbol{R}_2^T = \dot{\theta}_2 \begin{bmatrix} 0 & -1 & 0 \\ 1 & 0 & 0 \\ 0 & 0 & 0 \end{bmatrix}.$$

These give $\boldsymbol{\Omega}_1 = \boldsymbol{\omega}_1 = (0,0,\dot{\theta}_1)$ and $\boldsymbol{\Omega}_2 = \boldsymbol{\omega}_2 = (0,0,\dot{\theta}_2)$, and have the same physical interpretation as in the previous example.

It is then straightforward to compute

$$\begin{aligned}
\lambda_{\text{body}} \circ T\Pi_1((\theta_1,\theta_2),(\dot{\theta}_1,\dot{\theta}_2)) &= (-r_1\dot{\theta}_1 \sin\theta_1, r_1\dot{\theta}_1\cos\theta_1, 0), \\
\lambda_{\text{body}} \circ T\Pi_2((\theta_1,\theta_2),(\dot{\theta}_1,\dot{\theta}_2)) &= (-\ell_1\dot{\theta}_1 \sin\theta_1 - r_2\dot{\theta}_2\sin\theta_2, \\
&\qquad \ell_1\dot{\theta}_1\cos\theta_1 + r_2\dot{\theta}_2\cos\theta_2, 0), \\
\rho_{\text{body}} \circ T\Pi_1((\theta_1,\theta_2),(\dot{\theta}_1,\dot{\theta}_2)) &= (0,0,\dot{\theta}_1), \\
\rho_{\text{body}} \circ T\Pi_2((\theta_1,\theta_2),(\dot{\theta}_1,\dot{\theta}_2)) &= (0,0,\dot{\theta}_2).
\end{aligned}$$

In this example, again since the motion is planar, the above expressions agree with their spatial counterparts, so we do not reproduce these. Similarly, we will not display the explicit formulae for $\iota_{\text{body}}|\mathsf{TQ}$ and $\iota_{\text{spatial}}|\mathsf{TQ}$, since these are easily written using the above computations. •

Finally, we look at the rolling disk.

Example 4.21 (Rolling disk cont'd). For the rolling disk we have $\mathsf{Q} = \mathbb{R}^2 \times \mathbb{S}^1 \times \mathbb{S}^1$, and we use the coordinates (x,y,θ,ϕ) defined in Example 4.9. Following the preceding two examples, coordinates for the tangent bundle are written as $((x,y,\theta,\phi),(\dot{x},\dot{y},\dot{\theta},\dot{\phi}))$. From the coordinate expression for the forward kinematic map in Example 4.15, we compute

$$\widehat{\boldsymbol{\Omega}} = \boldsymbol{R}^T\dot{\boldsymbol{R}} = \begin{bmatrix} 0 & \dot{\phi} & \dot{\theta}\cos\phi \\ -\dot{\phi} & 0 & \dot{\theta}\sin\phi \\ -\dot{\theta}\cos\phi & -\dot{\theta}\sin\phi & 0 \end{bmatrix}$$

and

$$\widehat{\boldsymbol{\omega}} = \dot{\boldsymbol{R}}\boldsymbol{R}^T = \begin{bmatrix} 0 & -\dot{\theta} & \dot{\phi}\cos\theta \\ \dot{\theta} & 0 & \dot{\phi}\sin\theta \\ -\dot{\phi}\cos\theta & -\dot{\phi}\sin\theta & 0 \end{bmatrix}.$$

Equivalently we have $\boldsymbol{\Omega} = (-\dot{\theta}\sin\phi, \dot{\theta}\cos\phi, -\dot{\phi})$ and $\boldsymbol{\omega} = (-\dot{\phi}\sin\theta, \dot{\phi}\cos\theta, \dot{\theta})$.

This then gives

$$\begin{aligned}
\lambda_{\text{body}} \circ T\Pi_1((x,y,\theta,\phi),(\dot{x},\dot{y},\dot{\theta},\dot{\phi})) &= (\dot{x},\dot{y},0), \\
\lambda_{\text{spatial}} \circ T\Pi_1((x,y,\theta,\phi),(\dot{x},\dot{y},\dot{\theta},\dot{\phi})) &= (\dot{x},\dot{y},0), \\
\rho_{\text{body}} \circ T\Pi_1((x,y,\theta,\phi),(\dot{x},\dot{y},\dot{\theta},\dot{\phi})) &= (-\dot{\theta}\sin\phi, \dot{\theta}\cos\phi, -\dot{\phi}), \\
\rho_{\text{spatial}} \circ T\Pi_1((x,y,\theta,\phi),(\dot{x},\dot{y},\dot{\theta},\dot{\phi})) &= (-\dot{\phi}\sin\theta, \dot{\phi}\cos\theta, \dot{\theta}).
\end{aligned}$$

From these expressions, one can also write explicit expressions for ι_{body} and ι_{spatial}, although we do not display the results of this transcription. •

4.2 The kinetic energy metric

In this section we assume that a configuration manifold has been found, and that perhaps coordinates have been chosen. We then define the kinetic energy for the system as a function on the tangent bundle of the configuration manifold. To define the kinetic energy for physical systems, we first introduce the inertia tensor for a rigid body. Using this object, along with the data from the preceding section, we systematically obtain an expression for the kinetic energy of an interconnected mechanical system.

4.2.1 Rigid bodies

To discuss in a precise and usefully general way the inertial properties of a rigid body, one actually should use some basic measure theory. We present the necessary background in Appendix A. A little measure theory is also useful when we come to control theory. However, for readers not wanting to get too deeply into the sort of (necessary for rigor but) fine points that necessitate measure theory, it is possible to proceed without them, making necessary adjustments. For example, in this section, if one wishes to exclude such degenerate rigid bodies as point masses and lines with no thickness, then one may escape without the measure theory. For those who are comfortable with measure theory, whenever a μ is encountered below, it represents a finite Borel measure, meaning that it is a measure defined on Borel sets in $\mathbb{R}^3$, and having the property that the measure of $\mathbb{R}^3$ is finite. For those who wish to avoid the measure theory, whenever a μ is encountered in the sequel, they may replace it with $\rho \, \mathrm{d}V$, where $\rho \colon \mathcal{B} \to \mathbb{R}$ is the "mass density" and where $\mathrm{d}V$ is the standard volume element in $\mathbb{R}^3$. This simplified interpretation, however, does not allow a point mass to be thought of as a rigid body.

A ***rigid body*** is a pair $(\mathcal{B}, \mu)$, where $\mathcal{B} \subset \mathbb{R}^3$ is compact, and μ is a finite Borel measure on $\mathbb{R}^3$ with support equal to $\mathcal{B}$ called the ***mass distribution*** for the body. We write

$$\mu(\mathcal{B}) = \int_{\mathcal{B}} \mathrm{d}\mu$$

as the ***mass*** of the body. In the computations that follow, we shall denote by $\boldsymbol{\chi}$ a point in the body frame. Thus $\boldsymbol{\chi}$ is the vector of components of a point in space measured relative to the coordinates defined by the body frame $(O_{\mathrm{body}}, \{\boldsymbol{b}_1, \boldsymbol{b}_2, \boldsymbol{b}_3\})$. In cases where we perform spatial integration, $\boldsymbol{\chi}$ is the independent variable.

Here are some typical rigid bodies.

Examples 4.22. 1. A particle of mass m is a special case of a rigid body. In this case we suppose the particle to be situated at a point $\boldsymbol{\chi}_0 \in \mathbb{R}^3$. The measure μ is that which assigns to any Borel set in $\mathbb{R}^3$ containing the point $\boldsymbol{\chi}_0$ the value m, and to all other Borel sets the value 0.

2. A slightly more interesting case consists of a rigid body that is contained in a line segment

$$\ell_{\chi_1,\chi_2} = \{\tau(\chi_2 - \chi_1)/\|\chi_2 - \chi_1\|_{\mathbb{R}^3} + \chi_1 \mid \tau \in [0, \|\chi_2 - \chi_1\|_{\mathbb{R}^3}]\},$$

for some $\chi_1, \chi_2 \in \mathbb{R}^3$. An example of such a situation would consist of two particles constrained so as to be equidistant. Also, an infinitely thin rod will fall into this category. In all cases, the mass distribution μ has support in ℓ_{χ_1,χ_2}.

3. The above two cases are degenerate examples of rigid bodies. The typical nondegenerate situation is when $\mathcal{B}$ occupies a subset of $\mathbb{R}^3$ that cannot be contained in any line. There is not much one can say in this general case. •

The ***center of mass*** of a body $(\mathcal{B}, \mu)$ is the point

$$\chi_c = \frac{1}{\mu(\mathcal{B})}\Big(\int_{\mathcal{B}} \chi \, d\mu\Big).$$

Let us look at our examples of rigid bodies to see how to determine their centers of mass.

Examples 4.23. 1. *Example 4.22–1 cont'd:* In this case we had a particle of mass m at $\chi_0 \in \mathbb{R}^3$. Since the support of μ must be $\{\chi_0\}$, the center of mass is

$$\chi_c = \mu(\mathcal{B})^{-1} \int_{\mathcal{B}} \chi \, d\mu = m^{-1} m \chi_0 = \chi_0.$$

Therefore, unsurprisingly, the center of mass for a point mass is that point at which the mass is located.

2. *Example 4.22–2 and Example 4.22–3 cont'd:* For the more general rigid bodies consisting of a line or the most general rigid body, there is not much one can say about the center of mass, except to apply the definition. One intuitively clear fact that is true is that $\chi_c \in \mathrm{int}_{\mathrm{aff}(\mathcal{B})}\mathrm{conv}(\mathcal{B})$. The proof of this fact we leave to the reader (Exercise E4.10). •

The properties of a rigid body are characterized by three things: (1) its mass, (2) its center of mass, and (3) its inertia tensor. It remains to define the latter. Let $(\mathcal{B}, \mu)$ be a rigid body and let $\chi_0 \in \mathbb{R}^3$. We define the ***inertia tensor about*** χ_0 of $(\mathcal{B}, \mu)$ to be the linear map $\mathbb{I}_{\chi_0} \in L(\mathbb{R}^3; \mathbb{R}^3)$ defined by

$$\mathbb{I}_{\chi_0}(v) = \int_{\mathcal{B}} (\chi - \chi_0) \times (v \times (\chi - \chi_0)) \, d\mu.$$

Here $\times$ is the usual vector cross-product in $\mathbb{R}^3$ [see e.g., Marsden and Tromba 2004]. We denote the inertia tensor about the center of mass of $(\mathcal{B}, \mu)$ by $\mathbb{I}_c$. A basic property of the inertia tensor is the following.

Proposition 4.24 (Symmetry of inertia tensor). *The inertia tensor* $\mathbb{I}_{\chi_0}$ *of a rigid body* $(\mathcal{B}, \mu)$ *is symmetric and positive-semidefinite with respect to the inner product* $\mathbb{G}_{\mathbb{R}^3}$.

Proof. Using the vector identity

$$\mathbb{G}_{\mathbb{R}^3}(\boldsymbol{u}, \boldsymbol{v} \times \boldsymbol{w}) = \mathbb{G}_{\mathbb{R}^3}(\boldsymbol{w}, \boldsymbol{u} \times \boldsymbol{v}), \tag{4.6}$$

we prove the symmetry of $\mathbb{I}_{\chi_0}$ by means of

$$\begin{aligned} \mathbb{G}_{\mathbb{R}^3}(\mathbb{I}_{\chi_0}(\boldsymbol{v}_1), \boldsymbol{v}_2) &= \int_{\mathcal{B}} \mathbb{G}_{\mathbb{R}^3}((\boldsymbol{\chi} - \boldsymbol{\chi}_0) \times (\boldsymbol{v}_1 \times (\boldsymbol{\chi} - \boldsymbol{\chi}_0)), \boldsymbol{v}_2)\, \mathrm{d}\mu \\ &= \int_{\mathcal{B}} \mathbb{G}_{\mathbb{R}^3}(\boldsymbol{v}_1 \times (\boldsymbol{\chi} - \boldsymbol{\chi}_0), \boldsymbol{v}_2 \times (\boldsymbol{\chi} - \boldsymbol{\chi}_0))\, \mathrm{d}\mu \\ &= \int_{\mathcal{B}} \mathbb{G}_{\mathbb{R}^3}(\boldsymbol{v}_1, (\boldsymbol{\chi} - \boldsymbol{\chi}_0) \times (\boldsymbol{v}_2 \times (\boldsymbol{\chi} - \boldsymbol{\chi}_0)))\, \mathrm{d}\mu \\ &= \mathbb{G}_{\mathbb{R}^3}(\boldsymbol{v}_1, \mathbb{I}_{\chi_0}(\boldsymbol{v}_2)). \end{aligned} \tag{4.7}$$

That $\mathbb{I}_{\chi_0}$ is positive-semidefinite follows directly from (4.7). ■

The eigenvalues of the inertia tensor are, therefore, guaranteed to be real (Proposition 2.39). If $\{J_1, J_2, J_3\}$ are the eigenvalues of $\mathbb{I}_c$, we call these the ***principal inertias*** of $(\mathcal{B}, \mu)$. If $\{\boldsymbol{u}_1, \boldsymbol{u}_2, \boldsymbol{u}_3\}$ are orthonormal eigenvectors associated with these eigenvalues, we call these the ***principal axes*** of $(\mathcal{B}, \mu)$.

To complete our discussion of rigid bodies, we show that, as far as the inertia tensor is concerned, there are three possible types of rigid body.

Examples 4.25. 1. *Example 4.22–1 cont'd:* Let us compute the inertia tensor of a particle of mass m, thought of as a rigid body. Thus $\mathcal{B} = \{\boldsymbol{\chi}_c\}$. Since the particle is located at its center of mass, by Example 4.23–1 we readily compute

$$\mathbb{I}_{\chi_0}(\boldsymbol{v}) = m(\boldsymbol{\chi}_c - \boldsymbol{\chi}_0) \times (\boldsymbol{v} \times (\boldsymbol{\chi}_c - \boldsymbol{\chi}_0)).$$

The matrix representation of $\mathbb{I}_{\chi_0}$ with respect to an orthonormal basis $\{\boldsymbol{v}_1, \boldsymbol{v}_2, \boldsymbol{v}_3\}$ for $\mathbb{R}^3$ is

$$m \begin{bmatrix} (a^2)^2 + (a^3)^2 & -a^1a^2 & -a^1a^3 \\ -a^1a^2 & (a^1)^2 + (a^3)^2 & -a^2a^3 \\ -a^1a^3 & -a^2a^3 & (a^1)^2 + (a^2)^2 \end{bmatrix}, \tag{4.8}$$

where (a^1, a^2, a^3) are the components of $\boldsymbol{a} \triangleq \boldsymbol{\chi}_c - \boldsymbol{\chi}_0$ in the basis $\{\boldsymbol{v}_1, \boldsymbol{v}_2, \boldsymbol{v}_3\}$. One may verify that the eigenvalues of this matrix are $\{m\|\boldsymbol{a}\|^2_{\mathbb{R}^3}, m\|\boldsymbol{a}\|^2_{\mathbb{R}^3}, 0\}$ with the corresponding orthonormal eigenvectors $\{\boldsymbol{u}_1, \boldsymbol{u}_2, \boldsymbol{a}/\|\boldsymbol{a}\|_{\mathbb{R}^3}\}$, where $\{\boldsymbol{u}_1, \boldsymbol{u}_2\}$ is any orthonormal basis for $\mathrm{span}_{\mathbb{R}}\{\boldsymbol{a}\}^{\perp}$, the $\mathbb{G}_{\mathbb{R}^3}$-orthogonal complement to $\mathrm{span}_{\mathbb{R}}\{\boldsymbol{a}\}$. Thus, for a particle at $\boldsymbol{\chi}_c$, there is no inertia about the axis $\boldsymbol{\chi}_c - \boldsymbol{\chi}_0$, as is reflected by the zero eigenvalue for $\mathbb{I}_{\chi_0}$ with eigenvector $\boldsymbol{\chi}_c - \boldsymbol{\chi}_0$. What is more, if we choose $\boldsymbol{\chi}_0 = \boldsymbol{\chi}_c$, then $\mathbb{I}_{\chi_0} = 0$, reflecting the fact that the inertia tensor about the center of mass is zero.

2. *Example 4.22–2 cont'd:* Let us consider the slightly less degenerate case for which μ is a mass distribution whose support is contained in the line ℓ_{χ_1,χ_2}. It is not as easy to give a formula for the inertia tensor as it is for the point mass. We, therefore, restrict ourselves to stating some properties of the inertia tensor.
   (a) If $\mathcal{B}$ is not a point, then $\mathbb{I}_{\chi_0}$ is never identically zero, no matter what the reference point χ_0. This is in contrast to the point mass example, where, if $\chi_0 = \chi_c$, then the inertia tensor is zero.
   (b) If $\boldsymbol{a} = \chi_2 - \chi_1$, then $\boldsymbol{a}/\|\boldsymbol{a}\|_{\mathbb{R}^3}$ is a principal axis corresponding to a zero eigenvalue.
   (c) The nonzero eigenvalues of $\mathbb{I}_c$ are equal, and their corresponding two principal axes are $\mathbb{G}_{\mathbb{R}^3}$-orthogonal to $\boldsymbol{a}$.
   (d) If the value of the nonzero eigenvalues of $\mathbb{I}_c$ is J, then the mass, center of mass, and inertia tensor of $(\mathcal{B}, \mu)$ is the same as that for the rigid body consisting of two point masses of mass $\frac{1}{2}m$ placed at a distance $r = \sqrt{2\frac{J}{m}}$ from χ_c along the line ℓ_{χ_1,χ_2}. Thus a rigid body contained in a line can be replaced inertially with two point masses.
3. *Example 4.22–3 cont'd:* The remaining case to consider is that in which the mass distribution μ corresponding to $\mathcal{B} \subset \mathbb{R}^3$ has support that cannot be contained in a line. In this case, it can easily be shown that $\mathbb{I}_c$ is an isomorphism. Let m be the mass of $\mathcal{B}$ and let $\{J_1, J_2, J_3\}$ be the three principal moments of inertia, with $\{\boldsymbol{u}_1, \boldsymbol{u}_2, \boldsymbol{u}_3\}$ the corresponding principal axes, and suppose that these vectors are orthonormal. Let us indicate some of the properties of this most general inertia tensor that will be useful in the sequel.
   (a) If the body is contained in a two-dimensional plane, then any unit vector $\mathbb{G}_{\mathbb{R}^3}$-orthogonal to the plane is a principal axis (see Exercise E4.12). This will come up frequently in our discussion, as we shall deal with a large number of planar rigid bodies.
   (b) It is common to encounter bodies that have symmetry. To properly describe this phenomenon, define the ***inertial ellipsoid*** for $(\mathcal{B}, \mu)$ to be the ellipsoid in $\mathbb{R}^3$ given by
$$E(\mathcal{B}) = \{x^1\boldsymbol{u}_1 + x^2\boldsymbol{u}_2 + x^3\boldsymbol{u}_3 \mid J_1(x^1)^2 + J_2(x^2)^2 + J_3(x^3)^2 = 1\},$$
   provided that none of the principal inertias vanish. An ***axis of symmetry*** for $(\mathcal{B}, \mu)$ is a unit vector $\boldsymbol{u} \in \mathbb{R}^3$ with the property that, for every rotation matrix $\boldsymbol{R} \in \mathsf{O}(3)$ for which $\boldsymbol{R}\boldsymbol{u} = \boldsymbol{u}$, it also holds that $\boldsymbol{R}(E(\mathcal{B})) = E(\mathcal{B})$. That is to say, an axis of symmetry is a vector about which rotations leave the inertial ellipsoid unchanged. One can then verify that, if $\boldsymbol{u}$ is an axis of symmetry, then it is also a principal axis, and the principal inertias for the remaining two principal axes are equal.

(c) Define a new rigid body consisting of six points

$$\{r_1\boldsymbol{u}_1\} \cup \{-r_1\boldsymbol{u}_1\} \cup \{r_2\boldsymbol{u}_2\} \cup \{-r_2\boldsymbol{u}_2\} \cup \{r_3\boldsymbol{u}_3\} \cup \{-r_3\boldsymbol{u}_3\},$$

$r_i = \sqrt{6\frac{J_i}{m}}$, $i \in \{1,2,3\}$, and place a mass distribution on this new body by asking that the mass at each of the six points be $\frac{1}{6}m$. This body has the same mass, center of mass, and inertia tensor as $(\mathcal{B},\mu)$. Thus a general rigid body can be inertially replaced with six point masses. •

4.2.2 The kinetic energy of a single rigid body

Now we determine the kinetic energy for a rigid body $(\mathcal{B},\mu)$ whose inertia tensor about the center of mass is $\mathbb{I}_c$. For a mechanical system comprised of a collection of particles and rigid bodies, the total kinetic energy will be the sum of that for each component. Therefore, by being able to compute the kinetic energy for a single rigid body, we are in actuality not losing any generality, as we shall see in the next section.

In Section 4.2.1 we denoted a point in the body, measured in the body frame, by χ. Now we wish to consider rigid bodies in motion, and we often wish to measure motion relative to the spatial reference frame $\Sigma_{\text{spatial}} = (O_{\text{spatial}}, \{\boldsymbol{s}_1, \boldsymbol{s}_2, \boldsymbol{s}_3\})$. We shall denote by $\boldsymbol{x}$ a point measured in the spatial reference frame. We will generalize the fact that the kinetic energy of a particle of mass m, moving in $\mathbb{R}^3$ with velocity $\dot{\boldsymbol{x}}(t)$, is $\frac{1}{2}m\|\dot{\boldsymbol{x}}(t)\|^2_{\mathbb{R}^3}$. To do this, we need to understand what it means for a rigid body to move in space. Along with our spatial reference frame, we fix to the body, at its center of mass, another orthonormal reference frame $\Sigma_{\text{body}} = (O_{\text{body}}, \{\boldsymbol{b}_1, \boldsymbol{b}_2, \boldsymbol{b}_3\})$. As we have seen, the frames are related by the vector $\boldsymbol{r} = O_{\text{body}} - O_{\text{spatial}}$ and by their relative orientation. As in Section 4.1.1, the orientations are related by the matrix $\boldsymbol{R} \in \mathsf{SO}(3)$ whose ith column is comprised of the components of $\boldsymbol{b}_i$ relative to the basis $\{\boldsymbol{s}_1, \boldsymbol{s}_2, \boldsymbol{s}_3\}$. Thus, as the body $\mathcal{B}$ moves, this movement is described by a curve $t \mapsto (\boldsymbol{R}(t), \boldsymbol{r}(t)) \in \mathsf{SO}(3) \times \mathbb{R}^3$. We suppose this curve to be differentiable for the purposes of our present discussion. Furthermore, a point $\chi \in \mathcal{B}$ at time t will be located at $\boldsymbol{x}(t) = \boldsymbol{r}(t) + \boldsymbol{R}(t)\chi$. The following list of assumptions characterizes the preceding discussion. These assumptions are important, since, if they do not hold, the ensuing discussion becomes more complicated.

Assumptions 4.26 (Properties of frames).

(i) The origin O_{body} of the body frame Σ_{body} is located at the center of mass of the body.

(ii) The rotation matrix $\boldsymbol{R}(t) \in \mathsf{SO}(3)$, therefore, measures the rotation of the body about its center of mass, and the vector $\boldsymbol{r}(t)$, therefore, measures the distance from the origin of the spatial frame Σ_{spatial} to the center of mass of the body.

(iii) The bases $\{\boldsymbol{s}_1, \boldsymbol{s}_2, \boldsymbol{s}_3\}$ and $\{\boldsymbol{b}_1, \boldsymbol{b}_2, \boldsymbol{b}_3\}$ share the same orientation (usually right-handed). •

The ***kinetic energy*** of the body $(\mathcal{B}, \mu)$ at time t is given by

$$\mathrm{KE}(t) = \frac{1}{2}\int_{\mathcal{B}} \|\dot{\boldsymbol{r}}(t) + \dot{\boldsymbol{R}}(t)\boldsymbol{\chi}\|^2_{\mathbb{R}^3}\,\mathrm{d}\mu,$$

i.e., the integral over the body of the "infinitesimal kinetic energies." To provide a simplification of this expression, we recall from Section 4.1.5 the definition of the body angular velocity, which we reproduce here for convenience:

$$\widehat{\boldsymbol{\Omega}}(t) = \boldsymbol{R}^T(t)\dot{\boldsymbol{R}}(t).$$

Recall that, in (4.5), we indicated how one can assign to $\widehat{\boldsymbol{\Omega}}(t)$ a vector $\boldsymbol{\Omega}(t) \in \mathbb{R}^3$.

Now we provide a simplified expression for the kinetic energy of a rigid body using the body angular velocity.

Proposition 4.27 (Decomposition of kinetic energy). *Let a rigid body* $(\mathcal{B}, \mu)$ *undergo a motion specified by a differentiable curve* $t \mapsto (\boldsymbol{R}(t), \boldsymbol{r}(t))$ *as above. If*

$$\mathrm{KE}_{\mathrm{tran}}(t) = \tfrac{1}{2}\mu(\mathcal{B})\|\dot{\boldsymbol{r}}(t)\|^2_{\mathbb{R}^3}, \quad \mathrm{KE}_{\mathrm{rot}}(t) = \tfrac{1}{2}\mathbb{G}_{\mathbb{R}^3}(\mathbb{I}_c(\boldsymbol{\Omega}(t)), \boldsymbol{\Omega}(t)),$$

then $\mathrm{KE}(t) = \mathrm{KE}_{\mathrm{tran}}(t) + \mathrm{KE}_{\mathrm{rot}}(t)$.

Proof. We compute

$$\begin{aligned}\mathrm{KE}(t) &= \frac{1}{2}\int_{\mathcal{B}} \|\dot{\boldsymbol{r}}(t)\|^2_{\mathbb{R}^3}\,\mathrm{d}\mu + \frac{1}{2}\int_{\mathcal{B}} \|\dot{\boldsymbol{R}}(t)\boldsymbol{\chi}\|^2_{\mathbb{R}^3}\,\mathrm{d}\mu + \int_{\mathcal{B}} \mathbb{G}_{\mathbb{R}^3}(\dot{\boldsymbol{r}}(t), \dot{\boldsymbol{R}}(t)\boldsymbol{\chi})\,\mathrm{d}\mu\\ &= \tfrac{1}{2}\mu(\mathcal{B})\|\dot{\boldsymbol{r}}(t)\|^2_{\mathbb{R}^3} + \frac{1}{2}\int_{\mathcal{B}} \|\dot{\boldsymbol{R}}(t)\boldsymbol{\chi}\|^2_{\mathbb{R}^3}\,\mathrm{d}\mu,\end{aligned}$$

using in the second line the fact that, by Assumption 4.26(i), $\int_{\mathcal{B}} \boldsymbol{\chi}\,\mathrm{d}\mu = \boldsymbol{0}$. Thus the result will follow if we can show that

$$\int_{\mathcal{B}} \|\dot{\boldsymbol{R}}(t)\boldsymbol{\chi}\|^2_{\mathbb{R}^3}\,\mathrm{d}\mu = \mathbb{G}_{\mathbb{R}^3}(\mathbb{I}_c(\boldsymbol{\Omega}(t)), \boldsymbol{\Omega}(t)).$$

To this end we compute

$$\begin{aligned}\int_{\mathcal{B}} \|\dot{\boldsymbol{R}}(t)\boldsymbol{\chi}\|^2_{\mathbb{R}^3}\,\mathrm{d}\mu &= \int_{\mathcal{B}} \|\boldsymbol{R}(t)(\boldsymbol{\Omega}(t) \times \boldsymbol{\chi})\|^2_{\mathbb{R}^3}\,\mathrm{d}\mu = \int_{\mathcal{B}} \|\boldsymbol{\Omega}(t) \times \boldsymbol{\chi}\|^2_{\mathbb{R}^3}\,\mathrm{d}\mu\\ &= \int_{\mathcal{B}} \mathbb{G}_{\mathbb{R}^3}(\boldsymbol{\chi} \times (\boldsymbol{\Omega}(t) \times \boldsymbol{\chi}), \boldsymbol{\Omega}(t))\,\mathrm{d}\mu\\ &= \mathbb{G}_{\mathbb{R}^3}(\mathbb{I}_c(\boldsymbol{\Omega}(t)), \boldsymbol{\Omega}(t)),\end{aligned}$$

where we have used the vector identity (4.6) and Exercise E4.11, noting that $\boldsymbol{\chi}_c = \boldsymbol{0}$. ■

Remarks 4.28. 1. The additive decomposition of the kinetic energy into a translational and rotational part is a direct consequence of Assumptions 4.26. If these assumptions are not made, then this sort of decomposition will not generally occur. This is related to the fact, sometimes taught in elementary courses on dynamics, that when using Newton's and Euler's laws to sum forces and moments, one should only sum moments about either a stationary point or a center of mass. For this reason, if it turns out that the rigid motion $t \mapsto \boldsymbol{r}(t) + \boldsymbol{R}(t)\boldsymbol{\chi}$ fixes a point $\boldsymbol{\chi}_0$, one can also represent the kinetic energy as the rotational kinetic energy using $\mathbb{I}_{\boldsymbol{\chi}_0}$.

2. The development in this section allows particles as rigid bodies. The inertia tensor for a particle about its center of mass is zero. Therefore, under Assumption 4.26(i), the rotational kinetic energy of a particle is zero. For this reason, the choice of the curve $t \mapsto \boldsymbol{R}(t)$ describing the "rotation" of a particle is immaterial. •

Thus far, our discussion has been geared towards rigid body dynamics for bodies moving in space. We shall also have occasion to consider rigid bodies moving in a fluid, in which case the dynamics must take into account the fluid properties. The reader is referred to Exercise E4.25, and the comments preceding that exercise, for a discussion of this.

4.2.3 From kinetic energy to a Riemannian metric

Now we use the constructions of the preceding section to produce a Riemannian metric on the configuration manifold Q of an interconnected mechanical system. We suppose our mechanical system to be comprised of rigid bodies $(\mathcal{B}_1, \mu_1), \dots, (\mathcal{B}_k, \mu_k)$. This is completely general, since our definition of a rigid body includes particles and other degenerate rigid bodies as special cases. We also fix a spatial reference frame $\Sigma_{\text{spatial}} = (O_{\text{spatial}}, \{\boldsymbol{s}_1, \boldsymbol{s}_2, \boldsymbol{s}_3\})$. The positions of the centers of mass and the orientations of the bodies about their centers of mass are then denoted by $\boldsymbol{r}_1, \dots, \boldsymbol{r}_k \in \mathbb{R}^3$ and $\boldsymbol{R}_1, \dots, \boldsymbol{R}_k \in \mathsf{SO}(3)$, with these having exactly the meaning as in the preceding section. We will show how, for each fixed $a \in \{1, \dots, k\}$, to define a positive-semidefinite $(0,2)$-tensor field on the configuration manifold Q. After this has been done for each $a \in \{1, \dots, k\}$, the Riemannian metric is constructed by summing these positive-semidefinite $(0,2)$-tensor fields.

We let $\gamma\colon I \to \mathsf{Q}$ be a differentiable curve at q_0. This induces a differentiable curve $\gamma_a = \Pi_a \circ \gamma\colon I \to \mathsf{SO}(3) \times \mathbb{R}^3$ at $(\boldsymbol{R}_{a,0}, \boldsymbol{r}_{a,0}) \triangleq \Pi_a(q_0)$. The curve γ_a then defines, as in the preceding section, a corresponding kinetic energy for the ath body. In this way, to the tangent vector $v_0 = [\gamma]_{q_0} \in \mathsf{T}_{q_0}\mathsf{Q}$, we associate the nonnegative number $\mathrm{KE}_a(v_0)$, which is the kinetic energy of the ath body at time 0 along the curve γ_a. Doing this for every tangent vector in TQ defines a function that we write as $\mathrm{KE}_a\colon \mathsf{TQ} \to \mathbb{R}$. Let us record the most useful feature of this function.

Lemma 4.29. *For each* $a \in \{1, \dots, k\}$, *there exists a* C^∞, *positive-semidefinite* $(0,2)$*-tensor field* B_a *on* Q *such that* $\mathrm{KE}_a(v_q) = \frac{1}{2}B_a(v_q, v_q)$.

Proof. Let $\Pi_a(q) = (\boldsymbol{R}_a(q), \boldsymbol{r}_a(q))$. We have

$$T_q\Pi_a(v_q) = (T_q\boldsymbol{R}_a(v_q), T_q\boldsymbol{r}_a(v_q)) \in \mathsf{T}_{\Pi_a(q)}(\mathsf{SO}(3) \times \mathbb{R}^3).$$

By the definition of kinetic energy given in the preceding section, and by Proposition 4.27, we have

$$\mathrm{KE}_a(v_q) = \tfrac{1}{2}\mu_a(\mathcal{B}_a)\|T_q\boldsymbol{r}_a(v_q)\|^2_{\mathbb{R}^3} + \tfrac{1}{2}\mathbb{G}_{\mathbb{R}^3}(\mathbb{I}_{a,c}(\boldsymbol{R}_a^T(q)\cdot T_q\boldsymbol{R}_a(v_q)), \boldsymbol{R}_a^T(q)\cdot T_q\boldsymbol{R}_a(v_q)).$$

This expression is clearly nonnegative, bilinear in v_q, and C^∞. We then define $B_a(q)$ by reference to Exercise E2.12. ■

By repeating the above procedure for each $a \in \{1, \dots, k\}$, we then define a positive-semidefinite $(0,2)$-tensor field on Q, denoted by $\mathbb{G} = B_1 + \dots + B_k$, and called the ***kinetic energy metric*** for the system. If the model is reasonable, then $\mathbb{G}$ will actually be a Riemannian metric. It is possible that $\mathbb{G}$ would not be positive-definite if there are configurations allowed by Q that have no inertia associated with them. For example, if one allows a rotational degree-of-freedom for a particle, then $\mathbb{G}$ as constructed using the above procedure will not be positive-definite. We shall subsequently suppose that Q has been chosen so that $\mathbb{G}$ is positive-definite.

The matter of computing the components of $\mathbb{G}$ in a set of coordinates is easy. The following lemma is proved by a direct computation.

Lemma 4.30. *Let* $\mathbb{G}$ *be the* C^∞, $(0,2)$*-tensor field described above, let* $(q^1, \dots, q^n)$ *be coordinates for* Q *in a chart* $(\mathcal{U}, \phi)$, *and let* $((q^1, \dots, q^n), (v^1, \dots, v^n))$ *be the corresponding natural coordinates for* TQ. *Then the components of* $\mathbb{G}$ *in the chart are given by*

$$\mathbb{G}_{ij} = \sum_{a=1}^{k} \frac{\partial^2 \mathrm{KE}_a}{\partial v^i \partial v^j}, \qquad i,j \in \{1, \dots, n\},$$

where $\mathrm{KE}_1, \dots, \mathrm{KE}_k$ *are the functions defined preceding Lemma 4.29.*

Let us illustrate the procedure described above by means of our three running examples.

Example 4.31 (Planar rigid body cont'd). We first note that the system is planar, in that all of its components (in this case, the single rigid body) move in a fixed plane P. We choose spatial and body reference frames $\Sigma_{\mathsf{spatial}}$ and Σ_{body} as in Example 4.3. We assume, for simplicity, that the rigid body $(\mathcal{B}, \mu)$ has the property that $\mathcal{B} \subset P$. As is shown in Exercise E4.12, this implies that $\boldsymbol{s}_3$ is a principal axis for $(\mathcal{B}, \mu)$. Let us denote by J the eigenvalue associated with this eigenvector. Let us also write $m = \mu(\mathcal{B})$.

Now it is a direct computation to produce the expressions for the kinetic energy, using the expressions for the body velocities determined in Example 4.19. First, the translational kinetic energy is simply $\mathrm{KE}_{\mathrm{tran}} = \frac{1}{2}m(\dot{x}^2+\dot{y}^2)$. Proposition 4.27 gives the expression for the rotational kinetic energy using body angular velocity as

$$\mathrm{KE}_{\mathrm{rot}} = \frac{1}{2}\mathbb{G}_{\mathbb{R}^3}(\mathbb{I}_c(0,0,\dot{\theta}),(0,0,\dot{\theta})).$$

Since $(0,0,\dot{\theta})$ is an eigenvector of $\mathbb{I}_c$ with eigenvalue J, it then follows that $\mathrm{KE}_{\mathrm{rot}} = \frac{1}{2}J\dot{\theta}^2$. Thus the total kinetic energy of the system is

$$\mathrm{KE} = \tfrac{1}{2}J\dot{\theta}^2 + \tfrac{1}{2}m(\dot{x}^2+\dot{y}^2).$$

One can now use Lemma 4.30 to derive the matrix representation of the kinetic energy metric in this case to be

$$[\mathbb{G}] = \begin{bmatrix} J & 0 & 0 \\ 0 & m & 0 \\ 0 & 0 & m \end{bmatrix}.$$

Equivalently, we may write

$$\mathbb{G} = J\mathrm{d}\theta\otimes\mathrm{d}\theta + m(\mathrm{d}x\otimes\mathrm{d}x + \mathrm{d}y\otimes\mathrm{d}y).$$

Of course, this can all be done more directly, but this construction is nonetheless illustrative of the procedure we describe. •

Example 4.32 (Two-link manipulator cont'd). We use the spatial and body reference frames of Example 4.4. We also assume that the bodies are contained in the plane of motion so that s_3 is a principal axis for both links. The eigenvalues associated with this eigenvector are denoted by J_1 and J_2, where the subscripts "1" and "2" have the same meaning as they did for the coordinates (θ_1,θ_2). The mass of the ath link is m_a, $a\in\{1,2\}$, and the length of the ath link is ℓ_a, $a\in\{1,2\}$. We suppose that the links have a uniform mass distribution so that their centers of mass are located at their midpoints. We also assume that the joints for the first link are located at the extremities of the link. When modeling this system in practice, these simplifying assumptions may not be valid, but suitable modifications are easily made.

We recall from Example 4.20 the coordinate expressions for the body velocities of the two bodies. The computation of the rotational kinetic energy then goes just like in Example 4.3, and gives the total rotational kinetic energy as $\mathrm{KE}_{\mathrm{rot}} = \frac{1}{2}J_1\dot{\theta}_1^2 + \frac{1}{2}J_2\dot{\theta}_2^2$. The translational kinetic energy is computed straightforwardly, if slightly tediously, as

$$\mathrm{KE}_{\mathrm{tran}} = \tfrac{1}{8}\ell_1^2(m_1+4m_2)\dot{\theta}_1^2 + \tfrac{1}{8}m_2\ell_2^2\dot{\theta}_2^2 + \tfrac{1}{2}m_2\ell_1\ell_2\cos(\theta_1-\theta_2)\dot{\theta}_1\dot{\theta}_2.$$

Thus the total kinetic energy is

$$\mathrm{KE} = \tfrac{1}{8}(m_1 + 4m_2)\ell_1^2\dot{\theta}_1^2 + \tfrac{1}{8}m_2\ell_2^2\dot{\theta}_2^2 + \tfrac{1}{2}m_2\ell_1\ell_2\cos(\theta_1 - \theta_2)\dot{\theta}_1\dot{\theta}_2 + \tfrac{1}{2}J_1\dot{\theta}_1^2 + \tfrac{1}{2}J_2\dot{\theta}_2^2,$$

so giving the matrix representation of the kinetic energy metric as

$$[\mathbb{G}] = \begin{bmatrix} J_1 + \frac{1}{4}(m_1 + 4m_2)\ell_1^2 & \frac{1}{2}m_2\ell_1\ell_2\cos(\theta_1 - \theta_2) \\ \frac{1}{2}m_2\ell_1\ell_2\cos(\theta_1 - \theta_2) & J_2 + \frac{1}{4}m_2\ell_2^2 \end{bmatrix}.$$

We may equivalently write

$$\mathbb{G} = \big(J_1 + \tfrac{1}{4}(m_1 + 4m_2)\ell_1^2\big)\mathrm{d}\theta_1 \otimes \mathrm{d}\theta_1 + \big(J_2 + \tfrac{1}{4}m_2\ell_2^2\big)\mathrm{d}\theta_2 \otimes \mathrm{d}\theta_2 + \tfrac{1}{2}m_2\ell_1\ell_2\cos(\theta_1 - \theta_2)(\mathrm{d}\theta_1 \otimes \mathrm{d}\theta_2 + \mathrm{d}\theta_2 \otimes \mathrm{d}\theta_1).$$

This agrees, of course, with what we gave in Section 1.2. Note that even in this simple example, the calculations get a little messy. •

Example 4.33 (Rolling disk cont'd). We take the same $\Sigma_{\mathrm{spatial}}$ and Σ_{body} as described in Example 4.5. The motion of the body is not planar, and we therefore need to make some assumptions about the inertia tensor of the disk. A reasonable assumption is that the body axis $\boldsymbol{b}_3$ is an axis of symmetry, and, therefore, a principal axis, for the disk. Let us denote the corresponding principal inertia by J_{roll}. Therefore, the other two principal moments of inertia are equal, and let us denote these by J_{spin}. Any vector $\mathbb{G}_{\mathbb{R}^3}$-orthogonal to $\boldsymbol{b}_3$ is an eigenvector for the eigenvalue J_{spin}. We denote by m the mass of the disk. For simplicity, we assume that the disk is uniform (or at least rotationally symmetric) so the center of mass is at its center of rotation about the $\boldsymbol{b}_3$-axis. With these assumptions, the inertia tensor about the center of mass has, relative to the basis $\{\boldsymbol{b}_1, \boldsymbol{b}_2, \boldsymbol{b}_3\}$, the matrix representation

$$[\mathbb{I}_c] = \begin{bmatrix} J_{\mathrm{spin}} & 0 & 0 \\ 0 & J_{\mathrm{spin}} & 0 \\ 0 & 0 & J_{\mathrm{roll}} \end{bmatrix}.$$

From Example 4.21, we recall the expressions for the body velocities. The translational kinetic energy is then easily computed to be $\mathrm{KE}_{\mathrm{tran}} = \frac{1}{2}m(\dot{x}^2+\dot{y}^2)$. To compute the rotational kinetic energy we need the inertia tensor. Using Proposition 4.27 we can straightforwardly compute the rotational kinetic energy to be $\mathrm{KE}_{\mathrm{rot}} = \frac{1}{2}J_{\mathrm{spin}}\dot{\theta}^2 + \frac{1}{2}J_{\mathrm{roll}}\dot{\phi}^2$. The total kinetic energy is then

$$\mathrm{KE} = \tfrac{1}{2}m(\dot{x}^2 + \dot{y}^2) + \tfrac{1}{2}J_{\mathrm{spin}}\dot{\theta}^2 + \tfrac{1}{2}J_{\mathrm{roll}}\dot{\phi}^2,$$

from which we deduce from Lemma 4.30 that the matrix representation for the kinetic energy metric is

$$[\mathbb{G}] = \begin{bmatrix} m & 0 & 0 & 0 \\ 0 & m & 0 & 0 \\ 0 & 0 & J_{\mathrm{spin}} & 0 \\ 0 & 0 & 0 & J_{\mathrm{roll}} \end{bmatrix}.$$

This can also be expressed as

$$\mathbb{G} = m(\mathrm{d}x \otimes \mathrm{d}x + \mathrm{d}y \otimes \mathrm{d}y) + J_{\mathrm{spin}}\mathrm{d}\theta \otimes \mathrm{d}\theta + J_{\mathrm{roll}}\mathrm{d}\phi \otimes \mathrm{d}\phi.$$

This is as we saw in Section 1.3. •

Remarks 4.34. 1. In this section we have been very systematic about how to derive a Riemannian metric on Q, assuming as a starting point a physical system consisting of an interconnection of particles and rigid bodies. In the general development that follows, we shall not normally make the assumption that the Riemannian metric comes about in this way. Indeed, the general setting is that of an arbitrary differentiable manifold Q that possesses a Riemannian metric $\mathbb{G}$. Although this is far more general than the physical backdrop that we have developed to this point in the chapter, the examples we consider always bring us back to the mechanical situation, in which we have particles and rigid bodies. There is, however, some clarity gained by reverting to the abstract setup when engaging in discussions of general properties of the systems we consider.

2. Although we normally inherit a Riemannian metric from the physics of the problems we deal with, the positing of a Riemannian manifold $(\mathsf{Q}, \mathbb{G})$ raises the question of whether, for a given manifold Q, there even *exists* a Riemannian metric on Q. It turns out that if Q is C^∞, Hausdorff, and second-countable, then it possesses a Riemannian metric if and only if it is paracompact. In applications, it can be expected that Q will have this property. •

4.3 The Euler–Lagrange equations

It is convenient to introduce the equations of motion at this juncture, even though we do not have at our disposal all the ingredients (we are missing forces and constraints) that will appear in the final description of the most general form of the governing equations. We first talk about a general variational principle that gives rise to the Euler–Lagrange equations. This variational principle tells us where these equations "come from." A bald introduction of the Euler–Lagrange equations from thin air comes across as exceedingly unmotivated, and the variational principle at least provides some context. However, for our purposes, no doubt the most compelling property of the Euler–Lagrange equations is that they agree with the Newton–Euler equations on the intersection of the domain on which they are defined. This is discussed briefly in Section 4.3.5. After talking about the variational principle in a general setting, we see how it applies in the presence of a Riemannian metric on the configuration manifold in Section 4.3.4.

4.3.1 A problem in the calculus of variations

Many important ideas in mechanics have a variational basis. The calculus of variations is an important and complex subject, and we shall only be able to deal with it superficially. Nevertheless, the notion that the laws of nature act in such a way as to extremize some function is an important one. There are many introductory accounts of the calculus of variations, including the texts [Gelfand and Fomin 1961, Lanczos 1949, Wan 1995]. More advanced treatments include [Bliss 1946, Bolza 1961, Carathéodory 1935, Troutman 1996], and the two volumes of Giaquinta and Hildebrandt [1996]. A historical account may be found in [Goldstine 1980].

Let Q be a C^r-differentiable manifold, $r \in \mathbb{N} \cup \{\infty\} \cup \{\omega\}$. A ***$C^r$-Lagrangian*** is a C^r-function L on $\mathbb{R} \times \mathsf{TQ}$. Thus L is a function of time, position, and velocity. A Lagrangian $L\colon \mathbb{R} \times \mathsf{TQ} \to \mathbb{R}$ is ***time-independent*** if there exists a function $L_0\colon \mathsf{TQ} \to \mathbb{R}$ such that $L(t, v_q) = L_0(v_q)$. We will almost exclusively be dealing with time-independent Lagrangians, and we will identify a time-independent Lagrangian L with the function L_0 above. If $(\mathcal{U}, \phi)$ is a chart for Q with coordinates $(q^1, \dots, q^n)$, we might write $L(t, q^1, \dots, q^n, v^1, \dots, v^n)$, or $L(t, \boldsymbol{q}, \boldsymbol{v})$ for short, if we are working in coordinates. We will postpone physical examples of Lagrangians to Sections 4.3.4 and 4.4.4. In this section we will use the simple polar coordinate example that was used throughout Chapter 3.

We wish to minimize a certain function over a class of curves. To do this we need to say how to "vary" a curve.

Definition 4.35 (Variation). Let $\gamma\colon [a, b] \to \mathsf{Q}$ be a C^2-curve. A ***variation*** of γ is a C^2-map $\vartheta\colon J \times [a, b] \to \mathsf{Q}$ with the properties

(i) $J \subset \mathbb{R}$ is an interval for which $0 \in \operatorname{int}(J)$,
(ii) $\vartheta(0, t) = \gamma(t)$ for all $t \in [a, b]$,
(iii) $\vartheta(s, a) = \gamma(a)$ for all $s \in J$, and
(iv) $\vartheta(s, b) = \gamma(b)$ for all $s \in J$.

The ***infinitesimal variation*** associated with a variation ϑ is the vector field along γ given by

$$\delta\vartheta(t) = \frac{\mathrm{d}}{\mathrm{d}s}\Big|_{s=0} \vartheta(s, t) \in \mathsf{T}_{\gamma(t)}\mathsf{Q}. \qquad \bullet$$

Remark 4.36. A variation of γ gives rise to an infinitesimal variation. The converse construction would be, "Given a vector field $\xi\colon [a, b] \to \mathsf{TQ}$ along γ with $\xi(a) = 0_{\gamma(a)}$ and $\xi(b) = 0_{\gamma(b)}$, is there a variation ϑ such that $\delta\vartheta = \xi$?" The reader is asked to consider this question in Exercise E4.13. $\bullet$

The idea is that a variation of γ is a "wiggling" of γ, and an infinitesimal variation corresponding to a certain wiggle is a measure of the wiggle for small values of the "wiggle parameter." Some intuition is given in Figure 4.10. Note that the endpoints of a variation, as we have defined it, remain stationary, so the infinitesimal variations vanish at the endpoints.

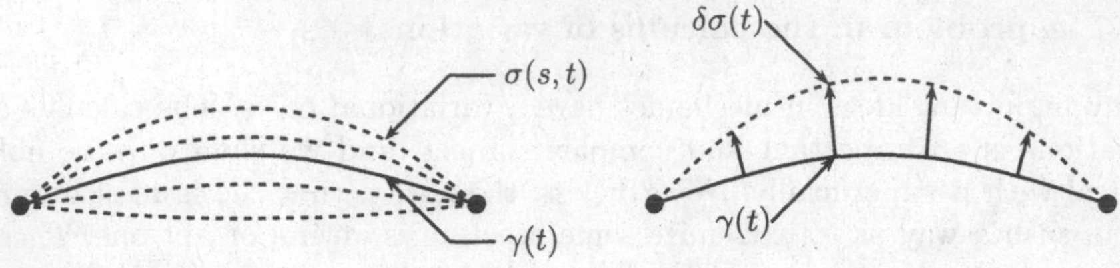

Figure 4.10. A variation (on the left) and an infinitesimal variation (on the right)

We let Q be a configuration manifold, let $a, b \in \mathbb{R}$ with $a < b$, and let $q_a, q_b \in \mathsf{Q}$. We let

$$C^2([a,b], q_a, q_b) = \{\gamma\colon [a,b] \to \mathsf{Q} \mid \gamma(a) = q_a, \gamma(b) = q_b, \gamma \text{ is of class } C^2\}$$

be the collection of twice continuously differentiable curves that are defined on $[a, b]$, and that start at q_a and end at q_b. Now we suppose that we have a Lagrangian on Q and define a function $A_L\colon C^2([a,b], q_a, q_b) \to \mathbb{R}$, called the ***action*** associated with L, by

$$A_L(\gamma) = \int_a^b L(t, \gamma'(t))\,\mathrm{d}t.$$

We may now formally state the problem we consider in this section.

Problem 4.37 (Problem in the calculus of variations). Find those curves $\gamma_0 \in C^2([a,b], q_a, q_b)$ for which $A_L(\gamma_0) \leq A_L(\gamma)$ for every $\gamma \in C^2([a,b], q_a, q_b)$. Such curves will be said to ***minimize*** A_L. •

4.3.2 Necessary conditions for minimization—the Euler–Lagrange equations

The following condition gives what might be regarded as a first-order necessary condition for a curve $\gamma \in C^2([a,b], q_a, q_b)$ to be a minimizer for A_L. Thus it should be thought of as being analogous to the first derivative condition (i.e., that the derivative be zero) for determining the minimum of a function. The result was originally obtained by Lagrange [1788].

Theorem 4.38 (Necessary conditions for minimization). *If a curve $\gamma \in C^2([a,b], q_a, q_b)$ minimizes A_L, then, for any chart $(\mathcal{U}, \phi)$ with the property that the image of γ intersects $\mathcal{U}$, the local representative $t \mapsto \boldsymbol{q}(t)$ of γ in that chart must satisfy*

$$\frac{\mathrm{d}}{\mathrm{d}t}\Big(\frac{\partial L}{\partial v^i}\Big) - \frac{\partial L}{\partial q^i} = 0, \qquad i \in \{1, \dots, n\}, \tag{4.9}$$

at $(t, \boldsymbol{q}(t), \dot{\boldsymbol{q}}(t))$ for each t with the property that $\gamma(t) \in \mathcal{U}$.

Proof. See Section B.1. ■

The equations (4.9) are called the ***Euler–Lagrange equations*** for the Lagrangian L. To those unfamiliar with the Euler–Lagrange equations, it is often not clear how to write them down in practice, and indeed it is genuinely vague from the way they are written. This is clarified by expanding the first term in the equations to give

$$\frac{\partial^2 L}{\partial v^i \partial v^j}\ddot{q}^j + \frac{\partial^2 L}{\partial v^i \partial q^j}\dot{q}^j + \frac{\partial^2 L}{\partial v^i \partial t} - \frac{\partial L}{\partial q^i} = 0, \qquad i \in \{1, \dots, n\}.$$

Thus we see that the Euler–Lagrange equations are implicit second-order differential equations when written in coordinates. As implicit equations, they have all the associated problems, like possible lack of existence and uniqueness of solutions. However, we shall only be considering in this book Lagrangians for which the Euler–Lagrange equations are well-posed.[2] The common language is to call a curve in $C^2(q_a, q_b, [a,b])$ an ***extremal*** for the action A_L when it satisfies the Euler–Lagrange equations for L. This reflects the fact that the Euler–Lagrange equations are essentially analogous to the first derivative conditions in calculus for the minimization of a $\mathbb{R}$-valued function on $\mathbb{R}^n$. More precisely, as can be see in the proof of Theorem 4.38, curves γ that satisfy the Euler–Lagrange equations are exactly those that satisfy

$$\frac{\mathrm{d}}{\mathrm{d}s}\Big|_{s=0} \int_a^b L(t, \tfrac{\mathrm{d}}{\mathrm{d}t}\vartheta(s,t))\,\mathrm{d}t$$

for all variations ϑ of γ. This is referred to as ***Hamilton's Principle***.

Let us look at a simple example.

Example 4.39 (Polar coordinates cont'd). We work again with $\mathsf{Q} = \mathbb{R}^2$ and its two charts $(\mathcal{U}_1, \phi_1)$ and $(\mathcal{U}_2, \phi_2)$, representing Cartesian and polar coordinates, respectively. We define a Lagrangian on Q by defining it in the chart $(\mathcal{U}_1, \phi_1)$ to be

$$L(t, x, y, v_x, v_y) = \tfrac{1}{2}(v_x^2 + v_y^2).$$

Suppose that we have a curve $t \mapsto (x(t), y(t))$ that is a minimizer for A_L for some interval and with some endpoints. Then the Euler–Lagrange equations say that this curve must satisfy

[2] The reader can deduce, after a moment's thought, that the matter of existence of solutions to the Euler–Lagrange equations is intimately related to the invertibility of the matrix with components

$$\frac{\partial^2 L}{\partial v^i \partial v^j}, \qquad i, j \in \{1, \dots, n\}.$$

These issues are discussed by Abraham and Marsden [1978], for example.

$$\frac{\mathrm{d}}{\mathrm{d}t}\Big(\frac{\partial L}{\partial v_x}\Big) - \frac{\partial L}{\partial x} = \frac{\mathrm{d}}{\mathrm{d}t} v_x(t) = \ddot{x}(t) = 0,$$
$$\frac{\mathrm{d}}{\mathrm{d}t}\Big(\frac{\partial L}{\partial v_y}\Big) - \frac{\partial L}{\partial y} = \frac{\mathrm{d}}{\mathrm{d}t} v_y(t) = \ddot{y}(t) = 0.$$

Thus the Euler–Lagrange equations are the simple second-order differential equations

$$\ddot{x}(t) = 0, \qquad \ddot{y}(t) = 0,$$

and these have solution $x(t) = x(0) + \dot{x}(0)t$ and $y(t) = y(0) + \dot{y}(0)t$. These describe straight lines in the plane $\mathsf{Q} = \mathbb{R}^2$.

Now let us look at the same example in the polar coordinate chart $(\mathcal{U}_2, \phi_2)$. First we need to determine the Lagrangian in these coordinates. Using the transformation rule (3.5), we have

$$v_x = \frac{\partial x}{\partial r} v_r + \frac{\partial x}{\partial \theta} v_\theta = \cos\theta v_r - r\sin\theta v_\theta,$$
$$v_y = \frac{\partial y}{\partial r} v_r + \frac{\partial y}{\partial \theta} v_\theta = \sin\theta v_r + r\cos\theta v_\theta.$$

With these relations we readily compute $L = \frac{1}{2}(v_r^2 + r^2 v_\theta^2)$. The Euler–Lagrange equations are then

$$\frac{\mathrm{d}}{\mathrm{d}t}\Big(\frac{\partial L}{\partial v_r}\Big) - \frac{\partial L}{\partial r} = \frac{\mathrm{d}}{\mathrm{d}t} v_r(t) - r(t)v_\theta^2(t) = \ddot{r}(t) - r(t)\dot{\theta}^2(t) = 0,$$
$$\frac{\mathrm{d}}{\mathrm{d}t}\Big(\frac{\partial L}{\partial v_\theta}\Big) - \frac{\partial L}{\partial \theta} = \frac{\mathrm{d}}{\mathrm{d}t}(r^2(t)v_\theta(t)) = r^2(t)\ddot{\theta}(t) + 2r(t)\dot{r}(t)\dot{\theta}(t) = 0.$$

That is, the Euler–Lagrange equations are the two coupled second-order differential equations

$$\ddot{r}(t) - r(t)\dot{\theta}^2(t) = 0, \qquad r^2(t)\ddot{\theta}(t) + 2r(t)\dot{r}(t)\dot{\theta}(t) = 0.$$

These equations are less amenable to solution by inspection than their Cartesian coordinate counterparts. However, since we know the Cartesian solutions and the change of coordinate formulae, we may in principal write the solutions to the polar coordinate equations. •

4.3.3 The Euler–Lagrange equations and changes of coordinate

In the previous example, we wrote the Euler–Lagrange equations in two different sets of coordinates, and it seems reasonable that the two sets of equations should be equivalent. Let us explore exactly what is meant by this. First of all, we remark that the *problem* of minimizing A_L over curves in $C^2([a, b], q_a, q_b)$ obviously does not depend upon any choice of coordinates. Now, if the problem *statement* is independent on coordinate chart, then so too should be any statements relating to its solution. Thus we should hope that the Euler–Lagrange equations are in some sense coordinate-invariant. Indeed they are, and in just the manner of the following result.

Proposition 4.40 (Euler–Lagrange equations and change of coordinate). *Let* Q *be a configuration manifold with* $(q^1,\dots,q^n)$ *and* $(\tilde{q}^1,\dots,\tilde{q}^n)$ *coordinates for* Q *in overlapping charts* $(\mathcal{U},\phi)$ *and* $(\tilde{\mathcal{U}},\tilde{\phi})$*. If* L *is a Lagrangian on* Q*, then*

$$\frac{\mathrm{d}}{\mathrm{d}t}\Big(\frac{\partial L}{\partial \tilde{v}^i}\Big) - \frac{\partial L}{\partial \tilde{q}^i} = \frac{\partial q^j}{\partial \tilde{q}^i}\Big(\frac{\mathrm{d}}{\mathrm{d}t}\Big(\frac{\partial L}{\partial v^j}\Big) - \frac{\partial L}{\partial q^j}\Big).$$

In particular, the Euler–Lagrange equations are satisfied in the coordinates $(q^1,\dots,q^n)$ *if and only if they are satisfied in the coordinates* $(\tilde{q}^1,\dots,\tilde{q}^n)$*.*

Proof. By (3.5) we have

$$\frac{\partial v^j}{\partial \tilde{v}^i} = \frac{\partial q^j}{\partial \tilde{q}^i}, \qquad \frac{\partial v^j}{\partial \tilde{q}^i} = \frac{\partial^2 q^j}{\partial \tilde{q}^i \partial \tilde{q}^k}\tilde{v}^k.$$

Therefore, using the Chain Rule, we have

$$\begin{aligned}\frac{\partial L}{\partial \tilde{v}^i} &= \frac{\partial L}{\partial v^j}\frac{\partial v^j}{\partial \tilde{v}^i} + \frac{\partial L}{\partial q^j}\frac{\partial q^j}{\partial \tilde{v}^i} = \frac{\partial L}{\partial v^j}\frac{\partial q^j}{\partial \tilde{q}^i},\\ \frac{\partial L}{\partial \tilde{q}^i} &= \frac{\partial L}{\partial v^j}\frac{\partial v^j}{\partial \tilde{q}^i} + \frac{\partial L}{\partial q^j}\frac{\partial q^j}{\partial \tilde{q}^i} = \frac{\partial L}{\partial v^j}\frac{\partial^2 q^j}{\partial \tilde{q}^i \partial \tilde{q}^k}\tilde{v}^k + \frac{\partial L}{\partial q^j}\frac{\partial q^j}{\partial \tilde{q}^i}.\end{aligned}$$

Now note that

$$\frac{\mathrm{d}}{\mathrm{d}t}\Big(\frac{\partial L}{\partial \tilde{v}^i}\Big) = \frac{\mathrm{d}}{\mathrm{d}t}\Big(\frac{\partial L}{\partial v^j}\Big)\frac{\partial q^j}{\partial \tilde{q}^i} + \frac{\partial L}{\partial v^j}\frac{\partial^2 q^j}{\partial \tilde{q}^i \partial \tilde{q}^k}\dot{\tilde{q}}^k.$$

Therefore, simple subtraction verifies the first part of the proposition. The second assertion follows from the first, since the matrix with components $\frac{\partial q^j}{\partial \tilde{q}^i}$, $i,j \in \{1,\dots,n\}$, is invertible, it being the Jacobian of the overlap map. ■

Remark 4.41. Note that the components of the Euler–Lagrange equations transform just like the components of a covector field on Q. However, the Euler–Lagrange equations are *not* a covector field, because their coefficients depend upon time, velocity, and acceleration. •

Let us see how this plays out in our running example.

Example 4.42 (Polar coordinates cont'd). We have $\mathsf{Q} = \mathbb{R}^2$ with its Cartesian and polar coordinate charts. In Example 4.39 we had derived the Euler–Lagrange equations for a particular Lagrangian in both sets of coordinates. Let us verify that these equations are indeed related as in Proposition 4.40.

In Cartesian coordinates the Euler Lagrange equations were

$$\ddot{x} = 0, \qquad \ddot{y} = 0,$$

and in polar coordinates they were

$$\ddot{r} - r\dot{\theta}^2 = 0, \qquad r^2\ddot{\theta} + 2r\dot{r}\dot{\theta} = 0.$$

We compute

$$\begin{aligned}\frac{\partial x}{\partial r}\left(\frac{\mathrm{d}}{\mathrm{d}t}\left(\frac{\partial L}{\partial v_x}\right)-\frac{\partial L}{\partial x}\right)+\frac{\partial y}{\partial r}\left(\frac{\mathrm{d}}{\mathrm{d}t}\left(\frac{\partial L}{\partial v_y}\right)-\frac{\partial L}{\partial y}\right) &= \cos\theta(\ddot{x}) + \sin\theta(\ddot{y})\\ &= \cos\theta(\ddot{r}\cos\theta - 2\dot{r}\dot{\theta}\sin\theta - r\dot{\theta}^2\cos\theta - r\ddot{\theta}\sin\theta)\\ &\quad + \sin\theta(\ddot{r}\sin\theta + 2\dot{r}\dot{\theta}\cos\theta - r\dot{\theta}^2\sin\theta + r\ddot{\theta}\cos\theta)\\ &= \ddot{r} - r\dot{\theta}^2.\end{aligned}$$

Similarly we compute

$$\begin{aligned}\frac{\partial x}{\partial \theta}\left(\frac{\mathrm{d}}{\mathrm{d}t}\left(\frac{\partial L}{\partial v_x}\right)-\frac{\partial L}{\partial x}\right)+\frac{\partial y}{\partial \theta}\left(\frac{\mathrm{d}}{\mathrm{d}t}\left(\frac{\partial L}{\partial v_y}\right)-\frac{\partial L}{\partial y}\right) &= -r\sin\theta(\ddot{x}) + r\cos\theta(\ddot{y})\\ &= -r\sin\theta(\ddot{r}\cos\theta - 2\dot{r}\dot{\theta}\sin\theta - r\dot{\theta}^2\cos\theta - r\ddot{\theta}\sin\theta)\\ &\quad + r\cos\theta(\ddot{r}\sin\theta + 2\dot{r}\dot{\theta}\cos\theta - r\dot{\theta}^2\sin\theta + r\ddot{\theta}\cos\theta)\\ &= r^2\ddot{\theta} + 2r\dot{r}\dot{\theta}.\end{aligned}$$

Thus the components do indeed transform as they ought to, although a direct verification of this, even in a simple example, is not an entirely pleasant task. •

4.3.4 The Euler–Lagrange equations on a Riemannian manifold

In this section we shall consider a special class of Lagrangians, this class being the basis for all of the Lagrangians we consider in this book. The starting point for the discussion is a Riemannian manifold $(\mathsf{Q}, \mathbb{G})$. This generalizes the situation considered in Section 4.2, in which Q was the configuration manifold for a mechanical system and $\mathbb{G}$ represented the kinetic energy of the system. For $(\mathsf{Q}, \mathbb{G})$, we define a time-independent Lagrangian $L_{\mathbb{G}}$ on Q by $L_{\mathbb{G}}(v_q) = \frac{1}{2}\mathbb{G}(v_q, v_q)$. Thus, for a mechanical system, this is the kinetic energy, and we will adopt the convention of calling this "kinetic energy," even though there may not be an associated physical system for which Q is the configuration manifold.

The following result describes the solutions to the Euler–Lagrange equations for the Lagrangian $L_{\mathbb{G}}$. The reader may wish to recall the notion of the Levi-Civita affine connection associated with a Riemannian metric in Section 3.8.2. This is the moment in which the first connection is made between mechanics and affine differential geometry. The following result is discussed in any text on Riemannian geometry, e.g., see [Do Carmo 1992].

Proposition 4.43 (Euler–Lagrange equations on a Riemannian manifold). *Let $(\mathsf{Q}, \mathbb{G})$ be a C^∞-Riemannian manifold with $L_{\mathbb{G}}$ the associated Lagrangian. The solutions of the Euler–Lagrange equations corresponding to $L_{\mathbb{G}}$ are exactly the geodesics of $\overset{\mathbb{G}}{\nabla}$.*

Proof. Suppose that $\gamma\colon [a,b] \to \mathsf{Q}$ minimizes $A_{L_{\mathbb{G}}}$, and let $\vartheta\colon J \times [a,b] \to \mathsf{Q}$ be a variation of γ with infinitesimal variation $\delta\vartheta$. Then we must have $\frac{\mathrm{d}}{\mathrm{d}s}\big|_{s=0} A_{L_{\mathbb{G}}}(\vartheta_s) = 0$.

We write $\vartheta_s(t) = \vartheta(s,t)$ and define $S_\vartheta, T_\vartheta\colon J \times [a,b] \to \mathsf{TQ}$ by

$$S_\vartheta(s,t) = \frac{\mathrm{d}}{\mathrm{d}s}\vartheta(s,t), \quad T_\vartheta(s,t) = \frac{\mathrm{d}}{\mathrm{d}t}\vartheta(s,t).$$

Thus $t \mapsto S_\vartheta(s,t)$ should be thought of as a vector field along the curve ϑ_s, and $t \mapsto T_\vartheta(s,t)$ should be thought of as the tangent vector field of the curve ϑ_s. The following lemma records a useful property of these two tangent vector fields along the curve ϑ_s.

Lemma. *Let X_{S_ϑ} and X_{T_ϑ} be vector fields having the property that $X_{S_\vartheta} \circ \vartheta = S_\vartheta$ and $X_{T_\vartheta} \circ \vartheta = T_\vartheta$. Then $[X_{S_\vartheta}, X_{T_\vartheta}](\vartheta(s,t)) = 0$ for $(s,t) \in J \times [a,b]$.*

Proof. Let $f \in C^\infty(\mathsf{Q})$ so that $f \circ \vartheta\colon J \times [a,b] \to \mathbb{R}$. The definition of the Lie bracket and a coordinate calculation give

$$\begin{aligned}\mathscr{L}_{[X_{S_\vartheta}, X_{T_\vartheta}]} f \circ \vartheta(s,t) &= \mathscr{L}_{X_{S_\vartheta}} \mathscr{L}_{X_{T_\vartheta}} f \circ \vartheta(s,t) - \mathscr{L}_{X_{T_\vartheta}} \mathscr{L}_{X_{S_\vartheta}} f \circ \vartheta(s,t) \\ &= \frac{\mathrm{d}^2(f\circ\vartheta)}{\mathrm{d}s\mathrm{d}t} - \frac{\mathrm{d}^2(f\circ\vartheta)}{\mathrm{d}t\mathrm{d}s}.\end{aligned}$$

Using the fact that mixed partial derivatives agree for C^2-functions, the result now follows. ▼

Note that, since $\overset{\mathbb{G}}{\nabla}$ satisfies Theorem 3.104(ii), the lemma implies that $\overset{\mathbb{G}}{\nabla}_{S_\vartheta} T_\vartheta = \overset{\mathbb{G}}{\nabla}_{T_\vartheta} S_\vartheta$. (This really means that $\overset{\mathbb{G}}{\nabla}_{S_\vartheta} T_\vartheta(\vartheta(s,t)) = \overset{\mathbb{G}}{\nabla}_{T_\vartheta} S_\vartheta(\vartheta(s,t))$ for all $(s,t) \in J \times [a,b]$, just as in the lemma.) We also note that, since $\overset{\mathbb{G}}{\nabla}$ is a metric connection, we have

$$\begin{aligned}\frac{\mathrm{d}}{\mathrm{d}t}\mathbb{G}(S_\vartheta, T_\vartheta)(s,t) &= \mathscr{L}_{T_\vartheta}\mathbb{G}(S_\vartheta, T_\vartheta)(s,t) \\ &= \mathbb{G}(\overset{\mathbb{G}}{\nabla}_{T_\vartheta} S_\vartheta, T_\vartheta)(s,t) + \mathbb{G}(S_\vartheta, \overset{\mathbb{G}}{\nabla}_{T_\vartheta} T_\vartheta)(s,t)\end{aligned} \tag{4.10}$$

(see also Exercise E3.38), and

$$\frac{\mathrm{d}}{\mathrm{d}s}\mathbb{G}(T_\vartheta, T_\vartheta)(s,t) = \mathscr{L}_{S_\vartheta}\mathbb{G}(T_\vartheta, T_\vartheta)(s,t) = 2\mathbb{G}(\overset{\mathbb{G}}{\nabla}_{S_\vartheta} T_\vartheta, T_\vartheta)(s,t). \tag{4.11}$$

With these equalities in hand, we compute

$$\begin{aligned}\frac{\mathrm{d}}{\mathrm{d}s}\Big|_{s=0} A_{L_{\mathbb{G}}}(\vartheta_s) &= \frac{\mathrm{d}}{\mathrm{d}s}\Big|_{s=0} \frac{1}{2}\int_a^b \mathbb{G}(\vartheta_s'(t), \vartheta_s'(t))\,\mathrm{d}t \\ &= \frac{\mathrm{d}}{\mathrm{d}s}\Big|_{s=0} \frac{1}{2}\int_a^b \mathbb{G}(T_\vartheta(s,t), T_\vartheta(s,t))\,\mathrm{d}t\end{aligned}$$

$$
\begin{aligned}
&= \int_a^b \mathbb{G}(\overset{\mathbb{G}}{\nabla}_{S_\vartheta(s,t)} T_\vartheta(s,t), T_\vartheta(s,t))\,\mathrm{d}t\Big|_{s=0} \\
&= \int_a^b \mathbb{G}(\overset{\mathbb{G}}{\nabla}_{T_\vartheta(s,t)} S_\vartheta(s,t), T_\vartheta(s,t))\,\mathrm{d}t\Big|_{s=0} \\
&= \int_a^b \mathbb{G}(\overset{\mathbb{G}}{\nabla}_{\gamma'(t)} \delta\vartheta(t), \gamma'(t))\,\mathrm{d}t \\
&= \int_a^b \Big(\frac{\mathrm{d}}{\mathrm{d}t}\mathbb{G}(\delta\vartheta(t), \gamma'(t)) - \mathbb{G}(\delta\vartheta(t), \overset{\mathbb{G}}{\nabla}_{\gamma'(t)}\gamma'(t))\Big)\,\mathrm{d}t \\
&= \mathbb{G}(\delta\vartheta(t), \gamma'(t))\Big|_{t=a}^{t=b} - \int_a^b \mathbb{G}(\delta\vartheta(t), \overset{\mathbb{G}}{\nabla}_{\gamma'(t)}\gamma'(t))\,\mathrm{d}t \\
&= -\int_a^b \mathbb{G}(\delta\vartheta(t), \overset{\mathbb{G}}{\nabla}_{\gamma'(t)}\gamma'(t))\,\mathrm{d}t.
\end{aligned}
$$

Here we have used (4.11) in the third step, the lemma in the fourth step, (4.10) in the sixth step, and the vanishing of $\delta\vartheta$ at the endpoints in the last step. The expression

$$
\int_a^b \mathbb{G}(\delta\vartheta(t), \overset{\mathbb{G}}{\nabla}_{\gamma'(t)}\gamma'(t))\,\mathrm{d}t
$$

vanishes for every variation ϑ if and only if $\overset{\mathbb{G}}{\nabla}_{\gamma'(t)}\gamma'(t) = 0$. ■

Let us see how the description of the Euler–Lagrange equations plays out for our running example.

Example 4.44 (Polar coordinates cont'd). We take as our Riemannian manifold $(\mathsf{Q} = \mathbb{R}^2, \mathbb{G} = \mathbb{G}_{\mathbb{R}^2})$, as considered in Example 3.87. In Cartesian coordinates, the resulting Lagrangian on TQ is given by

$$
L_{\mathbb{G}}(x, y, v_x, v_y) = \tfrac{1}{2}(v_x^2 + v_y^2).
$$

The Euler–Lagrange equations were computed in Example 4.39:

$$
\ddot{x} = 0 \quad \ddot{y} = 0. \tag{4.12}
$$

In Example 3.105 we determined that the Christoffel symbols for $\overset{\mathbb{G}}{\nabla}$ in Cartesian coordinates are zero, giving the geodesic equations as

$$
\ddot{x} = 0, \qquad \ddot{y} = 0.
$$

These equations are obviously the same as the Euler–Lagrange equations (4.12).

Now let us check that the computations also work out in polar coordinates. In these coordinates we have

$$
L_{\mathbb{G}}(r, \theta, v_r, v_\theta) = \tfrac{1}{2}(v_r^2 + r^2 v_\theta^2).
$$

The Euler–Lagrange equations are as in Example 4.39:

$$\ddot{r} - r\dot{\theta}^2 = 0, \quad r^2\ddot{\theta} + 2r\dot{r}\dot{\theta} = 0. \tag{4.13}$$

In Example 3.105 we provided the Christoffel symbols in polar coordinates, and determined that the geodesic equations are

$$\ddot{r} - r\dot{\theta}^2 = 0, \quad \ddot{\theta} + \frac{2}{r}\dot{r}\dot{\theta} = 0.$$

These equations are clearly equivalent to, although not identical in form to, the equations (4.13). •

It is also straightforward to use our preceding discussion to provide the unforced equations of motion for the two unconstrained physical systems we have been looking at in this chapter.

Example 4.45 (Planar rigid body cont'd). For the planar body with $\mathsf{Q} = \mathbb{S}^1 \times \mathbb{R}^2$ and with coordinates (θ, x, y), the Riemannian metric has matrix representation

$$[\mathbb{G}] = \begin{bmatrix} J & 0 & 0 \\ 0 & m & 0 \\ 0 & 0 & m \end{bmatrix}.$$

Since the components of the Riemannian metric are constant, it follows that the Christoffel symbols for the corresponding Levi-Civita connection are zero. The equations of motion in the absence of external forces are then just the geodesic equations, which in this case are simply

$$\ddot{\theta} = 0, \quad \ddot{x} = 0, \quad \ddot{y} = 0. \tag{4.14}$$

The solutions of these equations correspond to a uniform linear motion of the center of mass, along with a constant speed rotation of the body about its center of mass. •

Example 4.46 (Two-link manipulator cont'd). The system we consider here is the two-link planar manipulator whose configuration manifold is $\mathsf{Q} = \mathbb{S}^1 \times \mathbb{S}^1$. In the natural coordinates (θ_1, θ_2) we had determined the Riemannian metric associated with the system's kinetic energy to have matrix representation

$$[\mathbb{G}] = \begin{bmatrix} J_1 + \frac{1}{4}(m_1 + 4m_2)\ell_1^2 & \frac{1}{2}m_2\ell_1\ell_2\cos(\theta_1 - \theta_2) \\ \frac{1}{2}m_2\ell_1\ell_2\cos(\theta_1 - \theta_2) & J_2 + \frac{1}{4}m_2\ell_2^2 \end{bmatrix}.$$

The nonzero Christoffel symbols may be computed to be

$$\overset{\mathbb{G}}{\Gamma}{}^1_{11} = \frac{m_2^2\ell_1^2\ell_2^2\sin(2(\theta_1 - \theta_2))}{8\det[\mathbb{G}]}, \quad \overset{\mathbb{G}}{\Gamma}{}^1_{22} = \frac{m_2\ell_1\ell_2(4J_2 + m_2\ell_2^2)\sin(\theta_1 - \theta_2)}{8\det[\mathbb{G}]},$$

$$\overset{\mathbb{G}}{\Gamma}{}^2_{11} = -\frac{m_2\ell_1\ell_2(4J_1 + (m_1 + 4m_2)\ell_1^2)\sin(\theta_1 - \theta_2)}{8\det[\mathbb{G}]},$$

$$\overset{\mathbb{G}}{\Gamma}{}^2_{22} = -\frac{m_2^1\ell_1^2\ell_2^2\sin(2(\theta_1 - \theta_2))}{8\det[\mathbb{G}]}.$$

With these Christoffel symbols, one can write the equations of motion in the absence of external forces:

$$\ddot{\theta}_1 + \frac{m_2^2\ell_1^2\ell_2^2\sin(2(\theta_1-\theta_2))}{8\det[\mathbb{G}]}\dot{\theta}_1^2 + \frac{m_2\ell_1\ell_2(4J_2+m_2\ell_2^2)\sin(\theta_1-\theta_2)}{8\det[\mathbb{G}]}\dot{\theta}_2^2 = 0,$$

$$\ddot{\theta}_2 - \frac{m_2\ell_1\ell_2(4J_1+(m_1+4m_2)\ell_1^2)\sin(\theta_1-\theta_2)}{8\det[\mathbb{G}]}\dot{\theta}_1^2 - \frac{m_2^2\ell_1^2\ell_2^2\sin(2(\theta_1-\theta_2))}{8\det[\mathbb{G}]}\dot{\theta}_2^2 = 0.$$

Not much can be said about the behavior of the solution of these equations. Indeed, it is generally not possible to say much about the dynamics of many of the systems we consider. However, the reader will see as we progress that much of the analysis and design does not actually rely on the possession of the equations of motion, never mind their explicit solution. Of course, if one wishes to perform simulations, then these equations must be constructed at some point. In any case, for this system we shall generally consider the presence of gravitational effects, in which case the above equations are as yet incomplete. •

At this point, we are not ready to provide the equations of motion for the rolling disk, the example begun in Example 4.5, since this system is one with constraints, and we have not yet seen how these appear in the equations of motion.

4.3.5 Physical interpretations

In this section we provide three interpretations of the quantity $\overset{\mathbb{G}}{\nabla}_{\gamma'(t)}\gamma'(t)$ that may be helpful.

Equivalence of Euler–Lagrange and Newton–Euler equations

For us, the most important interpretation of the geodesic equations $\overset{\mathbb{G}}{\nabla}_{\gamma'(t)}\gamma'(t) = 0$ is that they provide the equations of motion for an interconnected mechanical system subject to no external forces. To make this precise, we must say something about the nature of the equations of motion of a rigid body subject to forces and torques. Particle motion, and the translational motion of a rigid body, are governed by Newton's three Laws of Motion. These are as follows, quoting the 1729 translation of Newton's original Latin [Newton 1687].

First Law: Every body continues in its state of rest, or of uniform motion in a right line, unless it is compelled to change that state by forces impressed upon it.

Second Law: The change of motion is proportional to the motive force impressed; and is made in the direction of the right line in which that force is impressed.

Third Law: To every action there is always opposed an equal reaction: or, the mutual action of two bodies upon each other are always equal, and directed to contrary parts.

For rotational motion, one must additionally use Euler's Law of Conservation of Angular Momentum, which will be stated below.

We let $(\mathcal{B}, \mu)$ be a rigid body, with Σ_{spatial} a spatial frame and Σ_{body} a body frame located at the body's center of mass. Let us explain the meaning of the words "force $\boldsymbol{f} \in \mathbb{R}^3$ applied at a point $\boldsymbol{r}_0$ in the body," and "torque $\boldsymbol{\tau} \in \mathbb{R}^3$." The meaning of the force is the same as that in elementary mechanics. In particular, the components of $\boldsymbol{f}$ are given with respect to the spatial frame. The representation of torque as a vector is not as commonplace in elementary mechanics courses, so let us explain how to interpret this. As with forces, the components of torques are written with respect to the spatial frame. A torque $\boldsymbol{\tau} \in \mathbb{R}^3$ is applied to the body about the $\boldsymbol{\tau}$-axis, and has magnitude $\|\boldsymbol{\tau}\|_{\mathbb{R}^3}$. In this way, a torque has "direction." Now let $\boldsymbol{r}_0$ be the vector from O_{body} to the point in the body at which the force is applied, measured with respect to the spatial frame. A force $\boldsymbol{f}$ applied at the point $\boldsymbol{r}_0$ may be shown to produce a force $\boldsymbol{f}$ applied at the center of mass, along with a torque $\boldsymbol{r}_0 \times \boldsymbol{f}$. For this reason, we shall always assume, without loss of generality, that forces are applied at the center of mass.

Next we assume that the body is undergoing a motion, so defining a curve $t \mapsto (\boldsymbol{R}(t), \boldsymbol{r}(t)) \in \mathsf{SO}(3) \times \mathbb{R}^3$. When discussing kinetic energy, it was convenient to use body angular velocity. Here, it is convenient to use spatial angular velocity, which we recall from Section 4.1.5 is defined by $\widehat{\boldsymbol{\omega}}(t) = \dot{\boldsymbol{R}}(t)\boldsymbol{R}^T(t)$. We suppose that, at time t, the body is subject to a force $\boldsymbol{f}(t) \in \mathbb{R}^3$ applied at the center of mass, and to a torque $\boldsymbol{\tau}(t) \in \mathbb{R}^3$. The force $\boldsymbol{f}$ and torque $\boldsymbol{\tau}$ are written relative to a chosen spatial frame $\Sigma_{\text{spatial}} = (O_{\text{spatial}}, \{\boldsymbol{s}_1, \boldsymbol{s}_2, \boldsymbol{s}_3\})$.

With the above as setup, let us now turn to the equations of motion of a rigid body subject to a force and a torque. At various points in the text, we have made reference to the Newton–Euler equations governing the behavior of a mechanical system. Let us reproduce these for completeness. We let $m = \mu(\mathcal{B})$ be the mass of the body and $\mathbb{I}_c$ be the inertia tensor of the body about its center of mass. The ***spatial linear momentum*** for the motion is given by $\boldsymbol{p}(t) = m\boldsymbol{v}(t)$, where $\boldsymbol{v}(t) = \dot{\boldsymbol{r}}(t)$, and the ***spatial angular momentum*** for the motion is $\boldsymbol{\mu}(t) = \boldsymbol{R}(t)\mathbb{I}_c\boldsymbol{R}^T(t)\boldsymbol{\omega}(t)$. Newton's law of rate of change of linear momentum, and Euler's law of rate of change of angular momentum then read

$$\frac{\mathrm{d}}{\mathrm{d}t}\boldsymbol{p}(t) = \boldsymbol{f}(t), \quad \frac{\mathrm{d}}{\mathrm{d}t}\boldsymbol{\mu}(t) = \boldsymbol{\tau}(t).$$

It is these equations to which we refer when we say ***Newton–Euler equations***. In standard formulations of mechanics, these equations are manipulated to yield equations in terms of body and spatial angular velocities (e.g., [Murray, Li, and Sastry 1994] and cf. Example 5.48). Note that the second equation, in the absence of the torque $\boldsymbol{\tau}$, is Euler's law of conservation of momentum.

Let us now sketch how one goes from Newton–Euler equations for the separate rigid bodies $(\mathcal{B}_1, \mu_1), \dots, (\mathcal{B}_k, \mu_k)$ in an interconnected mechanical system to the Euler–Lagrange equations. The interconnections between the bodies in an interconnected mechanical system give rise to a force $\boldsymbol{g}_a$ and a torque $\boldsymbol{\sigma}_a$ on the ath body, $a \in \{1, \dots, k\}$. These forces and torques are subject to certain conditions. To state these conditions, we recall the notation $\lambda_{\text{spatial}} \colon \mathsf{T}(\mathsf{SO}(3) \times \mathbb{R}^3) \to \mathbb{R}^3$ and $\rho_{\text{spatial}} \colon \mathsf{T}(\mathsf{SO}(3) \times \mathbb{R}^3) \to \mathbb{R}^3$ from Section 4.1.5. Using this notation, the conditions on the internal interconnection forces are:

1. if the system is in configuration q, then the force $\boldsymbol{g}_a$ and the torque $\boldsymbol{\sigma}_a$ applied to the system in this configuration must satisfy
$$\langle\!\langle \boldsymbol{g}_a, \lambda_{\text{spatial}} \circ T\Pi_a(v_q) \rangle\!\rangle_{\mathbb{R}^3} + \langle\!\langle \boldsymbol{\sigma}_a, \rho_{\text{spatial}} \circ T\Pi_a(v_q) \rangle\!\rangle_{\mathbb{R}^3} = 0$$
for all $v_q \in \mathsf{T}_q\mathsf{Q}$;
2. the interconnection force and torque exerted by body a on body b is equal in magnitude, and opposite in direction, to the force and torque exerted by body b on body a.

The first condition says that the internal interconnection forces and torques are $\mathbb{G}_{\mathbb{R}^3}$-orthogonal to the submanifold of admissible directions allowed by motion in $\mathsf{Q} \subset \mathsf{Q}_{\text{free}}$. This will be made explicit in a more general setting in Section 4.5.1. The second condition above is Newton's Third Law of motion. In interpreting this condition, one should be careful that the force exerted by body a on body b may not be applied directly at the center of mass of body b. Thus Newton's Third Law may not apply to the forces when they are redefined to be applied at the center of mass.

In elementary Newtonian mechanics, the forces $\boldsymbol{g}_a$ and torques $\boldsymbol{\sigma}_a$, $a \in \{1, \dots, k\}$, are unknowns, and are determined as part of the solution to the problem. In the Euler–Lagrange framework, these internal forces are ignored, and the equations of motion are determined for all bodies at once, using the correspondence between configurations of the system and the configuration manifold Q. Indeed, one has the following important theorem, whose proof we omit, and refer the reader to the texts on mechanics mentioned in the introduction to the chapter.

Theorem 4.47 (Equivalence of Newton–Euler and Euler–Lagrange equations I). *Let $(\mathcal{B}_1, \mu_1), \dots, (\mathcal{B}_k, \mu_k)$ be an interconnected mechanical system with configuration manifold Q, and let $\mathbb{G}$ be the kinetic energy Riemannian metric constructed as in Section 4.2.3. For a curve γ on Q, the following statements are equivalent:*

(i) the curve $\Pi_a \circ \gamma$ on $\mathsf{SO}(3) \times \mathbb{R}^3$ describes a motion of the ath rigid body, $a \in \{1, \dots, k\}$, for the system according to the Newton–Euler equations, and with the forces and torques of interconnection as described above;

(ii) γ is a geodesic of $\overset{\mathbb{G}}{\nabla}$.

At subsequent points in the text, we will state extended versions of this theorem, valid in the presence of external forces and nonholonomic constraints (see Theorems 4.58 and 4.79).

Geometric acceleration

Our next interpretation of the quantity $\overset{\mathbb{G}}{\nabla}_{\gamma'(t)}\gamma'(t)$ is not mechanical, but is nonetheless interesting in its own right. Given a smooth curve $\gamma\colon I \to \mathsf{Q}$, we think of $\gamma'\colon I \to \mathsf{TQ}$ as being the velocity of γ. One can then ask how one might define the acceleration of γ. One might be tempted to adopt the definition of acceleration as that object which, when represented in coordinates $(q^1, \dots, q^n)$, is given by $(\ddot{q}^1(t), \dots, \ddot{q}^n(t))$. This *is* a valid definition, but, unfortunately (or interestingly), this quantity is not a tangent vector field along γ, as γ' is (see Exercise E4.14). The appropriate setting for talking about acceleration in this way is that of jet bundles [Saunders 1989]. If we require that the acceleration defined by γ be a vector field along γ, then we claim that, for an affine connection ∇ on Q, the quantity $t \mapsto \nabla_{\gamma'(t)}\gamma'(t)$ does the job. We call this the ***geometric acceleration*** with respect to ∇.

We justify this assertion with an example.

Example 4.48 (Polar coordinates cont'd). The $\mathsf{Q} = \mathbb{R}^2$ example we have been using, although simple, is sufficient to illustrate why one cannot regard acceleration as a tangent vector. On Q we consider a curve $\gamma\colon \mathbb{R} \to \mathsf{Q}$ defined in Cartesian coordinates by $\gamma(t) = (\cos t, \sin t)$. Thus $\text{image}(\gamma) = \mathbb{S}^1$. In Cartesian coordinates we then have $(x(t), y(t)) = (\cos t, \sin t)$, so that

$$(\dot{x}(t), \dot{y}(t)) = (-\sin t, \cos t), \quad (\ddot{x}(t), \ddot{y}(t)) = (-\cos t, -\sin t).$$

In polar coordinates the curve is represented by $(r(t), \theta(t)) = (1, t)$, but note that $r(t)$ and $\theta(t)$ are only defined for $t \in]-\pi, \pi[$. We readily compute

$$(\dot{r}(t), \dot{\theta}(t)) = (0, 1), \quad (\ddot{r}(t), \ddot{\theta}(t)) = (0, 0).$$

Therefore, the naïve acceleration along the curve is nonzero in Cartesian coordinates, but zero in polar coordinates. This precludes acceleration from being a tangent vector, since it cannot be both zero and nonzero.

Now let us look at the geometric acceleration represented by the Levi-Civita connection corresponding to the Riemannian metric $\mathbb{G}$ from Example 3.87. In Cartesian coordinates, the Christoffel symbols are zero so that we have

$$\overset{\mathbb{G}}{\nabla}_{\gamma'(t)}\gamma'(t) = \ddot{x}\frac{\partial}{\partial x} + \ddot{y}\frac{\partial}{\partial y} = -\cos t\frac{\partial}{\partial x} - \sin t\frac{\partial}{\partial y}.$$

In polar coordinates, the Christoffel symbols were computed in Example 3.87, and using these we have

$$\overset{\mathbb{G}}{\nabla}_{\gamma'(t)}\gamma'(t) = (\ddot{r} - r\dot{\theta}^2)\frac{\partial}{\partial r} + (\ddot{\theta} + \tfrac{2}{r}\dot{r}\dot{\theta})\frac{\partial}{\partial \theta} = -\frac{\partial}{\partial r}.$$

Thus, in both Cartesian and polar coordinates, the quantity $\overset{\mathbb{G}}{\nabla}_{\gamma'(t)}\gamma'(t)$ is a tangent vector with base point $\gamma(t)$, of unit length with respect to the standard Riemannian metric on $\mathbb{R}^2$, and pointing toward the origin (see Figure 4.11). Thus, at least in this example, the quantity $\nabla_{\gamma'(t)}\gamma'(t)$ corresponds to the

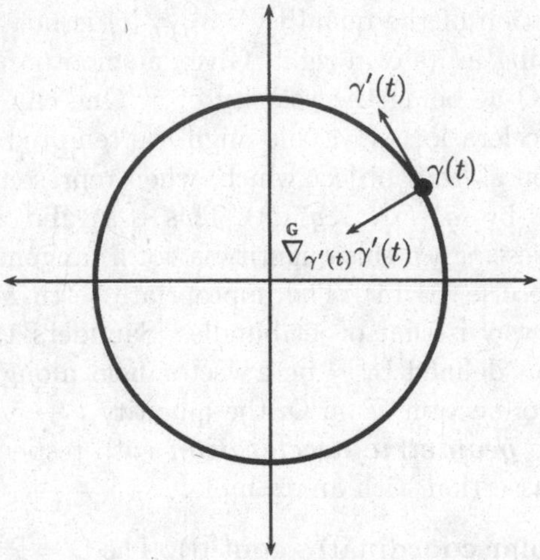

Figure 4.11. Velocity and geometric acceleration along a curve

quantity normally called acceleration. Indeed, in vector mechanics in $\mathbb{R}^3$, the reason one can get away with thinking of acceleration as a vector is that one usually works solely in Cartesian coordinates. Any attempt to use other coordinates will reveal the hidden role of the Levi-Civita affine connection corresponding to the standard Riemannian metric on $\mathbb{R}^3$. •

Remark 4.49. One often sees the Euler–Lagrange equations for the Lagrangian $L_{\mathbb{G}}(v_q)$ written in coordinates as

$$\boldsymbol{M}(\boldsymbol{q})\ddot{\boldsymbol{q}} + \boldsymbol{C}(\boldsymbol{q}, \dot{\boldsymbol{q}}) = \boldsymbol{0},$$

or in a form similar to this. Here $\boldsymbol{M}(\boldsymbol{q})$ is the matrix representation of $\mathbb{G}$ in the chosen coordinates, and the expression $\boldsymbol{C}(\boldsymbol{q}, \dot{\boldsymbol{q}})$ represents the contribution of the terms which we write in coordinates as $\overset{\mathbb{G}}{\Gamma}{}^k_{ij}\dot{q}^i\dot{q}^j$. This quantity is often referred to as "Coriolis forces," and sometimes there is a "Coriolis tensor" associated with the expression $\boldsymbol{C}(\boldsymbol{q}, \dot{\boldsymbol{q}})$. This language, however, is manifestly not coordinate-invariant. Indeed, neither of the quantities $\boldsymbol{M}(\boldsymbol{q})\ddot{\boldsymbol{q}}$ nor $\boldsymbol{C}(\boldsymbol{q}, \dot{\boldsymbol{q}})$ has coordinate-invariant meaning. Only together do they have meaning, as in this case they are the coordinate representation for $\mathbb{G}^\flat(\overset{\mathbb{G}}{\nabla}_{\gamma'(t)}\gamma'(t))$. •

Geodesics as minimizers of length

By our above interpretation, geodesics are zero acceleration curves. Furthermore, by their definition, they also satisfy the necessary conditions for minimization of the integral of the kinetic energy in the set of curves joining any two fixed points. Let us now provide yet another interpretation of geodesics. For a Riemannian manifold $(\mathsf{Q}, \mathbb{G})$ and a piecewise C^1-curve $\gamma\colon [a,b] \to \mathsf{Q}$,[3] define the ***Riemannian length*** of γ to be

$$\ell_{\mathbb{G}}(\gamma) = \int_a^b \sqrt{\mathbb{G}(\gamma'(t), \gamma'(t))}\, \mathrm{d}t.$$

One may show that this notion of length is independent of reparameterization, and so is actually describing $\mathrm{image}(\gamma)$ rather that the parameterization of $\mathrm{image}(\gamma)$ given by a particular γ. With this in mind, the following definition of the ***Riemannian distance*** between $q_1, q_2 \in \mathsf{Q}$ makes sense:

$$d_{\mathbb{G}}(q_1, q_2) = \inf\{\ell_{\mathbb{G}}(\gamma) \mid \gamma\colon [0,1] \to \mathsf{Q} \text{ is piecewise } C^1, \\ \text{and } \gamma(0) = q_1 \text{ and } \gamma(1) = q_2\}.$$

If there are no piecewise C^1-curves connecting q_1 and q_2 (which may be the case if Q is not connected), then take $d_{\mathbb{G}}(q_1, q_2) = +\infty$. It is possible to show that length minimizing curves are geodesics. For a proof of the following result, we refer to [Do Carmo 1992].

Theorem 4.50 (Minimizers of distance are geodesics). *Let* $(\mathsf{Q}, \mathbb{G})$ *be a* C^∞*-Riemannian manifold. If* $\gamma\colon [a,b] \to \mathsf{Q}$ *is a piecewise* C^1*-curve with the property that* $\ell_{\mathbb{G}}(\gamma) = d_{\mathbb{G}}(\gamma(a), \gamma(b))$*, then* γ *is a geodesic for* $\overset{\mathbb{G}}{\nabla}$.

4.4 Forces

If one wishes to talk about control theory for mechanical systems, as we do, then the natural "inputs" for these systems are forces. Furthermore, forces often come up in other ways, and we need to understand how these arise in our geometric setting. In a Newtonian point of view, the manner in which forces are treated has a natural physical basis, so there is no cause for confusion about what a force "is." In Lagrangian mechanics, since the origin of the theory is the variational principle discussed in Section 4.3.1, it is less clear what a force should be. We take recourse in the following motivational remarks.

We may think of the Euler–Lagrange equations for a Lagrangian L as representing the "natural" motion of the system, i.e., the motion of the system in the absence of external interaction. A force, on the other hand *is* an

[3] A continuous curve $\gamma\colon [a,b] \to \mathsf{Q}$ is ***piecewise*** C^1 if $[a,b]$ is a finite disjoint union of intervals, on each of which γ is C^1.

external interaction. If we believe that the effects of forces on a system are linear—meaning that the effect of two forces is the sum of their individual effects, and, if one scales a force, its effect on the system will be similarly scaled—this suggests that a force should appear linearly in the Euler–Lagrange equations. The question now is, "What is the proper geometric representation of a force?" To answer this, we consider Proposition 4.40, which tells us that the components of the Euler–Lagrange equations transform when changing coordinates like the components of a covector field. Correspondingly, we shall see that forces are indeed cotangent bundle-valued in our geometric setting.

Before we get to this, we begin with a general discussion of how forces appear in the Euler–Lagrange equations, starting with consideration of their Newtonian form.

4.4.1 From rigid body forces and torques to Lagrangian forces

We ask the reader to recall from Section 4.3.5 our discussion of forces, torques, and the Newton–Euler equations. In that section, the punchline was that the internal interconnection forces do not appear in the Euler–Lagrange setup. Now we wish to consider the effects not only of the internal interconnection forces, but also of external forces applied to the system (e.g., friction forces, or control forces). These forces *will* have to be accounted for in the Euler–Lagrange setup. Here we indicate how Newtonian forces and torques can be used to produce an object defined in terms of the configuration manifold of an interconnected mechanical system. We suppose our system to be comprised of rigid bodies $(\mathcal{B}_1, \mu_1), \dots, (\mathcal{B}_k, \mu_k)$. As usual, we fix a spatial frame $\Sigma_{\text{spatial}} = (O_{\text{spatial}}, \{\boldsymbol{s}_1, \boldsymbol{s}_2, \boldsymbol{s}_3\})$, and body frames for each body with origins at the bodies' centers of mass. We fix a body $(\mathcal{B}_a, \mu_a)$, $a \in \{1, \dots, k\}$, and suppose that it is subject to forces $\boldsymbol{f}_a$ and $\boldsymbol{\tau}_a$, which we call the Newtonian force and torque, respectively. We allow these forces to depend on the position $(\boldsymbol{R}_a, \boldsymbol{r}_a) \in \mathsf{SO}(3) \times \mathbb{R}^3$, on the velocities $(\dot{\boldsymbol{R}}_a, \dot{\boldsymbol{r}}_a) \in \mathsf{T}_{(\boldsymbol{R}_a, \boldsymbol{r}_a)}(\mathsf{SO}(3) \times \mathbb{R}^3)$, and on time. The force $\boldsymbol{f}_a$ and the torque $\boldsymbol{\tau}_a$ model only the applied external forces, since, essentially by virtue of Theorem 4.47, the internal interconnection forces will not play a role in the Lagrangian framework.

Our objective is to define, for each $v_q \in \mathsf{TQ}$, an element $F_{\boldsymbol{f}_a, \boldsymbol{\tau}_a}(t, v_q) \in \mathsf{T}_q^*\mathsf{Q}$ that models the effects of $\boldsymbol{f}_a$ and $\boldsymbol{\tau}_a$ in the Lagrangian framework. As in Section 4.1.4, we let $\Pi_a \colon \mathsf{Q} \to \mathsf{SO}(3) \times \mathbb{R}^3$ denote the forward kinematic map that assigns the position of body a to a point in Q. Our prescription for $F_{\boldsymbol{f}_a, \boldsymbol{\tau}_a}(t, v_q)$ is then the requirement that

$$\langle F_{\boldsymbol{f}_a, \boldsymbol{\tau}_a}(t, v_q); w_q \rangle = \langle\!\langle \boldsymbol{f}_a(t, T\Pi_a(v_q)), \lambda_{\text{spatial}} \circ T\Pi_a(w_q) \rangle\!\rangle_{\mathbb{R}^3} + \langle\!\langle \boldsymbol{\tau}_a(t, T\Pi_a(v_q)), \rho_{\text{spatial}} \circ T\Pi_a(w_q) \rangle\!\rangle_{\mathbb{R}^3} \quad (4.15)$$

hold for all $w_q \in \mathsf{T}_q\mathsf{Q}$, where λ_{spatial} and ρ_{spatial} are as defined in Section 4.1.5. We call $F_{\boldsymbol{f}_a, \boldsymbol{\tau}_a}$ the ***Lagrangian force*** associated with the Newtonian force and torque $\boldsymbol{f}_a$ and $\boldsymbol{\tau}_a$. By carrying out this procedure for all forces and torques applied to all bodies, we arrive at the ***total external force*** F, defined as a map from $\mathbb{R} \times \mathsf{TQ}$ to $\mathsf{T}^*\mathsf{Q}$ by

$$F(t, v_q) = \sum_{a=1}^{k} F_{\boldsymbol{f}_a, \boldsymbol{\tau}_a}(t, v_q).$$

The notation associated with the preceding general construction masks the fact that it is, in principle, simple to determine coordinate expressions for the total external force for an interconnected mechanical system. The steps can be summarized as follows.

Computing the total external force. For each body $(\mathcal{B}_a, \mu_a)$, and for each force and torque $\boldsymbol{f}_a$ and $\boldsymbol{\tau}_a$ applied to the body, do the following.

(i) In a set of coordinates for Q, write the map Π_a, so giving $(\boldsymbol{R}_a, \boldsymbol{r}_a) \in \mathsf{SO}(3) \times \mathbb{R}^3$ as functions of these coordinates.

(ii) For $v_q \in \mathsf{TQ}$, write $(\dot{\boldsymbol{R}}_a, \dot{\boldsymbol{r}}_a) = T\Pi_a(v_q)$ in terms of natural coordinates for TQ.

(iii) Compute the spatial translational velocity $\dot{\boldsymbol{r}}_a$ and the spatial angular velocity $\boldsymbol{\omega}_a = \dot{\boldsymbol{R}}_a \boldsymbol{R}_a^T$. Note that our above calculations express these as functions of the natural coordinates on TQ.

(iv) Using (4.15), define $F_{\boldsymbol{f}_a, \boldsymbol{\tau}_a}(t, v_q) \in \mathsf{T}^*_q\mathsf{Q}$. •

Next we abstract the above constructions to arbitrary configuration manifolds, not necessarily defined as above for an interconnected mechanical system.

4.4.2 Definitions and examples of forces in Lagrangian mechanics

With the above as backdrop, we may now formally define what we mean by a force. The reader may wish to recall the notion of a section along a curve as given in Section 3.4.1.

Definition 4.51 (Force). For $r \in \mathbb{N} \cup \{\infty\} \cup \{\omega\}$, a ***$C^r$-force*** on Q is a map $F\colon \mathbb{R} \times \mathsf{TQ} \to \mathsf{T}^*\mathsf{Q}$ with the property that F is a locally integrally class C^r bundle map over id_{Q} (see Definition 3.125 and Section A.2.1). A C^r-force F is

(i) ***time-independent*** if there exists a C^r-fiber bundle map $F_0\colon \mathsf{TQ} \to \mathsf{T}^*\mathsf{Q}$ over id_{Q} with the property that $F(t, v_q) = F_0(v_q)$, and is

(ii) ***basic*** if there exists a C^r-covector field F_0 on Q such that $F(t, v_q) = F_0(q)$.

If $\gamma\colon I \to \mathsf{Q}$ is a C^r-curve, then a ***C^r-force along*** γ is a C^r-covector field $F\colon I \to \mathsf{T}^*\mathsf{Q}$ along γ. •

Remarks 4.52. 1. The rather general assumptions on the time-dependence of a force are mandated by certain parts of the control theoretic analysis in the second part of the book. Nothing essential will be lost conceptually by regarding a C^r-force as being a map $F\colon I \times \mathsf{TQ} \to \mathsf{T}^*\mathsf{Q}$ that depends (at least) continuously on time, and for which $F(t, v_q) \in \mathsf{T}^*_q\mathsf{Q}$ for each $(t, v_q) \in \mathbb{R} \times \mathsf{TQ}$.

2. If $F\colon \mathbb{R} \times \mathsf{TQ} \to \mathsf{T}^*\mathsf{Q}$ is a force, then this defines a force along a curve γ by $t \mapsto F(t, \gamma'(t))$. In this case we make a convenient abuse of notation by denoting by F both the force and the force along γ. Note that, if F is a locally integrally class C^r bundle map and γ is of class C^r, then their composition will only be a locally integrable covector field.
3. We generally allow a force to depend on time, position, and velocity, but not on higher-derivatives of position with respect to time. In practice, this is not a restriction, and it ensures that the resulting equations of motion will be well-posed, as we shall see below.
4. For problems in control theory, it is often useful to allow the dependence of a force on configuration and velocity to be discontinuous. While this is not allowed by the above formal definition, the extension to the more general situation is clear.
5. If $(q^1, \dots, q^n)$ are coordinates in a chart $(\mathcal{U}, \phi)$, then we write a force F as $F = F_i \mathrm{d}q^i$, for some functions $F_i\colon \mathbb{R} \times \mathsf{T}\mathcal{U} \to \mathbb{R}$, $i \in \{1, \dots, n\}$, called the ***components*** of the force F. •

Let us consider some examples of forces for the three examples introduced in Section 4.1. In each example, we apply the procedure outlined at the end of Section 4.4.1.

Example 4.53 (Planar rigid body cont'd). For this example, let us consider two possible forces, both illustrated in Figure 4.12. The first is a force

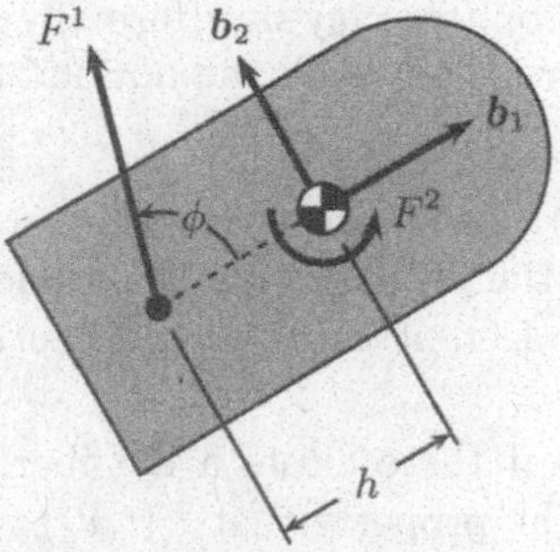

Figure 4.12. Forces on the planar body

F^1 that we apply to the body at a point that is distance $h > 0$ from the center of mass along the body $\boldsymbol{b}_1$-axis, and pointing in a direction forming an angle ϕ as shown in Figure 4.12. The second force is a torque F^2 applied about the center of mass. In the Newtonian framework, the first force gives the force $\boldsymbol{f}_1 = F(\cos(\theta + \phi), \sin(\theta + \phi), 0)$ applied to the center of mass, as well as a torque $\boldsymbol{\tau}_1 = F(0, 0, -h\sin\phi)$. Here F is a function of position and orientation of the body, velocity of the body, and time. The second force gives just the torque $\boldsymbol{\tau}_2 = \tau(0, 0, 1)$ in the Newtonian framework, where τ is again a function of position and orientation of the body, velocity of the body, and time. Next let w_q be a tangent vector in TQ with components (w_θ, w_x, w_y). To apply (4.15), we need to find the spatial linear and angular velocities corresponding to this tangent vector. These were computed in Example 4.19 to be $\boldsymbol{v} \triangleq \lambda_{\text{spatial}} \circ T\Pi_1(w_q) = (w_x, w_y, 0)$ and $\boldsymbol{\omega} \triangleq \rho_{\text{spatial}} \circ T\Pi_1(w_q) = (0, 0, w_\theta)$, respectively. We then have

$$\begin{aligned}\langle\langle \boldsymbol{f}_1, \boldsymbol{v}\rangle\rangle_{\mathbb{R}^3} &+ \langle\langle \boldsymbol{\tau}_1 + \boldsymbol{\tau}_2, \boldsymbol{\omega}\rangle\rangle_{\mathbb{R}^3} \\ &= F\cos(\theta + \phi)w_x + F\sin(\theta + \phi)w_y - Fh\sin\phi w_\theta + \tau w_\theta.\end{aligned}$$

We note that this expression is exactly $F^1(w_q) + F^2(w_q)$, where

$$F^1 = F\big(\cos(\theta + \phi)\mathrm{d}x + \sin(\theta + \phi)\mathrm{d}y - h\sin\phi\mathrm{d}\theta\big), \quad F^2 = \tau\mathrm{d}\theta.$$

The functions F and τ are now thought of as being defined on $\mathbb{R} \times \mathsf{TQ}$ via $T\Pi_1$.

At the moment, we think of these forces as being prescribed, and then asking, "How does the system behave when subjected to the prescribed forces?" When we come to looking at control problems, we shall think of forces as something we *design* in order to accomplish certain tasks. •

Example 4.54 (Two-link manipulator cont'd). For the two-link manipulator, we consider three forces as shown in Figure 4.13. The force F^a is a torque

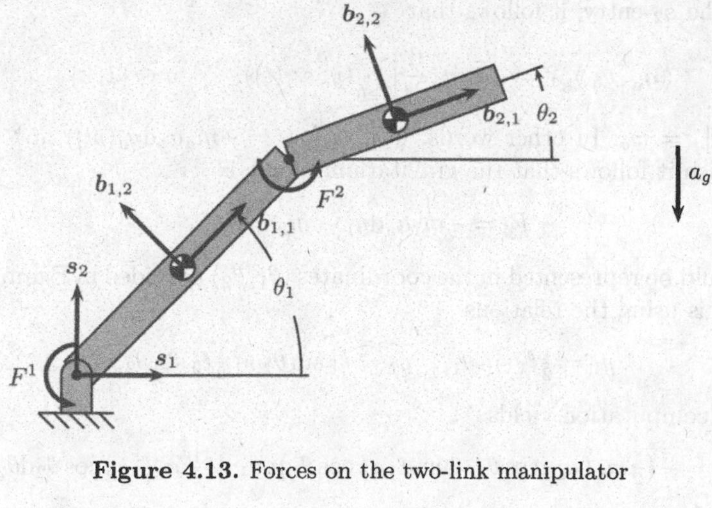

Figure 4.13. Forces on the two-link manipulator

applied at the base of link a, $a \in \{1, 2\}$. We also have gravity in this example that exerts a force on each of the links. The force F^1 is represented in the Newtonian setting by the torque $\boldsymbol{\tau}_{1,1} = \tau_1(0,0,1)$. The force F^2 is represented by the torque $\boldsymbol{\tau}_{2,1} = -\tau_2(0,0,1)$ applied to the first link, and the equal and opposite torque $\boldsymbol{\tau}_{2,2} = \tau_2(0,0,1)$ applied to the second link. The coefficients τ_1 and τ_2 are functions of position and velocity of the two links, and of time. Now let $w_q \in \mathsf{TQ}$ have components (w_1, w_2) in coordinates. Since we only have torques in this problem, we just need to compute the spatial angular velocities corresponding to w_q. Using the computations of Example 4.20, these are $\boldsymbol{\omega}_1 \triangleq \rho_{\text{spatial}} \circ T\Pi_1(w_q) = (0,0,w_1)$ and $\boldsymbol{\omega}_2 \triangleq \rho_{\text{spatial}} \circ T\Pi_2(w_q) = (0,0,w_2)$. We then have

$$\langle\langle \boldsymbol{\tau}_{1,1} + \boldsymbol{\tau}_{2,1}, \boldsymbol{\omega}_1 \rangle\rangle_{\mathbb{R}^3} + \langle\langle \boldsymbol{\tau}_{2,2}, \boldsymbol{\omega}_2 \rangle\rangle_{\mathbb{R}^3} = \tau_1 w_1 + \tau_2(w_2 - w_1).$$

This is exactly $F^1(w_q) + F^2(w_q)$, provided that we take

$$F^1 = \tau_1 \mathrm{d}\theta_1, \quad F^2 = \tau_2(\mathrm{d}\theta_2 - \mathrm{d}\theta_1).$$

As the reader may demonstrate in Exercise E4.26, the Newtonian force on the ath link resulting from the gravity has magnitude $m_a a_g$, $a \in \{1,2\}$, and acts in the direction $-\boldsymbol{s}_2$ as shown in Figure 4.3. Here, $a_g > 0$ is the acceleration due to gravity. Thus this force is represented in the Newtonian setting by the force $\boldsymbol{g}_1 = m_1 a_g(0,-1,0)$ on the first link, and the force $\boldsymbol{g}_2 = m_2 a_g(0,-1,0)$ on the second link. To determine the Lagrangian representation of the gravitational force, we proceed in a manner that is slightly different than that used for the torques above. Let $y_a \colon \mathsf{Q} \to \mathbb{R}$, $a \in \{1,2\}$, denote the function that is the height of the center of mass of link a, measured in the direction of $\boldsymbol{s}_2$ from O_{spatial}. As above, let $w_q \in \mathsf{TQ}$ have components (w_1, w_2). Let $\boldsymbol{v}_1 = \lambda_{\text{spatial}} \circ T\Pi_1(w_q)$ and $\boldsymbol{v}_2 = \lambda_{\text{spatial}} \circ T\Pi_2(w_q)$ be the spatial linear velocities associated with w_q. Since $\boldsymbol{g}_a$, $a \in \{1,2\}$, has a nonzero component only in the $\boldsymbol{s}_2$-entry, it follows that

$$\langle\langle \boldsymbol{g}_a, \boldsymbol{v}_a \rangle\rangle_{\mathbb{R}^3} = -m_a a_g \frac{\mathrm{d}}{\mathrm{d}t}\Big|_{t=0} (y_a \circ \gamma(t)), \qquad a \in \{1,2\},$$

where $[\gamma]_q = w_q$. In other words, $\langle\langle \boldsymbol{g}_a, \boldsymbol{v}_a \rangle\rangle_{\mathbb{R}^3} = -m_a a_g \mathrm{d}y_a(w_q)$, $a \in \{1,2\}$, from which it follows that the gravitational force is

$$F_g = -m_1 a_g \mathrm{d}y_1 - m_2 a_g \mathrm{d}y_2.$$

This should be represented in the coordinates (θ_1, θ_2) provided in Example 4.8. We do this using the relations

$$y_1 = \tfrac{1}{2}\ell_1 \sin\theta_1, \quad y_2 = \ell_1 \sin\theta_1 + \tfrac{1}{2}\ell_2 \sin\theta_2.$$

A direct computation yields

$$F_g = -\left(\tfrac{1}{2} m_1 \ell_1 a_g \cos\theta_1 + m_2 \ell_1 a_g \cos\theta_1\right)\mathrm{d}\theta_1 - \tfrac{1}{2} m_2 \ell_2 a_g \cos\theta_2 \mathrm{d}\theta_2.$$

This gravitational force, of course, cannot be controlled. This illustrates that one may have both forces one can control, and forces that are prescribed. •

Example 4.55 (Rolling disk cont'd). For the rolling disk, we consider two forces, one a torque F^1 that "spins" the disk, and the other a torque F^2 that "rolls" the disk (see Figure 4.14). The "spin" torque is represented in

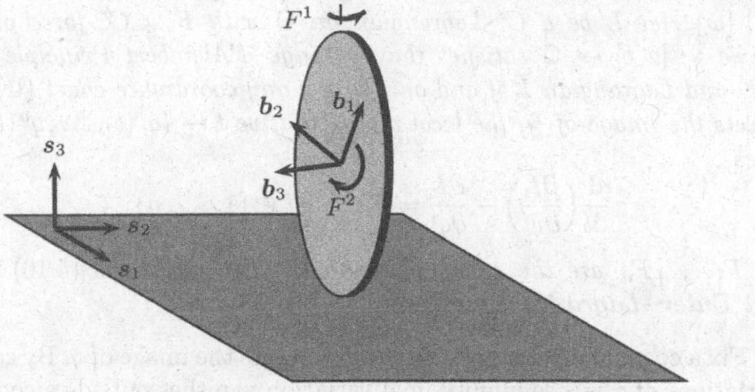

Figure 4.14. Forces on the rolling disk

the Newtonian setup by the torque $\boldsymbol{\tau}_1 = \tau_1(0,0,1)$, and the "roll" torque is represented by the torque $\boldsymbol{\tau}_2 = \tau_2(-\sin\theta, \cos\theta, 0)$. Here θ is the angle depicted in Figure 4.9. Let $w_q \in \mathsf{T}\mathsf{Q}$ have components $(w_x, w_y, w_\theta, w_\phi)$. From the computations in Example 4.21, the corresponding spatial angular velocity is $\boldsymbol{\omega} = (-\sin\theta w_\phi, \cos\theta w_\phi, w_\theta)$. Therefore, we compute

$$\langle\langle \boldsymbol{\tau}_1 + \boldsymbol{\tau}_2, \boldsymbol{\omega}\rangle\rangle_{\mathbb{R}^3} = \tau_1 w_\theta + \tau_2 w_\phi.$$

This is given by $F^1(w_q) + F^2(w_q)$, provided that

$$F^1 = \tau_1 \mathrm{d}\theta, \quad F^2 = \tau_2 \mathrm{d}\phi.$$

As usual, the functions τ_1 and τ_2 are defined on $\mathbb{R} \times \mathsf{T}\mathsf{Q}$. •

4.4.3 The Lagrange–d'Alembert Principle

The Lagrange–d'Alembert Principle tells us how a force F should appear in the Euler–Lagrange equations. This principle is discussed in most of the texts on mechanics mentioned in the introduction to the chapter.

Definition 4.56 (Lagrange–d'Alembert Principle). Let L be a C^r-Lagrangian on a C^r-manifold Q, and let F be a C^r-force, $r \in \mathbb{N} \cup \{\infty\} \cup \{\omega\}$. A C^2-curve $\gamma\colon [a,b] \to \mathsf{Q}$ satisfies the ***Lagrange–d'Alembert Principle*** for the force F and the Lagrangian L along γ if, for every variation $\vartheta\colon J \times [a,b] \to \mathsf{Q}$ of γ, we have

$$\left.\frac{\mathrm{d}}{\mathrm{d}s}\right|_{s=0} \int_a^b L(t, \tfrac{\mathrm{d}}{\mathrm{d}t}\vartheta(s,t))\,\mathrm{d}t + \int_a^b \langle F(t, \gamma'(t)); \delta\vartheta(t)\rangle\,\mathrm{d}t = 0.$$ •

The definition, of course, applies when F is defined not just along γ, since a globally defined force gives a force along a curve. Let us see what are the implications of the Lagrange–d'Alembert Principle for the equations describing the motion of a Lagrangian system in the presence of forces.

Proposition 4.57 (Forced Euler–Lagrange equations). *For $r \in \mathbb{N} \cup \{\infty\} \cup \{\omega\}$, let L be a C^r-Lagrangian on Q with F a C^r-force on Q. A C^2-curve $\gamma\colon [a,b] \to \mathsf{Q}$ satisfies the Lagrange–d'Alembert Principle for the force F and Lagrangian L if and only if, for any coordinate chart $(\mathcal{U}, \phi)$ that intersects the image of γ, the local representative $t \mapsto (q^1(t), \dots, q^n(t))$ of γ satisfies*

$$\frac{\mathrm{d}}{\mathrm{d}t}\Big(\frac{\partial L}{\partial v^i}\Big) - \frac{\partial L}{\partial q^i} = F_i, \qquad i \in \{1, \dots, n\}, \tag{4.16}$$

where $F_1, \dots, F_n$ are the components of F. The equations (4.16) are the ***forced Euler–Lagrange equations****.*

Proof. Fix a coordinate chart $(\mathcal{U}, \phi)$ that intersects the image of γ. By considering variations of γ whose infinitesimal variation vanishes outside a coordinate chart, we may suppose that $\gamma(t) \in \mathcal{U}$ for each t. Let ϑ be a variation of γ with ϑ_s the curve defined by $\vartheta_s = \vartheta(s,t)$. Then, by following the calculations of Theorem 4.38, we arrive at

$$\int_a^b \Big(\frac{\partial L}{\partial q^i} - \frac{\mathrm{d}}{\mathrm{d}t}\Big(\frac{\partial L}{\partial v^i}\Big) + F_i\Big) \frac{\mathrm{d}\vartheta_s^i(t)}{\mathrm{d}s}\Big|_{s=0} \mathrm{d}t = 0,$$

where $(s,t) \mapsto (\vartheta^1(s,t), \dots, \vartheta^n(s,t))$ is the local representative of ϑ. Since this must hold for every variation, the result follows. ■

Thus the manner in which forces are added to the Euler–Lagrange equations, governed by the Lagrange–d'Alembert Principle, is very simple: one merely puts the components of the force on the right-hand side of the Euler–Lagrange equations. What is not so simple is showing that the inclusion of forces as per the Lagrange–d'Alembert Principle agrees with the Newton–Euler equations for the motion of interconnected rigid bodies. We merely state the corresponding theorem, referring to texts such as mentioned in the chapter introduction for further discussion.

Theorem 4.58 (Equivalence of Newton–Euler and Euler–Lagrange equations II). *Let $(\mathcal{B}_1, \mu_1), \dots, (\mathcal{B}_k, \mu_k)$ be an interconnected mechanical system with configuration manifold Q, and let $\mathbb{G}$ be the kinetic energy Riemannian metric constructed as in Section 4.2.3. Suppose that the ath body, $a \in \{1, \dots, k\}$, is subject to an external force $\boldsymbol{f}_a$ applied at the center of mass, and to a torque $\boldsymbol{\tau}_a$, and let F be the total external force constructed as in Section 4.4.1. For a curve γ on Q, the following statements are equivalent:*

(i) the curve $\Pi_a \circ \gamma$ on $\mathsf{SO}(3) \times \mathbb{R}^3$ describes a motion of the ath rigid body, $a \in \{1, \dots, k\}$, for the system according to the Newton–Euler equations;

(ii) γ *satisfies the Lagrange–d'Alembert Principle for the force* F *and the Lagrangian* $L_{\mathbb{G}}$.

Let us see how this looks for Lagrangians defined using a Riemannian metric as in Section 4.3.4. This abstracts the interconnected mechanical system setup used in Theorem 4.58 to general Riemannian manifolds.

Proposition 4.59 (Lagrange–d'Alembert Principle on Riemannian manifolds). *Let* $r \in \mathbb{N} \cup \{\infty\} \cup \{\omega\}$ *and let* $(\mathsf{Q}, \mathbb{G})$ *be a* C^r*-Riemannian manifold. A curve* $\gamma\colon [a,b] \to \mathsf{Q}$ *satisfies the Lagrange–d'Alembert Principle for the* C^r*-force* F *and the* C^r*-Lagrangian* $L_{\mathbb{G}}$ *if and only if*

$$\overset{\mathbb{G}}{\nabla}_{\gamma'(t)}\gamma'(t) = \mathbb{G}^{\sharp}(F(t,\gamma'(t))). \tag{4.17}$$

Furthermore, given $(t_0, v_{q_0}) \in \mathbb{R}\times\mathsf{TQ}$, *there exists an interval* $J \subset \mathbb{R}$, *containing* t_0 *in its interior, and a locally absolutely differentiable curve* $\gamma\colon J \to \mathsf{Q}$, *such that* γ *satisfies* (4.17) *and such that* $\gamma'(t_0) = v_{q_0}$. *Moreover, if* $\tilde{J}$ *is another such interval and if* $\tilde{\gamma}\colon \tilde{J} \to \mathsf{Q}$ *is another such curve, then* $\gamma(t) = \tilde{\gamma}(t)$ *for* $t \in J \cap \tilde{J}$.

Proof. Proceeding as in the proof of Proposition 4.43, we ascertain that γ satisfies the Lagrange–d'Alembert Principle for the force F and the Lagrangian $L_{\mathbb{G}}$ if and only if

$$-\int_a^b \mathbb{G}(\overset{\mathbb{G}}{\nabla}_{\gamma'(t)}\gamma'(t), \delta\vartheta(t))\,\mathrm{d}t + \int_a^b \langle F(t,\gamma'(t)); \delta\vartheta(t)\rangle\,\mathrm{d}t = 0$$

for every variation ϑ. Rewriting this as

$$\int_a^b \Big(\mathbb{G}(\overset{\mathbb{G}}{\nabla}_{\gamma'(t)}\gamma'(t), \delta\vartheta(t)) - \mathbb{G}(\mathbb{G}^{\sharp}(F(t,\gamma'(t))), \delta\vartheta(t))\Big)\,\mathrm{d}t,$$

the first part of the result follows. The second assertion is equivalent to Corollary A.12. ■

4.4.4 Potential forces

A potential force is one resulting from potential energy. More usefully, given a ***potential function*** $V \in C^\infty(\mathsf{Q})$ on Q, its ***potential force*** is the basic force given by $F(t, v_q) = -\mathrm{d}V(q)$. In particular, potential forces are independent of time and velocity.[4] However, potential forces are more than simply independent of time and velocity. One should think of the potential function V as a "storage function," since, loosely speaking, by increasing its value one stores energy that is available to become kinetic energy.

[4] One can actually define potential forces to be time-dependent by allowing V to depend on time. Sometimes it is interesting to do this, but we shall not do so.

Potential forces are essentially regarded as "internal" forces. The meaning of this is perhaps best understood in terms of the following result, which tells us that potential forces can essentially be absorbed into the Lagrangian. Although the result actually holds for general Lagrangians, we state it here only for kinetic energy Lagrangians.

Proposition 4.60 (Euler–Lagrange equations with potential forces). *Let $(\mathsf{Q}, \mathbb{G})$ be a C^∞-Riemannian manifold and let $V \in C^\infty(\mathsf{Q})$. For the Lagrangian $L_{\mathbb{G}}$ and a curve $\gamma\colon [a,b] \to \mathsf{Q}$, the following statements are equivalent:*

(i) γ *satisfies the Lagrange–d'Alembert Principle for the force* $F(t, v_q) = -\mathrm{d}V(q)$ *and the Lagrangian* $L_{\mathbb{G}}$*;*

(ii) γ *is an extremal for the action corresponding to the Lagrangian* $L(t, v_q) = L_{\mathbb{G}}(v_q) - V(q)$*;*

(iii) γ *satisfies the differential equation* $\overset{\mathbb{G}}{\nabla}_{\gamma'(t)}\gamma'(t) = -\mathrm{grad}V(\gamma(t))$.

Proof. The equivalence of (i) and (iii) is a direct consequence of Proposition 4.59. We will show that (i) and (ii) are equivalent. From the proof of Proposition 4.59, we recall that (i) is equivalent to the equality

$$\int_a^b \Big(-\mathbb{G}(\overset{\mathbb{G}}{\nabla}_{\gamma'(t)}\gamma'(t), \delta\vartheta(t)) + \langle \mathrm{d}V(\gamma(t)); \delta\vartheta(t)\rangle\Big)\,\mathrm{d}t$$

holding for every variation ϑ of γ. The first term in the integrand corresponds to $\frac{\mathrm{d}}{\mathrm{d}s}\big|_{s=0} A_{L_{\mathbb{G}}}(\vartheta_s)$, as was shown in the proof of Proposition 4.43. The second term in the integrand we recognize as corresponding to $-\frac{\mathrm{d}}{\mathrm{d}s}\big|_{s=0} A_V(\vartheta_s)$. The proposition now follows by the linearity of the map $L \mapsto A_L$. ■

Remarks 4.61. 1. Mechanical systems with kinetic energy Lagrangian $L_{\mathbb{G}}$, and subject to a potential force $-\mathrm{d}V$ are very common in applications. We call such a system a ***C^r-simple mechanical system*** if all the data are C^r, $r \in \mathbb{N} \cup \{\infty\} \cup \{\omega\}$, and write it as $(\mathsf{Q}, \mathbb{G}, V)$. Of most interest to us is the case in which $r \in \{\infty, \omega\}$, and unless otherwise stated, a simple mechanical system is assumed to be C^∞. It will always be assumed that we use the time-independent Lagrangian formed by kinetic energy minus potential energy: $L(v_q) = \frac{1}{2}\mathbb{G}(v_q, v_q) - V(q)$.

2. We shall frequently wish to consider a simple mechanical system subject to a (possibly non-potential) force F. In this case we shall write the system as $(\mathsf{Q}, \mathbb{G}, V, F)$, and call at a ***$C^r$-forced simple mechanical system***, $r \in \mathbb{N} \cup \{\infty\} \cup \{\omega\}$. •

A commonly encountered potential force is a gravitational force. Let us show that the gravitational force introduced in Example 4.54 is a potential force.

Example 4.62 (Two-link manipulator cont'd). The potential function arising from the presence of gravity is proportional to the mass of the system multiplied by the height of the center of mass of the system (in the direction opposite that of gravity) above a reference. The constant of proportionality is the gravitational acceleration. Since a potential force is given by the differential of the potential function, the choice of reference is immaterial.

For the two-link manipulator, let us choose the origin of the spatial reference frame as reference. In this case the gravitational potential function is given by

$$V(\theta_1, \theta_2) = \tfrac{1}{2} m_1 a_g \ell_1 \sin\theta_1 + m_2 a_g \ell_1 \sin\theta_1 + \tfrac{1}{2} m_2 \ell_2 \sin\theta_2.$$

One then readily verifies that the force F_g of Example 4.54 is exactly $-\mathrm{d}V$. •

Note that, for a simple mechanical system $(\mathsf{Q}, \mathbb{G}, V)$, one has two natural energies, the kinetic energy defined by $\mathbb{G}$ and the potential energy defined by V. The natural Lagrangian is the difference between these two energies. It seems intuitively reasonable that the total energy, the sum of the two energies, should also be interesting.

Definition 4.63 (Energy). The ***energy*** for a simple mechanical system $(\mathsf{Q}, \mathbb{G}, V)$ is the function $E\colon \mathsf{TQ} \to \mathbb{R}$ given by $E(v_q) = \frac{1}{2}\mathbb{G}(v_q, v_q) + V(q)$. •

The following result gives a useful property of the energy, namely that it is conserved in the absence of non-potential external forces.

Proposition 4.64 (Energy along solutions of Euler–Lagrange equations). *Let $(\mathsf{Q}, \mathbb{G}, V, F)$ be a C^∞-forced simple mechanical system. If $\gamma\colon I \to \mathsf{Q}$ satisfies the Lagrange–d'Alembert Principle for the force F and the Lagrangian $L(v_q) = \frac{1}{2}\mathbb{G}(v_q, v_q) - V(q)$, then*

$$\frac{\mathrm{d}E(\gamma'(t))}{\mathrm{d}t} = \langle F(t, \gamma'(t)); \gamma'(t)\rangle .$$

In particular, if $F = 0$, then energy is constant along solutions of the Euler–Lagrange equations for the Lagrangian L.

Proof. We compute

$$\begin{aligned}\frac{\mathrm{d}E(\gamma'(t))}{\mathrm{d}t} &= \frac{1}{2}\frac{\mathrm{d}}{\mathrm{d}t}\mathbb{G}(\gamma'(t), \gamma'(t)) + \frac{\mathrm{d}}{\mathrm{d}t}V(\gamma(t)) \\ &= \mathbb{G}(\overset{\mathrm{G}}{\nabla}_{\gamma'(t)}\gamma'(t), \gamma'(t)) + \langle \mathrm{d}V(\gamma(t)); \gamma'(t)\rangle \\ &= -\,\mathbb{G}(\operatorname{grad}V(\gamma(t)), \gamma'(t)) + \mathbb{G}(\mathbb{G}^\sharp(F(t, \gamma'(t))), \gamma'(t)) \\ &\quad + \langle \mathrm{d}V(\gamma(t)); \gamma'(t)\rangle \\ &= \langle F(t, \gamma'(t)); \gamma'(t)\rangle .\end{aligned}$$

■

4.4.5 Dissipative forces

Dissipative forces are so named because they dissipate energy. For example, air resistance on a moving body is a dissipative force. As we shall see in Chapter 6, the dynamics of systems with dissipative forces tend to become increasingly simple as time increases. For now, we have the following definition.

Definition 4.65 (Dissipative force). A time-independent force $F\colon \mathsf{TQ} \to \mathsf{T}^*\mathsf{Q}$ is ***dissipative*** if $\langle F(v_q); v_q\rangle \leq 0$ for each $v_q \in \mathsf{TQ}$, and is ***strictly dissipative*** if it is dissipative and if $\langle F(v_q); v_q\rangle = 0$ only when $v_q \in Z(\mathsf{TQ})$. •

When we wish to denote a dissipative force, we shall use the notation F_{diss}. For example, if we write $(\mathsf{Q}, \mathbb{G}, V, F_{\mathrm{diss}})$, then we implicitly have in mind a simple mechanical system subject to a dissipative force.

Let us immediately show that, for simple mechanical systems, dissipative forces cause the energy to decrease, or at least not to increase.

Proposition 4.66 (Dissipative forces decrease energy). *Let* $(\mathsf{Q}, \mathbb{G}, V)$ *be a* C^∞*-simple mechanical system, let* F_{diss} *be a* C^∞*-dissipative force, and let* $\gamma\colon I \to \mathsf{Q}$ *be a curve satisfying*

$$\overset{\mathbb{G}}{\nabla}_{\gamma'(t)}\gamma'(t) = -\mathrm{grad}V(\gamma(t)) + \mathbb{G}^\sharp(F_{\mathrm{diss}}(\gamma'(t))).$$

Then, if E *is the energy for the system, the function* $t \mapsto E(\gamma'(t))$ *is nonincreasing.*

Proof. This follows directly from Proposition 4.64. ■

In applications, dissipative forces are often idealized as being linear in velocity. The following definition describes this type of force.

Definition 4.67 (Rayleigh dissipation). Let Q be a C^∞-manifold. A ***Rayleigh dissipation function*** is a class C^∞, symmetric, positive-semidefinite $(0, 2)$-tensor field R_{diss} on Q. If R_{diss} is positive-definite, then it is ***strict***. The dissipative force associated with a Rayleigh dissipation function is the map $-R^\flat_{\mathrm{diss}}\colon \mathsf{TQ} \to \mathsf{T}^*\mathsf{Q}$. •

4.5 Nonholonomic constraints

A final basic ingredient in our description of mechanical system is constraints on the velocities of a mechanical system. There is a fairly substantial literature on this subject, particularly from the analytical mechanics side, where much effort is expended in providing physical or metaphysical justifications for the equations of motion for these systems. Many classic texts, including those listed in the introduction to this chapter, give a thorough presentation of velocity constraints. The book of Pars [1965] also has a particularly detailed discussion of the related methodologies. A recent treatise in the classic style is that of Papastavridis [2002]. A geometric approach to problems in nonholonomic mechanics is taken by Bloch [2003] and Cortés [2002]. One of the reasons why there is such an expanse in the classic and recent literature concerning velocity constraints is that, when such constraints are present, there are many ways, most equivalent, of writing the equations of motion. We survey some of the most well-known methods in Remark 4.78 below. It is not uncommon for authors to identify strongly with one approach over another. We adopt a more pragmatic approach by simply indicating how velocity constraints are represented in our differential geometric framework, and how the governing equations are represented in ways that are conducive to our control-theoretic objectives.

4.5.1 From rigid body constraints to a distribution on Q

We have already seen in Section 4.2.3 how the physical concept of kinetic energy leads to a Riemannian metric, and in Section 4.4.1 how the physical concept of force and torque leads to the abstract notion of a force in Lagrangian mechanics. Now we indicate how constraints lead to a distribution on the configuration manifold.

As usual, we start with an interconnected mechanical system comprised of rigid bodies $(\mathcal{B}_1, \mu_1), \dots, (\mathcal{B}_k, \mu_k)$, and with configuration manifold Q. We fix $a \in \{1, \dots, k\}$, and let $(\boldsymbol{R}_{a,0}, \boldsymbol{r}_{a,0}) \in \mathsf{SO}(3) \times \mathbb{R}^3$ be a position and orientation of the ath body that is allowed by the interconnection, i.e., there is a point $q_0 \in \mathsf{Q}$ for which $\Pi_a(q_0) = (\boldsymbol{R}_{a,0}, \boldsymbol{r}_{a,0})$. Now let $t \mapsto (\boldsymbol{R}_a(t), \boldsymbol{r}_a(t))$ be a curve at $(\boldsymbol{R}_{a,0}, \boldsymbol{r}_{a,0})$, and suppose that, for each t sufficiently close to $t = 0$, $(\boldsymbol{R}_a(t), \boldsymbol{r}_a(t))$ is an admissible configuration of the body. We now impose the constraint that the spatial angular velocity and linear velocity of the ath body be restricted to lie in some subspace $\mathcal{C}_{a,q_0} \subset \mathbb{R}^3 \oplus \mathbb{R}^3$:

$$(\boldsymbol{\omega}_a(t), \dot{\boldsymbol{r}}_a(t)) \in \mathcal{C}_{a,q_0}.$$

Note that the dependence is on q_0, not just on $(\boldsymbol{R}_{a,0}, \boldsymbol{r}_{a,0})$, since the admissible velocities for the ath body may depend on the position of the other $k-1$ bodies. The subspaces $\mathcal{C}_{a,q_0}$ should be thought of as living in the fiber of the vector bundle $(\mathbb{R}^3 \oplus \mathbb{R}^3)^k_{\mathsf{Q}}$ that was defined in Section 4.1.5. Indeed, for $q_0 \in \mathsf{Q}$ we have the inclusion

$$\mathcal{C}_{q_0} \triangleq \mathcal{C}_{1,q_0} \oplus \cdots \oplus \mathcal{C}_{k,q_0} \subset (\mathbb{R}^3 \oplus \mathbb{R}^3) \oplus \cdots \oplus (\mathbb{R}^3 \oplus \mathbb{R}^3),$$

so specifying a subspace in each fiber of $(\mathbb{R}^3 \oplus \mathbb{R}^3)^k_{\mathsf{Q}}$. To regard this as a distribution on Q, we recall from Section 4.1.5 the vector bundle isomorphism ι_{spatial} from $\mathsf{TQ}_{\mathrm{free}}$ to $(\mathbb{R}^3 \oplus \mathbb{R}^3)^k_{\mathsf{Q}_{\mathrm{free}}}$. With this vector bundle isomorphism, we note that our constructions give the distribution $\mathcal{D}$ on Q defined by $\mathcal{D}_{q_0} = \iota^{-1}_{\mathrm{spatial}}(\mathcal{C}_{q_0})$. The subspaces in the distribution $\mathcal{D}$ should fit together nicely.

Assumption 4.68 (Smoothness of velocity constraints). The assignment $q_0 \mapsto \mathcal{D}_{q_0}$ has the property that its annihilator is a C^∞-codistribution on Q. •

We should now say a few words about how, in Newtonian mechanics, one obtains equations of motion for a system in the presence of constraints on velocities. When the system is in configuration q_0, one asks that the ath body be subject to a force $\boldsymbol{h}_a \in \mathbb{R}^3$ and a torque $\boldsymbol{\rho}_a \in \mathbb{R}^3$ whose purpose is to ensure that the constraints on the velocity are satisfied. Let us call these ***constraint forces*** and ***constraint torques***. The constraint forces and torques are required to satisfy the conditions:

1. if the system is in configuration q, then the force $\boldsymbol{h}_a$ and the torque $\boldsymbol{\rho}_a$ applied to the system in this configuration must satisfy
$$\langle\!\langle \boldsymbol{h}_a, \lambda_{\text{spatial}} \circ T\Pi_a(v_q)\rangle\!\rangle_{\mathbb{R}^3} + \langle\!\langle \boldsymbol{\rho}_a, \rho_{\text{spatial}} \circ T\Pi_a(v_q)\rangle\!\rangle_{\mathbb{R}^3} = 0$$
for all $v_q \in \mathcal{C}_q$, and
2. the constraint force and torque exerted by body a on body b is equal in magnitude, and opposite in direction to the constraint force and torque exerted by body b on body a.

The physical interpretations of these conditions are analogous to those for the similar conditions given in Section 4.3.5 on the forces and torques that model internal interconnection forces. What is different here is that the constraint forces $\boldsymbol{h}_a$ and torques $\boldsymbol{\rho}_a$, $a \in \{1, \dots, k\}$, do not disappear in the Euler–Lagrange framework, at least not in the same way that they disappear in the case of the forces and torques required to maintain the interconnections for an interconnected mechanical system. The exact manner in which these forces appear in the Euler–Lagrange equations is the subject of Section 4.5.3 below.

4.5.2 Definitions and basic properties

Now we abstract the previous construction for an interconnected mechanical system to the setting of a general configuration manifold. First we give the definition of the general type of constraint we consider.

Definition 4.69 (Linear velocity constraint). For $r \in \mathbb{N} \cup \{\infty\} \cup \{\omega\}$, let Q be a C^r-differentiable manifold. A ***C^r-linear velocity constraint***, or simply a ***C^r-constraint***, is a distribution $\mathcal{D}$ on Q with the property that $\operatorname{ann}(\mathcal{D})$ is a C^r-codistribution. A C^r-constraint $\mathcal{D}$ is ***regular*** if $\mathcal{D}$ is a regular distribution. A locally absolutely continuous curve $\gamma\colon I \to$ Q ***satisfies*** a constraint $\mathcal{D}$ if $\gamma'(t) \in \mathcal{D}_{\gamma(t)}$ for a.e. $t \in I$. •

Remarks 4.70. 1. On occasion, it is convenient to say "constraint distribution" in reference to a linear velocity constraint $\mathcal{D}$.

2. We will always be considering $r \in \{\infty, \omega\}$, and if the smoothness is left unstated, we assume that a linear velocity constraint is C^∞.
3. Note that we allow the rank of $\mathcal{D}$ to change. Moreover, we do so in such a way that the annihilator of $\mathcal{D}$ is C^r. As remarked upon in Section 3.7.4, this is different than asking that $\mathcal{D}$ be C^r. Only if $\mathcal{D}$ is regular are the two notions are equivalent (this is Proposition 3.98). A good question, and one entirely related to Assumption 4.68, is then, "Why do we ask that $\text{ann}(\mathcal{D})$ be regular, and not that $\mathcal{D}$ be regular?" The answer to this question is embedded in a detailed consideration of the physical modeling of constraints, something that we do not consider here. In Section 13.4 we will consider a system with a linear velocity constraint that is not regular. We shall do in this example as everyone does, and that is we will suppose that the singular points in Q are omitted from consideration. A paper where constraint distributions are considered whose rank is not locally constant is [Cortés, de León, de Diego, and Martínez 2001]. However, there appears to be work to be done in this area.
4. Our notion of a constraint is not very general. Possible generalizations include the following.
   (a) Constraints may be affine in velocity rather than linear. That is, rather than restricting velocities to a subspace $\mathcal{D}_q \subset \mathsf{T}_q\mathsf{Q}$ for each point in Q, one might restrict to an affine subspace, i.e., a subset of the form $\{v_q\} + \mathcal{D}_q$ for some vector $v_q \in \mathsf{T}_q\mathsf{Q}$. An example of a system exhibiting constraints of this type is a uniform ball rolling on a table rotating with constant speed [Bloch and Crouch 1995, Lewis and Murray 1995].
   (b) The constraints may be time-dependent. This arises in the rolling ball example above, by relaxing the assumption that the table rotates with uniform speed.
   (c) Some authors [de León, Marrero, and de Diego 1997, Lewis 2000c] consider quite general classes of constraints, depending possibly nonlinearly on velocity. There seems to be a lack of mechanical instances of these sorts of constraints (but see [Kitzka 1986]), although their geometry is quite interesting. •

Within the class of linear velocity constraints, there is an important dichotomy.

Definition 4.71 (Holonomic and nonholonomic constraints). A regular linear velocity constraint $\mathcal{D}$ is ***holonomic*** if $\mathcal{D}$ is integrable in the sense explained in Section 3.7.2. If a regular linear velocity constraint is not holonomic, it is ***nonholonomic***.[5] •

Remarks 4.72. 1. Essentially, if a linear velocity constraint $\mathcal{D}$ is holonomic, this means that all curves through a point $q \in \mathsf{Q}$ satisfying the constraint

[5] The etymology of the word "holonomic" is from the Greek, *holos*, meaning "whole," and *nomos*, meaning "principle."

must lie on the maximal integral manifold for $\mathcal{D}$ through q. In such an event, one could restrict consideration to maximal integral manifolds of $\mathcal{D}$. However, there are sometimes reasons for not doing this. For example, although $\mathcal{D}$ may be holonomic, the maximal integral manifold may not be a submanifold. Indeed, Theorem 3.95 only guarantees that integral manifolds will be immersed in Q. It is also possible that it may be easier to represent certain quantities on the larger configuration manifold Q rather than on a maximal integral manifold. We shall consider holonomic constraints in detail in Section 4.5.7, along with examples.

2. When we introduced the notion of an interconnected mechanical system, we were essentially subjecting the free mechanical system to a holonomic constraint, namely a holonomic constraint for which Q is a maximal integral manifold. In Section 4.5.7 we will justify our undeclared assumption that we can simply restrict the Lagrangian to $\mathsf{TQ} \subset \mathsf{TQ}_{\text{free}}$, and then apply the Euler–Lagrange formalism in the smaller configuration manifold.
3. For constraints that are nonholonomic, one may then wonder, "How nonholonomic are they?" This can be worded precisely in the language of Section 3.7.3 by considering the nature of the $\mathcal{D}$-orbits. To this end, we say that a regular linear velocity constraint $\mathcal{D}$ is ***totally nonholonomic*** if $\mathcal{O}(q, \mathcal{D}) = \mathsf{Q}$ for some (and so, for all) $q \in \mathsf{Q}$.
4. Note that our definitions of holonomic and nonholonomic are made only for regular linear velocity constraints. In cases where $\mathcal{D}$ is not regular, it is not clear what holonomic should mean; as mentioned at the end of Section 3.7.4, the notion of integrability for codistributions is not as direct as it is for distributions. •

We now provide an example to illustrate these concepts.

Example 4.73 (Rolling disk cont'd). We consider the rolling disk first considered in Section 1.3, and considered earlier in this chapter. Let us apply the formalism of Section 4.5.1 to derive the constraint arising from the condition that the disk roll without slipping on the plane. Suppose that the disk is in a configuration q, and let $\Pi_1(q) = (\boldsymbol{R}, \boldsymbol{r}) \in \mathsf{SO}(3) \times \mathbb{R}^3$ be determined as in Example 4.15. The point of contact of the disk with the plane is

$$\boldsymbol{x}_{\boldsymbol{R},\boldsymbol{r}} = \langle\!\langle \boldsymbol{r}, \boldsymbol{s}_1 \rangle\!\rangle_{\mathbb{R}^3} \boldsymbol{s}_1 + \langle\!\langle \boldsymbol{r}, \boldsymbol{s}_2 \rangle\!\rangle_{\mathbb{R}^3} \boldsymbol{s}_2,$$

measured in spatial coordinates. The same point, measured relative to the body frame, is denoted by $\boldsymbol{\chi}_{\boldsymbol{R},\boldsymbol{r}}$, and satisfies $\boldsymbol{r} + \boldsymbol{R}\boldsymbol{\chi}_{\boldsymbol{R},\boldsymbol{r}} = \boldsymbol{x}_{\boldsymbol{R},\boldsymbol{r}}$. Thus

$$\boldsymbol{\chi}_{\boldsymbol{R},\boldsymbol{r}} = \boldsymbol{R}^T(\boldsymbol{x}_{\boldsymbol{R},\boldsymbol{r}} - \boldsymbol{r}) = - \langle\!\langle \boldsymbol{r}, \boldsymbol{s}_3 \rangle\!\rangle_{\mathbb{R}^3} \boldsymbol{R}^T \boldsymbol{s}_3 = -\rho \boldsymbol{R}^T \boldsymbol{s}_3,$$

where ρ is the radius of the disk. Now suppose that the body is in motion, so defining a curve $t \mapsto (\boldsymbol{R}(t), \boldsymbol{r}(t)) \in \mathsf{SO}(3) \times \mathbb{R}^3$. The spatial velocity of a point $\boldsymbol{\chi}$ in the body, measure relative to the body frame, is $\dot{\boldsymbol{r}}(t) + \dot{\boldsymbol{R}}(t)\boldsymbol{\chi}$. Taking $\boldsymbol{\chi} = -\rho \boldsymbol{R}^T(t)\boldsymbol{s}_3$ we see that the velocity of the point of contact of the disk with the plane is $\dot{\boldsymbol{r}} - \rho \dot{\boldsymbol{R}}(t)\boldsymbol{R}^T(t)\boldsymbol{s}_3$. The constraint is that the velocity of this point should be zero, since it is fixed to the plane. In short, the constraint is that the spatial angular velocity and the linear velocity should satisfy

$$\dot{\boldsymbol{r}}(t) - \rho\boldsymbol{\omega}(t) \times \boldsymbol{s}_3 = \boldsymbol{0}.$$

In Example 4.21 we computed $\boldsymbol{\omega} = (-\dot{\phi}\sin\theta, \dot{\phi}\cos\theta, \dot{\theta})$, so giving, after a simple calculation, the constraint equations as

$$\dot{x} = \rho\dot{\phi}\cos\theta, \quad \dot{y} = \rho\dot{\phi}\sin\theta. \tag{4.18}$$

We will now demonstrate that the subset of $\mathsf{T}\mathsf{Q}$ defined by these conditions is an analytic distribution. We shall do this by writing down analytic local generators for this distribution. The relations (4.18) can be written in matrix form as

$$\begin{bmatrix} 1 & 0 & 0 & -\rho\cos\theta \\ 0 & 1 & 0 & -\rho\sin\theta \end{bmatrix} \begin{bmatrix} \dot{x} \\ \dot{y} \\ \dot{\theta} \\ \dot{\phi} \end{bmatrix} = \begin{bmatrix} 0 \\ 0 \end{bmatrix}.$$

A basis for the kernel of the coefficient matrix is given by the two vector fields

$$X_1 = \rho\cos\theta\frac{\partial}{\partial x} + \rho\sin\theta\frac{\partial}{\partial y} + \frac{\partial}{\partial\phi}, \quad X_2 = \frac{\partial}{\partial\theta},$$

defined on the domain of the coordinate chart. We therefore have

$$\mathcal{D}_q = \operatorname{span}_{\mathbb{R}}\{X_1(q), X_2(q)\},$$

showing that the relations (4.18) do define a linear velocity constraint, at least on the domain of definition of the coordinates. One can verify that the vector fields X_1 and X_2 can be extended to analytic vector fields on all of Q, so giving global generators for the constraint. This shows that $\mathcal{D}$ is a regular analytic linear velocity constraint.

Let us ascertain whether $\mathcal{D}$ is holonomic or nonholonomic. We compute

$$[X_1, X_2] = \rho\sin\theta\frac{\partial}{\partial x} - \rho\cos\theta\frac{\partial}{\partial y},$$
$$[X_2, [X_1, X_2]] = \rho\cos\theta\frac{\partial}{\partial x} + \rho\sin\theta\frac{\partial}{\partial y}.$$

We claim that the vector fields $\{X_1, X_2, [X_1, X_2], [X_2, [X_1, X_2]]\}$ are linearly independent on the entirety of the domain of the coordinate chart. To show this, we assemble the components of the four vector fields into the columns of a matrix, and compute the determinant to be $-\rho^2$, which is never zero. One can easily see that these four vector fields can be extended to be globally defined linearly independent vector fields. Thus, by Theorem 3.95 we can assert that, not only is $\mathcal{D}$ nonholonomic, but it is totally nonholonomic. •

Remarks 4.74. 1. In the preceding example we were able to find a global basis of everywhere linearly independent vector fields. Thus the constraint is regular. Here we have two occurrences that we should not generally expect to arise: a regular constraint and a globally defined set of generators. There are simple physical examples for which either or both of these fortunate circumstances do not arise.

2. The rolling disk is also totally nonholonomic. Unlike the properties discussed in the preceding remark, we will frequently consider examples for which this is the case. The reasons for this are control-theoretic rather than mechanical; unless the constraint is totally nonholonomic, the system will not be controllable (see Chapter 7). •

4.5.3 The Euler–Lagrange equations in the presence of constraints

Now we suppose that we have a Lagrangian $L: \mathbb{R} \times \mathsf{TQ} \to \mathbb{R}$ and a force $F: \mathbb{R} \times \mathsf{TQ} \to \mathsf{T}^*\mathsf{Q}$ to accompany a constraint $\mathcal{D}$. We wish to provide a means by which one can produce equations governing the motion of the system. There is not a unique way to do this, and indeed there are approaches that seem reasonable, but that are not correct. The key to the proper approach is the notion of what is meant by a constraint force.

Definition 4.75 (Constraint force). For a linear velocity constraint $\mathcal{D}$, a ***constraint force*** is a force taking values in the codistribution $\mathrm{ann}(\mathcal{D})$. If $\gamma: I \to \mathsf{Q}$ is a continuous curve, then a ***constraint force along*** γ is a covector field $\alpha: I \to \mathsf{T}^*\mathsf{Q}$ along γ with the property that $\alpha(t) \in \mathrm{ann}(\mathcal{D})_{\gamma(t)}$ for each $t \in I$. •

Remarks 4.76. 1. For a C^r-linear velocity constraint, the codistribution in which constraint forces take their values is a C^r-codistribution, even when $\mathcal{D}$ is not regular.

2. A physical motivation for the definition of constraint forces for linear velocity constraints is that constraint forces do no work[6] on curves satisfying the constraint. For more general constraints, such physical motivation is more difficult to provide. The reader should be aware of the common fallacy of concluding that, for a force F, the conditions of (1) doing no work on any curve satisfying the constraint and (2) being a constraint force, are not equivalent. It is true that (2) implies (1), but the converse is not true (see Exercise E4.28). •

We may now say what it means for a curve to be a trajectory for a Lagrangian system with a constraint.

---

[6] The work done by a force F along a differentiable curve γ is defined in the preamble to Exercise E4.27.

Definition 4.77 (Constrained Lagrangian system). For $r \in \mathbb{N} \cup \{\infty\} \cup \{\omega\}$, let Q be a C^r-differentiable manifold. A ***C^r-constrained Lagrangian system*** on Q is a triple $(L, F, \mathcal{D})$, where L is a Lagrangian on Q, F is a force on Q, and $\mathcal{D}$ is a constraint on Q, with all data being of class C^r. If $\mathcal{D}$ is regular, a curve $\gamma\colon [a,b] \to$ Q is a ***trajectory for the constrained Lagrangian system $(L, F, \mathcal{D})$*** if there exists a constraint force α along γ such that, together, γ and α have the following properties:

(i) γ satisfies the constraint $\mathcal{D}$;

(ii) γ satisfies the Lagrange–d'Alembert Principle for the force $F + \alpha$ and Lagrangian L. •

Remark 4.78. Definition 4.77 provides a method for computing equations of motion for Lagrangian systems with constraints. There are other methods available for determining equations of motion.

1. ***The Poincaré equations:*** This method is one that eliminates the need for the addition of the constraint force α by incorporating the equations of constraint into the equations of motion and by introducing "pseudo-velocities." This method is considered in Section 2.4 in Chapter 1 of Arnol'd [1988]. We consider this method in an affine connection setting in Section 4.5.6.
2. ***The Gibbs–Appell equations:*** These equations use not only "pseudo-velocities," but "pseudo-accelerations." In the original formulation of Appell [1900a,b] and Gibbs [1879], the treatment is for particles and rigid bodies. A generalization to arbitrary Lagrangian systems is provided by Lewis [1996].
3. ***Gauss's Principle of Least Constraint:*** This method is of a variational nature, although Gauss [1829] was able to cast the problem so that the variational methods are finite-dimensional. That this method bears a strong resemblance to the Gibbs–Appell equations is fleshed out in a general setting by Lewis [1996].
4. ***Hertz's Principle of Least Curvature:*** In the method of Hertz [1894], one defines a quantity called curvature, the minimization of which produces the desired equations of motion. When one looks at it a little carefully, the method is very much like Gauss's Principle of least constraint.
5. ***Kane's equations:*** Kane's equations (see [Kane and Levinson 1985]; the original papers are [Kane 1961, Kane and Wang 1965]) can be thought of as "Newton's equations in 'pseudo-velocities'." This method for deriving the Euler–Lagrange equations in the presence of constraints is quite popular in the multibody dynamics literature, as it is specially adapted to situations where the mechanical system is derived from a collection of rigid bodies (as opposed to an abstract Riemannian manifold). It bears resemblance to the traditional (as opposed to differential geometric) version of the Gibbs–Appell equations. Lesser [1992] gives a differential geometric interpretation of Kane's equations.

6. ***The "vakonomic method":*** The so-called vakonomic method for producing equations for systems with constraints is, unlike the other methods, a variational method. Indeed, it results from determining necessary conditions for minima among curves satisfying the constraints, in the same manner as Theorem 4.38 for the unconstrained case. However, as is shown with a simple counterexample by Lewis and Murray [1995], the resulting equations are not the same as those derived using the Definition 4.77 or, equivalently, by the other techniques listed above. What is more, it can be shown that the equations derived from Definition 4.77 are equivalent to Newtonian mechanics, in the case of an interconnected mechanical system with constraints. We state this as Theorem 4.79 below. This, therefore, renders the vakonomic method an unsuitable formulation for problems in constrained mechanics. Much has been written on this, and we refer the interested reader to the literature, [e.g., Bloch 2003, Bloch and Crouch 1995, Cardin and Favretti 1996, Cortés, de León, de Diego, and Martínez 2002, Gràcia, Marín-Solano, and Muñoz-Lecanda 2003, Kharlomov 1992, Kozlov 1992, Kupka and Oliva 2001, Lewis and Murray 1995, Zampieri 2000].

This is an incomplete list. However, with the exception of the "vakonomic method," all the methods we list, and most that we do not list, are equivalent. Depending on the problem one is looking at, and what one wishes to do, one or the other of the available methods may be the most suitable. •

Let us state the last of our theorems concerning the equivalence of the Newton–Euler equations and the Euler–Lagrange equations, now adding constraints to the mix.

Theorem 4.79 (Equivalence of Newton–Euler and Euler–Lagrange equations III). *Let $(\mathcal{B}_1, \mu_1), \dots, (\mathcal{B}_k, \mu_k)$ be an interconnected mechanical system with configuration manifold Q, and let $\mathbb{G}$ be the kinetic energy Riemannian metric constructed as in Section 4.2.3. Suppose that the ath body, $a \in \{1, \dots, k\}$, is subject to an external force $\boldsymbol{f}_a$ applied at the center of mass, and to a pure torque $\boldsymbol{\tau}_a$, let $F_{\boldsymbol{f}_a, \boldsymbol{\tau}_a}$ be the force constructed as in Section 4.4.1, and let F be the corresponding total external force. Suppose that the bodies are subject to constraints on their velocities, as discussed in Section 4.5.1, producing a regular distribution $\mathcal{D}$ on Q. For a curve γ on Q, the following statements are equivalent:*

(i) the curve $\Pi_a \circ \gamma$ on $\mathsf{SO}(3) \times \mathbb{R}^3$ describes a motion of the ath rigid body, $a \in \{1, \dots, k\}$, for the system according to the Newton–Euler equations;
(ii) γ is a trajectory for the constrained Lagrangian system $(L_{\mathbb{G}}, F, \mathcal{D})$.

Since we do not consider general Lagrangians in this book, we shall end here our discussion of constraints for arbitrary Lagrangians, and proceed to a discussion of simple mechanical systems subjected to linear velocity constraints. Readers interested in the more general discussions are referred to the sources in the literature mentioned previously in this section, and in the introductory remarks of the chapter.

4.5.4 Simple mechanical systems with constraints

We begin with a definition.

Definition 4.80 (Simple mechanical system with constraints). For $r \in \mathbb{N} \cup \{\infty\} \cup \{\omega\}$, a ***$C^r$-simple mechanical system with constraints*** is a 4-tuple $(\mathsf{Q}, \mathbb{G}, V, \mathcal{D})$, where

(i) Q is a manifold,
(ii) $\mathbb{G}$ is Riemannian metric on Q,
(iii) V is a function on Q, and
(iv) $\mathcal{D}$ is a linear velocity constraint,

with all data of class C^r. If, additionally, we have

(v) a C^r-force $F\colon \mathbb{R} \times \mathsf{TQ} \to \mathsf{T}^*\mathsf{Q}$,

then the 5-tuple $(\mathsf{Q}, \mathbb{G}, V, F, \mathcal{D})$ is a ***C^r-forced simple mechanical system with constraints***. •

As usual, the two most interesting cases will be $r \in \{\infty, \omega\}$. We wish to provide a formulation of the equations of motion for simple mechanical systems with constraints. In order to do so we define two bundle maps over id_{Q}, denoted by $P_{\mathcal{D}}, P_{\mathcal{D}}^{\perp}\colon \mathsf{TQ} \to \mathsf{TQ}$, by asking that, for each $v_q \in \mathsf{T}_q\mathsf{Q}$, we have

$$v_q = P_{\mathcal{D}}(v_q) \oplus P_{\mathcal{D}}^{\perp}(v_q) \quad \text{and} \quad P_{\mathcal{D}}(v_q) \in \mathcal{D}_q,\ P_{\mathcal{D}}^{\perp}(v_q) \in \mathcal{D}_q^{\perp},$$

where $\mathcal{D}_q^{\perp}$ is the $\mathbb{G}$-orthogonal complement to $\mathcal{D}_q$ in $\mathsf{T}_q\mathsf{Q}$. Thus $P_{\mathcal{D}}$ is the $\mathbb{G}$-orthogonal projection onto $\mathcal{D}_q$ and $P_{\mathcal{D}}^{\perp}$ is the $\mathbb{G}$-orthogonal projection onto $\mathcal{D}^{\perp}$. If $\mathcal{D}$ is a regular constraint, then $P_{\mathcal{D}}$ and $P_{\mathcal{D}}^{\perp}$ are C^r-vector bundle maps, or equivalently C^r-tensor fields of type $(1,1)$. However, if $\mathcal{D}$ is not a regular constraint, then these bundle maps will not generally even be continuous. Nonetheless, they can be defined, and have the differentiability of $\mathcal{D}$ on the open dense set on which $\mathcal{D}$ is regular.

Let us see how to compute these objects in our rolling disk example.

Example 4.81 (Rolling disk cont'd). For the rolling disk, we have $\mathsf{Q} = \mathbb{R}^2 \times \mathbb{S}^1 \times \mathbb{S}^1$, and we were using coordinates (x, y, θ, ϕ). The Riemannian metric in these coordinates, we recall, is

$$\mathbb{G} = m(\mathrm{d}x \otimes \mathrm{d}x + \mathrm{d}y \otimes \mathrm{d}y) + J_{\text{spin}}\mathrm{d}\theta \otimes \mathrm{d}\theta + J_{\text{roll}}\mathrm{d}\phi \otimes \mathrm{d}\phi.$$

Referring to (4.18), covector fields comprising local generators for $\mathrm{ann}(\mathcal{D})$ are

$$\alpha^1 = \mathrm{d}x - \rho\cos\theta\mathrm{d}\phi, \quad \alpha^2 = \mathrm{d}y - \rho\sin\theta\mathrm{d}\phi.$$

The vector fields $\mathbb{G}^{\sharp}(\alpha^1)$ and $\mathbb{G}^{\sharp}(\alpha^2)$ then comprise local generators for $\mathcal{D}^{\perp}$. A direct computation gives

$$\mathbb{G}^{\sharp}(\alpha^1) = \frac{1}{m}\frac{\partial}{\partial x} - \frac{\rho}{J_{\text{roll}}}\cos\theta\frac{\partial}{\partial\phi}, \quad \mathbb{G}^{\sharp}(\alpha^2) = \frac{1}{m}\frac{\partial}{\partial y} - \frac{\rho}{J_{\text{roll}}}\sin\theta\frac{\partial}{\partial\phi}.$$

Let us now turn to the computation of the orthogonal projection $P_{\mathcal{D}}^{\perp}$. This is actually not an entirely trivial object to compute, but let us go through the steps. From Example 4.73 we have the vector fields

$$X_1 = \rho\cos\theta\frac{\partial}{\partial x} + \rho\sin\theta\frac{\partial}{\partial y} + \frac{\partial}{\partial\phi}, \quad X_2 = \frac{\partial}{\partial\theta}$$

as local generators for $\mathcal{D}$. We also have the vector fields

$$X_3 = \frac{1}{m}\frac{\partial}{\partial x} - \frac{\rho}{J_{\text{roll}}}\cos\theta\frac{\partial}{\partial\phi}, \quad X_4 = \frac{1}{m}\frac{\partial}{\partial y} - \frac{\rho}{J_{\text{roll}}}\sin\theta\frac{\partial}{\partial\phi}$$

as local generators for $\mathcal{D}^{\perp}$. In the basis $\{X_1, X_2, X_3, X_4\}$, $P_{\mathcal{D}}^{\perp}$ has the matrix representation

$$\begin{bmatrix} 0 & 0 & 0 & 0 \\ 0 & 0 & 0 & 0 \\ 0 & 0 & 1 & 0 \\ 0 & 0 & 0 & 1 \end{bmatrix}.$$

Using change of basis formula (2.2), we may now compute the components of $P_{\mathcal{D}}^{\perp}$ relative to the basis $\{\frac{\partial}{\partial x}, \frac{\partial}{\partial y}, \frac{\partial}{\partial\theta}, \frac{\partial}{\partial\phi}\}$ as

$$[P_{\mathcal{D}}^{\perp}] = \frac{1}{m\rho^2 + J_{\text{roll}}}\begin{bmatrix} J_{\text{roll}} + m\rho^2\sin^2\theta & -m\rho^2\sin\theta\cos\theta & 0 & -\rho J_{\text{roll}}\cos\theta \\ -m\rho^2\sin\theta\cos\theta & J_{\text{roll}} + m\rho^2\cos^2\theta & 0 & -\rho J_{\text{roll}}\sin\theta \\ 0 & 0 & 0 & 0 \\ -m\rho\cos\theta & -m\rho\sin\theta & 0 & m\rho^2 \end{bmatrix}.$$

The matrix representation for $P_{\mathcal{D}}$ is then the above matrix subtracted from the 4×4 identity matrix. •

The following result now gives a description of the equations of motion for a simple mechanical control system with constraints.

Proposition 4.82 (Equations of motion for constrained systems). *Let $(\mathsf{Q}, \mathbb{G}, V, F, \mathcal{D})$ be a C^∞-forced simple mechanical system with regular constraints with $L = L_{\mathbb{G}} - V\circ\pi_{\mathsf{TQ}}$ the associated Lagrangian. For a curve $\gamma\colon I\to\mathsf{Q}$, the following statements are equivalent:*

(i) γ is a trajectory for the constrained Lagrangian system $(L, F, \mathcal{D})$ on Q;

(ii) there exists a vector field λ along γ with the property that $\lambda(t)\in\mathcal{D}^{\perp}_{\gamma(t)}$, and such that, together, λ and γ satisfy the equations

$$\overset{\mathbb{G}}{\nabla}_{\gamma'(t)}\gamma'(t) = -\operatorname{grad}V(\gamma(t)) + \mathbb{G}^\sharp(F(t,\gamma'(t))) + \lambda(t),$$
$$P_{\mathcal{D}}^{\perp}(\gamma'(t)) = 0.$$

Proof. The relation $P_{\mathcal{D}}^{\perp}(\gamma'(t)) = 0$ simply means that γ satisfies the constraint. Also, for a constraint force α along γ, γ satisfies the Lagrange–d'Alembert Principle for $F+\alpha$ if and only if

$$\overset{\mathbb{G}}{\nabla}_{\gamma'(t)}\gamma'(t) = -\mathrm{grad}V(\gamma(t)) + \mathbb{G}^\sharp(F(t,\gamma'(t))) + \mathbb{G}^\sharp(\alpha(t)).$$

The result now follows from Definition 4.77, since points in $\mathrm{ann}(\mathcal{D})$ are mapped bijectively onto $\mathcal{D}^\perp$ by $\mathbb{G}^\sharp$. ■

Let us consider Proposition 4.82 for our rolling disk example.

Example 4.83 (Rolling disk cont'd). We have yet to compute the Christoffel symbols for the Levi-Civita connection for the rolling disk, but this is easy, since the coefficients of the Riemannian metric in the coordinates (x, y, θ, ϕ) are constant, implying that all Christoffel symbols are zero. In Example 4.81 we determined that the vector fields

$$X_3 = \frac{1}{m}\frac{\partial}{\partial x} - \frac{\rho}{J_{\mathrm{roll}}}\cos\theta\frac{\partial}{\partial\phi}, \quad X_4 = \frac{1}{m}\frac{\partial}{\partial y} - \frac{\rho}{J_{\mathrm{roll}}}\sin\theta\frac{\partial}{\partial\phi}$$

were a basis for $\mathcal{D}^\perp$. Therefore, the vector field $\lambda(t)$ of Proposition 4.82 must be of the form $\lambda(t) = \lambda^1(t)X_3(\gamma(t)) + \lambda^2(t)X_4(\gamma(t))$ for some functions $\lambda^1, \lambda^2 \colon I \to \mathbb{R}$. Therefore, we may write the differential equation of Proposition 4.82 as

$$\ddot{x} = \frac{1}{m}\lambda^1, \quad \ddot{y} = \frac{1}{m}\lambda^2, \quad \ddot{\theta} = 0,$$
$$\ddot{\phi} = -\frac{\rho}{J_{\mathrm{roll}}}\cos\theta\lambda^1 - \frac{\rho}{J_{\mathrm{roll}}}\sin\theta\lambda^2.$$

To these differential equations, we must append the constraint equations, which, in Proposition 4.82 are the equations $P_{\mathcal{D}}^\perp(\gamma'(t)) = 0$. Since we have computed $P_{\mathcal{D}}^\perp$ in Example 4.81, we can in principle write down these equations. They are, in any case, simply equivalent to the equations

$$\dot{x} = \rho\dot{\phi}\cos\theta, \quad \dot{y} = \rho\dot{\phi}\sin\theta,$$

that we have already given for the constraints to be satisfied. •

4.5.5 The constrained connection

In this section we push the formulation of Proposition 4.82 a little further to arrive at a geometric description of the equations governing a simple mechanical system with constraints. We let $(\mathsf{Q}, \mathbb{G}, V, \mathcal{D})$ be a C^∞-simple mechanical system with constraints, and we assume that $\mathcal{D}$ is regular. Note that, in this case, the bundle maps $P_{\mathcal{D}}$ and $P_{\mathcal{D}}^\perp$ are C^∞-tensor fields of type $(1,1)$. Thus the following definition makes sense, recalling from (3.11) the notion of covariant differentiation of tensor fields.

Definition 4.84 (Constrained connection). Let $(\mathsf{Q}, \mathbb{G})$ be a C^∞-Riemannian manifold and let $\mathcal{D}$ be a regular C^∞-linear velocity constraint.

The corresponding ***constrained affine connection***, or simply the ***constrained connection***, is the affine connection $\overset{\mathcal{D}}{\nabla}$ on Q given by

$$\overset{\mathcal{D}}{\nabla}_X Y = \overset{\mathbb{G}}{\nabla}_X Y + (\overset{\mathbb{G}}{\nabla}_X P_{\mathcal{D}}^{\perp})(Y),$$ •

The geodesic spray associated with the affine connection $\overset{\mathcal{D}}{\nabla}$ we denote by $\overset{\mathcal{D}}{S}$. Before we get to the role of this affine connection in mechanics, let us record some of its differential geometric properties.

Proposition 4.85 (Properties of the constrained connection). *Let* $(\mathsf{Q}, \mathbb{G})$ *be a* C^∞*-Riemannian manifold, with* $\mathcal{D}$ *a* C^∞*-regular linear velocity constraint on* Q*. For the affine connection* $\overset{\mathcal{D}}{\nabla}$*, for* $X \in \Gamma^\infty(\mathsf{TQ})$*, and for* $Y \in \Gamma^\infty(\mathcal{D})$*, the following statements hold:*

(i) $\overset{\mathcal{D}}{\nabla}_X Y = P_{\mathcal{D}}(\overset{\mathbb{G}}{\nabla}_X Y)$ *and, therefore,* $\overset{\mathcal{D}}{\nabla}_X Y \in \Gamma^\infty(\mathcal{D})$*;*

(ii) $(\overset{\mathbb{G}}{\nabla}_X P_{\mathcal{D}}^{\perp})(Y) \in \Gamma^\infty(\mathcal{D}^{\perp})$*.*

Proof. If $Y \in \Gamma^\infty(\mathcal{D})$, then

$$P_{\mathcal{D}}^{\perp}(Y) = 0 \quad \Longrightarrow \quad (\overset{\mathbb{G}}{\nabla}_X P_{\mathcal{D}}^{\perp})(Y) + P_{\mathcal{D}}^{\perp}(\overset{\mathbb{G}}{\nabla}_X Y) = 0.$$

Hence $(\overset{\mathbb{G}}{\nabla}_X P_{\mathcal{D}}^{\perp})(Y) = -P_{\mathcal{D}}^{\perp}(\overset{\mathbb{G}}{\nabla}_X Y) \in \Gamma^\infty(\mathcal{D}^{\perp})$, and, therefore,

$$\overset{\mathcal{D}}{\nabla}_X Y = \overset{\mathbb{G}}{\nabla}_X Y + (\overset{\mathbb{G}}{\nabla}_X P_{\mathcal{D}}^{\perp})(Y) = \overset{\mathbb{G}}{\nabla}_X Y - P_{\mathcal{D}}^{\perp}(\overset{\mathbb{G}}{\nabla}_X Y) = P_{\mathcal{D}}(\overset{\mathbb{G}}{\nabla}_X Y).$$ ■

Remark 4.86. The condition (i) is of sufficient interest that we give it its own language. Given a regular C^∞-distribution $\mathcal{D}$ and a C^∞-affine connection ∇ on Q, ∇ ***restricts*** to $\mathcal{D}$ if $\nabla_X Y \in \Gamma^\infty(\mathcal{D})$ for every $Y \in \Gamma^\infty(\mathcal{D})$. With this terminology, condition (i) of the preceding result then says that $\overset{\mathcal{D}}{\nabla}$ restricts to $\mathcal{D}$. •

With the notion of a constrained connection, we may state the following result.

Theorem 4.87 (Equations of motion using constrained connection). *Let* $(\mathsf{Q}, \mathbb{G}, V, F, \mathcal{D})$ *be a* C^∞*-forced simple mechanical system with regular constraints, with* $L = L_{\mathbb{G}} - V \circ \pi_{\mathsf{TQ}}$ *the associated Lagrangian. For a curve* $\gamma\colon I \to \mathsf{Q}$*, the following statements are equivalent:*

(i) γ *is a trajectory for the constrained Lagrangian system* $(L, F, \mathcal{D})$ *on* Q*;*

(ii) the following conditions hold:

(a) $\gamma'(t_0) \in \mathcal{D}_{\gamma(t_0)}$ *for some* $t_0 \in I$*, and*

(b) $\overset{\mathcal{D}}{\nabla}_{\gamma'(t)}\gamma'(t) = -P_{\mathcal{D}}(\operatorname{grad} V(\gamma(t))) + P_{\mathcal{D}}(\mathbb{G}^\sharp(F(t, \gamma'(t))))$.

Furthermore, given $(t_0, v_{q_0}) \in \mathbb{R}\times\mathcal{D}$, there exists an interval $J \subset \mathbb{R}$ containing t_0 in its interior, and a locally absolutely differentiable curve $\gamma\colon J \to \mathsf{Q}$ such that γ satisfies (ii b) and such that $\gamma'(t_0) = v_{q_0}$. Moreover, if $\tilde{J}$ is another such interval and if $\tilde{\gamma}\colon \tilde{J} \to \mathsf{Q}$ is another such curve, then $\gamma(t) = \tilde{\gamma}(t)$ for $t \in J \cap \tilde{J}$.

Proof. First suppose that (i) holds. This is equivalent to the holding of the two equations of part (ii) of Proposition 4.82. In particular (ii a) above holds. Furthermore, we have $P_{\mathcal{D}}^{\perp}(\gamma'(t)) = 0$ for $t \in I$. Differentiating this relation, and solving for λ in the first of the equations from Proposition 4.82(ii), gives

$$\lambda(t) = -(\overset{\mathbb{G}}{\nabla}_{\gamma'(t)} P_{\mathcal{D}}^{\perp})(\gamma'(t)) - P_{\mathcal{D}}^{\perp}(\mathbb{G}^{\sharp}(F(t,\gamma'(t)))) + P_{\mathcal{D}}^{\perp}(\operatorname{grad}V(\gamma(t))). \quad (4.19)$$

Thus

$$\begin{aligned}\overset{\mathbb{G}}{\nabla}_{\gamma'(t)}\gamma'(t) = -(\overset{\mathbb{G}}{\nabla}_{\gamma'(t)} P_{\mathcal{D}}^{\perp})(\gamma'(t)) - P_{\mathcal{D}}^{\perp}\big(\mathbb{G}^{\sharp}(F(t,\gamma'(t)))\\ - \operatorname{grad}V(\gamma(t))\big) + \mathbb{G}^{\sharp}(F(t,\gamma'(t))) - \operatorname{grad}V(\gamma(t)),\end{aligned}$$

which is equivalent to (ii b) by the definition of $\overset{\mathcal{D}}{\nabla}$, and since $P_{\mathcal{D}} = \mathrm{id}_{\mathsf{TQ}} - P_{\mathcal{D}}^{\perp}$.

Now suppose that $\gamma'(t_0) \in \mathcal{D}_{\gamma(t_0)}$ for some $t_0 \in I$ and that γ satisfies (ii b). To show that γ satisfies the constraint, we need to show that the submanifold $\mathcal{D}$ of TQ is invariant under the flow of the vector field on TQ defined by part (ii b). This is most easily done using ideas established during the course of the proof of Theorem 3.108. In particular, from the proof of that result follows the fact that the geodesic spray $\overset{\mathcal{D}}{S}$ of $\overset{\mathcal{D}}{\nabla}$ is tangent to $\mathcal{D}$ by virtue of Proposition 4.85(i). Also note that the vector field

$$t \mapsto -P_{\mathcal{D}}(\operatorname{grad}V(\gamma(t))) + P_{\mathcal{D}}(\mathbb{G}^{\sharp}(F(t,\gamma'(t))))$$

along γ is $\mathcal{D}$-valued, so its vertical lift is tangent to $\mathcal{D}$. This fact is asserted during the course of the proof of Theorem 3.108, and can be directly proved easily. Now we note that part (ii b) is equivalent to the following equation on TQ:

$$\Upsilon'(t) = \overset{\mathcal{D}}{S}(\Upsilon(t)) + \operatorname{vlft}\big(-P_{\mathcal{D}}(\operatorname{grad}V(\gamma(t))) + P_{\mathcal{D}}(\mathbb{G}^{\sharp}(F(t,\Upsilon(t))))\big), \quad (4.20)$$

where $\Upsilon = \gamma'$. This sort of equation will come up in Section 4.6.3. The punchline is that all terms on the right-hand side of (4.20) are tangent to $\mathcal{D}$, and so $\Upsilon = \gamma'$ will also be tangent to $\mathcal{D}$. That is to say, γ satisfies the constraint, and so satisfies the second of the equations in part (ii) of Proposition 4.82. Also, γ satisfies the first of the equations in part (ii) of Proposition 4.82 with λ given by (4.19). Thus γ is a trajectory for the constrained Lagrangian system $(L, F, \mathcal{D})$ on Q.

The final assertion in the theorem follows from Corollary A.12. ■

The theorem suggests the possibility of generalizing the notion of a forced simple mechanical system with constraints to allow for a general affine connection. To this end, we need a notion of force that is naturally TQ-valued, rather than $\mathsf{T}^*\mathsf{Q}$-valued.

Definition 4.88 (Vector force). For $r \in \mathbb{N} \cup \{\infty\} \cup \{\omega\}$, a ***$C^r$-vector force*** is a map $Y : \mathbb{R} \times \mathsf{TQ} \to \mathsf{TQ}$ with the property that Y is a locally integrally class C^r bundle map over id_{Q} (see Definition 3.125 and Section A.2.1). A C^r-vector force Y is

(i) ***time-independent*** if there exists a C^r-fiber bundle map $Y_0 : \mathsf{TQ} \to \mathsf{TQ}$ over id_{Q} with the property that $Y(t, v_q) = Y_0(v_q)$, and is

(ii) ***basic*** if there exists a C^r-vector field Y_0 on Q such that $Y(t, v_q) = Y_0(q)$. •

We may now utilize this notion of a vector force to provide the following generalization of a forced simple mechanical system with constraints.

Definition 4.89 (Forced affine connection system). For $r \in \mathbb{N} \cup \{\infty\} \cup \{\omega\}$, a ***$C^r$-forced affine connection system*** is a 4-tuple $(\mathsf{Q}, \nabla, Y, \mathcal{D})$, where

(i) Q is a manifold,

(ii) ∇ is an affine connection on Q,

(iii) Y is a vector force, and

(iv) $\mathcal{D}$ is a regular distribution to which ∇ restricts, and for which Y is $\mathcal{D}$-valued,

with all data being class C^r. •

The equations governing a forced affine connection system are then

$$\nabla_{\gamma'(t)}\gamma'(t) = Y(t, \gamma'(t)).$$

Note that a simple mechanical system with regular constraints gives rise to a forced affine connection system $(\mathsf{Q}, \overset{\mathcal{D}}{\nabla}, P_{\mathcal{D}}(\mathbb{G}^\sharp \circ F - \pi^*_{\mathsf{T}^*\mathsf{Q}}\mathrm{d}V), \mathcal{D})$.

Theorem 4.87 has the following corollary, which is quite remarkable at a first glance. It is this corollary that enables us to seamlessly include systems with constraints into our approach for handling certain types of control problems.

Corollary 4.90 (Unforced, constrained motion is geodesic). *Let $(\mathsf{Q}, \mathbb{G}, 0, \mathcal{D})$ be a C^∞-simple mechanical system with regular constraints. The trajectories for the constrained Lagrangian system $(L_{\mathbb{G}}, 0, \mathcal{D})$ are exactly the geodesics of $\overset{\mathcal{D}}{\nabla}$ having some tangent vector (and, therefore, all tangent vectors) in $\mathcal{D}$.*

Remarks 4.91. 1. The idea that trajectories for constrained Lagrangian systems may be represented as the geodesics of an affine connection seems to originate with Synge [1928]. Other researchers have subsequently and independently rediscovered this interesting idea [Bloch and Crouch 1998, Cattaneo 1963, Cattaneo-Gasparini 1963, Vershik 1984]. The approach takes here is that of Lewis [1998].

2. In the case of a not necessarily regular constraint, the construction of $\overset{\mathcal{D}}{\nabla}$ is not currently understood.

3. In examples, the Christoffel symbols for $\overset{\mathcal{D}}{\nabla}$ are often extremely unwieldy, as is amply evidenced by the formulae in [Lewis 2000b]. In the next section we provide an alternative means of locally representing the equations of Theorem 4.87 that, in practice, is far superior to proceeding straightforward with the computation of Christoffel symbols. •

4.5.6 The Poincaré representation of the equations of motion

In this section we provide a means of locally representing the expression $\nabla_{\gamma'(t)}\gamma'(t)$, first for a general affine connection, then for the constrained connection $\overset{\mathcal{D}}{\nabla}$. These local representations will prove useful to us in subsequent sections when we need to produce equations of motion for systems with constraints. The essential idea here is known in the classic literature as "the Poincaré equations" [Arnol'd 1988].

Let ∇ be an affine connection on Q, let $\mathcal{U} \subset \mathsf{Q}$ be an open set, and suppose that there exists a collection of vector fields $\mathscr{X} = \{X_1, \dots, X_n\}$ on $\mathcal{U}$ that form a basis for each tangent space in $\mathcal{U}$. For this data, we define the ***generalized Christoffel symbols*** for ∇ with respect to $\mathscr{X}$ as the n^3 functions $\overset{\mathscr{X}}{\Gamma}{}^k_{ij} \colon \mathcal{U} \to \mathbb{R}$, $i, j, k \in \{1, \dots, n\}$, defined by

$$\nabla_{X_i} X_j = \overset{\mathscr{X}}{\Gamma}{}^k_{ij} X_k, \qquad i, j \in \{1, \dots, n\}.$$

In the special case that $(\mathcal{U}, \phi)$ is a chart and $X_i = \frac{\partial}{\partial q^i}$, $i \in \{1, \dots, n\}$, are the coordinate basis vector fields, then the generalized Christoffel symbols are the usual Christoffel symbols.

We now provide a local expression for $\nabla_{\gamma'(t)}\gamma'(t)$ using generalized Christoffel symbols.

Proposition 4.92 (Geometric acceleration in pseudo-velocities). *Let ∇ be a C^∞-affine connection on a manifold Q, let $\mathcal{U} \subset \mathsf{Q}$ be an open subset, and let $\mathscr{X} = \{X_1, \dots, X_n\}$ be vector fields on $\mathcal{U}$ that form a basis for each tangent space in $\mathcal{U}$. If $\gamma \colon I \to \mathcal{U}$ is a differentiable curve, define $v^k \colon I \to \mathbb{R}$, $k \in \{1, \dots, n\}$, by asking that $\gamma'(t) = v^k(t) X_k(\gamma(t))$. Then*

$$\nabla_{\gamma'(t)}\gamma'(t) = \big(\dot{v}^k(t) + \overset{\mathscr{X}}{\Gamma}{}^k_{ij}(\gamma(t)) v^i(t) v^j(t)\big) X_k(\gamma(t)).$$

Further suppose that $\mathcal{D}$ is a regular C^∞-distribution on $\mathcal{U}$, that, for each $q \in \mathcal{U}$, the vectors $\{X_1(q), \dots, X_{\mathrm{rank}(\mathcal{D})}(q)\}$ form a basis for $\mathcal{D}_q$, and that ∇ restricts to $\mathcal{D}$. Then, if $\gamma'(t) \in \mathcal{D}_{\gamma(t)}$ for each $t \in I$,

$$\nabla_{\gamma'(t)}\gamma'(t) = \big(\dot{v}^\delta(t) + \overset{\mathscr{X}}{\Gamma}{}^\delta_{\alpha\beta}(\gamma(t)) v^\alpha(t) v^\beta(t)\big) X_\delta(\gamma(t)), \quad \delta \in \{1, \dots, \mathrm{rank}(\mathcal{D})\},$$

and where the implied summation over α and β goes from 1 to $\mathrm{rank}(\mathcal{D})$.

Proof. The first assertion is a simple computation using properties of affine connections:

$$\begin{aligned}\nabla_{\gamma'(t)}\gamma'(t) &= \nabla_{\gamma'(t)}v^i(t)X_i(\gamma(t)) \\ &= v^i(t)\nabla_{\gamma'(t)}X_i(\gamma(t)) + \dot{v}^i(t)X_i(\gamma(t)) \\ &= \dot{v}^i(t)X_i(\gamma(t)) + v^i(t)v^j(t)\nabla_{X_j(\gamma(t))}X_i(\gamma(t)).\end{aligned}$$

In this computation we have used Exercise E3.37. Using the definition of the generalized Christoffel symbols and relabeling indices gives the result.

Now suppose that the second set of hypotheses hold. The second conclusion then holds, since $\gamma'(t) = v^\alpha(t)X_\alpha(\gamma(t))$, with the summation being over $\{1,\dots,\mathrm{rank}(\mathcal{D})\}$, and by the assumed property that ∇ restricts to $\mathcal{D}$. ■

Remarks 4.93. 1. Let Q, ∇, $\mathscr{X}$, γ, and v^k, $k \in \{1,\dots,n\}$, be as in the proposition. The equations

$$\dot{v}^k(t) + \overset{\mathscr{X}}{\Gamma}{}^k_{ij}(\gamma(t))v^i(t)v^j(t) = 0, \qquad k \in \{1,\dots,n\}, \tag{4.21}$$

are known as the ***Poincaré representation*** for the geodesic equations. They are a coordinate representation of the geodesic equation $\nabla_{\gamma'(t)}\gamma'(t) = 0$.

2. In the general treatment of the Poincaré representation, the functions v^k, $k \in \{1,\dots,n\}$, are called ***pseudo-velocities***. •

When ∇ is the constrained affine connection associated with a Riemannian metric $\mathbb{G}$ and a regular constraint $\mathcal{D}$, and when the vector fields $\mathscr{X}$ are $\mathbb{G}$-orthogonal, the generalized Christoffel symbols are easily computed.

Proposition 4.94 (Generalized Christoffel symbols in orthogonal basis). *Let $(\mathsf{Q},\mathbb{G})$ be a C^∞-Riemannian manifold, let $\mathcal{D}$ be a regular C^∞-constraint on Q, and let $\mathcal{U} \subset \mathsf{Q}$ be an open set on which the rank of $\mathcal{D}$ is constant. Let $\mathscr{X} = \{X_1,\dots,X_n\} \subset \Gamma^\infty(\mathsf{TQ})$ have the property that, for each $q \in \mathcal{U}$, $\{X_1(q),\dots,X_{\mathrm{rank}(\mathcal{D})}(q)\}$ is a $\mathbb{G}$-orthogonal basis for $\mathcal{D}_q$. Then the generalized Christoffel symbols for $\overset{\mathcal{D}}{\nabla}$ satisfy*

$$\overset{\mathscr{X}}{\Gamma}{}^\delta_{\alpha\beta}(q) = \frac{1}{\|X_\delta(q)\|^2_{\mathbb{G}}}\mathbb{G}(\overset{\mathbb{G}}{\nabla}_{X_\alpha}X_\beta(q), X_\delta(q)), \qquad \alpha,\beta,\delta \in \{1,\dots,\mathrm{rank}(\mathcal{D})\}.$$

Proof. By part (i) of Proposition 4.85 we have, for $\alpha,\beta,\delta \in \{1,\dots,\mathrm{rank}(\mathcal{D})\}$,

$$\begin{aligned}\|X_\delta\|^{-2}_{\mathbb{G}}\,\mathbb{G}(\overset{\mathcal{D}}{\nabla}_{X_\alpha}X_\beta, X_\delta) &= \|X_\delta\|^{-2}_{\mathbb{G}}\,\mathbb{G}(P_{\mathcal{D}}(\overset{\mathbb{G}}{\nabla}_{X_\alpha}X_\beta), X_\delta) \\ &= \|X_\delta\|^{-2}_{\mathbb{G}}\,\mathbb{G}(\overset{\mathbb{G}}{\nabla}_{X_\alpha}X_\beta, X_\delta).\end{aligned}$$

However, this last expression is simply the $\mathbb{G}$-orthogonal projection of $\overset{\mathbb{G}}{\nabla}_{X_\alpha}X_\beta$ onto the basis vector X_δ, and from this the result follows. ■

Remark 4.95. One of the useful features of the generalized Christoffel symbols in an orthogonal basis is that their computation requires neither the computation of $P_{\mathcal{D}}^{\perp}$, nor the covariant differentiation of that tensor. It is these computations that make the computation of the standard Christoffel symbols for $\overset{\mathcal{D}}{\nabla}$ so cumbersome, and which make the use of an orthogonal basis so inviting. The idea of using an orthogonal basis for velocities is classic. Our presentation follows the paper of Bullo and Žefran [2002]. •

Let us see how these results combine to give the equations of motion for the rolling disk. We refer the reader to Section 13.4 and Exercise E4.42 for more complex examples of the procedure for computing generalized Christoffel symbols in an orthogonal basis.

Example 4.96 (Rolling disk cont'd). Using the orthogonal local generators $\{X_1, X_2, X_3, X_4\}$ of Example 4.81, we determine, using Proposition 4.94, that the generalized Christoffel symbols are all identically zero. Therefore, the equations for the rolling disk, in the absence of external forces, are the equations (4.21), in addition to the equation $\gamma'(t) = v^{\alpha}(t)X_{\alpha}(\gamma(t))$. Translating this into coordinates gives

$$\dot{x} = \rho\dot{\phi}\cos\theta, \quad \dot{y} = \rho\dot{\phi}\sin\theta, \quad \ddot{\theta} = 0, \quad \ddot{\phi} = 0.$$

Apart from their obvious simplicity, the remarkable feature of these equations, as compared to those obtained in Example 4.83, is the absence of the components of the vector field λ. •

4.5.7 Special features of holonomic constraints

In the case in which the constraint is holonomic, the constrained connection reduces to a familiar object. The setting we consider is as follows. We let $(\mathsf{Q}, \mathbb{G})$ be a C^{∞}-Riemannian manifold and we let $\mathcal{D}$ be a holonomic C^{∞}-linear velocity constraint having a submanifold $\mathsf{Q}_0 \subset \mathsf{Q}$ as a maximal integral manifold. We note that $(\mathsf{Q}_0, \mathbb{G}_0 \triangleq i_{\mathsf{Q}_0}^{*}\mathbb{G})$ is also a Riemannian manifold.

Proposition 4.97 (Equations of motion for holonomic constraints). *Let* $(\mathsf{Q}, \mathbb{G})$, $\mathcal{D}$, *and* $(\mathsf{Q}_0, \mathbb{G}_0)$ *be as above. For vector fields* $X, Y \in \Gamma^{\infty}(\mathsf{TQ}_0)$, *let* $\tilde{X}, \tilde{Y} \in \Gamma^{\infty}(\mathsf{TQ})$ *have the property that they agree with* X *and* Y *when evaluated on* Q_0. *Then* $\overset{\mathcal{D}}{\nabla}_{\tilde{X}}\tilde{Y}(q) = \overset{\mathbb{G}_0}{\nabla}_X Y(q)$ *for all* $q \in \mathsf{Q}_0$.

Proof. In the proof, when we write vector fields on Q_0, we shall consider them to be extended to Q when necessary. One may verify, by a direct computation using a submanifold chart for Q_0, that the results do not depend on the choice of extension.[7]

[7] There is a question here about whether it is *possible* to extend a vector field from Q_0 to Q. This can be done when Q possesses a Riemannian metric, as is the case here. We refer to the discussion of tubular neighborhoods in [Hirsch 1976].

We must show that $\overset{\mathfrak{D}}{\nabla}$, restricted to $\Gamma^\infty(\mathsf{TQ}_0)$, satisfies Theorem 3.104(ii) and preserves the metric $\mathbb{G}_0$. We first show that $\overset{\mathfrak{D}}{\nabla}$ satisfies Theorem 3.104(ii). Let $X, Y \in \Gamma^\infty(\mathsf{TQ}_0)$. We have

$$\overset{\mathfrak{D}}{\nabla}_X Y - \overset{\mathfrak{D}}{\nabla}_Y X = \overset{\mathbb{G}}{\nabla}_X Y - \overset{\mathbb{G}}{\nabla}_Y X + (\overset{\mathbb{G}}{\nabla}_X P_{\mathfrak{D}}^{\perp})(Y) - (\overset{\mathbb{G}}{\nabla}_Y P_{\mathfrak{D}}^{\perp})(X).$$

By Proposition 4.85(ii) we see that $(\overset{\mathbb{G}}{\nabla}_X P_{\mathfrak{D}}^{\perp})(Y)$ and $(\overset{\mathbb{G}}{\nabla}_Y P_{\mathfrak{D}}^{\perp})(X)$ are orthogonal to Q_0, and by Proposition 4.85(i) we see that $\overset{\mathfrak{D}}{\nabla}_X Y, \overset{\mathfrak{D}}{\nabla}_Y X \in \Gamma^\infty(\mathsf{TQ}_0)$. Also, since $\overset{\mathbb{G}}{\nabla}$ satisfies Theorem 3.104(ii), and since $\mathfrak{D}$ is integrable,

$$\overset{\mathbb{G}}{\nabla}_X Y - \overset{\mathbb{G}}{\nabla}_Y X = [X, Y] \in \Gamma^\infty(\mathsf{TQ}_0).$$

Thus we obtain

$$\overset{\mathfrak{D}}{\nabla}_X Y - \overset{\mathfrak{D}}{\nabla}_Y X = [X, Y], \quad (\overset{\mathbb{G}}{\nabla}_X P_{\mathfrak{D}}^{\perp})(Y) = (\overset{\mathbb{G}}{\nabla}_Y P_{\mathfrak{D}}^{\perp})(X),$$

which shows in particular that $\overset{\mathfrak{D}}{\nabla}$ satisfies Theorem 3.104(ii).

Now we show that $\overset{\mathfrak{D}}{\nabla}$ preserves $\mathbb{G}_0$. Let $X, Y, Z \in \Gamma^\infty(\mathsf{TQ}_0)$. Since $\overset{\mathbb{G}}{\nabla}$ is a metric connection,

$$\mathscr{L}_X(\mathbb{G}(Y, Z)) = \mathbb{G}(\overset{\mathbb{G}}{\nabla}_X Y, Z) + \mathbb{G}(Y, \overset{\mathbb{G}}{\nabla}_X Z).$$

Now we have

$$\mathbb{G}(\overset{\mathfrak{D}}{\nabla}_X Y, Z) = \mathbb{G}(\overset{\mathbb{G}}{\nabla}_X Y, Z) + \mathbb{G}((\overset{\mathbb{G}}{\nabla}_X P_{\mathfrak{D}}^{\perp})(Y), Z) = \mathbb{G}(\overset{\mathbb{G}}{\nabla}_X Y, Z),$$

since $(\overset{\mathbb{G}}{\nabla}_X P_{\mathfrak{D}}^{\perp})(Y)$ is orthogonal to Q_0. Similarly, we may show that

$$\mathbb{G}(Y, \overset{\mathfrak{D}}{\nabla}_X Z) = \mathbb{G}(Y, \overset{\mathbb{G}}{\nabla}_X Z).$$

Therefore,

$$\mathscr{L}_X(\mathbb{G}(Y, Z)) = \mathbb{G}(\overset{\mathfrak{D}}{\nabla}_X Y, Z) + \mathbb{G}(Y, \overset{\mathfrak{D}}{\nabla}_X Z),$$

which, when restricted to Q_0, shows that $\overset{\mathfrak{D}}{\nabla}$ preserves $\mathbb{G}_0$. ∎

Remark 4.98. The outcome of the preceding result is that restriction to Q_0 commutes with the computation of the constrained equations on Q and the unconstrained Euler–Lagrange equations on Q_0. That is to say, consider the following two constructions.

1. On Q, treat the holonomic constraint in the way in which constraints are normally treated, so giving the equations for the solution through $v_{q_0} \in \mathsf{TQ}_0$ as in Theorem 4.87.

2. Pull-back the kinetic energy metric to Q_0, and then, on Q_0, compute the usual unconstrained Euler-Lagrange equations to give the equations for the solution through $v_{q_0} \in \mathsf{TQ}_0$.

These two approaches yield the same solution through v_{q_0}. This provides an *a posteriori* justification of restriction to $\mathsf{Q} \subset \mathsf{Q}_{\text{free}}$ when talking about interconnected mechanical systems. •

Example 4.99 (Geodesics on the 2-sphere). We consider the Riemannian manifold $(\mathsf{Q} = \mathbb{R}^3, \mathbb{G} = \mathbb{G}_{\mathbb{R}^3})$, and we define a linear velocity constraint $\mathcal{D}$ on Q by

$$\mathcal{D}_{\boldsymbol{x}} = \left\{ \boldsymbol{v} \in \mathsf{T}_{\boldsymbol{x}}\mathsf{Q} \simeq \mathbb{R}^3 \;\middle|\; \mathbb{G}(\boldsymbol{v}, \boldsymbol{x}) = 0 \right\}.$$

Thus the subspace $\mathcal{D}_{\boldsymbol{x}}$ is orthogonal to $\boldsymbol{x}$. It is clear that this distribution possesses $\mathsf{Q}_0 = \mathbb{S}^2$ as a maximal integral manifold. Let $i_{\mathbb{S}^2} : \mathbb{S}^2 \to \mathbb{R}^3$ be the inclusion. The Riemannian metric $\mathbb{G}_{\text{rnd}} = i_{\mathbb{S}^2}^* \mathbb{G}$ on $\mathbb{S}^2$ is called the ***round Riemannian metric***. We shall use Proposition 4.97 to provide the equations governing the geodesics of the round Riemannian metric. What Proposition 4.97, together with Theorem 4.87, tells us is that curve $t \mapsto \boldsymbol{x}(t) \in \mathbb{S}^2$ is a geodesic on $\mathbb{S}^2$ if and only if it satisfies the equations

$$\ddot{\boldsymbol{x}}(t) = \boldsymbol{\lambda}(t), \quad \mathbb{G}(\dot{\boldsymbol{x}}(t), \boldsymbol{x}(t)) = 0, \tag{4.22}$$

for some vector field $t \mapsto \boldsymbol{\lambda}(t)$ along $t \mapsto \boldsymbol{x}(t)$ having the property that $\boldsymbol{\lambda}(t)$ is orthogonal to $\mathbb{S}^2$ for each t. Differentiating the second of these equations with respect to t gives

$$\mathbb{G}(\ddot{\boldsymbol{x}}(t), \boldsymbol{x}(t)) + \|\dot{\boldsymbol{x}}(t)\|_{\mathbb{R}^2}^2 = 0.$$

Taking the inner product of the first of equations (4.22) with $\boldsymbol{x}(t)$ gives $\mathbb{G}(\boldsymbol{x}(t), \ddot{\boldsymbol{x}}(t)) = \mathbb{G}(\boldsymbol{x}(t), \boldsymbol{\lambda}(t))$. Since $\boldsymbol{\lambda}(t)$ is orthogonal to $\mathbb{S}^2$, this means that $\boldsymbol{\lambda}(t)$ is the orthogonal projection of $\ddot{\boldsymbol{x}}(t)$ onto the orthogonal complement. In other words, $\boldsymbol{\lambda}(t) = \mathbb{G}(\boldsymbol{x}(t), \ddot{\boldsymbol{x}}(t))\boldsymbol{x}(t)$. From this we ascertain that the geodesics of $\mathbb{G}_{\text{rnd}}$ satisfy

$$\ddot{\boldsymbol{x}}(t) + \|\dot{\boldsymbol{x}}(t)\|_{\mathbb{R}^2}^2 \boldsymbol{x}(t) = \boldsymbol{0}.$$

With these equations, one can directly verify that the geodesic with initial velocity $\boldsymbol{v} \in \mathsf{T}_{\boldsymbol{x}}\mathbb{S}^2 \setminus \{\boldsymbol{0}\}$ is given by

$$\boldsymbol{x}(t) = \cos(\|\boldsymbol{v}\|_{\mathbb{R}^2} t)\boldsymbol{x} + \sin(\|\boldsymbol{v}\|_{\mathbb{R}^2} t)\frac{\boldsymbol{v}}{\|\boldsymbol{v}\|_{\mathbb{R}^2}}.$$

This may be checked with a direct substitution into the differential equation. We recognize these geodesics as "great circles," and we depict the situation in Figure 4.15.

The above computations further verify that, if $t \mapsto \boldsymbol{x}(t)$ is any curve on $\mathbb{S}^2$, then we have

$$\overset{\mathbb{G}_{\text{rnd}}}{\nabla}_{\dot{\boldsymbol{x}}(t)} \dot{\boldsymbol{x}}(t) = \ddot{\boldsymbol{x}}(t) + \|\dot{\boldsymbol{x}}(t)\|_{\mathbb{R}^2}^2 \boldsymbol{x}(t).$$

We shall use this formula for the affine connection on $\mathbb{S}^2$ in subsequent development when we study control systems on the sphere. •

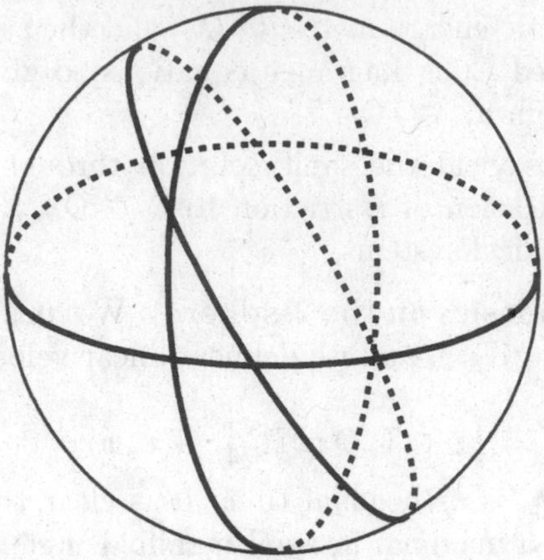

Figure 4.15. Geodesics for the round metric on $\mathbb{S}^2$

4.6 Simple mechanical control systems and their representations

Up to this point in the book, we have said little in a formal way about control theory. The objective in this section is to formulate the classes of mechanical control systems we study, and provide convenient means of representing them, both locally and globally. This section provides a summary of all notions we have presented so far and a reference point for the following chapters. It might be advisable to skim this chapter on a first reading, and then refer to it subsequently when we refer in the text to specific classes of mechanical (control) systems.

4.6.1 Control-affine systems

Before we discuss *mechanical* control systems, let us briefly say what we mean by a control system. The notion of a control-affine system is standard in the nonlinear control literature, and for more information we refer the reader to the texts [Agrachev and Sachkov 2004, Isidori 1995, Nijmeijer and van der Schaft 1990].

Definition 4.100 (Control-affine system). For $r \in \mathbb{N} \cup \{\infty\} \cup \{\omega\}$, a ***$C^r$-control-affine system*** is a triple $(\mathsf{M}, \mathscr{C} = \{f_0, f_1, \dots, f_m\}, U)$, where

(i) M is a C^r-manifold,
(ii) $\mathscr{C} = \{f_0, f_1, \dots, f_m\}$ are C^r-vector fields on M, and
(iii) $U \subset \mathbb{R}^m$. •

Let us immediately write the differential equations governing the system:

$$\gamma'(t) = f_0(\gamma(t)) + \sum_{a=1}^{m} u^a(t) f_a(\gamma(t)). \tag{4.23}$$

Here $u: I \to U \subset \mathbb{R}^m$ are the ***controls*** or ***inputs*** taking values in the ***control set*** U.[8] Controls are assumed to be locally integrable (see Appendix A). When the controls are specified, they give rise to the locally absolutely continuous ***trajectory*** $\gamma: I \to \mathsf{M}$ taking values in the ***state manifold*** M. The vector field f_0 is the ***drift vector field***, which specifies the dynamics of the system in the absence of controls. The vector fields $f_1, \ldots, f_m$ are the ***control vector fields*** or ***input vector fields***. A control-affine system for which the drift vector field is identically zero is called a ***driftless*** control system. It is not uncommon to see a driftless system be referred to as a "nonholonomic control system." However, we shall not use this terminology, since it creates an unnecessary coincidence of terminology with nonholonomic constraints in mechanics.

Remarks 4.101. In physical applications it is typically the case that control-affine systems have only a finite number of input vector fields. Furthermore, these input vector fields are typically part of the problem specification. For this reason, there is physically no limitation made by specifying a finite number of input vector fields. However, it is worth making some comments on the consequences of this assumption in terms of its mathematical generality.

1. When the distribution generated by the input vector fields is regular and has bounded rank, and when M is paracompact, then there is no loss of generality in assuming that the number of input vector fields is finite. That is to say, under these assumptions, if a control-affine system were to have infinite number of input vector fields, then it would be possible to choose a finite number of them that would generate the same input distribution. This is a consequence of Theorem 3.114.
2. When the control-affine system is analytic, and when one is interested only in local questions, then the assumption of a finite number of input vector fields is again made without loss of generality. Again, this means that, even if a system were to have an infinite number of input vector fields, provided they generate an analytic distribution, then there would be a finite number of input vector fields locally generating the same distribution. In this case, it is possible that the finite generators may not be included in the original set of input vector fields. This is the content of Theorem 3.112.
3. In cases different from the preceding two, the assumption that there be a finite number of input vector fields will generally be a restrictive assumption. •

A special sort of control-affine system is one that is linear. First let us state the formal definition, and then, after we do this, understand how this gives a control-affine system.

[8] Note that we shall adopt the convention of *not* writing controls with a bold symbol, despite the fact that they are $\mathbb{R}^m$-valued. This is somewhat at odds with the notation predominantly used in the book, and we adopt it for aesthetic reasons.

Definition 4.102 (Linear control system). A ***linear control system*** is a triple (V, A, B), where

(i) V is a finite-dimensional $\mathbb{R}$-vector space,
(ii) $A \in L(\mathsf{V}; \mathsf{V})$, and
(iii) $B \in L(\mathbb{R}^m; \mathsf{V})$ for some $m \in \mathbb{Z}_+$. •

Remark 4.103. Note that we omit from the data of a linear control system the control set U. One can certainly include this piece of data. However, in our dealings with linear control systems we will not have occasion to consider the issues arising from the nature of the control set. •

Given a linear control system (V, A, B), we define a control-affine system $(\mathsf{M}, \mathscr{C} = \{f_0, f_1, \dots, f_m\}, U)$ by $\mathsf{M} = \mathsf{V}$, $f_0(x) = A(x)$, $f_a(x) = B(e_a)$, $a \in \{1, \dots, m\}$, and $U = \mathbb{R}^m$. Note that, in making these definitions, we make the natural identification of $\mathsf{T}_x\mathsf{V}$ with V. The equations (4.23) for a linear control system then read

$$\dot{x}(t) = A(x(t)) + B(u(t)).$$

Linear systems are much studied in the control theory literature, and there are many basic references, including [Brockett 1970, Chen 1984, Kailath 1980, Sontag 1998, Wonham 1985]. On occasion, throughout the text, we will use linear control systems to illustrate how certain concepts appear for this particularly structured class of systems. Also, linear control systems come up in the process of linearization, and so, even for nonlinear systems, one can sometimes answer certain questions using linear theory. In Section 4.6.5 we shall consider linear mechanical control systems.

The control theoretic tasks we consider in this book include the following rather imprecisely stated problems.

1. ***Controllability:*** Given states $x_1, x_2 \in \mathsf{M}$, does there exist a control $u\colon [0, T] \to U$ such that the solution $\gamma\colon [0, T] \to \mathsf{M}$ to the initial value problem

$$\gamma'(t) = f_0(\gamma(t)) + \sum_{a=1}^{m} u^a(t) f_a(\gamma(t)), \quad \gamma(0) = x_1,$$

   has the property that $\gamma(T) = x_2$?
2. ***Motion planning:*** If the answer to the preceding question is, "Yes," can a simple expression for a control u be obtained? Often, the solution to this problem comes in the form of a precomputed function of time u, designed using some symbolic or numerical methodology. In this case the control is said to be ***open-loop*** in that, once it is implemented, no measurements or actions are taken to correct for possible inaccuracies and to account for the fact that the control was precomputed for an idealized model. An idealized model may in fact be a poor representation of the actual system.

3. ***Stabilization:*** Given a state $x_0 \in \mathsf{M}$, is it possible to design the controls as functions of the state, $u\colon \mathsf{M} \to U$, such that the resulting ***closed-loop system***,

$$\gamma'(t) = f_0(\gamma(t)) + \sum_{a=1}^{m} u^a(\gamma(t)) f_a(\gamma(t)),$$

renders the point x_0 stable in some sense? This means of designing the controls so that they are dependent on the state is known as ***closed-loop*** design. In applications, this type of control law has the following advantage. If the behavior of the system is measured, then the controls have a chance to react to the difference between what the system is doing and what it is supposed to be doing. This general philosophy is known as using ***feedback***.

In the stabilization problem, one might also stabilize to manifolds more general than points.

4. ***Trajectory tracking:*** This problem closely resembles the stabilization problem, except that, rather than stabilizing a point x_0, one wants to steer the system state along a reference trajectory $\gamma_{\mathrm{ref}}\colon I \to \mathsf{M}$.
5. ***Optimal control:*** For any of the preceding problems, one might also impose a cost on either or both of the control and the state, and ask whether it is possible to accomplish the objective while minimizing the cost. Optimal control is not considered in the text, but a chapter on optimal control will be included as an internet supplement. Topics in optimal control are also dealt with in the books of Bloch [2003] and [Agrachev and Sachkov 2004].

The above is an extremely incomplete list of interesting control problems. Problems not addressed include anything involving outputs (including state estimation and observers), disturbance rejection, model reduction, robust control, adaptive control, and model predictive control. We refer the reader to the texts [Isidori 1995, Khalil 2001, Sastry 1999, Sontag 1998], and the references contained therein to gain an introduction to parts of the broad subject of mathematical control theory.

4.6.2 Classes of simple mechanical control systems

Let us begin by defining the most general class of mechanical control system we will consider. That is to say, every system, and every control problem we consider in mechanics, will fall under the umbrella of this large class.

Definition 4.104 (General simple mechanical control system). For $r \in \mathbb{N} \cup \{\infty\} \cup \{\omega\}$, a ***$C^r$-general simple mechanical control system*** is a 7-tuple $(\mathsf{Q}, \mathbb{G}, V, F, \mathcal{D}, \mathscr{F}, U)$, where

(i) Q is a manifold (the ***configuration manifold***),
(ii) $\mathbb{G}$ is a Riemannian metric on Q (the ***kinetic energy metric***),

(iii) V is a function on Q (the ***potential function***),
(iv) F is a force on Q (the ***uncontrolled external force***, or simply ***external force***),
(v) $\mathcal{D}$ is a linear velocity constraint (the ***constraint distribution***),
(vi) $\mathscr{F} = \{F^1, \dots, F^m\}$ is a collection of covector fields on Q (the ***control forces***), and
(vii) $U \subset \mathbb{R}^m$ (the ***control set***),

with all data being of class C^r. If $\mathcal{D} = \mathsf{TQ}$ (i.e., the system is unconstrained), then the remaining data $(\mathsf{Q}, \mathbb{G}, V, F, \mathscr{F}, U)$ is called a ***C^r-forced simple mechanical control system***. The codistribution $\mathfrak{F}$ generated by $\mathscr{F}$ is the ***input codistribution***. The system is ***fully actuated at q_0*** if $\mathsf{T}^*_{q_0}\mathsf{Q} = \mathfrak{F}_{q_0}$, and is otherwise ***underactuated at q_0***. If the system is fully actuated at all points in Q, then it is ***fully actuated***. •

As usual, the cases in which $r \in \{\infty, \omega\}$ are most interesting for us, and unless we state otherwise, systems are assumed to be C^∞.

Let us make sure that we understand the physical relevance of each of the seven objects that comprise a general simple mechanical control system. Of course, Q is the configuration manifold, $\mathbb{G}$ is the kinetic energy metric, V is the potential function, and $\mathcal{D}$ is a linear velocity constraint (for systems with no constraints, take $\mathcal{D} = \mathsf{TQ}$). By F we mean to model any non-potential external forces on the system over which we exert no control. For example, uncontrolled dissipative forces fall into this category. The covector fields $\mathscr{F} = \{F^1, \dots, F^m\}$ represent forces we control and U is the set of controls we can access. Thus input forces to the system are linear combinations of the covector fields $F^1, \dots, F^m$, with coefficients being locally integrable U-valued functions of time. We refer the reader to Remark 4.101 for a discussion of the assumption that there are a finite number of input covector fields. By asking that the forces $\mathscr{F}$ be basic (in the language of Definition 4.51), we are losing some, but not a lot of, generality. The assumption is that directions in which one can apply forces depends only on configuration, and not on time or velocity. Although many examples fall into our setup, not all do.

The class of general simple mechanical control systems is very large, and, without making restrictions, one cannot hope to say much about these problems. Therefore, we typically make restrictions on the kinds of systems we consider, and the following is a list of the most commonly used subclasses of general simple mechanical control systems.

1. $(\mathsf{Q}, \mathbb{G}, V, \mathscr{F} = \{F^1, \dots, F^m\}, U)$: We shall frequently refer to systems described by this data as ***simple mechanical control systems***. This class is itself a rather large class, characterized by the fact that the models are quite idealized, since they necessarily exclude the effects of uncontrolled non-potential forces.
2. $(\mathsf{Q}, \mathbb{G}, V, \mathcal{D}, \mathscr{F} = \{F^1, \dots, F^m\}, U)$: The systems with this data we call ***simple mechanical control systems with constraints***. This class con-

tains as a subset the class of simple mechanical control systems by taking $\mathcal{D} = \mathsf{TQ}$.

3. $(\mathsf{Q}, \mathbb{G}, \mathscr{F} = \{F^1, \dots, F^m\}, U)$ and $(\mathsf{Q}, \mathbb{G}, \mathcal{D}, \mathscr{F} = \{F^1, \dots, F^m\}, U)$: We consider these two classes of system together, since they are both described essentially by the data of $(\mathsf{Q}, \nabla, \mathcal{D}, \mathscr{Y} = \{Y_1, \dots, Y_m\}, U)$, where
   (a) Q is a manifold,
   (b) ∇ is an affine connection on Q,
   (c) $\mathcal{D}$ is a linear velocity constraint,
   (d) $\mathscr{Y} = \{Y_1, \dots, Y_m\}$ are vector fields on Q, and
   (e) $U \subset \mathbb{R}^m$.

   Indeed, for systems described by the data $(\mathsf{Q}, \mathbb{G}, \mathscr{F} = \{F^1, \dots, F^m\}, U)$, we take $\mathsf{Q} = \mathsf{Q}$ (abuse of notation), $\nabla = \overset{\mathbb{G}}{\nabla}$, $\mathcal{D} = \mathsf{TQ}$, $Y_a = \mathbb{G}^\sharp(F^a)$, $a \in \{1, \dots, m\}$, and $U = U$ (abuse of notation). For systems described by the data $(\mathsf{Q}, \mathbb{G}, \mathcal{D}, \mathscr{F} = \{F^1, \dots, F^m\}, U)$, we take $\mathsf{Q} = \mathsf{Q}_{\mathcal{D}}$, where $\mathsf{Q}_{\mathcal{D}}$ is the set of regular points of $\mathcal{D}$, $\nabla = \overset{\mathcal{D}}{\nabla}$, $\mathcal{D} = \mathcal{D}|\mathsf{Q}_{\mathcal{D}}$ (abuse of notation), and $Y_a = P_{\mathcal{D}}(\mathbb{G}^\sharp(F^a))$, $a \in \{1, \dots, m\}$, and $U = U$ (abuse of notation).

Note that the forces in the preceding case all end up being represented, not as $\mathsf{T}^*\mathsf{Q}$-valued, but as TQ-valued, by an application of $\mathbb{G}^\sharp$ (and possibly followed by projection onto $\mathcal{D}$).

Motivated by the last of these classifications, we define an interesting class of control problems that extends to the control theory setting the notion of a forced affine connection system of Definition 4.89.

Definition 4.105 (Forced affine connection control system). For $r \in \mathbb{N} \cup \{\infty\} \cup \{\omega\}$, a ***$C^r$-forced affine connection control system*** is a 6-tuple $(\mathsf{Q}, \nabla, Y, \mathcal{D}, \mathscr{Y} = \{Y_1, \dots, Y_m\}, U)$, where

(i) Q is a manifold,
(ii) ∇ is a affine connection on Q,
(iii) Y is a vector force,
(iv) $\mathcal{D}$ is a regular linear velocity constraint having the property that ∇ restricts to $\mathcal{D}$,
(v) $\mathscr{Y} = \{Y_1, \dots, Y_m\}$ is a set of vector fields on Q taking values in $\mathcal{D}$, and
(vi) $U \subset \mathbb{R}^m$,

with all data assumed to be of class C^r. If Y is identically zero, the remaining components comprise a 5-tuple $(\mathsf{Q}, \nabla, \mathcal{D}, \mathscr{Y} = \{Y_1, \dots, Y_m\}, U)$, which is called a ***$C^r$-affine connection control system***. The distribution $\mathcal{Y}$ generated by $\mathscr{Y}$ is the ***input distribution***. •

Remark 4.106. Much of the time the control set U will not be of concern to us. For example, in the first treatment of many control problems, one often makes the idealizing assumption that $U = \mathbb{R}^m$. In such cases, we shall explicitly identify the control set as $\mathbb{R}^m$. On other occasions, we simply wish to place the emphasis on those properties of the system that are in some sense independent of the control set. In this latter case, we shall refer to the data minus the control set as a ***pre-control system***, with the appropriate prefix (e.g., "forced affine connection pre-control system" or "general simple mechanical pre-control system"). •

4.6.3 Global representations of equations of motion

Now we turn to writing the equations of motion for the subclasses of systems above. For a general simple mechanical control system $(\mathsf{Q}, \mathbb{G}, V, F, \mathcal{D}, \mathscr{F} = \{F^1, \dots, F^m\}, U)$, the equations of motion can be written as

$$\overset{\mathcal{D}}{\nabla}_{\gamma'(t)}\gamma'(t) = -P_{\mathcal{D}}(\mathrm{grad}V(\gamma(t))) + P_{\mathcal{D}}(\mathbb{G}^\sharp(F(t,\gamma'(t)))) + \sum_{a=1}^{m} u^a(t) P_{\mathcal{D}}(\mathbb{G}^\sharp(F^a(\gamma(t)))), \quad (4.24)$$

for a control $u\colon I \to \mathbb{R}^m$ that is locally integrable, and a locally absolutely differentiable curve $\gamma\colon I \to \mathsf{Q}$ satisfying $\gamma'(t) \in \mathcal{D}_{\gamma(t)}$ for some (and so for all) $t \in I$. This full expression will then simplify in the presence of various further assumptions on the problem data. For simplicity of exposition, we shall generally assume that the external force F is time-independent. The more general case can be easily handled by introducing the notion of time-dependence in the definition of control-affine systems.

We will also need a version of these equations on TQ. That is to say, we need to construct a control-affine system with state manifold TQ that corresponds to a general simple mechanical control system. To do this, we recall from Definition 3.73 the notion of vertical lift. By simply converting in coordinates the second-order equations (4.24) on Q to first-order equations on TQ, the system (4.24) can be written as

$$\Upsilon'(t) = \overset{\mathcal{D}}{S}(\Upsilon(t)) - \mathrm{vlft}(P_{\mathcal{D}}(\mathrm{grad}V))(\Upsilon(t)) + \mathrm{vlft}(P_{\mathcal{D}}(\mathbb{G}^\sharp(F)))(\Upsilon(t)) + \sum_{a=1}^{m} u^a(t)\mathrm{vlft}(P_{\mathcal{D}}(\mathbb{G}^\sharp(F^a)))(\Upsilon(t)), \quad (4.25)$$

for a locally absolutely continuous trajectory $\Upsilon\colon I \to \mathsf{TQ}$ satisfying $\Upsilon(t) \in \mathcal{D}$ for some (and so for all) $t \in I$. The reader is encouraged to understand why the equations (4.24) and (4.25) are in fact the same. This is not difficult, but it is essential in order to understand exactly how mechanical control systems are a subset of control-affine systems. Indeed, (4.25) shows that, to a general simple mechanical control system $(\mathsf{Q}, \mathbb{G}, V, F, \mathcal{D}, \mathscr{F}, U)$, we assign the control-affine system $(\mathsf{M}, \mathscr{C} = \{f_0, f_1, \dots, f_m\}, U)$, where

1. $\mathsf{M} = \mathsf{TQ}$,

2. $f_0 = \overset{\mathcal{D}}{S} - \mathrm{vlft}(P_{\mathcal{D}}(\mathrm{grad}V)) + \mathrm{vlft}(P_{\mathcal{D}}(\mathbb{G}^\sharp(F))$,
3. $f_a = \mathrm{vlft}(P_{\mathcal{D}}(\mathbb{G}^\sharp(F^a)))$, $a \in \{1, \dots, m\}$, and
4. $U = U$ (abuse of notation).

Let us introduce the same definitions for an affine connection control system $(\mathsf{Q}, \nabla, \mathcal{D}, \mathscr{Y} = \{Y_1, \dots, Y_m\}, U)$. In this case the governing equations are simply

$$\nabla_{\gamma'(t)}\gamma'(t) = \sum_{a=1}^{m} u^a(t) Y_a(\gamma(t)),$$

where $\gamma'(t) \in \mathcal{D}_{\gamma(t)}$ for some (and so for all) $t \in I$. As a system on TQ we have

$$\Upsilon'(t) = S(\Upsilon(t)) + \sum_{a=1}^{m} u^a(t)\mathrm{vlft}(Y_a)(\Upsilon(t)),$$

where S is the geodesic spray for ∇, and with $\Upsilon \colon I \to \mathsf{TQ}$ satisfying $\Upsilon(t) \in \mathcal{D}$ for some (and so for all) t. Therefore, to an affine connection control system $(\mathsf{Q}, \nabla, \mathscr{Y}, U)$ we assign a control-affine system $(\mathsf{M}, \mathscr{C} = \{f_0, f_1, \dots, f_m\}, U)$ with

1. $\mathsf{M} = \mathsf{TQ}$,
2. $f_0 = S$,
3. $f_a = \mathrm{vlft}(Y_a)$, $a \in \{1, \dots, m\}$, and
4. $U = U$ (abuse of notation).

In reality we are only interested in the system when restricted to the submanifold $\mathcal{D} \subset \mathsf{TQ}$. However, it is convenient to work with the system on all of TQ, taking care of the role played by $\mathcal{D}$ as it comes up in the various sorts of control problems that we consider.

4.6.4 Local representations of equations of motion

In order to analyze or perform control design for a given example, it is often convenient, or necessary, to write the governing equations in some concrete way. This is frequently done by choosing a coordinate chart and writing the equations. Here we merely provide various local expressions for the equations developed in the preceding section. These are a little messy in their most general form. The main point to note is that all objects are straightforwardly and concretely defined.

First let us consider the most general Poincaré representation for a general simple mechanical control system $(\mathsf{Q}, \mathbb{G}, V, F, \mathcal{D}, \mathscr{F} = \{F^1, \dots, F^m\}, U)$, under the assumption that $\mathcal{D}$ is regular, so that $\overset{\mathcal{D}}{\nabla}$ can be defined. We let $(\mathcal{U}, \phi)$ be a chart for Q with coordinates $(q^1, \dots, q^n)$. We also consider a collection $\mathscr{X} = \{X_1, \dots, X_n\}$ of vector fields forming a basis for $\mathsf{T}_q\mathsf{Q}$ for each $q \in \mathcal{U}$. A ***Poincaré representation*** of the equations (4.24) for a curve γ whose image intersects $\mathcal{U}$ is the collection of equations

$$\begin{aligned}\dot{q}^i &= X^i_j v^j, & i \in \{1,\dots,n\},\\ \dot{v}^i &= -\overset{\mathscr{X}}{\Gamma}{}^i_{jk} v^j v^k - A^i_l P^l_j \mathbb{G}^{jk}\frac{\partial V}{\partial q^k} + A^i_l P^l_j \mathbb{G}^{jk} F_k \\ &\quad + \sum_{a=1}^m u^a A^i_l P^l_j \mathbb{G}^{jk} F^a_k, & i \in \{1,\dots,n\}.\end{aligned}$$

Here

1. $(q^1,\dots,q^n)$ are the coordinates in the chart $(\mathcal{U},\phi)$,
2. $(v^1,\dots,v^n)$ are the fiber coordinates for $\mathsf{T}\mathcal{U}$ *in the basis* $\{X_1,\dots,X_n\}$ (note that it will not generally be the case that $v^i = \dot{q}^i$),
3. X^i_j, $i \in \{1,\dots,n\}$, are the components of X_j, $j \in \{1,\dots,n\}$, in the chart $(\mathcal{U},\phi)$,
4. $\overset{\mathscr{X}}{\Gamma}{}^i_{jk}$, $i,j,k \in \{1,\dots,n\}$, are the generalized Christoffel symbols for $\overset{\mathcal{D}}{\nabla}$ with respect to $\mathscr{X}$,
5. A^i_l, $i,l \in \{1,\dots,n\}$, satisfy $A^i_k X^k_j = \delta^i_j$, $i,j \in \{1,\dots,n\}$,
6. P^l_j, $l,j \in \{1,\dots,n\}$, are the components of $P_{\mathcal{D}}$ in the chart $(\mathcal{U},\phi)$,
7. $\mathbb{G}^{jk}$, $j,k \in \{1,\dots,n\}$, are the components of $\mathbb{G}^{-1}$ in the chart $(\mathcal{U},\phi)$,
8. F_k, $k \in \{1,\dots,n\}$, are the components of F in the chart $(\mathcal{U},\phi)$, and
9. F^a_k, $k \in \{1,\dots,n\}$, are the components of F^a, $a \in \{1,\dots,m\}$, in the chart $(\mathcal{U},\phi)$.

Note that a general Poincaré representation is a set of first-order differential equations. In the event that we take the vector fields $\mathscr{X}$ as the coordinate vector fields, $X_i = \frac{\partial}{\partial q^i}$, $i \in \{1,\dots,n\}$, then we call the corresponding Poincaré representation the ***natural representation***. This representation has the property that the equations may be also written in the second-order form

$$\ddot{q}^i + \overset{\mathcal{D}}{\Gamma}{}^i_{jk}\dot{q}^j\dot{q}^k = P^i_j \mathbb{G}^{jk}\frac{\partial V}{\partial q^k} + P^i_j \mathbb{G}^{jk} F_k + \sum_{a=1}^m u^a P^i_j \mathbb{G}^{jk} F^a_k, \qquad i \in \{1,\dots,n\}.$$

Here $\overset{\mathcal{D}}{\Gamma}{}^i_{jk}$, $i,j,k \in \{1,\dots,n\}$, are the Christoffel symbols for $\overset{\mathcal{D}}{\nabla}$. In this equation, everything represents the components of the corresponding object in the chart $(\mathcal{U},\phi)$.

Now we may specialize these equations further by assuming that $\mathcal{D}|\mathcal{U}$ has constant rank, and that $\{X_1,\dots,X_{\mathrm{rank}(\mathcal{D})}\}$ forms a basis for $\mathcal{D}_q$ at each point $q \in \mathcal{U}$. In this case, as we saw in Proposition 4.92, the Poincaré representation simplifies to

$$\begin{aligned}
\dot{q}^i &= X^i_\alpha v^\alpha, & i \in \{1,\dots,n\},\\
\dot{v}^\delta &= -\overset{\mathfrak{X}}{\Gamma}{}^\delta_{\alpha\beta} v^\alpha v^\beta - A^\delta_l P^l_j \mathbb{G}^{jk}\frac{\partial V}{\partial q^k} + A^\delta_l P^l_j \mathbb{G}^{jk} F_k\\
&\quad + \sum_{a=1}^m u^a A^\delta_l P^l_j \mathbb{G}^{jk} F^a_k, & \delta \in \{1,\dots,\mathrm{rank}(\mathcal{D})\},
\end{aligned}$$

where the implied summation over α and β goes from 1 to $\mathrm{rank}(\mathcal{D})$. In the case that $\mathcal{D} \subsetneq \mathsf{T}Q$, this is a simplification one will certainly wish to make, since the number of variables reduces to $2n - \mathrm{rank}(\mathcal{D})$, i.e., the minimum required to describe the dynamics of the system. The resulting equations we call a ***$\mathcal{D}$-adapted Poincaré representation***. Recall from Proposition 4.94 that, when the vector fields $\{X_1,\dots,X_{\mathrm{rank}(\mathcal{D})}\}$ are $\mathbb{G}$-orthogonal, the formulae for the generalized Christoffel symbols potentially simplify. Also, in this case the expression for $A^\delta_l P^l_j$, $\delta \in \{1,\dots,\mathrm{rank}(\mathcal{D})\}$, $j \in \{1,\dots,n\}$, simplifies to

$$A^\delta_l P^l_j = \frac{\mathbb{G}_{jk} X^k_\delta}{\|X_\delta\|^2_{\mathbb{G}}}, \qquad \delta \in \{1,\dots,\mathrm{rank}(\mathcal{D})\},\ j \in \{1,\dots,n\}.$$

The resulting simplification then renders the equations as

$$\begin{aligned}
\dot{q}^i &= X^i_\alpha v^\alpha, \qquad i \in \{1,\dots,n\},\\
\dot{v}^\delta &= -\overset{\mathfrak{X}}{\Gamma}{}^\delta_{\alpha\beta} v^\alpha v^\beta + \frac{1}{\|X_\delta\|^2_{\mathbb{G}}}\Big(-\langle \mathrm{d}V; X_\delta\rangle + \langle F; X_\delta\rangle + \sum_{a=1}^m u^a \langle F^a; X_\delta\rangle\Big),\\
&\quad \delta \in \{1,\dots,\mathrm{rank}(\mathcal{D})\}.
\end{aligned}$$

These equations we call an ***orthogonal Poincaré representation*** for the system.

Remarks 4.107. 1. While these coordinate representations are indeed somewhat imposing, the punchline is that they are implementable, for example, in a symbolic programming language. The authors have written a package in Mathematica®, and details for using this package are provided with the supplementary internet material for the book (see the preface for the URL).

2. The choice of local representation can be a key step in confronting any problem. By making a good or bad choice of local representation, certain control theoretic objectives may become more of less easy to achieve. •

4.6.5 Linear mechanical control systems

While this is certainly not a book on linear systems theory, it will be interesting and at times useful to consider linear mechanical systems. This is particularly true in the context of linearization, which we use to study stability in Chapter 6 and stabilization in Chapters 10 and 12.

First we look at linear mechanical systems in the uncontrolled context. This provides a specialization of the methodology in this chapter to a linear setting.

Definition 4.108 (Linear mechanical system). A ***linear mechanical system*** is a triple (V, M, K), where

(i) V is a finite-dimensional $\mathbb{R}$-vector space,
(ii) M is an inner product on V, and
(iii) $K \in \Sigma_2(\mathsf{V})$.

A ***forced linear mechanical system*** is a 4-tuple $(\mathsf{V}, M, K, (F_1, F_2))$, where (V, M, K) is a linear mechanical system and where

(iv) $F_1, F_2 \in L(\mathsf{V}; \mathsf{V}^*)$. •

To a linear mechanical system (V, M, K), we associated a linear map $A_{M,K}\colon \mathsf{V} \oplus \mathsf{V} \to \mathsf{V} \oplus \mathsf{V}$ by

$$A_{M,K} = \begin{bmatrix} 0 & \mathrm{id}_{\mathsf{V}} \\ -M^\sharp \circ K^\flat & 0 \end{bmatrix},$$

where the partitioning of the linear map corresponds to the natural partition of $\mathsf{V} \oplus \mathsf{V}$. This is called the ***state form*** for (V, M, K). Next consider adding forces to get the forced linear mechanical system $(\mathsf{V}, M, K, (F_1, F_2))$. This then gives rise to the linear differential equation

$$\begin{bmatrix} \dot{x}(t) \\ \dot{v}(t) \end{bmatrix} = \left(A_{M,K} + \begin{bmatrix} 0 & 0 \\ M^\sharp \circ F_1 & M^\sharp \circ F_2 \end{bmatrix} \right) \begin{bmatrix} x(t) \\ v(t) \end{bmatrix}$$

on $\mathsf{V} \oplus \mathsf{V}$. This may be written in second-order form as

$$\begin{aligned} & M^\flat(\ddot{x}(t)) + K^\flat(x(t)) = F_1(x(t)) + F_2(\dot{x}(t)) \\ \iff \quad & \ddot{x}(t) + M^\sharp \circ K^\flat(x(t)) = M^\sharp \circ F_1(x(t)) + M^\sharp \circ F_2(\dot{x}(t)). \end{aligned}$$

Note that the unforced equations are the Euler–Lagrange equations for the system with configuration manifold V and Lagrangian $L(x, v) = \frac{1}{2}M(v, v) - \frac{1}{2}K(x, x)$. Therefore, linear mechanical systems are examples of simple mechanical systems. In Section 6.2.1 we shall see that (forced) linear mechanical systems are linearizations of (forced) simple mechanical systems.

Next we turn to linear mechanical *control* systems.

Definition 4.109 (Linear mechanical control system). A ***general linear mechanical control system*** is a 5-tuple $\Sigma = (\mathsf{V}, M, K, (F_1, F_2), F)$, where

(i) V is a finite-dimensional $\mathbb{R}$-vector space,
(ii) M is an inner product on V,
(iii) $K \in \Sigma_2(\mathsf{V})$,
(iv) $F_1, F_2 \in L(\mathsf{V}; \mathsf{V}^*)$, and
(v) $F \in L(\mathbb{R}^m; \mathsf{V}^*)$.

If $F_1 = F_2 = 0$, then the remaining data (V, M, K, F) defines a ***linear mechanical control system***. •

Corresponding to a general linear mechanical control system $(\mathsf{V}, M, K, (F_1, F_2), F)$, we define a linear control system $(\mathsf{V} \oplus \mathsf{V}, A, B)$ by

$$A = \begin{bmatrix} 0 & \mathrm{id}_{\mathsf{V}} \\ -M^\sharp \circ K^\flat + M^\sharp \circ F_1 & M^\sharp \circ F_2 \end{bmatrix}, \quad B = \begin{bmatrix} 0 \\ M^\sharp \circ F \end{bmatrix}.$$

The linear control system $(\mathsf{V} \oplus \mathsf{V}, A, B)$ is called the ***state form*** for the general linear mechanical control system (V, M, K, F). The governing equations for a general linear mechanical control system in first-order state form are

$$\begin{bmatrix} \dot{x}(t) \\ \dot{v}(t) \end{bmatrix} = \begin{bmatrix} 0 & \mathrm{id}_{\mathsf{V}} \\ -M^\sharp \circ K^\flat + M^\sharp \circ F_1 & M^\sharp \circ F_2 \end{bmatrix} \begin{bmatrix} x(t) \\ v(t) \end{bmatrix} + \begin{bmatrix} 0 \\ M^\sharp \circ F \end{bmatrix} u(t).$$

In second-order form these are

$$\begin{aligned} & M^\flat(\ddot{x}(t)) + K^\flat(x(t)) = F_1(x(t)) + F_2(\dot{x}(t)) + F(u(t)) \\ \iff \quad & \ddot{x}(t) + M^\sharp \circ K^\flat(x(t)) = M^\sharp \circ F_1(x(t)) + M^\sharp \circ F_2(\dot{x}(t)) + M^\sharp \circ F(u(t)). \end{aligned}$$

Note that these are the equations for the general simple mechanical control system $(\mathsf{Q}, \mathbb{G}, V, F, \mathcal{D}, \mathscr{F}, U)$ with

1. $\mathsf{Q} = \mathsf{V}$,
2. $\mathbb{G}$ the constant Riemannian metric on V defined by M,
3. $V(x) = \frac{1}{2}K(x, x)$,
4. $F(x, v) = F_1(x) + F_2(v)$ (first half of abuse of notation with two F's),
5. $\mathcal{D} = \mathsf{V} \oplus \mathsf{V}$,
6. $F^a(x) = F(e_a)$, $a \in \{1, \dots, m\}$, (second half of abuse of notation with two F's), and
7. $U = \mathbb{R}^m$.

Thus general linear mechanical control systems are also general simple mechanical control systems. Furthermore, a linear mechanical control system (V, M, K, F) is also a simple mechanical control system $(\mathsf{Q}, \mathbb{G}, V, \mathscr{F}, \mathbb{R}^m)$ with the data as defined above.

4.6.6 Alternative formulations

The preceding discussion in this section provides a point of reference to all of the ways we shall use to represent mechanical control systems. In the literature, one can find other representations, and it is worth devoting some space to some of these to explain how they are related to our formulations. We do this in a fairly informal way.

A common piece of terminology arising in reference to mechanical control systems is "nonholonomic control systems." Most frequently this terminology is used in reference to what we call "driftless control systems," i.e., control-affine systems for which the drift vector field is zero; there are many instances of this in the literature [e.g., Astolfi 1996, Fantoni and Lozano 2001, Godhavn and Egeland 1997, Kolmanovsky and McClamroch 1995, Lin, Pongvuthithum, and Qian 2002, Liu 1997, Murray 1994]. In this context, "nonholonomic" is a somewhat unfortunate word choice. The connection between driftless systems and mechanical systems with nonholonomic constraints is not obvious. Indeed, it generally does not exist,[9] and one way to view the results in Chapter 8 is that they shed some light on cases for which such connections do exist. We advise the reader to be aware of making ill-advised links between mechanical control systems and nonholonomic control systems, in the case when the latter refers to driftless control systems. In this book we shall remain faithful to the mechanical meaning of the word nonholonomic.

Some authors [e.g., Aneke, Nijmeijer, and de Jager 2001, Ge, Zhendong, Lee, and Spong 2001, Lynch, Shiroma, Arai, and Tanie 2000, Oriolo and Nakamura 1991, Ostrowski 2000, Reyhanoglu, van der Schaft, McClamroch, and Kolmanovsky 1999], realizing the danger in confounding driftless systems with mechanical systems, adapt the language "nonholonomic control system" to the mechanical setting, referring to "systems with second-order nonholonomic constraints." The idea in this sort of formulation is that the control forces serve to constrain the accelerations on the system, whereas for "nonholonomic control systems," the velocities are constrained by the controls. This sort of language can be make precise; indeed, in our language such systems are called underactuated.

A major effort made in modeling and analysis of mechanical control systems in a Hamiltonian framework is the port-controlled Hamiltonian methodology. This is a quite general framework that allows modeling and interconnection of mechanical systems with constraints and general forces. The framework can also be used to model electrical systems, as well as infinite-dimensional systems, e.g., fluids. This work is reported, for example, in the papers [Dalsmo and van der Schaft 1998, van der Schaft 2000]. Some related modeling and control theoretic techniques are used in the recent book of Stramigioli [2001].

Exercises

E4.1 Consider a particle constrained to move on the surface of a cylinder.
- (a) What is the configuration manifold for the system, as a submanifold of Q_{free}?
- (b) What is a simple abstract model for the configuration manifold?
- (c) Find a set of coordinates for the configuration manifold, making sure that you state *exactly* how they coordinatize the manifold, and their range of validity.

[9] A bicycle is subject to nonholonomic constraints, yet it is not a driftless system; when peddling is stopped, a bicycle keeps moving.

(d) Is it possible to find a single set of coordinates that are valid on the entire configuration manifold?
(e) Is it possible to find vector fields that form a basis at each point of the configuration manifold?

E4.2 Consider a disk rolling on the inside of a circular track as in Figure E4.1.

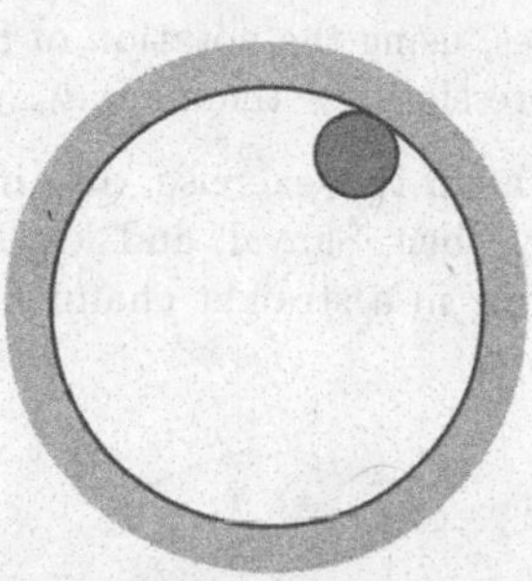

Figure E4.1. Disk in a circular track

Let r be the radius of the disk and let R be the radius of the circular track. Let $\mathsf{Q} \subset \mathsf{Q}_{\text{free}} = \mathsf{SO}(3) \times \mathbb{R}^3$ be the set of admissible configurations for the system.

(a) Show that, if $\frac{r}{R} \in \mathbb{Q}$, then Q is an embedded submanifold of Q_{free} that is diffeomorphic to $\mathbb{S}^1$.
(b) Show that, if $\frac{r}{R} \in \mathbb{R} \setminus \mathbb{Q}$, then Q is an immersed, but not embedded, submanifold of Q_{free}, and that the differentiable structure induced on Q as in Section 3.9.1 makes Q diffeomorphic to $\mathbb{R}$.

It turns out to be convenient for this example not to use $\mathbb{S}^1$ as a model for Q in the event that $\frac{r}{R} \in \mathbb{Q}$. Let us describe a more convenient model. For $k \in \mathbb{N}$, define an equivalence relation $\sim_k$ in $\mathbb{R}$ by $x \sim_k y$ if $x - y = 2\pi k j$ for some $j \in \mathbb{Z}$. Let us denote by $\mathbb{R}/(2\pi k)$ the set of equivalence classes.

(c) Suppose that $\frac{r}{R} \in \mathbb{Q}$. What is a natural (i.e., physically sensible) choice for k if we wish to think of the configuration manifold for the system as being $\mathbb{R}/(2\pi k)$?

In the exercises below, we shall use the natural coordinate x for $\mathbb{R}$ or for $\mathbb{R}/(2\pi k)$ in this example.

E4.3 Consider the pair of identical masses interconnected by three identical springs as shown in Figure E4.2. In modeling the system, we ignore the possibility

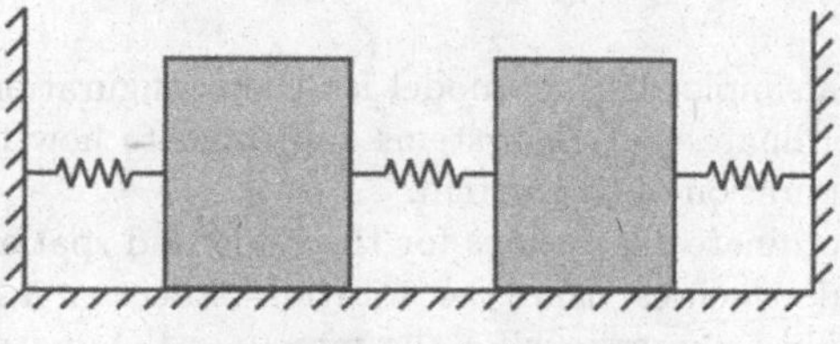

Figure E4.2. Coupled masses

of collisions, either between the masses, or with the walls.

(a) What is the configuration manifold for the system, as a submanifold of $\mathsf{Q}_{\mathrm{free}}$?
(b) Show that a simple abstract model for the configuration space is $\mathbb{R}^2$.
(c) Find coordinates for the system, and indicate how they are related to the configuration of the system.
(d) Write coordinate expressions for the body and spatial translational and rotational velocities, using the notation of Section 4.1.5.
(e) Write explicit expressions for the maps $\iota_{\mathrm{body}}|\mathsf{Q}$ and $\iota_{\mathrm{spatial}}|\mathsf{Q}$.

E4.4 The example we look at in this exercise, and in the subsequent related exercises, was studied by Shen, Sanyal, and McClamroch [2002]. The system consists of a ball moving in a straight channel on a rotating disk (see Fig-

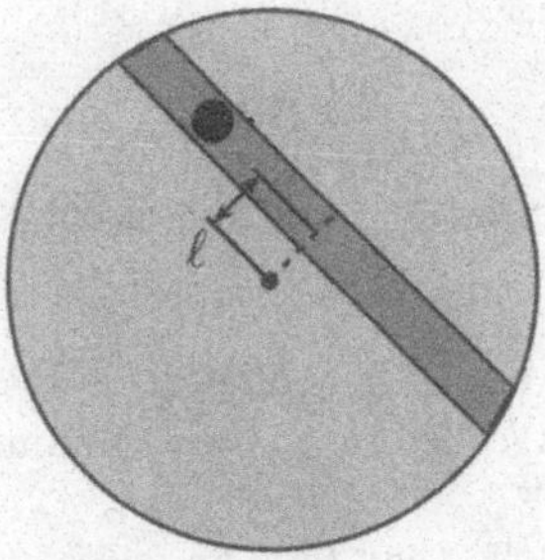

Figure E4.3. Ball in rotating channel

ure E4.3). The channel is offset from the center of rotation of the disk, and we simplify the model by thinking of the ball as a particle. Regard the channel as being infinitely long, so that you do not have to fuss with the boundaries.

(a) What is the configuration manifold for the system, as a submanifold of $\mathsf{Q}_{\mathrm{free}}$?
(b) What is a simple abstract model for the configuration manifold?
(c) Find coordinates for the system, and indicate how they are related to the configuration of the system.
(d) Write coordinate expressions for the body and spatial translational and rotational velocities, using the notation of Section 4.1.5.
(e) Write explicit expressions for the maps $\iota_{\mathrm{body}}|\mathsf{Q}$ and $\iota_{\mathrm{spatial}}|\mathsf{Q}$.

E4.5 Consider a pendulum swinging atop a cart constrained to move in a line (Figure E4.4).

(a) What is the configuration manifold of the system, as a submanifold of $\mathsf{Q}_{\mathrm{free}}$?
(b) What is a simple abstract model for the configuration manifold?
(c) Find coordinates for the system, and indicate how they are related to the configuration of the system.
(d) Write coordinate expressions for the body and spatial translational and rotational velocities, using the notation of Section 4.1.5.
(e) Write explicit expressions for the maps $\iota_{\mathrm{body}}|\mathsf{Q}$ and $\iota_{\mathrm{spatial}}|\mathsf{Q}$.

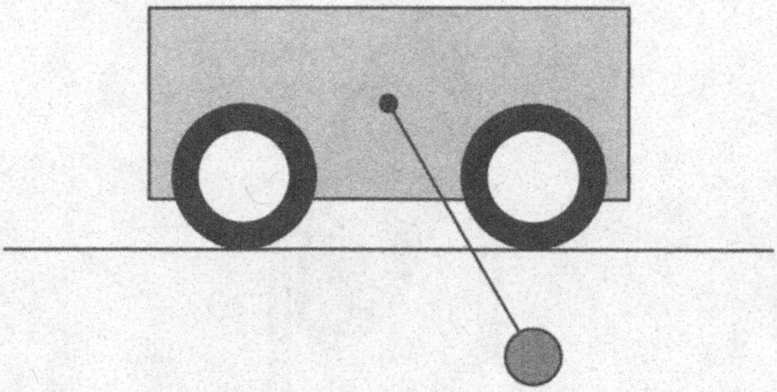

Figure E4.4. Pendulum on cart

E4.6 Consider a pendulum attached to a radial arm by a universal joint. The base of the radial arm is constrained to move in a line. See Figure E4.5.

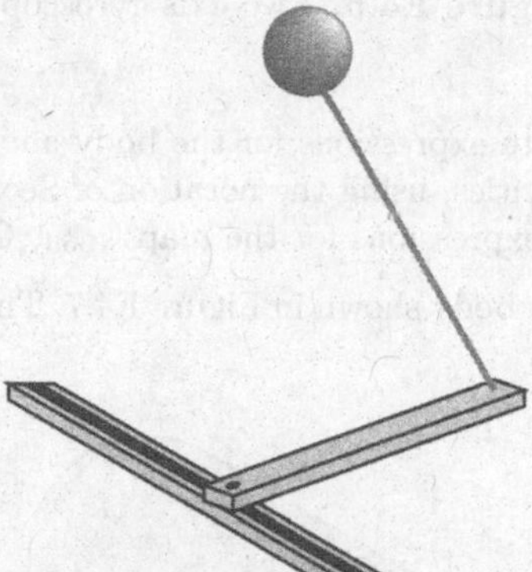

Figure E4.5. Pendulum on arm on linear track

(a) What is the configuration manifold of the system, as a submanifold of Q_{free}?
(b) What is a simple abstract model for the configuration manifold?
(c) Find coordinates for the system, and indicate how they are related to the configuration of the system.
(d) Write coordinate expressions for the body and spatial translational and rotational velocities, using the notation of Section 4.1.5.
(e) Write explicit expressions for the maps $\iota_{\text{body}}|\mathsf{Q}$ and $\iota_{\text{spatial}}|\mathsf{Q}$.

E4.7 Consider the two-axis gyroscope of Figure E4.6.
(a) What is the configuration manifold for the system, as a submanifold of Q_{free}?
(b) What is a simple abstract model for the configuration manifold?
(c) Find coordinates for the system, and indicate how they are related to the configuration of the system.

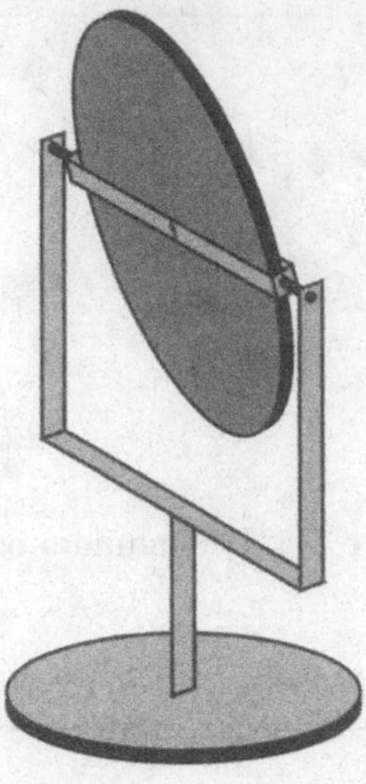

Figure E4.6. Two-axis gyroscope

(d) Write coordinate expressions for the body and spatial translational and rotational velocities, using the notation of Section 4.1.5.
(e) Write explicit expressions for the maps $\iota_{\text{body}}|\mathsf{Q}$ and $\iota_{\text{spatial}}|\mathsf{Q}$.

E4.8 Consider the planar body shown in Figure E4.7. This is similar to the system

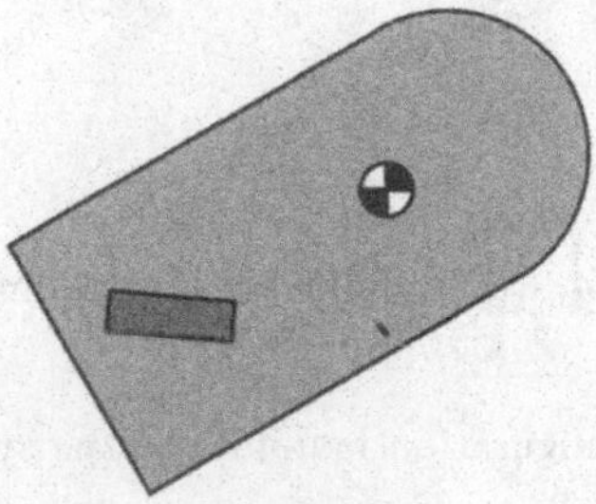

Figure E4.7. Planar rigid body with fan dynamics

considered in Section 1.1, but now the thrust fan has associated with it a degree-of-freedom, a mass, and an inertia.
(a) What is the configuration manifold for the system, as a submanifold of Q_{free}?
(b) What is a simple abstract model for the configuration manifold?
(c) Find coordinates for the system, and indicate how they are related to the configuration of the system.
(d) Write coordinate expressions for the body and spatial translational and rotational velocities, using the notation of Section 4.1.5.
(e) Write explicit expressions for the maps $\iota_{\text{body}}|\mathsf{Q}$ and $\iota_{\text{spatial}}|\mathsf{Q}$.

E4.9 Consider a car-like robot as depicted in Figure E4.8. It will not be necessary

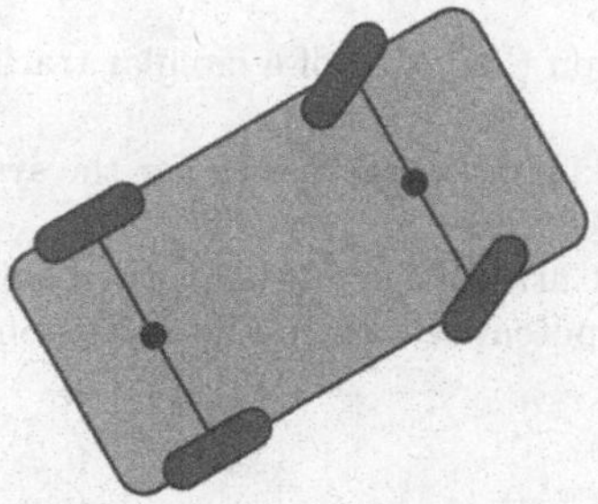

Figure E4.8. Car-like robot

to associate a degree of freedom to the back wheels, but it will be necessary to have a degree of freedom for the rotation about the vertical axis of the front (steering) wheels. Assume that the center of mass of the car is located at the midpoint of the rear axle.

(a) What is the configuration manifold for the system, as a submanifold of Q_{free}?
(b) What is a simple abstract model for the configuration manifold?
(c) Find coordinates for the system, and indicate how they are related to the configuration of the system.
(d) Write coordinate expressions for the body and spatial translational and rotational velocities, using the notation of Section 4.1.5.
(e) Write explicit expressions for the maps $\iota_{\text{body}}|\mathsf{Q}$ and $\iota_{\text{spatial}}|\mathsf{Q}$.

E4.10 For a rigid body $(\mathcal{B}, \mu)$, show that $\chi_c \in \text{int}_{\text{aff}(\mathcal{B})}\text{conv}(\mathcal{B})$.

E4.11 Prove the following formula for the inertia tensor about the center of mass:
$$\mathbb{I}_c(\boldsymbol{v}) = \int_{\mathcal{B}} \boldsymbol{\chi} \times (\boldsymbol{v} \times \boldsymbol{\chi})\, d\mu - \mu(\mathcal{B})(\boldsymbol{\chi}_c \times (\boldsymbol{v} \times \boldsymbol{\chi}_c)).$$

E4.12 Let $\boldsymbol{x}_0 \in \mathbb{R}^3$ and let $\mathsf{A} = \{\boldsymbol{v} + \boldsymbol{x}_0 \mid \boldsymbol{v} \in \mathsf{W}\}$ be an affine subspace associated with a two-dimensional subspace W, and let $(\mathcal{B}, \mu)$ be a rigid body with $\mathcal{B} \subset \mathsf{A}$. Show that the orthogonal complement of W is an eigenspace for $\mathbb{I}_c$.

E4.13 Let Q be a second-countable, paracompact manifold and let $\gamma\colon [a, b] \to \mathsf{Q}$ be a C^2-curve with $\xi\colon [a, b] \to \mathsf{TQ}$ a C^2-curve along γ satisfying $\xi(a) = 0_{\gamma(a)}$ and $\xi(b) = 0_{\gamma(b)}$. Show that there exists a variation ϑ of γ such that $\delta\vartheta = \xi$.
Hint: *By Remark 4.34–2, let $\mathbb{G}$ be a Riemannian metric on Q. For $q \in \mathsf{Q}$, let $\exp_q\colon \mathsf{T}_q\mathsf{Q} \to \mathsf{Q}$ be the map assigning to $v_q \in \mathsf{T}_q\mathsf{Q}$ the point $\gamma_{v_q}(1) \in \mathsf{Q}$, where γ_{v_q} is the unique geodesic satisfying $\gamma'_{v_q}(0) = v_q$. Then take $\vartheta(s, t) = \exp_{\gamma(t)}(s\xi(t))$.*

E4.14 Let $\gamma\colon I \to \mathsf{Q}$ be a C^2-curve. Explain why the object represented in coordinates by $t \mapsto (\ddot{q}^1(t), \ldots, \ddot{q}^n(t))$ does not define a vector field along γ.

E4.15 For the particle constrained to a cylinder problem of Exercise E4.1, if the particle has mass m, what is the kinetic energy for the system in the coordinates you specified in part (c).

E4.16 For the disk rolling on the inside of a circular track of Exercise E4.2, do the following.

(a) Determine the kinetic energy metric for the system in the natural set of coordinates.
(b) Determine the Christoffel symbols for the Levi-Civita affine connection.
(c) Determine the potential function if the system is subject to gravity as in Figure E4.9.

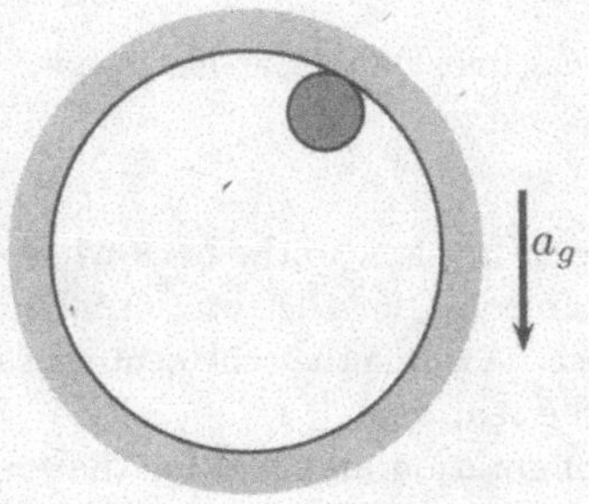

Figure E4.9. Gravity for the disk in a circular track

(d) Using this data, write the Euler–Lagrange equations for the system.

E4.17 For the coupled mass system of Exercise E4.3, do the following.

(a) Determine the kinetic energy metric for the system in your set of coordinates.
(b) Determine the Christoffel symbols for the Levi-Civita affine connection.
(c) Determine the potential function.
(d) Using this data, write the Euler–Lagrange equations for the system.

E4.18 For the ball in a rotating channel system of Exercise E4.4, do the following:

(a) Determine the kinetic energy metric for the system in your set of coordinates.
(b) Determine the Christoffel symbols for the Levi-Civita affine connection.
(c) Determine the potential function.
(d) Using this data, write the Euler–Lagrange equations for the system.

E4.19 For the pendulum/cart system of Exercise E4.5, do the following.

(a) Determine the kinetic energy metric for the system in your set of coordinates.
(b) Determine the Christoffel symbols for the Levi-Civita affine connection.
(c) Determine the potential function.
(d) Using this data, write the Euler–Lagrange equations for the system.

E4.20 For the spherical pendulum system on a track described in Exercise E4.6, do the following.

(a) Determine the kinetic energy metric for the system in your set of coordinates.
(b) Determine the Christoffel symbols for the Levi-Civita affine connection.
(c) Determine the potential function.
(d) Using this data, write the Euler–Lagrange equations for the system.

E4.21 For the two-axis gyro of Exercise E4.7, do the following.
(a) Determine the kinetic energy metric for the system in your set of coordinates.
(b) Determine the Christoffel symbols for the Levi-Civita affine connection.
(c) Determine the potential function.
(d) Using this data, write the Euler–Lagrange equations for the system.

E4.22 For the planar body with fan dynamics described in Exercise E4.8, do the following.
(a) Determine the kinetic energy metric for the system in your set of coordinates.
(b) Determine the Christoffel symbols for the Levi-Civita affine connection.
(c) Determine the potential function.
(d) Using this data, write the Euler–Lagrange equations for the system.

E4.23 For the car-like robot described in Exercise E4.9, do the following.
(a) Determine the kinetic energy metric for the system in your set of coordinates.
(b) Determine the Christoffel symbols for the Levi-Civita affine connection.
(c) Determine the potential function.

E4.24 Let $\mathbb{G}_{\mathrm{rnd}}$ be the round Riemannian metric on $\mathbb{S}^2$, and denote by $d_{\mathrm{rnd}}\colon \mathbb{S}^2 \times \mathbb{S}^2 \to \bar{\mathbb{R}}_+$ the corresponding Riemannian distance function.
(a) Show that $\mathrm{image}(d_{\mathrm{rnd}}) = [0, \pi]$.
(b) Show that if $d_{\mathrm{rnd}}(q_1, q_2) < \pi$, then there is a unique geodesic of minimum length joining q_1 and q_2.
(c) Show that if $d_{\mathrm{rnd}}(q_1, q_2) = \pi$, then there are infinitely many geodesics of minimum length joining q_1 and q_2.
(d) For $q_1, q_2 \in \mathbb{S}^2$, show that there are infinitely many geodesics joining q_1 and q_2 that do not minimize the distance between these points.

In the next exercise we consider the motion of a body in an ideal fluid, a problem that will arise in various examples and exercises in the text. The study of fluid dynamics and the modeling of fluid-body interactions is beyond the scope of this book. We refer the interested reader to the classic references [Kirchhoff 1869, Lamb 1895], the modern texts [Arnol'd and Khesin 1998, Chorin and Marsden 1994, Etkin 1995], and some sample recent developments in [Kelly and Murray 2000, Leonard 1997, Leonard and Marsden 1997, Radford 2003]. Borrowing from this body of literature, it is possible to state some straightforward and useful concepts. We make the following assumptions: (1) the rigid body is immersed in an infinitely large volume of fluid, (2) the fluid is "ideal," that is, in the language of Chorin and Marsden [1994], the fluid is incompressible, irrotational and inviscid, and (3) the fluid is at rest at infinity. Under these assumptions, the surprising result by Kirchhoff [1869], see also [Lamb 1895, Leonard 1997], is that the motion of rigid body immersed in fluid can be described as that of a rigid body whose inertia tensor includes additional terms corresponding to the inertial properties of the fluid. It is common terminology to refer to these additional terms as "added masses and inertias." If the rigid body has an ellipsoidal shape and uniform mass density, then closed-form expressions for the coefficients of the fluid-body inertia tensor are available [Leonard 1997]. In what follows, it will be convenient for us to assume in our models that the body is not subject to any potential forces. Physically, this corresponds to assuming that (1) the rigid body is neutrally buoyant, i.e., the buoyant force is equal in magnitude with, and opposite in direction to, the force of gravity, and (2) the center of buoyancy coincides with the center of mass. We summarize this discussion and our assumptions as follows:

Assumptions 4.110 (Assumptions for rigid body motion in a fluid).

(i) The body is immersed in an infinitely large volume of fluid.
(ii) The fluid is ideal.
(iii) The fluid is at rest at infinity.
(iv) The body has an ellipsoidal shape and uniform mass density.
(v) The body is neutrally buoyant, i.e., the buoyant force is equal in magnitude with, and opposite in direction to, the force of gravity.
(vi) The center of buoyancy (the point through which the buoyant force acts) coincides with the center of mass. •

E4.25 Consider a planar rigid body in ideal fluid satisfying Assumptions 4.110. The system dynamics are those of the simple mechanical system $(\mathbb{S}^1 \times \mathbb{R}^2, \mathbb{G}, V)$, where, in the chart for $\mathbb{S}^1 \times \mathbb{R}^2$ given in Example 4.7, the kinetic energy metric $\mathbb{G}$ is given by

$$[\mathbb{G}] = \begin{bmatrix} J & 0 & 0 \\ 0 & m_1(\cos\theta)^2 + m_2(\sin\theta)^2 & (m_1 - m_2)\cos\theta\sin\theta \\ 0 & (m_1 - m_2)\cos\theta\sin\theta & m_2(\cos\theta)^2 + m_1(\sin\theta)^2 \end{bmatrix}.$$

Here $J, m_1, m_2 \in \mathbb{R}_+$ are inertial parameters that include the influence of the fluid surrounding the vehicle. We take $V = 0$, consistent with Assumption 4.110(v).

(a) Determine the Christoffel symbols for the Levi-Civita affine connection.
(b) Using this data, write the Euler–Lagrange equations for the system.

E4.26 Let $(\mathcal{B}, \mu)$ be a rigid body in a gravitational field with gravitational acceleration a_g in the direction of the unit vector $\boldsymbol{g} \in \mathbb{R}^3$.

(a) Show that the gravitational force (in the Newtonian sense) is given by $\mu(\mathcal{B})a_g\boldsymbol{g}$ acting at the center of mass of the body.

Now choose a spatial frame $(O_{\text{spatial}}, \{\boldsymbol{s}_1, \boldsymbol{s}_2, \boldsymbol{s}_3\})$ and a body frame $(O_{\text{body}}, \{\boldsymbol{b}_1, \boldsymbol{b}_2, \boldsymbol{b}_3\})$. The configuration of the body is then described by $\mathsf{Q} = \mathsf{SO}(3) \times \mathbb{R}^3$, assuming the frames have the same orientation. Do not necessarily choose O_{body} to be situated at the center of mass of $(\mathcal{B}, \mu)$.

(b) Write the force exerted by gravity as a covector field on Q.

The next exercise alludes to an interesting feature of potential forces, namely that they do no work on closed curves. In Newtonian mechanics, work is "force×distance," this making sense for constant velocity motions. In the Lagrangian setting, if $\gamma\colon I \to \mathsf{Q}$ is a differentiable curve, and if F is a force, the ***work*** done by F along γ is

$$W(F,\gamma) = \int_I \langle F(t,\gamma'(t)); \gamma'(t)\rangle \, dt.$$

Note that the converse of part (a) of the following exercise is also true. Thus a force is a potential force if and only if the work done by the force along any closed differentiable curve is zero.

E4.27 Let $F\colon \mathbb{R} \times \mathsf{TQ} \to \mathsf{T}^*\mathsf{Q}$ be a potential force on a manifold Q.

(a) Show that, for any differentiable closed curve $\gamma\colon [0,1] \to \mathsf{Q}$ (i.e., any such curve γ for which $\gamma(0) = \gamma(1)$), the work done by F along γ is zero.

(b) Using part (a), show that the forces τ^1 and τ^2 for the two-link manipulator of Example 4.54 are not potential forces.

A ***gyroscopic force*** on a manifold Q is a force of the form $F(t, v_q) = -C^\flat(v_q)$, where C is a skew-symmetric $(0,2)$-tensor field on Q. A discussion of gyroscopic forces in control theory may be found in [Wang and Krishnaprasad 1992].

E4.28 Let $(\mathsf{Q}, \mathbb{G}, V, F)$ be a forced simple mechanical system with $F = -C^\flat$ a gyroscopic force.

(a) Using the definition of work given prior to Exercise E4.27, show that the gyroscopic force F along any curve is zero.

(b) Show that the energy $E(v_q) = \frac{1}{2}\mathbb{G}(v_q, v_q) - V(q)$ is constant along any curve satisfying the Lagrange–d'Alembert Principle for F with Lagrangian $L(v_q) = \frac{1}{2}\mathbb{G}(v_q, v_q) - V(q)$.

E4.29 This is a continuation of the coupled masses problem of Exercise E4.3.

(a) Suppose that there is a dashpot[10] between the leftmost mass and the wall. Write the force exerted by the dashpot as a covector field on Q.

(b) Repeat part (a) for a dashpot placed between the masses.

(c) Suppose an external force is applied to the leftmost mass; see Figure E4.10. Write this force as a covector field on Q.

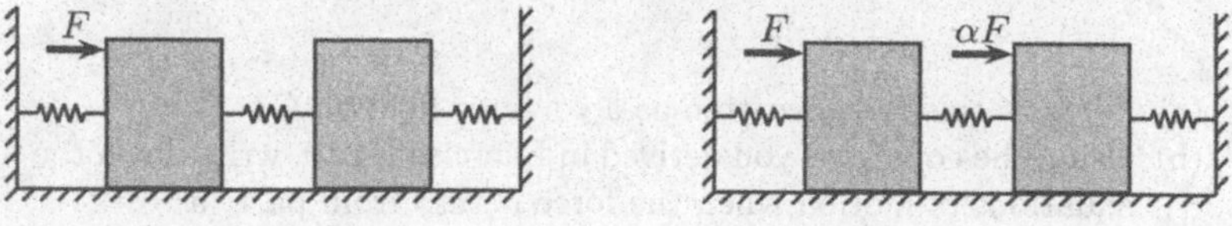

Figure E4.10. Force on the coupled masses for part (c) (left) and part (d) (right)

(d) Repeat part (c) for the case of a force on the system for which the force on the rightmost mass is equal to a proportionality constant α times the force on the leftmost mass.

E4.30 This is a continuation of the ball in a rotating channel problem of Exercise E4.4. Suppose that a torque is applied to the disk as in Figure E4.11.

[10] Recall that a ***dashpot*** is a device that exerts a force in the direction opposite to the direction it is being compressed or extended, and proportional to the velocity with which it is being compressed or extended.

Figure E4.11. Force for the ball in a rotating channel

(a) What is this force, written as a covector field on Q?
(b) Using the equations you derived in Exercise E4.18, write down the forced equations of motion when the force is that from part (a).

E4.31 This is a continuation of pendulum/cart problem of Exercise E4.5. Suppose that a force is applied to the base of the pendulum (see Figure E4.12).

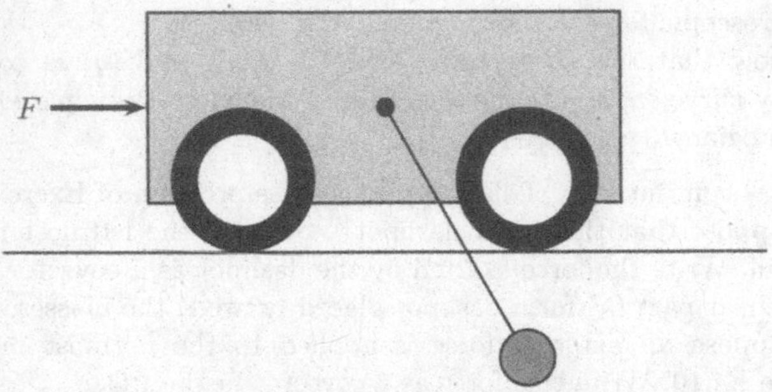

Figure E4.12. Force for the pendulum on a cart

(a) What is this force, written as a covector field on Q?
(b) Using the equations you derived in Exercise E4.19, write down the forced equations of motion when the force is that from part (a).

E4.32 This is a continuation of the spherical pendulum on a track considered in Exercise E4.6. Suppose that two forces are applied to the system, one a torque at the base of the arm to rotate the arm, and the other a linear force moving the base along the track (see Figure E4.13).
(a) What are these forces, written as covector fields on Q?
(b) Using the equations you derived in Exercise E4.20, write down the forced equations of motion when the force is a linear combination of the two forces from part (a).

E4.33 This is a continuation of the two-axis gyroscope problem of Exercise E4.7. Suppose that a torque is applied that rotates the gyro's frame (see Figure E4.14).

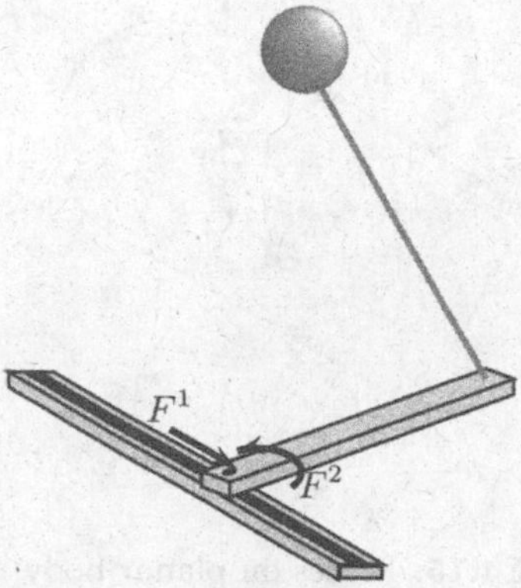

Figure E4.13. Forces for the spherical pendulum

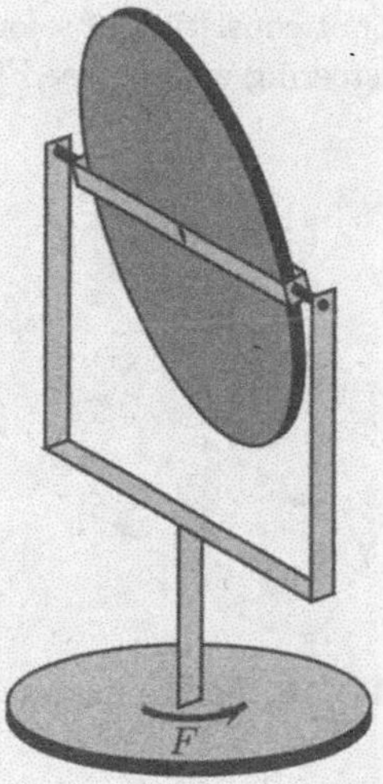

Figure E4.14. Force on the two-axis gyro

(a) What is this force, written as a covector field on Q?
(b) Using the equations you derived in Exercise E4.21, write down the forced equations of motion when the force is that from part (a).

E4.34 This is a continuation of the planar body problem with fan dynamics of Exercise E4.8. Consider the two forces on the planar body consisting of a torque that rotates the fan relative to the body, and a thrust force applied by the fan pointing in the direction the fan is pointing (see Figure E4.15).
(a) In the coordinates of Exercise E4.8, determine covector fields that represent the above forces.
(b) Using the equations you derived in Exercise E4.22, write down the forced equations of motion when the force is a linear combination of the two forces from part (a).

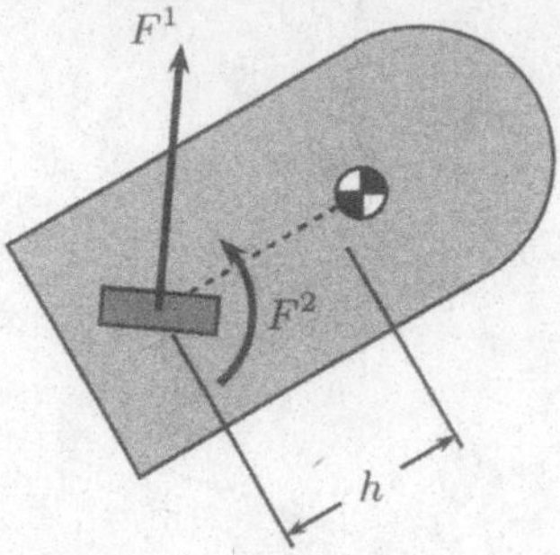

Figure E4.15. Forces on planar body with fan

E4.35 This is a continuation of the car-like robot problem of Exercise E4.8. Consider the two forces on the robot consisting of a force from the drive wheels and a torque that rotates the steering wheels (see Figure E4.16). In the coordinates

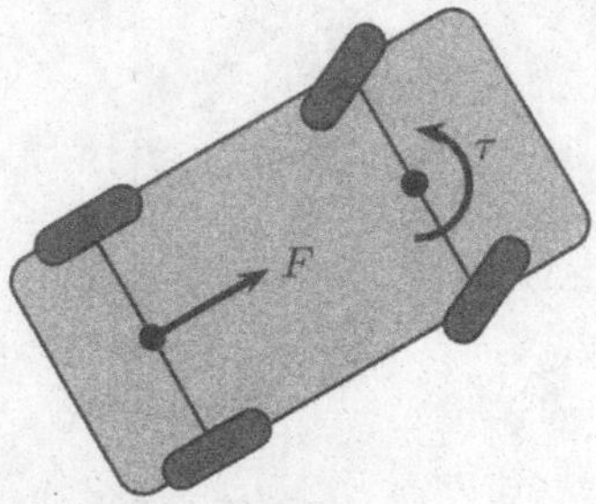

Figure E4.16. Forces on car-like robot

of Exercise E4.9, determine covector fields that represent the above forces.

E4.36 Prove that the total energy is conserved along the solution of a simple mechanical system with constraints.

Hint: *Consider the constrained connection* $\overset{\mathcal{D}}{\nabla}$*, and show that it is "compatible" with the kinetic energy metric when restricted to the constraint* $\mathcal{D}$*. (Part of the problem is to show what "compatible" means.)*

E4.37 *Mathematica®-assisted project:* In Figure E4.17 we show a three-link manipulator. In this exercise you will produce the elements of the mechanical model for the system. The expressions for the various quantities become quite lengthy, so Mathematica® (or equivalent) is recommended.

(a) Determine the configuration manifold for the system, and choose coordinates that have simple physical meaning.
(b) Determine the kinetic energy metric for the system in your set of coordinates. You should make the same modeling assumptions as were made for the two-link manipulator:

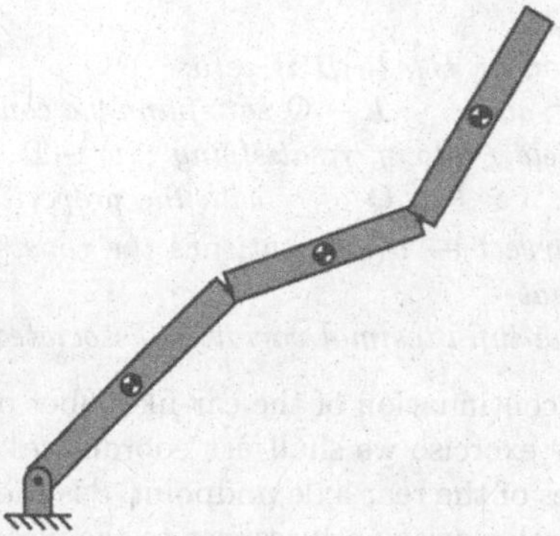

Figure E4.17. Three-link manipulator

1. number the links 1, 2, and 3, starting from the fixed link;
2. assume the motion is planar;
3. assume the centers of mass are at the midpoints of the links;
4. denote the lengths of the links by ℓ_a, $a \in \{1, 2, 3\}$;
5. denote the masses by m_a, $a \in \{1, 2, 3\}$;
6. denote the moments of inertia about the centers of mass by J_a, $a \in \{1, 2, 3\}$.

(c) Determine the Christoffel symbols for the Levi-Civita affine connection.
(d) Determine the potential function.

As forces, we will consider various combinations of torques applied at each of the three joints.

(e) Write the three input forces for the system as covector fields on Q.
(f) Write the equations of motion for the system.

E4.38 Let $\mathcal{U} \subset \mathsf{Q}$ be an open submanifold and let $\{X_1, \dots, X_n\}$ be vector fields on $\mathcal{U}$ for which $\{X_1(q), \dots, X_n(q)\}$ is a basis of $\mathsf{T}_q\mathsf{Q}$ for each $q \in \mathcal{U}$. Let $\overset{\mathscr{X}}{\Gamma}{}^i_{jk} \colon \mathcal{U} \to \mathbb{R}$, $i, j, k \in \{1, \dots, n\}$ be n^3 functions of class C^∞. Show that there exists a unique C^∞-affine connection ∇ on $\mathcal{U}$ for which $\overset{\mathscr{X}}{\Gamma}{}^i_{jk}$, $i, j, k \in \{1, \dots, n\}$, are the generalized Christoffel symbols for ∇ with respect to $\mathscr{X}$.

E4.39 Let ∇ be a torsion-free affine connection (see Definition 3.126 for the definition of torsion) on Q and let $\mathcal{D}$ be a regular distribution to which ∇ restricts. That is, if $X \in \Gamma^\infty(\mathsf{TQ})$ and $Y \in \Gamma^\infty(\mathcal{D})$, then $\nabla_X Y \in \Gamma^\infty(\mathcal{D})$. Show that $\mathcal{D}$ is integrable.

E4.40 Prove the following result.

Proposition. *Let $\mathcal{D}$ be a regular C^∞-distribution on* Q*, let* $\mathbb{G}$ *be a Riemannian metric on* Q*, and define a skew-symmetric $(0, 2)$-tensor on* Q *by*

$$B_{\mathcal{D}}^{\mathbb{G}}(X, Y) = (\overset{\mathbb{G}}{\nabla}_X P_{\mathcal{D}}^{\perp})(Y) - (\overset{\mathbb{G}}{\nabla}_Y P_{\mathcal{D}}^{\perp})(X).$$

Then the following are equivalent:

(i) *$\mathcal{D}$ is integrable;*
(ii) *the restriction of $B^{\mathbb{G}}_{\mathcal{D}}$ to $\mathcal{D}$ is zero;*
(iii) *for every C^2-curve $\gamma\colon I \to \mathsf{Q}$ satisfying the constraint $\mathcal{D}$, and for every C^2-vector field ξ along γ satisfying $\xi(t) \in \mathcal{D}_{\gamma(t)}$, $t \in I$, there exists a variation $\vartheta\colon J \times I \to \mathsf{Q}$ of γ with the properties that*
  (a) *the curve $t \mapsto \vartheta(s,t)$ satisfies the constraint $\mathcal{D}$ for each $s \in J$, and that*
  (b) *ξ is the infinitesimal variation associated with ϑ.*

E4.41 This exercise is a continuation of the car-like robot of Exercises E4.9, E4.23, and E4.35. In this exercise we shall use coordinates (x, y, θ, ϕ), where (x, y) are the coordinates of the rear axle midpoint, θ is the orientation of the body (measured counterclockwise with respect to the horizontal spatial axis), and ϕ is the steering angle.
(a) Show that if the system is subject to the constraint that the wheels roll without slipping, then the resulting constraint $\mathcal{D}$ is generated by the vector fields

$$X_1 = \rho\ell \sin\phi \cos\theta \frac{\partial}{\partial x} + \rho\ell \sin\phi \sin\theta \frac{\partial}{\partial y} + \cos\phi \frac{\partial}{\partial \theta}, \quad X_2 = \frac{\partial}{\partial \phi},$$

where ρ is the driving wheel radius and ℓ is the distance from rear to front axle.
(b) Find a $\mathbb{G}$-orthogonal basis for $\mathcal{D}$ and use these to write down the orthogonal Poincaré representation for the equations of motion.

E4.42 *Mathematica®-assisted project:* In this exercise we study the roller racer system and characterize it as a simple mechanical control system with constraints. This device was originally investigated by Krishnaprasad and Tsakiris [1995] and later in [Krishnaprasad and Tsakiris 2001] and [Bullo and Žefran 2002]. The roller racer, depicted in Figure E4.18, is a planar sys-

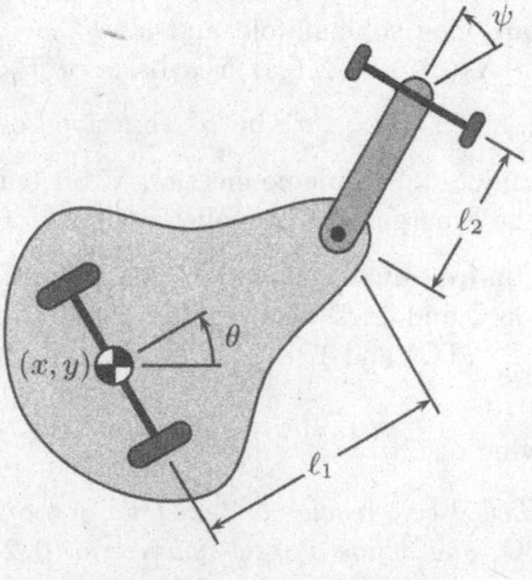

Figure E4.18. Roller racer

tem composed of two links (the main body and the front link) with wheels on both links and a control torque applied at the joint connecting the two links. The configuration manifold is $\mathbb{R}^2 \times \mathbb{S}^1 \times \mathbb{S}^1$; as coordinate chart we select (x, y, θ, ψ), where (x, y) denotes the position of the center of mass of the main body, θ is the angle of the main body relative to the horizontal axis, and ψ is the relative angle between the main body and the front link. Let m be the mass of the main body and let J_1 (resp. J_2) be the moment of inertia of the main body (resp. the front link) about its center of mass. For simplicity, we assume that the front link has negligible mass. The control force is $F = \mathrm{d}\psi$.

(a) Determine the kinetic energy metric and verify that all Christoffel symbols for the associated Levi-Civita affine connection vanish.

(b) Find an open dense subset $\mathcal{U}$ of Q on which the vector fields

$$X_1 = \cos\theta \frac{\partial}{\partial x} + \sin\theta \frac{\partial}{\partial y} + \left(\frac{\sin\psi}{\ell_2 + \ell_1 \cos\psi}\right) \frac{\partial}{\partial \theta},$$
$$X_2' = -\left(\frac{\ell_2}{\ell_2 + \ell_1 \cos\psi}\right) \frac{\partial}{\partial \theta} + \frac{\partial}{\partial \psi}$$

form a basis for the constraint distribution.

(c) For $X_2 = X_2' - \frac{\langle\langle X_2', X_1 \rangle\rangle}{\langle\langle X_1, X_1 \rangle\rangle} X_1$, consider the orthogonal basis $\{X_1, X_2\}$ for the constraint, valid on the subset $\mathcal{U}$ determined in part (b). Compute the generalized Christoffel symbols with respect to $\{X_1, X_2\}$ for the constrained affine connection of the roller racer.

(d) Using this data, write the orthogonal Poincaré representation of the equations of motion for the roller racer.

5

Lie groups, systems on groups, and symmetries

The previous chapter provides a general framework for modeling mechanical control systems on manifolds. Frequently however, there is additional structure in a mechanical system that can be exploited for analysis and control. This chapter discusses an important class of manifolds, called Lie groups, that arise naturally in rigid body kinematics, as well as the properties of mechanical systems defined on Lie groups or possessing Lie group symmetries.

Lie groups have a very rich geometric structure and are worthy of careful study in their own right. It is therefore not surprising that the literature on the subject is vast. Let us mention a few references related to the material we present. Rigid body kinematics is discussed in [Brockett 1990, Craig 1989, Murray, Li, and Sastry 1994, Selig 1996, Spong and Vidyasagar 1989], and the development of Lie theory from matrix Lie groups is discussed in [Crampin 1986, Howe 1983, 1984]. Textbooks and monographs on Lie groups and matrix groups include [Chevalley 1946, Curtis 1979, Helgason 1978, Sattinger and Weaver 1986, Varadarajan 1984, Warner 1971]. The invariant affine connection approach to modeling mechanical systems on Lie groups is discussed in [Arnol'd 1978, Bullo, Leonard, and Lewis 2000]. References on systems on groups, group actions, and symmetries include [Abraham and Marsden 1978, Bloch 2003, Hermann 1968, Kobayashi and Nomizu 1963, Marsden 1992, Marsden and Ratiu 1999]. Finally, other references that have affected the presentation of this chapter are the work in [Brockett 1972] on systems on groups, and in [Lamb 1895, Leonard 1997] on underwater rigid body dynamics.

The outline of this chapter is as follows. In Section 5.1, we study rigid body kinematics in the language of matrix groups and matrix algebras, and characterize the groups of rotations and rigid displacements in two- and three-dimensional Euclidean space. We then consider in Section 5.2 the abstract notions of a Lie group and of a Lie algebra. Roughly speaking, a Lie group is a smooth manifold equipped with a group operation and a Lie algebra is a vector space equipped with a bracket operation. Via the notion of a one-parameter subgroup generator, we present an explicit construction of matrix Lie algebras. The differential geometric structure of Lie groups is extremely rich and is connected to the notion of invariance under the canonical left action. Our treatment discusses invariance notions for vector fields, Riemannian metrics, and affine connections; see Section 5.3. Next, we investigate the equations of motion for rigid bodies. The structure and the properties of these equations are linked to the structure of the configuration Lie groups of these systems. We present the classic Euler and Kirchhoff equations for rigid body motion in space and in an ideal fluid, respectively. In Section 5.4, we introduce the notions of group actions and of invariance under a group action. Mechanical systems invariant under a group action are said to have a symmetry and are known to satisfy the classic Noether Conservation Law. As an illustration of this, we derive the angular momentum conservation law for a rigid body with a fixed center of mass. Finally, in Section 5.5, we discuss principal fiber bundles, an infinitesimal equivalent of principal fiber bundles, and present a Riemannian geometric treatment of reduction.

Since our presentation of reduction theory is admittedly limited, we provide a few additional pointers to the interested reader. The theory of reduction and symmetries in mechanical systems has been the subject of significant research. Early work on Hamiltonian reduction includes [Marsden and Weinstein 1974, Meyer 1973]. The theory is now exposed very elegantly in [Abraham and Marsden 1978] and [Marsden and Ratiu 1999]. Various approaches to reduction theory for systems with nonholonomic constraints, and with free and proper symmetries, are developed in [Bates and Śniatycki 1993, Bloch, Krishnaprasad, Marsden, and Murray 1996, Marle 1995, Śniatycki 1998, van der Schaft and Maschke 1994]. Reduction using Poisson structures is discussed in [Koon and Marsden 1998]. A modern comprehensive approach to Lagrangian reduction is developed in [Cendra, Marsden, and Ratiu 2001a]. Modern texts on reduction include [Abraham and Marsden 1978, Bloch 2003, Marsden and Ratiu 1999]. Finally, it is also possible to perform reduction when the symmetry action is not free and proper; we refer the reader to [Ortega and Ratiu 2004] and references therein.

5.1 Rigid body kinematics

We start by reviewing and extending our discussion in Chapter 4 about the motion of rigid bodies. We study in some detail (1) the properties of the orthogonal group and of the Euclidean group, (2) the notions of body velocity and spatial velocity, and (3) the matrix exponential map and its restriction to the Euclidean group. We refer the readers to [Brockett 1990, Murray, Li, and Sastry 1994] for related concepts and also warn them that, since this section reviews some concepts from Chapter 4, there is some overlap with the treatment in that chapter.

5.1.1 Rigid body transformations

Consider a rigid body $\mathcal{B}$ moving in three-dimensional Euclidean space, see Figure 5.1, and introduce the reference frame $\Sigma_{\text{spatial}} = (O_{\text{spatial}}, \{\boldsymbol{s}_1, \boldsymbol{s}_2, \boldsymbol{s}_3\})$

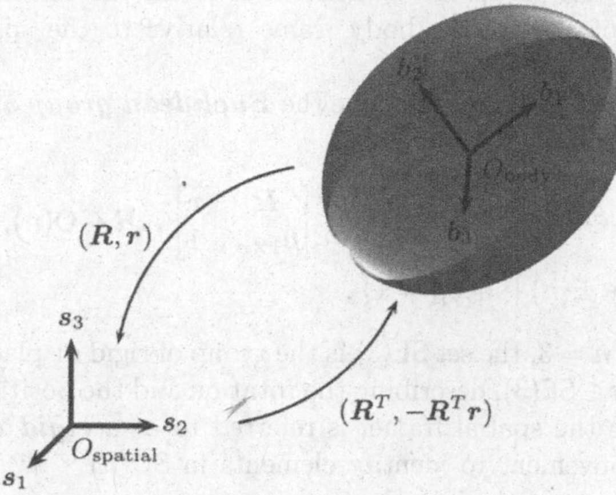

Figure 5.1. Change of reference frame transformations

fixed in space and the reference frame $\Sigma_{\text{body}} = (O_{\text{body}}, \{\boldsymbol{b}_1, \boldsymbol{b}_2, \boldsymbol{b}_3\})$ fixed to the body. Assume that $\{\boldsymbol{s}_1, \boldsymbol{s}_2, \boldsymbol{s}_3\}$ and $\{\boldsymbol{b}_1, \boldsymbol{b}_2, \boldsymbol{b}_3\}$ have the same orientation. Let the vector $\boldsymbol{r} \in \mathbb{R}^3$ connect the origin O_{spatial} of the frame Σ_{spatial} to the origin O_{body} of the frame Σ_{body} (i.e., $\boldsymbol{r} = O_{\text{spatial}} - O_{\text{body}}$), and let the matrix $\boldsymbol{R} \in \mathbb{R}^{3\times 3}$ have as its ath column the components of $\boldsymbol{b}_a$ relative to the basis $\{\boldsymbol{s}_1, \boldsymbol{s}_2, \boldsymbol{s}_3\}$.

The coordinates of a point in space can be measured by an observer fixed either in the spatial frame or in the body frame. We refer to these two vectors respectively as $\boldsymbol{\chi}_{\text{s}}$ and $\boldsymbol{\chi}_{\text{b}}$, noting that this notation differs from that used in Sections 4.2.1 and 4.2.2. The vectors $\boldsymbol{\chi}_{\text{s}}$ and $\boldsymbol{\chi}_{\text{b}}$ are related through the affine transformation

$$\boldsymbol{\chi}_{\text{s}} = \boldsymbol{R}\boldsymbol{\chi}_{\text{b}} + \boldsymbol{r}. \tag{5.1}$$

It is often convenient to write this transformation in the following equivalent form. For $\boldsymbol{x} \in \mathbb{R}^3$, we call $\bar{\boldsymbol{x}} = (\boldsymbol{x}, 1) \in \mathbb{R}^4$ the ***homogeneous coordinates*** of $\boldsymbol{x}$. We can now rewrite the transformation (5.1) and its inverse as

$$\bar{\boldsymbol{\chi}}_{\text{s}} = \begin{bmatrix} \boldsymbol{R} & \boldsymbol{r} \\ \boldsymbol{0}_{1\times 3} & 1 \end{bmatrix} \bar{\boldsymbol{\chi}}_{\text{b}}, \quad \bar{\boldsymbol{\chi}}_{\text{b}} = \begin{bmatrix} \boldsymbol{R}^T & -\boldsymbol{R}^T\boldsymbol{r} \\ \boldsymbol{0}_{1\times 3} & 1 \end{bmatrix} \bar{\boldsymbol{\chi}}_{\text{s}}, \tag{5.2}$$

where we recall that $\boldsymbol{0}_{m\times n}$ is the zero matrix in $\mathbb{R}^{m\times n}$. The two transformations are illustrated graphically in Figure 5.1.

Motivated by this kinematic analysis, let us recall from Exercise E3.17 the orthogonal group $\mathsf{O}(n)$ and the special orthogonal group $\mathsf{SO}(n)$ defined by

$$\mathsf{O}(n) = \{\boldsymbol{R} \in \mathbb{R}^{n\times n} \mid \boldsymbol{R}\boldsymbol{R}^T = \boldsymbol{I}_n\}, \quad \mathsf{SO}(n) = \{\boldsymbol{R} \in \mathsf{O}(n) \mid \det \boldsymbol{R} = 1\},$$

respectively. (We postpone the definition of group to Section 5.2.) For $n = 3$, the set $\mathsf{SO}(3)$ is the group of rotations in $\mathbb{R}^3$, and the matrix $\boldsymbol{R} \in \mathsf{SO}(3)$, describing the rotation of the body frame relative to the spatial frame, is referred to as a ***rotation matrix***.

Next, motivated by (5.2), we define the ***Euclidean group*** and the ***special Euclidean group*** by

$$\mathsf{E}(n) = \left\{\boldsymbol{g} \in \mathbb{R}^{(n+1)\times(n+1)} \;\middle|\; \boldsymbol{g} = \begin{bmatrix} \boldsymbol{R} & \boldsymbol{r} \\ \boldsymbol{0}_{1\times n} & 1 \end{bmatrix}, \ \boldsymbol{R} \in \mathsf{O}(n), \ \boldsymbol{r} \in \mathbb{R}^n \right\},$$
$$\mathsf{SE}(n) = \{\boldsymbol{g} \in \mathsf{E}(n) \mid \det \boldsymbol{g} = 1\}.$$

respectively. For $n = 3$, the set $\mathsf{SE}(3)$ is the group of rigid displacements in $\mathbb{R}^3$, and the matrix $\boldsymbol{g} \in \mathsf{SE}(3)$, describing the rotation and the position of the body frame relative to the spatial frame, is referred to as a ***rigid displacement matrix***. It is convenient to identify elements in $\mathsf{SO}(n) \times \mathbb{R}^n$ with elements and $\mathsf{SE}(n)$ by means of the bijection

$$(\boldsymbol{R}, \boldsymbol{r}) \mapsto \begin{bmatrix} \boldsymbol{R} & \boldsymbol{r} \\ \boldsymbol{0}_{1\times n} & 1 \end{bmatrix}.$$

Remarks 5.1. 1. Equation (5.1) states that elements of $\mathsf{SE}(3)$, when representing the displacement of a rigid body relative to a spatial frame, are transformations of reference frame.

2. The sets of matrices $\mathsf{O}(n)$, $\mathsf{SO}(n)$, $\mathsf{E}(n)$, and $\mathsf{SE}(n)$ are closed under the operation of matrix inversion and of matrix multiplication. This structure leads to numerous properties that form the basis for the material in this chapter. •

In what follows, let us examine the elements of $\mathsf{O}(n)$ and $\mathsf{E}(n)$ by considering them as linear and affine maps on $\mathbb{R}^n$, respectively. One can see that orthogonal matrices, i.e., elements of $\mathsf{O}(n)$, preserve the length of vectors and, geometrically, include rotations and reflections. Similarly, elements of $\mathsf{E}(n)$ include rotations, translations, and reflections. It is common to disregard reflections by imposing the positive determinant restriction on elements of $\mathsf{O}(n)$. When the orthogonal matrix represents the rotation of a body frame relative to a spatial frame, the positive determinant restriction amounts to the assumption that $\{\boldsymbol{s}_1, \boldsymbol{s}_2, \boldsymbol{s}_3\}$ and $\{\boldsymbol{b}_1, \boldsymbol{b}_2, \boldsymbol{b}_3\}$ have the same orientation.

Let us be more precise with regards to the property of preserving distances between points. We shall say that a map $\phi\colon \mathbb{R}^n \to \mathbb{R}^n$ is an ***isometry*** of $\mathbb{R}^n$ if $\|\phi(\boldsymbol{y}) - \phi(\boldsymbol{z})\|_{\mathbb{R}^n} = \|\boldsymbol{y} - \boldsymbol{z}\|_{\mathbb{R}^n}$ for all $\boldsymbol{y}, \boldsymbol{z} \in \mathbb{R}^n$. In what follows, we identify elements $\boldsymbol{g} = (\boldsymbol{R}, \boldsymbol{r}) \in \mathsf{E}(n)$ with the corresponding maps $\phi_{\boldsymbol{g}}\colon \mathbb{R}^n \to \mathbb{R}^n$ given by $\boldsymbol{y} \mapsto \boldsymbol{R}\boldsymbol{y} + \boldsymbol{r}$.

Proposition 5.2 (Isometries of $\mathbb{R}^n$). *A map $\phi\colon \mathbb{R}^n \to \mathbb{R}^n$ is an isometry if and only if there exists $\boldsymbol{g} \in \mathsf{E}(n)$ such that $\phi = \phi_{\boldsymbol{g}}$.*

Proof. For $\boldsymbol{g} \in \mathsf{E}(n)$, it is straightforward to show that $\phi_{\boldsymbol{g}}$ is an isometry. Next, suppose that ϕ is an isometry that fixes $\boldsymbol{0} \in \mathbb{R}^n$. Recall that the norm on an inner product space satisfies the parallelogram law:

$$\|\boldsymbol{y}+\boldsymbol{z}\|_{\mathbb{R}^n}^2 + \|\boldsymbol{y}-\boldsymbol{z}\|_{\mathbb{R}^n}^2 = 2\big(\|\boldsymbol{y}\|_{\mathbb{R}^n}^2 + \|\boldsymbol{z}\|_{\mathbb{R}^n}^2\big).$$

Using this equality, and the fact that ϕ is an isometry fixing $\boldsymbol{0}$, we compute

$$\begin{aligned}\|\phi(\boldsymbol{y})+\phi(\boldsymbol{z})\|_{\mathbb{R}^n}^2 &= 2\|\phi(\boldsymbol{y})\|_{\mathbb{R}^n}^2 + 2\|\phi(\boldsymbol{z})\|_{\mathbb{R}^n}^2 - \|\phi(\boldsymbol{y})-\phi(\boldsymbol{z})\|_{\mathbb{R}^n}^2 \\ &= 2\|\boldsymbol{y}\|_{\mathbb{R}^n}^2 + 2\|\boldsymbol{z}\|_{\mathbb{R}^n}^2 - \|\boldsymbol{y}-\boldsymbol{z}\|_{\mathbb{R}^n}^2 = \|\boldsymbol{y}+\boldsymbol{z}\|_{\mathbb{R}^n}^2.\end{aligned} \tag{5.3}$$

By the polarization identity, see Exercise E2.12, we obtain

$$\mathbb{G}_{\mathbb{R}^n}(\boldsymbol{y},\boldsymbol{z}) = \tfrac{1}{2}\left(\|\boldsymbol{y}+\boldsymbol{z}\|_{\mathbb{R}^n}^2 - \|\boldsymbol{y}\|_{\mathbb{R}^n}^2 - \|\boldsymbol{z}\|_{\mathbb{R}^n}^2\right)$$

for every $\boldsymbol{y},\boldsymbol{z} \in \mathbb{R}^n$. In particular, using (5.3) and the fact that ϕ is an isometry fixing $\boldsymbol{0}$, we compute

$$\begin{aligned}\mathbb{G}_{\mathbb{R}^n}(\phi(\boldsymbol{y}),\phi(\boldsymbol{z})) &= \tfrac{1}{2}\left(\|\phi(\boldsymbol{y})+\phi(\boldsymbol{z})\|_{\mathbb{R}^n}^2 - \|\phi(\boldsymbol{y})\|_{\mathbb{R}^n}^2 - \|\phi(\boldsymbol{z})\|_{\mathbb{R}^n}^2\right) \\ &= \tfrac{1}{2}\left(\|\boldsymbol{y}+\boldsymbol{z}\|_{\mathbb{R}^n}^2 - \|\boldsymbol{y}\|_{\mathbb{R}^n}^2 - \|\boldsymbol{z}\|_{\mathbb{R}^n}^2\right) = \mathbb{G}_{\mathbb{R}^n}(\boldsymbol{y},\boldsymbol{z}).\end{aligned}$$

We now claim that this implies that ϕ is a linear map. Indeed, let $\{\boldsymbol{e}_1,\dots,\boldsymbol{e}_n\}$ be the standard orthonormal basis for $\mathbb{R}^n$ and let $(y^1,\dots,y^n)$ be the components of $\boldsymbol{y} \in \mathbb{R}^n$ in this basis (thus $y^i = \mathbb{G}_{\mathbb{R}^n}(\boldsymbol{y},\boldsymbol{e}_i)$, $i \in \{1,\dots,n\}$). Since $\mathbb{G}_{\mathbb{R}^n}(\phi(\boldsymbol{e}_i),\phi(\boldsymbol{e}_j)) = \mathbb{G}_{\mathbb{R}^n}(\boldsymbol{e}_i,\boldsymbol{e}_j)$, for $i,j \in \{1,\dots,n\}$, the vectors $\{\phi(\boldsymbol{e}_1),\dots,\phi(\boldsymbol{e}_n)\}$ form an orthonormal basis for $\mathbb{R}^n$. The components of $\phi(\boldsymbol{y})$ in this basis are given by $\mathbb{G}_{\mathbb{R}^n}(\phi(\boldsymbol{y}),\phi(\boldsymbol{e}_i))$, $i \in \{1,\dots,n\}$. But, since ϕ preserves $\mathbb{G}_{\mathbb{R}^n}$, this means that the components of $\phi(\boldsymbol{y})$ are precisely $(y^1,\dots,y^n)$. That is,

$$\phi\Big(\sum_{i=1}^n y^i\boldsymbol{e}_i\Big) = \sum_{i=1}^n y^i\phi(\boldsymbol{e}_i).$$

Therefore, if ϕ fixes $\boldsymbol{0} \in \mathbb{R}^n$, then ϕ is linear and there exists $\boldsymbol{R} \in \mathsf{O}(n)$ such that $\phi = \phi_{(\boldsymbol{R},\boldsymbol{0})}$. Now, suppose that ϕ fixes not $\boldsymbol{0}$, but some other point $\boldsymbol{y}_0 \in \mathbb{R}^n$. Then we have $\phi_{(\boldsymbol{0},-\boldsymbol{y}_0)} \circ \phi \circ \phi_{(\boldsymbol{0},\boldsymbol{y}_0)}(\boldsymbol{0}) = \boldsymbol{0}$, which implies that $\phi_{(\boldsymbol{0},-\boldsymbol{y}_0)} \circ \phi \circ \phi_{(\boldsymbol{0},\boldsymbol{y}_0)} = \phi_{(\boldsymbol{R},\boldsymbol{0})}$ for some $\boldsymbol{R} \in \mathsf{O}(n)$. Since $\phi_{\boldsymbol{g}_1} \circ \phi_{\boldsymbol{g}_2} = \phi_{\boldsymbol{g}_1\boldsymbol{g}_2}$ for $\boldsymbol{g}_1,\boldsymbol{g}_2 \in \mathsf{E}(n)$, it follows that $\phi = \phi_{\boldsymbol{g}}$ for $\boldsymbol{g} \in \mathsf{E}(n)$. Finally, suppose that ϕ maps $\boldsymbol{y}_1 \in \mathbb{R}^n$ to $\boldsymbol{y}_2 \in \mathbb{R}^n$. In this most general case, letting $\boldsymbol{y}_0 = \boldsymbol{y}_1 - \boldsymbol{y}_2$, we have $\phi_{(\boldsymbol{0},-\boldsymbol{y}_0)} \circ \phi(\boldsymbol{y}_1) = \boldsymbol{y}_1$ and so $\phi_{(\boldsymbol{0},-\boldsymbol{y}_0)} \circ \phi = \phi_{\boldsymbol{g}}$ for some $\boldsymbol{g} \in \mathsf{SE}(n)$. ■

This result shows that the Euclidean group $\mathsf{E}(n)$ is the isometry group of $\mathbb{R}^n$, i.e., the set of all isometries of $\mathbb{R}^n$. In this sense, we shall say that rotations and displacements are ***rigid body transformations***.

5.1.2 Infinitesimal rigid body transformations

In this section we study the velocity of a rigid body motion. We discuss differential equations involving linear velocity and angular velocity. We start by reproducing a few notions from Section 4.1.4 and by studying them in more detail. For $n \in \mathbb{N}$, let $\mathfrak{so}(n)$ be the vector space of skew-symmetric matrices in $\mathbb{R}^{n\times n}$:

$$\mathfrak{so}(n) = \{\boldsymbol{S} \in \mathbb{R}^{n\times n} \mid \boldsymbol{S}^T = -\boldsymbol{S}\}.$$

Let $\times$ denote the vector cross-product on $\mathbb{R}^3$ and define the linear map $\widehat{\cdot}\colon \mathbb{R}^3 \to \mathfrak{so}(3)$ by $\widehat{\boldsymbol{\omega}}\boldsymbol{y} = \boldsymbol{\omega} \times \boldsymbol{y}$ for all $\boldsymbol{\omega}, \boldsymbol{y} \in \mathbb{R}^3$. Since the linear map $\widehat{\cdot}$ is an isomorphism of vector spaces, we write $\mathfrak{so}(3) \simeq \mathbb{R}^3$. Also, for later reference, it is convenient to let $\cdot^\vee\colon \mathfrak{so}(3) \to \mathbb{R}^3$ denote the inverse isomorphism. For $\boldsymbol{\omega} = (\omega^1, \omega^2, \omega^3)$, we write

$$\widehat{\boldsymbol{\omega}} = \begin{bmatrix} 0 & -\omega_3 & \omega_2 \\ \omega_3 & 0 & -\omega_1 \\ -\omega_2 & \omega_1 & 0 \end{bmatrix}, \quad \begin{bmatrix} 0 & -\omega_3 & \omega_2 \\ \omega_3 & 0 & -\omega_1 \\ -\omega_2 & \omega_1 & 0 \end{bmatrix}^\vee = \boldsymbol{\omega}.$$

These notions can be extended as follows. We define

$$\mathfrak{se}(n) = \left\{ \begin{bmatrix} \boldsymbol{S} & \boldsymbol{v} \\ \boldsymbol{0}_{1\times n} & 0 \end{bmatrix} \in \mathbb{R}^{(n+1)\times(n+1)} \;\middle|\; \boldsymbol{S} \in \mathfrak{so}(n),\ \boldsymbol{v} \in \mathbb{R}^n \right\}.$$

Analogously to the $\mathfrak{so}(3)$ case, it is convenient to introduce, with slight abuse of notation, the linear map $\widehat{\cdot}\colon \mathbb{R}^3 \oplus \mathbb{R}^3 \to \mathfrak{se}(3)$ given by

$$\widehat{\boldsymbol{\xi}} = \begin{bmatrix} \widehat{\boldsymbol{\omega}} & \boldsymbol{v} \\ \boldsymbol{0}_{1\times 3} & 0 \end{bmatrix}, \tag{5.4}$$

for $\boldsymbol{\xi} = (\boldsymbol{\omega}, \boldsymbol{v}) \in \mathbb{R}^3 \oplus \mathbb{R}^3$. Since $\widehat{\cdot}\colon \mathbb{R}^3 \oplus \mathbb{R}^3 \to \mathfrak{se}(3)$ is an isomorphism of vector spaces, we write $\mathfrak{se}(3) \simeq \mathbb{R}^3 \oplus \mathbb{R}^3$ and we let $\cdot^\vee\colon \mathfrak{se}(3) \to \mathbb{R}^3 \oplus \mathbb{R}^3$ denote the corresponding inverse isomorphism. The elements of $\mathfrak{se}(3)$ are often referred to as ***twists***.

Let us now return to the study of rigid body motion and expand on the discussion begun in Section 4.2.2. Given a spatial frame $\Sigma_{\mathsf{spatial}}$ and a body frame Σ_{body}, the movement of the rigid body $\mathcal{B}$ is described by a curve $t \mapsto (\boldsymbol{R}(t), \boldsymbol{r}(t)) \in \mathsf{SO}(3) \times \mathbb{R}^3$, equivalently written as $t \mapsto \boldsymbol{g}(t) \in \mathsf{SE}(3)$. We suppose this curve to be differentiable on $\mathbb{R}$. We recall from Section 4.1.5 the definition of the body angular velocity and of the spatial angular velocity, which we reproduce here for convenience. We call ***spatial angular velocity*** the curve $\boldsymbol{\omega}\colon \mathbb{R} \to \mathbb{R}^3$ by

$$\widehat{\boldsymbol{\omega}}(t) = \dot{\boldsymbol{R}}(t)\boldsymbol{R}^T(t),$$

and we call ***body angular velocity*** the curve $\boldsymbol{\Omega}\colon \mathbb{R} \to \mathbb{R}^3$ given by

$$\widehat{\boldsymbol{\Omega}}(t) = \boldsymbol{R}^T(t)\dot{\boldsymbol{R}}(t).$$

Recall that these definitions are well-posed because of Lemma 4.17. Let us provide here a generalization, whose proof is left to the reader as Exercise E5.2.

Lemma 5.3. *Let $\boldsymbol{g}_0 \in \mathsf{SE}(3)$. Consider the smooth curves $\boldsymbol{A}\colon \mathbb{R} \to \mathbb{R}^{4\times 4}$ and $\boldsymbol{g}\colon \mathbb{R} \to \mathbb{R}^{4\times 4}$, related by the matrix differential equation*

$$\dot{\boldsymbol{g}}(t) = \boldsymbol{g}(t)\boldsymbol{A}(t), \quad \boldsymbol{g}(0) = \boldsymbol{g}_0.$$

Then the curve $\boldsymbol{g}$ takes values in $\mathsf{SE}(3)$ if and only if the curve $\boldsymbol{A}$ takes values in $\mathfrak{se}(3)$. Furthermore, the same conclusion holds if the matrix differential equation is instead $\dot{\boldsymbol{g}}(t) = \boldsymbol{A}(t)\boldsymbol{g}(t); \boldsymbol{g}(0) = \boldsymbol{g}_0$.

We can now introduce notions of spatial and body velocity for the joint rotation and translation of a rigid body. Given the curve $\boldsymbol{g}\colon \mathbb{R} \to \mathsf{SE}(3)$, we call ***spatial velocity*** the curve $\boldsymbol{\xi}_{\mathrm{s}}\colon \mathbb{R} \to \mathbb{R}^3 \oplus \mathbb{R}^3$ given by

$$\widehat{\boldsymbol{\xi}}_{\mathrm{s}}(t) = \dot{\boldsymbol{g}}(t)\boldsymbol{g}^{-1}(t),$$

and we call ***body velocity*** the curve $\boldsymbol{\xi}_{\mathrm{b}}\colon \mathbb{R} \to \mathbb{R}^3 \oplus \mathbb{R}^3$ given by

$$\widehat{\boldsymbol{\xi}}_{\mathrm{b}}(t) = \boldsymbol{g}^{-1}(t)\dot{\boldsymbol{g}}(t).$$

Remarks 5.4. 1. For a rigid body trajectory $t \mapsto (\boldsymbol{R}(t), \boldsymbol{r}(t))$ with spatial angular velocity $t \mapsto \boldsymbol{\omega}(t)$ and body angular velocity $t \mapsto \boldsymbol{\Omega}(t)$, equivalent definitions of spatial velocity and of body velocity are:

$$\boldsymbol{\xi}_{\mathrm{s}}(t) = (\boldsymbol{\omega}(t), \dot{\boldsymbol{r}}(t) + \boldsymbol{r}(t) \times \boldsymbol{\omega}(t)),$$
$$\boldsymbol{\xi}_{\mathrm{b}}(t) = (\boldsymbol{\Omega}(t), -\boldsymbol{R}^T(t)\dot{\boldsymbol{r}}(t)).$$

2. Consider a point moving along a trajectory in space and let $t \mapsto \boldsymbol{\chi}_{\mathrm{s}}(t)$ and $t \mapsto \boldsymbol{\chi}_{\mathrm{b}}(t)$ be its coordinates with respect to $\Sigma_{\mathsf{spatial}}$ and Σ_{body}, respectively. By differentiating (5.2) with respect to time, we obtain

$$\begin{aligned}\frac{\mathrm{d}}{\mathrm{d}t}\bar{\boldsymbol{\chi}}_{\mathrm{s}}(t) &= \boldsymbol{g}(t)\Big(\frac{\mathrm{d}}{\mathrm{d}t}\bar{\boldsymbol{\chi}}_{\mathrm{b}}(t) + \widehat{\boldsymbol{\xi}}_{\mathrm{b}}(t)\bar{\boldsymbol{\chi}}_{\mathrm{b}}(t)\Big)\\ &= \widehat{\boldsymbol{\xi}}_{\mathrm{s}}(t)\bar{\boldsymbol{\chi}}_{\mathrm{s}}(t) + \boldsymbol{g}(t)\frac{\mathrm{d}}{\mathrm{d}t}\bar{\boldsymbol{\chi}}_{\mathrm{b}}(t). \end{aligned} \tag{5.5}$$

Analogous relationships can be written for the time-derivative of $t \mapsto \bar{\boldsymbol{\chi}}_{\mathrm{b}}(t)$.

3. Spatial velocity and body velocity represent the same physical quantity, only expressed in a different reference frame. Given a rigid body with configuration $\boldsymbol{g} \in \mathsf{SE}(3)$, the transformation from body to spatial reference frame for twists is called the ***adjoint map*** $\mathrm{Ad}_{\boldsymbol{g}}\colon \mathfrak{se}(3) \to \mathfrak{se}(3)$ and is given by

$$\mathrm{Ad}_{\boldsymbol{g}}\,\widehat{\boldsymbol{\eta}} = \boldsymbol{g}\widehat{\boldsymbol{\eta}}\boldsymbol{g}^{-1}. \tag{5.6}$$

Using the identification $\mathfrak{se}(3) \simeq \mathbb{R}^3 \oplus \mathbb{R}^3$, one can show that this definition is well-posed and that $\mathrm{Ad}_{\boldsymbol{g}}\colon \mathbb{R}^3 \oplus \mathbb{R}^3 \to \mathbb{R}^3 \oplus \mathbb{R}^3$ can be written as

$$\mathrm{Ad}_{(\boldsymbol{R},\boldsymbol{r})} = \begin{bmatrix} \boldsymbol{R} & 0_{3\times 3} \\ \widehat{\boldsymbol{r}}\boldsymbol{R} & \boldsymbol{R} \end{bmatrix}. \tag{5.7}$$

For any rigid body trajectory $t \mapsto \boldsymbol{g}(t)$ with spatial velocity $t \mapsto \boldsymbol{\xi}_{\mathrm{s}}(t)$ and body velocity $t \mapsto \boldsymbol{\xi}_{\mathrm{s}}(t)$, we have $\widehat{\boldsymbol{\xi}}_{\mathrm{s}}(t) = \mathrm{Ad}_{\boldsymbol{g}(t)}\,\widehat{\boldsymbol{\xi}}_{\mathrm{b}}(t)$. •

5.1.3 Rigid body transformations as exponentials of twists

As illustrated in the previous section, elements of $\mathfrak{so}(3)$ and $\mathfrak{se}(3)$ describe infinitesimal rigid body rotations and infinitesimal rigid body transformations, respectively. In this section we show that, by means of the matrix exponential map, the vector spaces $\mathfrak{so}(3)$ and $\mathfrak{se}(3)$ provide an alternative way of representing elements in $\mathsf{SO}(3)$ and $\mathsf{SE}(3)$.

We start by introducing the exponential map on matrices. We define the ***matrix exponential*** $\exp\colon \mathbb{R}^{n\times n} \to \mathbb{R}^{n\times n}$ as

$$\exp \boldsymbol{A} = \sum_{k=0}^{+\infty} \frac{\boldsymbol{A}^k}{k!}.$$

The reader is invited to demonstrate the convergence of this series, and other properties of the matrix exponential, in Exercise E5.3. This operation plays a key role in the study of linear time-independent differential equations. For example, one can easily prove the following facts.

Proposition 5.5. *The curve* $\boldsymbol{x}\colon \mathbb{R} \to \mathbb{R}^n$ *given by* $t \mapsto \exp(\boldsymbol{A}t)\boldsymbol{x}_0$ *is the solution to the initial value problem* $\dot{\boldsymbol{x}}(t) = \boldsymbol{A}\boldsymbol{x}(t); \boldsymbol{x}(0) = \boldsymbol{x}_0$. *Similarly, the curve* $\boldsymbol{g}\colon \mathbb{R} \to \mathbb{R}^{n\times n}$ *given by* $t \mapsto \boldsymbol{g}_0 \exp(\boldsymbol{A}t)$ *is the solution to the initial value problem* $\dot{\boldsymbol{g}}(t) = \boldsymbol{g}(t)\boldsymbol{A}; \boldsymbol{g}(0) = \boldsymbol{g}_0$.

Proof. The reader is invited to prove this result in Exercise E5.4. ■

Consider a point fixed in the body frame, i.e., a point whose coordinates χ_{b} in the body frame are constant. Assume that the rigid body $\mathcal{B}$ is moving with constant spatial velocity $\widehat{\boldsymbol{\xi}}_{\mathrm{s}}$. Therefore, we write (5.5) as

$$\frac{\mathrm{d}}{\mathrm{d}t}\bar{\chi}_{\mathrm{s}}(t) = \widehat{\boldsymbol{\xi}}_{\mathrm{s}}\,\bar{\chi}_{\mathrm{s}}(t).$$

Since this is a linear time-independent differential equation, its solution is

$$\bar{\chi}_{\mathrm{s}}(t) = \exp\left(t\widehat{\boldsymbol{\xi}}_{\mathrm{s}}\right)\bar{\chi}_{\mathrm{s}}(0).$$

Therefore, one should expect the matrix exponential of a twist to be an element of $\mathsf{SE}(3)$, that is, a rigid body transformation. The intuition from this discussion is formalized as follows.

Proposition 5.6 (Exponential of a twist). *For* $\boldsymbol{\omega}, \boldsymbol{v} \in \mathbb{R}^3$,

$$\exp\widehat{\boldsymbol{\omega}} = \begin{cases} \boldsymbol{I}_3, & \boldsymbol{\omega} = \boldsymbol{0}, \\ \boldsymbol{I}_3 + \dfrac{\sin\|\boldsymbol{\omega}\|_{\mathbb{R}^3}}{\|\boldsymbol{\omega}\|_{\mathbb{R}^3}}\widehat{\boldsymbol{\omega}} + \dfrac{1-\cos\|\boldsymbol{\omega}\|_{\mathbb{R}^3}}{\|\boldsymbol{\omega}\|^2_{\mathbb{R}^3}}\widehat{\boldsymbol{\omega}}^2, & \boldsymbol{\omega} \neq \boldsymbol{0}, \end{cases} \tag{5.8}$$

$$\exp\widehat{(\boldsymbol{\omega}, \boldsymbol{v})} = \begin{bmatrix} \exp\widehat{\boldsymbol{\omega}} & \boldsymbol{B}(\boldsymbol{\omega})\boldsymbol{v} \\ \boldsymbol{0}_{1\times 3} & 1 \end{bmatrix},$$

where

$$\boldsymbol{B}(\boldsymbol{\omega}) = \begin{cases} \boldsymbol{I}_3, & \boldsymbol{\omega} = \boldsymbol{0}, \\ \boldsymbol{I}_3 + \left(\dfrac{1 - \cos\|\boldsymbol{\omega}\|_{\mathbb{R}^3}}{\|\boldsymbol{\omega}\|_{\mathbb{R}^3}}\right)\dfrac{\widehat{\boldsymbol{\omega}}}{\|\boldsymbol{\omega}\|_{\mathbb{R}^3}} + \left(1 - \dfrac{\sin\|\boldsymbol{\omega}\|_{\mathbb{R}^3}}{\|\boldsymbol{\omega}\|_{\mathbb{R}^3}}\right)\dfrac{\widehat{\boldsymbol{\omega}}^2}{\|\boldsymbol{\omega}\|^2_{\mathbb{R}^3}}, & \boldsymbol{\omega} \neq \boldsymbol{0}. \end{cases}$$

Furthermore, $\exp\widehat{\boldsymbol{\omega}} \in \mathsf{SO}(3)$ *and* $\exp\widehat{(\boldsymbol{\omega}, \boldsymbol{v})} \in \mathsf{SE}(3)$.

An immediate consequence of these results is that the matrix exponential satisfies $\exp(\mathfrak{so}(3)) \subset \mathsf{SO}(3)$ and $\exp(\mathfrak{se}(3)) \subset \mathsf{SE}(3)$. Equation (5.8) is referred to as ***Rodrigues' Formula***.

Proof of Proposition 5.6. We consider the $\mathsf{SO}(3)$ case. The result is trivial for $\boldsymbol{\omega} = \boldsymbol{0}$, so we will assume that $\boldsymbol{\omega} \neq \boldsymbol{0}$. First, we prove inductively that $\widehat{\boldsymbol{\omega}}^{2k+1} = (-1)^k \|\boldsymbol{\omega}\|^{2k}_{\mathbb{R}^3}\widehat{\boldsymbol{\omega}}$ for all $k \in \mathbb{N}$. For $k = 0$, the statement is trivial. For $k = 1$, one can verify that $\widehat{\boldsymbol{\omega}}^3 = -\|\boldsymbol{\omega}\|^2_{\mathbb{R}^3}\widehat{\boldsymbol{\omega}}$ either directly by expanding $\boldsymbol{\omega}^3$ in terms of the component of $\boldsymbol{\omega}$, or by recalling the properties of the vector cross-product and showing that $\widehat{\boldsymbol{\omega}}^2 = \boldsymbol{\omega}\boldsymbol{\omega}^T - \|\boldsymbol{\omega}\|^2_{\mathbb{R}^3}\boldsymbol{I}_3$. Next, we assume the statement correct for $k-1$ and prove it for k. We have

$$\begin{aligned} \widehat{\boldsymbol{\omega}}^{2k+1} &= \widehat{\boldsymbol{\omega}}^{2(k-1)}\widehat{\boldsymbol{\omega}}^3 = (-1)^{k-1}\|\boldsymbol{\omega}\|^{2(k-1)}_{\mathbb{R}^3}\widehat{\boldsymbol{\omega}}^3 \\ &= (-1)^{k-1}\|\boldsymbol{\omega}\|^{2k-2}_{\mathbb{R}^3}(-\|\boldsymbol{\omega}\|^2_{\mathbb{R}^3})\widehat{\boldsymbol{\omega}}. \end{aligned}$$

This concludes the inductive proof. Now, we exploit the equality $\widehat{\boldsymbol{\omega}}^{2k-1} = (-1)^{(k-1)}\|\boldsymbol{\omega}\|^{2k-2}_{\mathbb{R}^3}\widehat{\boldsymbol{\omega}}$ for all $k > 0$ to compute

$$\begin{aligned} \exp\widehat{\boldsymbol{\omega}} &= \sum_{k=0}^{+\infty}\frac{\widehat{\boldsymbol{\omega}}^k}{k!} = \boldsymbol{I}_3 + \sum_{k=1}^{+\infty}\frac{\widehat{\boldsymbol{\omega}}^{2k-1}}{(2k-1)!} + \sum_{k=1}^{+\infty}\frac{\widehat{\boldsymbol{\omega}}^{2k}}{(2k)!} \\ &= \boldsymbol{I}_3 - \frac{\widehat{\boldsymbol{\omega}}}{\|\boldsymbol{\omega}\|_{\mathbb{R}^3}}\sum_{k=1}^{+\infty}\frac{(-1)^k\|\boldsymbol{\omega}\|^{2k-1}_{\mathbb{R}^3}}{(2k-1)!} - \frac{\widehat{\boldsymbol{\omega}}^2}{\|\boldsymbol{\omega}\|^2_{\mathbb{R}^3}}\sum_{k=1}^{+\infty}\frac{(-1)^k\|\boldsymbol{\omega}\|^{2k}_{\mathbb{R}^3}}{(2k)!}. \end{aligned}$$

The result follows from the Taylor series for the sine and cosine functions: $\sin x = \sum_{n=0}^{+\infty}\frac{(-1)^n x^{2n+1}}{(2n+1)!}$, and $\cos x = \sum_{n=0}^{+\infty}\frac{(-1)^n x^{2n}}{(2n)!}$. To show that $\exp\widehat{\boldsymbol{\omega}} \in \mathsf{SO}(3)$, it suffices to notice that $\exp(-\widehat{\boldsymbol{\omega}}) = (\exp\widehat{\boldsymbol{\omega}})^T$ and that $\exp(-\widehat{\boldsymbol{\omega}}) = (\exp\widehat{\boldsymbol{\omega}})^{-1}$ by Exercise E5.3(b). Exercise E5.5 invites the reader to provide the proof for the $\mathsf{SE}(3)$ statement. ■

Propositions 5.5 and 5.6 establish that $\exp(\boldsymbol{A})$, for $\boldsymbol{A} \in \mathfrak{se}(3)$, describes the position at time $t = 1$ of a rigid body whose position at time $t = 0$ is the identity matrix and whose spatial velocity and body velocity are constant and equal to $\boldsymbol{A}$ (see Exercise E5.3(b)).

5.1.4 Coordinate systems on the group of rigid displacements

The previous sections have illustrated how numerous computations can be performed directly on $\mathsf{SE}(3)$ by manipulating matrices. Next, we introduce coordinate charts on SO(3) and SE(3). Although we will not have occasion in the book to use coordinates for SO(3) or SE(3), it is, at times, useful to be able to do so.

Exponential coordinates

As discussed in Exercise E5.3(e), the matrix exponential is a diffeomorphism from a neighborhood of $\mathbf{0}_{n\times n}$ to a neighborhood of $\boldsymbol{I}_n$. Therefore, a local inverse map exists; we characterize this inverse map for the SO(3) and SE(3) settings.

Proposition 5.7 (Matrix logarithm). *The matrix exponential is a C^ω-diffeomorphism between $\mathcal{U}_{\mathfrak{so}(3)} = \{\widehat{\omega} \in \mathfrak{so}(3) \mid \omega \in \mathbb{R}^3, \|\omega\|_{\mathbb{R}^3} < \pi\}$ and $\mathcal{U}_{\mathsf{SO}(3)} = \{\boldsymbol{R} \in \mathsf{SO}(3) \mid \operatorname{tr}(\boldsymbol{R}) \neq -1\}$; the inverse map $\log\colon \mathcal{U}_{\mathsf{SO}(3)} \to \mathcal{U}_{\mathfrak{so}(3)}$ is given by*

$$\log(\boldsymbol{R}) = \begin{cases} \mathbf{0}_{3\times 3}, & \boldsymbol{R} = \boldsymbol{I}_3, \\ \dfrac{\phi(\boldsymbol{R})}{2\sin(\phi(\boldsymbol{R}))}(\boldsymbol{R} - \boldsymbol{R}^T), & \boldsymbol{R} \neq \boldsymbol{I}_3, \end{cases} \tag{5.9}$$

where $\phi\colon \mathcal{U}_{\mathsf{SO}(3)} \to [0,\pi[$ is defined by $\phi(\boldsymbol{R}) = \arccos(\frac{1}{2}(\operatorname{tr}(\boldsymbol{R}) - 1))$.

Similarly, the matrix exponential is a C^ω-diffeomorphism between $\mathcal{U}_{\mathfrak{se}(3)} = \{\widehat{(\omega, v)} \in \mathfrak{se}(3) \mid \widehat{\omega} \in \mathcal{U}_{\mathfrak{so}(3)}, v \in \mathbb{R}^3\}$ and $\mathcal{U}_{\mathsf{SE}(3)} = \mathcal{U}_{\mathsf{SO}(3)} \times \mathbb{R}^3$; the inverse map $\log\colon \mathcal{U}_{\mathsf{SE}(3)} \to \mathcal{U}_{\mathfrak{se}(3)}$ is given by

$$\log(\boldsymbol{R}, \boldsymbol{r}) = \begin{bmatrix} \log(\boldsymbol{R}) & \boldsymbol{B}^{-1}(\log(\boldsymbol{R})^\vee)\boldsymbol{r} \\ \mathbf{0}_{1\times 3} & 0 \end{bmatrix},$$

where, for $\psi \in \mathbb{R}^3$, $\|\psi\|_{\mathbb{R}^3} < \pi$,

$$\boldsymbol{B}^{-1}(\psi) = \begin{cases} \boldsymbol{I}_3, & \psi = \mathbf{0}, \\ \boldsymbol{I}_3 - \dfrac{1}{2}\widehat{\psi} + \left(1 - \dfrac{\|\psi\|_{\mathbb{R}^3}}{2}\cot\left(\dfrac{\|\psi\|_{\mathbb{R}^3}}{2}\right)\right)\dfrac{\widehat{\psi}^2}{\|\psi\|_{\mathbb{R}^3}^2}, & \psi \neq \mathbf{0}. \end{cases}$$

The maps $\log\colon \mathcal{U}_{\mathsf{SO}(3)} \to \mathcal{U}_{\mathfrak{so}(3)}$ and $\log\colon \mathcal{U}_{\mathsf{SE}(3)} \to \mathcal{U}_{\mathfrak{se}(3)}$ are referred to as the ***matrix logarithm*** on SO(3) and SE(3), respectively.

Proof of Proposition 5.7. We work out the SO(3) case and invite the reader to prove the corresponding SE(3) statement in Exercise E5.6. We start by proving that the matrix logarithm in (5.9) is the inverse to the matrix exponential in (5.8). It suffices to prove that $\widehat{\omega} = \log(\exp(\widehat{\omega}))$ for all $0 < \|\omega\|_{\mathbb{R}^3} < \pi$. Let us first compute $\phi(\boldsymbol{R})$ for $\boldsymbol{R} = \exp(\widehat{\omega})$:

$$\begin{aligned} \cos(\phi(\boldsymbol{R})) &= \tfrac{1}{2}\operatorname{tr}(\exp(\widehat{\omega})) - \tfrac{1}{2} \\ &= \frac{1}{2}\operatorname{tr}\left(\boldsymbol{I}_3 + \frac{\sin\|\omega\|_{\mathbb{R}^3}}{\|\omega\|_{\mathbb{R}^3}}\widehat{\omega} + \frac{1 - \cos\|\omega\|_{\mathbb{R}^3}}{\|\omega\|_{\mathbb{R}^3}^2}\widehat{\omega}^2\right) - \frac{1}{2} \\ &= \frac{1}{2}\left(3 + \frac{1 - \cos\|\omega\|_{\mathbb{R}^3}}{\|\omega\|_{\mathbb{R}^3}^2}\operatorname{tr}(\widehat{\omega}^2)\right) - \frac{1}{2} = \cos\|\omega\|_{\mathbb{R}^3}, \end{aligned}$$

where we have used the equality $\operatorname{tr}(\widehat{\omega}^2) = -2\|\omega\|^2_{\mathbb{R}^3}$. Because $\|\omega\|_{\mathbb{R}^3} < \pi$, it follows that $\phi(\boldsymbol{R})$ is equal to $\|\omega\|_{\mathbb{R}^3}$. Next, we are ready to examine

$$\begin{aligned}\log(\exp(\widehat{\omega})) &= \frac{\phi(\boldsymbol{R})}{2\sin\phi(\boldsymbol{R})}(\boldsymbol{R}-\boldsymbol{R}^T)\Big|_{\boldsymbol{R}=\boldsymbol{I}_3+\frac{\sin\|\omega\|_{\mathbb{R}^3}}{\|\omega\|_{\mathbb{R}^3}}\widehat{\omega}+\frac{1-\cos\|\omega\|_{\mathbb{R}^3}}{\|\omega\|^2_{\mathbb{R}^3}}\widehat{\omega}^2} \\ &= \frac{\|\omega\|_{\mathbb{R}^3}}{2\sin\|\omega\|_{\mathbb{R}^3}}\left(2\frac{\sin\|\omega\|_{\mathbb{R}^3}}{\|\omega\|_{\mathbb{R}^3}}\widehat{\omega}\right) = \widehat{\omega},\end{aligned}$$

where we have used the equality $\widehat{\omega}^2 = (\widehat{\omega}^2)^T$. This completes the proof that $\log\colon \mathcal{U}_{\mathsf{SO}(3)} \to \mathcal{U}_{\mathfrak{so}(3)}$ is the inverse map to $\exp\colon \mathcal{U}_{\mathfrak{so}(3)} \to \mathcal{U}_{\mathsf{SO}(3)}$. Next, we prove that these maps are analytic. First, we note that the matrix exponential is analytic, see Exercise E5.3(c), and its restriction to $\mathsf{SO}(3)$ is analytic because $\mathsf{SO}(3)$ is an analytic submanifold. Second, we note that the matrix logarithm is analytic in a neighborhood of any point $\boldsymbol{R} \neq \boldsymbol{I}_3$ with $\operatorname{tr}(\boldsymbol{R}) < \pi$ because it is a composition of analytic maps. Finally, one can show that the matrix logarithm is an analytic map also in a neighborhood of $\boldsymbol{I}_3$ because the function $\mathbb{R} \ni x \mapsto \frac{x}{\sin x} \in \mathbb{R}$ is analytic in a neighborhood of 0. ■

The matrix logarithm provides us, therefore, with local charts for $\mathsf{SO}(3)$ and $\mathsf{SE}(3)$. In the neighborhood $\mathcal{U}_{\mathsf{SO}(3)}$ containing $\boldsymbol{I}_3 \in \mathsf{SO}(3)$, we call $\log(\boldsymbol{R})^\vee \in B_\pi(\boldsymbol{0}) \subset \mathbb{R}^3$ the ***exponential coordinates*** of the rotation matrix $\boldsymbol{R}$. In the neighborhood $\mathcal{U}_{\mathsf{SE}(3)}$ containing $\boldsymbol{I}_4 \in \mathsf{SE}(3)$, we call $\log(\boldsymbol{g})^\vee \in B_\pi(\boldsymbol{0}) \times \mathbb{R}^3 \subset \mathbb{R}^3 \times \mathbb{R}^3$ the ***exponential coordinates*** of the rigid displacement $\boldsymbol{g}$.

Note that the exponential coordinates $\log(\boldsymbol{R})^\vee$ for a rotation matrix $\boldsymbol{R}$ are related to the notion of angle and axis of rotation. Specifically, the vector $(\boldsymbol{R}-\boldsymbol{R}^T)^\vee \in \mathbb{R}^3$ is parallel to the axis of rotation of $\boldsymbol{R}$, and the magnitude of the logarithm of $\boldsymbol{R}$ is proportional to the rotation angle. We refer to [Murray, Li, and Sastry 1994, Chapter 2] for a discussion of these issues.

Euler angles

Let $\{\boldsymbol{e}_1, \boldsymbol{e}_2, \boldsymbol{e}_3\}$ be the standard basis for $\mathbb{R}^3$ and recall our definition of the arctangent function $\operatorname{atan}\colon \mathbb{R}^2 \setminus \{(0,0)\} \to]-\pi,\pi]$. Consider the map

$$\operatorname{Eul}\colon\ ]-\pi,\pi] \times\ ]-\tfrac{\pi}{2},\tfrac{\pi}{2}[\ \times\ ]-\pi,\pi] \to \{\boldsymbol{R} \in \mathsf{SO}(3) \mid R_{31} \neq \pm 1\}$$

defined by

$$(\alpha,\beta,\gamma) \mapsto \exp(\alpha\,\widehat{\boldsymbol{e}}_3)\exp(\beta\,\widehat{\boldsymbol{e}}_2)\exp(\gamma\,\widehat{\boldsymbol{e}}_1),$$

where $\exp(\alpha\,\widehat{\boldsymbol{e}}_3)\exp(\beta\,\widehat{\boldsymbol{e}}_2)\exp(\gamma\,\widehat{\boldsymbol{e}}_1)$ admits the explicit expression

$$\begin{bmatrix} \cos\alpha\cos\beta & \cos\alpha\sin\beta\sin\gamma - \cos\gamma\sin\alpha & \cos\alpha\cos\gamma\sin\beta + \sin\alpha\sin\gamma \\ \cos\beta\sin\alpha & \cos\alpha\cos\gamma + \sin\alpha\sin\beta\sin\gamma & \cos\gamma\sin\alpha\sin\beta - \cos\alpha\sin\gamma \\ -\sin\beta & \cos\beta\sin\gamma & \cos\beta\cos\gamma \end{bmatrix}.$$

One can prove that Eul is a bijection with inverse map $\mathrm{Eul}^{-1}\colon]-\pi,\pi]\times]-\frac{\pi}{2},\frac{\pi}{2}[\times]-\pi,\pi]\to\{\boldsymbol{R}\in\mathsf{SO}(3)\mid R_{31}\neq\pm1\}$ given by

$$\mathrm{Eul}^{-1}(\boldsymbol{R})=\big(\mathrm{atan}(R_{21},R_{11}),\mathrm{atan}(-R_{31},\sqrt{R_{11}^2+R_{21}^2}),\mathrm{atan}(R_{32},R_{33})\big).$$

Because these maps are smooth, Eul restricted to $]-\pi,\pi[\times]-\frac{\pi}{2},\frac{\pi}{2}[\times]-\pi,\pi[$ is a diffeomorphism. Note that it is possible to extend Eul to the domain $]-\pi,\pi]\times[-\frac{\pi}{2},\frac{\pi}{2}]\times]-\pi,\pi]$ so that it becomes a surjection onto $\mathsf{SO}(3)$. However, the extended map is not a homeomorphism.

The quantities $(\alpha,\beta,\gamma)=\mathrm{Eul}^{-1}(\boldsymbol{R})$ are referred to as the ***ZYX Euler angles*** for $\boldsymbol{R}\in\mathsf{SO}(3)$. This local chart of $\mathsf{SO}(3)$ is one set of Euler angles, other possibilities are possible, for example, by interchanging the order of composition of the matrix exponentials in the definition of Eul. These and related topics are discussed in robotics texts, e.g., [Craig 1989, Murray, Li, and Sastry 1994, Spong and Vidyasagar 1989].

5.2 Lie groups and Lie algebras

We present an account of Lie group theory inspired by the treatments in [Curtis 1979, Sattinger and Weaver 1986, Warner 1971]. The key concepts in this section are the notion of Lie group and its associated Lie algebra of left invariant vector fields. We discuss in some detail the case of matrix Lie group and matrix Lie algebras.

5.2.1 Groups

Definition 5.8 (Group). A set G endowed with a binary operation, denoted by $\mathsf{G}\times\mathsf{G}\ni(a,b)\mapsto a\star b\in\mathsf{G}$, is a ***group*** if:

(i) $a\star(b\star c)=(a\star b)\star c$ for all $a,b,c\in\mathsf{G}$ (***associativity***);
(ii) there exists $e\in\mathsf{G}$ such that $a\star e=e\star a=a$ for all $a\in\mathsf{G}$ (***identity element***);
(iii) there exists $a^{-1}\in\mathsf{G}$ such that $a\star a^{-1}=a^{-1}\star a=e$ for all $a\in\mathsf{G}$ (***inverse elements***).

Sometimes it is convenient to write $(\mathsf{G},\star)$ to emphasize the operation with respect to which G is a group. Other times it will be convenient to omit the symbol $\star$. •

A ***topological group*** is a group that is also a topological space for which the group operation and the inverse operation are continuous. A ***Lie group*** is a topological group that is also a manifold, and in which the group and the inverse operations are smooth. A group is ***Abelian*** (or commutative) if the group operation is commutative, i.e., if $a\star b=b\star a$ for all $a,b\in\mathsf{G}$. There are interesting examples of both Abelian and non-Abelian groups.

Examples 5.9. 1. For $m \in \mathbb{N}$, the set of permutations of $\{1, \dots, m\}$ is a group called the permutation group and denoted by S_m (see Section 2.1). This group is not Abelian for $m \geq 3$ and is composed of a finite number of elements.

2. $(\mathbb{Z}, +)$, $(\mathbb{Q} \setminus \{0\}, \cdot)$, and $(\mathbb{R}_+, \cdot)$ are examples of Abelian groups.
3. A vector space V with the operation of vector addition is an Abelian group.
4. A ring is a group in two ways, first with respect to addition, and second with respect to multiplication. Addition is commutative, and thus the group is Abelian in this case. The group formed using multiplication is Abelian if and only if the ring is commutative.
5. The set $\mathsf{GL}(n; \mathbb{R})$ of invertible $n \times n$ matrices with real entries is a Lie group with respect to the operation of matrix multiplication. It is called the ***real general linear group***, or simply, the ***general linear group***). The identity element is $\boldsymbol{I}_n$ and the inverse element to $\boldsymbol{A} \in \mathsf{GL}(n; \mathbb{R})$ is $\boldsymbol{A}^{-1}$. For all $n > 1$, the group $\mathsf{GL}(n; \mathbb{R})$ is non-Abelian.
6. The n-torus $\mathbb{T}^n$ is an Abelian Lie group with respect to the sum operation $([\boldsymbol{x}], [\boldsymbol{y}]) \mapsto [\boldsymbol{x} + \boldsymbol{y}]$, where $[\boldsymbol{x}]$ is the equivalence class $\{\boldsymbol{z} \in \mathbb{R}^n \mid \boldsymbol{z} - \boldsymbol{x} \in \mathbb{Z}^n\}$. We denote this group $(\mathbb{T}^n, +)$.
7. Let M be a smooth manifold and let $\mathrm{Diff}(\mathsf{M})$ be the set of smooth diffeomorphisms of M. The set $\mathrm{Diff}(\mathsf{M})$ is a group with respect to the operation of composition of maps. •

Let $(\mathsf{G}, \star)$ and $(\mathsf{H}, \bullet)$ be two groups. A ***group homomorphism*** is a map $\rho\colon \mathsf{G} \to \mathsf{H}$ that satisfies $\rho(a \star b) = \rho(a) \bullet \rho(b)$ for all $a, b \in \mathsf{G}$. A ***Lie group homomorphism*** is a smooth group homomorphism between Lie groups. One can show that a group homomorphism $\rho\colon \mathsf{G} \to \mathsf{H}$ maps the identity element of G into the identity element of H, and inverse elements in G into inverse elements in H, see Exercise E5.10. If the group homomorphism ρ is a bijection, then ρ is called an ***isomorphism***. In that case, the two groups G and H are said to be ***isomorphic*** and one writes $\mathsf{G} \simeq \mathsf{H}$.

Examples 5.10. 1. The map $t \mapsto \mathrm{e}^t$ is a Lie group isomorphism between $(\mathbb{R}, +)$ and $(\mathbb{R}_+, \cdot)$, because $\mathrm{e}^{t_1 + t_2} = \mathrm{e}^{t_1} \cdot \mathrm{e}^{t_2}$.

2. $(\mathbb{T}^1, +)$ is isomorphic to $\mathsf{SO}(2)$ by way of $[x] \mapsto \begin{bmatrix} \cos(2\pi x) & -\sin(2\pi x) \\ \sin(2\pi x) & \cos(2\pi x) \end{bmatrix}$. Exercise E5.11 invites the reader to explore other groups isomorphic to $(\mathbb{T}^1, +)$.
3. The set
$$\mathcal{L} = \left\{ L_a = \begin{bmatrix} 1 & a \\ 0 & 1 \end{bmatrix} \;\middle|\; a \in \mathbb{R} \right\}$$
is a group and is isomorphic to $(\mathbb{R}, +)$ because of the equality $L_a \cdot L_b = L_{a+b}$.
4. The determinant is a group homomorphism between $\mathsf{GL}(n; \mathbb{R})$ and $(\mathbb{R} \setminus \{0\}, \cdot)$.

5. The map sgn: $S_m \to \{-1, 1\}$ defined in Section 2.1 is a group homomorphism.
6. The adjoint map $g \mapsto \mathrm{Ad}_g$ defined in (5.6) is a group homomorphism of $\mathsf{SE}(3)$ into the set of invertible linear maps on $\mathfrak{se}(3)$. •

Remark 5.11. Group homomorphisms are analogous to linear maps between vector spaces. Linear maps and group homomorphisms respect the structure of vector spaces and groups, respectively. •

Definition 5.12 (Subgroup). Let G be a group.

(i) A ***subgroup*** H of G is a subset of G that is a group with respect to the group operation in G.

Let G be a Lie group.

(ii) A ***Lie subgroup*** of G is a subgroup $\mathsf{H} \subset \mathsf{G}$ for which the inclusion $i_{\mathsf{H}}: \mathsf{H} \to \mathsf{G}$ is an injective immersion.

(iii) A Lie subgroup H of G that is a submanifold of G is a ***regular Lie subgroup***. •

In other words, a subgroup H of a group G is closed under the group operation, is closed under the inverse operation, and contains the identity element. With these definitions it is also possible to define the following concept.

Definition 5.13 (Matrix Lie group). A ***matrix Lie group*** is a subgroup of $\mathsf{GL}(n; \mathbb{R})$. •

Note that a matrix Lie group is an immersed submanifold of $\mathsf{GL}(n; \mathbb{R})$ and it is an embedded submanifold of $\mathsf{GL}(n; \mathbb{R})$ if and only if it is topologically closed; see Exercise E5.13. Matrix Lie groups find important applications in mathematical physics; see [Curtis 1979, Sattinger and Weaver 1986, Weyl 1939].

Remark 5.14. Sometimes it is convenient to consider the following equivalent definition of a subgroup. The pair (K, ρ) is a subgroup of G if K is a group and ρ is an injective group homomorphism between K and G. In other words, one can define the notion of subgroup up to a group isomorphism. •

Let us give some examples to illustrate these concepts.

Examples 5.15. 1. The group $(\mathbb{Z}, +)$ is a subgroup of $(\mathbb{R}, +)$, and $(\mathbb{Q} \setminus \{0\}, \cdot)$ is a subgroup of $(\mathbb{R} \setminus \{0\}, \cdot)$.

2. The group of rotations $\mathsf{SO}(3)$ and the group of rigid displacements $\mathsf{SE}(3)$ are matrix Lie groups.
3. The groups $\mathsf{SO}(3) \subset \mathsf{GL}(3; \mathbb{R})$ and $(\mathbb{R}^3, +)$ are isomorphic to subgroups of the group of rigid displacements $\mathsf{SE}(3) \subset \mathsf{GL}(4; \mathbb{R})$ by way of the inclusions

$$\mathsf{SO}(3) \ni \boldsymbol{R} \mapsto \begin{bmatrix} \boldsymbol{R} & 0_{3\times 1} \\ 0_{1\times 3} & 1 \end{bmatrix} \in \mathsf{SE}(3), \quad \mathbb{R}^3 \ni \boldsymbol{v} \mapsto \begin{bmatrix} 0_{3\times 3} & \boldsymbol{v} \\ 0_{1\times 3} & 1 \end{bmatrix} \in \mathsf{SE}(3).$$

4. Other examples of matrix Lie groups are (1) the set of diagonal matrices, (2) the set of matrices with unit determinant, and (3) the set of matrices $\boldsymbol{A} \in \mathsf{GL}(n;\mathbb{R})$ such that $A_{ii} = 1$, for $i \in \{1,\dots,n\}$, and $A_{ij} = 0$, for $i,j \in \{1,\dots,n\}$ and $i < j$. •

5.2.2 From one-parameter subgroups to matrix Lie algebras

Let us focus our attention on matrix Lie groups and perform an instructive construction. We start by defining a particular class of subgroups.

Definition 5.16 (One-parameter subgroup). A ***one-parameter subgroup*** of a Lie group G is the image of a smooth group homomorphism $\rho\colon \mathbb{R} \to \mathsf{G}$. A ***matrix one-parameter subgroup*** is a one-parameter subgroup of $\mathsf{GL}(n;\mathbb{R})$. •

In what follows, it is convenient to identify a one-parameter subgroup with the group homomorphism defining it.

Lemma 5.17 (Matrix one-parameter subgroups). *Every matrix one-parameter subgroup $\rho\colon \mathbb{R} \to \mathsf{GL}(n;\mathbb{R})$ is of the form $t \mapsto \exp(t\boldsymbol{A})$, where*

$$\boldsymbol{A} = \frac{\mathrm{d}\rho(t)}{\mathrm{d}t}\bigg|_{t=0} \in \mathbb{R}^{n\times n}.$$

Proof. Because ρ is a group homomorphism, we know that $\rho(t+s) = \rho(t)\rho(s)$. Differentiating with respect to s we have

$$\frac{\mathrm{d}\rho(t+s)}{\mathrm{d}s} = \rho(t)\frac{\mathrm{d}\rho(s)}{\mathrm{d}s}.$$

Evaluating this equality at $s = 0$ and using the definition of $\boldsymbol{A}$, we obtain $\frac{\mathrm{d}}{\mathrm{d}t}\rho(t) = \rho(t)\boldsymbol{A}$. Given the initial condition $\rho(0) = \boldsymbol{I}_n$, the unique solution is $\rho(t) = \exp(t\boldsymbol{A})$. ■

Let us give an example to illustrate the notion of a matrix one-parameter subgroup.

Example 5.18. Consider the group of rotations $\mathsf{SO}(3)$. The one-parameter subgroups of $\mathsf{SO}(3)$ describing rotations about the three coordinate axes are the images of the maps

$$t \mapsto \begin{bmatrix} 1 & 0 & 0 \\ 0 & \cos(t) & -\sin(t) \\ 0 & \sin(t) & \cos(t) \end{bmatrix}, \quad t \mapsto \begin{bmatrix} \cos(t) & 0 & -\sin(t) \\ 0 & 1 & 0 \\ \sin(t) & 0 & \cos(t) \end{bmatrix},$$

$$t \mapsto \begin{bmatrix} \cos(t) & -\sin(t) & 0 \\ \sin(t) & \cos(t) & 0 \\ 0 & 0 & 1 \end{bmatrix}.$$

The corresponding matrix one-parameter subgroup generators are computed by differentiating with respect to time and setting $t = 0$:

$$\widehat{e}_1 = \begin{bmatrix} 0 & 0 & 0 \\ 0 & 0 & -1 \\ 0 & 1 & 0 \end{bmatrix}, \quad \widehat{e}_2 = \begin{bmatrix} 0 & 0 & -1 \\ 0 & 0 & 0 \\ 1 & 0 & 0 \end{bmatrix}, \quad \widehat{e}_3 = \begin{bmatrix} 0 & -1 & 0 \\ 1 & 0 & 0 \\ 0 & 0 & 0 \end{bmatrix}. \qquad \bullet$$

We refer the reader to Exercise E5.13 for an interesting characterization of matrix one-parameter subgroups.

Definition 5.19 (One-parameter subgroup generator). Let G be a matrix Lie group. The matrix $\boldsymbol{A} \in \mathbb{R}^{n\times n}$ is a ***one-parameter subgroup generator*** of G if $\mathbb{R} \ni t \mapsto \exp(t\boldsymbol{A}) \in \mathbb{R}^{n\times n}$ takes values in G. We let $\mathcal{G}(\mathsf{G})$ denote the set of one-parameter subgroup generators of G. •

In other words, $\boldsymbol{A} \in \mathcal{G}(\mathsf{G})$ is the generator of the matrix one-parameter subgroup $t \mapsto \exp(t\boldsymbol{A}) \subset \mathsf{G}$. To characterize the set $\mathcal{G}(\mathsf{G})$, let us introduce the ***matrix commutator*** $[\cdot,\cdot]\colon \mathbb{R}^{n\times n} \times \mathbb{R}^{n\times n} \to \mathbb{R}^{n\times n}$ given by $[\boldsymbol{A},\boldsymbol{B}] = \boldsymbol{A}\boldsymbol{B} - \boldsymbol{B}\boldsymbol{A}$. We can now illustrate the structure of $\mathcal{G}(\mathsf{G})$.

Theorem 5.20. *The set of one-parameter subgroup generators $\mathcal{G}(\mathsf{G})$ of the matrix Lie group $\mathsf{G} \subset \mathsf{GL}(n;\mathbb{R})$ is a subspace of $\mathbb{R}^{n\times n}$ closed under the matrix commutator.*

Proof. It is immediate to see that, if $\boldsymbol{A} \in \mathcal{G}(\mathsf{G})$, then $\lambda\boldsymbol{A} \in \mathcal{G}(\mathsf{G})$ for all $\lambda \in \mathbb{R}$. The fact that $\mathcal{G}(\mathsf{G})$ is closed under the operations of matrix sum and matrix commutator is a consequence of the equalities

$$\exp(\boldsymbol{A}+\boldsymbol{B}) = \lim_{n\to+\infty} \left(\exp\left(\frac{\boldsymbol{A}}{n}\right)\exp\left(\frac{\boldsymbol{B}}{n}\right)\right)^n,$$

$$\exp([\boldsymbol{A},\boldsymbol{B}]) = \lim_{n\to+\infty} \left(\exp\left(\frac{\boldsymbol{A}}{n}\right)\exp\left(\frac{\boldsymbol{B}}{n}\right)\exp\left(\frac{-\boldsymbol{A}}{n}\right)\exp\left(\frac{-\boldsymbol{B}}{n}\right)\right)^{n^2},$$

for all $\boldsymbol{A},\boldsymbol{B}$ in $\mathbb{R}^{n\times n}$. In turn, these equalities follow from the Taylor expansions

$$\exp(t\boldsymbol{A})\exp(t\boldsymbol{B}) = \exp(t(\boldsymbol{A}+\boldsymbol{B}) + o(t^2)),$$
$$\exp(t\boldsymbol{A})\exp(t\boldsymbol{B})\exp(-t\boldsymbol{A})\exp(-t\boldsymbol{B}) = \exp(t^2[\boldsymbol{A},\boldsymbol{B}] + o(t^3)),$$

where, for $k \in \mathbb{N}$, the quantities $o(t^k)$ are smooth $\mathbb{R}^{n\times n}$-valued functions of t such that $\frac{o(t^k)}{t^k} \to \boldsymbol{0}_{n\times n}$ as $t \to 0$. We leave the proof of these Taylor expansions as an exercise to the interested reader in Exercise E5.15. ■

Motivated by this result, it is convenient to introduce the following notion.

Definition 5.21 (Matrix Lie algebra). A ***matrix Lie algebra*** is a subspace of $\mathbb{R}^{n\times n}$ closed under the operation of matrix commutator. •

To illustrate Theorem 5.20, let us compute some interesting sets of one-parameters subgroup generators and verify that they indeed are matrix Lie algebras. Proposition 5.6 directly implies that $\mathcal{G}(\mathsf{SO}(3)) = \mathfrak{so}(3)$ and that $\mathcal{G}(\mathsf{SE}(3)) = \mathfrak{se}(3)$. Indeed, it is easy to verify that $\mathfrak{so}(3)$ and $\mathfrak{se}(3)$ are vector spaces. Furthermore, for all $\boldsymbol{A}, \boldsymbol{B} \in \mathfrak{so}(3)$, we have

$$[\boldsymbol{A}, \boldsymbol{B}]^T = (\boldsymbol{AB} - \boldsymbol{BA})^T = (-\boldsymbol{B})(-\boldsymbol{A}) - (-\boldsymbol{A})(-\boldsymbol{B}) = -[\boldsymbol{A}, \boldsymbol{B}],$$

thus verifying that $\mathfrak{so}(3)$ is closed under matrix commutator. The same property can be directly verified for $\mathfrak{se}(3)$. As an additional example, let us consider the general linear group $\mathsf{GL}(n;\mathbb{R})$ and introduce the symbol $\mathfrak{gl}(n;\mathbb{R})$ for $\mathcal{G}(\mathsf{GL}(n;\mathbb{R}))$. Because $\exp(\boldsymbol{A}) \in \mathsf{GL}(n;\mathbb{R})$ for all $\boldsymbol{A} \in \mathbb{R}^{n\times n}$, it is clear that $\mathfrak{gl}(n;\mathbb{R}) = \mathbb{R}^{n\times n}$.

Remark 5.22. A consequence of Theorem 5.20 is that each matrix Lie group admits a corresponding matrix Lie algebra. Remarkably, the following stronger result is true. There is a one-to-one correspondence between connected k-dimensional subgroups $\mathsf{G} \subset \mathsf{GL}(n;\mathbb{R})$ and k-dimensional matrix Lie algebras $\mathcal{G}(\mathsf{G}) \subset \mathbb{R}^{n\times n}$. This is an important result in the study of Lie groups and Lie algebras and we refer the interested reader to [Chevalley 1946, Warner 1971] for more details. •

5.2.3 Lie algebras

Matrix Lie algebras are just an example of a more general object that we characterize as follows; see [Jacobson 1962, Serre 1992].

Definition 5.23 (Lie algebra). A ***Lie algebra*** V is a $\mathbb{R}$-vector space endowed with a bilinear operation $[\cdot,\cdot]\colon \mathsf{V}\times\mathsf{V} \to \mathsf{V}$ called the ***bracket*** satisfying
(i) anti-commutativity, i.e., $[\xi,\eta] = -[\eta,\xi]$ for all $\xi,\eta \in \mathsf{V}$, and
(ii) the Jacobi identity, i.e., $[\xi,[\eta,\zeta]]+[\eta,[\zeta,\xi]]+[\zeta,[\xi,\eta]] = 0$ for all $\xi,\eta,\zeta \in \mathsf{V}$. •

We adopt the standard convention that elements of a Lie algebra are denoted by Greek letters. Sometimes it is convenient to write $(\mathsf{V},[\cdot,\cdot]_{\mathsf{V}})$ to emphasize the operation with respect to which V is a Lie algebra.

Examples 5.24. 1. $\mathbb{R}^3$ with the operation of vector cross-product is a Lie algebra.
2. The matrix commutator satisfies the Jacobi identity (see Exercise E5.16), and, therefore, every matrix Lie algebra is a Lie algebra.
3. The set of linear maps $L(\mathsf{V};\mathsf{V})$ on a vector space V is a Lie algebra using the commutator $(A,B) \mapsto [A,B] \triangleq A\circ B - B\circ A$.
4. The vector space $\Gamma^\infty(\mathsf{TM})$ of all smooth vector fields on a smooth manifold M is a Lie algebra with respect to the operation of Lie bracket. •

Let $(\mathsf{V}, [\cdot,\cdot]_{\mathsf{V}})$ and $(\mathsf{U}, [\cdot,\cdot]_{\mathsf{U}})$ be two Lie algebras. A ***Lie algebra homomorphism*** is a linear map $\rho\colon \mathsf{V} \to \mathsf{U}$ that satisfies $\rho([\xi,\eta]_{\mathsf{V}}) = [\rho(\xi),\rho(\eta)]_{\mathsf{U}}$ for all $\xi,\eta \in \mathsf{V}$. If the Lie algebra homomorphism ρ is a bijection, then ρ is called a ***Lie algebra isomorphism***. In that case, the two Lie algebras V and U are said to be ***isomorphic*** and one writes $\mathsf{V} \simeq \mathsf{U}$.

Examples 5.25. 1. Consider the Lie algebras $\mathbb{R}^3$ with the operation of vector cross-product and $\mathfrak{so}(3)$ with the matrix commutator. One can verify that, for all $\boldsymbol{x}$ and $\boldsymbol{y} \in \mathbb{R}^3$,

$$[\widehat{\boldsymbol{x}}, \widehat{\boldsymbol{y}}] = \widehat{\boldsymbol{x}}\widehat{\boldsymbol{y}} - \widehat{\boldsymbol{y}}\widehat{\boldsymbol{x}} = \widehat{\boldsymbol{x} \times \boldsymbol{y}}.$$

This equality proves that the map $\widehat{\cdot}\colon \mathbb{R}^3 \to \mathfrak{so}(3)$ is a Lie algebra isomorphism between $\mathfrak{so}(3)$ and $\mathbb{R}^3$.

2. Consider the three vector fields X_1, X_2, and X_3 on $\mathbb{R}^3$ defined by

$$X_k = x^j \frac{\partial}{\partial x^i} - x^i \frac{\partial}{\partial x^j},$$

where $(i,j,k) \in \{(1,2,3),(2,3,1),(3,1,2)\}$. One can verify that $[X_1,X_2] = X_3$, $[X_2,X_3] = X_1$ and $[X_3,X_1] = X_2$; see Exercise E3.24. Accordingly, the vector space $\operatorname{span}_{\mathbb{R}}\{X_1,X_2,X_3\}$ is a Lie algebra. Furthermore, one can show that $\operatorname{span}_{\mathbb{R}}\{X_1,X_2,X_3\}$ is isomorphic to $(\mathbb{R}^3,\times)$ and the Lie algebra isomorphism is the map defined by $c^k X_k \mapsto c^k \boldsymbol{e}_k$. •

A ***Lie subalgebra*** U of the Lie algebra V is a nonempty subset of V that is a Lie algebra with respect to the bracket operation in V. In other words, a Lie subalgebra U is a subspace closed under the bracket operation. The Lie subalgebras of $\mathfrak{gl}(n;\mathbb{R})$ are the matrix Lie algebras. Given $S \subset \mathsf{V}$, the ***Lie algebra generated by S*** is the smallest Lie subalgebra of V containing S. The Lie algebra generated by S can be shown to be the subspace generated by vectors of the form

$$[\xi_1,[\xi_2,\dots,[\xi_{k-1},\xi_k]]], \qquad k \in \mathbb{N},\ \xi_1,\dots,\xi_k \in S.$$

An ***ideal*** of a Lie algebra V is a nonempty set $\mathsf{U} \subset \mathsf{V}$ with the property that $[\xi,\eta] \in \mathsf{U}$ if either $\xi \in \mathsf{U}$ or $\eta \in \mathsf{U}$. It is clear that every ideal is a Lie subalgebra. Given $S \subset \mathsf{V}$, the ***ideal generated by S*** is the smallest ideal of V containing S. One may readily verify that the ideal generated by S consists of the subspace generated by vectors of the form

$$[\xi_1,[\xi_2,\dots,[\xi_k,\eta]]], \qquad k \in \mathbb{Z}_+,\ \xi_1,\dots,\xi_k \in \mathsf{V},\ \eta \in S.$$

Next, let us provide some additional notation and concepts related to the bracket. Let V be a Lie algebra. For $\xi \in \mathsf{V}$, we define the linear map $\operatorname{ad}_\xi\colon \mathsf{V} \to \mathsf{V}$ by

$$\operatorname{ad}_\xi \eta = [\xi,\eta].$$

We call ad_ξ the ***adjoint operator*** corresponding to $\xi \in \mathsf{V}$. It is possible to compose the adjoint operation; we define $\mathrm{ad}_\xi^0 = \mathrm{id}_\mathsf{V}$ and $\mathrm{ad}_\xi^k = \mathrm{ad}_\xi \circ \mathrm{ad}_\xi^{k-1}$ for all $k \in \mathbb{N}$. If $\{e_1, \dots, e_n\}$ is a basis for a Lie algebra V, then there must exist coefficients $c_{ij}^k \in \mathbb{R}$, $i, j, k \in \{1, \dots, n\}$, called the ***structure constants***, defined by

$$[e_i, e_j] = c_{ij}^k e_k, \qquad i, j \in \{1, \dots, n\}. \tag{5.10}$$

From anti-commutativity and from the Jacobi identity, the structure constants inherit the properties

$$\begin{aligned} c_{ij}^k + c_{ji}^k &= 0, & i, j &\in \{1, \dots, n\}, \\ c_{ij}^l c_{lk}^m + c_{ki}^l c_{lj}^m + c_{jk}^l c_{li}^m &= 0, & i, j, k &\in \{1, \dots, n\}. \end{aligned} \tag{5.11}$$

Examples 5.26. Let us explicitly provide the adjoint operators for the matrix Lie algebras $\mathfrak{so}(3)$ and $\mathfrak{se}(3)$. To simplify matters, we rely on the vector space identifications $\mathfrak{so}(3) \simeq \mathbb{R}^3$ and $\mathfrak{se}(3) \simeq \mathbb{R}^3 \oplus \mathbb{R}^3$.

1. Given $\boldsymbol{y} \in \mathbb{R}^3$, the adjoint operator on $\mathbb{R}^3$ is $\boldsymbol{x} \mapsto \mathrm{ad}_{\boldsymbol{y}} \boldsymbol{x} = \widehat{\boldsymbol{y}}\boldsymbol{x}$, for all $\boldsymbol{x} \in \mathbb{R}^3$.
2. Given $(\boldsymbol{x}, \boldsymbol{y}) \in \mathbb{R}^3 \oplus \mathbb{R}^3$ and correspondingly $\widehat{(\boldsymbol{x}, \boldsymbol{y})} \in \mathfrak{se}(3)$, the adjoint operator on $\mathfrak{se}(3) \simeq \mathbb{R}^3 \oplus \mathbb{R}^3$ is given in matrix form by

$$\mathrm{ad}_{(\boldsymbol{x},\boldsymbol{y})} = \begin{bmatrix} \widehat{\boldsymbol{x}} & \boldsymbol{0}_{3\times 3} \\ \widehat{\boldsymbol{y}} & \widehat{\boldsymbol{x}} \end{bmatrix}. \tag{5.12}$$

This gives a linear map on $\mathfrak{se}(3)$ under the identification with $\mathbb{R}^3 \oplus \mathbb{R}^3$. •

Finally, we consider the dual space to a Lie algebra. Given $\xi \in \mathsf{V}$, we shall find useful the dual map $\mathrm{ad}_\xi^* \colon \mathsf{V}^* \to \mathsf{V}^*$ defined by $\langle \mathrm{ad}_\xi^* \alpha; \eta \rangle = \langle \alpha; [\xi, \eta] \rangle$ for all $\alpha \in \mathsf{V}^*$. As usual, given a basis for the vector space V, the matrix representation $[\mathrm{ad}_\xi^*]$ is simply the transpose of $[\mathrm{ad}_\xi]$.

5.2.4 The Lie algebra of a Lie group

As we discussed in Section 5.2.2, every matrix Lie group admits a matrix Lie algebra, i.e., the set of its one-parameter subgroup generators. In this section we show how every Lie group is endowed with an important Lie algebra.

Let $(\mathsf{G}, \star)$ be a Lie group. For $g \in \mathsf{G}$, the ***left translation map*** is the map

$$\begin{aligned} L_g \colon \mathsf{G} &\to \mathsf{G} \\ h &\mapsto g \star h. \end{aligned}$$

The map L_g is smooth, its smooth inverse is $L_{g^{-1}}$, and L_g is therefore a diffeomorphism of G. For $g \in \mathsf{G}$, the tangent map $T_e L_g \colon \mathsf{T}_e\mathsf{G} \to \mathsf{T}_g\mathsf{G}$ is a natural isomorphism between $\mathsf{T}_e\mathsf{G}$ and $\mathsf{T}_g\mathsf{G}$. Thus, we have an isomorphism between each tangent space and the tangent space at the identity. This isomorphism induces a vector bundle isomorphism from TG to $\mathsf{G} \times \mathsf{T}_e\mathsf{G}$ given by

$$v_g \to (g, T_g L_{g^{-1}}(v_g)),$$

for all $v_g \in \mathsf{TG}$. It is convenient to record this result by denoting the equivalences $\mathsf{T}_g\mathsf{G} \simeq \mathsf{T}_e\mathsf{G}$ and $\mathsf{TG} \simeq \mathsf{G} \times \mathsf{T}_e\mathsf{G}$.

Next, let us study the action of the left translation map on functions and vector fields. A function $f\colon \mathsf{G} \to \mathbb{R}$ is ***left-invariant*** if $L_g^* f = f$ for all $g \in \mathsf{G}$, or equivalently if $f(h) = f(g \star h)$ for all $g, h \in \mathsf{G}$. Clearly, left-invariant functions satisfy $f(g) = f(g^{-1} \star g) = f(e)$ for all $g \in \mathsf{G}$. Therefore, they are constant functions identified by their value at the identity.

A vector field X on G is ***left-invariant*** if $L_g^* X = X$ for all $g \in \mathsf{G}$, or equivalently if $X(g \star h) = T_h L_g(X(h))$ for all $g, h \in \mathsf{G}$. Left-invariant vector fields are identified by their value at the identity via the equality $X(g) = T_e L_g(X(e))$. It is convenient to let ξ_{L} denote the left-invariant vector field on G defined by $\xi_{\mathrm{L}}(e) = \xi \in \mathsf{T}_e\mathsf{G}$. In other words, we let ξ_{L} be the vector field $g \mapsto T_e L_g(\xi)$. Left-invariant vector fields enjoy the following remarkable properties.

Theorem 5.27 (Properties of left-invariant vector fields). *For a Lie group* G*, the following statements hold:*

(i) left-invariant vector fields are smooth;

(ii) the set of left-invariant vector fields of G*, denoted by* $\mathfrak{L}(\mathsf{G})$*, is a Lie subalgebra of* $\Gamma^\infty(\mathsf{TG})$*;*

(iii) the vector spaces $\mathsf{T}_e\mathsf{G}$ *and* $\mathfrak{L}(\mathsf{G})$ *are isomorphic by means of the vector space isomorphism* $\mathsf{T}_e\mathsf{G} \ni X_e \mapsto (X_e)_{\mathrm{L}} \in \mathfrak{L}(\mathsf{G})$*, which has inverse given by* $\mathfrak{L}(\mathsf{G}) \ni X \mapsto X(e) \in \mathsf{T}_e\mathsf{G}$*. Accordingly,* $\dim(\mathsf{G}) = \dim(\mathsf{T}_e\mathsf{G}) = \dim(\mathfrak{L}(\mathsf{G}))$.

Proof. We refer to [Warner 1971] for the proof of part (i). Part (ii) is a consequence of the fact that Lie differentiation and pull-back commute; see Section 3.6.2. Indeed, for all left-invariant $X, Y \in \Gamma^\infty(\mathsf{TG})$ and $g \in \mathsf{G}$, we have $L_g^*[X,Y] = [L_g^* X, L_g^* Y] = [X,Y]$. Part (iii) is straightforward. ■

With these results in mind, it is useful to now introduce a new Lie algebra.

Definition 5.28 (Lie algebra of a Lie group). The ***Lie algebra*** $\mathfrak{g}$ ***of a Lie group*** G is the tangent space at the identity $\mathsf{T}_e\mathsf{G}$ with the bracket $[\xi, \eta] = [\xi_{\mathrm{L}}, \eta_{\mathrm{L}}](e)$. •

By construction, the vector space isomorphism in Theorem 5.27(iii) is an isomorphism of Lie algebras, so we may write $\mathfrak{g} \simeq \mathfrak{L}(\mathsf{G})$. Indeed, the Lie algebra of a Lie group is sometimes *defined* to be the set of left-invariant vector fields; see [Warner 1971].

Remarks 5.29. The properties of a Lie group are intimately connected to the character of its Lie algebra. We refer to [Chevalley 1946, Varadarajan 1984] for a comprehensive treatment. Let us just briefly mention two important results.

1. Let G be a Lie group and $\mathfrak{g}$ be its Lie algebra. If $\mathfrak{h}$ be a Lie subalgebra of $\mathfrak{g}$, then there is a unique connected Lie subgroup H with Lie algebra $\mathfrak{h}$.
2. Let G and H be Lie groups with corresponding Lie algebras $\mathfrak{g}$ and $\mathfrak{h}$. If $\varphi\colon \mathsf{G} \to \mathsf{H}$ is a Lie group homomorphism, then its tangent map at the identity, $T_e\varphi\colon \mathsf{T}_e\mathsf{G} \simeq \mathfrak{g} \to \mathsf{T}_e\mathsf{H} \simeq \mathfrak{h}$, is a Lie algebra homomorphism. Conversely, let $\psi\colon \mathfrak{g} \to \mathfrak{h}$ be a Lie algebra homomorphism. If G is simply connected, then there exists a unique group homomorphism $\varphi\colon \mathsf{G} \to \mathsf{H}$ such that $T_e\varphi = \psi$. •

Let us now study the integral curves of left-invariant vector fields on Lie groups.

Definition 5.30 (Exponential map). The ***exponential map*** $\exp\colon \mathfrak{g} \to \mathsf{G}$ is given by

$$\exp(\xi) = \Phi_1^{\xi_L}(e). \quad \bullet$$

Let us give an example to illustrate the notion of exponential map.

Example 5.31. Let us consider the Abelian group $(\mathbb{T}^n, +)$; see Example 3.61. Given coordinates $(x^1, \dots, x^n)$, left-invariant vector fields are of the form $c_i \frac{\partial}{\partial x^i}$, for $(c_1, \dots, c_n) \in \mathbb{R}^n$, and the Lie algebra of $(\mathbb{T}^n, +)$ is therefore isomorphic to $(\mathbb{R}^n, [\cdot,\cdot]_0)$, where $[\boldsymbol{x}, \boldsymbol{y}]_0 = \boldsymbol{0}$ for all $\boldsymbol{x}, \boldsymbol{y} \in \mathbb{R}^n$. The exponential map on $(\mathbb{T}^n, +)$ admits the local trivial expression $(c_1, \dots, c_n) \mapsto [(c_1, \dots, c_n)]$. •

We can now characterize the exponential map as follows.

Theorem 5.32 (Properties of exponential map). *Let G be a Lie group and let $\xi \in \mathfrak{g}$. The following statements hold:*

(i) the integral curve $\mathbb{R} \ni t \mapsto \Phi_t^{\xi_L}(e) = \exp(t\xi)$ is a one-parameter subgroup of G;
(ii) for all $g \in \mathsf{G}$ and $t \in \mathbb{R}$, $\Phi_t^{\xi_L}(g) = L_g \circ \exp(\xi t)$ and, as a consequence, left-invariant vector fields are complete;
(iii) $\exp\colon \mathfrak{g} \to \mathsf{G}$ is a local C^∞-diffeomorphism and $T_0 \exp = \mathrm{id}_{\mathfrak{g}}$.

We refer to [Warner 1971] for the proof and include some comments on these results.

Remarks 5.33. 1. Because the exponential map is a local diffeomorphism, it induces a coordinate chart in a neighborhood of $e \in \mathsf{G}$. The coordinates obtained in this way are called ***exponential coordinates of the first kind***. Next, suppose that $\{e_1, \dots, e_n\}$ is a basis for $\mathfrak{g}$. Let $\phi\colon \mathfrak{g} \to \mathsf{G}$ be the map defined by $\phi(\xi) = \exp(\xi^1 e_1) \star \dots \star \exp(\xi^n e_n)$, where $\xi = \xi^i e_i$. One can show that ϕ is a diffeomorphism onto its image when restricted to a neighborhood of $0 \in \mathfrak{g}$. The coordinates obtained in this way for the corresponding neighborhood of $e \in \mathsf{G}$ are called ***exponential coordinates of the second kind***.

2. For $g \in \mathsf{G}$, the ***right translation map*** is the map
$$\begin{aligned} R_g \colon \mathsf{G} &\to \mathsf{G} \\ h &\mapsto h \star g. \end{aligned}$$
We let ξ_{R} denote the right-invariant vector field on G determined by $\xi_{\mathrm{R}}(e) = \xi \in \mathsf{T}_e\mathsf{G}$, that is, the vector field $g \mapsto T_eR_g(\xi)$. We leave it for the reader to characterize their properties in a parallel way to what we have done for left-invariant vector fields. For example, one can show that, for all $g \in \mathsf{G}$ and $t \in \mathbb{R}$
$$\Phi_t^{\xi_{\mathrm{R}}}(g) = R_g \circ \exp(\xi t).$$
3. Given $g \in \mathsf{G}$, define the ***conjugation map*** $I_g \colon \mathsf{G} \to \mathsf{G}$ according to $h \mapsto g \star h \star g^{-1}$. Because I_g is a group isomorphism that leaves $e \in \mathsf{G}$ fixed, its tangent map T_eI_g is a Lie algebra isomorphism of $\mathfrak{g}$. It is convenient to refer to this tangent map as the ***adjoint map***, and denote it by $\mathrm{Ad}_g \colon \mathfrak{g} \to \mathfrak{g}$. In other words, we define $\mathrm{Ad}_g = T_eI_g = T_{g^{-1}}L_g \circ T_eR_{g^{-1}}$. The adjoint map enjoys two remarkable properties. For $\xi, \eta \in \mathfrak{g}$, we have
$$\begin{aligned} I_g(\exp(\xi)) &= \exp(\mathrm{Ad}_g(\xi)), \\ \left.\frac{\mathrm{d}}{\mathrm{d}t}\right|_{t=0} \mathrm{Ad}_{\exp(t\xi)}(\eta) &= \mathrm{ad}_\xi\, \eta. \end{aligned}$$
The reader is invited to provide the proof of these equations and of related generalizations in Exercise E5.19. Note that the adjoint map is interpreted in rigid body kinematics as a change of reference frame transformation; see Section 5.1.2. For matrix Lie groups, I_g is a similarity transformation; see Section 2.2.3.
4. We briefly introduce and discuss left-invariant covector fields on a Lie group G. For all $g \in \mathsf{G}$, the dual map $T_e^*L_g \colon \mathsf{T}_g^*\mathsf{G} \to \mathsf{T}_e^*\mathsf{G}$ is an isomorphism between cotangent spaces, and allows us to write $\mathsf{T}^*\mathsf{G} \simeq \mathsf{G} \times \mathsf{T}_e^*\mathsf{G}$. We shall also write $\mathfrak{g}^*$ for the cotangent space $\mathsf{T}_e^*\mathsf{G}$. A covector field α on G is ***left-invariant*** if $L_g^*\alpha = \alpha$ for all $g \in \mathsf{G}$, or equivalently if
$$\alpha(g) = T_g^*L_{g^{-1}}(\alpha(e)),$$
for all $g \in \mathsf{G}$. Left-invariant covector fields enjoy similar properties to the ones in Theorem 5.27 for left-invariant vector fields: (1) the set of left-invariant covector fields is isomorphic to $\mathfrak{g}^*$, (2) left-invariant covector fields are smooth, and (3) the pairing between a left-invariant covector field and a left-invariant vector field is a constant function. •

5.2.5 The Lie algebra of a matrix Lie group

Here we show how the Lie algebra of a matrix Lie group coincides with the set of one-parameter subgroup generators of the group. In other words, we analyze left-invariant vector fields on a matrix Lie group and relate them to one-parameter subgroup generators.

Let $\mathsf{G} \subset \mathsf{GL}(n;\mathbb{R})$ be a matrix Lie group, and let $\boldsymbol{g}, \boldsymbol{h} \in \mathsf{G}$. On a matrix Lie group, left translation is left matrix multiplication, i.e., $L_{\boldsymbol{g}}$ maps the matrix $\boldsymbol{h}$ into the matrix $\boldsymbol{g} \cdot \boldsymbol{h} = \boldsymbol{gh}$. Given a matrix Lie group $\mathsf{G} \subset \mathsf{GL}(n;\mathbb{R})$, both its Lie algebra $\mathsf{T}_{\boldsymbol{I}_n}\mathsf{G}$ and its set of one-parameters subgroup generators $\mathcal{G}(\mathsf{G})$ are subsets of $\mathfrak{gl}(n;\mathbb{R})$. Because $\mathsf{G} \subset \mathsf{GL}(n;\mathbb{R})$, its tangent bundle TG is a subset of $\mathsf{T}(\mathsf{GL}(n;\mathbb{R}))$. Because $\mathsf{GL}(n;\mathbb{R})$ is an open subset of $\mathfrak{gl}(n;\mathbb{R})$, there is a natural way of writing $\mathsf{TG} \simeq \mathsf{GL}(n;\mathbb{R}) \times \mathfrak{gl}(n;\mathbb{R})$ and, therefore, of writing a tangent vector on G as a pair of matrices.

Remark 5.34. Before proceeding, let us clarify that there are two different ways to write matrix representations for the tangent bundle of a matrix Lie group. Firstly, on *any* Lie group G, the left translation map provides an isomorphism $\mathsf{T}_{\boldsymbol{g}}\mathsf{G} \simeq \mathsf{T}_e\mathsf{G}$. Accordingly, a tangent vector on G is identified by its base point, i.e., a group element, and its left translation to the identity, i.e., a Lie algebra element. Secondly, on a *matrix* Lie group $\mathsf{G} \subset \mathsf{GL}(n;\mathbb{R})$, a tangent vector and its base point are *naturally* a matrix pair, by virtue of the fact that $\mathsf{GL}(n;\mathbb{R})$ is an open subset of the vector space $\mathbb{R}^{n\times n}$, and that the tangent space of $\mathsf{GL}(n;\mathbb{R})$ at each base point is naturally isomorphic to $\mathbb{R}^{n\times n} \simeq \mathfrak{gl}(n;\mathbb{R})$. In what follows, we shall write tangent vectors in their natural matrix representations without resorting to left translations to the identity. •

We write an element of the tangent space at the identity $\mathfrak{g} = \mathsf{T}_{\boldsymbol{I}_n}\mathsf{G}$ as a matrix $\boldsymbol{A} \in \mathfrak{gl}(n;\mathbb{R})$ or as a pair $(\boldsymbol{I}_n, \boldsymbol{A})$ when we want to be specific about the base point. The tangent map for left matrix multiplication is also left matrix multiplication; that is, for all $\boldsymbol{g} \in \mathsf{G}$, we have

$$T_{\boldsymbol{I}_n} L_{\boldsymbol{g}}(\boldsymbol{A}) = \boldsymbol{gA} \in \mathsf{T}_{\boldsymbol{g}}\mathsf{G}.$$

If $\{\boldsymbol{A}_1, \dots, \boldsymbol{A}_n\}$ is a basis for $\mathfrak{g}$, then the tangent space $\mathsf{T}_{\boldsymbol{g}}\mathsf{G}$ at $\boldsymbol{g} \in \mathsf{G}$ is the subspace of $\mathfrak{gl}(n;\mathbb{R})$ given by

$$\mathsf{T}_{\boldsymbol{g}}\mathsf{G} = \operatorname{span}_{\mathbb{R}} \{\boldsymbol{gA}_1, \dots, \boldsymbol{gA}_n\}.$$

Given $\boldsymbol{A} \in \mathfrak{g} \subset \mathfrak{gl}(n;\mathbb{R})$, the left-invariant vector field $\boldsymbol{A}_{\mathrm{L}}$ on G corresponding to $\boldsymbol{A}$ is $\boldsymbol{g} \mapsto (\boldsymbol{g}, \boldsymbol{gA})$.

Proposition 5.35. *For a matrix Lie group $\mathsf{G} \subset \mathsf{GL}(n;\mathbb{R})$, we have the following statements:*

(i) the exponential map $\exp\colon \mathfrak{g} \to \mathsf{G}$ *coincides with the restriction to* $\mathfrak{g} \subset \mathfrak{gl}(n;\mathbb{R})$ *of the matrix exponential map* $\exp\colon \mathfrak{gl}(n;\mathbb{R}) \to \mathsf{GL}(n;\mathbb{R})$;

(ii) $\mathcal{G}(\mathsf{G})$ *is identical to the tangent space* $\mathsf{T}_{\boldsymbol{I}_n}\mathsf{G}$, *and* $\mathcal{G}(\mathsf{G})$ *is isomorphic to* $\mathfrak{g}$ *as a Lie algebra.*

Proof. Given $\boldsymbol{A} \in \mathfrak{g} \subset \mathfrak{gl}(n;\mathbb{R})$, the curve $\mathbb{R} \ni t \mapsto \exp(t\boldsymbol{A})$ is the solution to the differential equation defined by the left-invariant vector field $\boldsymbol{g} \mapsto (\boldsymbol{g}, \boldsymbol{g}\boldsymbol{A})$ with initial condition $\boldsymbol{I}_n$; this was discussed in Section 5.1.3. This proves (i).

Next, let us prove that $\mathcal{G}(\mathsf{G}) = \mathsf{T}_{\boldsymbol{I}_n}\mathsf{G}$. For $\boldsymbol{A} \in \mathfrak{g}$, the exponential map $t \mapsto \exp(\boldsymbol{A}t)$ takes values in G by Theorem 5.32(i). Hence, $\boldsymbol{A}$ is a one-parameter subgroup generator and $\mathsf{T}_{\boldsymbol{I}_n}\mathsf{G} \subset \mathcal{G}(\mathsf{G})$. Conversely, $\mathcal{G}(\mathsf{G}) \subset \mathsf{T}_{\boldsymbol{I}_n}\mathsf{G}$ because one-parameter subgroup generators are, by construction, tangent vectors at the identity. This proves that $\mathcal{G}(\mathsf{G})$ is equal to $\mathsf{T}_{\boldsymbol{I}_n}\mathsf{G}$. To establish the Lie algebra isomorphism, let us be specific and write the two Lie algebras as $(\mathcal{L}(\mathsf{G}), [\cdot,\cdot]_{\mathcal{L}(\mathsf{G})})$ and $(\mathcal{G}(\mathsf{G}), [\cdot,\cdot]_{\mathcal{G}(\mathsf{G})})$, where $[\cdot,\cdot]_{\mathcal{L}(\mathsf{G})}$ is the Lie bracket and $[\cdot,\cdot]_{\mathcal{G}(\mathsf{G})}$ is the matrix commutator. Then we need to prove that

$$[\boldsymbol{A}_{\mathrm{L}}, \boldsymbol{B}_{\mathrm{L}}]_{\mathcal{L}(\mathsf{G})} = \left([\boldsymbol{A}, \boldsymbol{B}]_{\mathcal{G}(\mathsf{G})}\right)_{\mathrm{L}}.$$

This is easily seen in coordinates as follows. Let $\boldsymbol{g}_i^j$, $i, j \in \{1, \dots, n\}$, be standard coordinates for $\mathsf{GL}(n;\mathbb{R})$. We compute

$$\begin{aligned}
([\boldsymbol{A}_{\mathrm{L}}, \boldsymbol{B}_{\mathrm{L}}]_{\mathcal{L}(\mathsf{G})}(\boldsymbol{g}))_i^j &= \frac{\partial(\boldsymbol{g}\boldsymbol{B})_i^j}{\partial \boldsymbol{g}_k^l}(\boldsymbol{g}\boldsymbol{A})_k^l - \frac{\partial(\boldsymbol{g}\boldsymbol{A})_i^j}{\partial \boldsymbol{g}_k^l}(\boldsymbol{g}\boldsymbol{B})_k^l \\
&= (\boldsymbol{g}\boldsymbol{A}\boldsymbol{B} - \boldsymbol{g}\boldsymbol{B}\boldsymbol{A})_i^j \\
&= \left([\boldsymbol{A}, \boldsymbol{B}]_{\mathcal{G}(\mathsf{G})}\right)_{\mathrm{L}}(\boldsymbol{g})_i^j,
\end{aligned}$$

because $\dfrac{\partial(\boldsymbol{g}_m^j \boldsymbol{B}_i^m)}{\partial \boldsymbol{g}_k^l}(\boldsymbol{g}\boldsymbol{A})_k^l = \delta_l^j \delta_m^k \boldsymbol{B}_i^m (\boldsymbol{g}\boldsymbol{A})_k^l = (\boldsymbol{g}\boldsymbol{A})_k^j \boldsymbol{B}_i^k$. ■

Let us now work out in some detail an example to illustrate these concepts.

Example 5.36 (The Lie algebra of SO(3)). The general notions above have an instructive interpretation on the group of rotations $\mathsf{SO}(3)$. A parallel development can be performed for left-invariant vector fields on $\mathsf{SE}(3)$ and elements of $\mathfrak{se}(3)$.

1. The Lie algebra of $\mathsf{SO}(3)$ is the set of skew-symmetric matrices $\mathfrak{so}(3)$. For $\boldsymbol{\omega} \in \mathbb{R}^3$, consider the left-invariant vector field $\boldsymbol{\omega}_{\mathrm{L}}$ on $\mathsf{SO}(3)$ whose value at the identity is $\widehat{\boldsymbol{\omega}} \in \mathfrak{so}(3)$. In other words, $\boldsymbol{\omega}_{\mathrm{L}}$ is the map $\mathsf{SO}(3) \ni \boldsymbol{R} \mapsto (\boldsymbol{R}, \boldsymbol{R}\widehat{\boldsymbol{\omega}}) \in \mathsf{T}_{\boldsymbol{R}}\mathsf{SO}(3)$. This is indeed a left-invariant vector field because
$$\boldsymbol{\omega}_{\mathrm{L}}(\boldsymbol{R}_2\boldsymbol{R}_1) = (\boldsymbol{R}_2\boldsymbol{R}_1)\widehat{\boldsymbol{\omega}} = \boldsymbol{R}_2(\boldsymbol{R}_1\widehat{\boldsymbol{\omega}}) = \boldsymbol{R}_2\boldsymbol{\omega}_{\mathrm{L}}(\boldsymbol{R}_1),$$
for all $\boldsymbol{R}_1, \boldsymbol{R}_2 \in \mathsf{SO}(3)$.
2. That the Lie bracket of left-invariant vector fields is left-invariant is illustrated as follows. For all $\boldsymbol{\omega}, \boldsymbol{\eta} \in \mathbb{R}^3$, we have
$$[\boldsymbol{\omega}_{\mathrm{L}}(\boldsymbol{R}), \boldsymbol{\eta}_{\mathrm{L}}(\boldsymbol{R})] = \boldsymbol{R}[\widehat{\boldsymbol{\omega}}, \widehat{\boldsymbol{\eta}}] = (\boldsymbol{\omega} \times \boldsymbol{\eta})_{\mathrm{L}}(\boldsymbol{R}).$$

3. Regarding the flow of a left-invariant vector field on $\mathsf{SO}(3)$, for $\boldsymbol{\omega} \in \mathbb{R}^3$, let $\mathbb{R} \ni t \mapsto \gamma(t) \in \mathsf{SO}(3)$ be an integral curve of $\boldsymbol{\omega}_{\mathrm{L}}$. That is, let $\gamma'(t) = \gamma(t)\widehat{\boldsymbol{\omega}}$ and, therefore, $\gamma(t) = \gamma(0)\exp(t\widehat{\boldsymbol{\omega}})$. As discussed in Section 5.1.3, the curve γ describes the rotation of a rigid body with constant body velocity $\boldsymbol{\omega}$. In the section below on mechanical systems on Lie groups, we shall present more details on the relationship between left-invariant vector fields and body angular velocity.
4. Regarding exponential coordinate charts of the first and second kind, these coincide with the ones described in Section 5.1.4. In particular, Proposition 5.7 characterizes the logarithm map on $\mathsf{SO}(3)$ and the corresponding exponential coordinates of the first kind, whereas the Euler angles parametrization of $\mathsf{SO}(3)$ is a set of exponential coordinates of the second kind. •

5.3 Metrics, connections, and systems on Lie groups

In this section we investigate a class of mechanical control systems whose configuration manifold is a Lie group and whose data is left-invariant. Typically these are vehicle systems defined on the group of rigid displacements $\mathsf{SE}(3)$ or on one of its Lie subgroups. We start with the preliminary notions of left-invariant Riemannian metrics and left-invariant affine connections.

5.3.1 Invariant metrics and connections

Left-invariant Riemannian metrics and affine connections on Lie groups are described, for example, in [Sattinger and Weaver 1986], and are used in the study of ideal fluids, see [Arnol'd 1966] and [Arnol'd 1978, Appendix 1 and 2].

Definition 5.37 (Left-invariant Riemannian metric). A Riemannian metric $\mathbb{G}$ on a Lie group $(\mathsf{G}, \star)$ is ***left-invariant*** if $L_g^*\mathbb{G} = \mathbb{G}$ for all $g \in \mathsf{G}$, or equivalently if

$$\mathbb{G}(g) \cdot (X_g, Y_g) = \mathbb{G}(h \star g) \cdot (T_g L_h(X_g), T_g L_h(Y_g)),$$

for all $g, h \in \mathsf{G}$ and $X_g, Y_g \in \mathsf{T}_g\mathsf{G}$. •

The importance of this notion is illustrated by the following result.

Theorem 5.38 (Properties of left-invariant Riemannian metrics). *For a Lie group* G*, the following statements hold.*

(i) An inner product $\mathbb{I}$ *on* $\mathfrak{g}$ *determines a smooth left-invariant Riemannian metric* $\mathbb{G}_{\mathbb{I}}$ *on* G *via left translation, i.e., by means of*

$$\mathbb{G}_{\mathbb{I}}(g) \cdot (X_g, Y_g) = \mathbb{I}\left(T_g L_{g^{-1}}(X_g), T_g L_{g^{-1}}(Y_g)\right), \tag{5.13}$$

for all $g \in \mathsf{G}$ *and* $X_g, Y_g \in \mathsf{T}_g\mathsf{G}$. *Furthermore, any left-invariant Riemannian metric* $\mathbb{G}$ *on* G *is determined via left translation by its value at the identity* $\mathbb{G}(e)$.

(ii) A Riemannian metric is left-invariant if and only if the inner product of any two left-invariant vector fields is a left-invariant function, i.e., a constant function.

Proof. If $\mathbb{I}$ is an inner product on $\mathfrak{g}$, then the $\mathbb{G}_{\mathbb{I}}$ map defined according to (5.13) is clearly bilinear, symmetric, and positive definite. Because the tangent map of the left translation map is smooth, $\mathbb{G}_{\mathbb{I}}$ is smooth. For any $g, h \in \mathsf{G}$ and $X_g, Y_g \in \mathsf{T}_g\mathsf{G}$, we compute

$$\begin{aligned}
&\mathbb{G}_{\mathbb{I}}(h \star g) \cdot (T_gL_h(X_g), T_gL_h(Y_g)) \\
&\qquad = \mathbb{I}\left(T_{h\star g}L_{(h\star g)^{-1}}(T_gL_h(X_g)), T_{h\star g}L_{(h\star g)^{-1}}(T_gL_h(Y_g))\right) \\
&\qquad = \mathbb{I}\left(T_gL_{g^{-1}}(X_g), T_gL_{g^{-1}}(Y_g)\right) \\
&\qquad = \mathbb{G}_{\mathbb{I}}(g) \cdot (X_g, Y_g),
\end{aligned}$$

because $L_{(h\star g)^{-1}} \circ L_h = L_{g^{-1}}$. This shows that $\mathbb{G}_{\mathbb{I}}$ defined according to (5.13) is left-invariant. Furthermore, if $\mathbb{G}$ is a left-invariant Riemannian metric, then $\mathbb{G}(g) \cdot (X_g, Y_g) = \mathbb{G}(e) \cdot (T_gL_{g^{-1}}(X_g), T_gL_{g^{-1}}(Y_g))$. This completes the proof of (i).

Next, let us prove (ii). Let $\mathbb{G}$ be a left-invariant Riemannian metric. If X and Y are left-invariant vector fields on G, then, for all $g \in \mathsf{G}$, we have

$$\begin{aligned}
\mathbb{G}(g) \cdot (X(g), Y(g)) &= \mathbb{G}(g^{-1} \star g) \cdot (T_gL_{g^{-1}}(X(g)), T_gL_{g^{-1}}(Y(g))) \\
&= \mathbb{G}(e) \cdot (X(e), Y(e)),
\end{aligned}$$

which is a constant value, independent of $g \in \mathsf{G}$. Conversely, assume that $g \mapsto \mathbb{G}(g) \cdot (X(g), Y(g))$ is left-invariant for left-invariant vector fields X and Y. Then, for all $g, h \in \mathsf{G}$ we have

$$\begin{aligned}
\mathbb{G}(g) \cdot (X(g), Y(g)) &= \mathbb{G}(h \star g) \cdot (X(h \star g), Y(h \star g)) \\
&= \mathbb{G}(h \star g) \cdot (T_gL_h(X_g), T_gL_h(Y_g)).
\end{aligned}$$

This shows that $\mathbb{G}$ is left-invariant. ∎

The preceding theorem shows how left-invariant Riemannian metrics on a Lie group and inner products on its Lie algebra are closely related. The next theorem characterizes the consequences of this fact for the associated Levi-Civita affine connection.

Definition 5.39 (Left-invariant affine connection). An affine connection ∇ on a Lie group G is ***left-invariant*** if

$$L_g^*(\nabla_X Y) = \nabla_{L_g^* X} L_g^* Y$$

for any vector fields $X, Y \in \Gamma^\infty(\mathsf{TG})$ and any $g \in \mathsf{G}$. •

Given $\xi \in \mathfrak{g}$, recall that $\xi_{\mathrm{L}} \in \Gamma^{\infty}(\mathsf{TG})$ is the left-invariant vector field $g \mapsto T_eL_g(\xi)$.

Theorem 5.40 (Properties of left-invariant affine connections). *For a Lie group* G, *the following statements hold.*

(i) An affine connection on G *is left-invariant if and only if the covariant derivative of any two left-invariant vector fields on* G *is a left-invariant vector field on* G.

(ii) Given a left-invariant affine connection ∇, *there exists a unique bilinear map* $B\colon \mathfrak{g} \times \mathfrak{g} \to \mathfrak{g}$ *such that, for all* $\xi, \eta \in \mathfrak{g}$,

$$\nabla_{\xi_{\mathrm{L}}} \eta_{\mathrm{L}} = (B(\xi, \eta))_{\mathrm{L}}. \tag{5.14}$$

Conversely, given a bilinear map $B\colon \mathfrak{g} \times \mathfrak{g} \to \mathfrak{g}$, *there exists a unique left-invariant affine connection* ∇ *satisfying* (5.14) *on the set of left-invariant vector fields.*

(iii) Let $\mathbb{I}$ *be an inner product on* $\mathfrak{g}$ *and let* $\mathbb{G}_{\mathbb{I}}$ *be its associated left-invariant Riemannian metric. The Levi-Civita connection* $\overset{\mathbb{G}_{\mathbb{I}}}{\nabla}$ *induced by* $\mathbb{G}_{\mathbb{I}}$ *is left-invariant and the corresponding bilinear map, denoted by* $\overset{\mathfrak{g}}{\nabla}\colon \mathfrak{g} \times \mathfrak{g} \to \mathfrak{g}$, *is given by*

$$\overset{\mathfrak{g}}{\nabla}_{\xi}\eta = \frac{1}{2}[\xi, \eta] - \frac{1}{2}\mathbb{I}^{\sharp}\big(\operatorname{ad}^*_{\xi} \mathbb{I}^{\flat}(\eta) + \operatorname{ad}^*_{\eta} \mathbb{I}^{\flat}(\xi)\big). \tag{5.15}$$

Note that part (ii) implies that a left-invariant affine connection on G determines, and is uniquely determined by, a bilinear map $B\colon \mathfrak{g} \times \mathfrak{g} \to \mathfrak{g}$; we refer to this bilinear map B as the ***restriction*** of the left-invariant affine connection ∇ to $\mathfrak{g}$. Furthermore, part (ii) also implies that a left-invariant affine connection on G is smooth.

Proof of Theorem 5.40. If X, Y are left-invariant vector fields and ∇ is a left-invariant affine connection, then $L_g^*(\nabla_X Y) = \nabla_{L_g^* X} L_g^* Y = \nabla_X Y$, so that $\nabla_X Y$ is left-invariant. Conversely, assume that, for all left-invariant vector fields X, Y, $\nabla_X Y$ is a left-invariant vector field. Then $L_g^*(\nabla_X Y) = \nabla_X Y = \nabla_{L_g^* X} L_g^* Y$, so that ∇ is left-invariant. This proves part (i).

Given a left-invariant affine connection ∇, define $B\colon \mathfrak{g} \times \mathfrak{g} \to \mathfrak{g}$ by means of $(\xi, \eta) \mapsto (\nabla_{\xi_{\mathrm{L}}} \eta_{\mathrm{L}})(e) \in \mathfrak{g}$. This map is bilinear because ∇ is $\mathbb{R}$-bilinear. Next, consider the vector fields $(B(\xi, \eta))_{\mathrm{L}}$ and $\nabla_{\xi_{\mathrm{L}}} \eta_{\mathrm{L}}$. Because they are identical at the identity and because they are both left-invariant, they must be equal everywhere on G. This implies equation (5.14). Conversely, consider a bilinear $B\colon \mathfrak{g} \times \mathfrak{g} \to \mathfrak{g}$ and a basis $\{e_1, \dots, e_n\}$ for $\mathfrak{g}$. We claim that equation (5.14) uniquely determines a left-invariant affine connection on G. Indeed, equation (5.14) provides a procedure for computing the covariant derivative between any pair of left-invariant vector fields. We can therefore compute the generalized Christoffel symbols of such a connection with respect to the basis $\{(e_1)_{\mathrm{L}}, \dots, (e_n)_{\mathrm{L}}\}$. An application of the result in Exercise E4.38 concludes the proof of part (ii).

Regarding part (iii), recall from Theorem 3.104 that $\overset{\mathbb{G}_{\mathbb{I}}}{\nabla}$ is determined by equation (3.12) which we reproduce here for convenience:

$$\begin{aligned}\mathbb{G}_{\mathbb{I}}(\overset{\mathbb{G}_{\mathbb{I}}}{\nabla}_X Y, Z) = \tfrac{1}{2}\big(\mathscr{L}_X(\mathbb{G}_{\mathbb{I}}(Y,Z)) + \mathscr{L}_Y(\mathbb{G}_{\mathbb{I}}(Z,X)) - \mathscr{L}_Z(\mathbb{G}_{\mathbb{I}}(X,Y))\\ + \mathbb{G}_{\mathbb{I}}([X,Y],Z) - \mathbb{G}_{\mathbb{I}}([X,Z],Y) - \mathbb{G}_{\mathbb{I}}([Y,Z],X)\big),\end{aligned} \quad (5.16)$$

for $X, Y, Z \in \Gamma^\infty(\mathsf{TG})$. Now, take $X = \xi_{\mathrm{L}}$, $Y = \eta_{\mathrm{L}}$, and $Z = \zeta_{\mathrm{L}}$ for some $\xi, \eta, \zeta \in \mathfrak{g}$. We need to show that $\overset{\mathbb{G}_{\mathbb{I}}}{\nabla}_{\xi_{\mathrm{L}}}\eta_{\mathrm{L}}$ is the left-invariant vector field with value $\overset{\mathfrak{g}}{\nabla}_\xi \eta$ at the identity. The first, second, and third terms on the right-hand side of equation (5.16) vanish because the left-invariant inner product of left-invariant vector fields is a constant function, and the Lie derivative of a constant function is zero. Because the remaining terms contain left-invariant inner products between left-invariant vector fields, we can write

$$2\mathbb{G}_{\mathbb{I}}(\overset{\mathbb{G}_{\mathbb{I}}}{\nabla}_{\xi_{\mathrm{L}}}\eta_{\mathrm{L}}, \zeta_{\mathrm{L}}) = \mathbb{I}([\xi,\eta],\zeta) - \mathbb{I}([\xi,\zeta],\eta) - \mathbb{I}([\eta,\zeta],\xi).$$

Therefore, the left-hand side $\mathbb{G}_{\mathbb{I}}(\overset{\mathbb{G}_{\mathbb{I}}}{\nabla}_{\xi_{\mathrm{L}}}\eta_{\mathrm{L}}, \zeta_{\mathrm{L}})$ is a left-invariant function on G. Because $\mathbb{G}_{\mathbb{I}}$ and ζ_{L} are left-invariant, Theorem 5.38(ii) implies that $\overset{\mathbb{G}_{\mathbb{I}}}{\nabla}_{\xi_{\mathrm{L}}}\eta_{\mathrm{L}}$ is a left-invariant vector field. Now, part (i) implies that $\overset{\mathbb{G}_{\mathbb{I}}}{\nabla}$ is a left-invariant affine connection because the covariant derivative with respect to $\overset{\mathbb{G}_{\mathbb{I}}}{\nabla}$ of any two $\xi_{\mathrm{L}}, \eta_{\mathrm{L}}$ is a left-invariant vector field. Let $\overset{\mathfrak{g}}{\nabla}$ be the bilinear map on $\mathfrak{g}$ associated with $\overset{\mathbb{G}_{\mathbb{I}}}{\nabla}$ because of the statement (ii). In summary, we have

$$\mathbb{G}_{\mathbb{I}}(\overset{\mathbb{G}_{\mathbb{I}}}{\nabla}_{\xi_{\mathrm{L}}}\eta_{\mathrm{L}}, \zeta_{\mathrm{L}}) = \mathbb{I}(\overset{\mathfrak{g}}{\nabla}_\xi \eta, \zeta).$$

Next, we write

$$2\langle \mathbb{I}^\flat(\zeta); \overset{\mathfrak{g}}{\nabla}_\xi \eta\rangle = \langle \mathbb{I}^\flat(\zeta); [\xi,\eta]\rangle - \langle \mathbb{I}^\flat(\eta); [\xi,\zeta]\rangle - \langle \mathbb{I}^\flat(\xi); [\eta,\zeta]\rangle,$$

and, in terms of dual adjoint operators,

$$\begin{aligned}2\langle \mathbb{I}^\flat(\zeta); \overset{\mathfrak{g}}{\nabla}_\xi \eta\rangle &= \langle \mathbb{I}^\flat(\zeta); [\xi,\eta]\rangle - \langle \mathrm{ad}_\xi^* \mathbb{I}^\flat(\eta); \zeta\rangle - \langle \mathrm{ad}_\eta^* \mathbb{I}^\flat(\xi); \zeta\rangle\\ &= \langle \mathbb{I}^\flat(\zeta); [\xi,\eta]\rangle - \langle \mathbb{I}^\flat(\zeta); \mathbb{I}^\sharp(\mathrm{ad}_\xi^* \mathbb{I}^\flat(\eta))\rangle - \langle \mathbb{I}^\flat(\zeta); \mathbb{I}^\sharp(\mathrm{ad}_\eta^* \mathbb{I}^\flat(\xi))\rangle\\ &= \langle \mathbb{I}^\flat(\zeta); [\xi,\eta] - \mathbb{I}^\sharp(\mathrm{ad}_\xi^* \mathbb{I}^\flat(\eta)) - \mathbb{I}^\sharp(\mathrm{ad}_\eta^* \mathbb{I}^\flat(\xi))\rangle\,.\end{aligned}$$

This proves equation (5.15) in part (iii) because this equality must hold for all $\zeta \in \mathfrak{g}$, and because $\mathbb{I}$ is nondegenerate. ■

We conclude this section by remarking that Theorem 5.40(iii) provides the means to compute the generalized Christoffel symbols of $\overset{\mathbb{G}_{\mathbb{I}}}{\nabla}$ with respect to a basis of left-invariant vector fields; the reader is invited to perform these computations in Exercise E5.21.

5.3.2 Simple mechanical control systems on Lie groups

We are finally ready to introduce an interesting class of mechanical systems. These are systems whose configuration space is a Lie group and whose constitutive elements, such as kinetic energy metric and forces, are left-invariant.

Definition 5.41 (Simple mechanical (control) system on a Lie group). A ***simple mechanical control system on a Lie group*** is a 4-tuple $(\mathsf{G}, \mathbb{I}, \{f^1, \dots, f^m\}, U)$, where

(i) G is an n-dimensional Lie group, defining the configuration space,
(ii) $\mathbb{I}$ is an inner product on the Lie algebra $\mathfrak{g}$, defining the kinetic energy metric via left translation,
(iii) $\{f^1, \dots, f^m\}$ is a collection of covectors in $\mathfrak{g}^*$, defining a collection of left-invariant control forces, and
(iv) $U \subset \mathbb{R}^m$ is the control set.

A ***simple mechanical system on a Lie group*** is a pair $(\mathsf{G}, \mathbb{I})$. •

Remarks 5.42. 1. A simple mechanical control system on a Lie group $(\mathsf{G}, \mathbb{I}, \{f^1, \dots, f^m\}, U)$ defines a C^∞-simple mechanical control system as follows. The configuration manifold is the Lie group G. The kinetic energy metric is the Riemannian metric $\mathbb{G}_{\mathbb{I}}$ induced by the inner product $\mathbb{I}$. The potential function is zero. The control forces are the left-invariant covector fields $g \mapsto T_g^* L_{g^{-1}}(f^a)$, $a \in \{1, \dots, m\}$. The control set is U. The kinetic energy of the resulting simple mechanical control system is the function $\mathrm{KE}\colon \mathsf{TG} \to \mathbb{R}$ given by

$$\mathrm{KE}(v_g) = \tfrac{1}{2}\mathbb{G}_{\mathbb{I}}(g) \cdot (v_g, v_g).$$

2. A left-invariant function on G is a constant function and, so long as we are interested in simple mechanical systems comprised only of left-invariant quantities, we neglect potential function terms in Definition 5.41.
3. Simple mechanical systems on the group of rigid displacements $\mathsf{SE}(3)$ or on one of its Lie subgroups are often equipped with control forces that, in their Newtonian representation, are constant when expressed in the body reference frame. We call such forces ***body-fixed*** and model them, in our Lagrangian framework, as left-invariant covector fields. We present examples of mechanical systems with body-fixed forces in Section 5.3.3.
4. We shall call the inner product $\mathbb{I}$ the ***kinetic energy tensor***. By Theorem 5.38, the kinetic energy tensor induces a left-invariant kinetic energy metric $\mathbb{G}_{\mathbb{I}}$ on G. For a rigid body fixed at a point, the configuration manifold is $\mathsf{SO}(3)$ and the kinetic energy tensor is related to the inertia tensor $\mathbb{I}_c$ defined in Section 4.2.2. For this system (see also Example 5.48 below), the relationship between $\mathbb{I}_c$ and $\mathbb{I}$ is $\mathbb{I}(\widehat{\boldsymbol{\Omega}}_1, \widehat{\boldsymbol{\Omega}}_2) = \langle\langle \mathbb{I}_c(\boldsymbol{\Omega}_1), \boldsymbol{\Omega}_2 \rangle\rangle_{\mathbb{R}^3}$. •

In the study of simple mechanical control systems on Lie groups, it is convenient to write the velocity of a controlled trajectory as a Lie algebra element. We do this by means of the following notion.

Definition 5.43 (Body velocity). Let $\gamma\colon I \to \mathsf{G}$ be differentiable. The ***body velocity*** of γ is the curve $v_\gamma\colon I \to \mathfrak{g}$ defined by $t \mapsto T_{\gamma(t)}L_{\gamma(t)^{-1}}(\gamma'(t))$. •

Remarks 5.44. 1. One can easily see that, if v_γ is the body velocity of γ, then the system's kinetic energy along γ is

$$\mathrm{KE}(\gamma'(t)) = \tfrac{1}{2}\mathbb{I}(v_\gamma(t), v_\gamma(t)).$$

2. For $\mathsf{G} = \mathsf{SE}(3)$, this notion of body velocity agrees with that introduced in Section 5.1 within the context of rigid body kinematics. In other words, for systems defined on the rigid displacement group $\mathsf{SE}(3)$ or on one of its Lie subgroups, writing the velocity curve γ' in terms of the body velocity curve on the Lie algebra $\mathfrak{se}(3)$, or on one of its Lie subalgebras, corresponds to writing the velocity in the body reference frame.
3. The notion of body velocity has a second important interpretation. Given a basis $\{e_1, \dots, e_n\}$ for $\mathfrak{g}$, we can write

$$\gamma'(t) = T_eL_{\gamma(t)}(v_\gamma(t)) = T_eL_{\gamma(t)}\Big(\sum_{i=1}^{n} v_\gamma^i(t)e_i\Big) = \sum_{i=1}^{n} v_\gamma^i(t)(e_i)_{\mathrm{L}}(\gamma(t)), \tag{5.17}$$

where the tangent vectors $(e_i)_{\mathrm{L}}(g) = T_eL_g(e_i)$, $i \in \{1, \dots, n\}$, form a basis for $\mathsf{T}_g\mathsf{G}$ for each $g \in \mathsf{G}$. Therefore, in the language introduced in Section 4.6.4, the components of the body velocity are pseudo-velocities. As the next result illustrates, the notion of body velocity is indeed useful in obtaining Poincaré representations of the Euler–Lagrange equations. •

We are now ready to characterize the equations of motion for systems on Lie groups.

Theorem 5.45 (Euler–Poincaré equations). *Consider a simple mechanical control system on a Lie group* $(\mathsf{G}, \mathbb{I}, \{f^1, \dots, f^m\}, U)$. *Assume that the control inputs* $u^a\colon I \to \mathbb{R}$, $a \in \{1, \dots, m\}$, *are locally integrable. The locally absolutely differentiable curve* $\gamma\colon I \to \mathsf{G}$ *satisfies the governing equations*

$$\overset{\mathsf{G}_{\mathbb{I}}}{\nabla}_{\gamma'(t)}\gamma'(t) = \sum_{a=1}^{m} u^a(t)\mathbb{G}_{\mathbb{I}}^\sharp(T^*_{\gamma(t)}L_{\gamma^{-1}(t)}(f^a))$$

if and only if there exists a locally absolutely continuous curve $v\colon I \to \mathfrak{g}$ *such that*

$$\gamma'(t) = T_eL_{\gamma(t)}(v(t)), \tag{5.18}$$

$$v'(t) - \mathbb{I}^\sharp\big(\operatorname{ad}^*_{v(t)}\mathbb{I}^\flat(v(t))\big) = \sum_{a=1}^{m} u^a(t)\mathbb{I}^\sharp(f^a). \tag{5.19}$$

Equation (5.19) is referred to as the ***controlled Euler–Poincaré equation.***

Proof of Theorem 5.45. The proof is an application of the analysis in Section 4.5.6; indeed, equation (5.19) is a Poincaré representation of the governing equations. Let $\{e_1, \dots, e_n\}$ be a basis for $\mathfrak{g}$ and write $v(t) = v^i(t)e_i$. From (5.17) and from the first result in Proposition 4.92, we have

$$\begin{aligned}\overset{\mathbb{G}_{\mathbb{I}}}{\nabla}_{\gamma'(t)}\gamma'(t) &= \Big(\frac{\mathrm{d}}{\mathrm{d}t}v^i(t)\Big)(e_i)_{\mathrm{L}}(\gamma(t)) + v^j(t)v^k(t)\Big(\overset{\mathbb{G}_{\mathbb{I}}}{\nabla}_{(e_j)_{\mathrm{L}}}(e_k)_{\mathrm{L}}\Big)(\gamma(t))\\ &= T_eL_{\gamma(t)}\Big(\frac{\mathrm{d}}{\mathrm{d}t}v^i(t)e_i + v^j(t)v^k(t)\overset{\mathfrak{g}}{\nabla}_{e_j}e_k\Big)\\ &= T_eL_{\gamma(t)}\Big(\frac{\mathrm{d}}{\mathrm{d}t}v(t) + \overset{\mathfrak{g}}{\nabla}_{v(t)}v(t)\Big).\end{aligned}$$

Finally, we note that, for all $g \in \mathsf{G}$, the left-invariant Riemannian metric $\mathbb{G}_{\mathbb{I}}$ satisfies $T_gL_{g^{-1}} \circ \mathbb{G}_{\mathbb{I}}^{\sharp} \circ T_g^*L_{g^{-1}}(f^a) = \mathbb{I}^{\sharp}(f^a)$. The result follows from (5.15). ■

This concludes our treatment of simple mechanical control systems on Lie groups. To present a useful generalization, let us say that a distribution $\mathcal{D}$ on the Lie group G is ***left-invariant*** if $\mathcal{D}_g = T_eL_g(\mathcal{D}_e)$ for all $g \in \mathsf{G}$. Note that left-invariant distributions are automatically C^∞, regular, and finitely generated.

Definition 5.46 (Affine connection control system on a Lie group). An ***affine connection control system on a Lie group*** is a 5-tuple $(\mathsf{G}, \nabla, \mathcal{D}, \mathscr{Y} = \{Y_1, \dots, Y_m\}, U)$, where

(i) G is a Lie group,
(ii) ∇ is a left-invariant affine connection on G,
(iii) $\mathcal{D}$ is a left-invariant distribution on G having the property that ∇ restricts to $\mathcal{D}$,
(iv) $\mathscr{Y} = \{Y_1, \dots, Y_m\}$ is a set of left-invariant vector fields on G taking values in $\mathcal{D}$, and
(v) $U \subset \mathbb{R}^m$ is the control set. •

Note that an equivalent way of describing an affine connection control system on a Lie group is by the restriction of the constitutive elements ∇, $\mathcal{D}$, and $\mathscr{Y}$ to Lie algebra $\mathfrak{g}$.

5.3.3 Planar and three-dimensional rigid bodies as systems on Lie groups

In what follows, we work out a few examples of mechanical control systems on matrix Lie groups. We provide some basic constructions for the matrix Lie groups $\mathsf{SE}(2)$, $\mathsf{SO}(3)$, and $\mathsf{SE}(3)$, and we present models for planar bodies, satellites, and underwater vehicles.

Example 5.47 (Planar rigid body). Let us revisit the planar rigid body that we studied in Section 1.1 and Chapter 4; see Section 4.1, Section 4.2, and

Section 4.4. In these previous discussions, we set the configuration manifold of this system to be $\mathbb{S}^1 \times \mathbb{R}^2$. Here we shall study the planar rigid body as a simple mechanical control system on the Lie group of planar rigid displacements $\mathsf{SE}(2)$. In this section, we also aim to present coordinate expressions for an example of an invariant affine connection on $\mathsf{SE}(2)$. Let us therefore present some basic constructions for $\mathsf{SE}(2)$ and for its Lie algebra $\mathfrak{se}(2)$.

On the Lie group $\mathsf{SE}(2)$, we consider coordinates (θ, x, y) as introduced in Section 4.1. For the planar rigid body example, the coordinates (x, y) are the position of the center of mass of the planar rigid body, and θ is the angle the body forms with the horizontal. In matrix representations we let $\boldsymbol{g} \in \mathsf{SE}(2)$ denote a configuration and write

$$\boldsymbol{g} = \begin{bmatrix} \cos\theta & -\sin\theta & x \\ \sin\theta & \cos\theta & y \\ 0 & 0 & 1 \end{bmatrix}, \quad \boldsymbol{g}^{-1} = \begin{bmatrix} \cos\theta & \sin\theta & -x\cos\theta - y\sin\theta \\ -\sin\theta & \cos\theta & x\sin\theta - y\cos\theta \\ 0 & 0 & 1 \end{bmatrix}.$$

To study the Lie algebra $\mathfrak{se}(2)$, it is convenient to introduce the isomorphism $\widehat{\cdot}\colon \mathbb{R}^3 \to \mathfrak{se}(2)$ given by

$$\widehat{(v^1, v^2, v^3)} = \begin{bmatrix} 0 & -v^1 & v^2 \\ v^1 & 0 & v^3 \\ 0 & 0 & 0 \end{bmatrix}.$$

Accordingly, if $\{\boldsymbol{e}_1, \boldsymbol{e}_2, \boldsymbol{e}_3\}$ is the standard basis for $\mathbb{R}^3$, then $\{\widehat{\boldsymbol{e}}_1, \widehat{\boldsymbol{e}}_2, \widehat{\boldsymbol{e}}_3\}$ is a basis for $\mathfrak{se}(2)$. Given $\boldsymbol{v} = (v^1, v^2, v^3) \in \mathbb{R}^3$ and $\boldsymbol{g} \in \mathsf{SE}(2)$ with coordinates (θ, x, y), the adjoint operator $\mathrm{ad}_{\widehat{\boldsymbol{v}}} \in L(\mathfrak{se}(2); \mathfrak{se}(2))$ and the adjoint map $\mathrm{Ad}_{\boldsymbol{g}} \in L(\mathfrak{se}(2); \mathfrak{se}(2))$ admit the matrix representations

$$[\mathrm{ad}_{\widehat{\boldsymbol{v}}}] = \begin{bmatrix} 0 & 0 & 0 \\ v^3 & 0 & -v^1 \\ -v^2 & v^1 & 0 \end{bmatrix}, \quad [\mathrm{Ad}_{\boldsymbol{g}}] = \begin{bmatrix} 1 & 0 & 0 \\ y & \cos\theta & -\sin\theta \\ -x & \sin\theta & \cos\theta \end{bmatrix}. \tag{5.20}$$

The isomorphism $\mathfrak{se}(2) \simeq \mathbb{R}^3$ induces a natural isomorphism between the dual spaces $\mathfrak{se}(2)^*$ and $(\mathbb{R}^3)^*$

$$(\mathbb{R}^3)^* \ni \boldsymbol{f} = f_1\boldsymbol{e}^1 + f_2\boldsymbol{e}^2 + f_3\boldsymbol{e}^3 \mapsto \widehat{\boldsymbol{f}} = f_1\widehat{\boldsymbol{e}}^1 + f_2\widehat{\boldsymbol{e}}^2 + f_3\widehat{\boldsymbol{e}}^3 \in \mathfrak{se}(2)^*,$$

where $\{\widehat{\boldsymbol{e}}^1, \widehat{\boldsymbol{e}}^2, \widehat{\boldsymbol{e}}^3\}$ is the dual basis on $\mathfrak{se}(2)^*$. With this isomorphism, we have $\langle\widehat{\boldsymbol{f}}; \widehat{\boldsymbol{v}}\rangle = \langle\boldsymbol{f}; \boldsymbol{v}\rangle = f_i v^i$, for all $\widehat{\boldsymbol{v}} \in \mathfrak{se}(2)$ and $\widehat{\boldsymbol{f}} \in \mathfrak{se}(2)^*$. Given $\boldsymbol{v} \in \mathbb{R}^3$, the dual adjoint operator $\mathrm{ad}^*_{\widehat{\boldsymbol{v}}} \in L(\mathfrak{se}(2)^*; \mathfrak{se}(2)^*)$ admits the matrix representation $[\mathrm{ad}^*_{\widehat{\boldsymbol{v}}}] = [\mathrm{ad}_{\widehat{\boldsymbol{v}}}]^T$.

Next, let us provide $\mathfrak{se}(2)$ with an inner product. Given $J, m_1, m_2 \in \mathbb{R}_+$, we define $\mathbb{I} = J\widehat{\boldsymbol{e}}^1 \otimes \widehat{\boldsymbol{e}}^1 + m_1\widehat{\boldsymbol{e}}^2 \otimes \widehat{\boldsymbol{e}}^2 + m_2\widehat{\boldsymbol{e}}^3 \otimes \widehat{\boldsymbol{e}}^3$. The reader is invited to verify in Exercise E5.24 that this inner product corresponds to the kinetic energy tensor of a planar rigid body in ideal fluid. When $m_1 = m_2$, this inner product corresponds to the kinetic energy tensor of the usual planar rigid body

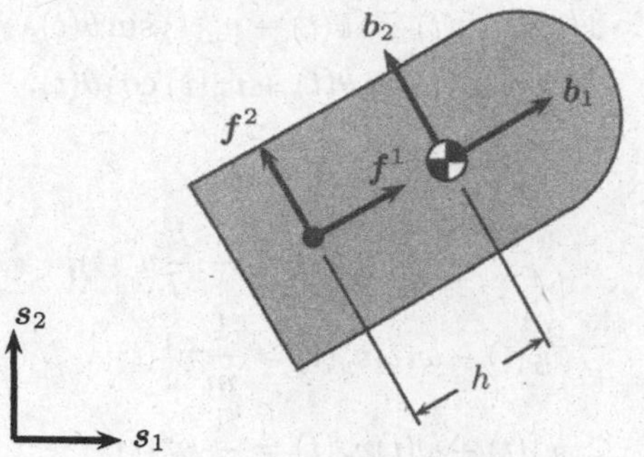

Figure 5.2. A planar rigid body a variable-direction thruster

as discussed in Chapter 4. The matrix representation of $\mathbb{I}$, and, therefore, of $\mathbb{I}^\flat$, is

$$\begin{bmatrix} J & 0 & 0 \\ 0 & m_1 & 0 \\ 0 & 0 & m_2 \end{bmatrix}.$$

This inner product on $\mathfrak{se}(2)$ induces a Riemannian metric and an affine connection on $\mathsf{SE}(2)$. Using equation (5.15), some straightforward computations show that the restriction $\overset{\mathfrak{se}(2)}{\nabla} : \mathfrak{se}(2) \times \mathfrak{se}(2) \to \mathfrak{se}(2)$ of the affine connection associated with $\mathbb{I}$ is, for $\boldsymbol{v} = (v^1, v^2, v^3) \in \mathbb{R}^3$ and $\boldsymbol{w} = (w^1, w^2, w^3) \in \mathbb{R}^3$,

$$\begin{aligned} \overset{\mathfrak{se}(2)}{\nabla}_{\widehat{\boldsymbol{v}}} \widehat{\boldsymbol{w}} = & \frac{(m_2 - m_1)(v^3 w^2 + v^2 w^3)}{2J} \widehat{\boldsymbol{e}}_1 \\ & + \frac{(m_1 - m_2) v^3 w^1 - (m_1 + m_2) v^1 w^3}{2m_1} \widehat{\boldsymbol{e}}_2 \\ & + \frac{(m_1 - m_2) v^2 w^1 + (m_1 + m_2) v^1 w^2}{2m_2} \widehat{\boldsymbol{e}}_3. \end{aligned} \tag{5.21}$$

Let us now return to the planar rigid body with kinetic energy $\frac{1}{2}J\omega^2 + \frac{1}{2}m(v_x^2 + v_y^2)$, where m is the mass of the body, J is the moment of inertia about the center of mass, $(\theta, x, y) \in \mathsf{SE}(2)$ is the system configuration, and (ω, v_x, v_y) is its body velocity. The control input to the system consists of a variable-direction force applied at a distance h from the center of mass; see Figure 5.2. This body-fixed force can be resolved into two components, one along the body $\boldsymbol{b}_1$-axis and the other along the body $\boldsymbol{b}_2$-axis, thus giving the system two inputs that we denote by f^1 and f^2. In summary, the planar body with a variable-direction thruster is a simple mechanical control system on a Lie group described by the triple $(\mathsf{SE}(2), \mathbb{I}, \{f^1, f^2\})$. In coordinates the equations of motion (5.18) and (5.19) read

$$\begin{aligned}\dot{\theta}(t) &= \omega(t),\\ \dot{x}(t) &= v_x(t)\cos\theta(t) - v_y(t)\sin\theta(t),\\ \dot{y}(t) &= v_x(t)\sin\theta(t) + v_y(t)\cos\theta(t),\end{aligned}$$

and

$$\begin{aligned}\dot{\omega}(t) &= -\frac{h}{J}u^2(t),\\ \dot{v}_x(t) - \omega(t)v_y(t) &= \frac{1}{m}u^1(t),\\ \dot{v}_y(t) + \omega(t)v_x(t) &= \frac{1}{m}u^2(t).\end{aligned}$$

These equations are a Poincaré representation of the equations of motion for the planar rigid body, and are equivalent to the natural representation in equation (4.14). •

Example 5.48 (Rigid body fixed at a point (cont'd)). As a second example, we consider a rigid body $(\mathcal{B}, \mu)$ fixed at a point $\boldsymbol{\chi}_0$ discussed in Sections 4.1 and 4.2. The resultant model is that of a simple mechanical control system on the Lie group $\mathsf{SO}(3)$, and with kinetic energy metric induced by the inertia tensor $\mathbb{I}_{\boldsymbol{\chi}_0}$ of the rigid body $(\mathcal{B}, \mu)$ about $\boldsymbol{\chi}_0$. This model is often used to describe the attitude dynamics of a satellite.

It is convenient to start by investigating the Lie group $\mathsf{SO}(3)$ and its Lie algebra $\mathfrak{so}(3)$. Recall the isomorphism $\widehat{\cdot}\colon \mathbb{R}^3 \to \mathfrak{so}(3)$ and its inverse $\cdot^{\vee}\colon \mathfrak{so}(3) \to \mathbb{R}^3$ introduced in Section 5.1.2. Given $\boldsymbol{\omega} \in \mathbb{R}^3$ and $\boldsymbol{R} \in \mathsf{SO}(3)$, the matrix representations of $\mathrm{ad}_{\widehat{\omega}} \in L(\mathfrak{so}(3); \mathfrak{so}(3))$ and $\mathrm{Ad}_{\boldsymbol{R}} \in L(\mathfrak{so}(3); \mathfrak{so}(3))$ are:

$$[\mathrm{ad}_{\widehat{\omega}}] = \widehat{\boldsymbol{\omega}}, \quad [\mathrm{Ad}_{\boldsymbol{R}}] = \boldsymbol{R}. \tag{5.22}$$

Here the matrix representations are with respect to the basis for $\mathfrak{so}(3)$ induced from $\widehat{\cdot}$ by the standard basis $\{\boldsymbol{e}_1, \boldsymbol{e}_2, \boldsymbol{e}_3\}$ for $\mathbb{R}^3$. Via the dual basis $\{\widehat{\boldsymbol{e}}^1, \widehat{\boldsymbol{e}}^2, \widehat{\boldsymbol{e}}^3\}$, we identify $\widehat{\boldsymbol{t}} = t_i\widehat{\boldsymbol{e}}^i \in \mathfrak{so}(3)^*$ with $\boldsymbol{t} = (t_1, t_2, t_3) \in \mathbb{R}^3$, and we write $\langle \widehat{\boldsymbol{t}}; \widehat{\boldsymbol{\omega}} \rangle = \langle \boldsymbol{t}; \boldsymbol{\omega} \rangle = t_i\omega^i$, for all $\widehat{\boldsymbol{\omega}} \in \mathfrak{so}(3)$. Given $\boldsymbol{\omega} \in \mathbb{R}^3$, the dual adjoint operator $\mathrm{ad}^*_{\widehat{\omega}} \in L(\mathfrak{so}(3)^*; \mathfrak{so}(3)^*)$ admits the matrix representation $[\mathrm{ad}^*_{\widehat{\omega}}] = -\widehat{\boldsymbol{\omega}}$.

Next, let us provide $\mathfrak{so}(3)$ with an inner product. As discussed in Section 4.2.1, the inertial properties of the rigid body $(\mathcal{B}, \mu)$ about $\boldsymbol{\chi}_0$ are characterized by its principal inertias and principal axes. In what follows, let us assume that the principal axes of $(\mathcal{B}, \mu)$ are aligned with the axes of the body reference frame. Accordingly, the matrix representation of $\mathbb{I}$, and, therefore, of $\mathbb{I}^\flat$, is

$$[\mathbb{I}] = [\mathbb{I}^\flat] = \begin{bmatrix} J_1 & 0 & 0 \\ 0 & J_2 & 0 \\ 0 & 0 & J_3 \end{bmatrix}.$$

This inner product on $\mathfrak{so}(3)$ induces a Riemannian metric and, therefore, an affine connection on $\mathsf{SO}(3)$. Using equation (5.15), some straightforward computations show that the restriction $\overset{\mathfrak{so}(3)}{\nabla} : \mathfrak{so}(3) \times \mathfrak{so}(3) \to \mathfrak{so}(3)$ of the affine connection associated with $\mathbb{I}$ is

$$\left(\overset{\mathfrak{so}(3)}{\nabla}_{\widehat{v}} \widehat{w}\right)^{\vee} = \frac{1}{2} v \times w + \frac{1}{2} [\mathbb{I}]^{-1} \left(v \times ([\mathbb{I}] w) + w \times ([\mathbb{I}] v)\right),$$

or, for $v = (v^1, v^2, v^3)$ and $w = (w^1, w^2, w^3) \in \mathbb{R}^3$,

$$\begin{aligned}\overset{\mathfrak{so}(3)}{\nabla}_{\widehat{v}} \widehat{w} = & \frac{(J_1 - J_2 + J_3) v^2 w^3 - (J_1 + J_2 - J_3) v^3 w^2}{2J_1} \widehat{e}_1 \\ & + \frac{(J_1 + J_2 - J_3) v^3 w^1 + (J_1 - J_2 - J_3) v^1 w^3}{2J_2} \widehat{e}_2 \\ & + \frac{(-J_1 + J_2 + J_3) v^1 w^2 - (J_1 - J_2 + J_3) v^2 w^1}{2J_3} \widehat{e}_3. \end{aligned} \quad (5.23)$$

Finally, let us assume that the rigid body is subject to three body-fixed control torques aligned with the three principal axes. Accordingly, the rigid body fixed at a point is a simple mechanical control system on a Lie group described by the triple $(\mathsf{SO}(3), \mathbb{I}, \{e^1, e^2, e^3\})$. The kinematic equation governing the trajectory $t \mapsto \boldsymbol{R}(t)$ is

$$\dot{\boldsymbol{R}}(t) = \boldsymbol{R}(t) \widehat{\boldsymbol{\Omega}}(t),$$

where $t \mapsto \boldsymbol{\Omega}(t) \in \mathbb{R}^3$ is the body angular velocity. The Euler–Poincaré equations describing the trajectory $t \mapsto \boldsymbol{\Omega}(t)$ are

$$\dot{\boldsymbol{\Omega}}(t) + [\mathbb{I}]^{-1}(\boldsymbol{\Omega}(t) \times [\mathbb{I}]\boldsymbol{\Omega}(t)) = [\mathbb{I}]^{-1} \boldsymbol{\tau}(t), \quad (5.24)$$

for any torque $t \mapsto \boldsymbol{\tau}(t) = (\tau_1(t), \tau_2(t), \tau_3(t))$ acting on the body. Equations (5.24) are referred to as the ***controlled Euler equations***. •

Example 5.49 (Rigid body in ideal fluid). As a final example we consider a six degree-of-freedom rigid body system and study its attitude and translation in three-dimensional Euclidean space. An early reference on the control of this system is [Fjellstad and Fossen 1994]. For this system, the configuration manifold is the group of rigid displacement $\mathsf{SE}(3)$. The vehicle's configuration is described by the attitude $\boldsymbol{R}$ and the position $\boldsymbol{r}$ of a body frame with respect to an inertial frame. Additionally, let $(\boldsymbol{\Omega}, \boldsymbol{V}) \in \mathbb{R}^3 \oplus \mathbb{R}^3 \simeq \mathfrak{se}(3)$ denote the body velocity. Accordingly, the kinematic equations are

$$\begin{aligned} \dot{\boldsymbol{R}}(t) &= \boldsymbol{R}(t) \widehat{\boldsymbol{\Omega}}(t), \\ \dot{\boldsymbol{r}}(t) &= \boldsymbol{R}(t) \boldsymbol{V}(t). \end{aligned} \quad (5.25)$$

Before proceeding, let us present some basic constructions for the Lie algebra $\mathfrak{se}(3)$, expanding on the treatment in Section 5.1. From equation (5.4), (5.7), and (5.12) let us recall that the isomorphism $\widehat{\cdot} : \mathbb{R}^3 \oplus \mathbb{R}^3 \to \mathfrak{se}(3)$ is

$$\widehat{(\boldsymbol{\omega}, \boldsymbol{v})} = \begin{bmatrix} \widehat{\boldsymbol{\omega}} & \boldsymbol{v} \\ \mathbf{0}_{1\times 3} & 0 \end{bmatrix},$$

and the matrix representations of the adjoint map and adjoint operator on $\mathfrak{se}(3)$ are

$$[\mathrm{Ad}_{(\boldsymbol{R},\boldsymbol{r})}] = \begin{bmatrix} \boldsymbol{R} & \mathbf{0}_{3\times 3} \\ \widehat{\boldsymbol{r}}\boldsymbol{R} & \boldsymbol{R} \end{bmatrix}, \quad [\mathrm{ad}_{(\boldsymbol{\omega},\boldsymbol{v})}] = \begin{bmatrix} \widehat{\boldsymbol{\omega}} & \mathbf{0}_{3\times 3} \\ \widehat{\boldsymbol{v}} & \widehat{\boldsymbol{\omega}} \end{bmatrix}. \tag{5.26}$$

Next, we assume that the rigid body is immersed in an ideal fluid and, under Assumptions 4.110 discussed in the preamble to Exercise E4.25, we introduce the kinetic energy tensor of the fluid-body system. If the underwater vehicle is an ellipsoidal body with uniformly distributed mass, then the kinetic energy of the body-fluid system is $\frac{1}{2}\mathbb{J}(\boldsymbol{\Omega},\boldsymbol{\Omega})+\frac{1}{2}\mathbb{M}(\boldsymbol{V},\boldsymbol{V})$, where the tensors $\mathbb{M}$ and $\mathbb{J}$ are inner products on $\mathbb{R}^3$ that take into account the inertial properties of the body-fluid system. The kinetic energy induces the inner product $\mathbb{I}$ on the Lie algebra $\mathfrak{se}(3) \simeq \mathbb{R}^3 \oplus \mathbb{R}^3$ by means of $\mathbb{I}((\boldsymbol{\omega}_1,\boldsymbol{v}_1),(\boldsymbol{\omega}_2,\boldsymbol{v}_2)) = \mathbb{J}(\boldsymbol{\omega}_1,\boldsymbol{\omega}_2) + \mathbb{M}(\boldsymbol{v}_1,\boldsymbol{v}_2)$. The matrix representation of $\mathbb{I}$, and, therefore, of $\mathbb{I}^\flat$, is

$$[\mathbb{I}] = [\mathbb{I}^\flat] = \begin{bmatrix} [\mathbb{J}] & \mathbf{0}_{3\times 3} \\ \mathbf{0}_{3\times 3} & [\mathbb{M}] \end{bmatrix}.$$

This inner product on $\mathfrak{se}(3)$ induces a Riemannian metric and an affine connection on $\mathsf{SE}(3)$. Using equation (5.15), some straightforward computations show that the restriction $\overset{\mathfrak{se}(3)}{\nabla} : \mathfrak{se}(3) \times \mathfrak{se}(3) \to \mathfrak{se}(3)$ of the affine connection associated with $\mathbb{I}$ is

$$\left(\overset{\mathfrak{se}(3)}{\nabla}_{\widehat{(\boldsymbol{\omega}_1,\boldsymbol{v}_1)}}\widehat{(\boldsymbol{\omega}_2,\boldsymbol{v}_2)}\right)^\vee = \frac{1}{2}\begin{bmatrix} \boldsymbol{\omega}_1 \times \boldsymbol{\omega}_2 + [\mathbb{J}]^{-1}(\boldsymbol{\omega}_1 \times [\mathbb{J}]\boldsymbol{\omega}_2 + \boldsymbol{v}_1 \times [\mathbb{M}]\boldsymbol{v}_2 + \boldsymbol{\omega}_2 \times [\mathbb{J}]\boldsymbol{\omega}_1 + \boldsymbol{v}_2 \times [\mathbb{M}]\boldsymbol{v}_1) \\ \boldsymbol{v}_1 \times \boldsymbol{\omega}_2 + \boldsymbol{\omega}_1 \times \boldsymbol{v}_2 + [\mathbb{M}]^{-1}(\boldsymbol{\omega}_1 \times [\mathbb{M}]\boldsymbol{v}_2 + \boldsymbol{\omega}_2 \times [\mathbb{M}]\boldsymbol{v}_1) \end{bmatrix}. \tag{5.27}$$

In summary the rigid body in ideal fluid is a simple mechanical system on a Lie group described by the pair $(\mathsf{SE}(3),\mathbb{I})$. The Euler–Poincaré equations describing the trajectory of the body velocity are

$$\begin{aligned} \dot{\boldsymbol{\Omega}}(t) + [\mathbb{J}]^{-1}(\boldsymbol{\Omega}(t) \times [\mathbb{J}]\boldsymbol{\Omega}(t) + \boldsymbol{V}(t) \times [\mathbb{M}]\boldsymbol{V}(t)) &= [\mathbb{J}]^{-1}\boldsymbol{f}_{\boldsymbol{\Omega}}(t), \\ \dot{\boldsymbol{V}}(t) + [\mathbb{M}]^{-1}(\boldsymbol{V}(t) \times [\mathbb{M}]\boldsymbol{\Omega}(t)) &= [\mathbb{M}]^{-1}\boldsymbol{f}_{\boldsymbol{V}}(t), \end{aligned} \tag{5.28}$$

for any generalized force $t \mapsto \boldsymbol{f}(t) = (\boldsymbol{f}_{\boldsymbol{\Omega}}(t), \boldsymbol{f}_{\boldsymbol{V}}(t)) \in \mathbb{R}^3 \oplus \mathbb{R}^3 \simeq \mathfrak{se}(3)^*$ acting on the body. These equations are referred to as the ***controlled Kirchhoff equations***. For example, we can assume that the vehicle is endowed with a body-fixed variable-direction thruster applied at a point a distance h from the center of mass, as depicted in Figure 5.49. Then $\boldsymbol{f}(t) = \sum_{a=1}^3 u_a(t)\boldsymbol{f}_a$, where $\boldsymbol{f}_1 = \boldsymbol{e}_4$, $\boldsymbol{f}_2 = -h\boldsymbol{e}_3 + \boldsymbol{e}_5$, and $\boldsymbol{f}_3 = h\boldsymbol{e}_2 + \boldsymbol{e}_6$, and $\{\boldsymbol{e}_1, \dots, \boldsymbol{e}_6\}$ is the standard basis on $\mathbb{R}^6 \simeq \mathbb{R}^3 \oplus \mathbb{R}^3$. •

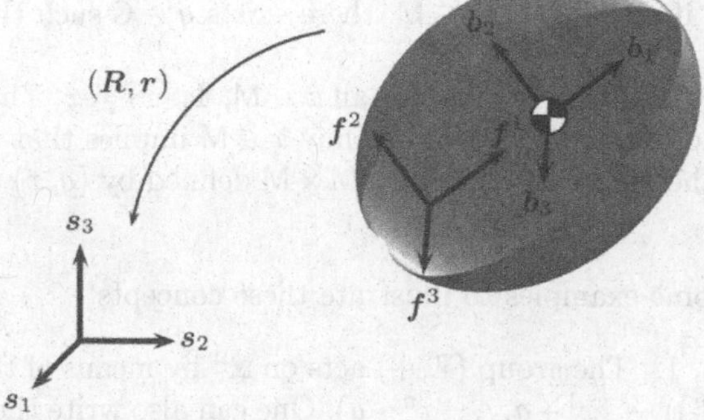

Figure 5.3. Rigid body in SE(3) with three forces applied at a point a distance h from the center of mass

5.4 Group actions, isometries, and symmetries

In this section we present the notion of symmetries and some aspects of reduction theory from a Riemannian geometric point of view. We refer the reader to [Kobayashi and Nomizu 1963, Marsden 1992, Marsden and Ratiu 1999] for more complete treatments of these subjects.

5.4.1 Group actions and infinitesimal generators

We begin with a definition.

Definition 5.50 (Group action). Let M be a manifold and let $(\mathsf{G}, \star)$ be a group. A ***left action of* G *on* M** (or ***left* G*-action on* M**) is a map $\Phi \colon \mathsf{G} \times \mathsf{M} \to \mathsf{M}$ satisfying

$$\Phi(e, x) = x, \quad \Phi(g_1, \Phi(g_2, x)) = \Phi(g_1 \star g_2, x),$$

for all $x \in \mathsf{M}$. Similarly, a ***right action of* G *on* M** (or ***right* G*-action on* M**) is a map $\Phi \colon \mathsf{M} \times \mathsf{G} \to \mathsf{M}$ satisfying

$$\Phi(x, e) = x, \quad \Phi(\Phi(x, g_2), g_1) = \Phi(x, g_2 \star g_1),$$

for all $x \in \mathsf{M}$. A left or right action Φ is ***class C^r*** if Φ is a C^r-map and is ***smooth*** if Φ is of class C^∞. Given a left action Φ and an element $g \in \mathsf{G}$, we let $\Phi_g \colon \mathsf{M} \to \mathsf{M}$ denote the map $x \mapsto \Phi(g, x)$. •

In other words, a left action is a group homomorphism from G to the group Diff(M) of diffeomorphisms of M. Given a left (resp. right) action Φ of G on M, we say that G ***acts*** on the left (resp. on the right) on M via Φ.

Definition 5.51 (Properties of group actions). We say that an action Φ of G on M is

(i) ***faithful*** if, for all $g \in \mathsf{G} \setminus \{e\}$, there exists $x \in \mathsf{M}$ such that $\Phi_g(x) \neq x$;
(ii) ***transitive*** if, for all $x_1, x_2 \in \mathsf{M}$, there exists $g \in \mathsf{G}$ such that $\Phi_g(x_1) = x_2$;
(iii) ***free*** if, for all $g \in \mathsf{G} \setminus \{e\}$ and for all $x \in \mathsf{M}$, $\Phi_g(x) \neq x$. Thus, an action is free if and only if $\Phi_g(x) = x$ for any $x \in \mathsf{M}$ implies that $g = e$;
(iv) ***proper*** if the map from $\mathsf{G} \times \mathsf{M}$ to $\mathsf{M} \times \mathsf{M}$ defined by $(g, x) \mapsto (x, \Phi_g(x))$ is proper. •

Let us give some examples to illustrate these concepts.

Examples 5.52. 1. The group $(\mathbb{R}, +)$ acts on $\mathbb{R}^n$ by means of the left action $(a, (x^1, \dots, x^n)) \mapsto (x^1 + a, \dots, x^n + a)$. One can also write the right action $((x^1, \dots, x^n), a) \mapsto (x^1 + a, \dots, x^n + a)$. These actions are smooth, faithful, free, and proper, but not transitive.
2. The group $\mathsf{SO}(3)$ acts on the left on $\mathbb{R}^3$ via $(\boldsymbol{R}, \boldsymbol{x}) \mapsto \boldsymbol{R}\boldsymbol{x}$. This action is smooth and faithful, but neither free nor transitive. The group $\mathsf{SE}(3)$ acts on the left on $\mathbb{R}^3$ via $((\boldsymbol{R}, \boldsymbol{r}), \boldsymbol{x}) \mapsto \boldsymbol{R}\boldsymbol{x} + \boldsymbol{r}$. This action is smooth, faithful, and transitive, but not free.
3. A Lie group $(\mathsf{G}, \star)$ acts on itself in three smooth ways. The left translation $(g, h) \mapsto L_g(h) = g \star h$ and the conjugation map $(g, h) \mapsto I_g(h) = g \star h \star g^{-1}$ are left actions; the right translation $(h, g) \mapsto R_g(h) = h \star g$ is a right action. Left and right translation are free and transitive; the conjugation map is not free because $I_g(e) = e$ for all $g \in \mathsf{G}$.
4. If X is a smooth complete vector field on a manifold M, then $(t, x) \mapsto \Phi_t^X(x)$ is a smooth left action of $\mathbb{R}$ on M. Note the following confluence of notation: if we denote this left $\mathbb{R}$-action on M by Φ, then $\Phi_t = \Phi_t^X$, for all $t \in \mathbb{R}$. •

When the left action Φ of G on M is smooth, the map $\Phi_g \colon \mathsf{M} \to \mathsf{M}$, $g \in \mathsf{G}$, is a smooth diffeomorphism of M. We shall study the relationship between a left action $\Phi \colon \mathsf{G} \times \mathsf{M} \to \mathsf{M}$ and geometric objects on M. Although not necessary, we shall restrict our study to smooth objects and emphasize this assumption whenever convenient.

Definition 5.53 (Invariance under group action). Let Φ be a smooth left action of G on M.

(i) A function $f \colon \mathsf{M} \to \mathbb{R}$ is ***Φ-invariant*** if, for all $g \in \mathsf{G}$, $f \circ \Phi_g = f$.
(ii) A tensor field t on M is ***Φ-invariant*** if, for all $g \in \mathsf{G}$, $\Phi_g^* t = t$.
(iii) A distribution $\mathcal{D}$ on M is ***Φ-invariant*** if $T_x\Phi_g(\mathcal{D}_x) = \mathcal{D}_{\Phi_g(x)}$ for all $x \in \mathsf{M}$ and $g \in \mathsf{G}$.
(iv) An affine connection ∇ on M is ***Φ-invariant*** if

$$\Phi_g^*(\nabla_X Y) = \nabla_{\Phi_g^* X} \Phi_g^* Y,$$

for all $g \in \mathsf{G}$ and $X, Y \in \Gamma^\infty(\mathsf{TM})$.

(v) A fiber bundle map $F\colon \mathsf{TM} \to \mathsf{T}^*\mathsf{M}$ over id_{M} is ***Φ-equivariant*** if

$$\Phi_g^*(F(X)) = F(\Phi_g^* X),$$

for all $g \in \mathsf{G}$ and $X \in \Gamma^\infty(\mathsf{TM})$. •

As an example of the importance of invariance, let us briefly review a basic fact about integral curves of Φ-invariant vector fields; e.g., see Exercise E3.28.

Lemma 5.54. *Let Φ be a smooth left action of G on M. Let $X \in \Gamma^\infty(\mathsf{TM})$ be Φ-invariant, and let $\gamma\colon I \to \mathsf{M}$ be an integral curve of X. Then the curve $\Phi_g \circ \gamma\colon I \to \mathsf{M}$ is an integral curve of X for all $g \in \mathsf{G}$.*

Proof. We compute

$$\frac{\mathrm{d}}{\mathrm{d}t}\Phi_g \circ \gamma(t) = T_{\gamma(t)}\Phi_g \circ \gamma'(t) = T_{\gamma(t)}\Phi_g \circ X(\gamma(t)) = X \circ \Phi_g(\gamma(t)). \qquad \blacksquare$$

Next, we study the infinitesimal counterpart of group actions. In what follows, we assume that G is a Lie group with Lie algebra $\mathfrak{g}$, and we assume that the left action Φ of G on M is smooth.

Definition 5.55 (Infinitesimal generator). Let Φ be a smooth left action of a Lie group G on M. The ***infinitesimal generator corresponding to $\boldsymbol{\xi} \in \mathfrak{g}$***, is the smooth vector field ξ_{M} on M defined by

$$\xi_{\mathsf{M}}(x) = \left.\frac{\mathrm{d}}{\mathrm{d}t}\right|_{t=0} \Phi(\exp(t\xi), x). \qquad \bullet$$

To illustrate this concept let us give some examples, whose proof is left to the reader.

Examples 5.56. 1. Consider the action $\mathbb{R} \times \mathbb{R}^n \mapsto \mathbb{R}^n$ given by $(a, (x^1, \dots, x^n)) \mapsto (x^1 + a, \dots, x^n + a)$. The infinitesimal generator corresponding to $a \in \mathbb{R}$ is

$$a_{\mathbb{R}^n} = \frac{\partial}{\partial x^1} + \cdots + \frac{\partial}{\partial x^n}.$$

2. Example 5.25–2 introduces the three vector fields X_1, X_2, and X_3 on $\mathbb{R}^3$ defined by

$$X_k = x^j \frac{\partial}{\partial x^i} - x^i \frac{\partial}{\partial x^j},$$

where $(i, j, k) \in \{(1, 2, 3), (2, 3, 1), (3, 1, 2)\}$. Recall that $\{\widehat{\mathbf{e}}_1, \widehat{\mathbf{e}}_2, \widehat{\mathbf{e}}_3\}$ is a basis for $\mathfrak{so}(3)$. For $k \in \{1, 2, 3\}$, the vector field X_k is the infinitesimal generator $(\widehat{\mathbf{e}}_k)_{\mathbb{R}^3}$ of the left action $\mathsf{SO}(3) \times \mathbb{R}^3 \to \mathbb{R}^3$ given by $(\boldsymbol{R}, \boldsymbol{x}) \mapsto \boldsymbol{R}\boldsymbol{x}$.

3. Consider the three actions of G on itself. Let $\xi, \eta \in \mathfrak{g}$. The infinitesimal generator corresponding to ξ for the left action of G on itself is the right-invariant vector field ξ_{R}. The infinitesimal generator corresponding to ξ for the right action of G on itself is the left-invariant vector field ξ_{L}. Finally, the infinitesimal generator corresponding to ξ for the conjugation map is the sum $\xi_{\mathrm{L}} + \xi_{\mathrm{R}}$.
4. If X is a complete vector field on the manifold M, then X is the infinitesimal generator of the action $\mathbb{R} \times \mathsf{M} \to \mathsf{M}$ given by $(t, x) \mapsto \Phi_t^X(x)$. •

Let us illustrate some of the properties of infinitesimal generators.

Proposition 5.57 (Properties of infinitesimal generators). *Let Φ be a smooth, free, and proper left action of a Lie group G on M. The following statements hold:*

(i) for $\xi \in \mathfrak{g}$, the integral curve of $\xi_{\mathsf{M}} \in \Gamma^\infty(\mathsf{TM})$ through $x \in \mathsf{M}$ is $t \mapsto \Phi(\exp(t\xi), x)$;

(ii) the set of infinitesimal generators is a Lie algebra, and the map from $\mathfrak{g}$ to $\Gamma^\infty(\mathsf{TM})$ given by $\xi \mapsto \xi_{\mathsf{M}}$ is linear and satisfies $[\xi, \eta]_{\mathsf{M}} = -[\xi_{\mathsf{M}}, \eta_{\mathsf{M}}]$ for all $\xi, \eta \in \mathfrak{g}$.

Proof. By definition of left action and of infinitesimal generator,

$$\begin{aligned}
\left.\frac{\mathrm{d}}{\mathrm{d}t}\right|_{t=t_0} \Phi(\exp(t\xi), x) &= \left.\frac{\mathrm{d}}{\mathrm{d}t}\right|_{t=t_0} \Phi(\exp((t-t_0)\xi), \Phi(\exp(t_0\xi), x)) \\
&= \left.\frac{\mathrm{d}}{\mathrm{d}s}\right|_{s=0} \Phi(\exp(s\xi), \Phi(\exp(t_0\xi), x)) \\
&= \xi_{\mathsf{M}}(\Phi(\exp(t_0\xi), x)).
\end{aligned}$$

This proves part (i). Next, for $\xi, \eta \in \mathfrak{g}$, we apply Proposition 3.69 to infinitesimal generators on M:

$$\begin{aligned}
[\xi_{\mathsf{M}}, \eta_{\mathsf{M}}](x) &= \left.\frac{\mathrm{d}}{\mathrm{d}t}\right|_{t=0} \Phi^{-\eta_{\mathsf{M}}}_{\sqrt{t}} \circ \Phi^{-\xi_{\mathsf{M}}}_{\sqrt{t}} \circ \Phi^{\eta_{\mathsf{M}}}_{\sqrt{t}} \circ \Phi^{\xi_{\mathsf{M}}}_{\sqrt{t}}(x) \\
&= \left.\frac{\mathrm{d}}{\mathrm{d}t}\right|_{t=0} \Phi\big(\exp(-\sqrt{t}\eta) \star \exp(-\sqrt{t}\xi) \star \exp(\sqrt{t}\eta) \star \exp(\sqrt{t}\xi), x\big).
\end{aligned}$$

Recall from Theorem 5.32(ii) that $\Phi_t^{\xi_{\mathrm{L}}}(g) = g \star \exp(\xi t)$ for all $g \in \mathsf{G}$. Therefore,

$$\begin{aligned}
&\exp(-\sqrt{t}\eta) \star \exp(-\sqrt{t}\xi) \star \exp(\sqrt{t}\eta) \star \exp(\sqrt{t}\xi) \\
&\qquad = \Phi^{\xi_{\mathrm{L}}}_{\sqrt{t}} \circ \Phi^{\eta_{\mathrm{L}}}_{\sqrt{t}} \circ \Phi^{-\xi_{\mathrm{L}}}_{\sqrt{t}} \circ \Phi^{-\eta_{\mathrm{L}}}_{\sqrt{t}}(e).
\end{aligned}$$

Fact (ii) follows from again applying Proposition 3.69 to left-invariant vector fields on G:

$$\left.\frac{\mathrm{d}}{\mathrm{d}t}\right|_{t=0} \Phi^{\xi_{\mathrm{L}}}_{\sqrt{t}} \circ \Phi^{\eta_{\mathrm{L}}}_{\sqrt{t}} \circ \Phi^{-\xi_{\mathrm{L}}}_{\sqrt{t}} \circ \Phi^{-\eta_{\mathrm{L}}}_{\sqrt{t}}(e) = [-\eta, -\xi] = -[\xi, \eta].$$

■

Given a group action Φ, the notion of Φ-invariance has an infinitesimal counterpart for smooth objects. Let us first define appropriate notions of invariance with respect to a vector field.

Definition 5.58 (Invariance under vector field). Let X be a smooth vector field on M.

(i) A function $f\colon \mathsf{M} \to \mathbb{R}$ is ***X-invariant*** if $\mathscr{L}_X f = 0$.
(ii) A tensor field t on M is ***X-invariant*** if $\mathscr{L}_X t = 0$.
(iii) A distribution $\mathcal{D}$ on M is ***X-invariant*** if $\mathscr{L}_X Y \in \Gamma^\infty(\mathcal{D})$ for all $Y \in \Gamma^\infty(\mathcal{D})$.
(iv) An affine connection ∇ on M is ***X-invariant*** if, for all $Y, Z \in \Gamma^\infty(\mathsf{TM})$,

$$\mathscr{L}_X(\nabla_Y Z) = \nabla_{\mathscr{L}_X Y} Z + \nabla_X(\mathscr{L}_Y Z).$$

(v) A fiber bundle map $F\colon \mathsf{TM} \to \mathsf{T}^*\mathsf{M}$ over id_{M} is ***X-equivariant*** if, for all $Y \in \Gamma^\infty(\mathsf{TM})$,

$$\mathscr{L}_X(F(Y)) = F(\mathscr{L}_X Y).$$ •

Now we can establish the following equivalences. It is convenient to adopt the following convention: we shall say that an object is invariant under the action Φ or under the vector field X if that object is Φ-invariant or X-invariant, respectively.

Theorem 5.59 (Invariance under actions and infinitesimal generators I). *Let let Φ be a smooth left G-action on M and X be a smooth complete vector field on M. Assume that all following data is smooth. A function, tensor field, regular distribution, or affine connection on M is:*

(i) Φ-invariant only if it is invariant under all infinitesimal generators of Φ;
(ii) Φ-invariant if it is invariant under all infinitesimal generators of Φ and if G is connected;
(iii) X-invariant if and only if it is invariant under the $\mathbb{R}$-action $(t, q) \mapsto \Phi_t^X(q)$.

The same three statements hold for a fiber bundle map on TM if, in the statements, "invariant" is replaced with "equivariant."

Proof. Let us prove fact (i) for a tensor field τ on M. We need to show that $\Phi^*\tau = \tau$ implies $\mathscr{L}_{\xi_{\mathsf{M}}}\tau = 0$, for all $\xi \in \mathfrak{g}$. By Propositions 3.85 and 5.57 we compute

$$(\Phi_t^{\xi_{\mathsf{M}}})^* \mathscr{L}_{\xi_{\mathsf{M}}}\tau = \frac{\mathrm{d}}{\mathrm{d}t}(\Phi_t^{\xi_{\mathsf{M}}})^*\tau = \frac{\mathrm{d}}{\mathrm{d}t}(\Phi_{\exp(t\xi)})^*\tau = \frac{\mathrm{d}}{\mathrm{d}t}\tau = 0.$$

Analogous proofs can be provided for regular distributions and affine connections. The proof of the converse statement (ii) is less direct and appears difficult to find in the literature. We present only a brief outline in what follows. If $\mathscr{L}_{\xi_{\mathsf{M}}}\tau = 0$, then $(\Phi_{\exp(t\xi)})^*\tau = \tau$ for all $t \in \mathbb{R}$. This means that τ is invariant on all one-parameter subgroups of G. In order to show that τ is invariant under Φ_g for all $g \in \mathsf{G}$, it suffices to show that g is the product of a finite number of elements of the form $\exp(t\xi_k)$, for $\xi_k \in \mathfrak{g}$. This is a consequence of the surjectivity of the exponential map in a neighborhood of the origin (see Theorem 5.32(iii)) and of the connectivity of G (see [Warner 1971, Proposition 3.18]). Fact (iii) is a consequence of facts (i) and (ii). ■

Finally, it is useful to consider the tangent map to a group action and its infinitesimal version. As usual, let Φ be a smooth left action of G on M. The ***lifted action*** $\Phi^T: \mathsf{G} \times \mathsf{TM} \to \mathsf{TM}$ is the left action of G on TM defined by

$$\Phi^T(g, v_x) = T_x\Phi_g(v_x).$$

It is immediate to verify that Φ^T is indeed a left action. As usual, we let $\Phi_g^T(v_x) = \Phi^T(g, v_x)$, for all $g \in \mathsf{G}$ and $v_x \in \mathsf{TM}$. For all $g \in \mathsf{G}$, the lifted action satisfies the following commutative diagram:

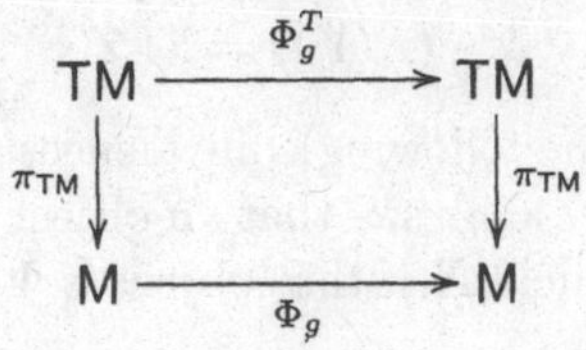

Additionally, we can state the following result.

Lemma 5.60. *Let Φ be a smooth left action of a Lie group G on M. For all $\xi \in \mathfrak{g}$, the infinitesimal generator $\xi_{\mathsf{TM}} \in \Gamma^\infty(\mathsf{TTM})$ of the lifted action $\Phi^T: \mathsf{G} \times \mathsf{TM} \to \mathsf{TM}$ is equal to the tangent lift $\xi_{\mathsf{M}}^T \in \Gamma^\infty(\mathsf{TTM})$ of the corresponding infinitesimal generator of $\Phi: \mathsf{G} \times \mathsf{M} \to \mathsf{M}$.*

Proof. The reader is invited to prove this result in Exercise E5.28. ■

5.4.2 Isometries

We begin introducing the notion of isometry that naturally arises in the study of Riemannian manifolds.

Definition 5.61 (Isometry). Let $(\mathsf{M}, \mathbb{G})$ and $(\tilde{\mathsf{M}}, \tilde{\mathbb{G}})$ be Riemannian manifolds.

(i) An ***isometry*** is a smooth diffeomorphism $\varphi: \mathsf{M} \to \tilde{\mathsf{M}}$ that satisfies $\varphi^*\tilde{\mathbb{G}} = \mathbb{G}$.

(ii) If $\varphi: \mathsf{M} \to \tilde{\mathsf{M}}$ is an isometry, then M and $\tilde{\mathsf{M}}$ are said to be ***isometric*** and one writes $\mathsf{M} \simeq \tilde{\mathsf{M}}$.

(iii) An isometry $\varphi: \mathsf{M} \to \mathsf{M}$ is an ***isometry of*** M. The ***isometry group of*** M is the set of isometries of M. •

Note that the set of isometries of a Riemannian manifold $(\mathsf{M}, \mathbb{G})$ is a group because the composition of isometries and the inverse of an isometry are again isometries of M.

It can be shown that the isometry group of a manifold M is always isomorphic to a finite-dimensional Lie group acting smoothly on M; see [Kobayashi and Nomizu 1963, Theorem VI.3.4]. We provide an example of this fact in Proposition 5.2 where $\mathsf{E}(3)$ is shown to be the isometry group of $(\mathbb{R}^3, \mathbb{G}_{\mathbb{R}^3})$. Let us be more precise on what we mean by a group action that consist of isometries.

Definition 5.62 (Infinitesimal isometry). Let $(\mathsf{M}, \mathbb{G})$ be a Riemannian manifold.

(i) A smooth left action Φ of a Lie group G on M is ***isometric*** if $\Phi_g^*\mathbb{G} = \mathbb{G}$ for all $g \in \mathsf{G}$. In this case, we say that G acts on M by isometries.

(ii) A smooth vector field X is an ***infinitesimal isometry*** (or, a ***Killing vector field***) of $(\mathsf{M}, \mathbb{G})$ if $\mathscr{L}_X\mathbb{G} = 0$. •

As in the previous subsection, let us study the relationship between isometric group actions and their infinitesimal counterparts. The following statements are direct consequences of Theorem 5.59.

Theorem 5.63 (Invariance under actions and infinitesimal generators II). *Let* $(\mathsf{M}, \mathbb{G})$ *be a* C^∞*-Riemannian manifold. Let* Φ *be a smooth left* G*-action on* M *and* X *be a smooth complete vector field on* M*. The following statements hold:*

(i) if Φ *is an isometry of* $(\mathsf{M}, \mathbb{G})$*, then all its infinitesimal generators are complete infinitesimal isometries of* $(\mathsf{M}, \mathbb{G})$*. Conversely, if all the infinitesimal generators of* Φ *are complete infinitesimal isometries of* $(\mathsf{M}, \mathbb{G})$ *and if* G *is connected, then* Φ *is an isometry of* $(\mathsf{M}, \mathbb{G})$*;*

(ii) X *is an infinitesimal isometry of* $(\mathsf{M}, \mathbb{G})$ *if and only if the* $\mathbb{R}$*-action* $(t, q) \mapsto \Phi_t^X(q)$ *is an isometry of* $(\mathsf{M}, \mathbb{G})$*.*

Note that if the Riemannian manifold $(\mathsf{M}, \mathbb{G})$ is complete, then any infinitesimal isometry is a complete vector field.

Given $X \in \Gamma^\infty(\mathsf{TM})$, recall that the covariant differential $\overset{\mathbb{G}}{\nabla} X : \Gamma^\infty(\mathsf{TM}) \to \Gamma^\infty(\mathsf{TM})$ is the $(1,1)$-tensor field given by $Y \mapsto \overset{\mathbb{G}}{\nabla}_Y X$; see Section 3.8. Let us now present some properties of infinitesimal isometries.

Proposition 5.64 (Characterization of infinitesimal isometry). *Let* $(\mathsf{M}, \mathbb{G})$ *be a Riemannian manifold. The following statements hold:*

(i) the smooth vector field X *is an infinitesimal isometry if and only if the covariant differential* $\overset{\mathbb{G}}{\nabla} X$ *is skew-symmetric with respect to* $\mathbb{G}$*; that is, for all* $Y, Z \in \Gamma^\infty(\mathsf{TM})$*,*

$$\mathbb{G}(Y, \overset{\mathbb{G}}{\nabla}_Z X) + \mathbb{G}(Z, \overset{\mathbb{G}}{\nabla}_Y X) = 0; \tag{5.29}$$

(ii) if the vector field X is an infinitesimal isometry, then the smooth function $x \mapsto \|X\|^2_{\mathbb{G}}(x)$ is X-invariant and satisfies

$$\tfrac{1}{2}\mathrm{grad}\,\|X\|^2_{\mathbb{G}} = -\overset{\mathbb{G}}{\nabla}_X X. \tag{5.30}$$

Proof. Regarding part (i), we compute

$$\begin{aligned}\mathbb{G}(Y, \overset{\mathbb{G}}{\nabla}_Z X) + \mathbb{G}(Z, \overset{\mathbb{G}}{\nabla}_Y X) &= \mathbb{G}(Y, \overset{\mathbb{G}}{\nabla}_X Z + [Z, X]) + \mathbb{G}(Z, \overset{\mathbb{G}}{\nabla}_X Y + [Y, X]) \\ &= \mathscr{L}_X(\mathbb{G}(Y, Z)) - \mathbb{G}(Y, [X, Z]) - \mathbb{G}(Z, [X, Y]) \\ &= (\mathscr{L}_X \mathbb{G})(Y, Z).\end{aligned}$$

The result follows since this equality holds for all $Y, Z \in \Gamma^\infty(\mathsf{TM})$.

Regarding part (ii), both facts are immediate consequences of the skew-symmetry (5.29) of the $(1,1)$-tensor field $\overset{\mathbb{G}}{\nabla} X$. To prove equation (5.30), let $Z \in \Gamma^\infty(\mathsf{TM})$ and compute

$$\langle\langle Z, \mathrm{grad}\,\|X\|^2_{\mathbb{G}} \rangle\rangle = \mathscr{L}_Z \|X\|^2_{\mathbb{G}} = 2\langle\langle X, \overset{\mathbb{G}}{\nabla}_Z X \rangle\rangle = -2\langle\langle Z, \overset{\mathbb{G}}{\nabla}_X X \rangle\rangle.$$

To show that $x \mapsto \|X\|^2_{\mathbb{G}}(x)$ is X-invariant, note that

$$\mathscr{L}_X \|X\|^2_{\mathbb{G}} = -2\langle\langle X, \overset{\mathbb{G}}{\nabla}_X X \rangle\rangle = 0. \qquad \blacksquare$$

Remark 5.65. Let $\{X_k\}_{k \in \{1,\dots,l\}}$ be infinitesimal isometries of $(\mathsf{M}, \mathbb{G})$. It is interesting to note that $C^\infty(\mathsf{M})$-linear combinations of infinitesimal isometries are, in general, not infinitesimal isometries. Let us illustrate why this is the case. For $f^k \in C^\infty(\mathsf{M})$, $k \in \{1,\dots,l\}$, we compute

$$\overset{\mathbb{G}}{\nabla}(f^k X_k) = f^k \overset{\mathbb{G}}{\nabla} X_k + X_k \otimes \mathrm{d}f^k,$$

where, according to the definition of tensor product,

$$\begin{aligned} X_k \otimes \mathrm{d}f^k \colon \Gamma^\infty(\mathsf{TM}) &\to \Gamma^\infty(\mathsf{TM}) \\ Z &\mapsto \langle \mathrm{d}f^k; Z \rangle X_k. \end{aligned}$$

From this calculation we deduce that the vector field $f^k X_k$ is an infinitesimal isometry only if the $(1,1)$-tensor field $X_k \otimes \mathrm{d}f^k$ is skew-symmetric. A sufficient condition is for $f^k \in C^\infty(\mathsf{M})$, $k \in \{1,\dots,l\}$ to be constant functions. In general, however, $f^k X_k$ is not an infinitesimal isometry. •

5.4.3 Symmetries and conservation laws

In this section we study the relationship between group actions and mechanical control systems. It is instructive to consider actions that are isometries for the kinetic energy metric and that leave the potential function and the constraints invariant. We focus on the notion of infinitesimal symmetry and show how the corresponding momentum map gives rise to conserved quantities for the mechanical system.

Definition 5.66 (Symmetry). Let $\Sigma = (\mathsf{Q}, \mathbb{G}, V, F, \mathcal{D}, \{F^1, \dots, F^m\}, U)$ be a C^∞-general simple mechanical control system.

(i) A smooth left action Φ of a Lie group G on Q is a ***symmetry*** of Σ if Φ is an isometry of $(\mathsf{Q}, \mathbb{G})$, if V, $\mathcal{D}$, and $\{F^1, \dots, F^m\}$ are Φ-invariant, and if F is Φ-equivariant.

(ii) A smooth vector field X on Q is an ***infinitesimal symmetry*** of Σ if X is an infinitesimal isometry of $(\mathsf{Q}, \mathbb{G})$, if V, $\mathcal{D}$, and $\{F^1, \dots, F^m\}$ are X-invariant, and if F is X-equivariant. •

As in the previous subsections, let us study the relationship between symmetries and their infinitesimal counterparts. The following statements are direct consequences of Theorems 5.59 and 5.63.

Theorem 5.67 (Invariance under actions and infinitesimal generators III). *Let $\Sigma = (\mathsf{Q}, \mathbb{G}, V, F, \mathcal{D}, \{F^1, \dots, F^m\}, U)$ be a C^∞-general simple mechanical control system with a regular constraint $\mathcal{D}$. Let Φ be a smooth left G-action on Q and X be a smooth complete vector field on Q. The following statements hold:*

(i) if Φ is a symmetry of Σ, then all its infinitesimal generators are complete infinitesimal symmetries of Σ. Conversely, if all the infinitesimal generators of Φ are complete infinitesimal symmetries of Σ and if G is connected, then Φ is a symmetry of Σ;

(ii) X is an infinitesimal symmetry of Σ if and only if the $\mathbb{R}$-action $(t, q) \mapsto \Phi^X_t(q)$ is a symmetry of Σ.

Remarkably, symmetries of mechanical control systems give rise, under certain conditions, to conserved quantities. A well-known example of this fact is the conservation law for the spatial angular momentum of a rigid body. To formalize this fact we introduce the notion of momentum map for systems without constraints.

Definition 5.68 (Momentum map). Let $(\mathsf{Q}, \mathbb{G}, V, F, \{F^1, \dots, F^m\}, U)$ be a C^∞-forced simple mechanical control system.

(i) Given a symmetry $\Phi\colon \mathsf{G} \times \mathsf{Q} \to \mathsf{Q}$, the ***$\Phi$-momentum map*** $J_\Phi\colon \mathsf{TQ} \to \mathfrak{g}^*$ is defined by
$$\langle J_\Phi(v_q); \xi\rangle = \langle\langle \xi_{\mathsf{Q}}(q), v_q\rangle\rangle\,,$$
for all $\xi \in \mathfrak{g}$.

(ii) Given an infinitesimal symmetry X, the ***X-momentum map*** $J_X\colon \mathsf{TQ} \to \mathbb{R}$ is
$$J_X(v_q) = \langle\langle X(q), v_q\rangle\rangle\,.$$ •

Note that both momentum maps are smooth. Next, it is convenient to restrict our attention to systems *without* control forces. Recall that the forced simple mechanical system $(\mathsf{Q}, \mathbb{G}, V, F)$ gives rise to the following governing equations for the trajectory $\gamma\colon I \to \mathsf{Q}$:

$$\overset{\mathbb{G}}{\nabla}_{\gamma'(t)}\gamma'(t) = -\operatorname{grad}V(\gamma(t)) + \mathbb{G}^\sharp(F(t, \gamma(t))). \tag{5.31}$$

Theorem 5.69 (Noether Conservation Law). *Let* $(\mathsf{Q}, \mathbb{G}, V, F)$ *be a* C^∞*-forced simple mechanical system and let* $\gamma\colon I \to \mathsf{Q}$ *be a solution to the equations of motion* (5.31)*. The following statements hold:*

(i) if $\Phi\colon \mathsf{G} \times \mathsf{Q} \to \mathsf{Q}$ *is a symmetry, then, for all* $\xi \in \mathfrak{g}$*,*

$$\frac{\mathrm{d}}{\mathrm{d}t} \langle J_\Phi(\gamma'(t)); \xi\rangle = \langle F(t, \gamma'(t)); \xi_{\mathsf{Q}}(\gamma(t))\rangle\,; \tag{5.32}$$

(ii) if X *is an infinitesimal symmetry, then*

$$\frac{\mathrm{d}}{\mathrm{d}t} J_X(\gamma'(t)) = \langle F(t, \gamma'(t)); X(\gamma(t))\rangle\,. \tag{5.33}$$

Proof. We first prove part (ii). Because $\overset{\mathbb{G}}{\nabla}$ is a Levi-Civita affine connection (see Exercise E3.39), we have

$$\begin{aligned}\frac{\mathrm{d}}{\mathrm{d}t} J_X(\gamma'(t)) &= \frac{\mathrm{d}}{\mathrm{d}t} \langle\langle \gamma'(t), X(\gamma(t))\rangle\rangle \\ &= \langle\langle \overset{\mathbb{G}}{\nabla}_{\gamma'(t)}\gamma'(t), X(\gamma(t))\rangle\rangle + \langle\langle \gamma'(t), \overset{\mathbb{G}}{\nabla}_{\gamma'(t)} X(\gamma(t))\rangle\rangle \\ &= \langle\langle -\mathrm{grad} V(\gamma(t)) + \mathbb{G}^\sharp(F(t, \gamma'(t))), X(\gamma(t))\rangle\rangle \\ &\quad + \langle\langle \gamma'(t), \overset{\mathbb{G}}{\nabla}_{\gamma'(t)} X(\gamma(t))\rangle\rangle.\end{aligned}$$

Fact (ii) follows from the equality $\mathscr{L}_X V = 0$ and from the skew-symmetry of $\overset{\mathbb{G}}{\nabla} X$. For all $\xi \in \mathfrak{g}$, we compute $J_{\xi_{\mathsf{Q}}}(\gamma'(t)) = \langle\langle \xi_{\mathsf{Q}}(\gamma(t)), \gamma'(t)\rangle\rangle = \langle J_\Phi(\gamma'(t)); \xi\rangle$ and part (i) follows. ■

In other words, for a simple mechanical system with no external force, no constraints, and no control forces, the momentum map associated with an infinitesimal symmetry X or a symmetry Φ is a constant function along a trajectory γ of the simple mechanical system. In this sense we refer to the momentum map as a ***constant of motion***. For mechanical systems with constraints, the existence of symmetries does not, in general, give rise to conserved quantities. We refer the readers to [Bloch 2003, Bloch, Krishnaprasad, Marsden, and Murray 1996, Cendra, Marsden, and Ratiu 2001b], and invite them to explore this fact in Exercise E5.29.

Finally, we investigate those mechanical quantities that are invariant under isometries. We refer the reader to [Abraham and Marsden 1978, Kobayashi and Nomizu 1963, Lewis 1998] for related discussions.

Theorem 5.70 (Invariance under actions and infinitesimal generators IV). *Let* $\Sigma = (\mathsf{Q}, \mathbb{G}, V, F, \mathcal{D}, \{F^1, \dots, F^m\}, U)$ *be a* C^∞*-general simple mechanical control system with a regular constraint* $\mathcal{D}$*. The following statements hold.*

(i) If Φ *is a symmetry of* Σ*, then the Levi-Civita affine connection associated with* $(\mathsf{Q}, \mathbb{G})$*, and the constrained affine connection associated with*

$(\mathsf{Q}, \mathbb{G}, \mathcal{D})$, *are* Φ*-invariant, the energy function is* Φ^T*-invariant, and the* Φ*-momentum map satisfies* $\mathrm{Ad}_g^* J_\Phi(v_q) = J_\Phi(T_q\Phi_g(v_q))$ *for all* $g \in \mathsf{G}$ *and* $v_q \in \mathsf{TQ}$.

(ii) If X *is an infinitesimal symmetry of* Σ*, then the Levi-Civita affine connection associated with* $(\mathsf{Q}, \mathbb{G})$*, and the constrained affine connection associated with* $(\mathsf{Q}, \mathbb{G}, \mathcal{D})$*, are* X*-invariant, and the energy function and the* X*-momentum map are* X^T*-invariant.*

Proof. These statements are direct consequences of Theorems 5.59, 5.63, and 5.67. The statements regarding the Levi-Civita affine connection are proved via (3.12). ■

5.4.4 Examples of mechanical systems with symmetries

In this section we apply the results on symmetries and momentum maps to a few example systems.

Example 5.71 (Robotic leg). This example is depicted in Figure 5.4. The

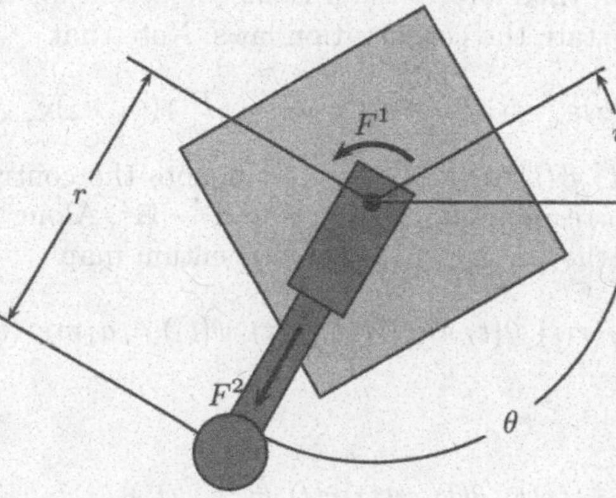

Figure 5.4. The robotic leg

main body rotates about a fixed point, and at that fixed point is attached an extensible leg with a point mass on its tip. We assume that the direction of the gravitational field is orthogonal to the plane of motion of the system. There are two control forces: a torque applied at the point of rotation that controls the *relative* angle between the body and the leg, and a force that extends the leg. This system seems to have been first studied, at least in the form we consider, by Li, Montgomery, and Raibert [1989].

Let us simply present the data resulting from an application of the methodology outlined in Chapter 4. We will represent this system as an affine connection control system. The configuration manifold for the system is $\mathsf{Q} = \mathbb{R}_+ \times \mathbb{S}^1 \times \mathbb{S}^1$, and the coordinates we use are (r, θ, ψ) as indicated in Figure 5.4. The Riemannian metric for the system is

$$\mathbb{G} = m(\mathrm{d}r \otimes \mathrm{d}r + r^2 \mathrm{d}\theta \otimes \mathrm{d}\theta) + J\mathrm{d}\psi \otimes \mathrm{d}\psi,$$

where m is the mass of the particle on the end of the extensible leg, and J is the moment of inertia of the base rigid body about the pivot point. The input forces are $F^1 = \mathrm{d}\theta - \mathrm{d}\psi$ and $F^2 = \mathrm{d}r$. Therefore, the robotic leg is described by the simple mechanical control system $(\mathbb{R}_+ \times \mathbb{S}^1 \times \mathbb{S}^1, \mathbb{G}, V = 0, \{F^1, F^2\})$.

Let us consider the left action $\Phi\colon (\mathsf{SO}(2) \times \mathsf{SO}(2)) \times (\mathbb{R}_+ \times \mathbb{S}^1 \times \mathbb{S}^1) \to (\mathbb{R}_+ \times \mathbb{S}^1 \times \mathbb{S}^1)$ given by $((\boldsymbol{R}_1, \boldsymbol{R}_2), (r, \boldsymbol{x}, \boldsymbol{y})) \mapsto (r, \boldsymbol{R}_1\boldsymbol{x}, \boldsymbol{R}_2\boldsymbol{y})$. Note the natural isomorphism of Lie algebras $\mathfrak{so}(2) \simeq \mathbb{R}$. Let $(a_1, a_2) \in \mathbb{R}^2$ be an arbitrary element of $\mathfrak{so}(2) \times \mathfrak{so}(2)$, and compute the corresponding infinitesimal generator in the coordinates (r, θ_1, θ_2) to be

$$(a_1, a_2)_{\mathbb{R}_+ \times \mathbb{S}^1 \times \mathbb{S}^1} = a_1 \frac{\partial}{\partial \theta} + a_2 \frac{\partial}{\partial \psi}.$$

One can compute $\mathscr{L}_{(a_1,a_2)_{\mathbb{R}_+ \times \mathbb{S}^1 \times \mathbb{S}^1}} \mathbb{G} = 0$ and $\mathscr{L}_{(a_1,a_2)_{\mathbb{R}_+ \times \mathbb{S}^1 \times \mathbb{S}^1}} F^i = 0$, $i \in \{1, 2\}$. This implies that $(a_1, a_2)_{\mathbb{R}_+ \times \mathbb{S}^1 \times \mathbb{S}^1}$ is an infinitesimal isometry for all $(a_1, a_2) \in \mathbb{R}^2$, and, therefore, Φ is an isometric action for the robotic leg.

Next, we can state the conservation laws. Note that

$$\langle F^1; (a_1, a_2)_{\mathbb{R}_+ \times \mathbb{S}^1 \times \mathbb{S}^1} \rangle = a_1 - a_2, \quad \langle F^2; (a_1, a_2)_{\mathbb{R}_+ \times \mathbb{S}^1 \times \mathbb{S}^1} \rangle = 0.$$

Let $t \mapsto (r(t), \theta(t), \psi(t)) \in \mathbb{R}_+ \times \mathbb{S}^1 \times \mathbb{S}^1$ denote the controlled trajectory of the robotic leg corresponding to control $u\colon I \to \mathbb{R}^2$. Along this trajectory, for all $(a_1, a_2) \in \mathbb{R}^2$, the $(a_1, a_2)_{\mathbb{R}_+ \times \mathbb{S}^1 \times \mathbb{S}^1}$-momentum map

$$J_{(a_1,a_2)_{\mathbb{R}_+ \times \mathbb{S}^1 \times \mathbb{S}^1}}(r(t), \theta(t), \psi(t), \dot{r}(t), \dot{\theta}(t), \dot{\psi}(t)) = a_1 m r^2(t) \dot{\theta}(t) + a_2 J \dot{\psi}(t)$$

satisfies

$$\begin{aligned}\frac{\mathrm{d}}{\mathrm{d}t} & J_{(a_1,a_2)_{\mathbb{R}_+ \times \mathbb{S}^1 \times \mathbb{S}^1}}(r(t), \theta(t), \psi(t), \dot{r}(t), \dot{\theta}(t), \dot{\psi}(t)) \\ &= \sum_{a=1}^{2} \langle F^a; (a_1, a_2)_{\mathbb{R}_+ \times \mathbb{S}^1 \times \mathbb{S}^1} \rangle u^a(t) = (a_1 - a_2) u^1(t).\end{aligned}$$ •

Example 5.72 (Spherical pendulum with gravity). We consider a spherical pendulum with mass m; see Figure 5.5. This is a simple mechanical control system on the two-dimensional sphere $\mathbb{S}^2$. Its kinetic energy metric is $m\mathbb{G}_{\mathrm{rnd}}$, where $\mathbb{G}_{\mathrm{rnd}}$ is the round Riemannian metric on $\mathbb{S}^2$ induced by the standard Riemannian metric $\mathbb{G}_{\mathbb{R}^3}$ on $\mathbb{R}^3$; see Example 4.99. Assuming that gravity is aligned with the third reference axis $\boldsymbol{s}_3$, the potential function $V\colon \mathbb{S}^2 \to \mathbb{R}$ is $\boldsymbol{x} \mapsto m a_g \langle\langle \boldsymbol{s}_3, \boldsymbol{x} \rangle\rangle_{\mathbb{R}^2}$, where a_g is the acceleration due to gravity. Therefore, the spherical pendulum with gravity is described by the simple mechanical system $(\mathbb{S}^2, m\mathbb{G}_{\mathrm{rnd}}, V)$.

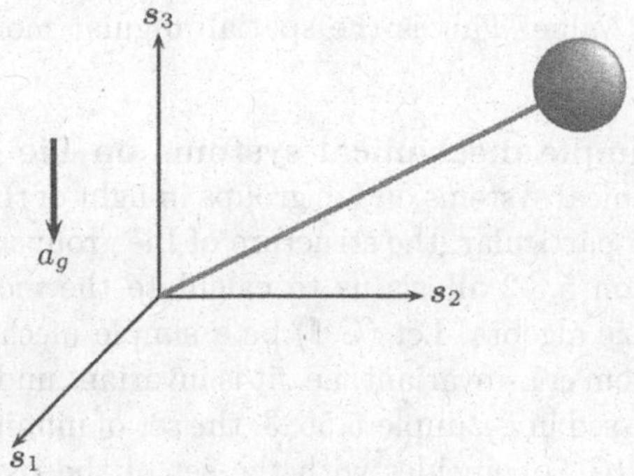

Figure 5.5. Spherical pendulum in a gravitational field

Let $\Phi\colon \mathsf{SO}(3) \times \mathbb{R}^3 \to \mathbb{R}^3$ be the standard $\mathsf{SO}(3)$ left action on $\mathbb{R}^3$; see Example 5.52–2. This action induces by restriction an action of $\mathsf{SO}(3)$ on $\mathbb{S}^2$ that we denote $\Phi|\mathbb{S}^2\colon \mathsf{SO}(3) \times \mathbb{S}^2 \to \mathbb{S}^2$. It is now possible to show that $\Phi|\mathbb{S}^2$ is a symmetry for $(\mathbb{S}^2, m\mathbb{G}_{\mathrm{rnd}})$. Recall that $\Phi\colon \mathsf{SO}(3) \times \mathbb{R}^3 \to \mathbb{R}^3$ is an isometry, and recall that $\mathbb{G}_{\mathrm{rnd}} = i_{\mathbb{S}^2}^* \mathbb{G}_{\mathbb{R}^3}$, where $i_{\mathbb{S}^2}\colon \mathbb{S}^2 \to \mathbb{R}^3$ is the inclusion map. For $\boldsymbol{R} \in \mathsf{SO}(3)$, we compute

$$(\Phi_{\boldsymbol{R}}|\mathbb{S}^2)^* \mathbb{G}_{\mathrm{rnd}} = (\Phi_{\boldsymbol{R}}|\mathbb{S}^2)^* i_{\mathbb{S}^2}^* \mathbb{G}_{\mathbb{R}^3} = (i_{\mathbb{S}^2} \circ \Phi_{\boldsymbol{R}}|\mathbb{S}^2)^* \mathbb{G}_{\mathbb{R}^3},$$

and, because the map $i_{\mathbb{S}^2} \circ \Phi_{\boldsymbol{R}}|\mathbb{S}^2\colon \mathbb{S}^2 \to \mathbb{R}^3$ is equal to $\Phi_{\boldsymbol{R}} \circ i_{\mathbb{S}^2}\colon \mathbb{S}^2 \to \mathbb{R}^3$, we conclude that

$$(i_{\mathbb{S}^2} \circ \Phi_{\boldsymbol{R}}|\mathbb{S}^2)^* \mathbb{G}_{\mathbb{R}^3} = (\Phi_{\boldsymbol{R}} \circ i_{\mathbb{S}^2})^* \mathbb{G}_{\mathbb{R}^3} = (i_{\mathbb{S}^2})^* \Phi_{\boldsymbol{R}}^* \mathbb{G}_{\mathbb{R}^3} = (i_{\mathbb{S}^2})^* \mathbb{G}_{\mathbb{R}^3} = \mathbb{G}_{\mathrm{rnd}}.$$

This shows that $\Phi|\mathbb{S}^2$ is indeed a symmetry for $(\mathbb{S}^2, m\mathbb{G}_{\mathrm{rnd}})$; Exercise E5.34 asks the reader to verify this fact in spherical coordinates.

Next, one can show that the potential function V is invariant under rotations about the vertical axis and not invariant under any other rotation. Accordingly, the action $\Phi|\mathbb{S}^2$ is not a symmetry for the simple mechanical system $(\mathbb{S}^2, m\mathbb{G}_{\mathrm{rnd}}, V)$. Let us instead consider the left action of the subgroup $\{\exp(t\widehat{\boldsymbol{s}}_3) \in \mathsf{SO}(3) \mid t \in [0, 2\pi[\,\} \simeq \mathsf{SO}(2)$ on $\mathbb{S}^2$ given by left multiplication. Given the corresponding Lie algebra element $\widehat{\boldsymbol{s}}_3 \in \mathfrak{so}(3)$, the infinitesimal symmetry is

$$(\widehat{\boldsymbol{s}}_3)_{\mathbb{S}^2}(\boldsymbol{x}) = x^1 \frac{\partial}{\partial x^2} - x^2 \frac{\partial}{\partial x^1},$$

for $\boldsymbol{x} = (x^1, x^2, x^3) \in \mathbb{S}^2 \subset \mathbb{R}^3$. This result is consistent with Example 5.56–2. Finally, we can state the conservation law induced by the infinitesimal symmetry $(\widehat{\boldsymbol{s}}_3)_{\mathbb{S}^2}$. Along a trajectory $t \mapsto (\gamma^1(t), \gamma^2(t), \gamma^3(t)) \in \mathbb{S}^2$ of the spherical pendulum with gravity, the momentum map

$$J_{(\widehat{\boldsymbol{s}}_3)_{\mathbb{S}^2}}(\gamma'(t)) = \gamma^1(t)\dot{\gamma}^2(t) - \gamma^2(t)\dot{\gamma}^1(t)$$

assumes a constant value. This is the spatial angular momentum about the s_3-axis. •

Example 5.73 (Simple mechanical systems on Lie groups). Here we study simple mechanical systems on Lie groups in light of the general results in Proposition 5.70. In particular, the structure of Lie groups and of Lie algebras introduced in Section 5.3.2 allows us to calculate the momentum maps via operations on the Lie algebra. Let $(\mathsf{G}, \mathbb{I})$ be a simple mechanical system on a Lie group. This system is L-invariant, i.e., it is invariant under the left action of G on itself. As discussed in Example 5.56–3, the set of infinitesimal symmetries of the action L on $(\mathsf{G}, \mathbb{I})$ coincides with the set of the right-invariant vector fields. For $v_g \in \mathsf{T}_g\mathsf{G}$ and $\xi \in \mathfrak{g}$, we compute the ξ_{R}-momentum map and the L-momentum map as

$$\begin{aligned} J_{\xi_{\mathrm{R}}}(v_g) &= \langle\!\langle \xi_{\mathrm{R}}(g), v_g \rangle\!\rangle = \langle \mathbb{I}^\flat(\mathrm{Ad}_{g^{-1}}(\xi)); T_gL_g^{-1}(v_g)\rangle, \\ J_L(v_g) &= \mathrm{Ad}^*_{g^{-1}}(\mathbb{I}^\flat(T_gL_g^{-1}(v_g))). \end{aligned}$$

In particular, along the trajectory $t \mapsto \gamma(t) \in \mathsf{G}$, introducing the body velocity $t \mapsto v(t) = T_{\gamma(t)}L_{\gamma(t)^{-1}}(\gamma'(t)) \in \mathfrak{g}$, the momentum maps satisfy

$$\begin{aligned} J_{\xi_{\mathrm{R}}}(\gamma'(t)) &= \langle \mathrm{Ad}^*_{\gamma^{-1}(t)}(\mathbb{I}^\flat(v(t))); \xi \rangle, \\ J_L(\gamma'(t)) &= \mathrm{Ad}^*_{\gamma^{-1}(t)}(\mathbb{I}^\flat(v(t))). \end{aligned}$$ •

Example 5.74 (Rigid body fixed at a point (cont'd)). Finally, let us apply the concepts in Example 5.73 to a rigid body fixed at a point. The pair $(\mathsf{SO}(3), \mathbb{I})$ is the corresponding simple mechanical system on a Lie group, as described in Example 5.48. The governing equations in $t \mapsto (\boldsymbol{R}(t), \boldsymbol{\Omega}(t))$ are

$$\dot{\boldsymbol{R}}(t) = \boldsymbol{R}(t)\widehat{\boldsymbol{\Omega}}(t), \quad \dot{\boldsymbol{\Omega}}(t) + [\mathbb{I}]^{-1}(\boldsymbol{\Omega}(t) \times [\mathbb{I}]\boldsymbol{\Omega}(t)) = 0.$$

Along trajectories $t \mapsto (\boldsymbol{R}(t), \boldsymbol{\Omega}(t))$, the L-momentum map

$$t \mapsto \mathrm{Ad}^*_{\boldsymbol{R}^{-1}(t)}([\mathbb{I}](\boldsymbol{\Omega}(t))) = \boldsymbol{R}(t)[\mathbb{I}]\boldsymbol{\Omega}(t) \in \mathbb{R}^3$$

is a conserved quantity. This is the total spatial angular momentum for the body. •

5.5 Principal bundles and reduction

We refer the reader to [Abraham and Marsden 1978, Cendra, Marsden, and Ratiu 2001b, Kobayashi and Nomizu 1963, Marsden 1992, Marsden and Ratiu 1999] for detailed discussions of principal fiber bundles and reduction. A prerequisite for the material in this section is Section 3.9.5.

5.5.1 Principal fiber bundles

Let G be a Lie group with Lie algebra $\mathfrak{g}$, and let Φ be a smooth left action of G on M. The left action Φ of G on M induces an equivalence relation in M whereby $x_1 \sim x_2$ if $x_1 = \Phi(g, x_2)$ for some $g \in \mathsf{G}$. For $x_0 \in \mathsf{M}$, the ***Φ-orbit*** through x_0 is the subset of M given by

$$\mathcal{O}(x_0, \Phi) = \{\, \Phi(g, x_0) \mid \ g \in \mathsf{G}\}.$$

(This notion is analogous to our definition of orbits for distributions in Section 3.7.3.) In other words, the set $\mathcal{O}(x_0, \Phi)$ is the equivalence class $\{x \in \mathsf{M} \mid \ x \sim x_0\}$. We shall denote by M/G the set of all Φ-orbits, which we call the ***quotient space*** defined by the group action Φ.

Examples 5.75. 1. Consider the left action $(a, (x^1, \dots, x^n)) \mapsto (x^1 + a, \dots, x^n + a)$ of $(\mathbb{R}, +)$ on $\mathbb{R}^n$. The orbit of this action through $(x^1, \dots, x^n) \in \mathbb{R}^n$ is uniquely specified by the point $(0, x^2 - x^1, \dots, x^n - x^1)$ on the orbit. The quotient space $\mathbb{R}^n/\mathbb{R}$ is therefore naturally diffeomorphic to $\mathbb{R}^{n-1}$.
2. The group $(\mathbb{Z}^n, +)$ acts on the left on $\mathbb{R}^n$ by $((k^1, \dots, k^n), (x^1, \dots, x^n)) \mapsto (x^1 + k^1, \dots, x^n + k^n)$. The orbit of this action through $(x^1, \dots, x^n) \in \mathbb{R}^n$ is $(x^1 + k^1, \dots, x^n + k^n)$ for $k^i \in \mathbb{Z}$. The quotient set is $\mathbb{R}^n/\mathbb{Z}^n \simeq \mathbb{T}^n$; see Example 3.26.
3. Consider the left action $\mathsf{SO}(3) \times \mathbb{R}^3 \to \mathbb{R}^3$ given by $(\boldsymbol{R}, \boldsymbol{x}) \mapsto \boldsymbol{R}\boldsymbol{x}$. The orbit of this action through $\boldsymbol{x}$ is the sphere of radius $\|\boldsymbol{x}\|_{\mathbb{R}^3}$. The quotient set is $\mathbb{R}^3/\mathsf{SO}(3) \simeq \bar{\mathbb{R}}_+$. •

Definition 5.76. A ***C^∞-principal fiber bundle*** is a 4-tuple $(\mathsf{M}, \mathsf{G}, \Phi, \mathsf{B})$, where
1. M and B are manifolds, and G is a Lie group,
2. Φ is a smooth, free, and proper left action of G on M for which $\mathsf{B} = \mathsf{M}/\mathsf{G}$, and
3. there exists an atlas $\mathscr{A} = \{(\mathcal{U}_a, \phi_a)\}_{a \in A}$ for B such that, for each $a \in A$, there is a C^∞-diffeomorphism $\psi_a \colon \pi^{-1}(\mathcal{U}_a) \to \mathcal{U}_a \times \mathsf{G}$ given by $\psi_a(x) = (\pi(x), \varphi_a(x))$, for which $\varphi_a \colon \pi^{-1}(\mathcal{U}_a) \to \mathsf{G}$ satisfies $\varphi_a(\Phi(g, x)) = g \star \varphi_a(x)$ for all $x \in \pi^{-1}(\mathcal{U}_a)$ and $g \in \mathsf{G}$.

We call M the ***total space***, B the ***base space***, and G the ***structure group***. •

In a principal fiber bundle $(\mathsf{M}, \mathsf{G}, \Phi, \mathsf{B})$, the projection $\pi \colon \mathsf{M} \to \mathsf{B}$ is a surjective submersion. Note that this nomenclature is appropriate since a C^∞-principal fiber bundle is a C^∞-locally trivial fiber bundle as defined in Section 3.4. We refer to [Kobayashi and Nomizu 1963] for a detailed discussion of principal fiber bundles.

Let us give some examples to illustrate these concepts.

Examples 5.77. 1. Let ϕ be the left action of $(\mathbb{R}, +)$ on $\mathbb{R}^n$ given by $(a, (x^1, \dots, x^n)) \mapsto (x^1 + a, \dots, x^n + a)$. Because this action is free and proper, the quotient space $\mathbb{R}^n/\mathbb{R} \simeq \mathbb{R}^{n-1}$ is a manifold. The 4-tuple $(\mathbb{R}^n, \mathbb{R}, \phi, \mathbb{R}^n/\mathbb{R})$ is a principal fiber bundle.

2. Let $\mathrm{Rot}_{\boldsymbol{e}_1}$ be the left action of $\mathsf{SO}(2)$ on $\mathsf{SO}(3)$ given by

$$\mathrm{Rot}_{\boldsymbol{e}_1}\left(\begin{bmatrix}\cos(\theta) & -\sin(\theta)\\ \sin(\theta) & \cos(\theta)\end{bmatrix}, \boldsymbol{R}\right) = \begin{bmatrix}1 & 0 & 0\\ 0 & \cos(\theta) & -\sin(\theta)\\ 0 & \sin(\theta) & \cos(\theta)\end{bmatrix} \boldsymbol{R}.$$

In the quotient set $\mathsf{SO}(3)/\mathsf{SO}(2)$, two rotations are equivalent when they are equal up to a left multiplication by a rotation about $\boldsymbol{e}_1$. One can show that $\mathsf{SO}(3)/\mathsf{SO}(2)$ is diffeomorphic to $\mathbb{S}^2$ via the diffeomorphism that maps the $\mathrm{Rot}_{\boldsymbol{e}_1}$-orbit through $\boldsymbol{R}$ to the vector $\boldsymbol{R}^T\boldsymbol{e}_1 \in \mathbb{S}^2$. The action $\mathrm{Rot}_{\boldsymbol{e}_1}$ is smooth, free, and proper. Therefore, the quotient space $\mathsf{SO}(3)/\mathsf{SO}(2)$ is a manifold and $(\mathsf{SO}(3), \mathsf{SO}(2), \mathrm{Rot}_{\boldsymbol{e}_1}, \mathsf{SO}(3)/\mathsf{SO}(2))$ is a principal fiber bundle. This bundle is related to the Hopf fibration; see [Abraham, Marsden, and Ratiu 1988].

3. Let H be a regular Lie subgroup of the Lie group G. The subgroup H acts on the left on G in the natural way; $L^{\mathsf{H}} \colon \mathsf{H} \times \mathsf{G} \to \mathsf{G}$ defined by $(h, g) \mapsto L_h(g) = hg$. This action is free and proper. We can write the quotient space as $\mathsf{G}/\mathsf{H} = \{\mathsf{H}g \mid g \in \mathsf{G}\}$, where $\mathsf{H}g = \{hg \mid h \in \mathsf{H}\}$. The 4-tuple $(\mathsf{G}, \mathsf{H}, L^{\mathsf{H}}, \mathsf{G}/\mathsf{H})$ is a principal fiber bundle. •

5.5.2 Reduction by an infinitesimal isometry

In this section we shall investigate the dynamics of a C^∞-simple mechanical system with a complete infinitesimal symmetry. As mentioned in the introduction, our treatment is limited in scope; for a more comprehensive treatment we refer the reader to the related work in [Cendra, Marsden, and Ratiu 2001a].

We begin with some preliminary concepts. Given a complete vector field X on Q, the ***X-orbit*** through q_0 is the set $\mathcal{O}(q_0, X) = \{\Phi^X_t(q_0) \mid t \in \mathbb{R}\}$. Note that the X-orbit through q_0 is equal to the orbit through q_0 of the distribution generated by X and to the orbit through q_0 of the left $\mathbb{R}$-action $(t, q) \mapsto \Phi^X_t(q)$.

Assumption 5.78. Let X be a complete infinitesimal symmetry for a C^∞-simple mechanical system $(\mathsf{Q}, \mathbb{G}, V)$. We assume that the set of X-orbits is a manifold, say B, and that the projection $\pi_{\mathsf{B}} \colon \mathsf{Q} \to \mathsf{B}$ is a surjective submersion. •

The following result gives an example of when these assumptions are valid.

Lemma 5.79. *Let $(\mathsf{Q}, \mathbb{G}, V)$ be a C^∞-simple mechanical system, let X be a complete vector field on Q, and let $q_0 \in \mathsf{Q}$ have the property that the integral curve of X through q_0 is an embedding. Then there exists an X-invariant* *neighborhood* $\mathcal{U}$ *of* $\mathcal{O}(q_0, X)$ *such that the set of* X*-orbits contained in* $\mathcal{U}$ *is a manifold, and the projection from* $\mathcal{U}$ *to this set of* X*-orbits is a surjective submersion.*

Note that, in order for Assumption 5.78 to hold, it is not sufficient for X to be the infinitesimal generator of a smooth, free, and proper action; the reader is invited to explore this issue in Exercise E5.32. In this section we investigate

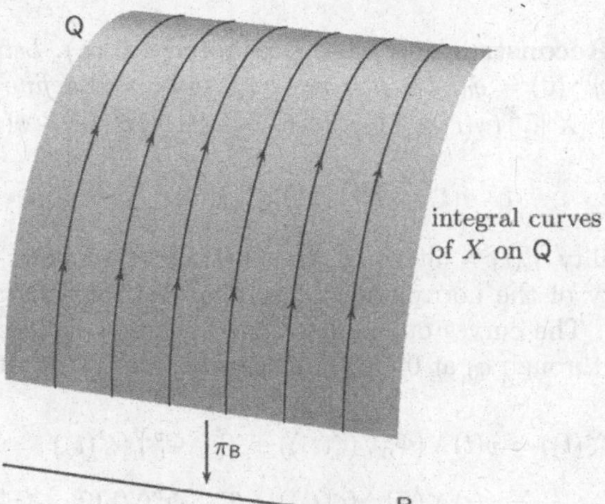

Figure 5.6. The bundle of X-orbits

the consequences of this structure on mechanical control systems.

Now we introduce some useful definitions. The ***vertical distribution*** VQ is the distribution on Q generated by the vector field X. The ***horizontal distribution*** HQ is the $\mathbb{G}$-orthogonal complement of VQ so that $\mathsf{TQ} = \mathsf{HQ} \oplus \mathsf{VQ}$. The vertical and horizontal distributions are smooth, regular, and locally finitely generated because of Assumption 5.78.

At each $q \in \mathsf{Q}$, the linear map $T_q\pi_{\mathsf{B}} \colon \mathsf{T}_q\mathsf{Q} \to \mathsf{T}_{\pi_{\mathsf{B}}(q)}\mathsf{B}$ is a surjection. Therefore, the map $T_q\pi_{\mathsf{B}}|\mathsf{H}_q\mathsf{Q} \colon \mathsf{H}_q\mathsf{Q} \to \mathsf{T}_{\pi_{\mathsf{B}}(q)}\mathsf{B}$ is a linear isomorphism. The ***horizontal lift*** of the tangent vector $v_b \in \mathsf{T}_b\mathsf{B}$ at point $q \in \mathsf{Q}$ is the tangent vector in $\mathsf{T}_q\mathsf{Q}$ given by

$$\mathrm{hlft}_q(v_b) = \left(T_q\pi_{\mathsf{B}}|\mathsf{H}_q\mathsf{Q}\right)^{-1}(v_b). \tag{5.34}$$

Furthermore, the ***horizontal lift*** of the vector field $Y \in \Gamma^\infty(\mathsf{TB})$ is the vector field $\mathrm{hlft}(Y) \in \Gamma^\infty(\mathsf{TQ})$ defined by $\mathrm{hlft}(Y)(q) = \mathrm{hlft}_q(Y(\pi_{\mathsf{B}}(q)))$. This vector field is X-invariant and takes values in HQ.

Lemma 5.80 (Horizontal lift of curves). *Given* $q_0 \in \mathsf{Q}$, *an interval* I *containing* 0, *and a* C^1*-curve* $\eta \colon I \to \mathsf{B}$, *there exists a unique* C^1*-curve* $\mathrm{hlft}_{q_0}(\eta) \colon I \to \mathsf{Q}$ *such that*

(i) $\pi_{\mathsf{B}} \circ \mathrm{hlft}_{q_0}(\eta) = \eta$,
(ii) $\mathrm{hlft}_{q_0}(\eta)'(t) \in \mathsf{HQ}$ *for all* $t \in I$, *and*
(iii) $\mathrm{hlft}_{q_0}(\eta)(0) = q_0$.

Proof. Such a curve $\zeta\colon I \to \mathsf{Q}$ needs to satisfy $\zeta'(t) = \mathrm{hlft}_{\zeta(t)}(\eta'(t))$ and $\zeta(0) = q_0$. The result follows from the existence and uniqueness of solutions to ordinary differential equations. ■

Recall the notion of X-momentum map from Definition 5.68.

Lemma 5.81 (Reconstruction from projected curve). *Let the* C^1*-curve* $\gamma\colon I \to \mathsf{Q}$ *satisfy* $\gamma(0) = q_0$. *Let* $\eta = \pi_{\mathsf{B}} \circ \gamma\colon I \to \mathsf{B}$, *and define* $v\colon I \to \mathbb{R}$ *by* $v(t) = J_X(\gamma'(t)) \|X\|_{\mathbb{G}}^{-2}(\gamma(t))$. *Then* $\gamma'(t) = \mathrm{hlft}_{\gamma(t)}(\eta'(t)) + v(t)X(\gamma(t))$ *for all* $t \in I$ *and*

$$\gamma(t) = \Phi_{0,t}^{vX}(\mathrm{hlft}_{q_0}(\eta)(t)). \tag{5.35}$$

Proof. The equality $\gamma'(t) = \mathrm{hlft}_{\gamma(t)}(\eta'(t)) + v(t)X(\gamma(t))$ follows directly from the orthogonality of the horizontal and vertical distributions and from the definition of J_X. The curves on the left-hand and right-hand sides of equation (5.35) pass through q_0 at $0 \in I$. Furthermore, for $\zeta(t) = \mathrm{hlft}_{q_0}(\eta)(t)$,

$$\begin{aligned}
\frac{\mathrm{d}}{\mathrm{d}t}\Phi_{0,t}^{vX}(\zeta(t)) &= v(t)X(\Phi_{0,t}^{vX}(\zeta(t))) + T_{\zeta(t)}\Phi_{0,t}^{vX}(\zeta'(t)) \\
&= v(t)X(\Phi_{0,t}^{vX}(\zeta(t))) + T_{\zeta(t)}\Phi_{0,t}^{vX}(\mathrm{hlft}_{\zeta(t)}(\eta'(t))) \\
&= v(t)X(\Phi_{0,t}^{vX}(\zeta(t))) + \mathrm{hlft}_{\Phi_{0,t}^{vX}(\zeta(t))}(\eta'(t)).
\end{aligned}$$

The result now follows from the existence and uniqueness of solutions to ordinary differential equations. ■

Next, we note that is possible to project certain X-invariant objects from Q onto B in natural ways. For example, given the Riemannian metric $\mathbb{G}$ on Q, we define the Riemannian metric $\mathbb{G}_{\mathsf{B}}$ on B, called the ***projected metric***, by

$$\mathbb{G}_{\mathsf{B}}(b)(v_b, w_b) = \mathbb{G}(\mathrm{hlft}(v_b), \mathrm{hlft}(w_b)),$$

for $b \in \mathsf{B}$ and $v_b, w_b \in \mathsf{T}_b\mathsf{B}$. Let $\overset{\mathbb{G}_{\mathsf{B}}}{\nabla}$ be the Levi-Civita affine connection on the Riemannian manifold $(\mathsf{B}, \mathbb{G}_{\mathsf{B}})$ and, given a function $f\colon \mathsf{B} \to \mathbb{R}$, let $\mathrm{grad}_{\mathsf{B}}f$ be its gradient with respect to the metric $\mathbb{G}_{\mathsf{B}}$. Given an X-invariant function $h\colon \mathsf{Q} \to \mathbb{R}$, let $h_{\mathsf{B}}\colon \mathsf{B} \to \mathbb{R}$ be its projection defined by $h_{\mathsf{B}}([q]) = h(q)$. One can then show that

$$T\pi_{\mathsf{B}} \circ \mathrm{grad}h = \mathrm{grad}_{\mathsf{B}} h_{\mathsf{B}}. \tag{5.36}$$

We will also find useful the following result.

Lemma 5.82. *For all* $Y, Z \in \Gamma^\infty(\mathsf{TB})$, $T\pi_{\mathsf{B}}(\overset{\mathbb{G}}{\nabla}_{\mathrm{hlft}(Y)}\mathrm{hlft}(Z)) = \overset{\mathbb{G}_{\mathsf{B}}}{\nabla}_Y Z$.

Proof. The reader is invited to prove this result in Exercise E5.33. ■

Let us now return to the study of a simple mechanical system $(\mathsf{Q}, \mathbb{G}, V)$ with infinitesimal symmetry X. We need to introduce two final concepts. Firstly, for $\lambda \in \mathbb{R}$, we define the ***effective potential function*** $V^{\text{eff}}_{X,\lambda} \colon \mathsf{Q} \to \mathbb{R}$ by

$$V^{\text{eff}}_{X,\lambda}(q) = V(q) - \frac{\lambda^2}{2} \|X\|^2_{\mathbb{G}}(q),$$

and the ***amended potential function*** $V^{\text{amd}}_{X,\lambda} \colon \mathsf{Q} \to \mathbb{R}$ by

$$V^{\text{amd}}_{X,\lambda}(q) = V(q) + \frac{\lambda^2}{2} \|X\|^{-2}_{\mathbb{G}}(q).$$

By Proposition 5.64, the effective potential and the amended potential are X-invariant functions on Q. Second, we define the ***gyroscopic tensor*** C_X as the $(1,1)$-tensor field on B such that, for any $q \in \pi_{\mathsf{B}}^{-1}(b)$ and $v_b \in \mathsf{T}_b\mathsf{B}$

$$C_X(v_b) = -2T_q\pi_{\mathsf{B}}\Big(\overset{\mathbb{G}}{\nabla}_{\text{hlft}_q(v_b)}X(q)\Big).$$

One can show that C_X is well-defined in the sense that the choice of q is immaterial, and that C_X is skew-symmetric with respect to $\mathbb{G}_{\mathsf{B}}$; that is, for all $v_b, w_B \in \mathsf{T}_b\mathsf{B}$,

$$\mathbb{G}_{\mathsf{B}}(v_b, C_X(w_b)) + \mathbb{G}_{\mathsf{B}}(w_b, C_X(v_b)) = 0.$$

The skew-symmetry of C_X is an immediate consequence of the skew-symmetry of $\overset{\mathbb{G}}{\nabla}X$. We are finally ready to state the main result of this section.

Theorem 5.83 (Reduced dynamics). *Let $(\mathsf{Q}, \mathbb{G}, V)$ be a C^∞-simple mechanical system with a complete infinitesimal symmetry X satisfying Assumption 5.78. The following statements about the C^1-curves $\gamma \colon I \to \mathsf{Q}$, $\eta \colon I \to \mathsf{B}$, $v \colon I \to \mathbb{R}$, and $\mu \colon I \to \mathbb{R}$ are equivalent:*

(i) γ satisfies

$$\begin{aligned}\overset{\mathbb{G}}{\nabla}_{\gamma'(t)}\gamma'(t) &= -\operatorname{grad}V(\gamma(t)),\\ \gamma'(0) &= v_{q_0} \in \mathsf{T}_{q_0}\mathsf{Q},\end{aligned}$$

and, in turn, η, μ, and v are defined by $\eta(t) = \pi_{\mathsf{B}} \circ \gamma(t)$, $\mu(t) = J_X \circ \gamma'(t)$, and $v(t) = (J_X \circ \gamma'(t))(\|X\|^{-2}_{\mathbb{G}} \circ \gamma(t))$, respectively;

(ii) η and v together satisfy

$$\begin{aligned}\overset{\mathbb{G}_{\mathsf{B}}}{\nabla}_{\eta'(t)}\eta'(t) &= -\operatorname{grad}_{\mathsf{B}}\big(V^{\text{eff}}_{X,v(t)}\big)_{\mathsf{B}}(\eta(t)) + v(t)C_X(\eta'(t)),\\ \dot v(t) &= -\frac{v(t)}{(\|X\|^2_{\mathbb{G}})_{\mathsf{B}}(\eta(t))}\langle \mathrm{d}(\|X\|^2_{\mathbb{G}})_{\mathsf{B}}(\eta(t)); \eta'(t)\rangle,\\ \eta'(0) = T_{q_0}\pi_{\mathsf{B}}(v_{q_0}), &\quad v(0) = J_X(v_{q_0})\,\|X\|^{-2}_{\mathbb{G}}(q_0),\end{aligned}$$

and, in turn, γ and μ are defined by $\gamma(t) = \Phi^{vX}_{0,t}(\text{hlft}_{q_0}(\eta)(t))$, and $\mu(t) = v(t)((\|X\|^2_{\mathbb{G}})_{\mathsf{B}} \circ \eta(t))$, respectively;

(iii) η and μ together satisfy

$$\begin{aligned}\overset{\mathsf{G}_\mathsf{B}}{\nabla}_{\eta'(t)}\eta'(t) &= -\operatorname{grad}_\mathsf{B}\left(V^{\mathrm{amd}}_{X,\mu(t)}\right)_\mathsf{B}(\eta(t)) + \frac{\mu(t)}{(\|X\|^2_\mathsf{G})_\mathsf{B}(\eta(t))}C_X(\eta'(t)),\\ \dot{\mu}(t) &= 0,\\ \eta'(0) &= T_{q_0}\pi_\mathsf{B}(v_{q_0}), \quad \mu(0) = J_X(v_{q_0}),\end{aligned}$$

and, in turn, v and γ are defined by $v(t) = \mu(t)((\|X\|^{-2}_\mathsf{G})_\mathsf{B}\circ\eta(t))$ and $\gamma(t) = \Phi^{vX}_{0,t}(\mathrm{hlft}_{q_0}(\eta)(t))$, respectively.

Proof. Let us prove that fact (i) implies, and is implied by, fact (ii). Let $\{Y_1,\dots,Y_{n-1}\}$ be a family of vector fields that forms a basis for each tangent space in an open set $\mathcal{U}\subset\mathsf{B}$. We write $\gamma'(t) = u^k(t)\mathrm{hlft}(Y_k)(\gamma(t)) + v(t)X(\gamma(t))$ for some functions $u^k\colon I\to\mathbb{R}$, $k\in\{1,\dots,n-1\}$, and compute

$$\begin{aligned}\overset{\mathsf{G}}{\nabla}_{\gamma'(t)}\gamma'(t) = {}& \dot{u}^k(t)\mathrm{hlft}(Y_k)(\gamma(t)) + \dot{v}(t)X(\gamma(t))\\ &+ u^k(t)\overset{\mathsf{G}}{\nabla}_{\gamma'(t)}\mathrm{hlft}(Y_k)(\gamma(t)) + v(t)\overset{\mathsf{G}}{\nabla}_{\gamma'(t)}X(\gamma(t)).\end{aligned}$$

The equality $\eta = \pi_\mathsf{B}\circ\gamma$ implies that $\eta'(t) = \mathsf{T}_{\gamma(t)}\pi_\mathsf{B}(\gamma'(t))$, which in coordinates reads $\dot{\eta}^k(t) = u^k(t)$, $k\in\{1,\dots,n-1\}$. Now we project the previous equation onto TB to obtain

$$\begin{aligned}&T_{\gamma(t)}\pi_\mathsf{B}\left(\overset{\mathsf{G}}{\nabla}_{\gamma'(t)}\gamma'(t)\right)\\ &\quad= \ddot{\eta}^k(t)Y_k(t) + \dot{\eta}^k(t)T_{\gamma(t)}\pi_\mathsf{B}\left(\overset{\mathsf{G}}{\nabla}_{\gamma'(t)}\mathrm{hlft}(Y_k)(\gamma(t))\right)\\ &\qquad+ v(t)T_{\gamma(t)}\pi_\mathsf{B}\left(\overset{\mathsf{G}}{\nabla}_{\gamma'(t)}X(\gamma(t))\right)\\ &\quad= \ddot{\eta}^k(t)Y_k(\gamma(t))\\ &\qquad+ \dot{\eta}^k(t)T_{\gamma(t)}\pi_\mathsf{B}\left(\dot{\eta}^j(t)\overset{\mathsf{G}}{\nabla}_{\mathrm{hlft}(Y_j)}\mathrm{hlft}(Y_k)(\gamma(t)) + v(t)\overset{\mathsf{G}}{\nabla}_X\mathrm{hlft}(Y_k)(\gamma(t))\right)\\ &\qquad+ v(t)T_{\gamma(t)}\pi_\mathsf{B}\left(\dot{\eta}^j(t)\overset{\mathsf{G}}{\nabla}_{\mathrm{hlft}(Y_j)}X(\gamma(t)) + v(t)\overset{\mathsf{G}}{\nabla}_XX(\gamma(t))\right)\\ &\quad= \left(\ddot{\eta}^k(t)Y_k(t) + \dot{\eta}^k(t)\dot{\eta}^j(t)\overset{\mathsf{G}_\mathsf{B}}{\nabla}_{Y_j}Y_k(\eta(t))\right)\\ &\qquad+ 2v(t)\dot{\eta}^k(t)T_{\gamma(t)}\pi_\mathsf{B}\left(\overset{\mathsf{G}}{\nabla}_{\mathrm{hlft}(Y_k)}X(\gamma(t))\right) + v^2(t)T_{\gamma(t)}\pi_\mathsf{B}\left(\overset{\mathsf{G}}{\nabla}_XX(\gamma(t))\right),\end{aligned}$$

where we have used Lemma 5.82 and the X-invariance of $\mathrm{hlft}(Y_k)$, $k\in\{1,\dots,n-1\}$, i.e., $\overset{\mathsf{G}}{\nabla}_{\mathrm{hlft}(Y_k)}X = \overset{\mathsf{G}}{\nabla}_{\mathrm{hlft}(X)}Y_k$. From Proposition 5.64, equation (5.36), and the definition of C_X, we obtain

$$\begin{aligned}T_{\gamma(t)}\pi_\mathsf{B}(\overset{\mathsf{G}}{\nabla}_{\gamma'(t)}\gamma'(t)) = {}& \overset{\mathsf{G}_\mathsf{B}}{\nabla}_{\eta'(t)}\eta'(t) - v(t)C_X(\eta'(t))\\ &- v^2(t)\operatorname{grad}_\mathsf{B}\left(\tfrac{1}{2}\|X\|^2_\mathsf{G}\right)_\mathsf{B}(\eta(t)).\end{aligned}$$

Next, note that $\mu = J_X \circ \gamma' = v\,(\|X\|^2_{\mathbb{G}})_{\mathsf{B}} \circ \eta$. Therefore, the vertical component of γ' satisfies

$$\frac{\mathrm{d}}{\mathrm{d}t}\mu(t) = \dot{v}(t)(\|X\|^2_{\mathbb{G}})_{\mathsf{B}}(\eta(t)) + v(t)\langle \mathrm{d}(\|X\|^2_{\mathbb{G}})_{\mathsf{B}}(\eta(t)); \eta'(t)\rangle.$$

These computations show the equivalence of statements (i) and (ii). The equivalence between statements (ii) and (iii) follows, after some bookkeeping, from the equality

$$\mathrm{grad}_{\mathsf{B}}\left(\|X\|^2_{\mathbb{G}} \right)_{\mathsf{B}} = -\left(\|X\|^4_{\mathbb{G}} \right)_{\mathsf{B}} \mathrm{grad}_{\mathsf{B}}\left(\|X\|^{-2}_{\mathbb{G}} \right)_{\mathsf{B}}. \qquad \blacksquare$$

Remark 5.84 (Forced simple mechanical systems). It is immediate to extend the results in the theorem to the setting of forced simple mechanical systems. The uncontrolled external force F will in general appear in both the horizontal and the vertical equations, so that the equations in Theorem 5.83(ii) read

$$\begin{aligned} \overset{\mathbb{G}_{\mathsf{B}}}{\nabla}_{\eta'(t)}\eta'(t) &= -\mathrm{grad}_{\mathsf{B}}\left(V^{\mathrm{eff}}_{X,v}\right)_{\mathsf{B}}(\eta(t)) + v(t)C_X(\eta'(t)) + T\pi_{\mathsf{B}}\left(\mathbb{G}^\sharp(F(t,\gamma'(t)))\right), \\ \dot{v}(t) &= -\frac{v(t)\langle \mathrm{d}(\|X\|^2_{\mathbb{G}})_{\mathsf{B}}(\eta(t)); \eta'(t)\rangle}{(\|X\|^2_{\mathbb{G}})_{\mathsf{B}}(\eta(t))} + \frac{1}{(\|X\|^2_{\mathbb{G}})_{\mathsf{B}}(\eta(t))}\langle F(t,\gamma'(t)); X(\gamma(t))\rangle. \end{aligned}$$

•

We conclude this section by illustrating the results of Theorem 5.83 in an example system.

Example 5.85 (Spherical pendulum with gravity). This example is a continuation of Example 5.72. We consider the simple mechanical system $(\mathbb{S}^2, m\mathbb{G}_{\mathrm{rnd}}, V)$ depicted in Figure 5.5. The infinitesimal symmetry X is the infinitesimal generator $(\widehat{\boldsymbol{s}}_3)_{\mathbb{S}^2}$, that we write as

$$X(\boldsymbol{x}) = x^1 \frac{\partial}{\partial x^2} - x^2 \frac{\partial}{\partial x^1},$$

for $\boldsymbol{x} = (x^1, x^2, x^3) \in \mathbb{S}^2$. The X-orbit $\mathcal{O}(\boldsymbol{x}, X)$ through a point $\boldsymbol{x} \in \mathbb{S}^2$ is the intersection of $\mathbb{S}^2$ with a horizontal plane at height $x^3 = \langle\langle \boldsymbol{s}_3, \boldsymbol{x}\rangle\rangle_{\mathbb{R}^3}$ (see Figure 5.7). In other words, we write $\mathcal{O}(\boldsymbol{x}, X) = \left\{ \boldsymbol{y} \in \mathbb{S}^2 \mid y^3 = x^3 \right\}$.

Note that this infinitesimal symmetry X corresponds to a left action of $\mathsf{SO}(2)$ on $\mathbb{S}^2$ that is not free, and the quotient space $\mathbb{S}^2/\mathsf{SO}(2)$ is not a principal fiber bundle. Therefore, the vector field X on $\mathbb{S}^2$ fails to satisfy Assumption 5.78 and it is not possible to apply Theorem 5.83 directly. To deal with this obstacle, we make the preliminary assumption that the trajectory of the simple mechanical system $(\mathbb{S}^2, m\mathbb{G}_{\mathrm{rnd}}, V)$ is restricted to the manifold $\mathbb{S}^2 \setminus \{(0,0,-1),(0,0,1)\}$. We then define

$$\mathsf{B} = \left\{ \mathcal{O}(\boldsymbol{x}, X) \mid \boldsymbol{x} \in \mathbb{S}^2 \setminus \{(0,0,-1),(0,0,1)\} \right\}.$$

Note that B is a manifold, it is diffeomorphic to the interval $]-1,1[$, and it possesses a global coordinate chart (B, ϕ) defined by $\phi(\mathcal{O}(\boldsymbol{x}, X)) = x^3$. In this

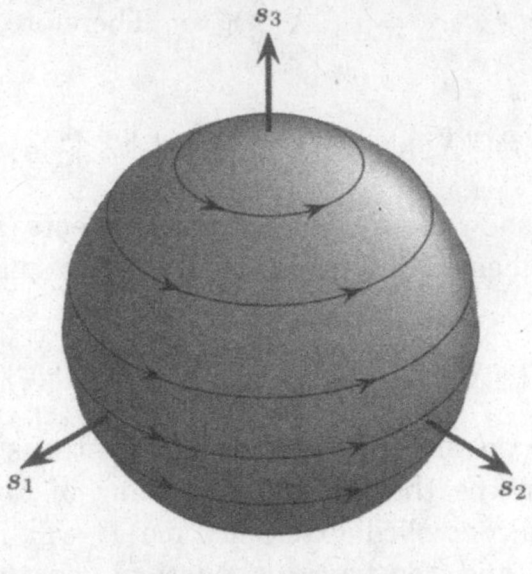

Figure 5.7. Integral curves of X

coordinate chart, the projection $\pi_{\mathsf{B}}\colon \mathbb{S}^2 \setminus \{(0,0,-1),(0,0,1)\} \to \mathsf{B}$ admits the expression

$$\pi_{\mathsf{B}}(\boldsymbol{x}) = x^3$$

and it is a surjective submersion. This shows that the vector field X restricted to $\mathbb{S}^2 \setminus \{(0,0,-1),(0,0,1)\}$ satisfies Assumption 5.78.

We now proceed with some basic calculations. For $\boldsymbol{x} \in \mathbb{S}^2 \setminus \{(0,0,-1),(0,0,1)\}$, the tangent map of the projection satisfies $T_{\boldsymbol{x}}\pi_{\mathsf{B}}(\boldsymbol{v}) = v^3$ for each $\boldsymbol{v} = (v^1, v^2, v^3) \in \mathsf{T}_{\boldsymbol{x}}(\mathbb{S}^2 \setminus \{(0,0,-1),(0,0,1)\})$. By definition, the vertical distribution is generated by X. Some straightforward computations show that the generator for the horizontal distribution is the vector field $Y \in \Gamma^\infty(\mathsf{T}(\mathbb{S}^2 \setminus \{(0,0,-1),(0,0,1)\}))$ defined by

$$Y(\boldsymbol{x}) = x^1 x^3 \frac{\partial}{\partial x^1} + x^2 x^3 \frac{\partial}{\partial x^2} + ((x^3)^2 - 1)\frac{\partial}{\partial x^3}.$$

According to equation (5.34), the horizontal lift of a vector $v \in \mathsf{TB}$ is $\mathrm{hlft}_{\boldsymbol{x}}(v) = v\frac{1}{(x^3)^2-1}Y(\boldsymbol{x})$, and, in turn, one can compute

$$(m\mathbb{G}_{\mathrm{rnd}})_{\mathsf{B}}(x^3) = \frac{m}{1-(x^3)^2}\mathrm{d}x^3 \otimes \mathrm{d}x^3.$$

The corresponding Levi-Civita affine connection has the Christoffel symbol $x^3 \mapsto \frac{x^3}{1-(x^3)^2}$. Furthermore, we compute

$$\begin{aligned}
(\|X\|_{\mathbb{G}}^2)_{\mathsf{B}}(x^3) &= 1-(x^3)^2,\\
(V_{X,\lambda}^{\mathrm{eff}})_{\mathsf{B}}(x^3) &= m a_g x^3 - \frac{\lambda^2}{2}(1-(x^3)^2),\\
(V_{X,\lambda}^{\mathrm{amd}})_{\mathsf{B}}(x^3) &= m a_g x^3 + \frac{\lambda^2}{2(1-(x^3)^2)},
\end{aligned}$$

where a_g is the acceleration due to gravity. Finally, we note that the gyroscopic tensor C_X must vanish because of its skew-symmetry and the fact that $\dim(\mathsf{B}) = 1$.

We are now ready to write the reduced equations. If $t \mapsto \gamma(t) = (\gamma^1(t), \gamma^2(t), \gamma^3(t)) \in \mathbb{S}^2 \setminus \{(0,0,-1), (0,0,1)\}$ is a trajectory of $(\mathbb{S}^2, m\mathbb{G}_{\mathrm{rnd}}, V)$, we compute its projections as in Theorem 5.83:

$$
\begin{aligned}
t &\mapsto \eta(t) = \pi_{\mathsf{B}}(\gamma(t)) = \gamma^3(t), \\
t &\mapsto \mu(t) = \gamma^1(t)\dot{\gamma}^2(t) - \gamma^2(t)\dot{\gamma}^1(t), \\
t &\mapsto v(t) = \frac{\mu(t)}{1 - (\gamma^3)^2(t)}.
\end{aligned}
$$

The projections (η, v) satisfy

$$
\begin{aligned}
\ddot{\eta}(t) + \frac{\eta(t)\dot{\eta}(t)^2}{1 - \eta(t)^2} &= -\frac{1 - \eta(t)^2}{m}\left(m a_g + v(t)^2 \eta(t)\right), \\
\dot{v}(t) &= 2\frac{v(t)\eta(t)}{1 - \eta(t)^2}\dot{\eta}(t),
\end{aligned}
$$

and the projections (η, μ) satisfy

$$
\begin{aligned}
\ddot{\eta}(t) + \frac{\eta(t)\dot{\eta}(t)^2}{1 - \eta(t)^2} &= -\left((1 - \eta(t)^2)a_g + \mu(t)^2 \frac{\eta(t)}{m(1 - \eta(t)^2)}\right), \\
\dot{\mu}(t) &= 0.
\end{aligned}
$$

•

Exercises

E5.1 Prove that $\mathsf{SO}(n)$, and $\mathsf{SE}(n)$ are closed with respect to the operations of matrix inversion and matrix multiplication.

E5.2 Prove Lemma 5.3.

E5.3 Prove the following statements concerning the matrix exponential $\exp\colon \mathbb{R}^{n\times n} \to \mathbb{R}^{n\times n}$ defined in Section 5.1.3.

(a) The definition of matrix exponential is well-posed in the sense that the series converges uniformly and absolutely in any compact subset of $\mathbb{R}^{n\times n}$.
Hint: *Show that, for some appropriate norm,* $\|\exp(\boldsymbol{A})\| \leq \exp(\|\boldsymbol{A}\|)$.

(b) If $\boldsymbol{A}, \boldsymbol{B} \in \mathbb{R}^{n\times n}$ satisfy $\boldsymbol{AB} = \boldsymbol{BA}$, then

$$
\begin{aligned}
\exp(\boldsymbol{A})\boldsymbol{B} &= \exp(\boldsymbol{A})\boldsymbol{B}, \\
\exp(\boldsymbol{A})\exp(\boldsymbol{B}) &= \exp(\boldsymbol{B})\exp(\boldsymbol{A}) = \exp(\boldsymbol{A} + \boldsymbol{B}).
\end{aligned}
$$

(c) The matrix exponential is neither injective nor surjective.
Hint: *Consider* $\boldsymbol{A} = \lambda\left[\begin{smallmatrix} 0 & 1 \\ -1 & 0 \end{smallmatrix}\right]$ *for appropriate* λ.

(d) The matrix exponential is analytic in a neighborhood of $\boldsymbol{0}_{n\times n}$.
Hint: *What is the Taylor expansion of* exp*?*

(e) The matrix exponential is a diffeomorphism from a neighborhood of $\mathbf{0}_{n\times n}$ to a neighborhood of $\boldsymbol{I}_n$.

E5.4 Prove Proposition 5.5.

E5.5 Complete the proof of Proposition 5.6; that is, prove the statement for the group $\mathsf{SE}(3)$.

E5.6 Complete the proof of Proposition 5.7; that is, prove the statement for the group $\mathsf{SE}(3)$.

E5.7 Compute the matrix exponential and matrix logarithm for the group of planar displacements $\mathsf{SE}(2)$. In particular, show that, for $\omega \neq 0$,

$$\exp\left(\begin{bmatrix} 0 & -\omega & v^1 \\ \omega & 0 & v^2 \\ 0 & 0 & 0 \end{bmatrix}\right) = \begin{bmatrix} \cos\omega & -\sin\omega & \frac{\sin\omega}{\omega}v^1 - \frac{1-\cos\omega}{\omega}v^2 \\ \sin\omega & \cos\omega & \frac{1-\cos\omega}{\omega}v^1 + \frac{\sin\omega}{\omega}v^2 \\ 0 & 0 & 1 \end{bmatrix}.$$

E5.8 Recall that $\widehat{\cdot}\colon \mathbb{R}^3 \to \mathfrak{so}(3)$ is a Lie algebra isomorphism. For what inner product $\langle\!\langle\cdot,\cdot\rangle\!\rangle$ on $\mathfrak{so}(3)$, is it true that $\langle\!\langle \boldsymbol{v}, \boldsymbol{w}\rangle\!\rangle_{\mathbb{R}^3} = \langle\!\langle \widehat{\boldsymbol{v}}, \widehat{\boldsymbol{w}}\rangle\!\rangle$, for all $\boldsymbol{v}, \boldsymbol{w} \in \mathbb{R}^3$?
***Hint:** See Exercise E2.8.*

E5.9 Prove that $(\mathbb{Z}, +)$, $(\mathbb{Q}\setminus\{0\}, \cdot)$, and $(\mathbb{R}_+, \cdot)$ are Abelian groups. What is the identity element and inverse operation in each case?

E5.10 Show that a group homomorphism $\rho\colon \mathsf{G} \to \mathsf{H}$ maps the identity element of G into the identity element of H, and inverse elements in G into inverse elements in H.

E5.11 This exercise is a continuation of Exercise E3.14. Show that the following diffeomorphic manifolds are isomorphic as Lie groups:
1. $\mathbb{T}^1 = \mathbb{R}/\sim$, where $\sim$ is the equivalence relation $x \sim y$ if $x - y \in \mathbb{Z}$;
2. $\mathbb{S}^1 = \{\, (x,y) \in \mathbb{R}^2 \mid x^2 + y^2 = 1\}$;
3. $\mathsf{SO}(2) = \{\boldsymbol{R} \in \mathsf{GL}(2;\mathbb{R}) \mid \boldsymbol{R}^T\boldsymbol{R} = \boldsymbol{I}_2,\ \det\boldsymbol{R} = 1\}$;
4. $\mathbb{S}^1_{\mathbb{C}} = \{z \in \mathbb{C} \mid |z| = 1\}$.

E5.12 Show that the ***special linear group*** $\mathsf{SL}(n;\mathbb{R}) = \{\, \boldsymbol{A} \in \mathbb{R}^{n\times n} \mid \det\boldsymbol{A} = 1\}$ is a matrix Lie group.

E5.13 Prove the following statements.
(a) Every matrix one-parameter subgroup is isomorphic to either $(\mathbb{R}, +)$ or $(\mathbb{T}^1, +)$.
(b) A matrix one-parameter subgroup is isomorphic to $\mathbb{T}^1$ if and only if its generator $\boldsymbol{A}$ can be converted to a diagonal matrix by a complex similarity transformation and the eigenvalues of $\boldsymbol{A}$ are purely imaginary and linearly dependent over $\mathbb{Q}$.
***Hint:** Express **A** in Jordan normal form.*
(c) The matrix one-parameter subgroup generated by

$$\boldsymbol{A} = \begin{bmatrix} 0 & 1 & 0 & 0 \\ -1 & 0 & 0 & 0 \\ 0 & 0 & 0 & \sqrt{2} \\ 0 & 0 & -\sqrt{2} & 0 \end{bmatrix}$$

is an immersed, but not embedded, submanifold of $\mathsf{GL}(n;\mathbb{R})$.

(d) Every one-parameter subgroup of $\mathsf{SE}(3)$ is an embedding.

E5.14 Characterize the Lie algebra elements of $\mathbb{T}^2$ that give rise to one-parameter subgroups that are not regular Lie subgroups of $\mathbb{T}^2$.

E5.15 Complete the proof of Theorem 5.20, i.e., show that, for $\boldsymbol{A}, \boldsymbol{B} \in \mathbb{R}^{n\times n}$,

$$\begin{aligned}\exp(t\boldsymbol{A})\exp(t\boldsymbol{B}) &= \exp(t(\boldsymbol{A}+\boldsymbol{B}) + o(t^2)),\\ \exp(t\boldsymbol{A})\exp(t\boldsymbol{B})\exp(-t\boldsymbol{A})\exp(-t\boldsymbol{B}) &= \exp(t^2[\boldsymbol{A},\boldsymbol{B}] + o(t^3)).\end{aligned}$$

E5.16 Show that the matrix commutator satisfies the Jacobi identity.

E5.17 Let $c^k_{ij} \in \mathbb{R}$, $i, j, k \in \{1, \dots, n\}$, be the structure constants for a Lie algebra V. Show they satisfy the properties in equation (5.11).

E5.18 Let G be an Abelian Lie group with $\mathfrak{g}$ its Lie algebra.
(a) Show that $[\xi, \eta] = 0$ for all $\xi, \eta \in \mathfrak{g}$.
Now we claim that the converse to part (a) is not true. Define $\mathsf{G} = \mathbb{R} \times S_m$, for $m \geq 3$, and make G a group by $(a_1, \sigma_1) \star (a_2, \sigma_2) = (a_1 + a_2, \sigma_1 \circ \sigma_2)$. Prove the following statements.
(b) G is a one-dimensional Lie group.
(c) The Lie algebra $\mathfrak{g}$ of G is isomorphic to the Lie algebra $(\mathbb{R}, [\cdot,\cdot]_0)$, where $[x, y]_0 = 0$ for all $x, y \in \mathbb{R}$.
(d) G is not Abelian.

E5.19 Let G be a Lie group with $\mathfrak{g}$ its Lie algebra. Let $g \in \mathsf{G}$ and $\xi, \eta \in \mathfrak{g}$. Consider the conjugation map $I_g\colon \mathsf{G} \to \mathsf{G}$, the adjoint map $\mathrm{Ad}_g\colon \mathfrak{g} \to \mathfrak{g}$ and adjoint operator $\mathrm{ad}_\xi\colon \mathfrak{g} \to \mathfrak{g}$.
(a) Prove the following identities:

$$I_g(\exp(\xi)) = \exp(\mathrm{Ad}_g(\xi)), \quad \left.\frac{\mathrm{d}}{\mathrm{d}t}\right|_{t=0} \mathrm{Ad}_{\exp(t\xi)}(\eta) = \mathrm{ad}_\xi\,\eta.$$

(b) Let $\gamma\colon I \to \mathsf{G}$, and let $v\colon I \to \mathfrak{g}$ satisfy $\gamma'(t) = T_eL_{\gamma(t)}(v(t))$. Show that

$$\begin{aligned}\frac{\mathrm{d}}{\mathrm{d}t}\gamma^{-1}(t) &= -\,T_eR_{\gamma(t)}(v(t)),\\ \frac{\mathrm{d}}{\mathrm{d}t}\mathrm{Ad}_{\gamma(t)}(\eta) &= \mathrm{Ad}_{\gamma(t)}[v(t), \eta],\\ \frac{\mathrm{d}}{\mathrm{d}t}\mathrm{Ad}_{\gamma^{-1}(t)}(\eta) &= [\mathrm{Ad}_{\gamma^{-1}(t)}(\eta), v(t)].\end{aligned}$$

(c) Make sense of and prove the following equalities:

$$\begin{aligned}\mathscr{L}_{\xi_{\mathrm{L}}}\mathrm{Ad}_g(\eta) &= \mathrm{Ad}_g[\xi, \eta];\\ \mathscr{L}_{\xi_{\mathrm{L}}}\mathrm{Ad}_{g^{-1}}(\eta) &= [\mathrm{Ad}_{g^{-1}}(\eta), \xi].\end{aligned}$$

E5.20 Let G be a Lie group with $\mathfrak{g}$ its Lie algebra.
(a) Show that $[\xi_{\mathrm{L}}, \eta_{\mathrm{R}}] = 0$ for all $\xi, \eta \in \mathfrak{g}$.
(b) Show that the set of right-invariant vector fields is equal to the set of all vector fields that commute with all left-invariant vector fields.

E5.21 Let G be a Lie group with $\mathfrak{g}$ its Lie algebra. Let $\{e_1, \dots, e_n\}$ be a basis for $\mathfrak{g}$ with $c^k_{ij} \in \mathbb{R}$, $i, j, k \in \{1, \dots, n\}$, the structure constants for $\mathfrak{g}$. Let $\mathbb{I}_{ij}$ be the components of an inner product $\mathbb{I}$ on $\mathfrak{g}$. Define $\gamma_{ij}^k \in \mathbb{R}$, $i, j, k \in \{1, \dots, n\}$, by means of

$$\overset{\mathfrak{g}}{\nabla}_{e_i} e_j = \gamma_{ij}^k e_k.$$

Prove the following statements:

(a) the constants γ_{ij}^k, $i, j, k \in \{1, \dots, n\}$, are the generalized Christoffel symbols of $\overset{G_{\mathbb{I}}}{\nabla}$ with respect to some family of vector fields;

(b) the constants γ_{ij}^k, $i, j, k \in \{1, \dots, n\}$, satisfy

$$\gamma_{ij}^k = \tfrac{1}{2}\mathbb{I}^{km}\left(c_{ij}^l \mathbb{I}_{lm} + c_{mi}^l \mathbb{I}_{lj} + c_{mj}^l \mathbb{I}_{li}\right),$$

and, if the basis $\{e_1, \dots, e_n\}$ is orthonormal,

$$\gamma_{ij}^k = \tfrac{1}{2}(c_{ij}^k + c_{ki}^j - c_{jk}^i).$$

E5.22 Verify the expressions (5.20), (5.22), and (5.26) for the adjoint operator and adjoint map on the Lie algebras $\mathfrak{se}(2)$, $\mathfrak{so}(3)$, and $\mathfrak{se}(3)$.

E5.23 Show that the bilinear maps in equations (5.21), (5.23), and (5.27) are the restriction of the Levi-Civita affine connections on the Lie algebras $\mathfrak{se}(2)$, $\mathfrak{so}(3)$, and $\mathfrak{se}(3)$ discussed in Example 5.47, Example 5.48, and Example 5.49, respectively.

E5.24 Consider the planar rigid body in an ideal fluid discussed in Exercise E4.25. Let (ω, v_x, v_y) be the body velocity. Prove the following statements.

(a) The kinetic energy is $\frac{1}{2}J\omega^2 + \frac{1}{2}m_1 v_x^2 + m_2 v_y^2$, so that the kinetic energy tensor of this simple mechanical system on $\mathsf{SE}(2)$ admits the matrix representation

$$\begin{bmatrix} J & 0 & 0 \\ 0 & m_1 & 0 \\ 0 & 0 & m_2 \end{bmatrix}.$$

(b) The Euler–Poincaré equations are

$$\begin{aligned} \dot{\omega} + \frac{m_2 - m_1}{J} v_x v_y &= 0, \\ \dot{v}_x - \frac{m_2}{m_1}\omega v_y &= 0, \\ \dot{v}_y + \frac{m_1}{m_2}\omega v_x &= 0. \end{aligned}$$

E5.25 A satellite with internal rotors is a rigid body with fixed center of mass equipped with internal rotors aligned with the principal axes of the satellite.

(a) Show that the configuration manifold $\mathsf{SO}(3) \times \mathbb{S}^1 \times \mathbb{S}^1 \times \mathbb{S}^1$ is a Lie group. Determine its Lie algebra and the adjoint operator on it.

(b) Let $\mathbb{I}_{\text{sat-rot}} = \sum_{i=1}^3 J_i \widehat{e}^i \otimes \widehat{e}^i$ be the inertia of the satellite-rotors system with the rotors locked, and let $J_{\text{rot},k}$ be the inertia of kth rotor about its spin axis. Compute the kinetic energy of the satellite with internal rotors.

(c) Compute the corresponding left-invariant affine connection.

(d) Compute the Euler–Poincaré equations.

E5.26 Consider a Lie group G with Lie algebra $\mathfrak{g}$. Assume that the Riemannian metric $\mathbb{G}$ on G is simultaneously left-invariant and right-invariant. Prove the following statements:

(a) $\mathbb{G}(e)([\xi,\eta],\zeta) = -\mathbb{G}(e)(\xi,[\zeta,\eta])$, for all $\xi,\eta,\zeta \in \mathfrak{g}$;
Hint: *Differentiate* $\mathbb{G}(e)(\mathrm{Ad}_{\exp(t\eta)}\,\xi, \mathrm{Ad}_{\exp(t\eta)}\,\zeta)$ *with respect to time.*

(b) $\overset{\mathbb{G}}{\nabla}_{\xi_L}\eta_L = \frac{1}{2}[\xi,\eta]_L$, for all $\xi,\eta \in \mathfrak{g}$;

(c) the geodesics on G for the metric $\mathbb{G}$ through the identity e are one-parameter subgroups.

E5.27 The Killing form K of a Lie algebra V is a $(0,2)$-tensor on V defined by

$$(\xi,\eta) \mapsto K(\xi,\eta) = \mathrm{tr}(\mathrm{ad}_\xi \circ \mathrm{ad}_\eta),$$

where ad_ξ and ad_η are linear transformations of V. Let $\{e_1,\dots,e_n\}$ be a basis for V, and let c^k_{ij}, $i,j,k \in \{1,\dots,n\}$, be the corresponding structure constants.

(a) Show that $K_{ij} = c^r_{is}c^s_{jr}$, $i,j \in \{1,\dots,n\}$.

(b) Show that, for $\mathfrak{so}(3)$ with the standard basis inherited from $\mathbb{R}^3$, the Killing form satisfies $K_{ij} = -2\delta_{ij}$.

E5.28 Prove Lemma 5.60.

E5.29 Let $\Sigma = (\mathsf{Q},\mathbb{G},V,F,\mathcal{D})$ be a C^∞-forced simple mechanical system with a regular constraint $\mathcal{D}$. Let X be an infinitesimal symmetry for Σ.

(a) Give conditions under which the momentum $J_X \circ \gamma'$ is conserved along any trajectory γ of Σ.

Then, let $X_k \in \Gamma^\infty(\mathsf{TQ})$, $k \in \{1,\dots,l\}$, be infinitesimal symmetries for the simple mechanical system $(\mathsf{Q},\mathbb{G},V)$.

(b) Show that, if there exist $f^k \in C^\infty(\mathsf{Q})$, $k \in \{1,\dots,l\}$, such that $f^kX_k \in \Gamma^\infty(\mathcal{D})$, then any trajectory γ of the constrained system Σ satisfies

$$\begin{aligned}\frac{\mathrm{d}}{\mathrm{d}t}J_{f^kX_k(\gamma(t))}&(\gamma'(t))\\ &= \Big(\frac{\mathrm{d}}{\mathrm{d}t}f^k(\gamma(t))\Big)J_{X_k}(\gamma'(t)) + \langle F(t,\gamma'(t)); f^kX_k(\gamma(t))\rangle.\end{aligned}$$

Here, J_X is the X-momentum map as defined in Definition 5.68.

(c) Comment on how this result relates to Remark 5.65 and to Theorem 5.69.

E5.30 Consider the robotic leg of Example 5.71 and the left action $\Phi\colon \mathsf{SO}(2) \times (\mathbb{R}_+ \times \mathbb{S}^1 \times \mathbb{S}^1) \to (\mathbb{R}_+ \times \mathbb{S}^1 \times \mathbb{S}^1)$ given by $(\boldsymbol{R},(r,\boldsymbol{x},\boldsymbol{y})) \mapsto (r,\boldsymbol{R}\boldsymbol{x},\boldsymbol{R}\boldsymbol{y})$. Prove the following statements.

(a) Φ is a symmetry.

(b) Along trajectories of the controlled robotic leg, the momentum associated with Φ is a conserved quantity for all locally integrable control inputs u^1 and u^2.

E5.31 Let $\Phi\colon \mathsf{G}\times\mathsf{Q} \to \mathsf{Q}$ be a smooth left action of the Lie group G on the manifold Q and let ∇ be a Φ-invariant affine connection on Q. Let $\mathscr{X} = \{X_1,\dots,X_n\}$ be a family of Φ-invariant vector fields that forms a basis for each tangent space in an open set $\mathcal{U} \subset \mathsf{Q}$.

(a) Show that the generalized Christoffel symbols of ∇ with respect to $\mathscr{X}$ are Φ-invariant functions.

We now consider the robotic leg discussed in Example 5.71 and Exercise E5.30. Recall the left action $\Phi\colon (\mathsf{SO}(2)\times\mathsf{SO}(2))\times(\mathbb{R}_+\times\mathbb{S}^1\times\mathbb{S}^1)\to(\mathbb{R}_+\times\mathbb{S}^1\times\mathbb{S}^1)$. Show that:

(b) $\{\frac{\partial}{\partial r},\frac{\partial}{\partial\theta},\frac{\partial}{\partial\phi}\}$ is a basis of Φ-invariant vector fields on $\mathbb{R}_+\times\mathbb{S}^1\times\mathbb{S}^1$, and that

(c) the Christoffel symbols for the Levi-Civita affine connection of the robotic leg are Φ-invariant.

E5.32 Find an example of a smooth, free, proper group action on a manifold for which the integral curves of infinitesimal generators are not embedded submanifolds.
Hint: *See Example 3.44–1.*

E5.33 Prove Lemma 5.82.
Hint: *Apply the definition of Levi-Civita affine connection in equation* (3.12) *to* $\overset{\mathbb{G}_\mathsf{B}}{\nabla}_Y Z$*, and horizontally lift the vector fields on both sides.*

E5.34 Consider the spherical pendulum system of Examples 4.99 and 5.72. As in Example 5.72, let $\Phi|\mathbb{S}^2$ be the left $\mathsf{SO}(3)$-action on $\mathbb{S}^2$ induced by the standard left $\mathsf{SO}(3)$-action on $\mathbb{R}^3$. Performing computations in the spherical coordinates (θ,ϕ), see Exercise E1.6 and Figure E1.1, verify that

(a) $\Phi|\mathbb{S}^2$ is an isometry, and that

(b) for all $\boldsymbol{x}\in\mathbb{R}^3$, the infinitesimal generator $\widehat{\boldsymbol{x}}_{\mathbb{S}^2}$ is an infinitesimal isometry.

E5.35 Consider the pendulum on a cart system of Exercises E4.5 and E4.19, with configuration manifold $\mathsf{Q}=\mathbb{R}\times\mathbb{S}^1$ and with natural coordinates (x,θ).

(a) Show that $X=\frac{\partial}{\partial x}$ is an infinitesimal isometry for the system, that the set B of X-orbits is a manifold diffeomorphic to $\mathbb{S}^1$, and that the projection $\pi\colon\mathsf{Q}\to\mathsf{B}$ is a surjective submersion.

(b) Find the corresponding horizontal distribution HQ.

(c) What is the projected metric $\mathbb{G}_\mathsf{B}$?

(d) What is the effective potential function?

(e) What is the amended potential function?

(f) What is the gyroscopic tensor?

(g) If γ is a trajectory of the unforced (but with the potential force) equations of motion for the system, what is the differential equation governing $\pi\circ\gamma$?

Part II

Analysis of mechanical control systems

6

Stability

The analysis of stability of mechanical systems is a classic topic in mechanics and dynamics that has affected a number of mathematical and engineering disciplines. In his classic work, Lagrange [1788] investigated the stability of mechanical systems at local minima of the potential function using energy arguments. In a work widely recognized to be one of the first on control theory, Maxwell [1868] analyzed the stability of certain mechanical governors using linearization. At about the same time, Thomson and Tait [1867] studied the asymptotic stability of mechanical systems subject to dissipative forces also via linear methods. Finally, Lyapunov [1892] developed the key elements of a stability notion and of stability criteria applicable to a broad class of nonlinear systems; this work laid the foundations for modern stability theory. We present the Lyapunov Stability Critera in Theorem 6.14. So-called invariance principles were later developed to establish stability properties of dynamical systems on the basis of weaker requirements than those required by Lyapunov's original criteria. Early work on invariance principles in stability is due to Barbashin and Krasovskiĭ [1952]; LaSalle presented his Invariance Principle in [LaSalle 1968]. Recent influential works on stability include [LaSalle and Lefschetz 1962], [Hahn 1963, 1967] and [Chetaev 1955]. Nowadays, stability theory is a cornerstone of dynamical systems and control theory; examples of modern treatments in nonlinear control monographs include [Khalil 2001, Sastry 1999, Sontag 1998], and dynamical systems references include [Arnol'd 1992, Guckenheimer and Holmes 1990, Hirsch and Smale 1974, Merkin 1997].

This chapter presents an overview of classic stability theory; we first discuss linearization and linear methods, and later the Lyapunov Stability Critera and the LaSalle Invariance Principle. The main subject of the chapter is the stability analysis of equilibria of simple mechanical systems subject to dissipation. We develop both classic approaches to stability analysis: a linearized analysis based on the spectrum of the linearization of the mechanical system; and a nonlinear Lyapunov analysis based on the properties of the energy function. Via the nonlinear Lyapunov analysis we obtain the classic Lagrange–Dirichlet Criterion for simple mechanical systems: if the potential function has positive-definite Hessian at a critical configuration, then that configuration is a stable equilibrium. Stronger statements can be made for systems subject to strict dissipative forces: any isolated local minimum point of the potential function is a locally asymptotically stable equilibrium configuration. In performing the nonlinear Lyapunov analysis, we discuss the relationship between the asymptotic behavior of mechanical systems with dissipation and gradient dynamical system; e.g., see [Koditschek 1989, Terjéki 1987]. To characterize exponential convergence we introduce a notion of quadratic functions based on Morse theory; e.g., see [Hirsch 1976, Milnor 1963]. Throughout the chapter, we focus our analysis on the setting of smooth mechanical systems and smooth Lyapunov functions.

We also consider existence and stability problems for relative equilibria. This family of trajectories is of great interest in theory and applications, as they provide a rich family of easily understood solutions of the mechanical system. For systems defined on the group of rigid displacements $\mathsf{SE}(3)$ or one of its Lie subgroups, relative equilibria are translations of one-parameter subgroups. The notion of relative equilibria is discussed in more detail in [Marsden 1992], the Energy-Momentum Method to analyze their stability is discussed in [Marsden 1992, Simo, Lewis, and Marsden 1991], and other references on stability of relative equilibria include [Bloch 2003, Leonard and Marsden 1997, Ortega and Ratiu 1999, Patrick 1992].

Let us conclude this introduction by mentioning some topics that are not discussed in this chapter. Concerning Lyapunov theory, we do not discuss (1) the existence of a Lyapunov function for a stable or locally asymptotically stable system (these are so-called converse Lyapunov theorems), or (2) criteria to establish instability based on the existence of an appropriate function with similar, but opposite monotonicity properties, to a Lyapunov function. We refer the reader to [Chetaev 1955, Khalil 2001, Krasovskiĭ 1963, Massera 1956, 1958, Merkin 1997] for discussions of converse Lyapunov and instability theorems. An important topic in the stability of equilibria that we do not discuss is Center Manifold Theory. We refer to [Guckenheimer and Holmes 1990] for a discussion of this. An additional topic that we do not discuss in this book is passivity. A brief discussion about the passivity properties of mechanical systems with inputs and outputs, and about their relationships with stability and stabilization, is postponed to Section 10.5.4. Finally, in the stability analysis for mechanical systems, we do not consider systems subject to constraints or to arbitrary external forces. There has not been a lot of work done in this area, but let us outline some of it. Some results on existence and stability of equilibrium configurations and relative equilibria for systems with constraints are obtained in [Krasinskiĭ 1988, Muñoz-Lecanda and Yániz 2003, Wang and McClamroch 1994, Zenkov, Bloch, and Marsden 1998], and references contained therein. The linearization of systems with constraints is considered by Neĭmark and Fufaev [1972] (see also [DeMarco 1999, Krasinskiĭ 1988]). Other work on the stability of mechanical systems with constraints is that of Wang and McClamroch [1994]).

6.1 An overview of stability theory for dynamical systems

In this section we provide a short introduction to stability theory for dynamical systems. After defining some basic concepts, we present the classic linearization analysis for stability. We then present the Lyapunov Stability Criteria and the LaSalle Invariance Principle. We conclude with some useful concepts from Morse theory.

6.1.1 Stability notions

Let M be a smooth manifold, let X be a smooth vector field on M, and consider the differential equation

$$\gamma'(t) = X(\gamma(t)), \tag{6.1}$$

where $\gamma\colon I \to \mathsf{M}$ is a curve on M.

Definition 6.1 (Equilibrium point). Given $X \in \Gamma^\infty(\mathsf{M})$, a point $x_0 \in \mathsf{M}$ is an ***equilibrium point for X*** if the trivial curve $\gamma\colon \mathbb{R} \to \mathsf{M}$, given by $\gamma(t) = x_0$ for all $t \in \mathbb{R}$, is an integral curve for X. •

Remarks 6.2. 1. It is immediate to see that x_0 is an equilibrium point for X if and only if $X(x_0) = 0_{x_0}$.

2. The expressions "fixed point" or "critical point" are sometimes used for equilibrium point. •

We shall say that a curve $\gamma\colon \bar{\mathbb{R}}_+ \to \mathsf{M}$ ***approaches*** a set $S \subset \mathsf{M}$ as $t \to +\infty$ if, for any neighborhood $\mathcal{U}$ of S, there exists a time $T \in \mathbb{R}_+$ such that, for all $t > T$, we have $\gamma(t) \in \mathcal{U}$. In such a case, we shall write $\gamma(t) \to S$ as $t \to +\infty$. If the set S consists of a single point x_0, then it is customary to write $\lim_{t\to+\infty} \gamma(t) = x_0$, and to say that γ ***converges*** to x_0 as $t \to +\infty$.

Given these preliminary concepts, we can now introduce various stability notions for an equilibrium point.

Definition 6.3 (Stability notions). Let X be a vector field on the manifold M. An equilibrium point x_0 for X is

(i) ***stable*** (or ***Lyapunov stable***) if, for any neighborhood $\mathcal{U}$ of x_0, there exists a neighborhood $\mathcal{W}$ of x_0 such that for all $x \in \mathcal{W}$, the integral curve $t \mapsto \Phi_t^X(x)$ takes values in $\mathcal{U}$, is

(ii) ***unstable*** if it is not stable, is

(iii) ***locally asymptotically stable*** if it is stable and if there exists a neighborhood $\mathcal{U}$ of x_0 such that, for all $x \in \mathcal{U}$, the integral curve $t \mapsto \Phi_t^X(x)$ converges to x_0, and is

(iv) ***globally asymptotically stable*** if it is Lyapunov stable and if the integral curve $t \mapsto \Phi_t^X(x)$ converges to x_0 for all $x \in \mathsf{M}$. •

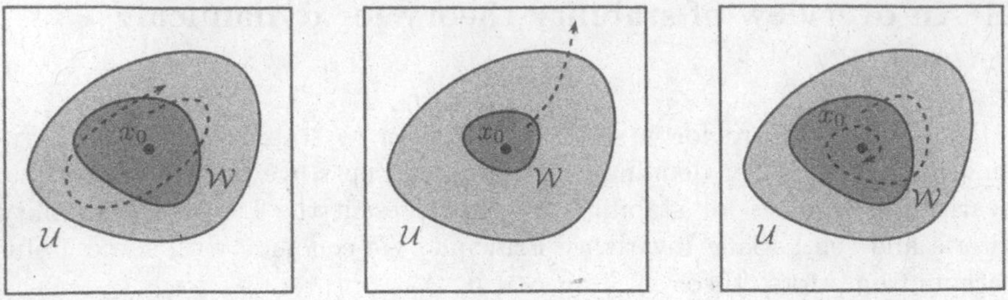

Figure 6.1. An interpretation of stability, instability, and local asymptotic stability

Figure 6.1 illustrates the difference between some of these concepts. If x_0 is an unstable equilibrium point, for all neighborhoods $\mathcal{U}$ and $\mathcal{W}$ of x_0, then there exists $x \in \mathcal{W}$ such that the integral curve $t \mapsto \Phi_t^X(x)$ does not remain in $\mathcal{U}$. Note that parts (i), (iii) and (iv) implicitly require the existence of a neighborhood $\mathcal{W}$ such that the integral curves of X through any point in $\mathcal{W}$ can be defined for all positive times. This will be true, for example, if the vector field X is positively complete.

Examples 6.4. We shall study the stability properties of some example vector fields to illustrate the tests we present in the following sections. It is convenient to consider vector fields defined by polynomial differential equations of first- and second-order on $\mathsf{M} = \mathbb{R}$. To make things even simpler, we shall restrict our attention to low-order polynomial functions. Specifically, we shall consider the four differential equations

$$\ddot{x} = -x, \quad \dot{x} = x - x^3, \quad \dot{x} = -x + x^2, \quad \dot{x} = -x, \tag{6.2}$$

and, by means of a linear analysis, we shall show that the origin 0 is stable for $\ddot{x} = -x$ (but not asymptotically stable), unstable for $\dot{x} = x - x^3$, locally asymptotically stable for $\dot{x} = -x + x^2$ (but not globally), and globally asymptotically stable for $\dot{x} = -x$. Furthermore, by means of so-called Lyapunov and LaSalle analysis, we shall establish the local asymptotic stability of the origin for

$$\dot{x} = -x^3, \quad \ddot{x} = -x^3 - \dot{x}, \tag{6.3}$$

for which a linear analysis is inconclusive. •

We conclude this section by generalizing some stability notions to the setting of integral curves that are periodic functions of time. We first review some basic nomenclature. Given a set S, a map $f\colon \mathbb{R} \to S$ is ***periodic*** if there exists $T \in \mathbb{R}_+$, called the ***period of*** $\boldsymbol{f}$, such that $f(t+T) = f(t)$ for all $t \in \mathbb{R}$. If f is periodic with period T, we say that f is $\boldsymbol{T}$***-periodic***.

Definition 6.5 (Periodic trajectory). Let X be a vector field on the manifold M. A ***periodic trajectory for*** $\boldsymbol{X}$ is a periodic integral curve of X. A periodic trajectory γ for X is

(i) ***stable*** if, for any neighborhood $\mathcal{U}$ of image(γ), there exists a neighborhood $\mathcal{W}$ of image(γ) such that, for all $x \in \mathcal{W}$, the integral curve $t \mapsto \Phi_t^X(x)$ takes values in $\mathcal{U}$, and is

(ii) ***locally asymptotically stable*** if it is stable and if there exists a neighborhood $\mathcal{U}$ of image(γ) such that, for all $x \in \mathcal{U}$, the integral curve $t \mapsto \Phi_t^X(x)$ approaches the set image(γ). •

6.1.2 Linearization and linear stability analysis

Classically, the way in which one initiates a stability investigation is via linear methods. Thus one begins with a reference equilibrium point and linearizes about it. The methods for investigating stability of linearizations are well understood, and in some instances stability of the linearized system can be used to infer local stability of the system. In this section we review the notions of linearized stability for general vector fields.

We let M be a differentiable manifold, X be a smooth vector field on M, and $x_0 \in \mathsf{M}$ be an equilibrium point for X. Recall from Proposition 3.75 the linearization of X at x_0 that we had denoted by $A_X(x_0) \in L(\mathsf{T}_{x_0}\mathsf{M}; \mathsf{T}_{x_0}\mathsf{M})$. In a chart $(\mathcal{U}, \phi)$ with coordinates $(x^1, \dots, x^n)$, the matrix representation of $A_X(x_0)$ in the natural basis is

$$[A_X(x_0)] = \begin{bmatrix} \frac{\partial X^1}{\partial x^1}(x_0) & \cdots & \frac{\partial X^1}{\partial x^n}(x_0) \\ \vdots & \ddots & \vdots \\ \frac{\partial X^n}{\partial x^1}(x_0) & \cdots & \frac{\partial X^n}{\partial x^n}(x_0) \end{bmatrix},$$

i.e., the Jacobian of the components of X.

In what follows, it is convenient to first consider the situation in which $A \colon \mathsf{V} \to \mathsf{V}$ is a linear map and in which the differential equation of interest is $\dot{x}(t) = A(x(t))$. The following result gives the properties of the stability of this system in terms of the eigenvalues and eigenvectors of A.

Theorem 6.6 (Stability of linear systems). *Let V be a finite-dimensional $\mathbb{R}$-vector space, let $A \in L(\mathsf{V}; \mathsf{V})$, and consider the differential equation*

$$\dot{x}(t) = A(x(t)) \tag{6.4}$$

with equilibrium point $0 \in \mathsf{V}$. The following statements hold:

(i) 0 is unstable if $\mathrm{spec}(A) \cap \mathbb{C}_+ \neq \emptyset$;

(ii) 0 is globally asymptotically stable if $\mathrm{spec}(A) \subset \mathbb{C}_-$;

(iii) 0 is stable if $\mathrm{spec}(A) \subset \overline{\mathbb{C}}_-$ *and if, for every* $\lambda \in \mathrm{spec}(A) \cap \mathrm{i}\mathbb{R}$, *we have* $m_a(\lambda) = m_g(\lambda)$;

(iv) 0 is unstable if $\mathrm{spec}(A) \subset \overline{\mathbb{C}}_-$ *and if, for some* $\lambda \in \mathrm{spec}(A) \cap \mathrm{i}\mathbb{R}$, *we have* $m_a(\lambda) > m_g(\lambda)$.

The result is easily proved if one knows certain properties of the matrix exponential; see [Arnol'd 1992, Hirsch and Smale 1974].

Given that x_0 is an equilibrium point for a vector field X, one can ask how the stability of x_0 for X is related to the stability of the origin for the linearization. The following result gives the only general implications one can make; see [Lyapunov 1892] or [Chetaev 1955] for a proof.

Proposition 6.7 (Stability from linear stability). *Let $x_0 \in \mathsf{M}$ be an equilibrium point for a vector field X and let $A_X(x_0) \in L(\mathsf{T}_{x_0}\mathsf{M}; \mathsf{T}_{x_0}\mathsf{M})$ denote the linearization of X at x_0. The following statements hold:*

(i) *if* $\operatorname{spec}(A_X(x_0)) \subset \mathbb{C}_-$, *then x_0 is locally asymptotically stable;*
(ii) *if* $\operatorname{spec}(A_X(x_0)) \cap \mathbb{C}_+ \neq \emptyset$, *then x_0 is unstable.*

Based on this result we make the following definitions.

Definition 6.8 (More stability notions). Let X be a vector field with $x_0 \in \mathsf{M}$ an equilibrium point for X. Then x_0 is

(i) ***linearly stable*** if 0 is stable for the linearization $A_X(x_0)$, is
(ii) ***linearly asymptotically stable*** if 0 is asymptotically stable for the linearization $A_X(x_0)$, is
(iii) ***linearly unstable*** if it is not linearly stable, is
(iv) ***spectrally stable*** if $\operatorname{spec}(A_X(x_0)) \subset \overline{\mathbb{C}}_-$, and is
(v) ***spectrally unstable*** if it is not spectrally stable. •

With these definitions, the following implications hold:

$$\text{linear asymptotic stability} \implies \text{local asymptotic stability}$$
$$\text{spectral instability} \implies \text{instability.}$$

One cannot generally say more than this, as is illustrated with a few simple examples.

Examples 6.9 (Example 6.4 cont'd). Let us consider the four examples in (6.2). Their respective linearizations at the origin read

$$\ddot{x} = -x, \quad \dot{x} = x, \quad \dot{x} = -x, \quad \dot{x} = -x.$$

Based on the eigenvalues of these linear systems and on Definition 6.8, we see that the origin 0 is, respectively, linearly stable (but not asymptotically), spectrally unstable, linearly asymptotically stable, and linearly asymptotically stable.

Next, let us draw conclusions about the original systems. First, notice that the first and fourth examples are linear systems, so that we can immediately conclude that the origin is stable for the first example system $\ddot{x} = -x$, and globally asymptotically stable for the fourth example system $\dot{x} = -x$. Second, Proposition 6.7 implies that the origin is unstable for the second example system $\dot{x} = x - x^3$, and locally asymptotically stable for the third example system $\dot{x} = -x + x^2$.

We conclude by noticing how the linear analysis described in this section is inconclusive when applied to the systems in (6.3). Indeed, their respective linearizations are $\dot{x} = 0$, and $\ddot{x} = -\dot{x}$, implying that, for both systems, the origin is linearly stable but not linearly asymptotically stable. In this case, one cannot, without further analysis, draw any conclusions concerning the behavior of the nonlinear system. •

6.1.3 Lyapunov Stability Criteria and LaSalle Invariance Principle

In this section we present a nonlinear analysis of stability based on the Lyapunov Stability Criteria and on the LaSalle Invariance Principle. We shall show how the stability notions introduced in Definition 6.3 can be characterized in terms of the existence of a so-called Lyapunov function. Let us start by introducing the required concepts.

Definition 6.10 (Locally positive-definite function). A function $\psi\colon \mathsf{M} \to \mathbb{R}$ is ***locally positive-definite about*** $\boldsymbol{x_0}$ (respectively, ***locally positive-semidefinite about*** $\boldsymbol{x_0}$) if $\psi(x_0) = 0$ and if there exists a neighborhood $\mathcal{U}$ of x_0 such that $\psi(x) > 0$ (respectively, $\psi(x) \geq 0$) for all $x \in \mathcal{U} \setminus \{x_0\}$. In this case we also say that ψ is positive-definite about x_0 on $\mathcal{U}$. •

We shall also talk about locally negative-definite and locally negative-semidefinite functions defined in the obvious way.

Definition 6.11 (Sublevel set). Let $L \in \mathbb{R}$, $\psi \in C^\infty(\mathsf{M})$, and $x_0 \in \mathsf{M}$.

(i) Let $\psi^{-1}(\leq L)$ denote the ***L-sublevel set of*** $\boldsymbol{\psi}$, that is, the set $\psi^{-1}(]-\infty, L]) = \{x \in \mathsf{M} \mid \psi(x) \leq L\}$.

(ii) Let $\psi^{-1}(\leq L, x_0)$ denote the connected component of the L-sublevel set of ψ containing x_0, if $\psi(x_0) \leq L$, or the empty set, if $\psi(x_0) > L$. •

Note that, for $L > \psi(x_0)$, a continuity argument immediately shows that $\psi^{-1}(\leq L)$ contains a neighborhood of x_0. Additionally, one can see that, in general, the L-sublevel set $\psi^{-1}(\leq L)$ is composed of multiple connected components. We illustrate the situation in Figure 6.2.

The following result will be useful in our subsequent analysis.

Lemma 6.12. *Let $\psi\colon \mathsf{M} \to \mathbb{R}$ be smooth and locally positive-definite about $x_0 \in \mathsf{M}$. For any neighborhood $\mathcal{U}$ of x_0, there exists $\alpha \in \mathbb{R}_+$ such that $\psi^{-1}(\leq\alpha, x_0)$ is a compact subset of $\mathcal{U}$.*

Proof. Let $\mathcal{U}$ be a neighborhood of x_0, and let $\mathcal{V}$ be a neighborhood of x_0 on which the function ψ is positive-definite. Let $(\mathcal{W}, \phi)$ be a chart with the property that $\phi(x_0) = \mathbf{0}$ and with coordinates $(x^1, \dots, x^n)$; we will identify sets and functions on M with their representative in the chart. There exists a sufficiently small $r \in \mathbb{R}_+$ such that $\bar{B}_r(x_0) \subset (\mathcal{W} \cap \mathcal{U} \cap \mathcal{V})$, where we let $\bar{B}_r(x_0) = \{\boldsymbol{x} \in \mathbb{R}^n \mid \|\boldsymbol{x}\|_{\mathbb{R}^n} \leq r\}$. Next, let $\alpha = \frac{1}{2}\min\{\psi(x) \mid x \in \mathrm{bd}(\bar{B}_r(x_0))\}$. Note that $\alpha > 0$ because ψ is locally positive-definite about x_0 on $\mathcal{V}$. Also note that

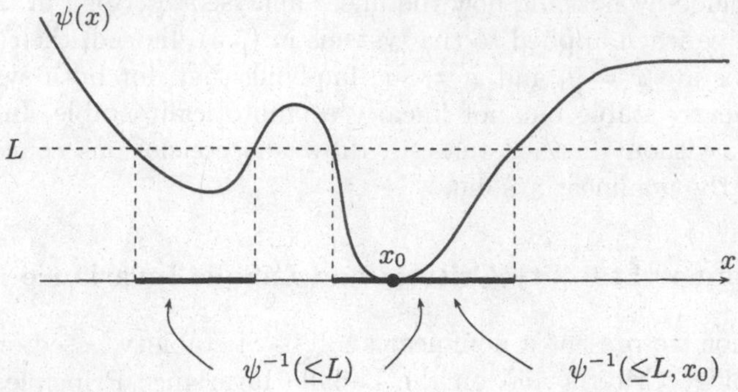

Figure 6.2. A sublevel set of $\psi\colon \mathbb{R} \to \mathbb{R}$ and its connected components

$\psi^{-1}(\leq\alpha, x_0) \cap \mathrm{bd}(\bar{B}_r(x_0)) = \emptyset$, because $\psi|\mathrm{bd}(\bar{B}_r(x_0))$ assumes values greater than or equal to 2α. Because $\psi^{-1}(\leq\alpha, x_0)$ is connected, it cannot be the union of two disjoint components and, therefore, its intersection with $\mathbb{R} \setminus \bar{B}_r(x_0)$ must be empty. This proves that $\psi^{-1}(\leq\alpha, x_0) \subset \bar{B}_r(x_0) \subset \mathcal{U}$; see Figure 6.3. Finally, $\psi^{-1}(\leq\alpha, x_0)$ is closed because, for continuous functions, the preimage

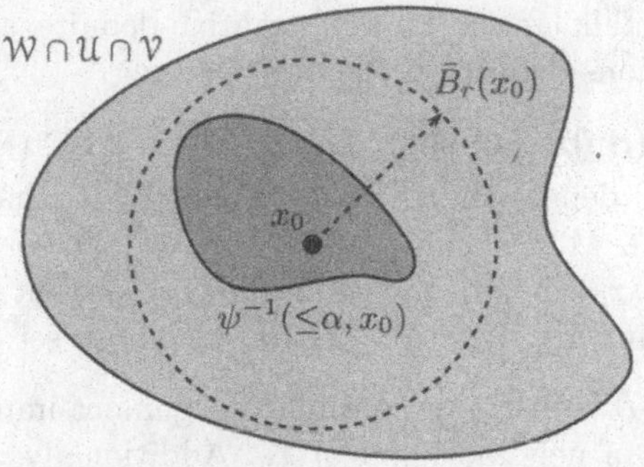

Figure 6.3. The open set $\mathcal{W} \cap \mathcal{U} \cap \mathcal{V}$ contains the closed ball $\bar{B}_r(x_0)$ that contains the sublevel set $\psi^{-1}(\leq\alpha, x_0)$ that contains x_0.

of a closed set is closed. ■

Now we define the important concept of a Lyapunov function.

Definition 6.13 (Lyapunov function). Given $X \in \Gamma^\infty(\mathsf{M})$, a smooth function $\psi\colon \mathsf{M} \to \mathbb{R}$ is a ***Lyapunov function for X about x_0*** if ψ is locally positive-definite about x_0, and if $\mathscr{L}_X\psi$ is locally negative-semidefinite about x_0. •

In other words, ψ is a Lyapunov function if $\psi(x_0) = 0$ and if there exists a neighborhood $\mathcal{U}$ of x_0 such that, for all $x \in \mathcal{U} \setminus x_0$, $\psi(x) > 0$ and $\mathscr{L}_X\psi(x) \leq 0$.

Given these notions, we can state the following stability criteria.

Theorem 6.14 (Lyapunov Stability Criteria). *Let $x_0 \in \mathsf{M}$ be an equilibrium point for the smooth vector field X.*

(i) If there exists a Lyapunov function ψ for X about x_0, then the equilibrium point x_0 is stable.

(ii) If there exists a Lyapunov function ψ for X about x_0 and if $\mathscr{L}_X\psi$ is locally negative-definite about x_0, then the equilibrium point x_0 is locally asymptotically stable.

Proof. We prove part (i) in three steps. First, let $\mathcal{U}$ be a neighborhood of x_0, and let $\mathcal{V}$ be a neighborhood of x_0 on which ψ is positive-definite and $\mathscr{L}_X\psi$ is negative-semidefinite. Lemma 6.12 provides $\alpha \in \mathbb{R}_+$ such that $\psi^{-1}(\leq\alpha, x_0) \subset \mathcal{U} \cap \mathcal{V}$. Second, because $x_0 \in \operatorname{int}(\psi^{-1}(\leq\alpha, x_0))$, there exists a neighborhood $\mathcal{W}$ of x_0 contained in $\psi^{-1}(\leq\alpha, x_0)$. Recall from Proposition 3.85 that

$$\frac{\mathrm{d}}{\mathrm{d}t}\psi(\Phi_t^X(x)) = \mathscr{L}_X\psi(\Phi_t^X(x)).$$

Accordingly, $\mathscr{L}_X\psi$ being negative semidefinite along with a continuity argument implies that the value of ψ along integral curves of X is nonincreasing. Third, we have the following chain of implications

$$\begin{aligned} x \in \mathcal{W} &\implies \psi(x) \leq \alpha \implies \psi(\Phi_t^X(x)) \leq \alpha \\ &\implies \Phi_t^X(x) \in \psi^{-1}(\leq\alpha), \quad t \in \bar{\mathbb{R}}_+. \end{aligned}$$

By continuity, $\Phi_t^X(x)$ must remain in $\psi^{-1}(\leq\alpha, x_0) \subset \mathcal{U}$. The proof of part (i) follows from the next lemma.

Lemma. *Let X be a C^∞-vector field on M, and let $A \subset \mathsf{M}$ be compact and positively X-invariant (see Definition 6.16). Then $\sigma_+(X, x) = +\infty$ for all $x \in A$.*

Proof. For $x \in A$, assume that $\sigma_+(X, x) = b \in \mathbb{R}_+$. Take $\{t_j\}_{j\in\mathbb{N}}$ in $[0, b[$ with the property that $t_j \to b$ as $j \to +\infty$. Because A is compact, by Remark 3.12–4 there exist a subsequence $\{t_{j_k}\}_{k\in\mathbb{N}}$ and a point $y \in \mathsf{M}$ such that $\lim_{k\to+\infty}\Phi_{t_{j_k}}^X(x) = y$. By Theorem A.11, there exists $\varepsilon > 0$ such that $]-\varepsilon, \varepsilon[\subset]\sigma_-(X, y), \sigma_+(X, y)[$. By continuity of the flow, there exist $N \in \mathbb{N}$ and $\tau \in]-\varepsilon, 0[$ such that $\Phi_{t_{j_N}}^X(x) = \Phi_\tau^X(y)$. In turn, this implies that $\sigma_+(X, x) \geq b + \varepsilon$ and this is a contradiction. ▼

Next, assume that $\mathscr{L}_X\psi$ is locally negative-definite. Let $\mathcal{U}$, α, $\mathcal{V}$, and $\mathcal{W}$ be defined as before, except now suppose that $\mathscr{L}_X\psi < 0$ on $\mathcal{V}$. For all $x \in \mathcal{W}$, the map $t \mapsto \psi(\Phi_t^X(x))$ assumes positive values and is nonincreasing. Therefore, for all $x \in \mathcal{W}$,

$$\lim_{t\to+\infty}\psi(\Phi_t^X(x)) = \beta \in \bar{\mathbb{R}}_+.$$

Clearly, $\beta \leq \psi(x)$. In what follows, we prove that $\beta = 0$ by contradiction. The set $\psi^{-1}(\leq\psi(x), x_0) \setminus \mathrm{int}(\psi^{-1}(\leq\beta, x_0))$ is compact and, assuming $\beta > 0$, does not contain x_0. On this compact set the smooth function $\mathscr{L}_X\psi$ assumes a maximum, say $-\delta < 0$ for some $\delta \in \mathbb{R}_+$. Therefore,

$$\psi(\Phi_t^X(x)) = \psi(x) + \int_0^t \mathscr{L}_X\psi(\Phi_t^X(x))\mathrm{d}t \leq \psi(x) - \delta t.$$

For sufficiently large t, the right-hand side is negative. This contradiction proves that $\beta = 0$. Now, we are ready to prove the convergence result. For any neighborhood $\mathcal{S}$ of x_0, there exists $\rho \in \mathbb{R}_+$ such that $\psi^{-1}(\leq\rho, x_0) \subset \mathcal{S}$. Because $\lim_{t\to+\infty} \psi(\Phi_t^X(x)) = 0$, there exists $T \in \mathbb{R}_+$ such that $t > T$ implies $\psi(\Phi_t^X(x)) < \rho$ and, therefore, $\Phi_t^X(x) \in \mathcal{S}$. This proves part (ii). ■

The Lyapunov Stability Criteria are very powerful in that they apply to general nonlinear differential equations on manifolds. For example, the asymptotic stability criteria presented in Theorem 6.6 can be shown to be a consequence of Theorem 6.14; see Exercise E6.1. The main limitation of the criteria is that they are not constructive, in the sense that the theorem does not suggest how to find Lyapunov functions. Unless the vector field X has special structure, no general methodology is available. Let us identify some Lyapunov functions for some interesting examples.

Examples 6.15 (Example 6.4 cont'd). Let us derive for a second time the stability results for the examples in (6.2) and consider then the systems in (6.3).

1. For the third and fourth example systems $\dot{x} = -x + x^2$ and $\dot{x} = -x$, we consider the candidate Lyapunov function $\frac{1}{2}x^2$. This function is locally positive-definite about the origin, and its Lie derivative along $(-x+x^2)\frac{\partial}{\partial x}$ and $-x\frac{\partial}{\partial x}$ is $-x^2 + x^3$ and $-x^2$, respectively. Both Lie derivatives are locally negative-definite, since they assume strictly negative values for all $0 < |x| < 1$. According to Theorem 6.14, the origin is locally asymptotically stable.
2. For the first example system $\ddot{x} = -x$, we consider the candidate Lyapunov function $\frac{1}{2}(x^2 + \dot{x}^2)$. This function is locally positive-definite about the origin and it can be shown to be constant along the flow of the differential equation (equivalently, its Lie derivative along the vector field representing the differential equation vanishes). According to Theorem 6.14, the origin is stable.
3. From the analysis in Example 6.9, we know that no Lyapunov function exists for the example system $\dot{x} = x - x^3$.
4. For the example system $\dot{x} = -x^3$, we consider the candidate Lyapunov function $\frac{1}{2}x^2$. It is locally positive-definite about the origin and its Lie derivative is $-x^4$, which is clearly locally negative-definite. According to Theorem 6.14, the origin is locally asymptotically stable.

5. We rewrite the system $\ddot{x} = -x^3 - \dot{x}$ in first-order form as $\dot{x}_1 = x_2$; $\dot{x}_2 = -x_1^3 - x_2$. We consider the candidate Lyapunov function $\frac{1}{4}x_1^4 + \frac{1}{2}x_2^2$. Its Lie derivative along $x_2\frac{\partial}{\partial x_1} + (-x_1^3 - x_2)\frac{\partial}{\partial x_2}$ is $-x_2^2$, which can be seen to be negative-semidefinite only. Therefore, Theorem 6.14 only allows us to conclude that the origin is stable. •

The analysis of the last example system is inconclusive as to whether the origin is a stable or an asymptotically stable equilibrium point. Indeed, the last example is designed to illustrate a drawback to the application of the Lyapunov analysis in Theorem 6.14. The characterization of stability via Lyapunov functions is strengthened by analyzing in more detail the properties of certain sets.

Definition 6.16 (Invariant set). Let X be a smooth vector field on M.

(i) A set $A \subset \mathsf{M}$ is ***X-invariant*** (or ***invariant for X***) if, for all $x \in A$, the integral curve $\mathbb{R} \ni t \mapsto \Phi_t^X(x)$ takes values in A.

(ii) A set $A \subset \mathsf{M}$ is ***positively X-invariant*** if, for all $x \in A$, the integral curve $\bar{\mathbb{R}}_+ \ni t \mapsto \Phi_t^X(x)$ takes values in A. •

Definition 6.17 (Positive limit set). Let X be a smooth vector field on M and let $x \in \mathsf{M}$. The ***positive limit set of x for X***, denoted by $\Omega(X, x)$, is the set of $y \in \mathsf{M}$ for which there exists a sequence $\{t_k\}_{k\in\mathbb{N}} \subset \mathbb{R}$ such that

(i) $t_k < t_{k+1}$, for $k \in \mathbb{N}$,

(ii) $\lim_{k\to+\infty} t_k = +\infty$, and

(iii) $\lim_{k\to+\infty} \Phi_{t_k}^X(x) = y$. •

Positive limit sets have many interesting properties.

Proposition 6.18. *Let $A \subset \mathsf{M}$ be compact and positively X-invariant. If $x \in A$, then $\Omega(X, x)$ is a nonempty, compact, and positively X-invariant subset of A. Furthermore, the curve $t \mapsto \Phi_t^X(x)$ approaches $\Omega(X, x)$ as $t \to +\infty$.*

Proof. Let $\{t_k\}_{k\in\mathbb{N}} \subset \mathbb{R}_+$ satisfy $t_k < t_{k+1}$, for $k \in \mathbb{N}$, and $\lim_{k\to+\infty} t_k = +\infty$. The sequence $\{\Phi_{t_k}^X(x)\}_{k\in\mathbb{N}} \subset A$ has a convergence subsequence by the Bolzano–Weierstrass Theorem; see Remark 3.12–4. Without loss of generality, let $y \in \Omega(X, x)$ and let $\{t_k\}_{k\in\mathbb{N}}$ satisfy $\lim_{k\to+\infty} \Phi_{t_k}^X(x) = y$. This proves that $\Omega(X, x)$ is a nonempty subset of A.

For $z \in A \setminus \Omega(X, x)$, there exists a neighborhood $\mathcal{U}$ of z and $T \in \mathbb{R}_+$ such that $\mathcal{U} \cap \{\Phi_t^X(x) \mid t \geq T\} = \emptyset$. Therefore, $A \setminus \Omega(X, x)$ is open and its complement $\Omega(X, x)$ is closed. For $\tau \in \mathbb{R}_+$, consider the sequence $\{\tau + t_k\}_{k\in\mathbb{N}}$ and compute

$$\lim_{k\to+\infty} \Phi_{\tau+t_k}^X(x) = \lim_{k\to+\infty} \Phi_\tau^X \circ \Phi_{t_k}^X(x) = \Phi_\tau^X(y).$$

Hence $\Phi_\tau^X(y) \in \Omega(X, x)$. This proves that $\Omega(X, x)$ is positively X-invariant.

It is left to show that $\Phi_t^X(x) \to \Omega(X, x)$ as $t \to +\infty$. Assume by contradiction that there exist a neighborhood $\mathcal{U}$ of $\Omega(X, x)$ and a sequence $\{\tau_k\}_{k\in\mathbb{N}}$

satisfying $\tau_k < \tau_{k+1}$, for $k \in \mathbb{N}$, $\lim_{k\to+\infty} \tau_k = +\infty$, and $\Phi^X_{\tau_k}(x) \notin \mathcal{U}$. By the Bolzano–Weierstrass Theorem, the sequence $\Phi^X_{\tau_k}(x) \in A$ admits a subsequent convergence with limit $x^* \in A \setminus \mathcal{U}$. However, x^* must also belong to $\Omega(X, x)$. ■

We are now ready to present the LaSalle Invariance Principle on the asymptotic behavior of the integral curves of vector fields. The original reference for this work is [LaSalle 1968].

Theorem 6.19 (LaSalle Invariance Principle). *For $X \in \Gamma^\infty(\mathsf{M})$, let $A \subset \mathsf{M}$ be compact and positively X-invariant. Let $\psi \in C^\infty(\mathsf{M})$ satisfy $\mathscr{L}_X\psi(x) \leq 0$ for all $x \in A$, and let B be the largest positively X-invariant set contained in $\{x \in A \mid \mathscr{L}_X\psi(x) = 0\}$. Then the following statements hold:*

(i) each integral curve of X with initial condition in A approaches B as $t \to +\infty$;

(ii) if B consists of a finite number of isolated points, then each integral curve of X with initial condition in A converges to a point of B as $t \to +\infty$.

Proof. The function $\psi|A$ is bounded from below, because it is continuous on the compact set A. For $x \in A$, let $\gamma(t) = \Phi^X_t(x)$. The function $t \mapsto \psi \circ \gamma(t)$ is nonincreasing (i.e., $t_1 > t_2 \implies \psi(\gamma(t_1)) \leq \psi(\gamma(t_2))$) and bounded from below. Therefore, $\lim_{t\to+\infty} \psi(\gamma(t))$ exists and is equal to, say, $a \in \mathbb{R}$. Now, let $y \in \Omega(X, x)$ and let $\{t_k\}_{k\in\mathbb{N}}$ satisfy $\lim_{k\to+\infty} \gamma(t_k) = y$. By continuity of ψ, $a = \lim_{k\to+\infty} \psi \circ \Phi^X_{t_k}(x) = \psi(y)$. This proves that $\psi(y) = a$ for all $y \in \Omega(X, x)$.

Because $\Omega(X, x)$ is positively X-invariant, $\Phi^X_t(y) \in \Omega(X, x)$ for all $t \in \bar{\mathbb{R}}_+$ and $y \in \Omega(X, x)$. Therefore, $\psi \circ \Phi^X_t(y) = a$ for all $t \in \bar{\mathbb{R}}_+$ and in particular $\mathscr{L}_X\psi(y) = 0$. Now, because $\mathscr{L}_X\psi(y) = 0$ for all $y \in \Omega(X, x)$, we know that $\Omega(X, x) \subset \{x \in A \mid \mathscr{L}_X\psi(x) = 0\}$. This implies that $\Omega(X, x) \subset B$, and in turn fact (i). The reader is invited to prove fact (ii) in Exercise E6.3. ■

It is also useful to state this result in the following form.

Corollary 6.20 (Barbashin and Krasovskiĭ 1952). *Let $x_0 \in \mathsf{M}$ be an equilibrium point for the smooth vector field X. Assume there exists a smooth Lyapunov function ψ for X about x_0 and let $\mathcal{U}$ be a neighborhood of x_0 on which ψ is positive-definite and $\mathscr{L}_X\psi$ is negative-semidefinite. Let $C = \{x \in \mathcal{U} \mid \mathscr{L}_X\psi(x) = 0\}$. If $\{x_0\}$ is the only positively X-invariant set in C, then x_0 is locally asymptotically stable.*

Note that requiring that the only positively X-invariant set in C is $\{x_0\}$ is equivalent to requiring that no integral curve of X takes values in C, other than the trivial integral curve $t \mapsto x_0$.

Example 6.21 (Example 6.4 cont'd). Let us continue the analysis of the example system $\dot{x}_1 = x_2$; $\dot{x}_2 = -x_1^3 - x_2$ with corresponding vector field

$X = x_2 \frac{\partial}{\partial x_1} + (-x_1^3 - x_2)\frac{\partial}{\partial x_2}$. The Lyapunov function $\psi = \frac{1}{4}x_1^4 + \frac{1}{2}x_2^2$ has Lie derivative $\mathscr{L}_X\psi = -x_2^2$. To apply Corollary 6.20, we examine the set

$$C = \{(x_1, x_2) \in \mathbb{R}^2 \mid -x_2^2 = 0\} = \{(x_1, 0) \mid x_1 \in \mathbb{R}\}.$$

From any point in $C \setminus \{(0,0)\}$, the flow of X leaves the set C and, therefore, $t \mapsto (0,0)$ is the only integral curve of X taking value in C. According to Corollary 6.20, the origin is locally asymptotically stable. •

Remark 6.22 (Domain of attraction). Consider the problem of describing the region of stability associated with an asymptotically stable equilibrium point x_0. The ***domain of attraction of*** $\boldsymbol{x_0}$ is the largest set $D(X, x_0)$ such that, for all $x \in D(X, x_0)$, $\Phi_t^X(x)$ converges to x_0 as $t \to +\infty$, that is,

$$D(X, x_0) = \{x \in \mathsf{M} \mid \lim_{t\to+\infty} \Phi_t^X(x) = x_0\}.$$

In general it is impossible to characterize in closed form the domain of attraction of an equilibrium point of a nonlinear system. It can be shown that $D(X, x_0)$ is ***contractible*** to $\{x_0\}$, by which it is meant that there exists a continuous map $h\colon [0,1] \times D(X, x_0) \to D(X, x_0)$ with the property that $h(0, x) = x$ and $h(1, x) = x_0$ for all $x \in D(X, x_0)$. This is proved in [Sontag 1998]. This property of $D(X, x_0)$ holds even if X is only continuous. For smooth vector fields, it can be further shown that $D(X, x_0)$ is diffeomorphic to $\mathbb{R}^k$ for some $k \in \mathbb{Z}_+$; see [Sontag 1999]. As we shall see in Section 10.1, this places restrictions on the manner in which systems can be stabilized using even continuous feedback.

However, if X possesses a Lyapunov function ψ about x_0, then this can be used to assert some properties of $D(X, x_0)$. Indeed, assume there exists a neighborhood $\mathcal{U}$ of x_0 such that $\psi(x)$ is positive and $\mathscr{L}_X\psi(x)$ is negative for all $x \in \mathcal{U} \setminus \{x_0\}$. For all $L \in \mathbb{R}$ such that $\psi^{-1}(\leq L, x_0) \subset \mathcal{U}$, the domain of attraction of x_0 contains the L-sublevel set of the Lyapunov function, that is,

$$D(X, x_0) \supset \psi^{-1}(\leq L, x_0). \qquad \bullet$$

6.1.4 Elements of Morse theory

Here we introduce some basic concepts and results from Morse theory. In particular, we consider scalar functions defined on manifolds and we characterize their behavior in neighborhoods of their critical points. We shall say that a point $x_0 \in \mathsf{M}$ is a ***critical zero*** for a smooth function $\psi\colon \mathsf{M} \to \mathbb{R}$ if $\psi(x_0) = 0$ and $\mathrm{d}\psi(x_0) = 0_{x_0}$.

Let $(\mathsf{M}, \mathbb{G})$ be a Riemannian manifold and let $\overset{\mathbb{G}}{\nabla}$ be the corresponding Levi-Civita affine connection. The ***Hessian*** $\operatorname{Hess}\psi$ of ψ is the symmetric $(0,2)$-tensor field on M defined by

$$\operatorname{Hess}\psi(x) \cdot (v_x, w_x) = \langle\langle v_x, \overset{\mathbb{G}}{\nabla}_{w_x} \operatorname{grad}\psi \rangle\rangle,$$

for $v_x, w_x \in \mathsf{T}_x\mathsf{M}$. It is important to remark that $\operatorname{Hess}\psi$ depends on $\mathbb{G}$, and that $\operatorname{Hess}\psi$ is defined everywhere on M (not only at critical points of ψ). However, given a critical point x_0 for ψ, and given a chart $(\mathcal{U}, \phi)$ about x_0 with coordinates $(x^1, \dots, x^n)$, the Hessian at x_0 has components

$$(\operatorname{Hess}\psi(x_0))_{ij} = \frac{\partial^2 \psi}{\partial x^i \partial x^j}(x_0),$$

which are clearly independent of the choice of Riemannian metric. We leave it to the reader to prove these and related facts; see Exercise E6.6.

The Hessian of a smooth function at a critical point provides useful information about the function's local behavior in a neighborhood of the point. Let us recall the discussion in Section 2.3 on the notion of nondegenerate symmetric bilinear maps. A critical point x_0 for ψ is ***nondegenerate*** if the Hessian of ψ at x_0 is nondegenerate. Let $(x^1, \dots, x^n)$ be coordinates for M corresponding to a chart $(\mathcal{U}, \phi)$. Without loss of generality, let $\phi(x_0) = \mathbf{0} \in \mathbb{R}^n$. The local representative of the function ψ admits the following Taylor expansion about the critical nondegenerate zero x_0:

$$\psi(x^1, \dots, x^n) = \tfrac{1}{2}(\operatorname{Hess}\psi(x_0))_{ij} x^i x^j + o\big(\|(x^1, \dots, x^n)\|^3_{\mathbb{R}^n}\big). \tag{6.5}$$

This result can be stated in stronger terms as follows.

Lemma 6.23 (Morse Lemma). *Let the smooth function $\psi\colon \mathsf{M} \to \mathbb{R}$ have a nondegenerate critical zero x_0. Let k be the index of the symmetric bilinear map* $\operatorname{Hess}\psi(x_0)$. *Then there exists a chart $(\mathcal{U}, \phi)$ with coordinates $(x^1, \dots, x^n)$ such that $\phi(x_0) = \mathbf{0}$ and such that the local representative of ψ is*

$$(x^1, \dots, x^n) \mapsto \sum_{i=1}^{n-k-1} (x^i)^2 - \sum_{i=n-k}^{n} (x^i)^2.$$

For a concise proof of Morse Lemma, we refer to [Abraham and Marsden 1978]. Motivated by this result, we shall call $\psi\colon \mathsf{M} \to \mathbb{R}$ a ***Morse function*** if all of its critical points are nondegenerate.

An immediate consequence of this discussion is the observation that our definition of positive-definite functions does not necessarily imply that the function has a positive-definite Hessian.

Corollary 6.24. *If x_0 is a critical zero for the smooth function $\psi\colon \mathsf{M} \to \mathbb{R}$ and if the Hessian of ψ is positive-definite at x_0, then ψ is locally positive-definite about x_0. Conversely, if ψ is locally positive-definite about x_0, then x_0 is a critical zero for ψ and the Hessian of ψ is positive-semidefinite at x_0.*

We conclude this section by emphasizing the coordinate-free nature of the concepts we have introduced.

Remark 6.25. If the Hessian of a function at a critical point is nondegenerate or positive-definite in a coordinate chart, then it is so in every admissible coordinate chart. In other words, having a nondegenerate or positive-definite Hessian is a coordinate-invariant property. This is true because the differentiable structure of the manifold M restricts the admissible coordinate charts according to the overlap condition; see Definition 3.24 and Remark 3.27. This is not true on coordinate charts belonging to non-equivalent atlases; see Exercise E6.7. •

6.1.5 Exponential convergence

In this section we provide a coordinate-invariant notion of exponential convergence of a curve γ to a point x_0 of a Riemannian manifold $(\mathsf{M}, \mathbb{G})$. This will allow us to define a notion of exponential convergence for systems on manifolds, that resembles, but is weaker than, the notion of exponential stability for linear systems. Let us start by defining the notion of exponential convergence for functions of time.

A function $f\colon \bar{\mathbb{R}}_+ \to \mathbb{R}^n$ with $\lim_{t\to+\infty} f(t) = \mathbf{0}$ is said to converge to $\mathbf{0} \in \mathbb{R}^n$ ***exponentially fast***, or with ***exponential convergence rate***, if there exist a nonnegative time t_0 and positive constants α and β, such that for all $t > t_0$

$$\|f(t)\|_{\mathbb{R}^n} \leq \alpha e^{-\beta t}. \tag{6.6}$$

In what follows, we extend this definition to the setting of manifolds.

Proposition 6.26 (Exponential convergence). *Let $x_0 \in \mathsf{M}$ and consider a curve $\gamma\colon \bar{\mathbb{R}}_+ \to \mathsf{M}$ with $\lim_{t\to+\infty} \gamma(t) = x_0$. The following statements are equivalent:*

(i) there exists a smooth function $\psi\colon \mathsf{M} \to \mathbb{R}$ with critical zero x_0 and with $\operatorname{Hess}\psi(x_0)$ positive-definite, such that $\psi \circ \gamma\colon t \mapsto \psi(\gamma(t))$ converges to zero exponentially fast;

(ii) given any smooth function $\psi\colon \mathsf{M} \to \mathbb{R}$ with critical zero x_0 and with $\operatorname{Hess}\psi(x_0)$ positive-definite, $\psi \circ \gamma\colon t \mapsto \psi(\gamma(t))$ converges to zero exponentially fast;

(iii) there exists a chart $(\mathcal{U}, \phi)$ such that, for all $i \in \{1, \dots, n\}$, $t \mapsto \gamma^i(t)$ converges to zero exponentially fast, where $t \mapsto (\gamma^1(t), \dots, \gamma^n(t))$ is the local representative of γ;

(iv) given any chart $(\mathcal{U}, \phi)$ with $x_0 \in \mathcal{U}$ and $\phi(x_0) = \mathbf{0}$, $t \mapsto \gamma^i(t)$ converges to zero exponentially fast for all components γ^i of γ in $(\mathcal{U}, \phi)$.

If any, and therefore all, of these facts holds, we shall say that γ converges to x_0 with ***exponential convergence rate*** *or* ***exponentially fast***.

Proof. Clearly, part (iv) implies part (iii), and part (ii) implies part (i).

Next, we show that part (i) implies (iv). Assume $\psi \circ \gamma\colon t \mapsto \psi(\gamma(t))$ converges to zero exponentially fast with $\psi\colon \mathsf{M} \to \mathbb{R}$ having the property that x_0 is a critical zero and $\operatorname{Hess}\psi(x_0)$ is positive-definite. Let $(x^1, \dots, x^n)$ be coordinates corresponding to an arbitrary chart $(\mathcal{U}, \phi)$. Without loss of generality, let $\phi(x_0) = \mathbf{0} \in \mathbb{R}^n$. The Taylor expansion in equation (6.5) and Exercise E2.18

implies the existence of a neighborhood $\mathcal{W} \subset \mathcal{U}$ of the coordinate chart origin, such that, for any two constants $0 < b_1 < 1 < b_2$ and for all $(x^1, \dots, x^n)$ in $\mathcal{W}$,

$$\frac{b_1}{2}\lambda_{\min}([\operatorname{Hess}\psi(x_0)])\sum_{i=1}^{n}(x^i)^2 < \psi(x^1, \dots, x^n) < \frac{b_2}{2}\lambda_{\max}([\operatorname{Hess}\psi(x_0)])\sum_{i=1}^{n}(x^i)^2. \quad (6.7)$$

In other words, the function ψ can be bounded from above and from below by constants multiplying the square of the Euclidean norm of the coordinate vector $(x^1, \dots, x^n)$. Since γ converges to x_0, we know that there exists t_0 such that $\phi(\gamma(t)) \in \mathcal{W}$ for all $t > t_0$. Accordingly, for $t > t_0$,

$$\psi(\gamma^1(t), \dots, \gamma^n(t)) > \frac{b_1}{2}\lambda_{\min}([\operatorname{Hess}\psi(x_0)])\sum_{i=1}^{n}(\gamma^i(t))^2.$$

Since $t \mapsto \psi(\gamma^1(t), \dots, \gamma^n(t))$ converges exponentially fast, then $t \mapsto \sum_{i=1}^{n}(\gamma^i(t))^2$ converges exponentially fast, and, therefore, also $t \mapsto \gamma^i(t)$ must also do so for all $i \in \{1, \dots, n\}$.

Finally, we show that part (iii) implies (ii). We know that in a chart $(\mathcal{U}, \phi)$, all coordinate curves $t \mapsto \gamma^i(t)$ converge to zero exponentially fast. However, in the chart, the bounds in equation (6.7) still hold. Therefore, in a neighborhood of origin of the coordinate chart, we have

$$\psi(\gamma^1(t), \dots, \gamma^n(t)) < \frac{b_2}{2}\lambda_{\max}([\operatorname{Hess}\psi(x_0)])\sum_{i=1}^{n}(\gamma^i(t))^2.$$

Since $t \mapsto \gamma^i(t)$ converges exponentially fast for all $i \in \{1, \dots, n\}$, the curve $t \mapsto \sum_{i=1}^{n}(\gamma^i(t))^2$ converges exponentially fast. In turn, also $t \mapsto \psi(\gamma^1(t), \dots, \gamma^n(t))$ converges exponentially fast. ■

Remark 6.27. 1. For a linear system on $\mathbb{R}^n$, the origin is an exponentially stable equilibrium point if there exist $\alpha, \beta \in \mathbb{R}_+$ such that the integral curve from initial condition $\boldsymbol{x}_0 \in \mathbb{R}^n$ is bounded from above by $\alpha\|\boldsymbol{x}_0\|_{\mathbb{R}^n} \mathrm{e}^{-\beta t}$. For a dynamical system on a manifold M, we replace this notion with a weaker one. Namely, assuming that x_0 is a locally asymptotically stable equilibrium point, we investigate when solutions converge to x_0 exponentially fast (see (6.6)). The difference is that we do not require a dependency of the form $\alpha\|\boldsymbol{x}_0\|_{\mathbb{R}^n}$ in our notion of exponential convergence.

2. Note that for systems on a manifold, coordinate charts have to belong to the same atlas, and the notion of indefinite Hessian is independent of coordinate charts. However, if more general coordinate changes are allowed, then the results in [Grüne, Sontag, and Wirth 1999] show how exponential convergence is not independent of coordinate chart transformations.

In particular, local asymptotic stability is equivalent to local exponential stability on n-dimensional manifolds, for $n \in \mathbb{N} \setminus \{4,5\}$. The omission of the exceptional cases $n \in \{4,5\}$ is related to the presently outstanding Poincaré Conjecture; we refer the reader to Exercise E6.4 and to the relevant references [Dubrovin, Fomenko, and Novikov 1990, Milnor 1965a,b, 2003, Wilson, Jr. 1967]. •

6.1.6 Quadratic functions

Here we introduce a notion of quadratic function on a Riemannian manifold. This notion will later be useful to provide sufficient conditions for exponential convergence. From Section 3.6.2, recall that, on the Riemannian manifold $(\mathsf{M}, \mathbb{G})$, there is an induced pointwise norm $\|\cdot\|_{\mathbb{G}}$ on tensor fields on M.

Definition 6.28 (Quadratic function). Let $(\mathsf{M}, \mathbb{G})$ be a Riemannian manifold, let $A \subset \mathsf{M}$, and let $\psi\colon \mathsf{M} \to \mathbb{R}$ have a critical zero $x_0 \in A$. The function ψ is ***quadratic on A***, if there exist two constants $0 < b_1 \leq b_2$ such that

$$0 < b_1 \|\mathrm{d}\psi\|_{\mathbb{G}}^2 (x) \leq \psi(x) \leq b_2 \|\mathrm{d}\psi\|_{\mathbb{G}}^2 (x), \tag{6.8}$$

for all $x \in A \setminus \{x_0\}$. •

To provide a characterization of when a function is quadratic, and in particular of the set on which it is quadratic, we present the following consequence of Lemma 6.23.

Corollary 6.29. *Let the smooth function* $\psi\colon \mathsf{M} \to \mathbb{R}$ *have a critical zero* x_0 *with positive-definite* $\operatorname{Hess}\psi(x_0)$*. Then there exists* $\varepsilon > 0$ *such that* $\psi^{-1}(\leq\varepsilon, x_0)$ *is compact and contains no critical point other than* x_0*.*

Building on this corollary, we can define two scalar quantities that characterize the sublevel sets of a function ψ about a critical zero x_0 with positive-definite $\operatorname{Hess}\psi(x_0)$. We set

$$\begin{aligned} L_{\mathrm{cpt}}(\psi, x_0) &= \sup\{ L \in \mathbb{R} \mid \psi^{-1}(\leq L, x_0) \text{ is compact}\}, \\ L_{\mathrm{reg}}(\psi, x_0) &= \sup\{ L \in \mathbb{R} \mid x \in \psi^{-1}(\leq L, x_0) \setminus \{x_0\} \implies \mathrm{d}\psi(x) \neq 0\}, \end{aligned}$$

and we let $L_{\mathrm{reg,cpt}}(\psi, x_0) = \min\{L_{\mathrm{cpt}}(\psi, x_0), L_{\mathrm{reg}}(\psi, x_0)\}$. In other words, $L_{\mathrm{reg,cpt}}(\psi, x_0)$ bounds from above the largest L such that the connected component of the L-sublevel set of ψ is compact and contains no critical point other than x_0. We illustrate the situation in Figure 6.4.

Proposition 6.30. *Let* $(\mathsf{M}, \mathbb{G})$ *be a Riemannian manifold. Let the smooth function* $\psi\colon \mathsf{M} \to \mathbb{R}$ *have a critical zero* x_0 *with* $\operatorname{Hess}\psi(x_0)$ *positive-definite. Then* ψ *is quadratic on* $\psi^{-1}(\leq L, x_0)$ *for all* $L \in]0, L_{\mathrm{reg,cpt}}(\psi, x_0)[$*.*

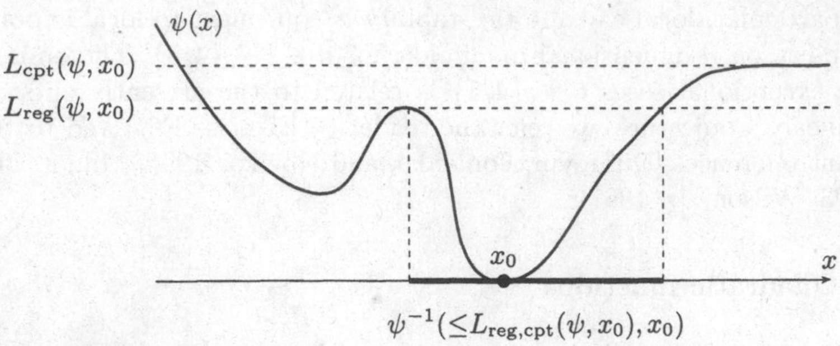

Figure 6.4. Example of $L_{\mathrm{reg,cpt}}(\psi, x_0) = \min\{L_{\mathrm{cpt}}(\psi, x_0), L_{\mathrm{reg}}(\psi, x_0)\}$, and of associated connected component of sublevel set

Proof. As in Lemma 6.23, choose a chart $(\mathcal{U}, \phi)$ with coordinates $(x^1, \dots, x^n)$ such that $\psi(x) = \sum_{i=1}^n (x^i)^2$. Because $\mathcal{U}$ is open, there exists $\varepsilon > 0$ such that $\psi^{-1}(\leq \varepsilon, x_0) = \{x \in \mathsf{M} \mid \|\phi(x)\|_{\mathbb{R}^n}^2 \leq \varepsilon\} \subset \mathcal{U}$. Let $[\mathbb{G}^{-1}]$ be the matrix representation of the tensor $\mathbb{G}^{-1}$ in the coordinate chart $(\mathcal{U}, \phi)$. On the compact set $\{x \in \mathcal{U} \mid \|\phi(x)\|_{\mathbb{R}^n}^2 \leq \varepsilon\}$, let ℓ_1, ℓ_2 be the minimum and maximum eigenvalues of $[\mathbb{G}^{-1}]$, and compute

$$\|\mathrm{d}\psi\|_{\mathbb{G}}^2(x) = \left\| 2 \sum_{i=1}^n x^i \mathrm{d}x^i \right\|_{\mathbb{G}}^2 = 4 \sum_{i,j=1}^n \mathbb{G}^{ij}(x) x^i x^j,$$
$$\implies \quad 4\ell_1 \psi(x) \leq \|\mathrm{d}\psi\|_{\mathbb{G}}^2(x) \leq 4\ell_2 \psi(x).$$

Therefore, equation (6.8) holds on $\psi^{-1}(\leq \varepsilon, x_0)$ for $b_1 = (4\ell_2)^{-1}$ and $b_2 = (4\ell_1)^{-1}$.

To complete the proof we need to prove the bound (6.8) on the set $\mathsf{N} = \psi^{-1}(\leq L, x_0) \setminus \psi^{-1}(\leq \varepsilon, x_0)$. It suffices to prove that the functions ψ and $\|\mathrm{d}\psi\|_{\mathbb{G}}$ are bounded from above and below by strictly positive constants on N. We prove this fact in three steps. First, note that the two functions ψ and $\|\mathrm{d}\psi\|_{\mathbb{G}}$ are bounded from above because N is compact. Second, since $L < L_{\mathrm{reg,cpt}}(\psi, x_0)$, we know that $\mathrm{d}\psi(x) \neq 0$ for all $x \in \psi^{-1}(\leq L, x_0)$, and, therefore, the minimum of $\|\mathrm{d}\psi\|_{\mathbb{G}}$ is strictly positive. Thirdly and finally, we need to prove that ψ is bounded from below by a strictly positive constant on N. Note that N is compact, $\mathrm{int}(\mathsf{N})$, $\psi|\mathsf{N}\colon \mathsf{N} \to [\varepsilon, L]$ has no critical points, and $\psi|\mathsf{N}$ on the boundary of N assumes the values $\{\varepsilon, L\}$. Under these assumptions, the Regular Interval Theorem [Hirsch 1976, page 153] guarantees that all level surfaces of $\psi|\mathsf{N}$ are diffeomorphic. Now, suppose that there exists a point x^* in N such that $\psi(x^*) = 0$. Then $\psi^{-1}(\leq \varepsilon, x^*)$ is a subset of $\psi^{-1}(\leq L, x^*) = \psi^{-1}(\leq L, x_0)$ because $\psi^{-1}(\leq L, x_0)$ is a connected set by definition. We have shown that the ε-level set of $\psi|\mathsf{N}$ contains at least two connected components, whereas the L-level set of $\psi|\mathsf{N}$ contains a single connected component. This is a contradiction because these two sets are diffeomorphic.

Therefore, there cannot exist a point $x^* \in \mathsf{N}$ with $\psi(x^*) = 0$ and ψ is bounded from below by a strictly positive constant. ■

Remark 6.31. Some global properties of the function ψ might allow one to immediately draw conclusions on $L_{\mathrm{cpt}}(\psi, x_0)$ or $L_{\mathrm{reg}}(\psi, x_0)$. Let us mention two such cases. First, recall that the function ψ is proper if the preimage of any compact set is compact. It is then clear that, if ψ is proper and bounded from below, then $L_{\mathrm{cpt}}(\psi, x_0) = +\infty$. Second, if the function ψ has a single critical point, then $L_{\mathrm{reg}}(\psi, x_0) = +\infty$. •

6.2 Stability analysis for equilibrium configurations of mechanical systems

Starting in this section, and for the remainder of this chapter, we consider a forced simple mechanical system $\Sigma = (\mathsf{Q}, \mathbb{G}, V, F)$ with time-independent external force F, and with governing equations

$$\overset{\mathbb{G}}{\nabla}_{\gamma'(t)}\gamma'(t) = -\mathrm{grad}V(\gamma(t)) + \mathbb{G}^\sharp(F(\gamma'(t))). \tag{6.9}$$

Note that we do not consider the stability of equilibria for systems with constraints; we refer to the introduction of the chapter for some references on this.

A point $q_0 \in \mathsf{Q}$ is an ***equilibrium configuration*** for Σ if the constant curve $t \mapsto q_0$ satisfies (6.9). An equilibrium configuration q_0 gives rise to the equilibrium point 0_{q_0} for the simple mechanical system, thought of as a vector field on TQ. The stability notions of Definition 6.3 can be made for the equilibrium configuration q_0 by reference to the equilibrium point 0_{q_0}.

Lemma 6.32 (Characterization of equilibrium configurations). *Consider a forced simple mechanical system* $(\mathsf{Q}, \mathbb{G}, V, F)$ *with time-independent external force* F*, and let* $\gamma\colon I \to \mathsf{Q}$ *be a solution to* (6.9). *Then* $q_0 \in \mathsf{Q}$ *is an equilibrium configuration if and only if* $\mathrm{d}V(q_0) = F(0_{q_0})$. *If the force* F *has the property that* $F(0_q) = 0_q$ *for each* $q \in \mathsf{Q}$*, then the equilibrium configurations are exactly the critical points of* V.

To study the stability properties of equilibrium configurations, we perform two stability analyses. First, we study the linearization of a mechanical control system about an equilibrium. We do this by studying the tangent space $\mathsf{T}_{0_{q_0}}\mathsf{TQ}$ and the spectrum of the linearized flow. Second, we present a Lyapunov stability analysis based on the use of energy as a Lyapunov function.

6.2.1 Linearization of simple mechanical systems

In this section we indicate how one can linearize a forced simple mechanical system about a controlled equilibrium, and so arrive at a forced linear mechanical system. We refer to Section 4.6.5 for the definitions concerning forced linear mechanical systems. We rewrite (6.9) in first-order form

$$\Upsilon'(t) = \overset{\mathbb{G}}{S}(\Upsilon(t)) - \operatorname{vlft}(\operatorname{grad}V)(\Upsilon(t)) + \operatorname{vlft}(\mathbb{G}^\sharp(F))(\Upsilon(t)), \qquad (6.10)$$

which has 0_{q_0} as an equilibrium point. This is then a system like that described in Section 6.1.2, and so can be linearized in the standard way. However, the mechanical structure leads to a linearization with a special form. First we describe the vector space $\mathsf{T}_{0_{q_0}}\mathsf{TQ}$ in which the linearization lives.

Lemma 6.33. *There is a natural isomorphism of* $\mathsf{T}_{0_{q_0}}\mathsf{TQ}$ *with* $\mathsf{T}_{q_0}\mathsf{Q} \oplus \mathsf{T}_{q_0}\mathsf{Q}$.

Proof. Passing through the point 0_{q_0} are two submanifolds of TQ, the zero section $Z(\mathsf{TQ})$ and the tangent space $\mathsf{T}_{q_0}\mathsf{Q}$. Clearly $\mathsf{T}_{0_{q_0}}(Z(\mathsf{TQ}))$ is naturally isomorphic to $\mathsf{T}_{q_0}\mathsf{Q}$, since the map $q \mapsto 0_q$ is a diffeomorphism of Q with $Z(\mathsf{TQ})$. Also, the map $v_{q_0} \mapsto \operatorname{vlft}_{0_{q_0}}(v_{q_0})$ is an isomorphism of $\mathsf{T}_{q_0}\mathsf{Q}$ with $\mathsf{T}_{0_{q_0}}(\mathsf{T}_{q_0}\mathsf{Q})$. Since $\mathsf{T}_{0_{q_0}}(Z(\mathsf{TQ})) \cap \mathsf{T}_{0_{q_0}}(\mathsf{T}_{q_0}\mathsf{Q}) = \{0_{q_0}\}$, the result follows. ■

We shall adopt the convention that the first component in the decomposition $\mathsf{T}_{0_{q_0}}\mathsf{TQ} \simeq \mathsf{T}_{q_0}\mathsf{Q} \oplus \mathsf{T}_{q_0}\mathsf{Q}$ is tangent to the zero section at 0_{q_0} and that the second component is tangent to the fiber of TQ at 0_{q_0}. It is common to say that the first component is "horizontal" and that the second is "vertical."

To linearize (6.10), we linearize separately the three terms on the right-hand side. We shall record the intermediate steps as three lemmas, with Proposition 6.37 summarizing these steps.

Lemma 6.34. *Let* $(\mathsf{Q}, \mathbb{G})$ *be a Riemannian manifold and let* $\overset{\mathbb{G}}{S}$ *be the geodesic spray of the associated Levi-Civita affine connection. Then* $\overset{\mathbb{G}}{S}{}^T(0_{q_0})(v_1 \oplus v_2) = v_2 \oplus 0_{q_0}$, *for* $v_1 \oplus v_2 \in \mathsf{T}_{q_0}\mathsf{Q} \oplus \mathsf{T}_{q_0}\mathsf{Q} \simeq \mathsf{T}_{0_{q_0}}\mathsf{TQ}$.

Proof. This follows directly from the coordinate expressions for $\overset{\mathbb{G}}{S}$ and for the tangent lift of a vector field; see Sections 3.5.3 and 3.8.3. ■

To provide the formula for the linearization of $\operatorname{vlft}(\operatorname{grad}V)$, we use the notion of the Hessian. Recall from Section 6.1.6 and from Exercise E6.6 that if q_0 is a critical point for V, then the second derivative of V at q_0 is independent of coordinates, and is denoted by $\operatorname{Hess}V(q_0) \in \Sigma_2(\mathsf{T}_{q_0}\mathsf{Q})$.

Lemma 6.35. *If* q_0 *is a critical point for* V, *then*

$$(\operatorname{vlft}(\mathbb{G}^\sharp \circ \mathrm{d}V))^T(0_{q_0})(v_1 \oplus v_2) = 0_{q_0} \oplus (\mathbb{G}(q_0)^\sharp \circ \operatorname{Hess}V(q_0)^\flat(v_1)).$$

Proof. Again, this is easily proved in coordinates. ■

To consider the linearization of a general time-independent force $F \colon \mathsf{TQ} \to \mathsf{T}^*\mathsf{Q}$, we proceed as follows. We will make the assumption that $F(Z(\mathsf{TQ})) = Z(\mathsf{T}^*\mathsf{Q})$; this assumption will be satisfied in the cases we consider below. Given $v_{q_0} \in \mathsf{T}_{q_0}\mathsf{Q}$, let $\gamma_1 \colon [-\varepsilon, \varepsilon] \to Z(\mathsf{TQ})$ be a differentiable curve for which $\gamma_1(0) = 0_{q_0}$ and $\gamma_1'(0) = v_{q_0} \oplus 0_{q_0}$. By the properties of F, the curve $t \mapsto F(\gamma_1(t))$ takes values in $Z(\mathsf{T}^*\mathsf{Q}) \simeq \mathsf{Q}$. Note that $\frac{\mathrm{d}}{\mathrm{d}t}\big|_{t=0} F(\gamma_1(t)) \in \mathsf{T}_{0_{q_0}}\mathsf{T}^*\mathsf{Q}$. Furthermore, one can verify that $\mathsf{T}_{0_{q_0}}\mathsf{T}^*\mathsf{Q} \simeq \mathsf{T}_{q_0}\mathsf{Q} \oplus \mathsf{T}^*_{q_0}\mathsf{T}^*\mathsf{Q}$, entirely along the lines of Lemma 6.34. Since $t \mapsto F(\gamma_1(t))$ is a curve in $Z(\mathsf{T}^*\mathsf{Q})$, we have that $\frac{\mathrm{d}}{\mathrm{d}t}\big|_{t=0} F(\gamma_1(t)) = v_{q_0} \oplus \mathrm{d}_1 F(q_0) \cdot v_{q_0}$. This defines $\mathrm{d}_1 F(q_0) \in L(\mathsf{T}_{q_0}\mathsf{Q}; \mathsf{T}^*_{q_0}\mathsf{Q})$. Next, given $v_{q_0} \in \mathsf{T}_{q_0}\mathsf{Q}$, let $\gamma_2 \colon [-\varepsilon, \varepsilon] \to \mathsf{T}_{q_0}\mathsf{Q}$ be a differentiable curve in TQ for which $\gamma_2(0) = 0_{q_0}$ and $\gamma_2'(0) = 0_{q_0} \oplus v_{q_0}$. Since F is a bundle map, the curve $t \mapsto F(\gamma_2(t))$ takes values in $\mathsf{T}^*_{q_0}\mathsf{Q}$. We then define $\mathrm{d}_2 F(q_0) \in L(\mathsf{T}_{q_0}; \mathsf{T}^*_{q_0}\mathsf{Q})$ by

$$\mathrm{d}_2 F(q_0) \cdot v_{q_0} = \frac{\mathrm{d}}{\mathrm{d}t}\Big|_{t=0} F(\gamma_2(t)).$$

In coordinates $(q^1, \dots, q^n)$, the components of $\mathrm{d}_1 F(q_0)$ and $\mathrm{d}_2 F(q_0)$ in the standard basis are

$$\frac{\partial F_i}{\partial q^j}(0_{q_0}), \qquad i, j \in \{1, \dots, n\},$$
$$\frac{\partial F_i}{\partial v^j}(0_{q_0}), \qquad i, j \in \{1, \dots, n\},$$

respectively.

With these definitions, the following result follows by direct computation.

Lemma 6.36. *Let $F \colon \mathsf{TQ} \to \mathsf{T}^*\mathsf{Q}$ be a time-independent force satisfying $F(Z(\mathsf{TQ})) = Z(\mathsf{T}^*\mathsf{Q})$. Then*

$$(\mathrm{vlft}(\mathbb{G}^\sharp \circ F))^T(0_{q_0})(v_1 \oplus v_2) = 0_{q_0} \oplus \left(\mathbb{G}(q_0)^\sharp \circ \mathrm{d}_1 F(q_0) \cdot v_1 + \mathbb{G}(q_0)^\sharp \circ \mathrm{d}_2 F(q_0) \cdot v_2\right).$$

Assembling the preceding computations together, we obtain the following result.

Proposition 6.37 (Linearization of forced simple mechanical system). *Let $\Sigma = (\mathsf{Q}, \mathbb{G}, V, F)$ be a C^∞-forced simple mechanical system and assume that*

(i) $\mathrm{d}V(q_0) = 0$,
(ii) F is time-independent, and
(iii) $F(Z(\mathsf{TQ})) = Z(\mathsf{T}^*\mathsf{Q})$.

Then the linearization of (6.10) *at 0_{q_0} in the decomposition $\mathsf{T}_{0_{q_0}}\mathsf{TQ} \simeq \mathsf{T}_{q_0}\mathsf{Q} \oplus \mathsf{T}_{q_0}\mathsf{Q}$ is given by*

$$A_\Sigma(q_0) = \begin{bmatrix} 0 & \mathrm{id}_{\mathsf{T}_{q_0}\mathsf{Q}} \\ -\mathbb{G}(q_0)^\sharp \circ \mathrm{Hess}\, V(q_0)^\flat + \mathbb{G}(q_0)^\sharp \circ \mathrm{d}_1 F(q_0) & \mathbb{G}(q_0)^\sharp \circ \mathrm{d}_2 F(q_0) \end{bmatrix}.$$

That is to say, the linearization is given by the forced linear mechanical system $(\mathsf{T}_{q_0}\mathsf{Q}, \mathbb{G}(q_0), \mathrm{Hess}\, V(q_0), (\mathrm{d}_1 F(q_0), \mathrm{d}_2 F(q_0)))$.

Remark 6.38. It is also possible to consider linearization for forced affine connection systems, and in so doing allow the consideration of constraints. This topic is discussed in [DeMarco 1999, Krasinskiĭ 1988, Neĭmark and Fufaev 1972, Wang and McClamroch 1994]. The reader is asked to produce the form of the linearization in Exercise E6.12. The issues surrounding the stability of these more general systems is somewhat more complicated than the setup we consider here. •

6.2.2 Linear stability analysis for unforced systems

Next we consider the stability analysis for unforced linear mechanical systems, and then we provide the implications that can be made, using the stability of the linearization, for the stability of an unforced simple mechanical system.

We first describe the eigenvalues of an unforced linear mechanical system.

Proposition 6.39 (Eigenvalues of linear mechanical systems). *Let* $\Sigma = (\mathsf{V}, M, K)$ *be a linear mechanical system. The following statements hold:*

(i) the eigenvalues of $M^\sharp \circ K^\flat$ *are real;*

(ii) if $\lambda \in \mathbb{C}$ *is an eigenvalue of* $A_{M,K}$ *of algebraic multiplicity* k*, then there exists an eigenvalue* $\ell \in \mathbb{R}$ *of* $M^\sharp \circ K^\flat$ *of algebraic multiplicity* $\frac{k}{2}$ *such that* $\lambda = \pm\sqrt{-\ell}$*.*

Proof. The first assertion follows, since $M^\sharp \circ K^\flat$ is symmetric with respect to the inner product M, as may be checked by direct computation.

For the second assertion, note that $\lambda \in \mathbb{C}$ is an eigenvalue of $A_{M,K}$ if and only if

$$\det \begin{bmatrix} -\lambda \mathrm{id}_{\mathsf{V}} & \mathrm{id}_{\mathsf{V}} \\ -M^\sharp \circ K^\flat & -\lambda \mathrm{id}_{\mathsf{V}} \end{bmatrix} = \det(\lambda^2 \mathrm{id}_{\mathsf{V}} + M^\sharp \circ K^\flat) = 0.$$

This shows that λ is an eigenvalue for $A_{M,K}$ if and only if $-\lambda^2$ is an eigenvalue for $M^\sharp \circ K^\flat$. The assertions concerning algebraic multiplicity are also evident from the form of the characteristic polynomial of $A_{M,K}$. ■

The following characterization of the eigenvalues of $M^\sharp \circ K^\flat$ follows directly from Exercise E2.22.

Corollary 6.40. *For a linear mechanical system* $\Sigma = (\mathsf{V}, M, K)$ *with* $\dim(\mathsf{V}) = n$*, the following statements hold:*

(i) the number of zero eigenvalues of $M^\sharp \circ K^\flat$ *is* $n - \mathrm{rank}(\mathrm{Hess}\, V(q_0))$*;*

(ii) the number of negative eigenvalues of $M^\sharp \circ K^\flat$ *is* $\mathrm{ind}(\mathrm{Hess}\, V(q_0))$*;*

(iii) the number of positive eigenvalues of $M^\sharp \circ K^\flat$ *is* $\mathrm{rank}(\mathrm{Hess}\, V(q_0)) - \mathrm{ind}(\mathrm{Hess}\, V(q_0))$.

The next result uses the preceding eigenvalue characterization to derive the conditions for linear stability of an equilibrium configuration.

Proposition 6.41 (Stability of linear mechanical systems). *Let* $\Sigma = (\mathsf{V}, M, K)$ *be a linear mechanical system with equilibrium configuration* $0 \in \mathsf{V}$*. The following statements hold:*

(i) 0 *is spectrally stable if and only if* $\mathrm{ind}(K) = 0$*;*

(ii) 0 *is stable if and only if* $\mathrm{rank}(K) = n$ *and* $\mathrm{ind}(K) = 0$*;*

(iii) 0 *is unstable if* $\mathrm{ind}(K) > 0$.

Proof. (i) This follows, since there are no eigenvalues of $A_{M,K}$ in $\mathbb{C}_+$ if and only if $M^\sharp \circ K^\flat$ has no negative eigenvalues. From part (iii) of Corollary 6.40 the result then follows.

(ii) For linear stability, we must have all eigenvalues of $A_{M,K}$ on the imaginary axis. This means that all eigenvalues of $M^\sharp \circ K^\flat$ must be nonnegative. We shall show that they must further be strictly positive, and that this is in fact necessary and sufficient for linear stability. First suppose that all eigenvalues of $M^\sharp \circ K^\flat$ are strictly positive. The only problem that can arise is when we have eigenvalues whose algebraic and geometric multiplicities are not equal. To resolve this, let $\{e_1, \dots, e_n\}$ be a basis in which the matrix of $M^\sharp \circ K^\flat$ is diagonal, with the diagonal entries therefore being the eigenvalues. Let $\ell \in \mathbb{R}$ be an eigenvalue of $M^\sharp \circ K^\flat$ and suppose that the basis $\{e_1, \dots, e_n\}$ is arranged so that the matrix for $M^\sharp \circ K^\flat$ has the form

$$\begin{bmatrix} \ell \boldsymbol{I}_k & \boldsymbol{0} \\ \boldsymbol{0} & \boldsymbol{D} \end{bmatrix},$$

where $\boldsymbol{D}$ is a diagonal matrix with diagonal entries d_i, $i \in \{1, \dots, n-k\}$, none of whose entries are ℓ. Corresponding to the eigenvalue ℓ of $M^\sharp \circ K^\flat$ are the eigenvalues $\pm\sqrt{-\ell}$ for $A_{M,K}$.

Let us look at the eigenvalue $\lambda = \sqrt{-\ell}$. In the basis we have for $\mathsf{V} \oplus \mathsf{V}$, the matrix for $A_{M,K} - \lambda \mathrm{id}_{\mathsf{V}^2}$ has the form

$$\begin{bmatrix} -\lambda \boldsymbol{I}_k & \boldsymbol{0} & \boldsymbol{I}_k & \boldsymbol{0} \\ \boldsymbol{0} & -\lambda \boldsymbol{I}_{n-k} & \boldsymbol{0} & \boldsymbol{I}_{n-k} \\ -\ell \boldsymbol{I}_k & \boldsymbol{0} & -\lambda \boldsymbol{I}_k & \boldsymbol{0} \\ \boldsymbol{0} & -\boldsymbol{D} & \boldsymbol{0} & -\lambda \boldsymbol{I}_{n-k} \end{bmatrix}.$$

First suppose that $\ell \neq 0$. By performing the row operations $R_{n+i} + \sqrt{-\ell} R_i$, $i \in \{1, \dots, k\}$, we obtain the matrix

$$\begin{bmatrix} -\lambda \boldsymbol{I}_k & \boldsymbol{0} & \boldsymbol{I}_k & \boldsymbol{0} \\ \boldsymbol{0} & -\lambda \boldsymbol{I}_{n-k} & \boldsymbol{0} & \boldsymbol{I}_{n-k} \\ \boldsymbol{0} & \boldsymbol{0} & \boldsymbol{0} & \boldsymbol{0} \\ \boldsymbol{0} & -\boldsymbol{D} & \boldsymbol{0} & -\lambda \boldsymbol{I}_{n-k} \end{bmatrix}.$$

Now, performing the row operations $R_{n+k+i} - \frac{d_i}{\sqrt{-\ell}} R_{k+i}$, $i \in \{1, \dots, n-k\}$, we obtain the matrix

$$\begin{bmatrix} -\lambda \boldsymbol{I}_k & \boldsymbol{0} & \boldsymbol{I}_k & \boldsymbol{0} \\ \boldsymbol{0} & -\lambda \boldsymbol{I}_{n-k} & \boldsymbol{0} & \boldsymbol{I}_{n-k} \\ \boldsymbol{0} & \boldsymbol{0} & \boldsymbol{0} & \boldsymbol{0} \\ \boldsymbol{0} & \boldsymbol{0} & \boldsymbol{0} & -\lambda \boldsymbol{I}_{n-k} - \frac{1}{\sqrt{-\ell}} \boldsymbol{D} \end{bmatrix}.$$

Thus we see that $\dim(\ker(A_{M,K} - \lambda \mathrm{id}_{\mathsf{V}^2})) = k$. Thus we have shown that nonzero imaginary eigenvalues for $M^\sharp \circ K^\flat$ give rise to eigenvalues for $A_{M,K}$

whose geometric and algebraic multiplicities are equal. By Theorem 6.6, this proves that q_0 is linearly stable if all eigenvalues of $M^\sharp \circ K^\flat$ are positive.

Finally, suppose that ℓ is a zero eigenvalue for $M^\sharp \circ K^\flat$ of algebraic multiplicity k. By Proposition 6.39, this gives rise to a zero eigenvalue of $A_{M,K}$ of algebraic multiplicity $2k$. We choose a basis $\{e_1, \dots, e_n\}$ for V as above so that the matrix for $A_{M,K}$ has the form

$$\begin{bmatrix} \mathbf{0} & \mathbf{0} & \boldsymbol{I}_k & \mathbf{0} \\ \mathbf{0} & \mathbf{0} & \mathbf{0} & \boldsymbol{I}_{n-k} \\ \mathbf{0} & \mathbf{0} & \mathbf{0} & \mathbf{0} \\ \mathbf{0} & \boldsymbol{D} & \mathbf{0} & \mathbf{0} \end{bmatrix}.$$

The rank of this matrix is k, which shows that zero eigenvalues of the linearization of a simple mechanical system *never* have equal algebraic and geometric multiplicities. Therefore, by Theorem 6.6, if $M^\sharp \circ K^\flat$ has a zero eigenvalue, q_0 is linearly unstable. This part of the proposition now follows from Corollary 6.40(i).

(iii) Recall from Corollary 6.40(iii) that $A_{M,K}$ has a nonzero real eigenvalue if and only if $\operatorname{ind}(K(q_0)) > 0$. The result now follows from Theorem 6.6. ■

6.2.3 Linear stability analysis for systems subject to Rayleigh dissipation

Now we consider the special case of a forced simple mechanical system, where the force is determined by a Rayleigh dissipation function. We first consider the case of a linear mechanical system, and then apply these results to an equilibrium configuration for a simple mechanical system.

For a linear mechanical system (V, M, K), a ***linear Rayleigh dissipation function*** is a symmetric, positive-semidefinite $(0,2)$-tensor R on V. This gives rise to the forced linear mechanical system $(\mathsf{V}, M, K, (0, -R^\flat))$, which we say is subject to a ***linear dissipative force***. The governing equations are

$$M^\flat(\ddot{x}(t)) + R^\flat(\dot{x}(t)) + K^\flat(x(t)) = 0.$$

The following result gives some statements regarding the stability of this system. The form of the asymptotic stability part of the following result is due to Walker and Schmitendorf [1973] and Miller and Michel [1980], and relies on the notation $\langle \mathscr{L}, \mathsf{U} \rangle$ introduced on Page 29 for a subset $\mathscr{L} \subset L(\mathsf{V}; \mathsf{V})$ and a subspace $\mathsf{U} \subset \mathsf{V}$.

Theorem 6.42 (Stability of linear mechanical systems with dissipation). *Let $\Sigma = (\mathsf{V}, M, K, (0, -R^\flat))$ be a linear mechanical system subject to a linear dissipative force. For the equilibrium configuration $0 \in \mathsf{V}$, the following statements hold:*

(i) Σ is unstable if K is not positive-semidefinite;

(ii) if R is positive-semidefinite, then Σ is stable if and only if K is positive-semidefinite and $\ker(K^\flat) \cap \ker(R^\flat) = \{0\}$;

(iii) if R is positive-semidefinite, then Σ is asymptotically stable if and only if K is positive-definite and $\langle M^\sharp \circ K^\flat, \mathrm{image}(M^\sharp \circ R^\flat)\rangle = \mathsf{V}$.

Proof. The $\mathbb{R}$-inner product M on V may be extended to a $\mathbb{C}$-inner product $M_{\mathbb{C}}$ on $\mathsf{V}_{\mathbb{C}}$ by

$$M_{\mathbb{C}}((u_1, v_1), (u_2, v_2)) = \big(M(u_1, u_2) - M(v_1, v_2)\big) + \mathrm{i}\big(M(u_1, v_2) + M(u_2, v_1)\big).$$

Similarly, we may extend K and R to $(0,2)$-tensors $K_{\mathbb{C}}$ and $R_{\mathbb{C}}$ on $\mathsf{V}_{\mathbb{C}}$ that satisfy all the properties of an inner product with the exception of positive-definiteness, i.e., to Hermitian $(0,2)$-tensors on $\mathsf{V}_{\mathbb{C}}$ as in Remark 2.41. One can then check that the resulting $\mathbb{C}$-linear map $M_{\mathbb{C}}^\sharp \circ K_{\mathbb{C}}^\flat$ on $\mathsf{V}_{\mathbb{C}}$ satisfies

$$M_{\mathbb{C}}(M_{\mathbb{C}}^\sharp \circ K_{\mathbb{C}}^\flat(v_1), v_2) = M_{\mathbb{C}}(v_1, M_{\mathbb{C}}^\sharp \circ K_{\mathbb{C}}^\flat(v_2)).$$

It follows that the eigenvalues of $M_{\mathbb{C}}^\sharp \circ K_{\mathbb{C}}^\flat$ will be real, and equal to the eigenvalues of $M^\sharp \circ K^\flat$, cf. Exercise E2.15.

Let $A_{M,K,R} \colon \mathsf{V} \oplus \mathsf{V} \to \mathsf{V} \oplus \mathsf{V}$ be the linear map

$$A_{M,K,R} = \begin{bmatrix} 0 & \mathrm{id}_{\mathsf{V}} \\ -M^\sharp \circ K^\flat & -M^\sharp \circ R^\flat \end{bmatrix},$$

and let $A_{M,K,R}^{\mathbb{C}}$ denote its extension to $\mathsf{V}_{\mathbb{C}} \oplus \mathsf{V}_{\mathbb{C}}$. The eigenvalues of $A_{M,K,R}$ satisfy

$$\begin{bmatrix} v_2 \\ -M_{\mathbb{C}}^\sharp \circ K_{\mathbb{C}}^\flat(v_1) - M_{\mathbb{C}}^\sharp \circ R_{\mathbb{C}}^\flat(v_2) \end{bmatrix} = \begin{bmatrix} \lambda v_1 \\ \lambda v_2 \end{bmatrix},$$

for some nonzero $v_1 \oplus v_2 \in \mathsf{V}_{\mathbb{C}} \oplus \mathsf{V}_{\mathbb{C}}$. We then compute

$$\begin{aligned} M_{\mathbb{C}}(\bar\lambda v_1, \lambda v_1) &= M_{\mathbb{C}}(\bar\lambda v_1, v_2) = M_{\mathbb{C}}(v_1, \lambda v_2) \\ &= M_{\mathbb{C}}(v_1, -M_{\mathbb{C}}^\sharp \circ K_{\mathbb{C}}^\flat(v_1) - M_{\mathbb{C}}^\sharp \circ R_{\mathbb{C}}^\flat(v_2)) \\ &= -M_{\mathbb{C}}(v_1, M_{\mathbb{C}}^\sharp \circ K_{\mathbb{C}}^\flat(v_1)) - \lambda M_{\mathbb{C}}(v_1, M_{\mathbb{C}}^\sharp \circ R_{\mathbb{C}}^\flat(v_1)), \end{aligned}$$

where symmetry of $M_{\mathbb{C}}^\sharp \circ R_{\mathbb{C}}^\flat$ is used in the last step. We also have $M_{\mathbb{C}}(\bar\lambda v_1, \lambda v_1) = \lambda^2 M_{\mathbb{C}}(v_1, v_1)$, thus giving

$$M_{\mathbb{C}}(v_1, v_1)\lambda^2 + R_{\mathbb{C}}(v_1, v_1)\lambda + K_{\mathbb{C}}(v_1, v_1) = 0. \tag{6.11}$$

Now note that if $v_1 = 0$, then we also have $v_2 = 0$. Since eigenvectors are necessarily nonzero, this implies that $v_1 \neq 0$. Thus λ satisfies the equation

$$a\lambda^2 + b\lambda + c = 0, \tag{6.12}$$

where $a > 0$, since $v_1 \neq 0$.

The following lemma describes those eigenvalues for the system on the imaginary axis.

Lemma. *Let λ be an imaginary eigenvalue for $A_{M,K,R}$. Then λ is an eigenvalue for $A_{M,K}$ and*

(i) if $\lambda = 0$, then $m_a(\lambda) = \dim(\ker(K^\flat)) + \dim(\ker(R^\flat))$ and $m_g(\lambda) = \dim(\ker(K^\flat))$, and

(ii) if $\lambda \neq 0$, then $m_a(\lambda) = m_g(\lambda)$.

Proof. From (6.11) it is clear that imaginary eigenvalues of $A_{M,K,R}$ are also eigenvalues of $A_{M,K}$.

(i) Let $\{e_1, \dots, e_n\}$ be an M-orthonormal basis for V with the properties

1. $\{e_1, \dots, e_k\}$ forms a basis for $\ker(K^\flat_{\mathbb{C}}) \cap \ker(R^\flat_{\mathbb{C}})$, and
2. $\{e_1, \dots, e_{k+l}\}$ forms a basis for $\ker(K^\flat)$.

Then the matrix representation for $A_{M,K,R}$ in this basis has the form

$$\begin{bmatrix} \mathbf{0} & \mathbf{0} & \boldsymbol{I}_{k+l} & \mathbf{0} \\ \mathbf{0} & \mathbf{0} & \mathbf{0} & \boldsymbol{I}_{n-k-l} \\ \mathbf{0} & \mathbf{0} & \boldsymbol{R}_{11} & \boldsymbol{R}_{12} \\ \mathbf{0} & \boldsymbol{K} & \boldsymbol{R}_{21} & \boldsymbol{R}_{22} \end{bmatrix}$$

for suitably-sized matrices $\boldsymbol{K}_1$ and $\boldsymbol{R}_{ij}$, $i,j \in \{1,2\}$, and with $\boldsymbol{M}_1$ invertible. We then immediately see that $\dim(\ker(A_{M,K,R})) = \dim(\ker(K^\flat))$, so giving $m_g(0) = \dim(\ker(K^\flat))$, as claimed. Furthermore, corresponding to each vector $e_1, \dots, e_k$, there are two zero roots of (6.12), and corresponding to each vector $e_{k+1}, \dots, e_{k+l}$ there is one zero root of (6.12). This gives $m_a(\lambda) = \dim(\ker(K^\flat)) + \dim(\ker(R^\flat))$, as claimed.

(ii) Suppose that $\lambda = \sqrt{-\ell}$ for a positive eigenvalue ℓ of $M^\sharp \circ K^\flat$. Let $\{e_1, \dots, e_k\}$ be an M-orthonormal basis for the corresponding eigenspace of $M^\sharp \circ K^\flat$, and extend this to an M-orthonormal basis $\{e_1, \dots, e_n\}$ for V consisting of eigenvectors for $M^\sharp \circ K^\flat$. Note that, since λ is an eigenvalue of $A_{M,K,R}$, we must have $R^\flat(e_j) = 0$, $j \in \{1, \dots, k\}$. Therefore, the matrix representative for $A^{\mathbb{C}}_{M,K,R} - \lambda_{\mathsf{V}_{\mathbb{C}} \oplus \mathsf{V}_{\mathbb{C}}}$ in this basis is

$$\begin{bmatrix} -\lambda \boldsymbol{I}_k & \mathbf{0} & \boldsymbol{I}_k & \mathbf{0} \\ \mathbf{0} & -\lambda \boldsymbol{I}_{n-k} & \mathbf{0} & \boldsymbol{I}_{n-k} \\ \mathbf{0} & \mathbf{0} & -\lambda \boldsymbol{I}_k & \mathbf{0} \\ \mathbf{0} & \boldsymbol{D} & \mathbf{0} & \boldsymbol{R}_{22} - \lambda \boldsymbol{I}_{n-k} \end{bmatrix},$$

for some symmetric matrix $\boldsymbol{R}_{22}$ and some diagonal matrix $\boldsymbol{D}$. The same sequence of row operations as used in the proof of part (ii) of Proposition 6.41 gives the same conclusion, namely that $m_g(\lambda) = m_a(\lambda)$. ▼

(i) This part of the result may be proved with a Lyapunov-type *instability* result. We do not develop this here, but refer to [Krasovskiĭ 1963, Merkin 1997], particularly Theorem 6.6 of the latter reference.

(ii) From part (i) we know that K must be positive-semidefinite. From (6.12) it then follows that $\operatorname{spec}(A_{M,K,R}) \subset \overline{\mathbb{C}}_-$. Therefore, the system will be stable if and only if all eigenvalues on the imaginary axis have equal algebraic and geometric multiplicities. This part of the result now follows immediately from the lemma.

(iii) Suppose that the system is asymptotically stable. Since the system is stable, K must be positive-semidefinite. From the lemma, K must be positive definite for asymptotic stability, since otherwise $A_{M,K,R}$ will have zero eigenvalues. Conversely, if K is positive-definite and R is positive-semidefinite, it follows that all roots of (6.12) will lie in $\overline{\mathbb{C}}_-$. Furthermore, since K is positive-definite, it follows that there are no zero eigenvalues for $A_{M,K,R}$. From the lemma we then know that the system is linearly stable, and so the system will therefore be asymptotically stable if and only if there are no periodic trajectories corresponding to the imaginary eigenvalues. From our computations of the lemma, we know that the system is stable, but not asymptotically stable, if and only if $\ker(R^\flat)$ has nontrivial intersection, say W, with an eigenspace U_ℓ, where ℓ is a positive eigenvalue for $M^\sharp \circ K^\flat$. Since R is symmetric, this is equivalent to asking that W be M-orthogonal to $\operatorname{image}(M^\sharp \circ R^\flat)$ (Exercise E2.17). Note that $\mathsf{W}^\perp$ is then $M^\sharp \circ K^\flat$-invariant, since K is symmetric, and contains $\operatorname{image}(M^\sharp \circ R^\flat)$. Thus we have argued that there are imaginary eigenvalues of $A_{M,K,R}$ if and only if there is some nontrivial subspace whose orthogonal complement is $M^\sharp \circ K^\flat$-invariant and contains $\operatorname{image}(M^\sharp \circ R^\flat)$. By definition of $\langle M^\sharp \circ K^\flat, \operatorname{image}(M^\sharp \circ R^\flat)\rangle$ this part of the result follows. ∎

Remarks 6.43. 1. The preceding result may be improved by a detailed analysis of the symmetric linear maps $M^\sharp \circ K^\flat$ and $M^\sharp \circ R^\flat$. For example, it is possible to relax the assumption of the positive-semidefiniteness of R (cf. Exercise E6.5). However, this is the realm of classic dynamics, and we do not wish to discuss these matters in detail. We refer to references such as [Chetaev 1955, Merkin 1997].

2. The instability result of part (iii) holds in the presence of gyroscopic forces. The reader is asked to explore this in Exercise E6.10. •

Now we extend the preceding result to the case in which one can infer stability properties of an equilibrium configuration for a simple mechanical system with dissipation from its linearization. Note that if $\Sigma = (\mathsf{Q}, \mathbb{G}, V, F = -R^\flat_{\text{diss}})$ is a simple mechanical system subjected to the force associated with a Rayleigh dissipation function R_{diss}, then one can readily verify that $\mathrm{d}_1 F(q_0) = 0$ and $\mathrm{d}_2 F(q_0) = -R^\flat_{\text{diss}}(q_0)$. Therefore, the linearization is a linear mechanical system subject to a linear dissipative force. Now, using Proposition 6.7 in combination with Theorem 6.42, one can readily deduce the following result.

Proposition 6.44. *Let* $(\mathsf{Q}, \mathbb{G}, V, -R^\flat_{\text{diss}})$ *be a* C^∞*-simple mechanical system subject to the force associated with a Rayleigh dissipation function* R_{diss} *and let* $(\mathsf{T}_{q_0}\mathsf{Q}, \mathbb{G}(q_0), \operatorname{Hess} V(q_0), (0, -R^\flat_{\text{diss}}(q_0)))$ *be the linearization at the equilibrium configuration* q_0*. Then* q_0 *is*

(i) unstable if $\operatorname{Hess} V(q_0)$ *is not positive-semidefinite, is*

(ii) stable if $\operatorname{Hess} V(q_0)$ *is positive-definite and* $R_{\text{diss}}(q_0)$ *is positive-semidefinite, and is*

(iii) locally asymptotically stable if $\operatorname{Hess} V(q_0)$ *is positive-definite and* $\langle \mathbb{G}(q_0)^\sharp \circ (\operatorname{Hess} V(q_0)^\flat), \operatorname{image}(\mathbb{G}(q_0)^\sharp \circ R_{\text{diss}}(q_0)^\flat)\rangle = \mathsf{T}_{q_0}\mathsf{Q}$.

6.2.4 Lyapunov stability analysis

In this section we present criteria for stability of equilibrium configurations based on the use of the energy as a Lyapunov function. We shall restrict our analysis to simple mechanical systems with dissipation $(\mathsf{Q}, \mathbb{G}, V, F_{\text{diss}})$ with equation of motion

$$\overset{\mathbb{G}}{\nabla}_{\gamma'(t)}\gamma'(t) = -\operatorname{grad}V(\gamma(t)) + \mathbb{G}^\sharp(F_{\text{diss}}(\gamma'(t))). \tag{6.13}$$

The main result in this section is the following.

Theorem 6.45 (Lyapunov Stability Criteria for equilibrium configurations). *Consider a* C^∞*-simple mechanical system with dissipation* $\Sigma = (\mathsf{Q}, \mathbb{G}, V, F_{\text{diss}})$. *Let* $q_0 \in \mathsf{Q}$ *be a critical point for* V, *so that* q_0 *is an equilibrium configuration for* Σ, *and let* $\gamma\colon I \to \mathsf{Q}$ *be a solution to* (6.13). *Then the equilibrium configuration* q_0 *is*

(i) stable if V *is locally positive-definite about* q_0, *and is*

(ii) locally asymptotically stable if q_0 *is an isolated local minimum for* V *and* F_{diss} *is strictly dissipative.*

Furthermore, assume $\operatorname{Hess} V(q_0)$ *is positive-definite, assume* F_{diss} *is a strict Rayleigh dissipation force, and assume there exist* $L < L_{\text{reg,cpt}}(V, q_0)$ *and* $t_0 \in \bar{\mathbb{R}}_+$ *such that* $\gamma'(t_0) \in E^{-1}(\leq L, 0_{q_0})$. *Then*

(iii) the solution γ *is defined for* $t \in I \cup [t_0, +\infty[$, *and the curve* $\gamma'\colon I \cup [t_0, +\infty[\to \mathsf{TQ}$ *converges to* 0_{q_0} *exponentially fast.*

Proof. To prove (i), let us consider the energy function $E\colon \mathsf{TQ} \to \mathbb{R}$ defined in Definition 4.63:

$$E(v_q) = \tfrac{1}{2}\,\|v_q\|_{\mathbb{G}}^2 + V(q).$$

Without loss of generality, let us assume that $V(q_0) = 0$. We claim that the energy E is a Lyapunov function at the point 0_{q_0} for the simple mechanical system with dissipation, and that accordingly, the equilibrium configuration q_0 is stable. To prove the claim, we make three observations. First, we note that $E(0_{q_0}) = 0$. Second, we recall from Proposition 4.64 that the energy is nonincreasing along the solution to the equation of motion (6.13). Finally, we show that there exists a neighborhood $\mathcal{U} \subset \mathsf{TQ}$ of 0_{q_0} such that, for all $v_q \in \mathcal{U} \setminus \{0_{q_0}\}$, we have $E(v_q) > 0$. Note that $\|v_q\|_{\mathbb{G}}^2 > 0$ for all $v_q \neq 0_q$. Also, since V is locally positive-definite about q_0, there exists a neighborhood $\mathcal{W} \subset \mathsf{Q}$ of q_0 such that $V(q) > V(q_0)$ for all $q \in \mathcal{W} \setminus \{q_0\}$. Hence, for all $v_q \in \mathsf{T}\mathcal{W} \setminus \{0_{q_0}\}$, we have $E(v_q) > 0$. This proves (i).

To prove (ii), we invoke LaSalle Invariance Principle as presented in Corollary 6.20. Since q_0 is an isolated local minimum for V, there exists a neighborhood $\mathcal{W} \subset \mathsf{Q}$ of q_0 such that $V(q) > V(q_0)$ for all $q \in \mathcal{W} \setminus \{q_0\}$ and $\mathcal{W}$ contains no critical point other than q_0. Next, we define the set $A = \{v_q \in \mathsf{T}\mathcal{W} \mid \langle F_{\mathrm{diss}}(v_q); v_q\rangle = 0\}$. A continuity argument shows that, because F_{diss} is strictly dissipative, $A = \{0_q \in \mathsf{T}\mathcal{W}\}$. We claim that the only positively invariant set in A is $\{0_{q_0}\}$. To show this, assume $0_{q_1} \neq 0_{q_0}$ is a point in the largest positively invariant set in A. Since $\mathrm{grad}V(q_1) \neq 0_{q_1}$, the solution to equation (6.13) will leave A and from this our claim follows. Therefore, Corollary 6.20 guarantees that q_0 is locally asymptotically stable.

Finally, we prove (iii). Since we assume the dissipative force arises from a Rayleigh dissipation function, we write $F_{\mathrm{diss}} = -R^{\flat}_{\mathrm{diss}}$. Let ε be a positive constant, and define $E_\varepsilon \colon \mathsf{TQ} \to \mathbb{R}$ by

$$E_\varepsilon(v_q) = E(v_q) + \varepsilon \langle \mathrm{d}V(q); v_q\rangle .$$

Along the trajectory γ, we have $\frac{\mathrm{d}}{\mathrm{d}t}V(\gamma(t)) = \langle \mathrm{d}V(\gamma(t)); \gamma'(t)\rangle$. Furthermore, we compute

$$\begin{aligned}
&\frac{\mathrm{d}^2}{dt^2}V(\gamma(t)) \\
&= \frac{\mathrm{d}}{\mathrm{d}t}\langle \mathrm{d}V(\gamma(t)); \gamma'(t)\rangle = \langle \overset{\mathbb{G}}{\nabla}_{\gamma'(t)}\mathrm{d}V; \gamma'(t)\rangle + \langle \mathrm{d}V(\gamma(t)); \overset{\mathbb{G}}{\nabla}_{\gamma'(t)}\gamma'(t)\rangle \\
&= \mathrm{Hess}\, V(\gamma(t)) \cdot (\gamma'(t), \gamma'(t)) + \langle \mathrm{d}V(\gamma(t)); \mathbb{G}^\sharp(-\mathrm{d}V(\gamma(t)) - R_{\mathrm{diss}}(\gamma'(t)))\rangle \\
&= \mathrm{Hess}\, V(\gamma(t)) \cdot (\gamma'(t), \gamma'(t)) - \|\mathrm{d}V(\gamma(t))\|^2_{\mathbb{G}} - \langle R_{\mathrm{diss}}(\gamma'(t)); \mathrm{grad}V(\gamma(t))\rangle.
\end{aligned}$$

Next, we aim to bound from above the time derivatives of $E_\varepsilon \circ \gamma'$ and bound from below $E_\varepsilon \circ \gamma'$. To compute such bounds, let us discuss some compactness properties. Because E decreases along the trajectory γ, the assumption that $E(\gamma'(t_0)) \leq L$ implies that $E(\gamma'(t)) < L$ for all subsequent $t \in [t_0, +\infty[$. Therefore, γ' remains in $E^{-1}(\leq L, 0_{q_0})$ for all $t \in [t_0, +\infty[$. Since $E(v_q) < L$ implies $V(q) < L$, we infer that γ remains in $V^{-1}(\leq L, q_0)$ for all $t \in [t_0, +\infty[$, and this sublevel set is compact because $L < L_{\mathrm{reg,cpt}}(V, q_0)$ by assumption. There are two consequences of this compactness result. First, by the lemma in the proof of Theorem 6.14, the trajectory γ can be extended to the interval $[t_0, +\infty[$. Second, the smooth tensor fields R_{diss}, $\mathrm{d}V$, and $\mathrm{Hess}\, V$ have bounded norm inside $V^{-1}(\leq L, q_0)$, and V satisfies the bound in equation (6.8). Exploiting this second consequence, we bound from below the function E_ε:

$$\begin{aligned}
E_\varepsilon(\gamma'(t)) &= V(\gamma(t)) + \tfrac{1}{2}\|\gamma'(t)\|^2_{\mathbb{G}} + \varepsilon \langle \mathrm{d}V(\gamma(t)); \gamma'(t)\rangle \\
&\geq b_1 \|\mathrm{d}V(\gamma(t))\|^2_{\mathbb{G}} + \tfrac{1}{2}\|\gamma'(t)\|^2_{\mathbb{G}} - \varepsilon \|\mathrm{d}V(\gamma(t))\|_{\mathbb{G}} \|\gamma'(t)\|_{\mathbb{G}} \\
&= \begin{bmatrix} \|\mathrm{d}V(\gamma(t))\|_{\mathbb{G}} \\ \|\gamma'(t)\|_{\mathbb{G}} \end{bmatrix}^T \boldsymbol{P}_\varepsilon \begin{bmatrix} \|\mathrm{d}V(\gamma(t))\|_{\mathbb{G}} \\ \|\gamma'(t)\|_{\mathbb{G}} \end{bmatrix},
\end{aligned}$$

for

$$P_\varepsilon \triangleq \begin{bmatrix} b_1 & -\frac{\varepsilon}{2} \\ -\frac{\varepsilon}{2} & \frac{1}{2} \end{bmatrix} \in \mathbb{R}^{2\times 2}.$$

Similarly, we bound from above the time derivative of E_ε:

$$\begin{aligned}
&\frac{\mathrm{d}}{\mathrm{d}t} E_\varepsilon(\gamma'(t)) \\
&\quad = -\langle R_{\mathrm{diss}}(\gamma'(t)); \gamma'(t)\rangle + \varepsilon \operatorname{Hess} V(\gamma(t)) \cdot (\gamma'(t), \gamma'(t)) \\
&\qquad - \varepsilon \|\mathrm{d}V(\gamma(t))\|_{\mathbb{G}}^2 - \varepsilon \langle\!\langle R_{\mathrm{diss}}(\gamma'(t)), \mathrm{d}V(\gamma(t))\rangle\!\rangle \\
&\quad \leq -\inf \left\{ \|R_{\mathrm{diss}}(q)\|_{\mathbb{G}} \mid q \in V^{-1}(\leq L, q_0) \right\} \|\gamma'(t)\|_{\mathbb{G}}^2 \\
&\qquad - \varepsilon \sup \left\{ \|\operatorname{Hess} V(q)\|_{\mathbb{G}} \mid q \in V^{-1}(\leq L, q_0) \right\} \|\gamma'(t)\|_{\mathbb{G}}^2 - \varepsilon \|\mathrm{d}V(\gamma(t))\|_{\mathbb{G}}^2 \\
&\qquad + \varepsilon \sup \left\{ \|R_{\mathrm{diss}}(q)\|_{\mathbb{G}} \mid q \in V^{-1}(\leq L, q_0) \right\} \|\gamma'(t)\|_{\mathbb{G}} \|\mathrm{d}V(\gamma(t))\|_{\mathbb{G}} \\
&\quad \leq -\begin{bmatrix} \|\mathrm{d}V(\gamma(t))\|_{\mathbb{G}} \\ \|\gamma'(t)\|_{\mathbb{G}} \end{bmatrix}^T Q_\varepsilon \begin{bmatrix} \|\mathrm{d}V(\gamma(t))\|_{\mathbb{G}} \\ \|\gamma'(t)\|_{\mathbb{G}} \end{bmatrix},
\end{aligned}$$

where $Q_\varepsilon \in \mathbb{R}^{2\times 2}$ is given by

$$\begin{aligned}
(Q_\varepsilon)_{11} &= \varepsilon, \\
(Q_\varepsilon)_{12} = (Q_\varepsilon)_{21} &= \frac{\varepsilon}{2} \sup \left\{ \|R_{\mathrm{diss}}(q)\|_{\mathbb{G}} \mid q \in V^{-1}(\leq L, q_0) \right\}, \\
(Q_\varepsilon)_{22} &= \inf \left\{ \|R_{\mathrm{diss}}(q)\|_{\mathbb{G}} \mid q \in V^{-1}(\leq L, q_0) \right\} \\
&\quad - \varepsilon \sup \left\{ \|\operatorname{Hess} V(q)\|_{\mathbb{G}} \mid q \in V^{-1}(\leq L, q_0) \right\}.
\end{aligned}$$

Now, one can show that there exists an $\bar{\varepsilon} > 0$ such that, for all $0 < \varepsilon < \bar{\varepsilon}$, the trace and the determinant of the matrices P_ε and Q_ε are positive, so that the matrices P_ε and Q_ε are positive-definite. Furthermore, two applications of Exercise E2.18 show that there exist positive constants λ_1, λ_2 such that

$$\begin{aligned}
E(\gamma'(t)) &\leq \lambda_1 E_\varepsilon(\gamma'(t)), \\
\frac{\mathrm{d}}{\mathrm{d}t} E_\varepsilon(\gamma'(t)) &\leq -\lambda_2 E_\varepsilon(\gamma'(t)),
\end{aligned}$$

therefore proving that there exist $\alpha, \beta \in \mathbb{R}_+$ independent of γ, such that $E(\gamma'(t)) \leq \alpha E(\gamma'(t_0)) \mathrm{e}^{-\beta(t-t_0)}$ for all $t \in [t_0, +\infty[$. To conclude, we note that E has a positive-definite Hessian at 0_{q_0}. ■

Remarks 6.46. 1. In part (ii) we assume the potential function to have an isolated local minimum at q_0. This assumption strictly implies that the potential function is locally positive-definite about the critical point q_0; the converse is only true for analytic functions. The reader can explore this fact in Exercise E6.17.

2. In a simple mechanical system $(\mathsf{Q}, \mathbb{G}, V)$, if the potential function V has positive-definite Hessian at a critical point q_0, then the equilibrium configuration q_0 is stable. This consequence of part (ii) is known as the ***Lagrange–Dirichlet Criterion***; see the original work in [Lagrange 1788] and the treatments in [Bloch 2003, Marsden 1992].

3. The proof of part (iii) relies on a Lyapunov function with strictly negative-definite Lie derivative. This function is designed by modifying the total energy of the system with the addition of a term that depends on a parameter ε. This idea is sometimes referred to as "Chetaev's trick," and we refer the reader to the example involving gyroscopic forces on page 100 of [Chetaev 1961]. Early references on this procedure in the control literature include [Arimoto and Miyazaki 1984, Koditschek 1988].
4. To prove part (iii), the assumption that F_{diss} is a strict Rayleigh dissipation force can be replaced by the weaker statement that there exists a strict Rayleigh dissipation function R_{diss} such that, for all $v_q \in \mathsf{TQ}$,
$$\langle F_{\mathrm{diss}}(v_q); v_q \rangle \leq -\langle R_{\mathrm{diss}}(v_q); v_q \rangle .$$
We also mention that it is also possible to obtain asymptotic and exponential convergence without requiring F_{diss} to be strictly dissipative. We do not pursue this here, except to refer the reader to Proposition 6.44 and to the notion of "zero-state detectability" in [van der Schaft 1999].
5. Part (iii) contains an estimate for the domain of attraction of the equilibrium point 0_{q_0} of the simple mechanical system with dissipation $\Sigma = (\mathsf{Q}, \mathbb{G}, V, F_{\mathrm{diss}})$. As discussed in Remark 6.22, this domain of attraction contains the L-sublevel set of the energy function $E^{-1}(\leq L, 0_{q_0})$ for all $L < L_{\mathrm{reg,cpt}}(V, q_0)$.
6. It is instructive to compare the results presented in Theorem 6.45 via a Lyapunov analysis with the results previously obtained via linear analysis in Proposition 6.41 and Theorem 6.42. Let us examine the three parts of Theorem 6.45, one at a time. Part (i) provides a sufficient condition for stability, and there is no equivalent statement that can be obtained via a linearized analysis. Part (ii) extends the result obtained in Theorem 6.42 to (1) positive-definite potential functions that do not have positive-definite Hessian, and (2) general dissipative forces as opposed to Rayleigh dissipation forces. Part (iii) presents an estimate of the domain of attraction of 0_{q_0} and a statement on exponential convergence; both results were not available from the previous linear analysis.
7. A corollary to Theorem 6.45 is that, if an equilibrium configuration of a simple mechanical system (with or without dissipation) is linearly stable, then it is stable. Furthermore, asymptotic stability of the linearization implies exponential stability of the equilibrium configuration.
The converse of the first of these statements is generally false, as the following example shows. We take $\mathsf{Q} = \mathbb{R}$ with x the usual coordinate. We take $\mathbb{G} = \mathrm{d}x \otimes \mathrm{d}x$ and $V(x) = \frac{1}{2}\alpha x^2 + \frac{1}{4}\beta x^4$ for $\alpha, \beta \in \mathbb{R}$. The governing equations for the resulting simple mechanical system are
$$\ddot{x} + \alpha x + \beta x^3 = 0. \tag{6.14}$$
The linearization at the equilibrium point $q_0 = 0$ is then $(\mathbb{R}, \boldsymbol{e}^1 \otimes \boldsymbol{e}^1, \alpha \boldsymbol{e}^1 \otimes \boldsymbol{e}^1)$, where $\boldsymbol{e}_1$ is the standard basis vector in $\mathbb{R}$ (i.e., the vector 1). The

linearization at q_0 is therefore stable if and only if $\alpha > 0$. The system itself is stable at q_0 if and only if either (1) $\alpha > 0$ or (2) $\alpha = 0$ and $\beta > 0$. These assertions are intuitively reasonable from (6.14), and follow rigorously from the treatment in the present section. •

6.2.5 Global stability analysis

Using the LaSalle Invariance Principle, we characterize some global stability properties of simple mechanical systems with dissipation.

Theorem 6.47 (Global stability of systems with dissipation). *Consider a C^∞-simple mechanical system with dissipation $(\mathsf{Q}, \mathbb{G}, V, F_{\text{diss}})$. Assume that F_{diss} is strictly dissipative and that there exists $L \in \mathbb{R}$ such that*

(i) $V^{-1}(\leq L)$ is compact and

(ii) V has a finite number of critical points in $V^{-1}(\leq L)$.

Then every solution $\gamma \colon \bar{\mathbb{R}}_+ \to \mathsf{Q}$ to (6.13) with initial condition $\gamma'(0) \in E^{-1}(\leq L)$ converges as $t \to +\infty$ to an equilibrium configuration in $V^{-1}(\leq L)$.

Proof. First, let us introduce the useful shorthands, $A_1 = V^{-1}(\leq L) \subset \mathsf{Q}$ and $A_2 = E^{-1}(\leq L) \subset \mathsf{TQ}$. Since the sublevel set A_1 of V is compact, V is bounded from below. Since $E(v_q) < L$ implies that $V(q) < L$, we know that $v_q \in A_2$ implies that $q \in A_1$. Therefore, we have $A_2 = \cup_{q \in A_1} \pi_{\mathsf{TQ}}^{-1}(q) \cap A_2$. Also we note that, for each $q \in A_1$, the set $A_2 \cap \pi_{\mathsf{TQ}}^{-1}(q)$ is compact.

Next, we prove that A_2 is compact. Let $\{\mathcal{U}_c\}_{c \in C}$ be an open cover for A_2 equipped with the subspace topology. We need to show that there exists a finite subset $\{c_1, \dots, c_k\} \subset C$ such that $A_2 = \cup_{j=1}^k \mathcal{U}_{c_j}$. Let $q \in A_1$. Since the set $A_2 \cap \pi_{\mathsf{TQ}}^{-1}(q)$ is compact, given the open cover $\{\mathcal{U}_c \cap \pi_{\mathsf{TQ}}^{-1}(q)\}_{c \in C}$ of $A_2 \cap \pi_{\mathsf{TQ}}^{-1}(q)$, there exists a finite subset $\{c_1(q), \dots, c_k(q)\} \subset C$ such that $A_2 \cap \pi_{\mathsf{TQ}}^{-1}(q) = \cup_{j=1}^k \mathcal{U}_{c_j(q)}$. Now, note that $\mathcal{O}_q = \cap_{j=1}^k \pi_{\mathsf{TQ}}(\mathcal{U}_{c_j(q)})$ is an open neighborhood of q. In turn, this implies that $\mathscr{O} = \{\mathcal{O}_q\}_{q \in A_1}$ is an open cover of the compact set A_1. Therefore, there exits a finite subset $\{q_1, \dots, q_m\} \subset A_1$ such that $A_1 = \cup_{i=1}^m \mathcal{O}_{q_i}$. In summary we have

$$A_2 = \cup_{q \in A_1} (A_2 \cap \pi_{\mathsf{TQ}}^{-1}(q)) = \cup_{q \in A_1} \cup_{j=1}^k \mathcal{U}_{c_j(q)} = \cup_{i=1}^m \cup_{j=1}^k \mathcal{U}_{c_j(q_i)}.$$

This concludes the proof that A_2 is compact.

Since the energy is nonincreasing along the solutions of (6.13), A_2 is positively invariant for $(\mathsf{Q}, \mathbb{G}, V, F_{\text{diss}})$. Let $B \subset A_2$ be the largest positively invariant set for $(\mathsf{Q}, \mathbb{G}, V, F_{\text{diss}})$ contained in $A \triangleq \{v_q \in A_2 \mid \langle F_{\text{diss}}(v_q); v_q \rangle = 0\} = \{0_q \in A_2 \mid q \in A_1\}$. We claim that $B = \{0_{q_0} \in A_2 \mid q_0 \in A_1 \text{ and } \operatorname{grad} V(q_0) = 0\}$. To show this, assume $0_{q_1} \notin B$ is a point in the largest positively invariant set in A. Since $\operatorname{grad} V(q_1) \neq 0_{q_1}$, the solution to equation (6.13) will leave A and from this our claim follows.

In summary, we have established that A_2 is positively invariant for $(\mathsf{Q}, \mathbb{G}, V, F_{\text{diss}})$ and is compact, and that E is nonincreasing. Under these conditions, the LaSalle Invariance Principle implies that every solution to (6.13) with initial condition $v_q \in A_2$ approaches B as $t \to +\infty$. Furthermore, since we assumed the set $\{q_0 \in A_1 \mid \text{grad}V(q_0) = 0\}$ to consist of a finite number of configurations, every solution to (6.13) with initial condition $v_q \in A_2$ will converge to a point in B. ■

Remarks 6.48. 1. Theorem 6.47 has the following interpretation. The global asymptotic behavior of the simple mechanical system with dissipation $(\mathsf{Q}, \mathbb{G}, V, F_{\text{diss}})$ is related to the global asymptotic behavior of the differential equation

$$\gamma'(t) = -\text{grad}V(\gamma(t)). \tag{6.15}$$

Indeed, integral curves of both dynamical systems converge to critical points of V, if the sublevel sets of V are compact. This fact is referred to as the ***lifting property of dissipative mechanical systems***; see [Koditschek 1991, Terjéki 1987]. The differential equation (6.15) is analyzed in Exercise E6.8.

2. According to the result in the theorem, local stability properties can be extended to global stability properties if the potential function has a single critical point. Unfortunately, it turns out that this is not always possible, and that certain configuration manifolds do not possess smooth functions with a single critical point. This is an important topological obstruction to obtaining global stability results in analysis and control design settings.
3. The compactness assumption on the sublevel sets of V prevents the possibility that the integral curves become unbounded even while the energy function is decreasing; see the example system in Example 6.49. •

6.2.6 Examples illustrating configuration stability results

In this section we apply the results of Section 6.2 to a few examples.

Example 6.49 (One degree-of-freedom point mass). Consider a particle of mass m with configuration described by the point $x \in \mathbb{R}$. We assume the particle is subject to a force derived from a potential function V depicted in Figure 6.5. From the figure we note that V has three critical points: $x_{c,1}$ and $x_{c,3}$ are local minima and $x_{c,2}$ is a local maximum. We shall assume that the Hessian of V at the three critical points is sign definite, i.e., positive-definite at $x_{c,1}$ and $x_{c,3}$, and negative-definite at $x_{c,2}$.

Before proceeding with the analysis, let us identify some critical values of V that characterize compact sublevel sets and domains of attraction, as defined in Section 6.1.6. Figure 6.6 illustrates how $L_{\text{reg,cpt}}(V, x_{c,1})$ and $L_{\text{reg,cpt}}(V, x_{c,3})$ are equal and it includes an example configuration x_0 with the property that $V(x_0) > L_{\text{cpt}}(V, x_{c,3})$.

We can now apply the methods of Section 6.2 to analyze the behavior of the point mass and the stability of its equilibrium configurations. We shall consider both the setting of a system subject, and not subject, to a strict dissipative force.

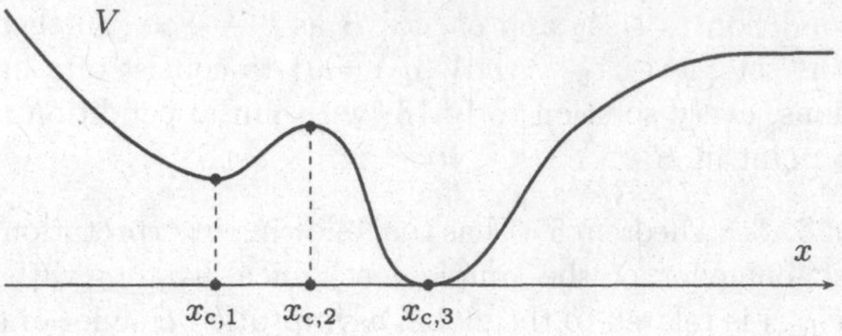

Figure 6.5. Potential function V for the point mass: $x_{c,1}$, $x_{c,2}$, and $x_{c,3}$ are critical points.

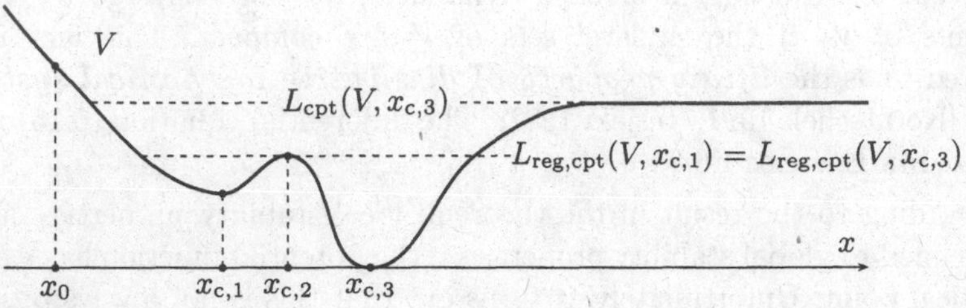

Figure 6.6. Potential function values that affect the topology of sublevel sets.

1. *Local asymptotic stability of the equilibria* $x_{c,1}, x_{c,3}$*:* The local minima $x_{c,1}$ and $x_{c,3}$ are stable for the system without dissipation (see Theorem 6.45(i)) and are locally asymptotically stable for system with dissipation (see Theorem 6.45(ii)). The local maximum $x_{c,2}$ is unstable in both cases (see Proposition 6.41(iii) and Theorem 6.42(iii)).
2. *Domain of attraction with exponential convergence for the equilibria* $x_{c,1}, x_{c,3}$*:* Assume the point mass is subject to a strict Rayleigh dissipation function. According to Theorem 6.45(iii), from all initial conditions inside $E^{-1}(\leq L, 0_{x_{c,1}})$, for any $L < L_{reg,cpt}(V, x_{c,1})$, the point mass converges to $x_{c,1}$ exponentially fast. An identical statement holds for $x_{c,3}$.
3. *Global behavior:* Assume the point mass is subject to a strictly dissipative force. Given $L \in \mathbb{R}$ with the property that $V^{-1}(\leq L)$ is compact, and given any initial condition inside $E^{-1}(\leq L)$, Theorem 6.47 guarantees that the corresponding integral curve will converge to one of the equilibrium configurations $x_{c,1}$, $x_{c,2}$, or $x_{c,3}$.

   Finally, consider the integral curve γ starting from rest at initial configuration x_0. Depending on the magnitude of the dissipative force, γ could converge to any of the critical points $x_{c,1}, x_{c,2}$, and $x_{c,3}$, or, in case of appropriately vanishing dissipation forces, γ could escape to infinity along the horizontal asymptote of V. •

Example 6.50 (Two-link manipulator subject to gravity). We consider the example of two interconnected links subject to gravity and with the base of the first link fixed spatially; see Figure 6.7. This example was first considered

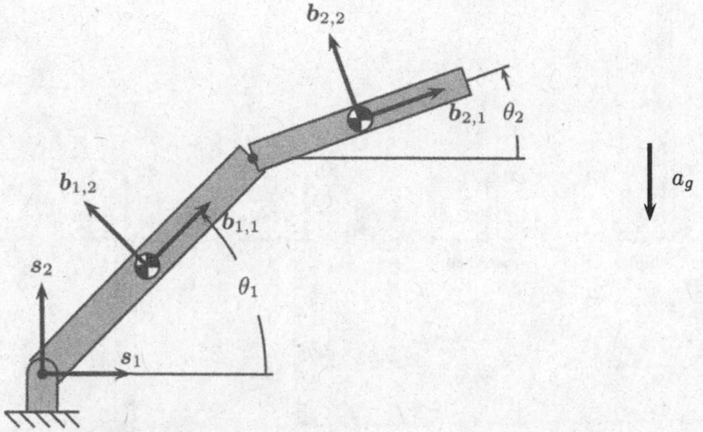

Figure 6.7. A two-link manipulator subject to gravity

in Section 1.2, and more carefully analyzed in Chapter 4. The configuration manifold is $\mathbb{S}^1 \times \mathbb{S}^1$, and the coordinates (θ_1, θ_2) are the two angles describing the orientation of the two links with respect to the horizontal axis. We refer to Example 4.32 for the manipulator's kinetic energy metric. The potential function V due to gravity is given in Example 4.62.

We perform both the linear and the Lyapunov stability analyses. We start by computing the equilibrium configurations of the manipulator. The differential of V is

$$\mathrm{d}V = \left(\tfrac{1}{2}m_1a_g\ell_1 + m_2a_g\ell_1\right)\cos\theta_1\mathrm{d}\theta_1 + \left(\tfrac{1}{2}m_2a_g\ell_2\right)\cos\theta_2\mathrm{d}\theta_2,$$

and, therefore, the equilibrium configurations of the manipulator are $q_{\mathrm{up/up}} = (\frac{\pi}{2}, \frac{\pi}{2})$, $q_{\mathrm{up/down}} = (\frac{\pi}{2}, -\frac{\pi}{2})$, $q_{\mathrm{down/up}} = (-\frac{\pi}{2}, \frac{\pi}{2})$, and $q_{\mathrm{down/down}} = (-\frac{\pi}{2}, -\frac{\pi}{2})$. These four configuration are illustrated in Figure 6.8. The Hessian of V is

$$[\mathrm{Hess}\, V] = \begin{bmatrix} -\left(\frac{1}{2}m_1a_g\ell_1 + m_2a_g\ell_1\right)\sin\theta_1 & 0 \\ 0 & -\left(\frac{1}{2}m_2a_g\ell_2\right)\sin\theta_2 \end{bmatrix}.$$

One can easily see that $\mathrm{Hess}\, V$ is negative-definite at $q_{\mathrm{up/up}} = (\frac{\pi}{2}, \frac{\pi}{2})$, indefinite at $q_{\mathrm{down/up}} = (-\frac{\pi}{2}, \frac{\pi}{2})$ and at $q_{\mathrm{up/down}} = (\frac{\pi}{2}, -\frac{\pi}{2})$, and positive-definite at $q_{\mathrm{down/down}} = (-\frac{\pi}{2}, -\frac{\pi}{2})$.

We can now apply the methods of Section 6.2 to the manipulator.

1. *Local asymptotic stability of the equilibrium* $q_{\mathrm{down/down}}$: Assuming the system is not subject to a dissipative force, Proposition 6.41 states that this equilibrium configuration is linearly stable, and Theorem 6.45 states that this equilibrium configuration is stable. Assuming the system is subject to

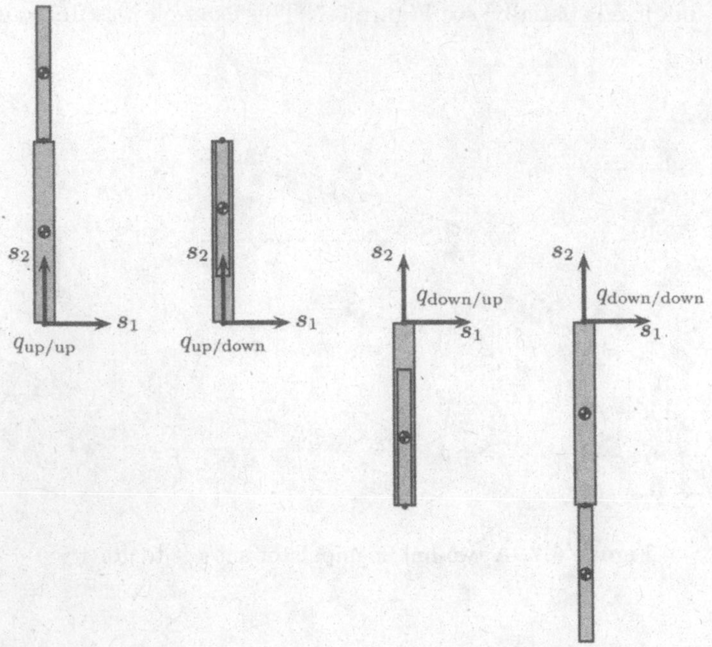

Figure 6.8. The four equilibrium configurations for a two-link manipulator subject to gravity. The equilibrium configuration $q_{\text{down/down}}$ is the only stable one.

a strict dissipative force, Theorem 6.45 states that this equilibrium configuration is locally asymptotically stable.

2. *Domain of attraction with exponential convergence for the equilibrium* $q_{\text{down/down}}$*:* Assuming the system is subject to a strict Rayleigh dissipative force, Theorem 6.45 establishes a local domain of attraction with exponential convergence. To characterize this domain, we compute the minimum of the potential function at the three unstable equilibrium configurations. Because of the equality

$$V(q_{\text{down/up}}) = -\tfrac{1}{2}m_1 a_g \ell_1 - m_2 a_g(\ell_1 - \tfrac{1}{2}\ell_2) = -V(q_{\text{up/down}}),$$

one can directly verify that

$$L_{\text{reg}}(V, q_{\text{down/down}}) = -\left|\tfrac{1}{2}m_1 a_g \ell_1 + m_2 a_g(\ell_1 - \tfrac{1}{2}\ell_2)\right|.$$

From all initial conditions inside $E^{-1}(\leq L, 0_{q_{\text{down/down}}})$ and for any $L < L_{\text{reg}}(V, q_{\text{down/down}})$, the manipulator's configuration converges to $q_{\text{down/down}}$ exponentially fast.

3. *Instability of the other three equilibrium configurations:* Because $\text{ind}(\text{Hess}\, V(q_0)) > 0$, Proposition 6.41 and Theorem 6.42 state that these equilibria are unstable. This fact is independent of whether the system is subject to a dissipative force or not.

4. *Global behavior:* Assume the manipulator is subject to a strictly dissipative force. Note that the configuration manifold is compact, so that the compactness assumption in the statement of Theorem 6.47 is automatically satisfied. Therefore, we know that, given any initial condition in $\mathsf{T}(\mathbb{S}^1 \times \mathbb{S}^1)$, the resulting trajectory will converge to one of the four equilibrium configurations. •

6.3 Relative equilibria and their stability

In this section we study a class of steady motions called relative equilibria that arise in systems with symmetries; see Section 5.4. A classic example of such motion is the steady rotation of a rigid body. We provide conditions for the existence and stability of relative equilibria in terms of certain modified potential functions. Prerequisites for the material in this section are Sections 3.9.5 and 5.5. We refer the interested reader to the texts [Bloch 2003, Marsden 1992, Marsden and Ratiu 1999] for further discussions on the stability of relative equilibria.

6.3.1 Existence and stability definitions

Let us establish the basic setup for all the following definitions and results. We consider a C^∞-forced simple mechanical system $\Sigma = (\mathsf{Q}, \mathbb{G}, V, F)$ with a time-independent external force F and with the following governing equations:

$$\overset{\mathbb{G}}{\nabla}_{\gamma'(t)}\gamma'(t) = -\text{grad}V(\gamma(t)) + \mathbb{G}^\sharp(F(\gamma'(t))). \tag{6.16}$$

We assume for simplicity that all solutions to this equation can be defined on all of $\mathbb{R}$.

Definition 6.51 (Relative equilibrium). Let X be a complete infinitesimal symmetry for Σ and let $\chi\colon \mathbb{R} \to \mathsf{Q}$ be a maximal integral curve of X.

(i) The curve χ is a ***relative equilibrium*** for Σ if it is a solution to the equation of motion (6.16).

(ii) The relative equilibrium χ is ***regular*** if χ is an embedding. •

Note that a relative equilibrium is a smooth curve because an infinitesimal symmetry is a smooth vector field. Also note that, if a relative equilibrium χ is regular, then Lemma 5.79 guarantees that, in a neighborhood $\mathcal{U}$ of image(χ), the set of orbits of X is a manifold, say B, and that the projection $\pi_{\mathsf{B}}\colon \mathcal{U} \to \mathsf{B}$ is a surjective submersion. These properties are the required assumptions for the reduction procedure described in Section 5.5.2.

Definition 6.52 (Base and fiber stability). Let X be a complete infinitesimal symmetry for Σ, let $\chi\colon \mathbb{R} \to \mathsf{Q}$ be a regular relative equilibrium, and let $b_0 = \pi_{\mathsf{B}}(\chi(0))$.

(i) The relative equilibrium χ is ***base stable*** if, for all neighborhoods $\mathcal{U}$ of 0_{b_0}, there exists a neighborhood $\mathcal{W}$ of $\chi'(0)$ such that each solution $\gamma\colon \mathbb{R} \to \mathsf{Q}$ of equation (6.16), with initial condition $\gamma'(0) \in \mathcal{W}$, satisfies $T\pi_{\mathsf{B}}(\gamma(t)) \in \mathcal{U}$ for all $t \in \mathbb{R}$.

(ii) The relative equilibrium χ is ***locally asymptotically base stable*** if it is base stable, and if there exists a neighborhood $\mathcal{V}$ of $\chi'(0)$ such that each solution $\gamma\colon \mathbb{R} \to \mathsf{Q}$ of equation (6.16), with initial condition $\gamma'(0) \in \mathcal{V}$, satisfies $\lim_{t\to+\infty} T_{\gamma(t)}\pi_{\mathsf{B}}(\gamma'(t)) = 0_{b_0}$.

(iii) The relative equilibrium χ is ***fiber stable*** if, for all neighborhoods $\mathcal{U}_{\mathbb{R}}$ of 1, there exists a neighborhood $\mathcal{W}$ of $\chi'(0)$ such that each solution $\gamma\colon \mathbb{R} \to \mathsf{Q}$ of equation (6.16), with initial condition $\gamma'(0) \in \mathcal{W}$, satisfies $J_X(\gamma'(t))\,\|X\|_{\mathbb{G}}^{-2}(\gamma(t)) \in \mathcal{U}_{\mathbb{R}}$ for all $t \in \mathbb{R}$.

(iv) The relative equilibrium χ is ***locally asymptotically fiber stable*** if it is fiber stable and if there exists a neighborhood $\mathcal{V}$ of $\chi'(0)$ such that each solution $\gamma\colon \mathbb{R} \to \mathsf{Q}$ of equation (6.16), with initial condition $\gamma'(0) \in \mathcal{V}$, satisfies $\lim_{t\to+\infty} J_X(\gamma'(t))\,\|X\|_{\mathbb{G}}^{-2}(\gamma(t)) = 1$. •

To describe base stability consider an invariant tube around the regular relative equilibrium χ. Base stability means that trajectories starting in the tube remain in the tube at subsequent times; see Figure 6.9. Fiber stability means that trajectories starting with nearby initial velocities will maintain nearby fiber velocities at subsequent times. Even if a system is base and fiber stable, trajectories with nearby initial conditions may diverge.

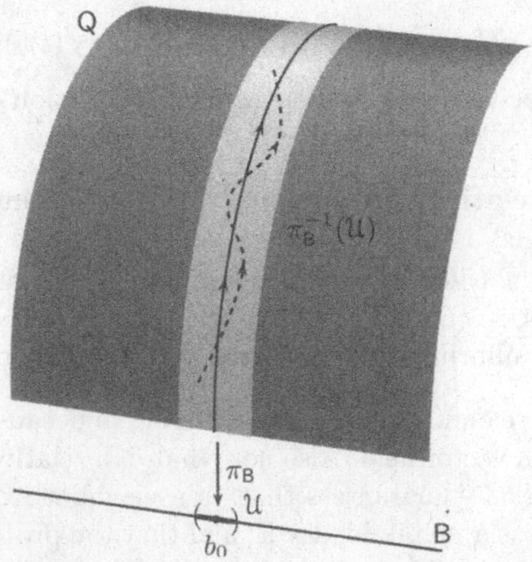

Figure 6.9. Base stability of relative equilibria

6.3.2 Lyapunov stability analysis

Next, we endeavor to provide existence and stability criteria for relative equilibria. We start by introducing a few appropriate concepts. From Section 5.5.2, recall the effective potential function $V^{\text{eff}}_{X,\lambda}\colon \mathsf{Q} \to \mathbb{R}$ and its role in Theorem 5.83. In what follows, it will be convenient to adopt the more expeditious notation

$$V_X(q) = V^{\text{eff}}_{X,1}(q) = V(q) - \tfrac{1}{2}\|X\|^2_{\mathbb{G}}(q).$$

It is now possible to introduce a modified notion of the energy function that will play the role of Lyapunov function in assessing stability of relative equilibria. Recall that the energy function and the momentum map are both constants of motion for simple mechanical systems with symmetry. A third useful constant of motion is the difference between the energy function and the momentum map.

Definition 6.53 (Effective energy). Let X be a complete infinitesimal symmetry for Σ, and let $J_X\colon \mathsf{TQ} \to \mathbb{R}$ be the associated momentum map. The ***effective energy*** $E_X\colon \mathsf{TQ} \to \mathbb{R}$ is the function

$$E_X(v_q) = E(v_q) - J_X(v_q). \qquad \bullet$$

Let us also introduce the appropriate notions of dissipative forces along a relative equilibrium.

Definition 6.54. Let $X \in \Gamma^\infty(\mathsf{TQ})$. A time-independent force $F\colon \mathsf{TQ} \to \mathsf{T}^*\mathsf{Q}$ is ***dissipative about X*** if $\langle F(v_q); v_q - X(q)\rangle \leq 0$ for each $v_q \in \mathsf{TQ}$, and is ***strictly dissipative about X*** if it is dissipative about X and if $\langle F(v_q); v_q - X(q)\rangle = 0$ only when $v_q = X(q)$. $\bullet$

We can now collect some basic statements about these concepts.

Lemma 6.55. *Consider the C^∞-forced simple mechanical system $(\mathsf{Q}, \mathbb{G}, V, F)$ with infinitesimal symmetry X. Then the following statements are true:*

(i) for all $v_q \in \mathsf{TQ}$

$$E_X(v_q) = V_X(q) + \tfrac{1}{2}\|v_q - X(q)\|^2_{\mathbb{G}}; \tag{6.17}$$

(ii) if F is dissipative about X, and if $\gamma\colon I \to \mathsf{Q}$ is a trajectory of equation (6.16), then the function $t \mapsto E_X(\gamma'(t))$ is nonincreasing.

Proof. Equation (6.17) is a consequence of the following "completing the square" computation:

$$\begin{aligned} E(v_q) - J_X(v_q) &= V(q) + \tfrac{1}{2}\|v_q\|^2_{\mathbb{G}} - \mathbb{G}(v_q, X(q)) \\ &= V(q) - \tfrac{1}{2}\|X\|^2_{\mathbb{G}}(q) + \tfrac{1}{2}\|X\|^2_{\mathbb{G}}(q) + \tfrac{1}{2}\|v_q\|^2_{\mathbb{G}} - \langle\langle v_q, X(q)\rangle\rangle \\ &= V_X(q) + \tfrac{1}{2}\|v_q - X(q)\|^2_{\mathbb{G}}. \end{aligned}$$

To prove that $t \mapsto E_X(\gamma'(t))$ is nonincreasing, recall that the vector field on TQ describing the dynamics of $(\mathsf{Q}, \mathbb{G}, V, F)$ is $\overset{\mathbb{G}}{S} - \mathrm{vlft}(\mathrm{grad}V) + \mathrm{vlft}(\mathbb{G}^\sharp \circ F)$. Because E_X is a constant of motion for $(\mathsf{Q}, \mathbb{G}, V)$, we know that

$$\mathscr{L}_{\overset{\mathbb{G}}{S} - \mathrm{vlft}(\mathrm{grad}V)} E_X = 0.$$

Next, from equation (6.17) we compute

$$\mathscr{L}_{\mathrm{vlft}(\mathbb{G}^\sharp \circ F)} E_X(v_q) = \langle F(v_q); v_q - X(q) \rangle \leq 0. \quad \blacksquare$$

Finally, we can provide criteria for the existence and stability of relative equilibria in terms of the existence of critical points of V_X, and of the character of the Hessian of V_X at critical points. These results parallel those presented for equilibrium configurations in Theorem 6.45. Recall that, under Assumption 5.78, $(V_X)_{\mathsf{B}}$ is the projection of V_X onto the base space B.

Theorem 6.56 (Existence and stability of relative equilibria). *Consider a C^∞-forced simple mechanical system $\Sigma = (\mathsf{Q}, \mathbb{G}, V, F)$ and let X be a complete infinitesimal symmetry for Σ. Assume F is dissipative about X and assume $F(X(q)) = 0_q$ for each $q \in \mathsf{Q}$. Then the following statements are true.*

(i) A maximal integral curve $\chi\colon \mathbb{R} \to \mathsf{Q}$ of X is a relative equilibrium for Σ if and only if V_X has a critical point at $\chi(t_0)$ for some (hence for all) $t_0 \in \mathbb{R}$.

(ii) A regular relative equilibrium $\chi\colon \mathbb{R} \to \mathsf{Q}$ is base and fiber stable if $(V_X)_{\mathsf{B}}$ is locally positive-definite about $\pi_{\mathsf{B}}(\chi(0))$.

(iii) A regular relative equilibrium χ is locally asymptotically base stable and locally asymptotically fiber stable if $(V_X)_{\mathsf{B}}$ has an isolated local minimum at $\pi_{\mathsf{B}}(\chi(0))$, and if F is strictly dissipative about X.

Proof. Let us prove part (i) in a direct way. Assume that χ is a maximal integral curve of X. The curve χ satisfies the equations of motion (6.16) if and only if

$$\begin{aligned} 0 &= \overset{\mathbb{G}}{\nabla}_{\chi'(t)} \chi'(t) + \mathrm{grad}V(\chi(t)) - F(\chi'(t)) \\ &= \overset{\mathbb{G}}{\nabla}_{X(\chi(t))} X(\chi(t)) + \mathrm{grad}V(\chi(t)) - F(X(\chi(t))) \\ &= \mathrm{grad}\big(-\tfrac{1}{2}\|X\|_{\mathbb{G}}^2 + V\big)(\chi(t)) \\ &= \mathrm{grad}V_X(\chi(t)). \end{aligned}$$

Therefore, χ is a relative equilibrium if and only if $\mathrm{grad}V_X(\chi(t)) = 0$ for all $t \in \mathbb{R}$. Because V_X is X-invariant, $\mathrm{grad}V_X(\chi(t)) = 0$ for all $t \in \mathbb{R}$ if and only if V_X has a critical point at $\chi(t_0)$ for some (hence for all) $t_0 \in \mathbb{R}$. This concludes the proof of part (i).

To prove parts (ii) and (iii), we use the regularity of the relative equilibrium χ to invoke Theorem 5.83. We recall the equivalent representation of the dynamics of Σ obtained in Theorem 5.83(ii). If $\gamma\colon \mathbb{R} \to \mathsf{Q}$ is a solution to (6.16), then the curve $\eta = \pi_{\mathsf{B}} \circ \gamma\colon \mathbb{R} \to \mathsf{B}$, and the curve $v\colon \mathbb{R} \to \mathbb{R}$ given by $v(t) = J_X(\gamma'(t)) \, \|X\|_{\mathbb{G}}^{-2} \, (\gamma(t))$, together satisfy

$$\begin{aligned}
\overset{\mathbb{G}_{\mathsf{B}}}{\nabla}_{\eta'(t)}\eta'(t) &= -\operatorname{grad}_{\mathsf{B}}\big(V_{X,v}^{\mathrm{eff}}\big)_{\mathsf{B}}(\eta(t)) + v(t)C_X(\eta'(t)) + T\pi_{\mathsf{B}}\left(\mathbb{G}^\sharp \circ F(\gamma'(t))\right), \\
\dot{v}(t) &= -\frac{v(t)\langle \mathrm{d}(\|X\|_{\mathbb{G}}^2)_{\mathsf{B}}(\eta(t)); \eta'(t)\rangle}{(\|X\|_{\mathbb{G}}^2)_{\mathsf{B}}(\eta(t))} + \frac{1}{(\|X\|_{\mathbb{G}}^2)_{\mathsf{B}}(\eta(t))}\langle F(\gamma'(t)); X(\gamma(t))\rangle .
\end{aligned} \tag{6.18}$$

From the curves η and v, the curve $\gamma\colon \mathbb{R} \to \mathsf{Q}$ is computed to be $\gamma(t) = \Phi_{0,t}^{vX}(\mathrm{hlft}_{q_0}(\eta)(t))$.

Let $\Sigma_{\mathsf{TB}\times\mathbb{R}}$ be reduced dynamical system on $\mathsf{TB} \times \mathbb{R}$ defined by equations (6.18). Given this characterization, the curve $\chi\colon \mathbb{R} \to \mathsf{Q}$ is a relative equilibrium for Σ if and only if its projection $t \mapsto (\pi_{\mathsf{B}}(\chi(0)), 1)$ is a solution to equations (6.18); that is, an equilibrium point for $\Sigma_{\mathsf{TB}\times\mathbb{R}}$.

Now we note that E_X is a function invariant under the flow of X^T (see Proposition 5.70). Therefore, E_X can be projected onto the reduced space $\mathsf{TB} \times \mathbb{R}$. Given the decomposition $\mathsf{TQ} = \mathsf{HQ} \oplus \mathsf{VQ}$, we write $v_q \in \mathsf{TQ}$ as $v_q = \mathrm{hlft}_q(w_b) + sX(q)$ for $(w_b, s) \in \mathsf{TB} \times \mathbb{R}$. Accordingly

$$\begin{aligned}
(E_X)_{\mathsf{TB}\times\mathbb{R}}(w_b, s) &= E_X(\mathrm{hlft}_q(w_b) + sX(q)) \\
&= (V_X)_{\mathsf{B}}(b) + \tfrac{1}{2}\mathbb{G}_{\mathsf{B}}(w_b, w_b) + \tfrac{1}{2}(\|X\|_{\mathbb{G}}^2)_{\mathsf{B}}(b)(s-1)^2.
\end{aligned}$$

Because E_X is a nonincreasing function along the trajectories of Σ, its projection $(E_X)_{\mathsf{TB}\times\mathbb{R}}(w_b, s)$ is a nonincreasing function along the trajectories of $\Sigma_{\mathsf{TB}\times\mathbb{R}}$. If $(V_X)_{\mathsf{B}}$ is locally positive-definite about $\pi_{\mathsf{B}}(\chi(0))$, then

$$(w_b, s) \mapsto (E_X)_{\mathsf{TB}\times\mathbb{R}}(w_b, s) - V_X(\chi(0))$$

is a Lyapunov function for the dynamical system $\Sigma_{\mathsf{TB}\times\mathbb{R}}$ about the point $(0_{\pi_{\mathsf{B}}(\chi(0))}, 1) \in \mathsf{TB} \times \mathbb{R}$. This proves part (ii), that is, the relative equilibrium $\chi\colon \mathbb{R} \to \mathsf{Q}$ is base and fiber stable.

To prove (iii), we mirror the proof of Theorem 6.45(ii) and invoke the LaSalle Invariance Principle, as presented in Corollary 6.20, on the dynamical system $\Sigma_{\mathsf{TB}\times\mathbb{R}}$. We now assume that F is strictly dissipative about X and that F is X-invariant. Since $\pi_{\mathsf{B}}(\chi(0)) \in \mathsf{B}$ is an isolated local minimum for $\left(V_X\right)_{\mathsf{B}}$, there exists a neighborhood $\mathcal{W} \subset \mathsf{B}$ of $\pi_{\mathsf{B}}(\chi(0))$ such that $\left(V_X\right)_{\mathsf{B}}(b) > \left(V_X\right)_{\mathsf{B}}(b_0)$ for all $b \in \mathcal{W} \setminus \{\pi_{\mathsf{B}}(\chi(0))\}$, and $\mathcal{W}$ contains no critical point other than $\pi_{\mathsf{B}}(\chi(0))$. Next, we compute

$$\begin{aligned}
A &= \{\, (w_b, s) \in \mathsf{T}\mathcal{W} \times \mathbb{R} \mid \langle F(v_q); v_q\rangle = 0 \text{ for } v_q = \mathrm{hlft}_q(w_b) + sX(q)\} \\
&= \{\, (w_b, s) \in \mathsf{T}\mathcal{W} \times \mathbb{R} \mid v_q = X(q) \text{ for } v_q = \mathrm{hlft}_q(w_b) + sX(q)\} \\
&= \{(0_b, 1) \in \mathsf{T}\mathcal{W} \times \mathbb{R}\}.
\end{aligned}$$

On A, the equations (6.18) reduce to $0 = -\mathrm{grad}_{\mathsf{B}}(V_X)_{\mathsf{B}}(\eta)$. Therefore, no trajectory of (6.18) takes values in A, other than the trivial $t \mapsto (\pi_{\mathsf{B}}(\chi(0)), 1)$. Under these conditions, Corollary 6.20 guarantees that $(\pi_{\mathsf{B}}(\chi(0)), 1)$ is locally asymptotically stable. ■

Theorem 6.56(ii) can be stated directly in terms of the effective potential function rather than in terms of its projection. The helpful equivalence is stated as follows.

Lemma 6.57. *Let $X \in \Gamma^\infty(\mathsf{TQ})$ satisfy Assumption 5.78. Let $\psi \in C^\infty(\mathsf{Q})$ be X-invariant and let $\psi(q_0) = 0$ for $q_0 \in \mathsf{Q}$. Then the following statements are true:*

(i) $\pi_{\mathsf{B}}(q_0)$ is a critical point for ψ_{B} if and only if q_0 is a critical point for ψ;
(ii) ψ_{B} is locally positive-definite about $\pi_{\mathsf{B}}(q_0)$ if and only if there exists a neighborhood $\mathcal{W}$ of q_0 with the properties that
  (a) $q \in \mathcal{W}$ implies $\psi(q) \geq 0$, and
  (b) $\psi(q) = 0$ only if q belongs to the integral curve of X through q_0;
(iii) Assume q_0 is a critical point for ψ. Then $\mathrm{Hess}\,\psi_{\mathsf{B}}(\pi_{\mathsf{B}}(q_0))$ is positive-definite if and only if $\mathrm{Hess}\,\psi(q_0)$ is positive-definite on any complement to $\mathrm{span}_{\mathbb{R}}\{X(q_0)\}$ in $\mathsf{T}_{q_0}\mathsf{Q}$.

Proof of Lemma 6.57. According to Theorem 3.68, there exists a chart $(\mathcal{U}, \phi)$ with coordinates $(q^1, \dots, q^n)$, with $\phi(q_0) = \mathbf{0}$, and $X = \frac{\partial}{\partial q^1}$. Because ψ is X-invariant, the coordinate representation of ψ does not depend on q^1 and therefore its Taylor expansion is

$$\psi(q^1, \dots, q^n) = \sum_{i=2}^{n} (\mathrm{d}\psi(q_0))_i q^i + \frac{1}{2} \sum_{i,j=2}^{n} (\mathrm{Hess}\,\psi(q_0))_{ij} q^i q^j + o(\|(q^1, \dots, q^n)\|^3_{\mathbb{R}^n}).$$

The integral curve of X through q_0 is an open subset of the axis $\{(q^1, 0, \dots, 0) \mid q^1 \in \mathbb{R}\}$, and by construction $\psi_{\mathsf{B}}(q^2, \dots, q^n) = \psi(0, q^2, \dots, q^n)$. Statements (i) and (ii) are consequences of these facts. Part (iii) follows by noting that $\mathrm{Hess}\,\psi(q_0)$ is positive-definite on any complement to $\mathrm{span}_{\mathbb{R}}\{(1, 0, \dots, 0)\}$ if and only if the $(n-1)$ eigenvalues of the $(n-1) \times (n-1)$ matrix with components $(\mathrm{Hess}\,\psi(q_0))_{ij}$, $i, j \in \{2, \dots, n\}$, are strictly positive. ■

Remarks 6.58. 1. *Effective versus amended potential functions:* Theorem 6.56(i) and (ii) can be stated in terms of the amended potential function instead of the effective potential function; see Exercise E6.20. For part (iii), it is convenient to rely on the effective potential function rather than the amended potential function.

2. *Similarities with equilibrium configurations:* The results of Theorem 6.56 are similar to those of Lemma 6.32 and Theorem 6.45 characterizing the

existence and stability of equilibrium configurations. Roughly speaking, the effective energy E_X plays the same role in analyzing a relative equilibria as the energy function E plays in analyzing equilibrium configurations. In the study of relative equilibria, the effective potential function plays a role parallel to the one the potential function plays in studying equilibrium configurations. Furthermore, the second term in the equation (6.17) characterizing the effective energy is a positive-definite function in the "velocity error" $(v_q - X(q))$.

3. *Comparison with the Energy-Momentum Method:* The stability criterion in Theorem 6.56 is only sufficient: it turns out that, because of the Noether Conservation Law (Theorem 5.69), it is not necessary to require positive-definitness of Hess V_X on every complement to $\operatorname{span}_{\mathbb{R}}\{X(q_0)\}$ in $\mathsf{T}_{q_0}\mathsf{Q}$. The Energy-Momentum Method described in [Marsden 1992, Marsden and Ratiu 1999, Simo, Lewis, and Marsden 1991] provides a sharper, more detailed analysis by taking this into account. •

6.3.3 Examples illustrating existence and stability of relative equilibria

In this section we apply the results of Section 6.3 to a few examples.

Example 6.59 (Two-link manipulator with no gravity). We consider here the same example as in Example 6.50; see Figure 6.7 for an illustration of the example system. However, here we assume that the manipulator lies in a horizontal plane so that it is not subject to any gravitational force, so we can take $V = 0$. This is the only difference with the previous treatment. Recall that the configuration manifold is $\mathbb{T}^2$, the coordinate chart (θ_1, θ_2) describes the orientation of the two links, and from Example 4.32 the kinetic energy metric has matrix representation

$$[\mathbb{G}] = \begin{bmatrix} J_1 + \frac{1}{4}(m_1 + 4m_2)\ell_1^2 & \frac{1}{2}m_2\ell_1\ell_2\cos(\theta_1 - \theta_2) \\ \frac{1}{2}m_2\ell_1\ell_2\cos(\theta_1 - \theta_2) & J_2 + \frac{1}{4}m_2\ell_2^2 \end{bmatrix}.$$

We assume no force is applied to the system.

For $\omega \in \mathbb{R} \setminus \{0\}$, consider the infinitesimal generator

$$X = \omega\Big(\frac{\partial}{\partial\theta_1} + \frac{\partial}{\partial\theta_2}\Big)$$

corresponding to the action $\mathbb{R} \times \mathbb{T}^2 \to \mathbb{T}^2$ given by $(t, (\theta_1, \theta_2)) \mapsto (\theta_1 + \omega t, \theta_2 + \omega t)$. In other words, the maximal integral curves of X are counterclockwise (for $\omega > 0$) and clockwise (for $\omega < 0$) rotations of both links at constant relative angle. Using the formulae in Proposition 3.84, one can directly verify that $\mathscr{L}_X\mathbb{G} = 0$. Accordingly, X is an infinitesimal symmetry for the two-link manipulator with no gravity. It is instructive to compute the momentum map induced by X as

$$J_X(\theta_1, \theta_2, v_1, v_2) = \left(J_1 + \tfrac{1}{4}(m_1 + 4m_2)\ell_1^2 + \tfrac{1}{2}m_2\ell_1\ell_2\cos(\theta_1 - \theta_2)\right)\omega v_1 + \left(\tfrac{1}{2}m_2\ell_1\ell_2\cos(\theta_1 - \theta_2) + J_2 + \tfrac{1}{4}m_2\ell_2^2\right)\omega v_2.$$

To study the existence of relative equilibria related to the infinitesimal symmetry X, we proceed according to Theorem 6.56 and search for critical points of the effective potential function. We compute

$$\begin{aligned} V_X(\theta_1, \theta_2) &= -\tfrac{1}{2}\left\|X\right\|_{\mathbb{G}}^2(\theta_1, \theta_2) \\ &= -\omega^2\Big(J_1 + J_2 + \tfrac{1}{4}(m_1\ell_1^2 + 4m_2\ell_1^2 + m_2\ell_2^2)\Big) - \omega^2 m_2\ell_1\ell_2\cos(\theta_1 - \theta_2), \end{aligned}$$

and in turn

$$\mathrm{d}V_X(\theta_1, \theta_2) = \omega^2 m_2\ell_1\ell_2\left(\sin(\theta_1 - \theta_2)\mathrm{d}\theta_1 - \sin(\theta_1 - \theta_2)\mathrm{d}\theta_2\right).$$

Setting $\mathrm{d}V_X = 0$, we have $\sin(\theta_1 - \theta_2) = 0$, and, therefore, any configuration with $\theta_1 = \theta_2$ or $\theta_1 = \theta_2 + \pi$ is a critical point of V_X. These two sets of critical points correspond to counterclockwise and clockwise rotations of both links when the two links are aligned or folded.

To study the stability of the two relative equilibria, we proceed according to Theorem 6.56 and investigate the Hessian of the effective potential function. If (θ_1, θ_2) is a critical point for V_X, then

$$[\operatorname{Hess} V_X](\theta_1, \theta_2) = \omega^2 m_2\ell_1\ell_2 \begin{bmatrix} \cos(\theta_1 - \theta_2) & -\cos(\theta_1 - \theta_2) \\ -\cos(\theta_1 - \theta_2) & \cos(\theta_1 - \theta_2) \end{bmatrix}.$$

Next, we evaluate the Hessian at the two configurations $(\theta_1, \theta_2) = (0, 0)$ and $(\theta_1, \theta_2) = (0, \pi)$, representative of the two relative equilibria. We have

$$[\operatorname{Hess} V_X](0, 0) = -[\operatorname{Hess} V_X](0, \pi) = \omega^2 m_2\ell_1\ell_2 \begin{bmatrix} 1 & -1 \\ -1 & 1 \end{bmatrix}.$$

Clearly, the matrix representations of the Hessian at the two critical points have an eigenvalue equal to zero. This is consistent with the fact that V_X is X-invariant. The second eigenvalue of $[\operatorname{Hess} V_X](0, 0)$ is $2\omega^2 m_2\ell_1\ell_2$, whereas the second eigenvalue of $[\operatorname{Hess} V_X](0, \pi)$ is $-2\omega^2 m_2\ell_1\ell_2$. According to Theorem 6.56(ii), for $\omega > 0$ and $\omega < 0$, the relative equilibrium corresponding to $(\theta_1, \theta_2) = (0, 0)$ is base and fiber stable. The theorem is inconclusive about the relative equilibrium corresponding to $(\theta_1, \theta_2) = (0, \pi)$. One could use the sharper Energy-Momentum Method to analyze the stability properties of this relative equilibrium. •

Example 6.60 (Planar rigid body in ideal fluid). We consider a planar rigid body in an ideal fluid according to Assumptions 4.110. This is the system discussed in Exercises E4.25 and E5.24. As usual, the configuration manifold is $\mathsf{SE}(2) \simeq \mathbb{S}^1 \times \mathbb{R}^2$ and a coordinate chart is given by the position (x, y) of the center of mass and the angle θ the body forms with the horizontal axis. In what follows, it is convenient to define the chart domain to be $(\mathcal{U}, \phi)$, with

$$\mathcal{U} = (\mathbb{S}^1 \setminus \{(-\tfrac{\sqrt{2}}{2}, -\tfrac{\sqrt{2}}{2})\}) \times \mathbb{R}^2,$$

with $\phi(\chi, \eta, x, y) = (\theta(\chi, \eta), x, y)$, and where $\theta(\chi, \eta)$ takes values in $]-\frac{3\pi}{4}, \frac{5\pi}{4}[$. Note that this chart domain includes the values $\theta = k\frac{\pi}{2}$ for $k \in \{-1, 0, 1, 2\}$.

We refer the reader to Exercise E4.25 for the matrix representation of the kinetic energy metric in this chart. No potential function is present; $V = 0$.

For all $v_0 \in \mathbb{R}\setminus\{0\}$, the vector field $X = v_0 \frac{\partial}{\partial x}$ is an infinitesimal symmetry for the planar underwater body, and the corresponding momentum is

$$\begin{aligned} &J_X(\theta, x, y, v_\theta, v_x, v_y) \\ &\qquad = \left(m_1(\cos\theta)^2 + v_0 m_2(\sin\theta)^2\right) v_0 v_x + \left((m_1 - m_2)\sin\theta\cos\theta\right) v_0 v_y. \end{aligned}$$

The maximal integral curves of X correspond to translations of the planar body along the spatial x-axis.

To study the existence of relative equilibria related to the infinitesimal symmetry X, we proceed according to Theorem 6.56 and search for critical points of the effective potential function. We compute

$$V_X(\theta, x, y) = -\frac{v_0^2}{2}\left(m_1(\cos\theta)^2 + m_2(\sin\theta)^2\right).$$

Note that V_X is independent of x and y. Therefore, it suffices to investigate

$$\frac{\partial V_X}{\partial\theta} = 0 \quad\Longleftrightarrow\quad \sin 2\theta = 0 \quad\Longleftrightarrow\quad \theta \in \{-\tfrac{\pi}{2}, 0, \tfrac{\pi}{2}, \pi\}.$$

Hence, the maximal integral curve of X through $(\frac{\pi}{2}k, x, y)$, for all $k \in \{-1, 0, 1, 2\}$, $(x, y) \in \mathbb{R}^2$, are relative equilibria. Because V_X depends only on the variable θ, its Hessian cannot be positive-definite on any two-dimensional subspace. Therefore, we cannot apply Theorem 6.56 to assess stability of these relative equilibria.

As mentioned in Remark 6.58–3, we refer the reader to [Leonard and Marsden 1997] for a more detailed analysis using the Energy-Momentum Method for the stability of relative equilibria of rigid bodies in ideal fluid. •

6.3.4 Relative equilibria for simple mechanical systems on Lie groups

Here we specialize the results of Theorem 6.56 to the setting of simple mechanical systems on Lie groups. In particular we investigate simplified tests for the existence of relative equilibria based on the structure of Lie groups and Lie algebras introduced in Section 5.3.2.

Proposition 6.61 (Relative equilibrium for system on a Lie group). *Let* $(\mathsf{G}, \mathbb{I})$ *be a simple mechanical system on a Lie group. Let* $g_0 \in \mathsf{G}$ *and* $\xi \in \mathfrak{g}$. *Then the following statements are true:*

(i) the curve $t \mapsto g_0 \exp(t\xi)$ *is a relative equilibrium corresponding to the infinitesimal symmetry* $(\mathrm{Ad}_{g_0} \xi)_{\mathrm{R}}$ *if and only if*

$$\mathrm{ad}^*_{\xi} \mathbb{I}^\flat(\xi) = 0; \tag{6.19}$$

(ii) the regular relative equilibrium $t \mapsto g_0 \exp(t\xi)$ *is base and fiber stable if the symmetric bilinear map on* $\mathfrak{g}$ *given by*

$$(\eta, \zeta) \mapsto -\langle \mathbb{I}^\flat([\xi, \eta]); [\xi, \zeta] \rangle - \tfrac{1}{2} \langle \mathbb{I}^\flat(\xi); \mathrm{ad}_\eta \mathrm{ad}_\zeta \xi + \mathrm{ad}_\zeta \mathrm{ad}_\eta \xi \rangle$$

is positive-definite on any complement to $\mathrm{span}_{\mathbb{R}} \{\xi\}$ *in* $\mathfrak{g}$.

Part (i) can be equivalently stated as follows: if $\nu \in \mathfrak{g}$ and $g_0 \in \mathsf{G}$ satisfy $\mathrm{ad}^*_{\mathrm{Ad}_{g_0^{-1}}(\nu)} \mathbb{I}^\flat(\mathrm{Ad}_{g_0^{-1}}(\nu)) = 0$, then $t \mapsto \exp(t\nu) g_0$ is a relative equilibrium generated by the infinitesimal symmetry ν_{R}.

Proof of Proposition 6.61. Recall from Section 5.3.2 and Example 5.73 that a simple mechanical system on a Lie group $(\mathsf{G}, \mathbb{I})$ is L-invariant and its infinitesimal symmetries are the right-invariant vector fields $X = \nu_{\mathrm{R}}$ for $\nu \in \mathfrak{g}$. Also, recall from Remark 5.33–2 that the maximal integral curve of ν_{R} through $g_0 \in \mathsf{G}$ is $t \mapsto \exp(t\nu) g_0$. The set of orbits of ν_{R} is the quotient space $\mathsf{G}/\{\exp(t\nu) \mid t \in \mathbb{R}\}$. From Exercise E5.19(a) we write $\exp(t\nu) g_0 = g_0 \exp(t\,\mathrm{Ad}_{g_0^{-1}}(\nu))$, i.e., $t \mapsto \exp(t\nu) g_0$ is the integral curve of $(\mathrm{Ad}_{g_0^{-1}}(\nu))_{\mathrm{L}}$ through $g_0 \in \mathsf{G}$.

The effective potential function $V_{\nu_{\mathrm{R}}} : \mathsf{G} \to \mathbb{R}$ corresponding to the complete infinitesimal symmetry ν_{R} is

$$V_{\nu_{\mathrm{R}}}(g) = -\tfrac{1}{2} \|\nu_{\mathrm{R}}(g)\|^2_{\mathbb{G}_{\mathbb{I}}} = -\tfrac{1}{2} \|\mathrm{Ad}_{g^{-1}}(\nu)\|^2_{\mathbb{I}}.$$

According to Theorem 6.56(i), the curve $t \mapsto \exp(\nu t) g_0$ is a relative equilibrium for $(\mathsf{G}, \mathbb{I})$ if and only if $\mathrm{d}V_{\nu_{\mathrm{R}}}(g_0) = 0$. Let $\eta \in \mathfrak{g}$, recall the equalities presented in Exercise E5.19(c), and compute

$$\begin{aligned}
\mathscr{L}_{\eta_{\mathrm{L}}} V_{\nu_{\mathrm{R}}}(g) &= -\left\langle \mathbb{I}^\flat(\mathrm{Ad}_{g^{-1}}(\nu)); \mathscr{L}_{\eta_{\mathrm{L}}} \mathrm{Ad}_{g^{-1}}(\nu) \right\rangle \\
&= -\left\langle \mathbb{I}^\flat(\mathrm{Ad}_{g^{-1}}(\nu)); [\mathrm{Ad}_{g^{-1}}(\nu), \eta] \right\rangle \\
&= -\left\langle \mathrm{ad}^*_{\mathrm{Ad}_{g^{-1}}(\nu)} \mathbb{I}^\flat(\mathrm{Ad}_{g^{-1}}(\nu)); \eta \right\rangle.
\end{aligned}$$

Therefore, $\mathrm{d}V_{\nu_{\mathrm{R}}}(g_0) = 0$ if and only if $\mathrm{ad}^*_{\mathrm{Ad}_{g_0^{-1}}(\nu)} \mathbb{I}^\flat(\mathrm{Ad}_{g_0^{-1}}(\nu)) = 0$. In summary, if $\xi = \mathrm{Ad}_{g_0^{-1}}(\nu) \in \mathfrak{g}$ satisfies (6.19), then $t \mapsto g_0 \exp(t\,\mathrm{Ad}_{g_0^{-1}}(\nu)) = g_0 \exp(t\xi)$ is a relative equilibrium. The corresponding infinitesimal symmetry is $\nu_{\mathrm{R}} = (\mathrm{Ad}_{g_0} \xi)_{\mathrm{R}}$. This proves fact (i).

For $\zeta \in \mathfrak{g}$, we compute the second Lie derivative of $V_{\nu_{\mathrm{R}}}$ as follows:

$$\mathscr{L}_{\zeta_{\mathrm{L}}}\mathscr{L}_{\eta_{\mathrm{L}}}V_{\nu_{\mathrm{R}}}(g) = -\left\langle \mathbb{I}^\flat([\mathrm{Ad}_{g^{-1}}(\nu),\zeta]);[\mathrm{Ad}_{g^{-1}}(\nu),\eta]\right\rangle \\ -\left\langle \mathbb{I}^\flat(\mathrm{Ad}_{g^{-1}}(\nu));[[\mathrm{Ad}_{g^{-1}}(\nu),\zeta],\eta]\right\rangle.$$

At the critical point $g_0 \in \mathsf{G}$ such that $\mathrm{Ad}_{g_0^{-1}}(\nu) = \xi$ satisfies equation (6.19), we compute the Hessian of $V_{\nu_{\mathrm{R}}}$ by symmetry considerations to be

$$\mathrm{Hess}\, V_{\nu_{\mathrm{R}}}(g_0)(T_eL_{g_0}(\zeta), T_eL_{g_0}(\eta)) \\ = -\langle \mathbb{I}^\flat([\xi,\zeta]);[\xi,\eta]\rangle - \tfrac{1}{2}\langle \mathbb{I}^\flat(\xi); \mathrm{ad}_\eta\,\mathrm{ad}_\zeta\,\xi + \mathrm{ad}_\zeta\,\mathrm{ad}_\eta\,\xi\rangle.$$

Let us now assume that this symmetric bilinear map is positive-definite on a complement to $\mathrm{span}_{\mathbb{R}}\{\xi\}$ in $\mathfrak{g}$. At $g_0 \in \mathsf{G}$ such that $\mathrm{Ad}_{g_0^{-1}}(\nu) = \xi$, the subspace $\mathrm{span}_{\mathbb{R}}\{\xi,\eta,\zeta\} \subset \mathfrak{g}$ has maximal dimension if and only if the subspace $\mathrm{span}_{\mathbb{R}}\{\nu_{\mathrm{R}}(g_0), \eta_{\mathrm{L}}(g_0), \zeta_{\mathrm{L}}(g_0)\} \subset \mathsf{T}_{g_0}\mathsf{G}$ has maximal dimension. If the Hessian of $V_{\nu_{\mathrm{R}}}$ at g_0 is positive-definite on $\mathrm{span}_{\mathbb{R}}\{\eta_{\mathrm{L}}(g_0), \zeta_{\mathrm{L}}(g_0)\} \subset \mathsf{T}_{g_0}\mathsf{G}$, then, by Lemma 6.57(iii), the Hessian of the projection of $V_{\nu_{\mathrm{R}}}$ onto the manifold $\mathsf{G}/\{\exp(t\nu) \mid t \in \mathbb{R}\}$ is positive-definite. In turn, Lemma 6.57(iii) and Theorem 6.56(ii) imply that the relative equilibrium $t \mapsto g_0\exp(t\xi)$ is base and fiber stable. ■

Example 6.62 (Rigid body fixed at a point (cont'd)). We consider the same example system as in Example 5.48. Let $\{J_1, J_2, J_3\}$ be the three principal inertias of the rigid body and let $\mathbb{I}$ be the inertia tensor. The simple mechanical system on a Lie group is therefore $(\mathsf{SO}(3), \mathbb{I})$. As illustrated in Example 5.48, it is convenient to rely on the identifications $\mathbb{R}^3 \simeq \mathfrak{so}(3) \simeq (\mathbb{R}^3)^* \simeq \mathfrak{so}(3)^*$.

Let $\boldsymbol{\xi} \in \mathbb{R}^3$ and $\boldsymbol{R}_0 \in \mathsf{SO}(3)$. According to Proposition 6.61(i), the curve $t \mapsto \boldsymbol{R}_0\exp(t\widehat{\boldsymbol{\xi}})$ is a relative equilibrium if and only if

$$\boldsymbol{\xi} \times ([\mathbb{I}]\boldsymbol{\xi}) = 0,$$

that is, if and only if $\boldsymbol{\xi}$ and $[\mathbb{I}]\boldsymbol{\xi}$ are aligned. In other words, rotation at constant body velocity $\boldsymbol{\xi}$ is a relative equilibrium for the rigid body system $(\mathsf{SO}(3), \mathbb{I})$ if and only if $\boldsymbol{\xi}$ is a scalar multiple of one of the principal axes.

Next, let us proceed according to Proposition 6.61(ii). Note that the curve $t \mapsto \boldsymbol{R}_0\exp(t\widehat{\boldsymbol{\xi}})$ is an embedding for all $\boldsymbol{\xi} \in \mathbb{R}^3$, see Exercise E5.13(d). We consider the symmetric bilinear map on $\mathbb{R}^3$ given by

$$(\boldsymbol{\eta},\boldsymbol{\zeta}) \mapsto -(\boldsymbol{\xi}\times\boldsymbol{\eta})^T[\mathbb{I}](\boldsymbol{\xi}\times\boldsymbol{\zeta}) - \frac{1}{2}\boldsymbol{\xi}^T[\mathbb{I}](\boldsymbol{\zeta}\times(\boldsymbol{\eta}\times\boldsymbol{\xi}) + \boldsymbol{\zeta}\times(\boldsymbol{\eta}\times\boldsymbol{\xi})). \quad (6.20)$$

We study whether this symmetric bilinear map is positive definite on a complement to $\mathrm{span}_{\mathbb{R}}\{\boldsymbol{\xi}\}$. Without loss of generality, we let the orthonormal vectors $\{\boldsymbol{e}_1, \boldsymbol{e}_2, \boldsymbol{e}_3\}$ be the principal axes of the rigid body, and we set $\boldsymbol{\xi} = \omega\boldsymbol{e}_1$ for $\omega \in \mathbb{R}\setminus\{0\}$. Next, we compute a 2×2 matrix representation of the symmetric bilinear map restricted to $\mathrm{span}_{\mathbb{R}}\{\boldsymbol{e}_2, \boldsymbol{e}_3\}$. In other words, we let $(\boldsymbol{\eta}, \boldsymbol{\zeta}) \in \{(\boldsymbol{e}_2, \boldsymbol{e}_2), (\boldsymbol{e}_2, \boldsymbol{e}_3), (\boldsymbol{e}_3, \boldsymbol{e}_2), (\boldsymbol{e}_3, \boldsymbol{e}_3)\}$. One can easily verify that this matrix representation is a diagonal matrix with entries $\{\omega^2(J_1 - J_3), \omega^2(J_1 - J_2)\}$, see Exercise E6.21. In summary, rotation at constant body angular velocity $\boldsymbol{\xi}$ is base and fiber stable if $\boldsymbol{\xi}$ is aligned with the principal axis corresponding to the largest principal inertia.

As mentioned in Remark 6.58–3, we refer the reader to [Marsden 1992, Marsden and Ratiu 1999] for a more detailed analysis through the Energy-Momentum Method of the stability of spinning rigid bodies. For example, using the more detailed analysis one can show that rotation at constant body angular velocity $\boldsymbol{\xi}$ is base stable if $\boldsymbol{\xi}$ is aligned with the principal axis corresponding to the smallest principal inertia, and it is not base stable if $\boldsymbol{\xi}$ is aligned with the principal axis corresponding to the intermediate principal inertia. •

Exercises

E6.1 Given a matrix $\boldsymbol{A} \in \mathbb{R}^{n \times n}$, prove that the origin $\boldsymbol{0} \in \mathbb{R}^n$ is globally asymptotically stable for the vector field $\boldsymbol{x} \mapsto \boldsymbol{A}\boldsymbol{x}$ if and only if there exist symmetric positive-definite matrices $\boldsymbol{P}, \boldsymbol{Q} \in \mathbb{R}^{n \times n}$ such that $\boldsymbol{A}^T\boldsymbol{P} + \boldsymbol{P}\boldsymbol{A} = -\boldsymbol{Q}$. The latter equation is referred to as the ***Lyapunov equation***.
Hint: *Given a symmetric, positive-definite matrix $\boldsymbol{Q} \in \mathbb{R}^{n \times n}$, define $\boldsymbol{P} = -\int_0^{+\infty} \exp(t\boldsymbol{A}^T)\boldsymbol{Q}\exp(t\boldsymbol{A})\mathrm{d}t$.*

E6.2 Let X be a C^r-vector field on $\mathbb{S}^1$ and let $x_0 \in \mathbb{S}^1$ be a locally asymptotically equilibrium point for X.
(a) Show that for $r = 0$, x_0 is not globally asymptotically stable.
(b) Show that for $r \geq 1$ the domain of attraction for x_0 is diffeomorphic to $\mathbb{R}$.

E6.3 Complete the proof of LaSalle Invariance Principle in Theorem 6.19 by showing that fact (ii) in the theorem statement is true.

In the next exercise we investigate the character of the level sets of a Lyapunov function. This investigation relies on some ideas from algebraic topology. We refer the reader to [Munkres 1984] for details. At the end of Section 3.1.1 we defined what it meant for two loops to be homotopic. We generalize this here to maps between topological spaces. Continuous maps $f, g\colon S \to T$ between topological spaces S and T are ***homotopic*** if there exists a continuous map $h\colon S \times [0,1] \to T$ with the properties that $h(0,x) = f(x)$ for all $x \in S$ and that $h(1,x) = g(x)$ for all $x \in S$. Topological spaces S and T are ***homotopic*** if there are continuous maps $f\colon S \to T$ and $g\colon T \to S$ such that $f \circ g$ is homotopic to id_T and $g \circ f$ is homotopic to id_S. A ***deformation retract*** of a topological space S is a subset $D \subset S$ for which there exists a continuous map $h\colon S \times [0,1] \to S$ with the property that $h(x,0) = x$ for all $x \in S$, that $h(x,1) \in D$ for all $x \in S$, and that $h(1,x) = x$ for all $x \in D$.

E6.4 The first few problems deal with the preceding definitions.
(a) Show that the relation "$S \sim T$ if S and T are homotopic" is an equivalence relation in the set of topological spaces.

(b) Show that if D is a deformation retract of S, then D and S are homotopic.
(c) Conclude that if D_1 and D_2 are two deformation retracts of S, then D_1 and D_2 are homotopic.

Now we see how the preceding discussion relates to Lyapunov functions. For simplicity, we suppose that $\mathsf{M} = \mathbb{R}^n$, and that $\psi\colon \mathbb{R}^n \to \mathbb{R}$ is a proper smooth function with the properties that $\mathbf{0}$ is its only critical point and that $\psi(x) > 0$ for all $x \in \mathbb{R}^n \setminus \{\mathbf{0}\}$. On $\mathbb{R}^n \setminus \{\mathbf{0}\}$ consider the vector field

$$X_\psi(\boldsymbol{x}) = -\frac{\operatorname{grad}\psi(\boldsymbol{x})}{\|\operatorname{grad}\psi(\boldsymbol{x})\|_{\mathbb{R}^n}^2},$$

where "grad" refers to the gradient with respect to the standard Riemannian metric on $\mathbb{R}^n$. We comment that the properness of ψ ensures that X_ψ is complete.

(d) Show that for $c > 0$ the level set $\psi^{-1}(c)$ is a deformation retract of $\mathbb{R}^n \setminus \{\mathbf{0}\}$.
Hint: *Consider the map from $\mathbb{R}^n \setminus \{\mathbf{0}\}$ to $\psi^{-1}(c)$ defined by*

$$f(\boldsymbol{x}) = \Phi^{X_\psi}_{c-\psi(\boldsymbol{x})}(\boldsymbol{x}).$$

(e) Show that $\mathbb{S}^{n-1}$ is a deformation retract of $\mathbb{R}^n \setminus \{\mathbf{0}\}$.
(f) Conclude that $\psi^{-1}(c)$ and $\mathbb{S}^{n-1}$ are homotopic.

The ***Poincaré Conjecture*** states that a compact k-dimensional manifold that is homotopic to $\mathbb{S}^k$ is diffeomorphic to $\mathbb{S}^k$. The Poincaré Conjecture is known to be true for $k \notin \{3,4\}$, is known to be true for $k = 4$ provided one replaces "diffeomorphic" with "homeomorphic," and is currently undecided when $k = 3$ (although this may be on the verge of changing at the time of publication of this book).

(g) Show that when, $n \notin \{4,5\}$, the level sets of ψ in the preceding discussion are diffeomorphic to $(n-1)$-dimensional spheres.

E6.5 In this exercise we consider the coupled mass system of Exercises E4.3 and E4.17. Note that this is a linear mechanical system (provided you chose your coordinates in the "natural" way in Exercise E4.3).
(a) Determine whether the origin is a stable equilibrium point.

Now we consider the addition of dissipative forces as in parts (a) and (b) of Exercise E4.29.

(b) For the dissipation of Exercise E4.29(a), determine whether the origin is asymptotically stable.
(c) For the dissipation of Exercise E4.29(b), determine whether the origin is asymptotically stable.
(d) Provide a physical interpretation of your answers from parts (b) and (c).

Now suppose that the system is subject to both dissipative forces and that the dissipative factor for the dissipation of part (b) of Exercise E4.29 is negative. Thus the dashpot does not dissipate energy, it increases energy.

(e) Show that, even with this negative dissipative, the origin can be asymptotically stable, and provide the condition on the dissipative constant that ensures asymptotic stability. Provide a physical interpretation for your answer.

E6.6 Let $(\mathsf{M}, \mathbb{G})$ be a Riemannian manifold with Levi-Civita connection $\overset{\mathbb{G}}{\nabla}$, and let $f\colon \mathsf{M} \to \mathbb{R}$ be smooth. Prove the following facts:

(a) the Hessian of f satisfies

$$\operatorname{Hess} f(x) \cdot (v_x, w_x) = (\mathscr{L}_W \mathscr{L}_V - \mathscr{L}_{\overset{\mathbb{G}}{\nabla}_W V}) f(x), \qquad \text{(E6.1)}$$

where the vector fields V and W are smooth extensions to a neighborhood of x of the tangent vectors $v_x, w_x \in \mathsf{T}_x\mathsf{M}$;

(b) equation (E6.1) is independent of the extensions V and W;

(c) $\operatorname{Hess} f$ is a symmetric $(0,2)$-tensor field;

(d) at a critical point x_0 for f, the Hessian has components

$$(\operatorname{Hess} f(x_0))_{ij} = \frac{\partial^2 f}{\partial x^i \partial x^j}(x_0),$$

in a chart $(\mathcal{U}, \phi)$ with coordinates $(x^1, \dots, x^n)$;

(e) at a point $x \in \mathsf{M}$, the Hessian has components

$$(\operatorname{Hess} f(x))_{ij} = \frac{\partial^2 f}{\partial x^i \partial x^j}(x) - \Gamma^k_{ij}(x) \frac{\partial f}{\partial x^k}(x).$$

E6.7 Exercise E3.10 presents two atlases on the set $\mathbb{R}$,

$$\mathscr{A}_1 = \{(\mathcal{U}_1 = \mathbb{R}, \phi_1 = \mathrm{id}_{\mathbb{R}})\},$$
$$\mathscr{A}_2 = \{(\mathcal{U}_2 = \mathbb{R}, \phi_2 = \{x \mapsto x^3\})\},$$

and defines M_1 and M_2 as the differentiable manifolds whose differentiable structures are generated by $\mathscr{A}_1$ and $\mathscr{A}_2$, respectively. Prove the following statements:

(a) The function $\mathsf{M}_1 \ni x \mapsto x^2$ has positive-definite Hessian at 0, and the function $\mathsf{M}_2 \ni x \mapsto x^2$ is not differentiable at 0.

(b) The function $\mathsf{M}_1 \ni x \mapsto x^6$ has degenerate Hessian at 0, and the function $\mathsf{M}_2 \ni x \mapsto x^6$ has positive-definite Hessian at 0.

E6.8 Let $(\mathsf{Q}, \mathbb{G})$ be a Riemannian manifold. Given a smooth function $V\colon \mathsf{Q} \to \mathbb{R}$ and $L \in \mathbb{R}$, assume the sublevel set $V^{-1}(\leq L)$ is compact. Let $\gamma\colon \bar{\mathbb{R}}_+ \to \mathsf{Q}$ be an integral curve for $-\operatorname{grad} V$:

$$\gamma'(t) = -\operatorname{grad} V(\gamma(t)).$$

The solution to this differential equation is sometimes referred to as the ***negative gradient flow*** associated with $(\mathsf{Q}, \mathbb{G}, V)$.

(a) Prove that, for every $q \in V^{-1}(\leq L)$, the integral curve with initial condition $\gamma(0) = q$ converges to a critical point of V.

Furthermore, assume that there exists $t_0 \in \bar{\mathbb{R}}_+$ such that $\gamma(t_0) \in V^{-1}(\leq L, q_0)$, where q_0 is a critical point with positive-definite $\operatorname{Hess} V(q_0)$ and $L < L_{\mathrm{reg,cpt}}(V, q_0)$.

(b) Prove exponential convergence of γ to q_0.

E6.9 Let $x_0 = (0,0) \in \mathbb{R}^2$, and compute $L_{\mathrm{cpt}}(\psi, x_0)$ and $L_{\mathrm{reg}}(\psi, x_0)$ for the functions $\psi\colon \mathbb{R}^2 \to \mathbb{R}$ given by

(a) $\psi(x_1, x_2) = x_1^2 + \frac{x_2^2}{1+x_1^4} + x_1^3$, and

(b) $\psi(x_1, x_2) = x_1^2 + \frac{1}{2-\cos x_2}$.

In the next exercise we consider the effect of linear gyroscopic forces. A forced linear mechanical system $(\mathsf{V}, M, K, (F_1, F_2))$ is subject to a ***linear gyroscopic force*** if F_2 is skew-symmetric (making the identification of $L(\mathsf{V}; \mathsf{V}^*)$ with $T_2^0(\mathsf{V})$) and if $F_1 = 0$. The following exercise serves as an illustration of the following facts.

1. Sometimes an unstable linear mechanical system can be stabilized by the addition of a linear gyroscopic force.
2. Even when an unstable linear mechanical system can be stabilized by the addition of a gyroscopic force, it is *always* destabilized by the further addition of a linear dissipative force.

For a discussion, we refer to [Merkin 1997, Section 6.5]. See also the discussion in [Bloch, Krishnaprasad, Marsden, and Ratiu 1993] as concerns the stability of relative equilibria.

E6.10 On $\mathsf{V} = \mathbb{R}^2$ consider the linear mechanical system with

$$M = e^1 \otimes e^1 + e^2 \otimes e^2, \quad K = k_1 e^1 \otimes e^1 + k_2 e^2 \otimes e^2,$$

where $k_1, k_2 \in \mathbb{R}$.

(a) Show that, for $k_1, k_2 < 0$, the equilibrium configuration $(0, 0)$ is unstable.

Let C be the skew-symmetric $(0, 2)$-tensor

$$C = g e^1 \otimes e^2 - g e^2 \otimes e^1,$$

with $g \in \mathbb{R}$.

(b) Show that, given $k_1, k_2 < 0$, there exists $g \in \mathbb{R}$ for which the system subject to the linear gyroscopic force defined by C has $(0, 0)$ as a stable equilibrium configuration.

***Hint:** The system is linear, so stability can be deduced from the characteristic polynomial.*

Now define the symmetric $(0, 2)$-tensor

$$R = d e^1 \otimes e^1 + d e^2 \otimes e^2,$$

where $d \in \mathbb{R}$.

(c) Let $k_1, k_2 < 0$ and let g be chosen as in part (b) so that $(0, 0)$ is stable. Show that for any $d > 0$, the further addition of the linear dissipative force defined by R renders the equilibrium configuration $(0, 0)$ unstable.

In the next exercise we consider the effect of so-called "linear nonconservative forces." A forced linear mechanical system $(\mathsf{V}, M, K, (F_1, F_2))$ is subject to a ***linear nonconservative force*** if F_1 is skew-symmetric (making the identification of $L(\mathsf{V}; \mathsf{V}^*)$ with $T_2^0(\mathsf{V})$) and if $F_2 = 0$. The following exercise serves as an illustration of the following facts.

1. If K is negative-definite, then the system cannot be stabilized by the addition of a linear nonconservative force.
2. The preceding statement remains true even in the presence of arbitrary dissipation.

For a discussion, we refer to [Merkin 1997, Section 6.8].

E6.11 On $\mathsf{V} = \mathbb{R}^2$ consider the linear mechanical system with

$$M = e^1 \otimes e^1 + e^2 \otimes e^2, \quad K = k_1 e^1 \otimes e^1 + k_2 e^2 \otimes e^2,$$

where $k_1, k_2 \in \mathbb{R}$.

(a) Show that, for $k_1, k_2 < 0$, the equilibrium configuration $(0,0)$ is unstable.

Let N be the skew-symmetric $(0,2)$-tensor defined by

$$N = ce^1 \otimes e^2 - ce^2 \otimes e^1,$$

for $c \in \mathbb{R}$.

(b) Show that, for any $c \in \mathbb{R}$, the system subject to the linear nonconservative force defined by N still has $(0,0)$ as an unstable equilibrium configuration.

Hint: *The system is linear, so stability can be deduced from the characteristic polynomial.*

Now define the symmetric $(0,2)$-tensor

$$R = de^1 \otimes e^1 + de^2 \otimes e^2,$$

where $d \in \mathbb{R}$.

(c) For $k_1, k_2 < 0$, $c \in \mathbb{R}$, and $d > 0$, show that the further addition of the linear dissipative force defined by R cannot render the equilibrium configuration $(0,0)$ stable.

In Section 6.2.1 we considered the linearization of equilibria for forced simple mechanical systems. In the next exercise we will extend this to forced affine connection systems as defined in Definition 4.89. Using the developments of Section 4.5.5, this allows us to consider equilibria for forced simple mechanical systems with constraints.

E6.12 Let $\Sigma = (\mathsf{Q}, \nabla, Y, \mathcal{D})$ be a forced affine connection system, supposing for simplicity that Y is time-independent. A point $q_0 \in \mathsf{Q}$ is an ***equilibrium configuration*** for Σ if $Y(0_{q_0}) = 0_{q_0}$.

(a) Follow the arguments giving Proposition 6.37 to show that the linearization of Σ about an equilibrium configuration q_0 is governed by a second-order linear differential equation

$$\ddot{x}(t) + A_2(\dot{x}(t)) + A_1(x(t)) = 0 \tag{E6.2}$$

for some $A_1, A_2 \in L(\mathsf{T}_{q_0}\mathsf{Q}; \mathsf{T}_{q_0}\mathsf{Q})$. Give coordinate expressions for A_1 and A_2.

Now let $(\mathsf{Q}, \mathbb{G}, V, F, \mathcal{D})$ be a simple mechanical control system with regular constraints which then defines a forced affine connection system $(\mathsf{Q}, \overset{\mathcal{D}}{\nabla}, P_{\mathcal{D}} \circ (\mathbb{G}^\sharp \circ F - \operatorname{grad} V), \mathcal{D})$.

(b) Find the explicit form for A_1 and A_2 in this case. In particular, show that $A_2 = 0$ if F is basic.

(c) Argue that, in this case, the linearized equations (E6.2) can naturally be written as

$$M^\flat(\ddot{x}(t)) + (R^\flat + C^\flat)(\dot{x}(t)) + (K^\flat + N^\flat)(x(t)) = 0,$$

where $M, R, K \in T_2^0(\mathsf{T}_{q_0}\mathsf{Q})$ are symmetric and $C, N \in T_2^0(\mathsf{T}_{q_0}\mathsf{Q})$ are skew-symmetric.

Note that this is then a natural situation where one encounters a linear nonconservative force.

E6.13 Let $\mathsf{Q} = \mathbb{R}^2$. Since Q is a vector space, it makes sense to ask whether a Lagrangian on Q is that corresponding to a linear mechanical system. For each of the Lagrangians given below, answer the following questions:

1. Is the Lagrangian a Lagrangian for a linear mechanical system?
2. Is the Lagrangian a Lagrangian for a simple mechanical system? If so, write the Riemannian metric and the potential function.
3. Find all equilibrium points for the system.
4. Find the linearization about each of the equilibrium points.
5. Is the linearization stable? State which of the results from this chapter you applied to obtain your answer.
6. Is it possible, using the results given in this chapter, to determine whether the equilibrium is stable under the actual dynamics of the system? If it is possible, state the result which is applicable. If it is not possible,
   (a) state why none of the results in this chapter apply, and
   (b) see if you can figure out using your ingenuity whether the equilibrium point is stable (you can play with the equations, or try numerical simulation).

Here are the Lagrangians.

(a) $L(x, y, v_x, v_y) = \frac{1}{2}(v_x^2 + (1 + x^2 + y^2)v_y^2) - \frac{1}{2}(x^2 - y^2)$.
(b) $L(x, y, v_x, v_y) = \frac{1}{2}(v_x^2 + v_y^2) - \frac{1}{2}(x^2 + 4y^2)$.
(c) $L(x, y, v_x, v_y) = \frac{1}{2}(v_x^2 + v_y^2) - \frac{1}{2}(x^2 + y^4)$.
(d) $L(x, y, v_x, v_y) = \frac{1}{2}(v_x^2 + v_y^2) - \frac{1}{2}(x^4 - y^4)$.
(e) $L(x, y, v_x, v_y) = \frac{1}{2}(v_x^2 + 2v_y^2) - \frac{1}{2}(x^2 + \frac{1}{2}x^4 + y^2)$.
(f) $L(x, y, v_x, v_y) = \frac{1}{2}(v_x^2 + v_y^2) - \frac{1}{2}(y^2 - x^2 - \frac{1}{2}x^4)$.
(g) $L(x, y, v_x, v_y) = \frac{1}{2}(2v_x^2 + v_y^2) - \frac{1}{2}(2x^2 + xy + 3y^2 - xy^2)$.
(h) $L(x, y, v_x, v_y) = \frac{1}{2}(v_x^2 + 2v_y^2) - \frac{1}{2}xy$.
(i) $L(x, y, v_x, v_y) = \frac{1}{2}(v_x^2 + v_y^2) - \frac{1}{2}(x + y^2)$.

E6.14 This exercise is a continuation of the pendulum/cart system considered in Exercises E4.5, E4.19, and E4.31.

(a) Determine all equilibrium configurations for the system.
(b) For each equilibrium configuration q_0, compute $\mathbb{G}(q_0)$ and $\operatorname{Hess} V(q_0)$.
(c) For each equilibrium configuration q_0, determine whether it is linearly stable.
(d) For each equilibrium configuration q_0, determine whether it is stable.

E6.15 This exercise is a continuation of the spherical pendulum on a track system considered in Exercises E4.6 and E4.20.

(a) Determine all equilibrium configurations for the system.
(b) For each equilibrium configuration q_0, compute $\mathbb{G}(q_0)$ and $\operatorname{Hess} V(q_0)$.
(c) For each equilibrium configuration q_0, determine whether it is linearly stable.
(d) For each equilibrium configuration q_0, determine whether it is stable.

E6.16 This exercise is a continuation of the two-axis gyroscope considered in Exercises E4.7 and E4.21.
(a) Determine all equilibrium configurations for the system.
(b) For each equilibrium configuration q_0, compute $\mathbb{G}(q_0)$ and $\operatorname{Hess} V(q_0)$.
(c) For each equilibrium configuration q_0, determine whether it is linearly stable.
(d) For each equilibrium configuration q_0, determine whether it is stable.

E6.17 Let $f: \mathbb{R} \to \mathbb{R}$ be a C^∞-function.
(a) Show that if f has an isolated local minimum, then it is locally positive-definite about the minimum.
For $a, b \in \mathbb{R}$, we now consider the function $f: \mathbb{R} \to \mathbb{R}$ given by

$$f(x) = \begin{cases} \exp\left(-\frac{1}{x^2}\right)\left(\sin\left(\frac{a}{x^2}\right) + b\right), & x \neq 0, \\ 0, & x = 0. \end{cases}$$

Do the following:
(b) show that f is C^∞;
(c) show that f is locally positive-definite about 0 for $b > 1$;
(d) determine for what values of a and b the function f has a sequence of local minima converging to 0.
Using Theorem 3.109, answer the following question.
(e) Let M be a C^ω-manifold and $f: \mathsf{M} \to \mathbb{R}$ a C^ω-function. Show that f is locally positive-definite about x_0 if and only if x_0 is an isolated local minimum for f.

E6.18 This exercise is a continuation of the pendulum/cart system considered in Exercises E5.35 and E6.14.
(a) Find all relative equilibria for the system and describe the corresponding physical motions.
(b) Use your conclusions to complete part (d) of Exercise E6.14.

E6.19 Consider the spherical pendulum with gravity discussed in Examples 4.99, 5.72, and Exercise E5.34. Find all relative equilibria of the system and describe the corresponding physical motions.

E6.20 From Section 5.5.2, recall the definition of effective and amended potential functions. Using the same assumptions as Theorem 6.56 show that
(a) q_0 is a critical point for $V^{\mathrm{eff}}_{X,1}$ if and only if it is a critical point for $V^{\mathrm{amd}}_{X,\mu_0}$, where $\mu_0 = \|X\|^2_{\mathbb{G}}(q_0)$, and that
(b) if q_0 is a critical point for both functions and if $\mu_0 = \|X\|^2_{\mathbb{G}}(q_0)$, then $\operatorname{Hess} V^{\mathrm{eff}}_{X,1}(q_0)$ is positive-definite if and only if $\operatorname{Hess} V^{\mathrm{amd}}_{X,\mu_0}(q_0)$ is positive-definite.

E6.21 Compute the matrix representation of the symmetric $(0,2)$-tensor defined in equation (6.20). Compute its restriction to the subspace $\operatorname{span}_{\mathbb{R}}\{e_2, e_3\}$ for $\nu = \omega e_1$.

7

Controllability

One of the fundamental problems in control theory is that of controllability. Indeed, many design methodologies rely on some hypotheses that concern controllability. The problem of controllability is essentially one of describing the nature of the set of states reachable from an initial state. In the development of this theory, there are two properties that arise as being important. The first is the property of "accessibility," which means that the reachable set has a nonempty interior. The treatment of accessibility we present follows the approach of the fundamental paper of Sussmann and Jurdjevic [1972]. Results of a related nature are those of Krener [1974] and Hermann and Krener [1977]. The property of "controllability" extends accessibility by further asking whether the initial state lies in the interior of the reachable set. The matter of providing general conditions for determining controllability is currently unresolved, although there have been many deep and insightful contributions. While we cannot hope to provide anything close to a complete overview of the literature, we will mention some work that is commensurate with the approach that we take here. Sussmann has made various important contributions to controllability, starting with the paper [Sussmann 1978]. In the paper [Sussmann 1983], a Lie series approach was developed for the controllability of control-affine systems, and this approach culminated in the quite general results of [Sussmann 1987], which incorporated the ideas of Crouch and Byrnes [1986] concerning input symmetries. The Lie series methods rely on the notion that a system can be well approximated by a "nilpotent approximation." Contributions to this sort of approach have been made, for example, by Hermes [1982a,b, 1991], Kawski [1988b, 1998], and Kawski and Sussmann [1997]. A recent paper by Bianchini and Kawski [2003] indicates that there may well be some limitations to the approach of using nilpotent approximations to determine conditions for controllability. A related approach is the construction of "control variations," and this is explained, for example, in the papers of Bianchini and Stefani [1993] and Kawski [1988a]. Another approach to local controllability is that taken by Agrachev and Gamkrelidze [1993a,b], based on the chronological calculus of the same authors [Agrachev and Gamkrelidze 1978, Agrachev and Sachkov 2004]. The fact that some of the very basic properties of the reachable set for a nonlinear control system are yet to be understood is the subject of the open problems paper by Agrachev [1999]. Sontag [1988] and Kawski [1990] show that a general answer to the controllability problem will be computationally difficult. Nonetheless, the problem of controllability is so fundamental that there continues to be much work in the area.

The controllability problem for mechanical systems has a somewhat shorter history. As far as we are aware, the first paper dealing explicitly with controllability in a mechanical setting is that of Bonnard [1984]. Other early work is that of Bloch, Reyhanoglu, and McClamroch [1992] where systems with complementary input and constraint force distributions are considered. In [Bloch and Crouch 1992] global controllability was investigated for systems with symmetries, using the results of San Martin and Crouch [1984]. The approach we follow here is that initiated in the paper of Lewis and Murray [1997a] (see [Lewis and Murray 1999] for a more gentle treatment). This approach has been followed up in other papers [Hirschorn and Lewis 2001, Lewis 1997, 2000b, Lewis and Murray 1997b]. These methods have been employed by other authors in various applications [Lynch 1999, Lynch and Mason 1999, Manikonda and Krishnaprasad 2002, Shen 2002, Shen, McClamroch, and Bloch 2004, Shen, Sanyal, and McClamroch 2002]. Other treatments of nonlinear controllability, in textbook form, include [Agrachev and Sachkov 2004, Isidori 1995, Jurdjevic 1997, Nijmeijer and van der Schaft 1990] for accessibility, and [Bloch 2003] for accessibility and controllability. Some global controllability results are given in [Agrachev and Sachkov 2004].

In this chapter we shall review certain aspects of the general theory of controllability that are currently known to have direct application to mechanical systems. These general results are then applied to classes of mechanical systems, and controllability conclusions are drawn for a collection of examples.

7.1 An overview of controllability for control-affine systems

We now provide a quick overview of the results in the existing literature that will be useful for developing an understanding of the controllability of mechanical systems. The treatment here is incomplete, reflecting the currently incomplete understanding of controllability. In this section, $\Sigma = (\mathsf{M}, \mathscr{C} = \{f_0, f_1, \dots, f_m\}, U)$ will denote a control-affine system, and it will be assumed to be at least C^∞. We will also frequently require Σ to be analytic, and will be careful to say when this is required. We will also allow in the general discussion quite general control sets. A useful property of a control set is that $\mathbf{0} \in \operatorname{int}(\operatorname{conv}(U))$, in which case we shall say that U is ***proper***.

We should remark that the approach we take to accessibility is not an entirely standard one. We start by presenting, in some detail in Section 7.1.3, the results of Sussmann and Jurdjevic [1972]. These are then used to prove the standard accessibility results for control-affine systems in Section 7.1.4. This approach allows us to concentrate on some of the simpler, but control theoretically interesting, aspects of control-affine systems. A more standard direct approach to accessibility theory may be found in the texts [Isidori 1995, Nijmeijer and van der Schaft 1990].

7.1.1 Reachable sets

A control-affine system $\Sigma = (\mathsf{M}, \mathscr{C} = \{f_0, f_1, \dots, f_m\}, U)$ has associated with it the differential equations

$$\gamma'(t) = f_0(\gamma(t)) + \sum_{a=1}^{m} u^a(t) f_a(\gamma(t)). \tag{7.1}$$

A ***controlled trajectory*** of Σ is a pair (γ, u), where $u\colon I \to U$ is locally integrable, and where $\gamma\colon I \to \mathsf{M}$ is the corresponding locally absolutely continuous curve satisfying (7.1) (we refer to Section A.2 for a discussion of existence and uniqueness of solutions for (7.1)). The set of controlled trajectories for Σ is denoted by $\mathrm{Ctraj}(\Sigma)$, and the set of controlled trajectories for Σ defined on $[0, T]$ is denoted by $\mathrm{Ctraj}(\Sigma, T)$. We may now define the various sets of points that can be reached by trajectories of a control-affine system. For $x_0 \in \mathsf{M}$, we define

$$\mathcal{R}_\Sigma(x_0, T) = \{\gamma(T) \mid (\gamma, u) \in \mathrm{Ctraj}(\Sigma, T),\ \gamma(0) = x_0\}.$$

This is the set of states reachable in time exactly T. We are also interested in the set of states reachable in time at most T, and this is simply given by $\mathcal{R}_\Sigma(x_0, \leq T) = \bigcup_{t \in [0,T]} \mathcal{R}_\Sigma(x_0, t)$. The set of states reachable from x_0 we denote by $\mathcal{R}_\Sigma(x_0)$. Note that $x_0 \in \mathcal{R}_\Sigma(x_0, \leq T)$ for any $T > 0$.

Let us give some examples to illustrates these concepts.

Examples 7.1. 1. We take $\mathsf{M} = \mathbb{R}^2$, $m = 1$, $U = \mathbb{R}$, and the control system

$$\dot{x} = u, \quad \dot{y} = x^2.$$

Thus $f_0 = x^2 \frac{\partial}{\partial y}$ and $f_1 = \frac{\partial}{\partial x}$. Since the controls are allowed to be arbitrarily large, one can show that $\mathcal{R}_\Sigma((0,0), T) = \{(x, y) \mid y > 0\} \cup \{(0,0)\}$ for any $T > 0$.

2. Next we modify the preceding example by considering the control set $U = [-1, 1]$. In this case the character of the reachable sets is shown in Figure 7.1. One can show that the left and right boundaries for $\mathcal{R}_\Sigma((0,0))$ are given by the graph of the function $y(x) = \frac{1}{3}|x|^3$, and that the upper boundary for $\mathcal{R}_\Sigma((0,0), T)$ is given by the graph of the function

$$y(x) = -\frac{|x|^3}{4} + \frac{T|x|^2}{4} + \frac{T^2|x|}{4} + \frac{T^3}{12}.$$

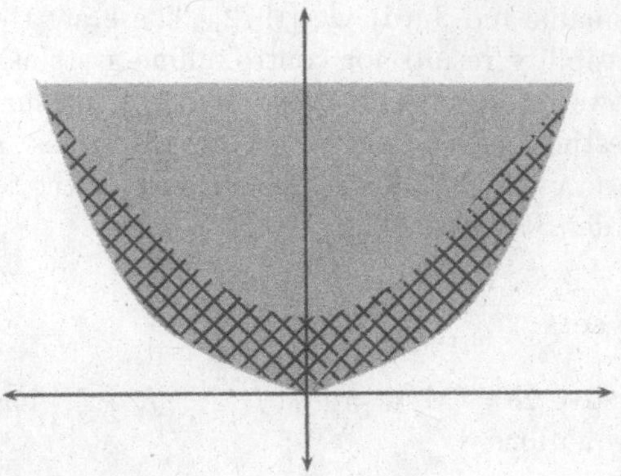

Figure 7.1. Reachable sets for $\dot{x} = u$, $\dot{y} = x^2$ with $U = [-1, 1]$: the shaded area represents $\mathcal{R}_\Sigma((0,0))$ and the hatched area represents $\mathcal{R}_\Sigma((0,0), T)$ for some $T > 0$.

3. We take $\mathsf{M} = \mathbb{R}^2$, $m = 1$, $U = \mathbb{R}$, and consider the control system

$$\dot{x} = u, \quad \dot{y} = 1.$$

Thus $f_0 = \frac{\partial}{\partial y}$ and $f_1 = \frac{\partial}{\partial x}$. In Figure 7.2 we show the reachable sets. For

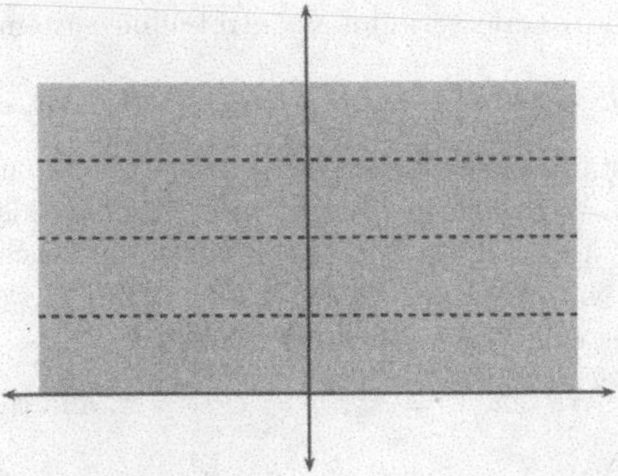

Figure 7.2. Reachable sets for $\dot{x} = u$, $\dot{y} = 1$ with $U = \mathbb{R}$: the shaded region represents $\mathcal{R}_\Sigma((0,0))$ and the dashed lines represent $\mathcal{R}_\Sigma((0,0), T)$ for various T's.

this example, we see that $\mathcal{R}_\Sigma((0,0), T) \subsetneq \mathcal{R}_\Sigma((0,0), \leq T)$, in contrast to the preceding examples.

4. We take $\mathsf{M} = \mathbb{R} \times \mathbb{S}^1$ with coordinates (x, θ), $m = 1$, $U = [-1, 1]$, and defined by the control system

$$\dot{x} = u, \quad \dot{\theta} = 1.$$

The reachable sets are shown in Figure 7.3, thinking of $\mathbb{R} \times \mathbb{S}^1$ as being $\mathbb{R} \times [0, 1]$ with the edges identified. •

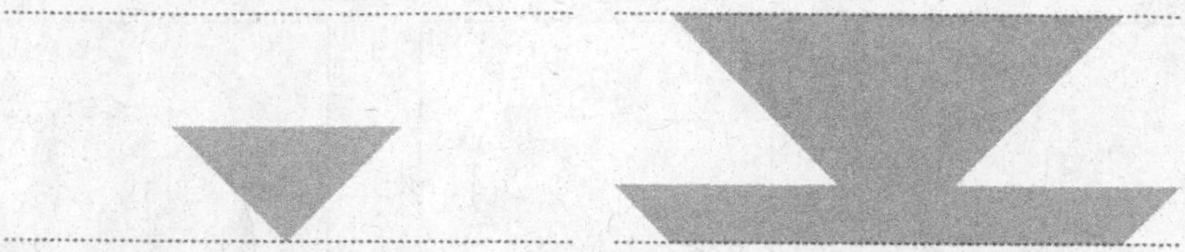

Figure 7.3. Reachable set for $\dot{x} = u$, $\dot{\theta} = 1$ with $U = [-1, 1]$: small time (left) and larger time (right).

7.1.2 Notions of controllability

Having introduced reachable sets, we may now turn to describing various notions of controllability. There are many ways to do this, and our presentation covers only some of those that will be interesting when we come to discussing mechanical systems.

Definition 7.2 (Controllability notions). Let $\Sigma = (\mathsf{M}, \mathscr{C} = \{f_0, f_1, \dots, f_m\}, U)$ be a C^∞-control-affine system and let $x_0 \in \mathsf{M}$.

(i) Σ is ***accessible*** from x_0 if there exists $T > 0$ such that $\mathrm{int}(\mathcal{R}_\Sigma(x_0, \leq t)) \neq \emptyset$ for $t \in]0, T]$.

(ii) Σ is ***controllable*** from x_0 if, for each $x \in \mathsf{M}$, there exists a $T > 0$ and $(\gamma, u) \in \mathrm{Ctraj}(\Sigma, T)$ such that $\gamma(0) = x_0$ and $\gamma(T) = x$.

(iii) Σ is ***small-time locally controllable*** (***STLC***) from x_0 if there exists $T > 0$ such that $x_0 \in \mathrm{int}(\mathcal{R}_\Sigma(x_0, \leq t))$ for each $t \in]0, T]$.

(iv) The pair $(\mathsf{M}, \mathscr{C})$ is ***properly small-time locally controllable*** (***properly STLC***) if $(\mathsf{M}, \mathscr{C}, U)$ is STLC for every proper control set U.

(v) The pair $(\mathsf{M}, \mathscr{C})$ is ***small-time locally uncontrollable*** (***STLUC***) if $(\mathsf{M}, \mathscr{C}, U)$ is not STLC for every compact control set U. •

Remarks 7.3. 1. Of these controllability notions, those of accessibility, small-time local controllability, proper small-time local controllability, and small-time local uncontrollability are most important in this book. In Figure 7.4 we depict the difference between some of these concepts.

2. We will not say anything concerning the notion of STLUC for general control-affine systems. For single-input systems, there are a few conditions for STLUC [Stefani 1986, Sussmann 1983]. Stefani [1988] also gives a necessary condition for local controllability for multi-input systems. A recent second-order necessary condition using the methods like those presented in Chapter 8 is given by Hirschorn and Lewis [2002]. •

Let us look at the examples we gave for reachable sets, and determine their controllability character.

Examples 7.4 (Example 7.1 cont'd).

1. In the first example, $\mathcal{R}_\Sigma((0,0), T)$ is the upper half-plane which has a nonempty interior. Thus this system is accessible from $(0, 0)$. It is, however, neither STLC nor controllable.

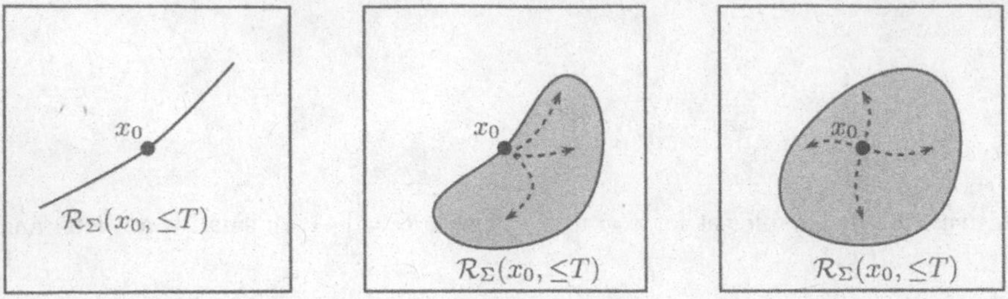

Figure 7.4. An interpretation of local accessibility and controllability, depicting lack of accessibility (left), accessibility but not local controllability (middle), and local controllability (right)

2. This example is the same as the preceding one with the exception that the controls are restricted. From Figure 7.1, we see that this restriction of the control set does not alter the property that the system is accessible from $(0, 0)$. Of course, it is not STLC, since the system with unbounded controls is not STLC from $(0, 0)$. The system is also not controllable from $(0, 0)$.
3. From Figure 7.2, we see that the system is accessible from $(0, 0)$. It is neither STLC nor controllable from $(0, 0)$.
4. As with the preceding example, the system is accessible from $(0, 0)$. Similarly, is it not STLC from $(0, 0)$. However, the system *is* controllable from $(0, 0)$. Here we see that the global nature of the state manifold M affects the controllability of the system. This is one of the reasons why we restrict our attention to local controllability in this book. •

7.1.3 The Sussmann and Jurdjevic theory of attainability

We review the approach of Sussmann and Jurdjevic [1972]. Readers may wish to refer at this point to the discussion of Chow's Theorem in Section 3.7.3. For our present discussion, we assume that M is an analytic manifold and that $\mathscr{V} \subset \Gamma^\omega(\mathsf{TM})$ is an arbitrary family of complete analytic vector fields. We note that, for a control-affine system $\Sigma = (\mathsf{M}, \mathscr{C}, U)$, there is an associated family of vector fields

$$\mathscr{V}_\Sigma = \left\{ f_0 + \sum_{a=1}^{m} u^a f_a \;\middle|\; u \in U \right\}.$$

After our discussion of *attainability* in this section we will use this family of vector fields to give conditions for *accessibility*. The reader wishing to simply see the accessibility results may skip ahead to Section 7.1.4.

We first note that the set $\mathrm{Diff}(\mathsf{M})$ of all diffeomorphisms of M is a group with multiplication defined by composition of diffeomorphisms (see Section 5.2.1 for definitions concerning groups). Given a subset $\mathscr{M} \subset \mathrm{Diff}(\mathsf{M})$, the subgroup of $\mathrm{Diff}(\mathsf{M})$ ***generated*** by $\mathscr{M}$ is the smallest subgroup of $\mathrm{Diff}(\mathsf{M})$

containing $\mathscr{M}$. For a family $\mathscr{V}$ of vector fields on M, we denote by $\mathrm{Diff}(\mathscr{V})$ the subgroup of diffeomorphisms of M generated by elements of the form

$$\Phi^{X_1}_{t_1} \circ \cdots \circ \Phi^{X_k}_{t_k}, \qquad t_1, \dots, t_k \in \mathbb{R},\ X_1, \dots, X_k \in \mathscr{V},\ k \in \mathbb{N}.$$

Analogously to our definition of orbits for distributions in Section 3.7.3, the ***$\mathscr{V}$-orbit*** through x_0 is the set

$$\mathcal{O}(x_0, \mathscr{V}) = \{\, \phi(x_0) \mid \ \phi \in \mathrm{Diff}(\mathscr{V})\} .$$

Now our presentation departs in spirit from that for distributions in Section 3.7.3, due that the fact that in control theory, time usually only goes forward. With this in mind, we let $\mathrm{Diff}^+(\mathscr{V})$ be the semi-group[1] of diffeomorphisms generated by elements of the form

$$\Phi^{X_1}_{t_1} \circ \cdots \circ \Phi^{X_k}_{t_k}, \qquad t_1, \dots, t_k \geq 0,\ X_1, \dots, X_k \in \mathscr{V},\ k \in \mathbb{N}.$$

This semi-group defines a subset of $\mathcal{O}(x_0, \mathscr{V})$ given by

$$\mathcal{O}^+(x_0, \mathscr{V}) = \{\, \phi(x_0) \mid \ \phi \in \mathrm{Diff}^+(\mathscr{V})\} .$$

A family $\mathscr{V}$ of vector fields is ***attainable*** from x_0 if $\mathrm{int}(\mathcal{O}^+(x_0, \mathscr{V})) \neq \emptyset$. This definition obviously closely mirrors the definition of accessibility.

Let us first describe the orbits $\mathcal{O}(x_0, \mathscr{V})$. This description is provided in varying degrees of generality by many authors [Hermann 1960, 1962, Matsuda 1968, Nagano 1966, Sussmann 1973], and generalizes Theorem 3.95. Let us denote by $\mathrm{Lie}^{(\infty)}(\mathscr{V})$ the smallest involutive distribution on M for which $\mathscr{V} \subset \Gamma^\omega(\mathrm{Lie}^{(\infty)}(\mathscr{V}))$. One may verify that this distribution is well-defined and is itself analytic (cf. Lemma 3.94).

Theorem 7.5 (Orbit Theorem). *If $\mathscr{V}$ is a family of complete analytic vector fields on M and $x_0 \in \mathsf{M}$, then the following statements are true:*

(i) $\mathcal{O}(x_0, \mathscr{V})$ is an analytic immersed submanifold;

(ii) for each $x \in \mathcal{O}(x_0, \mathscr{V})$, $\mathsf{T}_x(\mathcal{O}(x_0, \mathscr{V})) = \mathrm{Lie}^{(\infty)}(\mathscr{V})_x$;

(iii) M is the disjoint union of all orbits of $\mathscr{V}$.

Now let us turn to attainability. The results here are from the landmark paper of Sussmann and Jurdjevic [1972]. First note that by Theorem 7.5 it is evident that, if $\mathscr{V}$ is attainable from x_0, then $\mathrm{Lie}^{(\infty)}(\mathscr{V})_{x_0} = \mathsf{T}_{x_0}\mathsf{M}$. This condition is sufficient as well. The following result also provides an important description of the topological character of the attainable set.

Theorem 7.6 (Attainability). *Let $\mathscr{V}$ be an family of analytic vector fields on M, and let $x_0 \in \mathsf{M}$. The family $\mathscr{V}$ is attainable from x_0 if and only if $\mathrm{Lie}^{(\infty)}(\mathscr{V})_{x_0} = \mathsf{T}_{x_0}\mathsf{M}$. Furthermore, the interior of $\mathcal{O}^+(x_0, \mathscr{V})$ is dense in $\mathcal{O}^+(x_0, \mathscr{V})$.*

[1] A ***semi-group*** is a set with an associative product. There need not be an identity, and elements need not have inverses. The semi-group ***generated*** by a subset $\mathscr{M}$ of $\mathrm{Diff}(\mathsf{M})$ is the smallest semi-group containing $\mathscr{M}$.

Remarks 7.7. 1. In the second statement of the theorem, one actually needs to carefully specify the topology on $\mathcal{O}(x_0, \mathscr{V})$. The correct topology is the ***orbit topology***, which, by definition, is the manifold topology associated with the differentiable structure on the immersed submanifold $\mathcal{O}(x_0, \mathscr{V})$. This is discussed in Section 3.9.1, and we also refer to [Jurdjevic 1997].

2. The final assertion of the theorem is important, since it tells us that the character of the set $\mathcal{O}^+(x_0, \mathscr{V})$ is not too strange. For example, it rules out situations like that represented by Figure 7.5, where there are thin subsets

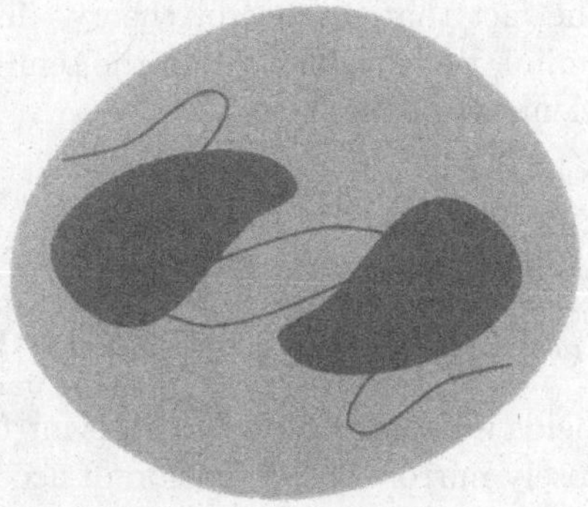

Figure 7.5. This cannot be the picture for $\mathcal{O}^+(x_0, \mathscr{V})$. That is, if the lightly shaded region represents M, then $\mathcal{O}^+(x_0, \mathscr{V})$ cannot have the character of the dark shaded region, since the interior of the dark shaded region is not dense in itself.

branching off a nice open set. It should be remarked that in the case when the vector fields are merely C^∞, the strong conclusions of Theorem 7.6 no longer hold. •

7.1.4 From attainability to accessibility

Most of the hard work for accessibility is contained in the attainability results from the preceding section. What we will do is explicitly provide the connection, and in so doing, arrive at fairly easily computable conditions for accessibility.

For a control-affine system $\Sigma = (\mathsf{M}, \mathscr{C} = \{f_0, f_1, \dots, f_m\}, U)$, we have previously defined the family of vector fields

$$\mathscr{V}_\Sigma = \left\{ f_0 + \sum_{a=1}^{m} u^a f_a \;\middle|\; u \in U \right\}.$$

To this family of vector fields, all of the machinery of Section 7.1.3 can be applied. However, we wish to see exactly how $\mathscr{V}_\Sigma$ is related to $\mathscr{C}$. Key to this is the following lemma.

Lemma 7.8. *Let $\Sigma = (\mathsf{M}, \mathscr{C}, U)$ be a C^∞-control-affine system and suppose that*

(i) $\mathbf{0} \in \mathrm{conv}(U)$ *and*
(ii) $\mathrm{aff}(U) = \mathbb{R}^m$.

Then $\mathrm{span}_{\mathbb{R}} \{\mathscr{C}\} = \mathrm{span}_{\mathbb{R}} \{\mathscr{V}_\Sigma\}$.

Proof. By definition of $\mathrm{span}_{\mathbb{R}} \{\mathscr{V}_\Sigma\}$, the inclusion $\mathrm{span}_{\mathbb{R}} \{\mathscr{V}_\Sigma\} \subset \mathrm{span}_{\mathbb{R}} \{\mathscr{C}\}$ holds. By (i) there exist $\lambda_1, \dots, \lambda_k \in \bar{\mathbb{R}}_+$ and $u_1, \dots, u_k \in U$ such that

$$\sum_{j=1}^{k} \lambda_j = 1, \quad \mathbf{0} = \sum_{j=1}^{k} \lambda_j u_j.$$

Therefore,

$$\sum_{j=1}^{k} \lambda_j \Big(f_0 + \sum_{a=1}^{m} u_j^a f_a \Big) = f_0 + \sum_{j=1}^{k} \sum_{a=1}^{m} \lambda_j u_j^a f_a = f_0.$$

Thus $f_0 \in \mathrm{span}_{\mathbb{R}} \{\mathscr{V}_\Sigma\}$.

Similarly, by (ii), for each $a \in \{1, \dots, m\}$, there exist $\lambda_1, \dots, \lambda_k \in \mathbb{R}$ and $u_1, \dots, u_k \in U$ such that

$$\sum_{j=1}^{k} \lambda_j = 1, \quad e_a = \sum_{j=1}^{k} \lambda_j u_j.$$

Therefore,

$$\sum_{j=1}^{k} \lambda_j \Big(f_0 + \sum_{a=1}^{m} u_j^a f_a \Big) = f_0 + \sum_{j=1}^{k} \sum_{b=1}^{m} \lambda_j u_j^b f_b = f_0 + f_a.$$

Thus $f_0 + f_a \in \mathrm{span}_{\mathbb{R}} \{\mathscr{V}_\Sigma\}$, showing that $f_a \in \mathrm{span}_{\mathbb{R}} \{\mathscr{V}_\Sigma\}$, $a \in \{1, \dots, m\}$. Thus we have shown that the inclusion $\mathrm{span}_{\mathbb{R}} \{\mathscr{C}\} \subset \mathrm{span}_{\mathbb{R}} \{\mathscr{V}_\Sigma\}$ also holds. ■

With this result as motivation, let us call a subset $U \subset \mathbb{R}^m$ ***almost proper*** if it has properties (i) and (ii) of the lemma. Note that if U is proper, then it is almost proper. Now we may state the following result, derived from Theorems 7.5 and 7.6, characterizing accessibility.

Theorem 7.9 (Accessibility). *Let $\Sigma = (\mathsf{M}, \mathscr{C}, U)$ be an analytic control-affine system with U almost proper. Then Σ is accessible from x_0 if and only if* $\mathrm{Lie}^{(\infty)}(\mathscr{C})_{x_0} = \mathsf{T}_{x_0}\mathsf{M}$.

Proof. Suppose that Σ is accessible. Note that the vector fields $f_0, f_1, \dots, f_m$ are tangent to $\mathcal{O}(x_0, \mathscr{C})$. Since a controlled trajectory (γ, u) has the property that γ is a locally absolutely continuous curve for which

$$\gamma'(t) \in \operatorname{span}_{\mathbb{R}}\{f_0(\gamma(t)), f_1(\gamma(t)), \dots, f_m(\gamma(t))\} \quad \text{a.e.},$$

it follows that if $\gamma(0) = x_0$, then $\gamma(t) \in \mathcal{O}(x_0, \mathscr{C})$. Thus $\mathcal{R}_\Sigma(x_0) \subset \mathcal{O}(x_0, \mathscr{C})$. In particular, if Σ is locally accessible, we must have $\mathsf{T}_{x_0}(\mathcal{O}(x_0, \mathscr{C})) = \mathsf{T}_{x_0}\mathsf{M}$. From Theorem 7.5 this implies that $\mathrm{Lie}^{(\infty)}(\mathscr{C})_{x_0} = \mathsf{T}_{x_0}\mathsf{M}$.

Now suppose that $\mathrm{Lie}^{(\infty)}(\mathscr{C})_{x_0} = \mathsf{T}_{x_0}\mathsf{M}$. By Lemma 7.8 this implies that $\mathrm{Lie}^{(\infty)}(\mathscr{V}_\Sigma)_{x_0} = \mathsf{T}_{x_0}\mathsf{M}$. Now, since piecewise constant controls $u\colon [0,T] \to U$ are locally integrable, it follows that $\mathcal{O}^+(x_0, \mathscr{V}_\Sigma) \subset \mathcal{R}_\Sigma(x_0)$. From Theorem 7.6 this means that $\operatorname{int}(\mathcal{R}_\Sigma(x_0)) \neq \emptyset$. From the second assertion of Theorem 7.6 we also conclude that $\operatorname{int}(\mathcal{R}_\Sigma(x_0, \leq T)) \neq \emptyset$ for $T > 0$, and thus Σ is accessible. ■

Remark 7.10. Note that for C^∞-systems, the condition that $\mathrm{Lie}^{(\infty)}(\mathscr{C})_{x_0} = \mathsf{T}_{x_0}\mathsf{M}$ is only sufficient for accessibility. There are C^∞-systems that are accessible, but that violate this condition. However, as is always the case for objects that are C^∞ but not analytic, such examples are a little peculiar. In the context of orbits for distributions, this is considered in Exercise E3.33. •

Let us illustrate Theorem 7.9 for the examples originally introduced in Section 7.1.1.

Examples 7.11 (Example 7.1 cont'd).

1. We take $\mathsf{M} = \mathbb{R}^2$, $f_0 = x^2\frac{\partial}{\partial y}$, and $f_1 = \frac{\partial}{\partial x}$. We compute $[f_1, [f_0, f_1]] = -2\frac{\partial}{\partial y}$. Therefore, the vector fields $\{f_1, [f_1, [f_0, f_1]]\}$ form a global basis for TM. This shows that $\mathrm{Lie}^{(\infty)}(\mathscr{C}) = \mathsf{TM}$. We may then say that, as long as the control set U is almost proper, the system will be accessible from $(0,0)$. The two cases in Example 7.1 had $U \in \{\mathbb{R}, [-1,1]\}$. Both of these control sets are almost proper, so our conclusions hold for these cases. This is consistent with the explicit descriptions we gave for the reachable sets in these cases.
2. Here we take $\mathsf{M} = \mathbb{R}^2$, $f_0 = \frac{\partial}{\partial y}$, and $f_1 = \frac{\partial}{\partial x}$. In this case we immediately have $\mathrm{Lie}^{(\infty)}(\mathscr{C}) = \mathsf{TM}$. This system is, therefore, accessible from $(0,0)$, provided that the control set is almost proper. The control set $U = \mathbb{R}$ is almost proper, and we do indeed see that our conclusions are consistent with Figure 7.2.
3. Our final example had $\mathsf{M} = \mathbb{R} \times \mathbb{S}^1$, $f_0 = \frac{\partial}{\partial \theta}$, and $f_1 = \frac{\partial}{\partial x}$. We again immediately conclude that, as long as U is almost proper, the system is accessible from $(0,0)$. This is consistent with the reachable set as depicted in Figure 7.3. •

Let us apply these ideas to a linear control system. We note that since a linear control system is a control-affine system, all notions and characterizations of controllability are relevant.

Example 7.12 (Accessibility of linear control systems). We refer the reader to Definition 4.102 for a recollection of the notion of a linear control system, which is a triple (V, A, B) defining the control equations

$$\dot{x}(t) = A(x(t)) + B(u(t)).$$

For the accessibility of a linear control system, using the computations of Exercise E3.23, we may deduce that the nonzero Lie brackets for the system are those of the form

$$\mathrm{ad}_{f_0}^k f_a \triangleq \underbrace{[f_0, [f_0, \cdots, [f_0, f_a]]]}_{k \text{ times}}, \qquad a \in \{1, \dots, m\}.$$

What is more, these brackets have the explicit expression $\mathrm{ad}_{f_0}^k f_a = (-1)^k A^k(b_a)$, $a \in \{1, \dots, m\}$, where $b_a = B(e_a)$. Thus we have

$$\mathrm{Lie}^{(\infty)}(\mathscr{C})_x = \mathrm{span}_{\mathbb{R}}\Big\{\{A(x)\} \cup \Big(\bigcup_{k=0}^{+\infty} \bigcup_{a=1}^{m} \{A^k(b_a)\}\Big)\Big\}.$$

By the Cayley–Hamilton Theorem, the linear maps A^k, $k \geq n$, are linear combinations of $\mathrm{id}_{\mathsf{V}}, A, \dots, A^{n-1}$. Provided that the control set is almost proper, the condition for accessibility from the origin in V reduces to checking whether the linear map from $\oplus_{a=1}^m \mathbb{R}^m$ to V given by the block matrix

$$\left[\ B \mid A \circ B \mid \cdots \mid A^{n-1} \circ B\ \right]$$

has maximal rank.

The condition we derive here is called the ***Kalman Rank Condition*** [Kalman, Ho, and Narendra 1963], and is derived using linear algebraic methods in linear systems theory. We refer to Exercise E7.4 for an alternative version of the Kalman Rank Condition. •

Note that we have shown in this section that there are computable (at least in terms of differentiations and linear algebra) necessary and sufficient conditions for accessibility, at least for analytic systems with a sufficiently nice control set.[2] As we shall see, things are not so well-developed for small-time local controllability.

7.1.5 Some results on small-time local controllability

We now give some conditions from the literature concerning small-time local controllability of a control-affine system $\Sigma = (\mathsf{M}, \mathscr{C} = \{f_0, f_1, \dots, f_m\}, U)$. Before we begin, let us make some introductory comments concerning the nature

[2] The "computability" of the conditions we give can be made precise; Sontag [1988] shows that the problem of determining accessibility is solvable in polynomial time.

of the control set U. Sussmann [1987, Proposition 2.3] shows that provided that $\mathrm{aff}(U) = \mathbb{R}^m$, $\Sigma = (\mathsf{M}, \mathscr{C}, U)$ is STLC from x_0 if $\tilde{\Sigma} = (\mathsf{M}, \mathscr{C}, \mathrm{cl}(\mathrm{conv}(U)))$ is STLC from x_0. For this reason, the control set may be assumed closed and convex, if this is necessary. Sussmann [1978] also shows that if

$$f_0(x_0) \notin \mathrm{conv}\Big\{ f_0(x_0) + \sum_{a=1}^{m} u^a f_a(x_0) \;\Big|\; u \in U \Big\},$$

then Σ is not STLC from x_0. Therefore, we must assume that this relation does not hold. In fact, we will assume something more, namely that

$$f_0(x_0) \in \mathrm{int}\Big(\mathrm{conv}\Big\{ f_0(x_0) + \sum_{a=1}^{m} u^a f_a(x_0) \;\Big|\; u \in U \Big\}\Big). \tag{7.2}$$

By noting as above that U may without loss of generality be assumed to be convex, one may see that the drift vector field f_0 and the control set U can be redefined such that the following assumption holds (see Exercise E7.1).

Assumption 7.13 (Vanishing drift at initial state). In this section we assume that $f_0(x_0) = 0_{x_0}$ and that U is proper. •

Let us begin the technical development in this section with a motivating example.

Example 7.14 (Example 7.1 cont'd). We resume the example in which $\mathsf{M} = \mathbb{R}^2$, and with the governing equations

$$\dot{x} = u, \quad \dot{y} = x^2.$$

For simplicity, let us take the control set $U = \mathbb{R}$. We have already seen that this system is accessible from $(0, 0)$, both "by hand" and by checking the hypotheses of Theorem 7.9. "By hand," we also saw that the system was *not* STLC from $(0, 0)$. Let us now begin to understand why this is so. The following argument is by no means precise, but captures the essence of the theory that predicts the lack of controllability of this example.

In Example 7.11 we saw that the vector fields $\{f_1, [f_1, [f_0, f_1]]\}$ form a global basis for $\mathrm{Lie}^{(\infty)}(\mathscr{C})_{(0,0)}$. What is more, one can check that the bracket $[f_0, f_1]$ vanishes at $(0, 0)$. Thus the basis $\{f_1(0, 0), [f_1, [f_0, f_1]](0, 0)\}$ for $\mathrm{Lie}^{(\infty)}(\mathscr{C})_{(0,0)}$ is the "lowest-order" collection of brackets possible that comprise such a basis. It turns out that the reason why this example is not STLC from $(0, 0)$ is the presence of the bracket $[f_1, [f_0, f_1]]$ in the basis. While this bracket gives an "accessible direction," and so adds a dimension to the reachable set, it does so in a directionally biased manner. To wit, note that since the control vector field f_1 appears twice in $[f_1, [f_0, f_1]]$, one cannot change the direction one can flow in the direction of this bracket, since changing the sign of the control will not alter the sign of the bracket. This is in contrast with a bracket in which the control vector field appears linearly (or more generally, with an odd number of terms), as with $[f_0, f_1]$. The bracket $[f_1, [f_0, f_1]]$ is an example of what is known as an "obstruction" to local controllability. •

The preceding example allows us to provide an imprecise description of how a lack of controllability might arise, and what we shall present in the remainder of this section is the result of a dedicated effort by numerous researchers to understand this, and similar phenomenon. During the course of our development it will be helpful to carry along another example to illustrate the use of the notation.

Example 7.15. The example we take is a somewhat general one that shows both the power and limitations of the theory. We take $\mathsf{M} = \mathbb{R}^3$ with the control-affine system defined by the equations

$$\dot{x} = u^1, \quad \dot{y} = u^2, \quad \dot{z} = \tfrac{1}{2}ax^2 + bxy + \tfrac{1}{2}cy^2,$$

for $a, b, c \in \mathbb{R}$. Thus

$$f_0 = \left(\tfrac{1}{2}ax^2 + bxy + \tfrac{1}{2}cy^2\right)\frac{\partial}{\partial z}, \quad f_1 = \frac{\partial}{\partial x}, \quad f_2 = \frac{\partial}{\partial y}.$$

We then compute

$$[f_0, f_1] = -(ax + by)\frac{\partial}{\partial z}, \quad [f_0, f_2] = -(by + cy)\frac{\partial}{\partial z}, \quad [f_1, f_2] = 0,$$
$$[f_1, [f_0, f_1]] = -a\frac{\partial}{\partial z}, \quad [f_2, [f_0, f_2]] = -c\frac{\partial}{\partial z},$$
$$[f_1, [f_0, f_2]] = [f_2, [f_0, f_1]] = -b\frac{\partial}{\partial z}.$$

Provided that $a^2 + b^2 + c^2 \neq 0$, we see that the system is accessible from $x_0 = (0,0,0)$. In this section, we shall investigate conditions on a, b, and c for which the system can be shown to be STLC from x_0. For the moment, we point out that this system possesses obstructions to controllability in the sense that we encountered in Example 7.14. However, what can happen in this example, that cannot happen in Example 7.14, is that the obstructions can be "neutralized." •

Now we proceed with the formulation of the main results. The development that follows is essentially that of Bianchini and Stefani [1993]. The main result we state in this section is Theorem 7.20. To state this result requires the introduction of a significant amount of notation that the uninitiated reader can expect to devote some time to understanding. Corollary 7.24 follows from Theorem 7.20 (and was originally proved as Theorem 7.3 in [Sussmann 1987]), and is somewhat easier to apply, although it is less general.

A heuristic brief on free Lie algebras. To get started in a coherent way, we need to say at least some things about free Lie algebras. We refer to [Jacobson 1962, Serre 1992] for details, and give only a heuristic outline here. The

free Lie algebra in indeterminates $\boldsymbol{\xi} = \{\xi_1, \dots, \xi_k\}$ is the $\mathbb{R}$-vector space $\mathrm{Lie}(\boldsymbol{\xi})$ generated by the indeterminates $\xi_1, \dots, \xi_k$ and their formal brackets, with the relations of skew-symmetry and the Jacobi identity enforced. Thus we mandate that, for example, $[\xi_\alpha, \xi_\beta] = -[\xi_\beta, \xi_\alpha]$, $\alpha, \beta \in \{1, \dots, k\}$, and that

$$[\xi_\alpha, [\xi_\beta, \xi_\delta]] + [\xi_\delta, [\xi_\alpha, \xi_\beta]] + [\xi_\beta, [\xi_\delta, \xi_\alpha]] = 0, \qquad \alpha, \beta, \delta \in \{1, \dots, k\}.$$

By a ***bracket*** in $\mathrm{Lie}(\boldsymbol{\xi})$ we mean an element that cannot be written as a sum of other element. Thus $\xi_\alpha + [\xi_\beta, \xi_\delta]$ is not a bracket, but both ξ_α and $[\xi_\beta, \xi_\delta]$ are brackets. We denote by $\mathrm{Br}(\boldsymbol{\xi}) \subset \mathrm{Lie}(\boldsymbol{\xi})$ the set of brackets. A bracket can be broken into its ***components***, by which we mean all of its sub-brackets. For example, the components of $[[\xi_{\alpha_1}, \xi_{\alpha_2}], [\xi_{\alpha_3}, \xi_{\alpha_4}]]$ are

$$[[\xi_{\alpha_1}, \xi_{\alpha_2}]; [\xi_{\alpha_3}, \xi_{\alpha_4}]],\ [\xi_{\alpha_1}, \xi_{\alpha_2}],\ [\xi_{\alpha_3}, \xi_{\alpha_4}],\ \xi_{\alpha_1},\ \xi_{\alpha_2},\ \xi_{\alpha_3},\ \xi_{\alpha_4}.$$

Except for the relations of skew-symmetry and Jacobi identity, brackets are linearly independent, meaning that $\mathrm{Lie}(\boldsymbol{\xi})$ is an infinite-dimensional $\mathbb{R}$-vector space. For a bracket $B \in \mathrm{Br}(\boldsymbol{\xi})$, we write $|B|_\alpha$ as the number of times the indeterminate ξ_α appears in B. Thus, for example, $|[\xi_1, [\xi_2, \xi_1]]|_1 = 2$. The ***degree*** of a bracket is given by $\deg(B) = \sum_{\alpha=1}^{k} |B|_\alpha$.

For a subset of the indeterminates, suppose without loss of generality the subset $\tilde{\boldsymbol{\xi}} = \{\xi_{l+1}, \dots, \xi_k\}$, the ***ideal*** of $\mathrm{Lie}(\boldsymbol{\xi})$ generated by $\tilde{\boldsymbol{\xi}}$ is denoted by $\mathrm{Ideal}(\tilde{\boldsymbol{\xi}})$ (see Section 5.2.3 for the definition of an ideal). One may verify that $\mathrm{Ideal}(\tilde{\boldsymbol{\xi}})$ is the subspace of $\mathrm{Lie}(\boldsymbol{\xi})$ generated by brackets of the form

$$[\xi_{\alpha_{r-1}}, [\xi_{\alpha_{r-2}}, \cdots, [\xi_{\alpha_1}, \xi_\alpha]]], \\ \alpha_1, \dots, \alpha_{r-1} \in \{0, 1, \dots, k\},\ \alpha \in \{l+1, \dots, k\},\ r \in \mathbb{N}.$$

If, corresponding to the indeterminates $\xi_1, \dots, \xi_k$, there are vector fields $\mathscr{X} = \{X_1, \dots, X_k\} \in \Gamma^\infty(\mathsf{T}\mathsf{M})$, we ***evaluate*** a bracket $B \in \mathrm{Lie}(\boldsymbol{\xi})$ by replacing in the expression for B each occurrence of the indeterminate ξ_α with the vector field X_α, $\alpha \in \{1, \dots, k\}$. We denote the resulting vector field by $\mathrm{Ev}^{\mathscr{X}}(B)$. For $x_0 \in \mathsf{M}$, we denote by $\mathrm{Ev}^{\mathscr{X}}_{x_0} : \mathrm{Lie}(\boldsymbol{\xi}) \to \mathsf{T}_{x_0}M$ the map that evaluates at x_0 the vector field $\mathrm{Ev}^{\mathscr{X}}(B)$. The reason for our use of free Lie algebras is that it enables us to define the notion of degree in a precise way. Suppose, for example, that we have a family of vector fields $\{X_1, X_2, X_3\}$ for which $[X_1, [X_2, X_1]] = X_3$. The expression $|[X_1, [X_2, X_1]]|_\alpha$, $\alpha \in \{1, 2, 3\}$ is then ambiguous. For example, we could have either $|[X_1, [X_2, X_1]]|_3 = 0$ or $|[X_1, [X_2, X_1]]|_3 = 1$. •

We now consider a manifold M with a family $\mathscr{C} = \{f_0, f_1, \dots, f_m\}$ of C^∞-vector fields. We denote by $\boldsymbol{\xi} = \{\xi_0, \xi_1, \dots, \xi_m\}$ a set of corresponding indeterminates with $\mathrm{Lie}(\boldsymbol{\xi})$ the free Lie algebra with these indeterminates. By $\mathrm{Ideal}_0(\boldsymbol{\xi})$ we denote the ideal of $\mathrm{Lie}(\boldsymbol{\xi})$ generated by the indeterminates $\{\xi_1, \dots, \xi_m\}$. An ***admissible weight*** for $\boldsymbol{\xi}$ is an $(m+1)$-tuple $\boldsymbol{w} = (w_0, w_1, \dots, w_m)$ of nonnegative real numbers satisfying $w_a \geq w_0$,

$a \in \{1, \dots, m\}$. The set of admissible weights for $\boldsymbol{\xi}$ is denoted by $\mathrm{wgt}(\boldsymbol{\xi})$. For an admissible weight $\boldsymbol{w}$ and a bracket $B \in \mathrm{Br}(\boldsymbol{\xi})$, the ***w*-weight** of B is

$$\|B\|_{\boldsymbol{w}} = \sum_{a=0}^{m} w_a \, |B|_a .$$

An element L in the vector space $\mathrm{Lie}(\boldsymbol{\xi})$ is ***w*-homogeneous** if it is a linear combination of brackets of the same $\boldsymbol{w}$-weight. For an admissible weight $\boldsymbol{w}$, if L is $\boldsymbol{w}$-homogeneous, then we define $\|L\|_{\boldsymbol{w}}$ to be the $\boldsymbol{w}$-weight of one, and therefore all, of its summands.

We begin our illustration of these concepts using Example 7.15.

Example 7.16 (Example 7.15 cont'd). For the example, we use the indeterminates $\boldsymbol{\xi} = \{\xi_0, \xi_1, \xi_2\}$. Let us consider for the moment the admissible weight $\boldsymbol{w} = (1, 1, 3)$. Then $\|[\xi_1, [\xi_0, \xi_1]]\|_{\boldsymbol{w}} = 3$ and $\|\xi_2\|_{\boldsymbol{w}} = 3$. Therefore, $\xi_2 + [\xi_1, [\xi_0, \xi_1]]$ is $\boldsymbol{w}$-homogeneous. •

For an admissible weight $\boldsymbol{w}$ and $k \in \mathbb{N}$, let us define

$$\mathcal{V}_{\mathscr{C}}^{\boldsymbol{w},k}(x_0) = \mathrm{span}_{\mathbb{R}}\{\mathrm{Ev}_{x_0}^{\mathscr{C}}(B) | \; B \in \mathrm{Ideal}_0(\boldsymbol{\xi}) \cap \mathrm{Br}(\boldsymbol{\xi}), \; \|B\|_{\boldsymbol{w}} \leq k\}.$$

We adopt the convention that $\mathcal{V}_{\mathscr{C}}^{\boldsymbol{w},0}(x_0) = \{0_{x_0}\}$. If $L \in \mathrm{Lie}(\boldsymbol{\xi})$ is $\boldsymbol{w}$-homogeneous, it is ***w*-neutralized** by $\mathscr{C}$ at x_0 if there exists an integer $k < \|L\|_{\boldsymbol{w}}$ such that $\mathrm{Ev}_{x_0}^{\mathscr{C}}(L) \in \mathcal{V}_{\mathscr{C}}^{\boldsymbol{w},k}(x_0)$.

Example 7.17 (Example 7.15 cont'd). In our example with $m = 2$, $\boldsymbol{w} = (1, 1, 1)$, and $\boldsymbol{x}_0 = (0, 0, 0)$, some bookkeeping, using the computations presented in Example 7.15, shows that

$$\begin{aligned}
\mathcal{V}_{\mathscr{C}}^{\boldsymbol{w},1}(x_0) &= \mathrm{span}_{\mathbb{R}}\{\mathrm{Ev}_{x_0}^{\mathscr{C}}(B) | \; B \in \{\xi_1, \xi_2\}\} \\
&= \mathrm{span}_{\mathbb{R}}\{(1,0,0), (0,1,0)\}, \\
\mathcal{V}_{\mathscr{C}}^{\boldsymbol{w},2}(x_0) &= \mathrm{span}_{\mathbb{R}}\{\mathrm{Ev}_{x_0}^{\mathscr{C}}(B) | \; B \in \{\xi_1, \xi_2, [\xi_0, \xi_1], [\xi_0, \xi_2], [\xi_1, \xi_2]\}\} \\
&= \mathrm{span}_{\mathbb{R}}\{(1,0,0), (0,1,0)\}, \\
\mathcal{V}_{\mathscr{C}}^{\boldsymbol{w},3}(x_0) &= \mathrm{span}_{\mathbb{R}}\{\mathrm{Ev}_{x_0}^{\mathscr{C}}(B) | \; B \in \{\xi_1, \xi_2, [\xi_0, \xi_1], [\xi_0, \xi_2], [\xi_1, \xi_2], \\
&\qquad [\xi_0, [\xi_0, \xi_1]], [\xi_1, [\xi_0, \xi_1]], [\xi_2, [\xi_0, \xi_1]], \\
&\qquad [\xi_0, [\xi_0, \xi_2]], [\xi_1, [\xi_0, \xi_2]], [\xi_2, [\xi_0, \xi_2]], \\
&\qquad [\xi_0, [\xi_1, \xi_2]], [\xi_1, [\xi_1, \xi_2]], [\xi_2, [\xi_1, \xi_2]]\}\} \\
&= \mathrm{span}_{\mathbb{R}}\{(1,0,0), (0,1,0), (0,0,a^2+b^2+c^2)\}.
\end{aligned}$$

Concerning neutralization, let us make some observations based on specific choices for a, b, and c. Throughout, we take $\boldsymbol{w} = (1, 1, 1)$.

1. The bracket $[\xi_1, [\xi_0, \xi_1]]$ is $\boldsymbol{w}$-neutralized if and only if $a = 0$.
2. The bracket $[\xi_2, [\xi_0, \xi_2]]$ is $\boldsymbol{w}$-neutralized if and only if $c = 0$.

3. Let $\boldsymbol{w}$-homogeneous member of $\mathrm{Lie}(\boldsymbol{\xi})$, $[\xi_1, [\xi_0, \xi_1]] + [\xi_2, [\xi_0, \xi_2]]$, is $\boldsymbol{w}$-neutralized if and only if $a + c = 0$. •

Now we come to an important concept, that of an obstruction to small-time local controllability. Define a subspace of $\mathrm{Ideal}_0(\boldsymbol{\xi})$ by

$$\mathcal{B}(\boldsymbol{\xi}) = \mathrm{span}_{\mathbb{R}}\{B \in \mathrm{Br}(\boldsymbol{\xi}) \cap \mathrm{Ideal}_0(\boldsymbol{\xi}) \mid |B|_0 \text{ is odd and } |B|_a \text{ is even, } a \in \{1, \dots, m\}\}.$$

Recall from Section 2.1 that S_m denotes the permutation group on m symbols. To each $\sigma \in S_m$, we associate a bijection of $\mathrm{wgt}(\boldsymbol{\xi})$, which we also denote by σ for convenience. This bijection is defined by

$$\sigma(w_0, w_1, \dots, w_m) = (w_0, w_{\sigma(1)}, \dots, w_{\sigma(m)}).$$

For $\boldsymbol{w} \in \mathrm{wgt}(\boldsymbol{\xi})$, we denote by $S_m^{\boldsymbol{w}}$ those permutations $\sigma \in S_m$ for which $\sigma(\boldsymbol{w}) = \boldsymbol{w}$. For a bracket $B \in \mathrm{Lie}(\boldsymbol{\xi})$, $\sigma(B)$ denotes the bracket obtained by replacing ξ_a with $\xi_{\sigma(a)}$, $a \in \{1, \dots, m\}$, and by leaving ξ_0 unchanged. If $L \in \mathrm{Lie}(\boldsymbol{\xi})$, then $\sigma(L)$ is defined by applying σ to each of the brackets comprising L. For $\boldsymbol{w} \in \mathrm{wgt}(\boldsymbol{\xi})$, we then define

$$\mathcal{B}^{\boldsymbol{w}}(\boldsymbol{\xi}) = \{L \in \mathcal{B}(\boldsymbol{\xi}) \mid \sigma(L) = L \text{ for all } \sigma \in S_m^{\boldsymbol{w}}\},$$

and then we denote by $\mathrm{Lie}^{(\infty)}(\xi_0, \mathcal{B}^{\boldsymbol{w}}(\boldsymbol{\xi}))$ the Lie algebra generated by $\{\xi_0\} \cup \mathcal{B}^{\boldsymbol{w}}(\boldsymbol{\xi})$. Finally, for $\boldsymbol{w} \in \mathrm{wgt}(\boldsymbol{\xi})$, a ***w*-obstruction** is a member of the subspace

$$\mathcal{B}_0^{\boldsymbol{w}}(\boldsymbol{\xi}) = \mathrm{Lie}^{(\infty)}(\xi_0, \mathcal{B}^{\boldsymbol{w}}(\boldsymbol{\xi})) \cap \mathrm{Ideal}_0(\boldsymbol{\xi}).$$

Example 7.18 (Example 7.15 cont'd). Let us resume with our example in which $m = 2$. If $\boldsymbol{w} = (1, 1, 1)$, then $[\xi_1, [\xi_0, \xi_1]] + [\xi_2, [\xi_0, \xi_2]]$ is in $\mathcal{B}^{\boldsymbol{w}}(\boldsymbol{\xi})$, but individually $[\xi_1, [\xi_0, \xi_1]]$ and $[\xi_2, [\xi_0, \xi_2]]$ are not. If $\boldsymbol{w} = (1, 1, 3)$, then $[\xi_1, [\xi_0, \xi_1]] + [\xi_2, [\xi_0, \xi_2]]$ is not in $\mathcal{B}^{\boldsymbol{w}}(\boldsymbol{\xi})$ but $[\xi_1, [\xi_0, \xi_1]]$ and $[\xi_2, [\xi_0, \xi_2]]$ separately are. •

Define a map $\rho_{\mathscr{C}}^{x_0} : \mathrm{wgt}(\boldsymbol{\xi}) \to \mathbb{N}$ by asking that $\rho_{\mathscr{C}}^{x_0}(\boldsymbol{w})$ be the largest integer for which all $\boldsymbol{w}$-obstructions L satisfying $\|L\|_{\boldsymbol{w}} \leq \rho_{\mathscr{C}}^{x_0}(\boldsymbol{w})$ are $\boldsymbol{w}$-neutralized by $\mathscr{C}$ at x_0. For $\boldsymbol{w} \in \mathrm{wgt}(\boldsymbol{\xi})$ and $x_0 \in \mathsf{M}$, let

$$\mathcal{V}_{\mathscr{C}}^{\boldsymbol{w}}(x_0) = \mathrm{span}\{A_{f_0}(x_0)^j(X) \mid X \in \mathcal{V}_{\mathscr{C}}^{\boldsymbol{w}, \rho_{\mathscr{C}}^{x_0}(\boldsymbol{w})}(x_0),\ j \in \mathbb{Z}_+\}.$$

Note that, under our assumption that $f_0(x_0) = 0_{x_0}$, the map $A_{f_0}(x_0) \in L(\mathsf{T}_{x_0}\mathsf{M}; \mathsf{T}_{x_0}\mathsf{M})$ is as defined in Proposition 3.75.

Example 7.19 (Example 7.15 cont'd). For our example, let us consider $\boldsymbol{w} = (1, 1, 1)$ and $x_0 = (0, 0, 0)$. As we saw in Examples 7.17 and 7.18, any obstruction L satisfying $\|L\|_{\boldsymbol{w}} \leq 3$ is $\boldsymbol{w}$-neutralized if and only if $a + c = 0$. Therefore, $\rho_{\mathscr{C}}^{x_0}(1, 1, 1) \geq 3$ if $a + c = 0$, and $\rho_{\mathscr{C}}^{x_0}(1, 1, 1) = 2$ if $a + c \neq 0$. •

We are now ready to state the general result of Bianchini and Stefani [1993].

Theorem 7.20 (Sufficient conditions for STLC). *Let* M *be a* C^∞*-manifold with* $\mathscr{C} = \{f_0, f_1, \dots, f_m\}$ *a family of* C^∞*-vector fields on* M*, and suppose that* $f_0(x_0) = 0_{x_0}$*. Let* $\boldsymbol{\xi} = \{\xi_0, \xi_1, \dots, \xi_m\}$ *be indeterminates. If*

$$\sum_{\boldsymbol{w}\in\mathrm{wgt}(\boldsymbol{\xi})} \mathcal{V}_{\mathscr{C}}^{\boldsymbol{w}}(x_0) = \mathsf{T}_{x_0}\mathsf{M},$$

then $(\mathsf{M}, \mathscr{C})$ *is properly STLC from* x_0*.*

Let us next see how this theorem *fails* to apply to Example 7.14, which we know is not STLC from $(0,0)$.

Example 7.21 (Example 7.14 cont'd). The bracket $L = [\xi_1, [\xi_0, \xi_1]]$ is an obstruction for this system. However, it cannot be $\boldsymbol{w}$-neutralized no matter how we choose $\boldsymbol{w}$. Indeed, if $\boldsymbol{w} = (w_0, w_1)$, then $\|L\|_{\boldsymbol{w}} = w_0 + 2w_1$. If $B \in \mathrm{Ideal}_0(\boldsymbol{\xi})$ is a bracket satisfying $\|B\|_{\boldsymbol{w}} < \|L\|_{\boldsymbol{w}}$, then $B \in \{\xi_1, \mathrm{ad}_{\xi_0}\xi_1, \dots, \mathrm{ad}_{\xi_0}^k \xi_1\}$, where k satisfies $kw_0 + w_1 < w_0 + 2w_1$. All brackets $\mathrm{ad}_{\xi_0}\xi_1, \dots, \mathrm{ad}_{\xi_0}^k \xi_1$ evaluate to zero, and $\mathrm{Ev}_{x_0}^{\mathscr{C}}(L) \notin \mathrm{span}_{\mathbb{R}}\{\mathrm{Ev}_{x_0}^{\mathscr{C}}(\xi_1)\}$. Therefore, we cannot conclude that the system is STLC using Theorem 7.20, and indeed the system is not STLC. •

Let us consider our running example for this section, and see what Theorem 7.20 tells us about the controllability of the system.

Example 7.22 (Example 7.15 cont'd). For this example, we claim that the hypotheses of Theorem 7.20 are satisfied if and only if $a + c = 0$. Certainly, the hypotheses are satisfied when $a + c = 0$ since, in this case, $\mathcal{V}_{\mathscr{C}}^{(1,1,1),3}(x_0) = \mathsf{T}_{x_0}\mathbb{R}^3$. Now suppose that $a + c \neq 0$. We take an admissible weight $\boldsymbol{w} = (w_0, w_1, w_2)$, and consider three cases.

1. $w_1 < w_2$: We look at the two $\boldsymbol{w}$-obstructions $L_1 = [\xi_1, [\xi_0, \xi_1]]$ and $L_2 = [\xi_2, [\xi_0, \xi_2]]$. We have $\|L_1\|_{\boldsymbol{w}} = w_0 + 2w_1$. The brackets in $\mathrm{Ideal}_0(\boldsymbol{\xi})$ whose $\boldsymbol{w}$-weight is less than that of L_1 are $\{\xi_1\}$ with the possibility of including $\{\xi_2\}$, along with $\mathrm{ad}_{\xi_0}^k$ applied to any of these brackets for sufficiently small k. In any event we see that, if $a \neq 0$, then L_1 cannot be $\boldsymbol{w}$-neutralized by $\mathscr{C}$ at $x_0 = (0,0,0)$. A similar argument shows that, if $c \neq 0$, then L_2 cannot be $\boldsymbol{w}$-neutralized by $\mathscr{C}$ at $x_0 = (0,0,0)$. Since both a and c cannot be zero, it follows that L_1 and/or L_2 cannot be $\boldsymbol{w}$-neutralized.
2. $w_1 > w_2$: The argument here mirrors the preceding one to show that the $\boldsymbol{w}$-obstructions $L_1 = [\xi_1, [\xi_0, \xi_1]]$ and/or $L_2 = [\xi_2, [\xi_0, \xi_2]]$ cannot be $\boldsymbol{w}$-neutralized by $\mathscr{C}$ at x_0.
3. $w_1 = w_2$: In this case, we consider the $\boldsymbol{w}$-obstruction $L = [\xi_1, [\xi_0, \xi_1]] + [\xi_2, [\xi_0, \xi_2]]$ whose $\boldsymbol{w}$-weight is $w_0 + 2w_1$. The brackets in $\mathrm{Ideal}_0(\boldsymbol{\xi})$ whose $\boldsymbol{w}$-weight is less than that of L are $\{\xi_1, \xi_2, [\xi_0, \xi_1], [\xi_0, \xi_2], [\xi_1, \xi_2]\}$, possibly

along with $\mathrm{ad}_{\xi_0}^k$ of some of these brackets for sufficiently small k. One can readily check that the evaluation of L does not lie in the span of any of these brackets at x_0, so L cannot be $\boldsymbol{w}$-neutralized by $\mathscr{C}$ at x_0.

The preceding arguments show that the hypotheses of Theorem 7.20 apply only in the rare case that $a+c=0$. However, this leaves open the question of whether the system can still be STLC, even when $a+c\neq 0$. It turns out that it can. Let us look at a concrete instance of this. We take $a=4$, $b=0$, and $c=-2$. Thus the system does not satisfy the hypotheses of Theorem 7.20. Let us now consider a new family of vector fields $\tilde{\mathscr{C}}=\{\tilde{f}_0,\tilde{f}_1,\tilde{f}_2\}$ defined by

$$\tilde{f}_0 = f_0, \quad \tilde{f}_1 = f_1 + \sqrt{2} f_2, \quad \tilde{f}_2 = f_1 - \sqrt{2} f_2.$$

The control system defined by these vector fields is essentially the same, since all we have done is make a change of basis for the input vector fields. We do not expect that the controllability properties of the system will be affected by such a change of basis. More precisely, we expect that, and it is actually true that, $(\mathsf{M},\mathscr{C})$ is properly STLC from x_0 if and only if $(\mathsf{M},\tilde{\mathscr{C}})$ is properly STLC from x_0. We compute

$$[\tilde{f}_0,\tilde{f}_1] = (2\sqrt{2}y - 4x)\frac{\partial}{\partial z}, \quad [\tilde{f}_0,\tilde{f}_2] = -(2\sqrt{2}y + 4x)\frac{\partial}{\partial z}, \quad [\tilde{f}_1,\tilde{f}_2] = 0,$$
$$[\tilde{f}_1,[\tilde{f}_0,\tilde{f}_1]] = [\tilde{f}_2,[\tilde{f}_0,\tilde{f}_2]] = 0, \quad [\tilde{f}_1,[\tilde{f}_0,\tilde{f}_2]] = [\tilde{f}_2,[\tilde{f}_0,\tilde{f}_1]] = -8\frac{\partial}{\partial z}.$$

Here both possible obstructions, $[\xi_1,[\xi_0,\xi_1]]$ and $[\xi_2,[\xi_0,\xi_2]]$, evaluate to zero, and so will be neutralizable with weight, for example, $\boldsymbol{w}=(1,1,1)$. Furthermore, with this $\boldsymbol{w}$, $\mathcal{V}_{\tilde{\mathscr{C}}}^{\boldsymbol{w},3} = \mathsf{T}_{x_0}\mathbb{R}^3$, so we conclude that $(\mathsf{M},\tilde{\mathscr{C}})$ is properly STLC from x_0 by Theorem 7.20.

The punchline is that, although the pair $(\mathsf{M},\mathscr{C})$ does not satisfy the hypotheses of Theorem 7.20, it is transformable to a system that does, and so is itself properly STLC. Furthermore, one can show that this system is STLC from x_0 if and only if $ac-b^2<0$ (note that the condition $a+c=0$ is a special case of this more general condition). The way in which one can see this is related to the treatment of low-order controllability in Chapter 8; see [Hirschorn and Lewis 2002] for the control-affine case. •

Remark 7.23. The reason that Theorem 7.20, general though it is, fails to apply to the system of Example 7.15—even for choices of the parameters a, b, and c that render the system STLC—is that, by making a correspondence between the indeterminates $\boldsymbol{\xi}=\{\xi_0,\xi_1,\ldots,\xi_m\}$ and the system vector fields $\mathscr{C}=\{f_0,f_1,\ldots,f_m\}$, one is making a choice that is not necessarily natural. That is to say, Theorem 7.20 is not "feedback-invariant" in the sense that if one changes from a given set of vector fields $\mathscr{C}=\{f_0,f_1,\ldots,f_m\}$ to another set $\tilde{\mathscr{C}}=\{\tilde{f}_0,\tilde{f}_1,\ldots,\tilde{f}_{\tilde{m}}\}$ giving "the same" control equations, then the controllability conditions of Theorem 7.20 are not invariant under this equivalence. This is a drawback of any approach that entails a specific choice of drift and input vector fields. This is addressed in the paper of Hirschorn and Lewis [2002], where control systems are considered in such a way that one does not make a choice of generators, and so one is led to state results in such a way that the generators are suppressed. Hirschorn and Lewis [2002] consider only "second-order" conditions for controllability. A general generator-independent theory of controllability has yet to be achieved. •

Theorem 7.20 also has a corollary due to Sussmann [1987, Theorem 7.3].[3] To state this result, it is convenient to introduce some notation. A bracket $B \in \mathrm{Br}(\boldsymbol{\xi})$ is ***bad*** if $|B|_0$ is odd and if $|B|_a$ is even for all $a \in \{1, \dots, m\}$. $B \in \mathrm{Br}(\boldsymbol{\xi})$ is ***good*** if it is not bad. For $\theta \in [1, +\infty[\cup \{+\infty\}$, the ***$\theta$-degree*** of a bracket B is defined by

$$\deg_\theta(B) = \begin{cases} \frac{1}{\theta}|B|_0 + \sum_{a=1}^m |B|_a, & \theta \in [1, +\infty[, \\ \sum_{a=1}^m |B|_a, & \theta = +\infty. \end{cases}$$

For a bracket $B \in \mathrm{Br}(\boldsymbol{\xi})$, define

$$\beta(B) = \sum_{\sigma \in S_m} \sigma(B).$$

We now state the corollary, leaving the proof as Exercise E7.3. In Exercise E7.5 the reader may show that the hypotheses of the corollary are strictly stronger than those of Theorem 7.40.

Corollary 7.24. *Let* M *be a* C^∞*-manifold with* $\mathscr{C} = \{f_0, f_1, \dots, f_m\}$ *a family of* C^∞*-vector fields on* M*, and suppose that* $f_0(x_0) = 0_{x_0}$*. Let* $\boldsymbol{\xi} = \{\xi_0, \xi_1, \dots, \xi_m\}$ *be indeterminates. Suppose that, for every bad bracket* B*, there exist good brackets* $C_1, \dots, C_k$ *and* $\theta \in [1, +\infty[\cup \{+\infty\}$ *such that*

(i) $\mathrm{Ev}_{x_0}^{\mathscr{C}}(\beta(B)) \in \mathrm{span}_{\mathbb{R}}\{\mathrm{Ev}_{x_0}^{\mathscr{C}}(C_j) \mid j \in \{1, \dots, k\}\}$*, and*
(ii) $\deg_\theta(C_j) < \deg_\theta(B)$*,* $j \in \{1, \dots, k\}$*.*

Then $(\mathsf{M}, \mathscr{C})$ *is properly STLC from* x_0 *if* $\mathrm{Lie}^{(\infty)}(\mathscr{C})_{x_0} = \mathsf{T}_{x_0}\mathsf{M}$*.*

Before we leave our general discussion of controllability, let us show how some classic results follow from Theorem 7.20. First we show that a local version of Chow's Theorem, Theorem 3.95, follows from Theorem 7.20.

Corollary 7.25 (Controllability of driftless systems). *Let* $\mathscr{C} = \{f_0, f_1, \dots, f_m\}$ *be* C^∞*-vector fields on* M *with* f_0 *identically zero. The pair* $(\mathsf{M}, \mathscr{C})$ *is properly STLC from* x_0 *if* $\mathrm{Lie}^{(\infty)}(\mathscr{C})_{x_0} = \mathsf{T}_{x_0}\mathsf{Q}$*.*

Proof. This follows from Corollary 7.24, since all bad brackets vanish identically by virtue of the drift vector field being zero. ■

[3] The result we state is actually a corollary to a more general result in the paper [Sussmann 1987].

Next we show that accessibility for linear control systems is equivalent to small-time local controllability from the origin for these systems.

Corollary 7.26 (Controllability of linear control systems). *Let* $\Sigma = (\mathsf{V}, A, B)$ *be a linear control system and suppose that the controls take values in a proper control set* U. *Then* Σ *is STLC from* 0 *if and only if it is accessible from* 0.

Proof. We apply Corollary 7.24 with $\theta = +\infty$. The system is accessible if and only if the constant vector fields

$$\mathrm{ad}_{f_0}^k f_a = (-1)^k A^k b_a, \qquad a \in \{1, \dots, m\},\ k \in \{0, 1, \dots, n-1\},$$

generate V. The θ-degree of the brackets yielding these vector fields is 1. Furthermore, all of these brackets are good. Any bad bracket will have θ-degree at least 2 (in fact for linear control systems, all bad brackets are identically zero). Therefore, any bad bracket must lie in the span of good brackets of lower θ-degree. Thus the result follows from Corollary 7.24. ■

Our next result has to do with inferring the controllability of a system from the controllability of its linearization. To do this, we need to define what we mean by linearization.

Definition 7.27 (Linearization of a control-affine system). The ***linearization*** at x_0 of a control-affine system $\Sigma = (\mathsf{M}, \mathscr{C} = \{f_0, f_1, \dots, f_m\}, U)$, satisfying Assumption 7.13, is the linear control system (V, A, B) of the sort introduced in Definition 4.102, with

$$\mathsf{V} = \mathsf{T}_{x_0}\mathsf{M}, \quad A = A_\Sigma(x_0), \quad B(u) = B_\Sigma(x_0)(u) \triangleq \sum_{a=1}^m u^a f_a(x_0),$$

where $A_\Sigma(x_0)$ is the linearization of f_0 at x_0, as introduced in Proposition 3.75. •

With this notation at hand, we have the following result.

Corollary 7.28. *A* C^∞*-control-affine system* $\Sigma = (\mathsf{M}, \mathscr{C}, U)$, *satisfying Assumption 7.13, is STLC from* x_0 *if its linearization is STLC from* 0.

Proof. If $\mathscr{C} = \{f_0, f_1, \dots, f_m\}$, then write $A = A_\Sigma(x_0)$ and $b_a = f_a(x_0)$, $a \in \{1, \dots, m\}$. Direct computations in coordinates yield

$$\mathrm{ad}_{f_0} f_a(x_0) = (-1)^k A_{f_0}^k(b_a), \qquad a \in \{1, \dots, m\},\ k \in \mathbb{Z}_+.$$

Now the argument used to prove Corollary 7.26 may be applied directly. Note that while the bad brackets may not be identically zero as they are for the linearization, the hypotheses of Corollary 7.24 still apply, since, by taking $\theta = +\infty$, all bad brackets have θ-degree at least two. ■

7.2 Controllability definitions for mechanical control systems

In this section we begin to apply the discussion of the preceding section to arrive at controllability results for mechanical control systems. Mechanical control systems have more structure than control-affine systems, some of which leads to more refined notions of controllability. In particular, mechanical systems have a configuration manifold Q to go along with the state manifold TQ. Thus it makes sense to talk about controllability in terms only of configurations. For the purpose of stating definitions, we consider a general simple mechanical control system $\Sigma = (\mathsf{Q}, \mathbb{G}, V, F, \mathcal{D}, \mathscr{F} = \{F^1, \dots, F^m\}, U)$ with a time-independent external force F, and with governing equations (4.24), which we reproduce here for reference:

$$\overset{\mathcal{D}}{\nabla}_{\gamma'(t)}\gamma'(t) = -P_{\mathcal{D}}(\mathrm{grad}V(\gamma(t))) + P_{\mathcal{D}}(\mathbb{G}^\sharp(F(\gamma'(t)))) \\ + \sum_{a=1}^{m} u^a(t) P_{\mathcal{D}}(\mathbb{G}^\sharp(F^a(\gamma(t)))). \quad (7.3)$$

A ***controlled trajectory*** for Σ is a pair (γ, u), where $u\colon I \to U$ is locally integrable, and where $\gamma\colon I \to \mathsf{Q}$, satisfying $\gamma'(t_0) \in \mathcal{D}_{\gamma(0t)}$ for some $t_0 \in I$ (and so for all $t \in I$), is the resulting solution of (7.3). Note that Corollary A.12 ensures that γ' is locally absolutely differentiable. We denote by $\mathrm{Ctraj}(\Sigma)$ the controlled trajectories for Σ, and by $\mathrm{Ctraj}(\Sigma, T)$ the controlled trajectories defined on $[0, T]$. Note that there is potential confusion here between this notation and that of Section 7.1.1, since here we are dealing with trajectories in the configuration manifold Q, not the state manifold TQ. However, the intended meaning should be clear from context.

We will only consider controllability from states with zero velocity. The reachable sets of interest, therefore, take the following form:

$$\mathcal{R}_{\Sigma,\mathsf{TQ}}(q_0, T) = \{\gamma'(T) \mid (\gamma, u) \in \mathrm{Ctraj}(\Sigma, T),\ \gamma'(0) = 0_{q_0}\},$$
$$\mathcal{R}_{\Sigma,\mathsf{Q}}(q_0, T) = \{\gamma(T) \mid (\gamma, u) \in \mathrm{Ctraj}(\Sigma, T),\ \gamma'(0) = 0_{q_0}\},$$
$$\mathcal{R}_{\Sigma,\mathsf{TQ}}(q_0, \leq T) = \bigcup_{t\in[0,T]} \mathcal{R}_{\Sigma,\mathsf{TQ}}(q_0, t),$$
$$\mathcal{R}_{\Sigma,\mathsf{Q}}(q_0, \leq T) = \bigcup_{t\in[0,T]} \mathcal{R}_{\Sigma,\mathsf{Q}}(q_0, t).$$

Controllability from states with nonzero velocity is not currently well understood, but preliminary results can be found in the papers [Martínez and Cortés 2003, Žefran, Bullo, and Radford 1999].

We now have the following controllability definitions.

Definition 7.29 (Controllability notions for mechanical control systems). Let $\Sigma = (\mathsf{Q}, \mathbb{G}, V, F, \mathcal{D}, \mathscr{F}, U)$ be a general simple mechanical control system with F time-independent, let $\Sigma_{\text{pre}} = (\mathsf{Q}, \mathbb{G}, V, F, \mathcal{D}, \mathscr{F})$ be the corresponding general simple mechanical pre-control system, and let $q_0 \in \mathsf{Q}$.

(i) Σ_{pre} is ***accessible*** from q_0 if, for every almost proper control set U, there exists $T > 0$ such that $\text{int}_{\mathcal{D}}(\mathcal{R}_{\Sigma,\mathsf{TQ}}(q_0, \leq t)) \neq \emptyset$ for $t \in]0, T]$.

(ii) Σ_{pre} is ***configuration accessible*** from q_0 if, for every almost proper control set U, there exists $T > 0$ such that $\text{int}(\mathcal{R}_{\Sigma,\mathsf{Q}}(q_0, \leq t)) \neq \emptyset$ for $t \in]0, T]$.

(iii) Σ is ***small-time locally controllable*** (***STLC***) from q_0 if there exists $T > 0$ such that $0_{q_0} \in \text{int}_{\mathcal{D}}(\mathcal{R}_{\Sigma,\mathsf{TQ}}(q_0, \leq t))$ for $t \in]0, T]$.

   (a) Σ_{pre} is ***properly small-time locally controllable*** (***properly STLC***) from q_0 if Σ is STLC from q_0 for every proper control set U.

   (b) Σ_{pre} is ***small-time locally uncontrollable*** (***STLUC***) from q_0 if Σ is not STLC from q_0 for every compact control set U.

(iv) Σ is ***small-time locally configuration controllable*** (***STLCC***) from q_0 if there exists $T > 0$ such that $q_0 \in \text{int}(\mathcal{R}_{\Sigma,\mathsf{Q}}(q_0, \leq t))$ for $t \in]0, T]$.

   (a) Σ_{pre} is ***properly small-time locally configuration controllable*** (***properly STLCC***) from q_0 if Σ is STLCC from q_0 for every proper control set U.

   (b) Σ_{pre} is ***small-time locally configuration uncontrollable*** (***STLCUC***) from q_0 if Σ is not STLCC from q_0 for every compact control set U. •

Remarks 7.30. 1. In these definitions we are careful to distinguish between those properties having to do with the system data minus the control set, and those having to do with the system data including the control set. To this end, note that our definitions of accessibility and configuration accessibility are sensible given Theorem 7.9. Indeed, that theorem ensures that, if the set of states (resp. configurations) reachable using a particular proper control set has nonempty interior, then the set of states (resp. configurations) reachable using *any* almost proper control set will have nonempty interior. Thus accessibility is really a property not involving the control set. This is not true of small-time local controllability, however.

2. The definitions above apply to the most general class of mechanical control system we consider in the book. We will not be able to produce much in the way of controllability *results* for such a large class. Thus, when we state and prove results in the next section, we will restrict our attention to subclasses of systems such as outlined in Section 4.6. In such cases, the preceding definitions still hold, of course, with the corresponding simplifications in the data defining the system.

3. The definitions for controllability apply verbatim to a forced affine connection control system $(\mathsf{Q}, \nabla, Y, \mathcal{D}, \mathscr{Y}, U)$. We shall take advantage of this in Sections 7.3.2 and 7.3.3.

4. While it is true that if Σ is accessible (resp. STLC) from q_0, then it is configuration accessible (resp. STLCC) from q_0, the converse is not necessarily true, and we shall encounter simple examples for which this is the case.
5. Note that the notions of accessibility and small-time local controllability we give use the interior relative to $\mathcal{D} \subset \mathsf{T}Q$. This is only sensible, since states not in the constraint distribution are, by definition, not reachable. •

7.3 Controllability results for mechanical control systems

Let us now produce some results for the various notions of controllability, restricting to subcases of general simple mechanical control systems described in Section 4.6, as suits the results that are currently available. The results we give can be broken down as follows.

1. Linear and linearization results are given in Section 7.3.1.
2. In Section 7.3.2 we characterize the accessibility of affine connection control systems. This analysis includes simple mechanical systems, possibly with linear velocity constraints, that are subject to no external forces. This means that potential forces are not included in this analysis. The analysis here is quite complete in that we are able to provide a nice geometric interpretation of the accessibility results.
3. Sufficient conditions for local controllability of affine connection control systems are given in Section 7.3.3, using Theorem 7.20.
4. In internet supplements, the analysis of Sections 7.3.2 and 7.3.3 is extended to systems with certain types of external forces.
   (a) One extension is to systems with a basic external force, considered first in the work of Lewis and Murray [1997a]. These results allow the inclusion of potential forces in the analysis for both accessibility and local controllability. The cost is that the straightforward geometry, seen in the accessibility results for affine connection control systems, disappears. An interesting open problem is that of providing a nice geometric description in this more general setting. We comment that although the accessibility characterization of Lewis and Murray [1997a] is correct, the controllability result, given as Theorem 5.15 in the paper, is only valid for low-order. It is, however, correct for systems with no external force (in this case it specializes to Corollary 7.41).
   (b) Another extension considered in the supplementary material is to systems with isotropic Rayleigh dissipation. These results were originally recorded in [Cortés, Martínez, and Bullo 2003].
   (c) The third controllability supplement deals with controllability for systems on Lie groups and principal fiber bundles. The Lie group case

can be worked out by the reader as Exercise E7.20, and is presented by Bullo, Leonard, and Lewis [2000]. The principal bundle results have not been published, although related results may be found in [Cortés, Martínez, Ostrowski, and Zhang 2002].

Examples are postponed to Section 7.4. The results we give in this section are of a general nature. In Chapter 8 we consider low-order controllability results, as these have a rather different character, and are closely related to design issues that are interesting in their own right (see Chapter 13).

7.3.1 Linearization results

We first deal with linear controllability theory for simple mechanical control systems. We recall from Section 4.6.5 that a linear mechanical control system is a 4-tuple (V, M, K, F), where V is a finite-dimensional $\mathbb{R}$-vector space, M is an inner product on V, $K \in \Sigma_2(\mathsf{V})$, and $F \in L(\mathbb{R}^m; \mathsf{V}^*)$. The following result characterizes the controllability of linear mechanical control systems. This result appears in the paper of Hughes and Skelton [1980].

Theorem 7.31 (Controllability of linear mechanical control systems). *Let $\Sigma = (\mathsf{V}, M, K, F)$ be a linear mechanical control system with $\dim(\mathsf{V}) = n$, and consider the linear map from $(\mathbb{R}^m)^n$ to V represented by the block linear map*

$$\left[\, M^\sharp \circ F \;\middle|\; M^\sharp \circ K^\flat \circ (M^\sharp \circ F) \;\middle|\; \cdots \;\middle|\; (M^\sharp \circ K^\flat)^{n-1} \circ (M^\sharp \circ F) \,\right]. \tag{7.4}$$

Then the following three statements are equivalent:

(i) Σ is accessible from $0 \oplus 0$;
(ii) Σ is properly STLC from $0 \oplus 0$;
(iii) the linear map in (7.4) has maximal rank.

Proof. The equivalence of (i) and (ii) is exactly Corollary 7.26. Thus we show the equivalence of these to (iii). For brevity and convenience, let us write $A_0 = -M^\sharp \circ K^\flat$ and $B_0 = M^\sharp \circ F$. A straightforward computation shows that the Kalman Rank Condition of Example 7.12, for accessibility from $0 \oplus 0$ of the state form for Σ, asks that the linear map

$$\left[\begin{array}{c|c|c|c|c|c|c} 0 & B_0 & 0 & A_0 \circ B_0 & \cdots & 0 & A_0^{n-1} \circ B_0 \\ B_0 & 0 & A_0 \circ B_0 & 0 & \cdots & A_0^{n-1} \circ B_0 & 0 \end{array}\right]$$

have rank $2n$. This clearly holds if and only if the linear map

$$\left[\, B_0 \;\middle|\; A_0 \circ B_0 \;\middle|\; \cdots \;\middle|\; A_0^{n-1} \circ B_0 \,\right]$$

has rank n, which is the result. ■

The main use we shall make of the preceding result is in terms of linearization. This requires that we prescribe the manner in which one linearizes a simple mechanical control system $\Sigma = (\mathsf{Q}, \mathbb{G}, V, \mathscr{F})$. We consider a point $q_0 \in \mathsf{Q}$ that is an equilibrium configuration for the simple mechanical system $(\mathsf{Q}, \mathbb{G}, V)$. That is to say, q_0 is a critical point for V. It is possible to consider more general equilibria where some control force is used to maintain the equilibrium. This generality is discussed in Section 10.2, and we refer to Exercise E7.7 for the consideration of the effects of this on linear controllability.

Let us formally define what we mean by linearization in our current context.

Definition 7.32 (Linearization of mechanical control systems). The ***configuration linearization*** of a simple mechanical control system $\Sigma = (\mathsf{Q}, \mathbb{G}, V, \mathscr{F} = \{F^1, \dots, F^m\}, \mathbb{R}^m)$, at an equilibrium configuration q_0 for $(\mathsf{Q}, \mathbb{G}, V)$, is the linear mechanical control system $(\mathsf{T}_{q_0}\mathsf{Q}, \mathbb{G}(q_0), K_\Sigma(q_0), F_\Sigma(q_0))$, where $K_\Sigma(q_0) \in \Sigma_2(\mathsf{T}_{q_0}\mathsf{Q})$ and $F_\Sigma(q_0) \in L(\mathbb{R}^m; \mathsf{T}^*_{q_0}\mathsf{Q})$ are defined by

$$K_\Sigma(q_0) = \operatorname{Hess} V(q_0), \quad F_\Sigma(q_0) \cdot u = \sum_{a=1}^m u^a F^a(q_0).$$

The state form for $(\mathsf{T}_{q_0}\mathsf{Q}, \mathbb{G}(q_0), K_\Sigma(q_0), F_\Sigma(q_0))$ is called the ***state linearization*** and is denoted by $(\mathsf{T}_{q_0}\mathsf{Q} \oplus \mathsf{T}_{q_0}\mathsf{Q}, A_\Sigma(q_0), B_\Sigma(q_0))$. •

Remark 7.33. Following the computations leading up to Proposition 6.37, one can easily show that the state form of the linearization of a simple mechanical control system is exactly the linearization of the system, thought of as a first-order system with state manifold TQ. •

Of course, Corollary 7.28 still holds in the mechanical context. That is to say, if the linearization of a simple mechanical control system at q_0 is STLC from q_0, then the system itself will be STLC from q_0. The following corollary indicates why it is that underactuated mechanical systems with no potential function are such a challenging class of control problems. This was pointed out in the paper of Oriolo and Nakamura [1991].

Corollary 7.34 (Controllability of underactuated systems without external forces). *If $\Sigma = (\mathsf{Q}, \mathbb{G}, V = 0, \mathscr{F}, U)$ is underactuated at q_0, then its linearization about 0_{q_0} is not accessible from the origin.*

Proof. We use the notation of the proof of Theorem 7.31. Since $A_0 = 0$, the linearization is controllable if and only if the matrix

$$\begin{bmatrix} 0 & B_0 \\ B_0 & 0 \end{bmatrix}$$

has rank $2n$, or equivalently if and only if $\operatorname{rank}(B_0) = n$. However, if Σ is underactuated at q_0, then $\operatorname{rank}(B_0) < n$. ■

Remark 7.35. One of the consequences of Corollary 7.34 is that the problem of designing stabilizing control laws is difficult for underactuated systems without a potential function. Indeed, from a theorem of Brockett [1983] (stated as Theorem 10.8), it follows that for underactuated simple mechanical control systems with no external forces, potential or otherwise, it is not possible to design a differentiable feedback control law for which the closed-loop system renders q_0 asymptotically stable. We shall further see in Section 10.1 (see also Exercise E10.7) that it is not possible to stabilize the system with *continuous* state feedback. Therefore, a great many techniques for stabilization, including those discussed in Chapters 10 and 11, are not applicable to the stabilization problem for these systems. •

7.3.2 Accessibility of affine connection control systems

Corollary 7.34 indicates that systems without a potential function are difficult to control, even in terms of simply analyzing their controllability. Let us, therefore, turn our attention to this class of system, and admit constraints back into the picture. Thus we now consider an affine connection control system $\Sigma = (\mathsf{Q}, \nabla, \mathcal{D}, \mathscr{Y} = \{Y_1, \dots, Y_m\}, U)$, assuming that $\mathcal{D}$ has locally constant rank. For these systems, we will produce a complete characterization of their accessibility, and an incomplete, but still broadly applicable, characterization of their controllability. It is convenient to assign to Σ a control-affine system, as described in Section 4.6. Thus we take $\Sigma_{\mathsf{TQ}} = (\mathsf{M} = \mathsf{TQ}, \mathscr{C}_\Sigma = \{f_0 = S, f_1 = \mathrm{vlft}(Y_1), \dots, f_m = \mathrm{vlft}(Y_m)\}, U)$, where S is the geodesic spray for ∇. Of course, we are always interested in the restriction of the system to $\mathcal{D} \subset \mathsf{TQ}$. However, this is taken care of automatically, since (1) we consider initial states with zero velocity, (2) ∇ restricts to $\mathcal{D}$, and (3) the vector fields in $\mathscr{Y}$ are $\mathcal{D}$-valued (this is a consequence of Theorem 4.87).

By Theorem 7.9, we know that the accessibility of Σ from q_0 is characterized by the involutive closure at 0_{q_0} of the distribution generated by the drift vector field and the control vector fields for the control-affine system Σ_{TQ}. Let us provide a few examples of Lie brackets of the vector fields $\mathscr{C}_\Sigma = \{S, \mathrm{vlft}(Y_1), \dots, \mathrm{vlft}(Y_m)\}$, so as to motivate our main result. To present these calculations, recall from Lemma 6.34 that there is a natural decomposition $\mathsf{T}_{0_q}\mathsf{TQ} \simeq \mathsf{T}_q\mathsf{Q} \oplus \mathsf{T}_q\mathsf{Q}$. We take the convention that the first component in this decomposition describes the tangent space to $Z(\mathsf{TQ})$ at 0_q, and the second component describes the tangent space to the fiber at 0_q. Intuitively, the first component is "horizontal" and the second component is "vertical." Now, by direct calculation in coordinates, one may verify the following formulae, recalling from Section 3.8.4 the definition of the symmetric product:

$$\begin{aligned} 2[S, \mathrm{vlft}(Y_a)](0_{q_0}) &= -Y_a(q_0) \oplus 0_{q_0}, && a \in \{1, \dots, m\}, \\ [\mathrm{vlft}(Y_a), [S, \mathrm{vlft}(Y_b)]] &= \mathrm{vlft}(\langle Y_a : Y_b \rangle), && a, b \in \{1, \dots, m\}, \\ [[S, \mathrm{vlft}(Y_a)], [S, \mathrm{vlft}(Y_b)]](0_q) &= [Y_a, Y_b](q_0) \oplus 0_{q_0}, && a, b \in \{1, \dots, m\}. \end{aligned} \tag{7.5}$$

These calculations are indicative of two facts: (1) the vertical component of the involutive closure for $\mathscr{C}_\Sigma$ is specified by iterated symmetric products, and (2) the horizontal component of the same involutive closure is specified by iterated Lie brackets of iterated symmetric products. In particular, our simple calculations reveal the importance of the symmetric product.

Now let us provide the formal notation needed to state our main result. We take $r \in \{\infty\} \cup \{\omega\}$. We denote by $\mathcal{Y}$ the C^r-distribution generated by the vector fields $\mathscr{Y}$, and inductively define a sequence $\mathrm{Sym}^{(k)}(\mathcal{Y})$, $k \in \mathbb{Z}_+$, of C^r-distributions by $\mathrm{Sym}^{(0)}(\mathcal{Y})_q = \mathcal{Y}_q$, and, for $k \geq 1$,

$$\begin{aligned} \mathrm{Sym}^{(k)}(\mathcal{Y})_q = \mathrm{Sym}^{(k-1)}(\mathcal{Y})_q + \mathrm{span}_{\mathbb{R}}\{\langle X : Y \rangle (q) | \\ X \in \Gamma^\infty(\mathrm{Sym}^{(k_1)}(\mathcal{Y})),\ Y \in \Gamma^\infty(\mathrm{Sym}^{(k_2)}(\mathcal{Y})),\ k_1 + k_2 = k - 1\}. \end{aligned}$$

As with the constructions of the sequence of distributions $\mathrm{Lie}^{(k)}(\mathcal{D})$, $k \in \mathbb{Z}_+$, in Section 3.7.3, these constructions do define C^r-distributions. One can also prove an analogue of Lemma 3.94, replacing the Lie bracket with the symmetric product. This gives the existence of the smallest C^r-distribution $\mathrm{Sym}^{(\infty)}(\mathcal{Y})$ with the property that $\mathcal{Y} \subset \mathrm{Sym}^{(\infty)}(\mathcal{Y})$, and that $\langle X : Y \rangle \in \Gamma^\infty(\mathrm{Sym}^{(\infty)}(\mathcal{Y}))$ for each $X, Y \in \Gamma^\infty(\mathrm{Sym}^{(\infty)}(\mathcal{Y}))$. As with the involutive closure, one must exercise care when relating the distribution $\mathrm{Sym}^{(\infty)}(\mathcal{Y})$ with the sequence of distributions $\mathrm{Sym}^{(k)}(\mathcal{Y})$, $k \in \mathbb{Z}_+$. We refer to Sections 3.7.2 and 3.9.4 for a discussion in the context of the Lie bracket rather than the symmetric product.

With these definitions, we now describe the tangent spaces to the reachable sets in TQ for Σ_{TQ}.

Theorem 7.36 (Accessibility of affine connection control systems). *Let $\Sigma = (\mathsf{Q}, \nabla, \mathcal{D}, \mathscr{Y} = \{Y_1, \dots, Y_m\})$ be an analytic affine connection pre-control system. If $\mathscr{C}_\Sigma = \{S, \mathrm{vlft}(Y_1), \dots, \mathrm{vlft}(Y_m)\}$, then, for $q_0 \in \mathsf{Q}$,*

$$\mathrm{Lie}^{(\infty)}(\mathscr{C}_\Sigma)_{0_{q_0}} \simeq \mathrm{Lie}^{(\infty)}(\mathrm{Sym}^{(\infty)}(\mathcal{Y}))_{q_0} \oplus \mathrm{Sym}^{(\infty)}(\mathcal{Y})_{q_0} \subset \mathsf{T}_{q_0}\mathsf{Q} \oplus \mathsf{T}_{q_0}\mathsf{Q}.$$

In particular, the following statements hold:

(i) Σ is accessible from q_0 if and only if $\mathrm{Sym}^{(\infty)}(\mathcal{Y})_{q_0} = \mathcal{D}_{q_0}$ and $\mathrm{Lie}^{(\infty)}(\mathcal{D})_{q_0} = \mathsf{T}_{q_0}\mathsf{Q}$;

(ii) Σ is configuration accessible from q_0 if and only if $\mathrm{Lie}^{(\infty)}(\mathrm{Sym}^{(\infty)}(\mathcal{Y}))_{q_0} = \mathsf{T}_{q_0}\mathsf{Q}$.

Proof. See Section B.2. ■

Remark 7.37. The conditions of Theorem 7.36 for accessibility and configuration accessibility are sufficient, but not necessary, if the system is C^∞ but not analytic. •

This characterization of the tangent spaces to the $\mathscr{C}_\Sigma$-orbits through 0_{q_0} leads to an elegant characterization of the orbits themselves. This characterization relies on the notion of geodesic invariance introduced in Definition 3.107, and characterized by Theorem 3.108. Suppose our affine connection control system is analytic and note that the C^ω-distribution $\mathrm{Lie}^{(\infty)}(\mathrm{Sym}^{(\infty)}(\mathcal{Y}))$ is involutive, so, through any point $q_0 \in \mathsf{Q}$, there passes a unique maximal integral manifold, which we denote by $\Lambda_{\mathcal{Y},q_0}$. Thus $\Lambda_{\mathcal{Y},q_0}$ is an immersed analytic submanifold of Q, so if $\mathcal{U}$ is a sufficiently small connected neighborhood of q_0, there exists a unique connected component of $\Lambda_{\mathcal{Y},q_0} \cap \mathcal{U}$ containing q_0 that is an analytic embedded submanifold of $\mathcal{U}$, and which we denote by $\Lambda^{\mathcal{U}}_{\mathcal{Y},q_0}$ (cf. the discussion in Section 3.9.1). On $\Lambda^{\mathcal{U}}_{\mathcal{Y},q_0}$ we have a distribution $\mathcal{S}^{\mathcal{U}}_{\mathcal{Y},q_0}$ defined by $(\mathcal{S}^{\mathcal{U}}_{\mathcal{Y},q_0})_q = \mathrm{Sym}^{(\infty)}(\mathcal{Y})_q$. Since $\mathrm{Lie}^{(\infty)}(\mathscr{C}_\Sigma)$ is an involutive distribution by Theorem 7.5, $\mathcal{S}^{\mathcal{U}}_{\mathcal{Y},q_0}$ does indeed have constant rank on $\Lambda^{\mathcal{U}}_{\mathcal{Y},q_0}$. Therefore, $\mathcal{S}^{\mathcal{U}}_{\mathcal{Y},q_0}$ is a submanifold of $\mathsf{T}(\Lambda^{\mathcal{U}}_{\mathcal{Y},q_0})$. Since the preceding construction can be performed around every point on $\Lambda_{\mathcal{Y},q_0}$, we have produced an analytic immersed submanifold $\mathcal{S}_{\mathcal{Y},q_0}$ of $\mathcal{D}$ that can be thought of as a vector bundle over the analytic immersed submanifold $\Lambda_{\mathcal{Y},q_0}$. This is summarized by the following result.

Theorem 7.38 (Geometric characterization of accessibility for affine connection control systems). *Let* $\Sigma = (\mathsf{Q}, \nabla, \mathcal{D}, \mathscr{Y} = \{Y_1, \dots, Y_m\}, U)$ *be an analytic affine connection control system with* $\mathscr{C}_\Sigma = \{S, \mathrm{vlft}(Y_1), \dots, \mathrm{vlft}(Y_m)\}$. *Let* $\Lambda_{\mathcal{Y},q_0}$ *be the maximal integral manifold through* $q_0 \in \mathsf{Q}$ *for the involutive distribution* $\mathrm{Lie}^{(\infty)}(\mathrm{Sym}^{(\infty)}(\mathcal{Y}))$. *The following statements hold:*

(i) the $\mathscr{C}_\Sigma$*-orbit through* $0_{q_0} \in \mathcal{D}$ *is the smallest geodesically invariant subbundle (denoted by* $\mathcal{S}_{\mathcal{Y},q_0}$ *above) of* $\mathsf{T}(\Lambda_{\mathcal{Y},q_0})$ *containing* $\mathcal{Y}|\Lambda_{\mathcal{Y},q_0}$*;*

(ii) if U *is almost proper, then* $\mathrm{int}_{\mathcal{S}_{\mathcal{Y},q_0}}(\mathcal{R}_{\Sigma,\mathsf{TQ}}(q_0)) \neq \emptyset$ *and* $\mathrm{int}_{\mathcal{S}_{\mathcal{Y},q_0}}(\mathcal{R}_{\Sigma,\mathsf{TQ}}(q_0))$ *is dense in* $\mathcal{R}_{\Sigma,\mathsf{TQ}}(q_0) \subset \mathcal{S}_{\mathcal{Y},q_0}$.

Remark 7.39. As was mentioned in Remark 7.7–1, one has to use the orbit topology on $\mathcal{S}_{\mathcal{Y},q_0}$ in the second part of the theorem. •

7.3.3 Controllability of affine connection control systems

Our next results have to do with small-time local configuration controllability of affine connection control systems. Thus the results here are specializations of the results of Section 7.1.5 to affine connection control systems. The results of Lewis and Murray [1997a] utilized the result of Sussmann [1987] that we stated as Corollary 7.24. The development we give will be more general, relying on the results of Bianchini and Stefani [1993], stated as our Theorem 7.20.

A heuristic brief on free symmetric algebras. A ***symmetric algebra*** is a $\mathbb{R}$-vector space with a product, denoted by $\langle \cdot : \cdot \rangle$, satisfying $\langle v_1 : v_2 \rangle = \langle v_2 : v_1 \rangle$. Note that if ∇ is an affine connection on Q, then the corresponding symmetric product $\langle \cdot : \cdot \rangle$ makes $\Gamma^\infty(\mathsf{TQ})$ into a symmetric algebra.

Just as we defined the free Lie algebra with a certain collection of indeterminates, we can do this for symmetric algebras. We let $\boldsymbol{\eta} = \{\eta_1, \dots, \eta_k\}$ be indeterminates, and consider iterated $\mathbb{R}$-bilinear products in these indeterminates, asking only that the relation $\langle \eta_\alpha : \eta_\beta \rangle = \langle \eta_\beta : \eta_\alpha \rangle$ be satisfied. The result is an infinite-dimensional symmetric algebra, denoted by $\mathrm{Sym}(\boldsymbol{\eta})$, and called the ***free symmetric algebra*** in indeterminates $\boldsymbol{\eta}$. A ***product*** in $\mathrm{Sym}(\boldsymbol{\eta})$ is a member of $\mathrm{Sym}(\boldsymbol{\eta})$ that is not a sum of two other elements. We denote by $\mathrm{Pr}(\boldsymbol{\eta})$ the set of products, noting that $\mathrm{Pr}(\boldsymbol{\eta})$ is to $\mathrm{Sym}(\boldsymbol{\eta})$ what $\mathrm{Br}(\boldsymbol{\xi})$ is to $\mathrm{Lie}(\boldsymbol{\xi})$. We also define $|P|_\alpha$ to be the number of times the indeterminate η_α appears in P. The ***degree*** of $P \in \mathrm{Pr}(\boldsymbol{\eta})$ is $\deg(P) = \sum_{\alpha=1}^m |P|_\alpha$.

Given vector fields $\mathscr{Y} = \{Y_1, \dots, Y_k\}$ and an affine connection ∇ on Q, we may also evaluate elements of $\mathrm{Sym}(\boldsymbol{\eta})$, just as we did for $\mathrm{Lie}(\boldsymbol{\xi})$. In particular, we have a map $\mathrm{Ev}^{\mathscr{Y}}_{q_0} : \mathrm{Sym}(\boldsymbol{\eta}) \to \mathsf{T}_{q_0}\mathsf{Q}$ that assigns to $P \in \mathrm{Sym}(\boldsymbol{\eta})$ the evaluation at q_0 of the corresponding vector field defined by replacing η_α with Y_α, and replacing the symmetric product in $\mathrm{Sym}(\boldsymbol{\eta})$ with that on vector fields defined by ∇. •

Now we go through the same sort of development as we did in Section 7.1.5 to arrive at the analogue of Theorem 7.20 for affine connection control systems. However, the development is sufficiently different from that in Section 7.1.5 that we must go through it again. We suppose that we have a manifold Q, with a C^∞-affine connection ∇ and a family $\mathscr{Y} = \{Y_1, \dots, Y_m\}$ of C^∞-vector fields. Associated with these vector fields are indeterminates $\boldsymbol{\eta} = \{\eta_1, \dots, \eta_m\}$ and their free symmetric algebra $\mathrm{Sym}(\boldsymbol{\eta})$. An ***admissible weight*** for $\boldsymbol{\eta}$ is an m-tuple $(w_1, \dots, w_m)$ of nonnegative real numbers, and the set of admissible weights is denoted by $\mathrm{wgt}(\boldsymbol{\eta})$. For an admissible weight $\boldsymbol{w}$ and for $P \in \mathrm{Pr}(\boldsymbol{\eta})$, the ***$\boldsymbol{w}$-weight*** of P is

$$\|P\|_{\boldsymbol{w}} = \sum_{a=1}^{m} w_a \, |P|_a .$$

A member S of $\mathrm{Sym}(\boldsymbol{\eta})$ is ***$\boldsymbol{w}$-homogeneous*** if it is a sum of products of the same $\boldsymbol{w}$-weight. If S is $\boldsymbol{w}$-homogeneous, then we define $\|S\|_{\boldsymbol{w}}$ to be the $\boldsymbol{w}$-weight of one, and therefore all, of its summands. For an admissible weight $\boldsymbol{w}$ and for $k \in \mathbb{N}$, we write

$$\mathcal{V}^{\boldsymbol{w},k}_{\mathscr{Y}}(q_0) = \mathrm{span}_{\mathbb{R}} \{ \mathrm{Ev}^{\mathscr{Y}}_{q_0}(P) \mid \ P \in \mathrm{Pr}(\boldsymbol{\eta}), \ \|P\|_{\boldsymbol{w}} \le k \}.$$

If $S \in \mathrm{Sym}(\boldsymbol{\eta})$ is $\boldsymbol{w}$-homogeneous, then it is ***$\boldsymbol{w}$-neutralized*** by $\mathscr{Y}$ at q_0 if there exists an integer $k < \|S\|_{\boldsymbol{w}}$ such that $\mathrm{Ev}^{\mathscr{Y}}_{q_0}(S) \in \mathcal{V}^{\boldsymbol{w},k}_{\mathscr{Y}}(q_0)$. Define a subspace $\mathcal{P}(\boldsymbol{\eta})$ of $\mathrm{Sym}(\boldsymbol{\eta})$ by

$$\mathcal{P}(\boldsymbol{\eta}) = \mathrm{span}_{\mathbb{R}} \{ P \in \mathrm{Pr}(\boldsymbol{\eta}) \mid \ |P|_a \text{ is even}, \ a \in \{1, \dots, m\} \} .$$

For $\sigma \in S_m$ and $(w_1, \dots, w_m) \in \mathrm{wgt}(\boldsymbol{\eta})$, we define

$$\sigma(w_1, \dots, w_m) = (w_{\sigma(1)}, \dots, w_{\sigma(m)}),$$

and we denote by $S_m^{\boldsymbol{w}}$ those permutations fixing $\boldsymbol{w} \in \mathrm{wgt}(\boldsymbol{\eta})$. For $P \in \mathrm{Pr}(\boldsymbol{\eta})$ and $\sigma \in S_m$, $\sigma(P)$ denotes the product obtained by replacing η_a with $\eta_{\sigma(a)}$, $a \in \{1, \dots, m\}$, and for $S \in \mathrm{Sym}(\boldsymbol{\eta})$, $\sigma(S)$ is defined by applying σ to each summand of S. For $\boldsymbol{w} \in \mathrm{wgt}(\boldsymbol{\eta})$, we define

$$\mathcal{P}_0^{\boldsymbol{w}}(\boldsymbol{\eta}) = \{ S \in \mathcal{P}(\boldsymbol{\eta}) \mid \sigma(S) = S \text{ for all } \sigma \in S_m^{\boldsymbol{w}} \}$$

as the set of ***w*-obstructions** in $\mathrm{Sym}(\boldsymbol{\eta})$. We define $\rho_{\mathscr{Y}}^{q_0} \colon \mathrm{wgt}(\boldsymbol{\eta}) \to \mathbb{N}$ by asking that $\rho_{\mathscr{Y}}^{q_0}(\boldsymbol{w})$ be the largest integer for which all $\boldsymbol{w}$-obstructions S satisfying $\|S\|_{\boldsymbol{w}} \leq \rho_{\mathscr{Y}}^{q_0}(\boldsymbol{w})$ are $\boldsymbol{w}$-neutralized by $\mathscr{Y}$ at q_0. We then define

$$\mathcal{V}_{\mathscr{Y}}^{\boldsymbol{w}}(q_0) = \mathcal{V}^{\boldsymbol{w}, \rho_{\mathscr{Y}}^{q_0}(\boldsymbol{w})}(q_0).$$

We now have the following analogue of Theorem 7.20.

Theorem 7.40 (Sufficient conditions for STLCC). *Let* $\Sigma = (\mathsf{Q}, \nabla, \mathcal{D}, \mathscr{Y})$ *be an analytic affine connection pre-control system and let* $\boldsymbol{\eta} = \{\eta_1, \dots, \eta_m\}$ *be indeterminates. If*

$$\sum_{\boldsymbol{w} \in \mathrm{wgt}(\boldsymbol{\eta})} \mathcal{V}_{\mathscr{Y}}^{\boldsymbol{w}}(q_0) = \mathrm{Sym}^{(\infty)}(\mathcal{Y})_{q_0},$$

then the following statements hold:

(i) if $\mathrm{Sym}^{(\infty)}(\mathcal{Y})_{q_0} = \mathcal{D}_{q_0}$ *and* $\mathrm{Lie}^{(\infty)}(\mathcal{D})_{q_0} = \mathsf{T}_{q_0}\mathsf{Q}$*, then* Σ *is properly STLC from* q_0*;*

(ii) if $\mathrm{Lie}^{(\infty)}(\mathrm{Sym}^{(\infty)}(\mathcal{Y}))_{q_0} = \mathsf{T}_{q_0}\mathsf{Q}$*, then* Σ *is properly STLCC from* q_0.

Proof. The proof relies on an understanding of the proof of Theorem 7.36. From the proof of that theorem, particularly Lemma B.3(i) and Lemma B.6, we see that obstructions for affine connection control systems occur in brackets that, when evaluated, lie in $0_{q_0} \oplus \mathrm{Sym}^{(\infty)}(\mathcal{Y})_{q_0}$. We shall show that a collection of such obstructions, sufficient to allow us to generate $0_{q_0} \oplus \mathrm{Sym}^{(\infty)}(\mathcal{Y})_{q_0}$, can be neutralized. The hypotheses of the theorem assert that we can find products $P_1, \dots, P_k \in \mathrm{Pr}(\boldsymbol{\eta})$ and weights $\boldsymbol{w}_1, \dots, \boldsymbol{w}_k \in \mathrm{wgt}(\boldsymbol{\eta})$ such that $\|P_j\|_{\boldsymbol{w}_j} < \rho_{\mathscr{Y}}^{q_0}(\boldsymbol{w}_j)$, $j \in \{1, \dots, k\}$, and such that

$$\mathrm{span}_{\mathbb{R}} \{ \mathrm{Ev}_{q_0}^{\mathscr{Y}}(P_j) \mid j \in \{1, \dots, k\} \} = \mathrm{Sym}^{(\infty)}(\mathcal{Y})_{q_0}. \tag{7.6}$$

Now let $\boldsymbol{\xi} = \{\xi_0, \xi_1, \dots, \xi_m\}$ correspond, as in the proof of Theorem 7.36, to $\mathscr{C}_\Sigma = \{S, \mathrm{vlft}(Y_1), \dots, \mathrm{vlft}(Y_m)\}$. Define weights $\boldsymbol{w}_{0,j} \in \mathrm{wgt}(\boldsymbol{\xi})$ by $\boldsymbol{w}_{0,j} = (0, \boldsymbol{w}_j)$, $j \in \{1, \dots, k\}$. Fix $j \in \{1, \dots, k\}$. Let $L \in \mathrm{Lie}(\boldsymbol{\xi})$ be a $\boldsymbol{w}_{0,j}$-obstruction. Then, by the proof of Theorem 7.36,

$$\mathrm{Ev}^{\mathscr{C}_\Sigma}_{0_{q_0}}(L) = \sum_{l=1}^{r} \mathrm{vlft}\big(\mathrm{Ev}^{\mathscr{Y}}_{q_0}(\tilde{P}_l)\big),$$

where $\|\tilde{P}_l\|_{\boldsymbol{w}_j} = \|L\|_{\boldsymbol{w}_{0,j}}$. The hypotheses of the theorem ensure that the $\boldsymbol{w}_j$-obstruction $\tilde{P}_1 + \cdots + \tilde{P}_r$ can be $\boldsymbol{w}_j$-neutralized. By understanding the computations of the proof of Theorem 7.36 that allow one to relate products with weight $\boldsymbol{w}_j$ and brackets of weight $\boldsymbol{w}_{0,j}$, one can easily see that the $\boldsymbol{w}_{0,j}$-obstruction L can be $\boldsymbol{w}_{0,j}$-neutralized.

Now let $B_1, \dots, B_k \in \mathrm{Br}(\boldsymbol{\xi})$ be the natural brackets having the properties that $\|B_j\|_{\boldsymbol{w}_{0,j}} = \|P_j\|_{\boldsymbol{w}_j}$, $j \in \{1, \dots, k\}$, and that $\mathrm{Ev}^{\mathscr{C}_\Sigma}_{0_{q_0}}(B_j) = \mathrm{vlft}\big(\mathrm{Ev}^{\mathscr{Y}}_{q_0}(P_j)\big)$, $j \in \{1, \dots, k\}$. That such brackets exist is a direct consequence of the computations in the proof of Theorem 7.36. Our argument above shows that $\mathrm{Ev}^{\mathscr{C}_\Sigma}_{0_{q_0}}(B_j) \in \mathcal{V}^{\boldsymbol{w}_{0,j}}_{\mathscr{C}_\Sigma}$. By (7.6) we then have

$$0_{q_0} \oplus \mathrm{Sym}^{(\infty)}(\mathscr{Y})_{q_0} \subset \sum_{\boldsymbol{w} \in \mathrm{wgt}(\boldsymbol{\xi})} \mathcal{V}^{\boldsymbol{w}}_{\mathscr{C}_\Sigma}.$$

Now we wish to show that

$$\mathrm{Lie}^{(\infty)}(\mathrm{Sym}^{(\infty)}(\mathscr{Y})_{q_0}) \oplus 0_{q_0} \subset \sum_{\boldsymbol{w} \in \mathrm{wgt}(\boldsymbol{\xi})} \mathcal{V}^{\boldsymbol{w}}_{\mathscr{C}_\Sigma}.$$

We choose the products $P_1, \dots, P_k \in \mathrm{Pr}(\boldsymbol{\eta})$ as above, and we let $B_1, \dots, B_k \in \mathrm{Br}(\boldsymbol{\xi})$ be the corresponding brackets, again just as defined above. We also use the weights $\boldsymbol{w}_1, \dots, \boldsymbol{w}_k$ as above, with their associated weights $\boldsymbol{w}_{0,j} = (0, \boldsymbol{w}_j)$, $j \in \{1, \dots, k\}$. The vector fields $X_j \triangleq \mathrm{Ev}^{\mathscr{Y}}(P_j)$, $j \in \{1, \dots, k\}$, may be used to generate $\mathrm{Lie}^{(\infty)}(\mathrm{Sym}^{(\infty)}(\mathscr{Y}))$, since $\mathrm{Sym}^{(\infty)}(\mathscr{Y})$ is regular when restricted to integral manifolds of $\mathrm{Lie}^{(\infty)}(\mathrm{Sym}^{(\infty)}(\mathscr{Y}))$. Consider a Lie bracket

$$[X_{\alpha_1}, [X_{\alpha_2}, \dots, [X_{\alpha_{r-1}}, X_{\alpha_r}]]], \qquad \alpha_1, \dots, \alpha_r \in \{1, \dots, k\},\ r \in \mathbb{N}.$$

By Exercise E7.2, we know that such brackets generate $\mathrm{Lie}^{(\infty)}(\mathrm{Sym}^{(\infty)}(\mathscr{Y}))$. What is more, consulting the proof of Theorem 7.36, we know that

$$[X_{\alpha_1}, [X_{\alpha_2}, \dots, [X_{\alpha_{r-1}}, X_{\alpha_r}]]](q) \oplus 0_q = \mathrm{Ev}^{\mathscr{C}_\Sigma}_q(B), \qquad q \in \mathsf{Q},$$

where $B \in \mathrm{Br}(\boldsymbol{\xi})$ has the property that

$$|B|_0 = |B|_a, \quad |B|_a = \sum_{j=1}^{k} |P|_a, \qquad a \in \{1, \dots, m\}.$$

In particular, $\|B\|_{\boldsymbol{w}_{0,j}} \geq \|P_j\|_{\boldsymbol{w}_j}$ for $j \in \{\alpha_1, \dots, \alpha_r\}$. From our computations above, all $\boldsymbol{w}_{0,j}$-obstructions of $\boldsymbol{w}_{0,j}$-weight at most that of B are $\boldsymbol{w}_{0,j}$-neutralized for $j \in \{1, \dots, k\}$. Thus $\mathrm{Ev}^{\mathscr{C}_\Sigma}_{0_{q_0}}(B) \in \mathcal{V}^{\boldsymbol{w}_{0,j}}_{\mathscr{C}_\Sigma}$. From this, the result now follows. ■

Just as the theorem of Sussmann [1987], stated as Corollary 7.24, follows from Theorem 7.20, there is a result for affine connection control systems, following from Theorem 7.40 and adapted from the statement of Sussmann. This result is often easier to apply in examples, although it is not as general. Call a product $P \in \mathrm{Pr}(\boldsymbol{\eta})$ ***bad*** if $|P|_a$ is even for each $a \in \{1, \dots, m\}$, and ***good*** otherwise. For $P \in \mathrm{Pr}(\boldsymbol{\eta})$, define

$$\rho(P) = \sum_{\sigma \in S_m} \sigma(P).$$

We leave the proof of the following result as Exercise E7.10. In Exercise E7.12 the reader can show that the corollary is strictly weaker than Theorem 7.40.

Corollary 7.41. *Let $\Sigma = (\mathsf{Q}, \nabla, \mathcal{D}, \mathscr{Y})$ be an analytic affine connection pre-control system and let $\boldsymbol{\eta} = \{\eta_1, \dots, \eta_m\}$ be indeterminates. Suppose that, for every bad product P, there exist good products $Q_1, \dots, Q_k$, for which $\deg(Q_j) < \deg(P)$, $j \in \{1, \dots, k\}$, and such that*

$$\mathrm{Ev}^{\mathscr{Y}}_{q_0}(\rho(P)) \in \mathrm{span}_{\mathbb{R}}\{\mathrm{Ev}^{\mathscr{Y}}_{q_0}(Q_j) |\ j \in \{1, \dots, k\}\}.$$

Then the following statements hold:

(i) if $\mathrm{Sym}^{(\infty)}(\mathcal{Y})_{q_0} = \mathcal{D}_{q_0}$ and $\mathrm{Lie}^{(\infty)}(\mathrm{Sym}^{(\infty)}(\mathcal{D}))_{q_0} = \mathsf{T}_{q_0}\mathsf{Q}$, then Σ is properly STLC from q_0;

(ii) if $\mathrm{Lie}^{(\infty)}(\mathrm{Sym}^{(\infty)}(\mathcal{Y})) = \mathsf{T}_{q_0}\mathsf{Q}$, then Σ is properly STLCC from q_0.

7.4 Examples illustrating controllability results

In this section we apply the results of Section 7.3 to a few examples. The examples we consider in this section all illustrate the application of the results of Sections 7.3.2 and 7.3.3. The linear and linearization results of Section 7.3.1 are relatively straightforward to apply, and the reader can work out examples of this in Exercises E7.6, E7.16, E7.17, and E7.18.

7.4.1 Robotic leg

This example was first studied in Example 5.71, and is depicted in Figure 7.6. The controllability of the system, as we present it here, follows Lewis and Murray [1997a].

In Example 5.71 the configuration manifold for the system was determined to be $\mathsf{Q} = \mathbb{R}_+ \times \mathbb{S}^1 \times \mathbb{S}^1$, and the system's Riemannian metric was presented as

$$\mathbb{G} = m(\mathrm{d}r \otimes \mathrm{d}r + r^2 \mathrm{d}\theta \otimes \mathrm{d}\theta) + J \mathrm{d}\psi \otimes \mathrm{d}\psi,$$

where we use the coordinates (r, θ, ψ) indicated in Figure 7.6, and where m is the mass of the particle on the end of the extensible leg, and J is the moment

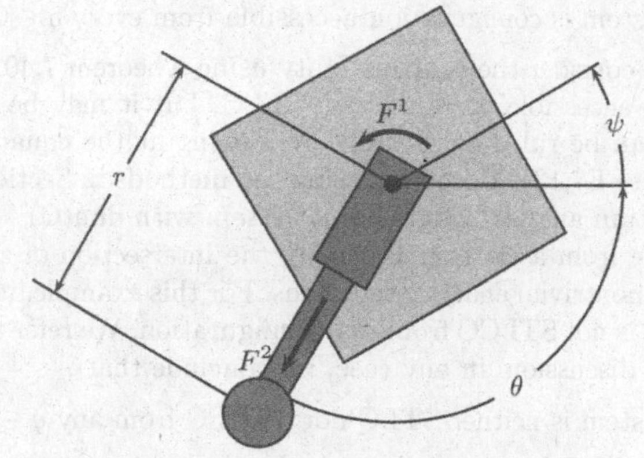

Figure 7.6. The robotic leg

of inertia of the base rigid body about the pivot point. The nonzero Christoffel symbols for the associated affine connection were computed to be

$$\Gamma^r_{\theta\theta} = -r, \quad \Gamma^\theta_{r\theta} = \Gamma^\theta_{\theta r} = \frac{1}{r}.$$

The input forces are $F^1 = \mathrm{d}\theta - \mathrm{d}\psi$ and $F^2 = \mathrm{d}r$, giving the input vector fields

$$Y_1 = \frac{1}{mr^2}\frac{\partial}{\partial\theta} - \frac{1}{J}\frac{\partial}{\partial\psi}, \quad Y_2 = \frac{1}{m}\frac{\partial}{\partial r}.$$

There are no constraints on the system, so we take $\mathcal{D} = \mathsf{TQ}$, giving $\overset{\mathcal{D}}{\nabla} = \overset{\mathrm{G}}{\nabla}$. Thus the system is an affine connection control system, and the results of Sections 7.3.2 and 7.3.3 are applicable. We shall concern ourselves with controllability properties that do not depend on the control set (i.e., we take the control set to be proper, but otherwise arbitrary).

Let us first compute some symmetric products and Lie brackets. We have

$$\langle Y_1 : Y_1\rangle = -\frac{2}{m^2r^3}\frac{\partial}{\partial r}, \quad \langle Y_1 : Y_2\rangle = 0, \quad \langle Y_2 : Y_2\rangle = 0,$$
$$[Y_1, Y_2] = \frac{2}{m^2r^3}\frac{\partial}{\partial\theta}, \quad [Y_1, \langle Y_1 : Y_1\rangle] = -\frac{4}{m^3r^6}\frac{\partial}{\partial\theta}.$$

Now let us consider all different possible combinations of inputs for the system.

1. Y_1 *only:* We first observe that $\langle Y_1 : Y_1\rangle$ is collinear with Y_2. Since $\langle Y_1 : Y_2\rangle = 0$, it follows that $\{Y_1, \langle Y_1 : Y_1\rangle\}$ are actually generators for $\mathrm{Sym}^{(\infty)}(\mathscr{Y})$. We also note that $\{Y_1, \langle Y_1 : Y_1\rangle, [Y_1, \langle Y_1 : Y_1\rangle]\}$ are generators for TQ, and, therefore, also for $\mathrm{Lie}^{(\infty)}(\mathrm{Sym}^{(\infty)}(\mathscr{Y}))$. Thus the accessibility properties of the system in this case are determined from Theorem 7.36:
   (a) the system is not accessible from q for any $q \in \mathsf{Q}$;

(b) the system is configuration accessible from every $q \in \mathsf{Q}$.

Now let us consider the controllability using Theorem 7.40. Since the system is not accessible, it cannot be STLC, but it may be STLCC. This, however, can be ruled out merely by looking at the equations of motion (see Exercise E7.13). Moreover, using the methods in Section 8.2 it can be shown that an analytic single-input system with $\dim(\mathsf{Q}) > 1$ can only be controllable from a set that is locally the intersection of zeros of a finite number of nontrivial analytic functions. For this example, it turns out that the system is not STLCC from *any* configuration. We refer to Remark 8.13 for further discussion. In any case, we conclude that

(c) the system is neither STLC nor STLCC from any $q \in \mathsf{Q}$.

2. Y_2 *only:* Since $\langle Y_2 : Y_2 \rangle = 0$, there is no way to generate anything in either $\mathrm{Sym}^{(\infty)}(\mathscr{Y})$ or $\mathrm{Lie}^{(\infty)}(\mathrm{Sym}^{(\infty)}(\mathscr{Y}))$ other than Y_2 itself. Therefore, following Theorem 7.36, we make the following conclusions about the system with this input:

   (a) the system is neither accessible nor configuration accessible from q for any $q \in \mathsf{Q}$.

   This, of course, also rules out any sort of controllability as well.
3. Y_1 *and* Y_2: Since $\langle Y_1 : Y_2 \rangle = \langle Y_2 : Y_2 \rangle = 0$, $\mathrm{Sym}^{(\infty)}(\mathscr{Y})$ is generated by $\{Y_1, \langle Y_1 : Y_1 \rangle\}$. We also have $\{Y_1, Y_2, [Y_1, Y_2]\}$ as generators for TQ, and, therefore, for $\mathrm{Lie}^{(\infty)}(\mathrm{Sym}^{(\infty)}(\mathscr{Y}))$. From this we conclude the following about the accessibility of the system, using Theorem 7.36:

   (a) the system is not accessible from q for any $q \in \mathsf{Q}$;

   (b) the system is configuration accessible from every $q \in \mathsf{Q}$.

   To determine the controllability properties, we use the corollary to Theorem 7.40 stated in Corollary 7.41. We see that the product $\langle \eta_1 : \eta_1 \rangle + \langle \eta_2 : \eta_2 \rangle$, when evaluated, is collinear with the evaluation of η_2. Since $\mathrm{Sym}^{(\infty)}(\mathscr{Y})$ is generated by $\{Y_1, \langle Y_1 : Y_1 \rangle\}$, we can conclude that *all* bad symmetric products evaluate to linear combinations of lower degree good symmetric products. Thus we make the following conclusions about the controllability of the system:

   (c) the system is STLCC, but not STLC, from each $q \in \mathsf{Q}$.

7.4.2 Planar body with variable-direction thruster

We consider here the example originally studied in Section 1.1, and also throughout Chapter 4. This example seems to have first been introduced in the control literature by Lewis and Murray [1997a]. The system is reproduced in Figure 7.7 for convenience. We recall that the configuration manifold for the system is $\mathsf{Q} = \mathbb{S}^1 \times \mathbb{R}^2$, that coordinates were denoted by (θ, x, y), and that the Riemannian metric for the system is

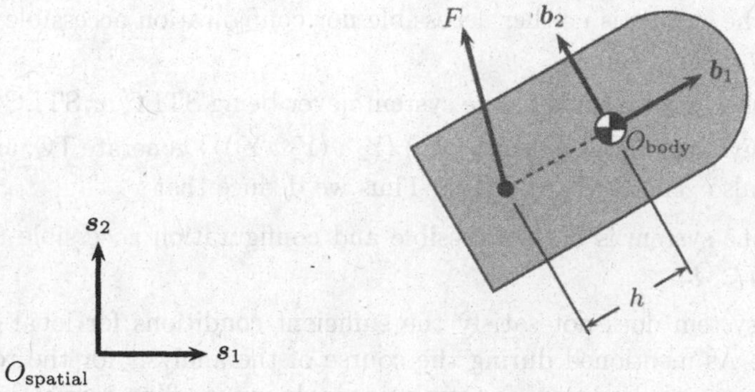

Figure 7.7. A planar body with a variable-direction thruster

$$\mathbb{G} = J\mathrm{d}\theta \otimes \mathrm{d}\theta + m(\mathrm{d}x \otimes \mathrm{d}x + \mathrm{d}y \otimes \mathrm{d}y),$$

where m is the mass of the body and J is its moment of inertia about its center of mass. The input for the system are taken to be a force, as in Example 5.47, applied to the body at a point a distance h from the center of mass, along the body $\boldsymbol{b}_1$-axis. The force can be resolved into two components, one in the body $\boldsymbol{b}_1$-direction, and the other in the body $\boldsymbol{b}_2$-direction. Here we will not take into account the symmetry of the problem, so we consider these forces to be specified by covector fields F^1 and F^2 on Q. One readily ascertains that

$$F^1 = \cos\theta \mathrm{d}x + \sin\theta \mathrm{d}y, \quad F^2 = -h\mathrm{d}\theta - \sin\theta \mathrm{d}x + \cos\theta \mathrm{d}y,$$

giving

$$Y_1 = \frac{\cos\theta}{m}\frac{\partial}{\partial x} + \frac{\sin\theta}{m}\frac{\partial}{\partial y}, \quad Y_2 = -\frac{h}{J}\frac{\partial}{\partial\theta} - \frac{\sin\theta}{m}\frac{\partial}{\partial x} + \frac{\cos\theta}{m}\frac{\partial}{\partial y}.$$

This system is again an affine connection control system, and so we can use Theorems 7.36 and 7.40, and Corollary 7.41 to evaluate its accessibility and controllability properties.

Some symmetric products and Lie brackets are

$$\langle Y_1 : Y_1\rangle = 0, \quad \langle Y_1 : Y_2\rangle = \frac{h\sin\theta}{mJ}\frac{\partial}{\partial x} - \frac{h\cos\theta}{mJ}\frac{\partial}{\partial y},$$

$$\langle Y_2 : Y_2\rangle = \frac{2h\cos\theta}{mJ}\frac{\partial}{\partial x} + \frac{2h\sin\theta}{mJ}\frac{\partial}{\partial y}, \quad [Y_1, Y_2] = -\frac{h\sin\theta}{mJ}\frac{\partial}{\partial x} + \frac{h\cos\theta}{mJ}\frac{\partial}{\partial y},$$

$$\langle Y_2 : \langle Y_2 : Y_2\rangle\rangle = \frac{2h^2\sin\theta}{mJ^2}\frac{\partial}{\partial x} - \frac{2h^2\cos\theta}{mJ^2}\frac{\partial}{\partial y}.$$

From this, we make the following deductions concerning the accessibility and controllability properties of the system.

1. Y_1 *only:* In this case, since $\langle Y_1 : Y_1\rangle = 0$, it is not possible to generate anything in $\mathrm{Sym}^{(\infty)}(\mathscr{Y})$ and $\mathrm{Lie}^{(\infty)}(\mathrm{Sym}^{(\infty)}(\mathscr{Y}))$ except Y_1. Thus we conclude that

(a) the system is neither accessible nor configuration accessible from any $q \in \mathsf{Q}$.

Clearly this also leads to the system never being STLC or STLCC as well.

2. Y_2 *only:* In this case $\{Y_2, \langle Y_2 : Y_2\rangle\, \langle Y_2 : \langle Y_2 : Y_2\rangle\rangle\}$ generate TQ, and, therefore, also generate $\mathrm{Sym}^{(\infty)}(\mathscr{Y})$. Thus we deduce that

   (a) the system is both accessible and configuration accessible from each $q \in \mathsf{Q}$.

   The system does not satisfy the sufficient conditions for local controllability. As mentioned during the course of the analysis for the robotic leg in the previous section, for analytic single-input affine connection control systems, controllability occurs from at most a very small set of configurations. For the planar rigid body, this set may be shown to be empty, so that we have:

   (b) the system is neither STLC nor STLCC from each $q \in \mathsf{Q}$.

   Whereas for the robotic leg example it is easy to see that the system is not STLCC by looking at the equations of motion, for the planar body, this is not the case. Lewis [1997] gives some of the properties of the reachable set for this example.

3. Y_1 *and* Y_2*:* Since the system is accessible and configuration accessible with just the input Y_1, we conclude that

   (a) the system is both accessible and configuration accessible from each $q \in \mathsf{Q}$.

   For controllability, we use Corollary 7.41. The bad product $\langle \eta_1 : \eta_1\rangle + \langle \eta_2 : \eta_2\rangle$ evaluates to a vector field that is collinear with Y_1. Thus this obstruction is neutralized, and we conclude that

   (b) the system is both STLC and STLCC from each $q \in \mathsf{Q}$.

7.4.3 Rolling disk

In this section we resume looking at the upright rolling disk example that was introduced in Section 1.3, and was also used throughout Chapter 4. This classic example seems to have first been studied in the control literature by Bloch, Reyhanoglu, and McClamroch [1992]. For convenience, we reproduce a figure of the system in Figure 7.8. We recall that the configuration manifold is $\mathsf{Q} = \mathbb{R}^2 \times \mathbb{S}^1 \times \mathbb{S}^1$, and that we used coordinates (x, y, θ, ϕ) as indicated in Figure 7.8. This is a system with constraints, and so, to compute symmetric products, we need to use the constrained connection of Section 4.5.5. In particular, we use the generalized Christoffel symbols of Section 4.5.6. We recall from Example 4.73 that the vector fields $\mathscr{X} = \{X_1, X_2\}$ given by

$$X_1 = \rho\cos\theta\frac{\partial}{\partial x} + \rho\sin\theta\frac{\partial}{\partial y} + \frac{\partial}{\partial \phi}, \quad X_2 = \frac{\partial}{\partial \theta}$$

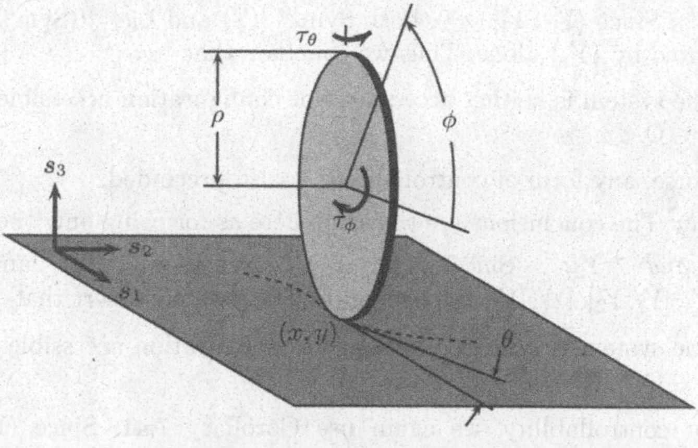

Figure 7.8. The upright rolling disk

formed a $\mathbb{G}$-orthogonal set of generators for the constraint distribution $\mathcal{D}$. Relative to these generators, the generalized Christoffel symbols that are of use to us are $\overset{\mathscr{X}}{\Gamma}{}^k_{ij} = 0$, $i, j, k \in \{1, 2\}$. The two input forces for the system are $F^1 = \mathrm{d}\theta$ and $F^2 = \mathrm{d}\phi$, giving rise to the input vector fields

$$Y_1 = P_{\mathcal{D}}(\mathbb{G}^\sharp(F^1)) = \frac{1}{J_{\mathrm{spin}}} \frac{\partial}{\partial \theta},$$
$$Y_2 = P_{\mathcal{D}}(\mathbb{G}^\sharp(F^2)) = \frac{1}{m\rho^2 + J_{\mathrm{roll}}} \Big(\rho \cos\theta \frac{\partial}{\partial x} + \rho \sin\theta \frac{\partial}{\partial y} + \frac{\partial}{\partial \phi} \Big).$$

Here we use the expression for $P_{\mathcal{D}}$ given in Example 4.81. The data thus described renders the system an affine connection control system, and so we use the results of Sections 7.3.2 and 7.3.3.

To use the generalized Christoffel symbols, we need to write Y_1 and Y_2 as linear combinations of X_1 and X_2, and this is easily done to get

$$Y_1 = \frac{1}{J_{\mathrm{spin}}} X_2, \quad Y_2 = \frac{1}{m\rho^2 + J_{\mathrm{roll}}} X_1.$$

With this data, we compute the symmetric products and Lie brackets

$$\langle Y_1 : Y_1 \rangle = 0, \quad \langle Y_1 : Y_2 \rangle = 0, \quad \langle Y_2 : Y_2 \rangle = 0,$$
$$[Y_1, Y_2] = \frac{1}{J_{\mathrm{spin}}(m\rho^2 + J_{\mathrm{roll}})} \Big(-\rho \sin\theta \frac{\partial}{\partial x} + \rho \cos\theta \frac{\partial}{\partial y} \Big),$$
$$[Y_1, [Y_1, Y_2]] = -\frac{1}{J_{\mathrm{spin}}^2(m\rho^2 + J_{\mathrm{roll}})} \Big(-\rho \cos\theta \frac{\partial}{\partial x} - \rho \sin\theta \frac{\partial}{\partial y} \Big).$$

Now we may draw the following conclusions concerning the controllability properties of the system.

1. Y_1 *only:* Since $\langle Y_1 : Y_1\rangle = 0$, both $\mathrm{Sym}^{(\infty)}(\mathscr{Y})$ and $\mathrm{Lie}^{(\infty)}(\mathrm{Sym}^{(\infty)}(\mathscr{Y}))$ are generated by $\{Y_1\}$ alone. Thus we conclude that

   (a) the system is neither accessible not configuration accessible from any $q \in \mathsf{Q}$.

   Of course, any form of controllability is also precluded.
2. Y_2 *only:* The conclusions are the same here as for using only the Y_1 input.
3. Y_1 *and* Y_2*:* Since $\{Y_1, Y_2\}$ generate $\mathcal{D}$ and since $\{Y_1, Y_2, [Y_1, Y_2], [Y_1, [Y_1, Y_2]]\}$ generate TQ, we may assert that

   (a) the system is both accessible and configuration accessible from each $q \in \mathsf{Q}$.

   As for controllability, we again use Corollary 7.41. Since $\langle Y_1 : Y_1\rangle = \langle Y_2 : Y_2\rangle = 0$, the bad product $\langle \eta_1 : \eta_1\rangle + \langle \eta_2 : \eta_2\rangle$ evaluates to zero. Thus we may assert that

   (b) the system is both STLC and STLCC from each $q \in \mathsf{Q}$.

Exercises

E7.1 Show that Assumption 7.13 holds, without loss of generality, provided that (7.2) holds and that the control set is convex.

E7.2 Let $\boldsymbol{\xi} = \{\xi_1, \dots, \xi_k\}$. Show that all brackets in the Lie algebra $\mathrm{Lie}(\boldsymbol{\xi})$ are $\mathbb{R}$-linear combinations of those of the form

$$[\xi_{\alpha_1}, [\xi_{\alpha_2}, \cdots, [\xi_{\alpha_{l-1}}, \xi_{\alpha_l}]]], \qquad \alpha_1, \dots, \alpha_l \in \{1, \dots, k\}.$$

E7.3 Prove Corollary 7.24 using Theorem 7.20.

E7.4 Let (V, A, B) be a linear control system, and recall from Example 7.12 that the system is STLC from 0 if and only if the rank of the linear map

$$C(A, B) = [\ B \mid A \circ B \mid \cdots \mid A^{n-1} \circ B\ ]$$

from $\oplus_{a=1}^{m} \mathbb{R}^m$ to V is maximal. Show that $\mathrm{image}(C(A, B)) = \langle A, \mathrm{image}(B)\rangle$.

Hint: *Use the Cayley–Hamilton Theorem.*

Readers wishing to learn more about the approach to linear control systems represented by the preceding exercise are referred to the book of Wonham [1985].

E7.5 Consider the control-affine system $\Sigma = (\mathsf{M}, \mathscr{C} = \{f_0, f_1, f_2\}, \mathbb{R}^2)$ with $\mathsf{M} = \mathbb{R}^3$ and with

$$f_0 = (\tfrac{1}{2}x^2 - \tfrac{1}{24}y^4)\frac{\partial}{\partial z}, \quad f_1 = \frac{\partial}{\partial x}, \quad f_2 = \frac{\partial}{\partial y}.$$

For this system, taking $x_0 = (0, 0, 0)$, do the following.

(a) Show that the hypotheses of Corollary 7.24 are not satisfied for this system, and so this corollary cannot be used to deduce that the system is STLC from x_0.
Hint: Consider the bad bracket $[f_1, [f_0, f_1]]$.
(b) Show that the hypotheses of Theorem 7.20 are satisfied, so that the system *is* STLC from x_0.

E7.6 In this exercise we consider the coupled masses whose data is described in Exercises E4.3, E4.17, and E4.29. The equilibrium configurations for the system were determined in Exercise E6.5. Note that this is a linear mechanical control system (provided you chose your coordinates in the "natural" way in Exercise E4.3).
(a) For the input configuration of Exercise E4.29(c), determine whether the system is STLC from $0 \oplus 0$.
(b) For the input configuration of Exercise E4.29(d), determine for which values of α the system is STLC from $0 \oplus 0$. Provide a physical interpretation of your answer.

E7.7 Consider a simple mechanical control system $\Sigma = (\mathsf{Q}, \mathbb{G}, V, \mathscr{F}, \mathbb{R}^m)$ and let $q_0 \in \mathsf{Q}$ have the property that $\mathrm{d}V(q_0) \in \mathrm{span}_{\mathbb{R}}\{F^1(q_0), \dots, F^m(q_0)\}$. If $u_0 \in \mathbb{R}^m$ satisfies

$$\sum_{a=1}^{m} u_0^a F^a(q_0) = \mathrm{d}V(q_0),$$

then in Section 10.2 we shall call (q_0, u_0) a ***controlled equilibrium configuration***.
(a) Show that the linearization of Σ at the controlled equilibrium (q_0, u_0) is a forced linear mechanical control system of the form $(\mathsf{V} = \mathsf{T}_{q_0}\mathsf{Q}, M = \mathbb{G}(q_0), K = K_\Sigma(q_0), (F_1 = F_{\Sigma,1}(q_0), 0), F = F_\Sigma(q_0))$. Find the explicit coordinate formulae for $K_\Sigma(q_0)$ and $F_{\Sigma,1}(q_0)$.
Hint: *Think of the system as a control-affine system with the drift modified by the term corresponding to the external force* $\sum_{a=1}^{m} u_0^a F^a$ *required to maintain the equilibrium.*
(b) Show that Theorem 7.31, by a suitable modification, applies to general linear mechanical systems of the form $(\mathsf{V}, M, K, (F_1, 0), F)$.
(c) State the theorem for controllability of the linearization of Σ about the controlled equilibrium (q_0, u_0).

E7.8 Prove the formulae in (7.5).

The next exercise is a followup on Exercise E6.12, where the linearization of a forced affine connection system is considered.

E7.9 We consider a forced affine connection control system $\Sigma = (\mathsf{Q}, \nabla, Y, \mathcal{D}, \mathscr{Y}, \mathbb{R}^m)$ and we suppose that Y is basic. Consider a general equilibrium point for which $Y(q_0) \in \mathrm{span}_{\mathbb{R}}\{Y_1(q_0), \dots, Y_m(q_0)\}$ with $u_0 \in \mathbb{R}^m$ satisfying

$$Y(q_0) = \sum_{a=1}^{m} u_0^a Y_a(q_0).$$

Do the following.

(a) Show that, by redefining Y appropriately, we may suppose, without loss of generality, that $u_0 = 0$. (This is in contrast with the case of a simple mechanical system as we saw in Exercise E7.7.)

Thus we may, without loss of generality, suppose that $Y(q_0) = 0_{q_0}$ at an equilibrium point.

(b) Show that the linearization of the system about an equilibrium point q_0 is governed by a second-order differential equation

$$\ddot{x}(t) + A_1(x(t)) = B_1(u(t))$$

for $A_1 \in L(\mathsf{T}_{q_0}\mathsf{Q}; \mathsf{T}_{q_0}\mathsf{Q})$ and $B_1 \in L(\mathbb{R}^m; \mathsf{T}_{q_0}\mathsf{Q})$. Give formulae for A_1 and B_1.

(c) Is it possible for the linearization to be controllable if $\mathcal{D}_{q_0} \subsetneq \mathsf{T}_{q_0}\mathsf{Q}$?

E7.10 Prove Corollary 7.41 using Theorem 7.40.

The next exercise deals with a notion of controllability that is sometimes useful, but was not discussed explicitly in the text. Let $\Sigma = (\mathsf{Q}, \mathbb{G}, V, \mathscr{F}, \mathbb{R}^m)$ be a simple mechanical control system and let q_0 be an equilibrium configuration for $(\mathsf{Q}, \mathbb{G}, V)$. The system Σ is ***equilibrium controllable*** from q_0 if, for any equilibrium point $q \in \mathsf{Q}$ for $(\mathsf{Q}, \mathbb{G}, V)$, there exists $(\gamma, u) \in \mathrm{Ctraj}(\Sigma, T)$ such that $\gamma'(0) = 0_{q_0}$ and $\gamma'(T) = 0_q$.

E7.11 Let $\Sigma = (\mathsf{Q}, \nabla, \mathcal{D}, \mathscr{Y}, U)$ be an affine connection control system, and suppose that the hypotheses of Theorem 7.40 hold at each point $q \in \mathsf{Q}$. Show that, if Q is connected, then Σ is equilibrium controllable from all points in Q.
Hint: *Take a look at the last part of the proof of Theorem 13.2.*

E7.12 Consider the affine connection control system $\Sigma = (\mathsf{Q}, \nabla, \mathscr{Y} = \{Y_1, Y_2\}, \mathbb{R}^2)$ with $\mathsf{Q} = \mathbb{R}^3$, $Y_1 = \frac{\partial}{\partial x}$, $Y_2 = \frac{\partial}{\partial y}$, and where ∇ is the affine connection whose nonzero Christoffel symbols are $\Gamma^z_{xx} = \frac{1}{2}$ and $\Gamma^z_{yy} = -\frac{1}{4}y^2$. For this system and with $q_0 = (0, 0, 0)$, do the following.

(a) Show that the hypotheses of Corollary 7.41 are not satisfied for this system, and so this corollary cannot be used to deduce that the system is STLCC from q_0.
Hint: *Consider the bad product* $\langle Y_1 : Y_1 \rangle$.

(b) Show that the hypotheses of Theorem 7.40 are satisfied, so that the system *is* STLC from q_0.

E7.13 Show "by hand" that the robotic leg example of Section 7.4.1 is not STLCC if one uses only the input u^1.

E7.14 Consider the planar rigid body of Section 7.4.2, but now take as inputs the fixed-direction force F^2 as in Section 7.4.2, along with a torque $F^3 = \mathrm{d}\theta$ applied at the center of mass.

(a) Find the corresponding control vector field Y_3.

(b) Show that the system is accessible from every $q \in \mathsf{Q}$.

(c) Show that the system does not satisfy the hypotheses of Theorem 7.40.

Now consider a change of input by taking

$$\tilde{F}^2 = -\sin\theta \mathrm{d}x + \cos\theta \mathrm{d}y = F^2 + hF^3, \quad \tilde{F}^3 = F^3.$$

Since $\operatorname{span}_{\mathbb{R}}\{F^2(q), F^3(q)\} = \operatorname{span}_{\mathbb{R}}\{\tilde{F}^2(q), \tilde{F}^3(q)\}$, the controllability properties of the system might be expected to be (and indeed are) the same.

(d) Show that the system with the modified forces satisfies the hypotheses of Theorem 7.40, and so is STLC from each $q \in \mathsf{Q}$.

E7.15 In this exercise, we consider the ball in a rotating channel of Exercises E4.4, E4.18, and E4.30. We consider the system as an affine connection control system, and apply the results of Sections 7.3.2 and 7.3.3.

(a) Determine the accessibility properties of the system. If these vary as a function of the initial point in the configuration manifold, then spell out this dependence.

(b) Determine the small-time local controllability properties of the system. If these vary as a function of the initial point in the configuration manifold, then spell out this dependence.

E7.16 In this exercise we consider the pendulum/cart system whose data is described in Exercises E4.5 and E4.19. The equilibrium configurations for the system were determined in Exercise E6.14.

We first consider the input force of Exercise E4.31.

(a) Determine the configuration linearization of the system at the equilibria.

(b) Is the linearization STLC from $0 \oplus 0$?

(c) From the controllability of the linearization, what can you deduce about the controllability of the full system from the equilibrium configurations?

Now we take a different input force, a torque applied to the base of the pendulum arm.

(d) What is the covector field F that defines this input?

(e) Determine the configuration linearization of the system at the equilibria.

(f) Is the linearization STLC from $0 \oplus 0$?

(g) From the controllability of the linearization, what can you deduce about the controllability of the full system from the equilibrium configurations?

E7.17 In this exercise we consider the spherical pendulum on a track whose data is described in Exercises E4.6, E4.20, and E4.32. The equilibrium configurations for the system were determined in Exercise E6.15.

(a) Determine the configuration linearization of the system at the equilibria.

(b) Is the linearization STLC from $0 \oplus 0$?

(c) From the controllability of the linearization, what can you deduce about the controllability of the full system from the equilibrium configurations?

E7.18 In this exercise we consider the two-axis gyro whose data is described in Exercises E4.7, E4.21, and E4.33. The equilibrium configurations for the system were determined in Exercise E6.16.

(a) Determine the configuration linearization of the system at the equilibria.

(b) Is the linearization STLC from $0 \oplus 0$?

(c) From the controllability of the linearization, what can you deduce about the controllability of the full system from the equilibrium configurations?

E7.19 Let G be a Lie group, $\mathfrak{g}$ be its Lie algebra, $\mathbb{I}$ be an inner product on $\mathfrak{g}$, and $\overset{\mathsf{G}_{\mathbb{I}}}{\nabla}$ be the corresponding Levi-Civita affine connection.

(a) Show that the symmetric product, with respect to $\overset{\mathbb{G}_{\mathbb{I}}}{\nabla}$, of left-invariant vector fields is left-invariant.

(b) Write the symmetric product with respect to $\overset{\mathbb{G}_{\mathbb{I}}}{\nabla}$ of two left-invariant vector fields by way of the symmetric bilinear map $\langle\cdot:\cdot\rangle_{\mathfrak{g}}:\mathfrak{g}\times\mathfrak{g}\to\mathfrak{g}$ given by

$$\langle\xi:\eta\rangle_{\mathfrak{g}} = -\mathbb{I}^{\sharp}\big(\operatorname{ad}^*_{\xi}\mathbb{I}^{\flat}(\eta) + \operatorname{ad}^*_{\eta}\mathbb{I}^{\flat}(\xi)\big).$$

E7.20 Consider an affine connection control system defined on a Lie group as characterized in Section 5.3.2. The object of this exercise is to perform the appropriate symmetric product computations on the Lie algebra of the Lie group.
(a) State and prove a version of Theorem 7.40 for affine connection control systems defined on a Lie group.
(b) What are the consequences of left-invariance?

E7.21 Consider the planar rigid body of Section 7.4.2 and, following Example 5.47, regard it as a simple mechanical control system on a Lie group. Recover the results in Section 7.4.2 and in Exercise E7.14, using the version of Theorem 7.40 worked out in Exercise E7.20.

E7.22 Consider the rigid body of Example 5.48, fixed at a point, and consider two control torques aligned with the first and second major axes. In other words, consider the simple mechanical system on a Lie group given by $(\mathsf{SO}(3),\mathbb{I},\{e^1,e^2\})$. Characterize the controllability properties of this system, using the version of Theorem 7.40 in Exercise E7.20.

For the following exercise, you will need the following ideas. Let $\mathcal{D}$ be a C^∞-distribution on Q, let X_0 be a vector field on Q, and let ∇ be an affine connection on Q. The vector field X_0 is ***$\mathcal{D}$-parallel*** if $\nabla_X X_0\in\Gamma^\infty(\mathcal{D})$ for all $X\in\Gamma^\infty(\mathcal{D})$.

E7.23 Let $X_0\in\Gamma^\infty(\mathsf{TQ})$ have the property that $X(q_0)=0_{q_0}$.
(a) Show that the map $\mathsf{T}_{q_0}\mathsf{Q}\ni v_{q_0}\mapsto\langle X_0:V\rangle(q_0)\in\mathsf{T}_{q_0}\mathsf{Q}$, where V is an arbitrary vector field satisfying $V(q_0)=v_{q_0}$, is well-defined and linear.
Write the preceding linear map as $\operatorname{sym}_{X_0}(q_0)\in L(\mathsf{T}_{q_0}\mathsf{Q};\mathsf{T}_{q_0}\mathsf{Q})$.
(b) Prove the following result of Bullo and Lewis [2002].

Proposition. *Let $\Sigma=(\mathsf{Q},\mathbb{G},V,\mathscr{F},\mathbb{R}^m)$ be a simple mechanical control system with $\dim(\mathsf{Q})=n$, let q_0 be a critical point for V, and consider the linearization of Σ at q_0 as given in Definition 7.32. Let $\mathscr{Y}=\{\mathbb{G}^\sharp(F^1),\dots,\mathbb{G}^\sharp(F^m)\}$. The following statements are equivalent:*

(i) the linearization of Σ at q_0 is controllable;

(ii) the linear map from $\oplus_{i=1}^n\mathbb{R}^m$ to $\mathsf{T}_{q_0}\mathsf{Q}$ represented by the block matrix

$$\big[B_\Sigma(q_0)\mid\mathbb{G}(q_0)^\sharp\circ K_\Sigma(q_0)^\flat\circ B_\Sigma(q_0)\mid\cdots\mid(\mathbb{G}(q_0)^\sharp\circ K_\Sigma(q_0)^\flat)^{n-1}\circ B_\Sigma(q_0)\big],$$

has maximal rank;

(iii) the smallest $\operatorname{sym}_{\operatorname{grad}V}(q_0)$-invariant subspace of $\mathsf{T}_{q_0}\mathsf{Q}$ containing $\mathscr{Y}_{q_0}$ is $\mathsf{T}_{q_0}\mathsf{Q}$;

(iv) *the smallest* $\nabla(\mathrm{grad}V)(q_0)$*-invariant subspace of* $\mathsf{T}_{q_0}\mathsf{Q}$ *containing* $\mathcal{Y}_{q_0}$ *is* $\mathsf{T}_{q_0}Q$;

(v) *if* $\mathcal{P}(\mathrm{grad}V, \mathscr{Y})$ *is the smallest distribution containing* $\mathcal{Y}$ *for which* $\mathrm{grad}V$ *is* $\mathcal{P}(\mathrm{grad}V, \mathscr{Y})$*-parallel, then* $\mathcal{P}(\mathrm{grad}V, \mathscr{Y})_{q_0} = \mathsf{T}_{q_0}Q$.

8

Low-order controllability and kinematic reduction

The discussions of controllability in Chapter 7 are of a quite general nature. However, as we saw in Exercise E7.14 (see also Example 7.22 for a control-affine example), even these very general results can fail to provide complete characterizations of the controllability of a system, even in rather simple examples. In this chapter we provide low-order controllability results that are quite sharp. These are extensions of results initially due of Hirschorn and Lewis [2001], and considered in the context of motion planning by Bullo and Lewis [2003b]. One of the interesting features of these results is that their hypotheses are "feedback-invariant," a notion that we do not discuss here; see [Lewis 2000a] for an introduction in the setting of affine connection control systems. Interestingly, these controllability results are related to a notion called "kinematic controllability" by Bullo and Lynch [2001], and considered here in Section 8.3. While the controllability results of Section 8.2 have more restrictive hypotheses than those of Chapter 7, it turns out that the restricted class of systems are those for which it is possible to develop some simplified design methodologies for motion planning. These methodologies are presented by way of some examples in Chapter 13. The emphasis in this chapter is on understanding the sometimes subtle relationships between various concepts; the design issues and serious consideration of examples are postponed to Chapter 13.

Mathematically, the ideas in this chapter are united by a certain vector-valued quadratic form. The usefulness of vector-valued, symmetric, bilinear maps in control has been observed in previous work, most commonly in optimal control [e.g., Agrachev 1990a]. Controllability for a class of control systems was considered by Melody, Başar, and Bullo [2003] using quadratic forms, although only $\mathbb{R}$-valued. Related results for control-affine systems are those of Basto-Gonçalves [1998]. Recent results of Hirschorn and Lewis [2002] extend those of Basto-Gonçalves, and, like the results in this chapter, are of a completely geometric character. However, for general control-affine systems, the development is significantly more complicated. A presentation of some control problems utilizing vector-valued quadratic forms appears in the paper of

Bullo, Cortés, Lewis, and Martínez [2004], along with a discussion of some related computational issues. A paper containing a discussion of vector-valued quadratic forms in a manner related to our interests is that of Matveev [1999]. Related papers include [Agrachev 1990b, Agrachev and Gamkrelidze 1991].

8.1 Vector-valued quadratic forms

An essential ingredient in our development in this chapter is a vector-valued quadratic form. The properties of such an object, as we shall use it, are not part of standard linear algebra. Therefore, in this section we present an overview of some of these properties. At the end of the section we indicate how to define a vector-valued quadratic form associated to an affine connection control system.

8.1.1 Basic definitions and properties

Let V and U be finite-dimensional $\mathbb{R}$-vector spaces and let $\Sigma_2(\mathsf{V};\mathsf{U})$ denote the set of symmetric $\mathbb{R}$-bilinear maps from $\mathsf{V}\times\mathsf{V}$ to U. For $B\in\Sigma_2(\mathsf{V};\mathsf{U})$, we define $Q_B\colon\mathsf{V}\to\mathsf{U}$ by $Q_B(v)=B(v,v)$. Just as in the scalar-valued case, we call Q_B the ***quadratic form*** associated with B, and note that Exercise E2.12 also applies to the vector-valued case, so that an element $B\in\Sigma_2(\mathsf{V};\mathsf{U})$ is uniquely determined by Q_B. This justifies our calling B a ***vector-valued quadratic form***, accepting a slight abuse of language.

For $\lambda\in\mathsf{U}^*$, we define $\lambda B\colon\mathsf{V}\times\mathsf{V}\to\mathbb{R}$ by $\lambda B(v_1,v_2)=\langle\lambda;B(v_1,v_2)\rangle$.

Definition 8.1 (Definiteness of vector-valued quadratic forms). Let $B\in\Sigma_2(\mathsf{V};\mathsf{U})$.

(i) B is ***definite*** if there exists $\lambda\in\mathsf{U}^*$ such that λB is positive-definite.
(ii) B is ***semidefinite*** if there exists $\lambda\in\mathsf{U}^*\setminus\{0\}$ such that λB is positive-semidefinite.
(iii) B is ***strongly semidefinite*** if there exists $\lambda\in\mathsf{U}^*\setminus\{0\}$ such that λB is nonzero and positive-semidefinite.
(iv) B is ***indefinite*** if, for each $\lambda\in\mathsf{U}^*\setminus\{0\}$, λB is neither positive nor negative-semidefinite.
(v) B is ***essentially indefinite*** if, for each $\lambda\in\mathsf{U}^*$, λB is either
  (a) zero or
  (b) neither positive nor negative-semidefinite. •

Some of the notions in the definition overlap. However, there are two that are mutually exclusive, and provide us with the characterizations of interest.

Lemma 8.2. *If V and U are finite-dimensional $\mathbb{R}$-vector spaces and if $B\in\Sigma_2(\mathsf{V};\mathsf{U})$, then B satisfies exactly one of the following conditions, provided that $\mathsf{V}\neq\{0\}$:*

(i) B is strongly semidefinite;

(ii) B is essentially indefinite.

Furthermore, if B is definite, then it is strongly semidefinite (i.e., it satisfies condition (i), as well as (ii)).

Proof. First suppose that $\mathsf{U} = \{0\}$. Then $\lambda B = 0$ for all $\lambda \in \mathsf{U}^*$, which means that B is essentially indefinite. It is clear that B cannot be strongly semidefinite since, in this case, $Q_{\lambda B}^{-1}(0) = \mathsf{V} \neq \{0\}$ for all $\lambda \in \mathsf{U}^*$, from which it follows that B cannot be definite.

Now suppose that $\mathsf{U} \neq \{0\}$. Let $\lambda \in \mathsf{U}^* \setminus \{0\}$. Then either

1. $\lambda B = 0$,
2. λB is nonzero and positive-semidefinite,
3. λB is nonzero and negative-semidefinite, or
4. λB is nonzero and neither positive nor negative-semidefinite.

Furthermore, these four properties of λB are mutually exclusive. Therefore, it must be that either properties 1 and 4 hold for all $\lambda \in \mathsf{U}^*$ or that properties 2 and 3 hold for some $\lambda \in \mathsf{U}^*$. This is exactly the first assertion of the lemma. The second assertion follows directly from the definitions. ■

Based on the preceding classification of vector-valued quadratic forms, we have the following useful result.

Lemma 8.3. *Let V and U be finite-dimensional $\mathbb{R}$-vector spaces with $B \in \Sigma_2(\mathsf{V};\mathsf{U})$. Suppose that $\mathsf{V} \neq \{0\}$. Then the following statements hold:*

(i) if $\mathsf{U} = \{0\}$, then B is essentially indefinite;

(ii) if $\mathsf{U} \neq \{0\}$, then B is essentially indefinite if and only if

$$0 \in \operatorname{int}_{\operatorname{aff}(\operatorname{image}(Q_B))}(\operatorname{conv}(\operatorname{image}(Q_B))); \tag{8.1}$$

(iii) if $\mathsf{U} \neq \{0\}$, then B is definite if and only if there exists a codimension one subspace P of U, such that

(a) image(Q_B) *lies on one side of* P[1] *and*

(b) image$(Q_B) \cap \mathsf{P} = \{0\}$.

Proof. Part (i) follows directly from the definitions.

Next, note that when $\mathsf{U} \neq \{0\}$, (8.1) holds if and only if there is no codimension one subspace $\mathsf{P} \subset \mathsf{U}$ for which image(Q_B) lies on one side of P. Equivalently, (8.1) holds if and only if there is no $\lambda \in \mathsf{U}^* \setminus \{0\}$ such that $\langle\lambda; Q_B(v)\rangle \geq 0$ for every $v \in \mathsf{V}$, and such that there is $v \in \mathsf{V}$ for which $\langle\lambda; Q_B(v)\rangle > 0$.

(ii) Assume that (8.1) does not hold. Then there exists $\lambda \in \mathsf{U}^* \setminus \{0\}$ such that $\langle\lambda; Q_B(v)\rangle \geq 0$ for every $v \in \mathsf{V}$, and such that there is at least one $v \in \mathsf{V}$ such that $\langle\lambda; Q_B(v)\rangle > 0$. Thus λB is positive-semidefinite and

[1] This is made precise as follows. Let $\lambda \in \mathsf{U}^*$ have the property that $\mathsf{P} = \ker(\lambda)$. Then image$(Q_B)$ ***lies on one side*** of P if either image$(Q_B) \subset \{u \in \mathsf{U} \mid \lambda(u) \geq 0\}$ or image$(Q_B) \subset \{u \in \mathsf{U} \mid \lambda(u) \leq 0\}$.

nonzero, so B is not essentially indefinite. Now assume that B is not essentially indefinite. Then there exists $\lambda \in \mathsf{U}^*$ such that λB is nonzero and positive-semidefinite. Then $\langle \lambda; Q_B(v) \rangle \geq 0$ for all $v \in \mathsf{V}$, and there exists $v \in \mathsf{V}$ such that $\langle \lambda; Q_B(v) \rangle > 0$. This proves (ii).

(iii) Suppose that there is a codimension one subspace P having properties (iii a) and (iii b). If λ has the property that $\mathsf{P} = \ker(\lambda)$, then, by changing the sign of λ if necessary, we have $\langle \lambda; Q_B(v) \rangle \geq 0$ for all $v \in \mathsf{V}$ by property (iii a). Property (iii b) ensures that $\langle \lambda; Q_B(v) \rangle = 0$ only if $v = 0$. Thus λB is positive-definite. The preceding argument may be reversed to show that if λB is positive-definite, then $\mathsf{P} = \ker(\lambda)$ has properties (iii a) and (iii b). ■

The question of whether it is possible to ascertain, in a computationally efficient manner, whether a given vector-valued symmetric bilinear map is essentially indefinite is unresolved. Designing an algorithm to ascertain when a general vector-valued symmetric bilinear map is essentially indefinite can be shown to be an NP-complete problem (see [Sipser 1996] as an introductory reference for computational complexity).[2] This is given some consideration by Bullo, Cortés, Lewis, and Martínez [2004]. In Exercise E8.1 the reader can see how the rather abstract notions of definiteness can be more obviously phrased using standard linear algebra concepts.

The following result gives some properties of $\mathbb{R}$-valued quadratic forms that will be useful in our discussion.

Lemma 8.4 (Indefinite $\mathbb{R}$-valued quadratic forms). *Let V be a finite-dimensional $\mathbb{R}$-vector space and let $B \in \Sigma_2(\mathsf{V}; \mathbb{R})$. For a basis $\mathscr{V} = \{v_1, \dots, v_n\}$ for V, let $[B]_{\mathscr{V}}$ be the $n \times n$ matrix representation of B. The following statements are equivalent:*

(i) there exists a basis $\mathscr{V}$ for V for which the sum of the diagonal entries in the matrix $[B]_{\mathscr{V}}$ is zero;
(ii) there exists a basis $\mathscr{V}$ for V for which the diagonal entries in the matrix $[B]_{\mathscr{V}}$ are all zero;
(iii) B is essentially indefinite.

Proof. See Section B.3. ■

8.1.2 Vector-valued quadratic forms and affine connection control systems

Let $\Sigma = (\mathsf{Q}, \nabla, \mathcal{D}, \mathscr{Y}, U)$ be a C^∞-affine connection control system and let $q \in \mathsf{Q}$. Denote by $\mathcal{Y}$ the C^∞-distribution generated by $\mathscr{Y}$. Note that we do not *a priori* ask that $\mathcal{Y}$ have locally constant rank. The discussion here makes use of the idea of a quotient space, for which we refer the reader to Definition 2.10.

[2] A solution to the problem of determining the essential indefiniteness of a vector-valued symmetric bilinear map implies a solution to the problem of maximizing a quadratic function over the unit hypercube, and this is known to be NP-complete.

If $\mathsf{S}_q \subset \mathsf{T}_q\mathsf{Q}$ is a subspace, we define $B_{\mathcal{Y}_q}(\mathsf{S}_q)\colon \mathcal{Y}_q \times \mathcal{Y}_q \to \mathsf{T}_q\mathsf{Q}/\mathsf{S}_q$ as the $\mathsf{T}_q\mathsf{Q}/\mathsf{S}_q$-valued symmetric, bilinear map on $\mathcal{Y}_q$ given by

$$B_{\mathcal{Y}_q}(\mathsf{S}_q)(u,v) = \pi_{\mathsf{S}_q}(\langle U : V\rangle(q)), \tag{8.2}$$

where U and V are vector fields extending $u, v \in \mathcal{Y}_q$, and where $\pi_{\mathsf{S}_q}\colon \mathsf{T}_q\mathsf{Q} \to \mathsf{T}_q\mathsf{Q}/\mathsf{S}_q$ is the canonical projection. Note that $B_{\mathcal{Y}_q}(\mathsf{S}_q)$ is not necessarily well-defined, meaning that it may depend on the extensions U and V. The following lemma gives a condition that ensures the well-definedness of $B_{\mathcal{Y}_q}(\mathsf{S}_q)$.

Lemma 8.5. *If $\mathcal{Y}_q \subset \mathsf{S}_q$, then $B_{\mathcal{Y}_q}(\mathsf{S}_q)$ is well-defined.*

Proof. We need to show that the definition in (8.2) does not depend on the extensions U and V of u and v. This will follow if $\pi_{\mathsf{S}_q}(\langle U : V\rangle(q))$ depends only on the values of U and V at q, and not on their derivatives. Let $\phi, \psi \in C^\infty(\mathsf{Q})$ and compute

$$\langle \phi U : \psi V\rangle = \phi\psi\,\langle U : V\rangle + (\mathscr{L}_U \psi)V + (\mathscr{L}_V \phi)U.$$

Therefore, we have $\pi_{\mathsf{S}_q}(\langle \phi U : \psi V\rangle(q)) = \phi(q)\psi(q)\pi_{\mathsf{S}_q}(\langle U : V\rangle(q))$, showing that $\pi_{\mathsf{S}_q}(\langle U : V\rangle(q))$ does not depend on the derivatives of U and V at q, and so the result follows. ■

Remark 8.6. As the reader may show in Exercise E2.5, $(\mathsf{T}_q\mathsf{Q}/\mathsf{S}_q)^*$ is naturally isomorphic to $\operatorname{ann}(\mathsf{S}_q)$. This allows a more concrete representation for the $\mathbb{R}$-valued symmetric bilinear maps $\lambda B_{\mathcal{Y}_q}(\mathsf{S}_q)$ used to determine the definiteness properties of $B_{\mathcal{Y}_q}(\mathsf{S}_q)$. Indeed, for $u, v \in \mathsf{S}_q$ and $\lambda \in \operatorname{ann}(\mathsf{S}_q) \simeq (\mathsf{T}_q\mathsf{Q}/\mathsf{S}_q)^*$, we have

$$\lambda B_{\mathcal{Y}_q}(\mathsf{S}_q)(u,v) = \langle \lambda; \langle U : V\rangle(q)\rangle,$$

where, as usual, U and V are vector fields extending u and v, respectively. •

Note that our definition of $B_{\mathcal{Y}_q}(\mathsf{S}_q)$ above is made at a fixed point in Q. We will also wish to consider a global notion in Section 8.3. In this case we need to assume that $\mathcal{Y}$ has locally constant rank so that $\mathsf{TQ}/\mathcal{Y}$ is a vector bundle over Q. Taking $\mathsf{S}_q = \mathcal{Y}_q$ for each $q \in \mathsf{Q}$, we apply the construction of (8.2) for each $q \in \mathsf{Q}$ to get a map $B_{\mathcal{Y}}$ that assigns to each $q \in \mathsf{Q}$ an element $B_{\mathcal{Y}}(q) \in \Sigma_2(\mathcal{Y}_q; \mathsf{T}_q\mathsf{Q}/\mathcal{Y}_q)$.

8.2 Low-order controllability results

In this section we state the controllability results of this chapter. The results we give are those of Bullo and Lewis [2003b]. Slightly weaker versions of these results were first reported by Hirschorn and Lewis [2001]. While the results are comparatively easy to state, complete proofs would require a significant diversion to introduce the necessary concepts. For this reason, our proofs are not quite complete, and we refer to the relevant references for details. In particular, we use some of the development of Hirschorn and Lewis [2002], which in turn uses some results from the papers [Bianchini and Stefani 1993, Sussmann 1987].

8.2.1 Constructions concerning vanishing input vector fields

We let $\Sigma = (\mathsf{Q}, \nabla, \mathcal{D}, \mathscr{Y}, U)$ be a C^∞-affine connection control system and we let $q_0 \in \mathsf{Q}$. It may be that q_0 is not a regular point for the C^∞-distribution $\mathcal{Y}$ generated by $\mathscr{Y}$. In this case, there may be vector fields in the collection $\mathscr{Y}$ that vanish at q_0. More generally, it may be the case that the input vector fields are otherwise not linearly independent at q_0. For example, if one wishes to globally define a control system for which the input distribution $\mathcal{Y}$ has locally constant rank, but is not trivializable, then one will necessarily have to choose more input vector fields than $\operatorname{rank}(\mathcal{Y})$ to generate $\mathcal{Y}$, implying that the input vector fields will never be linearly independent. It will be convenient to organize the vector fields in $\mathscr{Y}$ in a manner consistent with these possibilities. The following result gives a useful way of doing this.

Lemma 8.7. *Let $\Sigma = (\mathsf{Q}, \nabla, \mathcal{D}, \mathscr{Y} = \{Y_1, \dots, Y_m\}, U)$ be a C^∞-affine connection control system with $q_0 \in \mathsf{Q}$. There exists $\boldsymbol{T} \in \mathsf{GL}(m; \mathbb{R})$ with the property that, if $\tilde{Y}_a = T^b_a Y_b$, $a \in \{1, \dots, m\}$, then*

(i) $\{\tilde{Y}_1(q_0), \dots, \tilde{Y}_k(q_0)\}$ is a basis for $\mathcal{Y}_{q_0}$ and

(ii) the vector fields $\tilde{Y}_{k+1}, \dots, \tilde{Y}_m$ vanish at q_0.

Proof. We let $k = \dim(\mathcal{Y}_{q_0})$. Since $\mathscr{Y}$ generates $\mathcal{Y}$, we may find $\boldsymbol{R} \in \mathsf{GL}(m; \mathbb{R})$ with the property that, if $X_a = R^b_a Y_b$, $a \in \{1, \dots, m\}$, then $\{X_1, \dots, X_k\}$ form a basis for $\mathcal{Y}_{q_0}$. Now let $L_{q_0} \colon \mathbb{R}^m \to \mathcal{Y}_{q_0}$ be defined by $L_{q_0}(\boldsymbol{u}) = u^a X_a(q_0)$. Let $\boldsymbol{u}_{k+1}, \dots, \boldsymbol{u}_m \in \mathbb{R}^m$ be a basis for $\ker(L_{q_0})$ and define $\boldsymbol{S} \in \mathsf{GL}(m; \mathbb{R})$ by

$$\boldsymbol{S} = \left[\ \boldsymbol{e}_1 \mid \cdots \mid \boldsymbol{e}_k \mid \boldsymbol{u}_{k+1} \mid \cdots \mid \boldsymbol{u}_m\ \right].$$

It is readily checked that $\boldsymbol{S}$ is invertible. It is then clear that, if we take $\tilde{Y}_a = S^b_a X_b$, $a \in \{1, \dots, m\}$, then $\{\tilde{Y}_1(q_0), \dots, \tilde{Y}_k(q_0)\}$ form a basis for $\mathcal{Y}_{q_0}$, and that $\tilde{Y}_{k+1}, \dots, \tilde{Y}_m$ vanish at q_0. Now we take $\boldsymbol{T} = \boldsymbol{R}\boldsymbol{S}$. ■

Remarks 8.8. 1. If the vector fields $\mathscr{Y}$ are linearly independent at q_0, then one may take $\boldsymbol{T} = \boldsymbol{I}_m$ in the lemma.

2. If we take $\boldsymbol{T} \in \mathsf{GL}(m; \mathbb{R})$ and $\tilde{\mathscr{Y}} = \{\tilde{Y}_1, \dots, \tilde{Y}_m\}$ as in the lemma, and if we define $\tilde{U} = \left\{\boldsymbol{T}^{-1}\boldsymbol{u} \mid\ \boldsymbol{u} \in U\right\}$, this gives an affine connection control system $\tilde{\Sigma} = (\mathsf{Q}, \nabla, \mathcal{D}, \tilde{\mathscr{Y}}, \tilde{U})$. Clearly the controlled trajectories for Σ and $\tilde{\Sigma}$ agree, so we can, without loss of generality, assume that the input vector fields for an affine connection control system satisfy conditions (i) and (ii). Input vector fields satisfying these conditions at q_0 will be said to be ***adapted at*** $\boldsymbol{q_0}$. •

Let $X, Y \in \Gamma^\infty(\mathsf{TQ})$. If $X(q_0) = 0_{q_0}$, then the expression $\langle X : Y \rangle (q_0)$ may be verified (in coordinates, for example) to depend only on the value of Y at q_0. That is to say, we may define a linear map $\operatorname{sym}_X \colon \mathsf{T}_{q_0}\mathsf{Q} \to \mathsf{T}_{q_0}\mathsf{Q}$ by $v \mapsto \langle X : V \rangle (q_0)$, where V is any extension of $v \in \mathsf{T}_{q_0}\mathsf{Q}$ (see Exercise E7.23(a)). If $\mathscr{Y}$ is adapted at q_0, then we denote by $Z_{q_0}(\mathscr{Y})$ the set of linear maps

$\{\mathrm{sym}_{Y_1}, \ldots, \mathrm{sym}_{Y_m}\}$. Recalling from Page 29 the notation associated with the smallest subspace containing a given subspace and invariant under a family of linear maps, the subspace $\langle Z_{q_0}(\mathscr{Y}), \mathsf{S}_{q_0}\rangle$ will be of interest to us. As a point of reference for understanding this notation in the current context, the reader may wish to consult Exercise E7.4 for an alternative statement of the Kalman Rank Condition that employs the ideas of invariant subspaces.

8.2.2 First-order controllability results

Recall our notation of $\mathrm{Sym}^{(k)}(\mathcal{Y})_q$, $k \in \mathbb{Z}_+ \cup \infty$, from Section 7.3.2. Let us first state a sufficient condition for controllability.

Theorem 8.9 (Low-order sufficient condition for STLCC). *Let $\Sigma = (\mathsf{Q}, \nabla, \mathcal{D}, \mathscr{Y})$ be a C^∞-affine connection pre-control system with the vector fields $\mathscr{Y}$ adapted at q_0, and suppose that*

(i) $\mathrm{Sym}^{(2)}(\mathcal{Y})_{q_0} = \mathrm{Sym}^{(\infty)}(\mathcal{Y})$ *and that*

(ii) $B_{\mathcal{Y}_{q_0}}(\langle Z_{q_0}(\mathscr{Y}), \mathcal{Y}_{q_0}\rangle)$ *is essentially indefinite.*

Then Σ is properly STLC from q_0 if it is accessible from q_0 and is properly STLCC from q_0 if it is configuration accessible from q_0.

Proof. The proof will essentially follow from Corollary 7.41. However, the extension to allow singular points for the input distribution $\mathcal{Y}$ does not follow directly from Corollary 7.41, but requires some manipulations with the variational cone that we will not go through here. The idea, in essence, is that if an input vector field vanishes at the reference point, then directions generated by symmetric products using these vector fields come "for free." Since these symmetric products are simply applications of a linear map, this explains the presence of the invariant subspace characterizations of the tangent space to the reachable set. We refer to [Hirschorn and Lewis 2002, Lemma 7.2] for the details behind this, noting that the discussion in that paper builds on concepts presented in [Bianchini and Stefani 1993, Sussmann 1987]. The outcome of these discussions, once they are specialized to our setting, is the following result.

Lemma. *Let $(\mathsf{Q}, \nabla, \mathcal{D}, \mathscr{Y} = \{Y_1, \ldots, Y_m\})$ be a C^∞-affine connection pre-control system for which $\mathscr{Y}$ is adapted at $q_0 \in \mathsf{Q}$. Assume the following:*

(i) $\mathrm{Sym}^{(\infty)}(\mathcal{Y}) = \langle Z_{q_0}(\mathscr{Y}), \mathrm{Sym}^{(2)}(\mathcal{Y})_{q_0}\rangle$;

(ii) there exist $\tilde{m} \geq m$ and a maximal rank matrix $\boldsymbol{T} \in \mathbb{R}^{m \times \tilde{m}}$ such that, if $\tilde{Y}_\alpha = T^a_\alpha Y_a$, then

$$\sum_{a=1}^{\tilde{m}} \langle \tilde{Y}_\alpha : \tilde{Y}_\alpha\rangle(q_0) \in \langle Z_{q_0}(\mathscr{Y}), \mathcal{Y}_{q_0}\rangle.$$

Then $(\mathsf{Q}, \nabla, \mathcal{D}, \mathscr{Y})$ is properly STLC from q_0 if it is accessible from q_0 and is properly STLCC from q_0 if it is configuration accessible from q_0.

We shall show that, if the hypotheses of Theorem 8.9 are satisfied at q_0, then the hypotheses of the lemma are satisfied for some possibly different collection of input vector fields. From this, the conclusion of Theorem 8.9 will follow.

For brevity, let us write $\mathsf{S}_{q_0} = \langle Z_{q_0}(\mathscr{Y}), \mathcal{Y}_{q_0}\rangle$ and $B = B_{\mathcal{Y}_{q_0}}(\mathsf{S}_{q_0})$. First we need to find an appropriate collection of input vector fields. Choose $v_1, \dots, v_l \in \mathcal{Y}_{q_0}$ so that $0_{q_0} + \mathsf{S}_{q_0} \in \mathrm{Sym}^{(2)}(\mathcal{Y})_{q_0}/\mathsf{S}_{q_0}$ lies in the interior of the convex hull of the vectors $B(v_1, v_1), \dots, B(v_l, v_l)$. That this is possible is guaranteed by the hypotheses of Theorem 8.9 and by Lemma 8.3. If necessary, add vectors $v_{l+1}, \dots, v_{\tilde{k}}$ so that the vectors $v_1, \dots, v_{\tilde{k}}$ span $\mathcal{Y}_{q_0}$. It now follows that the vectors $B(v_1, v_1), \dots, B(v_{\tilde{k}}, v_{\tilde{k}})$ contain $0_{q_0} + \mathsf{S}_{q_0} \in \mathrm{Sym}^{(2)}(\mathcal{Y})_{q_0}/\mathsf{S}_{q_0}$ in the interior of their convex hull. Thus the vectors $v_1, \dots, v_{\tilde{k}}$ may be rescaled by strictly positive constants (for simplicity, let us denote the rescaled vectors also by $v_1, \dots, v_{\tilde{k}}$) so that

$$\sum_{a=1}^{\tilde{k}} B(v_a, v_a) = 0_{q_0} + \mathsf{S}_{q_0} \in \mathrm{Sym}^{(2)}(\mathcal{Y})_{q_0}/\mathsf{S}_{q_0}. \tag{8.3}$$

It is now possible to define vector fields $\tilde{\mathscr{Y}} = \{\tilde{Y}_1, \dots, \tilde{Y}_{\tilde{m}}\}$ so that if $\dim(\mathcal{Y}_{q_0}) = k$, then

1. $\tilde{Y}_{\tilde{k}+a} = Y_{k+a}$, $a \in \{1, \dots, m-k\}$, and
2. $\tilde{Y}_\alpha = \sum_{a=1}^{k} \tilde{T}^a_\alpha Y_a$, $\alpha \in \{1, \dots, \tilde{k}\}$, for a maximal rank matrix $\tilde{\boldsymbol{T}} \in \mathbb{R}^{k\times\tilde{k}}$.

Clearly this then implies the existence of a maximal rank matrix $\boldsymbol{T} \in \mathbb{R}^{m\times\tilde{m}}$ so that $\tilde{Y}_\alpha = T^a_\alpha Y_a$, $\alpha \in \{1, \dots, \tilde{m}\}$. From (8.3) it immediately follows that $(\mathsf{Q}, \nabla, \mathcal{D}, \mathscr{Y})$ satisfies the hypotheses of the Lemma, and so Theorem 8.9 follows. ■

A necessary condition for controllability is the following.

Theorem 8.10 (Low-order necessary condition for STLCC). *Let $\Sigma = (\mathsf{Q}, \nabla, \mathcal{D}, \mathscr{Y})$ be an analytic affine connection pre-control system and suppose that*

(i) q_0 is a regular point for the distribution $\mathcal{Y}$ and that
(ii) $B_{\mathcal{Y}_{q_0}}(\mathcal{Y}_{q_0})$ is definite and nonzero.

Then Σ is STLCUC from q_0.

Proof. We work locally. Therefore, we may assume that the sets of vector fields $\{Y_1, \dots, Y_m\}$ are linearly independent in a neighborhood of q_0. First we show that the system is not STLC from q_0 for any compact control set, using calculations of Hirschorn and Lewis [2002]. We will not provide here a self-contained justification for all of our computations, since they take considerable space, but we refer to the paper [Hirschorn and Lewis 2002]. The calculation uses the Chen–Fliess–Sussmann series [Chen 1957, Fliess 1981, Sussmann 1983]. For an analytic control-affine system with governing equations

$$\gamma'(t) = f_0(\gamma(t)) + \sum_{a=1}^{m} u^a(t) f_a(\gamma(t)), \qquad \gamma(t) \in \mathsf{M},$$

on a manifold M with a compact control set, and for an analytic function ϕ, the Chen–Fliess–Sussmann series gives the following formula for the value of ϕ along a controlled trajectory (γ, u):

$$\phi(\gamma(t)) = \sum_J U_J(t) f_J \phi(\gamma(0)).$$

The sum is over multi-indices $J = (a_1, \dots, a_k)$ in $\{0, 1, \dots, m\}$,

$$U_J(t) = \int_0^t u_{a_k}(t_k) \int_0^{t_k} u_{a_{k-1}}(t_{k-1}) \dots \int_0^{t_2} u_{a_1}(t_1)\, \mathrm{d}t_1 \dots \mathrm{d}t_{k-1}\, \mathrm{d}t_k.$$

and

$$f_J \phi = \mathscr{L}_{f_{a_1}} \mathscr{L}_{f_{a_2}} \cdots \mathscr{L}_{f_{a_k}} \phi.$$

We adopt the convention that $u_0 = 1$. We also regard an affine connection control system as a control-affine system in the usual manner by taking f_0 to be the geodesic spray for ∇ and $f_1, \dots, f_m$ to be the vertical lifts of $Y_1, \dots, Y_m$ (cf. Section 4.6.3).

The function we evaluate is defined as follows. We let λ be an analytic covector field defined in a neighborhood of q_0 with the following properties:

1. λ annihilates the distribution $\mathcal{Y}$;
2. $\lambda(q_0) B_{\mathcal{Y}_{q_0}} | \mathcal{Y}_{q_0}$ is negative-definite.

By a linear transformation applied to the inputs, one can ensure that the input vector fields diagonalize $\lambda(q_0) B_{\mathcal{Y}_{q_0}}$ with the diagonal entries being -1. We assume this input transformation to have been made. We then define a function ϕ_λ on TQ by $\phi_\lambda(v_q) = \lambda(q) \cdot v_q$, and we also define

$$\Phi_\lambda^+ = \{v_q \in \mathsf{TQ} \mid \phi_\lambda(v_q) > 0\}, \quad \Phi_\lambda^- = \{v_q \in \mathsf{TQ} \mid \phi_\lambda(v_q) < 0\}.$$

Note that, in any neighborhood $\mathcal{V}$ of 0_{q_0} in TQ, the sets $\mathcal{V} \cap \Phi_\lambda^-$ and $\mathcal{V} \cap \Phi_\lambda^+$ will be nonempty, since ϕ_λ is linear on the fibers of TQ. Therefore, we can show that $(\mathsf{Q}, \nabla, \mathcal{D}, \mathscr{Y})$ is STLUC from q_0 by showing that ϕ_λ has constant sign along any controlled trajectory. One may directly verify that ϕ_λ has the following properties:

1. $\mathscr{L}_{f_a} \phi_\lambda$, $a \in \{1, \dots, m\}$, is zero in a neighborhood of 0_{q_0};
2. $\mathscr{L}_{\mathrm{ad}_{f_0}^k f_a} \phi_\lambda(0_{q_0}) = 0$, $a \in \{1, \dots, m\}$, $k \in \mathbb{N}$;
3. $\mathscr{L}_{[f_a, [f_0, f_a]]} \phi_\lambda(0_{q_0}) = -1$, $a \in \{1, \dots, m\}$ (this and the next fact use the formula $[f_a, [f_0, f_b]] = \mathrm{vlft}(\langle Y_a : Y_b \rangle)$, $a, b \in \{1, \dots, m\}$);
4. $\mathscr{L}_{[f_a, [f_0, f_b]]} \phi_\lambda(0_{q_0}) = 0$, $a, b \in \{1, \dots, m\}$, $a \neq b$.

For an input $u\colon [0,T] \to U$ and $t \in [0,T]$, let us define

$$\|u\|_{2,t} = \max\left\{ \left(\int_0^t |u_a(t)|^2 \right)^{1/2} \;\middle|\; a \in \{1,\dots,m\} \right\}.$$

The calculations of Hirschorn and Lewis [2002] now immediately give the following inequality for $\phi_\lambda(\gamma'(t))$ along a controlled trajectory (γ,u) for an affine connection control system like that under consideration here:

$$\phi_\lambda(\gamma'(t)) \geq \tfrac{1}{2}(\|u\|_{2,t})^2 - |E(t)|.$$

Here $t \mapsto E(t)$ is a function that Hirschorn and Lewis [2002] show satisfies a bound $|E(t)| \leq tE_0(\|u\|_{2,t})^2$, for some $E_0 > 0$. For t sufficiently small, this shows that $\phi_\lambda(\gamma'(t))$ has constant sign. This shows that $(\mathsf{Q},\nabla,\mathcal{D},\mathscr{Y})$ is STLUC from q_0.

Now let us show that our above constructions also preclude the system from being locally *configuration* controllable. Choose a chart $(\mathcal{U},\chi)$ for Q around q_0 with the following properties: (1) $\chi(q_0) = \mathbf{0}$ and (2) $\mathrm{d}q^n(q_0) = \lambda(q_0)$. Let us define a function ψ_λ on the coordinate domain $\mathcal{U}$ by $\psi_\lambda(q) = q^n$ so that the sets

$$\Psi_\lambda^+ = \{q \in \mathsf{Q} \mid \psi_\lambda(q) > 0\}, \quad \Psi_\lambda^- = \{q \in \mathsf{Q} \mid \psi_\lambda(q) < 0\}$$

each intersect any neighborhood of $q_0 \in \mathsf{Q}$. Along any nonstationary trajectory $t \mapsto \gamma(t)$ we have

$$\left.\frac{\mathrm{d}\psi_\lambda(\gamma(t))}{\mathrm{d}t}\right|_{t=0} = \mathrm{d}\psi_\lambda(\gamma'(0)) = \phi_\lambda(\gamma'(0)) < 0.$$

Since $\psi_\lambda(q_0) = 0$, this means that, for sufficiently small positive t, $\psi_\lambda(\gamma(t)) < 0$. Since both γ and γ' are continuous, this shows that the points in Ψ_λ^+ are not reachable in small time, and so Σ is not locally configuration controllable. ■

8.2.3 Examples and discussion

Let us give an example that illustrates how Theorem 8.9 is an improvement over Theorem 7.40.

Example 8.11 (Exercise E7.14 cont'd). We consider the controllability of the planar rigid body of Section 7.4.2, but we take as input forces

$$F^1 = -h\mathrm{d}\theta - \sin\theta \mathrm{d}x + \cos\theta \mathrm{d}y, \quad F^2 = \mathrm{d}\theta.$$

(We have changed our numbering from that of Exercise E7.14.) These give the input vector fields

$$Y_1 = -\frac{h}{J}\frac{\partial}{\partial\theta} - \frac{\sin\theta}{m}\frac{\partial}{\partial x} + \frac{\cos\theta}{m}\frac{\partial}{\partial y}, \quad Y_2 = \frac{1}{J}\frac{\partial}{\partial\theta}.$$

These vector fields are linearly independent at all points in Q, so they are necessarily adapted at each point in Q. We therefore also have $\langle Z_{q_0}(\mathscr{Y}), \mathcal{Y}_{q_0}\rangle = \mathcal{Y}_{q_0}$. Since $\dim(\mathcal{Y}_{q_0}) = 2$ and $\dim(\mathsf{T}_{q_0}\mathsf{Q}) = 3$, the quotient $\mathsf{T}_{q_0}\mathsf{Q}/\mathcal{Y}_{q_0}$ is isomorphic to $\mathbb{R}$, and the isomorphism, by Exercise E2.1, is established by choosing a vector field complementary to the input distribution at each point. An example of such a vector field is

$$X = \cos\theta\frac{\partial}{\partial x} + \sin\theta\frac{\partial}{\partial y}.$$

We now compute

$$\langle Y_1 : Y_1\rangle = \frac{2h}{mJ}X, \quad \langle Y_1 : Y_2\rangle = \frac{1}{mJ}X, \quad \langle Y_2 : Y_2\rangle = 0.$$

We therefore conclude that the matrix representation of $B_{\mathcal{Y}_{q_0}}(\langle Z_{q_0}(\mathscr{Y}), \mathcal{Y}_{q_0}\rangle)$ relative to the basis $\{Y_1(q_0), Y_2(q_0)\}$ for $\mathcal{Y}_{q_0}$ and the basis $X(q_0) + \langle Z_{q_0}(\mathscr{Y}), \mathcal{Y}_{q_0}\rangle$ for $\mathsf{T}_{q_0}\mathsf{Q}/\langle Z_{q_0}(\mathscr{Y}), \mathcal{Y}_{q_0}\rangle$ is

$$[B_{\mathcal{Y}_{q_0}}(\langle Z_{q_0}(\mathscr{Y}), \mathcal{Y}_{q_0}\rangle)] = \begin{bmatrix} \frac{2h}{mJ} & \frac{1}{mJ} \\ \frac{1}{mJ} & 0 \end{bmatrix}.$$

By Exercise E2.22, $B_{\mathcal{Y}_{q_0}}(\langle Z_{q_0}(\mathscr{Y}), \mathcal{Y}_{q_0}\rangle)$ is essentially indefinite if and only if the two eigenvalues of the above matrix representation have different sign. This in turn will be the case if and only if the determinant of the matrix is negative. This, however, is readily seen to be the case, and so we conclude from Theorem 8.9 that the system is properly STLC from each $q_0 \in \mathsf{Q}$. Note that we were able to ascertain this without having to change the inputs as we did in Exercise E7.14. This reflects the fact that the essential indefiniteness condition is a basis free statement of the sufficient condition of Corollary 7.41. •

Now let us consider an example for which one of the input vector fields vanishes, so that we may see how this is handled by Theorem 8.9.

Example 8.12. The system we consider here is not physical, so that the concepts can be illustrated on as simple a system as possible. We take $\mathsf{Q} = \mathbb{R}^2$ with standard Cartesian coordinates (x, y), and with ∇ the affine connection defined by asking that its nonzero Christoffel symbols be

$$\Gamma^y_{xx} = 1.$$

The input vector fields we take to be

$$Y_1 = \frac{\partial}{\partial x}, \quad Y_2 = x\frac{\partial}{\partial y}.$$

The corresponding equations governing the affine connection control system are

$$\ddot{x} = u_1, \quad \ddot{y} = -\dot{x}^2 + xu_2.$$

Note that the input vector fields are adapted at $q_0 = (0,0)$. With respect to the standard basis for $\mathsf{T}_{q_0}\mathsf{Q}$, the matrix representation for sym_{Y_2} is readily computed to be

$$[\mathrm{sym}_{Y_2}] = \begin{bmatrix} 0 & 0 \\ 1 & 0 \end{bmatrix}.$$

One can see then that $\langle Z_{q_0}(\mathscr{Y}), \mathcal{Y}_{q_0}\rangle = \mathsf{T}_{q_0}\mathsf{Q}$, immediately implying that the system is properly STLC from q_0. •

Remark 8.13. The ideas in this section allow one to characterize in a fairly thorough way the controllability properties of systems with small numbers of inputs. For example, one can show that single-input analytic affine connection control systems are controllable only from a set of points forming an analytic set (i.e., a set that is defined locally by the intersection of the zeros of a finite number of analytic functions). The single-input case is considered by Lewis [1997], although the result proved there is not quite correct, in that an assumption concerning the regularity of the distribution $\mathrm{Sym}^{(1)}(\mathcal{Y})$ is neglected. What we have stated here is correct. The two-input case is more complicated, and is described, along with more details on the single-input case, by Tyner and Lewis [2004b]. •

8.3 Reductions of affine connection control systems

The controllability results of Section 8.2 turn out to apply to a great many examples. That is to say, many interesting physical examples may be shown to be controllable or uncontrollable using these results. What is not obvious is that many of these systems are describable, in some sense, by a driftless system, by which we mean a control-affine system $(\mathsf{Q}, \mathscr{C} = \{f_0, f_1, \dots, f_m\}, U)$ for which the drift vector field f_0 is identically zero. This effectively simplifies the system, making certain control design tasks, especially motion planning, considerably simpler. In this section we introduce the framework for discussing these simplifications. The work in this section was motivated by work in [Arai, Tanie, and Shiroma 1998, Lynch, Shiroma, Arai, and Tanie 2000]. However, the geometric approach was initiated by Bullo and Lynch [2001], and the approach here follows Bullo and Lewis [2003b].

8.3.1 Inputs for dynamic and kinematic systems

The thrust of this section is that we will compare the controlled trajectories of second-order systems on Q with those of first-order systems on Q. This requires being a little careful about the class of inputs used for the two types of systems. Let us first give reasons why one should differentiate between inputs for systems using forces as inputs, and inputs for systems using velocities as inputs.

Example 8.14. To illustrate this concept, let us consider the simplest mechanical system: a particle moving on the line under the influence of a control force. The control system thus has two states, configuration x and velocity v, and the governing equations are

$$\dot{x} = v, \quad \dot{v} = \frac{u}{m}, \tag{8.4}$$

where u is the input force and $m > 0$ is the mass of the particle. Supposing that u is locally integrable, and that we start the system at $t = 0$, the equation (8.4) is equivalent to the control system $\dot{x} = \tilde{u}$, where

$$\tilde{u}(t) = \frac{1}{m}\int_0^t u(s)\,ds.$$

Thus the system is reducible to a system with a *locally absolutely continuous* velocity input $\tilde{u}$. However, the trajectories of the problem $\dot{x} = \tilde{u}$, where $\tilde{u}$ is only locally integrable are not realizable as trajectories of (8.4) with bounded inputs. There is a physical interpretation of this: locally integrable inputs for velocity imply that one can cause the velocity of the system to change instantaneously. However, the system's inertia will not allow this motion to take place with bounded forces. •

The above example shows that, for kinematic and dynamic problems, one must consider different classes of inputs. In this section we shall fix a class of inputs for mechanical systems (i.e., those that are dynamic), and from these derive what must be the corresponding inputs for the kinematic system (in cases in which there is even a correspondence between the two). Unsurprisingly, given the example above, the inputs for the kinematic system are roughly the integrals of those for the mechanical system.

For the mechanical system, we shall consider inputs $u\colon [0,T] \to \mathbb{R}^m$ that are locally integrable. Let us denote such inputs by $\mathscr{U}^m_{\mathrm{dyn}}$. It will turn out that the corresponding inputs to use for kinematic systems are locally absolutely continuous functions $\tilde{u}\colon [0,T] \to \mathbb{R}^{\tilde{m}}$, which we denote by $\mathscr{U}^{\tilde{m}}_{\mathrm{kin}}$. Note that $\mathscr{U}^m_{\mathrm{kin}} \subset \mathscr{U}^m_{\mathrm{dyn}}$.

An idea that will come up in this section is that of the reparameterization of a curve.

Definition 8.15 (Reparameterization). Given an interval $I \subset \mathbb{R}$, a ***reparameterization of I*** is a map $\tau\colon J \to I$ with the properties

(i) τ is surjective,
(ii) τ is either strictly increasing (i.e., $\tau(t_1) > \tau(t_2)$ if $t_1 > t_2$) or strictly decreasing (i.e., $\tau(t_1) < \tau(t_2)$ if $t_1 > t_2$), and
(iii) τ is locally absolutely continuous.

If $\gamma\colon I \to \mathsf{Q}$ is a locally absolutely continuous curve, a ***reparameterization of*** $\boldsymbol{\gamma}$ is a curve $\gamma \circ \tau\colon J \to \mathsf{Q}$, where τ is a reparameterization of I. •

Since $(\gamma \circ \tau)'(t) = \tau'(t)\gamma'(\tau(t))$, the image of a reparameterized curve is the same as that of the original curve. What changes is how quickly or slowly one moves along the curve. The following lemma is easily proved, and is implicitly assumed in some parts of the ensuing discussion, and also in parts of Chapter 13.

Lemma 8.16. *Let $X \in \Gamma^\infty(\mathsf{TQ})$ and let $\gamma\colon I \to \mathsf{Q}$ be the integral curve of X at q_0. For a curve $\tilde{\gamma}\colon J \to \mathsf{Q}$, the following statements are equivalent:*

(i) $\tilde{\gamma} = \gamma \circ \tau$ for a reparameterization τ of γ;

(ii) there exists a locally integrable function $u\colon J \to \mathbb{R}$ such that $\tilde{\gamma}$ is an integral curve of the time-dependent vector field $(t, q) \mapsto u(t)X(q)$ at (t_0, q_0), for some $t_0 \in J$.

8.3.2 Kinematic reductions

In Section 4.6.1 we said that a driftless system is a control-affine system for which the drift vector field is zero. Since the ideas in this section rely heavily on driftless systems, let us introduce some associated elementary language and notation. In this section, all data are required to be C^∞ unless otherwise stated.

In order to emphasize the difference between the two kinds of systems we are comparing, we shall denote an affine connection control system by $\Sigma_{\mathrm{dyn}} = (\mathsf{Q}, \nabla, \mathcal{D}, \mathscr{Y}, \mathbb{R}^m)$. A ***driftless system*** is a triple $\Sigma_{\mathrm{kin}} = (\mathsf{Q}, \mathscr{X} = \{X_1, \dots, X_{\tilde{m}}\}, U)$, where $X_1, \dots, X_{\tilde{m}} \in \Gamma^\infty(\mathsf{TQ})$ and $U \subset \mathbb{R}^{\tilde{m}}$. The associated control system is then

$$\gamma'(t) = \sum_{\alpha=1}^{\tilde{m}} \tilde{u}^\alpha(t) X_\alpha(\gamma(t)), \tag{8.5}$$

so that a ***controlled trajectory*** is a pair $(\gamma, \tilde{u})$, where

1. $\gamma\colon I \to \mathsf{Q}$ and $\tilde{u}\colon I \to U$ are both defined on the same interval $I \subset \mathbb{R}$,
2. $\tilde{u} \in \mathscr{U}_{\mathrm{kin}}^{\tilde{m}}$, and
3. $(\gamma, \tilde{u})$ together satisfy (8.5).

A driftless system $(\mathsf{Q}, \mathscr{X}, U)$ is ***STLC from q_0*** if the set of points reachable from q_0 contains q_0 in its interior, and a pair $(\mathsf{Q}, \mathscr{X})$ is ***properly STLC from q_0*** if $(\mathsf{Q}, \mathscr{X}, U)$ is STLC from q_0 for every proper U. We refer the reader to Corollary 7.25 for conditions concerning the controllability of driftless systems; familiarity with this result is assumed in this section.

Let us define what we mean by a kinematic reduction. In the following definition, and in the remainder of this chapter, we use the notation, already introduced above, of denoting the distribution generated by $\mathscr{Y}$ (resp. $\mathscr{X}$) by $\mathcal{Y}$ (resp. $\mathcal{X}$).

Definition 8.17 (Kinematic reduction). Let $\Sigma_{\mathrm{dyn}} = (\mathsf{Q}, \nabla, \mathcal{D}, \mathscr{Y}, \mathbb{R}^m)$ be a C^∞-affine connection control system with $\mathcal{Y}$ having locally constant rank. A driftless system $\Sigma_{\mathrm{kin}} = (\mathsf{Q}, \mathscr{X}, \mathbb{R}^{\tilde{m}})$ is a ***kinematic reduction*** of Σ_{dyn} if

(i) $\mathcal{X}$ is a locally constant rank subbundle of $\mathcal{D}$ and if
(ii) for every controlled trajectory (γ, u_{kin}) for Σ_{kin} with $u_{\text{kin}} \in \mathscr{U}_{\text{kin}}^{\tilde{m}}$, there exists $u_{\text{dyn}} \in \mathscr{U}_{\text{dyn}}^{m}$ such that (γ, u_{dyn}) is a controlled trajectory for Σ_{dyn}.

The ***rank*** of the kinematic reduction Σ_{kin} at q is the rank of $\mathcal{X}$ at q. •

Thus kinematic reductions are driftless systems whose controlled trajectories, at least for controls in $\mathscr{U}_{\text{kin}}^{\tilde{m}}$, are also controlled trajectories of Σ_{dyn}, possibly up to reparameterization. Let us characterize kinematic reductions. To do so, recall that with our locally constant rank assumptions, given an affine connection ∇ and a family of vector fields $\mathscr{Y} = \{Y_1, \dots, Y_m\}$ on Q, we may globally define $B_{\mathcal{Y}}$ as at the end of Section 8.1.2. This also allows us to define a map $Q_{B_{\mathcal{Y}}} : \Gamma^\infty(\mathsf{TQ}) \to \Gamma^\infty(\mathsf{TQ}/\mathcal{Y})$ by

$$Q_{B_{\mathcal{Y}}}(X)(q) = B_{\mathcal{Y}}(q)(X(q), X(q)).$$

With this notation, we have the following result.

Theorem 8.18 (Characterization of kinematic reductions). *Let $\Sigma_{\text{dyn}} = (\mathsf{Q}, \nabla, \mathcal{D}, \mathscr{Y} = \{Y_1, \dots, Y_m\}, \mathbb{R}^m)$ be a C^∞-affine connection control system with $\mathcal{Y}$ of locally constant rank and let $\Sigma_{\text{kin}} = (\mathsf{Q}, \mathscr{X} = \{X_1, \dots, X_{\tilde{m}}\}, \mathbb{R}^{\tilde{m}})$ be a driftless system with $\mathcal{X}$ of locally constant rank. The following statements are equivalent:*
(i) Σ_{kin} is a kinematic reduction of Σ_{dyn};
(ii) $\mathrm{Sym}^{(1)}(\mathcal{X}) \subset \mathcal{Y}$;
(iii) $\mathcal{X} \subset \mathcal{Y}$ and $Q_{B_{\mathcal{Y}}}|\mathcal{X} = 0$.

Proof. (i) $\Longrightarrow$ (ii) Let $X \in \Gamma^\infty(\mathcal{X})$ so that $X = \phi^\alpha X_\alpha$ for some $\phi^1, \dots, \phi^{\tilde{m}} \in C^\infty(\mathsf{Q})$ (cf. Exercise E3.42). For $q_0 \in \mathsf{Q}$, define controls $\tilde{u}_1, \tilde{u}_2 \in \mathscr{U}_{\text{kin}}^{\tilde{m}}$ by $\tilde{u}_1(t) = (\phi^1(q_0), \dots, \phi^{\tilde{m}}(q_0))$ and $\tilde{u}_2(t) = (1+t)\tilde{u}_1(t)$. Thus $\tilde{u}_1$ is a constant control and $\tilde{u}_2$ is a linear function of t. Let $(\gamma_1, \tilde{u}_1)$ and $(\gamma_2, \tilde{u}_2)$ be the corresponding controlled trajectories of Σ_{kin} satisfying $\gamma_1(0) = \gamma_2(0) = q_0$. Thus $\gamma_i'(t) = \tilde{u}_i^\alpha(t) X_\alpha(\gamma_i(t))$, $i \in \{1,2\}$. (In this formula, and in what follows, we use the summation convention.) We compute

$$\begin{aligned}\nabla_{\gamma_1'(t)}\gamma_1'(t) &= \nabla_{\tilde{u}_1^\alpha(t) X_\alpha(\gamma_1(t))} \tilde{u}_1^\beta(t) X_\beta(\gamma_1(t)) \\ &= \tilde{u}_1^\alpha(t)\tilde{u}_1^\beta(t) \nabla_{X_\alpha(\gamma_1(t))} X_\beta(\gamma_1(t)) + \dot{\tilde{u}}_1^\beta(t) X_\beta(\gamma_1(t)).\end{aligned}$$

Evaluating this at $t = 0$ gives

$$\nabla_{\gamma_1'(t)}\gamma_1'(t)\big|_{t=0} = \tilde{u}_1^\alpha(0)\tilde{u}_1^\beta(0)\nabla_{X_\alpha} X_\beta(q_0) + \dot{\tilde{u}}_1^\beta(0) X_\beta(q_0) = \nabla_X X(q_0).$$

Similarly, for γ_2, we have

$$\nabla_{\gamma_2'(t)}\gamma_2'(t)\big|_{t=0} = \nabla_X X(q_0) + X(q_0).$$

Therefore, since Σ_{kin} is a kinematic reduction of Σ_{dyn}, we have $\nabla_X X(q_0), \nabla_X X(q_0) + X(q_0) \in \mathcal{Y}_{q_0}$, or simply $X, \nabla_X X \in \Gamma^\infty(\mathcal{Y})$, since the above constructions can be performed for all $X \in \Gamma^\infty(\mathcal{X})$ and $q_0 \in \mathsf{Q}$. Therefore, for $X, Y \in \Gamma^\infty(\mathcal{X})$ we have the polarization identity,

$$\langle X : Y \rangle = \tfrac{1}{2}\big(\langle X+Y : X+Y\rangle - \langle X : X\rangle - \langle Y : Y\rangle\big) \in \Gamma^\infty(\mathcal{Y}), \tag{8.6}$$

which gives (ii).

(ii) $\Longrightarrow$ (iii) From the definition of $B_{\mathcal{Y}}$ we readily see that $Q_{B_{\mathcal{Y}}}|\mathcal{X} = 0$ exactly means that $\langle X : X\rangle = 2\nabla_X X \in \Gamma^\infty(\mathcal{Y})$ for each $X \in \Gamma^\infty(\mathcal{X})$. From this observation, the current implication follows easily by employing the formula for $\langle X : Y\rangle$ in (8.6).

(iii) $\Longrightarrow$ (i) As in the preceding step, we saw that the condition $Q_{B_{\mathcal{Y}}}|\mathcal{X} = 0$ is equivalent to asserting that $\nabla_X X \in \Gamma^\infty(\mathcal{Y})$ for each $X \in \Gamma^\infty(\mathcal{X})$. By (8.6), this implies that $\langle X_\alpha : X_\beta\rangle \in \Gamma^\infty(\mathcal{Y})$ for $\alpha, \beta \in \{1, \dots, \tilde{m}\}$. Let $u_{\mathrm{kin}} \in \mathscr{U}^{\tilde{m}}_{\mathrm{kin}}$ and let $(\gamma, u_{\mathrm{kin}})$ be a corresponding controlled trajectory for Σ_{kin}. We then have

$$\nabla_{\gamma'(t)}\gamma'(t) = u^\alpha_{\mathrm{kin}}(t)u^\beta_{\mathrm{kin}}(t)\nabla_{X_\alpha(\gamma(t))}X_\beta(\gamma(t)) + \dot{u}^\alpha_{\mathrm{kin}}(t)X_\alpha(\gamma(t)).$$

We note that

$$u^\alpha_{\mathrm{kin}}(t)u^\beta_{\mathrm{kin}}(t)\nabla_{X_\alpha(\gamma(t))}X_\beta(\gamma(t)) = \tfrac{1}{2}u^\alpha_{\mathrm{kin}}(t)u^\beta_{\mathrm{kin}}(t)\big\langle X_\alpha(\gamma(t)) : X_\beta(\gamma(t))\big\rangle.$$

Since $X_\alpha, \langle X_\alpha : X_\beta\rangle \in \Gamma^\infty(\mathcal{Y})$, it now follows that $\nabla_{\gamma'(t)}\gamma'(t) \in \mathcal{Y}_{\gamma(t)}$, implying that there exists a control $u_{\mathrm{dyn}} \in \mathscr{U}^m_{\mathrm{dyn}}$ such that $(\gamma, u_{\mathrm{dyn}})$ is a controlled trajectory for Σ_{dyn}. ■

Of particular interest are kinematic reductions of rank-one: $(\mathsf{Q}, \{X_1\}, \mathbb{R})$. In this case, any vector field of the form $X = \phi X_1$, where $\phi \in C^\infty(\mathsf{Q})$ is nowhere vanishing, is called a ***decoupling vector field***. From Theorem 8.18 we have the following description of a decoupling vector field.

Corollary 8.19 (Characterization of decoupling vector fields). *A vector field $X \in \Gamma^\infty(\mathsf{TQ})$ is a decoupling vector field for $\Sigma_{\mathrm{dyn}} = (\mathsf{Q}, \nabla, \mathcal{D}, \mathscr{Y}, \mathbb{R}^m)$ if and only if $X, \nabla_X X \in \Gamma^\infty(\mathcal{Y})$.*

Remarks 8.20. 1. It is the notion of a decoupling vector field that was initially presented by Bullo and Lynch [2001], and which is generalized by the idea of a kinematic reduction. In the Bullo and Lynch definition, a decoupling vector field is a vector field whose integral curves, and any reparameterization of these integral curves, can be followed by a controlled trajectory of the dynamic system. That this definition is equivalent to ours is demonstrated easily, and we leave this to the reader as Exercise E8.3.

2. While in general, even when a kinematic reduction exists, it will not be easy to find, it turns out in practice that many examples exhibit kinematic reductions in a more or less obvious way. We shall see this in the examples below. Note that condition (iii) of Theorem 8.18 provides a set of algebraic equations that can, in principle, be solved to identify decoupling

vector fields. This is discussed by Bullo and Lynch [2001]. We comment that, in general, the resulting computational problem is intractable, for much the same reasons that the computational problem of checking the indefiniteness of a vector-valued quadratic form is intractable. •

Next, let us consider affine connection control systems endowed with multiple kinematic reductions. It is interesting to characterize when the concatenation of controlled trajectories of the kinematic reductions gives rise to a controlled trajectory for the affine connection control system. The following lemma follows immediately from the definition of a kinematic reduction.

Lemma 8.21. *Consider a C^∞-affine connection control system $\Sigma_{\text{dyn}} = (\mathsf{Q}, \nabla, \mathcal{D}, \mathscr{Y}, \mathbb{R}^m)$ with two kinematic reductions $\Sigma_{\text{kin},1} = (\mathsf{Q}, \mathscr{X}_1, \mathbb{R}^{m_1})$ and $\Sigma_{\text{kin},2} = (\mathsf{Q}, \mathscr{X}_2, \mathbb{R}^{m_2})$. For $i \in \{1,2\}$, let $(\gamma_i, u_{\text{kin},i})$ be a controlled trajectory for $\Sigma_{\text{kin,i}}$ defined on the interval $[0, T_i]$ with $u_{\text{kin},i} \in \mathscr{U}_{\text{kin}}^{m_i}$. There exists a control $u_{\text{dyn}} \in \mathscr{U}_{\text{dyn}}^m$ such that $(\gamma_1 * \gamma_2, u_{\text{dyn}})$ is a controlled trajectory for Σ_{dyn} if and only if $\gamma_1'(T_1) = \gamma_2'(0)$.*

Let us consider an example of a system where kinematic reductions appear in a more or less obvious manner.

Example 8.22 (Planar rigid body). We consider again the planar rigid body with a variable-direction thruster. We refer the reader to Section 7.4.2 for the description of the system data upon which we base our discussion here. We claim that the vector fields

$$X_1 = \cos\theta \frac{\partial}{\partial x} + \sin\theta \frac{\partial}{\partial y}, \qquad X_2 = -\frac{mh}{J}\frac{\partial}{\partial \theta} - \sin\theta\frac{\partial}{\partial x} + \cos\theta\frac{\partial}{\partial y}$$

are decoupling vector fields. Since $X_1 = mY_1$ and $X_2 = mY_2$, clearly they are sections of $\mathcal{Y}$. We also compute

$$\nabla_{X_1} X_1 = 0, \quad \nabla_{X_2} X_2 = \frac{mh\cos\theta}{J}\frac{\partial}{\partial x} + \frac{mh\sin\theta}{J}\frac{\partial}{\partial y}.$$

Therefore, $\nabla_{X_1} X_1, \nabla_{X_2} X_2 \in \Gamma^\infty(\mathcal{Y})$, showing that X_1 and X_2 are indeed decoupling vector fields.

Let us explore, within the confines of the example for the moment, the implications of the existence of these decoupling vector fields. Since X_1 and X_2 are decoupling vector fields, we may follow their integral curves. In Figure 8.1 we show sample motions of the body along a pair of integral curves. In actuality, one can follow not only the integral curves of the decoupling vector fields, but any reparameterization of these integral curves. With this in mind, here is a following possible methodology for moving the body around in the plane.

1. Given $q_1, q_2 \in \mathsf{Q}$, find a concatenation of the integral curves of X_1 and X_2 that connects q_1 with q_2. (This is possible if the involutive closure of the distribution generated by $\{X_1, X_2\}$ is TQ, which is in fact the case in this example.)

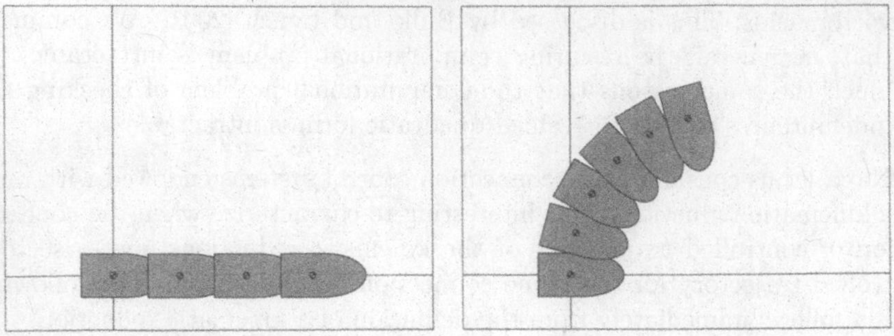

Figure 8.1. Decoupling motions for the planar rigid body: X_1 on the left and X_2 on the right

2. Reparameterize each segment of the preceding concatenated curve so that each segment has zero initial and final velocity.
3. The resulting reparameterized curve can be followed by controlled trajectories of Σ_{dyn}.

This procedure is further explored in Chapter 13. •

Motivated by this example we introduce the following definition.

Definition 8.23 (Kinematic controllability). A C^∞-affine connection control system $\Sigma_{\mathrm{dyn}} = (\mathsf{Q}, \nabla, \mathcal{D}, \mathscr{Y}, \mathbb{R}^m)$ is ***kinematically controllable*** from $q_0 \in \mathsf{Q}$ (***KC*** from $q_0 \in \mathsf{Q}$) if there exists a finite collection

$$\Sigma_{\mathrm{kin},1} = (\mathsf{Q}, \mathscr{X}_1, \mathbb{R}^{m_1}), \dots, \Sigma_{\mathrm{kin},k} = (\mathsf{Q}, \mathscr{X}_k, \mathbb{R}^{m_k})$$

of kinematic reductions for Σ_{dyn} such that $(\mathsf{Q}, \mathscr{X}_1 \cup \cdots \cup \mathscr{X}_k)$ is properly STLC from q_0. •

Remarks 8.24. 1. We recall from Corollary 7.25 that, for analytic systems, the condition that $(\mathsf{Q}, \mathscr{X}_1 \cup \cdots \cup \mathscr{X}_k)$ be properly STLC from q_0 is equivalent to the condition that $\mathrm{Lie}^{(\infty)}(\mathcal{X}_1 + \cdots + \mathcal{X}_k)_{q_0} = \mathsf{T}_{q_0}\mathsf{Q}$.

2. If an affine connection control system $\Sigma_{\mathrm{dyn}} = (\mathsf{Q}, \nabla, \mathcal{D}, \mathscr{Y}, \mathbb{R}^m)$ is kinematically controllable from q_0, then it is STLCC from q_0. This fact is proved in Proposition 8.31 below, and we refer to Section 8.4 for a discussion of the relationships between the various notions of controllability introduced in this section and in Section 7.2.
3. Suppose that the affine connection control system $\Sigma_{\mathrm{dyn}} = (\mathsf{Q}, \nabla, \mathcal{D}, \mathscr{Y}, \mathbb{R}^m)$ is kinematically controllable from all $q \in \mathsf{Q}$. A standard control problem is to find a controlled trajectory connecting two given configurations $q_1, q_2 \in \mathsf{Q}$, starting and ending with zero velocity. Lemma 8.21 says that this can be done for Σ_{dyn} by concatenating integral curves of decoupling vector fields for which each segment is reparameterized to start and end at zero velocity. This is the point of view of Bullo and Lynch [2001]. •

8.3.3 Maximally reducible systems

If $\Sigma_{\text{kin}} = (\mathsf{Q}, \mathscr{X}, \mathbb{R}^{\tilde{m}})$ is a kinematic reduction of $\Sigma_{\text{dyn}} = (\mathsf{Q}, \nabla, \mathcal{D}, \mathscr{Y}, \mathbb{R}^m)$, then, by definition, any controlled trajectory of Σ_{kin} may be followed by a controlled trajectory of Σ_{dyn}. In this section we wish to consider the possibility of the converse statement. The following definition, and the attendant Theorem 8.27 below, are due to Lewis [1999].

Definition 8.25 (Maximal reducibility). A C^∞-affine connection control system $\Sigma_{\text{dyn}} = (\mathsf{Q}, \nabla, \mathcal{D}, \mathscr{Y}, \mathbb{R}^m)$ with $\mathcal{Y}$ locally constant rank is ***maximally reducible*** to $\Sigma_{\text{kin}} = (\mathsf{Q}, \mathscr{X}, \mathbb{R}^{\tilde{m}})$ if Σ_{kin} is a kinematic reduction of Σ_{dyn} and if, for every controlled trajectory (γ, u_{dyn}) for Σ_{dyn} satisfying $\gamma'(0) \in \mathcal{X}_{\gamma(0)}$, there exists a control $u_{\text{kin}} \in \mathscr{U}_{\text{kin}}^{\tilde{m}}$ such that (γ, u_{kin}) is a controlled trajectory for Σ_{kin}. •

Before we proceed to characterize maximally reducible systems, let us illustrate that a system may not be maximally reducible to a given kinematic reduction.

Example 8.26 (Planar rigid body (cont'd)). We claim that the affine connection control system corresponding to the planar rigid body with a variable-direction thruster is not maximally reducible to either of the kinematic reductions $\Sigma_{\text{kin},1} = (\mathsf{Q}, \mathscr{X}_1 = \{X_1\}, \mathbb{R})$ and $\Sigma_{\text{kin},2} = (\mathsf{Q}, \mathscr{X}_2 = \{X_2\}, \mathbb{R})$ exhibited in Example 8.22. We shall exhibit this explicitly for $\Sigma_{\text{kin},1}$, and leave the other case to the reader.

Consider the control $t \mapsto u(t) = (0,1) \in \mathscr{U}_{\text{dyn}}^2$ along with the initial condition $\gamma'(0)$ represented in coordinates (θ, x, y) by $((0,0,0),(0,1,0))$. We have $\gamma'(0) \in \mathcal{X}_{1,\gamma(0)}$, where $\mathcal{X}_1$ is the regular distribution generated by the vector field X_1. If Σ_{dyn} is to be maximally reducible to $\Sigma_{\text{kin},1}$, then we should have $\gamma'(t) \in \mathcal{X}_{1,\gamma(t)}$ for each $t > 0$. To show that this is not the case, consider the governing equations for the system with the given control:

$$\ddot{\theta} = -\frac{h}{J}, \quad \ddot{x} = -\frac{\sin\theta}{m}, \quad \ddot{y} = \frac{\cos\theta}{m}.$$

Clearly the solution to this ordinary differential equation is not a reparameterization of the integral curve for X_1 through $\gamma(0)$, since the latter is given by $t \mapsto (0,t,0)$. Thus it cannot be that $\gamma'(t) \in \mathcal{X}_{1,\gamma(t)}$ for each $t > 0$. •

Now let us establish when an affine connection control system is in fact maximally reducible to *some* driftless system. Note that in the statement of the following theorem, the driftless systems to which Σ_{dyn} is maximally reducible are characterized sharply.

Theorem 8.27. *A C^∞-affine connection control system $\Sigma_{\text{dyn}} = (\mathsf{Q}, \nabla, \mathcal{D}, \mathscr{Y}, \mathbb{R}^m)$ with $\mathcal{Y}$ locally constant rank is maximally reducible to $\Sigma_{\text{kin}} = (\mathsf{Q}, \mathscr{X}, \mathbb{R}^{\tilde{m}})$ if and only if the following two conditions hold:*

(i) $\mathcal{X} = \mathcal{Y}$;

(ii) $\mathrm{Sym}^{(\infty)}(\mathcal{Y}) = \mathcal{Y}$.

Proof. It will be useful to recall from the proof of Theorem 3.108 that a vector field X is a section of a distribution $\mathcal{D}$ if and only if $\mathrm{vlft}(X)$ is tangent to $\mathcal{D} \subset \mathsf{TQ}$.

First suppose that Σ_{dyn} is maximally reducible to a driftless system Σ_{kin}. Let $\gamma\colon [0,T] \to \mathsf{Q}$ be a geodesic so that $(\gamma', 0)$ is a controlled trajectory for Σ_{dyn}. If we ask that $\gamma'(0) \in \mathcal{X}$, then Definition 8.25 implies that there exists $u_{\mathrm{kin}} \in \mathscr{U}_{\mathrm{kin}}^{\tilde{m}}$ such that $(\gamma, u_{\mathrm{kin}})$ is a controlled trajectory of Σ_{dyn}. Indeed, u_{kin} is defined by

$$\gamma'(t) = u_{\mathrm{kin}}^{\alpha}(t) X_{\alpha}(\gamma(t))$$

and so is smooth. Further, this implies that $\mathcal{X}$ is geodesically invariant. The remainder of this part of the proof will be directed towards showing that $\mathcal{X} = \mathcal{Y}$.

Let $\boldsymbol{e}_a$ be the ath standard basis vector for $\mathbb{R}^m$ and let $u_a\colon [0,T] \to \mathbb{R}^m$ be the control defined by $u_a(t) = \boldsymbol{e}_a$. If $\Upsilon\colon [0,T] \to \mathsf{TQ}$ is an integral curve for the vector field $S + \mathrm{vlft}(Y_a)$, then (Υ, u_a) is a controlled trajectory for $(\mathsf{TQ}, \mathscr{C}_{\Sigma_{\mathrm{dyn}}}, \mathbb{R}^m)$, the control-affine system defined by Σ_{dyn} (see Section 4.6). By Definition 8.25, Υ must be tangent to $\mathcal{X}$. Since $\mathcal{X}$ is geodesically invariant, S is tangent to $\mathcal{X}$, therefore, $\mathrm{vlft}(Y_a)$ must be tangent to $\mathcal{X}$. This implies that $\mathcal{Y} \subset \mathcal{X}$.

To show that $\mathcal{X} \subset \mathcal{Y}$ we employ the following lemma.

Lemma. *If a regular distribution $\mathcal{D}$ is geodesically invariant for an affine connection ∇, then, for each $q \in \mathsf{Q}$ and each $v_q \in \mathcal{D}_q$, there exist $T > 0$ and a smooth curve $\gamma\colon [0,T] \to \mathsf{Q}$ with the following properties:*

(i) $\gamma'(t) \in \mathcal{D}_{\gamma(t)}$ *for* $t \in]0,T]$;

(ii) $\nabla_{\gamma'(0)}\gamma'(0) = v_q$.

Proof. Let $(\mathcal{U}, \phi)$ be a normal coordinate chart [Kobayashi and Nomizu 1963, volume 1, Proposition 8.4] with $\phi(q) = \mathbf{0}$. In such a chart, the Christoffel symbols for ∇ satisfy $\Gamma^i_{jk}(\mathbf{0}) + \Gamma^i_{kj}(\mathbf{0}) = 0$, $i,j,k \in \{1,\dots,n\}$. Let $\tilde{T} > 0$ be as small as necessary and let $\tilde{\gamma}\colon [0,\tilde{T}] \to \mathsf{Q}$ be the geodesic satisfying $\tilde{\gamma}'(0) = v_q$. Let us denote the local representative of $\tilde{\gamma}$ in our normal coordinate chart by $t \mapsto (\tilde{q}^1(t), \dots, \tilde{q}^n(t))$. We must then have $\ddot{\tilde{q}}^i(0) = 0$, $i \in \{1,\dots,n\}$, since $\tilde{\gamma}$ is a geodesic and we are using normal coordinates. Since $\mathcal{D}$ is geodesically invariant, $\tilde{\gamma}'(t) \in \mathcal{D}_{\tilde{\gamma}(t)}$ for $t \in]0,\tilde{T}]$. Now define $\tau\colon [0,\tilde{T}] \to [0, \frac{1}{2}\tilde{T}^2]$ by $\tau(t) = \frac{1}{2}t^2$. Let $T = \frac{1}{2}\tilde{T}^2$, define $\gamma\colon [0,T] \to \mathsf{Q}$ by $\gamma = \tilde{\gamma} \circ \tau$, and denote by $t \mapsto (q^1(t), \dots, q^n(t))$ the local representative of γ. Then we have

$$\dot{q}^i(t) = t\dot{\tilde{q}}^i(t), \qquad i \in \{1,\dots,n\},$$
$$\ddot{q}^i(0) = \dot{\tilde{q}}^i(0), \qquad i \in \{1,\dots,n\}.$$

Since $\tilde{\gamma}'(0) = v_q$, the result follows. ▼

Now let $q \in \mathsf{Q}$ and $X \in \mathscr{X}_q$. Choose a curve $\gamma\colon [0,T] \to \mathsf{Q}$ as in the lemma. Define a smooth map $u_{\mathrm{kin}}\colon [0,T] \to \mathbb{R}^{\tilde{m}}$ by asking that it satisfy

$$\gamma'(t) = u^{\alpha}_{\mathrm{kin}}(t) X_{\alpha}(\gamma(t)).$$

Then $(\gamma, u_{\mathrm{kin}})$ is a controlled trajectory for Σ_{kin}. Therefore, by Definition 8.25, there exists a map $u_{\mathrm{dyn}}\colon [0,T] \to \mathbb{R}^m$ such that $(\gamma', u_{\mathrm{dyn}})$ is a controlled trajectory for $(\mathsf{TQ}, \mathscr{C}_{\Sigma_{\mathrm{dyn}}}, \mathbb{R}^m)$. Indeed, since γ' is smooth, u_{dyn} will also be smooth. Furthermore, we have

$$X = \nabla_{\gamma'(0)}\gamma'(0) = u^{a}_{\mathrm{dyn}}(0) Y_a(\gamma(0)).$$

This shows that $\mathscr{X} \subset \mathscr{Y}$, which completes the proof of the "only if" part of the theorem.

Now suppose that parts (i) and (ii) of the theorem hold. Let us work locally, so we may as well assume that the vector fields $\{Y_1, \ldots, Y_m\}$ and $\{X_1, \ldots, X_{\tilde{m}}\}$ are linearly independent (and so $\tilde{m} = m$). First, (ii) implies $\mathscr{Y}$ is an invariant submanifold for the system $(\mathsf{TQ}, \mathscr{C}_{\Sigma_{\mathrm{dyn}}}, \mathbb{R}^m)$, since $\mathrm{vlft}(Y_a)$, $a \in \{1, \ldots, m\}$, is tangent to $\mathscr{Y}$. If $(\Upsilon, u_{\mathrm{dyn}})$ is a controlled trajectory of $(\mathsf{TQ}, \mathscr{C}_{\Sigma_{\mathrm{dyn}}}, \mathbb{R}^m)$, then $\Upsilon\colon [0,T] \to \mathsf{TQ}$ is locally absolutely continuous, and so $\gamma \triangleq \pi_{\mathsf{TQ}} \circ \Upsilon$ is also locally absolutely continuous. In fact, $\Upsilon = \gamma'$ and so not only is γ locally absolutely continuous, but γ' is locally absolutely continuous. If we further suppose that $\gamma'(0) \in \mathscr{Y}_{\gamma(0)}$, then $\gamma'(t) \in \mathscr{Y}_{\gamma(t)}$ for $t \in [0,T]$. We may then define $u_{\mathrm{kin}}\colon [0,T] \to \mathbb{R}^{\tilde{m}}$ by

$$\gamma'(t) = u^{\alpha}_{\mathrm{kin}}(t) X_{\alpha}(\gamma(t)),$$

which uniquely defines u_{kin}, since $(\mathsf{TQ}, \mathscr{C}_{\Sigma_{\mathrm{dyn}}}, \mathbb{R}^m)$ leaves $\mathscr{Y}$, and hence $\mathscr{X}$, invariant. It remains to show that u_{kin} is locally absolutely continuous. To prove this, we work locally around $q_0 = \gamma(0)$ and extend the vector fields $\{X_1, \ldots, X_{\tilde{m}}\}$ to a set $\{X_1, \ldots, X_n\}$ of local generators for TQ. Thus the matrix with components $X^i_j(q)$, $i, j \in \{1, \ldots, n\}$, is invertible for q sufficiently near q_0, and we denote the components of the inverse matrix by $W^i_j(q)$, $i, j \in \{1, \ldots, n\}$. We also extend u_{kin} to take values in $\mathbb{R}^n$ by asking that the last $n - \tilde{m}$ entries be zero, and we denote the corresponding map by $w_{\mathrm{kin}}\colon [0,T] \to \mathbb{R}^n$. For t sufficiently near zero, we have

$$\begin{aligned} \dot{q}^i(t) &= w^j_{\mathrm{kin}}(t) X^i_j(q(t)), \qquad i \in \{1, \ldots, n\}, \\ \implies \quad w^i_{\mathrm{kin}}(t) &= W^i_j(q(t)) \dot{q}^j(t), \qquad i \in \{1, \ldots, n\}. \end{aligned}$$

Note that the terms in the right-hand side of the second equation are locally absolutely continuous. This then implies that w_{kin} is locally absolutely continuous, and so u_{kin} is locally absolutely continuous.

Finally, let $(\gamma, u_{\mathrm{kin}})$ be a controlled trajectory for Σ_{kin}. Thus $t \mapsto \gamma'(t)$ is locally absolutely continuous, and $t \mapsto \nabla_{\gamma'(t)}\gamma'(t)$ is locally integrable. Since $\mathscr{Y}$, and, therefore, $\mathscr{X}$, is geodesically invariant, $\nabla_{\gamma'(t)}\gamma'(t) \in \mathscr{Y}_{\gamma(t)}$ for $t \in [0,T]$. Thus we may write

$$\nabla_{\gamma'(t)}\gamma'(t) = u^a_{\text{dyn}}(t)Y_a(\gamma(t)),$$

which defines $u_{\text{dyn}}\colon [0,T] \to \mathbb{R}^m$. It remains to show that u is locally integrable. Again, we work locally around $q_0 = \gamma(0)$. We extend $\{Y_1,\dots,Y_m\}$ to a local basis $\{Y_1,\dots,Y_n\}$ of vector fields near q_0. We denote by $V^i_j(q)$, $i,j \in \{1,\dots,n\}$, the components of the inverse of the matrix with components $Y^i_j(q)$, $i,j \in \{1,\dots,n\}$. We then extend u_{dyn} to a map $w_{\text{dyn}}\colon [0,T] \to \mathbb{R}^m$ by asking that w_{dyn} be zero in its last $n-m$ components. We have

$$\begin{aligned} &\ddot{q}^i(t) + \Gamma^i_{jk}(q(t))\dot{q}^j(t)\dot{q}^k(t) = w^j_{\text{dyn}}(t)Y^i_j(q(t)), && i \in \{1,\dots,n\},\\ \implies\ & w^i_{\text{dyn}}(t) = V^i_j(\ddot{q}^j(t) + \Gamma^j_{kl}(q(t))\dot{q}^k(t)\dot{q}^l(t)), && i \in \{1,\dots,n\}. \end{aligned}$$

As in the previous step of the proof, using properties of locally integrable functions, we assert that u_{dyn} is locally integrable. This completes the proof. ■

Remark 8.28 (Maximal reducibility). Note that all driftless systems to which a given affine connection control system $\Sigma_{\text{dyn}} = (\mathsf{Q},\nabla,\mathcal{D},\mathscr{Y} = \{Y_1,\dots,Y_m\},\mathbb{R}^m)$ is maximally reducible are essentially the same, by which we mean that for two such driftless systems, $\Sigma_{\text{kin}} = (\mathsf{Q},\mathscr{X} = \{X_1,\dots,X_m\},\mathbb{R}^m)$ and $\tilde{\Sigma}_{\text{kin}} = (\mathsf{Q},\tilde{\mathscr{X}} = \{\tilde{X}_1,\dots,\tilde{X}_{\tilde{m}}\},\mathbb{R}^{\tilde{m}})$, we have $\mathcal{X} = \tilde{\mathcal{X}}$. Thus, without loss of generality, we may take $(\mathsf{Q},\{Y_1,\dots,Y_m\},\mathbb{R}^m)$ as the system to which Σ_{dyn} is maximally reducible. For this reason, it makes sense to simply say that Σ_{dyn} is simply ***maximally reducible*** if it is maximally reducible to *some* driftless system. •

Let us give an example of a system that is maximally reducible.

Example 8.29 (Robotic leg). We consider again the robotic leg example considered first in Example 5.71. A reference to the computations of Section 7.4.1 shows that $\langle Y_a : Y_b\rangle \in \Gamma^\infty(\mathcal{Y})$ for $a \in \{1,2\}$, thus showing that $\mathcal{Y}$ is closed under symmetric product. Thus the corresponding affine connection control system Σ_{dyn} is maximally reducible to $(\mathsf{Q},\{Y_1,Y_2\},\mathbb{R}^2)$. •

Since $\mathrm{Sym}^{(\infty)}(\mathcal{Y}) = \mathcal{Y}$ for an affine connection control system that is maximally reducible to a driftless system, by Theorem 7.40 such an affine connection control system, if analytic, is STLCC from $q \in \mathsf{Q}$ if and only if $\mathrm{Lie}^{(\infty)}(\mathcal{Y})_q = \mathsf{T}_q\mathsf{Q}$. Thus we make the following definition.

Definition 8.30. A maximally reducible affine connection control system $(\mathsf{Q},\nabla,\mathcal{D},\mathscr{Y},\mathbb{R}^m)$ is ***maximally reducibly kinematically controllable*** from $q_0 \in \mathsf{Q}$ (***MR-KC*** from $q_0 \in \mathsf{Q}$) if $(\mathsf{Q},\mathscr{Y})$ is properly STLC from q_0. •

8.4 The relationship between controllability and kinematic controllability

The appearance in Theorem 8.18 of the vector-valued quadratic form $B_{\mathcal{Y}}$ raises questions about how the notion of kinematic reductions are related to the low-order controllability results of Section 8.2. In this section we work out the proper relationships, and provide counterexamples to show that the relationships we provide are sharp.

8.4.1 Implications

Let $\Sigma_{\mathrm{dyn}} = (\mathsf{Q}, \nabla, \mathcal{D}, \mathscr{Y}, \mathbb{R}^m)$ be an affine connection control system. First let us list the various types of controllability we have at hand for Σ_{dyn} from a point $q_0 \in \mathsf{Q}$:

1. small-time local controllability (STLC) (Definition 7.29);
2. small-time local configuration controllability (STLCC) (Definition 7.29);
3. kinematic controllability (KC) (Definition 8.23);
4. maximally reducible kinematic controllability (MR-KC) (Definition 8.30).

The relationships between these concepts are demonstrated in Figure 8.2. Let

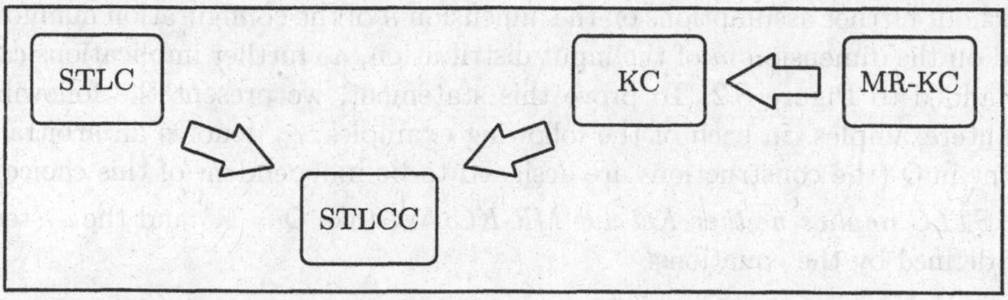

Figure 8.2. Relationships between various forms of controllability for affine connection control systems

us show that these implications do indeed hold.

Proposition 8.31. *For an analytic affine connection control system* $\Sigma_{\mathrm{dyn}} = (\mathsf{Q}, \nabla, \mathcal{D}, \mathscr{Y}, \mathbb{R}^m)$ *and for* $q_0 \in \mathsf{Q}$*, the implications of Figure 8.2 hold.*

Proof. The implications STLC $\implies$ STLCC and MR-KC $\implies$ KC follow directly from the definitions of the various notions of controllability involved. Thus we need only show that KC $\implies$ STLCC. We let

$$\Sigma_{\mathrm{kin},1} = (\mathsf{Q}, \mathscr{X}_1, \mathbb{R}^{m_1}), \dots, \Sigma_{\mathrm{kin},k} = (\mathsf{Q}, \mathscr{X}_k, \mathbb{R}^{m_k})$$

be a collection of kinematic reductions for which $\mathrm{Lie}^{(\infty)}(\mathcal{X}_1 + \cdots + \mathcal{X}_k)_{q_0} = \mathsf{T}_{q_0}\mathsf{Q}$. Let $\mathscr{X} = \mathscr{X}_1 \cup \cdots \cup \mathscr{X}_k$. Note that, since $\mathcal{X}_j \subset \mathcal{Y}$, $j \in \{1, \dots, k\}$, Σ_{dyn} is STLCC from q_0 if $(\mathsf{Q}, \nabla, \mathcal{D}, \mathscr{X})$ is properly STLCC from q_0. Select vector fields $X_{a_1}, \dots, X_{a_l}$ from the family $\mathscr{X}$ so that $\{X_{a_1}(q_0), \dots, X_{a_l}(q_0)\}$ is a basis for $\mathcal{X}_{q_0}$. For brevity, let us denote by $B \in \Sigma_2(\mathcal{Y}_{q_0}; \mathsf{T}_{q_0}\mathsf{Q}/\mathcal{Y}_{q_0})$ the vector-valued quadratic form $B_{\mathcal{Y}}(q_0)$. By Theorem 8.18, we know that

$Q_B|\mathcal{X}_{j,q_0} = 0$, $j \in \{1,\dots,k\}$. It therefore follows that, for each $\lambda \in \operatorname{ann}(\mathcal{Y}_{q_0})$, $\lambda B(X_{a_j}(q_0), X_{a_j}(q_0)) = 0$, $j \in \{1,\dots,l\}$. From Lemma 8.4 this means that λB is essentially indefinite, and since this holds for every $\lambda \in \operatorname{ann}(\mathcal{Y}_{q_0})$, B is itself essentially indefinite. Therefore, by Theorem 8.9, $(\mathsf{Q}, \nabla, \mathcal{D}, \mathscr{X})$ is properly STLCC if $\operatorname{Lie}^{(\infty)}(\mathcal{X})_{q_0} = \mathsf{T}_{q_0}\mathsf{Q}$. The result now follows directly. ■

Remark 8.32. All implications in Figure 8.2 are local. There are implications for global notions of controllability that follow from the local notions, but we do not consider this in a systematic way, since the understanding of this is, as yet, poor. •

Although it is generally not true that STLCC $\Longrightarrow$ KC, as we shall make clear in the next section, there are some cases in which the converse implication can be made. This is contained in the paper of Tyner and Lewis [2004b], and a special case is considered in Exercise E8.13.

8.4.2 Counterexamples

Without further assumptions on the dimension n of the configuration manifold and on the dimension m of the input distribution, no further implications can be added to Figure 8.2. To prove this statement, we present the following counterexamples. In each of the following examples, q_0 denotes an arbitrary point in Q (the constructions are designed to be independent of this choice).

1. *STLC implies neither KC nor MR-KC:* We take $\mathsf{Q} = \mathbb{R}^3$ and the system defined by the equations:
$$\ddot{x}^1 = u^1, \quad \ddot{x}^2 = u^2, \quad \ddot{x}^3 = \dot{x}^1\dot{x}^2.$$
The input vector fields are $Y_1 = \frac{\partial}{\partial x^1}$, $Y_2 = \frac{\partial}{\partial x^2}$. This system is STLC, since $\langle Y_1 : Y_2\rangle = -\frac{\partial}{\partial x^3}$. It is not KC, since Y_1 and Y_2 are the only decoupling vector fields (note that $\langle Y_1 : Y_1\rangle = \langle Y_2 : Y_2\rangle = 0$) but $[Y_1, Y_2] = 0$. Additionally, the system is not maximally reducible, since the input distribution is not geodesically invariant.
2. *MR-KC does not imply STLC:* We take $\mathsf{Q} = \mathbb{R}^3$ and let $\mathscr{X} = \{X_1, X_2, X_3\}$ be defined by
$$X_1 = \frac{\partial}{\partial x^1}, \quad X_2 = \cos(x^1)\frac{\partial}{\partial x^2} + \sin(x^1)\frac{\partial}{\partial x^3},$$
$$X_3 = -\sin(x^1)\frac{\partial}{\partial x^2} + \cos(x^1)\frac{\partial}{\partial x^3},$$
noting that $\mathscr{X}$ generates TQ. The system is then defined by the following Poincaré representation:
$$\begin{aligned} \dot{x}^1 &= v^1, & \dot{v}^1 &= u^1,\\ \dot{x}^2 &= \cos(x^1)v^2 - \sin(x^1)v^3, & \dot{v}^2 &= u^2,\\ \dot{x}^3 &= \sin(x^1)v^2 + \cos(x^1)v^3, & \dot{v}^3 &= 0. \end{aligned}$$

The input vector fields are $Y_1 = \frac{\partial}{\partial x^1}$ and $Y_2 = \cos(x^1)\frac{\partial}{\partial x^2} + \sin(x^1)\frac{\partial}{\partial x^3}$. This system is not STLC, since $\mathrm{Sym}^{(\infty)}(\mathscr{Y})_{q_0} = \mathrm{span}_{\mathbb{R}}\{Y_1(q_0), Y_2(q_0)\}$ for each $q_0 \in \mathsf{Q}$. In particular, along any solution of this mechanical control system starting from rest, $v^3(t) = 0$ for all time t. However, both input vector fields are decoupling and $\mathrm{Lie}^{(\infty)}(\mathscr{Y})$ is maximal rank. Hence the system is MR-KC, but not STLC.

3. *KC implies neither MR-KC nor STLC:* We take $\mathsf{Q} = \mathbb{R}^4$ and take a family of vector fields $\mathscr{X} = \{X_1, X_2, X_3, X_4\}$ with the following properties:

   (a) $\mathscr{X}$ generates TQ;

   (b) $\mathrm{Lie}^{(\infty)}(\tilde{\mathfrak{X}}) = \mathsf{TQ}$, where $\tilde{\mathscr{X}} = \{X_1, X_2\}$.

   It is a simple exercise to construct such a family of vector fields (Exercise E8.15). Now we consider the system defined by the following Poincaré representation:

$$\gamma'(t) = \sum_{i=1}^{4} X_i v^i, \qquad \begin{aligned} \dot{v}^1 &= u^1, \\ \dot{v}^2 &= u^2, \\ \dot{v}^3 &= v^1 v^2, \\ \dot{v}^4 &= \beta (v^3)^2, \end{aligned}$$

   for $\beta \in \mathbb{R}$. The input vector fields are thus $Y_a = X_a$, $a \in \{1, 2\}$. All generalized Christoffel symbols vanish except for $\overset{\mathscr{X}}{\Gamma}{}^3_{12} = \overset{\mathscr{X}}{\Gamma}{}^3_{21} = -\frac{1}{2}$, and $\overset{\mathscr{X}}{\Gamma}{}^4_{33} = -\beta$. We see that the vector fields X_1 and X_2 are decoupling, so the system is KC by property (b) of the vector fields $\mathscr{X}$. The system is not maximally reducible, since $\mathrm{Sym}^{(\infty)}(\tilde{\mathfrak{X}})$ is at least dimension 3. If $\beta = 0$, the system is not accessible. If $\beta \neq 0$, the system is accessible but not STLC.

4. *STLCC implies neither STLC nor KC nor MR-KC:* We take $\mathsf{Q} = \mathbb{R}^4$ and consider a family $\mathscr{X} = \{X_1, X_2, X_3, X_4\}$ of vector fields on Q with the following properties:

   (a) $\mathscr{X}$ generates TQ;

   (b) $\mathrm{Lie}^{(\infty)}(\tilde{\mathfrak{X}}) = \tilde{\mathfrak{X}}$, where $\tilde{\mathscr{X}} = \{X_1, X_2\}$;

   (c) $\mathrm{Lie}^{(\infty)}(\overline{\mathfrak{X}}) = \mathsf{TQ}$, where $\overline{\mathscr{X}} = \{X_1, X_2, X_3\}$.

   Again, one can show that such a family of vector fields exists (Exercise E8.16). The system is then specified by the following Poincaré representation:

$$\gamma'(t) = \sum_{i=1}^{4} X_i v^i, \qquad \begin{aligned} \dot{v}^1 &= u^1, \\ \dot{v}^2 &= u^2, \\ \dot{v}^3 &= v^1 v^2, \\ \dot{v}^4 &= 0. \end{aligned}$$

As previously, the input vector fields $Y_1 = X_1$ and $Y_2 = X_2$ are decoupling. Note that in this case the system is not KC, since the involutive closure of the input distribution is not maximal rank by property (b) of $\mathscr{X}$. Note that $\langle X_1 : X_2 \rangle = X_3$, and that $\mathrm{Sym}^{(\infty)}(\mathcal{X})_{q_0} = \mathrm{span}_{\mathbb{R}}\{X_1(q_0), X_2(q_0), X_3(q_0)\}$ for each $q_0 \in \mathsf{Q}$; therefore, the system is neither maximally reducible, nor STLC. It is STLCC, since $\mathrm{Lie}^{(\infty)}(\mathrm{Sym}^{(\infty)}(\mathcal{X}))$ is maximal rank, and since $B_{\mathscr{Y}}(q_0)$ is essentially indefinite.

Exercises

E8.1 Let V and U be finite-dimensional $\mathbb{R}$-vector spaces and let $B \in \Sigma_2(\mathsf{V}; \mathsf{U})$. Let $\{v_1, \dots, v_n\}$ and $\{u_1, \dots, u_m\}$ be bases for V and U. Define m symmetric $n \times n$ matrices $\boldsymbol{B}^1, \dots, \boldsymbol{B}^m$ by

$$B(v_i, v_j) = B^a_{ij} u_a, \qquad i, j \in \{1, \dots, n\}.$$

For $\boldsymbol{\lambda} \in \mathbb{R}^m$, define

$$\boldsymbol{B}_{\boldsymbol{\lambda}} = \sum_{a=1}^{m} \lambda^a \boldsymbol{B}^a.$$

With this notation, show the following.

(a) B is definite if and only if there exists $\boldsymbol{\lambda} \in \mathbb{R}^m$ such that $\mathrm{spec}(\boldsymbol{B}_{\boldsymbol{\lambda}}) \subset \mathbb{R}_+$.

(b) B is semidefinite if and only if there exists $\boldsymbol{\lambda} \in \mathbb{R}^m$ such that $\mathrm{spec}(\boldsymbol{B}_{\boldsymbol{\lambda}}) \subset \bar{\mathbb{R}}_+$.

(c) B is strongly semidefinite if and only if there exists $\boldsymbol{\lambda} \in \mathbb{R}^m$ such that $\mathrm{spec}(\boldsymbol{B}_{\boldsymbol{\lambda}}) \subset \bar{\mathbb{R}}_+$ and $\mathrm{spec}(\boldsymbol{B}_{\boldsymbol{\lambda}}) \cap \mathbb{R}_+ \neq \emptyset$.

(d) B is indefinite if and only if, for each $\boldsymbol{\lambda} \in \mathbb{R}^m$, $\mathrm{spec}(\boldsymbol{B}_{\boldsymbol{\lambda}}) \cap \mathbb{R}_+ \neq \emptyset$ and $\mathrm{spec}(\boldsymbol{B}_{\boldsymbol{\lambda}}) \cap (-\mathbb{R}_+) \neq \emptyset$.

(e) B is essentially indefinite if and only if, for each $\boldsymbol{\lambda} \in \mathbb{R}^m$, either (1) $\boldsymbol{B}_{\boldsymbol{\lambda}} = \boldsymbol{0}$ or (2) $\mathrm{spec}(\boldsymbol{B}_{\boldsymbol{\lambda}}) \cap \mathbb{R}_+ \neq \emptyset$ and $\mathrm{spec}(\boldsymbol{B}_{\boldsymbol{\lambda}}) \cap (-\mathbb{R}_+) \neq \emptyset$.

E8.2 This exercise is a continuation of the ball in a rotating channel system, whose controllability was considered in Exercise E7.15. In that exercise, you will have shown that the system is accessible from all configurations, but the results of Section 7.3.3 were not adequate to fully determine the controllability properties of the system. You will now finish the analysis for controllability.

(a) Determine the set of points in Q from which the system is STLC.

(b) Determine the set of points in Q from which the system is not STLCC.

E8.3 Show that $X \in \Gamma^\infty(\mathsf{TQ})$ is a decoupling vector field for $\Sigma = (\mathsf{Q}, \nabla, \mathcal{D}, \mathscr{Y}, \mathbb{R}^m)$ if and only if, for every integral curve $\gamma \colon I \to \mathsf{Q}$, and for every reparameterization $\tau \colon J \to I$ of γ, there exists a control $u_{\mathrm{dyn}} \in \mathscr{U}^m_{\mathrm{dyn}}$ such that $(\gamma, u_{\mathrm{dyn}})$ is a controlled trajectory for Σ.

E8.4 Consider the planar rigid body of Example 8.22, and let Σ_{dyn} denote the corresponding affine connection control system. Show that Σ_{dyn} is not maximally reducible to the kinematic reduction $\Sigma_{\mathrm{kin},2} = (\mathsf{Q}, \mathscr{X}_2 = \{X_2\}, \mathbb{R})$ constructed in Example 8.22.

E8.5 Let $(\mathsf{Q}, \mathbb{G}, 0, \mathcal{D}, \mathscr{F}, U)$ be a simple mechanical control system with regular constraints and with a zero potential function. Denote by $\mathcal{F}$ the codistribution generated by $\mathscr{F}$. Let $(\mathsf{Q}, \overset{\mathcal{D}}{\nabla}, \mathcal{D}, \mathscr{Y}, U)$ be the corresponding affine connection control system (thus $Y_a = P_{\mathcal{D}}(\mathbb{G}^\sharp(F^a))$, $a \in \{1, \dots, m\}$). Show that, if $\operatorname{ann}(\mathcal{D}) \oplus \mathcal{F} = \mathsf{T}^*\mathsf{Q}$, then $(\mathsf{Q}, \overset{\mathcal{D}}{\nabla}, \mathcal{D}, \mathscr{Y}, U)$ is maximally reducible.

E8.6 For the affine connection control systems below on $\mathbb{R}^n$ (with coordinates denoted by $(x^1, \dots, x^n)$), answer the following questions.

1. Write the differential equations governing the system.
2. Does the system satisfy the sufficient conditions of Theorem 8.9 in order to be properly STLCC from $q_0 = \mathbf{0}$?
3. Does the system satisfy the necessary conditions of Theorem 8.10 in order to be STLCUC from $q_0 = \mathbf{0}$?
4. Find as many decoupling vector fields as you can.
5. Is the system kinematically controllable?
6. Is the system maximally reducible?
7. Is the system maximally reducibly kinematically controllable?

Here are the systems, specified by giving n, the nonzero Christoffel symbols, and the control vector fields.

(a) $n = 3$, $\Gamma^3_{12} = 1$, $Y_1 = \frac{\partial}{\partial x^1}$, and $Y_2 = \frac{\partial}{\partial x^2}$.
(b) $n = 4$, $\Gamma^3_{11} = 1$, $\Gamma^3_{22} = 1$, $\Gamma^4_{12} = 1$, $Y_1 = \frac{\partial}{\partial x^1}$, and $Y_2 = \frac{\partial}{\partial x^2}$.
(c) $n = 6$, $\Gamma^4_{12} = 1$, $\Gamma^5_{13} = 1$, $\Gamma^6_{23} = 1$, $Y_1 = \frac{\partial}{\partial x^1}$, $Y_2 = \frac{\partial}{\partial x^2}$, and $Y_3 = \frac{\partial}{\partial x^3}$.
(d) $n = 3$, $\Gamma^3_{11} = 1$, $\Gamma^3_{22} = -y^2$, $Y_1 = \frac{\partial}{\partial x^1}$, and $Y_2 = \frac{\partial}{\partial x^2}$.
(e) $n = 2$, $\Gamma^2_{12} = 1$, and $Y_1 = \frac{\partial}{\partial x^1}$.

E8.7 Consider the car-like robot of Exercises E4.9, E4.23, E4.35, and E4.41.
(a) Show that this system satisfies the conditions in Exercise E8.5.
(b) Find two decoupling vector fields for the system.
(c) How many decoupling vector fields are there in total?

E8.8 This exercise is a continuation of the planar body system with fan dynamics of Exercises E4.8, E4.22, and E4.34. For the system as shown in Figure E8.1, show that there is a decoupling vector field, and ascertain what it is.

The presence of only one decoupling vector field for the system in the preceding exercise has negative consequences for the controllability of the system. This is discussed systematically by Tyner and Lewis [2004a]. The computations are a bit cumbersome to present here, either as an exercise or as an example.

E8.9 Consider the planar rigid body of Section 7.4.2 and, following Example 5.47, regard it as a simple mechanical control system on a Lie group. Recover the results in Example 8.22 by performing the appropriate symmetric product computations on the Lie algebra $\mathfrak{se}(2)$.

E8.10 Consider the rigid body of Example 5.48 fixed at a point and consider two control torques aligned with the first and second major axis. In other words, consider the triple $(\mathsf{SO}(3), \mathbb{I}, \{e^1, e^2\})$.
(a) Find two decoupling vector fields for the system.

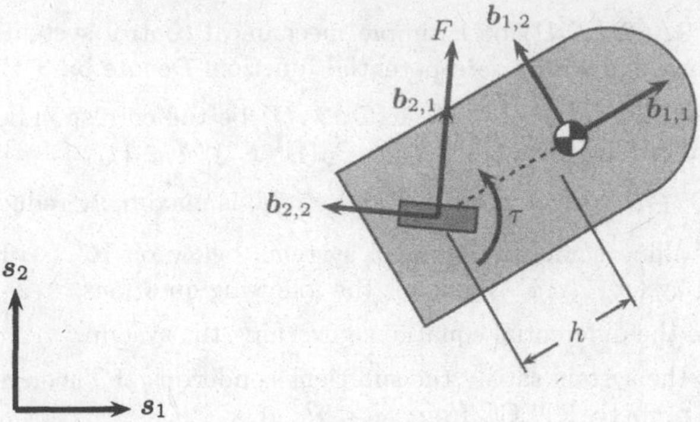

Figure E8.1. Planar body with fan and input forces

(b) Verify that the system is kinematically controllable.
(c) Is the system maximally reducible?

E8.11 Consider an affine connection control system $\Sigma_{\mathrm{dyn}} = (\mathsf{Q}, \nabla, \mathcal{D}, \mathscr{Y}, \mathbb{R}^m)$ for which $\dim(\mathcal{Y}_q) = \dim(\mathsf{Q}) - 1$ for each $q \in \mathsf{Q}$ and for which $\mathrm{Sym}^{(1)}(\mathcal{Y})$ has constant rank. Show that, if Σ_{dyn} possesses m decoupling vector fields, then it is either maximally reducible to some driftless system, or it is STLC from each $q \in \mathsf{Q}$.

E8.12 Let $(\mathsf{Q}, \nabla, \mathcal{D}, \mathscr{Y}, \mathbb{R}^m)$ be an affine connection control system for which $\mathrm{rank}(\mathcal{Y}_q) = \dim(\mathsf{Q}) - 1$ for each $q \in \mathsf{Q}$. Show that the following two statements are equivalent:

1. for each $q \in \mathsf{Q}$, there exists a neighborhood $\mathcal{U}$ of q such that $\mathcal{Y}|\mathcal{U}$ is generated by decoupling vector fields;
2. $B_{\mathcal{Y}_q}(\mathcal{Y}_q)$ is essentially indefinite for each $q \in \mathsf{Q}$.

E8.13 Let $\Sigma_{\mathrm{dyn}} = (\mathsf{Q}, \nabla, \mathcal{D}, \mathscr{Y}, U)$ be an analytic affine connection control system and suppose that the distribution $\mathcal{Y}$ generated by the vector fields $\mathscr{Y}$ has constant rank $\dim(\mathsf{Q}) - 1$.

(a) Show that Σ_{dyn} is kinematically controllable from q_0 if and only if (1) $B_{\mathcal{Y}}(\mathcal{Y}_{q_0})$ is essentially indefinite and (2) $\mathrm{Lie}^{(\infty)}(\mathcal{Y})_{q_0} = \mathsf{T}_{q_0}\mathsf{Q}$.

Next we illustrate that the assumption that the distribution $\mathcal{Y}$ has codimension one is essential in part (a). We consider the affine connection control system $(\mathsf{Q}, \nabla, \mathcal{D}, \mathscr{Y}, U)$ with

1. $\mathsf{Q} = \mathbb{R}^4$ (we let (x^1, x^2, x^3, x^4) denote the standard coordinates),
2. ∇ is defined by the nonzero Christoffel symbols
$$\Gamma^3_{11} = \tfrac{1}{2}, \quad \Gamma^3_{22} = -\tfrac{1}{2}, \quad \Gamma^4_{12} = \Gamma^4_{21} = \tfrac{1}{2},$$
3. $\mathcal{D} = \mathsf{TQ}$,
4. $Y_1 = \frac{\partial}{\partial x^1}$ and $Y_2 = \frac{\partial}{\partial x^2}$, and
5. $U = \mathbb{R}^2$.

Also define $X_1 = \frac{\partial}{\partial x^3}$ and $X_2 = \frac{\partial}{\partial x^4}$. Define two symmetric 2×2 matrices $\boldsymbol{B}^1$ and $\boldsymbol{B}^2$ by

$$\langle Y_a : Y_b \rangle = B^\alpha_{ab} X_\alpha, \qquad a, b \in \{1, 2\}.$$

Answer the following questions.

(b) Show that, for each $q_0 \in \mathsf{Q}$, $B_{\mathcal{Y}}(\mathcal{Y}_{q_0})$ is essentially indefinite.

(c) Use the conclusion of part (b) to show that the hypotheses of Theorem 8.9 are satisfied, and so the system is STLCC from any $q_0 \in \mathsf{Q}$.

(d) Show that the system possesses no kinematic reductions.
Hint: *It suffices to show that the system possesses no decoupling vector fields. Note that if X is a decoupling vector field, then it must take values in $\mathcal{Y}$. If $\tilde{X}$ is a complement to X in $\mathcal{Y}$ (thus $\{X, \tilde{X}\}$ are generators for $\mathcal{Y}$), then it must be that $B_{\mathcal{Y}}(X, X) = 0$. Show that this cannot happen, using the matrices $\boldsymbol{B}^1$ and $\boldsymbol{B}^2$.*

E8.14 *Mathematica®-assisted project:* This exercise is a continuation of the three-link manipulator problem of Exercise E4.37. The computations outlined here are presented in [Bullo and Lynch 2001]. For convenience, we say that an ***actuator configuration*** is a triple (a_1, a_2, a_3) of elements from the set $\{0, 1\}$, with $a_i = 0$ implying no actuation at joint i, $i \in \{1, 2, 3\}$, and $a_i = 1$ implying actuation at joint i, $i \in \{1, 2, 3\}$. In this exercise we assume coordinates $(\theta_1, \theta_2, \theta_3)$ which measure the angles of the links in a spatial frame.

(a) *Actuator configuration* $(0, 1, 1)$*:* Show that the system is maximally reducible.

To examine the $(1, 1, 0)$ actuator configuration requires some setup. Define a map $\psi\colon \mathsf{Q} \to \mathbb{R}^2$ by asking that $\psi(q)$ denotes the location of the end of link 2 when the system is in configuration q. Note that, in our coordinates $(\theta_1, \theta_2, \theta_3)$, ψ is a function only of θ_1 and θ_2.

(b) Show that $T_q\psi$ is surjective except for configurations q where the links 1 and 2 are collinear.

Let S denote the set of points $q \in \mathsf{Q}$ where $T_q\psi$ is not surjective.

(c) Show that, for $(x, y) \in \operatorname{image}(\psi(\mathsf{Q} \setminus \mathsf{S}))$, the set $\psi^{-1}(x, y)$ consists of two submanifolds, each diffeomorphic to $\mathbb{S}^1$.
Hint: *The idea is that there are two possible values of θ_1 and θ_2 that map to the same value (x, y), and that θ_3 is free, thus giving two copies of $\mathbb{S}^1$.*

(d) Argue that $\mathsf{Q} \setminus \mathsf{S}$ has two connected components, denoted by Q_1 and Q_2, and that on each connected component the map ψ is a submersion.

(e) Show, therefore, that $q \mapsto (\psi(q), \theta_3(q))$ is a valid coordinate chart with chart domain Q_1 or Q_2.

With this as backdrop, we now have the following problem.

(f) *Actuator configuration* $(1, 1, 0)$*:* Show that the vector fields

$$X_1 = \cos\theta_3 \frac{\partial}{\partial x} + \sin\theta_3 \frac{\partial}{\partial y},$$
$$X_2 = \sin\theta_3 \frac{\partial}{\partial x} - \cos\theta_3 \frac{\partial}{\partial y} + \frac{m_3 \ell_3}{J_3 + m_3 \ell_3^2} \frac{\partial}{\partial \theta_3}$$

are decoupling vector fields on each connected component Q_1 and Q_2 of $\mathsf{Q} \setminus \mathsf{S}$. Show that the system is furthermore kinematically controllable on each connected component.

(g) *Actuator configuration* $(1, 0, 1)$*:* For this actuator configuration, it is convenient to use relative coordinates $\theta_{2,1}$ and $\theta_{3,1}$ for the orientations of the second and third link. Thus $\theta_{2,1}$ is the orientation of the body frame for the second link relative to that for the first link, and $\theta_{3,2}$ is the orientation of the body frame for the third link relative to that of the second link.

With these coordinates, show that the vector fields

$$\begin{aligned} X_1 &= -\mathbb{G}_{23}\frac{\partial}{\partial\theta_{2,1}} + \mathbb{G}_{22}\frac{\partial}{\partial\theta_{3,2}}, \\ X_2 &= 2\Big(\mathbb{G}_{23}\frac{\partial\mathbb{G}_{12}}{\partial\theta_{2,1}} - \mathbb{G}_{22}\frac{\partial\mathbb{G}_{13}}{\partial\theta_{2,1}}\Big)\frac{\partial}{\partial\theta_1} \\ &\quad + \Big(2\mathbb{G}_{12}\frac{\partial\mathbb{G}_{13}}{\partial\theta_{2,1}} - \mathbb{G}_{23}\frac{\partial\mathbb{G}_{11}}{\partial\theta_{2,1}}\Big)\frac{\partial}{\partial\theta_{2,1}} \\ &\quad + \Big(\mathbb{G}_{22}\frac{\partial\mathbb{G}_{11}}{\partial\theta_{2,1}} - 2\mathbb{G}_{12}\frac{\partial\mathbb{G}_{12}}{\partial\theta_{2,1}}\Big)\frac{\partial}{\partial\theta_{3,2}} \end{aligned}$$

are decoupling vector fields, and that the system is kinematically controllable.

E8.15 Show that there exists a family of vector fields $\mathscr{X} = \{X_1, X_2, X_3, X_4\}$ on $\mathsf{Q} = \mathbb{R}^4$ with the properties

1. $\mathscr{X}$ generates TQ and
2. $\mathrm{Lie}^{(\infty)}(\tilde{\mathfrak{X}}) = \mathsf{TQ}$ if $\tilde{\mathscr{X}} = \{X_1, X_2\}$.

E8.16 Show that there exists a family of vector fields $\mathscr{X} = \{X_1, X_2, X_3, X_4\}$ on $\mathsf{Q} = \mathbb{R}^4$ with the properties

1. $\mathscr{X}$ generates TQ,
2. $\mathrm{Lie}^{(\infty)}(\tilde{\mathfrak{X}}) = \tilde{\mathfrak{X}}$ if $\tilde{\mathscr{X}} = \{X_1, X_2\}$, and
3. $\mathrm{Lie}^{(\infty)}(\overline{\mathfrak{X}}) = \mathsf{TQ}$, where $\overline{\mathscr{X}} = \{X_1, X_2, X_3\}$.

9

Perturbation analysis

In this chapter we apply perturbation methods to analyze the trajectories of forced affine connection control systems. We consider two types of control signals: periodic, large-amplitude, high-frequency signals, which we refer to as oscillatory; and small-amplitude signals. For both classes of signals, we perform a perturbation analysis that predicts, with some specified level of accuracy, the behavior of the resulting forced affine connection system.

The subject matter in this chapter builds on classic averaging and oscillation theory. Early studies include the works by Clairaut, Laplace, and Lagrange. Clairaut [1754] began the study of perturbation methods to analyze the solution of differential equations. Laplace [1799-1825] developed some elements of the method in his celestial mechanics study of the Sun-Jupiter-Saturn configuration. Lagrange [1788] provided the first detailed study of the averaging method, discussing, for example, the notion of standard averaging form and the variation of constants formula. These works were extended and developed, for example, by Poincaré [1892-1899], Jacobi [1884], and others. We refer the reader to [Bogoliubov and Mitropolsky 1961] for a classic text on nonlinear oscillations, to [Guckenheimer and Holmes 1990, Sanders and Verhulst 1985] for a modern dynamical system approach and for a historical overview of the development of averaging theory, and to [Bloch 2003, Khalil 2001] for a control theoretical view. In recent years, averaging and perturbation theory has been applied to animal and robotic locomotion [Golubitsky, Stewart, Buono, and Collins 1999, Hirose 1993], switching and power conversion circuits [Sanders, Noworolski, Liu, and Verghese 1991], for example.

In what follows, we provide an admittedly incomplete list of references that are related to our presentation. First of all, our analysis is related to the study of differential equations subject to periodic, high-frequency, large-amplitude forcing terms; see [Kurzweil and Jarnik 1988, Liu 1997, Liu and Sussmann 1999]. Roughly speaking, these works characterize the trajectory of the original system by showing that they converge to the trajectories of an appropriate averaged system. A second set of related results deals with the analysis of high-frequency vibrations in mechanical and other classes of systems.

Historical work here includes that of Hill [1886] and Mathieu [1868], while more recent papers are [Baillieul 1993, 1995, Kapitsa 1965, Levi 1998, 1999]. Finally, the material in this chapter is related to classic works on series expansions and on averaging results for small-amplitude inputs. Some early work on series expansions for the trajectories of nonlinear dynamical systems includes the Magnus and Chen series [Chen 1957, Magnus 1954]. Volterra series were then studied by Brockett [1976], Gilbert [1977], and Lesiak and Krener [1978]. A comprehensive treatment of series expansions for time-dependent vector fields is developed by Agrachev and Gamkrelidze [1978]; see also [Agrachev and Sachkov 2004]. Fliess [1981] and Sussmann [1986] later provided a comprehensive treatment of what is now known as the Chen–Fliess–Sussmann series. Other related works include [Kawski and Sussmann 1997, Leonard and Krishnaprasad 1995, Sarychev 2001, Vela 2003].

The outline of this chapter is the following. Section 9.1 presents a general approach to averaging for dynamical systems subject to oscillatory controls. We introduce the variation of constants formula, the classic first-order averaging theorem, and a fairly complete averaging analysis for systems subject to oscillatory inputs. These results are then applied in Section 9.2 to the setting of affine connection control systems. The objective of the analysis is the characterization of the behavior of simple mechanical systems forced by large-amplitude, high-frequency inputs. The averaged system is shown to be an affine connection system subject to an appropriate vector force. This vector force is a linear combination of symmetric products of control vector fields. We also introduce the notion of the averaged potential to characterize the averaged behavior of simple mechanical systems with potential control forces. In Section 9.3 we present a series expansion describing the controlled trajectories of a mechanical system starting at rest and subject to a small-amplitude force. The treatment we present is adapted from the papers [Bullo 2001, 2002, Martínez 2002, Martínez, Cortés, and Bullo 2003a]. We shall mostly consider systems of class C^∞, but we will need to state some results relying on weaker smoothness properties.

Finally, let us remark that our study of oscillatory controls and averaging in mechanical *control* systems differs substantially from the classic approach to averaging of Hamiltonian systems; see, for example, Chapter 4 in [Guckenheimer and Holmes 1990]. In that setting, the nominal mechanical system is assumed to be integrable. In our setting, it is the Lagrangian dynamics that plays the role of the perturbation to the dominant large-amplitude, high-frequency forcing.

9.1 An overview of averaging theory for oscillatory control systems

We present here some preliminary tools, study averaging theory, and consider systems subject to oscillatory controls. We state the first-order averaging theorem following the presentation in [Sanders and Verhulst 1985]. The variation of constants formula originates with Lagrange's work, see [Sanders and Verhulst 1985, page 183], and is presented here in a coordinate-invariant setting inspired by Agrachev and Gamkrelidze [1978]; see also [Agrachev and Sachkov 2004].

9.1.1 Iterated integrals and their averages

A ***multi-index in*** $\{1,\dots,m\}$ is an array $(a_1,\dots,a_k)$ of arbitrary length $k \in \mathbb{N}$, whose entries $a_1,\dots,a_k$ take values in $\{1,\dots,m\}$. Let $\mathcal{I}_{\{1,\dots,m\}}$ be the set of all multi-indices in $\{1,\dots,m\}$. Let $u_a\colon \bar{\mathbb{R}}_+ \to \mathbb{R}$, $a \in \{1,\dots,m\}$, be essentially bounded and measurable functions, and define their ***iterated integrals*** $U_J\colon \bar{\mathbb{R}}_+ \to \mathbb{R}$, $J \in \mathcal{I}_{\{1,\dots,m\}}$, by

$$U_{(a_1,\dots,a_k)}(t) = \int_0^t u_{a_k}(t_k) \int_0^{t_k} u_{a_{k-1}}(t_{k-1}) \cdots \int_0^{t_2} u_{a_1}(t_1)\mathrm{d}t_1 \cdots \mathrm{d}t_{k-1}\mathrm{d}t_k.$$

For $k_1,\dots,k_m \in \mathbb{Z}_+$, let $C_{k_1,\dots,k_m}$ be the set of m-tuples $(s_1,\dots,s_m)$ with the following properties: for $a \in \{1,\dots,m\}$, s_a is a subset of $\{1,\dots,k_1 + \cdots + k_m\}$, s_a has k_a elements, and the union of s_a, for $a \in \{1,\dots,m\}$, is $\{1,\dots,k_1+\cdots+k_m\}$. The number of m-tuples in $C_{k_1,\dots,k_m}$ is the ***multinomial coefficient***

$$\binom{k_1+\cdots+k_m}{k_1,\dots,k_m} = \frac{(k_1+\cdots+k_m)!}{k_1!\cdots k_m!}.$$

To each m-tuple $(s_1,\dots,s_m) \in C_{k_1,\dots,k_m}$, we associate a multi-index in $\{1,\dots,m\}$ of length $k_1+\cdots+k_m$, denoted by $J(s_1,\dots,s_m)$, as follows: for $a \in \{1,\dots,m\}$, place the index a in the k_a entries in $J(s_1,\dots,s_m)$ given by the set of integers $s_a \subset \{1,\dots,k_1+\cdots+k_m\}$. For example, let $m=3$, $k_1=2$, $k_2=3$, and $k_3=4$; then a sample triple in $C_{2,3,4}$ is $(\{2,4\},\{1,9,8\},\{5,6,3,7\})$, and the corresponding multi-index is $(2,1,3,1,3,3,3,2,2)$.

Let $u_a\colon \bar{\mathbb{R}}_+ \to \mathbb{R}$, $a \in \{1,\dots,m\}$, be essentially bounded and measurable functions, and define their ***multinomial iterated integrals*** $\mathsf{U}_{k_1,\dots,k_m}\colon \bar{\mathbb{R}}_+ \to \mathbb{R}$, $k_1,\dots,k_m \in \mathbb{Z}$, by

$$\mathsf{U}_{k_1,\dots,k_m}(t) = \begin{cases} 0, & k_a < 0, \text{ for some } a,\\ 1, & k_1 = \cdots = k_m = 0,\\ \displaystyle\sum_{(s_1,\dots,s_m)\in C_{k_1,\dots,k_m}} U_{J(s_1,\dots,s_m)}(t), & \text{otherwise.}\end{cases}$$

In other words, the multinomial iterated integral $\mathsf{U}_{k_1,\dots,k_m}$ is the sum of all iterated integrals containing k_a copies of u_a, for $a \in \{1,\dots,m\}$, in all possible orders. With this characterization, one can show that

$$\mathsf{U}_{k_1,\dots,k_m}(t) = \int_0^t \Big(u_1(s)\mathsf{U}_{k_1-1,\dots,k_m}(s) + \cdots + u_m(s)\mathsf{U}_{k_1,\dots,k_m-1}(s)\Big)\mathrm{d}s.$$

For $T \in \mathbb{R}_+$, a function $u\colon \bar{\mathbb{R}}_+ \to \mathbb{R}$ is ***T-periodic*** if $u(t+T) = u(t)$ for all $t \in \bar{\mathbb{R}}_+$. A T-periodic function $u\colon \bar{\mathbb{R}}_+ \to \mathbb{R}$ is ***zero-mean*** if $\int_0^T u(t)\mathrm{d}t = 0$. When studying periodic functions, it will sometimes be convenient to use the following notions: let $\lfloor \tau \rfloor$ denote the greatest integer less than or equal to $\tau \in \bar{\mathbb{R}}_+$, and let $\tau \bmod T \triangleq \tau - \lfloor \frac{\tau}{T} \rfloor T$.

Proposition 9.1 (Periodicity of the multinomial iterated integrals). *Let $u_a\colon \bar{\mathbb{R}}_+ \to \mathbb{R}$, $a \in \{1,\dots,m\}$, be essentially bounded and measurable. Their multinomial iterated integrals $\mathsf{U}_{k_1,\dots,k_m}\colon \bar{\mathbb{R}}_+ \to \mathbb{R}$, $k_1,\dots,k_m \in \mathbb{Z}_+$, have the following properties:*

(i) $\mathsf{U}_{k_1,\dots,k_m}(t) = \dfrac{1}{k_1!\cdots k_m!}\left(\displaystyle\int_0^t u_1(\tau)\mathrm{d}\tau\right)^{k_1}\cdots\left(\displaystyle\int_0^t u_m(\tau)\mathrm{d}\tau\right)^{k_m}$;

(ii) $\mathsf{U}_{k_1,\dots,k_m}$ *is T-periodic if and only if $u_1,\dots,u_m$ are T-periodic and zero-mean.*

Proof. We prove part (i) by induction on $k = k_1+\dots+k_m$. For $k=1$, we have $k_a = \delta_{ab}$, for some $b \in \{1,\dots,m\}$. Then $\mathsf{U}_{k_1,\dots,k_m}(t) = \int_0^t u_b(\tau)\mathrm{d}\tau = U_{(b)}(t)$. Assume that the claim is true for $k-1$, and let us prove it for k. Using the induction hypothesis, the time derivative of the right-hand side of the equation in part (i) can be written as

$$\frac{1}{k_1!\cdots k_m!}\Big(k_1u_1(t)U_{(1)}^{k_1-1}(t)\cdots U_{(m)}^{k_m}(t)+\dots+k_m u_m(t)U_{(1)}^{k_1}(t)\cdots U_{(m)}^{k_m-1}(t)\Big)$$
$$= u_1(t)\mathsf{U}_{k_1-1,\dots,k_m}(t)+\dots+u_m(t)\mathsf{U}_{k_1,\dots,k_m-1}(t).$$

The claim now follows by integrating with respect to time. Next, we consider part (ii). Since

$$\int_0^{T+t} u_a(\tau)\mathrm{d}\tau = \int_0^T u_a(\tau)\mathrm{d}\tau + \int_T^{T+t} u_a(\tau)\mathrm{d}\tau,$$

the functions $\int_0^t u_a(\tau)\mathrm{d}\tau$ are T-periodic if u_a are zero-mean and T-periodic. Furthermore, the u_a are T-periodic if their time integrals are T-periodic, and they are zero-mean if $\int_0^T u_a(\tau)\mathrm{d}\tau = 0$. ■

Remark 9.2. In the single-input case, the iterated integrals and the multinomial iterated integrals of the function $u_1 = u\colon \bar{\mathbb{R}}_+ \to \mathbb{R}$ are equal, in the sense that $\mathsf{U}_k(t) = \sum_{\alpha\in C_k} U_{J(\alpha)}(t) = U_{(1,\dots,1)}(t)$. •

Definition 9.3 (Average, zero-mean, ∞-norm). Let $u\colon \bar{\mathbb{R}}_+ \to \mathbb{R}$ be essentially bounded, measurable, and T-periodic.

(i) The ***average of u***, denoted by $\overline{u}$, is

$$\overline{u} = \frac{1}{T}\int_0^T u(\tau)\mathrm{d}\tau.$$

(ii) The function u is ***zero-mean*** if $\overline{u} = 0$.
(iii) The ***∞-norm of u***, denoted by $\|u\|_\infty$, is

$$\|u\|_\infty = \operatorname{ess\,sup} \{|u(t)| \mid t \in [0, T]\}. \quad \bullet$$

We can now state the following result.

Proposition 9.4 (Bounds on the multinomial iterated integrals). *Let $u_a : \overline{\mathbb{R}}_+ \to \mathbb{R}$, $a \in \{1, \dots, m\}$, be essentially bounded, measurable, T-periodic, and zero-mean. Their multinomial iterated integrals $\mathsf{U}_{k_1,\dots,k_m} : \overline{\mathbb{R}}_+ \to \mathbb{R}$, $k_1, \dots, k_m \in \mathbb{Z}_+$, have the following properties:*

$$\|\mathsf{U}_{k_1,\dots,k_m}\|_\infty \leq \frac{T^{k_1+\cdots+k_m}}{k_1! \cdots k_m!} \|u_1\|_\infty^{k_1} \cdots \|u_m\|_\infty^{k_m},$$

$$|\overline{\mathsf{U}}_{k_1,\dots,k_m}| \leq \frac{T^{k_1+\cdots+k_m}}{k_1! \cdots k_m!(1 + \sum_{b=1}^m k_b)} \|u_1\|_\infty^{k_1} \cdots \|u_m\|_\infty^{k_m}.$$

Proof. Under the given assumptions, the functions $\mathsf{U}_{k_1,\dots,k_m}$, $k_1, \dots, k_m \in \mathbb{Z}_+$, are T-periodic. For $0 \leq t \leq T$,

$$|\mathsf{U}_{k_1,\dots,k_m}(t)| \leq \frac{1}{k_1! \cdots k_m!} \left(\int_0^t |u_1(\tau)| \mathrm{d}\tau \right)^{k_1} \cdots \left(\int_0^t |u_m(\tau)| \mathrm{d}\tau \right)^{k_m}$$

$$\leq \frac{1}{k_1! \cdots k_m!} t^{k_1} \|u_1\|_\infty^{k_1} \cdots t^{k_m} \|u_1\|_\infty^{k_m},$$

which gives the first bound. The second bound is proven via the chain of inequalities

$$|\overline{\mathsf{U}}_{k_1,\dots,k_m}| \leq \frac{1}{T} \int_0^T |\mathsf{U}_{k_1,\dots,k_m}(t)| \mathrm{d}t$$

$$\leq \frac{\|u_1\|_\infty^{k_1} \cdots \|u_1\|_\infty^{k_m}}{k_1! \cdots k_m!} \frac{1}{T} \int_0^T t^{k_1+\cdots+k_m} \mathrm{d}t. \quad \blacksquare$$

Example 9.5. Consider the functions $u_a(t) = c_a \cos \omega t$, for $\omega \in \mathbb{N}$, $c_a \in \mathbb{R}$, and $a \in \{1, \dots, m\}$. Then

$$\mathsf{U}_{k_1,\dots,k_m}(t) = \frac{c_1^{k_1} \cdots c_m^{k_m}}{k_1! \cdots k_m!} \left(\frac{1}{\omega} \sin \omega t \right)^{k_1+\cdots+k_m}.$$

Let $k = \sum_{b=1}^m k_b$. From the identity $4^m \int_0^{2\pi} (\sin t)^{2m} \mathrm{d}t = 2\pi \binom{2m}{m}$ in [Spiegel and Liu 1998], the averages are

$$\overline{\mathsf{U}}_{k_1,\dots,k_m} = \begin{cases} 0, & \text{if } k \text{ is odd}, \\ \dfrac{c_1^{k_1} \cdots c_m^{k_m}}{k_1! \cdots k_m!} \left(\dfrac{1}{2\omega} \right)^k \dbinom{k}{\frac{k}{2}}, & \text{if } k \text{ is even}. \end{cases} \quad \bullet$$

9.1.2 Norms for objects defined on complex neighborhoods

In this section we introduce a norm on the set of analytic vector fields on a compact subset of a manifold M. We also provide bounds for partial derivatives of analytic functions. The bounds are not coordinate-invariant, that is, the bounds apply to the coordinate representation of the vector field of interest, and they depend on the specific selection of coordinate chart. However, if the bounds hold in one set of coordinates, they will also hold in any other. Accordingly, the treatment here assumes that M is a bounded open subset of $\mathbb{R}^n$; it is convenient to adopt the convention of *not* writing points in M with a bold symbol. We refer to [Hörmander 1973] for a standard treatment on complex analysis and to [Agrachev and Sachkov 2004] for a discussion of norms and seminorms on function spaces on manifolds.

To establish the estimates we require, we extend the analysis from $\mathsf{M} \subset \mathbb{R}^n$ to $\mathbb{C}^n$, by thinking of $\mathbb{R}^n$ as a subset of $\mathbb{C}^n$ via the inclusion

$$(x^1, \dots, x^n) \mapsto (x^1 + \mathrm{i}0, \dots, x^n + \mathrm{i}0).$$

Points in $\mathbb{C}^n$ will typically be denoted by z, whereas points in $\mathbb{R}^n$ will typically be denoted by x. The ***complex open ball*** of radius ρ centered at $x_0 \in \mathsf{M} \subset \mathbb{R}^n$ is given by

$$B_\rho^{\mathbb{C}}(x_0) = \{ z \in \mathbb{C}^n \mid \|z - x_0\|_{\mathbb{C}^n} < \rho \}.$$

Let f be an analytic function on M that admits a bounded analytic continuation to $B_\rho^{\mathbb{C}}(x_0)$, for some $\rho \in \mathbb{R}_+$. The ***∞-norm*** of f on $B_\rho^{\mathbb{C}}(x_0)$ is

$$\|f\|_{x_0,\rho} = \sup \{ |f(z)| \mid z \in B_\rho^{\mathbb{C}}(x_0) \},$$

where, for simplicity, f denotes both the function on M and its analytic continuation. Given a time-dependent vector field $Y \in \Gamma^\omega(\mathbb{R}; \mathsf{TM})$, we let $Y_t(x) = Y(t, x)$, and let Y_t^i be the ith component in the standard basis on $\mathbb{R}^n$. Assuming that every component function Y_t^i, for almost every $t \in [0, T]$, admits a bounded analytic continuation to $B_\rho^{\mathbb{C}}(x_0)$, we define ***∞-norm*** of Y on $B_\rho^{\mathbb{C}}(x_0)$ and $[0, T]$

$$\|Y\|_{x_0,\rho,T} = \sup \{ \operatorname{ess\,sup} \{ \|Y_t^i\|_{x_0,\rho} | \ t \in [0, T] \} \mid i \in \{1, \dots, n\} \}.$$

If Y is a time-independent vector field, i.e., $Y \in \Gamma^\omega(\mathsf{TM})$, we shall denote its ∞-norm on $B_\rho^{\mathbb{C}}(x_0)$ by $\|Y\|_{x_0,\rho}$.

Next, we examine the norm of high-order partial derivatives of functions and vector fields. Let $f \in C^\omega(\mathsf{M})$ and $Y \in \Gamma^\omega(\mathbb{R}; \mathsf{M})$ satisfy the same smoothness properties as before. For $\rho \in \mathbb{R}_+$ and for $m \in \mathbb{N}$, let us define the symbols

$$\|\partial^m f\|_{x_0,\rho} \triangleq \sup \left\{ \left\| \frac{\partial^m f}{\partial (x^1)^{i_1} \cdots \partial (x^n)^{i_n}} \right\|_{x_0,\rho} \;\middle|\; i_1, \dots, i_n \in \mathbb{Z}_+, \sum_{j=1}^n i_j = m \right\},$$

$$\|\partial^m Y\|_{x_0,\rho,T} \triangleq \sup \{ \operatorname{ess\,sup} \{ \|\partial^m Y_t^i\|_{x_0,\rho} | \ t \in [0, T] \} \mid i \in \{1, \dots, n\} \}.$$

Here $(x^1, \dots, x^n)$ are coordinates for M. Recall that the Cauchy Integral Representation of analytic functions leads to bounds on high-order derivatives of analytic functions in terms of the norm of the functions themselves; see the so-called Cauchy estimates in [Hörmander 1973, Theorem 2.2.7]. These estimates have the form, for $\rho' \in]0, \rho[$,

$$\|\partial^m f\|_{x_0,\rho'} \leq m!(\rho - \rho')^{-m}\|f\|_{x_0,\rho},$$
$$\|\partial^m Y\|_{x_0,\rho',T} \leq m!(\rho - \rho')^{-m}\|Y\|_{x_0,\rho,T}.$$

Given an analytic affine connection ∇ on the manifold M with Christoffel symbols $\Gamma^i_{jk} \colon \mathsf{M} \to \mathbb{R}$, $i, j, k \in \{1, \dots, n\}$, we introduce the notation:

$$\|\Gamma\|_{x_0,\rho} \triangleq \sup\left\{\|\Gamma^i_{jk}\|_{x_0,\rho} \mid i, j, k \in \{1, \dots, n\}\right\},$$
$$\|\partial^m\Gamma\|_{x_0,\rho} \triangleq \sup\left\{\|\partial^m\Gamma^i_{jk}\|_{x_0,\rho} \mid i, j, k \in \{1, \dots, n\}\right\}.$$

Again, for $\rho' \in]0, \rho[$, one can show that

$$\|\partial^m\Gamma\|_{x_0,\rho'} \leq m!(\rho - \rho')^{-m}\|\Gamma\|_{x_0,\rho}.$$

9.1.3 The variation of constants formula

In this section we introduce a set of tools that facilitate averaging analysis. We present the variation of constants formula as a means of characterizing the flows of certain time-dependent vector fields. Before we proceed, it is useful to review some properties of time-dependent vector fields. For a more complete treatment of these topics we refer to the work on chronological calculus developed by Agrachev and Gamkrelidze [1978] and documented by Agrachev and Sachkov [2004].

Recall that, for $r \in \mathbb{N} \cup \{\infty\} \cup \{\omega\}$, $\Gamma^r(\mathbb{R}; \mathsf{TM})$ is the set of C^r-time-dependent vector fields on M (see Definition 3.71). Also recall that, for $X \in \Gamma^r(\mathbb{R}; \mathsf{TM})$ and $\tau \in \mathbb{R}$, $X_\tau \in \Gamma^r(\mathsf{TM})$ is the time-independent vector field $x \mapsto X(\tau, x)$ obtained by freezing time at τ. For $X, Y \in \Gamma^{r+1}(\mathbb{R}; \mathsf{TM})$, the C^r-time-dependent vector field $[X, Y]$ defined by

$$[X, Y](t, x) = [X_t, Y_t](x)$$

is the ***(fixed-time) Lie bracket*** of X and Y. Later, it will also be convenient to define fixed-time symmetric products of time-dependent vector fields; see Section 9.3.

In what follows, we study the integral curves of the vector field $X + Y$, for $X, Y \in \Gamma^\infty(\mathbb{R}; \mathsf{TM})$. In our development, we treat X as a ***perturbation vector field*** to the ***nominal vector field*** Y. That is to say, we loosely speaking regard X as a "small" addition to Y. The relationship between the flow of $X + Y$ and the flows of X and Y is then given by the following result, recalling from Definition 3.81 the notion of pull-back of a vector field along a diffeomorphism.

Proposition 9.6 (Variation of constants formula). *For $r \in \mathbb{N} \cup \{\infty\} \cup \{\omega\}$, let $X \in \Gamma^r(\mathbb{R}; \mathsf{TM})$ and $Y \in \Gamma^{r+1}(\mathbb{R}; \mathsf{TM})$. For $x_0 \in \mathsf{M}$, $T \in \mathbb{R}_+$ and $t \in [0, T]$, assume that $\Phi^Y_{0,t}$ is a C^{r+1}-diffeomorphism from a neighborhood $\mathcal{U}$ of x_0 onto its image. Define $\Delta \in \Gamma^r(\mathbb{R}; \mathsf{T}\mathcal{U})$ by means of $\Delta(\tau, x) = ((\Phi^Y_{0,\tau})^* X_\tau)(x)$. Then*

$$\Phi^{X+Y}_{0,t}(x_0) = \Phi^Y_{0,t} \circ \Phi^\Delta_{0,t}(x_0), \tag{9.1}$$

for all $t \in [0, T]$ for which the three integral curves exist.

In other words, if the curves $\gamma, \zeta \colon [0, T] \to \mathsf{M}$ satisfy

$$\begin{aligned} \gamma'(t) &= X(t, \gamma(t)) + Y(t, \gamma(t)), & \gamma(0) &= x_0, \\ \zeta'(t) &= \big((\Phi^Y_{0,t})^* X_t\big)(\zeta(t)), & \zeta(0) &= x_0, \end{aligned} \tag{9.2}$$

then their final value $\gamma(T)$ and $\zeta(T)$ satisfy $\gamma(T) = \Phi^Y_{0,t}(\zeta(T))$. We illustrate formula (9.1) in Figure 9.1.

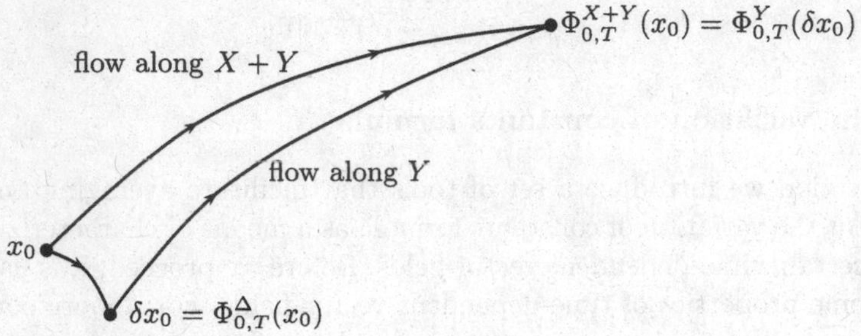

Figure 9.1. A depiction of the variation of constants formula

Proof of Proposition 9.6. Let γ and ζ be the integral curves defined in (9.2). Define $\eta \colon [0, T] \to \mathsf{M}$ by $\eta(t) = \Phi^Y_{0,t}(\zeta(t))$, and compute

$$\begin{aligned} \zeta'(t) &= \big((\Phi^Y_{0,t})^* X\big)(t, \zeta(t)) = \big(T_{\Phi^Y_{0,t}(\zeta(t))} \big(\Phi^Y_{0,t}\big)^{-1} \circ X_t \circ \Phi^Y_{0,t}\big)(\zeta(t)) \\ &= \big(T_{\zeta(t)} \Phi^Y_{0,t}\big)^{-1} \circ X(t, \eta(t)), \end{aligned}$$

so that

$$\begin{aligned} \eta'(t) &= \frac{\mathrm{d}}{\mathrm{d}t} \big(\Phi^Y_{0,t}(\zeta(t))\big) = Y\left(t, \Phi^Y_{0,t}(\zeta(t))\right) + \big(T_{\zeta(t)} \Phi^Y_{0,t}\big)\left(\zeta'(t)\right) \\ &= Y\left(t, \eta(t)\right) + X(t, \eta(t)). \end{aligned}$$

Therefore, η obeys the same differential equation as γ. Since it is also clear that $\gamma(0) = \eta(0)$, the curves γ and η must be equal. ■

Remark 9.7. Exercise E9.3 invites the reader to provide a second version of the variations of constants formula (9.1), in which the nominal flow $\Phi^Y_{0,t}$ precedes the perturbation effect. •

Example 9.8 (Trajectories of a linear control system). Consider the initial value problem on $\mathbb{R}^n$:

$$\dot{\boldsymbol{x}}(t) = \boldsymbol{A}\boldsymbol{x}(t) + \boldsymbol{B}u(t), \quad \boldsymbol{x}(0) = \boldsymbol{x}_0,$$

for $\boldsymbol{A} \in \mathbb{R}^{n\times n}$, $\boldsymbol{B} \in \mathbb{R}^{n\times m}$, and $u\colon \mathbb{R} \to \mathbb{R}^m$. We shall compute the solution to this initial value problem using variations of constants formula (9.1). Let $Y(\boldsymbol{y}) = \boldsymbol{A}\boldsymbol{y}$ be the nominal vector field and $X(t,\boldsymbol{y}) = \boldsymbol{B}u(t)$ be the perturbation vector field. By Proposition 5.5, we compute $\Phi_t^Y(\boldsymbol{y}) = \exp(t\boldsymbol{A})\boldsymbol{y}$. By Definition 3.81, we compute

$$\left((\Phi_t^Y)^* X\right)(t,\boldsymbol{y}) = \exp(-t\boldsymbol{A})\boldsymbol{B}u(t).$$

If $\boldsymbol{z}\colon \mathbb{R} \to \mathbb{R}^n$ satifies

$$\dot{\boldsymbol{z}}(t) = \exp(-t\boldsymbol{A})\boldsymbol{B}u(t), \quad \boldsymbol{z}(t) = \boldsymbol{x}_0,$$

then Proposition 9.6 states that $\boldsymbol{x}(t) = \Phi_t^Y(\boldsymbol{z}(t))$. Now, note that, if $\boldsymbol{z}\colon \mathbb{R} \to \mathbb{R}^n$ satifies $\dot{\boldsymbol{z}}(t) = Z(t,\boldsymbol{z}(t))$ for some vector field Z on $\mathbb{R}^n$, then $\boldsymbol{z}(t) = \boldsymbol{z}(0) + \int_0^t Z(\tau,\boldsymbol{z}(\tau))\mathrm{d}\tau$. In summary, we obtain

$$\boldsymbol{x}(t) = \exp(t\boldsymbol{A})\boldsymbol{x}_0 + \int_0^t \exp((t-\tau)\boldsymbol{A})\boldsymbol{B}u(\tau)\mathrm{d}\tau.$$

This equality is sometimes referred to as Cauchy's formula. •

Motivated by the previous result, we now obtain a series expansion representation for the pull-back of a vector field X along the flow map of a vector field Y.

Proposition 9.9. *Let $X, Y \in \Gamma^\infty(\mathbb{R};\mathsf{TM})$. If the series*

$$\sum_{k=1}^{+\infty} \int_0^t \cdots \int_0^{s_{k-1}} \left(\mathrm{ad}_{Y_{s_k}} \cdots \mathrm{ad}_{Y_{s_1}} X_t(x)\right) \mathrm{d}s_k \cdots \mathrm{d}s_1$$

contains only a finite number of non-vanishing terms, then

$$\begin{aligned}\left((\Phi_{0,t}^Y)^* X\right)&(t,x)\\ &= X(t,x) + \sum_{k=1}^{+\infty} \int_0^t \cdots \int_0^{s_{k-1}} \left(\mathrm{ad}_{Y_{s_k}} \cdots \mathrm{ad}_{Y_{s_1}} X_t(x)\right) \mathrm{d}s_k \cdots \mathrm{d}s_1. \quad (9.3)\end{aligned}$$

Proof. Let $\tau \in [0,T]$. One can show, e.g., [Abraham, Marsden, and Ratiu 1988, Theorem 4.2.31], that the result in Proposition 3.85 can be extended to the following:

$$\frac{\mathrm{d}}{\mathrm{d}t}\left((\Phi_{0,t}^Y)^* X_\tau\right)(x) = (\Phi_{0,t}^Y)^* [Y_t, X_\tau](x).$$

At fixed $x \in \mathsf{M}$, we integrate the previous equation from time 0 to t to obtain

$$\left(\left(\Phi^Y_{0,t}\right)^* X_\tau\right)(x) = X_\tau(x) + \int_0^t (\Phi^Y_{0,s})^*[Y_s, X_\tau](s,x)\mathrm{d}s.$$

The expansion in equation (9.3) follows from iteratively applying the previous equality and, finally, setting $\tau = t$. Convergence is automatically guaranteed by the presence of only a finite number of terms. ■

Remarks 9.10. 1. If X and Y are time-independent vector fields, Proposition 9.9 is sometimes referred to as the Infinitesimal Campbell–Baker–Hausdorff Formula, see [Isidori 1995]. In this case, equation (9.3) reads

$$(\Phi^Y_t)^* X = \sum_{k=0}^{+\infty} \frac{t^k}{k!} \operatorname{ad}^k_Y X.$$

This vector field is sometimes referred to as the "pull-back" or the "adjoint" system, e.g., see [Hermes 1991].

2. The convergence properties for the series expansion in (9.3) are difficult to characterize in the general case of an infinite number of non-vanishing terms. We refer the interested reader to the sufficient conditions for local convergence in [Agrachev and Gamkrelidze 1978, Proposition 2.1 and 3.1], and to a related discussion in [Varadarajan 1984] on the convergence of the Campbell–Baker–Hausdorff Formula. •

Next, let us introduce a convenient notion.

Definition 9.11 (Parameter- and time-dependent vector field). Let $r \in \mathbb{N} \cup \{\infty\} \cup \{\omega\}$.

(i) A ***C^r-parameter- and time-dependent vector field*** is a map $X : \mathbb{R} \times \mathbb{R} \times \mathsf{M} \to \mathsf{TM}$ with the property that, for each $\lambda \in \mathbb{R}$, the map $(\tau, x) \mapsto X(\lambda, \tau, x)$ is a C^r-time-dependent vector field.

(ii) A ***C^r-parameter-dependent vector field*** is a map $X : \mathbb{R} \times \mathsf{M} \to \mathsf{TM}$ with the property that, for each $\lambda \in \mathbb{R}$, the map $x \mapsto X(\lambda, x)$ is a C^r-vector field. •

Remarks 9.12. 1. Given a C^r-parameter- and time-dependent vector field X, it is convenient to introduce the shorthands $X_{\lambda,\tau} \in \Gamma^r(\mathsf{TM})$ and $X_\lambda \in \Gamma^r(\mathbb{R}; \mathsf{TM})$ by means of $X(\lambda, \tau, x) = X_\lambda(\tau, x) = X_{\lambda,\tau}(x)$.

2. If $\lambda \mapsto X(\lambda, \tau, x)$ is of class C^r for all $(\tau, x) \in \mathbb{R} \times \mathsf{M}$, then the map $\lambda \mapsto \Phi^{X_\lambda}_{0,\tau}(x)$ is of class C^{r+1}; see [Abraham, Marsden, and Ratiu 1988] for a related discussion. •

This notion allows us to present an application of the variation of constants formula that will be useful later in the chapter. Recall that a family of vector fields is commutative if all pairwise Lie brackets vanish. The next result is Proposition 2.7 in [Agrachev and Sachkov 2004].

Proposition 9.13 (Flow of a commutative time-dependent vector field). *Let $Y \in \Gamma^1(\mathbb{R}; \mathsf{TM})$ be of class C^0. Define the C^1-parameter-dependent vector field X by $X(\lambda, x) = \int_0^\lambda Y(\tau, x)\mathrm{d}\tau$, and assume that*

(i) the family of vector fields Y_t, $t \in [0, T]$, is commutative, and that

(ii) for $x_0 \in \mathsf{M}$ and $\lambda \in [0, T]$, the flow $\Phi_1^{X_\lambda}$ is a C^1-diffeomorphism from a neighborhood $\mathcal{U}$ of x_0 onto its image.

Then, for $t \in [0, T]$,

$$\Phi_{0,t}^Y(x_0) = \Phi_1^{X_\lambda}(x_0)\big|_{\lambda=t}.$$

In other words, if Y is a time-dependent vector field satisfying the commutativity assumption (i), then its integral curve can be computed by, first, integrating Y with respect to time and, second, computing the integral curve of the resulting time-independent vector field. The key assumption here is commutativity.

Proof of Proposition 9.13. Because $\Phi_{0,0}^Y(x_0) = \Phi_1^{X_\lambda}(x_0)\big|_{\lambda=0} = x_0$, it suffices to show that the curves $t \mapsto \Phi_{0,t}^Y(x_0)$ and $t \mapsto \Phi_1^{X_\lambda}(x_0)\big|_{\lambda=t}$ satisfy the same differential equation. Clearly

$$\frac{\mathrm{d}}{\mathrm{d}t}\Phi_{0,t}^Y(x_0) = Y_t \circ \Phi_{0,t}^Y(x_0).$$

To differentiate the second curve, we treat X as a time-independent vector-field depending on the parameter λ. Because of (ii), we may apply the result in Exercise E9.4:

$$\frac{\mathrm{d}}{\mathrm{d}\lambda}\Phi_1^{X_\lambda}(x_0) = \int_0^1 \Big((\Phi_{\tau-1}^{X_\lambda})^* \frac{\partial X_\lambda}{\partial \lambda}\Big)\mathrm{d}\tau \circ \Phi_1^{X_\lambda}(x_0).$$

By construction we know $\frac{\partial X_\lambda}{\partial \lambda} = Y_\lambda$. Assumption (i) implies $[X_t, Y_\tau] = 0$ for all $(\tau, t) \in [0, T] \times [0, T]$. Therefore, we can apply Proposition 9.9 to infer that $(\Phi_s^{X_\lambda})^* Y_\lambda = Y_\lambda$, for $s \in \mathbb{R}$. In turn, this proves that

$$\frac{\mathrm{d}}{\mathrm{d}\lambda}\Phi_1^{X_\lambda}(x_0) = Y_\lambda \circ \Phi_1^{X_\lambda}(x_0).$$ ■

Let us comment on this result by presenting an immediate consequence.

Remark 9.14. A time-dependent vector field is ***zero-mean*** if it is T-periodic and zero-mean in the time argument, for some $T \in \mathbb{R}_+$. Let $Y \in \Gamma^1(\mathbb{R}; \mathsf{TM})$ have the same properties as in the proposition. If, additionally, Y is zero-mean, then any integral curve of Y is a T-periodic function of time. •

9.1.4 First-order averaging

In this section we present some basic results on averaging theory and their coordinate-invariant interpretation. Because all results are local, we restrict our attention to a bounded open subset M of $\mathbb{R}^n$. For this section, we will write a typical point in M as x rather than $\boldsymbol{x}$.

Let M be a bounded open neighborhood of $x_0 \in \mathbb{R}^n$, let $X \in \Gamma^1(\bar{\mathbb{R}}_+; \mathsf{TM})$ be a time-dependent vector field, and for $\varepsilon \in \mathbb{R}_+$, consider the initial value problem in ***standard form***:

$$\gamma'(t) = \varepsilon X(t, \gamma(t)), \quad \gamma(0) = x_0. \tag{9.4}$$

When X is T-periodic in time, we define the ***averaged vector field*** $\overline{X} \in \Gamma^1(\mathsf{TM})$ by

$$\overline{X}(x) = \frac{1}{T} \int_0^T X(\tau, x) \mathrm{d}\tau.$$

In what follows, an open ball of radius $O(\varepsilon)$ centered at x_0 is an open ball centered at x_0 whose radius is a function of ε that is $O(\varepsilon)$. From [Sanders and Verhulst 1985], we have the following result.

Theorem 9.15 (First-order averaging). *Let M be a bounded open neighborhood of $x_0 \in \mathbb{R}^n$, and let $X \in \Gamma^1(\bar{\mathbb{R}}_+; \mathsf{TM})$ be a map of class C^0 and be T-periodic in time. Let $\varepsilon_0, t_0 \in \mathbb{R}_+$ be such that $\Phi_t^{\varepsilon\overline{X}}(x_0) \in \mathsf{M}$ for all $\varepsilon \in]0, \varepsilon_0[$ and for all $t \in [0, \frac{t_0}{\varepsilon}]$. Then, for all $t \in [0, \frac{t_0}{\varepsilon}]$,*

(i) $\Phi_{0,t}^{\varepsilon X}(x_0) = \Phi_t^{\varepsilon\overline{X}}(x_0) + O(\varepsilon)$.

Additionally, if $x^* \in \mathsf{M}$ is a linearly asymptotically stable equilibrium point for $\overline{X}$, then

(ii) there exists $\rho \in \mathbb{R}_+$ such that if $\|x_0 - x^*\|_{\mathbb{R}^n} < \rho$, then the estimate in part (i) holds for all $t \in \mathbb{R}_+$; and

(iii) there exists $\varepsilon_1 \in \mathbb{R}_+$ such that, for all $\varepsilon \in]0, \varepsilon_1[$, the vector field εX has a unique T-periodic trajectory that is locally asymptotically stable and that takes values in an open ball of radius $O(\varepsilon)$ centered at x^*.

In other words, if we define the ***averaged trajectory*** $t \mapsto \eta(t) \in \mathsf{M}$ by

$$\eta'(t) = \varepsilon \overline{X}(\eta(t)), \quad \eta(0) = x_0, \tag{9.5}$$

and if γ is the solution to the initial value problem in standard form (9.4), then we know that $\gamma(t) - \eta(t) = O(\varepsilon)$ for all $t \in [0, \frac{t_0}{\varepsilon}]$, for some $t_0 \in \mathbb{R}_+$.

Proof of Theorem 9.15. We refer the reader to [Sanders and Verhulst 1985] for a detailed proof of Theorem 9.15. In what follows, we outline a proof of the equality in part (i) that relies on the additional assumption that $X \in \Gamma^2(\bar{\mathbb{R}}_+; \mathsf{TM})$.

Define the C^2-parameter-dependent vector field $Y(\lambda, x) = \int_0^\lambda \big(X_\tau(x) - \overline{X}(x)\big)\mathrm{d}\tau$. Exercise E9.4 implies that

$$\frac{\partial}{\partial \lambda} \Phi_1^{\varepsilon Y_\lambda}(x) = \varepsilon \int_0^1 \big(\Phi_{s-1}^{\varepsilon Y_\lambda}\big)^* \big(X_\lambda(x) - \overline{X}(x)\big) \mathrm{d}s.$$

Define the curve $t \mapsto \xi(t) \in \mathsf{M}$ by

$$\xi(t) = \Phi_1^{\varepsilon Y_t} \circ \Phi_t^{\varepsilon \overline{X}}(x_0).$$

Note that $Y_\lambda = Y_{\lambda \bmod T}$ implies that $\Phi_1^{\varepsilon Y_\lambda}(x) = x + O(\varepsilon)$, for all $\lambda \in \mathbb{R}$. Therefore, to prove part (i), it suffices to show that $t \mapsto \Phi_{0,t}^{\varepsilon X}(x_0)$ differs from ξ by an $O(\varepsilon)$ term over a time interval of length of order $O(\frac{1}{\varepsilon})$. Compute the tangent vector field to ξ as

$$\xi'(t) = \varepsilon \left(\int_0^1 (\Phi_{s-1}^{\varepsilon Y_t})^* (X_t - \overline{X}) \mathrm{d}s + (\Phi_{-1}^{\varepsilon Y_t})^* \overline{X} \right) \circ \xi(t) \triangleq \varepsilon V_{\varepsilon,t} \circ \xi(t).$$

Here V is a C^1-parameter- and time-dependent vector field with the property that $V_0 = X$. This implies that there exists a C^1-parameter- and time-dependent vector field W such that $V_\varepsilon = X + \varepsilon W_\varepsilon$. Finally, it suffices to analyze the two curves:

$$\gamma'(t) = \varepsilon X(t, \gamma(t)), \qquad \gamma(0) = x_0,$$
$$\xi'(t) = \varepsilon X(t, \xi(t)) + \varepsilon^2 W_\varepsilon(t, \xi(t)), \qquad \xi(0) = x_0.$$

A time-scaling argument and an application of the smoothness properties of the flow of a parameter-dependent vector field, see Remark 9.12–2, show that these curves differ by an $O(\varepsilon)$ term over a time interval of $O(\frac{1}{\varepsilon})$ length. ■

Examples 9.16. 1. Consider the initial value problem on $\mathbb{R}$:

$$\dot{x}(t) = \varepsilon a (\cos t)^2 x(t), \quad x(0) = x_0,$$

for $\varepsilon \in \mathbb{R}_+$ and for $a \in \{-1, +1\}$. One can verify directly that the nominal solution is

$$x(t) = x_0 \mathrm{e}^{\varepsilon a (t + \cos t \sin t)/2}.$$

Using $\frac{1}{2\pi} \int_0^{2\pi} (\cos t)^2 \mathrm{d}t = \frac{1}{2}$, we compute the averaged trajectory as the curve $t \mapsto y(t) = x_0 \mathrm{e}^{\varepsilon a t/2}$ solution to

$$\dot{y}(t) = \tfrac{1}{2} \varepsilon a y(t), \quad y(0) = x_0.$$

The following statement is a direct consequence of the expressions for $t \mapsto x(t)$ and $t \mapsto y(t)$: for $\varepsilon_0 \in \mathbb{R}_+$, there exists $K \in \mathbb{R}_+$ such that, for all $t \in \mathbb{R}$, $\varepsilon \in [0, \varepsilon_0]$ and $a \in \{-1, +1\}$,

$$|x(t) - y(t)| = |x_0 \mathrm{e}^{\varepsilon a t/2}||1 - \mathrm{e}^{\varepsilon a (\cos t \sin t)/2}| \leq \varepsilon K |x_0 \mathrm{e}^{\varepsilon a t/2}|. \tag{9.6}$$

One can verify the following statements either directly from (9.6) or by invoking Theorem 9.15. For $\varepsilon \in [0, \varepsilon_0]$ and $a \in \{-1, +1\}$, the curves x and y differ by a function that is $O(\varepsilon)$ on any interval $[0, t_0]$, for t_0 independent of ε. For $a = -1$, the origin is an asymptotically stable equilibrium for the averaged system and, therefore, (1) x and y differ by a function that is $O(\varepsilon)$ on $\mathbb{R}$, and (2) the nominal system has a unique 2π-periodic trajectory with the properties in Theorem 9.15(iii) (the trivial trajectory $t \mapsto 0$ in this case).

2. Let us now provide a generalization of the result in part 1. Consider the initial value problem on $\mathbb{R}^n$:

$$\dot{\boldsymbol{x}}(t) = \varepsilon(\cos t)^2 \boldsymbol{A}\boldsymbol{x}(t), \quad \boldsymbol{x}(0) = \boldsymbol{x}_0,$$

for $\varepsilon \in \mathbb{R}_+$, and $\boldsymbol{A} \in \mathbb{R}^{n\times n}$. We compute the averaged trajectory as the solution to

$$\dot{\boldsymbol{y}}(t) = \tfrac{1}{2}\varepsilon \boldsymbol{A}\boldsymbol{x}(t), \quad \boldsymbol{y}(0) = \boldsymbol{x}_0.$$

Then, there exist $\varepsilon_0, t_0 \in \mathbb{R}_+$ such that for all $\varepsilon \in]0, \varepsilon_0[$ and for all $t \in [0, \frac{t_0}{\varepsilon}]$, the nominal trajectory $t \mapsto \boldsymbol{x}(t)$ and the averaged trajectory $t \mapsto \exp(\frac{1}{2}\varepsilon t\boldsymbol{A})\boldsymbol{x}_0$ differ by a function that is $O(\varepsilon)$. •

Remark 9.17. Many extensions of Theorem 9.15 are possible, and we briefly give one. Consider the initial value problem:

$$\gamma'(t) = X\Big(\frac{t}{\varepsilon}, \gamma(t)\Big), \quad \gamma(0) = x_0,$$

where $X \in \Gamma^1(\bar{\mathbb{R}}_+; \mathsf{TM})$ is a map of class C^0 and T-periodic in time. A time scaling argument shows that this differential equation can be written in the same form as equation (9.5). Accordingly, Theorem 9.15 implies that, if $\Phi_t^{\overline{X}}(x_0) \in \mathsf{M}$ for all $t \in [0, t_0]$, then $\gamma(t) - \eta(t) = O(\varepsilon)$ for all $t \in [0, t_0]$. Remarkably, this estimate now *only* holds for finite time, independent on ε. •

9.1.5 Averaging of systems subject to oscillatory inputs

In this subsection we study the averaged behavior of systems subject to oscillatory inputs, i.e., inputs of large amplitude and high frequency. We begin with a basic model.

As above, let M be an open bounded subset of $\mathbb{R}^n$. Let $X\colon \bar{\mathbb{R}}_+ \times \mathsf{M} \to \mathsf{TM}$ and $Y\colon \bar{\mathbb{R}}_+ \times \bar{\mathbb{R}}_+ \times \mathsf{M} \to \mathsf{TM}$ have the property that $(t, x) \mapsto X(t, x)$ and $(t, x) \mapsto Y(\frac{t}{\varepsilon}, t, x)$ are time-dependent vector fields, for all $\varepsilon \in \mathbb{R}_+$. For $T_0 \in \mathbb{R}_+$, let $\gamma\colon [0, T_0] \to \mathsf{M}$ satisfy

$$\gamma'(t) = X(t, \gamma(t)) + \frac{1}{\varepsilon}Y\Big(\frac{t}{\varepsilon}, t, \gamma(t)\Big), \quad \gamma(0) = x_0 \in \mathsf{M}. \tag{9.7}$$

We refer to (9.7) as an initial value problem in ***standard oscillatory form***.

We regard Y as a vector field depending on time in two different ways; it is sometimes useful to refer to t as the "slow time-scale" and $\frac{t}{\varepsilon}$ as the "fast time-scale." Typically, it makes sense to require the vector field Y to be a periodic zero-mean function of the fast time-scale and to perform averaging with respect to the fast time-scale. It is convenient to provide formal definitions of these concepts.

Definition 9.18 (Two-time-scales vector field). Let $r \in \mathbb{N} \cup \{\infty\} \cup \{\omega\}$.

(i) A ***two-time-scales vector field of class C^r*** is a map $Y : \mathbb{R}\times\mathbb{R}\times\mathsf{M} \to \mathsf{TM}$ with the following properties:
   (a) for $\varepsilon \in \mathbb{R}_+$, $(t,x) \mapsto Y(\frac{t}{\varepsilon}, t, x)$ is a C^r-time-dependent vector field;
   (b) the map $(\tau, t, x) \mapsto Y(\tau, t, x)$ is of class C^0.

(ii) A two-time-scales vector field Y is ***zero-mean*** if it is T-periodic in the first time argument, for some $T \in \mathbb{R}_+$, and zero-mean in the first time argument.

(iii) We let $\Gamma^r(\mathbb{R}\times\mathbb{R};\mathsf{TM})$ and $\Gamma^r_T(\mathbb{R};\mathsf{TM})$ denote the set of two-time-scales vector fields of class C^r and the set of zero-mean two-time-scales vector fields of class C^r, respectively. •

Remark 9.19. It will also be convenient to use the following two generalizations. Let $\mathcal{D}$ be a regular C^r-distribution on M and let $I_1 \times I_2 \subset \mathbb{R}\times\mathbb{R}$. We let $\Gamma^r(I_1\times I_2;\mathcal{D})$ denote the set of two-time-scales vector fields taking values in $\mathcal{D}$ and whose time arguments take values in $I_1 \times I_2$. •

We can now return to the system of interest in equation (9.7) and specify the assumptions under which we will study it.

Assumptions 9.20. In this section we shall assume that:

(i) $X \in \Gamma^1(\bar{\mathbb{R}}_+;\mathsf{TM})$ is of class C^1;

(ii) $Y \in \Gamma^2_T(\bar{\mathbb{R}}_+;\mathsf{TM})$ has the property that the maps $(t,x) \mapsto Y(\tau,t,x)$, for $\tau \in \mathbb{R}$, are of class C^1;

(iii) the vector fields $x \mapsto Y(\tau,t,x)$, for $(\tau,t) \in \bar{\mathbb{R}}_+ \times \bar{\mathbb{R}}_+$, are commutative. •

It is convenient to introduce the following shorthands. For $Y \in \Gamma^2(\bar{\mathbb{R}}_+ \times \bar{\mathbb{R}}_+;\mathsf{TM})$, $\varepsilon \in \mathbb{R}_+$ and $t_0 \in \bar{\mathbb{R}}_+$, we define the time-dependent vector fields $Y_{1,\varepsilon}$ and Y_{2,t_0} in $\Gamma^2(\bar{\mathbb{R}}_+;\mathsf{TM})$ by means of

$$Y_{1,\varepsilon}(t,x) = Y\Big(\frac{t}{\varepsilon}, t, x\Big), \quad \text{and} \quad Y_{2,t_0}(\tau,x) = Y(\tau,t_0,x), \tag{9.8}$$

respectively. We now introduce vector fields on an extended manifold. Let $\mathsf{M}_{\mathrm{e}} = \bar{\mathbb{R}}_+ \times \mathsf{M}$ and define the vector fields

$$\begin{aligned} X_{\mathrm{e}} \colon \mathsf{M}_{\mathrm{e}} &\to \mathsf{TM}_{\mathrm{e}} \\ (t,x) &\mapsto (1, X(t,x)) \end{aligned} \quad \text{and} \quad \begin{aligned} Y_{\mathrm{e}} \colon \bar{\mathbb{R}}_+ \times \mathsf{M}_{\mathrm{e}} &\to \mathsf{TM}_{\mathrm{e}} \\ (\tau,(t,x)) &\mapsto (0, Y(\tau,t,x)). \end{aligned}$$

The time-independent vector field X_{e} is referred to as the "suspension of X" in [Abraham, Marsden, and Ratiu 1988]. We now establish useful relationship between flows in M and M_{e}.

Lemma 9.21. *For $X \in \Gamma^1(\bar{\mathbb{R}}_+;\mathsf{TM})$, $Y \in \Gamma^2_T(\bar{\mathbb{R}}_+;\mathsf{TM})$, $x \in \mathsf{M}$, and $\tau \in \bar{\mathbb{R}}_+$, we have*

$$\Phi^{\varepsilon X_{\mathrm{e}}+Y_{\mathrm{e}}}_{0,\frac{t}{\varepsilon}}(0,x) = \big(t, \Phi^{X+\frac{1}{\varepsilon}Y_{1,\varepsilon}}_{0,t}(x)\big),$$

$$\Phi^{Y_{\mathrm{e}}}_{0,\tau}(t,x) = \big(t, \Phi^{Y_{2,\lambda}}_{0,\tau}(x)\big)\big|_{\lambda=t},$$

for all $t \in [0,T_0]$ for which the integral curves exist.

By definition, equation (9.7) is equivalent to

$$\gamma(t) = \Phi_{0,t}^{X+\frac{1}{\varepsilon}Y_{1,\varepsilon}}(x_0).$$

Lemma 9.21 states that, if $\eta\colon [0, \frac{T_0}{\varepsilon}] \to \mathsf{M}_{\mathrm{e}}$ satisfies

$$\eta'(\tau) = \varepsilon X_{\mathrm{e}}(\eta(\tau)) + Y_{\mathrm{e}}(\tau, \eta(\tau)), \quad \eta(0) = (0, x) \in \mathsf{M}_{\mathrm{e}}, \tag{9.9}$$

then, for $t \in [0, T_0]$, the solution γ to equation (9.7) satisfies $(t, \gamma(t)) = \eta(\frac{t}{\varepsilon})$. Loosely speaking, t is the "slow" time-scale and τ is the "fast" time-scale.

Proof of Lemma 9.21. In coordinates we write $\eta(\tau) = (\eta_1(\tau), \eta_2(\tau)) \in \mathbb{R} \times \mathsf{M}$. Then equation (9.9) reads

$$\begin{aligned}
\eta_1'(\tau) &= \varepsilon \quad \Longrightarrow \quad \eta_1(\tau) = \varepsilon\tau, \\
\eta_2'(\tau) &= \varepsilon X(\eta_1(\tau), \eta_2(\tau)) + Y(\tau, \eta_1(\tau), \eta_2(\tau)) \\
&= \varepsilon X(\varepsilon\tau, \eta_2(\tau)) + Y_{1,\varepsilon}(\varepsilon\tau, \eta_2(\tau)).
\end{aligned}$$

When $\tau = \tau(t) = \frac{t}{\varepsilon}$, we have that $\eta_1 \circ \tau(t) = t$ and $\eta_2 \circ \tau$ satisfies equation (9.7). To show the second statement, let $\chi = (\chi_1, \chi_2)\colon [0, \frac{T_0}{\varepsilon}] \to \mathsf{M}_{\mathrm{e}}$ solve

$$\chi'(\tau) = Y_{\mathrm{e}}(\tau, \chi(\tau)), \quad \chi(0) = (t, x) \in \mathsf{M}_{\mathrm{e}},$$

and write

$$\begin{aligned}
\chi_1'(\tau) &= 0 \quad \Longrightarrow \quad \chi_1(\tau) = t, \\
\chi_2'(\tau) &= Y(\tau, \chi_1(\tau), \chi_2(\tau)) = Y_{2,\lambda}(\tau, \chi_2(\tau))\big|_{\lambda=t}.
\end{aligned}$$

■

Let us now return to the analysis of equation (9.7) by means of the equivalent equation (9.9). We shall use the variation of constants formula (9.1), regarding Y_{e} as the nominal vector field and X_{e} as the perturbation. To do this, we define the pull-back vector field F_{e} as the time-dependent vector field on M_{e} given by

$$F_{\mathrm{e}}(\tau, x_{\mathrm{e}}) = \Big((\Phi_{0,\tau}^{Y_{\mathrm{e}}})^* X_{\mathrm{e}}\Big)(x_{\mathrm{e}})$$

for $(\tau, x_{\mathrm{e}}) \in \bar{\mathbb{R}}_+ \times \mathsf{M}_{\mathrm{e}}$.

Note that Assumptions 9.20 implies that $X_{\mathrm{e}} \in \Gamma^1(\mathsf{TM}_{\mathrm{e}})$, that $Y_{\mathrm{e}} \in \Gamma^1(\bar{\mathbb{R}}_+; \mathsf{TM}_{\mathrm{e}})$, and that Y_{e} is of class C^2 in the state x. Additionally, the flow $\Phi_{0,\tau}^{Y_{\mathrm{e}}}\colon \mathsf{M}_{\mathrm{e}} \to \mathsf{M}_{\mathrm{e}}$ is of class C^2 in the extended space $x_{\mathrm{e}} = (t, x)$ because of Remark 9.12–2. Therefore, we know that $F_{\mathrm{e}} \in \Gamma^1(\bar{\mathbb{R}}_+; \mathsf{TM}_{\mathrm{e}})$.

Note that F_{e} is of the form

$$F_{\mathrm{e}}(\tau, x_{\mathrm{e}}) = (1, F(\tau, x_{\mathrm{e}})), \tag{9.10}$$

for some map $F\colon \bar{\mathbb{R}}_+ \times \mathsf{M}_{\mathrm{e}} \to \mathsf{TM}$. Because of Proposition 9.13 and Assumptions 9.20 (see Remark 9.14), the flow $\tau \mapsto \Phi_{0,\tau}^{Y_{\mathrm{e}}}$, and thus the vector field F_{e}, are T-periodic. We can therefore define the averaged vector field $\overline{F}_{\mathrm{e}} \in \Gamma^1(\mathsf{TM}_{\mathrm{e}})$, and note that $\overline{F}_{\mathrm{e}}(x_{\mathrm{e}}) = (1, \overline{F}(x_{\mathrm{e}}))$, for some $\overline{F} \in \Gamma^1(\bar{\mathbb{R}}_+; \mathsf{TM})$.

We are finally ready to state the main averaging theorems for initial value problems in standard oscillatory form.

Theorem 9.22 (Averaging in systems with oscillatory inputs). *Let* M *be an open bounded subset of* $\mathbb{R}^n$*, and let the vector fields* X *and* Y *satisfy Assumptions 9.20. Assume that* $T_0 \in \mathbb{R}_+$ *has the property that* $\xi\colon [0, T_0] \to \mathsf{M}$ *satisfies*

$$\xi'(t) = \overline{F}(t, \xi(t)), \quad \xi(0) = x_0.$$

Then there exists $\varepsilon_0 \in \mathbb{R}_+$ *such that, for all* $\varepsilon \in]0, \varepsilon_0[$*, the solution* γ *to the initial value problem (9.7) is defined on* $[0, T_0]$ *and satisfies*

$$\gamma(t) = \Phi^{Y_{2,\lambda}}_{0,(\frac{t}{\varepsilon} \bmod T)}(\xi(t))\Big|_{\lambda=t} + O(\varepsilon). \tag{9.11}$$

Next, assume that X *is time-independent and that* Y *does not depend on the slow time-scale* t *in the sense that, with a slight abuse of notation,* $X(t,x) = X(x)$ *and* $Y(\tau, t, x) = Y(\tau, x)$*. If* $x^* \in \mathsf{M}$ *is a linearly asymptotically stable equilibrium point for* $\overline{F}$*, then*

(i) there exists $\rho \in \mathbb{R}_+$ *such that, if* $\|x_0 - x^*\|_{\mathbb{R}^n} < \rho$*, the estimate (9.11) holds for all* $t \in \mathbb{R}_+$*, and*

(ii) there exists $\varepsilon_1 \in \mathbb{R}_+$ *such that, for all* $\varepsilon \in]0, \varepsilon_1[$*, the vector field* $(t, x) \mapsto X(x) + \frac{1}{\varepsilon} Y(\frac{t}{\varepsilon}, x)$ *has a unique* εT*-periodic locally asymptotically stable trajectory that takes values in an open ball of radius* $O(1)$ *centered at* x^**.*

Proof. First, note that $(\tau, x_\mathrm{e}) \mapsto Y_\mathrm{e}(\tau, x_\mathrm{e})$ depends continuously on τ and so does $(\tau, x_\mathrm{e}) \mapsto \Phi^{Y_\mathrm{e}}_{0,\tau}(x_\mathrm{e})$. Therefore, F_e depends continuously on time. We can now compute

$$\begin{aligned}(t, \Phi^{X + \frac{1}{\varepsilon} Y_{1,\varepsilon}}_{0,t}(x)) &= \Phi^{\varepsilon X_\mathrm{e} + Y_\mathrm{e}}_{0,\frac{t}{\varepsilon}}(0, x) = \Phi^{Y_\mathrm{e}}_{0,\frac{t}{\varepsilon}} \circ \Phi^{\varepsilon F_\mathrm{e}}_{0,\frac{t}{\varepsilon}}(0, x) \\ &= \Phi^{Y_\mathrm{e}}_{0,\frac{t}{\varepsilon}} \circ \left(\Phi^{\varepsilon \overline{F}_\mathrm{e}}_{0,\frac{t}{\varepsilon}}(0, x) + O(\varepsilon) \right),\end{aligned}$$

where we have used the first equivalence results in Lemma 9.21, the variations of constants formula in Proposition 9.6, and the first-order averaging result in Theorem 9.15. Because of the periodicity of the flow of Y_e, and by continuity, we have

$$\begin{aligned}(t, \Phi^{X + \frac{1}{\varepsilon} Y_{1,\varepsilon}}_{0,t}(x)) &= \Phi^{Y_\mathrm{e}}_{0,(\frac{t}{\varepsilon} \bmod T)} \circ \left(\Phi^{\varepsilon \overline{F}_\mathrm{e}}_{0,\frac{t}{\varepsilon}}(0, x) + O(\varepsilon) \right) \\ &= \Phi^{Y_\mathrm{e}}_{0,(\frac{t}{\varepsilon} \bmod T)} \circ \Phi^{\varepsilon \overline{F}_\mathrm{e}}_{0,\frac{t}{\varepsilon}}(0, x) + O(\varepsilon).\end{aligned}$$

Let $\pi_\mathsf{M}\colon \mathsf{M}_\mathrm{e} = [0, T] \times \mathsf{M} \to \mathsf{M}$ be the projection onto the second factor. Using the second equivalence result in Lemma 9.21, and projecting onto the second factor, we have

$$\begin{aligned}\Phi^{X + \frac{1}{\varepsilon} Y_{1,\varepsilon}}_{0,t}(x) &= \pi_\mathsf{M} \circ \Phi^{Y_\mathrm{e}}_{0,(\frac{t}{\varepsilon} \bmod T)} \circ \Phi^{\varepsilon \overline{F}_\mathrm{e}}_{0,\frac{t}{\varepsilon}}(0, x) + O(\varepsilon) \\ &= \Phi^{Y_{2,\lambda}}_{0,(\frac{t}{\varepsilon} \bmod T)} \circ \pi_\mathsf{M} \circ \Phi^{\varepsilon \overline{F}_\mathrm{e}}_{0,\frac{t}{\varepsilon}}(0, x)\Big|_{\lambda=t} + O(\varepsilon).\end{aligned}$$

The estimate (9.11) follows from noting that

$$\Phi^{\varepsilon \overline{F}_{\mathrm{e}}}_{0,\frac{t}{\varepsilon}}(0,x) = \Phi^{\overline{F}_{\mathrm{e}}}_{0,t}(0,x) = \left(t, \Phi^{\overline{F}}_{0,t}(x)\right).$$

Next, assume that X and Y do not depend on the slow time-scale t. Under this assumption, we have that $(\tau, x) \mapsto F(\tau, x) = (\Phi^Y_{0,\tau})^* X(x)$ is a T-periodic vector field on M, and that

$$\Phi^{X+\frac{1}{\varepsilon}Y_{1,\varepsilon}}_{0,t}(x) = \Phi^{\varepsilon X+Y}_{0,\frac{t}{\varepsilon}}(x) = \Phi^Y_{0,\frac{t}{\varepsilon}} \circ \Phi^{\varepsilon F}_{0,\frac{t}{\varepsilon}}(x). \tag{9.12}$$

If $x^* \in \mathsf{M}$ is a linearly asymptotically stable equilibrium point for $\overline{F}$, then, by Theorem 9.15(ii), there exists $\rho \in \mathbb{R}_+$ such that, if $\|x_0 - x^*\|_{\mathbb{R}^n} < \rho$, then the estimate (9.11) holds for all $t \in \mathbb{R}_+$. Furthermore, by Theorem 9.15(iii), there exists $\varepsilon_1 \in \mathbb{R}_+$ such that, for all $\varepsilon \in \,]0, \varepsilon_1[$, the vector field εF has a unique T-periodic trajectory that is locally asymptotically stable and that takes values in an open ball of radius $O(\varepsilon)$ centered at x^*. This fact, equation (9.12), and the observation that $\Phi^Y_{0,\frac{t}{\varepsilon}}$ is εT-periodic, together imply that $X + \frac{1}{\varepsilon}Y_{1,\varepsilon}$ has a unique εT-periodic trajectory that is locally asymptotically stable and that takes values in an open ball of radius $O(1)$ centered at x^*. ■

Let us now apply this averaging result to some simple examples. For simplicity, we will pay special attention to two-time-scales vector fields Y of the form $Y(\tau, t, x) = u(\tau)Y(x)$.

Examples 9.23. 1. Consider the initial value problem on $\mathbb{R}^n$:

$$\dot{\boldsymbol{x}}(t) = \boldsymbol{a} + \frac{1}{\varepsilon} u\Big(\frac{t}{\varepsilon}\Big)\boldsymbol{b}, \quad \boldsymbol{x}(0) = \boldsymbol{x}_0,$$

for $\boldsymbol{a} \in \mathbb{R}^n$, $\boldsymbol{b} \in \mathbb{R}^n$, and $u\colon \mathbb{R} \to \mathbb{R}$ of class C^0, T-periodic, and zero-mean. We first compute $t \mapsto \boldsymbol{x}(t)$ directly by integration:

$$\boldsymbol{x}(t) = \boldsymbol{x}_0 + \frac{1}{\varepsilon}\int_0^t \left(\boldsymbol{a} + \boldsymbol{b}u\Big(\frac{\tau}{\varepsilon}\Big)\right)\mathrm{d}\tau = \boldsymbol{x}_0 + \boldsymbol{a}t + U_{(1)}\Big(\frac{t}{\varepsilon}\Big)\boldsymbol{b}, \tag{9.13}$$

where, from Section 9.1.1, we recall that $U_{(1)}(t) = \int_0^t u(s)\mathrm{d}s$. We also note that $U_{(1)}(t) = U_{(1)}(t \mod T)$ because u is zero-mean. Next, we compute an approximation of $t \mapsto \boldsymbol{x}(t)$ via Theorem 9.22. We set $Y(\tau, \boldsymbol{y}) = u(\tau)\boldsymbol{b}$ and $X(\boldsymbol{y}) = \boldsymbol{a}$. We compute the intermediate quantities:

$$\begin{aligned}
&\Phi^{u\boldsymbol{b}}_{0,\tau}(\boldsymbol{y}) = \boldsymbol{y} + U_{(1)}(\tau \mod T)\boldsymbol{b},\\
&F(\tau, \boldsymbol{y}) = \left((\Phi^Y_{0,\tau})^* X\right)(\boldsymbol{y}) = \boldsymbol{a}, \quad \overline{F}(\boldsymbol{y}) = \boldsymbol{a}.
\end{aligned}$$

Now we consider the initial value problem $\dot{\boldsymbol{z}}(t) = \overline{F}(\boldsymbol{z}(t))$; $\boldsymbol{z}(0) = \boldsymbol{x}_0$, and compute its solution as $\boldsymbol{z}(t) = \boldsymbol{x}_0 + t\boldsymbol{a}$. Theorem 9.22 states that there exists $T_0 \in \mathbb{R}_+$ independent of ε, such that, for all $t \in [0, T_0]$, we have

$$\boldsymbol{x}(t) = \boldsymbol{z}(t) + U_{(1)}\big(\tfrac{t}{\varepsilon} \bmod T\big)\boldsymbol{b} + O(\varepsilon).$$

Indeed, it is clear by direct inspection of (9.13) that $\boldsymbol{x}(t)$ is equal to $\boldsymbol{z}(t) + U_{(1)}\big(\tfrac{t}{\varepsilon} \bmod T\big)\boldsymbol{b}$ for all $t \in \mathbb{R}$.

2. Consider the initial value problem on $\mathbb{R}^n$:

$$\dot{\boldsymbol{x}}(t) = \boldsymbol{A}\boldsymbol{x}(t) + \frac{1}{\varepsilon}u\Big(\frac{t}{\varepsilon}\Big)\boldsymbol{b}, \quad \boldsymbol{x}(0) = \boldsymbol{x}_0,$$

for $\boldsymbol{A} \in \mathbb{R}^{n\times n}$, $\boldsymbol{b} \in \mathbb{R}^n$, and $u \colon \mathbb{R} \to \mathbb{R}$ of class C^0, T-periodic, and zero-mean. We compute $t \mapsto \boldsymbol{x}(t)$ via the formula in Example 9.8 as:

$$\boldsymbol{x}(t) = \exp(t\boldsymbol{A})\boldsymbol{x}_0 + \frac{1}{\varepsilon}\int_0^t \exp((t-\tau)\boldsymbol{A})\boldsymbol{b}u\Big(\frac{\tau}{\varepsilon}\Big)\mathrm{d}\tau.$$

We manipulate this expression via an integration by parts and other elementary steps to obtain:

$$\begin{aligned}\boldsymbol{x}(t) = \exp(t\boldsymbol{A})\boldsymbol{x}_0 + U_{(1)}\big(\tfrac{t}{\varepsilon} \bmod T\big)\boldsymbol{b} + \overline{U}_{(1)}(\exp(t\boldsymbol{A}) - \boldsymbol{I}_n)\boldsymbol{b}\\ + \boldsymbol{A}\int_0^t \exp((t-\tau)\boldsymbol{A})\boldsymbol{b}\big(U_{(1)}(\tfrac{\tau}{\varepsilon} \bmod T) - \overline{U}_{(1)}\big)\mathrm{d}\tau. \quad (9.14)\end{aligned}$$

Next, we compute an approximation of the exact solution (9.14) via Theorem 9.22. We set $Y(\tau, \boldsymbol{y}) = u(\tau)\boldsymbol{b}$ and $X(\boldsymbol{y}) = \boldsymbol{A}\boldsymbol{y}$. We compute the intermediate quantities:

$$\begin{aligned}\Phi^{u\boldsymbol{b}}_{0,\tau}(\boldsymbol{y}) &= \boldsymbol{y} + U_{(1)}(\tau \bmod T)\boldsymbol{b},\\ F(\tau, \boldsymbol{y}) &= \big((\Phi^{Y}_{0,\tau})^* X\big)(\tau, \boldsymbol{y}) = \boldsymbol{A}\boldsymbol{y} + \boldsymbol{A}\boldsymbol{b}U_{(1)}(\tau \bmod T),\\ \overline{F}(\boldsymbol{y}) &= \boldsymbol{A}\boldsymbol{y} + \overline{U}_{(1)}\boldsymbol{A}\boldsymbol{b}.\end{aligned}$$

We consider the initial value problem $\dot{\boldsymbol{z}}(t) = \overline{F}(\boldsymbol{z}(t)); \boldsymbol{z}(0) = \boldsymbol{x}_0$, and compute its solution as

$$\boldsymbol{z}(t) = \exp(t\boldsymbol{A})\boldsymbol{x}_0 + \overline{U}_{(1)}\big(\exp(t\boldsymbol{A}) - \boldsymbol{I}_n\big)\boldsymbol{b}.$$

Theorem 9.22 states that there exists $T_0 \in \mathbb{R}_+$ independent of ε, such that, for all $t \in [0, T_0]$, we have

$$\boldsymbol{x}(t) = \boldsymbol{z}(t) + U_{(1)}\big(\tfrac{t}{\varepsilon} \bmod T\big)\boldsymbol{b} + O(\varepsilon).$$

Indeed, this equality can be verified directly from (9.14) by showing that the last term is a function that is $O(\varepsilon)$. •

9.1.6 Series expansion results for averaging

In this section we consider control-affine systems subject to oscillatory controls; that is, systems in the form

$$\gamma'(t) = X(t, \gamma(t)) + \frac{1}{\varepsilon} \sum_{a=1}^{m} u^a\Big(\frac{t}{\varepsilon}, t\Big) Y_a(\gamma(t)). \qquad (9.15)$$

This is a variation on the initial value problem in standard oscillatory form. Let us collect the required assumptions on the control inputs for the averaging analysis of this class of systems.

Assumption 9.24. We shall assume that:

(i) the functions $u^a : \bar{\mathbb{R}}_+ \times \bar{\mathbb{R}}_+ \to \mathbb{R}$, $a \in \{1, \dots, m\}$, are of class C^0, of class C^1 in their second argument, and T-periodic and zero-mean in their first argument;

(ii) for $(a_1, \dots, a_k) \in \mathfrak{I}_{\{1,\dots,m\}}$, $U_{(a_1,\dots,a_k)} : \bar{\mathbb{R}}_+ \times \bar{\mathbb{R}}_+ \to \mathbb{R}$ and $\overline{U}_{(a_1,\dots,a_k)} : \bar{\mathbb{R}}_+ \to \mathbb{R}$ are the iterated integrals of u^a, $a \in \{1, \dots, m\}$, and their averages, with respect to the first argument;

(iii) for $k_1, \dots, k_m \in \mathbb{Z}_+$, $\mathsf{U}_{k_1,\dots,k_m} : \bar{\mathbb{R}}_+ \times \bar{\mathbb{R}}_+ \to \mathbb{R}$ and $\overline{\mathsf{U}}_{k_1,\dots,k_m} : \bar{\mathbb{R}}_+ \to \mathbb{R}$ are the multinomial iterated integrals of u^a, $a \in \{1, \dots, m\}$, and their averages, with respect to the first argument. •

The following theorem is the second main analysis result for oscillatory inputs.

Theorem 9.25 (Series expansion representation). *Let* M *be an open bounded subset of* $\mathbb{R}^n$, *let* $X \in \Gamma^1(\bar{\mathbb{R}}_+; \mathsf{TM})$ *be of class* C^1, *let* $Y_1, \dots, Y_m \in \Gamma^2(\mathsf{TM})$ *commute, and let* $u^a : \bar{\mathbb{R}}_+ \times \bar{\mathbb{R}}_+ \to \mathbb{R}$, $a \in \{1, \dots, m\}$, *satisfy Assumption 9.24.*

Then, for all $x \in \mathsf{M}$ *and* $(\tau, t) \in \bar{\mathbb{R}}_+ \times \bar{\mathbb{R}}_+$, *the pull-back vector field* $F_{\mathrm{e}} = (1, F)$ *in equation* (9.10) *satisfies the following series expansions, whenever only a finite number of terms do not vanish:*

$$\begin{aligned} F(\tau, t, x) &= X(t, x) + \sum_{k=1}^{+\infty} \sum_{\substack{(a_1,\dots,a_k) \\ \in \mathfrak{I}_{\{1,\dots,m\}}}} U_{(a_1,\dots,a_k)}(\tau, t)\, \mathrm{ad}_{Y_{a_1}} \cdots \mathrm{ad}_{Y_{a_k}} X_t(x) \\ &\quad - \sum_{a=1}^{m} \frac{\partial U_{(a)}}{\partial t}(\tau, t) Y_a(x) \\ &= \sum_{k_1,\dots,k_m=0}^{+\infty} \mathsf{U}_{k_1,\dots,k_m}(\tau, t)\, \mathrm{ad}_{Y_1}^{k_1} \cdots \mathrm{ad}_{Y_m}^{k_m} X_t(x) - \sum_{a=1}^{m} \frac{\partial U_{(a)}}{\partial t}(\tau, t) Y_a(x). \end{aligned} \qquad (9.16)$$

Furthermore, the average vector field $\overline{F}_{\mathrm{e}}$ *satisfies analogous series expansions, where the iterated integrals and the multinomial iterated integrals are replaced by their respective averages.*

Proof. Note that the smoothness requirements in Assumptions 9.20 are satisfied, and therefore Theorem 9.22 applies. We first prove the result for the single-input case, $m = 1$. With abuse of notation, we write $Y(\tau, t, x) =$

$u(\tau,t)Y(x)$ and compute $\left(\Phi_{0,\tau}^{Y_e}\right)^* X_e$ by means of Proposition 9.6. From Remark 9.2, recall that $\mathsf{U}_k = U_{(1,\dots,1)}$. Now, the expressions for the first two Lie brackets in terms of the original vector fields $(t,x) \mapsto X(t,x)$ and $(s,t,x) \mapsto u(s,t)Y(x)$ are

$$\mathrm{ad}_{(Y_e)_{s_1}} X_e(t,x) = \Big(u(s_1,t)\,\mathrm{ad}_Y X(t,x) - \frac{\partial u}{\partial t}(s_1,t)Y(x), 0\Big),$$

and

$$\begin{aligned}
&\mathrm{ad}_{(Y_e)_{s_2}} \mathrm{ad}_{(Y_e)_{s_1}} X_e(t,x)\\
&\quad = \Big(u(s_2,t)u(s_1,t)\,\mathrm{ad}_Y^2 X(t,x) - \frac{\partial u}{\partial t}(s_1,t)u(s_2,t)\,\mathrm{ad}_Y Y(x), 0\Big)\\
&\quad = \big(u(s_2,t)u(s_1,t)\,\mathrm{ad}_Y^2 X(t,x), 0\big).
\end{aligned}$$

Following an inductive argument, one can see that, for $k \geq 2$,

$$\mathrm{ad}_{(Y_e)_{s_k}} \cdots \mathrm{ad}_{(Y_e)_{s_1}} X_e = \big(0, u(s_k,t)\cdots u(s_1,t)\,\mathrm{ad}_Y^k X(t,x)\big).$$

In turn, this leads to the desired result

$$F(\tau,t,x) = X(t,x) + \sum_{k=1}^{+\infty} \mathsf{U}_k(\tau,t)\,\mathrm{ad}_Y^k X(t,x) - \frac{\partial U_{(1)}}{\partial t}(\tau,t)Y(x).$$

In the multi-input case, the first series expansion in (9.16) can be deduced in the same way. As for the second series expansion in (9.16), the fact that $Y_1,\dots,Y_m$ commute implies that

$$\mathrm{ad}_{Y_{a_{\sigma(1)}}} \cdots \mathrm{ad}_{Y_{a_{\sigma(k)}}} X = \mathrm{ad}_{Y_{a_1}} \cdots \mathrm{ad}_{Y_{a_k}} X,$$

for any $\sigma \in S_k$. Then, for each $k \in \mathbb{N}$,

$$\begin{aligned}
&\sum_{(a_1,\dots,a_k)\in\mathcal{J}\{1,\dots,m\}} U_{(a_1,\dots,a_k)}(\tau,t)\,\mathrm{ad}_{Y_{a_1}} \cdots \mathrm{ad}_{Y_{a_k}} X(t,x)\\
&\qquad = \sum_{\substack{k_1,\dots,k_m\geq 0\\ k_1+\dots+k_m=k}} \mathsf{U}_{k_1,\dots,k_m}(\tau,t)\,\mathrm{ad}_{Y_1}^{k_1}\cdots\mathrm{ad}_{Y_m}^{k_m} X(t,x). \qquad \blacksquare
\end{aligned}$$

Remark 9.26. It is possible to establish convergence properties for the series expansions (9.16) by means of the tools introduced in Section 9.1.2. Assume that all vector fields $X, Y_1,\dots,Y_m$ are analytic and admit bounded analytic continuations to $B_\rho^{\mathbb{C}}(x)$ for some $\rho \in \mathbb{R}_+$ and for all $x \in \mathsf{M}$. Additionally, assume that, for all $x \in \mathsf{M}$, there exists $\rho' \in]0,\rho[$ such that

$$4T\sum_{a=1}^{m} \|u^a\|_\infty \, \|Y_a\|_{x,\rho} < \rho - \rho'. \tag{9.17}$$

where $\|u^a\|_\infty = \sup\{|u^a(\tau,t)| \mid (t,\tau)\in\bar{\mathbb{R}}_+\times\bar{\mathbb{R}}_+\}$. Then, from Proposition 9.4 and Proposition 3.1 in [Agrachev and Gamkrelidze 1978], one can show that, for all $x\in\mathsf{M}$,

$$\begin{aligned}\big\|\mathsf{U}_{k_1,\dots,k_m}(\tau,t)\,\mathrm{ad}_{Y_1}^{k_1}\cdots\mathrm{ad}_{Y_m}^{k_m}X\big\|_{x,\rho',T}\\ &\le\left(\frac{4T}{\rho-\rho'}\right)^k\|u^1\|_\infty^{k_1}\cdots\|u^m\|_\infty^{k_m}\,\|Y_1\|_{x,\rho}^{k_1}\cdots\|Y_m\|_{x,\rho}^{k_m}\,\|X\|_{x,\rho}\\ &\le\Big(\frac{4T}{\rho-\rho'}\Big)^k\Big(\sum_{a=1}^m\|u^a\|_\infty\,\|Y_a\|_{x,\rho}\Big)^k\,\|X\|_{x,\rho},\end{aligned}$$

where $k=\sum_{b=1}^m k_b$. The bound (9.17) may now be shown to imply that the series expansions (9.16) converge absolutely and uniformly for all $x\in\mathsf{M}$ and $(t,\tau)\in\bar{\mathbb{R}}_+\times\bar{\mathbb{R}}_+$. •

We are finally ready to provide an example of the averaging procedure via series expansions for systems subject to oscillatory inputs.

Example 9.27. Consider the initial value problem on $\mathbb{R}^n$

$$\dot{\boldsymbol{x}}(t)=\boldsymbol{A}\boldsymbol{x}(t)+\frac{1}{\varepsilon}u\left(\frac{t}{\varepsilon}\right)\boldsymbol{B}\boldsymbol{x}(t),\quad \boldsymbol{x}(0)=\boldsymbol{x}_0,$$

for $\boldsymbol{A},\boldsymbol{B}\in\mathbb{R}^{n\times n}$ and $u\colon\mathbb{R}\to\mathbb{R}$ of class C^0, T-periodic, and zero-mean. If $X_{\boldsymbol{A}}$ and $X_{\boldsymbol{B}}$ are the linear vector fields defined by $\boldsymbol{x}\mapsto\boldsymbol{A}\boldsymbol{x}$ and $\boldsymbol{x}\mapsto\boldsymbol{B}\boldsymbol{x}$, respectively, their Lie bracket satisfies

$$\mathrm{ad}_{X_{\boldsymbol{B}}}X_{\boldsymbol{A}}(\boldsymbol{x})=(\mathrm{ad}_{\boldsymbol{A}}\boldsymbol{B})\boldsymbol{x},$$

where, as in Section 5.2.2, the operation of matrix commutation is $(\boldsymbol{A},\boldsymbol{B})\mapsto\mathrm{ad}_{\boldsymbol{A}}\boldsymbol{B}=\boldsymbol{A}\boldsymbol{B}-\boldsymbol{B}\boldsymbol{A}$. We compute the intermediate quantities:

$$\begin{aligned}\Phi^{uX_{\boldsymbol{B}}}_{0,(\frac{t}{\varepsilon}\bmod T)}(\boldsymbol{y})&=\exp\big(U_{(1)}(\tfrac{t}{\varepsilon}\bmod T)\boldsymbol{B}\big)\boldsymbol{y},\\ F(\tau,\boldsymbol{y})&=\big((\Phi^{uX_{\boldsymbol{B}}}_{0,\tau})^*X_{\boldsymbol{A}}\big)(\boldsymbol{y})\\ &=\exp\big(-U_{(1)}(\tfrac{t}{\varepsilon}\bmod T)\boldsymbol{B}\big)\boldsymbol{A}\exp\big(U_{(1)}(\tfrac{t}{\varepsilon}\bmod T)\boldsymbol{B}\big)\boldsymbol{y}.\end{aligned}$$

It is not clear how to obtain a simple expression for the average of F. We therefore take the approach advocated in Theorem 9.25 and write

$$\overline{F}(\boldsymbol{y})=\Big(\sum_{k=0}^{+\infty}(-1)^k\overline{\mathsf{U}}_k\,\mathrm{ad}_{\boldsymbol{B}}^k\boldsymbol{A}\Big)\boldsymbol{y}.$$

Note that $\overline{F}$ is a linear vector field. Let $t\mapsto\boldsymbol{z}(t)$ be the solution to the initial value problem $\dot{\boldsymbol{z}}(t)=\overline{F}(\boldsymbol{z}(t));\boldsymbol{z}(0)=\boldsymbol{x}_0$. Theorem 9.22 states that there exists $T_0\in\mathbb{R}_+$ independent of ε, such that, for all $t\in[0,T_0]$, we have

$$\boldsymbol{x}(t) = \exp\Big(U_{(1)}\big(\tfrac{t}{\varepsilon} \bmod T\big)\boldsymbol{B}\Big) \exp\Big(t\sum_{k=0}^{+\infty}(-1)^k \overline{\mathrm{U}}_k \operatorname{ad}_{\boldsymbol{B}}^k \boldsymbol{A}\Big)\boldsymbol{x}_0 + O(\varepsilon).$$

We conclude by noting that, by the Cayley–Hamilton Theorem, the series for $\overline{F}$ reduces to a finite sum in the powers of $\operatorname{ad}_{\boldsymbol{B}}$. •

9.2 Averaging of affine connection systems subject to oscillatory controls

In this section we study affine connection control systems subject to oscillatory controls. We show how the averaged system is again an affine connection control system, described by the same affine connection and subject to an appropriate vector force. We start by reviewing and developing some preliminary concepts.

9.2.1 The homogeneity properties of affine connection control systems

We discuss the Lie algebraic and homogeneous structure of affine connection control systems; see related ideas in [Bullo 2001, Lewis and Murray 1997a, Sontag and Sussmann 1986]. We here rely on the notion of geometric homogeneity as described by Kawski [1995] and the notion of the Liouville vector field as in [Libermann and Marle 1987]. Roughly speaking, geometric homogeneity corresponds to the existence of an infinitesimal symmetry in the equations of motion. For control systems described by an affine connection, the symmetry is invariance under affine time-scaling transformations. Let us recall some notation from Section 4.6.

Let $(\mathsf{Q}, \nabla, Y, \mathcal{D}, \mathscr{Y} = \{Y_1, \dots, Y_m\}, U)$ be a C^∞-forced affine connection control system with governing equation

$$\nabla_{\gamma'(t)}\gamma'(t) = Y(t, \gamma(t)) + \sum_{a=1}^{m} u^a(t) Y_a(\gamma(t)),$$

where $\gamma\colon I \to \mathsf{Q}$ and $\gamma'(t) \in \mathcal{D}_{\gamma(t)}$ for some (and so for all) t. This equation is written in first-order form on TQ as

$$\Upsilon'(t) = S(\Upsilon(t)) + \operatorname{vlft}(Y)(t, \Upsilon(t)) + \sum_{a=1}^{m} u^a(t) \operatorname{vlft}(Y_a)(\Upsilon(t)),$$

where S is the geodesic spray for ∇, and where $\Upsilon\colon I \to \mathsf{TQ}$ satisfies $\Upsilon(t) \in \mathcal{D}$ for some (and so for all) t. Here $\operatorname{vlft}(Y)(t, v_q) = \operatorname{vlft}(Y_t)(v_q)$. We let $(\mathcal{U}, \phi)$ be a chart for Q with coordinates $(q^1, \dots, q^n)$, we consider a collection $\mathscr{X} = \{X_1, \dots, X_n\}$ of vector fields forming a basis for $\mathsf{T}_q\mathsf{Q}$ for each $q \in \mathcal{U}$, and we write the governing equations in Poincaré representation for a curve γ whose image intersects $\mathcal{U}$ as

$$\begin{aligned}\dot{q}^i &= X^i_j v^j, \\ \dot{v}^i &= -\overset{\mathscr{X}}{\Gamma}{}^i_{jk} v^i v^j + A^i_l Y^l + \sum_{a=1}^m u^a A^i_l Y^l_a,\end{aligned} \qquad i \in \{1, \dots, n\}.$$

Loosely speaking, one fundamental structure of these governing equations in first-order form is the polynomial dependence of the vector fields S and $\mathrm{vlft}(Y_a)$ on the velocity variables $\{v^1, \dots, v^n\}$. This structure is reflected, in coordinate-free terms, in the Lie bracket computations involving S and $\mathrm{vlft}(Y_a)$, as we now show.

Given two vector fields X and X_0 and an integer m, we say that X is ***X_0-homogeneous of degree j*** if

$$[X_0, X] = jX.$$

For control systems described by an affine connection, we define the ***Liouville vector field*** on $\mathsf{T}\mathsf{Q}$ by

$$V_{\mathrm{L}}(v_q) = v^i \frac{\partial}{\partial v^i}, \tag{9.18}$$

for $v_q = v^i \frac{\partial}{\partial q^i} \in \mathsf{T}\mathsf{Q}$. See Exercise E9.6 for a coordinate-invariant definition of V_{L}. For $j \in \mathbb{Z}$, let $\mathcal{P}_j$ be the set of vector fields on $\mathsf{T}\mathsf{Q}$ that are V_{L}-homogeneous of degree j. With these definitions we can now state the following result.

Lemma 9.28 (Homogeneity of the geodesic spray and of vertically lifted vector fields). *Let $(\mathsf{Q}, \nabla, Y, \mathcal{D}, \mathscr{Y} = \{Y_1, \dots, Y_m\}, U)$ be a C^∞-forced affine connection control system. Then*

$$S \in \mathcal{P}_1, \quad \textit{and} \quad \mathrm{vlft}(Y_a) \in \mathcal{P}_{-1}, \quad a \in \{1, \dots, m\}.$$

Proof. It is a straightforward calculation in coordinates that $[V_{\mathrm{L}}, S] = (+1)S$, and that $[V_{\mathrm{L}}, \mathrm{vlft}(Y_a)] = (-1)\mathrm{vlft}(Y_a)$, for $a \in \{1, \dots, m\}$. ■

Regarding the vector force Y, if it arises from a potential function, then its vertical lift belongs to $\mathcal{P}_{-1}$, and if it arises from a Rayleigh dissipation function, then its vertical lift belongs to $\mathcal{P}_0$. We say that a time-dependent vector force $Y \colon \mathbb{R} \times \mathsf{T}\mathsf{Q} \to \mathsf{T}\mathsf{Q}$ is an ***affine bundle map*** if, for each $t \in \mathbb{R}$ and $v_q \in \mathsf{T}\mathsf{Q}$, we have $Y(t, v_q) = Y_0(t, q) + Y_1(t, v_q)$, where Y_0 is a vector field and Y_1 is a $(1,1)$-tensor field on Q. If Y is an affine bundle map, then the vertical lift of Y belongs to $\mathcal{P}_0 + \mathcal{P}_{-1}$.

The sets $\{\mathcal{P}_j\}_{j \in \mathbb{Z}}$ of V_{L}-homogeneous vector fields enjoy the following useful properties.

Proposition 9.29 (Properties of homogeneous vector fields). *The following statements hold:*

(i) for $i, j \in \mathbb{Z}$, $[\mathcal{P}_i, \mathcal{P}_j] \subset \mathcal{P}_{i+j}$, i.e., the Lie bracket between a vector field in $\mathcal{P}_i$ and a vector field in $\mathcal{P}_j$ belongs to $\mathcal{P}_{i+j}$;
(ii) for $k \leq -2$, $\mathcal{P}_k = \{0\}$;
(iii) for $k \geq 1$ and for $X \in \mathcal{P}_k$, $X(0_q) = 0_q$;
(iv) every $X \in \mathcal{P}_{-1}$ is the vertical lift of a vector field on Q.

Proof. The reader is invited to prove these properties in Exercise E9.7. ■

Remarks 9.30. 1. We illustrate these properties in Figure 9.2. The (i, j)th position in the table in figure contains Lie brackets with i copies of $\mathrm{verlift}(Y_a)$ and j copies of S; the corresponding homogeneous degree is $j - i$. All Lie brackets to the right of the line corresponding to $\mathcal{P}_{-1}$ exactly vanish, and all Lie brackets to the left of the line corresponding to $\mathcal{P}_{-1}$ vanish when evaluated at 0_q.

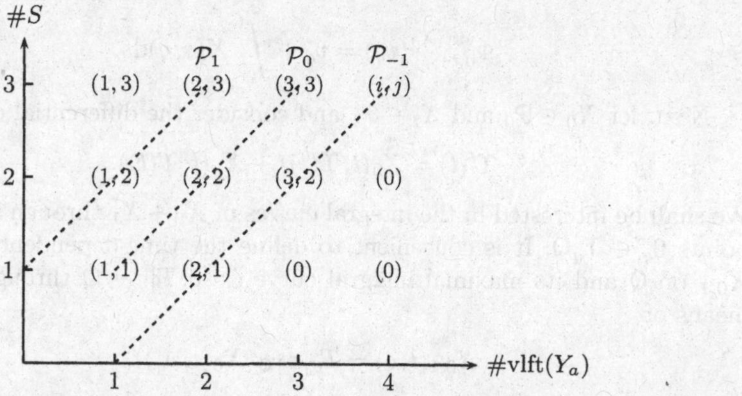

Figure 9.2. Table of Lie brackets between the drift vector field S and an input vector field $\mathrm{verlift}(Y_a)$

2. It is helpful to provide an interpretation of $\mathcal{P}_i$ in coordinates. Let $((q^1, \dots, q^n), (v^1, \dots, v^n))$ be coordinates on TQ, associated with coordinates $(q^1, \dots, q^n)$ on Q. Let $\mathcal{H}_i$ be the set of real-valued functions that are arbitrary functions of $(q^1, \dots, q^n)$ and that are homogeneous polynomials in $\{v^1, \dots, v^n\}$ of degree i. Then, $\mathcal{P}_i$ is the set of vector fields on TQ whose first n components belong to $\mathcal{H}_i$ and whose second n components belong to $\mathcal{H}_{i+1}$.

3. Given the properties in Proposition 9.29, we investigate the Lie brackets between the vector fields S and $\mathrm{vlft}(Y_a)$, $a \in \{1, \dots, m\}$. A few useful brackets are:

$$[S, \mathrm{vlft}(Y_a)] \in \mathcal{P}_0, \quad [\mathrm{vlft}(Y_a), \mathrm{vlft}(Y_b)] = 0,$$
$$[S, [S, \mathrm{vlft}(Y_a)]] \in \mathcal{P}_1, \quad [\mathrm{vlft}(Y_b), [S, \mathrm{vlft}(Y_a)]] \in \mathcal{P}_{-1}.$$

These results are consistent with the calculations we performed in Section 7.3.2, including, for example, the statement that

$$\mathrm{vlft}(\langle X : Y\rangle) = [\mathrm{vlft}(X), [S, \mathrm{vlft}(Y)]],$$

for any vector fields X and Y on Q. •

9.2.2 Flows for homogeneous vector fields

Here we compute the flows of certain homogeneous vector fields. In particular, significant simplifications take place in the two cases we consider. First, let X be a time-dependent vector field on Q, and consider the following differential equation for the curve $\Upsilon\colon I \to \mathsf{TQ}$:

$$\Upsilon'(t) = \mathrm{vlft}(X)(t, \Upsilon(t)). \tag{9.19}$$

For all $v_q \in \mathsf{T}_q\mathsf{Q}$, we have

$$\Phi_{0,t}^{\mathrm{vlft}(X)}(v_q) = v_q + \int_0^t X(s,q)\mathrm{d}s. \tag{9.20}$$

Next, let $X_0 \in \mathcal{P}_0$ and $X_1 \in \mathcal{P}_1$ and consider the differential equation

$$\Upsilon'(t) = X_0(t, \Upsilon(t)) + X_1(t, \Upsilon(t)). \tag{9.21}$$

We shall be interested in the integral curves of $X_0 + X_1$ through zero velocity points $0_q \in \mathsf{T}_q\mathsf{Q}$. It is convenient to define the time-dependent vector field $X_{0,1}$ on Q and its maximal integral curve $\zeta\colon [0, T] \to \mathsf{Q}$ through $q \in \mathsf{Q}$ by means of

$$X_{0,1}(t,q) = T_{v_q}\pi_{\mathsf{TQ}}(X_0(t, v_q)),$$
$$\zeta(t) = \Phi_{0,t}^{X_{0,1}}(q),$$

where $\pi_{\mathsf{TQ}}\colon \mathsf{TQ} \to \mathsf{Q}$ is the standard projection map. Note that the vector field $X_{0,1}$ is well-defined in the sense that its definition does not depend on the choice of $v_q \in \mathsf{T}_q\mathsf{Q}$. In coordinates, this vector field consists of the first n components of the vector field X_0 on TQ. It can be seen that

$$\Phi_{0,t}^{X_0+X_1}(0_q) = 0_{\zeta(t)}.$$

That is to say, $Z(\mathsf{Q})$ is an invariant manifold for the flow of $X_0 + X_1$, and the flow on Q is that of $X_{0,1}$. This assertion is easily proved in coordinates.

9.2.3 Averaging analysis

In this subsection we will state the main averaging result for affine connection control systems subject to oscillatory controls. We begin with a preliminary observation. Let $u = (u^1, \dots, u^m)\colon \bar{\mathbb{R}}_+ \times \bar{\mathbb{R}}_+ \to \mathbb{R}^m$ satisfy Assumption 9.24. Define the curve $\boldsymbol{\Lambda}\colon \bar{\mathbb{R}}_+ \to \mathbb{R}^{m\times m}$ with components

$$\Lambda_{ab}(t) = \tfrac{1}{2}\big(\overline{U_{(a)}U_{(b)}}(t) - \overline{U}_{(a)}(t)\overline{U}_{(b)}(t)\big), \qquad a, b \in \{1, \dots, m\}. \tag{9.22}$$

Lemma 9.31. *For each $t \in \bar{\mathbb{R}}_+$, the matrix $\mathbf{\Lambda}(t)$ is symmetric and positive-semidefinite. For each $t_0 \in \bar{\mathbb{R}}_+$ with the property that the functions $\tau \mapsto u^a(\tau, t_0)$, $a \in \{1, \dots, m\}$, are non-zero, the matrix $\mathbf{\Lambda}(t_0)$ is positive-definite.*

Proof. Recall from Exercise E2.14 the vector space of square-integrable functions $L_2([0,T])$ and the inner product $\langle\!\langle \cdot, \cdot \rangle\!\rangle_{L_2([a,b])}$. For all $s, t \in \bar{\mathbb{R}}_+$ and for all $\boldsymbol{\eta} = (\eta^1, \dots, \eta^m) \in \mathbb{R}^m$, let $U_{\boldsymbol{\eta},t}(s) = U_{\boldsymbol{\eta}}(s,t) = \sum_{a=1}^m \eta^a U_{(a)}(s,t)$ and compute

$$\begin{aligned}
2\boldsymbol{\eta}^T \mathbf{\Lambda}(t)\boldsymbol{\eta} &= 2\sum_{a,b=1}^m \Lambda_{ab}(t)\eta^a\eta^b \\
&= \frac{1}{T}\int_0^T U_{\boldsymbol{\eta}}(s,t)^2 \mathrm{d}s - \Big(\frac{1}{T}\int_0^T U_{\boldsymbol{\eta}}(s,t)\mathrm{d}s\Big)^2 \\
&= \frac{1}{T}\langle\!\langle U_{\boldsymbol{\eta},t}, U_{\boldsymbol{\eta},t}\rangle\!\rangle_{L_2([0,T])} - \frac{1}{T^2}\langle\!\langle U_{\boldsymbol{\eta},t}, 1\rangle\!\rangle^2_{L_2([0,T])},
\end{aligned}$$

where the $L_2([0,T])$-inner product is computed with respect to the first time variable, and where 1 stands for the constant function $s \mapsto 1$. From the Cauchy–Schwarz Inequality in Theorem 2.36 we know that $|\langle\!\langle U_{\boldsymbol{\eta},t}, 1\rangle\!\rangle_{L_2([0,T])}| \le \|U_{\boldsymbol{\eta},t}\|_{L_2([0,T])}\|1\|_{L_2([0,T])} = \sqrt{T}\|U_{\boldsymbol{\eta},t}\|_{L_2([0,T])}$, and this implies that $\boldsymbol{\eta}^T\mathbf{\Lambda}(t)\boldsymbol{\eta}$ is nonnegative. Here we have used $\langle\!\langle 1, 1\rangle\!\rangle_{L_2([0,T])} = T$. Hence, the matrix $\mathbf{\Lambda}(t)$ is positive-semidefinite.

For all $t_0 \in \bar{\mathbb{R}}_+$ such that all $\tau \mapsto u^a(\tau, t_0)$, $a \in \{1, \dots, m\}$, are non-zero, and for all non-zero $\boldsymbol{\eta} \in \mathbb{R}^m$, the function $s \mapsto U_{\boldsymbol{\eta},t}(s)$ is not a multiple of the trivial function $s \mapsto 1$, because $U_{\boldsymbol{\eta}}(0,t) \neq 0$ and because $U_{\boldsymbol{\eta}}(0,t) = U_{\boldsymbol{\eta}}(T,t) = 0$ by assumption. Therefore, the Cauchy–Schwarz Inequality holds with a strict sign. This implies that the matrix $\mathbf{\Lambda}(t_0)$ is positive-definite. ■

We are now ready to state one of the main results of this chapter.

Theorem 9.32 (Averaging affine connection systems subject to oscillatory controls). *Consider the C^∞-forced affine connection control system $(\mathsf{Q}, \nabla; Y, \mathcal{D}, \mathscr{Y} = \{Y_1, \dots, Y_m\}, \mathbb{R}^m)$ with governing equation*

$$\nabla_{\gamma'(t)}\gamma'(t) = Y(t, \gamma'(t)) + \sum_{a=1}^m \frac{1}{\varepsilon}u^a\Big(\frac{t}{\varepsilon}, t\Big)Y_a(\gamma(t)), \tag{9.23}$$

with $\varepsilon \in \mathbb{R}_+$ and with initial condition $\gamma'(0) \in \mathcal{D}$. Assume that Q is a bounded open subset of $\mathbb{R}^n$ and that Y is an affine bundle map of class C^1 in time. Let $u = (u^1, \dots, u^m)\colon \bar{\mathbb{R}}_+ \times \bar{\mathbb{R}}_+ \to \mathbb{R}^m$ satisfy Assumption 9.24, and consider the differential equation

$$\nabla_{\xi'(t)}\xi'(t) = Y(t, \xi'(t)) - \sum_{a,b=1}^m \Lambda_{ab}(t)\,\langle Y_a : Y_b\rangle\,(\xi(t)) \tag{9.24}$$

with initial condition $\xi'(0) = \gamma'(0) + \sum_{a=1}^{m} \overline{U}_{(a)}(0) Y_a(\gamma(0))$. *Then there exists* $\varepsilon_0, t_0 \in \mathbb{R}_+$ *such that, for all* $t \in [0, t_0]$ *and for all* $\varepsilon \in]0, \varepsilon_0[$,

$$\begin{aligned} \gamma(t) &= \xi(t) + O(\varepsilon), \\ \gamma'(t) &= \xi'(t) + \sum_{a=1}^{m} \big(U_{(a)}(\tfrac{t}{\varepsilon} \bmod T, t) - \overline{U}_{(a)}(t)\big) Y_a(\xi(t)) + O(\varepsilon). \end{aligned} \tag{9.25}$$

Furthermore, assume that the inputs u^a, $a \in \{1, \dots, m\}$, *do not depend on the slow time-scale* t *and that* $q_0 \in \mathsf{Q}$ *is a linearly asymptotically stable equilibrium configuration for the forced affine connection system* (9.24). *Then*

(i) *there exists a neighborhood* $\mathcal{U}$ *of* 0_{q_0} *with the property that, if* $\gamma'(0) \in \mathcal{U}$, *then the approximations in* (9.25) *hold for all* $t \in \mathbb{R}_+$, *and*

(ii) *there exists* $\varepsilon_1 > 0$ *such that, for all* $\varepsilon \in]0, \varepsilon_1[$, *there exists a unique trajectory* $\gamma_0 \colon \mathbb{R} \to \mathsf{Q}$ *for the forced affine connection system* (9.23) *with the following properties:* γ_0 *is* εT*-periodic and locally asymptotically stable,* γ_0 *takes values in an open ball of radius* $O(\varepsilon)$ *centered at* q_0, *and* γ_0' *takes values in an open ball of radius* $O(1)$ *centered at* 0_{q_0}.

Proof. The proof is based on a careful application of Theorem 9.22. Let us start by characterizing the quantity

$$\begin{aligned} &\sum_{k=1}^{+\infty} \sum_{\substack{(a_1,\dots,a_k) \\ \in \mathcal{J}_{\{1,\dots,m\}}}} \overline{U}_{(a_1,\dots,a_k)}(t)\, \mathrm{ad}_{\mathrm{vlft}(Y_{a_1})} \cdots \mathrm{ad}_{\mathrm{vlft}(Y_{a_k})} (S + \mathrm{vlft}(Y)) \\ &\quad = \sum_{a=1}^{m} \overline{U}_{(a)}(t)\, \mathrm{ad}_{\mathrm{vlft}(Y_a)} (S + \mathrm{vlft}(Y)) \\ &\quad\quad + \sum_{a,b=1}^{m} \overline{U}_{(a,b)}(t)\, \mathrm{ad}_{\mathrm{vlft}(Y_a)}\, \mathrm{ad}_{\mathrm{vlft}(Y_b)} (S + \mathrm{vlft}(Y)) \\ &\quad = \sum_{a=1}^{m} \overline{U}_{(a)}(t)\, \mathrm{ad}_{\mathrm{vlft}(Y_a)} (S + \mathrm{vlft}(Y)) - \sum_{a,b=1}^{m} \overline{U}_{(a,b)}(t) \mathrm{vlft}(\langle Y_a : Y_b \rangle), \end{aligned}$$

where the simplifications take place according to the Lie bracket analysis in Section 9.2.1. Only a finite number of terms do not vanish in the series expansion for the pull-back vector field.

Next, for $\Upsilon \colon I \to \mathsf{TQ}$, we write equation (9.23) in first-order form:

$$\Upsilon'(t) = S(\Upsilon(t)) + \mathrm{vlft}(Y)(t, \Upsilon(t)) + \sum_{a=1}^{m} \frac{1}{\varepsilon} u^a\Big(\frac{t}{\varepsilon}, t\Big) \mathrm{vlft}(Y_a)(\Upsilon(t)).$$

Consistent with the notation in Theorem 9.22 and in (9.8), for $t_0 \in [0, T]$, we let

$$\Big(\sum_{a=1}^{m} u^a \mathrm{vlft}(Y_a)\Big)_{2,t_0} (\tau, v_q) = \sum_{a=1}^{m} (u^a \mathrm{vlft}(Y_a))(\tau, t_0, v_q).$$

We now apply Theorem 9.22 to obtain

$$\Upsilon(t) = \Phi_{0,(\frac{t}{\varepsilon} \bmod T)}^{(\sum_{a=1}^{m} u^a \mathrm{vlft}(Y_a))_{2,t}}(\Xi(t)) + O(\varepsilon), \tag{9.26}$$

where $\Xi\colon I \to \mathsf{TQ}$, and where Theorem 9.25 leads to

$$\begin{aligned}\Xi'(t) = (S + \mathrm{vlft}(Y))\,(t, \Xi(t)) + \sum_{a=1}^{m} \overline{U}_{(a)}(t)\,\mathrm{ad}_{\mathrm{vlft}(Y_a)}\,(S + \mathrm{vlft}(Y))\,(t, \Xi(t))\\ - \sum_{a,b=1}^{m} \overline{U}_{(a,b)}(t)\mathrm{vlft}(\langle Y_a : Y_b\rangle)(\Xi(t)) - \sum_{a=1}^{m} \frac{\mathrm{d}}{\mathrm{d}t}\overline{U}_{(a)}(t)\mathrm{vlft}(Y_a)(\Xi(t)).\end{aligned} \tag{9.27}$$

We claim that the curve $\xi = \pi_{\mathsf{TQ}} \circ \Xi\colon I \to \mathsf{Q}$ satisfies equation (9.24). In coordinates, let $\xi(t) = (\xi^1(t), \dots, \xi^n(t))$ and $\Xi(t) = (\xi^1(t), \dots, \xi^n(t), \Xi^1(t), \dots, \Xi^n(t))$. Equation (9.27) reads

$$\begin{aligned}\frac{\mathrm{d}}{\mathrm{d}t}\xi^i(t) &= \Xi^i(t) + \sum_{a=1}^{m} \overline{U}_{(a)}(t) Y_a^i(\xi(t)),\\ \frac{\mathrm{d}}{\mathrm{d}t}\Xi^i(t) &= -\Gamma^i_{jk}(\xi(t))\Xi^j(t)\Xi^k(t) + Y^i(t, \Xi(t))\\ &\quad + \sum_{a=1}^{m} \overline{U}_{(a)}(t)(\mathrm{ad}_{\mathrm{vlft}(Y_a)}\,(S + \mathrm{vlft}(Y))^i\,(t, \Xi(t))\\ &\quad - \sum_{a,b=1}^{m} \overline{U}_{(a,b)}(t)\,\langle Y_a : Y_b\rangle^i\,(\xi(t)) - \sum_{a=1}^{m} \frac{\mathrm{d}}{\mathrm{d}t}\overline{U}_{(a)}(t) Y_a^i(\xi(t)).\end{aligned}$$

If $Y^i(t, \Xi(t)) = Y_0^i(t, \xi(t)) + Y_{1,j}^i(t, \xi(t))\Xi^j(t)$, then

$$\begin{aligned}(\mathrm{ad}_{\mathrm{vlft}(Y_a)}\,\mathrm{vlft}(Y))^i(t, \Xi(t)) &= Y_{1,j}^i(t, \xi(t)) Y_a^j(\xi(t)),\\ (\mathrm{ad}_{\mathrm{vlft}(Y_a)}\,S)^i(t, \Xi(t)) &= -\Gamma^i_{jk}(\xi(t))(Y_a^j(\xi(t))\Xi^k(t) + Y_a^k(\xi(t))\Xi^j(t))\\ &\quad - \frac{\partial Y_a^i}{\partial q^k}(\xi(t))\Xi^k(t).\end{aligned}$$

Computing the second time derivative of $(\xi^1(t), \dots, \xi^n(t))$, we obtain

$$\frac{\mathrm{d}^2}{\mathrm{d}t^2}\xi^i(t) = \frac{\mathrm{d}}{\mathrm{d}t}\Xi^i(t) + \sum_{a=1}^{m} \Big(\frac{\mathrm{d}}{\mathrm{d}t}\overline{U}_{(a)}(t)\Big) Y_a^i(\xi(t)) + \sum_{a=1}^{m} \overline{U}_{(a)}(t) \frac{\partial Y_a^i}{\partial q^k}(\xi(t)) \frac{\mathrm{d}}{\mathrm{d}t}\xi^k(t).$$

Next, we substitute into this equation the above expression for $\frac{\mathrm{d}}{\mathrm{d}t}\Xi^i(t)$ as well as $\Xi^i(t) = \frac{\mathrm{d}}{\mathrm{d}t}\xi^i(t) - \sum_{a=1}^{m} \overline{U}_{(a)}(t) Y_a^i(\xi(t))$, whenever possible. After a few simplifications we obtain

$$\begin{aligned}\frac{\mathrm{d}^2}{\mathrm{d}t^2}\xi^i(t) = & -\Gamma^i_{jk}(\xi(t))\xi^j(t)\xi^k(t) + Y^i_0(t,\xi(t)) + Y^i_{1,j}(t,\xi(t))\frac{\mathrm{d}}{\mathrm{d}t}\xi^j(t) \\ & - \sum_{a,b=1}^m \overline{U}_{(a,b)}(t)\,\langle Y_a : Y_b\rangle^i\,(\xi(t)) \\ & + \frac{1}{2}\sum_{a,b=1}^m \overline{U}_{(a)}(t)\overline{U}_{(b)}(t)\,\langle Y_a : Y_b\rangle\,(\xi(t)),\end{aligned}$$

which is equation (9.24) by noting that

$$\sum_{a,b=1}^m \overline{U}_{(a,b)}(t)\,\langle Y_a : Y_b\rangle^i\,(\xi(t)) = \frac{1}{2}\sum_{a,b=1}^m (\overline{U}_{(a,b)}(t) + \overline{U}_{(b,a)}(t))\,\langle Y_a : Y_b\rangle^i\,(\xi(t)),$$

and that $\overline{U}_{(a,b)}(t) + \overline{U}_{(b,a)}(t) = \overline{U_{(a)}U_{(b)}}(t)$ because of the integration by parts equality

$$\begin{aligned}\int_0^T u^b(s_b,t)\int_0^{s_b} u^a(s_a,t)\mathrm{d}s_a\mathrm{d}s_b + \int_0^T u^a(s_a,t)\int_0^{s_a} u^b(s_b,t)\mathrm{d}s_b\mathrm{d}s_a \\ = \int_0^T u^b(s_b,t)\mathrm{d}s_b\int_0^T u^a(s_a,t)\mathrm{d}s_a.\end{aligned}$$

This proves that the curve $\xi = \pi_{\mathsf{TQ}}\circ\Xi\colon I \to \mathsf{Q}$ satisfies equation (9.24). Therefore, equation (9.26) leads to

$$\gamma'(t) = \Phi^{(\sum_{a=1}^m u^a \mathrm{vlft}(Y_a))_{2,t}}_{0,(\frac{t}{\varepsilon}\bmod T)}\Big(\xi'(t) - \sum_{a=1}^m \overline{U}_{(a)}(t)Y_a(\xi(t))\Big) + O(\varepsilon).$$

The flow is in the form (9.19) and, therefore,

$$\begin{aligned}\gamma'(t) = & \Big(\xi'(t) - \sum_{a=1}^m \overline{U}_{(a)}(t)Y_a(\xi(t))\Big) \\ & + \int_0^{(\frac{t}{\varepsilon}\bmod T)}\Big(\sum_{a=1}^m u^a(\tau,t)Y_a(\xi(t))\Big)\mathrm{d}\tau + O(\varepsilon) \\ = & \;\xi'(t) + \sum_{a=1}^m (U_{(a)}(\tfrac{t}{\varepsilon}\bmod T,t) - \overline{U}_{(a)}(t))Y_a(\xi(t)) + O(\varepsilon),\end{aligned}$$

which completes the proof of (9.25). We leave it to the reader to provide the straightforward proof of the statements in the second half of the theorem. ■

Remarks 9.33. 1. Justified by the approximations in the theorem, we call the initial value problem in equation (9.24) the ***averaged forced affine connection system*** of the initial value problem in equation (9.23).

2. While the amplitude of the oscillatory forces is order $\frac{1}{\varepsilon}$, the velocity trajectory $t \mapsto \gamma'(t)$ contains oscillations of order 1, and the configuration trajectory $t \mapsto \gamma(t)$ contains oscillations of order ε.
3. Related alternative treatments of averaging in mechanical systems subject to oscillatory controls are discussed in [Baillieul 1993, 1995, Baillieul and Lehman 1996, Kapitsa 1965, Levi 1998, 1999], see also [Bloch 2003]. Often, these treatments introduce a so-called averaged potential function as a means of characterizing the averaged mechanical system. We discuss this notion and the corresponding averaging result in the following subsection. •

9.2.4 Simple mechanical control systems with potential control forces

In what follows we assume that the forced system is not subject to constraints, and we answer the question "when is the averaged system again a simple mechanical system?" In other words, we present conditions that guarantee that the averaged force in (9.24) is a potential force. The answer to this question involves the relationships between various definitions of the symmetric product that go back to the early treatment by Crouch [1981].

Let $(\mathsf{Q}, \mathbb{G}, V, F_{\text{diss}}, \mathscr{F} = \{F^1, \dots, F^m\}, \mathbb{R}^m)$ be a C^∞-simple mechanical control system subject to the dissipative force F_{diss}. We shall assume that the control forces $\{F^1, \dots, F^m\}$ are potential forces. In other words, we assume that there exist functions $\phi^1, \dots, \phi^m \in C^\infty(\mathsf{Q})$ with the property that $F^a = \mathrm{d}\phi^a$ for $a \in \{1, \dots, m\}$. We refer the reader to Proposition 3.93 and to Remark 10.38 for discussions on the relationship between this assumption and the integrability properties of the codistribution generated by $\mathscr{F}$. Under these assumptions, the governing equations of motion for the forced simple mechanical system subject to oscillatory controls take the form

$$\overset{\mathbb{G}}{\nabla}_{\gamma'(t)}\gamma'(t) = -\operatorname{grad}V(\gamma(t)) + \mathbb{G}^\sharp(F_{\text{diss}}(\gamma'(t))) + \sum_{a=1}^{m} \frac{1}{\varepsilon} u^a\Big(\frac{t}{\varepsilon}\Big) \operatorname{grad}(\phi^a)(\gamma(t)), \quad (9.28)$$

where the functions $u^a\colon \bar{\mathbb{R}}_+ \to \mathbb{R}$, $a \in \{1, \dots, m\}$, are continuous, T-periodic, and zero-mean. Let $\mathbf{\Lambda} \in \mathbb{R}^{m\times m}$ have components

$$\Lambda_{ab} = \frac{1}{2}\big(\overline{U_{(a)}U_{(b)}} - \overline{U}_{(a)}\overline{U}_{(b)}\big). \quad (9.29)$$

For this class of systems, a remarkable simplifications take place in the averaging analysis. We start by introducing and characterizing a useful notion of product between functions.

Definition 9.34 (Beltrami bracket). Let $(\mathsf{Q}, \mathbb{G})$ be a Riemannian manifold. For $\phi, \varphi \in C^\infty(\mathsf{Q})$, the function $\langle \phi : \varphi \rangle \in C^\infty(\mathsf{Q})$ defined by

$$\langle \phi : \varphi \rangle = \mathbb{G}(\mathrm{grad}\phi, \mathrm{grad}\varphi), \tag{9.30}$$

is the ***Beltrami bracket*** of ϕ and φ. •

The following result, originally obtained by Crouch [1981], characterizes the relationship between the Beltrami bracket of functions and the symmetric product of vector fields.

Lemma 9.35. *Let $(\mathsf{Q}, \mathbb{G})$ be a Riemannian manifold. For $\phi, \varphi \in C^\infty(\mathsf{Q})$, the symmetric product $\langle \mathrm{grad}\phi : \mathrm{grad}\varphi \rangle \in \Gamma^\infty(\mathsf{TQ})$ is the gradient of $\langle \phi : \varphi \rangle \in C^\infty(\mathsf{Q})$, that is,*

$$\langle \mathrm{grad}\phi : \mathrm{grad}\varphi \rangle = \mathrm{grad}\, \langle \phi : \varphi \rangle .$$

Proof. The reader is invited to prove this result in Exercise E9.8. ■

In other words, Lemma 9.35 states that the set of gradient vector fields is closed under the operation of symmetric product. This result finds direct application to the averaging analysis in Theorem 9.32.

Theorem 9.36 (Averaged potential function). *Let $(\mathsf{Q}, \mathbb{G}, V, F_{\mathrm{diss}}, \mathscr{F} = \{F^1, \dots, F^m\}, \mathbb{R}^m)$ be a C^∞-simple mechanical control system subject to a dissipative force and subject to oscillatory controls as in equation (9.28). If $F^a = \mathrm{d}\phi^a$ for some $\phi^a \in C^\infty(\mathsf{Q})$, $a \in \{1, \dots, m\}$, then the averaged system is the C^∞-simple mechanical system with dissipation $(\mathsf{Q}, \mathbb{G}, V_{\mathrm{avg}}, F_{\mathrm{diss}})$, where the* ***averaged potential function*** *$V_{\mathrm{avg}} \colon \mathsf{Q} \to \mathbb{R}$ is defined by*

$$V_{\mathrm{avg}} = V + \sum_{a,b=1}^{m} \Lambda_{ab} \langle \phi^a : \phi^b \rangle . \tag{9.31}$$

Accordingly, the equations of motions for the averaged system are

$$\overset{\mathbb{G}}{\nabla}_{\xi'(t)} \xi'(t) = -\mathrm{grad} V_{\mathrm{avg}}(\xi(t)) + \mathbb{G}^\sharp(F_{\mathrm{diss}}(\xi'(t))) . \tag{9.32}$$

Furthermore, let $q_0 \in \mathsf{Q}$ have the property that $\mathrm{d}V_{\mathrm{avg}}(q_0)$ vanishes and $\mathrm{Hess}\, V_{\mathrm{avg}}(q_0)$ *is positive-definite. If F_{diss} is a strict Rayleigh dissipation force, then there exists $\varepsilon_1 > 0$ such that, for all $\varepsilon \in]0, \varepsilon_1[$, there exists a unique trajectory $\gamma_0 \colon \mathbb{R} \to \mathsf{Q}$ for the forced simple mechanical system (9.28) with the following properties: γ_0 is εT-periodic and locally asymptotically stable, γ_0 takes values in an open ball of radius $O(\varepsilon)$ centered at q_0, and γ_0' takes values in an open ball of radius $O(1)$ centered at 0_{q_0}.*

Proof. These results are a consequence of Theorem 9.32. The local asymptotic stability of the averaged system is a consequence of Proposition 6.44. ■

We illustrate these results in the following example and in later chapters on control design.

Example 9.37. Consider the smooth dynamical system on $\mathbb{R}^n$ given by

$$\ddot{\boldsymbol{x}}(t) = \frac{1}{\varepsilon} u\Big(\frac{t}{\varepsilon}\Big) Y(\boldsymbol{x}(t)),$$

for a periodic zero-mean function $u \colon \mathbb{R} \to \mathbb{R}$. Assume that $Y^a = \frac{\partial \varphi}{\partial x^a}$, $a \in \{1, \dots, m\}$, for some $\varphi \colon \mathbb{R}^n \to \mathbb{R}$, and let $\langle \cdot : \cdot \rangle$ denote the Beltrami bracket with respect to the standard metric in $\mathbb{R}^n$. Then the function

$$\langle \varphi : \varphi \rangle = \Big\| \Big(\frac{\partial \varphi}{\partial x^1}, \dots, \frac{\partial \varphi}{\partial x^n} \Big) \Big\|^2_{\mathbb{R}^n}$$

is the classic averaged potential function, or Kapitsa's potential; see [Baillieul 1993, Kapitsa 1965, Levi 1998, 1999]. It is easy to see that every isolated critical point of φ is a minimum of the Kapitsa's potential. Hence, every isolated equilibrium point of the original system is a Lyapunov stable equilibrium point for the averaged system. •

9.3 A series expansion for a controlled trajectory from rest

This section provides a series expansion describing the controlled trajectory of a forced affine connection system with initial velocity equal to zero. Series expansions for control-affine systems are discussed for example in the books [Agrachev and Sachkov 2004, Bloch 2003, Isidori 1995, Nijmeijer and van der Schaft 1990].

We consider a C^ω-forced affine connection system $(\mathsf{Q}, \nabla, Y, \mathcal{D})$. The analyticity assumption is strictly required here. We assume that the vector force $Y \colon \mathbb{R} \times \mathsf{TQ} \to \mathcal{D}$ is basic and, with a slight abuse of notation, we write $Y(t, v_q) = Y(t, q)$. Given this assumption, we consider the initial value problem

$$\nabla_{\gamma'(t)} \gamma'(t) = Y(t, \gamma(t)), \quad \gamma'(0) = 0_{q_0}. \tag{9.33}$$

For $X, Y \in \Gamma^{r+1}(\mathbb{R}; \mathsf{TQ})$, the C^r-time-dependent vector field $\langle X : Y \rangle$ defined by

$$\langle X : Y \rangle (t, q) = \langle X_t : Y_t \rangle (q)$$

is the ***(fixed-time) symmetric product*** of X and Y. Finally, define $\kappa \colon \bar{\mathbb{R}}_+ \to [0, \frac{\pi}{2}]$ implicitly as the unique solution to $\kappa(x) \tan(\kappa(x)) = x$. With these notions, we can state the main result in this section.

Theorem 9.38 (Trajectory of a forced affine connection system starting at rest). *Given a C^ω-forced affine connection system $(\mathsf{Q}, \nabla, Y, \mathcal{D})$, define recursively the time-dependent vector fields $V_k \in \Gamma^\infty(\mathbb{R}; \mathcal{D})$, $k \in \mathbb{N}$, by*

$$V_1(t,q) = \int_0^t Y(s,q)\mathrm{d}s, \tag{9.34}$$

$$V_k(t,q) = -\frac{1}{2}\sum_{j=1}^{k-1}\int_0^t \langle V_j(s,q) : V_{k-j}(s,q)\rangle \mathrm{d}s, \qquad k \geq 2. \tag{9.35}$$

Select a coordinate chart $(\mathcal{U}, \phi)$ *with* $q_0 \in \mathcal{U}$*, let* ρ *and* ρ' *be two positive constants with* $\rho > \rho'$ *and with* $B_\rho(\phi(q_0)) \subset \phi(\mathcal{U})$*, and assume that*

$$\|Y\|_{q_0,\rho,T} T^2 < M \triangleq \min\left\{\frac{2^{-4}(\rho-\rho')}{n(n+1)}, \frac{2^{-4}}{n(n+1)\|\Gamma\|_\rho}, \frac{\kappa^2(\rho' n^2 \|\Gamma\|_{\rho'})}{n^2\|\Gamma\|_{\rho'}}\right\}. \tag{9.36}$$

Then the solution γ *to the initial value problem* (9.33) *is defined on* $[0,T]$*, takes values in* $\mathcal{U}$*, and satisfies*

$$\gamma'(t) = \sum_{k=1}^{+\infty} V_k(t,\gamma(t)), \tag{9.37}$$

where the vector fields V_k*,* $k \in \mathbb{N}$ *satisfy the bound*

$$\|V_k\|_{q_0,\rho',t} \leq M^{1-k}\|Y\|^k_{q_0,\rho,t} t^{2k-1}, \tag{9.38}$$

and the series $(t,q) \mapsto \sum_{k=1}^{+\infty} V_k(t,q)$ *converges absolutely and uniformly for* $q \in B^{\mathbb{C}}_{\rho'}(q_0)$ *and for* $t \in [0,T]$*.*

Proof. See Section B.4. ■

Let us add some comments on the theorem.

Remarks 9.39. 1. While the series expansion is stated in a coordinate-invariant manner, its convergence properties rely on the introduction of a coordinate system. However, if the series converges in one set of coordinates, it will do so in any other, at least in a relatively compact neighborhood of the initial configuration.

2. Equation (9.37) is well-defined in the sense that, at fixed q, the integration is performed with respect to the time variable. Let us introduce a convenient abbreviated notation. Given a time-dependent vector field X on Q, we denote its definite time integral from time 0 to time t by:

$$\underline{X}(t,q) = \int_0^t X(s,q)\,\mathrm{d}s. \tag{9.39}$$

With this notation, the first few terms of the sequence $V_k \in \Gamma^\infty(\mathbb{R};\mathcal{D})$, $k \in \mathbb{N}$, are

$$V_1 = \underline{Y}, \quad V_2 = -\frac{1}{2}\underline{\langle V_1 : V_1\rangle}, \quad V_3 = -\underline{\langle V_2 : V_1\rangle},$$

$$V_4 = -\underline{\langle V_3 : V_1\rangle} - \frac{1}{2}\underline{\langle V_2 : V_2\rangle},$$

and expanding the computations

$$V_1 = \underline{Y}, \quad V_2 = -\frac{1}{2}\langle \underline{Y} : \underline{Y} \rangle, \quad V_3 = \frac{1}{2}\Big\langle \langle \underline{Y} : \underline{Y} \rangle : \underline{Y} \Big\rangle,$$

$$V_4 = -\frac{1}{2}\Big\langle \Big\langle \langle \underline{Y} : \underline{Y} \rangle : \underline{Y} \Big\rangle : \underline{Y} \Big\rangle - \frac{1}{8}\Big\langle \langle \underline{Y} : \underline{Y} \rangle : \langle \underline{Y} : \underline{Y} \rangle \Big\rangle.$$

For $\gamma(0) = q_0$, $\rho \in \mathbb{R}_+$, and $t \in [0, T]$, we can write

$$\begin{aligned}\gamma'(t) = \underline{Y}(t, \gamma(t)) - \frac{1}{2}\langle \underline{Y} : \underline{Y} \rangle(t, \gamma(t)) + \frac{1}{2}\Big\langle \langle \underline{Y} : \underline{Y} \rangle : \underline{Y} \Big\rangle(t, \gamma(t)) \\ - \frac{1}{2}\Big\langle \Big\langle \langle \underline{Y} : \underline{Y} \rangle : \underline{Y} \Big\rangle : \underline{Y} \Big\rangle(t, \gamma(t)) - \frac{1}{8}\Big\langle \langle \underline{Y} : \underline{Y} \rangle : \langle \underline{Y} : \underline{Y} \rangle \Big\rangle(t, \gamma(t)) \\ + O(\|Y\|^5_{q_0,\rho,T} t^9).\end{aligned}$$

3. It is interesting to consider the setting in which the control input is composed of multiple terms of varying magnitude. Let ε be a small positive constant, and consider a vector force of the form

$$Y(t, q, \varepsilon) = \varepsilon X_1(t, q) + \varepsilon^2 X_2(t, q) + \varepsilon^3 X_3(t, q), \qquad t \in [0, 1]$$

Accordingly, equation (9.37) is equivalent to

$$\begin{aligned}\gamma'(t) = \varepsilon \underline{X_1}(t, \gamma(t)) + \varepsilon^2 \Big(\underline{X_2} - \frac{1}{2}\langle \underline{X_1} : \underline{X_1} \rangle \Big)(t, \gamma(t)) \\ + \varepsilon^3 \Big(\underline{X_3} - \frac{1}{2}\langle \underline{X_1} : \underline{X_2} \rangle + \frac{1}{2}\Big\langle \langle \underline{X_1} : \underline{X_1} \rangle : \underline{X_1} \Big\rangle \Big)(t, \gamma(t)) + O(\varepsilon^4).\end{aligned}$$

4. The series expansion contains symmetric products of the input vector field evaluated at different times. The presence of the symmetric product should come as no surprise, given the notion of geodesic invariance and the controllability analysis in Chapter 7.
5. We refer the reader to [Martínez and Cortés 2003] for a series expansion for a forced affine connection system with non-zero initial velocity. •

Exercises

E9.1 Recall that $\lfloor t \rfloor$ is the greatest integer less than or equal to $t \in \mathbb{R}$. Define the square signal

$$s_{\mathrm{square}}(t) = \begin{cases} \frac{1}{2}, & 0 \leq t - \lfloor t \rfloor < \frac{1}{2}, \\ -\frac{1}{2}, & \frac{1}{2} \leq t - \lfloor t \rfloor < 1. \end{cases}$$

Compute iterated integrals, multinomial iterated integrals, and their averages for

(a) $u_a(t) = c_a \sin(\omega t)$, for $\omega \in \mathbb{N}$ and $c_a \in \mathbb{R}$, $a \in \{1, \dots, m\}$, and for

(b) $u_a(t) = c_a s_{\text{square}}(\omega t)$, for $\omega \in \mathbb{N}$ and $c_a \in \mathbb{R}$, $a \in \{1, \ldots, m\}$.

E9.2 This exercise is an extension of Example 9.8. Consider the differential equation on $\mathbb{R}^n$

$$\dot{\boldsymbol{x}}(t) = \boldsymbol{A}\boldsymbol{x}(t) + X(t, \boldsymbol{x}(t)), \quad \boldsymbol{x}(0) = \boldsymbol{x}_0,$$

where $\boldsymbol{A} \in \mathbb{R}^{n \times n}$ and X is a time-dependent vector field on $\mathbb{R}^n$. Show that the variations of constants formula in Proposition 9.6 implies that

$$\boldsymbol{x}(t) = \exp(t\boldsymbol{A})\boldsymbol{x}_0 + \int_0^t \exp((t-\tau)\boldsymbol{A}) X(\tau, \boldsymbol{x}(\tau)) \mathrm{d}\tau.$$

In the next exercises you will develop an alternative version of the variations of constants formula for time-dependent vector fields and two versions for parameter- and time-dependent vector fields; see Proposition 9.6. These relationships and related concepts are elements of a calculus for time-dependent vector fields and their flows called chronological calculus, see [Agrachev and Sachkov 2004].

E9.3 Prove the following result of Agrachev and Sachkov [2004].

Proposition. *For $r \in \mathbb{N} \cup \{\infty\} \cup \{\omega\}$, let $X \in \Gamma^r(\mathbb{R}; \mathsf{TM})$ and $Y \in \Gamma^{r+1}(\mathbb{R}; \mathsf{TM})$. For $x_0 \in \mathsf{M}$, $T \in \mathbb{R}_+$, and $s, t \in [0, T]$, assume that $\Phi^Y_{s,t}$ is a C^{r+1}-diffeomorphism from a neighborhood $\mathcal{U}$ of x_0 onto its image. Let $F\colon [0,T] \times [0,T] \times \mathcal{U} \to \mathsf{T}\mathcal{U}$ be the C^r-parameter- and time-dependent vector field defined by $F(\lambda, \tau, x) = ((\Phi^Y_{\lambda,\tau})^* X_\tau)(x)$. Then*

$$\Phi^{X+Y}_{0,t}(x_0) = \left(\Phi^{F_\lambda}_{0,t}\big|_{\lambda=t}\right) \circ \Phi^Y_{0,t}(x_0),$$

for all $t \in [0, T]$ for which the three integral curves exist.

E9.4 Prove the following result of Agrachev and Sachkov [2004].

Proposition. *Let the parameter- and time-dependent vector field Y be of class C^1. For $x_0 \in \mathsf{M}$, $\lambda \in \mathbb{R}$, and $s, t \in [0, T]$, assume that $\Phi^{Y_\lambda}_{s,t}$ is a C^1-diffeomorphism from a neighborhood $\mathcal{U}$ of x_0 onto its image. Then the curve $\lambda \mapsto \Phi^{Y_\lambda}_{0,t}(x_0)$ satisfies the equivalent expressions:*

$$\begin{aligned}\frac{\partial}{\partial \lambda} \Phi^{Y_\lambda}_{0,t}(x_0) &= (T_x \Phi^{Y_\lambda}_{0,t}) \circ \int_0^t \left((\Phi^{Y_\lambda}_{0,\tau})^* \frac{\partial Y_{\lambda,\tau}}{\partial \lambda}\right)(x_0) \mathrm{d}\tau \\ &= \int_0^t \left((\Phi^{Y_\lambda}_{t,\tau})^* \frac{\partial Y_{\lambda,\tau}}{\partial \lambda}\right) \mathrm{d}\tau \circ \Phi^{Y_\lambda}_{0,t}(x_0).\end{aligned}$$

E9.5 Consider the control-affine system $\Sigma = (\mathsf{M} = \mathbb{R}^n, \{f_0, f_1\}, U)$ with governing equation

$$\dot{\boldsymbol{x}}(t) = f_0(\boldsymbol{x}(t)) + \frac{1}{\varepsilon} u\left(\frac{t}{\varepsilon}\right) f_1(\boldsymbol{x}(t)),$$

where the components of the vector field $\boldsymbol{y} \mapsto f_0(\boldsymbol{y})$ are polynomials in $\boldsymbol{y}$ of finite degree, and where the vector field $\boldsymbol{y} \mapsto f_1(\boldsymbol{y})$ is constant. Show that the averaged system $\overline{F}$ is a finite sum of polynomial vector fields.

E9.6 For $t \in \mathbb{R}$, consider the map

$$\begin{aligned}\phi_t\colon \mathsf{TQ} &\to \mathsf{TQ} \\ v_q &\mapsto \mathrm{e}^t v_q,\end{aligned}$$

and define the Liouville vector field V_{L} on TQ according to

$$V_{\mathrm{L}}\colon \mathsf{TQ} \to \mathsf{T}(\mathsf{TQ})$$
$$v_q \mapsto \left.\frac{\mathrm{d}}{\mathrm{d}t}\right|_{t=0} \phi_t(v_q).$$

Show that

(a) V_{L} is a well-defined vector field on TQ,
(b) $\phi_t = \Phi_t^{V_{\mathrm{L}}}$, and
(c) V_{L} has the coordinate expression of (9.18).

E9.7 Prove the four results in Proposition 9.29.

E9.8 Prove Lemma 9.35.
Hint: *Use Exercise E3.40.*

E9.9 *Mathematica®-assisted project:* Consider the planar rigid body with a control force applied at a distance h from the center of mass that we resolve into two components F^1 and F^2; see Example 4.53. Consider the two controlled trajectories for the planar body starting from initial condition $(x(0), y(0), \theta(0)) = (0, 0, 0)$ and subject to the two oscillatory inputs $(0, \frac{1}{\varepsilon}\cos(\frac{t}{\varepsilon}))$ and $(\frac{1}{\varepsilon}\cos(\frac{t}{\varepsilon}), 0)$.

(a) Using Theorem 9.32, compute the controlled trajectories for the averaged system in order to approximately predict the controlled trajectories for the forced planar rigid body as $\varepsilon \to 0$.
(b) Verify your prediction in numerical simulations with the following sample parameter values: $J = 1$, $m = 1$, and $h = 1$.

E9.10 *Mathematica®-assisted project:* Consider the roller racer system studied in Exercise E4.42. Consider the controlled trajectory for the roller racer starting from initial condition $(x(0), y(0), \theta(0), \psi(0)) = (0, 0, 0, 0)$ and subject to the oscillatory input $\frac{1}{\varepsilon}\cos(\frac{t}{\varepsilon})$.

(a) Using Theorem 9.32, compute the controlled trajectory for the averaged system in order to approximately predict the controlled trajectory for the forced roller racer as $\varepsilon \to 0$.
(b) Verify your prediction in numerical simulations with the following sample parameter values: $J_1 = 2$, $J_2 = 1$, $m = 1$, $\ell_1 = 2$, and $\ell_2 = 1$.

E9.11 *Mathematica®-assisted project:* Consider the three-link manipulator discussed in Exercise E4.37. For $i \in \{1, 2, 3\}$, let θ_i denote the orientation of the ith link measured counterclockwise from the positive horizontal axis. Assume that the system is equipped with a single control force $\mathrm{d}\theta_1$. Consider the controlled trajectory for the three-link planar manipulator starting from initial condition $(\theta_1(0), \theta_2(0), \theta_3(0)) = (0, \frac{\pi}{4}, 0)$ and subject to the oscillatory input $\frac{1}{\varepsilon}\cos(\frac{t}{\varepsilon})$.

(a) Using Theorem 9.32, compute the controlled trajectory for the averaged system in order to approximately predict the controlled trajectory for the forced three-link planar manipulator as $\varepsilon \to 0$.
(b) Verify your prediction in numerical simulations with all lengths, masses, and moments of inertia set equal to 1.

Part III

A sampling of design methodologies

10

Linear and nonlinear potential shaping for stabilization

Our first chapter on control design is concerned with somewhat classic methods for stabilization. The techniques described here involve linearization methods and potential energy shaping methods. Although these techniques are somewhat limited in scope, they are valuable because they provide some insight into the problem of stabilization, both generally and for mechanical systems. Readers can then use the experience gained from this chapter to more easily read the literature on stabilization, some of which is discussed in Section 10.5.

The problem of stabilization consists, roughly, of finding controls that will steer the system to a desired state. For a mechanical system, this state is typically an equilibrium configuration. In Chapters 11 and 12 we will consider the stabilization problem for more general trajectories. In the history of control, the stabilization problem has been one of the cornerstones. Indeed, perhaps the first modern treatment of a control problem, that of Maxwell [1868] concerning flyball governors, involved stabilization. The development of negative feedback by Nyquist [1932] was also made in the context of the stabilization problem, and led to some important tools for the design of controllers that remain in practice today. The methods of Nyquist dealt with transfer functions, and other important contributions to these techniques were made by Bode [1945] and Evans [1948, 1950]. After the Second World War, much of the development in the theory for stabilization shifted to so-called state space methods, with one of the fundamental contributions being made by Kalman [1960a]. We shall have more to say about this in Section 10.5.1. For the moment, suffice it to say that there is a quite well-developed theory for stabilization of linear control systems. There are many references for this, some being [Brockett 1970, Chen 1984, Kailath 1980, Sontag 1998, Wonham 1985]. For nonlinear systems, the development of stabilization techniques presents a more complicated picture, and Section 10.1 is devoted to a presentation of some of the more important ideas. An excellent overview of the stabilization problem for nonlinear systems can be found in the book of Bacciotti [1992].

For mechanical systems, the history of the stabilization problem is shorter. A very early paper, dealing with dissipative linear control, is that of Gabrielian and Krasovskiĭ [1965]. One of the earliest papers is that of Takegaki and Arimoto [1981], where potential energy shaping and dissipative control injection for fully actuated systems were originally introduced. These methodologies were applied to the setting of underactuated systems in [van der Schaft 1986], and were further investigated in [Koditschek 1989, Ortega and Spong 1989, Paden and Panja 1988, Wen and Bayard 1988]. The idea of these results is that one wishes to design control laws that retain the mechanical structure of the problem; it is from this idea that the expression "energy shaping" is derived. Related works in [Khatib 1986, Rimon and Koditschek 1991, 1992] discuss the so-called potential function method for global stabilization problems in the presence of obstacles. Other early papers include [Crouch 1984, Krasinskiĭ 1988]. Stabilization to equilibrium *manifolds* using PD control was considered by Oriolo and Nakamura [1991], who also point out that mechanical systems without a potential function are difficult to stabilize, by virtue of their not possessing continuous stabilizing control laws.

An outline of the chapter is as follows. In Section 10.1 we provide an overview of some of the important problems and approaches in stabilization theory. Again, here the reader can find pointers to the stabilization literature. In Section 10.2 we define stabilization problems for mechanical systems, building on the general definitions in Section 10.1. Sections 10.3 and 10.4 provide two techniques for stabilizing equilibrium configurations for mechanical systems. In the first of these sections, we consider linear methods. First we indicate how to stabilize linear systems, and then how to apply stabilizing control laws for linearizations to nonlinear systems. This is the "classic" approach to stabilization theory. In Section 10.4 we consider general potential shaping, following Takegaki and Arimoto [1981] and van der Schaft [1986]. In Section 10.5 we incompletely review approaches to the stabilization of mechanical systems that are commonly encountered in the literature. In particular, some quite recent work on these problems is described in Sections 10.5.4 and 10.5.6. A reader, referring to the references in Section 10.5, and to the work cited in these papers, will obtain a brief introduction to existing approaches to stabilization theory for mechanical systems.

10.1 An overview of stabilization

In this section we provide a short introduction to the theory of stabilization of equilibria. After defining the problem, we point out some of the subtleties in the quest for a general theory of stabilization. As we shall see, there are many impediments to arriving at a complete theory. The final sections deal with stabilization techniques that rely on Lyapunov theory. These Lyapunov-style results will form the basis for the design strategies in this chapter.

10.1.1 Defining the problem

We begin with definitions.

Definition 10.1 (Feedback and stabilizability). Let $\Sigma = (\mathsf{M}, \mathscr{C} = \{f_0, f_1, \dots, f_m\}, U)$ be a C^∞-control-affine system.

(i) A ***controlled equilibrium point*** for Σ is a pair $(x_0, u_0) \in \mathsf{M} \times U$ with the property that

$$f_0(x_0) + \sum_{a=1}^{m} u_0^a f_a(x_0) = 0_{x_0}.$$

(ii) A ***state feedback*** (resp. ***time-dependent state feedback***) for Σ is a map $u \colon \mathsf{M} \to U$ (resp. $u \colon \mathbb{R}_+ \times \mathsf{M} \to U$).

(iii) For a state feedback (resp. time-dependent state feedback) u for Σ, the ***closed-loop system*** is the vector field (resp. time-dependent vector field) defined by

$$x \mapsto f_0(x) + \sum_{a=1}^{m} u^a(x) f_a(x) \quad \Big(\text{resp. } (t,x) \mapsto f_0(x) + \sum_{a=1}^{m} u^a(t,x) f_a(x)\Big).$$

(iv) For $r \in \mathbb{Z}_+ \cup \{\infty\} \cup \{\omega\}$, a state feedback (resp. time-dependent state feedback) is $\boldsymbol{C^r}$ if the corresponding closed-loop system is of class C^r.

(v) For $r \in \mathbb{Z}_+ \cup \{\infty\} \cup \{\omega\}$ and $x_0 \in \mathsf{M}$, a state feedback is ***almost $\boldsymbol{C^r}$ about $\boldsymbol{x_0}$*** if there exists a neighborhood $\mathcal{U}$ of x_0 such that the corresponding closed-loop system is C^r on $\mathcal{U} \setminus \{x_0\}$.

(vi) A controlled equilibrium point (x_0, u_0) is ***stabilizable by state feedback*** (resp. ***stabilizable by time-dependent state feedback***) if there exists a state feedback (resp. time-dependent state feedback) u for Σ with the property that the closed-loop system has x_0 as a stable equilibrium point.

(vii) A controlled equilibrium point (x_0, u_0) is ***locally asymptotically stabilizable by state feedback*** (resp. ***by time-dependent state feedback***) if there exists a state feedback (resp. time-dependent state feedback) and a neighborhood $\mathcal{U}$ of x_0 with the properties that
  (a) the closed-loop system leaves $\mathcal{U}$ invariant, and
  (b) the restriction of the closed-loop system to $\mathcal{U}$ possesses x_0 as an asymptotically stable equilibrium point.

(viii) A controlled equilibrium point (x_0, u_0) is ***globally asymptotically stabilizable by state feedback*** (resp. ***by time-dependent state feedback***) if in part (vii) one can take $\mathcal{U} = \mathsf{M}$. •

Remarks 10.2. 1. Given a controlled equilibrium point (x_0, u_0) for $\Sigma = (\mathsf{M}, \mathscr{C} = \{f_0, f_1, \dots, f_m\}, U)$, one can define a modified control-affine system $\Sigma_{(x_0,u_0)} = (\mathsf{M}, \{\tilde{f}_0, f_1, \dots, f_m\}, \tilde{U})$, where

$$\tilde{f}_0(x) = f_0(x) + \sum_{a=1}^{m} u_0^a f_a(x), \quad \tilde{U} = \{u - u_0 \mid u \in U\}. \tag{10.1}$$

For this new system, $0 \in \tilde{U}$ and $(x_0, 0)$ is a controlled equilibrium point. For this reason, we will often, and without loss of generality, suppose that the control at a controlled equilibrium point is zero.

2. We adopt a slightly different definition of C^r-state feedback than is usual. The usual definition is that a state feedback $u\colon \mathsf{M} \to U$ (resp. time-dependent state feedback $u\colon \bar{\mathbb{R}}_+ \times \mathsf{M} \to U$) is C^r if it is a C^r-map from M to $\mathbb{R}^m$ (resp. from $\bar{\mathbb{R}}_+ \times \mathsf{M}$ to $\mathbb{R}^m$). Clearly, our definition is a special case of the usual definition. Furthermore, if the vector fields $\{f_1, \dots, f_m\}$ are linearly independent at each $x \in \mathsf{M}$, then our definition and the usual definition are equivalent. However, if the vector fields are not linearly independent, then the equivalence of the definitions does not immediately follow. The picture here is quite complicated, particularly if the distribution generated by the control vector fields has singular points; see Section 3.9.4, and Exercises E3.42 and E10.1.
3. When stating theorems concerning design strategies for stabilization, it is convenient to be able to say, "a system is stabilized at (x_0, u_0) by the state feedback 'X'." Corresponding to our preceding remark, one should be careful what is meant by 'X.' We shall allow two possibilities. First, we say a controlled equilibrium point (x_0, u_0) is stabilized (in one of the senses of Definition 10.1) ***by the state feedback u*** if the closed-loop system for u is stable (in the appropriate sense). To state the other variant of this language, we denote by $\mathcal{C}$ the distribution generated by the vector fields $\{f_1, \dots, f_m\}$. Let f be a vector field of unspecified regularity taking values in $\mathcal{C}$. We then say that (x_0, u_0) is stabilized (in one of the senses of Definition 10.1) ***by the vector field f*** if the closed-loop system for u is stable (in the appropriate sense), where u is any state feedback satisfying $u^a f_a = f$.
4. For time-dependent state feedback, one sometimes one wishes to allow the dependence on time and state to have different regularity. This will come up in Chapter 12.
5. In our definitions, we distinguish between stabilizability and asymptotic stabilizability. It is not customary to do so in the control literature, where stabilizability is the only concept defined, and is the same as our asymptotic stabilizability. In the mechanical context, however, it is useful to be able to talk about stabilizability, without necessarily having asymptotic stabilizability.
6. Note that, in the definition of the closed-loop system, we made no assumptions about continuity or differentiability. As we shall see, this is actually essential for the stabilizability problem, since many interesting systems cannot be stabilized with even continuous state feedback. This lack of continuity leads to problems with what is meant by a solution for

the closed-loop system. A commonly used notion for an integral curve of a discontinuous vector field is that of Filippov [1960] (see [Filippov 1988] for a thorough account). However, as shown by Ryan [1994] and Coron and Rosier [1994], stabilizability of control-affine systems by discontinuous feedback in the sense of Filippov implies stabilizability by almost C^∞-feedback about the desired equilibrium point.[1] This limits the usefulness of the Filippov notion of solutions using discontinuous feedback. Somewhat recently, a new method for defining solutions for discontinuous systems has been proposed, and it is far more flexible than the Filippov method. This is discussed at length in the book of Clarke, Ledyaev, Stern, and Wolenski [1998]. The consequences of this theory for stabilization will be discussed below.

7. We shall not concern ourselves much with global stabilization. One of the reasons for this is that global stabilization on manifolds is generally difficult to achieve. For example, as was mentioned in Remark 6.22, the domain of attraction for an equilibrium point of a smooth vector field is diffeomorphic to Euclidean space. This means that if the state manifold has some interesting topology, then one will have to go with a nonsmooth control law (cf. Exercise E6.2), even when the system is locally asymptotically stabilizable by smooth feedback. Thus the character of global stabilization is rather different than that of local stabilization, at least if one wishes to establish a theory with a useful degree of generality. The reader can explore some issues concerning global stabilization in Exercise E10.2. •

10.1.2 Stabilization using linearization

An important dichotomy that arises in the theory of local stabilization occurs in terms of the linearization of the system. First let us recall a standard result from the theory of linear systems, for which the reader may refer to the texts mentioned in the introduction. We recall from Example 7.12 the notion of a linear control-affine system. The following result concerns stabilizability by *linear* feedback. Recall that P_A denotes the characteristic polynomial of $A \in L(\mathsf{V};\mathsf{V})$.

Theorem 10.3 (Stabilizability of linear systems). *For a linear control system* (V, A, B), *let* $\langle A, \mathrm{image}(B)\rangle$ *denote the smallest* A*-invariant subspace containing* $\mathrm{image}(B)$. *Let* $\bar{A}$ *be the linear map defined on* $\mathsf{V}/\langle A, \mathrm{image}(B)\rangle$ *by*

$$\bar{A}(x + \langle A, \mathrm{image}(B)\rangle) = A(x) + \langle A, \mathrm{image}(B)\rangle.$$

If P *is a monic polynomial of degree* $\dim(\langle A, \mathrm{image}(B)\rangle)$, *then there exists* $F \in L(\mathsf{V};\mathbb{R}^m)$ *for which* $P_{A+B\circ F} = PP_{\bar{A}}$. *In particular, there exists an*

[1] This is true for control-affine systems, but not necessarily true for the more general class of control system $\gamma'(t) = f(\gamma(t), u(t))$. For this more general class, what is true is that stabilizability by discontinuous feedback in the sense of Filippov implies stabilizability by C^0-feedback.

$F \in L(\mathsf{V};\mathbb{R}^m)$ such that $\operatorname{spec}(A + B \circ F) \subset \mathbb{C}_-$ *if and only if* $\operatorname{spec}(\bar{A}) \subset \mathbb{C}_-$. *In such an event, the linear system* (V, A, B) *is* ***asymptotically stabilizable***.

Remarks 10.4. 1. The feedback being used in the theorem is $u = F(x)$, and with this feedback the closed-loop system is governed by the linear differential equation

$$\dot{x}(t) = (A + B \circ F)(x(t)).$$

Thus linear feedback, unsurprisingly, results in linear closed-loop dynamics.

2. In words the theorem says, "The system is stabilizable if and only if all uncontrollable modes of A are asymptotically stable." •

Next let us consider not linear systems, but linearized systems. We recall from Definition 7.27 the notion of the linearization of a control-affine system $(\mathsf{M}, \{f_0, f_1, \dots, f_m\}, U)$ at a point x_0 for which $(x_0, 0)$ is a controlled equilibrium point. The linearization is the linear control system $(\mathsf{T}_{x_0}\mathsf{M}, A_\Sigma(x_0), B_\Sigma(x_0))$, where $A_\Sigma(x_0)$ is the linearization of f_0 at x_0, as defined by Proposition 3.75, and $B_\Sigma(x_0) \in L(\mathbb{R}^m; \mathsf{T}_{x_0}\mathsf{M})$ is defined by

$$B_\Sigma(x_0)(u) = \sum_{a=1}^{m} u^a f_a(x_0).$$

The following result characterizes the relationship between the stabilizability of the linearization and local stabilizability. The proof of the first part of the theorem is given as Exercise E10.5, and the second part is proved by Brockett [1983].

Theorem 10.5 (Stabilization from linear stabilization). *Let $(x_0, 0)$ be a controlled equilibrium point for a C^∞-control-affine system $\Sigma = (\mathsf{M}, \mathscr{C}, \mathbb{R}^m)$, and let $(\mathsf{T}_{x_0}\mathsf{M}, A_\Sigma(x_0), B_\Sigma(x_0))$ be the linearization. Let $\bar{A}_\Sigma(x_0)$ be defined as in Theorem 10.3. The following statements hold:*

(i) if $(\mathsf{T}_{x_0}\mathsf{M}, A_\Sigma(x_0), B_\Sigma(x_0))$ is asymptotically stabilizable, then $(x_0, 0)$ is locally asymptotically stabilizable by C^∞-state feedback;

(ii) if $\operatorname{spec}(\bar{A}_\Sigma(x_0)) \cap \mathbb{C}_+ \neq \emptyset$, then it is not possible to locally asymptotically stabilize $(x_0, 0)$ by C^1-state feedback having the value 0 at x_0.

Remark 10.6. Suppose that the linearization of a system is stabilizable, and that, therefore, the equilibrium point is locally asymptotically stabilizable. The use of the linear techniques to design stabilizing controllers is often referred to as ***stabilization by Lyapunov's first method***. We shall see what is meant by stabilization by Lyapunov's second method in Section 10.1.5. •

10.1.3 The gaps in linear stabilization theory

Note that Theorem 10.5 has a gap between the necessary and sufficient condition. Some questions arising from this gap are:

1. if $\mathrm{spec}(\bar{A}_\Sigma(x_0)) \cap \mathbb{C}_+ \neq \emptyset$, is it possible to find stabilizing feedback control laws that are C^0, or perhaps discontinuous?
2. what can be said about the stabilization problem when $\mathrm{spec}(\bar{A}_\Sigma(x_0)) \subset \overline{\mathbb{C}}_-$ and $\mathrm{spec}(\bar{A}_\Sigma(x_0)) \cap \mathrm{i}\mathbb{R} \neq \emptyset$?

In this section we give a brief survey of the existing results that fill out this gap. This is presently an active area of research, and we refer to a fairly recent (at the time of publication) paper of Sontag [1999] for additional discussion. A major omission of our survey is center manifold theory for stabilization. A treatment of this may be found in Section 10.3 of [Nijmeijer and van der Schaft 1990] or Chapter 4 of [Khalil 2001].

We begin with a couple of examples.

Example 10.7. 1. Consider the control-affine system with $\mathsf{M} = \mathbb{R}^2$, $\{f_0 = (y - x^3)\frac{\partial}{\partial y}, f_1 = \frac{\partial}{\partial x}\}$, and $U = \mathbb{R}$. The control equations are

$$\dot{x} = u, \quad \dot{y} = y - x^3,$$

and we consider the controlled equilibrium point $((0,0),0)$. Note that the linearization in the standard basis is given by

$$A_\Sigma(x_0) = \begin{bmatrix} 0 & 0 \\ 0 & 1 \end{bmatrix}, \quad B_\Sigma(x_0) = \begin{bmatrix} 1 \\ 0 \end{bmatrix}.$$

One can easily check that the linearization has an unstable uncontrollable mode, and so is not asymptotically stabilizable. Furthermore, part (ii) of Theorem 10.5 tells us that it is not possible to find a locally asymptotically stabilizing C^1-state feedback for this system. However, the continuous state feedback $u(x,y) = -ax + by^{1/3}$ can be verified to globally asymptotically stabilize the system, provided that $1 < a < b$. The stabilization of this system seems to have first been discussed by Kawski [1989], and generalizations of the system have been considered by many authors, including [Čelikovsky and Aranda-Bricaire 1999, Coron and Praly 1991, Qian and Lin 2001]. The punchline is that one cannot replace "C^1" with "C^0" in part (ii) of Theorem 10.5.

2. Consider the control-affine system with $\mathsf{M} = \mathbb{R}^2$, $f_0 = -x^3\frac{\partial}{\partial x}$, $f_1 = \frac{\partial}{\partial y}$, and $U = \mathbb{R}$. This system satisfies the hypotheses of neither part of Theorem 10.5, but it is quite clear that the system is globally asymptotically stabilizable by C^∞-state feedback.
3. Consider the control-affine system with $\mathsf{M} = \mathbb{R}^2$, $f_0 = x^3\frac{\partial}{\partial x}$, $f_1 = \frac{\partial}{\partial y}$, and $U = \mathbb{R}$. This system again satisfies the hypotheses of neither part of Theorem 10.5, but it is obvious that there is no stabilizing state feedback, no matter how weak are the assumptions placed on its smoothness. •

The second and third of the preceding three examples indicate that, when the hypotheses of part (ii) of Theorem 10.5 are satisfied, one must have additional hypotheses to establish the presence or absence of stabilizability. A well-known result along these lines is due to Brockett [1983].

Theorem 10.8 (Brockett's Necessary Condition). *If there exists a C^0-locally asymptotically stabilizing state feedback for the C^∞-control-affine system $\Sigma = (\mathsf{M}, \mathscr{C}, \mathbb{R}^m)$ at $(x_0, 0)$, then, for any coordinate chart $(\mathcal{U}, \phi)$ for which $x_0 \in \mathcal{U}$, the map*

$$\phi(\mathcal{U}) \times \mathbb{R}^m \ni (\boldsymbol{x}, \boldsymbol{u}) \mapsto f_{\boldsymbol{u}}(\boldsymbol{x}) \in \mathbb{R}^n$$

is surjective onto a neighborhood of the origin, where $\boldsymbol{x} \mapsto (\boldsymbol{x}, f_{\boldsymbol{u}}(\boldsymbol{x}))$ is the local representative for the vector field

$$x \mapsto f_0(x) + \sum_{a=1}^{m} u^a f_a(x),$$

for fixed $u \in \mathbb{R}^m$.

Remarks 10.9. 1. Brockett initially stated the result for C^1-feedback. The result, however, is valid for continuous feedback [Zabczyk 1989], even without the assumption of uniqueness of solutions for the closed-loop system [Orsi, Praly, and Mareels 2003].

2. Another often cited necessary condition for stabilizability is that of Coron [1990]. Coron's result relies on techniques from algebraic topology, as does Theorem 10.8. Both Brockett's result and Coron's result are not sharp, in that there are systems that do not satisfy the necessary conditions, but are nonetheless locally asymptotically stabilizable by C^0-state feedback. •

Examples 10.10. 1. Consider a driftless control-affine system $\Sigma = (\mathsf{M}, \{f_0, f_1, \dots, f_m\}, \mathbb{R}^m)$; thus $f_0 = 0$. First of all, note that every point in M is a controlled equilibrium point with the zero control. Second, it is easy to show (Exercise E10.6) that Theorem 10.8 implies the impossibility of stabilizing any of these controlled equilibrium points with C^0-state feedback. There are thus two possible avenues for stabilizing driftless systems: (1) using discontinuous state feedback, or (2) using time-dependent state feedback. Along the lines of the second of these approaches, Coron [1992] shows that, for driftless systems, it is possible to construct a time-dependent C^∞-locally asymptotically stabilizing state feedback. In terms of design, there has been a significant amount of work done on stabilization for driftless systems. An incomplete list of references is [Astolfi 1996, Kolmanovsky and McClamroch 1995, Lin, Pongvuthithum, and Qian 2002, M'Closkey and Murray 1997, Morin, Pomet, and Samson 1999, Pomet 1992]. In Section 13.1 we shall have more to say about driftless systems in the context of motion planning.

2. The reader may show in Exercise E10.7, using Theorem 10.8, that underactuated affine connection control systems cannot be locally asymptotically stabilized using C^0-state feedback. This was pointed out by Oriolo and Nakamura [1991]. •

10.1.4 Control-Lyapunov functions

In this section we consider an important tool in the development of the theory of stabilization.

Definition 10.11 (Control-Lyapunov function). Let $\Sigma = (\mathsf{M}, \mathscr{C}, \mathbb{R}^m)$ be a C^∞-control-affine system for which $(x_0, 0)$ is a controlled equilibrium point.

(i) A ***control-Lyapunov triple*** for Σ at x_0 is a triple $(\psi, \phi, \mathcal{U})$, where
   (a) $\mathcal{U}$ is a neighborhood of x_0,
   (b) $\psi \colon \mathcal{U} \to \bar{\mathbb{R}}_+$ is continuous, proper, and locally positive-definite about x_0,
   (c) $\phi \colon \mathcal{U} \to \bar{\mathbb{R}}_+$ is continuous and locally positive-definite about x_0, and
   (d) for each compact subset $K \subset \mathcal{U}$, there exists a compact subset $U \subset \mathbb{R}^m$ such that, for all $x \in K$, there exists $\boldsymbol{u}_x \in U$ such that

$$\mathrm{d}\psi(x) \cdot \Big(f_0(x) + \sum_{a=1}^m u_x^a f_a(x) \Big) < -\phi(x).$$

(ii) A ***control-Lyapunov function*** for Σ at x_0 is a continuous function $\psi \colon \mathsf{M} \to \mathbb{R}$ for which there exist a neighborhood $\mathcal{U}$ and a continuous function $\phi \colon \mathcal{U} \to \bar{\mathbb{R}}_+$ for which $(\psi|\mathcal{U}, \phi, \mathcal{U})$ is a control-Lyapunov triple. •

Remark 10.12. An immediate problem with the preceding definition is that ψ is permitted to be only continuous, so it is not clear what one might mean by $\mathrm{d}\psi$. This can be resolved by using nonsmooth analysis, as was initially pointed out by Sontag and Sussmann [1995]. •

Control-Lyapunov functions are useful in the theory of stabilization, since, when they can be found, they often point the way to the design of a stabilizing feedback control law. Indeed, the following control law of Sontag [1989] gives a locally asymptotically stabilizing state feedback for $\Sigma = (\mathsf{M}, \mathscr{C}, \mathbb{R}^m)$ at a controlled equilibrium point $(x_0, 0)$, given a C^∞-control-Lyapunov function ψ for Σ at x_0:

$$u(x) = \begin{cases} 0, & \mathrm{d}\psi(x) \cdot f_a(x) = 0,\ a \in \{1, \dots, m\}, \\ -\frac{\alpha(x) + \left((\alpha(x))^2 + (\beta^T(x)\beta(x))^2\right)^{1/2}}{\beta^T(x)\beta(x)}, & \text{otherwise,} \end{cases}$$

where

$$\alpha \colon \mathsf{M} \to \mathbb{R} \colon x \mapsto \mathrm{d}\psi(x) \cdot f_0(x),$$
$$\beta \colon \mathsf{M} \to \mathbb{R}^m \colon x \mapsto (\mathrm{d}\psi(x) \cdot f_1(x), \dots, \mathrm{d}\psi(x) \cdot f_m(x)).$$

The preceding construction notwithstanding, the following important result due to Artstein [1983] indicates that for many systems it will be necessary to consider nonsmooth control-Lyapunov functions.

Theorem 10.13 (Artstein's Theorem). *For a controlled equilibrium point* $(x_0, 0)$ *for a* C^∞*-control-affine system* Σ*, the following statements are equivalent:*

(i) x_0 *is locally asymptotically stabilizable using almost* C^∞*-state feedback;*
(ii) there exists a C^1*-control-Lyapunov function for* Σ *at* x_0*.*

Let us next state a theorem of Clarke, Ledyaev, Sontag, and Subotin [1997] that brings together the techniques of nonsmooth analysis and control-Lyapunov functions to prove a strong result concerning the stabilizability of a system. To state the theorem, we say that a control-affine system $\Sigma = (\mathsf{M}, \mathscr{C}, \mathbb{R}^m)$ is ***locally asymptotically controllable*** to $x_0 \in \mathsf{M}$ if there exists a neighborhood $\mathcal{U}$ of x_0 with the property that, for each $x \in \mathcal{U}$, there exists a map $u_x \colon \bar{\mathbb{R}}_+ \to \mathbb{R}^m$ for which the solution for the initial value problem

$$\gamma'(t) = f_0(\gamma(t)) + \sum_{a=1}^{m} u_x^a(t) f_a(\gamma(t)), \quad \gamma(0) = x,$$

has the property that $\lim_{t \to +\infty} \gamma(t) = x_0$.

Theorem 10.14. *Let* $(x_0, 0)$ *be a controlled equilibrium point for the* C^∞*-control-affine system* $\Sigma = (\mathsf{M}, \mathscr{C}, \mathbb{R}^m)$*. Then* $(x_0, 0)$ *is locally asymptotically stabilizable by state feedback if and only if* Σ *is locally asymptotically controllable to* x_0*.*

As is to be expected from our preceding discussion, the control cannot be expected to be continuous, and so one must carefully consider what is meant by a solution for the closed-loop system. This is not something we get into here, and we refer the reader to [Clarke, Ledyaev, Sontag, and Subotin 1997] for a description of the necessary machinery.

10.1.5 Lyapunov-based dissipative control

We next consider a specific design methodology based on the use of Lyapunov functions for stability analysis. Sometimes the method we describe in this section is called "$\mathscr{L}_g V$-control," this terminology referring to notation for control-affine systems that we do not use. Some early work along these lines is that of Jurdjevic and Quinn [1978]. This work was followed up in numerous papers e.g., Lee and Arapostathis [1988]. The ideas we use in this section lie in the background of much of the analysis we present in this chapter. A more detailed discussion may be found in [Nijmeijer and van der Schaft 1990, Section 10.2]. The general approach outlined in this section is often referred to as ***stabilization by Lyapunov's second method***.

An example motivates the approach.

Example 10.15. Let us consider the control-affine system $(\mathsf{M} = \mathbb{R}^2, \mathscr{C} = \{f_0 = y\frac{\partial}{\partial x}, f_1 = \frac{\partial}{\partial y}\}, \mathbb{R})$. The governing equations are then

$$\dot{x} = y, \quad \dot{y} = u.$$

We claim that $\psi(x,y) = \frac{1}{2}x^2 + \frac{1}{2}y^2$ is a control-Lyapunov function for Σ at $x_0 = (0,0)$. We compute

$$\mathscr{L}_{f_0}\psi(x,y) = xy, \quad \mathscr{L}_{f_1}\psi(x,y) = y.$$

We then note that, if we take $u_{\mathrm{clf}}(x,y) = -x$, then

$$\mathrm{d}\psi(x,y) \cdot (f_0(x,y) + u_{\mathrm{clf}}(x,y) f_1(x,y)) = 0.$$

This shows that ψ is indeed a control-Lyapunov function (the triple $(\psi, \phi, \mathbb{R}^2)$ is a control-Lyapunov triple if we take $\phi(x,y) = 0$).

Note that in this example ψ is a Lyapunov function for the closed-loop system with the state feedback u_{clf}. Thus x_0 is stable for the closed-loop system. However, since the derivative of the Lyapunov function is zero along the trajectories of the closed-loop system, x_0 is not asymptotically stable. To asymptotically stabilize x_0 we define the function

$$u_{\mathrm{diss}}(x,y) = -\mathscr{L}_{f_1}\psi(x,y) = -y,$$

noting that

$$\mathrm{d}\psi(x,y) \cdot (f_0(x,y) + (u_{\mathrm{clf}}(x,y) + u_{\mathrm{diss}}(x,y)) f_1(x,y)) = -y^2.$$

One may now apply the LaSalle Invariance Principle to conclude that the closed-loop system with the state feedback $u_{\mathrm{clf}} + u_{\mathrm{diss}}$ has x_0 as an asymptotically stable equilibrium point.

The ideas in this example are generalized in the ensuing discussion. •

Given a control-affine system $\Sigma = (\mathsf{M}, \mathscr{C}, \mathbb{R}^m)$, an equilibrium point x_0 for f_0, and a Lyapunov function $\psi\colon \mathsf{M} \to \mathbb{R}$ for f_0 at x_0, define the functions $u^a_{\mathrm{diss}}\colon \mathsf{M} \to \mathbb{R}$ by

$$u^a_{\mathrm{diss}} = -\mathscr{L}_{f_a}\psi, \qquad a \in \{1, \dots, m\}. \tag{10.2}$$

We call the control functions $(u^1_{\mathrm{diss}}, \dots, u^m_{\mathrm{diss}})$ ***dissipative feedback***. It is an immediate computation to show that

$$\mathscr{L}_{(f_0 + \sum_a u^a_{\mathrm{diss}} f_a)}\psi(x) = \mathscr{L}_{f_0}\psi(x) - \sum_{a=1}^{m} (\mathscr{L}_{f_a}\psi(x))^2 \leq 0,$$

for all x in a neighborhood of x_0, because $\mathscr{L}_{f_0}\psi$ is locally negative-semidefinite. Therefore, ψ is a Lyapunov function for the closed-loop system under dissipative feedback control.

Note that the Lie derivative of ψ along the closed-loop might be negative-definite about x_0 even if $\mathscr{L}_{f_0}\psi$ is only negative-semidefinite. In this sense, dissipative feedback control improves the stability properties of the closed-loop system as compared to the vector field f_0. Next, we present sufficient conditions for the asymptotic stability of x_0 for the closed-loop system under dissipative controls.

Theorem 10.16 (Dissipative control). *Let $\Sigma = (\mathsf{M}, \mathscr{C}, \mathbb{R}^m)$ be a C^∞-control-affine system, let x_0 be an equilibrium point for f_0, let $\psi\colon \mathsf{M} \to \mathbb{R}$ be a class C^∞ Lyapunov function for f_0 at x_0 for which x_0 is an isolated local minimum. Also define*

$$\mathcal{W} = \{x \in \mathcal{U} \mid \mathscr{L}_{f_0}\psi(x) = 0,\ \mathscr{L}_{f_a}\psi(x) = 0,\ a \in \{1, \dots, m\}\},$$

and suppose that $\{x_0\}$ is the largest f_0-invariant subset of $\mathcal{W}$. Then the equilibrium point x_0 is locally asymptotically stable for the closed-loop control system with dissipative feedback controls (10.2).

Proof. Because the Lie derivative of ψ along the closed-loop system is

$$\mathscr{L}_{f_0}\psi(x) - \sum_{a=1}^{m} (\mathscr{L}_{f_a}\psi(x))^2 \leq 0,$$

we know that ψ is a Lyapunov function for the closed-loop system at x_0. Note that, since $\mathscr{L}_{f_a}\psi(x) = 0$ for $x \in \mathcal{W}$ and $a \in \{1, \dots, m\}$, any subset of $\mathcal{W}$ that is invariant under the closed-loop system will also be invariant under f_0. Therefore, the largest subset of $\mathcal{W}$ invariant under the closed-loop system is contained in the largest subset of $\mathcal{W}$ invariant under f_0. From this observation, the result follows from the LaSalle Invariance Principle. ■

In the previous theorem, the hypothesis concerning the largest f_0-invariant subset of $\mathcal{W}$ is often impractical to verify. Therefore, one would like to have stronger, but checkable, conditions from which this hypothesis follows. One such condition is given by the following result.

Lemma 10.17. *Let Σ, x_0, ψ, $\mathcal{U}$, and $\mathcal{W}$ be as in Theorem 10.16. If*

$$\operatorname{span}_{\mathbb{R}}\{\operatorname{ad}_{f_0}^k f_a(x_0) \mid a \in \{1, \dots, m\},\ k \in \mathbb{Z}_+\} = \mathsf{T}_{x_0}\mathsf{M},$$

then $\{x_0\}$ is the largest f_0-invariant subset of $\mathcal{W}$.

Proof. Note that the hypotheses ensure that

$$\operatorname{span}_{\mathbb{R}}\{\operatorname{ad}_{f_0}^k f_a(x) \mid a \in \{1, \dots, m\},\ k \in \mathbb{Z}_+\} = \mathsf{T}_{x}\mathsf{M},$$

for x in a neighborhood of x_0. Since ψ is a Lyapunov function for f_0, it follows that, if $x \in \mathcal{W}$, then x is a maximum for $\mathscr{L}_{f_0}\psi$. Therefore, $\mathrm{d}(\mathscr{L}_{f_0}\psi)(x) = 0$

for $x \in \mathcal{W}$. Now let γ be the integral curve of f_0 through $x \in \mathcal{W}$. Note that $\gamma(t) \in \mathcal{W}$ for all t, since ψ is a Lyapunov function for f_0 and since $\psi^{-1}(0) \subset \mathcal{W}$. Now, for $a \in \{1, \dots, m\}$, define $\phi_a(t) = \mathscr{L}_{f_a}\psi(\gamma(t))$, noting that $\phi_a(t)$ is necessarily zero for all t. Therefore,

$$\frac{\mathrm{d}^k}{\mathrm{d}t^k}\bigg|_{t=0} \phi_a(t) = \mathscr{L}^k_{f_0}\mathscr{L}_{f_a}\psi(x) = 0,$$

for $k \in \mathbb{Z}_+$. Since $\mathrm{d}(\mathscr{L}_{f_0}\psi)(x) = 0$, we determine that $\mathscr{L}^k_{f_0}\mathscr{L}_{f_a}\psi(x) = \mathscr{L}_{\mathrm{ad}^k_{f_0} f_a}\psi(x)$, $k \in \mathbb{Z}_+$. By hypothesis, the only way that $\mathscr{L}_{\mathrm{ad}^k_{f_0} f_a}\psi(x) = 0$, $k \in \mathbb{Z}_+$ and $a \in \{1, \dots, m\}$, is that $\mathrm{d}\psi(x) = 0$. This can only happen if the set $\mathcal{W}$ consists of the single point x_0, since $\mathrm{d}\psi(x) = 0 \in \mathcal{U}$ implies that $x = x_0$. ■

Remarks 10.18. 1. As is shown in [Nijmeijer and van der Schaft 1990], the condition that $\{x_0\}$ be the largest f_0-invariant subset of $\mathcal{W}$ is also necessary in order that the controls (10.2) locally asymptotically stabilize x_0.
2. If $\boldsymbol{K}$ is a symmetric positive-definite matrix, then one can replace u_{diss} with $\boldsymbol{K}u_{\mathrm{diss}}$, and the conclusions of Theorem 10.16 will still hold.
3. Note that rank condition of Lemma 10.17 is exactly the controllability rank condition for the linearization at x_0, cf. Corollary 7.26. •

10.2 Stabilization problems for mechanical systems

Before we get into specific methods of stabilizing mechanical control systems, we should say a few words about what we mean by stabilization for mechanical systems. We give a quite general definition.

Definition 10.19 (Feedback for simple mechanical control systems). Let $\Sigma = (\mathsf{Q}, \mathbb{G}, V, F, \mathcal{D}, \mathscr{F}, U)$ be a C^∞-general simple mechanical control system for which F is time-independent.

(i) A ***controlled equilibrium configuration*** for Σ is a pair $(q_0, u_0) \in \mathsf{Q} \times U$ with the property that

$$-\mathrm{d}V(q_0) + F(0_{q_0}) + \sum_{a=1}^m u_0^a F^a(q_0) \in \mathrm{ann}(\mathcal{D}_{q_0}).$$

(ii) A ***state feedback*** (resp. ***time-dependent state feedback***) for Σ is a map $u\colon \mathsf{TQ} \to U$ (resp. $u\colon \bar{\mathbb{R}}_+ \times \mathsf{TQ} \to U$).

(iii) For a state feedback (resp. time-dependent state feedback) u for Σ, the ***closed-loop system*** is the forced simple mechanical system with constraints defined by the 5-tuple $\Sigma_{\mathrm{cl}} = (\mathsf{Q}, \mathbb{G}, V, F_{\mathrm{cl}}, \mathcal{D})$, where

$$F_{\mathrm{cl}}(v_q) = F(v_q) + \sum_{a=1}^m u^a(v_q)F^a(q),$$

$$\Big(\text{resp. } F_{\mathrm{cl}}(t, v_q) = F(v_q) + \sum_{a=1}^m u^a(t, v_q)F^a(q)\Big).$$

(iv) For $r \in \mathbb{Z}_+ \cup \{\infty\} \cup \{\omega\}$, a state feedback (resp. time-dependent state feedback) is ***C^r*** if the corresponding closed-loop system is of class C^r. •

With these definitions one can define all notions of stabilizability given by parts (vi)–(viii) of Definition 10.1, but replacing (x_0, u_0) with $(0_{q_0}, u_0)$. Thus we shall only consider the stabilization of states with zero velocity, which is certainly a natural thing to do. Although our definition allows the inclusion of constraints in the stabilization problem, all of the stabilization problems we address in this book will be unconstrained. Indeed, there has been very little work done on the stabilization problem for equilibria of constrained Lagrangian systems (but see [Krasinskiĭ 1988, McClamroch and Bloch 1988]), although this is certain to change in the near future. A second problem that we do not address in this book is that of stabilization of relative equilibria. We refer to [Bullo 2000, Jalnapurkar and Marsden 2000] for some work in this direction. For some preliminary results on stabilization of relative equilibria for systems with nonholonomic constraints, we refer to [Zenkov, Bloch, and Marsden 1998]. Also, an internet supplement for the book will address stabilization of relative equilibria, based on the stability results in Section 6.3, and following the ideas in the present chapter.

Remarks 10.20. 1. One of the problems one might wish to address in stabilization theory for mechanical systems is that of stabilization that preserves the mechanical structure. That is to say, one would like to, if possible, ensure that the closed-loop system has mechanical properties. This will happen, for example, if the closed-loop force F_{cl} is a potential force corresponding to a potential function V_{P}. Then the closed-loop system will be the simple mechanical system $(\mathsf{Q}, \mathbb{G}, V + V_{\text{P}})$. Of course, such a system will not possess locally asymptotically stable equilibria. However, one can then modify the feedback to include dissipation, and hope that this will then provide local asymptotic stability. This essentially describes the strategy for stabilization used in this chapter, as well as in Chapters 11 and 12.

2. Related to Remark 10.2–3, let us introduce some terminology. Let (q_0, u_0) be a controlled equilibrium configuration and denote by $\mathcal{F}$ the codistribution generated by the covector fields $\{F^1, \dots, F^m\}$. We say that a controlled equilibrium configuration (q_0, u_0) is stabilized (in one of the senses of Definition 10.1) ***by the state feedback u*** if the closed-loop system for u is stable (in the appropriate sense). Now let F be a covector field of unspecified regularity taking values in $\mathcal{F}$. We also say that Σ is stabilized at (q_0, u_0) (in one of the senses of Definition 10.1) ***by the force F*** if the closed-loop system for u is stable (in the appropriate sense), where u is any state feedback satisfying $\sum_{a=1}^{m} u^a F_a = F$. •

Remark 10.21. In the preceding definitions of stabilization problems for mechanical systems, we allowed an arbitrary controlled equilibrium configuration (q_0, u_0). In practice we shall often take $u_0 = 0$. As mentioned in

Remark 10.2–1, for control-affine systems this can be done without loss of generality. This is not necessarily so for mechanical systems, since the resulting redefined drift vector field (cf. (10.1)) may no longer be that for a mechanical system. In Exercise E10.12 the reader is asked to consider a general controlled equilibrium configuration in two cases.

1. If the force exerted to maintain the equilibrium annihilates the involutive closure of the input distribution (see the exercise for the precise statement), then it *is* possible to assume that $u_0 = 0$ without loss of generality, while remaining in the mechanical setting. In this special case, the results in this chapter can be applied after a suitable modification of the potential function. We leave to the reader the fairly obvious statements of the modified results.
2. In the more general situation, instead, the effects of u_0 being nonzero must be taken into account. The results in this chapter do not apply in this more general case. •

10.3 Stabilization using linear potential shaping

In Section 7.3.1 we had defined the linearization of a simple mechanical control system $\Sigma = (\mathsf{Q}, \mathbb{G}, V, \mathscr{F} = \{F^1, \dots, F^m\}, U)$ at an equilibrium configuration q_0. The linearization is the 4-tuple $(\mathsf{T}_{q_0}\mathsf{Q}, \mathbb{G}(q_0), K_\Sigma(q_0), F_\Sigma(q_0))$, where $K_\Sigma(q_0) = \operatorname{Hess} V(q_0)$ and where $F_\Sigma(q_0) \in L(\mathbb{R}^m; \mathsf{T}^*_{q_0}\mathsf{Q})$ is defined by

$$F_\Sigma(q_0) \cdot u = \sum_{a=1}^{m} u^a F^a(q_0).$$

In this section we shall discuss stabilization of linear control systems of the sort that arise from the linearization of a simple mechanical control system. This is followed by a description of how control laws stabilizing the linearization can be applied to the nonlinear system to achieve stabilization.

10.3.1 Linear PD control

Recall from Section 7.3.1 the notion of a linear mechanical control system, denoted by the 4-tuple $\Sigma = (\mathsf{V}, M, K, F)$, where V is a finite-dimensional $\mathbb{R}$-vector space, M is an inner product on V, $K \in \Sigma_2(\mathsf{V})$, and $F \in L(\mathbb{R}^m; \mathsf{V}^*)$. The governing differential equations for the system, in second-order form, are

$$M^\flat(\ddot{x}(t)) + K^\flat(x(t)) = F(u(t)).$$

For this type of system, we wish to consider a special form of linear feedback. The linear feedback has the property that the closed-loop system has useful mechanical properties. Specifically, we desire the closed-loop system to be a linear mechanical system subject to a linear dissipative force. This allows us to use the results of Sections 6.2.2 and 6.2.3 to ensure stability of the closed-loop system. The requirement that the closed-loop system have this form is quite restrictive in that there are many system that cannot be stabilized in this sense, but can be stabilized using more general forms of linear feedback. However, in the cases in which the feedback we consider here *is* applicable, it is useful because it provides an easily implemented methodology.

The following linear algebra lemma is helpful in providing a useful definition of the sort of control law we consider.

Lemma 10.22. *Let* U *and* V *be finite-dimensional* $\mathbb{R}$*-vector spaces, and let* $F \in L(\mathsf{U};\mathsf{V}^*)$*. The following statements hold:*

(i) if $C \in L(\mathsf{V};\mathsf{U})$ *has the property that* $F \circ C = F \circ G^\sharp \circ F^*$ *for* $G \in \Sigma_2(\mathsf{U}^*)$*, then there exists* $K \in \Sigma_2(\mathsf{V})$ *such that* $F \circ C = K^\flat$*;*

(ii) if $C \in L(\mathsf{V};\mathsf{U})$ *has the property that* $F \circ C = K^\flat$ *for* $K \in \Sigma_2(\mathsf{V})$*, then there exists* $G \in \Sigma_2(\mathsf{U}^*)$ *for which* $F \circ C = F \circ G^\sharp \circ F^*$*.*

Proof. The first part of the result follows by defining K by

$$K(u,v) = \langle F \circ G^\sharp \circ F^*(v); u\rangle.$$

For the second part of the lemma, note that if $K^\flat = F \circ C$ as desired, then $\operatorname{image}(K^\flat) \subset \operatorname{image}(F)$. Thus we can choose a basis $\{e_1,\dots,e_n\}$ for V with the following properties:

1. $\{e^1,\dots,e^k\}$ is a basis for $\operatorname{image}(K^\flat)$;
2. $\{e^1,\dots,e^k,e^{k+1},\dots,e^r\}$ is a basis for $\operatorname{image}(F)$;
3. $\{e_{k+1},\dots,e_n\}$ is a basis for $\ker(K^\flat)$.

The last property of the basis follows from Theorem 2.46. Now let $\{f_1,\dots,f_m\}$ be a basis for U with the properties that

4. $\{f_{r+1},\dots,f_m\}$ is a basis for $\ker(F)$ and that
5. $F(f_a) = e^a$, $a \in \{1,\dots,r\}$.

Relative to the partitions $\{e_1,\dots,e_k\} \cup \{e_{k+1},\dots,e_r\} \cup \{e_{r+1},\dots,e_n\}$ and $\{f_1,\dots,f_k\} \cup \{f_{k+1},\dots,f_r\} \cup \{f_{r+1},\dots,f_m\}$ of the bases for V and U, and the induced partition of a basis for V^*, the matrix representations for $K^\flat$ and F have the form

$$[K^\flat] = \begin{bmatrix} \boldsymbol{K} & \boldsymbol{0} & \boldsymbol{0} \\ \boldsymbol{0} & \boldsymbol{0} & \boldsymbol{0} \\ \boldsymbol{0} & \boldsymbol{0} & \boldsymbol{0} \end{bmatrix}, \quad [F] = \begin{bmatrix} \boldsymbol{I}_k & \boldsymbol{0} & \boldsymbol{0} \\ \boldsymbol{0} & \boldsymbol{I}_{r-k} & \boldsymbol{0} \\ \boldsymbol{0} & \boldsymbol{0} & \boldsymbol{0} \end{bmatrix}.$$

From this we ascertain that the matrix representation for C must have the form

$$[C] = \begin{bmatrix} \boldsymbol{K} & \boldsymbol{0} & \boldsymbol{0} \\ \boldsymbol{0} & \boldsymbol{0} & \boldsymbol{0} \\ \boldsymbol{C}_{31} & \boldsymbol{C}_{32} & \boldsymbol{C}_{33} \end{bmatrix},$$

where C_{31}, C_{32}, and C_{33} are arbitrary matrices of the appropriate size. The lemma follows by choosing G to have the matrix representation

$$[G] = \begin{bmatrix} \boldsymbol{K} & \mathbf{0} & \mathbf{0} \\ \mathbf{0} & \mathbf{0} & \mathbf{0} \\ \mathbf{0} & \mathbf{0} & \mathbf{0} \end{bmatrix}$$

in the induced partition of the basis for U^*. ■

The lemma tells us that a linear feedback for a linear mechanical control system is symmetric if and only if it is of the form $C = G^\sharp \circ F^*$ for $G \in \Sigma_2(\mathsf{U}^*)$. This then motivates the following definition, telling us that the special form assumed for the control laws does not suffer from loss of generality.

Definition 10.23 (PD control for linear mechanical systems). For a linear mechanical control system $\Sigma = (\mathsf{V}, M, K, F)$, a ***linear proportional-derivative (PD) control law*** is a linear state feedback given by

$$u(x, v) = -K_{\mathrm{P}}^\flat \circ F^*(x) - K_{\mathrm{D}}^\flat \circ F^*(v),$$

where $K_{\mathrm{P}}, K_{\mathrm{D}} \in \Sigma_2((\mathbb{R}^m)^*)$. If $K_{\mathrm{D}} = 0$, then u is a ***linear proportional control law***. •

Remark 10.24. It is evident that one cannot locally asymptotically stabilize a linear mechanical system using proportional control. This is just as evidently true for the nonlinear proportional control we consider in Section 10.4. The reader is asked to prove this in Exercise E10.8. •

10.3.2 Stabilization using linear PD control

Now that we understand what is meant by linear PD control, let us consider using it to stabilize a linear mechanical system. First we give some definitions to make precise some convenient language.

Definition 10.25 (Stabilizability of linear systems by PD control). A linear mechanical control system $\Sigma = (\mathsf{V}, M, K, F)$ is

(i) ***stabilizable by proportional control*** if there exists a linear proportional control law that stabilizes $(0, \mathbf{0}) \in \mathsf{V} \times \mathbb{R}^m$, and is

(ii) ***asymptotically stabilizable by proportional-derivative control*** if there exists a linear PD control law that asymptotically stabilizes $(0, \mathbf{0}) \in \mathsf{V} \times \mathbb{R}^m$. •

The following result characterizes stabilizing and asymptotically stabilizing PD control laws for linear mechanical control systems. The result implicitly assumes some identifications of spaces of tensors that the reader can prove in Exercise E2.23.

Theorem 10.26 (Stabilization by linear PD control). *Let* $\Sigma = (\mathsf{V}, M, K, F)$ *be a linear mechanical control system. Then the following statements hold:*

(i) Σ *is stabilizable by proportional control if and only if* $K|\mathrm{coann}(\mathrm{image}(F))$ *is positive-definite;*

(ii) Σ *is asymptotically stabilizable by proportional-derivative control if and only if*

(a) it is stabilizable by proportional control and

(b) Σ *is STLC from* $0 \oplus 0$*.*

Proof. We first prove a linear algebra lemma for Σ.

Lemma. *If* $K|\mathrm{coann}(\mathrm{image}(F))$ *is positive-definite, then there exists* $G \in \Sigma_2((\mathbb{R}^m)^*)$ *for which the symmetric bilinear map associated to* $K^\flat + F \circ G^\sharp \circ F^* \in L(\mathsf{V}; \mathsf{V}^*)$ *is positive-definite.*

Proof. Let

$$\mathsf{W} = \{ w \in \mathsf{V} \mid \; K(w, v) = 0, \; v \in \mathrm{coann}(\mathrm{image}(F))\} .$$

We claim that $\mathsf{V} = \mathsf{W} \oplus \mathrm{coann}(\mathrm{image}(F))$. First suppose that $v \in \mathsf{W} \cap \mathrm{coann}(\mathrm{image}(F))$. Then $K(v, v) = 0$, which implies that $v = 0$, since K is positive-definite on $\mathrm{coann}(\mathrm{image}(F))$. This shows that $\mathsf{W} \cap \mathrm{coann}(\mathrm{image}(F)) = \{0\}$. Further, by Theorem 2.46 one can easily see that $\dim(\mathsf{W}) = n - \dim(\mathrm{coann}(\mathrm{image}(F)))$, again, since $K|\mathrm{coann}(\mathrm{image}(F))$ is positive-definite. This means that $\mathsf{V} = \mathsf{W} \oplus \mathrm{coann}(\mathrm{image}(F))$, as desired. If we write $K^\flat + F \circ G^\sharp \circ F^*$ with respect to this decomposition, we get

$$\begin{bmatrix} \boldsymbol{K}_1 + \boldsymbol{F}\boldsymbol{G}_1\boldsymbol{F}^T & \mathbf{0} \\ \mathbf{0} & \boldsymbol{K}_2 \end{bmatrix} \tag{10.3}$$

with $\boldsymbol{K}_2$ being positive-definite. By choice of $\boldsymbol{G}_1$ the matrix $\boldsymbol{F}\boldsymbol{G}_1\boldsymbol{F}^T$ has maximal rank, and so can be arbitrarily specified. The result therefore follows. ▼

(i) The "if" implication is clear from the lemma. The "only if" part of the statement follows immediately from Proposition 6.41, since the closed-loop system is a linear mechanical system, and if the matrix $\boldsymbol{K}_2$ is not positive-definite in (10.3), then the closed-loop system either has a zero eigenvalue, or an eigenvalue in $\mathbb{C}_+$. In either event, the closed-loop system will not be stable.

(ii) By Theorem 6.42(iii), if a system is asymptotically stabilized by the linear PD control law

$$u(x, v) = -K_{\mathrm{P}}^\flat \circ F^*(x) - K_{\mathrm{D}}^\flat \circ F^*(v),$$

then the linear feedback obtained by setting $K_{\mathrm{D}} = 0$ must stabilize the system. Note by Theorem 7.31 and Exercise E7.4 that Σ is STLC from $0 \oplus 0$ if and only if the smallest $M^\sharp \circ K^\flat$-invariant subspace containing image$(M^\sharp \circ F)$ is V. Suppose that Σ is not STLC from $0 \oplus 0$ and let $\mathsf{W} = \langle M^\sharp \circ K^\flat, \mathrm{image}(M^\sharp \circ F)\rangle$. If $\mathsf{W}^\perp$ is the M-orthogonal complement to W in V, then the linear maps $M^\sharp \circ K^\flat$ and $M^\sharp \circ F$ can be written relative to the decomposition $\mathsf{W} \oplus \mathsf{W}^\perp$ as

$$\begin{bmatrix} K_1 & 0 \\ 0 & K_2 \end{bmatrix}, \quad \begin{bmatrix} F_1 \\ 0 \end{bmatrix},$$

respectively. From this it is clear that, by linear PD control, the best one can do is stabilize the closed-loop system, since the closed-loop system will always contain the dynamics

$$\ddot{x}_2(t) + K_2(x_2(t)) = 0,$$

and these are never asymptotically stable. This shows that the system must be STLC from $0 \oplus 0$ if it is to be asymptotically stabilizable.

We now show the converse. To do this we shall use Theorem 10.16, particularly Lemma 10.17. We let $\Sigma' = (\mathsf{V}, M, K_1, F)$ denote the system after it has been stabilized by the proportional control law $u_1(x, v) = -K_{\mathrm{P}}^\sharp \circ F^*(x)$. Thus K_1 is defined by $K_1^\flat = K^\flat + F \circ K_{\mathrm{P}}^\sharp \circ F^*$. Note that $\psi(x, v) = \frac{1}{2}M(v, v) + \frac{1}{2}K(x, x)$ is a Lyapunov function for the unforced system associated with Σ'. Let $F^a : \mathbb{R} \to \mathsf{V}^*$ be defined by requiring that

$$F(u^1, \dots, u^m) = \sum_{a=1}^{m} F^a(u^a).$$

A direct computation shows that, if f_a is the constant vector field

$$f_a(x, v) = \begin{bmatrix} 0 \\ M^\sharp \circ F^a \end{bmatrix}$$

on $\mathsf{V} \oplus \mathsf{V}$ (i.e., the ath input vector field for the system in first-order form), then we have

$$\mathscr{L}_{f_a}\psi(x, v) = F^a(1) \cdot v.$$

Note, therefore, that the dissipative feedback of Theorem 10.16, given by $u_2^a(x, v) = -F^a(1) \cdot v$, is then the dissipative control for the linear mechanical system. Indeed, one readily computes that

$$F \circ u_2(x, v) = -F \circ K_{\mathrm{D}}^\sharp \circ F^*(v),$$

where $K_{\mathrm{D}} \in \Sigma_2((\mathbb{R}^m)^*)$ satisfies $[K_{\mathrm{D}}] = \boldsymbol{I}_m$ in the standard basis for $(\mathbb{R}^m)^*$. Now, by Lemma 10.17 the control law u_2 asymptotically stabilizes 0 for Σ'. Therefore, the linear PD control law $u_1 + u_2$ asymptotically stabilizes 0 for Σ, as desired. ■

Remarks 10.27. 1. The closed-loop system for the state feedback of part (i) is a linear mechanical system $(\mathsf{V}, M, K_{\mathrm{cl}})$ with K_{cl} defined by

$$K^\flat_{\mathrm{cl}} = K^\flat + F \circ K^\sharp_{\mathrm{P}} \circ F^*,$$

for $K_{\mathrm{P}} \in \Sigma_2((\mathbb{R}^m)^*)$. The content of part (i) is that if the hypotheses are met, then one can always find a suitable K_{P} that stabilizes $0 \in \mathsf{V}$. There are many ways in which one can choose a suitable K_{P}, and the reader can explore this in Exercise E10.9.

2. The closed-loop system for the state feedback of part (ii) is a forced linear mechanical system $(\mathsf{V}, M, K_{\mathrm{cl}}, (0, -R_{\mathrm{cl}}))$ with K_{cl} and R_{cl} defined by

$$K^\flat_{\mathrm{cl}} = K^\flat + F \circ K^\sharp_{\mathrm{P}} \circ F^*, \quad R^\flat_{\mathrm{cl}} = F \circ K^\sharp_{\mathrm{D}} \circ F^*,$$

for $K_{\mathrm{P}}, K_{\mathrm{D}} \in \Sigma_2((\mathbb{R}^m)^*)$. The content of part (ii) is that if K_{P} is determined as in part (i) and if K_{D} is positive-definite, then $(0, 0)$ will be asymptotically stabilized.

3. Although we used Theorem 10.16 to prove the second part of the theorem, we could have also directly appealed to Theorem 6.42. However, it is instructive to see the relationship between general dissipative control and dissipative control for mechanical systems. The reader interested in the use of Theorem 6.42 to prove stabilization results is referred to the proof of Theorem 10.43.
4. As we mentioned in the introductory prelude to this section, stabilization by linear PD control is not the most general form of linear state feedback, even for a mechanical system. Indeed, the usual methods for designing a linear state feedback (e.g., LQR) will not produce a PD control law. The reader is asked to explore this in Exercises E10.10 and E10.11. In particular, in Exercise E10.11, we consider the linear energy shaping (not just potential energy shaping) results of Zenkov [2002]. •

It is also interesting to consider the case in which the system without controls has some dissipation. The next result follows fairly directly from Theorems 6.42 and 10.26.

Corollary 10.28. *Let* $(\mathsf{V}, M, K, (0, -R^\flat_{\mathrm{diss}}), F)$ *be a forced linear mechanical control system with governing equation*

$$M^\flat(\ddot{x}(t)) + R^\flat_{\mathrm{diss}}(\dot{x}(t)) + K^\flat(x(t)) = F(u(t)).$$

Then

(i) the system is stabilizable by proportional control if and only if $K|\mathrm{coann}(\mathrm{image}(F))$ *is positive-definite and*

(ii) the system is asymptotically stabilizable by proportional-derivative control if and only if

(a) it is stabilizable by proportional control and

(b) $\langle M^\sharp \circ K^\flat, \mathrm{image}(M^\sharp \circ R^\flat_{\mathrm{diss}}) + \mathrm{image}(M^\sharp \circ F)\rangle = \mathsf{V}$.

10.3.3 Implementing linear control laws on nonlinear systems

In practice, a common method for stabilizing an equilibrium configuration is to linearize, design a stabilizing controller for the linearization (assuming this is possible), and then implement the linear controller on the nonlinear system. In this section we consider the last stage in this operation. To implement a linear feedback on a nonlinear system requires a means of "transferring" the control law from the tangent space to the configuration manifold. In practice, this is done naturally when one chooses coordinates. Here we briefly describe a geometric interpretation of this procedure. Although this geometric interpretation is simple, it is not commonly described, and the problem needs explicit addressing in our geometric, coordinate-invariant approach.

The following definition provides the basic tool for implementing a linear control law on a nonlinear system.

Definition 10.29 (Near identity diffeomorphism). Let Q be a C^∞-differentiable manifold with $q_0 \in \mathsf{Q}$. A ***near identity diffeomorphism at*** $\boldsymbol{q_0}$ is a triple $(\chi, \mathcal{U}_0, \mathcal{U}_1)$, where

(i) $\mathcal{U}_0 \subset \mathsf{T}_{q_0}\mathsf{Q}$ is a neighborhood of 0_{q_0},
(ii) $\mathcal{U}_1 \subset \mathsf{Q}$ is a neighborhood of q_0, and
(iii) $\chi\colon \mathcal{U}_0 \to \mathcal{U}_1$ is a C^∞-diffeomorphism satisfying
 (a) $\chi(0_{q_0}) = q_0$ and
 (b) $T_{0_{q_0}}\chi = \mathrm{id}_{\mathsf{T}_{q_0}\mathsf{Q}}$ (where we make the natural identification of $\mathsf{T}_{0_{q_0}}(\mathsf{T}_{q_0}\mathsf{Q})$ with $\mathsf{T}_{q_0}\mathsf{Q}$). •

In practice, a near identity diffeomorphism arises from a coordinate chart as described by the following lemma. The lemma also tells us that this is the most general way to obtain such a diffeomorphism.

Lemma 10.30 (Characterization of near identity diffeomorphisms). *Let $(\mathcal{U}, \psi)$ be a coordinate chart for Q satisfying $\psi(q_0) = \mathbf{0} \in \mathbb{R}^n$. Then the triple $(\chi = \psi^{-1} \circ T_{q_0}\psi, \mathcal{U}_0 = \chi(\mathcal{U}), \mathcal{U}_1 = \mathcal{U})$ is a near identity diffeomorphism at q_0.*

Conversely, let $(\chi, \mathcal{U}_0, \mathcal{U}_1)$ be a near identity diffeomorphism at q_0 and let $L\colon \mathsf{T}_{q_0}\mathsf{Q} \to \mathbb{R}^n$ be an isomorphism. Then $(\mathcal{U} = \mathcal{U}_1, \psi = L \circ \chi^{-1})$ is a coordinate chart for Q satisfying $\psi(q_0) = \mathbf{0} \in \mathbb{R}^n$ and $\chi = \psi^{-1} \circ \mathsf{T}_{q_0}\psi$.

Proof. With $(\chi, \mathcal{U}_0, \mathcal{U}_1)$ as defined in the first part of the lemma, we compute

$$\chi(0_{q_0}) = \psi^{-1} \circ \mathsf{T}_{q_0}\psi(0_{q_0}) = \psi^{-1}(\mathbf{0}) = q_0,$$

and, using the Chain Rule,

$$\begin{aligned} T_{0_{q_0}}\chi(v) &= T_{0_{q_0}}(\psi^{-1} \circ T_{q_0}\psi)(v) \\ &= T_{T_{q_0}\psi(0_{q_0})}\psi^{-1} \circ T_{0_{q_0}}(T_{q_0}\psi)(v) \\ &= T_{\mathbf{0}}\psi^{-1} \circ T_{q_0}\psi(v) = T_{q_0}(\psi^{-1} \circ \psi)(v) = v, \end{aligned}$$

showing that $(\chi, \mathcal{U}_0, \mathcal{U}_1)$ is indeed a near identity diffeomorphism at q_0.

Now let $(\psi, \mathcal{U})$ be as defined in the second part of the lemma. We then have

$$\psi(q_0) = L \circ \chi^{-1}(q_0) = L(0_{q_0}) = \mathbf{0}$$

and

$$\psi^{-1} \circ T_{q_0}\psi = \chi \circ L^{-1} \circ T_{q_0}(L \circ \chi) = \chi \circ L^{-1} \circ T_{\chi(q_0)} L \circ T_{q_0}\chi = \chi,$$

thus giving the desired assertion. ∎

In Figure 10.1 we illustrate the idea behind a near identity diffeomorphism,

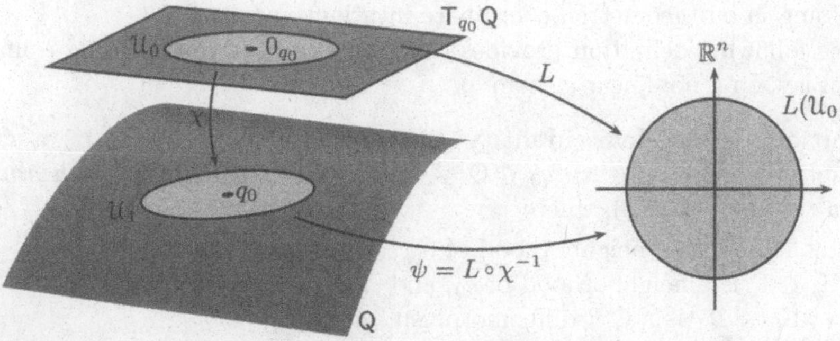

Figure 10.1. An illustration of Lemma 10.30

and its relationship with a coordinate chart as stated in Lemma 10.30.

With this understanding of the tool needed to apply a linear control law to a nonlinear system, let us formally define this process.

Definition 10.31 (Stabilizability of nonlinear systems by linear PD control). Let $(q_0, 0)$ be a controlled equilibrium configuration for a C^∞-simple mechanical control system $\Sigma = (\mathsf{Q}, \mathbb{G}, V, \mathscr{F}, \mathbb{R}^m)$, and let $\Sigma_{\text{lin}} = (\mathsf{T}_{q_0}\mathsf{Q}, \mathbb{G}(q_0), K_\Sigma(q_0), F_\Sigma(q_0))$ be the linearization at q_0. If

$$u_{\text{lin}}(x, v) = -K_{\text{P}}^\flat \circ F_\Sigma(q_0)^*(x) - K_{\text{D}}^\flat \circ F_\Sigma(q_0)^*(v)$$

is a linear PD control law for Σ_{lin}, then an ***implementation*** of u_{lin} for Σ is a state feedback for Σ on $\mathsf{T}\mathcal{U}_1$ given by $u_{\text{nonlin}} = u_{\text{lin}} \circ T\chi^{-1}$, where $(\chi, \mathcal{U}_0, \mathcal{U}_1)$ is a near identity diffeomorphism at q_0.

Furthermore, we say that Σ is

(i) ***stabilizable by linear proportional control*** at q_0 if there exist a linear proportional control law u_{lin} and an implementation of this control law for Σ which stabilizes $(q_0, 0)$, and is

(ii) ***locally asymptotically stabilizable by linear proportional-derivative control*** at q_0 if there exist a linear PD control law u_{lin} and an implementation of this control law for Σ which locally asymptotically stabilizes $(q_0, 0)$. •

The following result indicates that a near identity diffeomorphism can be used to effectively implement a stabilizing linear PD control law on a nonlinear simple mechanical system.

Theorem 10.32 (Stabilization of nonlinear systems by linear PD control). *Let $\Sigma = (\mathsf{Q}, \mathbb{G}, V, \mathscr{F}, \mathbb{R}^m)$ be a C^∞-simple mechanical control system, with $(q_0, 0)$ a controlled equilibrium configuration and $(\mathsf{T}_{q_0}\mathsf{Q}, \mathbb{G}(q_0), K_\Sigma(q_0), F_\Sigma(q_0))$ the linearization at q_0. For a near identity diffeomorphism $(\chi, \mathcal{U}_0, \mathcal{U}_1)$ at q_0 and a linear PD control law*

$$u_{\text{lin}}(x, v) = -K_{\text{P}}^\sharp \circ F_\Sigma(q_0)^*(x) - K_{\text{D}}^\sharp \circ F_\Sigma(q_0)^*(v)$$

for the linearization, define a state feedback for Σ on $\mathsf{T}\mathcal{U}_1$ by $u_{\text{nonlin}} = u_{\text{lin}} \circ T\chi^{-1}$. The following statements hold.

(i) If $K_\Sigma(q_0)|\text{coann}(\text{image}(F_\Sigma(q_0)))$ is positive-definite, then there exists $K_{\text{P}} \in \Sigma_2((\mathbb{R}^m)^*)$ such that $K_\Sigma(q_0) + F_\Sigma(q_0) \circ K_{\text{P}}^\flat \circ F_\Sigma(q_0)^* \in \Sigma_2(\mathsf{T}_{q_0}\mathsf{Q})$ is positive-definite. Furthermore $(q_0, 0)$ is stabilized by the state feedback u_{nonlin} with K_{P} so chosen and with K_{D} positive-semidefinite.

(ii) Suppose that $K_\Sigma(q_0)|\text{coann}(\text{image}(F_\Sigma(q_0)))$ is positive-definite, that K_{P} is chosen as in part (i), that

$$\langle \mathbb{G}(q_0)^\sharp \circ K_\Sigma(q_0)^\flat, \text{image}(\mathbb{G}(q_0)^\sharp \circ F_\Sigma(q_0)) \rangle = \mathsf{T}_{q_0}\mathsf{Q},$$

and that $K_{\text{D}} \in \Sigma_2((\mathbb{R}^m)^*)$ is positive-definite. Then $(q_0, 0)$ is locally asymptotically stabilized by the state feedback u_{nonlin} with K_{P} and K_{D} so chosen.

Proof. Using the near identity diffeomorphism, define a coordinate chart as in the second part of Lemma 10.30. One may now proceed directly, using the calculations that prove Proposition 6.37, to show in coordinates that the linearization of the closed-loop system is exactly the closed-loop system for the linearization. The result follows from Propositions 6.41 and 6.42 and Theorem 10.26. ■

Remarks 10.33. 1. Note that the assertion of part (ii) applies to *any* linear feedback that asymptotically stabilizes the linearization, not just to linear PD control. That is to say, if a linear feedback stabilizes the linearization, and if this linear feedback is implemented on the nonlinear system using a near identity diffeomorphism, then the resulting closed-loop nonlinear system will be locally asymptotically stable.

2. In the case in which the system already possesses some dissipation, it is possible to state an analogue to Corollary 10.28 in the setting of Theorem 10.32. Indeed, if the system is subject to the Rayleigh dissipative force defined by a positive-semidefinite symmetric $(0,2)$-tensor R_{diss}, then the condition in part (ii) of the theorem can be weakened to the requirement that $\langle \mathbb{G}(q_0)^\sharp \circ K_\Sigma(q_0)^\flat, \mathsf{W}_{q_0} \rangle = \mathsf{T}_{q_0}\mathsf{Q}$, where

$$\mathsf{W}_{q_0} = \text{image}(\mathbb{G}(q_0)^\sharp \circ R_{\text{diss}}(q_0)^\flat) + \text{image}(\mathbb{G}(q_0)^\sharp \circ F_\Sigma(q_0)).$$ •

The following corollary to Theorem 10.32 indicates that the definitions of stabilizability and local asymptotic stabilizability by linear PD control in Definition 10.31 are independent of a choice of implementation.

Corollary 10.34. *Let* $\Sigma = (\mathsf{Q}, \mathbb{G}, V, \mathscr{F}, \mathbb{R}^m)$ *be a simple mechanical control system, with* $(q_0, 0)$ *a controlled equilibrium configuration and* $(\mathsf{T}_{q_0}\mathsf{Q}, \mathbb{G}(q_0), K_\Sigma(q_0), F_\Sigma(q_0))$ *the linearization at* q_0*. The following statements hold:*

(i) Σ *is stabilizable by linear proportional control at* q_0 *if its linearization is stabilizable by linear proportional control;*

(ii) Σ *is locally asymptotically stabilizable by linear proportional-derivative control at* q_0 *if its linearization is asymptotically stabilizable by linear proportional-derivative control.*

Let us summarize the content of this section. We have the following two methods for stabilizing a nonlinear system at a controlled equilibrium configuration $(q_0, 0)$ using a linear control law.

Standard method

1. Choose a coordinate chart about q_0 and write the governing equations in these coordinates.

2. Linearize the equations using standard Jacobian linearization.

3. In these coordinates, design a linear PD control law that asymptotically stabilizes the origin for the linearization.

4. Using the identification provided by the coordinates of the tangent space $\mathsf{T}_{q_0}\mathsf{Q}$ with Q in a neighborhood of q_0, apply the linear control law to the nonlinear system.

Coordinate-invariant method

1. Perform the coordinate-invariant linearization.

2. Design a coordinate-invariant linear PD control that asymptotically stabilizes the origin for the linearization.

3. Choose a near identity diffeomorphism.

4. Implement the linear PD control law on the nonlinear system as in Definition 10.31.

5. Write the closed-loop equations in coordinates adapted to the near identity diffeomorphism as per Lemma 10.30.

The standard method outlined on the left is the usual construction one performs when designing a linear control law to stabilize a nonlinear system. The coordinate-invariant method on the right gives a construction that is independent of coordinates, at least until the last step, which exists only to compare it to the method on the left. Our discussion shows that both methods will produce a locally asymptotically stabilizing control law for the nonlinear system. Furthermore, the methods are "the same" in that

1. given a set of coordinates for the standard method, one may choose a suitable near identity diffeomorphism for the coordinate-invariant method such that the equations in Step 4 on the left and Step 5 on the right are the same, and

2. given a near identity diffeomorphism for the coordinate-invariant method, there exists a set of coordinates for the standard method such that the equations in Step 4 on the left and Step 5 on the right are the same.

10.3.4 Application to the two-link manipulator

Let us illustrate the value and the limitations of the methods in this section with an example. We consider the two-link manipulator looked at first in Section 1.2. The system has configuration manifold $\mathsf{Q} = \mathbb{S}^1 \times \mathbb{S}^1$, and we use standard coordinates (θ_1, θ_2), where $(\theta_1, \theta_2) = (0, 0)$ corresponds to both links being "horizontal" (see Figure 4.8). In Example 4.32 we computed the kinetic energy Riemannian metric to have matrix representation

$$[\mathbb{G}] = \begin{bmatrix} J_1 + \frac{1}{4}(m_1 + 4m_2)\ell_1^2 & \frac{1}{2}m_2\ell_1\ell_2\cos(\theta_1 - \theta_2) \\ \frac{1}{2}m_2\ell_1\ell_2\cos(\theta_1 - \theta_2) & J_2 + \frac{1}{4}m_2\ell_2^2 \end{bmatrix}.$$

In Example 6.50 we computed the potential function to be

$$V = \left(\tfrac{1}{2}m_1 a_g \ell_1 + m_2 a_g \ell_1\right)\sin\theta_1 + \left(\tfrac{1}{2}m_2 a_g \ell_2\right)\sin\theta_2,$$

so that the Hessian, to be evaluated at equilibrium configurations, has matrix representation

$$[\operatorname{Hess} V] = \begin{bmatrix} -\left(\frac{1}{2}m_1 a_g \ell_1 + m_2 a_g \ell_1\right)\sin\theta_1 & 0 \\ 0 & -\left(\frac{1}{2}m_2 a_g \ell_2\right)\sin\theta_2 \end{bmatrix}.$$

We will take as input a torque applied at the base of the first link. Thus we have the single input $F^1 = \mathrm{d}\theta_1$.

We wish to examine the possibility of stabilizing the system using linear PD control, implemented on the nonlinear system as per Theorem 10.32, in each of the four equilibrium configurations discussed in Example 6.50. Note that the linear PD control law we consider for the nonlinear system is that induced by the linear feedback using our chosen coordinates, as prescribed by Lemma 10.30.

1. $q_{\text{down/down}} = (-\frac{\pi}{2}, -\frac{\pi}{2})$: Since the linearization is actually stable in this configuration, it is certainly stabilizable using linear proportional control. Indeed, the closed-loop linear system "K-matrix" is

$$\begin{bmatrix} \left(\frac{1}{2}m_1 a_g \ell_1 + m_2 a_g \ell_1\right) + k & 0 \\ 0 & \left(\frac{1}{2}m_2 a_g \ell_2\right) \end{bmatrix},$$

for $k \in \mathbb{R}$. Therefore, provided that $k > -\left(\frac{1}{2}m_1 a_g \ell_1 + m_2 a_g \ell_1\right)$ the closed-loop system will be stable. As concerns local asymptotic stabilizability using PD control, according to Theorem 10.26 we need to check the controllability of the linearization. It is straightforward to check, using Theorem 7.31, that the linearization of the system is controllable at *all* equilibrium configurations. In particular, it follows that any equilibrium configuration of the system that can be stabilized using linear proportional control can also be locally asymptotically stabilized using linear PD control. Furthermore, any linear PD control law of the form

$$u = -k(\theta_1 - \tfrac{\pi}{2}) - d\dot{\theta}_1$$

will locally asymptotically stabilize $q_{\text{down/down}}$, provided that $k > -(\frac{1}{2}m_1 a_g \ell_1 + m_2 a_g \ell_1)$ and $d > 0$. Note that global asymptotic stabilization of the equilibrium configuration using smooth state feedback is not possible due to the topology of $\mathsf{Q} = \mathbb{S}^1 \times \mathbb{S}^1$.

2. $q_{\text{down/up}} = (-\frac{\pi}{2}, \frac{\pi}{2})$: In this case the system cannot be stabilized by linear proportional control, since $\operatorname{ann}(\operatorname{image}(F_\Sigma(q_{\text{down/up}}))) = \operatorname{span}_{\mathbb{R}}\{\frac{\partial}{\partial\theta_2}\}$, and $\operatorname{Hess} V(q_{\text{down/up}})$ is negative-definite on this subspace. However, it is nonetheless true that the system can be stabilized by linear state feedback, as indicated by Theorem 10.3. This indicates that linear PD control is somewhat limited in its applicability, and this certainly should be kept in mind.
3. $q_{\text{up/down}} = (\frac{\pi}{2}, -\frac{\pi}{2})$: In this case the system is stabilizable by linear proportional control, and so is locally asymptotically stabilizable by linear PD control. The closed-loop "K-matrix" is

$$\begin{bmatrix} -(\frac{1}{2}m_1 a_g \ell_1 + m_2 a_g \ell_1) + k & 0 \\ 0 & (\frac{1}{2}m_2 a_g \ell_2) \end{bmatrix},$$

for $k \in \mathbb{R}$. Thus we see that the system can be stabilized provided that we choose $k > \frac{1}{2}m_1 a_g \ell_1 + m_2 a_g \ell_1$. As in the stabilization of $q_{\text{down/down}}$, local asymptotic stabilization of $q_{\text{up/down}}$ will be achieved by any linear PD control law of the form

$$u = -k(\theta_1 - \tfrac{\pi}{2}) - d\dot{\theta}_1,$$

provided that $k > \frac{1}{2}m_1 a_g \ell_1 + m_2 a_g \ell_1$ and $d > 0$. Again, global asymptotic stabilization using smooth state feedback is not possible.
4. $q_{\text{up/up}} = (\frac{\pi}{2}, \frac{\pi}{2})$: This equilibrium is not stabilizable using linear PD control by virtue of the same reasoning as for $q_{\text{down/up}}$.

Remark 10.35. The example in this section shows the limitations of restricting to linear PD control when considering the use of linear methods to stabilize a system. The limitations arise, essentially, because linear PD control does not take advantage of the possible coupling arising in the system's kinetic energy Riemannian metric. Nevertheless, linear PD control has the advantage of retaining the mechanical structure of the system in the cases where it can be applied. An extension of the energy shaping method for linear mechanical control systems is given in Exercise E10.11, following [Zenkov 2002]. •

10.4 Stabilization using nonlinear potential shaping

In this section we consider PD control in the nonlinear setting. As we shall see, this sort of control is even more limited that linear PD control in its applicability. Nevertheless, it is interesting to understand the geometry of the problem, particularly in terms of relating nonlinear PD control to linear PD control. We refer to Section 10.5.6 for a discussion of some recent work that overcomes some of the limitations of potential shaping as discussed here.

10.4.1 Nonlinear PD control and potential energy shaping

The idea in nonlinear PD control is to "shape" the potential function using feedback. Of course, one cannot expect to be able to attain an arbitrary potential function using PD control, at least if the system is underactuated. In this section we investigate the structure of what, at the end of the section, we shall call proportional control. This amounts to understanding, as concretely as possible, the most general sort of potential function that one can attain via feedback. This requires a little buildup that we now provide.

We let $\Sigma = (\mathsf{Q}, \mathbb{G}, V, \mathscr{F}, \mathbb{R}^m)$ be a C^∞-simple mechanical control system with $(q_0, 0)$ a controlled equilibrium configuration. We denote by $\mathcal{F}$ the codistribution on Q with generators $\mathscr{F}$. We suppose that q_0 is a regular point for $\mathcal{F}$. In this case $\mathcal{D}_{\mathcal{F}} \triangleq \operatorname{coann}(\mathcal{F})$ is a regular distribution in a neighborhood of q_0. We let $\operatorname{Lie}^{(\infty)}(\mathcal{D}_{\mathcal{F}})$ denote the smallest involutive distribution containing $\mathcal{D}_{\mathcal{F}}$, and we suppose that q_0 is a regular point for $\operatorname{Lie}^{(\infty)}(\mathcal{D}_{\mathcal{F}})$. We shall say that $\mathcal{F}$ is ***totally regular*** at q_0 if q_0 is a regular point for both $\mathcal{F}$ and $\operatorname{Lie}^{(\infty)}(\mathcal{D}_{\mathcal{F}})$. We then define $\mathcal{F}^{(\infty)} = \operatorname{ann}(\operatorname{Lie}^{(\infty)}(\mathcal{D}_{\mathcal{F}}))$. We denote by $C^\infty(\mathsf{Q})_{\mathcal{F}}$ those C^∞-functions ϕ having the property that $\mathrm{d}\phi \in \Gamma^\infty(\mathcal{F}^{(\infty)})$.

With this language as backdrop, we state the following result.

Lemma 10.36. *Let $(q_0, 0)$ be a controlled equilibrium configuration for a C^∞-simple mechanical control system $\Sigma = (\mathsf{Q}, \mathbb{G}, V, \mathscr{F}, \mathbb{R}^m)$, and suppose that the codistribution $\mathcal{F}$ is totally regular at q_0. For $V_{\mathrm{P}} \in C^\infty(\mathsf{Q})$, the following statements are equivalent:*

(i) $\mathrm{d}V_{\mathrm{P}}(q) \in \mathcal{F}_q^{(\infty)}$ for each q in a neighborhood $\mathcal{U}$ of q_0;

(ii) there exists a C^∞- feedback $u\colon \mathcal{U} \to \mathbb{R}^m$, defined on a neighborhood $\mathcal{U}$ of q_0, such that

$$\sum_{a=1}^{m} u^a(q) F^a(q) = -\mathrm{d}V_{\mathrm{P}}(q), \qquad q \in \mathcal{U}.$$

Proof. (i) $\Longrightarrow$ (ii) Since $\mathrm{d}V_{\mathrm{P}} \in \Gamma^\infty(\mathcal{F}^{(\infty)})$, and since $\mathcal{F}^{(\infty)} \subset \mathcal{F}$, (ii) immediately follows.

(ii) $\Longrightarrow$ (i) Suppose that (ii) holds with some neighborhood $\mathcal{U}$ of q_0, and that $\mathrm{d}V_{\mathrm{P}}(\bar{q}) \notin \mathcal{F}_{\bar{q}}^{(\infty)}$ for some $\bar{q}$. Then there is a neighborhood $\mathcal{W}$

of $\bar{q}$ for which $\mathrm{d}V_{\mathrm{P}}(q) \notin \mathcal{F}_q^{(\infty)}$ for all $q \in \mathcal{W}$. By Proposition 3.93, choose coordinates $(q^1, \dots, q^n)$ around $\bar{q}$ with the property that $\mathcal{F}_q^{(\infty)} = \mathrm{span}_{\mathbb{R}} \{\mathrm{d}q^1(q), \dots, \mathrm{d}q^k(q)\}$ for $q \in \mathcal{W}$. By our assumption about $V_{\mathrm{P}}(\bar{q})$, we can also choose the coordinates such that $q^{k+1}(q) = V_{\mathrm{P}}(q)$ for $q \in \mathcal{W}$. Note that $\mathrm{span}_{\mathbb{R}} \{\mathrm{d}q^1(q), \dots, \mathrm{d}q^{k+1}(q)\} \subset \mathcal{F}_q$ for $q \in \mathcal{W}$, implying that $\mathcal{D}_{\mathcal{F},q} \subset \mathrm{span}_{\mathbb{R}}\{\frac{\partial}{\partial q^{k+2}}(q), \dots, \frac{\partial}{\partial q^n}(q)\}$ for $q \in \mathcal{W}$. This implies that $\mathrm{Lie}^{(\infty)}(\mathcal{D}_{\mathcal{F}})_q \subset \mathrm{span}_{\mathbb{R}}\{\frac{\partial}{\partial q^{k+2}}(q), \dots, \frac{\partial}{\partial q^n}(q)\}$ for $q \in \mathcal{W}$. This, however, violates the definition of the coordinates $(q^1, \dots, q^n)$, since in these coordinates we have $\mathrm{Lie}^{(\infty)}(\mathcal{D}_{\mathcal{F}})_q = \mathrm{span}_{\mathbb{R}}\{\frac{\partial}{\partial q^{k+1}}(q), \dots, \frac{\partial}{\partial q^n}(q)\}$ for $q \in \mathcal{W}$. ■

The preceding developments, along with Lemma 10.22, make sense of the following definition.

Definition 10.37 (PD control). Let $\Sigma = (\mathsf{Q}, \mathbb{G}, V, \mathscr{F}, \mathbb{R}^m)$ be a C^∞-simple mechanical control system with $(q_0, 0)$ a controlled equilibrium configuration, and suppose that $\mathcal{F}$ is totally regular at q_0. A ***proportional-derivative (PD)-control law*** at q_0 is a state feedback satisfying

$$F_\Sigma(q) \cdot u(v_q) = -\mathrm{d}V_{\mathrm{P}}(q) - F_\Sigma(q) \circ K_{\mathrm{D}}^\sharp \circ F_\Sigma(q)^*(v_q),$$

where $V_{\mathrm{P}} \in C^\infty(\mathsf{Q})_{\mathcal{F}}$, $K_{\mathrm{D}} \in \Sigma_2((\mathbb{R}^m)^*)$, and where $F_\Sigma(q)$ is as defined in Definition 7.32. If $K_{\mathrm{D}} = 0$, then u is a ***proportional control law***. •

Remarks 10.38. 1. Note that the closed-loop system for a simple mechanical control system $(\mathsf{Q}, \mathbb{G}, V, \mathscr{F}, \mathbb{R}^m)$, using PD control, is a forced simple mechanical system $(\mathsf{Q}, \mathbb{G}, V + V_{\mathrm{P}}, -R_{\mathrm{diss}})$, where the Rayleigh dissipative force R_{diss} is defined by

$$R_{\mathrm{diss}}(q) = F_\Sigma(q) \circ K_{\mathrm{D}}^\sharp \circ F_\Sigma(q)^*.$$

Since the net effect of the proportional control law is to change the potential function of the given system, this control strategy is often referred to as ***potential energy shaping***.

2. The preceding definition is global in nature, but has the disadvantage of being implicit about the nature of the function V_{P}. Let us indicate how this can be made more explicit, with the possible penalty of restricting our analysis to a neighborhood of q_0. We assume $\mathcal{F}$ is totally regular at q_0 so that we may define the codistribution $\mathcal{F}^{(\infty)}$ having q_0 as a regular point. By Proposition 3.93, it is possible to modify the generators $\mathscr{F} = \{F^1, \dots, F^m\}$ for $\mathcal{F}$ in such a way that there exist functions $\phi^1, \dots, \phi^k$, defined in a neighborhood of q_0, with $\{F^1 = \mathrm{d}\phi^1, \dots, F^k = \mathrm{d}\phi^k\}$ generators for $\mathcal{F}^{(\infty)}$. We may, without loss of generality, suppose that $\phi^a(q_0) = 0$, $a \in \{1, \dots, k\}$. Furthermore, these k generators may be assumed to be linearly independent in a neighborhood of q_0. In this case we say these generators are ***proportionally adapted at q_0***. Then the function V_{P} in Definition 10.37 takes the form $V_{\mathrm{P}}(q) = \Psi(\phi^1(q), \dots, \phi^k(q))$, for an arbitrary differentiable function $\Psi \colon \mathbb{R}^k \to \mathbb{R}$. •

The final observation we make in our discussion of the structure of nonlinear PD control is that the derivative control we define is actually dissipative control as we discussed in Section 10.1.5.

Proposition 10.39 (Derivative control is dissipative). *Let $\Sigma = (\mathsf{Q}, \mathbb{G}, V, \mathscr{F}, \mathbb{R}^m)$ be a C^∞-simple mechanical control system with $\Sigma_{\mathsf{TQ}} = (\mathsf{TQ}, \mathscr{C}_\Sigma = \{f_0, f_1, \dots, f_m\}, \mathbb{R}^m)$ the corresponding control-affine system (see the discussion in Section 7.3.2). Let $E \colon \mathsf{TQ} \to \mathbb{R}$ be the energy associated with Σ. Let K_{D}^{ab}, $a, b \in \{1, \dots, m\}$, be the components of $K_{\mathrm{D}} \in \Sigma_2((\mathbb{R}^m)^*)$ with respect to the standard basis for $(\mathbb{R}^m)^*$.*

Then the control law $u \colon \mathsf{TQ} \to \mathbb{R}^m$ defined by

$$u^a(v_q) = -\sum_{b=1}^{m} K_{\mathrm{D}}^{ab} \mathscr{L}_{f_b} E(v_q), \qquad a \in \{1, \dots, m\},$$

also satisfies

$$F_\Sigma(q) \cdot u(v_q) = -F_\Sigma(q) \circ K_{\mathrm{D}}^\sharp \circ F_\Sigma(q)^*(v_q).$$

That is to say, derivative control of a simple mechanical system is an example of dissipative control for the corresponding control-affine system.

Proof. This follows directly from the straightforward computation showing that $\mathscr{L}_{f_a} E(v_q) = \langle F_a(q); v_q \rangle$, using $f_a = \mathrm{vlft}(\mathbb{G}^\sharp(F^a))$ and $E(v_q) = \frac{1}{2}\mathbb{G}(v_q, v_q) + V(q)$. ■

10.4.2 Stabilization using nonlinear PD control

Now that we understand the character of the "P-part" of PD control, we are in a position to state conditions under which a system can be stabilized using PD control. It is convenient to first say just what this means.

Definition 10.40 (Stabilizability by PD control). Let $(q_0, 0)$ be a controlled equilibrium configuration for a C^∞-simple mechanical control system $\Sigma = (\mathsf{Q}, \mathbb{G}, V, \mathscr{F}, \mathbb{R}^m)$, and suppose that $\mathcal{F}$ is totally regular at q_0. Σ is

(i) ***stabilizable by proportional control*** if there exists $V_{\mathrm{P}} \in C^\infty(\mathsf{Q})_{\mathcal{F}}$ such that the simple mechanical system $(\mathsf{Q}, \mathbb{G}, V + V_{\mathrm{P}})$ has q_0 as a stable equilibrium configuration, and is

(ii) ***locally asymptotically stabilizable by proportional-derivative control*** if it is stabilizable by the proportional control defined by $V_{\mathrm{P}} \in C^\infty(\mathsf{Q})_{\mathcal{F}}$, and if there exists $K_{\mathrm{D}} \in \Sigma_2((\mathbb{R}^m)^*)$ such that the Rayleigh dissipation function R_{diss} defined by $R_{\mathrm{diss}}(q) = F_\Sigma(q) \circ K_{\mathrm{D}}^\sharp \circ F_\Sigma(q)^*$ ensures that the forced simple mechanical system $(\mathsf{Q}, \mathbb{G}, V + V_{\mathrm{P}}, -R_{\mathrm{diss}}^\flat)$ has q_0 as a locally asymptotically stable equilibrium configuration. •

Just as the discussion of Sections 6.2.2 and 6.2.3 suggests the manner in which one should stabilize linear mechanical systems using PD control, the discussion in Section 6.2.4 indicates how one should stabilize an equilibrium configuration for a simple mechanical system using PD control. Indeed, the general idea should be the following.

Problem 10.41 (Methodology for stabilization using PD control). Let Σ and $(q_0, 0)$ be as in Definition 10.40.

(i) To stabilize Σ at q_0, if possible, design V_P such that $V + V_P$ is locally positive-definite at q_0. By Theorem 6.45, q_0 is then a stable equilibrium configuration for the closed-loop system using proportional control.

(ii) Supposing one may find V_P as in (i), if possible, asymptotically stabilize q_0 using Rayleigh dissipation introduced by control. •

The preceding methodology involves two "if possible's." The remainder of this section will be devoted to providing sufficient conditions under which the methodology is possible, and to discussing the limitations of these sufficient conditions. We begin with a result that indicates when a linear PD control law, implemented on a nonlinear mechanical system, is a nonlinear PD control law. The outcome is that stabilization using PD control is possible strictly less frequently than linear PD control. Thus PD control is subject to the limitations of linear PD control, plus additional limitations.

Lemma 10.42. *Let $(q_0, 0)$ be a controlled equilibrium configuration for a C^∞-simple mechanical control system $\Sigma = (\mathsf{Q}, \mathbb{G}, V, \mathscr{F}, \mathbb{R}^m)$ and denote by $\mathcal{F}$ the codistribution generated by the covector fields $\mathscr{F}$. If q_0 is a regular point for $\mathcal{F}$, then the following statements are equivalent:*

(i) there exists a neighborhood $\mathcal{U}$ of q_0 on which $\operatorname{coann}(\mathcal{F})$ *is integrable;*

(ii) there exist

(a) a near identity diffeomorphism $(\chi, \mathcal{U}_0, \mathcal{U}_1)$ at q_0 and

(b) local generators $\tilde{\mathscr{F}}$ for $\mathcal{F}$ at q_0,

with the property that, if $u_{\mathrm{lin}}(x) = -G^\sharp \circ \tilde{F}_\Sigma(q_0)^*(x)$ is a linear proportional control law for the linearization of $\tilde{\Sigma} = (\mathsf{Q}, \mathbb{G}, V, \tilde{\mathscr{F}}, \mathbb{R}^m)$ at q_0, then the implementation $u_{\mathrm{nonlin}} = u_{\mathrm{lin}} \circ \chi^{-1}$ is a proportional control law for $\tilde{\Sigma}$ defined on $\mathsf{T}\mathcal{U}_1$.

Proof. First we note that, since q_0 is a regular point for $\mathcal{F}$, we can suppose that the covector fields $F^1, \dots, F^m$ in $\mathscr{F}$ are linearly independent at q_0, and, therefore, in a neighborhood of q_0.

(i) $\Longrightarrow$ (ii) Suppose that $\operatorname{coann}(\mathcal{F})$ is integrable in a neighborhood $\mathcal{U}$ of q_0 and let $u_{\mathrm{lin}}(x) = -G^\sharp \circ F_\Sigma(q_0)^*(x)$ be a linear proportional control law for the linearization of Σ at q_0. Choose coordinates $(q^1, \dots, q^n)$ for Q around q_0 with the property that $\operatorname{coann}(\mathcal{F}_q) = \operatorname{span}_{\mathbb{R}}\{\frac{\partial}{\partial q^{m+1}}(q), \dots, \frac{\partial}{\partial q^n}(q)\}$ for $q \in \mathcal{U}$ (if necessary, shrink $\mathcal{U}$). Change the basis covector fields to $\tilde{\mathscr{F}}$ with $\tilde{F}^a = \mathrm{d}q^a$, $a \in \{1, \dots, m\}$. With respect to the induced basis for $\mathsf{T}_{q_0}\mathsf{Q}$, the linear feedback will have the form

$$u_{\mathrm{lin}}(x) = -\boldsymbol{K} \begin{bmatrix} x^1 \\ \vdots \\ x^m \end{bmatrix},$$

for some $m \times m$ symmetric matrix $\boldsymbol{K}$. Using the near identity diffeomorphism induced by the coordinate chart as in Lemma 10.30, we then have

$$\sum_{a=1}^{m} u^a(q)\tilde{F}^a(q) = -\boldsymbol{K}\begin{bmatrix} q^1 \\ \vdots \\ q^m \end{bmatrix},$$

which is indeed a proportional control law for $\tilde{\Sigma} = (\mathsf{Q}, \mathbb{G}, V, \tilde{\mathscr{F}}, \mathbb{R}^m)$, as desired.

(ii) $\implies$ (i) Fix a near identity diffeomorphism $(\chi, \mathcal{U}_0, \mathcal{U}_1)$ and a set of local generators $\tilde{\mathscr{F}}$ for $\mathcal{F}$ at q_0 with the property that, for any linear proportional control law $u_{\text{lin}}(x) = -G^\sharp \circ \tilde{F}_\Sigma(q_0)^*(x)$, we have

$$\sum_{a=1}^{m} u^a_{\text{nonlin}}(q)\tilde{F}^a(q) = -\mathrm{d}V_{\text{nonlin}}(q),$$

for some smooth function V_{nonlin} defined in a neighborhood of q_0, and where $u_{\text{nonlin}} = u_{\text{lin}} \circ \chi$. Choose coordinates $(q^1, \dots, q^n)$ adapted to the near identity diffeomorphism as in Lemma 10.30, and with the property that $\mathrm{coann}(\mathcal{F}_{q_0}) = \mathrm{span}_{\mathbb{R}}\{\frac{\partial}{\partial q^{m+1}}(q_0), \dots, \frac{\partial}{\partial q^n}(q_0)\}$. In these coordinates we have

$$u_{\text{lin}}(x) = -\boldsymbol{K}\begin{bmatrix} x^1 \\ \vdots \\ x^m \end{bmatrix},$$

for some symmetric $m \times m$ matrix $\boldsymbol{K}$. This then gives

$$\sum_{a=1}^{m}\sum_{b=1}^{m} \boldsymbol{K}_{ab} q^b \tilde{F}^a_i = -\frac{\partial V_{\boldsymbol{K}}}{\partial q^i}, \qquad i \in \{1, \dots, n\},$$

for some smooth function $V_{\boldsymbol{K}}$ defined in a neighborhood of q_0. In particular, if we define $\boldsymbol{K}_a$, $a \in \{1, \dots, m\}$, to be the $m \times m$ diagonal matrix with zeros on the diagonal, except for a 1 in the ath position, then we have

$$q^a \tilde{F}^a = -\mathrm{d}V^a, \qquad a \in \{1, \dots, m\},$$

for suitable smooth functions $V^1, \dots, V^m$ defined on a neighborhood of q_0. For $a \in \{1, \dots, m\}$, define $\mathcal{U}_a = \{q \in \mathcal{U} \mid q^a \neq 0\}$ and let $\mathcal{U}_0 = \cap_{a=1}^m \mathcal{U}_a$. Note that $\mathcal{U}_0$ has the following properties:

1. $\mathcal{U}_0$ is an open dense subset of $\mathcal{U}$;
2. for each $q \in \mathcal{U}_0$, we have $\cap_{a=1}^m \ker(\tilde{F}^a(q)) = \cap_{a=1}^m \ker(\mathrm{d}V^a(q))$.

By Proposition 3.93, it follows that the distribution $\cap_{a=1}^m \ker(\tilde{F}^a)$ on $\mathcal{U}_0$ is integrable. We claim that the fact that $\mathcal{U}_0$ is open and dense in $\mathcal{U}$ implies that $\mathcal{D}_{\mathcal{F}} \triangleq \cap_{a=1}^m \ker(\tilde{F}^a)$ is integrable on $\mathcal{U}$. Indeed, let $\mathrm{Lie}^{(1)}(\mathcal{D}_{\mathcal{F}})$ be the distribution on $\mathcal{U}$ generated by vector fields taking values in $\mathcal{D}_{\mathcal{F}}$ and their Lie brackets (this notation was introduced in Section 3.7.3). This is a C^∞-distribution whose rank on $\mathcal{U}_0$ is $n - m$. We claim that its rank on $\mathcal{U} \setminus \mathcal{U}_0$ is also $n - m$. Indeed, if $\operatorname{rank}(\mathrm{Lie}^{(1)}(\mathcal{D}_{\mathcal{F}})_q) = p > n - m$ for $q \in \mathcal{U} \setminus \mathcal{U}_0$, then there exists C^∞-vector fields $X_1, \dots, X_p$ such that the set $\{X_1(q), \dots, X_p(q)\}$ forms a basis for $\mathrm{Lie}^{(1)}(\mathcal{D}_{\mathcal{F}})_q$. By continuity, it follows that the vector fields $X_1, \dots, X_p$ are linearly independent in a neighborhood of q, and this is not possible, since any such neighborhood will contain points in $\mathcal{U}_0$. ■

The lemma can be interpreted as saying that linear PD control can be regarded as nonlinear PD control if and only if the input codistribution is integrable near q_0 (in the sense that the distribution annihilating it is integrable). This amounts to saying that there are coordinates $(q^1, \dots, q^n)$ around q_0 for which the equations governing the system are

$$\begin{aligned} \mathbb{G}_{ak}\ddot{q}^k + \mathbb{G}_{ak}\overset{\mathbb{G}}{\Gamma}{}^k_{ij}\dot{q}^i\dot{q}^j &= -\frac{\partial V}{\partial q^a} + u^a, \qquad a \in \{1, \dots, m\}, \\ \mathbb{G}_{\alpha k}\ddot{q}^k + \mathbb{G}_{\alpha k}\overset{\mathbb{G}}{\Gamma}{}^k_{ij}\dot{q}^i\dot{q}^j &= -\frac{\partial V}{\partial q^\alpha}, \qquad \alpha \in \{m+1, \dots, n\}. \end{aligned}$$

The punchline is that only the first m degrees-of-freedom are actuated. Indeed, this is often a *starting point* for treatments of underactuated mechanical systems (e.g., [Auckly and Kapitanski 2002, Baillieul 1998, Nijmeijer and van der Schaft 1990, Olfati-Saber 2002, Oriolo and Nakamura 1991, Reyhanoglu, van der Schaft, McClamroch, and Kolmanovsky 1999, Spong 1998]). However, as we have seen in certain of our examples (e.g., the planar rigid body of Section 7.4.2 using only the input F^2, and the rigid body of Exercise E7.22 with one or two input torques), it is not always physically valid to assume that the input codistribution is integrable.

This allows the statement of the following sufficient condition for stabilizability by PD control. This result is similar to that arrived at by van der Schaft [1986] in a Hamiltonian setting.

Theorem 10.43 (Stabilization by PD control). *Let $\Sigma = (\mathsf{Q}, \mathbb{G}, V, \mathscr{F}, \mathbb{R}^m)$ be a C^∞-simple mechanical control system with $(q_0, 0)$ a controlled equilibrium configuration. If $\mathcal{F}$ is totally regular at q_0 and if $\mathscr{F} = \{\mathrm{d}\phi^1, \dots, \mathrm{d}\phi^k, F^{k+1}, \dots, F^m\}$ is proportionally adapted at q_0, then the following statements hold.*

(i) If $K_\Sigma(q_0)|\mathrm{coann}(\mathcal{F}^{(\infty)}_{q_0})$ is positive-definite, then there exists a positive-definite $K_{\mathrm{P}} \in \Sigma_2((\mathbb{R}^k)^*)$ such that $V_{\mathrm{cl}} = V + \frac{1}{2}\sum_{a,b=1}^k (K_{\mathrm{P}})_{ab}\phi^a\phi^b \in C^\infty(\mathsf{Q})$ is locally positive-definite about q_0. Furthermore, $(q_0, 0)$ is stabilized by the proportional control law

$$u(v_q) = -\sum_{a,b=1}^{k} (K_{\mathrm{P}})_{ab}\phi^b(q)e_a,$$

where $\{e_1, \dots, e_m\}$ is the standard basis for $\mathbb{R}^m$.

(ii) Suppose that $K_\Sigma(q_0)|\mathrm{coann}(\mathcal{F}^{(\infty)}_{q_0})$ *is positive-definite, that* K_P *is chosen as in part (i), that*

$$\langle \mathbb{G}(q_0)^\sharp \circ K_\Sigma(q_0)^\flat, \mathrm{image}(\mathbb{G}(q_0)^\sharp \circ F_\Sigma(q_0))\rangle = \mathsf{T}_{q_0}\mathsf{Q},$$

and that $K_\mathrm{D} \in \Sigma_2((\mathbb{R}^m)^*)$ *is positive-definite. Then* $(q_0, 0)$ *is locally asymptotically stabilized by the PD control law*

$$u(v_q) = -\sum_{a,b=1}^{k} (K_\mathrm{P})_{ab}\phi^b(q)e_a - K_\mathrm{D}^\sharp \circ F_\Sigma(q)(v_q).$$

Proof. The first part of the result follows from Theorem 10.26 and Lemma 10.42, using only the controls $u^1, \dots, u^k$. We claim that the second part of the theorem follows from Theorem 6.42. To show this, it suffices to show that

$$\begin{aligned}\langle \mathbb{G}(q_0)^\sharp \circ \mathrm{Hess}\, V(q_0)^\flat, &\mathrm{image}(\mathbb{G}(q_0)^\sharp \circ F_\Sigma(q_0))\rangle \\ &= \langle \mathbb{G}(q_0)^\sharp \circ \mathrm{Hess}(V + V_\mathrm{P})(q_0)^\flat, \mathrm{image}(\mathbb{G}(q_0)^\sharp \circ F_\Sigma(q_0))\rangle,\end{aligned}$$

where $V_\mathrm{P}(q) = \frac{1}{2}\sum_{a,b=1}^{k}(K_\mathrm{P})_{ab}\phi^a\phi^b$. This is the content of the following lemma.

Lemma. *Let* U *and* V *be finite-dimensional* $\mathbb{R}$*-vector spaces, and let* M *be an inner product on* V, $K_1, K_2 \in \Sigma_2(\mathsf{V})$, *and* $F \in L(\mathsf{U}; \mathsf{V}^*)$. *If* $\mathrm{image}(K_2^\flat) \subset \mathrm{image}(F)$, *then*

$$\langle M^\sharp \circ K_1^\flat, \mathrm{image}(M^\sharp \circ F)\rangle = \langle M^\sharp \circ (K_1 + K_2)^\flat, \mathrm{image}(M^\sharp \circ F)\rangle.$$

Proof. To simplify notation, and without loss of generality, suppose that $\mathsf{V} = \mathbb{R}^n$ and $\mathsf{U} = \mathbb{R}^m$. Let $\boldsymbol{M}$, $\boldsymbol{K}_1$, $\boldsymbol{K}_2$, and $\boldsymbol{F}$ be the matrix representations of, respectively, M, K_1, K_2, and F in the standard bases. By Exercise E7.4 we know that $\langle M^\sharp \circ K_1^\flat, \mathrm{image}(M^\sharp \circ F)\rangle$ is exactly the subspace spanned by the columns of the matrix

$$\left[\, \boldsymbol{M}^{-1}\boldsymbol{F} \mid (\boldsymbol{M}^{-1}\boldsymbol{K}_1)\boldsymbol{M}^{-1}\boldsymbol{F} \mid \cdots \mid (\boldsymbol{M}^{-1}\boldsymbol{K}_1)^{n-1}\boldsymbol{M}^{-1}\boldsymbol{F}\,\right], \tag{10.4}$$

and that $\langle M^\sharp \circ (K_1 + K_2)^\flat, \mathrm{image}(M^\sharp \circ F)\rangle$ is the subspace spanned by the columns of the matrix

$$\begin{aligned}\big[\, \boldsymbol{M}^{-1}\boldsymbol{F} \mid (\boldsymbol{M}^{-1}(\boldsymbol{K}_1 + \boldsymbol{K}_2))\boldsymbol{M}^{-1}\boldsymbol{F} \mid \\ \cdots \mid (\boldsymbol{M}^{-1}(\boldsymbol{K}_1 + \boldsymbol{K}_2))^{n-1}\boldsymbol{M}^{-1}\boldsymbol{F}\,\big].\end{aligned} \tag{10.5}$$

We claim that the span of the columns of the two matrices of equations (10.4) and (10.5) are equal. Since $\mathrm{image}(K_2^\flat) \subset \mathrm{image}(F)$, it follows that the span of the columns of the matrix $\boldsymbol{M}^{-1}(\boldsymbol{K}_1 + \boldsymbol{K}_2)\boldsymbol{M}^{-1}\boldsymbol{F}$ is contained in the span of the union of the columns of the matrices $\boldsymbol{M}^{-1}\boldsymbol{F}$ and $\boldsymbol{M}^{-1}\boldsymbol{K}_1\boldsymbol{M}^{-1}\boldsymbol{F}$. We now proceed inductively, supposing that the span of the columns of the matrix $(\boldsymbol{M}^{-1}(\boldsymbol{K}_1+\boldsymbol{K}_2))^k\boldsymbol{M}^{-1}\boldsymbol{F}$ is contained in the span of the union of the columns of the matrices $\boldsymbol{M}^{-1}\boldsymbol{F},\dots,(\boldsymbol{M}\boldsymbol{K}_1)^{k-1}\boldsymbol{M}^{-1}\boldsymbol{F}$. Then we have

$$\begin{aligned}&(\boldsymbol{M}^{-1}(\boldsymbol{K}_1+\boldsymbol{K}_2))^{k+1}\boldsymbol{M}^{-1}\boldsymbol{F}\\ &=\boldsymbol{M}^{-1}\boldsymbol{K}_1(\boldsymbol{M}^{-1}(\boldsymbol{K}_1+\boldsymbol{K}_2)^k\boldsymbol{M}^{-1}\boldsymbol{F}+\boldsymbol{M}^{-1}\boldsymbol{K}_2(\boldsymbol{M}^{-1}(\boldsymbol{K}_1+\boldsymbol{K}_2)^k\boldsymbol{M}^{-1}\boldsymbol{F}.\end{aligned}$$

The span of the columns of the first term on the right-hand side lies in the span of the union of the columns of the matrices $\boldsymbol{M}^{-1}\boldsymbol{F},\dots,(\boldsymbol{M}\boldsymbol{K}_1)^k\boldsymbol{M}^{-1}\boldsymbol{F}$, by the induction hypothesis, and the span of the columns of the second term on the right-hand side lies in the span of the columns of the matrix $\boldsymbol{M}^{-1}\boldsymbol{F}$. This shows that the span of the columns of the matrices of equations (10.4) and (10.5) are indeed the same. ▾

To apply the lemma to the current situation take $\mathsf{V}=\mathsf{T}_{q_0}\mathsf{Q}$, $\mathsf{U}=\mathbb{R}^m$, $M=\mathbb{G}(q_0)$, $K_1=\operatorname{Hess}V(q_0)$, $K_2=\operatorname{Hess}V_{\mathrm{P}}(q_0)$, and $F=F_\Sigma(q_0)$. That $\operatorname{image}(K_2^\flat)\subset\operatorname{image}(F)$ in this case follows, since $V_{\mathrm{P}}\in C^\infty(\mathsf{Q})_{\mathcal{F}}$. ■

Remarks 10.44. The hypotheses of the theorem can be easily weakened in two directions.

1. If the system already possesses natural dissipative forces defined by a Rayleigh dissipation function R_{diss}, then the condition in part (ii) can be weakened by asking that $\langle\mathbb{G}(q_0)^\sharp\circ K_\Sigma(q_0)^\flat,\mathsf{W}_{q_0}\rangle=\mathsf{T}_{q_0}\mathsf{Q}$, where
$$\mathsf{W}_{q_0}=\operatorname{image}(\mathbb{G}(q_0)^\sharp\circ R_{\mathrm{diss}}(q_0)^\flat)+\operatorname{image}(\mathbb{G}(q_0)^\sharp\circ F_\Sigma(q_0)),$$
just as in Remark 10.33–2.
2. Let us revisit the discussion initiated in Remark 10.21. It is possible, not just in the hypotheses of Theorem 10.43, but in all of the developments of this section, to generalize somewhat the assumption that the controlled equilibrium configuration has the form $(q_0,0)$, i.e., that the control required to maintain the equilibrium is zero. While we cannot allow arbitrary controlled equilibrium configurations, we can allow those of the form (q_0,u_0), where u_0 has the property that
$$\sum_{a=1}^m u_0^aF^a\in\Gamma^\infty(\mathcal{F}^{(\infty)}).$$
The reader is asked to prove this fact in Exercise E10.12, as well as to explore why general controlled equilibrium configurations cannot be handled directly using the techniques of this section. •

The following example, based on Remark 6.46–7, shows why the conditions in Theorem 10.43 are only sufficient.

Example 10.45. We take $\mathsf{Q} = \mathbb{R}^2$ with $\mathbb{G}$ the standard Riemannian metric on $\mathbb{R}^2$, $V(x,y) = \frac{1}{2}\alpha x^2 + \frac{1}{4}\beta x^4$, and $\mathscr{F} = \{F^1 = \mathrm{d}y\}$. One can readily check that the hypotheses of Theorem 10.43 for stabilization by proportional control are satisfied if and only if $\alpha > 0$. However, the system is actually stabilizable by proportional control if and only if either (1) $\alpha > 0$ or (2) $\alpha = 0$ and $\beta > 0$. •

10.4.3 A mathematical example

The example we consider in this section is not a physical one, since any physical example illustrating what we wish to illustrate will involve somewhat complicated computations. We take $\mathsf{Q} = \mathbb{R}^4$ with $\mathbb{G}$ the standard Riemannian metric on Q, $V(x^1,x^2,x^3,x^4) = \frac{1}{2}a_1(x^1)^2 + \frac{1}{2}a_2(x^2)^2 + \frac{1}{2}a_3(x^3)^2$, $a_1, a_2, a_3 \in \mathbb{R}$, and $\mathscr{F} = \{F^1 = \mathrm{d}x^4, F^2 = \mathrm{d}x^3 + x^1\mathrm{d}x^2\}$. Note that $(q_0, u_0) = ((0,0,0,0),(0,0))$ is a controlled equilibrium configuration.

Let us compute $\mathcal{F}^{(\infty)}$. We note that the vector fields

$$X_1 = x^1\frac{\partial}{\partial x^3} - \frac{\partial}{\partial x^2}, \quad X_2 = \frac{\partial}{\partial x^1}$$

are generators for $\mathcal{D}_{\mathcal{F}} = \mathrm{coann}(\mathcal{F})$. We then note that

$$[X_1, X_2] = -\frac{\partial}{\partial x^3},$$

and that $\{X_1, X_2, [X_1, X_2]\}$ are generators for $\mathrm{Lie}^{(\infty)}(\mathcal{D}_{\mathcal{F}})$. Furthermore, one can verify that $\{\frac{\partial}{\partial x^1}, \frac{\partial}{\partial x^2}, \frac{\partial}{\partial x^3}\}$ also form a set of generators for $\mathrm{Lie}^{(\infty)}(\mathcal{D}_{\mathcal{F}})$. We now make the following conclusions for the stabilizability of the system using PD control of various types.

1. The system is stabilizable using linear proportional control if and only if $a_1, a_2 > 0$.
2. The system is stabilizable using proportional control if $a_1, a_2, a_3 > 0$.
3. The system is not locally asymptotically stabilizable using PD control, either linear or otherwise, since the linearization is not controllable.

10.5 Notes on stabilization of mechanical systems

The techniques outlined in the first six subsections in this section apply only to systems with controllable (and therefore, stabilizable) linearizations at the equilibrium configuration whose stabilization is being sought. As indicated by Theorem 10.5, local stabilization of such systems is possible via C^∞-state feedback. Thus these systems are "easy" to stabilize. There are many other techniques that may be exploited to stabilize such equilibria, and in this section we briefly describe some of these, giving pointers to the literature. In the final subsection we say a few words about the situation in which stabilization by C^∞-state feedback is not possible.

10.5.1 General linear techniques

As was hinted at in various points during the presentation of Section 10.3, particularly during the investigation of the example, even though it can be impossible to locally asymptotically stabilize a nonlinear system using linear PD control, it may be possible to locally asymptotically stabilize the system via linearization using other techniques. Many such techniques exist in the framework of general control-affine systems. Some of these include the following.

1. ***Linear quadratic regulator*** (***LQR***) control is a classic technique for stabilization of a linear system, and has its roots in the work of Kalman [1960a]. The approach to the problem consists of the formulation of an optimal control problem on a semi-infinite time interval. It turns out that the solution to the optimal control problem comes in the form of a linear state feedback law that is guaranteed to produce an asymptotically stable closed-loop system [Brockett 1970, Goodwin, Graebe, and Salgado 2001, Sontag 1998]. The method works well in practice, at least locally, and is computationally tractable. For these reasons, LQR is commonly used in applications. Its drawbacks include (1) the defect, common to all state feedback schemes, that one requires full knowledge of the state to implement the control law, and (2) the defect of the control law, when implemented on the nonlinear system, being fundamentally local. The latter defect can be particularly debilitating, since a control law that is well designed for the linearization has no guarantee to work well when implemented on the nonlinear system.
   One place in which LQR methods may have some use in the context of the techniques described in this book is in the stabilization of a system along a nontrivial trajectory generated using a nonlinear control strategy. Indeed, it can turn out that a system that is not linearly controllable at its equilibria will nonetheless be linearly controllable along a nonequilibrium trajectory. This is discussed by Tyner [2002].
2. To overcome the first of the two listed problems for LQR control, that of the control law requiring full state knowledge, a common approach is to modify the control law to allow for only partial knowledge of the state. The most common approach to this is the so-called ***Kalman–Bucy filter*** [Kalman 1960b, Kalman and Bucy 1960], which has the advantage that it is implemented in a manner that is similar computationally to the LQR state feedback law. The use of the Kalman–Bucy filter in combination with LQR state feedback produces a control law called the ***linear quadratic Gaussian*** (***LQG***). This approach is described, for example, in the text of Goodwin, Graebe, and Salgado [2001]. This approach can work well in practice, but the implementation of the filter must often be done carefully, otherwise the resulting controller can be quite fragile, even locally. That is to say, the method overcomes one of the defects of LQR control, but can exacerbate the other.

3. During the past fifteen years, much work has been done in the area of robust control, with one popular design method being H^∞-control design. This technique has been developed quite fully for linear systems. An introductory book is that of Doyle, Francis, and Tannenbaum [1990], with more advanced books being those of Zhou [1996] and Dullerud and Paganini [1999]. The linear H^∞ techniques suffer from the same drawback of being valid only locally as was commented on for LQR and LQG.

10.5.2 Feedback linearization and partial feedback linearization

A common technique for stabilization consists of transforming a system to a linear system via a "feedback transformation." We shall not precisely describe what is meant by a feedback transformation, but will merely say that it consists of a change of coordinates, along with a state-dependent affine change of control. After a feedback transformation, a control-affine system remains a control-affine system. The idea in feedback linearization is that, although the system of interest may not be linear, it is only because it has been represented in the wrong coordinates, and with the wrong control vector fields. Systems that can be feedback transformed to a controllable linear system possess controllable linearizations. However, it is not the case that systems with controllable linearizations can always be feedback transformed to a linear system. Therefore, feedback linearization can only be applied to systems for which standard linearization techniques, as described in Section 10.1, already apply. However, the advantage of feedback linearization, when applicable, is that it may perform better globally than classic linearization. The problem of feedback linearization was first addressed systematically by Brockett [1978], and a complete characterization was first obtained by Jakubczyk and Respondek [1980]. For mechanical systems, Bedrossian and Spong [1995] characterize feedback linearization in terms of the curvature tensor associated with the system's kinetic energy metric; see also [Baillieul 1998].

The systems for which feedback linearization may be directly applied are typically uncommon. One approach is to employ so-called "partial feedback linearization," meaning that one essentially linearizes the system as much as possible. The precise meaning of this is described by Marino [1986]. The idea of stabilization by partial feedback linearization is related to the existence of a linearizing output with stable zero dynamics; see [Isidori 1995]. Spong [1998] considers partial feedback linearization in a mechanical setting, and we refer to Exercise E10.17 for a discussion of issues related to this. Other work in a related vein includes [Fantoni and Lozano 2001, Olfati-Saber 2002].

10.5.3 Backstepping

Backstepping is an approach that originated in the papers [Kanellakopoulos, Kokotovic, and Morse 1991, 1992]. The method is one for stabilization that relies on a "triangular structure" of the system in a certain set of coordinates.

The structure enables the use of recursive techniques to jointly design feedback control laws and corresponding Lyapunov functions. The backstepping idea has been extended in various directions, in particular in combination with adaptive control. A serious discussion of this would take us too far afield, so we instead refer to the book of Krstic, Kanellakopoulos, and Kokotovic [1995]. An early paper dealing with a mechanical application of backstepping is that of Seto and Baillieul [1994]. The idea of backstepping is useful, in that it provides a systematic design methodology, and one that allows for parameter uncertainties. The method is limited by the required triangular structure. For example, in some incarnations of backstepping, the triangular structure arises from an assumption that the system can be feedback linearized. Thus, such an assumption inherits the limitations of feedback linearization discussed in the preceding section. Backstepping is among the techniques utilized in the books [Fantoni and Lozano 2001, Kwatny and Blankenship 2001].

10.5.4 Passivity-based methods

The notion of passivity stems from the description of systems using loop gain and positivity by [Zames 1966], and of input/output dissipative systems by Willems [1972]. A passive system is equipped with an energy storage function whose behavior as a function of time is related to the amount of energy provided or extracted from the system via the input/output pair. In stabilization problems, this storage function is a natural candidate for a control-Lyapunov function. An early discussion of passivity for stabilization can be found in [Byrnes, Isidori, and Willems 1991], and more recent treatments include [Lozano, Brogliato, Egeland, and Maschke 2000, van der Schaft 1999]. We do not consider outputs anywhere in this book, and there are issues related to this that require additional explanation in order for this notion to make sense. Rather than do this, we refer to the books [Arimoto 1996, Ortega, Loria, Nicklasson, and Sira-Ramirez 1998]. Examples of geometric treatments of passivity methods include [Duindam, Stramigioli, and Scherpen 2003, Li and Horowitz 2001a,b, Muñoz-Lecanda and Yániz 2002].

Since they deal with the full nonlinear system, and not with the linearization at an equilibrium point, passivity-based methods can have the advantage that they provide a larger domain of attraction. A second feature of passivity-based methods is that they do not require accurate knowledge of the system data, since passivity is a structural property. On the other hand, it is worth mentioning that passivity methods are useful for stabilization problems only for systems for which local asymptotic stabilization is possible using C^∞-state feedback.

10.5.5 Sliding mode control

Sliding mode control seems to have originated in Russia, and its first thorough account appears in the book of the originator of the theory, Utkin [1992].

The sliding mode strategy may be summarized as follows. One first seeks a submanifold S of the state manifold M with the following properties.

1. S passes through the equilibrium point x_0.
2. At each point $x \in \mathsf{S}$, the tangent space $\mathsf{T}_x\mathsf{S}$ forms a complement in $\mathsf{T}_x\mathsf{M}$ to the subspace spanned by the control vector fields.
3. It turns out that the preceding condition guarantees that the solutions of the system can be forced, using the controls, to lie on S in a unique manner. One should choose S in such a way that the resulting dynamics render x_0 an asymptotically stable equilibrium point.
4. The control law is now this. Since the input vector fields are transverse to S (i.e., they are complementary to the tangent spaces to S) it is possible to steer points in a neighborhood of S to S in finite time. Once on S, the resulting dynamics have x_0 as an asymptotically stable equilibrium point.

One of the features of sliding mode control is that the closed-loop system is discontinuous. Thus one needs to be careful about the sense in which trajectories are defined (cf. Remark 10.2–6). It turns out that, for sliding mode control, this can be done using the notion of an integral curve for a discontinuous vector field defined by Filippov [1988]. One of the immediate consequences of this (cf. Remark 10.2–6) is that sliding mode control cannot be used to locally asymptotically stabilize an equilibrium that cannot be locally asymptotically stabilized using state feedback that is almost C^∞ about the equilibrium point (e.g., an equilibrium point for a driftless system).

Nevertheless, sliding mode control can be quite effective in practice, where it often exhibits good robustness qualities due to its aggressive control action. There is an enormous literature dealing with applications of sliding mode control.

10.5.6 Total energy shaping methods

We now describe two recent, quite similar, developments in the stabilization of equilibria that can be stabilized using C^∞-state feedback. The techniques both rely on Lyapunov theory and passivity methods. An essential idea of the approach is that one designs the controls so that the closed-loop system is mechanical. An early example of this approach is the paper of Bloch, Krishnaprasad, Marsden, and Sánchez de Alvarez [1992]. On the Lagrangian side, this idea is expanded upon in the so-called method of controlled Lagrangians originated in the work [Bloch, Chang, Leonard, and Marsden 2001, Bloch, Leonard, and Marsden 2000], with additional contributions along these lines made in the papers [Auckly and Kapitanski 2002, Auckly, Kapitanski, and White 2000, Hamberg 1999]. The original work was couched in the language of symmetry, although this is not so essential in the stabilization problem. Indeed, asymptotic stabilization typically requires one to break the symmetry. A self-contained overview of this technique for stabilization is given in Chapter 9 of [Bloch 2003].

A different, Hamiltonian approach, based on passivity methods, called Interconnection Damping Assignment–Passivity Based Control (IDA–PBC), was taken in the papers [Ortega, Spong, Gómez-Estern, and Blankenstein 2002, Ortega, van der Schaft, Maschke, and Escobar 2002]. The two approaches, Lagrangian and Hamiltonian, have been realized as being equivalent in some sense in the papers [Blankenstein, Ortega, and van der Schaft 2002, Chang, Bloch, Leonard, Marsden, and Woolsey 2002]. Another approach, again Hamiltonian in nature, and based on time-dependent canonical transformations, is that of Fujimoto and Sugie [1998, 2001a,b].

One way of looking at these techniques is that they extend our potential energy shaping discussion of Section 10.4 to the shaping also of kinetic energy. In this way, they produce a closed-loop system that still has mechanical structure. These techniques retain some of the limitations of mere potential energy shaping in that there are systems with controllable linearizations, but that cannot be stabilized with the use of energy shaping. We note, however, that *linear* energy shaping, including the shaping of the inner product defining the kinetic energy, can always be used to produce a stable closed-loop system, provided the linear mechanical system is controllable (see Exercise E10.11 and [Zenkov 2002]). The existence of stabilizing general energy shaping feedback is currently an active area of research, so we refer to the existing literature for more details.

As with passivity methods, one of the useful features of these techniques is that they allow an improvement in the size of the domain of attraction of the closed-loop system over what would be achievable using strictly linearization-based methods.

10.5.7 When stabilization by smooth feedback is not possible

The discussion in this chapter has revolved around systems that are easily stabilized, in the sense that they may be asymptotically stabilized by C^∞-state feedback. For systems that do not admit stabilization by C^∞, or perhaps even C^0, state feedback, the picture is much murkier, and the research literature has not provided a coherent picture of what "should" be done in this case. This is especially so when it comes to design related issues, like performance and robustness. We do not say much about this here. However, the stabilization and tracking techniques of Chapter 12 *are* applicable to systems for which stabilization is not possible with C^0-state feedback.

Exercises

E10.1 Let $(\mathsf{M}, \mathscr{C} = \{f_0, f_1, \dots, f_m\}, U)$ be a C^∞-control-affine system, and consider the following statements concerning a map $u\colon \mathsf{M} \to U$:

(i) $u\colon \mathsf{M} \to U$ has the property that the closed-loop system is C^r;
(ii) $u\colon \mathsf{M} \to U$ is a C^r-map from M to $\mathbb{R}^m$.

Answer the following questions.

(a) Show that, if the set $\{f_1(x), \dots, f_m(x)\}$ is linearly independent for each $x \in \mathsf{M}$, then (i) and (ii) are equivalent for any $r \in \mathbb{Z}_+$.

Take the special case of a three-input system on $\mathsf{M} = \mathbb{R}^2$, with $f_1 = \frac{\partial}{\partial x}$, $f_2 = \frac{\partial}{\partial y}$, $f_3 = \frac{\partial}{\partial x} + \frac{\partial}{\partial y}$, and $U = \mathbb{R}^3$.

(b) For $r \in \mathbb{Z}_+$, show that there exists a map $u\colon \mathsf{M} = \mathbb{R}^2 \to U = \mathbb{R}^3$ such that (i) holds, but (ii) does not.

(c) Show that, given a state feedback $u_1 : \mathsf{M} = \mathbb{R}^2 \to U = \mathbb{R}^3$ satisfying (i), there exists a (possibly different) state feedback $u_2 : \mathsf{M} = \mathbb{R}^2 \to U = \mathbb{R}^3$ satisfying (ii), and for which the closed-loop systems are the same.

Now take $\mathsf{M} = \mathbb{R}^2$, with $f_1 = x\frac{\partial}{\partial x}$ and $f_2 = \frac{\partial}{\partial x} + x\frac{\partial}{\partial y}$.

(d) Show that the vector field $f = \frac{\partial}{\partial x}$ has the property that $f(x, y) \in \operatorname{span}_{\mathbb{R}}\{f_1(x, y), f_2(x, y)\}$.

(e) Show that there is no C^0-map $u\colon \mathsf{M} \to U = \mathbb{R}^2$ for which $f = u^1 f_1 + u^2 f_2$.

(f) Can you think of generators $\{\tilde{f}_1, \tilde{f}_2\}$ for the distribution generated by $\{f_1, f_2\}$ which have the property that, if f is any vector field taking values in this distribution, then there exists a C^∞-map $u\colon \mathsf{M} \to U = \mathbb{R}^2$ with the property that $f = u^1 f_1 + u^2 f_2$.

In the next exercise, you will be asked to consider the matter of global stabilization. The exercise illustrates the following three points:

1. there are sometimes topological obstructions to continuous global stabilization, even though local stabilization may be trivial;
2. global stabilization is generally impossible to achieve with bounded controls, even in the absence of topological obstructions;
3. one can artificially fabricate global stabilization results from local stabilization results.

E10.2 Let $\mathsf{M} = \mathbb{S}^1$ and consider the control-affine system on M with zero drift vector field and with the single input vector field $f_1 = -y\frac{\partial}{\partial x} + x\frac{\partial}{\partial y}$. (Thus f_1 is the vector field, of unit length relative to the standard Riemannian metric, tangent to $\mathbb{S}^1 \subset \mathbb{R}^2$.)

(a) Show that the controlled equilibrium point $(x_0, u_0) = ((1, 0), 0)$ can be locally asymptotically stabilized using C^∞-state feedback.

(b) Show that it is not possible to globally asymptotically stabilize (x_0, u_0) using C^0-state feedback.

Hint: *Consider Exercise E6.2.*

Next consider the linear control system on $\mathsf{V} = \mathbb{R}$ defined by $A(x) = x$ and $B(u) = 1$.

(c) Show that the controlled equilibrium $(x_0, u_0) = (0, 0)$ can be locally asymptotically stabilized using linear state feedback.

Now suppose that controls are required to take values in a compact subset U of $\mathbb{R}$.

(d) Show that (x_0, u_0) cannot be globally asymptotically stabilized using any U-valued control.

Finally, let $\mathsf{M} = \mathbb{R}^n$ and consider a control-affine system $(\mathsf{M}, \{f_0, f_1, \dots, f_m\}, \mathbb{R}^m)$. Suppose that one can locally asymptotically stabilize the controlled equilibrium point $(\mathbf{0}, 0)$ with a C^r-state feedback $u\colon \mathsf{M} \to \mathbb{R}^m$, $r \geq 1$. Let ψ be a proper C^r-Lyapunov function for the closed-loop system for which $\mathbf{0}$ is the only critical point, let $\varepsilon > 0$, and suppose that the interior of the sublevel set $\psi^{-1}(\leq\varepsilon)$ is C^r-diffeomorphic to the open ball $B_1(\mathbf{0})$ with diffeomorphism $\phi_0\colon \operatorname{int}(\psi^{-1}(\leq\varepsilon)) \to B_1(\mathbf{0})$. (From Exercise E6.4, we know that this will be true, provided that the Poincaré Conjecture is true.)

(e) Show that the map $\phi_1\colon B_1(\mathbf{0}) \to \mathbb{R}^n$ defined by $\phi_1(\boldsymbol{x}) = \frac{\boldsymbol{x}}{1-\|\boldsymbol{x}\|^2_{\mathbb{R}^n}}$ is a C^∞-diffeomorphism.

(f) Conclude that the C^r-state feedback $u \circ \phi$ globally asymptotically stabilizes the control affine system $(\mathsf{M}, \{\phi_* f_0, \phi_* f_1, \dots, \phi_* f_m\}, \mathbb{R}^m)$, where $\phi = \phi_1 \circ \phi_0$.

E10.3 Consider the control-affine system $\Sigma = (\mathsf{M} = \mathbb{R}^2, \{f_0 = y\frac{\partial}{\partial x}, f_1 = \frac{\partial}{\partial y}\}, U)$ and consider the state feedback

$$u(x, y) = \begin{cases} -5, & x > -\frac{1}{2}\operatorname{sign}(y)y^2, \\ 5, & x < -\frac{1}{2}\operatorname{sign}(y)y^2, \\ -\operatorname{sign}(y), & x = -\frac{1}{2}\operatorname{sign}(y)y^2,\ y \neq 0, \\ 0, & (x, y) = 0. \end{cases}$$

Here sign: $\mathbb{R} \to \{-1, 0, 1\}$ is the function which gives the sign of a real number.

(a) Show that this state-feedback globally asymptotically stabilizes $(0, 0)$.
Hint: *The control law is discontinuous, so one must be careful in defining solutions. In this example, one can deal with this in an* ad hoc *manner by saying that when one reaches a point of discontinuity of the control law, one switches to the new control law defined at that point. For this example, this is an unambiguous procedure, as you will see.*

(b) Sketch the solution curves for the closed-loop system in $\mathbb{R}^2$.

E10.4 Find a control-Lyapunov function for the control-affine system on $\mathsf{M} = \mathbb{R}^2$ defined by the equations

$$\dot{x} = -y^3, \quad \dot{y} = u.$$

E10.5 Let $(x_0, 0)$ be a controlled equilibrium point for a control-affine system $\Sigma = (\mathsf{M}, \mathscr{C}, \mathbb{R}^m)$. Suppose that the linearization (A, B) at $(x_0, 0)$ is asymptotically stabilizable. Choose a coordinate chart $(\mathcal{U}, \phi)$ around x_0 satisfying $\phi(x_0) = \mathbf{0}$.

(a) Let $F \in L(\mathsf{T}_{x_0}\mathsf{M}; \mathbb{R}^m)$ have the property that $A + B \circ F$ has all eigenvalues in $\mathbb{C}_-$. Using the chart, indicate how to use the control for the linear system to naturally define a C^∞-control on $\mathcal{U}$ for Σ.

(b) Show that the linearization of the closed-loop system from part (a) is given by $A + B \circ F$.

(c) Conclude that x_0 is a locally asymptotically stable equilibrium point for the closed-loop system.

E10.6 Let $(\mathsf{M}, \mathscr{C} = \{f_0 = 0, f_1, \dots, f_m\}, \mathbb{R}^m)$ be an underactuated driftless system. For $x_0 \in \mathsf{M}$, use Theorem 10.8 to show that it is not possible to locally asymptotically stabilize $(x_0, 0)$ using C^0-state feedback.

E10.7 Let $(\mathsf{Q}, \nabla, F, \mathscr{Y}, \mathbb{R}^m)$ be an underactuated forced affine connection control system.

(a) Suppose that $F = 0$. For $q_0 \in \mathsf{Q}$, use Theorem 10.8 to show that it is not possible to locally asymptotically stabilize $(q_0, 0)$ using C^0-state feedback.

(b) Suppose that $F = -R^\flat_{\text{diss}}$ is a Rayleigh dissipative force. For $q_0 \in \mathsf{Q}$, use Theorem 10.8 to show that it is not possible to locally asymptotically stabilize $(q_0, 0)$ using C^0-state feedback.

E10.8 Show that it is not possible to locally asymptotically stabilize a simple mechanical control system using proportional control.

E10.9 Let $\boldsymbol{K}$ be a symmetric $n \times n$ matrix thought of as a symmetric $(0,2)$-tensor on $\mathbb{R}^n$.

(a) Show that, for $k \in \mathbb{R}_+$ sufficiently large, the matrix $\boldsymbol{K} + k\boldsymbol{I}_n$ is positive-definite.

Now suppose that $\boldsymbol{K}$ is positive-definite on a subspace $\mathsf{W} \subset \mathbb{R}^n$, i.e., $\boldsymbol{x}^T\boldsymbol{K}\boldsymbol{x} > 0$ for all $\boldsymbol{x} \in \mathsf{W} \setminus \{\boldsymbol{0}\}$.

(b) Show that there exists, and define, a diagonal matrix $\boldsymbol{D}$ with the following properties:

1. $\boldsymbol{x}^T\boldsymbol{D}\boldsymbol{x} = 0$ for all $\boldsymbol{x} \in \mathsf{W}$;
2. $\boldsymbol{K} + \boldsymbol{D}$ is positive-definite.

E10.10 Consider the linear mechanical control system (V, M, K, F) with $\mathsf{V} = \mathbb{R}^2$ and

$$M = \begin{bmatrix} 1 & 0 \\ 0 & 1 \end{bmatrix}, \quad K = \begin{bmatrix} 1 & 0 \\ 0 & 0 \end{bmatrix}, \quad F = \begin{bmatrix} 1 \\ -1 \end{bmatrix}.$$

Show that the linear state feedback

$$u = \begin{bmatrix} -8 & -1 & -8 & -4 \end{bmatrix} \begin{bmatrix} x \\ y \\ v_x \\ v_y \end{bmatrix}$$

asymptotically stabilizes the system, but is not a linear PD control law.

The next exercise deals with energy shaping, not just potential energy shaping, for linear mechanical systems. The ideas here follow Zenkov [2002].

E10.11 Let (V, M, K, F) be a controllable simple mechanical control system, and let $\{\lambda_1, \dots, \lambda_n\}$ be distinct real numbers. Show that there exists a linear feedback $F_{\text{cl}} \colon \mathsf{V} \to \mathbb{R}^m$, an inner product M_{cl} on V, and $K_{\text{cl}} \in \Sigma_2(\mathsf{V})$ such that

1. $\operatorname{spec}(M^\sharp_{\text{cl}} \circ K^\flat_{\text{cl}}) = \{\lambda_1, \dots, \lambda_n\}$, and
2. the dynamics

$$M^\flat(\ddot{x}(t)) + K^\flat(x(t)) = F \circ F_{\text{cl}}(x(t))$$

of the closed-loop system are those of the linear mechanical system $(\mathsf{V}, M_{\text{cl}}, K_{\text{cl}})$.

E10.12 Let $(\mathsf{Q}, \mathbb{G}, V, \mathscr{F}, \mathbb{R}^m)$ be a simple mechanical control system with (q_0, u_0) a controlled equilibrium configuration.

(a) Show that, if

$$\sum_{a=1}^{m} u_0^a F^a \in \Gamma^\infty(\mathfrak{F}^{(\infty)}), \tag{E10.6}$$

then there exists a simple mechanical control system $(\mathsf{Q}, \mathbb{G}, \tilde{V}, \mathscr{F}, \mathbb{R}^m)$ with the following properties:

1. $\mathrm{d}V - \mathrm{d}\tilde{V} = \sum_{a=1}^{m} u_0^a F^a$;
2. $(q_0, 0)$ is a controlled equilibrium configuration for $(\mathsf{Q}, \mathbb{G}, \tilde{V}, \mathscr{F}, \mathbb{R}^m)$.

We now illustrate with an example that, if (E10.6) is not satisfied, then we cannot, for the purposes of local asymptotic stabilization by PD control, take $u_0 = 0$. We consider the case in which $\mathsf{Q} = \mathbb{R}^3$, with $\mathbb{G}$ the standard Riemannian metric on $\mathbb{R}^3$, $V(x, y, z) = xy + z$, and $\mathscr{F} = \{F^1 = (1 + x)\mathrm{d}z + z\mathrm{d}x\}$.

(b) Show that $(q_0, u_0) = ((0, 0, 0), -1)$ is a controlled equilibrium configuration.

(c) Linearize the first-order system on TQ and show that it corresponds to a forced linear mechanical control system $(\mathbb{R}^3, M, K, (F_1, 0), 0)$, where M is the standard inner product on $\mathbb{R}^3$, $K = e^1 \otimes e^2 + e^2 \otimes e^1$, and F_1 is nonzero. Compute F_1.

The point here is that the linearization is not that corresponding to a simple mechanical system with a controlled equilibrium configuration of the form $(q_0, 0)$. Therefore, the theory in this chapter cannot be applied verbatim. We shall now show that this inapplicability of the theory has consequences for the stabilization problem. First consider the simple mechanical control system with the same data as above, except now take $\tilde{V}(x, y, z) = xy$.

(d) Show that $(q_0, 0)$ is now a controlled equilibrium configuration, and that its linearization is asymptotically stabilizable by PD control.

This shows that the difficulty with the example with the original potential function V arises exactly because of the nonzero control for the controlled equilibrium configuration. Suppose that we now go ahead and try to stabilize the system with the potential function V at the controlled equilibrium configuration (q_0, u_0) using potential control.

(e) Show that the system is not stabilizable using proportional state feedback by showing that the closed-loop system will always have eigenvalues in $\mathbb{C}_+$.

Hint: *One way to show this is to first show that any monic polynomial with roots in $\overline{\mathbb{C}}_-$ will have nonnegative coefficients. Now compute the characteristic polynomial for the linearized closed-loop system after feedback by linear proportional control.*

(f) Deduce that the system cannot be stabilized using (not necessarily linear) proportional control.

Finally, we show that the system, despite the difficulties involved in stabilizing it using proportional control, is actually quite well-behaved.

(g) Show that the linearization of the first-order system, with the potential function V, is controllable.

E10.13 This exercise is a continuation of the coupled mass system whose data was described in Exercises E4.3 and E4.17. First consider the input for the system as described in part (c) of Exercise E4.29.
(a) Is the origin stabilizable using linear proportional control?
(b) Is the origin locally asymptotically stabilizable using linear PD control?
Now consider the input for the system as described in part (c) of Exercise E4.29.
(c) Is the origin stabilizable using linear proportional control?
(d) Is the origin locally asymptotically stabilizable using linear PD control?

E10.14 This exercise is a continuation of the pendulum/cart system whose data was described in Exercises E4.5 and E4.19.
(a) For the control force specified in Exercise E4.31, show that it is not possible to stabilize the system using linear PD control at the "pendulum up" equilibria.
(b) Give the form of the general control force F that will permit the system to be locally asymptotically stabilized using linear PD control at the "pendulum down" equilibria.
Note that this system has a controllable linearization at all equilibrium configurations, so it is possible to locally asymptotically stabilize these equilibria using linear state feedback that is more general than linear PD control.
Now consider a general collection of input one-forms $\mathscr{F}$ for the system.
(c) Show that it is possible to stabilize the system using linear PD control in the "pendulum up" equilibria only if the system is fully actuated.

E10.15 Let $(\mathsf{Q}, \mathbb{G}, V, \mathscr{F}, \mathbb{R}^m)$ be a simple mechanical control system for which the covector fields $\mathscr{F}$ are linearly independent, and let R_{diss} be a Rayleigh dissipation function with the property that $\operatorname{image}(R_{\mathrm{diss}}(q)^\flat) \subset \mathfrak{F}_q$ for each $q \in \mathsf{Q}$.
(a) Show that there exists a map $K_{\mathrm{D}} \colon \mathsf{Q} \to \Sigma_2((\mathbb{R}^m)^*)$ with the property that, for all $v_q \in \mathsf{TQ}$,

$$R^\flat_{\mathrm{diss}}(v_q) = F_\Sigma(q) \circ K^\sharp_{\mathrm{D}}(q) \circ F_\Sigma(q)^*(v_q).$$

(b) Can the dissipative force $-R_{\mathrm{diss}}$ be implemented by state feedback? If so, give the form of the control law.

E10.16 This exercise is a variation on the planar rigid body introduced in Chapter 4. For this system, consider the following inputs:

1. F^1 and F^2 are the forces introduced in Section 7.4.2;
2. F^3 is a torque applied at the center of mass;
3. F^4 is a force applied in the direction of F^2, but applied at the center of mass.

For the following combinations of forces, describe (in coordinates, for example) the general nature of the potential function that can be achieved using Lemma 10.36:
(a) F^1 and F^2;
(b) F^1 and F^3;
(c) F^1 and F^4;
(d) F^2 and F^3;

(e) F^3 and F^4.

Now address

(f) which of the preceding five cases can be stabilized using the method of Theorem 10.43?

E10.17 Let $\mathsf{Q} = \mathbb{R}^{n-m} \times \mathbb{R}^m$, and let $\mathbb{G}$ be a Riemannian metric on Q whose matrix representative in standard coordinates $((x^1, \dots, x^{n-m}), (y^1, \dots, y^m))$ is given by

$$[\mathbb{G}] = \begin{bmatrix} \boldsymbol{M}_{11} & \boldsymbol{M}_{12} \\ \boldsymbol{M}_{12}^T & \boldsymbol{M}_{22} \end{bmatrix},$$

for smooth matrix functions $\boldsymbol{M}_{11} \colon \mathsf{Q} \to \mathbb{R}^{(n-m)\times(n-m)}$ and $\boldsymbol{M}_{22} \colon \mathsf{Q} \to \mathbb{R}^{m\times m}$ symmetric, and for $\boldsymbol{M}_{12} \colon \mathsf{Q} \to \mathbb{R}^{(n-m)\times m}$.

(a) Show that $\boldsymbol{M}_{11}$ is invertible.
Hint: *Recall that a symmetric matrix is positive-definite if and only if its principal minors are positive [Gantmacher 1958, Horn and Johnson 1990].*

Define $\tilde{\boldsymbol{M}}_{22} = \boldsymbol{M}_{22} - \boldsymbol{M}_{12}^T \boldsymbol{M}_{11}^{-1} \boldsymbol{M}_{12}$.

(b) Show that $\tilde{\boldsymbol{M}}_{22}$ is invertible.
Hint: *Define*

$$\boldsymbol{P} = \begin{bmatrix} -\boldsymbol{M}_{11}^{-1}\boldsymbol{M}_{12} \\ \boldsymbol{I}_m \end{bmatrix},$$

and note that $\tilde{\boldsymbol{M}}_{22} = \boldsymbol{P}^T[\mathbb{G}]\boldsymbol{P}$. Use the ranks of $\boldsymbol{P}$ and $[\mathbb{G}]$ to get the result.

Now let $V \colon \mathsf{Q} \to \mathbb{R}$ be a smooth potential function and let $F^a \in \Gamma^\infty(\mathsf{T}^*\mathsf{Q})$ be defined by $F^a = \mathrm{d}y^a$, $a \in \{1, \dots, m\}$. Do the following.

(c) Write the equations of motion for the simple mechanical control system $\Sigma = (\mathsf{Q}, \mathbb{G}, V, \{F^1, \dots, F^m\})$ in the form

$$\begin{aligned} \boldsymbol{M}_{11}(\boldsymbol{x},\boldsymbol{y})\ddot{\boldsymbol{x}} + \boldsymbol{M}_{12}(\boldsymbol{x},\boldsymbol{y})\ddot{\boldsymbol{y}} + \boldsymbol{C}_1(\boldsymbol{x},\boldsymbol{y},\dot{\boldsymbol{x}},\dot{\boldsymbol{y}}) + \boldsymbol{N}_1(\boldsymbol{x},\boldsymbol{y}) &= \boldsymbol{0}, \\ \boldsymbol{M}_{22}(\boldsymbol{x},\boldsymbol{y})\ddot{\boldsymbol{y}} + \boldsymbol{M}_{12}^T(\boldsymbol{x},\boldsymbol{y})\ddot{\boldsymbol{x}} + \boldsymbol{C}_2(\boldsymbol{x},\boldsymbol{y},\dot{\boldsymbol{x}},\dot{\boldsymbol{y}}) + \boldsymbol{N}_2(\boldsymbol{x},\boldsymbol{y}) &= \boldsymbol{u}. \end{aligned}$$

Find expressions for $\boldsymbol{C}_1$, $\boldsymbol{C}_2$, $\boldsymbol{N}_1$, and $\boldsymbol{N}_2$.

Define a new input $\boldsymbol{v}$ by

$$\begin{aligned} \boldsymbol{u} = \tilde{\boldsymbol{M}}_{22}(\boldsymbol{x},\boldsymbol{y})\boldsymbol{v} + \boldsymbol{C}_2(\boldsymbol{x},\boldsymbol{y},\dot{\boldsymbol{x}},\dot{\boldsymbol{y}}) - \boldsymbol{M}_{12}^T(\boldsymbol{x},\boldsymbol{y})\boldsymbol{M}_{11}^{-1}(\boldsymbol{x},\boldsymbol{y})\boldsymbol{C}_1(\boldsymbol{x},\boldsymbol{y},\dot{\boldsymbol{x}},\dot{\boldsymbol{y}}) \\ + \boldsymbol{N}_2(\boldsymbol{x},\boldsymbol{y}) - \boldsymbol{M}_{12}^T(\boldsymbol{x},\boldsymbol{y})\boldsymbol{M}_{11}^{-1}(\boldsymbol{x},\boldsymbol{y})\boldsymbol{N}_1(\boldsymbol{x},\boldsymbol{y}). \qquad \text{(E10.7)} \end{aligned}$$

Now do the following.

(d) Show that the system using this new input satisfies the equations

$$\begin{aligned} \boldsymbol{M}_{11}(\boldsymbol{x},\boldsymbol{y})\ddot{\boldsymbol{x}} + \boldsymbol{C}_1(\boldsymbol{x},\boldsymbol{y},\dot{\boldsymbol{x}},\dot{\boldsymbol{y}}) + \boldsymbol{N}_1(\boldsymbol{x},\boldsymbol{y}) &= -\boldsymbol{M}_{12}(\boldsymbol{x},\boldsymbol{y})\boldsymbol{v}, \\ \ddot{\boldsymbol{y}} &= \boldsymbol{v}. \end{aligned} \qquad \text{(E10.8)}$$

This procedure is called ***collocated partial feedback linearization*** by Spong [1998]. In the next exercise, you will show that collocated partial feedback linearization is independent of the ability to stabilize a system.

(e) Show that there exists a simple mechanical control system that can be partially feedback linearized using collocated partial feedback linearization, but that is not stabilizable using any kind of feedback.

Now, still following Spong [1998] and using the notation above, say that Σ is ***strongly inertially coupled*** if $\boldsymbol{M}_{12}(\boldsymbol{x}, \boldsymbol{y})$ has rank $n - m$ for all $(\boldsymbol{x}, \boldsymbol{y}) \in \mathbb{R}^{n-m} \times \mathbb{R}^m$.

(f) Show that, if Σ is strongly inertially coupled, then $m \geq n - m$.

(g) Show that, if Σ is strongly inertially coupled, then the matrix $\boldsymbol{M}_{12}\boldsymbol{M}_{12}^T$ is an invertible matrix function.

(h) Show that, if Σ is strongly inertially coupled, then the matrix $\boldsymbol{M}_{12}^\dagger \triangleq \boldsymbol{M}_{12}^T(\boldsymbol{M}_{12}\boldsymbol{M}_{12}^T)^{-1}$ is a right inverse for $\boldsymbol{M}_{12}$, i.e., that $\boldsymbol{M}_{12}\boldsymbol{M}_{12}^\dagger = \boldsymbol{I}_{n-m}$.

In the next part of the exercise, you will show that the notion of being strongly inertially coupled is not coordinate-invariant.

(i) Let V be a two-dimensional vector space and let $\Sigma = (\mathsf{V}, M, K, F)$ be a linear mechanical system with $\operatorname{rank}(F) = 1$. Show that there exists an isomorphism $L\colon \mathsf{V} \to \mathbb{R}^2 \simeq \mathbb{R} \times \mathbb{R}$ so that the linear mechanical system $(\mathbb{R}^2, L_*M, L_*K, (L^*)^{-1} \circ F)$ is strongly inertially coupled.

The next exercise shows that, given that the system is expressed in a coordinate system where strong inertial coupling holds, then one can adjust the equations (E10.8) by using another input.

(j) Suppose that Σ is strongly inertially coupled. Show that, if we define an input $\boldsymbol{w}$ by

$$\boldsymbol{v} = -\boldsymbol{M}_{12}^\dagger(\boldsymbol{M}_{11}\boldsymbol{w} + \boldsymbol{C}_1 + \boldsymbol{N}_1),$$

where $\boldsymbol{v}$ is as defined in (E10.7), then the system satisfies the equations

$$\begin{aligned} \ddot{\boldsymbol{x}} &= \boldsymbol{w}, \\ \ddot{\boldsymbol{y}} &= -\boldsymbol{M}_{12}^\dagger(\boldsymbol{x}, \boldsymbol{y})\big(\boldsymbol{M}_{11}(\boldsymbol{x}, \boldsymbol{y})\boldsymbol{w} + \boldsymbol{C}_1(\boldsymbol{x}, \boldsymbol{y}, \dot{\boldsymbol{x}}, \dot{\boldsymbol{y}}) + \boldsymbol{N}_1(\boldsymbol{x}, \boldsymbol{y})\big). \end{aligned}$$

The procedure by which these previous equations are determined is called ***non-collocated partial feedback linearization*** by Spong [1998]. In the next exercise, you will show that non-collocated partial feedback linearization is independent of the ability to stabilize a system.

(k) Using part (i) above, show that there exists a simple mechanical control system that can be partially feedback linearized using non-collocated partial feedback linearization, but that is not stabilizable using any kind of feedback.

11

Stabilization and tracking for fully actuated systems

In this chapter we design feedback control laws for simple mechanical control systems with dissipation. We assume that the system is fully actuated and that the control set is unbounded throughout the chapter. Our approach builds on the proportional-derivative control designs presented in the previous chapter, but we are able to obtain stronger results by exploiting the full actuation. We provide a comprehensive solution to the problems of stabilization of controlled equilibrium configurations and tracking of reference trajectories.

The material in this chapter is inspired by the work in the field of robotics dedicated to the control of fully actuated manipulators; see [Takegaki and Arimoto 1981] and the discussion in the introduction to Chapter 10. In the context of stabilization problems, our presentation focuses on the notion of a configuration error function that directly gives rise to a proportional control law. For fully actuated mechanical systems, it is also possible to provide a comprehensive solution to the problem of trajectory tracking. Early results on trajectory tracking for fully actuated robotic manipulators and mechanical systems are described, for example, in [Koditschek 1989, Paden and Panja 1988, Wen and Bayard 1988]. Our presentation is taken from [Bullo and Murray 1999]. Throughout the chapter, we restrict our control design to the setting of smooth mechanical systems.

The outline of the chapter is the following. Section 11.1 presents strategies for the stabilization of reference configurations and applies them to two example systems; this is a straightforward application of the PD control design methodology introduced in the previous chapter. Section 11.2 presents strategies for tracking of reference trajectories, and Section 11.3 contains two examples. The approach to trajectory tracking is based on the notions of tracking error function and velocity error function. Finally, Section 11.4 discusses stabilization and tracking results for systems on Lie groups. The attitude tracking problem for a fully actuated rigid body is the final example given in Section 11.4.3.

11.1 Configuration stabilization for fully actuated systems

In this section we consider the stabilization problem for equilibrium configurations of a fully actuated mechanical control system. We rely on the stability results in Section 6.2.4 to analyze the global stability of the closed-loop system induced by appropriate proportional control.

11.1.1 Stabilization via configuration error functions

We shall consider a fully actuated C^∞-simple mechanical control system with dissipation $(\mathsf{Q}, \mathbb{G}, V, F_{\mathrm{diss}}, \mathscr{F}, \mathbb{R}^m)$. Because the system is fully actuated, it is convenient to write its equation of motion as

$$\overset{\mathbb{G}}{\nabla}_{\gamma'(t)}\gamma'(t) = -\mathrm{grad}V(\gamma(t)) + \mathbb{G}^\sharp(F_{\mathrm{diss}}(\gamma'(t))) + \mathbb{G}^\sharp(F(t,\gamma'(t))), \qquad (11.1)$$

where the ***time-dependent state feedback force*** $F\colon \mathbb{R}\times\mathsf{TQ} \to \mathsf{T}^*\mathsf{Q}$ encodes the feedback control law. Given a time-dependent state feedback force F of class C^∞, one can define *a* time-dependent state feedback $u\colon \bar{\mathbb{R}}_+ \times \mathsf{TQ} \to \mathbb{R}^m$ of class C^∞ with the property that

$$F(t, v_q) = \sum_{a=1}^{m} F^a(q)u^a(t, v_q).$$

Remarks 11.1. 1. The choice of state feedback is in general not unique, since the covector fields in $\mathscr{F}$ might not be linearly independent at each $q \in \mathsf{Q}$.

2. The fact that this non-unique state feedback can be chosen of class C^∞ is a consequence of Theorem 3.114.
3. In what follows we shall also consider time-independent state feedback forces and correspondingly time-independent state feedbacks. •

Given a time-dependent state feedback F, the ***closed-loop system defined by F*** is the C^∞-forced simple mechanical system $(\mathsf{Q}, \mathbb{G}, V, F_{\mathrm{diss}} + F)$. The control objective for this section is the following.

Problem 11.2 (Configuration stabilization). Given a fully actuated simple mechanical control system with dissipation $(\mathsf{Q}, \mathbb{G}, V, F_{\mathrm{diss}}, \mathscr{F}, \mathbb{R}^m)$, and given a reference configuration $q_0 \in \mathsf{Q}$, design a time-independent state feedback force F such that q_0 is a locally asymptotically stable equilibrium configuration for the closed-loop system defined by F. •

We shall quickly provide a solution to this problem by exploiting the concepts introduced in the previous chapter and in Chapter 6. In particular, recall that Lemma 6.32 and Theorem 6.45 provide sufficient conditions for the existence and stability of equilibrium configurations in simple mechanical systems with dissipation. We start by introducing the notion of a configuration error function that will play the role of a potential function for the closed-loop system defined by an appropriate force.

Definition 11.3 (Configuration error function). A smooth function $\Psi\colon \mathsf{Q} \to \mathbb{R}$ is a ***configuration error function about $q_0 \in \mathsf{Q}$*** if it is proper, bounded from below, and if

(i) $\Psi(q_0) = 0$,
(ii) $\mathrm{d}\Psi(q_0) = 0_{q_0}$, and
(iii) Hess $\Psi(q_0)$ is positive-definite. •

Immediate consequences of the definition are that (1) a configuration error function Ψ is a locally positive-definite function about q_0, (2) the estimates obtained in Proposition 6.30 apply to Ψ, and (3) the sublevel sets of Ψ are compact.

We can now state the main result of this section.

Theorem 11.4 (Stabilization by PD control). *Consider a fully actuated C^∞-simple mechanical control system with dissipation $\Sigma = (\mathsf{Q}, \mathbb{G}, V, F_{\mathrm{diss}}, \mathscr{F}, \mathbb{R}^m)$. Let R_{diss} be a strict Rayleigh dissipation function and let $\Psi\colon \mathsf{Q} \to \mathbb{R}$ have the property that $V_{\mathrm{cl}} = (V + \Psi)$ is a configuration error function. Define the **proportional-derivative feedback force***

$$F_{\mathrm{PD}}(v_q) = -\mathrm{d}\Psi(q) - R^\flat_{\mathrm{diss}}(v_q). \tag{11.2}$$

Then the closed-loop system Σ_{cl} defined by F_{PD} has the following properties:

(i) Σ_{cl} is the C^∞-simple mechanical system with dissipation $(\mathsf{Q}, \mathbb{G}, V_{\mathrm{cl}}, F_{\mathrm{diss}} - R^\flat_{\mathrm{diss}})$ with energy function E_{cl} given by $v_q \mapsto \frac{1}{2}\|v_q\|^2_{\mathbb{G}} + V_{\mathrm{cl}}(q)$;
(ii) q_0 is a locally asymptotically stable equilibrium configuration for Σ_{cl};
(iii) if there exists $t_0 \in \bar{\mathbb{R}}_+$ such that the integral curve $\gamma\colon I \to \mathsf{Q}$ for Σ_{cl} satisfies $\gamma'(t_0) \in E^{-1}_{\mathrm{cl}}(\leq L, 0_{q_0})$ with $L < L_{\mathrm{reg}}(V_{\mathrm{cl}}, q_0)$, then γ is defined on $I \cup [t_0, +\infty[$ and γ' converges to 0_{q_0} exponentially fast;
(iv) given any initial condition $\gamma'(0) \in \mathsf{TQ}$, the resulting controlled trajectory takes values in the compact set $E^{-1}_{\mathrm{cl}}(\leq E_{\mathrm{cl}}(\gamma'(0)))$ and converges as $t \to +\infty$ to the set of critical points of V_{cl} in $V^{-1}_{\mathrm{cl}}(\leq E_{\mathrm{cl}}(\gamma'(0)))$.

The energy and the potential function of the closed-loop system defined by a proportional feedback force are called the ***closed-loop potential function*** and the ***closed-loop energy***, respectively.

Proof of Theorem 11.4. Part (i) is the statement in Remark 10.38–1. Part (ii) is a consequence of Theorem 6.45 (ii), since the closed-loop potential function V_{cl} has an isolated local minimum at q_0, and the closed-loop dissipation function $F_{\mathrm{diss}} - R^\flat_{\mathrm{diss}}$ is strictly dissipative. Part (iii) is a consequence of Theorem 6.45(iii), since V_{cl} is proper. Part (iv) is a consequence of Theorem 6.47, since V_{cl} is bounded from below and proper. ■

Let us add some comments on the theorem.

Remarks 11.5. 1. *Limitations on global configuration error functions:* In general, it is not possible to design globally stabilizing smooth feedback controllers for systems on manifolds because of topological properties of the configuration manifold; see Remark 10.2–7 and Exercise E6.2. The design of global error functions is discussed, for example, in [Koditschek 1989, Rimon and Koditschek 1991, 1992].

2. *Proportional and derivative control:* The proportional-derivative feedback force in equation (11.2) is the sum of a ***proportional feedback force*** $F_{\mathrm{P}} = -\mathrm{d}\Psi$ and a ***derivative feedback force*** $F_{\mathrm{D}} = -R^{\flat}_{\mathrm{diss}}$. •

In what follows, we apply the results of Section 11.1 to a few example simple mechanical control systems of the form $(\mathsf{Q}, \mathbb{G}, 0, \mathscr{F}, \mathbb{R}^m)$. Note that the design of configuration error functions depends only on the configuration manifold Q and on the target configuration q_0. In order to use Theorem 11.4, we shall compute all critical points and characterize the sublevel sets of Ψ.

11.1.2 PD control for a point mass in three-dimensional Euclidean space

We start by studying a simple example that we shall also refer to later. Consider a particle of mass m moving in three-dimensional Euclidean space subject to no gravitational field, no dissipative force, and with control forces along three independent reference axes. We let $\boldsymbol{x} = (x^1, x^2, x^3)$ denote the configuration of the particle. Specifically, this simple mechanical control system is described by the 4-tuple $(\mathbb{R}^3, m\mathbb{G}_{\mathbb{R}^3}, 0, \{\mathrm{d}x^1, \mathrm{d}x^2, \mathrm{d}x^3\})$. Relying on the identification $\mathsf{T}\mathbb{R}^3 \simeq \mathsf{T}^*\mathbb{R}^3 \simeq \mathbb{R}^3 \times \mathbb{R}^3$, the equations of motion (11.1) in the controlled trajectory $t \mapsto \boldsymbol{x}(t) \in \mathbb{R}^3$ are

$$\ddot{\boldsymbol{x}}(t) = \frac{1}{m} F(\boldsymbol{x}(t), \dot{\boldsymbol{x}}(t)), \tag{11.3}$$

where the feedback control force is a map $F\colon \mathbb{R}^3 \times \mathbb{R}^3 \to \mathbb{R}^3$.

We assume that the control objective is to locally asymptotically stabilize the particle to the configuration $\boldsymbol{r} \in \mathbb{R}^3$, and we design an appropriate configuration error function. Let $\boldsymbol{K}_{\mathrm{P}} \in \mathbb{R}^{3\times 3}$ be symmetric and positive-definite, define

$$\Psi_{\mathbb{R}^3}(\boldsymbol{x}) = \tfrac{1}{2}(\boldsymbol{x} - \boldsymbol{r})^T \boldsymbol{K}_{\mathrm{P}}(\boldsymbol{x} - \boldsymbol{r}) = \tfrac{1}{2}\left\|\boldsymbol{x} - \boldsymbol{r}\right\|^2_{\boldsymbol{K}_{\mathrm{P}}}, \tag{11.4}$$

and compute

$$\mathrm{d}\Psi_{\mathbb{R}^3}(\boldsymbol{x}) = \boldsymbol{K}_{\mathrm{P}}(\boldsymbol{x} - \boldsymbol{r}), \quad \operatorname{Hess}\Psi_{\mathbb{R}^3}(\boldsymbol{x}) = \boldsymbol{K}_{\mathrm{P}}.$$

The function $\Psi_{\mathbb{R}^3}\colon \mathbb{R}^3 \to \bar{\mathbb{R}}_+$ is smooth, proper, and bounded from below. Furthermore, $\mathrm{d}\Psi_{\mathbb{R}^3}(\boldsymbol{r})$ vanishes and $\operatorname{Hess}\Psi_{\mathbb{R}^3}(\boldsymbol{r})$ is positive-definite. According to Definition 11.3, $\Psi_{\mathbb{R}^3}$ is a configuration error function about $\boldsymbol{r} \in \mathbb{R}^3$.

Furthermore, $L_{\mathrm{reg}}(\Psi_{\mathbb{R}^3}, \boldsymbol{r}) = +\infty$ because $\Psi_{\mathbb{R}^3}$ is unbounded and has no critical point other than $\boldsymbol{r}$. Finally, we design a strict Rayleigh dissipation force as the map $\boldsymbol{v} \mapsto -\boldsymbol{K}_{\mathrm{D}}\boldsymbol{v}$ for a symmetric positive-definite $\boldsymbol{K}_{\mathrm{D}} \in \mathbb{R}^{3\times3}$.

According to Theorem 11.4, we define the proportional-derivative feedback control

$$F_{\mathrm{PD}}(\boldsymbol{x}, \boldsymbol{v}) = -\boldsymbol{K}_{\mathrm{P}}(\boldsymbol{x} - \boldsymbol{r}) - \boldsymbol{K}_{\mathrm{D}}\boldsymbol{v},$$

for $\boldsymbol{x}, \boldsymbol{v} \in \mathbb{R}^3$. We observe that the closed-loop system defined by F_{PD} has the stability properties described in Theorem 11.4: the configuration $\boldsymbol{r}$ is globally asymptotically stable with exponential convergence rate.

11.1.3 PD control for the spherical pendulum

As a second example system, we consider a spherical pendulum with mass m. This system on $\mathbb{S}^2$ is discussed in Examples 4.99 and 5.72. We assume that the control forces on the spherical pendulum generate $\mathsf{T}^*\mathbb{S}^2$, i.e., the system is fully actuated. As discussed in Remark 3.53, $\mathbb{S}^2$ is not parallelizable and a minimum of three inputs are required to achieve an arbitrary force globally. We denote these three covector fields by $\{F^1, F^2, F^3\}$, although we will not explicitly use them in our development. Also, we shall assume that the spherical pendulum is not subject to any potential force. In summary, this simple mechanical control system is described by the 4-tuple $(\mathbb{S}^2, m\mathbb{G}_{\mathrm{rnd}}, 0, \{F^1, F^2, F^3\})$.

As discussed in Example 4.99 we know that the affine connection $\overset{\mathbb{G}_{\mathrm{rnd}}}{\nabla}$ satisfies

$$\overset{\mathbb{G}_{\mathrm{rnd}}}{\nabla}_{\dot{\boldsymbol{x}}(t)}\dot{\boldsymbol{x}}(t) = \ddot{\boldsymbol{x}}(t) + \|\dot{\boldsymbol{x}}(t)\|^2_{\mathbb{R}^3}\boldsymbol{x}(t),$$

for all curves $t \mapsto \boldsymbol{x}(t) \in \mathbb{S}^2$. Relying on the identification $\mathsf{T}_{\boldsymbol{x}}\mathbb{S}^2 \simeq \mathsf{T}^*_{\boldsymbol{x}}\mathbb{S}^2 \simeq \operatorname{span}\{\boldsymbol{x}\}^{\perp}$, the equations of motion (11.1) for the controlled trajectory $t \mapsto \boldsymbol{x}(t) \in \mathbb{S}^2$ read

$$\overset{\mathbb{G}_{\mathrm{rnd}}}{\nabla}_{\dot{\boldsymbol{x}}(t)}\dot{\boldsymbol{x}}(t) = \frac{1}{m}F(\boldsymbol{x}(t), \dot{\boldsymbol{x}}(t)), \tag{11.5}$$

where the feedback control force $F\colon \mathbb{R}^3 \times \mathbb{R}^3 \to \mathbb{R}^3$ satisfies $F(\boldsymbol{x}, \boldsymbol{v}) \in \operatorname{span}\{\boldsymbol{x}\}^{\perp}$ for all $\boldsymbol{v} \in \mathbb{R}^3$. Finally, let us introduce some useful notation: for $\boldsymbol{x} \in \mathbb{S}^2$, let $P_{\boldsymbol{x}}\colon \mathbb{R}^3 \to \operatorname{span}\{\boldsymbol{x}\}^{\perp}$ denote the $\mathbb{G}_{\mathbb{R}^3}$-orthogonal projection onto $\operatorname{span}\{\boldsymbol{x}\}^{\perp}$. For $\boldsymbol{r} \in \mathbb{R}^3$, one can easily see that

$$P_{\boldsymbol{x}}(\boldsymbol{r}) = -\widehat{\boldsymbol{x}}^2\boldsymbol{r}. \tag{11.6}$$

Next, we design a configuration error function for systems defined on $\mathbb{S}^2$.

Lemma 11.6 (Configuration error function on $\mathbb{S}^2$). *For $\boldsymbol{r} \in \mathbb{S}^2$ and $k_{\mathrm{P}} \in \mathbb{R}_+$, define $\Psi_{\mathbb{S}^2}\colon \mathbb{S}^2 \to \bar{\mathbb{R}}_+$ by $\Psi_{\mathbb{S}^2}(\boldsymbol{x}) = k_{\mathrm{P}}(1 - \langle\langle \boldsymbol{r}, \boldsymbol{x}\rangle\rangle_{\mathbb{R}^3})$. Then*

(i) $\Psi_{\mathbb{S}^2}$ is a configuration error function about $\boldsymbol{r}$,
(ii) $\mathrm{d}\Psi_{\mathbb{S}^2}(\boldsymbol{x}) = k_{\mathrm{P}}\widehat{\boldsymbol{x}}^2\boldsymbol{r}$, and
(iii) the critical points of $\Psi_{\mathbb{S}^2}$ are $\{\boldsymbol{r}, -\boldsymbol{r}\}$, and $L_{\mathrm{reg}}(\Psi_{\mathbb{S}^2}, \boldsymbol{r}) = 2k_{\mathrm{P}}$.

Proof. Clearly, $\Psi_{\mathbb{S}^2}$ is smooth. $\Psi_{\mathbb{S}^2}$ is proper and bounded from below because $\mathbb{S}^2$ is compact. For $\boldsymbol{v} \in \mathsf{T}_{\boldsymbol{x}}\mathbb{S}^2 \simeq \operatorname{span}\{\boldsymbol{x}\}^{\perp}$, we have

$$\mathscr{L}_{\boldsymbol{v}}(1 - \langle\langle \boldsymbol{r}, \boldsymbol{x}\rangle\rangle_{\mathbb{R}^3}) = -\langle\langle \boldsymbol{r}, \boldsymbol{v}\rangle\rangle_{\mathbb{R}^3} = \langle\langle -P_{\boldsymbol{x}}(\boldsymbol{r}), \boldsymbol{v}\rangle\rangle_{\mathbb{R}^3}.$$

This proves (ii). It is immediate to compute that $\mathrm{d}\Psi_{\mathbb{S}^2}(\boldsymbol{x}) = 0$ if and only if $\boldsymbol{x} = \pm\boldsymbol{r}$. At the two critical points we have $\Psi_{\mathbb{S}^2}(\boldsymbol{r}) = 0$ and $\Psi_{\mathbb{S}^2}(-\boldsymbol{r}) = 2k_{\mathrm{P}}$. These fact prove (iii) (in other words, for all $L < 2k_{\mathrm{P}}$, the L-sublevel set of $\Psi_{\mathbb{S}^2}$ contains no critical point other than $\boldsymbol{r}$). To complete the proof of (i), we need to show that $\operatorname{Hess}\Psi_{\mathbb{S}^2}(\boldsymbol{r})$ is positive-definite. The reader can perform this computation in Exercise E11.1. ■

We are now ready to apply the results in Theorem 11.4 to systems on $\mathbb{S}^2$.

Lemma 11.7 (PD control on $\mathbb{S}^2$). *Consider the system in equation* (11.5). *Let $\boldsymbol{r} \in \mathbb{S}^2$, and let k_{P} and k_{D} be two positive constants. For $\boldsymbol{v} \in \mathsf{T}_{\boldsymbol{x}}\mathbb{S}^2$, define the proportional-derivative control law*

$$F_{\mathrm{PD}}(\boldsymbol{x}, \boldsymbol{v}) = -k_{\mathrm{P}}\widehat{\boldsymbol{x}}^2\boldsymbol{r} - k_{\mathrm{D}}\boldsymbol{v}.$$

The configuration $\boldsymbol{r}$ is locally asymptotically stable for the closed-loop system defined by F_{PD}, the controlled trajectories converge to either $\boldsymbol{r}$ or $-\boldsymbol{r}$, and the controlled trajectories converge to $\boldsymbol{r}$ with exponential convergence rate from all initial conditions $\boldsymbol{v} \in \mathsf{T}\mathbb{S}^2$ such that $2k_{\mathrm{P}}(1 + \langle\langle \boldsymbol{r}, \boldsymbol{x}\rangle\rangle_{\mathbb{R}^3}) > m\|\boldsymbol{v}\|^2_{\mathbb{R}^3}$.

Remark 11.8. This is an example for which global stabilization by PD control is not possible, due to the topological properties of $\mathbb{S}^2$. This does not quite follow from anything we have said thus far about $\mathbb{S}^2$, but follows from the fact that the Euler characteristic of $\mathbb{S}^2$ is 2. For more details on the Euler characteristic and related concepts, we refer the interested reader to [Munkres 1984]. •

11.2 Trajectory tracking for fully actuated systems

In this section we consider the trajectory tracking problem for a fully actuated simple mechanical control system. The extensions of the stabilizing proportional-derivative control to the tracking setting includes an appropriately defined feedforward control.

11.2.1 Time-dependent feedback control and the tracking problem

We concern ourselves with a fully actuated simple mechanical control system $\Sigma = (\mathsf{Q}, \mathbb{G}, V = 0, \mathscr{F}, \mathbb{R}^m)$ with equation of motion

$$\overset{\mathbb{G}}{\nabla}_{\gamma'(t)}\gamma'(t) = \mathbb{G}^{\sharp}(F(t, \gamma'(t))). \tag{11.7}$$

We define a first useful concept. A ***reference trajectory*** $\gamma_{\text{ref}}\colon \bar{\mathbb{R}}_+ \to \mathsf{Q}$ is a twice-differentiable curve with bounded velocity, in the sense that

$$\sup\left\{ \|\gamma'_{\text{ref}}(t)\|_{\mathbb{G}} \mid t \in \bar{\mathbb{R}}_+ \right\} < +\infty.$$

Roughly speaking, the control objective in this section is to design a time-dependent state feedback force such that the closed-loop controlled trajectory γ asymptotically tracks γ_{ref}. Although the notion of asymptotic tracking is undefined at this time, we can loosely say that the controlled trajectory asymptotically tracks a reference trajectory if the distance between the two curves approaches zero as time goes to infinity. We postpone a formal statement on this matter to Definition 11.18.

Remark 11.9. In the context of tracking problems, time-dependent state feedback forces have the control theoretic interpretation of *feedback plus feedforward* control forces in the sense that the time dependence in F is introduced by a dependence on the reference trajectory γ_{ref} and its derivatives. •

In the following sections, we study the geometric objects involved in the design of a tracking controller. These objects are extensions to the tracking problem of the tools introduced in Section 11.1 for the purpose of stabilization. To measure the distance between reference and actual configuration, we generalize the notion of a configuration error function. To measure the distance between reference and actual velocity, we introduce the notion of transport map. Since we shall require the notion of vector bundles and vector bundle maps, we refer the reader to Section 3.4.1 for the corresponding definitions.

11.2.2 Tracking error functions

To extend the definition of a configuration error function to the tracking setting, we consider functions on $\mathsf{Q}\times\mathsf{Q}$. A function $\Psi\colon \mathsf{Q}\times\mathsf{Q} \to \mathbb{R}$ is ***symmetric*** if $\Psi(q,r) = \Psi(r,q)$ for all $(q,r) \in \mathsf{Q}\times\mathsf{Q}$. For a symmetric $\Psi\colon \mathsf{Q}\times\mathsf{Q}\to\mathbb{R}$ and for $r \in \mathsf{Q}$, we define $\Psi_r\colon \mathsf{Q}\to\mathbb{R}$ by $\Psi_r(q) = \Psi(q,r)$.

To study function on $\mathsf{Q}\times\mathsf{Q}$, we introduce the two vector bundles $\pi_1\colon \mathsf{TQ}\times\mathsf{Q} \to \mathsf{Q}\times\mathsf{Q}$ and $\pi_2\colon \mathsf{Q}\times\mathsf{TQ} \to \mathsf{Q}\times\mathsf{Q}$. An element in $\mathsf{TQ}\times\mathsf{Q}$ is a pair (X_q, r), where $q, r \in \mathsf{Q}$, and $X_q \in \mathsf{T}_q\mathsf{Q}$. Similarly, (q, Y_r) belongs to $\mathsf{Q}\times\mathsf{TQ}$, where $Y_r \in \mathsf{T}_r\mathsf{Q}$. The vector bundle projections are

$$\pi_1(X_q, r) = (q,r), \quad \text{and} \quad \pi_2(q, Y_r) = (q,r).$$

Note the natural isomorphisms $\pi_2^{-1}(q,r) \simeq \mathsf{T}_r\mathsf{Q}$ and $\pi_1^{-1}(q,r) \simeq \mathsf{T}_q\mathsf{Q}$. We regard both vector bundles as subbundles of the tangent bundle of the product manifold $\mathsf{Q}\times\mathsf{Q}$, that is, as subbundles of $\pi_{\mathsf{T}(\mathsf{Q}\times\mathsf{Q})}\colon \mathsf{T}(\mathsf{Q}\times\mathsf{Q}) \to (\mathsf{Q}\times\mathsf{Q})$. Finally, we let $\mathsf{T}^*\mathsf{Q}\times\mathsf{Q}$ and $\mathsf{Q}\times\mathsf{T}^*\mathsf{Q}$ denote the corresponding dual vector bundles.

For a smooth, symmetric $\Psi\colon \mathsf{Q}\times\mathsf{Q}\to\mathbb{R}$, we define the ***differential with respect to first argument*** $\mathrm{d}_1\Psi \in \Gamma^\infty(\mathsf{T}^*\mathsf{Q}\times\mathsf{Q})$ and the ***differential with respect to second argument*** $\mathrm{d}_2\Psi \in \Gamma^\infty(\mathsf{Q}\times\mathsf{T}^*\mathsf{Q})$ by means of

$$d_1\Psi(q,r) = d\Psi_r(q) \in \mathsf{T}^*_q\mathsf{Q}, \quad \text{and} \quad d_2\Psi(q,r) = d\Psi_q(r) \in \mathsf{T}^*_r\mathsf{Q},$$

respectively. Similarly, we define the ***Hessian with respect to the first argument*** $\operatorname{Hess}_1 \Psi(q,r) = \operatorname{Hess} \Psi_r(q)$.

Definition 11.10 (Tracking error function). A smooth, symmetric function $\Psi\colon \mathsf{Q} \times \mathsf{Q} \to \mathbb{R}$ is a ***tracking error function*** if, for $r \in \mathsf{Q}$, the function $\Psi_r\colon \mathsf{Q} \to \mathbb{R}$ is a configuration error function about r. In other words, for all $r \in \mathsf{Q}$, Ψ_r is smooth, proper, and bounded from below, and Ψ satisfies

(i) $\Psi(r,r) = 0$,
(ii) $d_1\Psi(r,r) = 0_r$, and
(iii) $\operatorname{Hess}_1 \Psi(r,r)$ is positive-definite. •

Remark 11.11. A tracking error function Ψ is locally positive-definite about the set $\{(q,q) \in \mathsf{Q} \times \mathsf{Q} \mid q \in \mathsf{Q}\}$ in the following sense: for all (p,r) in a neighborhood of $\{(q,q) \in \mathsf{Q} \times \mathsf{Q} \mid q \in \mathsf{Q}\}$, we have $\Psi(p,r) \geq 0$ and $\Psi(p,r) = 0$ if only if $p = r$. •

For $r \in \mathsf{Q}$, the function $\Psi_r\colon q \mapsto \Psi(q,r)$ is a configuration error function, and Proposition 6.30 guarantees that Ψ_r is quadratic on $\Psi_r^{-1}(\leq L, r)$ for all $0 < L < L_{\text{reg}}(\Psi_r, r)$. Let us write explicitly the corresponding bound. For all $r \in \mathsf{Q}$ and for all $L \in]0, L_{\text{reg}}(\Psi_r, r)[$, there exist two constants $0 < b_1 \leq b_2$ such that, for all $q \in \Psi_r^{-1}(\leq L, r) \setminus \{r\}$, we have

$$0 < b_1 \|d_1\Psi(q,r)\|^2_{\mathbb{G}} \leq \Psi(q,r) \leq b_2 \|d_1\Psi(q,r)\|^2_{\mathbb{G}}.$$

Note that this bound is guaranteed to hold pointwise for all r, but *not uniformly*, on $r \in \mathsf{Q}$. It turns out that we shall require uniformity of this estimate. Therefore, let us define

$$L_{\text{reg}}(\Psi, \mathsf{Q}) = \inf \{ L_{\text{reg}}(\Psi_r, r) \mid r \in \mathsf{Q}\}.$$

Definition 11.12 (Uniformly quadratic function). A tracking error function Ψ is ***uniformly quadratic*** if $L_{\text{reg}}(\Psi, \mathsf{Q}) > 0$ and if, for all $L \in]0, L_{\text{reg}}(\Psi, \mathsf{Q})[$, there exist two constants $0 < b_1 \leq b_2$, independent of r, such that, for all $q \in \Psi_r^{-1}(\leq L, r) \setminus \{r\}$, we have

$$0 < b_1 \|d_1\Psi(q,r)\|^2_{\mathbb{G}} \leq \Psi(q,r) \leq b_2 \|d_1\Psi(q,r)\|^2_{\mathbb{G}}. \quad \bullet$$

11.2.3 Transport maps

Here we introduce a novel object that plays a critical role in control design for tracking problems. Let us start by discussing the need for a notion of velocity error, and, therefore, for an object that naturally compares velocity curves. Figure 11.1 illustrates how, assuming $\gamma(t_0) = q$ and $\gamma'_{\text{ref}}(t_0) = r$, the two velocity curves $\gamma'(t_0)$ and $\gamma'_{\text{ref}}(t_0)$ take values in different tangent spaces. To define a meaningful notion of error velocity, we need a formal manner of comparing tangent vectors living in $\mathsf{T}_r\mathsf{Q}$ and $\mathsf{T}_q\mathsf{Q}$, for arbitrary $(q,r) \in \mathsf{Q}\times\mathsf{Q}$. This is achieved via the notion of the transport map, which we now discuss.

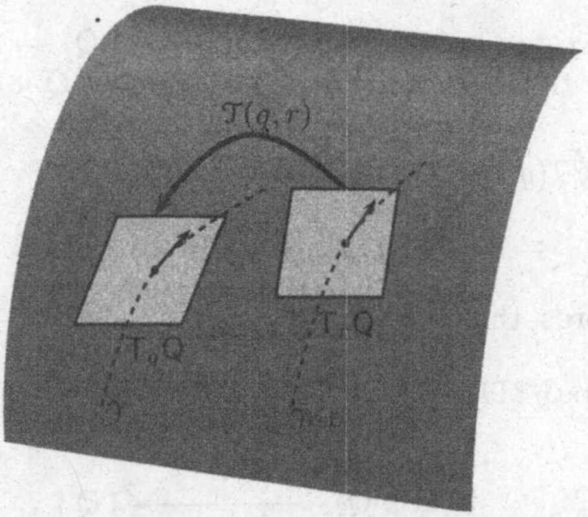

Figure 11.1. On the manifold Q we consider two curves γ and γ_{ref}; we assume that $\gamma(t_0) = q$ and $\gamma'_{\mathrm{ref}}(t_0) = r$. We introduce linear maps of the form $\mathcal{T}(q, r) \colon \mathsf{T}_r\mathsf{Q} \mapsto \mathsf{T}_q\mathsf{Q}$ in order to compare the two velocity curves γ' and γ'_{ref}.

Definition 11.13 (Transport map). A ***transport map*** is a smooth vector bundle map $\mathcal{T} \colon \mathsf{Q}\times\mathsf{TQ} \to \mathsf{TQ}\times\mathsf{Q}$ over $\mathrm{id}_{\mathsf{Q}\times\mathsf{Q}}$ with the property that $\mathcal{T}(q, X_q) = X_q$ for all $q \in \mathsf{Q}$ and $X_q \in \mathsf{T}_q\mathsf{Q}$. •

In equivalent words, the vector bundle map $\mathcal{T} \colon \mathsf{Q}\times\mathsf{TQ} \to \mathsf{TQ}\times\mathsf{Q}$ satisfies:

1. $\pi_1 \circ \mathcal{T} = \pi_2$, or equivalently, the following diagram commutes:

$$\begin{array}{ccc} \mathsf{Q}\times\mathsf{TQ} & \xrightarrow{\ \mathcal{T}\ } & \mathsf{TQ}\times\mathsf{Q} \\ {\scriptstyle \pi_2}\downarrow & & \downarrow{\scriptstyle \pi_1} \\ \mathsf{Q}\times\mathsf{Q} & \xrightarrow[\mathrm{id}_{\mathsf{Q}\times\mathsf{Q}}]{} & \mathsf{Q}\times\mathsf{Q} \end{array}$$

2. $\mathcal{T}|\pi_2^{-1}(q, r)$ is a linear map from $\pi_2^{-1}(q, r) \simeq \mathsf{T}_r\mathsf{Q}$ onto $\pi_1^{-1}(q, r) \simeq \mathsf{T}_q\mathsf{Q}$ with the property that $\mathcal{T}|\pi_2^{-1}(q, q) = \mathrm{id}_{\mathsf{T}_q\mathsf{Q}}$.

For simplicity, we shall write $\mathcal{T}|\pi_2^{-1}(q, r)$ as $\mathcal{T}(q, r) \colon \mathsf{T}_r\mathsf{Q} \to \mathsf{T}_q\mathsf{Q}$, so that, for all $(q, Y_r) \in \mathsf{Q}\times\mathsf{TQ}$, we write

$$\mathcal{T}(q, Y_r) = (\mathcal{T}(q, r) \cdot Y_r, r).$$

Note that, at fixed $r \in \mathsf{Q}$, the map $q \mapsto \mathcal{T}(q, r) \cdot Y_r$ is a smooth vector field on Q taking the value Y_r at $q = r$. Because this vector field has a well-defined covariant derivative $q \mapsto \overset{\mathrm{G}}{\nabla}_{X_q}(\mathcal{T}(q, r) \cdot Y_r)$, it is possible to define a notion of covariant derivative for the transport map $\mathcal{T}$.

Definition 11.14 (Covariant derivative of transport map). The ***covariant derivative*** of a transport map $\mathcal{T} \colon \mathsf{Q}\times\mathsf{TQ} \to \mathsf{TQ}\times\mathsf{Q}$ with respect to $\overset{G}{\nabla}$ is a smooth vector bundle map $\overset{G}{\nabla}\mathcal{T}\colon \mathsf{T}(\mathsf{Q}\times\mathsf{Q}) \to \mathsf{TQ}$ over the map $\mathrm{pr}_1\colon (q,r) \mapsto q$ with the property that, for all $(q,r) \in \mathsf{Q}\times\mathsf{Q}$,

$$\begin{aligned}\overset{G}{\nabla}\mathcal{T}(q,r)\colon\ &\mathsf{T}_{(q,r)}(\mathsf{Q}\times\mathsf{Q}) \to \mathsf{T}_q\mathsf{Q}\\ &(X_q, Y_r) \mapsto \overset{G}{\nabla}_{X_q}(\mathcal{T}(q,r)\cdot Y_r).\end{aligned}$$ •

In equivalent words, the vector bundle map $\mathcal{T}\colon \mathsf{Q}\times\mathsf{TQ} \to \mathsf{TQ}\times\mathsf{Q}$ satisfies:

1. $\pi_{\mathsf{TQ}} \circ \overset{G}{\nabla}\mathcal{T} = \pi_{\mathsf{T}(\mathsf{Q}\times\mathsf{Q})} \circ \mathrm{pr}_1$, or equivalently, the following diagram commutes

$$\begin{array}{ccc}\mathsf{T}(\mathsf{Q}\times\mathsf{Q}) & \xrightarrow{\ \overset{G}{\nabla}\mathcal{T}\ } & \mathsf{TQ}\\ \Big\downarrow{\scriptstyle \pi_{\mathsf{T}(\mathsf{Q}\times\mathsf{Q})}} & & \Big\downarrow{\scriptstyle \pi_{\mathsf{TQ}}}\\ \mathsf{Q}\times\mathsf{Q} & \xrightarrow[\ \mathrm{pr}_1\ ]{} & \mathsf{Q}\end{array}$$

2. for $(q,r) \in \mathsf{Q}\times\mathsf{Q}$, the map $\overset{G}{\nabla}\mathcal{T}|\pi_2^{-1}(q,r)$ is a linear map from $\pi_{\mathsf{T}(\mathsf{Q}\times\mathsf{Q})}^{-1}(q,r)$ onto $\pi_{\mathsf{TQ}}^{-1}(q)$.

Next, we present coordinate expressions for the transport map and its covariant derivative. Let $\frac{\partial}{\partial q^k}$, $k \in \{1,\dots,n\}$, be a basis for $\mathsf{T}_q\mathsf{Q}$ and let $\frac{\partial}{\partial r^\alpha}$, $\alpha \in \{1,\dots,n\}$, be a basis for $\mathsf{T}_r\mathsf{Q}$. Given a tangent vector $Y \in \mathsf{T}_r\mathsf{Q}$ with components Y^α, $\alpha \in \{1,\dots,n\}$, we then have

$$\mathcal{T}(q,r)(Y) = \mathcal{T}^k_\alpha Y^\alpha \frac{\partial}{\partial q^k},$$

and given a tangent vector $X \in \mathsf{T}_q\mathsf{Q}$ with components X^j, $j \in \{1,\dots,n\}$, we write

$$\overset{G}{\nabla}\mathcal{T}(Y,X) = (\overset{G}{\nabla}\mathcal{T})^k_{\alpha j} Y^\alpha X^j \frac{\partial}{\partial q^k},$$

where, if $\overset{G}{\Gamma}{}^k_{ij}$, $i,j,k \in \{1,\dots,n\}$, are the Christoffel symbols of $\overset{G}{\nabla}$, we then have

$$(\overset{G}{\nabla}\mathcal{T})^k_{\alpha j} = \frac{\partial \mathcal{T}^k_\alpha}{\partial q^j} + \overset{G}{\Gamma}{}^k_{ij}\mathcal{T}^i_\alpha. \tag{11.8}$$

11.2.4 Velocity error curves

Here, we bring together the notions of tracking error function and of transport map to define a third element of the control design.

Definition 11.15 (Compatibility of tracking error function and transport map). Let Ψ and $\mathcal{T}$ be a tracking error function and a transport map on the manifold Q, respectively. The pair $(\Psi, \mathcal{T})$ is ***compatible*** if, for all $(q,r) \in \mathsf{Q}\times\mathsf{Q}$,

$$d_2\Psi(q,r) = -\mathcal{T}(q,r)^*(d_1\Psi(q,r)), \tag{11.9}$$

where $\mathcal{T}(q,r)^* : \mathsf{T}^*_q\mathsf{Q} \to \mathsf{T}^*_r\mathsf{Q}$ is the dual of $\mathcal{T}(q,r)$. •

We shall interchangeably say that the pair $(\Psi, \mathcal{T})$ is compatible, that Ψ is compatible with $\mathcal{T}$, or that $\mathcal{T}$ is compatible with Ψ. An equivalent characterization of compatibility is the following equality. For all $(X_q, Y_r) \in \mathsf{T}_{(q,r)}(\mathsf{Q}\times\mathsf{Q})$, a compatible pair $(\Psi, \mathcal{T})$ satisfies

$$\begin{aligned}\mathscr{L}_{(X_q,Y_r)}\Psi(q,r) &= \langle d_1\Psi(q,r); X_q\rangle + \langle d_2\Psi(q,r); Y_r\rangle \\ &= \mathscr{L}_{(X_q-\mathcal{T}(q,r)\cdot Y_r,0)}\Psi(q,r).\end{aligned}$$

The compatibility condition (11.9) helps define a notion of error between velocity curves. The following result sheds light on its meaning.

Lemma 11.16 (Time derivative of tracking error function). *Let Ψ be a tracking error function and $\mathcal{T}$ be a transport map. The pair $(\Psi, \mathcal{T})$ is compatible if and only if, for any two C^1-curves $\gamma, \eta : \bar{\mathbb{R}}_+ \to \mathsf{Q}$, we have*

$$\frac{d}{dt}\Psi(\gamma(t),\eta(t)) = \langle d_1\Psi(\gamma(t),\eta(t)); \dot{e}(t)\rangle, \tag{11.10}$$

where $\dot{e} : \bar{\mathbb{R}}_+ \to \mathsf{TQ}$ is the vector field along γ given by

$$\dot{e}(t) = \gamma'(t) - \mathcal{T}(\gamma(t),\eta(t)) \cdot \eta'(t). \tag{11.11}$$

Proof. Let us first prove that compatibility between Ψ and $\mathcal{T}$ implies (11.10). Applying the compatibility condition (11.9), we have

$$\begin{aligned}\frac{d}{dt}\Psi(\gamma(t),\eta(t)) &= \langle d_1\Psi(\gamma(t),\eta(t)); \gamma'(t)\rangle + \langle d_2\Psi(\gamma(t),\eta(t)); \eta'(t)\rangle \\ &= \langle d_1\Psi(\gamma(t),\eta(t)); \gamma'(t)\rangle \\ &\quad + \big\langle -\mathcal{T}(\gamma(t),\eta(t))^*(d_1\Psi(\gamma(t),\eta(t))); \eta'(t)\big\rangle \\ &= \langle d_1\Psi(\gamma(t),\eta(t)); \gamma'(t) - \mathcal{T}(\gamma(t),\eta(t))\cdot\eta'(t)\rangle .\end{aligned}$$

Next, we prove that the holding of equality (11.10) for every pair of C^1-curves γ and η implies compatibility between Ψ and $\mathcal{T}$. Let γ be the trivial curve $t \mapsto q$. Because of equality (11.10), we can write

$$\langle d_2\Psi(q,\eta(t)); \eta'(t)\rangle = \frac{d}{dt}\Psi(q,\eta(t)) = \langle d_1\Psi(q,\eta(t)); -\mathcal{T}(q,\eta(t))\cdot\eta'(t)\rangle$$

and, therefore,

$$\langle d_2\Psi(q,\eta(t)); \eta'(t)\rangle = \langle -\mathcal{T}(q,\eta(t))^*(d_1\Psi(q,\eta(t))); \eta'(t)\rangle ,$$

for all C^1-curves $\eta : \bar{\mathbb{R}}_+ \to \mathsf{Q}$. ■

We refer to the vector field $\dot{e}\colon \bar{\mathbb{R}}_+ \to \mathsf{TQ}$ along γ defined in (11.11) as the ***velocity error between*** γ ***and*** η. Note the abuse of terminology: the velocity error curve $\dot{e}\colon \bar{\mathbb{R}}_+ \to \mathsf{TQ}$ is not necessarily the velocity curve of any curve on Q.

Remark 11.17. It is not obvious that compatible transport maps exist for any tracking error function on any manifold. Later in the chapter we will provide compatible pairs for a few instructive examples; however, the general question of existence of such compatible pairs remains open. •

11.2.5 Proportional-derivative and feedforward control

Let us start by introducing notions of stability and asymptotic stability for tracking problems on manifolds. Given a compatible pair $(\Psi, \mathcal{T})$, we define the ***error function*** $E_{\mathrm{cl}}\colon \mathsf{TQ} \times \mathsf{TQ} \to \mathbb{R}$ by means of

$$E_{\mathrm{cl}}(v_q, w_r) = \Psi(q,r) + \tfrac{1}{2}\left\|v_q - \mathcal{T}(q,r)\cdot w_r\right\|_{\mathbb{G}}^2. \tag{11.12}$$

Definition 11.18 (Stability of reference trajectory). Given a reference trajectory γ_{ref} and a controlled trajectory γ, let $t \mapsto E_{\mathrm{cl}}(t)$ denote $t \mapsto E_{\mathrm{cl}}(\gamma'(t), \gamma'_{\mathrm{ref}}(t))$. We say that

(i) the curve γ_{ref} is ***stable with respect to the error function*** $\boldsymbol{E_{\mathrm{cl}}}$ if there exists a neighborhood $\mathcal{U}$ of $\gamma'_{\mathrm{ref}}(0)$ such that, for all initial conditions $\gamma'(0) \in \mathcal{U}$, the error function $t \mapsto E_{\mathrm{cl}}(t)$ is nonincreasing;

(ii) the curve γ_{ref} is ***locally exponentially stable with respect to the error function*** $\boldsymbol{E_{\mathrm{cl}}}$ if it is stable with respect to E_{cl}, and if there exist a neighborhood $\mathcal{U}$ of $\gamma'_{\mathrm{ref}}(0)$ and two positive constants α, β such that $E_{\mathrm{cl}}(t) \le \alpha E_{\mathrm{cl}}(0)\mathrm{e}^{-\beta t}$, from all initial conditions $\gamma'(0) \in \mathcal{U}$ and for all $t \in \bar{\mathbb{R}}_+$. •

Finally, recall that we call a tensor field s on $\mathsf{Q} \times \mathsf{Q}$ ***uniformly bounded*** on $A \subset \mathsf{Q} \times \mathsf{Q}$ if there exists a constant M such that $\sup\{\|s(q,r)\|_{\mathbb{G}} \mid (q,r) \in A\} \le M$.

We are now ready to state the main result of this section.

Theorem 11.19 (Trajectory tracking using PD and feedforward control). *Consider a fully actuated C^∞-simple mechanical control system $(\mathsf{Q}, \mathbb{G}, V = 0, \mathscr{F}, \mathbb{R}^m)$ with equation of motion* (11.1) *in the controlled trajectory $\gamma\colon \bar{\mathbb{R}}_+ \to \mathsf{Q}$. Let $\gamma_{\mathrm{ref}}\colon \bar{\mathbb{R}}_+ \to \mathsf{Q}$ be a twice-differentiable reference trajectory with bounded velocity. Let Ψ be a tracking error function, $\mathcal{T}$ be a compatible transport map, and R_{diss} be a strict Rayleigh dissipation function. Define the proportional-derivative and feedforward control forces*

$$F_{\mathrm{PD}}(t, v_q) = -\,\mathrm{d}_1\Psi(q, \gamma_{\mathrm{ref}}(t)) - R^\flat_{\mathrm{diss}}(v_q - \mathcal{T}(q, \gamma_{\mathrm{ref}}(t))\cdot\gamma'_{\mathrm{ref}}(t)),$$

$$F_{\mathrm{FF}}(t, v_q) = \mathbb{G}^\flat\Big(\overset{\mathbb{G}}{\nabla}\mathcal{T}(q, \gamma_{\mathrm{ref}}(t))\cdot(v_q, \gamma'_{\mathrm{ref}}(t)) + \frac{\mathrm{d}}{\mathrm{d}t}\left(\mathcal{T}(q, \gamma_{\mathrm{ref}}(t))\cdot\gamma'_{\mathrm{ref}}(t)\right)\Big).$$

Then the closed-loop system defined by $F_{\mathrm{PD}} + F_{\mathrm{FF}}$ has the following properties:

(i) if there exists $t_0 \in \bar{\mathbb{R}}_+$ *such that* $\gamma'(t_0) = \gamma'_{\text{ref}}(t_0)$*, then* γ *and* γ_{ref} *are identical;*

(ii) the curve γ_{ref} *is stable with respect to the error function* E_{cl}*.*

Assume that Ψ *is uniformly quadratic, that at* $t_0 \in \bar{\mathbb{R}}_+$ *the energy level* $L = E_{\text{cl}}(t_0)$ *is strictly less that* $L_{\text{reg}}(\Psi, \mathsf{Q})$*, and that the controlled trajectory* γ *satisfies* $\gamma'(t_0) \in \Psi^{-1}_{\gamma_{\text{ref}}(t_0)}(\leq L, \gamma_{\text{ref}}(t_0))$*. Also, assume that* R_{diss} *is uniformly bounded from below, and that* $\text{Hess}_1 \Psi$*,* R_{diss}*, and* $\overset{\mathbb{G}}{\nabla}\mathcal{T}$ *are uniformly bounded from above on* $\Psi^{-1}(\leq L)$*. Then*

(iii) the trajectory γ_{ref} *is locally exponentially stable with respect to the error function* E_{cl}*.*

Proof. See Section B.5. ■

We conclude this section with some comments.

Remarks 11.20. 1. *Global versus local:* The design process and the stability results are global in the sense that the reference trajectory γ_{ref} is not restricted to any subset of Q. However, these results are only local with respect to the controlled trajectory γ in the sense that the error function $t \mapsto \Psi(\gamma(t), \gamma_{\text{ref}}(t))$ must remain smaller than the parameter L.

2. *On the design of error functions and transport maps:* Theorem 11.19 achieves stability and exponential stability with respect to the particular energy E_{cl} we synthesized. Therefore, the design of an error function and a transport map plays a central role in imposing performance requirements on the closed-loop system defined by the corresponding proportional-derivative and feedforward control forces. Similarly, the choice of $(\Psi, \mathcal{T})$ determines the control algorithm complexity. For example, one particular transport map might be desirable, since it generates an easily implementable velocity error and feedforward control. It is interesting to note that far more literature exists on the design of error functions than on the design of transport maps.

3. *On the design of error functions:* A useful feature for the design of an error function is the availability of matrix, rather than scalar, gains as a way of enforcing different weighting on different "error directions." A weighted error function for the matrix Lie group SO(3) is presented in Section 11.4.3.

4. *Compatibility condition:* The constraint on the set of admissible pairs $(\Psi, \mathcal{T})$ is given by the compatibility condition

$$d_2\Psi(q, r) = -\mathcal{T}(q, r)^*(d_1\Psi(q, r)).$$

This equation can be examined from two points of view. Given a transport map $\mathcal{T}$, a compatible error function can be computed via a set of partial differential equations. Alternatively, given an error function Ψ, a compatible transport map must satisfy (1) a pointwise underdetermined set of linear algebraic equations and (2) the smoothness requirement.

5. *The feedforward control force:* Let γ be the controlled trajectory for the closed-loop system defined by $F_{\mathrm{PD}} + F_{\mathrm{FF}}$. An intermediate result in the proof of Theorem 11.19, see Section B.5, is that

$$\overset{\mathbb{G}}{\nabla}_{\gamma'(t)}\dot{e}(t) = -\mathbb{G}^{\sharp}\big(F_{\mathrm{PD}}(t,\gamma'(t))\big).$$

This equality provides an interpretation of the feedforward control force F_{FF}. The closed-loop system defined by F_{FF} has the property that the quantity $\overset{\mathbb{G}}{\nabla}_{\gamma'(t)}\dot{e}(t)$ vanishes along the closed-loop trajectory γ; in particular, if $\dot{e}$ is zero at initial time, then it remains zero for all subsequent times.

6. *Simple mechanical control systems with dissipation:* Theorem 11.19 is stated for simple mechanical systems with $V = 0$ and no dissipative forces. It can be generalized to simple mechanical control systems with dissipation. The control strategy is based on compensating for these potential and dissipative forces by means of a feedback transformation. For fully actuated systems, this is always possible. •

11.3 Examples illustrating trajectory tracking results

In this section we present two example designs of tracking controllers for the same mechanical control systems as in Section 11.1, i.e., a point mass in three-dimensional Euclidean space and a spherical pendulum.

Before proceeding, let us summarize the design procedure. Clearly, we plan to design a tracking error function Ψ and a transport map $\mathcal{T}$ for each example system. Furthermore, in order to apply Theorem 11.19, we need to (1) show that Ψ is a uniformly quadratic configuration error function and compute the corresponding $L_{\mathrm{reg}}(\Psi, \mathrm{Q})$, (2) show that $\mathcal{T}$ is compatible and compute its covariant derivative, and (3) establish bounds on appropriate sublevel sets of Ψ for the maps $\mathrm{Hess}_1\,\Psi$, R_{diss}, and $\overset{\mathbb{G}}{\nabla}\mathcal{T}$.

11.3.1 PD and feedforward control for a point mass in three-dimensional Euclidean space

Consider the example system in Section 11.1.2 described by the 4-tuple $(\mathbb{R}^3, m\mathbb{G}_{\mathbb{R}^3}, 0, \{\mathrm{d}x^1, \mathrm{d}x^2, \mathrm{d}x^3\})$. It is convenient to reproduce here the governing equations (11.3) for the controlled trajectory $t \mapsto \boldsymbol{x}(t)$:

$$\ddot{\boldsymbol{x}}(t) = \frac{1}{m}F(\boldsymbol{x}(t), \dot{\boldsymbol{x}}(t)).$$

With a slight abuse of notation we define the tracking error function

$$\Psi_{\mathbb{R}^3}(\boldsymbol{x},\boldsymbol{r}) = \tfrac{1}{2}\,\|\boldsymbol{x}-\boldsymbol{r}\|^2_{\boldsymbol{K}_{\mathrm{P}}}$$

and compute

$$\mathrm{d}_1\Psi_{\mathbb{R}^3}(\boldsymbol{x},\boldsymbol{r}) = \boldsymbol{K}_{\mathrm{P}}(\boldsymbol{x}-\boldsymbol{r}), \quad \text{and} \quad \mathrm{d}_2\Psi_{\mathbb{R}^3}(\boldsymbol{x},\boldsymbol{r}) = -\boldsymbol{K}_{\mathrm{P}}(\boldsymbol{x}-\boldsymbol{r}).$$

Next we define a transport map $\mathcal{T}_{\mathbb{R}^3}\colon \mathbb{R}^3 \times \mathsf{T}\mathbb{R}^3 \to \mathsf{T}\mathbb{R}^3 \times \mathbb{R}^3$ over the identity. Because of the identification $\mathsf{T}\mathbb{R}^3 \simeq \mathbb{R}^3 \times \mathbb{R}^3$, any vector $\boldsymbol{v} \in \mathsf{T}_{\boldsymbol{r}}\mathbb{R}^3$ can be naturally thought of as belonging to $\mathsf{T}_{\boldsymbol{x}}\mathbb{R}^3$ for any other $\boldsymbol{x} \in \mathbb{R}^3$. We use this identification to define $\mathcal{T}_{\mathbb{R}^3}(\boldsymbol{x},\boldsymbol{r}) = \mathrm{id}_{\mathbb{R}^3}$.

Let $t \mapsto \boldsymbol{r}(t)$ be a twice-differentiable reference trajectory with bounded velocity. Let $\boldsymbol{K}_{\mathrm{D}} \in \mathbb{R}^{3\times 3}$ be a symmetric positive-definite matrix as in Section 11.1.2. For $\boldsymbol{x} \in \mathbb{R}^3$ and $\boldsymbol{v} \in \mathsf{T}_{\boldsymbol{x}}\mathbb{R}^3$, the closed-loop energy function is

$$E_{\mathrm{cl}}(t,\boldsymbol{x},\boldsymbol{v}) = \tfrac{1}{2}\,\|\boldsymbol{x}-\boldsymbol{r}\|^2_{\boldsymbol{K}_{\mathrm{P}}} + \tfrac{1}{2}m\|\boldsymbol{v}-\dot{\boldsymbol{r}}(t)\|^2_{\mathbb{R}^3},$$

and the proportional-derivative and feedforward control forces are

$$\begin{aligned} F_{\mathrm{PD}}(t,\boldsymbol{x},\boldsymbol{v}) &= -\boldsymbol{K}_{\mathrm{P}}(\boldsymbol{x}-\boldsymbol{r}(t)) - \boldsymbol{K}_{\mathrm{D}}\left(\boldsymbol{v}-\dot{\boldsymbol{r}}(t)\right), \\ F_{\mathrm{FF}}(t,\boldsymbol{x},\boldsymbol{v}) &= m\ddot{\boldsymbol{r}}(t). \end{aligned}$$

Lemma 11.21. *For the closed-loop system defined by $F_{\mathrm{PD}} + F_{\mathrm{FF}}$, the trajectory $t \mapsto \boldsymbol{r}(t)$ is locally exponentially stable with respect to E_{cl}.*

Proof. The result is a consequence of Theorem 11.19 after establishing the following facts. First, let us note that $\Psi_{\mathbb{R}^3}$ is uniformly quadratic because

$$\lambda_{\min}(\boldsymbol{K}_{\mathrm{P}})\Psi_{\mathbb{R}^3}(\boldsymbol{x},\boldsymbol{r}) \leq \|\boldsymbol{K}_{\mathrm{P}}(\boldsymbol{x}-\boldsymbol{r})\|^2_{\mathbb{R}^3} \leq \lambda_{\max}(\boldsymbol{K}_{\mathrm{P}})\Psi_{\mathbb{R}^3}(\boldsymbol{x},\boldsymbol{r})$$

for all $(\boldsymbol{x},\boldsymbol{r}) \in \mathbb{R}^3 \times \mathbb{R}^3$. Furthermore, we compute $L_{\mathrm{reg}}(\Psi_{\mathbb{R}^3},\mathbb{R}^3)$ to be $+\infty$ because, at fixed $\boldsymbol{r}$, $\Psi_{\mathbb{R}^3}$ has no critical points other $\boldsymbol{r}$, and its sublevel sets are compact. Second, we deduce that $\mathcal{T}_{\mathbb{R}^3}$ is compatible with $\Psi_{\mathbb{R}^3}$ from the equality $\mathrm{d}_2\Psi_{\mathbb{R}^3}(\boldsymbol{x},\boldsymbol{r}) = -\mathrm{d}_1\Psi_{\mathbb{R}^3}(\boldsymbol{x},\boldsymbol{r})$. Furthermore, we study the covariant derivative of the transport map computed with respect to the Levi-Civita connection $\overset{\mathbb{G}_{\mathbb{R}^3}}{\nabla}$ induced by $\mathbb{G}_{\mathbb{R}^3}$. From equation (11.8), all components of $\overset{\mathbb{G}_{\mathbb{R}^3}}{\nabla}\mathcal{T}_{\mathbb{R}^3}$ vanish because $(\mathcal{T}_{\mathbb{R}^3})^k_\alpha$ are constant functions and because all Christoffel symbols of $\overset{\mathbb{G}_{\mathbb{R}^3}}{\nabla}$ vanish. Accordingly, $\overset{\mathbb{G}_{\mathbb{R}^3}}{\nabla}\mathcal{T}_{\mathbb{R}^3}$ is the trivial function mapping any argument onto the zero vector. As the third and final step, we establish all the required bounds. This is immediate, since the constant tensors $\boldsymbol{K}_{\mathrm{D}}$ and $\mathrm{Hess}_1\,\Psi_{\mathbb{R}^3} = \boldsymbol{K}_{\mathrm{P}}$ are uniformly bounded from above and below. ■

We conclude by noting that, in addition to the local stability properties guaranteed by Theorem 11.19, the trajectory $t \mapsto \boldsymbol{r}(t)$ is exponentially stable from all initial conditions.

11.3.2 PD and feedforward control for the spherical pendulum

Consider the example system in Section 11.1.3 described by the 4-tuple $(\mathbb{S}^2, m\mathbb{G}_{\mathrm{rnd}}, 0, \{F^1, F^2, F^3\})$. It is convenient to reproduce here the governing equations (11.5) for the controlled trajectory $t \mapsto \boldsymbol{x}(t)$:

$$\overset{\mathrm{G_{rnd}}}{\nabla}_{\dot{\boldsymbol{x}}(t)}\dot{\boldsymbol{x}}(t) = \frac{1}{m}F(\boldsymbol{x}(t), \dot{\boldsymbol{x}}(t)).$$

With a slight abuse of notation we define the tracking error function

$$\Psi_{\mathbb{S}^2}(\boldsymbol{x}, \boldsymbol{r}) = k_{\mathrm{P}}(1 - \langle\!\langle \boldsymbol{r}, \boldsymbol{x}\rangle\!\rangle_{\mathbb{R}^3}),$$

and compute

$$\mathrm{d}_1\Psi_{\mathbb{S}^2}(\boldsymbol{x}, \boldsymbol{r}) = k_{\mathrm{P}}\widehat{\boldsymbol{x}}^2\boldsymbol{r}, \quad \text{and} \quad \mathrm{d}_2\Psi_{\mathbb{S}^2}(\boldsymbol{x}, \boldsymbol{r}) = k_{\mathrm{P}}\widehat{\boldsymbol{r}}^2\boldsymbol{x}.$$

Lemma 11.22. *Consider the transport map* $\mathcal{T}_{\mathbb{S}^2} : \mathbb{S}^2 \times \mathsf{T}\mathbb{S}^2 \to \mathsf{T}\mathbb{S}^2 \times \mathbb{S}^2$ *defined by* $\mathcal{T}_{\mathbb{S}^2}(\boldsymbol{x}, \boldsymbol{r}) = \langle\!\langle \boldsymbol{r}, \boldsymbol{x}\rangle\!\rangle_{\mathbb{R}^3} \boldsymbol{I}_3 + (\boldsymbol{r} \times \boldsymbol{x})\widehat{\ }$. *Then*
(i) $\mathcal{T}_{\mathbb{S}^2}$ *is compatible with* $\Psi_{\mathbb{S}^2}$, *and*
(ii) $\mathcal{T}_{\mathbb{S}^2}(\boldsymbol{x}, \boldsymbol{r}) \cdot \boldsymbol{v} = (\boldsymbol{r} \times \boldsymbol{v}) \times \boldsymbol{x}$ *for all* $\boldsymbol{v} \in \mathsf{T}_{\boldsymbol{r}}\mathbb{S}^2$.

We illustrate the transport map $\mathcal{T}_{\mathbb{S}^2}$ in Figure 11.2. Given a fixed reference velocity vector $\boldsymbol{v} = (0, \frac{1}{5}, 0) \in \mathsf{T}_{\boldsymbol{r}}\mathbb{S}^2$, for $\boldsymbol{r} = (0, 0, 1)$, we depict the vector field $\boldsymbol{x} \mapsto \mathcal{T}_{\mathbb{S}^2}(\boldsymbol{x}, \boldsymbol{r}) \cdot \boldsymbol{v}$.

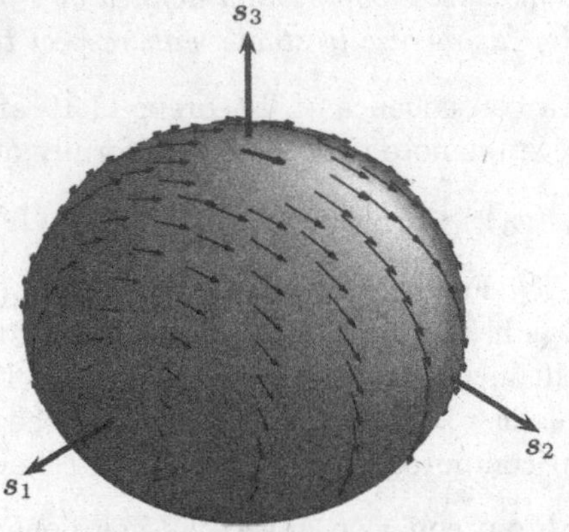

Figure 11.2. Transport map on $\mathbb{S}^2$

Proof of Lemma 11.22. Both facts are consequences of the equality $(\boldsymbol{x} \times \boldsymbol{y}) \times \boldsymbol{z} = (\boldsymbol{z}^T\boldsymbol{x})\boldsymbol{y} - (\boldsymbol{z}^T\boldsymbol{y})\boldsymbol{x}$ for all $\boldsymbol{x}, \boldsymbol{y}, \boldsymbol{z} \in \mathbb{R}^3$. First, we compute

$$\begin{aligned}
&\mathcal{T}_{\mathbb{S}^2}(\boldsymbol{x}, \boldsymbol{r})^*(\mathrm{d}_1\Psi_{\mathbb{S}^2}(\boldsymbol{x}, \boldsymbol{r}) \\
&\quad = -\left((\boldsymbol{r}^T\boldsymbol{x})\boldsymbol{I}_3 + \widehat{(\boldsymbol{r} \times \boldsymbol{x})}\right)^T \left(\boldsymbol{r} - (\boldsymbol{r}^T\boldsymbol{x})\boldsymbol{x}\right) \\
&\quad = -(\boldsymbol{r}^T\boldsymbol{x})\boldsymbol{r} + (\boldsymbol{r}^T\boldsymbol{x})^2\boldsymbol{x} - (\boldsymbol{r} - (\boldsymbol{r}^T\boldsymbol{x})\boldsymbol{x}) \times (\boldsymbol{r} \times \boldsymbol{x}) \\
&\quad = -(\boldsymbol{r}^T\boldsymbol{x})\boldsymbol{r} + (\boldsymbol{r}^T\boldsymbol{x})^2\boldsymbol{x} - \left((\boldsymbol{r} - (\boldsymbol{r}^T\boldsymbol{x})\boldsymbol{x})^T\boldsymbol{x}\right)\boldsymbol{r} - \left((\boldsymbol{r} - (\boldsymbol{r}^T\boldsymbol{x})\boldsymbol{x})^T\boldsymbol{r}\right)\boldsymbol{x} \\
&\quad = -(\boldsymbol{r}^T\boldsymbol{x})\boldsymbol{r} + (\boldsymbol{r}^T\boldsymbol{x})^2\boldsymbol{x} + (1 - (\boldsymbol{r}^T\boldsymbol{x})^2)\boldsymbol{x} \\
&\quad = \boldsymbol{x} - (\boldsymbol{r}^T\boldsymbol{x})\boldsymbol{r} = -\mathrm{d}_2\Psi_{\mathbb{S}^2}(\boldsymbol{x}, \boldsymbol{r}).
\end{aligned}$$

Next, for $\boldsymbol{v} \in \mathsf{T}_{\boldsymbol{r}}\mathbb{S}^2$ we have

$$\mathfrak{T}_{\mathbb{S}^2}(\boldsymbol{x}, \boldsymbol{r}) \cdot \boldsymbol{v} = (\boldsymbol{r}^T \boldsymbol{x})\boldsymbol{v} + (\boldsymbol{r} \times \boldsymbol{x}) \times \boldsymbol{v} = (\boldsymbol{r}^T \boldsymbol{x})\boldsymbol{v} - (\boldsymbol{v}^T \boldsymbol{x})\boldsymbol{r} = (\boldsymbol{r} \times \boldsymbol{v}) \times \boldsymbol{x}. \quad \blacksquare$$

Let $t \mapsto \boldsymbol{r}(t)$ be a twice-differentiable reference trajectory with bounded velocity. Let k_{D} be a positive constant as in Section 11.1.3. For $\boldsymbol{x} \in \mathbb{S}^2$ and $\boldsymbol{v} \in \mathsf{T}_{\boldsymbol{x}}\mathbb{S}^2$, the closed-loop energy function is

$$E_{\mathrm{cl}}(t, \boldsymbol{x}, \boldsymbol{v}) = k_{\mathrm{P}}(1 - \langle\langle \boldsymbol{r}, \boldsymbol{x} \rangle\rangle_{\mathbb{R}^3}) + \tfrac{1}{2}m\|\boldsymbol{v} - \big(\boldsymbol{r}(t) \times \dot{\boldsymbol{r}}(t)\big) \times \boldsymbol{x}\|^2_{\mathbb{R}^3},$$

and the proportional-derivative and feedforward control forces are

$$\begin{aligned} F_{\mathrm{PD}}(t, \boldsymbol{x}, \boldsymbol{v}) &= -k_{\mathrm{P}}\widehat{\boldsymbol{x}}^2 \boldsymbol{r}(t) - k_{\mathrm{D}}\left(\boldsymbol{v} - \big(\boldsymbol{r}(t) \times \dot{\boldsymbol{r}}(t)\big) \times \boldsymbol{x}\right), \\ F_{\mathrm{FF}}(t, \boldsymbol{x}, \boldsymbol{v}) &= m\,\langle\langle \boldsymbol{x}, \boldsymbol{r}(t) \times \dot{\boldsymbol{r}}(t) \rangle\rangle_{\mathbb{R}^3} (\boldsymbol{x} \times \boldsymbol{v}) + m\big(\boldsymbol{r}(t) \times \overset{\mathrm{G}_{\mathrm{rnd}}}{\nabla}_{\dot{\boldsymbol{r}}(t)} \dot{\boldsymbol{r}}(t)\big) \times \boldsymbol{x}. \end{aligned}$$

Lemma 11.23 (Tracking on $\mathbb{S}^2$). *For the closed-loop system defined by $F_{\mathrm{PD}} + F_{\mathrm{FF}}$, assume that the initial conditions $(\boldsymbol{x}(0), \dot{\boldsymbol{x}}(0))$ satisfy $\boldsymbol{x}(0) \neq -\boldsymbol{r}(0)$, and*

$$k_{\mathrm{P}} > \frac{m\|\dot{\boldsymbol{x}}(0) - \big(\boldsymbol{r}(0) \times \dot{\boldsymbol{r}}(0)\big) \times \boldsymbol{x}(0)\|^2_{\mathbb{R}^3}}{2\big(1 + \langle\langle \boldsymbol{x}(0), \boldsymbol{r}(0) \rangle\rangle_{\mathbb{R}^3}\big)}. \tag{11.13}$$

Then the resulting closed-loop trajectory $t \mapsto \boldsymbol{r}(t)$ is exponentially stable with respect to the error function E_{cl}.

Proof. The result is a consequence of Theorem 11.19 after establishing the following facts. First, we start by showing that $\Psi_{\mathbb{S}^2}$ is uniformly quadratic. Note that $L_{\mathrm{reg}}((\Psi_{\mathbb{S}^2})_{\boldsymbol{r}}, \boldsymbol{r}) = 2k_{\mathrm{P}}$ is independent of $\boldsymbol{r} \in \mathbb{S}^2$ and that, therefore,

$$L_{\mathrm{reg}}(\Psi_{\mathbb{S}^2}, \mathbb{S}^2) = \inf\left\{ L_{\mathrm{reg}}((\Psi_{\mathbb{S}^2})_{\boldsymbol{r}}, \boldsymbol{r}) \mid \boldsymbol{r} \in \mathbb{S}^2 \right\} = 2k_{\mathrm{P}}.$$

For $2k_{\mathrm{P}} > \varepsilon > 0$, we set $L = 2k_{\mathrm{P}} - \varepsilon$ and compute

$$\begin{aligned} \Psi^{-1}_{\mathbb{S}^2}(\leq L) &= \left\{ (\boldsymbol{x}, \boldsymbol{r}) \in \mathbb{S}^2 \times \mathbb{S}^2 \mid 0 \leq \Psi_{\mathbb{S}^2}(\boldsymbol{x}, \boldsymbol{r}) \leq 2k_{\mathrm{P}} - \varepsilon \right\} \\ &= \left\{ (\boldsymbol{x}, \boldsymbol{r}) \in \mathbb{S}^2 \times \mathbb{S}^2 \mid 2 \geq 1 + \boldsymbol{r}^T \boldsymbol{x} \geq \tfrac{\varepsilon}{k_{\mathrm{P}}} \right\}. \end{aligned}$$

On the other hand we can write

$$\|\mathrm{d}_1 \Psi_{\mathbb{S}^2}\|^2_{\mathbb{R}^3} = \|\boldsymbol{r} - (\boldsymbol{r}^T \boldsymbol{x})\boldsymbol{x}\|^2_{\mathbb{R}^3} = 1 - (\boldsymbol{r}^T \boldsymbol{x})^2 = (1 + \boldsymbol{r}^T \boldsymbol{x})\Psi_{\mathbb{S}^2}(\boldsymbol{x}, \boldsymbol{r})\tfrac{1}{k_{\mathrm{P}}},$$

where we used the equality $\widehat{\boldsymbol{x}}^2 \boldsymbol{r} = \boldsymbol{r} - (\boldsymbol{r}^T \boldsymbol{x})\boldsymbol{x}$. Since, for all $(\boldsymbol{x}, \boldsymbol{r}) \in \mathbb{S}^2 \times \mathbb{S}^2$ such that $\Psi_{\mathbb{S}^2}(\boldsymbol{x}, \boldsymbol{r}) = 0$, the bound in Definition 11.12 is verified, we can assume that $1 - \boldsymbol{r}^T \boldsymbol{x} \neq 0$. Therefore, we only need to check that there exist $0 < b_1 \leq b_2$ such that

$$\frac{b_1}{k_{\mathrm{P}}}(1 + \boldsymbol{r}^T \boldsymbol{x}) \leq 1 \leq \frac{b_2}{k_{\mathrm{P}}}(1 + \boldsymbol{r}^T \boldsymbol{x}).$$

For all $(\boldsymbol{x}, \boldsymbol{r}) \in \Psi^{-1}_{\mathbb{S}^2}(\leq L)$, this holds true with $b_1 = \frac{k_{\mathrm{P}}}{2}$ and $b_2 = \frac{k_{\mathrm{P}}}{\varepsilon}$.

As a second step, we compute the covariant derivative of the transport map and show that F_{FF} is designed as prescribed by Theorem 11.19. Note that

$$\mathcal{T}_{\mathbb{S}^2}(\boldsymbol{x}(t),\boldsymbol{r}(t))\cdot\dot{\boldsymbol{r}}(t) = (\boldsymbol{r}(t)\times\dot{\boldsymbol{r}}(t))\times\boldsymbol{x}(t).$$

For $\boldsymbol{v}\in\mathsf{T}_{\boldsymbol{r}}\mathbb{S}^2$, we compute the covariant derivative of the vector field $\boldsymbol{x}\mapsto(\boldsymbol{r}\times\boldsymbol{v})\times\boldsymbol{x}$ along $\boldsymbol{w}\in\mathsf{T}_{\boldsymbol{x}}\mathbb{S}^2$ by covariantly differentiating on $\mathbb{R}^3$, and projecting onto $\mathsf{T}_{\boldsymbol{x}}\mathbb{S}^2$; see Proposition 4.85. We have

$$\begin{aligned}\overset{\mathbb{G}_{\mathrm{rnd}}}{\nabla}_{\boldsymbol{w}}(\mathcal{T}_{\mathbb{S}^2}(\boldsymbol{x},\boldsymbol{r})\cdot(\boldsymbol{v})) &= P_{\boldsymbol{x}}\big(\overset{\mathbb{G}_{\mathbb{R}^3}}{\nabla}_{\boldsymbol{w}}(\boldsymbol{r}\times\boldsymbol{v})\times\boldsymbol{x}\big)\\ &= -\,\widehat{\boldsymbol{x}}^2((\boldsymbol{r}\times\boldsymbol{v})\times\boldsymbol{w})\\ &= \widehat{\boldsymbol{x}}\big(\boldsymbol{x}\times(\boldsymbol{w}\times(\boldsymbol{r}\times\boldsymbol{v})\big)\\ &= \widehat{\boldsymbol{x}}\big(\boldsymbol{x}^T(\boldsymbol{r}\times\boldsymbol{v})\boldsymbol{w}+\boldsymbol{x}^T\boldsymbol{w}(\boldsymbol{r}\times\boldsymbol{v})\big)\\ &= (\boldsymbol{x}^T(\boldsymbol{r}\times\boldsymbol{v}))(\boldsymbol{x}\times\boldsymbol{w}),\end{aligned}$$

where we have used the expression (11.6) for $P_{\boldsymbol{x}}$. We also compute

$$\begin{aligned}\frac{\mathrm{d}}{\mathrm{d}t}\big(\mathcal{T}_{\mathbb{S}^2}(\boldsymbol{x},\boldsymbol{r}(t))\cdot\dot{\boldsymbol{r}}(t)\big) &= \frac{\mathrm{d}}{\mathrm{d}t}((\boldsymbol{r}(t)\times\dot{\boldsymbol{r}}(t))\times\boldsymbol{x})\\ &= (\boldsymbol{r}(t)\times\ddot{\boldsymbol{r}}(t))\times\boldsymbol{x} = \big(\boldsymbol{r}(t)\times\overset{\mathbb{G}_{\mathrm{rnd}}}{\nabla}_{\dot{\boldsymbol{r}}(t)}\dot{\boldsymbol{r}}(t)\big)\times\boldsymbol{x}.\end{aligned}$$

As a third step, we establish the bounds required in Theorem 11.19. Because $\mathbb{S}^2$ is a compact manifold and the functions $\Psi_{\mathbb{S}^2}$ and $\mathcal{T}_{\mathbb{S}^2}$ are smooth, the Hessian of $\Psi_{\mathbb{S}^2}$ and the covariant derivative of $\mathcal{T}_{\mathbb{S}^2}$ are uniformly bounded from above on $\mathbb{S}^2\times\mathbb{S}^2$. Clearly, the dissipation function k_{D} is uniformly bounded from below and above on $\mathbb{S}^2$.

Finally, we invoke Theorem 11.19, and characterize the domain of attraction with exponential convergence rates by $E_{\mathrm{cl}}(0) < L_{\mathrm{reg}}(\Psi_{\mathbb{S}^2},\mathbb{S}^2) = 2k_{\mathrm{P}}$. After some straightforward manipulations, equation (11.13) turns out to be the required bound. ■

11.4 Stabilization and tracking on Lie groups

In this section we consider stabilization and tracking problems for simple mechanical control systems defined on Lie groups. This class includes rigid bodies moving in two or three-dimensional Euclidean space and endowed with body-fixed forces and left-invariant kinetic energy metric. As an example, we shall consider a satellite system defined on the group of rotations $\mathsf{SO}(3)$.

This class of mechanical system is studied in Section 5.3 and we shall utilize numerous concepts discussed in Chapter 5. We consider a fully actuated simple mechanical control system on a Lie group $(\mathsf{G},\mathbb{I},\mathscr{F}=\{f^1,\dots,f^n\},\mathbb{R}^n)$, where G is a Lie group, $\mathbb{I}$ is an inner product on the Lie algebra $\mathfrak{g}$, and $\{f^1,\dots,f^n\}$ is a basis for $\mathfrak{g}^*$. Let $\gamma\colon\bar{\mathbb{R}}_+\to\mathsf{G}$ be the controlled trajectory, and define the body velocity as the curve $t \mapsto v(t) = T_{\gamma(t)}L_{\gamma(t)^{-1}}(\gamma'(t))$ taking values in $\mathfrak{g}$. The controlled equations of motion are

$$\gamma'(t) = T_e L_{\gamma(t)}(v(t)),$$
$$v'(t) - \mathbb{I}^\sharp\left(\mathrm{ad}^*_{v(t)} \mathbb{I}^\flat(v(t))\right) = \mathbb{I}^\sharp(f(t, \gamma(t), v(t))),$$

where the time-dependent state feedback force $t \mapsto f(t, \gamma(t), v(t)) \in \mathfrak{g}^*$ uniquely defines a time-dependent state feedback $u\colon \bar{\mathbb{R}}_+ \times \mathsf{G} \times \mathfrak{g} \to \mathbb{R}^n$ with the property that

$$f(t, g, v) = \sum_{a=1}^{n} f^a u^a(t, g, v).$$

Remark 11.24. For systems defined on the rigid displacement group $\mathsf{SE}(3)$ or on one of its Lie subgroups, writing the velocity curve γ' in terms of the body velocity curve v corresponds to writing the velocity and control forces in the body reference frame. Similarly, writing the control law as a control force taking values in $\mathfrak{g}^*$ corresponds to writing the control force in the body reference frame. •

11.4.1 PD control on Lie groups

Let $\Psi_{\mathsf{G}}\colon \mathsf{G} \to \bar{\mathbb{R}}_+$ be a configuration error function on G about the identity element e. Let $R_{\mathrm{diss}}\colon \mathfrak{g} \times \mathfrak{g} \to \mathbb{R}$ be a symmetric positive-definite $(0,2)$-tensor on $\mathfrak{g}$. We design the proportional-derivative control force in the body frame as the map $f_{\mathrm{PD}}\colon \mathsf{G} \times \mathfrak{g} \to \mathfrak{g}^*$ given by

$$f_{\mathrm{PD}}(g, \xi) = -(T_e L_g)^* (\mathrm{d}\Psi_{\mathsf{G}}(g)) - R^\flat_{\mathrm{diss}}(\xi). \tag{11.14}$$

An immediate consequence of Theorem 11.4 is the following result.

Corollary 11.25 (Stabilization on Lie groups using PD control). *Consider the fully actuated simple mechanical control system on a Lie group $(\mathsf{G}, \mathbb{I}, \mathscr{F} = \{f^1, \dots, f^n\}, \mathbb{R}^n)$. The closed-loop system defined by f_{PD} has the following properties:*

(i) the closed-loop energy $E_{\mathrm{cl}}\colon \mathsf{TG} \to \mathbb{R}$ is

$$E_{\mathrm{cl}}(v_g) = \Psi_{\mathsf{G}}(g) + \tfrac{1}{2}\left\|T_g L_{g^{-1}}(v_g)\right\|^2_{\mathbb{I}};$$

(ii) $e \in \mathsf{G}$ is a locally asymptotically stable controlled equilibrium configuration;

(iii) if, for $t_0 \in \bar{\mathbb{R}}_+$, the controlled trajectory $\gamma\colon \bar{\mathbb{R}}_+ \to \mathsf{G}$ satisfies $\gamma'(t_0) \in E_{\mathrm{cl}}^{-1}(\leq L, (e, 0))$ for $L < L_{\mathrm{reg}}(\Psi_{\mathsf{G}}, e)$, then γ converges to $e \in \mathsf{G}$ exponentially fast, and the body velocity $v\colon \bar{\mathbb{R}}_+ \to \mathfrak{g}$ converges to zero exponentially fast;

(iv) *given any initial condition* $(\gamma(0), v(0)) \in \mathsf{G} \times \mathfrak{g}$, *the resulting controlled trajectory takes values in the compact set* $E_{\text{cl}}^{-1}(\leq E_{\text{cl}}(\gamma(0), v(0)))$ *and converges as* $t \to +\infty$ *to a critical point of* Ψ_{G} *inside* $\Psi_{\mathsf{G}}^{-1}(\leq E_{\text{cl}}(\gamma(0), v(0)))$.

Remarks 11.26. 1. Assume that the control objective is to locally asymptotically stabilize a specified configuration $g_0 \in \mathsf{G}$. An appropriate configuration error function can be designed by composing Ψ_{G} with the left translation $L_{g_0^{-1}}$. In other words, the function $g \mapsto \Psi_{\mathsf{G}}(g_0^{-1}g)$ is a configuration error function about g_0. Correspondingly, the proportional control is $f_{\mathrm{P}}(g) = -(T_e L_g)^* \left(\mathrm{d}\Psi_{\mathsf{G}}(g_0^{-1}g)\right)$.

2. It is often convenient to compute the covector $f_{\mathrm{P}}(g) = -(T_e L_g)^* \left(\mathrm{d}\Psi_{\mathsf{G}}(g)\right)$ by means of the equality

$$\mathscr{L}_{\xi_{\mathrm{L}}} \Psi_{\mathsf{G}}(g) = -\langle f_{\mathrm{P}}(g); \xi \rangle . \tag{11.15}$$

This computation can typically be performed without introducing coordinate charts. We shall see instances of this in the examples below. •

11.4.2 PD and feedforward control on Lie groups

To design a tracking error function about a reference configuration, we define the ***right group error function*** $g_{e,r} \colon \mathsf{G} \times \mathsf{G} \to \mathsf{G}$ by

$$g_{e,r}(g, g_{\text{ref}}) = g_{\text{ref}}^{-1} g.$$

We compose this function with the configuration error function Ψ_{G} to define the ***right tracking error function*** $\Psi_{\mathsf{G},r} \colon \mathsf{G} \times \mathsf{G} \to \mathbb{R}$ by

$$\Psi_{\mathsf{G},r}(g, g_{\text{ref}}) = \Psi_{\mathsf{G}}(g_{e,r}(g, g_{\text{ref}})) = \Psi_{\mathsf{G}}(g_{\text{ref}}^{-1} g). \tag{11.16}$$

Lemma 11.27. *Given a configuration error function* Ψ_{G} *on* G *about the identity* e, *the right tracking error function* $\Psi_{\mathsf{G},r}$ *is uniformly quadratic and* $L_{\text{reg}}(\Psi_{\mathsf{G},r}, \mathsf{G}) = L_{\text{reg}}(\Psi_{\mathsf{G}}, e)$.

Proof. From Proposition 6.30, we know that Ψ_{G} is quadratic on $\Psi_{\mathsf{G}}^{-1}(\leq L, e)$ for all $0 < L < L_{\text{reg}}(\Psi_{\mathsf{G}}, e)$. This means that, for all $0 < L < L_{\text{reg}}(\Psi_{\mathsf{G}}, e)$, there exist $0 < b_1 \leq b_2$ such that for all $h \in \Psi_{\mathsf{G}}^{-1}(\leq L, e) \setminus \{e\}$,

$$0 < b_1 \left\|\mathrm{d}\Psi_{\mathsf{G}}(h)\right\|_{\mathbb{G}_{\mathbb{I}}}^2 \leq \Psi_{\mathsf{G}}(h) \leq b_2 \left\|\mathrm{d}\Psi_{\mathsf{G}}(h)\right\|_{\mathbb{G}_{\mathbb{I}}}^2 .$$

Let $h = g_{\text{ref}}^{-1} g$, so that $\Psi_{\mathsf{G}}(h) = \Psi_{\mathsf{G},r}(g, g_{\text{ref}})$. Then, by (11.16), we have

$$g_{\text{ref}}^{-1} g \in \Psi_{\mathsf{G}}^{-1}(\leq L, e) \setminus \{e\} \quad \Longrightarrow \quad g \in (\Psi_{\mathsf{G},r})_{g_{\text{ref}}}^{-1}(\leq L, g_{\text{ref}}) \setminus \{g_{\text{ref}}\},$$

and then, by the left-invariance of $\mathbb{I}$ and by Lemma 11.28(i), we have

$$\|\mathrm{d}\Psi_{\mathsf{G}}(h)\|_{\mathbb{G}_{\mathbb{I}}}^2 = \|\mathrm{d}_1\Psi_{\mathsf{G},r}(g, g_{\mathrm{ref}})\|_{\mathbb{G}_{\mathbb{I}}}^2 .$$

Collecting these statements together, we have shown that, for all $L \in]0, L_{\mathrm{reg}}(\Psi_{\mathsf{G}}, e)[$, there exist $0 < b_1 \leq b_2$ such that, for all $g \in (\Psi_{\mathsf{G},r})_{g_{\mathrm{ref}}}^{-1}(\leq L, g_{\mathrm{ref}}) \setminus \{g_{\mathrm{ref}}\}$,

$$0 < b_1 \|\mathrm{d}_1\Psi_{\mathsf{G},r}(g, g_{\mathrm{ref}})\|_{\mathbb{G}_{\mathbb{I}}}^2 \leq \Psi_{\mathsf{G},r}(g, g_{\mathrm{ref}}) \leq b_2 \|\mathrm{d}_1\Psi_{\mathsf{G},r}(g, g_{\mathrm{ref}})\|_{\mathbb{G}_{\mathbb{I}}}^2 .$$

This proves that $\Psi_{\mathsf{G},r}$ is uniformly quadratic and that $L_{\mathrm{reg}}(\Psi_{\mathsf{G},r}, \mathsf{G}) = L_{\mathrm{reg}}(\Psi_{\mathsf{G}}, e)$. ■

Next, we define the ***right transport map*** $\mathcal{T}_r \colon \mathsf{G} \times \mathsf{TG} \to \mathsf{TG} \times \mathsf{G}$ by

$$\mathcal{T}_r(g, g_{\mathrm{ref}}) \cdot X_{g_{\mathrm{ref}}} = T_{g_{\mathrm{ref}}} R_{g_{\mathrm{ref}}^{-1} g}(X_{g_{\mathrm{ref}}}),$$

where $X_{g_{\mathrm{ref}}} \in \mathsf{T}_{g_{\mathrm{ref}}}\mathsf{G}$. Note that the definition is well-posed because $T_{g_{\mathrm{ref}}} R_{g_{\mathrm{ref}}^{-1} g} \colon \mathsf{T}_{g_{\mathrm{ref}}}\mathsf{G} \to \mathsf{T}_g\mathsf{G}$. It is also convenient to state the equivalent expressions:

$$\begin{aligned} T_{g_{\mathrm{ref}}} R_{g_{\mathrm{ref}}^{-1} g}(X_{g_{\mathrm{ref}}}) &= T_e L_g\big(\mathrm{Ad}_{g^{-1}g_{\mathrm{ref}}}\,(T_{g_{\mathrm{ref}}} L_{g_{\mathrm{ref}}^{-1}}(X_{g_{\mathrm{ref}}}))\big) \\ &= T_e L_g\left(\mathrm{Ad}_{g^{-1}g_{\mathrm{ref}}}\,(\xi_{\mathrm{ref}})\right), \end{aligned}$$

where $X_{g_{\mathrm{ref}}} = T_e L_{g_{\mathrm{ref}}}(\xi_{\mathrm{ref}})$, for $\xi_{\mathrm{ref}} \in \mathfrak{g}$.

Lemma 11.28. *The right tracking error function and transport map have the following properties:*

(i) $\mathrm{d}_1\Psi_{\mathsf{G},r}(g, g_{\mathrm{ref}}) = (T_g L_{g_{\mathrm{ref}}^{-1}})^*(\mathrm{d}\Psi_{\mathsf{G}}(g_{\mathrm{ref}}^{-1} g))$;

(ii) the pair $(\Psi_{\mathsf{G},r}, \mathcal{T}_r)$ *is compatible;*

(iii) the velocity error curve $t \mapsto \dot{e}_r(t)$ *between the curves* $t \mapsto \gamma(t) \in \mathsf{G}$ *and* $t \mapsto \gamma_{\mathrm{ref}}(t) \in \mathsf{G}$, *with body velocities* $t \mapsto v(t) = T_{\gamma(t)} L_{\gamma(t)^{-1}}(\gamma'(t)) \in \mathfrak{g}$ *and* $t \mapsto v_{\mathrm{ref}}(t) = T_{\gamma_{\mathrm{ref}}(t)} L_{\gamma_{\mathrm{ref}}(t)^{-1}}(\gamma_{\mathrm{ref}}'(t)) \in \mathfrak{g}$, *satisfies*

$$\dot{e}_r(t) = T_e L_{\gamma(t)}\left(v(t) - \mathrm{Ad}_{\gamma^{-1}(t)\gamma_{\mathrm{ref}}(t)}\left(v_{\mathrm{ref}}(t)\right)\right).$$

Proof. The proof relies on the characterization in Lemma 11.16. Let γ, γ_{ref}, v, and v_{ref} be as in (iii). We compute the velocity error curve between γ and γ_{ref} as follows:

$$\begin{aligned} \dot{e}_r(t) &= \gamma'(t) - \mathcal{T}_r(\gamma(t), \gamma_{\mathrm{ref}}(t)) \cdot \gamma_{\mathrm{ref}}'(t) \\ &= T_e L_{\gamma(t)}(v(t)) - T_e L_{\gamma(t)}\big(\mathrm{Ad}_{\gamma^{-1}(t)\gamma_{\mathrm{ref}}(t)}\,(T_{\gamma_{\mathrm{ref}}(t)} L_{\gamma_{\mathrm{ref}}^{-1}(t)}(\gamma_{\mathrm{ref}}'(t)))\big) \\ &= T_e L_{\gamma(t)}\left(v(t) - \mathrm{Ad}_{\gamma^{-1}(t)\gamma_{\mathrm{ref}}(t)}(v_{\mathrm{ref}}(t))\right). \end{aligned}$$

This proves part (iii). Next, we consider the curve $t \mapsto g_{e,r}(\gamma(t), \gamma_{\mathrm{ref}}(t)) = \gamma_{\mathrm{ref}}^{-1}(t)\gamma(t)$ and compute

$$\begin{aligned}
&\frac{\mathrm{d}}{\mathrm{d}t} g_{e,r}(\gamma(t), \gamma_{\mathrm{ref}}(t)) \\
&\quad = T_{\gamma(t)} L_{\gamma_{\mathrm{ref}}^{-1}(t)} (\gamma'(t)) + T_{\gamma_{\mathrm{ref}}^{-1}(t)} R_{\gamma(t)} \left((\gamma_{\mathrm{ref}}^{-1})'(t)\right) \\
&\quad = T_e L_{\gamma_{\mathrm{ref}}^{-1}(t)\gamma(t)}(v(t)) - T_e R_{\gamma_{\mathrm{ref}}^{-1}(t)\gamma(t)}(v_{\mathrm{ref}}(t)) \\
&\quad = T_e L_{\gamma_{\mathrm{ref}}^{-1}(t)\gamma(t)} \left(v(t) - T_{\gamma_{\mathrm{ref}}^{-1}(t)\gamma(t)} L_{\gamma(t)\gamma_{\mathrm{ref}}^{-1}(t)} (T_e R_{\gamma_{\mathrm{ref}}^{-1}(t)\gamma(t)}(v_{\mathrm{ref}}(t)))\right) \\
&\quad = T_{\gamma(t)} L_{\gamma_{\mathrm{ref}}^{-1}(t)} T_e L_{\gamma(t)} \left(v(t) - \mathrm{Ad}_{\gamma(t)\gamma_{\mathrm{ref}}^{-1}(t)}(v_{\mathrm{ref}}(t))\right),
\end{aligned}$$

where we have used the equality $(\gamma_{\mathrm{ref}}^{-1})'(t) = -T_e R_{\gamma_{\mathrm{ref}}^{-1}(t)}(v_{\mathrm{ref}}(t))$. Combining the two results, we have

$$\begin{aligned}
&\frac{\mathrm{d}}{\mathrm{d}t} \Psi_{\mathsf{G},r}(\gamma(t), \gamma_{\mathrm{ref}}(t)) \\
&\quad = \frac{\mathrm{d}}{\mathrm{d}t} \Psi_{\mathsf{G}}(\gamma_{\mathrm{ref}}^{-1}(t)\gamma(t)) \\
&\quad = \left\langle \mathrm{d}\Psi_{\mathsf{G}}(\gamma_{\mathrm{ref}}^{-1}(t)\gamma(t)); T_{\gamma(t)} L_{\gamma_{\mathrm{ref}}^{-1}(t)} T_e L_{\gamma(t)} (v(t) - \mathrm{Ad}_{\gamma(t)\gamma_{\mathrm{ref}}^{-1}(t)}(v_{\mathrm{ref}}(t)))\right\rangle \\
&\quad = \left\langle (T_{\gamma(t)} L_{\gamma_{\mathrm{ref}}^{-1}(t)})^*(\mathrm{d}\Psi_{\mathsf{G}}(\gamma_{\mathrm{ref}}^{-1}(t)\gamma(t))); \dot{e}_r(t)\right\rangle.
\end{aligned}$$

The last equality proves part (i) and part (ii). ■

We now state the main tracking theorem for systems on groups.

Theorem 11.29 (Trajectory tracking on Lie groups using PD and feedforward control). *Consider a fully actuated simple mechanical control system on a Lie group* $(\mathsf{G}, \mathbb{I}, \mathscr{F}, \mathbb{R}^n)$. *Consider a twice-differentiable reference trajectory* $t \mapsto \gamma_{\mathrm{ref}}(t) \in \mathsf{G}$ *with bounded reference body velocity* $t \mapsto v_{\mathrm{ref}}(t) = T_{\gamma_{\mathrm{ref}}(t)} L_{\gamma_{\mathrm{ref}}(t)^{-1}}(\gamma_{\mathrm{ref}}'(t)) \in \mathfrak{g}$.

Define proportional-derivative and feedforward control laws according to

$$\begin{aligned}
f_{\mathrm{PD}}(t, g, \xi) = &- (T_e L_{\gamma_{\mathrm{ref}}^{-1}(t)g})^*(\mathrm{d}\Psi_{\mathsf{G}}(\gamma_{\mathrm{ref}}^{-1}(t)g)) \\
&- R_{\mathrm{diss}}^{\flat}(\xi - \mathrm{Ad}_{g^{-1}\gamma_{\mathrm{ref}}(t)}(v_{\mathrm{ref}}(t))), \\
f_{\mathrm{FF}}(t, g, \xi) = \mathbb{I}^{\flat}\Big(&\overset{\mathfrak{g}}{\nabla}_{\xi}(\mathrm{Ad}_{g^{-1}\gamma_{\mathrm{ref}}(t)}(v_{\mathrm{ref}}(t))) + [\mathrm{Ad}_{g^{-1}\gamma_{\mathrm{ref}}(t)}(v_{\mathrm{ref}}(t)), \xi] \\
&+ \mathrm{Ad}_{g^{-1}\gamma_{\mathrm{ref}}(t)}(v_{\mathrm{ref}}'(t))\Big).
\end{aligned}$$

For the closed-loop system defined by $f_{\mathrm{PD}} + f_{\mathrm{FF}}$, *the curve* γ_{ref} *is stable with respect to the closed-loop energy function* $E_{\mathrm{cl}}\colon \bar{\mathbb{R}}_+ \times \mathsf{G} \times \mathfrak{g} \to \mathbb{R}$ *defined by*

$$E_{\mathrm{cl}}(t, g, \xi) = \Psi_{\mathsf{G}}(\gamma_{\mathrm{ref}}^{-1}(t)g) + \tfrac{1}{2}\left\|\xi - \mathrm{Ad}_{g^{-1}\gamma_{\mathrm{ref}}(t)}(v_{\mathrm{ref}}(t))\right\|_{\mathbb{I}}^2.$$

Assume that, at $t_0 \in \bar{\mathbb{R}}_+$, *the energy level* $E_{\mathrm{cl}}(t_0)$ *is strictly less that* $L_{\mathrm{reg}}(\Psi_{\mathsf{G}}, e)$, *and that* $\gamma_{\mathrm{ref}}(t_0)^{-1}\gamma(t_0)$ *lies in sublevel set* $\Psi_{\mathsf{G}}^{-1}(\leq E_{\mathrm{cl}}(t_0), e)$. *Then the curve* γ_{ref} *is locally exponentially stable with respect to the error function* E_{cl}.

Proof. See Section B.6. ■

In summary, the theorem suggests that the main step in the design of a tracking controller is the synthesis of a configuration error function Ψ_{G} and the computation of $L_{\text{reg}}(\Psi_{\mathsf{G}}, e)$. The right tracking error function and the right transport map lead directly to the expression for the tracking controller.

Remarks 11.30. 1. In the context of rigid body kinematics, the group element $g_{e,r} = g_{\text{ref}}^{-1} g$ describes the relative motion from the body frame to the reference frame. Note that it is not the only possible choice of group error. In particular, it is instructive to define a ***left group error function*** $g_{e,\ell}(g, g_{\text{ref}}) = g g_{\text{ref}}^{-1}$, and the corresponding ***left tracking error function*** $\Psi_{\mathsf{G},\ell}(g, g_{\text{ref}}) = \Psi_{\mathsf{G}}(g_{e,\ell}(g, g_{\text{ref}}))$. We then define the compatible ***left transport map***

$$\mathcal{T}_\ell(g, g_{\text{ref}}) \cdot X_{g_{\text{ref}}} = T_{g_{\text{ref}}} L_{g g_{\text{ref}}^{-1}}(X_{g_{\text{ref}}}).$$

Along the lines of Lemma 11.28, one can verify that the pair $(\Psi_{\mathsf{G},\ell}, \mathcal{T}_\ell)$ is compatible, that

$$\mathrm{d}_1 \Psi_{\mathsf{G},\ell}(g, g_{\text{ref}}) = (T_g R_{g_{\text{ref}}^{-1}})^*(\mathrm{d}\Psi_{\mathsf{G}}(g g_{\text{ref}}^{-1})),$$

and that, given the curves $t \mapsto \gamma(t)$ and $t \mapsto \gamma_{\text{ref}}(t)$ with body velocities $t \mapsto v(t)$ and $t \mapsto v_{\text{ref}}(t)$, respectively, the error velocity curve is

$$\dot{e}_\ell(t) = T_e L_{\gamma(t)}(v(t) - v_{\text{ref}}(t)).$$

The corresponding feedback and feedforward control are

$$\begin{aligned}(f_{\text{PD}})_\ell(t, g, \xi) = &- (T_e L_g)^* (T_g R_{\gamma_{\text{ref}}^{-1}(t)})^* (\mathrm{d}\Psi_{\mathsf{G}}(g \gamma_{\text{ref}}^{-1}(t))) \\ &- R_{\text{diss}}^\flat(\xi - v_{\text{ref}}(t)), \\ (f_{\text{FF}})_\ell(t, g, \xi) = &\, \mathbb{I}^\flat \left(\overset{\mathfrak{g}}{\nabla}_\xi v_{\text{ref}}(t) + v'_{\text{ref}}(t) \right).\end{aligned}$$

It is an instructive exercise for the reader to perform the following tasks: (1) show that $\Psi_{\mathsf{G},\ell}$ is a uniformly quadratic tracking error function and compute $L_{\text{reg}}(\Psi_{\mathsf{G},\ell}, \mathsf{Q})$, (2) show that $\mathcal{T}_\ell$ is compatible and compute its covariant derivative, and (3) establish bounds on appropriate sublevel sets of $\Psi_{\mathsf{G},\ell}$ for the maps $\text{Hess}_1 \Psi_{\mathsf{G},\ell}$, R_{diss}, and $\overset{\mathsf{G}}{\nabla} \mathcal{T}_\ell$. These steps guarantee that Theorem 11.19 applies to this "left" control design.

2. As discussed in Remark 11.26 for stabilization problems, it is convenient to identify a simple procedure to compute the proportional feedback control f_{P} when the group G is the matrix Lie group $\mathsf{SE}(3)$ or one of its Lie subgroups.

When designing right tracking error functions and right transport maps, the function $f_{\text{P}} \colon \mathsf{G} \to \mathfrak{g}^*$ is the same for stabilization or tracking purposes with the only distinction that the argument g is replaced by $g_{e,r}$. Therefore, the procedure outlined in equation (11.15) is still applicable.

An equivalent procedure is the following: (1) compute the time-derivative of the tracking error function evaluated on the actual and reference trajectory, and (2) factor out the body velocity error to obtain the proportional control. Specifically, one can compute the appropriate proportional feedback controls from

$$\frac{\mathrm{d}}{\mathrm{d}t}\Psi_{\mathrm{G},r}(\gamma(t),\gamma_{\mathrm{ref}}(t)) = \left\langle -(f_{\mathrm{P}})_r(t,\gamma(t),\gamma_{\mathrm{ref}}(t)); T_eL_{\gamma^{-1}(t)}(\dot{e}_r(t))\right\rangle,$$
$$\frac{\mathrm{d}}{\mathrm{d}t}\Psi_{\mathrm{G},\ell}(\gamma(t),\gamma_{\mathrm{ref}}(t)) = \left\langle -(f_{\mathrm{P}})_\ell(t,\gamma(t),\gamma_{\mathrm{ref}}(t)); T_eL_{\gamma^{-1}(t)}(\dot{e}_\ell(t))\right\rangle,$$

where $(f_{\mathrm{P}})_r$ and $(f_{\mathrm{P}})_\ell$ are the right and left proportional feedback controls, respectively.

3. Finally, it is interesting to investigate alternative expressions for the feed-forward control law. Let us define the modified control laws

$$(f_{\mathrm{FF}})_{r'}(t,g,\xi) = -\operatorname{ad}^*_\xi \mathbb{I}^\flat\left(\operatorname{Ad}_{g^{-1}\gamma_{\mathrm{ref}}(t)}(v_{\mathrm{ref}}(t))\right) + \mathbb{I}^\flat\left(\operatorname{Ad}_{g^{-1}\gamma_{\mathrm{ref}}(t)}(v'_{\mathrm{ref}}(t))\right),$$
$$(f_{\mathrm{FF}})_{\ell'}(t,g,\xi) = -\operatorname{ad}^*_{v_{\mathrm{ref}}(t)} \mathbb{I}^\flat(\xi) + \mathbb{I}^\flat(v'_{\mathrm{ref}}(t)).$$

We leave it to the reader to show that these expressions lead to a closed-loop system with the same properties as those discussed in Theorem 11.29. •

11.4.3 The attitude tracking problem for a fully actuated rigid body fixed at a point

In this section we discuss the attitude tracking problem for a fully actuated rigid body fixed at a point. This is a common model for satellites with three thrusters. The model is described in Examples 5.48, 5.74 and 6.62. Control problems for this classic example seem to have first been studied by Meyer [1971]. Let us briefly review some notation. The configuration manifold is the Lie group $\mathsf{SO}(3)$ and the inertia tensor is $\mathbb{I}$; see equation (5.23) in Example 5.48 for the restriction $\overset{\mathfrak{so}(3)}{\nabla} : \mathfrak{so}(3)\times\mathfrak{so}(3)\to\mathfrak{so}(3)$ of the affine connection associated with $\mathbb{I}$. We assume that the system is fully actuated and we let $\boldsymbol{f}$ denote the control force taking values on $\mathfrak{so}(3)^* \simeq \mathbb{R}^3$. We let $\boldsymbol{R}$ be the rigid body attitude, and $\boldsymbol{\Omega}$ be the body angular velocity. The equations of motion are

$$\begin{aligned}\dot{\boldsymbol{R}}(t) &= \boldsymbol{R}(t)\widehat{\boldsymbol{\Omega}}(t),\\ \dot{\boldsymbol{\Omega}}(t) &= [\mathbb{I}]^{-1}([\mathbb{I}]\boldsymbol{\Omega}(t)\times\boldsymbol{\Omega}(t)) + [\mathbb{I}]^{-1}\boldsymbol{f}.\end{aligned}\tag{11.17}$$

Configuration error function

Here we shall define and characterize an interesting error function on $\mathsf{SO}(3)$. An early study of this function in the context of mechanical control systems is that in [Koditschek 1989]. Let $\boldsymbol{K}_{\mathrm{P}} \in \mathbb{R}^{3\times 3}$ be symmetric positive-definite and define $\boldsymbol{G}_{\mathrm{P}} = \operatorname{tr}(\boldsymbol{K}_{\mathrm{P}})\boldsymbol{I}_3 - \boldsymbol{K}_{\mathrm{P}} \in \mathbb{R}^{3\times 3}$. The reader is asked to verify in Exercise E11.4 that

1. $\boldsymbol{G}_{\mathrm{P}}$ is symmetric,
2. the eigenvectors of $\boldsymbol{K}_{\mathrm{P}}$ and $\boldsymbol{G}_{\mathrm{P}}$ coincide,
3. if $\{k_1, k_2, k_3\}$ are the eigenvalues of $\boldsymbol{K}_{\mathrm{P}}$, then the eigenvalues of $\boldsymbol{G}_{\mathrm{P}}$ are $\{k_2 + k_3, k_1 + k_3, k_1 + k_2\}$, and
4. $\boldsymbol{K}_{\mathrm{P}} = \frac{1}{2}\operatorname{tr}(\boldsymbol{G}_{\mathrm{P}})\boldsymbol{I}_3 - \boldsymbol{G}_{\mathrm{P}}$.

In what follows, let $\{\boldsymbol{k}_1, \boldsymbol{k}_2, \boldsymbol{k}_3\}$ be an orthonormal collection of eigenvectors of $\boldsymbol{K}_{\mathrm{P}}$ and $\boldsymbol{G}_{\mathrm{P}}$ corresponding to the eigenvalues $\{k_1, k_2, k_3\}$ and $\{k_2+k_3, k_1+k_3, k_1+k_2\}$, respectively.

Proposition 11.31. *Consider the function* $\Psi_{\mathsf{SO}(3)} \colon \mathsf{SO}(3) \to \bar{\mathbb{R}}_+$ *given by*

$$\Psi_{\mathsf{SO}(3)}(\boldsymbol{R}) = \tfrac{1}{2}\operatorname{tr}\big(\boldsymbol{G}_{\mathrm{P}}(\boldsymbol{I}_3 - \boldsymbol{R})\big). \tag{11.18}$$

The following statements hold:

(i) $\Psi_{\mathsf{SO}(3)}$ *is a configuration error function about* $\boldsymbol{I}_3$*;*

(ii) with the identification $\mathfrak{so}(3)^* \simeq (\mathbb{R}^3)^* \simeq \mathbb{R}^3$*, the function* $\Psi_{\mathsf{SO}(3)}$ *satisfies* $(T_{\boldsymbol{I}_3}L_{\boldsymbol{R}})^*\left(\mathrm{d}\Psi_{\mathsf{SO}(3)}(\boldsymbol{R})\right) = \operatorname{skew}(\boldsymbol{G}_{\mathrm{P}}\boldsymbol{R})^\vee$*;*

(iii) the critical points of $\Psi_{\mathsf{SO}(3)}$ *are* $\boldsymbol{R} = \exp(l\pi\widehat{\boldsymbol{v}})$ *for* $l \in \{0, 1\}$*, where*

(a) $\boldsymbol{v} \in \{\boldsymbol{k}_1, \boldsymbol{k}_2, \boldsymbol{k}_3\}$*, if* $k_i \neq k_j$ *for* $i, j \in \{1, 2, 3\}$ *and* $i \neq j$*,*

(b) $\boldsymbol{v} \in \left\{\alpha\boldsymbol{k}_1 + \beta\boldsymbol{k}_2 \mid \alpha^2 + \beta^2 = 1\right\} \cup \{\boldsymbol{k}_3\}$*, if* $k_1 = k_2 \neq k_3$*, and*

(c) $\boldsymbol{v} \in \mathbb{S}^2$*, if* $k_1 = k_2 = k_3$*;*

(iv) $L_{\mathrm{reg}}(\Psi_{\mathsf{SO}(3)}, \boldsymbol{I}_3) = \lambda_{\min}(\boldsymbol{K}_{\mathrm{P}}) + \operatorname{tr}(\boldsymbol{K}_{\mathrm{P}})$.

Proof. Clearly, $\Psi_{\mathsf{SO}(3)}$ is smooth and $\Psi_{\mathsf{SO}(3)}(\boldsymbol{I}_3) = 0$. Because $\mathsf{SO}(3)$ is compact, $\Psi_{\mathsf{SO}(3)}$ is proper and bounded from below. To obtain the differential of $\Psi_{\mathsf{SO}(3)}$, let $t \mapsto \boldsymbol{R}(t)$ be an arbitrary curve and compute

$$\frac{\mathrm{d}}{\mathrm{d}t}\Psi_{\mathsf{SO}(3)}(\boldsymbol{R}(t)) = -\tfrac{1}{2}\operatorname{tr}\big(\boldsymbol{G}_{\mathrm{P}}\dot{\boldsymbol{R}}(t)\big) = -\tfrac{1}{2}\operatorname{tr}\big(\boldsymbol{G}_{\mathrm{P}}\boldsymbol{R}(t)\,\boldsymbol{R}^T(t)\dot{\boldsymbol{R}}(t)\big).$$

Recall from Exercise E2.8 that the subspaces of symmetric and skew-symmetric matrices are orthogonal with respect to the inner product $(\boldsymbol{A}, \boldsymbol{B}) \mapsto \operatorname{tr}(\boldsymbol{A}\boldsymbol{B}^T)$. Therefore,

$$\begin{aligned}\frac{\mathrm{d}}{\mathrm{d}t}\Psi_{\mathsf{SO}(3)}(\boldsymbol{R}(t)) &= -\tfrac{1}{2}\operatorname{tr}\Big(\big(\operatorname{skew}(\boldsymbol{G}_{\mathrm{P}}\boldsymbol{R}(t)) + \operatorname{sym}(\boldsymbol{G}_{\mathrm{P}}\boldsymbol{R}(t))\big)\,\big(\boldsymbol{R}^T(t)\dot{\boldsymbol{R}}(t)\big)\Big)\\ &= -\tfrac{1}{2}\operatorname{tr}\big(\operatorname{skew}(\boldsymbol{G}_{\mathrm{P}}\boldsymbol{R}(t))\,(\boldsymbol{R}^T(t)\dot{\boldsymbol{R}}(t))\big).\end{aligned}$$

From the equality $\operatorname{tr}(\widehat{\boldsymbol{x}}\widehat{\boldsymbol{y}}) = -2\boldsymbol{x}\cdot\boldsymbol{y}$ (see Exercise E11.4), we have

$$\frac{\mathrm{d}}{\mathrm{d}t}\Psi_{\mathsf{SO}(3)}(\boldsymbol{R}(t)) = \operatorname{skew}(\boldsymbol{G}_{\mathrm{P}}\boldsymbol{R}(t))^{\vee}\cdot(\boldsymbol{R}^T(t)\dot{\boldsymbol{R}}(t))^{\vee}.$$

This proves (ii).

The critical points of $\Psi_{\mathsf{SO}(3)}$ are the solutions $\boldsymbol{R}\in\mathsf{SO}(3)$ to the equation $\operatorname{skew}(\boldsymbol{G}_{\mathrm{P}}\boldsymbol{R}) = \boldsymbol{0}_{3\times3}$, or equivalently, to the equation $\boldsymbol{G}_{\mathrm{P}}\boldsymbol{R} = \boldsymbol{R}^T\boldsymbol{G}_{\mathrm{P}}$. We invite the reader to complete the proof of part (iii) in Exercise E11.5.

To show that $\Psi_{\mathsf{SO}(3)}$ is locally positive-definite about $\boldsymbol{I}_3$, let us study its Hessian. In a neighborhood of $\boldsymbol{I}_3$ we have $\operatorname{tr}(\boldsymbol{R}) \neq -1$ and we can define $\boldsymbol{x} = \log(\boldsymbol{R})^{\vee}$. From Rodrigues' Formula (5.8), we have

$$\begin{aligned}\Psi_{\mathsf{SO}(3)}(\exp\widehat{\boldsymbol{x}}) &= \tfrac{1}{2}\operatorname{tr}\Big(\boldsymbol{G}_{\mathrm{P}}\Big(-\frac{\sin\|\boldsymbol{x}\|_{\mathbb{R}^3}}{\|\boldsymbol{x}\|_{\mathbb{R}^3}}\widehat{\boldsymbol{x}} - \frac{1-\cos\|\boldsymbol{x}\|_{\mathbb{R}^3}}{\|\boldsymbol{x}\|^2_{\mathbb{R}^3}}\widehat{\boldsymbol{x}}^2\Big)\Big)\\ &= -\frac{1-\cos\|\boldsymbol{x}\|_{\mathbb{R}^3}}{2\|\boldsymbol{x}\|^2_{\mathbb{R}^3}}\operatorname{tr}\left(\boldsymbol{G}_{\mathrm{P}}\widehat{\boldsymbol{x}}^2\right)\\ &= \frac{1-\cos\|\boldsymbol{x}\|_{\mathbb{R}^3}}{2\|\boldsymbol{x}\|^2_{\mathbb{R}^3}}\,\boldsymbol{x}^T(\operatorname{tr}(\boldsymbol{K}_{\mathrm{P}})\boldsymbol{I}_3 + \boldsymbol{K}_{\mathrm{P}})\boldsymbol{x},\end{aligned}$$

where direct manipulations show that $\operatorname{tr}\left(\boldsymbol{G}_{\mathrm{P}}\widehat{\boldsymbol{x}}^2\right) = -\boldsymbol{x}^T(\operatorname{tr}(\boldsymbol{K}_{\mathrm{P}})\boldsymbol{I}_3 + \boldsymbol{K}_{\mathrm{P}})\boldsymbol{x}$; see Exercise E11.4. Because $\lim_{\boldsymbol{x}\to 0}\frac{1-\cos\|\boldsymbol{x}\|_{\mathbb{R}^3}}{\|\boldsymbol{x}\|^2_{\mathbb{R}^3}} = \frac{1}{2}$, it now follows that $\operatorname{Hess}\Psi_{\mathsf{SO}(3)}(\boldsymbol{I}_3) = \frac{1}{2}(\operatorname{tr}(\boldsymbol{K}_{\mathrm{P}})\boldsymbol{I}_3 + \boldsymbol{K}_{\mathrm{P}})$ is positive-definite. This proves (i).

Finally, we turn to (iv). Given the expressions for the critical points in part (iii), we compute

$$\Psi_{\mathsf{SO}(3)}(\exp(\pi\widehat{\boldsymbol{k}}_i)) = \frac{1-\cos\pi}{2}\,\boldsymbol{k}_i^T(\operatorname{tr}(\boldsymbol{K}_{\mathrm{P}})\boldsymbol{I}_3 + \boldsymbol{K}_{\mathrm{P}})\boldsymbol{k}_i = \operatorname{tr}(\boldsymbol{K}_{\mathrm{P}}) + k_i.$$

This shows that $L_{\mathrm{reg}}(\Psi_{\mathsf{SO}(3)}, \boldsymbol{I}_3) = \lambda_{\min}(\boldsymbol{K}_{\mathrm{P}}) + \operatorname{tr}(\boldsymbol{K}_{\mathrm{P}})$. ■

Tracking error functions and velocity errors

Next, we design tracking error functions on $\mathsf{SO}(3)$. Let $\boldsymbol{R}_{\mathrm{ref}}\colon \bar{\mathbb{R}}_+ \to \mathsf{SO}(3)$ denote a twice-differentiable reference trajectory and let $\widehat{\boldsymbol{\Omega}}_{\mathrm{ref}}\colon \bar{\mathbb{R}}_+ \to \mathfrak{so}(3)$ given by $t\mapsto \widehat{\boldsymbol{\Omega}}_{\mathrm{ref}}(t) = \boldsymbol{R}^T_{\mathrm{ref}}(t)\dot{\boldsymbol{R}}_{\mathrm{ref}}(t)$ be the reference body angular velocity. We define right and left tracking error functions by

$$\Psi_{\mathsf{SO}(3),r}(\boldsymbol{R},\boldsymbol{R}_{\mathrm{ref}}) = \Psi_{\mathsf{SO}(3)}(\boldsymbol{R}_{e,r}), \quad\text{and}\quad \Psi_{\mathsf{SO}(3),\ell}(\boldsymbol{R},\boldsymbol{R}_{\mathrm{ref}}) = \Psi_{\mathsf{SO}(3)}(\boldsymbol{R}_{e,\ell}),$$

where the right and left attitude errors are $\boldsymbol{R}_{e,r} = \boldsymbol{R}^T_{\mathrm{ref}}\boldsymbol{R}$ and $\boldsymbol{R}_{e,\ell} = \boldsymbol{R}\boldsymbol{R}^T_{\mathrm{ref}}$. Note that the matrix $\boldsymbol{R}_{e,r}$ describes the relative rotation from the body frame to the reference frame. One can see that $L_{\mathrm{reg}}(\Psi_{\mathsf{SO}(3),\ell},\mathsf{SO}(3)) = L_{\mathrm{reg}}(\Psi_{\mathsf{SO}(3),r},\mathsf{SO}(3)) = L_{\mathrm{reg}}(\Psi_{\mathsf{SO}(3)},\boldsymbol{I}_3) = \lambda_{\min}(\boldsymbol{K}_{\mathrm{P}})$.

The velocity error corresponding to left and right transport maps are

$$\dot{\boldsymbol{e}}_r = \dot{\boldsymbol{R}} - \dot{\boldsymbol{R}}_{\mathrm{ref}}(\boldsymbol{R}_{\mathrm{ref}}^T \boldsymbol{R}) = \boldsymbol{R}\widehat{\boldsymbol{\Omega}}_{\mathrm{e},r},$$
$$\dot{\boldsymbol{e}}_\ell = \dot{\boldsymbol{R}} - (\boldsymbol{R}\boldsymbol{R}_{\mathrm{ref}}^T)\dot{\boldsymbol{R}}_{\mathrm{ref}} = \boldsymbol{R}\widehat{\boldsymbol{\Omega}}_{\mathrm{e},\ell},$$

where we define right and left velocity errors (in the body frame) by

$$\boldsymbol{\Omega}_{\mathrm{e},r} = \boldsymbol{\Omega} - \boldsymbol{R}_{\mathrm{e},r}^T \boldsymbol{\Omega}_{\mathrm{ref}}, \quad \text{and} \quad \boldsymbol{\Omega}_{\mathrm{e},\ell} = \boldsymbol{\Omega} - \boldsymbol{\Omega}_{\mathrm{ref}}.$$

Remarks 11.32. 1. When the gain $\boldsymbol{K}_{\mathrm{P}}$ is a scalar multiple of the identity $k_{\mathrm{P}}\boldsymbol{I}_3$, one can show that the left and right error functions are equal, and all four possible pairs between $(\Psi_{\mathsf{SO}(3),r}, \Psi_{\mathsf{SO}(3),\ell})$ and $(\dot{\boldsymbol{e}}_{\mathrm{e},r}, \dot{\boldsymbol{e}}_{\mathrm{e},\ell})$ are compatible.

2. The pair $(\Psi_{\mathsf{SO}(3),r}, \dot{\boldsymbol{e}}_r)$ appears to be the most common choice in the literature; see for example [Egeland and Godhavn 1994, Koditschek 1989, Meyer 1971, Wen and Kreutz-Delgado 1991]. Left attitude and velocity error appear less frequently, although the design $(\Psi_{\mathsf{SO}(3),\ell}, \dot{\boldsymbol{e}}_\ell)$ leads to simple expressions for feedback and feedforward control; see [Luh, Walker, and Paul 1980]. Coordinate based approaches are also possible. In the flight control literature, Euler angles and their rates are sometimes used; see [Etkin 1995]. •

Control laws and simulations

Let us follow the procedure outlined in Remark 11.30 to compute right and left proportional control laws. We compute the time derivative of the two error functions:

$$\frac{\mathrm{d}}{\mathrm{d}t}\Psi_{\mathsf{SO}(3),r}(\boldsymbol{R}, \boldsymbol{R}_{\mathrm{ref}}) = (\mathrm{skew}(\boldsymbol{G}_{\mathrm{P}}\boldsymbol{R}_{\mathrm{e},r})^\vee) \cdot \boldsymbol{\Omega}_{\mathrm{e},r}, \tag{11.19}$$
$$\frac{\mathrm{d}}{\mathrm{d}t}\Psi_{\mathsf{SO}(3),\ell}(\boldsymbol{R}, \boldsymbol{R}_{\mathrm{ref}}) = \left(\boldsymbol{R}_{\mathrm{ref}}^T \,\mathrm{skew}(\boldsymbol{G}_{\mathrm{P}}\boldsymbol{R}_{\mathrm{e},\ell})^\vee\right) \cdot \boldsymbol{\Omega}_{\mathrm{e},\ell}. \tag{11.20}$$

As expected, equations (11.19) and (11.20) confirm that the right and left velocity error are compatible with the right and left configuration error functions, respectively.

We summarize the design process and the consequences of Theorem 11.29 as follows.

Lemma 11.33. *Consider the system* (11.17). *Assume that the reference body angular velocity* $\widehat{\boldsymbol{\Omega}}_{\mathrm{ref}}\colon \bar{\mathbb{R}}_+ \to \mathfrak{so}(3)$ *is bounded. Let* $\boldsymbol{K}_{\mathrm{P}}, \boldsymbol{K}_{\mathrm{D}} \in \mathbb{R}^{3\times 3}$ *be symmetric positive-definite and define* $\boldsymbol{G}_{\mathrm{P}} = \mathrm{tr}(\boldsymbol{K}_{\mathrm{P}})\boldsymbol{I}_3 - \boldsymbol{K}_{\mathrm{P}} \in \mathbb{R}^{3\times 3}$. *Corresponding to the choice of right and left attitude error, we define*

$$\begin{aligned} f_r = &- \mathrm{skew}(\boldsymbol{G}_{\mathrm{P}}\boldsymbol{R}_{\mathrm{e},r})^\vee - \boldsymbol{K}_{\mathrm{D}}\boldsymbol{\Omega}_{\mathrm{e},r} \\ &+ [\mathbb{I}]^{-1}\Big(\overset{\mathfrak{so}(3)}{\nabla}_{\boldsymbol{\Omega}}(\boldsymbol{R}_{\mathrm{e},r}^T\boldsymbol{\Omega}_{\mathrm{ref}}) + (\boldsymbol{R}_{\mathrm{e},r}^T\boldsymbol{\Omega}_{\mathrm{ref}}) \times \boldsymbol{\Omega} + \boldsymbol{R}_{\mathrm{e},r}^T\dot{\boldsymbol{\Omega}}_{\mathrm{ref}}\Big), \\ f_\ell = &- \boldsymbol{R}_{\mathrm{ref}}^T \,\mathrm{skew}(\boldsymbol{G}_{\mathrm{P}}\boldsymbol{R}_{\mathrm{e},\ell})^\vee - \boldsymbol{K}_{\mathrm{D}}\boldsymbol{\Omega}_{\mathrm{e},\ell} + [\mathbb{I}]^{-1}\Big(\overset{\mathfrak{so}(3)}{\nabla}_{\boldsymbol{\Omega}}\boldsymbol{\Omega}_{\mathrm{ref}}(t) + \dot{\boldsymbol{\Omega}}_{\mathrm{ref}}(t)\Big). \end{aligned}$$

Then, for the closed-loop system defined by f_r, the closed-loop energy $E_{cl,r}(t) = \Psi_{\mathsf{SO}(3)}(\boldsymbol{R}_{e,r}) + \frac{1}{2}\|\boldsymbol{\Omega}_{e,r}\|_{\mathbb{I}}^2$ converges exponentially to zero from all initial conditions $(\boldsymbol{R}(0), \boldsymbol{\Omega}(0))$ such that $E_{cl,r}(0) < \lambda_{\min}(\boldsymbol{K}_{\mathrm{P}})$. A corresponding result holds for the closed-loop system defined by f_{ℓ}.

Finally, we present a simulation of the control laws in Lemma 11.33. To illustrate the difference between the two velocity errors, we run simulations with the gain $\boldsymbol{K}_{\mathrm{P}}$ for the proportional-derivative control set to zero. The reference trajectory is a rotation about the vertical $\boldsymbol{s}_3$-axis performed on the time interval $[0, 10]$ sec; specifically, we set $t \mapsto \boldsymbol{R}_{\mathrm{ref}}(t) = \exp(2\pi(1-\cos\frac{\pi t}{10})\widehat{\boldsymbol{e}}_3) \in \mathsf{SO}(3)$; see Figure 11.3. Note $\boldsymbol{R}_{\mathrm{ref}}(0) = \boldsymbol{R}_{\mathrm{ref}}(10) = \boldsymbol{I}_3$. The initial attitude error is a

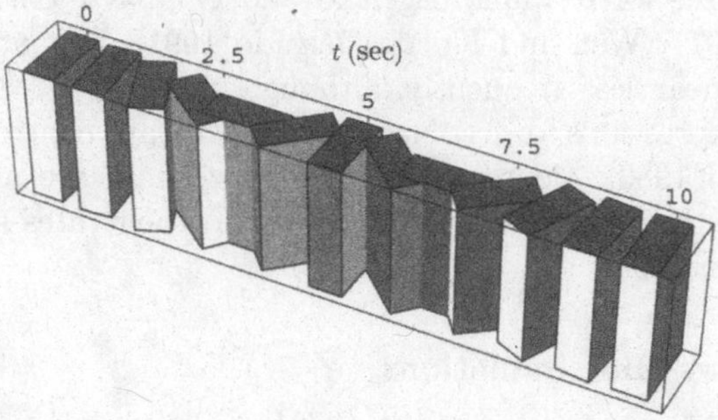

Figure 11.3. Reference trajectory on $\mathsf{SO}(3)$: the bricks orientation represents the rotation matrices and the bricks horizontal displacement represents time

rotation of $\frac{\pi}{4}$ rad about the $\boldsymbol{s}_1$-axis; both the body angular velocity and the reference body angular velocity are zero at time $t = 0$. One can see that the velocity error is therefore zero for all times. We note the different behavior of the two closed-loop simulations in Figure 11.4.

Exercises

E11.1 Consider the function $\Psi_{\mathbb{S}^2} : \mathbb{S}^2 \to \bar{\mathbb{R}}_+$ defined in Section 11.1.3. Prove that its Hessian is positive-definite at $\boldsymbol{r}$.
Hint: *Compute the coordinate representation of $\Psi_{\mathbb{S}^2}$ in the stereographic coordinates for $\mathbb{S}^2$ presented in Example 3.26.*

E11.2 Design an error function $\Psi_{\mathbb{S}^n}$ for $\mathbb{S}^n$ about the point $(1, 0, \dots, 0) \in \mathbb{R}^{n+1}$. Verify that your proposed $\Psi_{\mathbb{S}^n}$ satisfies all properties in Definition 11.3.

E11.3 In this exercise we shall consider a fully actuated simple mechanical control system on the Lie group $\mathsf{SE}(3)$.
(a) Design a configuration error function $\Psi_{\mathsf{SE}(3)}$ about $\boldsymbol{I}_4$ by allowing for as many tunable parameters as possible in your design. Furthermore,

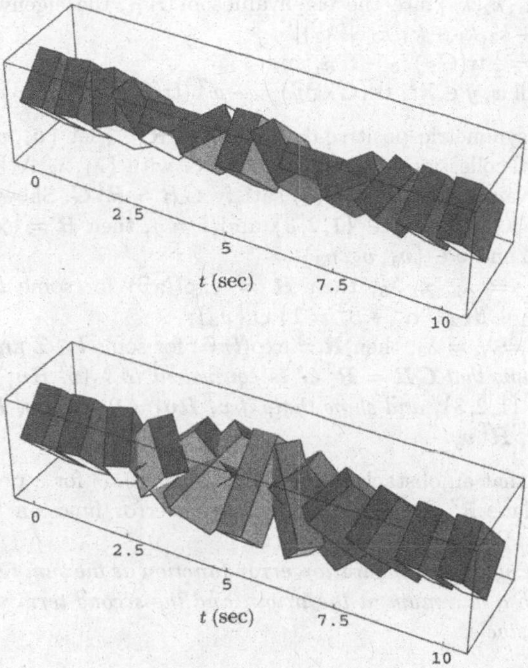

Figure 11.4. Controlled trajectories for f_ℓ (top) and f_r (bottom) with $\boldsymbol{K}_{\mathrm{P}} = 0$

compute the differential of $\Psi_{\mathsf{SE}(3)}$, express it as an element of $\mathfrak{se}(3)^*$ by left translation, and compute the parameter $L_{\mathrm{reg}}(\Psi_{\mathsf{SE}(3)}, \boldsymbol{I}_4)$.

(b) Assuming that $t \mapsto (\boldsymbol{R}_{\mathrm{ref}}(t), \boldsymbol{r}_{\mathrm{ref}}(t))$ is a reference trajectory, obtain coordinate expressions for the right and left group error functions $\boldsymbol{g}_{e,r}$ and $\boldsymbol{g}_{e,\ell}$.

(c) Obtain coordinate expressions for a compatible transport map and corresponding velocity error for tracking error functions induced by your design of $\Psi_{\mathsf{SE}(3)}$ and by the group error functions $\boldsymbol{g}_{e,r}$ and $\boldsymbol{g}_{e,\ell}$.

(d) Consider the two additional group error functions

$$\boldsymbol{g}_{e,1} = \left(\boldsymbol{R}_{\mathrm{ref}}^T \boldsymbol{R},\ \boldsymbol{p} - \boldsymbol{p}_{\mathrm{ref}}\right),$$
$$\boldsymbol{g}_{e,2} = \left(\boldsymbol{R}\boldsymbol{R}_{\mathrm{ref}}^T,\ \boldsymbol{R}^T \boldsymbol{p} - \boldsymbol{R}_{\mathrm{ref}}^T \boldsymbol{p}_{\mathrm{ref}}\right).$$

Compute a compatible transport map and the corresponding velocity error for tracking error functions based on $\boldsymbol{g}_{e,1}$ and $\boldsymbol{g}_{e,2}$.

(e) Write coordinate expressions for the proportional-derivative f_{PD} and feedforward f_{FF} control laws corresponding to the four group error functions $\boldsymbol{g}_{e,r}$, $\boldsymbol{g}_{e,\ell}$, $\boldsymbol{g}_{e,1}$, and $\boldsymbol{g}_{e,2}$.

E11.4 Let $\boldsymbol{K}_{\mathrm{P}} \in \mathbb{R}^{3\times 3}$ be symmetric positive-definite and define $\boldsymbol{G}_{\mathrm{P}} = \mathrm{tr}(\boldsymbol{K}_{\mathrm{P}})\boldsymbol{I}_3 - \boldsymbol{K}_{\mathrm{P}} \in \mathbb{R}^{3\times 3}$. Show that

(a) $\boldsymbol{G}_{\mathrm{P}}$ is symmetric,

(b) the eigenvectors of $\boldsymbol{K}_{\mathrm{P}}$ and $\boldsymbol{G}_{\mathrm{P}}$ coincide,

(c) if $\{k_1, k_2, k_3\}$ are the eigenvalues of $\boldsymbol{K}_{\mathrm{P}}$, the eigenvalues of $\boldsymbol{G}_{\mathrm{P}}$ are $\{k_2 + k_3, k_1 + k_3, k_1 + k_2\}$,
(d) $\boldsymbol{K}_{\mathrm{P}} = \frac{1}{2}\operatorname{tr}(\boldsymbol{G}_{\mathrm{P}})\boldsymbol{I}_3 - \boldsymbol{G}_{\mathrm{P}}$, and
(e) for all $\boldsymbol{x}, \boldsymbol{y} \in \mathbb{R}^3$, $\operatorname{tr}\left(\boldsymbol{G}_{\mathrm{P}}\widehat{\boldsymbol{x}}\widehat{\boldsymbol{y}}\right) = -\boldsymbol{x}^T(\operatorname{tr}(\boldsymbol{K}_{\mathrm{P}})\boldsymbol{I}_3 + \boldsymbol{K}_{\mathrm{P}})\boldsymbol{y}$.

E11.5 Given a symmetric positive-definite $\boldsymbol{G} \in \mathbb{R}^{3\times 3}$, let $\{\boldsymbol{v}_1, \boldsymbol{v}_2, \boldsymbol{v}_3\}$ be an orthonormal collection of eigenvectors of $\boldsymbol{G}$ with $\{\lambda_1, \lambda_2, \lambda_3\}$ the corresponding eigenvalues. Let $\boldsymbol{R} \in \mathsf{SO}(3)$ satisfy $\boldsymbol{G}\boldsymbol{R} = \boldsymbol{R}^T\boldsymbol{G}$. Show that:
(a) if $\lambda_i \neq \lambda_j$ for $i, j \in \{1, 2, 3\}$ and $i \neq j$, then $\boldsymbol{R} = \exp(l\pi\widehat{\boldsymbol{v}})$ for some $l \in \mathbb{Z}$ and $\boldsymbol{v} \in \{\boldsymbol{v}_1, \boldsymbol{v}_2, \boldsymbol{v}_3\}$;
(b) if $\lambda_1 = \lambda_2 \neq \lambda_3$, then $\boldsymbol{R} = \exp(l\pi\widehat{\boldsymbol{v}})$ for some $l \in \mathbb{Z}$ and $\boldsymbol{v} \in \left\{\alpha\boldsymbol{v}_1 + \beta\boldsymbol{v}_2 \mid \alpha^2 + \beta^2 = 1\right\} \cup \{\boldsymbol{v}_3\}$;
(c) if $\lambda_1 = \lambda_2 = \lambda_3$, then $\boldsymbol{R} = \exp(l\pi\widehat{\boldsymbol{v}})$ for some $l \in \mathbb{Z}$ and $\boldsymbol{v} \in \mathbb{S}^2$.
Hint: *Show that* $\boldsymbol{G}\boldsymbol{R} = \boldsymbol{R}^T\boldsymbol{G}$ *is equivalent to* $\lambda_i(\boldsymbol{v}_i^T\boldsymbol{R}\boldsymbol{v}_j) = (\boldsymbol{v}_i^T\boldsymbol{R}^T\boldsymbol{v}_j)\lambda_j$ *for* $i, j \in \{1, 2, 3\}$, *and show that, if* $\boldsymbol{v}_i^T\boldsymbol{R}\boldsymbol{v}_j \neq 0$, *then* $\boldsymbol{v}_i^T\boldsymbol{R}\boldsymbol{v}_j$ *has opposite sign to* $\boldsymbol{v}_i^T\boldsymbol{R}^T\boldsymbol{v}_j$.

E11.6 Suppose that an obstacle is placed at point $(0, 1)$ for a point mass moving in the plane $\mathbb{R}^2$. Design a configuration error function for the manifold $\mathbb{R}^2 \setminus \{(0, 1)\}$ about the origin $(0, 0)$.
Hint: *Design the configuration error function as the sum two terms, the first term with a minimum at the origin and the second term with a singularity at the obstacle.*

12

Stabilization and tracking using oscillatory controls

In this chapter we design oscillatory control laws to solve stabilization and tracking problems for forced affine connection control systems. The objective is to exploit the averaging analysis obtained in Chapter 9 for the purpose of control design. In particular, we shall present results on stabilization and tracking that are applicable to systems that are not linearly controllable. As in the perturbation analysis in Chapter 9, we shall consider smooth systems.

The use of time-dependent and periodic control signals in control theory is motivated by the limitations, discussed in Section 10.1, of smooth state-feedback. This has been studied by numerous researchers. Time-dependent stabilizing feedback laws for driftless systems are discussed, for example, in [Brockett 1983, Coron 1992, Morin, Pomet, and Samson 1999, Pomet 1992]. The use of oscillations in the context of nonlinear and geometric control is surveyed by Baillieul and Lehman [1996] and by Bloch [2003]. Two specific research areas are related to the material developed in this chapter. The first area deals with the design of oscillatory, i.e., large-amplitude, high-frequency controls for point stabilization in general nonlinear and mechanical control systems; see [Baillieul and Lehman 1996, Bellman, Bentsman, and Meerkov 1986, Bentsman 1987, Meerkov 1980, Meerkov and Tsitkin 1975]. The second research area is devoted to the design of oscillatory controls for trajectory planning in driftless systems [Lafferriere and Sussmann 1993, Liu 1997, Sussmann 1992], and for constructive controllability and approximate inversion [Brockett and Dai 1993, Cerven and Bullo 2003, Leonard and Krishnaprasad 1995, Morin and Samson 2003].

The outline of the chapter is as follows. In Section 12.1, we review the averaging results of Chapter 9 and cast them in terms of a certain symmetric bilinear map. We refer to the induced quadratic form as the averaging operator, for which we devise two closed-form inversion algorithms. Based on the results in Sections 12.2 and 12.3, we design oscillatory feedback laws for approximate stabilization and tracking purposes. We illustrate the results on some example systems, including a two-link planar manipulator, a controlled point mass, and a planar vertical takeoff and landing aircraft model. Some results in this chapter are adapted from the papers [Bullo 2002, Martínez, Cortés, and Bullo 2003a].

12.1 The design of oscillatory controls

In this section we present an application of the averaging analysis in Section 9.2 to the design of oscillatory controls. We consider a C^∞-forced affine connection control system $(\mathsf{Q}, \nabla, Y, \mathcal{D}, \mathscr{Y} = \{Y_1, \dots, Y_m\}, \mathbb{R}^m)$ with governing equation

$$\nabla_{\gamma'(t)}\gamma'(t) = Y(t, \gamma'(t)) + \sum_{a=1}^{m} \frac{1}{\varepsilon} u^a\Big(\frac{t}{\varepsilon}, t, \gamma(t)\Big) Y_a(\gamma(t)), \tag{12.1}$$

and with initial condition $\gamma'(0) \in \mathcal{D}$. Note that we allow the control inputs u^a, $a \in \{1, \dots, m\}$, to depend on the system configuration; this is therefore a more general averaging analysis problem than the one addressed by Theorem 9.32 in Section 9.2. Before proceeding, it is useful to state the following assumptions.

Assumptions 12.1. (i) The manifold Q is a bounded open subset of $\mathbb{R}^n$.
(ii) The vector force Y is an affine bundle map of class C^1 in time.
(iii) The distribution $\mathcal{Y}$ generated by $\mathscr{Y}$ is regular. •

12.1.1 The averaging operator

In this subsection we rephrase the results in Theorem 9.32 in a way that is independent of the generators $\{Y_1, \dots, Y_m\}$ for the input distribution $\mathcal{Y}$. This generalization will be helpful in the subsequent control design.

Recall from Definition 9.18 the notion of a two-time-scale vector field with zero mean. Also, recall that (1) the first and second time arguments of a two-time-scale vector field are referred to as the fast and slow time-scale, respectively, and that (2) a two-time-scale vector field has some specific smoothness properties spelled out in Definition 9.18. These notions allow us to formalize the averaging operation as follows.

Definition 12.2 (Averaging operator). Let ∇ be a C^∞-affine connection on Q.

(i) The ***averaging product*** is the map $\mathcal{A}_{[0,T]} : \Gamma^\infty_T(\mathbb{R}; \mathsf{TQ}) \times \Gamma^\infty_T(\mathbb{R}; \mathsf{TQ}) \to \Gamma^\infty(\mathbb{R}; \mathsf{TQ})$ defined by

$$\begin{aligned} &\mathcal{A}_{[0,T]}(V, W)(t, q) \\ &\quad = -\frac{1}{2T}\int_0^T \Big\langle \int_0^{\tau_1} V(\tau_2, t, q)\mathrm{d}\tau_2 : \int_0^{\tau_1} W(\tau_2, t, q)\mathrm{d}\tau_2 \Big\rangle \mathrm{d}\tau_1 \\ &\qquad + \frac{1}{2T^2}\Big\langle \int_0^T\int_0^{\tau_1} V(\tau_2, t, q)\mathrm{d}\tau_2\mathrm{d}\tau_1 : \int_0^T\int_0^{\tau_1} W(\tau_2, t, q)\mathrm{d}\tau_2\mathrm{d}\tau_1 \Big\rangle. \end{aligned}$$

(ii) The ***averaging operator*** is the map $Q_{\mathcal{A}_{[0,T]}}: \Gamma^\infty_T(\mathbb{R}; \mathsf{T}Q) \to \Gamma^\infty(\mathbb{R}; \mathsf{T}Q)$ defined by $Q_{\mathcal{A}_{[0,T]}}(V) = \mathcal{A}_{[0,T]}(V, V)$. •

Remarks 12.3. 1. The averaging analysis in Section 9.2 is performed for two-time-scales vector fields. The physical meaning is that the vector fields depend on a fast time-scale (the first time argument) as well as on a slow time-scale (the second time argument). The averaging operator $Q_{\mathcal{A}_{[0,T]}}$ involves computing time-integrals with respect to the fast time-scale argument.

2. The averaging product $\mathcal{A}_{[0,T]}$ is a symmetric $\mathbb{R}$-bilinear map, and the averaging operator $Q_{\mathcal{A}_{[0,T]}}$ is quadratic in the sense that $Q_{\mathcal{A}_{[0,T]}}(\lambda V) = \lambda^2 Q_{\mathcal{A}_{[0,T]}}(V)$ for all $\lambda \in \mathbb{R}$ and $V \in \Gamma^\infty_T(\mathbb{R}; \mathsf{T}Q)$.

3. Let $\Gamma^r_T(\mathsf{T}Q) \subset \Gamma^r(\mathbb{R}; \mathsf{T}Q)$ be the set of C^r-time-dependent vector fields that are continuous, T-periodic, and zero-mean in the time argument. Any element of $\Gamma^r_T(\mathsf{T}Q)$ can be regarded as a two-time-scale vector field that is independent of its second time argument (i.e., a two-time-scale vector field that does not depend on the slow time-scale). The averaging product and operator on $\Gamma^\infty_T(\mathbb{R}; \mathsf{T}Q)$ naturally induce corresponding maps on $\Gamma^\infty_T(\mathsf{T}Q)$. With a slight abuse of notation, we shall denote these maps with the same symbols $\mathcal{A}_{[0,T]}: \Gamma^\infty_T(\mathsf{T}Q) \times \Gamma^\infty_T(\mathsf{T}Q) \to \Gamma^\infty(\mathsf{T}Q)$, and $Q_{\mathcal{A}_{[0,T]}}: \Gamma^\infty_T(\mathsf{T}Q) \to \Gamma^\infty(\mathsf{T}Q)$. •

The averaging operator is related to an interesting inner product of functions. Let $L_{\mathcal{A}}([0,T])$ denote the set of functions from $\mathbb{R}$ to $\mathbb{R}$ that are continuous, T-periodic, and zero-mean. Consider the map $\langle\langle \cdot, \cdot \rangle\rangle_{L_{\mathcal{A}}([0,T])}: L_{\mathcal{A}}([0,T]) \times L_{\mathcal{A}}([0,T]) \to \mathbb{R}$ given by

$$\begin{aligned}\langle\langle f, g\rangle\rangle_{L_{\mathcal{A}}([0,T])} &= \frac{1}{2T}\int_0^T \left(\int_0^{\tau_1} f(\tau_2)\mathrm{d}\tau_2\right)\left(\int_0^{\tau_1} g(\tau_2)\mathrm{d}\tau_2\right)\mathrm{d}\tau_1 \\ &\quad - \frac{1}{2T^2}\left(\int_0^T\int_0^{\tau_1} f(\tau_2)\mathrm{d}\tau_2\mathrm{d}\tau_1\right)\left(\int_0^T\int_0^{\tau_1} g(\tau_2)\mathrm{d}\tau_2\mathrm{d}\tau_1\right). \quad (12.2)\end{aligned}$$

Exercise E12.1 asks the reader to show that the map $\langle\langle \cdot, \cdot \rangle\rangle_{L_{\mathcal{A}}([0,T])}$ is an inner product on the vector space $L_{\mathcal{A}}([0,T])$. This map is referred to as the $L_{\mathcal{A}}([0,T])$-inner product.

Lemma 12.4. *For $X, Y \in \Gamma^\infty(\mathsf{T}Q)$ and $f, g \in L_{\mathcal{A}}([0,T])$, define $fX, gY \in \Gamma^\infty_T(\mathsf{T}Q)$ by $(t,q) \mapsto f(t)X(q)$ and $(t,q) \mapsto g(t)Y(q)$, respectively. Then*

$$\mathcal{A}_{[0,T]}(fX, gY) = -\langle\langle f, g\rangle\rangle_{L_{\mathcal{A}}([0,T])} \langle X : Y\rangle .$$

Proof. The proof follows immediately from linearity of the integral. ■

The averaging operator allows us to obtain the following statement equivalent to Theorem 9.32.

Corollary 12.5 (Basis independent averaging). *Consider the C^∞-forced affine connection control system $(\mathsf{Q}, \nabla, Y, \mathcal{D}, \mathscr{Y} = \{Y_1, \dots, Y_m\}, \mathbb{R}^m)$ satisfying Assumptions 12.1. Let $W \in \Gamma_T^\infty(\mathbb{R}; \mathcal{Y})$ have the properties that $(t, q) \mapsto W(\tau, t, q)$, for $\tau \in \bar{\mathbb{R}}_+$, are of class C^1, and that $q \mapsto W(\tau, t, q)$, for $(\tau, t) \in \bar{\mathbb{R}}_+ \times \bar{\mathbb{R}}_+$, are commutative. For $\varepsilon \in \mathbb{R}_+$, write the governing equation in the form*

$$\nabla_{\gamma'(t)}\gamma'(t) = Y(t, \gamma'(t)) + \frac{1}{\varepsilon} W\Big(\frac{t}{\varepsilon}, t, \gamma(t)\Big).$$

Then the averaged forced affine connection system is

$$\nabla_{\xi'(t)}\xi'(t) = Y(t, \xi'(t)) + Q_{\mathcal{A}_{[0,T]}}(W)(t, \xi(t)) \tag{12.3}$$

with initial condition $\xi'(0) = \gamma'(0) + \frac{1}{T}\int_0^T \int_0^{\tau_1} W(\tau_2, 0, \gamma(0)) \mathrm{d}\tau_2 \mathrm{d}\tau_1$.

The trajectory of the averaged forced affine connection system approximates the trajectory of the original forced affine connection control system in the sense specified by Theorem 9.32. In particular, there exists $\varepsilon_0, t_0 \in \mathbb{R}_+$ such that, for all $t \in [0, t_0]$ and for all $\varepsilon \in]0, \varepsilon_0[$,

$$\begin{aligned}
\gamma(t) &= \xi(t) + O(\varepsilon), \\
\gamma'(t) &= \xi'(t) + \int_0^{\frac{t}{\varepsilon} \bmod T} W(\tau, t, \xi(t)) \mathrm{d}\tau - \frac{1}{T}\int_0^T \int_0^{\tau_1} W(\tau_2, t, \xi(t)) \mathrm{d}\tau_2 \mathrm{d}\tau_1 + O(\varepsilon).
\end{aligned}$$

Furthermore, stronger results hold if W does not depend on the slow timescale (i.e., $W \in \Gamma_T^\infty(\mathcal{Y})$) and if the averaged forced affine connection system possesses a linearly asymptotically stable equilibrium configuration; see Theorem 9.32.

Proof of Corollary 12.5. Because $q \mapsto W(\tau, t, q)$ takes values in $\Gamma^\infty(\mathcal{Y})$ for all $(\tau, t) \in \bar{\mathbb{R}}_+ \times \bar{\mathbb{R}}_+$, and because the C^∞-distribution $\mathcal{Y}$ is regular, there exist functions $(\tau, t) \mapsto u^a(\tau, t) \in \mathbb{R}$, $a \in \{1, \dots, m\}$, that satisfy $W(\tau, t, q) = \sum_{a=1}^m u^a(\tau, t) Y_a(q)$ and Assumption 9.24. This can be deduced essentially from Theorem 3.114 and Exercise E3.42. To apply Theorem 9.32 we compute the matrix $\mathbf{\Lambda}$ in equation (9.22). By the definition in (12.2),

$$\Lambda_{ab}(t) = \langle\langle u^a, u^b \rangle\rangle_{L_{\mathcal{A}}([0,T])}(t).$$

Here the $L_{\mathcal{A}}([0,T])$-inner product is computed with respect to the fast timescale. Using Lemma 12.4, we compute the vector force appearing in the right-hand side of the averaged system (9.24) as

$$\begin{aligned}
\sum_{a,b=1}^m \Lambda_{ab}(t) \langle Y_a : Y_b \rangle (q) &= \sum_{a,b=1}^m \langle\langle u^a, u^b \rangle\rangle_{L_{\mathcal{A}}([0,T])}(t) \langle Y_a : Y_b \rangle (q) \\
&= -Q_{\mathcal{A}_{[0,T]}}(W)(t, q).
\end{aligned}$$

■

12.1.2 Inverting the averaging operator

Based on the averaging operator, we now consider the following inversion problem that will play a key role in tackling the subsequent control design.

Problem 12.6 (Design of oscillatory controls). Let ∇ and $\mathscr{Y}$ be a C^∞-affine connection and a family of C^∞-vector fields on Q, respectively; let $\mathcal{Y}$ be the distribution generated by $\mathscr{Y}$.

(i) Given a time-dependent vector field $X \in \Gamma^\infty(\mathbb{R}; \mathsf{TQ})$, compute the vector fields $W_{X,\text{slow}} \in \Gamma^\infty(\mathbb{R}; \mathcal{Y})$ and $W_{X,\text{osc}} \in \Gamma^\infty_T(\mathbb{R}; \mathcal{Y})$ such that

$$W_{X,\text{slow}} + Q_{\mathcal{A}_{[0,T]}}(W_{X,\text{osc}}) = X. \tag{12.4}$$

(ii) If $X \in \Gamma^\infty(\mathsf{TQ})$, then compute $W_{X,\text{slow}}$ and $W_{X,\text{osc}}$ satisfying (12.4) as vector fields in $\Gamma^\infty(\mathcal{Y})$ and $\Gamma^\infty_T(\mathcal{Y})$, respectively. •

Remarks 12.7. 1. We refer to Problem 12.6 as an inversion problem in the following sense. Consider the ***augmented averaging operator*** $Q_{\text{aug}} : \Gamma^\infty(\mathbb{R}; \mathcal{Y}) \times \Gamma^\infty_T(\mathbb{R}; \mathcal{Y}) \to \Gamma^\infty(\mathbb{R}; \mathsf{TQ})$ defined by $(W_{\text{slow}}, W_{\text{osc}}) \mapsto W_{\text{slow}} + Q_{\mathcal{A}_{[0,T]}}(W_{\text{osc}})$. Problem 12.6(i) amounts to designing a right inverse for the map Q_{aug}. One should think of $W_{X,\text{slow}}$ and of $W_{X,\text{osc}}$ as being the time-dependent and two-time-scale components of the right inverse of Q_{aug} at X, respectively. An analog statement can be made for Problem 12.6(ii).

2. Problem 12.6 is trivial when the distribution $\mathcal{Y}$ is TQ, i.e., when there are no constraints on the design of $W_{X,\text{slow}}$ and $W_{X,\text{osc}}$. This problem is therefore most interesting when X does not take values in the distribution to which $W_{X,\text{slow}}$ and $W_{X,\text{osc}}$ belong. •

To tackle Problem 12.6, we present a few preliminary concepts. Recall from Section 7.3.2 the definition of the C^∞-distribution $\text{Sym}^{(1)}(\mathcal{Y})$:

$$\text{Sym}^{(1)}(\mathcal{Y})_q = \mathcal{Y}_q + \text{span}_{\mathbb{R}} \{\langle Y : Z\rangle (q) |\; Y \in \mathcal{Y}, Z \in \mathcal{Y}\}.$$

Let us consider the following assumptions on the pair $(\nabla, \mathscr{Y})$.

Assumptions 12.8. (i) The collection $\mathscr{Y} = \{Y_1, \dots, Y_m\}$ has the property that, for all $a \in \{1, \dots, m\}$, $\langle Y_a : Y_a\rangle \in \text{span}_{C^\infty(\mathsf{Q})} \{Y_1, \dots, Y_m\}$.

(ii) The distribution $\text{Sym}^{(1)}(\mathcal{Y})$ is regular. •

Note that Assumption 12.8(i) is equivalent to the existence of smooth functions $\sigma_a^b \in C^\infty(\mathsf{Q})$, $a, b \in \{1, \dots, m\}$, such that, for all $a \in \{1, \dots, m\}$,

$$\langle Y_a : Y_a\rangle = \sum_{b=1}^m \sigma_a^b Y_b. \tag{12.5}$$

Assumption 12.8(i) is also equivalent to $\mathscr{Y}$ being a family of decoupling vector fields for the affine connection control system $(\mathsf{Q}, \nabla, \mathsf{TQ}, \mathscr{Y}, \mathbb{R}^m)$; see Section 8.3.2. In what follows, we shall provide a constructive solution to Problem 12.6 for systems satisfying Assumptions 12.8.

Let us introduce two useful concepts. Firstly, for $T \in \mathbb{R}_+$ and $i \in \mathbb{N}$, define $\varphi_i : \mathbb{R} \to \mathbb{R}$ by

$$\varphi_i(t) = \frac{4\pi i}{T} \cos\left(\frac{2\pi i}{T} t\right). \tag{12.6}$$

Exercise E12.2 asks the reader to verify that $\{\varphi_i\}_{i\in\mathbb{N}}$ is an orthonormal family with respect to the inner product $\langle\!\langle \cdot, \cdot \rangle\!\rangle_{L_{\mathcal{A}}([0,T])}$. Secondly, define the ***lexicographic ordering*** of the set $\{ (a, b) \in \{1, \dots, m\}^2 \mid a < b\}$ as the bijective map lo: $\{ (a, b) \in \{1, \dots, m\}^2 \mid a < b\} \to \{1, \dots, \frac{1}{2}m(m-1)\}$ given by $\mathrm{lo}(a, b) = \sum_{j=1}^{a-1}(n - j) + (b - a)$. For example, if $m = 3$, then the lexicographic ordering is the bijection $\{(1, 2) \mapsto 1, (1, 3) \mapsto 2, (2, 3) \mapsto 3\}$. The functions $\{\varphi_i\}_{i\in\mathbb{N}}$, and the lexicographic ordering of $\{ (a, b) \in \{1, \dots, m\}^2 \mid a < b\}$ are related in the following way. For (a_1, b_1) and (a_2, b_2) in $\{ (a, b) \in \{1, \dots, m\}^2 \mid a < b\}$, we have

$$\left\langle\!\!\left\langle \varphi_{\mathrm{lo}(a_1,b_1)}, \varphi_{\mathrm{lo}(a_2,b_2)} \right\rangle\!\!\right\rangle_{L_{\mathcal{A}}([0,T])} = \delta_{a_1 a_2}\delta_{b_1 b_2}.$$

We are now ready to present an inversion algorithm for the augmented averaging operator.

Proposition 12.9 (Inversion with controllability assumption). *Let ∇ and $\mathscr{Y}$ be a C^∞-affine connection and a family of C^∞-vector fields on Q, respectively. Let $(\nabla, \mathscr{Y})$ satisfy Assumptions 12.8. Let $X \in \Gamma^\infty(\mathbb{R}; \mathrm{Sym}^{(1)}(\mathcal{Y}))$ be of class C^1, and define the functions $\eta^a, \eta^{bc} : \mathbb{R}\times\mathsf{Q} \to \mathbb{R}$, $a, b, c \in \{1, \dots, m\}$, $b < c$, as a solution of class C^1 to*

$$X(t, q) = \sum_{a=1}^{m} \eta^a(t, q) Y_a(q) + \sum_{\substack{b,c=1\\ b<c}}^{m} \eta^{bc}(t, q) \left\langle Y_b : Y_c \right\rangle (q). \tag{12.7}$$

Define $u_{X,\mathrm{slow}} : \mathbb{R} \times \mathsf{Q} \to \mathbb{R}^m$ and $u_{X,\mathrm{osc}} : \mathbb{R} \times \mathbb{R} \times \mathsf{Q} \to \mathbb{R}^m$ by

$$\begin{aligned} u^a_{X,\mathrm{slow}}(t, q) &= \eta^a(t, q) + \sum_{b=1}^{m} \Big(b - 1 + \sum_{i=b+1}^{m} \frac{(\eta^{bi}(t, q))^2}{4}\Big)\sigma^a_b(q) \\ &\quad + \sum_{b=a+1}^{m} \Big(\frac{1}{2}\eta^{ab}(\mathscr{L}_{Y_a}\eta^{ab}) - \mathscr{L}_{Y_b}\eta^{ab}\Big)(t, q), \\ u^a_{X,\mathrm{osc}}(\tau, t, q) &= \sum_{i=1}^{a-1} \varphi_{\mathrm{lo}(i,a)}(\tau) - \frac{1}{2} \sum_{i=a+1}^{m} \eta^{ai}(t, q)\varphi_{\mathrm{lo}(a,i)}(\tau). \end{aligned}$$

Define $W_{X,\mathrm{slow}} \in \Gamma^\infty(\mathbb{R};\mathscr{Y})$ *and* $W_{X,\mathrm{osc}} \in \Gamma^\infty_T(\mathbb{R};\mathscr{Y})$ *by* $W_{X,\mathrm{slow}}(t,q) = \sum_{a=1}^m u^a_{X,\mathrm{slow}}(t,q)Y_a(q)$ *and* $W_{X,\mathrm{osc}}(\tau,t,q) = \sum_{a=1}^m u^a_{X,\mathrm{osc}}(\tau,t,q)Y_a(q)$, *respectively. Then* $W_{X,\mathrm{slow}}$ *and* $W_{X,\mathrm{osc}}$ *solve Problem 12.6(i) and are of class* C^1.

Proof. See Section B.7. ■

Remark 12.10. Proposition 12.9 provides a solution to Problem 12.6(ii) as well. Indeed, if X is time-independent, the procedure in Proposition 12.9 leads to time-independent quantities $u_{X,\mathrm{slow}}$ and $W_{X,\mathrm{slow}}$, and to quantities $u_{X,\mathrm{osc}}$ and $W_{X,\mathrm{osc}}$ that depend on a single time-scale. •

Next, we illustrate this result on a simple example system. In the literature this model is sometimes referred to as a dynamic extension of Brockett's nonholonomic integrator [Brockett 1982].

Example 12.11 (Controlled point mass). We consider a point mass moving in $\mathbb{R}^3$, subject to a strict Rayleigh dissipation function and to two control forces. We let $\boldsymbol{x} = (x^1, x^2, x^3)$ be the coordinates of the point mass. The Rayleigh dissipation function we use is

$$R_{\mathrm{diss}} = -r_{\mathrm{d}}(\mathrm{d}x^1 \otimes \mathrm{d}x^1 + \mathrm{d}x^2 \otimes \mathrm{d}x^2 + \mathrm{d}x^3 \otimes \mathrm{d}x^3),$$

where $r_{\mathrm{d}} > 0$. We select the two control forces $F^1 = \mathrm{d}x^1 + x^2\mathrm{d}x^3$ and $F^2 = \mathrm{d}x^2 + x^1\mathrm{d}x^3$, and write the governing equations as

$$\begin{aligned}\ddot{x}^1 &= -\frac{r_{\mathrm{d}}}{m}\dot{x}^1 + \frac{1}{m}u^1,\\ \ddot{x}^2 &= -\frac{r_{\mathrm{d}}}{m}\dot{x}^2 + \frac{1}{m}u^2,\\ \ddot{x}^3 &= -\frac{r_{\mathrm{d}}}{m}\dot{x}^3 + \frac{1}{m}(u^1x^2 + u^2x^1).\end{aligned} \tag{12.8}$$

The input vector fields $Y_1 = \frac{1}{m}\left(\frac{\partial}{\partial x^1} + x^2\frac{\partial}{\partial x^3}\right)$ and $Y_2 = \frac{1}{m}\left(\frac{\partial}{\partial x^2} + x^1\frac{\partial}{\partial x^3}\right)$ satisfy

$$\langle Y_1 : Y_1\rangle = 0, \quad \langle Y_2 : Y_2\rangle = 0, \quad \langle Y_1 : Y_2\rangle = \frac{2}{m^2}\frac{\partial}{\partial x^3}. \tag{12.9}$$

It can now be easily seen that Assumptions 12.1 and 12.8 are satisfied. Because the input distribution $\mathscr{Y}$ satisfies $\mathrm{Sym}^{(1)}(\mathscr{Y})_{\boldsymbol{x}} = \mathsf{T}_{\boldsymbol{x}}\mathbb{R}^3$ for all $\boldsymbol{x} \in \mathbb{R}^3$, any vector field X can be written as $X = \eta^1 Y_1 + \eta^2 Y_2 + \eta^{12}\langle Y_1 : Y_2\rangle$ for appropriate smooth functions $\eta^1, \eta^2, \eta^{12} \colon \mathbb{R}^3 \to \mathbb{R}$. Next, we apply Proposition 12.9 to compute the pair $(u_{X,\mathrm{slow}}, u_{X,\mathrm{osc}})$ induced by X. We obtain

$$\begin{aligned}u^1_{X,\mathrm{slow}}(\boldsymbol{x}) &= \eta^1(\boldsymbol{x})\left(\tfrac{1}{2}\eta^{12}(\boldsymbol{x})(\mathscr{L}_{Y_1}\eta^{12})(\boldsymbol{x}) - \mathscr{L}_{Y_2}\eta^{12}(\boldsymbol{x})\right),\\ u^2_{X,\mathrm{slow}}(\boldsymbol{x}) &= \eta^2(\boldsymbol{x}),\\ u^1_{X,\mathrm{osc}}(\tau,\boldsymbol{x}) &= -\tfrac{1}{2}\eta^{12}(\boldsymbol{x})\varphi_1(\tau),\\ u^2_{X,\mathrm{osc}}(\tau,\boldsymbol{x}) &= \varphi_1(\tau),\end{aligned} \tag{12.10}$$

where $\varphi_1(t) = \frac{4\pi}{T}\cos\left(\frac{2\pi}{T}t\right)$, for $T \in \mathbb{R}_+$. •

Next, let us analyze Problem 12.6(ii) for input distributions that do not satisfy Assumption 12.8(i). Although a general inversion result as in Proposition 12.9 is not available, some statements can be made. We start with a preliminary result that relies on the natural identification between the set of real symmetric $m \times m$ matrices and $\Sigma_2((\mathbb{R}^m)^*)$.

Lemma 12.12. *Given a positive-semidefinite $K \in \Sigma_2((\mathbb{R}^m)^*)$, define $u_K : \bar{\mathbb{R}}_+ \to \mathbb{R}^m$ by*

$$u_K(t) = \sum_{i=1}^{m} \sqrt{\lambda_i} \boldsymbol{k}_i \varphi_i(t),$$

where $\{\boldsymbol{k}_i\}_{i\in\{1,\dots,m\}}$, is an orthonormal collection of eigenvectors of $[K]$ and $\lambda_i \in \bar{\mathbb{R}}_+$, $i \in \{1,\dots,m\}$, are the corresponding eigenvalues. Then, for each $a, b \in \{1,\dots,m\}$, we have $u_K^a \in L_{\mathcal{A}}([0,T])$ and $\langle\langle u_K^a, u_K^b \rangle\rangle_{L_{\mathcal{A}}([0,T])} = K^{ab}$.

Proof. That $u_K^a \in L_{\mathcal{A}}([0,T])$ is an obvious consequence of u_K^a being a linear combination of the functions $\{\varphi_i\}_{i\in\mathbb{N}}$. Furthermore,

$$\begin{aligned}\langle\langle u_K^a, u_K^b \rangle\rangle_{L_{\mathcal{A}}([0,T])} &= \sum_{i,j=1}^{m} \sqrt{\lambda_i\lambda_j} \boldsymbol{k}_i^a \boldsymbol{k}_j^b \langle\langle \varphi_i, \varphi_j \rangle\rangle_{L_{\mathcal{A}}([0,T])} \\ &= \sum_{i,j=1}^{m} \sqrt{\lambda_i\lambda_j} \boldsymbol{k}_i^a \boldsymbol{k}_j^b \delta_{ij} = \sum_{i=1}^{m} \lambda_i \boldsymbol{k}_i^a \boldsymbol{k}_i^b.\end{aligned}$$

The result follows from the equality $[K] = \sum_{i=1}^{m} \lambda_i \boldsymbol{k}_i \boldsymbol{k}_i^T$; see Exercise E2.16. ■

Next, we can obtain the following inversion result without the controllability assumptions in Proposition 12.9. Recall from Remark 12.3–3 that the averaging operator is well-defined on vector fields in $\Gamma_T^\infty(\mathsf{TQ})$.

Proposition 12.13 (Inversion without controllability assumption). *Let ∇ and $\mathscr{Y}$ be a C^∞-affine connection and a family of C^∞-vector fields on Q, respectively. Let $K \in \Sigma_2((\mathbb{R}^m)^*)$ be positive-semidefinite, and let $X \in \Gamma^\infty(\mathrm{Sym}^{(1)}(\mathcal{Y}))$ satisfy*

$$X = \sum_{a,b=1}^{m} K^{ab} \langle Y_a : Y_b \rangle .$$

Define $u_K : \bar{\mathbb{R}}_+ \to \mathbb{R}^m$ as in Lemma 12.12. Then the time-dependent vector field $(t,q) \mapsto \sum_{a=1}^{m} u_K^a(t) Y_a(q)$ belongs to $\Gamma_T^\infty(\mathsf{TQ})$ and satisfies

$$Q_{\mathcal{A}_{[0,T]}}\Big(\sum_{a=1}^{m} u_K^a Y_a\Big) = X.$$

Proof. This is an immediate consequence of Lemma 12.4. ■

12.2 Stabilization via oscillatory controls

In this section we define stability notions for general nonlinear systems and simple mechanical control systems subject to oscillatory controls. As usual, we let $\varepsilon \in \mathbb{R}_+$ be a parameter characterizing the amplitude and frequency of the oscillatory signal.

Definition 12.14 (Oscillatory feedback and vibrational stabilization). Let $\Sigma = (\mathsf{M}, \mathscr{C} = \{f_0, f_1, \dots, f_m\}, \mathbb{R}^m)$ be a C^∞-control-affine system with a controlled equilibrium point $(x_0, 0) \in \mathsf{M} \times \mathbb{R}^m$.

(i) An ***oscillatory state feedback*** for Σ is a pair $(u_{\mathrm{slow}}, u_{\mathrm{osc}})$, where $u_{\mathrm{slow}} \colon \mathsf{M} \to \mathbb{R}^m$ is a C^∞-state feedback and $u_{\mathrm{osc}} \colon \mathbb{R} \times \mathsf{M} \to \mathbb{R}^m$ is a time-dependent C^∞-state feedback that is a continuous map, and that is periodic, and zero-mean in time.

(ii) For an oscillatory state feedback $(u_{\mathrm{slow}}, u_{\mathrm{osc}})$ for Σ and for $\varepsilon \in \mathbb{R}_+$, the ***$(u_{\mathrm{slow}}, u_{\mathrm{osc}}, \varepsilon)$-closed-loop system*** is the C^∞-time-dependent vector field, that is of class C^0, defined by

$$(t, x) \mapsto f_0(x) + \sum_{a=1}^{m} \left(u^a_{\mathrm{slow}}(x) + \frac{1}{\varepsilon} u^a_{\mathrm{osc}}\left(\frac{t}{\varepsilon}, x\right)\right) f_a(x).$$

(iii) The controlled equilibrium point $(x_0, 0)$ is ***locally vibrationally stabilized*** by the oscillatory state feedback $(u_{\mathrm{slow}}, u_{\mathrm{osc}})$ if, for any neighborhood $\mathcal{U}$ of x_0, there exists $\varepsilon \in \mathbb{R}_+$ such that the $(u_{\mathrm{slow}}, u_{\mathrm{osc}}, \varepsilon)$-closed-loop system possesses a unique locally asymptotically stable periodic trajectory taking values in $\mathcal{U}$.

(iv) A controlled equilibrium point $(x_0, 0)$ is ***locally vibrationally stabilizable*** if there exists an oscillatory state feedback $(u_{\mathrm{slow}}, u_{\mathrm{osc}})$ that locally vibrationally stabilizes it. •

Remark 12.15. We refer the reader to [Bellman, Bentsman, and Meerkov 1986, Bentsman 1987, Meerkov 1980, Meerkov and Tsitkin 1975] for early control-theoretic treatments on vibrational stabilization theory. These works develop a collection of design techniques for certain classes of control-affine systems. •

Next, we adapt the preceding definition to mechanical systems.

Definition 12.16 (Oscillatory feedback and vibrational stabilization for simple mechanical control systems). Let $\Sigma = (\mathsf{Q}, \mathbb{G}, V, F, \mathcal{D}, \mathscr{F}, \mathbb{R}^m)$ be a C^∞-general simple mechanical control system with a controlled equilibrium configuration $(q_0, 0) \in \mathsf{Q} \times \mathbb{R}^m$.

(i) An ***oscillatory state feedback*** for Σ is a pair $(u_{\mathrm{slow}}, u_{\mathrm{osc}})$, where $u_{\mathrm{slow}} \colon \mathsf{TQ} \to \mathbb{R}^m$ is a C^∞-state feedback and $u_{\mathrm{osc}} \colon \mathbb{R} \times \mathsf{TQ} \to \mathbb{R}^m$ is a time-dependent C^∞-state feedback that is a continuous map, and is periodic and zero-mean in time.

(ii) For an oscillatory state feedback $(u_{\text{slow}}, u_{\text{osc}})$ for Σ and for $\varepsilon \in \mathbb{R}_+$, the ***$(u_{\text{slow}}, u_{\text{osc}}, \varepsilon)$-closed-loop system*** is the C^∞-forced simple mechanical system with constraints defined by the 5-tuple $\Sigma_{\text{cl},\varepsilon} = (\mathsf{Q}, \mathbb{G}, V, F_{\text{cl},\varepsilon}, \mathcal{D})$, where

$$F_{\text{cl},\varepsilon}(t, v_q) = F(v_q) + \sum_{a=1}^{m} \Big(u^a_{\text{slow}}(v_q) + \frac{1}{\varepsilon} u^a_{\text{osc}}\Big(\frac{t}{\varepsilon}, v_q\Big)\Big) F^a(q).$$

(iii) The controlled equilibrium configuration $(q_0, 0)$ is ***locally vibrationally stabilized*** by the oscillatory state feedback $(u_{\text{slow}}, u_{\text{osc}})$ if, for any neighborhood $\mathcal{U} \subset \mathsf{Q}$ of q_0, there exists $\varepsilon \in \mathbb{R}_+$ such that the $(u_{\text{slow}}, u_{\text{osc}}, \varepsilon)$-closed-loop system possesses a unique locally asymptotically stable periodic trajectory $\gamma^* \colon \bar{\mathbb{R}}_+ \to \mathsf{Q}$ taking values in $\mathcal{U}$.

(iv) A controlled equilibrium configuration $(q_0, 0)$ is ***locally vibrationally stabilizable*** if there exists an oscillatory state feedback $(u_{\text{slow}}, u_{\text{osc}})$ that locally vibrationally stabilizes it. •

In what follows, we shall tacitly assume that admissible oscillatory state feedbacks are smooth functions of the state in TQ.

Remarks 12.17. 1. It is important to point out the connection between the averaging analysis in Theorem 9.32 and the definition of local vibrational stabilization. In precise terms, the controlled equilibrium configuration $(q_0, 0)$ is locally vibrationally stabilized by $(u_{\text{slow}}, u_{\text{osc}})$ if it is a linearly asymptotically stable equilibrium configuration for the averaged mechanical system corresponding to the $(u_{\text{slow}}, u_{\text{osc}}, \varepsilon)$-closed-loop system.

2. If the controlled equilibrium *point* $(0_{q_0}, 0)$ is locally vibrationally stabilizable, then the controlled equilibrium *configuration* $(q_0, 0)$ is also locally vibrationally stabilizable. The converse is not necessarily true. The distinction is that, in the notion of a locally vibrationally stabilizable equilibrium configuration, no requirement is imposed on the velocity trajectory of the closed-loop system. •

12.2.1 Stabilization with the controllability assumption

A range of stabilization results can be obtained by combining oscillatory controls with linear PD control and nonlinear PD controls, and by assuming that the mechanical control system satisfies, or does not satisfy, Assumptions 12.8. We present some detailed results in this and in the next subsection, and leave it to the reader to deduce some of the possible extensions. The results in these subsections build on Theorem 10.32 on stabilization via linear PD control. In that theorem, a state feedback was designed to stabilize a controlled equilibrium configuration under an assumption on the Hessian of the potential function. Here we shall require only a weaker version of that stabilizability assumption, and we will design an oscillatory state feedback based on a controllability condition.

We begin by describing the setup. In this subsection we consider a C^∞-simple mechanical control system with dissipation, $\Sigma = (\mathsf{Q}, \mathbb{G}, V, -R^\flat_{\text{diss}}, \mathscr{F} = \{F^1, \dots, F^m\}, \mathbb{R}^m)$. We let $(q_0, 0)$ be a controlled equilibrium configuration, and $(\mathsf{T}_{q_0}\mathsf{Q}, \mathbb{G}(q_0), K_\Sigma(q_0), (0, -R^\flat_{\text{diss}}(q_0)), F_\Sigma(q_0))$ be the linearization of Σ at q_0. As usual, we define $\mathscr{Y} = \{Y_1, \dots, Y_m\}$ by means of $Y_a = \mathbb{G}^\sharp(F^a)$ for $a \in \{1, \dots, m\}$, and we let $\mathcal{Y}$ be the C^∞-distribution generated by the vector fields $\mathscr{Y}$. We will also assume that Σ satisfies Assumptions 12.1 and 12.8.

Now, we set about introducing some convenient notation. Let $\mathcal{U}$ be a neighborhood of q_0, and let $\{T^1, \dots, T^k\}$ be a family of C^∞-covector fields with the property that, for each $q \in \mathcal{U}$,

$$\mathrm{Sym}^{(1)}(\mathcal{Y})_q = \mathrm{span}_{\mathbb{R}}\left\{\mathbb{G}^\sharp(T^1)(q), \dots, \mathbb{G}^\sharp(T^k)(q)\right\}.$$

Observe that the first m of these k covector fields can be chosen to be $\{F^1, \dots, F^m\}$. Given these covector fields, we define the C^∞-simple mechanical control system with dissipation $\Sigma_{\text{ext}} = (\mathsf{Q}, \mathbb{G}, V, -R^\flat_{\text{diss}}, \{T^1, \dots, T^k\}, \mathbb{R}^k)$. We shall refer to Σ_{ext} as the ***extension of the simple mechanical control system*** $\boldsymbol{\Sigma}$, in the sense that it is the same system endowed with additional control inputs. Clearly, $(q_0, 0)$ is a controlled equilibrium configuration for Σ_{ext} and the linearization of Σ_{ext} at q_0 is $(\mathsf{T}_{q_0}\mathsf{Q}, \mathbb{G}(q_0), K_\Sigma(q_0), (0, -R^\flat_{\text{diss}}(q_0)), T_{\text{ext}}(q_0))$, where the map $T_{\text{ext}}(q_0) \in L(\mathbb{R}^k; \mathsf{T}^*_{q_0}\mathsf{Q})$ is given by

$$T_{\text{ext}}(q_0) \cdot u = \sum_{a=1}^{k} u^a T^a(q_0).$$

This construction is useful because of the following observation.

Remark 12.18. Because $(\nabla, \mathscr{Y})$ satisfy Assumptions 12.8, the inversion results in Proposition 12.9 are applicable. Because of Assumptions 12.1 and Corollary 12.5, any trajectory of the closed-loop system Σ_{ext}, subject to a state feedback of the form $u \colon \mathsf{Q} \to \mathbb{R}^k$, can be approximated by a closed-loop trajectory for the system Σ subject to an appropriate *oscillatory* state feedback. Therefore, our objective is to locally vibrationally stabilize $(q_0, 0)$ for the system Σ by designing a state feedback that locally asymptotically stabilizes $(q_0, 0)$ for the extension system Σ_{ext}. •

We are now ready for the main result in this subsection.

Theorem 12.19 (Vibrational stabilization with controllability assumption). *Let $\Sigma = (\mathsf{Q}, \mathbb{G}, V, -R^\flat_{\text{diss}}, \mathscr{F}, \mathbb{R}^m)$ be a C^∞-simple mechanical control system with dissipation satisfying Assumptions 12.1 and 12.8. Let $(q_0, 0)$ be a controlled equilibrium configuration, let $(\phi, \mathcal{U}_0, \mathcal{U}_1)$ be a near identity diffeomorphism at q_0, and let $K_{\mathrm{D}} \in \Sigma_2((\mathbb{R}^m)^*)$ be positive-semidefinite. Assume the following:*

(i) $K_\Sigma(q_0)|\mathrm{coann}(\mathrm{image}(T_{\mathrm{ext}}(q_0)))$ *is positive-definite so that there exists* $K_\mathrm{P} \in \Sigma_2((\mathbb{R}^k)^*)$ *for which* $K_\Sigma(q_0) + T_{\mathrm{ext}}(q_0) \circ K_\mathrm{P}^\sharp \circ T_{\mathrm{ext}}(q_0)^*$ *is positive-definite;*

(ii) $\langle \mathbb{G}(q_0)^\sharp \circ (K_\Sigma(q_0) + T_{\mathrm{ext}}(q_0) \circ K_\mathrm{P}^\sharp \circ T_{\mathrm{ext}}(q_0)^*)^\flat, \mathsf{W}_{q_0}\rangle = \mathsf{T}_{q_0}\mathsf{Q}$*, where*

$$\mathsf{W}_{q_0} = \mathrm{image}(\mathbb{G}(q_0)^\sharp \circ R_{\mathrm{diss}}(q_0)^\flat) + \mathrm{image}(\mathbb{G}(q_0)^\sharp \circ F_\Sigma(q_0)).$$

Define the derivative state feedback $u_\mathrm{D} \colon \mathsf{TQ} \to \mathbb{R}^m$ *and the proportional state feedback* $u_\mathrm{P} \colon \mathsf{Q} \to \mathbb{R}^k$ *for* Σ_{ext} *by*

$$u_\mathrm{D}(v_q) = -K_\mathrm{D}^\sharp \circ F_\Sigma(q)(v_q), \quad u_\mathrm{P}(q) = -K_\mathrm{P}^\sharp \circ T^*_{\Sigma_{\mathrm{ext}}}(q_0)(\phi^{-1}(q)).$$

Let $(u_{\mathrm{P,slow}}, u_{\mathrm{P,osc}})$ *be the oscillatory state feedback for* Σ *computed according to Proposition 12.9, so that*

$$\sum_{b=1}^k u_\mathrm{P}^b \mathbb{G}^\sharp(T^b) = \sum_{a=1}^m u_{\mathrm{P,slow}}^a Y_a + Q_{\mathcal{A}_{[0,T]}}\Big(\sum_{a=1}^m u_{\mathrm{P,osc}}^a Y_a\Big).$$

Then the oscillatory state feedback $(u_{\mathrm{P,slow}} + u_\mathrm{D}, u_{\mathrm{P,osc}})$ *locally vibrationally stabilizes* $(q_0, 0)$ *for* Σ.

Proof. Recall from Theorem 10.32(i) that, if $K_\Sigma(q_0)|\mathrm{coann}(\mathrm{image}(T_{\mathrm{ext}}(q_0)))$ is positive-definite, then there exists $K_\mathrm{P} \in \Sigma_2((\mathbb{R}^k)^*)$ such that $K_\Sigma(q_0) + T_{\mathrm{ext}}(q_0) \circ K_\mathrm{P} \circ T_{\mathrm{ext}}(q_0)^*$ is positive-definite and the linear proportional feedback u_P stabilizes the equilibrium configuration $(q_0, 0)$ for Σ_{ext}. Additionally, the vector field $q \mapsto \sum_{a=1}^k u_\mathrm{P}^a(q)\mathbb{G}^\sharp(T^a)(q)$ belongs to $\Gamma^\infty(\mathrm{Sym}^{(1)}(\mathscr{Y}))$ and, therefore, the results in Proposition 12.9 apply. The final statement is a consequence of Theorem 9.32 and Corollary 12.5: $(q_0, 0)$ being linearly asymptotically stable for the averaged system implies that $(q_0, 0)$ is locally vibrationally stabilized. ■

Remark 12.20. There exist a simple mechanical control system with dissipation, say Σ, and a controlled equilibrium configuration, say $(q_0, 0)$, with the following properties: (1) Σ does not have a controllable linearization at q_0, (2) Σ fails Brockett's Necessary Condition for the existence of a C^0-locally asymptotically stabilizing state feedback at q_0 (see Theorem 10.8), and (3) Σ satisfies the sufficient conditions for vibrational stabilization in Theorem 12.19. A system Σ and a controlled equilibrium configuration $(q_0, 0)$ with these properties are discussed in Example 12.21. Thus Theorem 12.19 genuinely extends the analysis of Section 10.3, and, furthermore, does so to a class of systems for which stabilization by state feedback is difficult. •

Example 12.21 (Controlled point mass (cont'd)). We consider the controlled point mass introduced in Example 12.11. Exercise E12.5 invites the reader to show that the controlled point mass system is not linearly controllable and does not satisfy Brockett's Necessary Condition. Here we design an oscillatory state feedback that locally vibrationally stabilizes the controlled equilibrium configuration $(\boldsymbol{x}_0, 0)$, for $\boldsymbol{x}_0 = (0,0,0)$. We start by characterizing the extension system Σ_{ext}. As described in Example 12.11, the family $\{Y_1, Y_2\}$ satisfies Assumptions 12.8. We choose $T^1 = F^1$, $T^2 = F^2$, and $T^3 = \mathbb{G}^\flat(\langle Y_1 : Y_2\rangle)$; that is,

$$T^1 = \mathrm{d}x^1 + x^2\mathrm{d}x^3, \quad T^2 = \mathrm{d}x^2 + x^1\mathrm{d}x^3, \quad T^3 = \frac{2}{m}\mathrm{d}x^3.$$

Because the extension system Σ_{ext} is fully actuated, it is easy to design a proportional state feedback to stabilize the controlled equilibrium configuration $(\boldsymbol{x}_0, 0)$. With the notation in Theorem 12.19, we set $u_{\mathrm{D}} = 0$, and, for $k_i \in \mathbb{R}_+$, $i \in \{1, 2, 3\}$,

$$u_{\mathrm{P}}(\boldsymbol{x}) = -(k_1 x^1, k_2 x^2, k_3 x^3).$$

One can verify that the controlled equilibrium configuration $(\boldsymbol{x}_0, 0)$ is locally asymptotically stable for the closed-loop system obtained by applying u_{P} to the extension system Σ_{ext}.

Next, we apply the algorithm in Proposition 12.9, cf. Example 12.11, to compute the oscillatory state feedback $(u_{\mathrm{P,slow}}, u_{\mathrm{P,osc}})$ induced by $X_{\mathrm{P}}(\boldsymbol{x}) = (-k_1x^1)Y_1(\boldsymbol{x}) + (-k_2x^2)Y_2(\boldsymbol{x}) + (-k_3x^3)\langle Y_1 : Y_2\rangle(\boldsymbol{x})$. After some straightforward calculation, we obtain

$$\begin{aligned} u^1_{\mathrm{P,slow}}(\boldsymbol{x}) &= \Big(\frac{k_3}{m} - k_1\Big)x^1 + \frac{k_3^2}{2m}x^2x^3, \\ u^2_{\mathrm{P,slow}}(\boldsymbol{x}) &= -k_2x^2, \\ u^1_{\mathrm{P,osc}}(\tau, \boldsymbol{x}) &= \frac{k_3}{2}x^3\varphi_1(\tau), \\ u^2_{\mathrm{P,osc}}(\tau, \boldsymbol{x}) &= \varphi_1(\tau), \end{aligned} \tag{12.11}$$

where $\varphi_1(t) = \frac{4\pi}{T}\cos\left(\frac{2\pi}{T}t\right)$, for $T \in \mathbb{R}_+$. Theorem 12.19 guarantees that the oscillatory state feedback $(u_{\mathrm{P,slow}}, u_{\mathrm{P,osc}})$ locally vibrationally stabilizes $(\boldsymbol{x}_0, 0)$ for Σ.

We run a simulation as follows. Regarding the system's physical parameters, we set $m = 1$ and $r_{\mathrm{d}} = \frac{1}{2}$. We design the control law parameters as $\varepsilon = 0.1$, $T = 2\pi$, $k_1 = k_2 = k_3 = 1$. At initial time, the point mass is at rest with configuration $\boldsymbol{x}(0) = (1, 1, 1)$. The results are shown in Figure 12.1. •

12.2.2 Stabilization without the controllability assumption

The results in this subsection build on Theorem 10.43 on stabilization via a nonlinear PD control law, i.e., via potential function shaping and dissipative control, and on Theorem 9.36 on averaging via the Beltrami bracket and the averaged potential function. For the notion of totally regular codistribution, we refer the reader to Section 10.4.1.

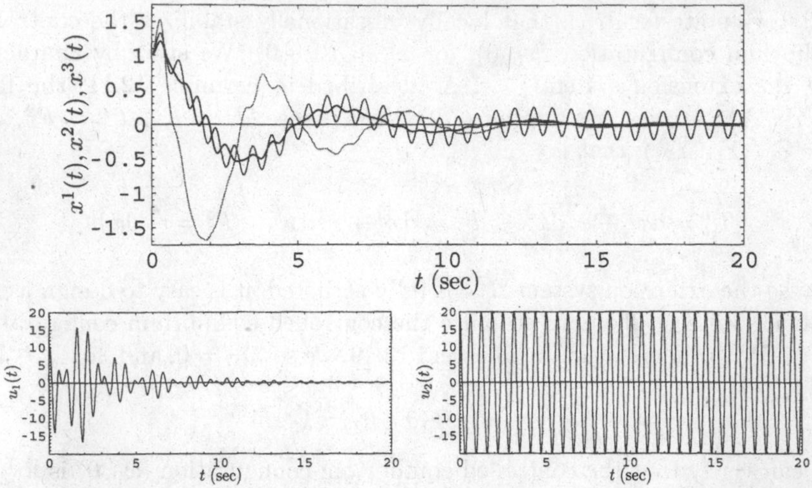

Figure 12.1. Vibrational stabilization of the point mass. The controlled trajectory is shown in the top figure, where the black line is x^1, the gray line is x^2, and the light gray line is x^3. The controls are shown in the bottom figures.

Theorem 12.22 (Vibrational stabilization without controllability assumption). *Let* $\Sigma = (\mathsf{Q}, \mathbb{G}, V, -R^{\flat}_{\text{diss}}, \mathscr{F}, \mathbb{R}^m)$ *be a* C^∞*-simple mechanical control system with dissipation. Let* $(q_0, 0)$ *be a controlled equilibrium configuration and let* $(\mathsf{T}_{q_0}\mathsf{Q}, \mathbb{G}(q_0), K_\Sigma(q_0), (0, -R^{\flat}_{\text{diss}}(q_0)), F_\Sigma(q_0))$ *be the linearization of* Σ *at* q_0*. Assume the following:*

(i) the codistribution generated by $\mathscr{F}$ *is totally regular at* q_0*, and* $\mathscr{F} = \{\mathrm{d}\phi^1, \dots, \mathrm{d}\phi^l, F^{l+1}, \dots, F^m\}$ *is proportionally adapted at* q_0*;*

(ii) there exist positive-definite $K_{\mathrm{P}}, K_{\mathrm{osc}} \in \Sigma_2((\mathbb{R}^l)^*)$ *such that the function* $V_{\text{avg,cl}} \colon \mathsf{Q} \to \mathbb{R}$*, defined by*

$$V_{\text{avg,cl}} = V + \frac{1}{2}\sum_{a,b=1}^{l} (K_{\mathrm{P}})^{ab}\phi^a\phi^b + \sum_{a,b=1}^{l} (K_{\mathrm{osc}})^{ab}\langle \phi^a : \phi^b\rangle,$$

has the property that $\mathrm{d}V_{\text{avg,cl}}(q_0) = 0_{q_0}$ *and* $\operatorname{Hess} V_{\text{avg,cl}}(q_0)$ *is positive-definite;*

(iii) $\langle \mathbb{G}(q_0)^\sharp \circ (\operatorname{Hess} V_{\text{avg,cl}})^\flat, \mathsf{W}_{q_0}\rangle = \mathsf{T}_{q_0}\mathsf{Q}$*, where*

$$\mathsf{W}_{q_0} = \operatorname{image}(\mathbb{G}(q_0)^\sharp \circ R_{\text{diss}}(q_0)^\flat) + \operatorname{image}(\mathbb{G}(q_0)^\sharp \circ F_\Sigma(q_0)).$$

Then, for all positive-semidefinite $K_{\mathrm{D}} \in \Sigma_2((\mathbb{R}^m)^*)$*, and for* $u_{K_{\mathrm{osc}}} \colon \bar{\mathbb{R}}_+ \to \mathbb{R}^m$ *computed as in Lemma 12.12, the oscillatory state feedback* $(u_{\text{slow}}, u_{\text{osc}})$*, defined by*

$$u_{\mathrm{slow}}(v_q) = -\sum_{a,b=1}^{l} (K_{\mathrm{P}})^{ab}\phi^b(q)e_a - K_{\mathrm{D}}^{\sharp}\circ F_{\Sigma}(q)(v_q),$$
$$u_{\mathrm{osc}}(\tau, v_q) = \sum_{a=1}^{l} u^a_{K_{\mathrm{osc}}}(\tau)e_a,$$

locally vibrationally stabilizes $(q_0, 0)$.

Proof. The input distribution $\mathscr{Y}$ is regular in a neighborhood of q_0 because $\mathscr{F}$ is totally regular. According to Theorem 9.36, Corollary 12.5, and Proposition 12.13, the averaged system is the simple mechanical system with dissipation $(\mathsf{Q}, \mathbb{G}, V_{\mathrm{avg,cl}}, -R^{\flat}_{\mathrm{diss}} - F_{\Sigma}\circ K_{\mathrm{D}}^{\sharp}\circ F^*_{\Sigma})$. As in Theorem 10.43, the local asymptotic stability of $(q_0, 0)$ for this system is a consequence of Theorem 6.42. The final statement is a consequence of Theorem 9.36: because $(q_0, 0)$ is linearly asymptotically stable for the averaged system, $(q_0, 0)$ is locally vibrationally stable. ■

Remark 12.23 (Averaged closed-loop potential function). We refer to the function $V_{\mathrm{avg,cl}}$ defined in part (ii) as the ***averaged closed-loop potential function***. It is useful to note that $V_{\mathrm{avg,cl}}$ depends linearly on the components of K_{P}, and K_{osc}. Therefore, the requirement that $\mathrm{d}V_{\mathrm{avg,cl}}(q_0)$ vanishes and that $\mathrm{Hess}\,V_{\mathrm{avg,cl}}(q_0)$ is positive-definite can be cast as a so-called linear matrix inequality in the free variables K_{P}, and K_{osc}. Efficient numerical methods are available to solve linear matrix inequalities, e.g., see [Boyd and Vandenberghe 2004]. •

We conclude this subsection with a simple example of stabilization via oscillatory controls.

Example 12.24 (Two-link planar manipulator (cont'd)). We consider the two-link planar manipulator looked at first in Section 1.2. The system has configuration manifold $\mathsf{Q} = \mathbb{S}^1 \times \mathbb{S}^1$. We choose the coordinate chart $(\mathcal{V}, \psi)$ defined in Example 4.8, that we reproduce for convenience

$$\begin{aligned}\mathcal{V} = \mathbb{S}^1\times\mathbb{S}^1 \setminus (\{((x_1,y_1),(x_2,y_2))\in\mathbb{S}^1\times\mathbb{S}^1 \mid x_1 = -1\}\\ \cup\{((x_1,y_1),(x_2,y_2))\in\mathbb{S}^1\times\mathbb{S}^1 \mid x_2 = -1\}),\\ \psi((x_1,y_1),(x_2,y_2)) = (\mathrm{atan}(x_1,y_1), \mathrm{atan}(x_2,y_2))\\ = (\theta_1,\theta_2) \in]-\pi,\pi[\times]-\pi,\pi[\subset\mathbb{R}^2.\end{aligned}$$

In what follows we restrict consideration to the mechanical system defined on $\mathcal{V}\subset\mathbb{S}^1\times\mathbb{S}^1$. We let (v_1, v_2) be the angular velocities of the two joints. The kinetic energy metric can be found, for example, in Example 4.32. Here, we assume that no potential function is present, and that the manipulator is subject to a strict Rayleigh dissipation force of the form $-r_{\mathrm{d}}(v_1\mathrm{d}\theta_1 + v_2\mathrm{d}\theta_2)$, for $r_{\mathrm{d}}\in\mathbb{R}_+$. We also assume that the system is equipped with a single control force; that is, a torque applied at the first joint. This control force is a potential force with respect to the function $\phi\colon \mathcal{V} \to \mathbb{R}$ given by $(\theta_1, \theta_2) \mapsto \theta_1$. (Note that this force is not a potential force on the entire configuration manifold Q.) The Beltrami bracket $\langle \phi : \phi \rangle$ is computed according to Lemma 9.35:

$$\langle \phi : \phi \rangle (\theta_1, \theta_2) = \frac{4(4J_2 + l_2^2 m_2)}{(4J_2 + l_2^2 m_2)(4J_1 + l_1^2(m_1 + 4m_2)) - 4l_1^2 l_2^2 m_2^2 \cos(\theta_1 - \theta_2)^2}.$$

We adopt the control law in Theorem 12.22. For k_{D}, k_{P}, and $k_{\mathrm{osc}} \in \mathbb{R}_+$, we apply Proposition 12.13 and compute

$$\begin{aligned} u_{\mathrm{slow}}(\theta_1, \theta_2, v_1, v_2) &= -k_{\mathrm{P}}\theta_1 - k_{\mathrm{D}} v_1, \\ u_{\mathrm{osc}}(\tau, \theta_1, \theta_2, v_1, v_2) &= \sqrt{k_{\mathrm{osc}}}\varphi_1(\tau), \end{aligned}$$

with $\varphi_1(t) = \frac{4\pi}{T}\cos\left(\frac{2\pi}{T}t\right)$, for $T \in \mathbb{R}_+$, as defined in (12.6). We write the averaged closed-loop potential function as

$$V_{\mathrm{avg,cl}}(\theta_1, \theta_2) = \frac{1}{2} k_{\mathrm{P}} \theta_1^2 + k_{\mathrm{osc}} \langle \phi : \phi \rangle (\theta_1, \theta_2). \tag{12.12}$$

Exercise E12.4 asks the reader to verify that, for any positive k_{P} and k_{osc}, the function $V_{\mathrm{avg,cl}}$ has two critical points with positive-definite Hessian at $(\theta_1, \theta_2) = (0, \pm\frac{\pi}{2})$. By Theorem 12.22, the two configurations $(\theta_1, \theta_2) = (0, \pm\frac{\pi}{2})$ are locally vibrationally stabilized by this oscillatory state feedback.

We run a simulation as follows. Regarding the system's physical parameters, we set $\ell_1 = 3$, $\ell_2 = 4$, $m_1 = J_1 = \ell_1^2$, $m_2 = J_2 = \ell_2^2$, and $r_{\mathrm{d}} = \frac{1}{10}$. We design the control law parameters as $T = 1$, $k_{\mathrm{D}} = \frac{1}{10}$, $k_{\mathrm{P}} = 5$, and $k_{\mathrm{osc}} = 50$. It turns out that, even for relatively large values of ε, the closed-loop system performs satisfactorily. Accordingly, we set $\varepsilon = 2$. At the initial time, the manipulator is at rest with configuration $(\theta_1(0), \theta_2(0)) = (0, \frac{\pi}{8})$. Using Theorem 6.45(iii), one can verify that, for the averaged closed-loop system, this initial condition belongs to the domain of attraction of the minimum $(\theta_1, \theta_2) = (0, \frac{\pi}{2})$. The results are shown in Figure 12.2.

We conclude the example by remarking that this vibrational stabilization result is similar to the classic vibrational stabilization of the inverted pendulum example, see [Baillieul and Lehman 1996], where the controlled variable is the speed of the joint connected to the second link, and where the joint itself is constrained to move vertically. •

12.3 Tracking via oscillatory controls

In this section we consider the following approximate tracking problem: given a reference curve $\gamma_{\mathrm{ref}}\colon I \to \mathsf{Q}$, design oscillatory controls laws such that, in the limit as $\varepsilon \to 0$, the solution to the closed-loop system approximates γ_{ref}. We assume that the interval I is of finite length, independent of the parameter

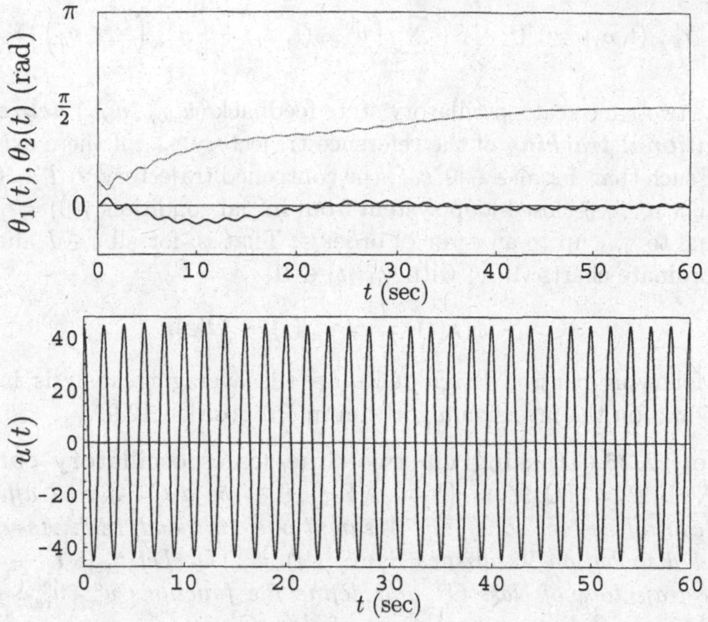

Figure 12.2. Vibrational stabilization of the two-link manipulator. The controlled trajectory is shown in the top figure, where the black line is θ^1, and the gray line is θ^2. The control is shown in the bottom figure.

ε. In this section we present the results for forced affine connection control systems rather than for general simple mechanical control systems. We start with some formal nomenclature.

Definition 12.25 (Vibrational tracking for forced affine connection control systems). Let $(\mathsf{Q}, \nabla, Y, \mathcal{D} = \mathsf{TQ}, \mathscr{Y} = \{Y_1, \ldots, Y_m\}, \mathbb{R}^m)$ be a C^∞-forced affine connection control system. Let I be a bounded interval containing 0, and let $\gamma_{\text{ref}}\colon I \to \mathsf{Q}$ be a curve of class C^2, called the ***reference trajectory***.

(i) A ***two-time-scales oscillatory state feedback*** for Σ is a pair $(u_{\text{slow}}, u_{\text{osc}})$, where $u_{\text{slow}}\colon I \times \mathsf{TQ} \to \mathbb{R}^m$ and $u_{\text{osc}}\colon \mathbb{R} \times I \times \mathsf{TQ} \to \mathbb{R}^m$ have the following properties:
   (a) u_{slow} is of class C^1 and of class C^∞ in its second argument;
   (b) u_{osc} is of class C^0, of class C^1 in its second argument, and of class C^∞ in its third argument;
   (c) u_{osc} is periodic and zero-mean in the first time argument.

(ii) For a two-time-scales oscillatory state feedback $(u_{\text{slow}}, u_{\text{osc}})$ for Σ and for $\varepsilon \in \mathbb{R}_+$, the ***$(u_{\text{slow}}, u_{\text{osc}}, \varepsilon)$-closed-loop system*** is the C^∞-forced affine connection system defined by the 4-tuple $(\mathsf{Q}, \nabla, Y_{\varepsilon\text{-cl}}, \mathcal{D} = \mathsf{TQ})$ where

$$Y_{\varepsilon\text{-cl}}(t, v_q) = Y(t, v_q) + \sum_{a=1}^{m} \left(u^a_{\text{slow}}(t, v_q) + \frac{1}{\varepsilon} u^a_{\text{osc}}\left(\frac{t}{\varepsilon}, t, v_q\right)\right) Y_a(q).$$

(iii) The two-time-scales oscillatory state feedback $(u_{\text{slow}}, u_{\text{osc}})$ ***achieves vibrational tracking*** of the reference trajectory γ_{ref}, if there exists $\varepsilon_0 \in \mathbb{R}_+$ such that, for all $\varepsilon \in]0, \varepsilon_0[$, the controlled trajectory $\gamma\colon I \to \mathsf{Q}$ of the $(u_{\text{slow}}, u_{\text{osc}}, \varepsilon)$-closed-loop system from initial condition $\gamma(0) = \gamma_{\text{ref}}(0)$ is equal to γ_{ref} up to an error of order ε. That is, for all $t \in I$ and for all coordinate charts $(\mathcal{U}, \phi)$ with $\gamma_{\text{ref}}(t) \in \mathcal{U}$,

$$\phi(\gamma(t)) = \phi(\gamma_{\text{ref}}(t)) + O(\varepsilon).$$ •

The following control design relies on the averaging analysis in Corollary 12.5 and on the inversion algorithm in Proposition 12.9.

Theorem 12.26 (Tracking via two-time-scales oscillatory controls). *Let $(\mathsf{Q}, \nabla, Y, \mathcal{D} = \mathsf{TQ}, \mathscr{Y} = \{Y_1, \dots, Y_m\}, \mathbb{R}^m)$ be a C^∞-forced affine connection control system satisfying Assumptions 12.1 and 12.8. Assume that the input distribution $\mathcal{Y}$ satisfies $\operatorname{Sym}^{(1)}(\mathcal{Y}) = \mathsf{TQ}$. Let $\gamma_{\text{ref}}\colon I \to \mathsf{Q}$ be a reference trajectory of class C^3 and define the functions $u^a_{\text{ref}}, u^{bc}_{\text{ref}}\colon I \to \mathbb{R}$, $a, b, c \in \{1, \dots, m\}, b < c$, as solutions of class C^1 to*

$$\nabla_{\gamma'_{\text{ref}}(t)} \gamma'_{\text{ref}}(t) - Y(t, \gamma'_{\text{ref}}(t)) = \sum_{a=1}^{m} u^a_{\text{ref}}(t) Y_a(\gamma_{\text{ref}}(t)) + \sum_{\substack{b,c=1\\ b<c}}^{m} u^{bc}_{\text{ref}}(t) \langle Y_b : Y_c \rangle (\gamma_{\text{ref}}(t)).$$

Define the two-time-scales oscillatory state feedback $(u_{X,\text{slow}}, u_{X,\text{osc}})$ by

$$u^a_{X,\text{slow}}(t, v_q) = u^a_{\text{ref}}(t) + \sum_{b=1}^{m} \left(b - 1 + \sum_{c=b+1}^{m} \frac{(u^{bc}_{\text{ref}}(t))^2}{4} \right) \sigma^a_b(q),$$

$$u^a_{X,\text{osc}}(\tau, t, v_q) = \sum_{c=1}^{a-1} \varphi_{\text{lo}(c,a)}(\tau) - \frac{1}{2} \sum_{c=a+1}^{m} u^{ac}_{\text{ref}}(t) \varphi_{\text{lo}(a,c)}(\tau).$$

Then $(u_{X,\text{slow}}, u_{X,\text{osc}})$ achieves vibrational tracking of γ_{ref}.

Proof. Let $X\colon I \to \mathsf{TQ}$ be the vector field along γ_{ref} defined by $X(t) = \nabla_{\gamma'_{\text{ref}}(t)} \gamma'_{\text{ref}}(t) - Y(t, \gamma'_{\text{ref}}(t))$. The existence of C^1-functions $u^a_{\text{ref}}, u^{bc}_{\text{ref}}\colon I \to \mathbb{R}$, $a, b, c \in \{1, \dots, m\}, b < c$, as specified in the statement follows from the regularity of $\operatorname{Sym}^{(1)}(\mathcal{Y})$. It is possible to extend X to a time-dependent vector field on Q via

$$X(t, q) = \sum_{a=1}^{m} u^a_{\text{ref}}(t) Y_a(q) + \sum_{\substack{b,c=1\\ b<c}}^{m} u^{bc}_{\text{ref}}(t) \langle Y_b : Y_c \rangle (q).$$

Clearly, X takes values in $\mathrm{Sym}^{(1)}(\mathcal{Y}) = \mathsf{T}\mathsf{Q}$. Therefore, Proposition 12.9 states that

$$X(t,q) = W_{X,\mathrm{slow}}(t,q) + Q_{\mathcal{A}_{[0,T]}}(W_{X,\mathrm{osc}})(t,q),$$

where $W_{X,\mathrm{slow}}(t,q) = \sum_{a=1}^{m} u^a_{X,\mathrm{slow}}(\tau,t,q)Y_a(q)$ and $W_{X,\mathrm{osc}}(\tau,t,q) = \sum_{a=1}^{m} u^a_{X,\mathrm{osc}}(\tau,t,q)Y_a(q)$, and where $u^a_{X,\mathrm{slow}}$ and $u^a_{X,\mathrm{osc}}$ are defined according to Proposition 12.9. Because of this construction, the solution ξ to the averaged system in equation (12.3) satisfies the same differential equation with the same initial conditions as the reference curve γ_{ref}. Therefore, $\xi = \gamma_{\mathrm{ref}}$. The result now follows from Corollary 12.5. ■

Next, we illustrate this result on two example systems taken from the paper [Martínez, Cortés, and Bullo 2003a].

Example 12.27 (Controlled point mass (cont'd)). We apply our two-time-scales oscillatory design methods to the controlled point mass of Examples 12.11 and 12.21; as shown in these exercises, this system satisfies Assumptions 12.1 and 12.8. Also, the input distribution $\mathcal{Y}$ satisfies $\mathrm{Sym}^{(1)}(\mathcal{Y}) = \mathsf{T}\mathbb{R}^3$. These are the assumptions in Theorem 12.26. Accordingly, we design control inputs to track a smooth reference trajectory, $t \mapsto (x^1_{\mathrm{ref}}(t), x^2_{\mathrm{ref}}(t), x^3_{\mathrm{ref}}(t))$, via

$$\begin{aligned} u^1(\varepsilon,t) &= u^1_{\mathrm{ref}}(t) - \frac{1}{2\varepsilon}\varphi_1\Big(\frac{t}{\varepsilon}\Big)u^{12}_{\mathrm{ref}}(t), \\ u^2(\varepsilon,t) &= u^2_{\mathrm{ref}}(t) + \frac{1}{\varepsilon}\varphi_1\Big(\frac{t}{\varepsilon}\Big), \end{aligned} \tag{12.13}$$

where $T = 1$, so that $\varphi_1(t) = 2\cos t$, and where $t \mapsto (u^1_{\mathrm{ref}}(t), u^2_{\mathrm{ref}}(t), u^{12}_{\mathrm{ref}}(t))$ are

$$\begin{aligned} u^1_{\mathrm{ref}}(t) &= m\ddot{x}^1_{\mathrm{ref}}(t) + r_{\mathrm{d}}\dot{x}^1_{\mathrm{ref}}(t), \\ u^2_{\mathrm{ref}}(t) &= m\ddot{x}^2_{\mathrm{ref}}(t) + r_{\mathrm{d}}\dot{x}^2_{\mathrm{ref}}(t), \\ u^{12}_{\mathrm{ref}}(t) &= \frac{m}{2}\Big(m\ddot{x}^3_{\mathrm{ref}}(t) + r_{\mathrm{d}}\dot{x}^3_{\mathrm{ref}}(t) - \big(u^1_{\mathrm{ref}}(t)x^2_{\mathrm{ref}}(t) + u^2_{\mathrm{ref}}(t)x^1_{\mathrm{ref}}(t)\big)\Big). \end{aligned}$$

The simulations are run with the parameters $m = 1$, $r_{\mathrm{d}} = \frac{1}{2}$, $\varepsilon = 0.01$, and with reference trajectory $t \mapsto (x^1_{\mathrm{ref}}(t), x^2_{\mathrm{ref}}(t), x^3_{\mathrm{ref}}(t)) = (\cos(4t)-1, \sin(2t), 0)$; see Figure 12.3. •

Example 12.28 (PVTOL aircraft). We consider a simple model of a planar vertical takeoff and landing (PVTOL) aircraft with Rayleigh dissipation forces; see Figure 12.4. This example was first studied in the control literature by Hauser, Sastry, and Meyer [1992]. This model is a variation of the usual planar rigid body example analyzed as a system on the Lie group $\mathsf{SE}(2)$ in Example 5.47. We parametrize the configuration and velocity of the aircraft via the state variables $(\theta, x, y, \omega, v_x, v_y)$. We let x and y be the horizontal and vertical displacement of the aircraft, respectively, and θ be its roll angle. The angular velocity is ω and the linear velocities in the body x-axis and y-axis

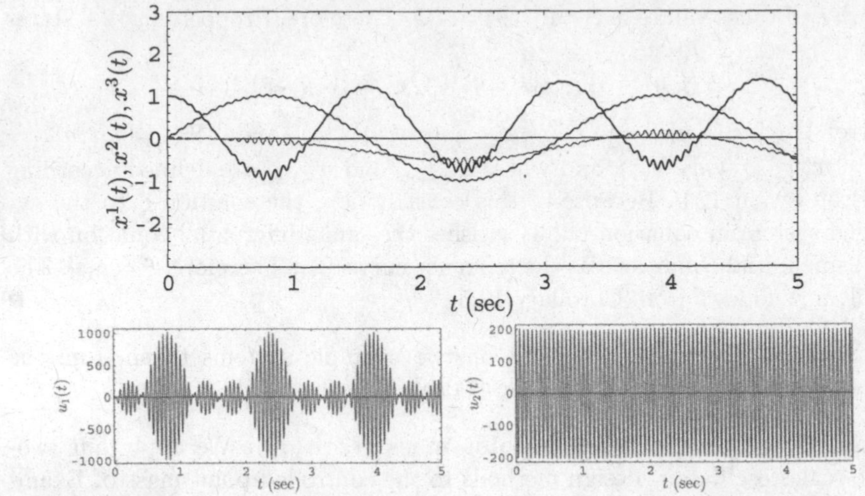

Figure 12.3. Vibrational tracking for the point mass. The controlled trajectory is shown in the top figure, where the black line is x^1, the gray line is x^2, and the light gray line is x^3. The controls are shown in the bottom figures.

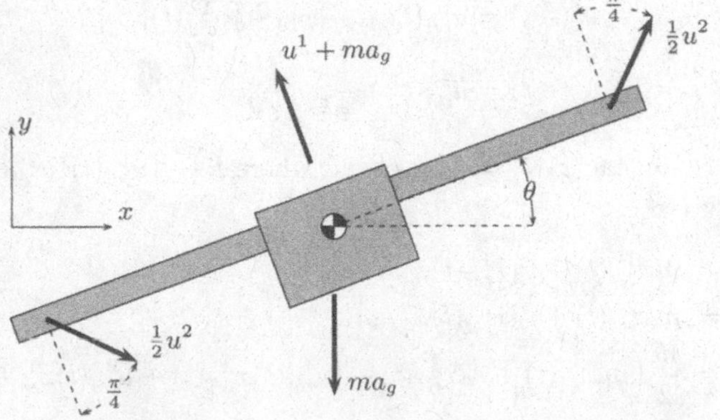

Figure 12.4. The PVTOL aircraft model

are v_x and v_y, respectively. The control u^1 corresponds to the body vertical force minus gravity, while u^2 corresponds to coupled forces on the tips of the wings with a net horizontal component. The other forces depend upon the constants $k_i \in \mathbb{R}_+$, $i \in \{1, 2, 3\}$, which parameterize a Rayleigh dissipation force, and a_g, the gravitational constant. The constant h is the distance from the center of mass to the wingtip, while m and J are mass and moment of inertia, respectively, of the body. Adapting the treatment in Example 5.47, the controlled Euler–Poincaré equations are

$$
\begin{aligned}
\dot{\theta} &= \omega, \\
\dot{x} &= v_x \cos\theta - v_y \sin\theta, \\
\dot{y} &= v_x \sin\theta + v_y \cos\theta, \\
\dot{\omega} &= -\frac{k_3}{J}\omega + \frac{h}{J}u^2, \\
\dot{v}_x - v_y\omega &= -\frac{k_1}{m}v_x - a_g \sin\theta + \frac{1}{m}u^2, \\
\dot{v}_y + v_x\omega &= -\frac{k_2}{m}v_y - a_g(\cos\theta - 1) + \frac{1}{m}u^1.
\end{aligned}
$$

Our objective is to design control inputs to track a reference trajectory $t \mapsto (\theta_{\text{ref}}(t), x_{\text{ref}}(t), y_{\text{ref}}(t))$ using the algorithm outlined in Theorem 12.26. Because the two input vector fields are left-invariant vector fields on $\mathsf{SE}(2)$, it is convenient to write them as elements of $\mathfrak{se}(2)$ using notation from Example 5.47:

$$
\boldsymbol{y}_1 = \frac{1}{m}\widehat{\boldsymbol{e}}_3, \quad \boldsymbol{y}_2 = \frac{h}{J}\widehat{\boldsymbol{e}}_1 + \frac{1}{m}\widehat{\boldsymbol{e}}_2.
$$

Because the Levi-Civita affine connection is left-invariant, we compute symmetric products via equation (5.21) as

$$
\langle \boldsymbol{y}_1 : \boldsymbol{y}_1 \rangle = 0, \quad \langle \boldsymbol{y}_2 : \boldsymbol{y}_2 \rangle = \frac{2h}{Jm}\widehat{\boldsymbol{e}}_3 = \frac{2h}{J}\boldsymbol{y}_1, \quad \langle \boldsymbol{y}_1 : \boldsymbol{y}_2 \rangle = -\frac{h}{Jm}\widehat{\boldsymbol{e}}_2. \tag{12.14}
$$

Therefore, this system satisfies Assumptions 12.1 and 12.8. In particular, equation (12.5) is satisfied by $\sigma_2^1 = \frac{2h}{J}$, with all other coefficients $\sigma_a^b \in \mathbb{R}$, $a, b \in \{1, 2\}$, being zero. Furthermore, we have $\operatorname{span}_{\mathbb{R}}\{\boldsymbol{y}_1, \boldsymbol{y}_2, \langle \boldsymbol{y}_1 : \boldsymbol{y}_2 \rangle\} = \mathfrak{se}(2)$.

Following the procedure detailed in Theorem 12.26, we now define functions $t \mapsto (u^1_{\text{ref}}(t), u^2_{\text{ref}}(t), u^{12}_{\text{ref}}(t))$ such that

$$
\begin{aligned}
\ddot{\theta}_{\text{ref}} + \frac{k_3}{J}\dot{\theta}_{\text{ref}} &= \frac{h}{J}u^2_{\text{ref}}, \\
(\dot{v}_x)_{\text{ref}} - (v_y)_{\text{ref}}\dot{\theta}_{\text{ref}} + \frac{k_1}{m}(v_x)_{\text{ref}} + a_g \sin\theta_{\text{ref}} &= \frac{1}{m}u^2_{\text{ref}} - \frac{h}{Jm}u^{12}_{\text{ref}}, \\
(\dot{v}_y)_{\text{ref}} + (v_x)_{\text{ref}}\dot{\theta}_{\text{ref}} + \frac{k_2}{m}(v_y)_{\text{ref}} + a_g(\cos\theta_{\text{ref}} - 1) &= \frac{1}{m}u^1_{\text{ref}},
\end{aligned}
$$

where $(v_x)_{\text{ref}} = \dot{x}_{\text{ref}}\cos\theta_{\text{ref}} + \dot{y}_{\text{ref}}\sin\theta_{\text{ref}}$ and $(v_y)_{\text{ref}} = -\dot{x}_{\text{ref}}\sin\theta_{\text{ref}} + \dot{y}_{\text{ref}}\cos\theta_{\text{ref}}$. After some careful bookkeeping, the final control laws are

$$
\begin{aligned}
u^1(\varepsilon, t) &= u^1_{\text{ref}}(t) - \frac{1}{2\varepsilon}u^{12}_{\text{ref}}(t)\varphi_1\left(\frac{t}{\varepsilon}\right) + \frac{2h}{J}, \\
u^2(\varepsilon, t) &= u^2_{\text{ref}}(t) + \frac{1}{\varepsilon}\varphi_1\left(\frac{t}{\varepsilon}\right),
\end{aligned} \tag{12.15}
$$

for $T = 2\pi$ so that $\varphi_1(t) = 2\cos t$. The simulations are run with parameters $m = 20$, $J = 10$, $h = 5$, $k_1 = 12$, $k_2 = 11$, $k_3 = 10$, $a_g = 9.8$, and with reference

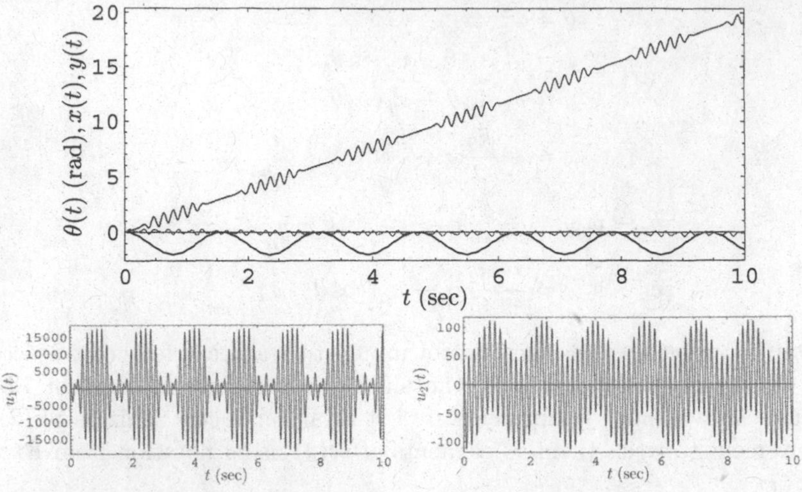

Figure 12.5. Vibrational tracking for the PVTOL aircraft model. The controlled trajectory is shown in the top figure, where the black line is θ, the gray line is x, and the light gray line is y. The controls are shown in the bottom figures.

trajectory $t \mapsto (\theta_{\text{ref}}(t), x_{\text{ref}}(t), x_{\text{ref}}(t)) = (\cos(4t) - 1, 2t, 0)$. As opposed to the stabilization problem in Example 12.24, accurate tracking performance for the PVTOL system is achieved only for relatively small values of ε. Figure 12.5 illustrates the behavior of the two-time-scales oscillatory controls at $\varepsilon = 0.025$. Although these large-amplitude, high-frequency signals are in line with the theoretical forecasts, the resulting control laws are likely not realizable in physical settings. •

Exercises

E12.1 Show that the map $(f, g) \mapsto \langle\langle f, g \rangle\rangle_{L_A([0,T])}$ defined in equation (12.2) is an inner product on the vector space $L_A([0, T])$.
Hint: *Refer to the proof of Lemma 9.31.*

E12.2 Show that $\{\varphi_i\}_{i \in \mathbb{N}}$, defined in equation (12.6), is an orthonormal family of vectors.

E12.3 Given a family of vector fields $Y_a \in \Gamma^\infty(\mathsf{TQ})$, $a \in \{1, \dots, m\}$, and a family of functions $f^a \in C^\infty(\mathsf{Q})$, $a \in \{1, \dots, m\}$, show that $\langle f^a Y_a : f^b Y_b \rangle = f^a f^b \langle Y_a : Y_b \rangle + 2(f^b \mathscr{L}_{Y_b} f^a) Y_a$.

E12.4 For the two-link planar manipulator in Example 12.24, verify that the function $V_{\text{avg,cl}} : \mathcal{V} \to \mathbb{R}$, defined in equation (12.12), has two critical points with positive-definite Hessian at $(\theta_1, \theta_2) = (0, \pm\frac{\pi}{2})$.

E12.5 Consider the controlled point mass system of Examples 12.11, 12.21, and 12.27. Let $\boldsymbol{x}_0 = (0, 0, 0) \in \mathbb{R}^3$.

(a) Show that this system does not have a controllable linearization at x_0.
(b) Show that this system does not satisfy Brockett's Necessary Condition for the existence of a C^0-locally asymptotically stabilizing state feedback at x_0.
(c) Verify the symmetric product computations in (12.9).
(d) Verify that the computations of $(u_{X,\text{slow}}, u_{X,\text{osc}})$ in (12.10) and of $(u_{P,\text{slow}}, u_{P,\text{osc}})$ in (12.11) are consistent with Proposition 12.9.
(e) Verify that the control laws in (12.13) are designed according to Theorem 12.26.

E12.6 Consider the oscillatory control laws designed in Example 12.28 for the PVTOL aircraft model.
(a) Verify the symmetric product computations in (12.14).
(b) Verify that the control law in (12.15) is designed according to Theorem 12.26.
(c) If the reference trajectory were the trivial curve $t \mapsto (\theta_{\text{ref}}(t), x_{\text{ref}}(t), y_{\text{ref}}(t)) = (0, 0, 0)$, what undesirable feature does the control law (12.15) exhibit?
(d) Change the order of the input vector fields, i.e., let $\boldsymbol{y}_2 = \frac{1}{m}\widehat{\mathbf{e}}_3$, and $\boldsymbol{y}_1 = \frac{h}{J}\widehat{\mathbf{e}}_1 + \frac{1}{m}\widehat{\mathbf{e}}_2$, and redesign the oscillatory control law. What are the terms that change and why?

13

Motion planning for underactuated systems

In Chapter 8 we saw that it was possible to come to a fairly complete understanding of low-order controllability of affine connection control systems, and that this sort of controllability was closely related to the notion of kinematic controllability. Implicit in these observations is that kinematic controllability is useful for motion planning for affine connection control systems. In this chapter we illustrate this by looking in detail at a couple of nontrivial examples. The material in this chapter has been developed only recently in the research literature. As mentioned in Chapter 8, the idea of a kinematic reduction is inspired by the work of Arai, Tanie, and Shiroma [1998], and Lynch, Shiroma, Arai, and Tanie [2000]. A first formal definition was given in [Bullo and Lynch 2001], while more complete work followed in the papers [Bullo and Lewis 2003a, Martínez, Cortés, and Bullo 2003b]. The kinematic reduction results of Section 8.3 provide a means of reducing the order of the dynamical systems being considered from two to one. The idea of lowering the complexity of representations of mechanical control systems can be related to numerous previous efforts, including work on hybrid models for motion control systems [Brockett 1993], oscillatory motion primitives [Bullo, Leonard, and Lewis 2000], consistent control abstractions [Pappas and Simić 2002], and maneuver automata [Frazzoli, Dahleh, and Feron 2003]. For a general introduction to motion planning, we refer to [Latombe 1991].

The material in the chapter is as follows. Motion planning for driftless systems is reviewed in Section 13.1. After considering the various types of natural motion planning algorithms for driftless systems, we provide a survey of the literature on this subject. There is no perfectly satisfactory methodology for motion planning for driftless systems, so this is the most significant limitation of the methods in this chapter. In Section 13.2 we present two methodologies for motion planning for affine connection control systems, one valid for kinematically controllable (KC) systems, and another valid for maximally reducibly kinematically controllable (MR-KC) systems. In Section 13.3 we look at the planar body and the robotic leg as systems that are KC and MR-KC, respectively, and we show that it is easy to devise *explicit* motion control algorithms for these systems, using the methods presented in Section 13.2. In Section 13.4 we provide motion planning algorithms for a certain mechanical system with nonholonomic constraints, known as the snakeboard, following Bullo and Lewis [2003a]. Since we have not to this point looked at the snakeboard, we begin by looking at the modeling of the snakeboard, and we then provide the motion control algorithms. From many points of view, the snakeboard is a nontrivial example. It provides an excellent illustration of the approach to modeling, analysis, and design presented in the book.

13.1 Motion planning for driftless systems

Some of the motion planning problems we consider for mechanical systems rely on the solution of a corresponding motion control problem for a driftless system. In this section we formulate the standard motion control problems for driftless systems, and provide a brief overview of what is a fairly extensive literature on solving, or approximately solving, these problems.

13.1.1 Definitions

Driftless systems comprise an interesting class of nonlinear control systems. On the one hand, they are challenging in that they cannot be stabilized using continuous state feedback (Exercise E10.6). However, in the class of nonlinear systems, they are in other ways fairly simple. For example, for analytic driftless systems, controllability and accessibility are equivalent (Corollary 7.25). Thus the matter of deciding controllability for driftless systems is quite simple. This raises the question of whether one can actually find controls that steer the system from one state to another. We state this as the following problem, recalling from Section 8.3.2 our somewhat formalized discussion of driftless systems. Note that we consider only the simplest form of motion planning, where it is desired to steer between two points, and where there are no additional constraints on the path that is to be constructed.

Problem 13.1 (Motion planning for driftless systems). Let $(\mathsf{M}, \mathscr{X}, U)$ be a C^∞-driftless system and let $\mathscr{U}$ be a collection of U-valued locally integrable inputs. The ***$\mathscr{U}$-motion planning problem*** is:

> Given $x_0, x_1 \in \mathsf{M}$, find $u \in \mathscr{U}$, defined on some interval $[0, T]$, so that the controlled trajectory (γ, u) with $\gamma(0) = x_0$ has the property that $\gamma(T) = x_1$. •

The nature of the controls $\mathscr{U}$ can change the methodology one uses to solve the $\mathscr{U}$-motion planning problem. Certain sets of controls will be of interest here, and we wish to assign to the corresponding motion planning problems special names. Let us list some common ones, and the only ones we will use.

1. Let $U = \{e_1, \dots, e_m, -e_1, \dots, -e_m\}$ and denote by $\mathscr{U}^m_{\text{prim}}$ the collection of piecewise constant U-valued controls. Then the $\mathscr{U}^m_{\text{prim}}$-motion planning problem is called the ***motion planning problem using primitives***. Note that, in this case, a controlled trajectory (γ, u) will have the property that γ is a concatenation of integral curves, possibly running backwards in time, of the vector fields $X_1, \dots, X_m$. One should think of each component in such a concatenation of being a "primitive" maneuver.
2. If $U = \mathbb{R}^m$ and $\mathscr{U} = \mathscr{U}^m_{\text{kin}}$, the set of locally absolutely continuous controls (see Section 8.3.1), then the $\mathscr{U}$-motion planning problem is called the ***motion planning problem with locally absolutely continuous inputs***. In Exercise E13.1 the reader is asked to show that the existence of a solution to the motion planning problem using primitives implies the existence of a solution to the motion planning problem using locally absolutely continuous controls.

By allowing the motion planning problem to depend on the set of controls, some doubt is raised as to whether the existence of solutions to the motion planning problem depends on the nature of the controls. Indeed, the local controllability result of Corollary 7.25 is for locally integrable inputs, and so it is not obvious that, if $\text{Lie}^{(\infty)}(\mathcal{X}) = \mathsf{TM}$, then the motion planning problem using primitives admits a solution. The following result addresses this issue, and at the same time shows that the local controllability result of Corollary 7.25 can be extended globally.

Theorem 13.2 (Motion planning using primitives). *Let* $(\mathsf{M}, \mathscr{X} = \{X_1, \dots, X_m\}, \mathbb{R}^m)$ *be a* C^∞*-driftless system with the vector fields* $X_1, \dots, X_m$ *being complete, and suppose that* $\text{Lie}^{(\infty)}(\mathcal{X}) = \mathsf{TM}$. *If* M *is connected, then, for each* $x_0, x_1 \in \mathsf{M}$*, there exist* $k \in \mathbb{N}$*,* $t_1, \dots, t_k \in \mathbb{R}$*, and* $a_1, \dots, a_k \in \{1, \dots, m\}$ *such that*

$$x_1 = \Phi^{X_{a_k}}_{t_k} \circ \dots \circ \Phi^{X_{a_1}}_{t_1}(x_0).$$

Proof. Let us first prove a local version of the result. Thus we shall prove that there is a neighborhood $\mathcal{U}$ of x_0 such that the theorem holds, provided that $x_1 \in \mathcal{U}$. Since $\text{Lie}^{(0)}(\mathcal{X})_{x_0} \neq \{0_{q_0}\}$, there exists $a_1 \in \{1, \dots, m\}$ such that $X_{a_1}(x_0) \neq 0_{x_0}$. For $T_1 > 0$, define

$$\mathsf{A}_1 = \left\{\Phi^{X_{a_1}}_{t_1}(x_0) \mid t_1 \in]0, T_1[\right\}.$$

We claim that there exist $a_2 \in \{1, \dots, m\}$ and $x \in \mathsf{A}_1$ such that $X_{a_2}(x) \notin \mathsf{T}_x\mathsf{A}_1$. Indeed, if this were not the case, by Theorem 3.90 it would follow that $\text{Lie}^{(\infty)}(\mathcal{X})|\mathsf{A}_1 = \mathsf{TA}_1$. Thus there exists $\tau_{11}, \tau_{12} \in]0, T_1[$ such that, if we redefine

$$\mathsf{A}_1 = \left\{\Phi^{X_{a_1}}_{t_1}(x_0) \mid t_1 \in]\tau_{11}, \tau_{12}[\right\},$$

then, for each $x \in \mathsf{A}_1$, we have $X_{a_2}(x) \notin \mathsf{T}_x\mathsf{A}_1$ for some $a_2 \in \{1, \dots, m\}$. We then define

$$\mathsf{A}_2 = \{\Phi_{t_2}^{X_{a_2}} \circ \Phi_{t_1}^{X_{a_1}}(x_0) \mid t_1 \in]\tau_{11}, \tau_{12}[,\ t_2 \in]0, T_2[\},$$

which is now a two-dimensional open submanifold. Proceeding in this way, we inductively define submanifolds $\mathsf{A}_1, \dots, \mathsf{A}_n$ of M of dimension $1, \dots, n$, respectively, with $n = \dim(\mathsf{M})$. Note that A_n is defined by

$$\mathsf{A}_n = \{\Phi_{t_n}^{X_{a_n}} \circ \cdots \circ \Phi_{t_1}^{X_{a_1}}(x_0) \mid t_j \in]\tau_{j1}, \tau_{j2}[,\ j \in \{1, \dots, n\}\}.$$

Now fix

$$(s_1, \dots, s_n) \in]\tau_{11}, \tau_{12}[\times \cdots \times]\tau_{n1}, \tau_{n2}[,$$

and consider the map

$$(t_1, \dots, t_n) \mapsto \Phi_{s_1}^{-X_{a_1}} \circ \cdots \circ \Phi_{s_n}^{-X_{a_n}} \circ \Phi_{t_n}^{X_{a_n}} \circ \cdots \circ \Phi_{t_1}^{X_{a_1}}(x_0). \tag{13.1}$$

This map can be defined for $(t_1, \dots, t_n)$ sufficiently close to $(s_1, \dots, s_n)$. Thus let $\mathcal{T}$ be a neighborhood of $(s_1, \dots, s_n) \in \mathbb{R}^n$. The image of the map

$$\mathcal{T} \ni (t_1, \dots, t_n) \mapsto \Phi_{t_n}^{X_{a_n}} \circ \cdots \circ \Phi_{t_1}^{X_{a_1}}(x_0)$$

is then a neighborhood of $\Phi_{s_n}^{X_{a_n}} \circ \cdots \circ \Phi_{s_1}^{X_{a_1}}(x_0)$, and let us denote this neighborhood by $\mathcal{U}$. The image of $\mathcal{T}$ under the map (13.1) is then

$$\Phi_{s_1}^{-X_{a_1}} \circ \cdots \circ \Phi_{s_n}^{-X_{a_n}}(\mathcal{U}),$$

which is evidently a neighborhood of x_0. This then gives the local version of the result.

For the global part of the proof, we proceed as follows. Since M is connected, it is path connected (Exercise E3.12), so there exists a continuous curve $\gamma\colon [0,1] \to \mathsf{M}$ such that $\gamma(0) = x_0$ and $\gamma(1) = x_1$. By our above argument, for each $s \in [0,1]$, there exist $a_1^s, \dots, a_{k_s}^s \in \{1, \dots, m\}$ such that the map

$$(t_1, \dots, t_{k_s}) \mapsto \Phi_{t_{k_s}}^{X_{a_{k_s}^s}} \circ \cdots \circ \Phi_{t_1}^{X_{a_1^s}}(\gamma(s)) \tag{13.2}$$

maps a neighborhood of $\mathbf{0} \in \mathbb{R}^{k_s}$ to a neighborhood of $\gamma(s) \in \mathsf{M}$. Thus there exists a nonempty interval $I_s \subset [0,1]$ around s, open in $[0,1]$, and having the property that, if $s' \in I_s$, then $\gamma(s')$ lies in the image of the map (13.2). Note that the collection of sets $\{I_s\}_{s\in[0,1]}$ comprises an open cover of $[0,1]$. Since $[0,1]$ is compact, there exists a finite subset $\{s_1, \dots, s_r\} \subset [0,1]$ satisfying

1. $0 = s_1 < s_2 < \cdots < s_{r-1} < s_r = 1$ and
2. $\{I_s\}_{s\in\{s_1,\dots,s_r\}}$ is an open cover of $[0,1]$.

By definition of the intervals I_s, for each $j \in \{1, \dots, r-1\}$, there exist $a_1^j, \dots, a_{k_j}^j \in \{1, \dots, m\}$ and $t_1^j, \dots, t_{k_j}^j \in \mathbb{R}$ such that

$$\Phi_{t_{k_j}^j}^{Y_{j,k_j}} \circ \cdots \circ \Phi_{t_1^j}^{Y_{j,1}}(\gamma(s_j)) \in \gamma(I_{s_{j+1}}),$$

where $Y_{j,l} = X_{a_l^j}$, $l \in \{1,\dots,k_j\}$. Also, for each $j \in \{2,\dots,r\}$, there exist $a_1^j,\dots,a_{k_j}^j \in \{1,\dots,m\}$ and $t_1^j,\dots,t_{k_j}^j \in \mathbb{R}$ such that

$$\Phi_{t_{k_j}^j}^{Y_{j,k_j}} \circ \cdots \circ \Phi_{t_1^j}^{Y_{j,1}}(\gamma(s_j)) \in \gamma(I_{s_{j-1}}).$$

Thus, following concatenated integral curves of the vector fields $X_1,\dots,X_m$, the system can be steered from a point reachable from $\gamma(s_j)$ to a point from which $\gamma(s_{j+1})$ can be reached. From this the theorem follows. ■

13.1.2 A brief literature survey of synthesis methods

Theorem 13.2 ensures that, for a given driftless system $(\mathsf{M}, \mathscr{X})$, the matter of existence of solutions to the motion planning problem is independent of the different types of controls that can be used. However, the *synthesis* problem, i.e., the problem of finding the controls that solve the motion planning problem, is definitely not the same. We shall first survey those papers that solve the motion planning problem using primitives. Then we consider the more extensive literature that deals with the motion planning problem with locally absolutely continuous inputs. We also refer the reader to [Murray, Li, and Sastry 1994, Chapter 8] and [Laumond, Sekhavat, and Lamiraux 1998] for surveys of this literature.

Motion planning using primitives

An approach taken by Lafferriere and Sussmann [1991, 1993] is to use nilpotent approximations. A driftless system is ***nilpotent*** when Lie brackets of the input vector fields of sufficiently large degree are identically zero. For systems having this property, it is possible to come up with algorithms that, provided one is able to determine the flow of the control vector fields, exactly solve the motion planning problem. For systems that are not nilpotent, one then can use nilpotent approximations to approximately solve the motion planning problem. An approach that considers only a finite set of controls is that of Bicchi, Marigo, and Piccoli [2002]. Note that by a finite set of controls, we mean not only that the controls take values in a finite subset of $\mathbb{R}^m$, but that we allow these controls to be defined only on a finite collection of time intervals. In this case, one can *never* get exact solutions to the motion planning problem, since the set of reachable points will be at best dense. However, in cases in which the reachable set *is* dense, fairly simple algorithms can be devised to approximately solve the problem. Some techniques for constructing nilpotent bases for distributions are described by Murray [1994].

Motion planning using locally absolutely continuous controls

The literature on the motion planning problem with locally absolutely continuous inputs is more extensive. In a coarse sense, many of these algorithms are based on Proposition 3.69. However, rather than using controls that are piecewise constant as are encountered in Proposition 3.69, controls are often taken to be periodic, often sinusoidal, and the synthesis problem is reduced to finding appropriately related frequencies and amplitudes for the periodic controls. Examples of this approach include the papers [Leonard and Krishnaprasad 1995, Liu 1997, Murray and Sastry 1993, Samson 1995, Tilbury, Murray, and Sastry 1995]. Another approach that does not constrain the inputs to take values in $\{e_1, \dots, e_m\}$ is that of Sontag [1995]. This technique relies on the idea of a "generic control." Such a control is one for which the linearization along the corresponding trajectory is controllable. This then allows one to use these trajectories as a basis for motion planning strategies. Chitour and Sussmann [1998] use continuation methods to provide solutions to the motion planning problem.

Flatness

Another approach that can sometimes be employed for motion planning is that of differential flatness, as introduced in a differential algebraic setting by Fliess, Lévine, Martin, and Rouchon [1995]. The notion is cast in a geometric setting using differential forms by van Nieuwstadt, Rathinam, and Murray [1998]. Roughly speaking, a system is ***differentially flat*** if there exist outputs, called ***flat outputs***, for which all states and inputs are determined by the outputs and a finite number of their derivatives. Thus one can reduce the motion planning problem to the sometimes more tractable problem of determining a suitable flat output. While it is not true that all driftless systems possess flat outputs, some driftless systems do, and when they do, the flat outputs can often simplify the motion planning problem. For a class of mechanical systems, flatness was characterized using Riemannian geometry by Rathinam and Murray [1998]. The use of flatness for motion planning of mobile robots is discussed by Lamiraux and Laumond [2000].

Numerical methods

When no analytical structure can be exploited, motion planning is tackled via numerical tools. Illustrative texts and surveys on this area include [Bertsekas 2000, Betts 1998, Stoer and Bulirsch 1992, von Stryk and Bulirsch 1992]. The objective of numerical trajectory design methods is often to, not only compute a feasible controlled trajectory, but also optimize the controlled trajectory with respect to, for example, energy consumption and/or trajectory duration. Various transcription techniques are used to turn the optimal control problem into a numerical optimization program. Depending on the specifics of the problem, transcription techniques include single shooting, multiple shooting, or collocation techniques. An important distinction is to be made between approaches that discretize the original optimal control problem, so-called "direct" methods, and "indirect" approaches, in which the necessary conditions for optimality are discretized (e.g., Hamilton's equations in the two-point boundary value formulation, or the Hamilton–Jacobi–Bellman equation in the dynamic programming formulation). Some recent research suggests that collocation is an effective direct transcription method, e.g., see [Betts 1998, Gopal and Biegler 1998, Hargraves and Paris 1987], and that successive quadratic programming is an effective nonlinear programming method, e.g., see [Barclay, Gill, and Rosen 1998, Panier and Tits 1993]. Finally, it is worth remarking that numerical approaches have a number of drawbacks. Gradient-based nonlinear optimization methods are inherently local and are not guaranteed to find a feasible solution. Dynamic programming methods may provide globally optimal solutions, but are computationally expensive.

We shall here not discuss motion planning problems for systems in environments with obstacles. Clearly, these problems are harder than problems without path constraints; from a computer science point of view, even simple formulations of motion planning problems are known to possess high complexity, e.g., see [Canny 1988, Reif 1979]. We refer the interested reader to the surveys [Halperin, Kavraki, and Latombe 1997, Latombe 1991, 1999, Laumond, Sekhavat, and Lamiraux 1998, Sharir 1997].

13.2 Motion planning for mechanical systems

Let us now turn to motion planning problems for mechanical systems. As in the preceding section, we begin with a precise statement of the problems we consider. We then discuss strategies that are applicable when a correspondence can be made between the mechanical system and a driftless system.

13.2.1 Definitions

Although one can certainly formulate motion control problems for the most general class of mechanical systems, we shall only consider affine connection control systems, since our results apply to that situation.

Problem 13.3 (Motion planning for mechanical systems). Let $(\mathsf{Q}, \nabla, \mathcal{D}, \mathscr{Y}, U)$ be a C^∞-affine connection control system and let $\mathscr{U}$ be a collection of U-valued locally integrable inputs. The ***$\mathscr{U}$-motion planning problem*** is:

Given $q_0, q_1 \in \mathsf{Q}$, find $u \in \mathscr{U}$, defined on some interval $[0, T]$, so that the controlled trajectory (γ, u) with $\gamma'(0) = 0_{q_0}$ has the property that $\gamma'(T) = 0_{q_1}$. •

Remark 13.4. Although one can consider more general motion planning problems (for example, by allowing trajectories to start and end with nonzero velocity), we restrict our attention to the simple case defined above. All motion planning algorithms presented in this chapter solve the problem stated above. Also, we use the terminology "$\mathscr{U}$-motion planning problem" for both driftless and mechanical systems. This will not be a cause of confusion, since it will be clear from context the type of system being considered. •

13.2.2 Kinematically controllable systems

In this section we consider an affine connection control system $\Sigma_{\text{dyn}} = (\mathsf{Q}, \nabla, \mathcal{D}, \mathscr{Y}, \mathbb{R}^m)$ that is kinematically controllable with decoupling vector fields $\{X_1, \dots, X_{\tilde{m}}\}$. Thus we suppose the system to possess $\tilde{m}$ rank-one kinematic reductions. Should the system possess kinematic reductions of higher rank, we simply choose a set of rank-one kinematic reductions whose span is the larger kinematic reduction. In this case, we expect there to be a relationship between the motion planning problem for Σ_{dyn} and that for the driftless system $\Sigma_{\text{kin}} = (\mathsf{Q}, \mathscr{X} = \{X_1, \dots, X_{\tilde{m}}\}, \mathbb{R}^{\tilde{m}})$. Indeed, since the vector fields $X_1, \dots, X_{\tilde{m}}$ are decoupling, their integral curves may be followed, with arbitrary reparameterization, by controlled trajectories of Σ_{dyn} (Exercise E8.3). In fact, with a given reparameterization, the controls needed for Σ_{dyn} to follow an integral curve of a decoupling vector field can be explicitly computed.

Proposition 13.5 (Controls for following decoupling vector fields). *Let $X \in \Gamma^\infty(\mathsf{TQ})$ be a decoupling vector field for the affine connection control system $\Sigma_{\text{dyn}} = (\mathsf{Q}, \nabla, \mathcal{D}, \mathscr{Y}, \mathbb{R}^m)$, let $\gamma \colon I \to \mathsf{Q}$ be an integral curve for X and let $\tau \colon J \to I$ be a reparameterization for γ that is of class C^2. Suppose that the vector fields $\mathscr{Y} = \{Y_1, \dots, Y_m\}$ are linearly independent at all points in Q. If $u \colon J \to \mathbb{R}^m$ is defined by*

$$u^a(t) Y_a(\gamma \circ \tau(t)) = (\tau'(t))^2 \nabla_X X(\gamma \circ \tau(t)) + \tau''(t) X(\gamma \circ \tau(t)),$$

then $(\gamma \circ \tau, u)$ is a controlled trajectory for Σ_{dyn}.

Proof. Since γ is an integral curve for X, $\gamma'(t) = X(\gamma(t))$. We therefore have

$$\begin{aligned}\nabla_{(\gamma\circ\tau)'(t)}(\gamma \circ \tau)'(t) &= (\tau'(t))^2 \nabla_{\gamma'(\tau(t))}\gamma'(\tau(t)) + \tau''(t)\gamma'(\tau(t)) \\ &= (\tau'(t))^2 \nabla_X X(\gamma \circ \tau(t)) + \tau''(t) X(\gamma \circ \tau(t)).\end{aligned}$$

Since X is a decoupling vector field, we have

$$(\tau'(t))^2 \nabla_X X(\gamma \circ \tau(t)) + \tau''(t) X(\gamma \circ \tau(t)) \in \mathcal{Y}_{\gamma \circ \tau(t)},$$

and the result follows. ■

This leads to the following strategy for solving the motion planning problem for affine connection control systems.

Proposition 13.6 (Motion planning for kinematically controllable systems). *Given:*

(i) a C^∞-affine connection control system $\Sigma_{\text{dyn}} = (\mathsf{Q}, \nabla, \mathcal{D}, \mathscr{Y}, \mathbb{R}^m)$ that is kinematically controllable with decoupling vector fields being the control vector fields for a driftless system $\Sigma_{\text{kin}} = (\mathsf{Q}, \mathscr{X} = \{X_1, \dots, X_{\tilde{m}}\}, \mathbb{R}^{\tilde{m}})$;

(ii) the set $\mathscr{U}^{\tilde{m}}_{\text{prim}}$ of piecewise constant controls taking values in the set $\tilde{U} = \{e_1, \dots, e_{\tilde{m}}, -e_1, \dots, -e_{\tilde{m}}\}$;

(iii) the set $\mathscr{U}^m_{\text{dyn}}$ of locally integrable controls;

(iv) $q_0, q_1 \in \mathsf{Q}$.

Do the following:

(v) solve the $\mathscr{U}^{\tilde{m}}_{\mathrm{prim}}$*-motion planning problem for* Σ_{kin} *by finding* $k \in \mathbb{N}$, $a_1, \dots, a_k \in \{1, \dots, \tilde{m}\}$, $t_1, \dots, t_k \in \mathbb{R}_+$, *and* $\epsilon_1, \dots, \epsilon_k \in \{-1, 1\}$ *such that*

$$\Phi^{\epsilon_k X_{a_k}}_{t_k} \circ \cdots \circ \Phi^{\epsilon_1 X_{a_1}}_{t_1}(q_0) = q_1$$

(this is possible by Theorem 13.2);

(vi) define $\bar{q}_0 = q_0$ *and* $\bar{q}_j = \Phi^{\epsilon_j X_{a_j}}_{t_j} \circ \cdots \circ \Phi^{\epsilon_1 X_{a_1}}_{t_1}(q_0)$, $j \in \{1, \dots, k\}$;

(vii) for each $j \in \{1, \dots, k\}$, *find a reparameterization* $\tau_j \colon [0, \bar{t}_j] \to [0, t_j]$ *satisfying* $\tau'(0) = \tau'(\bar{t}_j) = 0$;[1]

(viii) for each $j \in \{1, \dots, k\}$, *let* $u_j \colon [0, \bar{t}_j] \to \mathbb{R}^m$ *be defined as in Proposition 13.5, taking* γ *in the proposition to be the integral curve* $t \mapsto \Phi^{\epsilon_j X_{a_j}}_t(\bar{q}_{j-1})$, $t \in [0, t_j]$;

(ix) let $u \colon [0, T] \to \mathbb{R}^m$ *be the control in* $\mathscr{U}^m_{\mathrm{dyn}}$ *formed by concatenating the controls* $u_1, \dots, u_k$.

Then the controlled trajectory (γ, u) *for* Σ_{dyn} *satisfying* $\gamma'(0) = 0_{q_0}$ *has the property that* $\gamma(T) = 0_{q_1}$.

The idea then, is very simple. First, for the driftless system, one solves the motion planning problem using primitives. Then one follows each primitive with a curve for Σ_{dyn}, making sure to start and stop with zero velocity. In this way, when the end of one primitive is reached, the next one is ready to start. Of course, it is the first step in the strategy that is the difficult one. Nevertheless, the motion planning problem for a certain kind of affine connection control system has been successfully reduced to the motion planning problem for a driftless system. That this should even be possible is not obvious. Furthermore, we shall see in the examples in this chapter that sometimes one *can* explicitly solve the motion planning problem using primitives, and so arrive at simple, easily implementable, albeit open-loop, motion planning strategies.

13.2.3 Maximally reducible systems

In this section we consider an affine connection control system $\Sigma_{\mathrm{dyn}} = (\mathsf{Q}, \nabla, \mathcal{D}, \mathscr{Y}, \mathbb{R}^m)$ that is maximally reducible to the driftless system $\Sigma_{\mathrm{kin}} = (\mathsf{Q}, \mathscr{X} = \mathscr{Y}, \mathbb{R}^m)$. We make the assumption that $\mathrm{Lie}^{(\infty)}(\mathcal{Y}) = \mathsf{TQ}$. Since the vector fields $Y_1, \dots, Y_m$ are also decoupling vector fields, the strategy of the preceding section can be applied. However, one can also apply other strategies for maximally reducible systems that are not available for systems that are merely kinematically controllable. Indeed, maximal reducibility means that controlled trajectories of Σ_{kin} with locally absolutely continuous controls can be followed by controlled trajectories of Σ_{dyn}. This can serve to use strategies that do not have the "stop-start" character of the solution described by Proposition 13.6.

---

[1] By $\tau'(0)$ we mean $\lim_{\varepsilon \downarrow 0} \frac{\tau(\varepsilon) - \tau(0)}{\varepsilon}$ and by $\tau'(\bar{t}_j)$ we mean $\lim_{\varepsilon \downarrow 0} \frac{\tau(\bar{t}_j) - \tau(\bar{t}_j - \varepsilon)}{\varepsilon}$.

The following result indicates how to recover the appropriate controls for Σ_{dyn}.

Proposition 13.7 (Controls for maximally reducible systems). *Let $\Sigma_{\text{dyn}} = (\mathsf{Q}, \nabla, \mathcal{D}, \mathscr{Y}, \mathbb{R}^m)$ be a C^∞-affine connection control system that is maximally reducible to the driftless system $\Sigma_{\text{kin}} = (\mathsf{Q}, \mathscr{X} = \mathscr{Y}, \mathbb{R}^m)$. Suppose that the vector fields $\mathscr{Y} = \{Y_1, \dots, Y_m\}$ are linearly independent at all points in Q, and define C^∞-functions $\sigma^c_{ab} : Q \to \mathbb{R}$, $a, b, c \in \{1, \dots, m\}$, by*

$$\langle Y_a : Y_b \rangle = \sigma^c_{ab} Y_c,$$

which is possible by condition (ii) of Theorem 8.27. If $(\gamma, \tilde{u})$ is a controlled trajectory for the driftless system Σ_{kin}, and if we define the control $u \in \mathscr{U}^m_{\text{dyn}}$ by

$$u^c(t) = \dot{\tilde{u}}^c(t) + \tfrac{1}{2}\sigma^c_{ab}(\gamma(t))\tilde{u}^a(t)\tilde{u}^b(t), \qquad c \in \{1, \dots, m\},$$

then (γ, u) is a controlled trajectory for the affine connection control system Σ_{dyn}.

Proof. By definition $(\gamma, \tilde{u})$ satisfy $\gamma'(t) = \tilde{u}^a(t) Y_a(\gamma(t))$. Therefore,

$$\begin{aligned}
\nabla_{\gamma'(t)}\gamma'(t) &= \nabla_{\gamma'(t)}\big(\tilde{u}^b(t) Y_b(\gamma(t))\big) \\
&= \tilde{u}^b(t)\nabla_{\gamma'(t)} Y_b(\gamma(t)) + \dot{\tilde{u}}^b(t) Y_b(\gamma(t)) \\
&= \tilde{u}^b(t)\nabla_{\tilde{u}^a(t)Y_a(\gamma(t))} Y_b(\gamma(t)) + \dot{\tilde{u}}^b(t) Y_b(\gamma(t)) \\
&= \tilde{u}^a(t)\tilde{u}^b(t)\nabla_{Y_a(\gamma(t))} Y_b(\gamma(t)) + \dot{\tilde{u}}^b(t) Y_b(\gamma(t)) \\
&= \tilde{u}^a(t)\tilde{u}^b(t)\tfrac{1}{2}\big(\nabla_{Y_a} Y_b + \nabla_{Y_b} Y_a\big)(\gamma(t)) + \dot{\tilde{u}}^b(t) Y_b(\gamma(t)) \\
&= \big(\dot{\tilde{u}}^c(t) + \tfrac{1}{2}\sigma^c_{ab}(\gamma(t))\tilde{u}^a(t)\tilde{u}^b(t)\big) Y_c(t).
\end{aligned}$$

Now, if we define u as in the statement of the proposition, we have

$$\nabla_{\gamma'(t)}\gamma'(t) = u^c(t) Y_c(\gamma(t)),$$

as desired. ■

With this correspondence between controls for the driftless system and controls for the affine connection control system, we may easily state a strategy for solving the motion planning problem for maximally reducible affine connection control systems.

Proposition 13.8 (Motion planning for maximally reducible systems). *Given:*

(i) a C^∞-affine connection control system $\Sigma_{\text{dyn}} = (\mathsf{Q}, \nabla, \mathcal{D}, \mathscr{Y}, \mathbb{R}^m)$ that is maximally reducible to the driftless system $\Sigma_{\text{kin}} = (\mathsf{Q}, \mathscr{X} = \mathscr{Y}, \mathbb{R}^m)$;

(ii) the set $\mathscr{U}^m_{\text{kin}}$ of locally absolutely continuous controls;

(iii) the set $\mathscr{U}^m_{\text{dyn}}$ of locally integrable controls;

(iv) $q_0, q_1 \in \mathsf{Q}$.

Do the following:

(v) *solve the* $\mathscr{U}^m_{\mathrm{kin}}$*-motion planning problem, i.e., the motion planning problem with locally absolutely continuous controls, for* Σ_{kin} *by finding a control* $\bar{u}\colon [0,\bar{T}] \to \mathbb{R}^m$ *in* $\mathscr{U}^m_{\mathrm{kin}}$ *such that the controlled trajectory* $(\gamma, \bar{u})$ *for* Σ_{kin} *satisfying* $\gamma(0) = q_0$ *also satisfies* $\gamma(\bar{T}) = q_1$*;*

(vi) *find a reparameterization* $\tau\colon [0,T] \to [0,\bar{T}]$ *satisfying* $\tau'(0) = \tau'(T) = 0$*;*

(vii) *let* $\tilde{u}\colon [0,T] \to \mathbb{R}^m$ *be the control for which* $(\gamma \circ \tau, \tilde{u})$ *is a controlled trajectory for* Σ_{kin} *(cf. Lemma 8.16);*

(viii) *let* $u\colon [0,T] \to \mathbb{R}^m$ *be the control in* $\mathscr{U}^m_{\mathrm{dyn}}$ *defined as in Proposition 13.7.*

Then the controlled trajectory $(\gamma \circ \tau, u)$ *for* Σ_{dyn} *then satisfies* $\gamma'(0) = 0_{q_0}$ *and* $\gamma'(T) = 0_{q_1}$*.*

The idea here is a little simpler than for kinematically controllable systems. Here we need only perform a single reparameterization to ensure that, at the beginning and end of the motion, one has zero velocity. This is in contrast with motion planning for kinematically controllable systems, where each primitive component of the maneuver must begin and end with zero velocity. However, the major problem of finding the control $\bar{u}$ that solves the $\mathscr{U}^m_{\mathrm{kin}}$-motion planning problem still remains.

13.3 Motion planning for two simple systems

The two examples we consider in this section provide a simple, but effective illustration of the ideas behind motion planning using kinematic controllability and maximal reducible kinematic controllability.

13.3.1 Motion planning for the planar rigid body

The system we consider here has been developed in Section 1.1, throughout Chapter 4, and in Sections 5.3.3, 7.4.2, and 8.3.2 from various points of view. In Figure 13.1 we reproduce the system, and we recall that the system has configuration manifold $\mathsf{Q} = \mathsf{SE}(2)$, coordinates (θ, x, y), Riemannian metric

$$\mathbb{G} = J \mathrm{d}\theta \otimes \mathrm{d}\theta + m(\mathrm{d}x \otimes \mathrm{d}x + \mathrm{d}y \otimes \mathrm{d}y),$$

and input vector fields

$$Y_1 = \frac{\cos\theta}{m}\frac{\partial}{\partial x} + \frac{\sin\theta}{m}\frac{\partial}{\partial y}, \quad Y_2 = -\frac{h}{J}\frac{\partial}{\partial\theta} - \frac{\sin\theta}{m}\frac{\partial}{\partial x} + \frac{\cos\theta}{m}\frac{\partial}{\partial y}.$$

The system has decoupling vector fields

$$X_1 = \cos\theta\frac{\partial}{\partial x} + \sin\theta\frac{\partial}{\partial y}, \qquad X_2 = -\frac{mh}{J}\frac{\partial}{\partial\theta} - \sin\theta\frac{\partial}{\partial x} + \cos\theta\frac{\partial}{\partial y},$$

leading us to consider the driftless system $\Sigma_{\mathrm{kin}} = (\mathsf{Q}, \{X_1, X_2\}, \mathbb{R}^2)$.

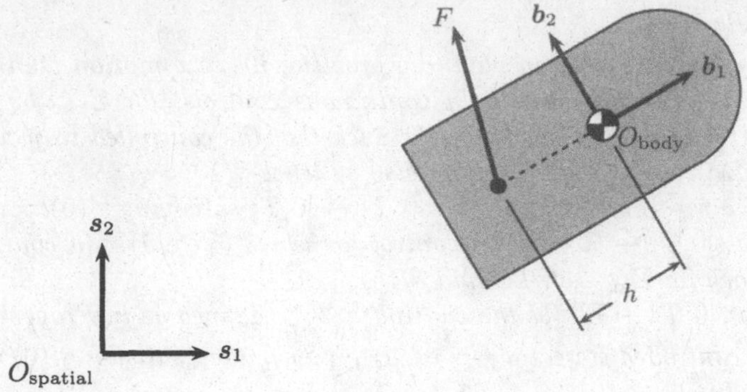

Figure 13.1. Planar body with a variable-direction thruster

While it is possible to write down a complete closed-form solution to the motion planning problem, it takes too much space to do this, and the explicit formulae are merely tedious. However, it is quite easy to describe the character of one of the many solutions to the problem. Since the system is a left-invariant system on SE(2), to ascertain how to steer from $g_0 = (\theta_0, x_0, y_0)$ to $g_1 = (\theta_1, x_1, y_1)$, one can simply use the controls that steer from $e = (0,0,0)$ to $g_0^{-1} g_1$. Thus we shall only indicate how to steer to a final point (θ_1, x_1, y_1) from $(\theta_0, x_0, y_0) = (0,0,0)$. The strategy is as follows.

1. Draw two circles C_0 and C_1 in the (x, y)-plane with radii $\frac{mh}{J}$ and centers $a_0 = (\frac{mh}{J}, 0)$ and $a_1 = (x_1 + \frac{mh}{J}\cos\theta_1, y_1 + \frac{mh}{J}\sin\theta_1)$, respectively.
2. Draw a vertical line ℓ_1 through a_1 so that it intersects the horizontal line ℓ_0 through a_0 and also intersects the circle C_1 in two places.
3. Let b_0 be the point on $C_0 \cap \ell_0$ that is furthest from $\ell_0 \cap \ell_1$ and let b_1 be the point on $C_1 \cap \ell_1$ that is closest to $\ell_0 \cap \ell_1$.
4. Draw a circle C_2 of radius $\frac{mh}{J}$ and center at $\ell_0 \cap \ell_1$.
5. Let c_0 be the point on $C_2 \cap \ell_0$ closest to b_0 and let c_1 be the point on $C_2 \cap \ell_1$ furthest from b_1.
6. Follow an integral curve for $\pm X_2$ to steer the center of mass of the body to b_0.
7. Follow an integral curve for $\pm X_1$ to steer the center of mass of the body to c_0.
8. Follow an integral curve for $\pm X_2$ to steer the center of mass of the body to c_1.
9. Follow an integral curve for $\pm X_1$ to steer the center of mass of the body to b_1.
10. Follow an integral curve for $\pm X_2$ to steer the center of mass of the body to (x_1, y_1).

The construction is illustrated in Figure 13.2. Note that although the motion

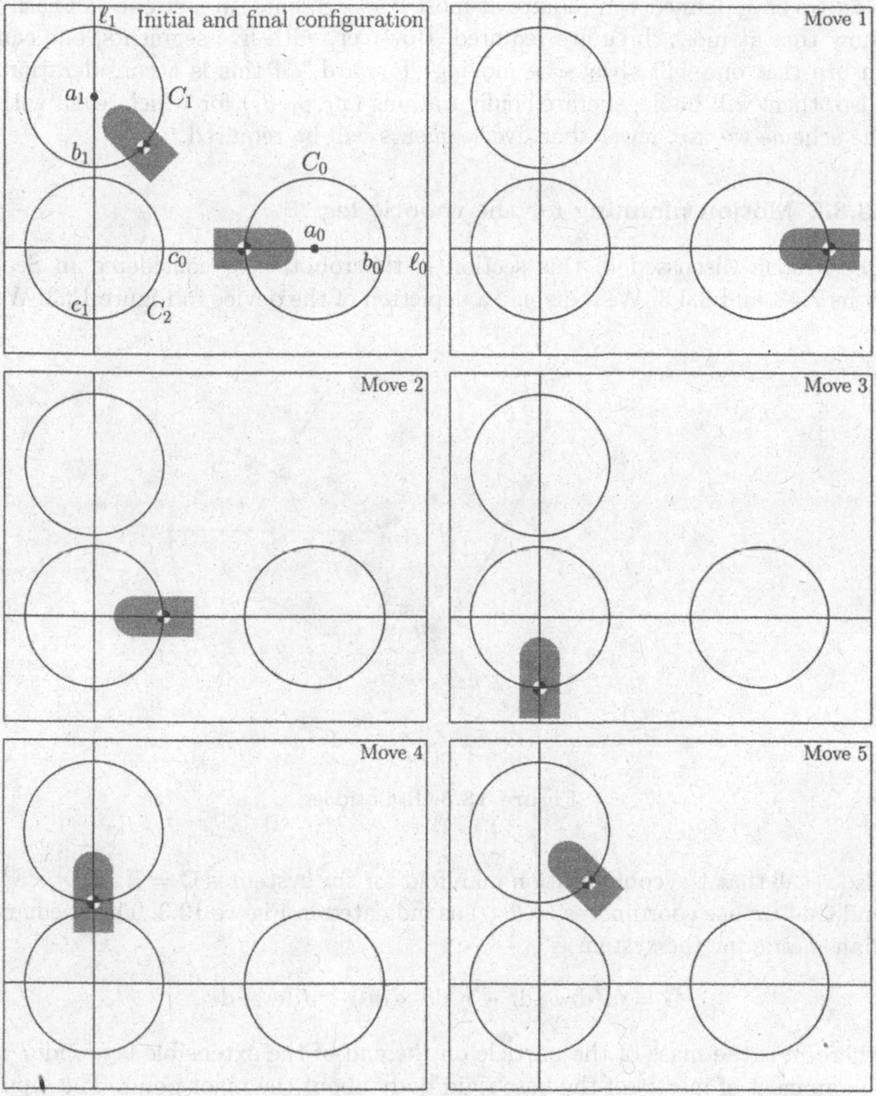

Figure 13.2. Motion primitives for the planar body motion planning problem. The initial and final configurations are shown in the top left figure, along with the illustration of the notation used to define the motion plan. The remaining five pictures indicate the position of the body at the end of each of the five segments defining the motion.

plan we have defined will require at most five segments, in fact one can easily show that at most three are required. However, with five segments, one can ensure that one will always be moving "forward," if this is a consideration. Also, there will be degenerate configurations (x_1, y_1, θ_1) for which, even with the scheme we use, fewer that five segments will be required.

13.3.2 Motion planning for the robotic leg

The system discussed in this section is the robotic leg considered in Sections 7.4.1 and 8.3.3. We redisplay a depiction of the device in Figure 13.3. We

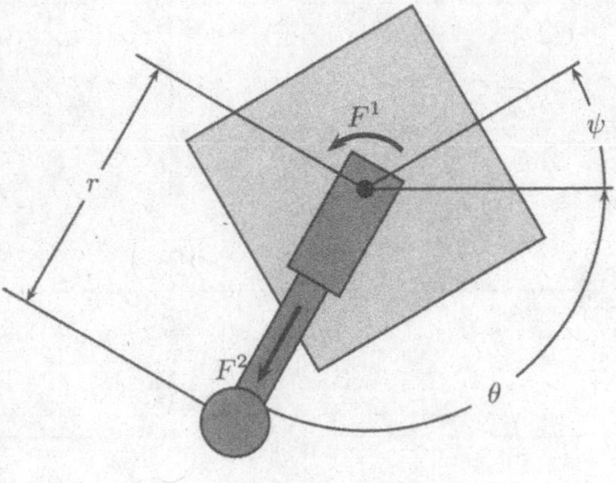

Figure 13.3. Robotic leg

also recall that the configuration manifold for the system is $\mathsf{Q} = \mathbb{R}_+ \times \mathbb{S}^1 \times \mathbb{S}^1$, and that we use coordinates (r, θ, ψ) as indicated in Figure 13.3. The Riemannian metric for the system is

$$\mathbb{G} = m(\mathrm{d}r \otimes \mathrm{d}r + r^2 \mathrm{d}\theta \otimes \mathrm{d}\theta) + J \mathrm{d}\psi \otimes \mathrm{d}\psi,$$

where m is the mass of the particle on the end of the extensible leg, and J is the moment of inertia of the base rigid body about the pivot point. The input vector fields are

$$Y_1 = \frac{1}{mr^2}\frac{\partial}{\partial\theta} - \frac{1}{J}\frac{\partial}{\partial\psi}, \quad Y_2 = \frac{1}{m}\frac{\partial}{\partial r}.$$

In Section 8.3.3 we showed that the system is maximally reducible. Therefore, we shall employ the solution of Proposition 13.8 to the motion planning problem for this system. One could also employ the strategy of Proposition 13.6 for kinematically controllable systems, but we do not do this here, in order to demonstrate the differences that can arise between systems that are maximally reducible and those that are not. The readers can consider using the approach of Proposition 13.6 in Exercise E13.4.

As with any motion planning strategy based on the method of Proposition 13.8, the most difficult part of the problem is finding the solution to the motion planning problem for the associated driftless system. The method we give here is that of Murray and Sastry [1993] (they regard the system as intrinsically a kinematic one). It is helpful to write the differential equations for the driftless system on Q with input vector fields $X_1 = Y_1$ and $X_2 = Y_2$:

$$\dot{r} = \frac{u^2}{m}, \quad \dot{\theta} = \frac{u^1}{mr^2}, \quad \dot{\psi} = -\frac{u^1}{J}. \tag{13.3}$$

The procedure relies on using both inputs at the same time (this is where maximal reducibility is important), and using sinusoidal signals of a certain frequency and phase relation. The following lemma records the most important idea.

Lemma 13.9. *Let* $c_1, c_2, \omega, r_0 \in \mathbb{R}$ *satisfy*

(i) $\omega, r_0 > 0$ *and*
(ii) $\left|\frac{c_2}{m\omega}\right| < r_0$.

The solution at time $T = \frac{2\pi}{\omega}$ *of the differential equation* (13.3) *with*

$$u^1(t) = c_1 \sin(\omega t), \quad u^2(t) = c_2 \cos(\omega t),$$

and with initial conditions $(r(0), \theta(0), \psi(0)) = (r_0, \theta_0, \psi_0)$ *is*

$$r(T) = r_0, \quad \theta(T) = -\frac{2\pi c_1 r_{\max}}{m\omega(r_0^2 - r_{\max}^2)^{\frac{3}{2}}}, \quad \psi(T) = \psi_0,$$

where $r_{\max} = \frac{c_2}{m\omega}$.

Proof. By direct integration we have

$$r(t) = r_0 + \frac{c_2}{m\omega}\sin(\omega t), \quad \psi(t) = \psi_0 - \frac{c_1}{J\omega}(1 - \cos(\omega t)).$$

It is then clear that $r(T) = r_0$ and $\psi(T) = \psi_0$. Also by direct integration we have

$$\theta(T) = \theta(0) + \int_0^T \frac{c_1 \sin(\omega t)}{mr(t)^2}\,dt = -\frac{2\pi c_1 r_{\max}}{m\omega(r_0^2 - r_{\max}^2)^{3/2}},$$

if $r_{\max} = \frac{c_2}{m\omega}$. ■

Based on the lemma, we offer the following solution to the motion planning problem for the robotic leg. We suppose that we begin at (r_0, θ_0, ψ_0) and wish to steer to (r_1, θ_1, ψ_1). To make the problem local in nature, suppose that we restrict the travel of the leg so that $r(t) \in]\min\{r_0, r_1\} - \bar{r}, \max\{r_0, r_1\} + \bar{r}[$ for some $\bar{r} > 0$, and we restrict the angular travel of the leg so that it does not deviate more than $\bar{\psi}$ beyond the initial angle ψ_0 or the final angle ψ_1. The following algorithm accomplishes this.

1. Using u^2, steer r from r_0 to r_1 while leaving θ and ψ fixed.

2. Using u^1, steer ψ from ψ_0 to ψ_1, while leaving r fixed and allowing θ to do whatever is prescribed by the control (this is described in the next step).
3. Let $\Delta\psi$ be the (signed) angle change in ψ incurred during the preceding maneuver. Then the change in θ is $-\frac{J}{mr_1^2}\Delta\psi$. Let $\tilde{\theta} = \theta_0 - \frac{J}{mr_1^2}\Delta\psi$.
4. Let $\Delta\theta$ be the (signed) angle required to steer from $\tilde{\theta}$ to θ_1 (choose the shortest distance on $\mathbb{S}^1$).
5. Choose $c_1, c_2, \omega \in \mathbb{R}$ and $N \in \mathbb{N}$ with the following properties:
   (a) $\omega > 0$;
   (b) $\left|\frac{c_2}{m\omega}\right| < \bar{r}$;
   (c) $\left|\frac{c_1}{m\omega}\right| < \bar{\psi}$;
   (d) $\Delta\theta = -\frac{2N\pi c_1 \frac{c_2}{m\omega}}{m\omega(r_1^2 - (\frac{c_2}{m\omega})^2)^{3/2}}$.
6. Apply the controls
$$u^1(t) = c_1 \sin(\omega t), \quad u^2(t) = c_2 \cos(\omega t)$$
for time $\frac{2\pi N}{\omega}$ to steer the system to (r_1, θ_1, ψ_1).

In Figure 13.4 we provide plots of the coordinates and the controls for a motion steering from $(r_0, \theta_0, \psi_0) = (1, 0, 0)$ to $(r_1, \theta_1, \psi_1) = (2, \frac{\pi}{3}, \frac{\pi}{5})$.

13.4 Motion planning for the snakeboard

The example in this section follows Bullo and Lewis [2003a], and represents a fairly complicated system that relies for its solution on many of the tools introduced in the book. For example, the system has nonholonomic constraints, and its representation benefits greatly from the development of generalized Christoffel symbols in Section 4.5.6. We begin by presenting these details for the snakeboard model.

13.4.1 Modeling

The snakeboard is a commercially available product that is a derivative of the standard skateboard. With this device, propulsion is achieved without contact with the ground, through a principle whereby the rider's angular momentum is converted to forward motion by the nonholonomic constraints. A depiction of the commercial product is given in Figure 13.5. The rider places one foot on each of the footpads, and these may be made to swivel. By swiveling the footpads and twisting their body appropriately, the rider can generate forward motion. Indeed, by combining foot movements with body movements in the right way, one can generate motions in the forward, lateral, and rotation directions. These "gaits" were investigated in the first paper on the snakeboard,

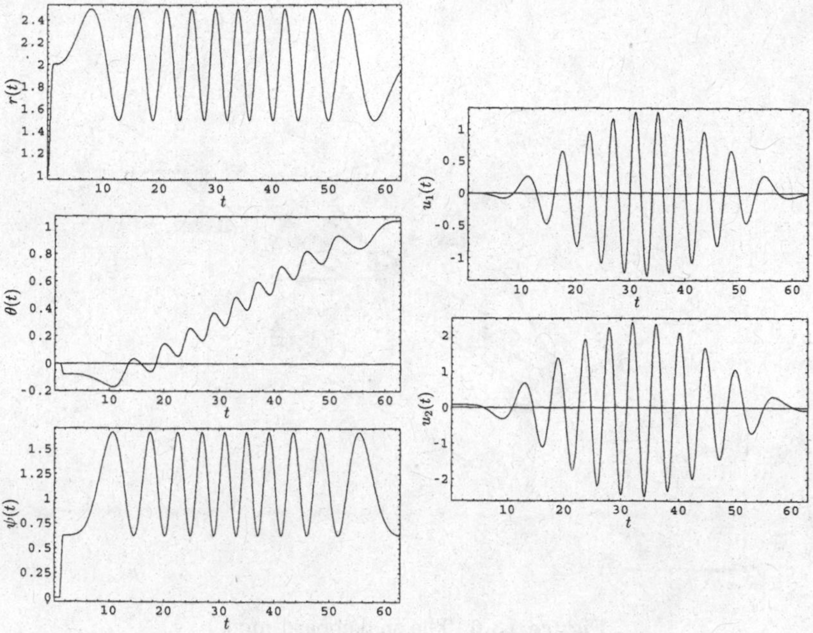

Figure 13.4. Motion planning for the robotic leg for a motion from $(1,0,0)$ to $(2,\frac{\pi}{3},\frac{\pi}{5})$: $r(t)$ (top left), $\theta(t)$ (middle left), $\psi(t)$ (bottom left), $u^1(t)$ (top right), and $u^2(t)$ (bottom right)

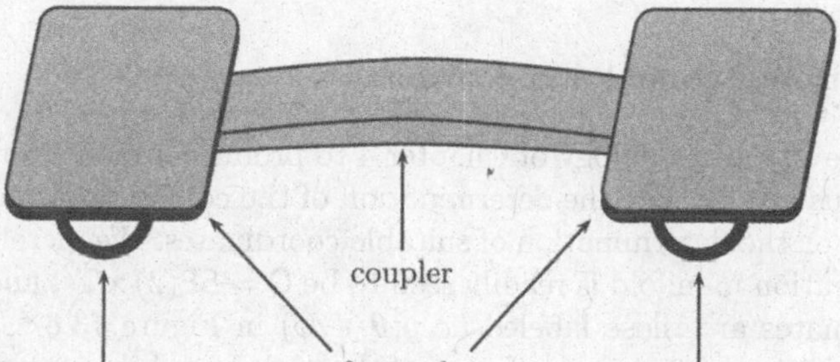

Figure 13.5. A depiction of the commercially available snakeboard

that of Lewis, Ostrowski, Murray, and Burdick [1994]. In that paper the mathematical model for the snakeboard was also introduced, the basis for which is the representation in Figure 13.6. See the papers [Marsden and Ostrowski 1996, Ostrowski and Burdick 1997] for further work on the snakeboard. In the model represented in Figure 13.6, we demand that the front and back axles of the snakeboard move in unison. In [Lewis, Ostrowski, Murray, and Burdick 1994] they were allowed to move independently. As concerns the motion planning we do here, the requirement that the axles move in unison does not pose

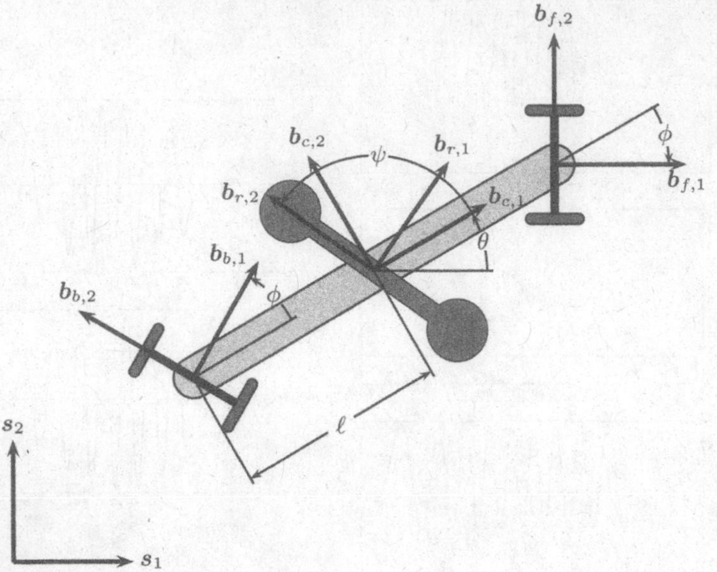

Figure 13.6. The snakeboard model

a problem (meaning, as we shall shortly see, that the snakeboard as shown in Figure 13.6 is kinematically controllable).

Configuration manifold and coordinates

We shall use the methodology of Chapter 4 to produce a model for the snakeboard. Let us not belabor the determination of the configuration manifold for the system, or the determination of suitable coordinates. We merely note that the configuration manifold is readily seen to be $\mathsf{Q} = \mathsf{SE}(2) \times \mathbb{T}^2$, and that suitable coordinates are those labeled $(x, y, \theta, \psi, \phi)$ in Figure 13.6.[2] Thus (x, y) labels the position of the center of mass of the snakeboard, which we assume by symmetry to lie exactly between the front and back wheels. The angle θ is that of a body frame attached to the coupler (denoted by $(O_{\mathrm{body}}, \{\boldsymbol{b}_{c,1}, \boldsymbol{b}_{c,2}, \boldsymbol{b}_{c,3}\})$ in Figure 13.6) relative to an inertial frame $(O_{\mathrm{spatial}}, \{\boldsymbol{s}_1, \boldsymbol{s}_2, \boldsymbol{s}_3\})$. The angle ψ is the angle of the rotor atop the coupler, measured relative to the coupler's body frame. The angle ϕ denotes the angle of the wheels, again measured relative to the body frame on the coupler.

[2] For the snakeboard, we write a typical point in $\mathsf{SE}(2)$ as (x, y, θ), as opposed to previous conventions where we wrote (θ, x, y). For the snakeboard, this has the advantage of collecting the three angular coordinates, θ, ψ, and ϕ, together in the coordinate list.

Kinetic energy and Riemannian metric

Now we derive the kinetic energy of the snakeboard. The snakeboard consists of four rigid body components: the coupler, the rotor, and two (assumed identical) wheels. We use the following physical parameters:

m_c	mass of coupler;
m_r	mass of rotor;
m_w	mass of each wheel assembly;
J_c	inertia of coupler about center of mass;
J_r	inertia of rotor about center of mass;
J_w	inertia of wheel assembly about center of mass;
ℓ	distance from coupler center of mass to wheel assembly.

With the given coordinates, the positions of the centers of mass of the coupler, the rotor, the back wheel, and the front wheel are

$$\begin{aligned}\boldsymbol{r}_c = (x, y, 0), \quad \boldsymbol{r}_r &= (x, y, 0),\\ \boldsymbol{r}_b &= (x - \ell\cos\theta, y - \ell\sin\theta, 0),\\ \boldsymbol{r}_f &= (x + \ell\cos\theta, y + \ell\sin\theta, 0),\end{aligned}$$

respectively. To compute the body angular velocities, we use the spatial reference frame as above, and frames $(O_c, \{\boldsymbol{b}_{c,1}, \boldsymbol{b}_{c,2}, \boldsymbol{b}_{c,3}\})$, $(O_r, \{\boldsymbol{b}_{r,1}, \boldsymbol{b}_{r,2}, \boldsymbol{b}_{r,3}\})$, $(O_b, \{\boldsymbol{b}_{b,1}, \boldsymbol{b}_{b,2}, \boldsymbol{b}_{b,3}\})$, and $(O_f, \{\boldsymbol{b}_{f,1}, \boldsymbol{b}_{f,2}, \boldsymbol{b}_{f,3}\})$ for the coupler, the rotor, the front wheel, and the back wheel, respectively. These frames are chosen so that the rotation matrices describing the orientation of the coupler, the rotor, the back wheel, and the front wheel are

$$\boldsymbol{R}_c = \begin{bmatrix}\cos\theta & -\sin\theta & 0\\ \sin\theta & \cos\theta & 0\\ 0 & 0 & 1\end{bmatrix}, \quad \boldsymbol{R}_r = \begin{bmatrix}\cos(\theta+\psi) & -\sin(\theta+\psi) & 0\\ \sin(\theta+\psi) & \cos(\theta+\psi) & 0\\ 0 & 0 & 1\end{bmatrix},$$

$$\boldsymbol{R}_b = \begin{bmatrix}\cos(\theta+\phi) & -\sin(\theta+\phi) & 0\\ \sin(\theta+\phi) & \cos(\theta+\phi) & 0\\ 0 & 0 & 1\end{bmatrix},$$

$$\boldsymbol{R}_f = \begin{bmatrix}\cos(\theta-\phi) & -\sin(\theta-\phi) & 0\\ \sin(\theta-\phi) & \cos(\theta-\phi) & 0\\ 0 & 0 & 1\end{bmatrix},$$

respectively. The inertia tensors for each of the components have $\boldsymbol{e}_3$ as an eigenvector, and the eigenvalues corresponding to this eigenvector are the moments of inertia denoted by J_c, J_r, and J_w above. Now one simply applies the procedure of Section 4.2.3 to ascertain that the Riemannian metric in the given coordinates is

$$\begin{aligned}\mathbb{G} = (m_c + m_r + 2m_w)(\mathrm{d}x \otimes \mathrm{d}x + \mathrm{d}y \otimes \mathrm{d}y) + (J_c + J_r + 2(J_w + m_w\ell^2))\mathrm{d}\theta \otimes \mathrm{d}\theta\\ + J_r\mathrm{d}\psi \otimes \mathrm{d}\psi + 2J_w\mathrm{d}\phi \otimes \mathrm{d}\phi + J_r(\mathrm{d}\theta \otimes \mathrm{d}\psi + \mathrm{d}\psi \otimes \mathrm{d}\theta). \quad (13.4)\end{aligned}$$

Remark 13.10. This Riemannian metric differs slightly from what one will find in most papers treating the snakeboard. This is because authors often simplify the parameters by lumping all masses into one mass, and by denoting the inertial contribution of the wheels simply by J_w, as opposed to $J_w + m_w\ell^2$ as we have done. Thus the expression (13.4) is equivalent to what one finds in the literature. Some authors [e.g., Ostrowski and Burdick 1997] make simplifying assumptions about the inertial parameters. •

Constraints

The constraints require that the wheels of the snakeboard roll without slipping. The location of the back and front wheels is given by $\boldsymbol{r}_b$ and $\boldsymbol{r}_f$, respectively. The constraint is that the component of the velocities of these points in the $\boldsymbol{b}_{b,2}$- and $\boldsymbol{b}_{f,2}$-direction, respectively, be zero. In terms of the problem data, these constraints read

$$\mathbb{G}_{\mathbb{R}^3}(\boldsymbol{R}_b\boldsymbol{s}_2, \dot{\boldsymbol{r}}_b) = 0, \quad \mathbb{G}_{\mathbb{R}^3}(\boldsymbol{R}_f\boldsymbol{s}_2, \dot{\boldsymbol{r}}_f) = 0.$$

Doing the computations gives

$$\begin{aligned} -\sin(\theta+\phi)\dot{x} + \cos(\theta+\phi)\dot{y} - \ell\cos\phi\dot{\theta} = 0,\\ -\sin(\theta-\phi)\dot{x} + \cos(\theta-\phi)\dot{y} + \ell\cos\phi\dot{\theta} = 0. \end{aligned}$$

Thus the constraints are specified as being the distribution $\mathcal{D}$ annihilated by the two covector fields

$$\begin{aligned} \alpha_b &= -\sin(\theta+\phi)\mathrm{d}x + \cos(\theta+\phi)\mathrm{d}y - \ell\cos\phi\mathrm{d}\theta,\\ \alpha_f &= -\sin(\theta-\phi)\mathrm{d}x + \cos(\theta-\phi)\mathrm{d}y + \ell\cos\phi\mathrm{d}\theta. \end{aligned} \tag{13.5}$$

Note that these two covector fields generate a C^ω-codistribution on Q. What is more, the covector fields in (13.5) become collinear when $\phi = \pm\frac{\pi}{2}$. Thus the constraint distribution $\mathcal{D}$ annihilated by these covector fields has constant rank and is analytic on the open dense subset

$$\mathsf{Q}_{\mathcal{D}} = \left\{(x, y, \theta, \psi, \phi) \;\middle|\; \phi \neq \pm\tfrac{\pi}{2}\right\}$$

of Q. However, the constraint is not regular, since its annihilator has singularities. Thus the snakeboard is an example of a system for which the constraint distribution is not a generalized subbundle, although its annihilator is. This is just as we discussed in Section 4.5. In the sequel, we will restrict our attention to $\mathsf{Q}_{\mathcal{D}}$, making the observation that, at this time, it is not clear how one handles the singularities on $\mathsf{Q} \setminus \mathsf{Q}_{\mathcal{D}}$.

Generalized Christoffel symbols

To write the equations of motion for the snakeboard, we shall use the orthogonal Poincaré representation of Section 4.6.4. To do this, we need a $\mathbb{G}$-orthogonal basis for $\mathcal{D}$. To write this basis, it is convenient to introduce the vector fields

$$V_1 = \cos\theta \frac{\partial}{\partial x} + \sin\theta \frac{\partial}{\partial y}, \quad V_2 = -\sin\theta \frac{\partial}{\partial x} + \cos\theta \frac{\partial}{\partial y}.$$

Note that V_1 and V_2 can be thought of as left-invariant vector fields on $\mathsf{SE}(2)$ that, at a point $(x, y, \theta) \in \mathsf{SE}(2)$, point along the body axes labeled $\boldsymbol{b}_{c,1}$ and $\boldsymbol{b}_{c,2}$ in Figure 13.6. One can then check by direct computation that the three vector fields

$$X_1 = \ell\cos\phi V_1 - \sin\phi \frac{\partial}{\partial\theta}, \quad X_2' = \frac{\partial}{\partial\psi}, \quad X_3 = \frac{\partial}{\partial\phi}$$

generate the distribution $\mathcal{D}$ on $\mathsf{Q}_{\mathcal{D}}$. Furthermore, a direct computation also shows that X_1 and X_3 are $\mathbb{G}$-orthogonal. If, for any nowhere zero function $c_1 : \mathsf{Q}_{\mathcal{D}} \to \mathbb{R}$, we define

$$X_2 = c_1^{-1} \left(\langle\langle X_1, X_1 \rangle\rangle X_2' - \langle\langle X_1, X_2' \rangle\rangle X_1 \right),$$

then one verifies directly that $\{X_1, X_2, X_3\}$ forms a $\mathbb{G}$-orthogonal basis for $\mathcal{D}$ on $\mathsf{Q}_{\mathcal{D}}$. A convenient choice for the function c_1 is

$$c_1(x, y, \theta, \psi, \phi) = (m_c + m_r + 2m_w)\ell^2 \cos^2\phi + (J_c + J_r + 2(J_w + m_w\ell^2)) \sin^2\phi. \tag{13.6}$$

For brevity, we shall simply write $c_1(\phi) = c_1(x, y, \theta, \psi, \phi)$. With this choice one computes

$$X_2 = \frac{J_r \ell \cos\phi \sin\phi}{c_1(\phi)} V_1 - \frac{J_r \sin^2\phi}{c_1(\phi)} \frac{\partial}{\partial\theta} + \frac{\partial}{\partial\psi}.$$

For brevity, let us define

$$a(\phi) = \frac{J_r \ell \cos\phi \sin\phi}{c_1(\phi)}, \quad b(\phi) = \frac{J_r \sin^2\phi}{c_1(\phi)}. \tag{13.7}$$

To summarize, we have the following conclusion,

Lemma 13.11 (Basis for $\mathcal{D}$). *The vector fields*

$$X_1 = \ell\cos\phi V_1 - \sin\phi \frac{\partial}{\partial\theta}, \quad X_2 = a(\phi)V_1 - b(\phi)\frac{\partial}{\partial\theta} + \frac{\partial}{\partial\psi}, \quad X_3 = \frac{\partial}{\partial\phi}$$

form a $\mathbb{G}$-orthogonal basis for $\mathcal{D}$ on $\mathsf{Q}_{\mathcal{D}}$.

Next we compute the generalized Christoffel symbols corresponding to the basis $\{X_1, X_2, X_3\}$ for the constraint distribution. The following result is a consequence of Proposition 4.94. We omit the tedious, but straightforward, computations for the sake of brevity.

Lemma 13.12 (Generalized Christoffel symbols). *For the vector fields $\mathscr{X} = \{X_1, X_2, X_3\}$ of Lemma 13.11, the nonzero generalized Christoffel symbols for $\overset{\mathcal{D}}{\nabla}$ are*

$$\overset{\mathfrak{X}}{\Gamma}{}^1_{13} = \frac{(J_c + J_r + 2J_w - (m_c + m_r)\ell^2)\cos\phi\sin\phi}{c_1(\phi)},$$

$$\overset{\mathfrak{X}}{\Gamma}{}^1_{23} = \frac{J_r(m_c + m_r + 2m_w)\ell^2\cos\phi}{c_1(\phi)^2},$$

$$\overset{\mathfrak{X}}{\Gamma}{}^2_{13} = -\frac{(m_c + m_r + 2m_w)\ell^2\cos\phi}{c_2(\phi)},$$

$$\overset{\mathfrak{X}}{\Gamma}{}^2_{23} = -\frac{J_r(m_c + m_r + 2m_w)\ell^2\cos\phi\sin\phi}{c_1(\phi)c_2(\phi)},$$

where

$$c_2(\phi) = (m_c + m_r + 2m_w)\ell^2\cos^2\phi + (J_c + 2(J_w + m_w\ell^2))\sin^2\phi.$$

Control forces

The two controls for the snakeboard are torques, one actuating the rotor and the other actuating the wheels. For the torque on the rotor, the Newtonian torque on the coupler is $\boldsymbol{\tau}_{\psi,c} = (0,0,-1)$ and the corresponding torque on the rotor is $\boldsymbol{\tau}_{\psi,r} = (0,0,1)$. The corresponding total external force, by an application of (4.15), is easily computed to be $F_\psi = \mathrm{d}\psi$. For the torque on the wheels, the Newtonian torques on the back and front wheels are $\boldsymbol{\tau}_{\phi,b} = (0,0,\frac{1}{2})$ and $\boldsymbol{\tau}_{\phi,f} = (0,0,-\frac{1}{2})$, respectively. We take the torque on each wheel to have magnitude $\frac{1}{2}$ since, when the total external force is written, it will be a sum of the forces on each wheel, and it is convenient for this to have magnitude 1. The corresponding torque on the body is zero, since the wheel torques have opposite orientations. The corresponding total external force is then computed, again using (4.15), to be $F_\phi = \mathrm{d}\phi$. To represent these forces in the orthogonal Poincaré representation, by Theorem 4.87 we must first convert them to vector fields with the metric $\mathbb{G}$, then orthogonally project them onto $\mathcal{D}$. We first compute

$$Y'_\psi = \mathbb{G}^\sharp(F_\psi) = \frac{1}{J_c + 2(J_w + m_w\ell^2)}\left(\frac{J_c + J_r + 2(J_w + m_w\ell^2)}{J_r}\frac{\partial}{\partial\psi} - \frac{\partial}{\partial\theta}\right),$$

$$Y'_\phi = \mathbb{G}^\sharp(F_\phi) = \frac{1}{2J_w}\frac{\partial}{\partial\phi}.$$

Note that Y'_ϕ is already a section of $\mathcal{D}$, so we may take $Y_\phi = Y'_\phi$. Then we use the formula

$$Y_\psi = P_{\mathcal{D}}(Y'_\psi) = \sum_{i=1}^{3}\frac{\langle\langle Y'_\psi, X_i\rangle\rangle}{\|X_i\|^2_{\mathbb{G}}}X_i \tag{13.8}$$

to give

$$Y_\psi = \frac{c_1(\phi)}{J_r c_2(\phi)}X_2, \quad Y_\phi = \frac{1}{2J_w}X_3.$$

Equations of motion

We are now ready to give the equations of motion as per the orthogonal Poincaré representation of Section 4.6.4. We use the pseudo-velocities $t \mapsto (v^1(t), v^2(t), v^3(t))$ defined by

$$\dot{q}^i(t)\frac{\partial}{\partial q^i} = v^\alpha(t) X_\alpha(\gamma(t)).$$

This then gives the equations

$$\begin{aligned}
\dot{x} &= \ell \cos\phi \cos\theta v^1 + \frac{J_r \ell \cos\phi \sin\phi \cos\theta}{c_1(\phi)}\dot{\psi},\\
\dot{y} &= \ell \cos\phi \sin\theta v^1 + \frac{J_r \ell \cos\phi \sin\phi \sin\theta}{c_1(\phi)}\dot{\psi},\\
\dot{\theta} &= -\sin\phi v^1 - \frac{J_r \sin^2\phi}{c_1(\phi)}\dot{\psi},\\
\dot{v}^1 &= -\frac{(J_c + J_r + 2J_w - (m_c + m_r)\ell^2)\cos\phi \sin\phi}{c_1(\phi)} v^1 \dot{\phi}\\
&\quad - \frac{J_r(m_c + m_r + 2m_w)\ell^2 \cos\phi}{c_1(\phi)^2}\dot{\psi}\dot{\phi},\\
\ddot{\psi} &= \frac{(m_c + m_r + 2m_w)\ell^2 \cos\phi}{c_2(\phi)} v^1 \dot{\phi}\\
&\quad + \frac{J_r(m_c + m_r + 2m_w)\ell^2 \cos\phi \sin\phi}{c_1(\phi) c_2(\phi)}\dot{\psi}\dot{\phi} + \frac{c_1(\phi)}{J_r c_2(\phi)} u_\psi,\\
\ddot{\phi} &= \frac{1}{2J_w} u_\phi.
\end{aligned}$$

Note that the pseudo-velocities v^2 and v^3 do not appear, since they turn out to simply be $v^2 = \dot{\psi}$ and $v^3 = \dot{\phi}$. In cases such as this, the pseudo-velocities are actual velocities, and we retain them in their more basic form.

Remark 13.13. The snakeboard provides an excellent illustration of the modeling methods of Chapter 4. While there are a few computations to go through, they are fairly rote, and the final equations are not excessively complicated. In the supplementary internet material for the book, we illustrate how to easily derive the snakeboard equations of motion using the Mathematica® software that accompanies this book. •

13.4.2 Motion planning on SE(2) for the snakeboard

In this section we show that the snakeboard is kinematically controllable, and investigate the consequences of this for motion planning.

Kinematic controllability

Let us first exhibit two decoupling vector fields for the snakeboard. From (13.8) we note that $X_2, X_3 \in \Gamma^\infty(\mathcal{Y})$. Furthermore, a computation gives

$$\overset{\mathfrak{D}}{\nabla}_{X_2} X_2 = \overset{\mathscr{X}}{\Gamma}{}^{\alpha}_{22} X_\alpha = 0, \quad \overset{\mathfrak{D}}{\nabla}_{X_3} X_3 = \overset{\mathscr{X}}{\Gamma}{}^{\alpha}_{33} X_\alpha = 0.$$

This shows that we also have $\overset{\mathfrak{D}}{\nabla}_{X_2} X_2, \overset{\mathfrak{D}}{\nabla}_{X_3} X_3 \in \Gamma^\infty(\mathcal{Y})$, and so X_2 and X_3 are in fact decoupling vector fields. In Figure 13.7 we show how the snakeboard

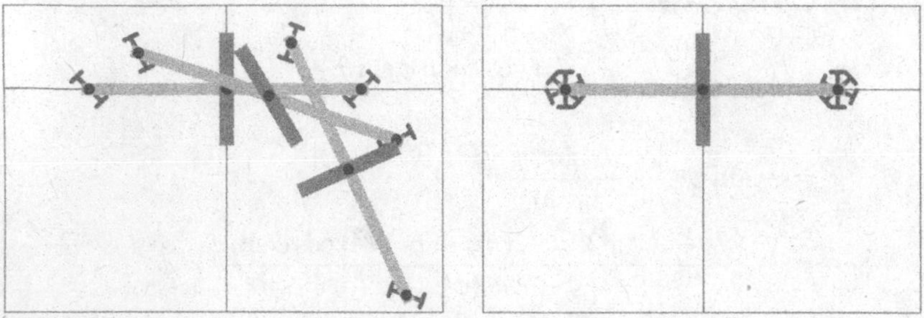

Figure 13.7. Snakeboard motion along X_2 (left) and X_3 (right)

appears as it moves along typical integral curves of X_2 and X_3. The motion of the snakeboard along X_3 is easily understood; the snakeboard remains in place and the front and back wheels rotate. The motion along X_2 is more complicated. Along integral curves of X_2, the snakeboard wheels are locked at a fixed angle ϕ_0, and the constraints are then holonomic, as shown in Figure 13.8. The system then conserves angular momentum, and this may be used to ascertain the relationship between the rotor angular velocity and the angular velocity of the snakeboard about the fixed point of rotation C in Figure 13.8. This is related to the nonholonomic momentum map of Bloch, Krishnaprasad, Marsden, and Murray [1996]. The reader may explore this in Exercise E13.5.

To check kinematic controllability of the snakeboard, we need to check that $\mathrm{Lie}^{(\infty)}(\mathcal{X}) = \mathsf{T}\mathsf{Q}_{\mathfrak{D}}$, where $\mathscr{X} = \{X_1, X_2\}$. The proof of the following result records the calculations necessary to check this.

Lemma 13.14 (Kinematic controllability). $\mathrm{Lie}^{(\infty)}(\mathcal{X}) = \mathsf{T}\mathsf{Q}_{\mathfrak{D}}$.

Proof. This will follow if and only if the involutive closure of $\tilde{\mathscr{X}} = \{c_1 X_2, X_3\}$ has maximal rank, with c_1 the function defined in (13.6). We compute

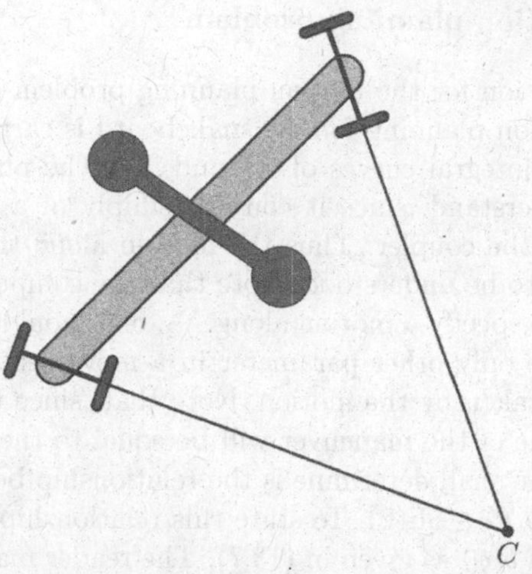

Figure 13.8. The motion of the snakeboard with locked wheels

$$
\begin{aligned}
[c_1X_2, X_3] = & - J_r\ell\cos(2\phi)V_1 + J_r\sin(2\phi)\frac{\partial}{\partial\theta} \\
& - (J_c + J_r + 2J_w - (m_c + m_r)\ell^2)\sin(2\phi)\frac{\partial}{\partial\psi} \\
[X_3, [c_1X_2, X_3]] = & \, 2J_r\ell\sin\phi\cos\phi V_1 + 2J_r\cos(2\phi)\frac{\partial}{\partial\theta} \\
& - 2(J_c + J_r + 2J_w - (m_c + m_r)\ell^2)\cos(2\phi)\frac{\partial}{\partial\psi} \\
[c_1X_2, [X_3, [c_1X_2, X_3]]] = & - J_r^2\ell\sin(2\phi)V_2 \\
[X_3, [c_1X_2, [X_3, [c_1X_2, X_3]]]] = & - 2J_r^2\ell\cos(2\phi)V_2.
\end{aligned}
$$

After assembling the components of the vector fields

$$\{c_1X_2, X_3, [c_1X_2, X_3], [X_3, [c_1X_2, X_3]], [X_3, [c_1X_2, [X_3, [c_1X_2, X_3]]]]\}$$

into the columns of a matrix, we compute the determinant of this matrix to be

$$-4J_r^4(m_c + m_r + 2m_w)\ell^4\cos(2\phi).$$

We see that this determinant is nonzero for $\phi \notin \{\pm\frac{\pi}{4}, \pm\frac{3\pi}{4}\}$. At these degenerate values of ϕ, the vector fields

$$\{c_1X_2, X_3, [c_1X_2, X_3], [X_3, [c_1X_2, X_3]], [c_1X_2, [X_3, [c_1X_2, X_3]]]\}$$

span the tangent space. This shows that the system is indeed kinematically controllable. ■

The general motion planning problem

Following the solution for the motion planning problem outlined in Proposition 13.6, the motion planning for the snakeboard is carried out by following concatenations of integral curves of X_2 and X_3. The physical motion along X_3 is easy to understand, since it consists simply of a reorientation of the wheels relative to the coupler. Thus the motion along the vector field X_2 is really what needs to be understood. Note that the components of X_2 depend on ϕ. Therefore, to specify a motion along X_2, one should first specify a wheel angle $\phi = \phi_0$. The only other parameter in a movement along X_3 will then be the total time taken by the motion. Note that, since the coefficient of $\frac{\partial}{\partial\psi}$ is 1 in X_2, the time of the maneuver will be equal to the change in the rotor angle $\Delta\psi$. What we shall determine is the relationship between $(\phi_0, \Delta\psi)$ and the motion in $(x, y, \theta) \in \mathsf{SE}(2)$. To state this relationship, we utilize the definitions of $a(\phi)$ and $b(\phi)$ as given in (13.7). The reader may also wish to recall the notation and computations concerning the Lie group $\mathsf{SE}(2)$, as presented in Example 5.47. We also recall the formula for the exponential map given in Exercise E5.7.

With the notation for $\mathsf{SE}(2)$ available, we have the following result.

Lemma 13.15. *Let $q_0 = (0, 0, 0, 0, \phi_0) \in \mathsf{Q}_{\mathfrak{D}}$ and let $(x, y, \theta, \Delta\psi, \phi_0) \in \mathsf{Q}_{\mathfrak{D}}$ be the point obtained by flowing along X_2 for time $\Delta\psi$ from q_0. Then*

$$\begin{bmatrix} \cos\theta & -\sin\theta & x \\ \sin\theta & \cos\theta & y \\ 0 & 0 & 1 \end{bmatrix} = \exp\left(\Delta\psi \begin{bmatrix} 0 & b(\phi_0) & a(\phi_0) \\ -b(\phi_0) & 0 & 0 \\ 0 & 0 & 0 \end{bmatrix}\right)$$
$$= \begin{bmatrix} \cos\big(b(\phi_0)\Delta\psi)\big) & \sin\big(b(\phi_0)\Delta\psi\big) & \ell\cot\phi_0\sin(b(\phi_0)\Delta\psi) \\ -\sin\big(b(\phi_0)\Delta\psi\big) & \cos\big(b(\phi_0)\Delta\psi\big) & \ell\cot\phi_0\big(\cos(b(\phi_0)\Delta\psi) - 1\big) \\ 0 & 0 & 1 \end{bmatrix}.$$

Proof. This is simply a matter of explicitly solving the differential equation associated with X_2 with ϕ specified as ϕ_0, and for time $\Delta\psi$. ■

It is evident from the definition of X_2 that, during a motion along X_2, one should have $\phi_0 \not\in \{0, \pi\}$, since in these degenerate configurations the rotor will simply move without changing (x, y, θ). The following lemma describes the possible values of the quantities $a(\phi)\Delta\psi$ and $b(\psi)\Delta\psi$ obtainable using wheel angles between $-\frac{\pi}{2}$ and $\frac{\pi}{2}$.

Lemma 13.16. *Let $\mathcal{S}, \mathcal{T} \subset \mathbb{R}^2$ be given by*

$$\mathcal{S} = \left(\left]-\tfrac{\pi}{2}, 0\right[\cup \left]0, \tfrac{\pi}{2}\right[\right) \times (\mathbb{R} \setminus \{0\}),$$
$$\mathcal{T} = \mathbb{R}^2 \setminus (\{(x, 0) \mid x \in \mathbb{R}\} \cup \{(0, y) \mid y \in \mathbb{R}\}).$$

The map $\mathcal{S} \ni (\phi, \Delta\psi) \mapsto (a(\phi)\Delta\psi, -b(\phi)\Delta\psi) \in \mathcal{T}$ is a C^∞-diffeomorphism.

Proof. The smoothness of the stated map is clear. We can also produce an explicit inverse for the map:

$$(\xi, \omega) \mapsto \Big(-\arctan\big(\tfrac{\omega \ell}{\xi}\big), -\frac{c_1\big(\arctan(\frac{\omega\ell}{\xi})\big)(\xi^2 + \ell^2\omega^2)}{J_r \ell^2 \omega}\Big).$$

This map is itself obviously smooth on the specified domain. ■

Combined, Lemmas 13.15 and 13.16 suggest that we should try to do motion planning in the coordinates (x, y, θ), using as parameters the translational and angular velocities $(\xi, \omega) = (a(\phi_0)\Delta\psi, b(\phi_0)\Delta\psi)$ constrained to take values in the set $\mathcal{T}$. For a physical snakeboard, it is reasonable to suppose that there will be restrictions on ψ and ϕ. For this reason, we make the assumption that $\psi \in [-\bar{\psi}, \bar{\psi}]$ and $\phi \in [-\bar{\phi}, \bar{\phi}]$ for some $\bar{\phi} \in]0, \frac{\pi}{2}[$. This provides a restriction that $|\Delta\psi| \leq 2\bar{\psi}$. This then defines a set

$$\bar{\mathcal{S}} = (]-\bar{\phi}, 0[\cup]0, \bar{\phi}[) \times (]-2\bar{\psi}, 2\bar{\psi}[),$$

and the map of Lemma 13.16, when restricted to this set, will have an image as shown in Figure 13.9. The angle χ in the figure is given by $\chi = -\arctan\big(\frac{\ell b(\bar{\phi})}{a(\bar{\phi})}\big)$.

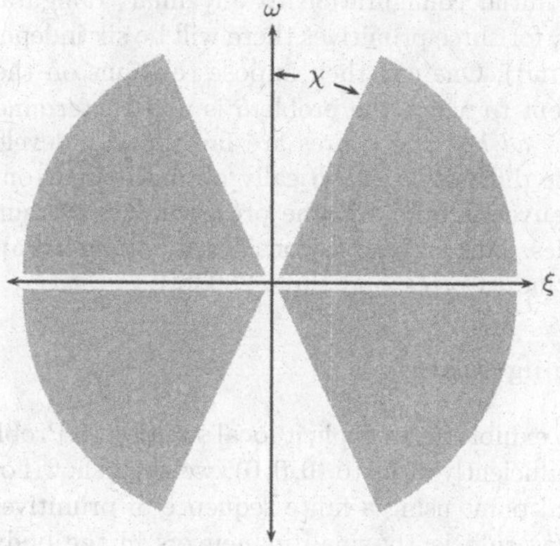

Figure 13.9. The gray region describes the set of admissible values for the forward and angular velocities (ξ, ω) with constraints on the wheel and rotor angles

Let us denote by $\bar{\mathcal{T}}$ the region of $\mathbb{R}^2$ shown in Figure 13.9. To summarize:

The set of (x, y, θ) reachable from $(0, 0, 0)$ using a single motion along X_2 is given by

$$\left\{ \left(\frac{\sin\omega}{\omega}\xi, \frac{1-\cos\omega}{\omega}\xi, \omega \right) \;\middle|\; (\xi,\omega) \in \bar{\mathfrak{T}} \right\}.$$

With this understanding of the integral curves of X_2 and X_3, we formally state the motion planning problem for the snakeboard.

Problem 13.17 (Snakeboard motion planning). Suppose that $q_{\mathrm{i}} = (x_{\mathrm{i}}, y_{\mathrm{i}}, \theta_{\mathrm{i}}, \psi_{\mathrm{i}}, \phi_{\mathrm{i}})$ and $q_{\mathrm{f}} = (x_{\mathrm{f}}, y_{\mathrm{f}}, \theta_{\mathrm{f}}, \psi_{\mathrm{f}}, \phi_{\mathrm{f}})$ are given. Find a finite collection of points $(\xi_1, \omega_1), \dots, (\xi_k, \omega_k) \in \bar{\mathfrak{T}}$ such that

$$\begin{bmatrix} \cos\theta_{\mathrm{i}} & -\sin\theta_{\mathrm{i}} & x_{\mathrm{i}} \\ \sin\theta_{\mathrm{i}} & \cos\theta_{\mathrm{i}} & y_{\mathrm{i}} \\ 0 & 0 & 1 \end{bmatrix} \circ \exp\left(\begin{bmatrix} 0 & -\omega_1 & \xi_1 \\ \omega_1 & 0 & 0 \\ 0 & 0 & 0 \end{bmatrix} \right) \circ \cdots$$
$$\circ \exp\left(\begin{bmatrix} 0 & -\omega_k & \xi_k \\ \omega_k & 0 & 0 \\ 0 & 0 & 0 \end{bmatrix} \right) = \begin{bmatrix} \cos\theta_{\mathrm{f}} & -\sin\theta_{\mathrm{f}} & x_{\mathrm{f}} \\ \sin\theta_{\mathrm{f}} & \cos\theta_{\mathrm{f}} & y_{\mathrm{f}} \\ 0 & 0 & 1 \end{bmatrix} \quad \bullet$$

Remarks 13.18. 1. Problem 13.17 reduces the motion planning problem for the snakeboard to a motion planning problem using primitives on $\mathsf{SE}(2)$.

2. Of course, in obtaining a solution to the problem, one will want to minimize k. In the absence of control constraints, it is possible to show that one can steer from an initial configuration to any final configuration with k at most 3. Indeed, for three primitives there will be six independent variables (three pairs (ξ, ω)). One can then impose relations on these variables to reduce the extent to which the problem is underdetermined. In this way, various sorts of *ad hoc* procedures are fairly easily developed. However, it appears quite difficult to analytically obtain bounds on the number of primitive maneuvers required in the presence of constraints on the rotor and wheel angles. This problem is considered by Iannitti and Lynch [2003] (see also [Martínez, Cortés, and Bullo 2003b]). •

A specific planning strategy

Let us now turn to exhibiting an explicit local solution to Problem 13.17. That is, for $(x_{\mathrm{f}}, y_{\mathrm{f}}, \theta_{\mathrm{f}})$ sufficiently close to $(0,0,0)$, we shall show how to steer from $(0,0,0)$ to the final point using a finite sequence of primitives. The strategy is first to show how sufficiently small maneuvers in the body $\boldsymbol{b}_{\mathrm{c},1}$ and $\boldsymbol{b}_{\mathrm{c},2}$-directions (see Figure 13.6) can be performed while leaving θ unchanged. Then we demonstrate how one may achieve a desired change in the angle θ, but with a translation by some predictable amount in x and y. Using the previously determined motions in the $\boldsymbol{b}_{\mathrm{c},1}$- and $\boldsymbol{b}_{\mathrm{c},2}$-directions, one can then achieve a change in θ while leaving x and y unchanged. This gives rise to a complete closed-form solution to the local motion planning problem. That is to say, for every initial configuration $q_{\mathrm{i}} \in \mathsf{Q}_{\mathfrak{D}}$, there exist a neighborhood $\mathcal{U}_{q_{\mathrm{i}}}$ and an explicit algorithm for steering the snakeboard from q_{i} to points in $\mathcal{U}_{q_{\mathrm{i}}}$. It is then possible to design a global motion planning algorithm by computing appropriate sequences of local displacements, and then using multiple invocations of the local planner (cf. the second part of the proof of Theorem 13.2).

Translation in the $\boldsymbol{b}_{c,1}$-direction

For fixed $\phi_0, \Delta\psi > 0$, we construct a sequence of three basic primitives as follows:

$$\exp\left(\Delta\psi\begin{bmatrix}0 & b(\phi_0) & a(\phi_0)\\ -b(\phi_0) & 0 & 0\\ 0 & 0 & 0\end{bmatrix}\right)$$
$$\circ\exp\left(-2\Delta\psi\begin{bmatrix}0 & b(-\phi_0) & a(-\phi_0)\\ -b(-\phi_0) & 0 & 0\\ 0 & 0 & 0\end{bmatrix}\right)$$
$$\circ\exp\left(\Delta\psi\begin{bmatrix}0 & b(\phi_0) & a(\phi_0)\\ -b(\phi_0) & 0 & 0\\ 0 & 0 & 0\end{bmatrix}\right) = \begin{bmatrix}1 & 0 & \Delta\xi\\ 0 & 1 & 0\\ 0 & 0 & 1\end{bmatrix},$$

where

$$\Delta\xi = 4\ell\cot\phi_0\sin\left(\frac{J_r\Delta\psi\sin^2\phi_0}{c_1(\phi_0)}\right).$$

Thus we can, for a fixed ϕ_0 and for $\Delta\xi$ sufficiently small, explicitly choose $\Delta\psi$ to translate the snakeboard in the body $\boldsymbol{b}_{c,1}$-direction by an amount $\Delta\xi$. In practice, the amount of the translation can be reasonably large. To obtain arbitrarily large translations, one can of course compose smaller ones as we have constructed here.

Translation in the $\boldsymbol{b}_{c,2}$-direction

The story here is much the same as for the $\boldsymbol{b}_{c,1}$-translation, except that we use the following sequence of four basic primitives:

$$\exp\left(\Delta\psi\begin{bmatrix}0 & b(\phi_0) & a(\phi_0)\\ -b(\phi_0) & 0 & 0\\ 0 & 0 & 0\end{bmatrix}\right)$$
$$\circ\exp\left(-\Delta\psi\begin{bmatrix}0 & b(-\phi_0) & a(-\phi_0)\\ -b(-\phi_0) & 0 & 0\\ 0 & 0 & 0\end{bmatrix}\right)$$
$$\circ\exp\left(-\Delta\psi\begin{bmatrix}0 & b(\phi_0) & a(\phi_0)\\ -b(\phi_0) & 0 & 0\\ 0 & 0 & 0\end{bmatrix}\right)$$
$$\circ\exp\left(\Delta\psi\begin{bmatrix}0 & b(-\phi_0) & a(-\phi_0)\\ -b(-\phi_0) & 0 & 0\\ 0 & 0 & 0\end{bmatrix}\right) = \begin{bmatrix}1 & 0 & 0\\ 0 & 1 & \Delta\eta\\ 0 & 0 & 1\end{bmatrix},$$

where

$$\Delta\eta = 4\ell \cot\phi_0 \Big(\cos\Big(\frac{J_r \Delta\psi \sin^2\phi_0}{c_1(\phi_0)}\Big) - 1\Big).$$

Thus, for $\Delta\eta$ sufficiently small and for a fixed ϕ_0, one can achieve a pure translation in the body $\boldsymbol{b}_{c,2}$-direction. In practice, and not surprisingly, the translation in the $\boldsymbol{b}_{c,2}$-direction is more costly to achieve than the corresponding $\boldsymbol{b}_{c,1}$-translation. Nevertheless, one can still compose sequences of small $\boldsymbol{b}_{c,2}$-translations to achieve any specified motion in that direction.

Rotation

Unlike our translations in the $\boldsymbol{b}_{c,1}$- and $\boldsymbol{b}_{c,2}$-directions, the motion in the θ-direction is achieved while incurring motion in the other variables as well. The single primitive we use is

$$\exp\left(\Delta\psi \begin{bmatrix} 0 & b(\phi_0) & a(\phi_0) \\ -b(\phi_0) & 0 & 0 \\ 0 & 0 & 0 \end{bmatrix}\right) = \begin{bmatrix} \cos(\Delta\theta) & -\sin(\Delta\theta) & \Delta x \\ \sin(\Delta\theta) & \cos(\Delta\theta) & \Delta y \\ 0 & 0 & 1 \end{bmatrix},$$

where

$$\Delta\theta = -\frac{J_r \Delta\psi \sin^2\phi_0}{c_1(\phi_0)},$$

$$\Delta x = \ell \cot\phi_0 \sin\Big(\frac{J_r \Delta\psi \sin^2\phi_0}{c_1(\phi_0)}\Big),$$

$$\Delta y = \ell \cot\phi_0 \Big(\cos\Big(\frac{J_r \Delta\psi \sin^2\phi_0}{c_1(\phi_0)}\Big) - 1\Big).$$

Although there is a possibly undesired translational displacement incurred, this can be corrected by using the above translational maneuvers, since the undesired displacements are explicitly determined. Thus it is possible to achieve pure rotation by following the single primitive above with two translational primitives.

13.4.3 Simulations

We shall present two simulations for the snakeboard. All simulations use the following parameter values: $m_c = \frac{1}{2}$, $m_r = \frac{3}{4}$, $m_w = \frac{1}{4}$, $J_c = \frac{1}{2}$, $J_r = \frac{3}{4}$, $J_w = \frac{1}{8}$, $\ell = \frac{1}{2}$. In Figure 13.10 we show a motion of the snakeboard from $(x, y, \theta) = (0, 0, 0)$ to $(x, y, \theta) = \big(\sqrt{2}, 2, \frac{\pi}{5}\big)$ using the basic translation and rotation primitives from our explicit solution to the local motion planning problem. The rotation by $\frac{\pi}{5}$ was broken into two smaller rotations to mollify the effects of the deviation in x and y. We note that, including final corrections to the wheel and rotor angles, the above sort of motion will involve a concatenation of twenty basic kinematic motions. While some of these twenty

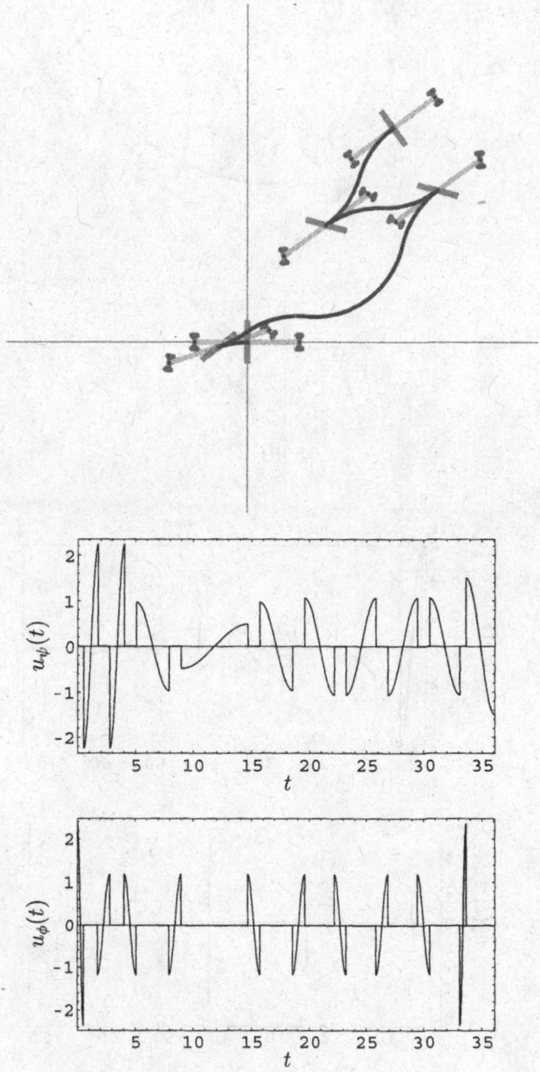

Figure 13.10. Snakeboard motion from $(0,0,0)$ to $\left(\sqrt{2}, 2, \frac{\pi}{5}\right)$ using concatenations of basic primitives. The controls are shown in the bottom figure.

motions are very simple, it is clear that this is generally a rather inefficient, albeit systematic, way of generating snakeboard motion.

In Figure 13.11 we show a three primitive motion, with the same initial and final configuration as in Figure 13.10, resulting from an *ad hoc* solution to Problem 13.17. There are many *ad hoc* solution techniques, and they amount to removing the redundancy in the problem to give the same number of equations as unknowns. Doing this in the right way will give an equation that

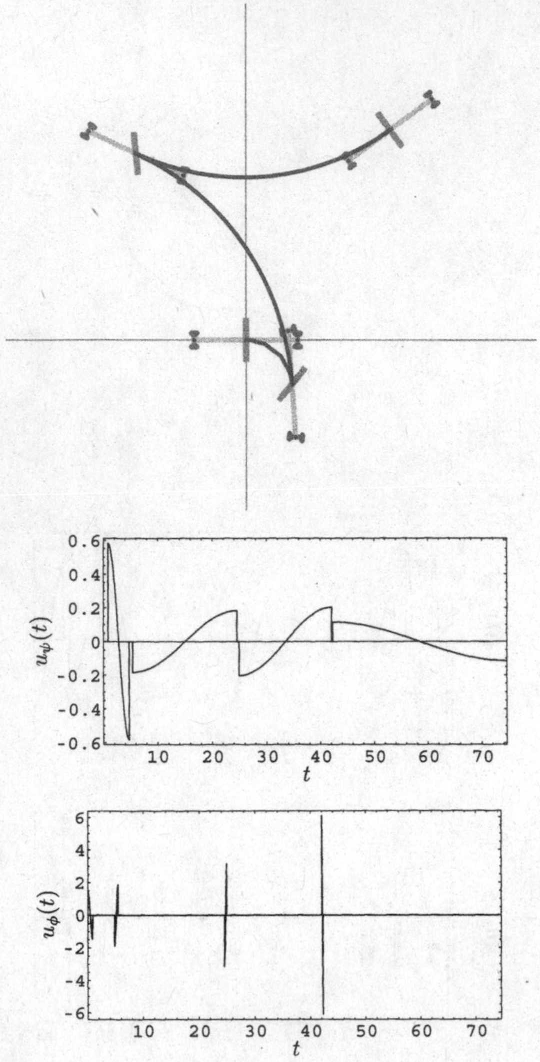

Figure 13.11. Snakeboard motion from $(0,0,0)$ to $\left(\sqrt{2}, 2, \frac{\pi}{5}\right)$ using an *ad hoc* solution to Problem 13.17. The controls are shown in the bottom figure.

can be solved explicitly. One way of carrying this out is indicated in Exercise E13.7. It should be noted that an *ad hoc* solution of this sort will not necessarily satisfy a given set of constraints on the wheel and rotor angles.

Remarks 13.19. 1. For the purposes of making more easily understood plots, we have ignored constraints of the rotor angle ψ. When constraints on this angle are enforced, the motions become comparatively smaller than those in Figures 13.10 and 13.11.

2. The controls that provide the motions shown in Figures 13.10 and 13.11 are simple in form, being merely cosines with an easily represented amplitude and frequency.
3. For an accurate comparison of the motion in Figure 13.11 with the motion of Figure 13.10, we should also count the motions involving positioning the wheels, and straightening out the rotor and wheels after the motion. In this case, there are eight segments of the motion in Figure 13.11, compared with twenty for the motion in Figure 13.10. •

Exercises

E13.1 Let $\Sigma = (\mathsf{M}, \mathscr{X}, U)$ be a driftless system. Show that if the motion planning problem using primitives is solvable for Σ, then so is the motion planning problem using locally absolutely continuous controls.

E13.2 Consider the planar rigid body for which the motion planning problem is solved in Section 13.3.1. For each of the decoupling vector fields X_1 and X_2, do the following:
(a) compute the integral curve through the point $(\theta, x, y) = (0, 0, 0)$;
(b) reparameterize the integral curve from part (a) on $[0, T]$, $T > 0$, so that it starts and ends with zero velocity;
(c) compute the control for the affine connection control system that will follow the reparameterized integral curve of part (b).

E13.3 Consider the robotic leg for which the motion planning problem is solved in Section 13.3.2. For each of the decoupling vector fields Y_1 and Y_2, do the following:
(a) compute the integral curve through the point $(r, \theta, \psi) = (r_0, 0, 0)$;
(b) reparameterize the integral curve from part (a) on $[0, T]$, $T > 0$, so that it starts and ends with zero velocity;
(c) compute the control for the affine connection control system that will follow the reparameterized integral curve of part (b).

E13.4 Solve the motion planning problem for the robotic leg using the solution of Proposition 13.6 for kinematically controllable systems.

E13.5 Consider the snakeboard with the wheels locked at angle $\phi_0 \notin \{-\frac{\pi}{2}, 0, \pi, \frac{\pi}{2}\}$.
(a) Show that the resulting constraints are holonomic, and that the maximal integral manifolds for the integrable constraints are diffeomorphic to $\mathbb{T}^2$.
Let Q_0 be the maximal integral manifold through the configuration of the snakeboard with coordinates $(x, y, \theta, \psi, \phi) = (0, 0, 0, 0, \phi_0)$.
(b) Show that one may use (θ, ψ) as coordinates for Q_0.
(c) Show that $\mathsf{TQ}_0 \subset \mathsf{TQ}$ is an invariant submanifold for the unforced snakeboard equations of motion. (This means that no forces need be exerted to keep the snakeboard on TQ_0 provided that it starts on TQ_0.)
(d) Show that the Riemannian metric for the system on Q_0 is
$$\mathbb{G}_0 = c_1(\phi_0) \csc^2 \phi_0 \mathrm{d}\theta \otimes \mathrm{d}\theta + J_r(\mathrm{d}\psi \otimes \mathrm{d}\psi + \mathrm{d}\theta \otimes \mathrm{d}\psi + \mathrm{d}\psi \otimes \mathrm{d}\theta).$$

Consider the group action of $\mathsf{SO}(2)$ on Q_0 given in coordinates by $\Phi(\alpha,(\theta,\psi))=(\theta+\alpha,\psi)$.

(e) Show that the group action leaves $\mathbb{G}_0$ invariant.

(f) Construct the momentum map for the group action (see Definition 5.68).

(g) With the snakeboard starting at rest at a point in Q_0 with the wheels locked, use conservation of momentum to derive a relationship between the change in angle of ψ with the change in angle of θ. Comment on how this is useful in motion planning.

E13.6 For each of the three basic motions defining the solution to the local path planning problem for the snakeboard (i.e., the translations in the $\boldsymbol{b}_1$ and $\boldsymbol{b}_2$-directions, and the rotation), sketch the motion of the snakeboard in the plane.

E13.7 In this problem we indicate how to obtain "by hand" a solution to the snakeboard motion planning problem that will involve at most three motions along the decoupling vector field X_2. This means that we need to show that, given $(x_{\mathrm{f}},y_{\mathrm{f}},\theta_{\mathrm{f}})$, there exists a solution $(\xi_1,\xi_2,\xi_3,\omega_1,\omega_2,\omega_3)$ of the equation

$$\exp\left(\begin{bmatrix}0&-\omega_1&\xi_1\\ \omega_1&0&0\\ 0&0&0\end{bmatrix}\right)\circ\exp\left(\begin{bmatrix}0&-\omega_2&\xi_2\\ \omega_2&0&0\\ 0&0&0\end{bmatrix}\right)\circ\exp\left(\begin{bmatrix}0&-\omega_3&\xi_3\\ \omega_3&0&0\\ 0&0&0\end{bmatrix}\right)=\begin{bmatrix}\cos\theta_{\mathrm{f}}&-\sin\theta_{\mathrm{f}}&x_{\mathrm{f}}\\ \sin\theta_{\mathrm{f}}&\cos\theta_{\mathrm{f}}&y_{\mathrm{f}}\\ 0&0&1\end{bmatrix}. \quad \text{(E13.1)}$$

Fix $\alpha\in\mathbb{R}\setminus\{0\}$.

(a) Show that, if we restrict ω_1, ω_2, and ω_3 to satisfy

$$\omega_1=-\alpha,\quad \omega_2=\omega_3=\tfrac{1}{2}(\theta_{\mathrm{f}}+\alpha),$$

then the rotational part of (E13.1) (i.e., the components $(1,1)$, $(1,2)$, $(2,1)$, and $(2,2)$ of the matrix equation) will be satisfied.

Now we wish to solve for (ξ_1,ξ_2,ξ_3). To do so we specify $\xi_1=\frac{1}{\alpha}$.

(b) Show that, with this constraint on ξ_1 and with the constraints on ω_1, ω_2, and ω_3 specified in part (a), the $(1,3)$ and $(2,3)$ components of the matrix equation (E13.1) form a linear equation for ξ_2 and ξ_3.

(c) Find conditions on $(x_{\mathrm{f}},y_{\mathrm{f}},\theta_{\mathrm{f}})$ that ensure that the linear equations of part (b) have a solution.

This was the method used to generate the motion of Figure 13.11, with $\alpha=\frac{3}{2}$.

(d) Reproduce Figure 13.11.

E13.8 *Mathematica®-assisted project:* Consider the kinematic reductions described for the three-link manipulator in Exercise E8.14.

(a) *Actuator configuration* $(0,1,1)$*:* For a general curve $\gamma\colon[0,T]\to\mathsf{Q}$ that satisfies $\gamma'(t)\in\mathscr{Y}_{\gamma(t)}$, determine a symbolic expression for the control required to follow γ, starting at $t=0$ at rest, and ending at $t=T$ at rest.

Hint: *First reparameterize γ to achieve the end conditions, then use Proposition 13.7.*

(b) *Actuator configuration* $(1, 1, 0)$*:* For each of the decoupling vector fields X_1 and X_2 from part (f) of Exercise E8.14, do the following:

1. find the integral curve of the vector field with initial condition $(\theta_1(0), \theta_2(0), \theta_3(0)) = (0, 0, 0)$;
2. find the control required to follow the integral curve for time T, starting and stopping with zero velocity.

   Hint: *Reparameterize the integral curve, and use Proposition 13.5.*

(c) *Actuator configuration* $(1, 0, 1)$*:* For each of the decoupling vector fields X_1 and X_2 from part (g) of Exercise E8.14, do the following:

1. find the integral curve of the vector field with initial condition $(\theta_1(0), \theta_2(0), \theta_3(0)) = (0, 0, 0)$;
2. find the control required to follow the integral curve for time T, starting and stopping with zero velocity.

   Hint: *Reparameterize the integral curve, and use Proposition 13.5.*

A

Time-dependent vector fields

In this chapter we review, without proofs, some results for differential equations that depend on time. A good concise reference for these results is Appendix C of [Sontag 1998]. Since in control theoretic applications one typically has a rather general (e.g., measurable) dependence on time, we first provide some background in measure theory. The treatment will be hardly adequate for a reader to really learn measure theory, and we refer the interested reader to the books [Cohn 1980, Halmos 1974a]. We comment that not much is lost conceptually if general measurable controls are replaced with something more easily understood, such as, for example, piecewise continuous controls.

A.1 Measure and integration

We begin by providing a brief discussion of some general measure theoretic concepts, since these arise not only in time-dependent ordinary differential equations, but (briefly) in our discussion of the inertia tensor. Then we define the Lebesgue measure on $\mathbb{R}$, and give some of its properties.

A.1.1 General measure theory

In measure theory, it is helpful to have on hand the extended real numbers. Let us denote by $[-\infty, +\infty]$ the union of the real numbers $\mathbb{R}$ with the two point sets with one element, $\{-\infty\}$ and $\{+\infty\}$. This set is ordered in the following way. Points in $]-\infty, +\infty[$ adopt the usual order on $\mathbb{R}$, and we declare that $-\infty < x$ and $x < +\infty$ for every $x \in]-\infty, +\infty[$. We also declare that $x + (+\infty) = +\infty$ and $x + (-\infty) = -\infty$ for every $x \in [-\infty, +\infty[$, and that $(+\infty) + (+\infty) = +\infty$, thus defining addition in $[-\infty, +\infty]$ (we ask that addition be commutative and we decree that $(-\infty) + (+\infty)$ makes no sense). Multiplication on $[-\infty, +\infty]$ is defined by $x \cdot (+\infty) = +\infty$ and $x \cdot (-\infty) = -\infty$ for $x > 0$, $x \cdot (+\infty) = -\infty$ and $x \cdot (-\infty) = +\infty$ for $x < 0$, $(+\infty) \cdot (+\infty) =$

$(-\infty)\cdot(-\infty)=+\infty$, $(-\infty)\cdot(+\infty)=-\infty$, and $0\cdot(+\infty)=0\cdot(-\infty)=0$. We declare multiplication to be commutative. Note that $[-\infty,+\infty]$ is *not* a ring.

A measure is applied to a certain class of subsets.

Definition A.1 (σ-algebra). A ***σ-algebra*** on a set S is a collection $\mathscr{A}$ of subsets of S with the properties

(i) $S\in\mathscr{A}$,
(ii) if $A\in\mathscr{A}$ then $S\setminus A\in\mathscr{A}$,
(iii) for each sequence $\{A_n\}_{n\in\mathbb{N}}$ in $\mathscr{A}$, $\cup_{n\in\mathbb{N}}A_n\in\mathscr{A}$, and
(iv) for each sequence $\{A_n\}_{n\in\mathbb{N}}$ in $\mathscr{A}$, $\cap_{n\in\mathbb{N}}A_n\in\mathscr{A}$.

A ***measurable space*** is a pair $(S,\mathscr{A})$, where $\mathscr{A}$ is a σ-algebra on S. •

Examples A.2. 1. If S is a set, the collection $\mathbf{2}^S$ of all subsets of S is a σ-algebra.

2. Given an arbitrary collection $\mathscr{S}$ of subsets of S, we may construct "the smallest σ-algebra containing $\mathscr{S}$." We do this as follows. Let $\mathscr{C}_{\mathscr{S}}$ be the collection of σ-algebras containing $\mathscr{S}$. Note that $\mathbf{2}^S$ is a σ-algebra containing $\mathscr{S}$, so that $\mathscr{C}_{\mathscr{S}}$ is not empty. One then ascertains that
$$\bigcap_{\mathscr{A}\in\mathscr{C}_{\mathscr{S}}}\mathscr{A}$$
is itself a σ-algebra. Obviously it is the smallest σ-algebra containing $\mathscr{S}$ by the very manner in which it is constructed. We call this σ-algebra that ***generated*** by $\mathscr{S}$.

3. We shall be interested in measures on $\mathbb{R}^n$. There is a natural σ-algebra on $\mathbb{R}^n$, which we call the ***Borel subsets***. This is the σ-algebra generated by the open sets. •

A measurable space awaits the definition of a measure.

Definition A.3 (Measure). If $(S,\mathscr{A})$ is a measurable space, then a ***measure*** on $(S,\mathscr{A})$ is a function $\mu\colon\mathscr{A}\to[0,+\infty]$ with the following properties:

(i) $\mu(\emptyset)=0$;
(ii) $\mu\Big(\bigcup_{n\in\mathbb{N}}A_n\Big)=\sum_{n=1}^{+\infty}\mu(A_n)$ for every collection $\{A_n\}_{n\in\mathbb{N}}$ of disjoint elements of $\mathscr{A}$. •

Now suppose that S has a topology $\mathscr{O}$. Let Z denote the union of all open sets in S that have zero measure. This union exists, and is an open set. Thus its complement is closed, and we call this complement the ***support*** of μ. Thus the support of a measure is the smallest closed set whose complement has zero measure.

A.1.2 Lebesgue measure

We now describe the measure on $\mathbb{R}$ that we will use in our discussion of time-dependence. We begin with a definition of the "size" of an arbitrary subset of $\mathbb{R}$.

Definition A.4 (Lebesgue outer measure). For $S \in \mathbf{2}^{\mathbb{R}}$, the ***Lebesgue outer measure*** of S is defined by

$$\lambda^*(S) = \inf\left\{ \sum_{n=1}^{+\infty} |b_n - a_n| \;\middle|\; S \subset \bigcup_{n\in\mathbb{N}}]a_n, b_n[\right\}. \qquad \bullet$$

It is possible to verify that the Lebesgue outer measure has the following properties:

1. $\lambda^*(\emptyset) = 0$;
2. if $S \subset T \subset \mathbb{R}$, then $\lambda^*(S) \leq \lambda^*(T)$;
3. $\lambda^*\Big(\bigcup_{n\in\mathbb{N}} S_n\Big) \leq \sum_{n=1}^{+\infty} \mu^*(S_n)$ for every collection $\{S_n\}_{n\in\mathbb{N}}$ of subsets of $\mathbb{R}$;
4. if $I \in \mathscr{I}$, then $\lambda^*(I)$ is the length of I.

Unfortunately, the Lebesgue outer measure is not a measure on $(\mathbb{R}, \mathbf{2}^{\mathbb{R}})$. However, it may be used to define a measure by restricting it to certain subsets of $\mathbb{R}$.

Definition A.5 (Lebesgue measure). Denote by $\mathscr{L}(\mathbb{R})$ the collection of subsets A of $\mathbb{R}$ for which

$$\lambda^*(S) = \lambda^*(S \cap A) + \lambda^*(S \cap (\mathbb{R} \setminus A)), \qquad S \in \mathbf{2}^{\mathbb{R}}.$$

A set in $\mathscr{L}(\mathbb{R})$ is ***Lebesgue measurable***. The map $\lambda\colon \mathscr{L}(\mathbb{R}) \to [0, +\infty]$ defined by $\lambda(A) = \lambda^*(A)$ is the ***Lebesgue measure***. A subset $Z \in \mathscr{L}(\mathbb{R})$ is a ***set of measure zero*** if $\lambda(Z) = 0$. •

Remarks A.6. 1. We will frequently be interested in subsets, not of all of $\mathbb{R}$, but of $I \in \mathscr{I}$. One can define measurable subsets of I by $\mathscr{L}(I) = \{A \cap I \mid A \in \mathscr{L}(\mathbb{R})\}$.

2. A property P holds ***almost everywhere (a.e.)*** on I, or for ***almost every*** $\boldsymbol{t \in I}$ **(*a.e.*** $\boldsymbol{t \in I}$**)** if the set of points for which P is not true has measure zero. •

One must, of course, verify that the Lebesgue measure is a measure, and for this we refer to the references given at the beginning of the chapter. It is generally not so easy to understand the character of Lebesgue measurable sets. On the other hand, it is not entirely simple to construct a set that is not measurable. What is true is that a Lebesgue measurable set is the union of a Borel set and a set of measure zero. Indeed, much of the difficult character of Lebesgue measurable sets occurs in sets of measure zero. For this reason, many ideas involving Lebesgue measure often agree with one's intuition about measure. Royden [1988] makes some comments along these lines.

In Section 4.2.1 we shall require the Lebesgue measure on $\mathbb{R}^3$. To construct the Lebesgue outer measure on $\mathbb{R}^n$, $n > 1$, one proceeds much as in the case $n = 1$, but using not intervals, but cubes in $\mathbb{R}^n$ as the sets with which to cover a subset in $\mathbb{R}^n$. The construction of Lebesgue measurable sets and then the Lebesgue measure on $\mathbb{R}^n$ proceeds just as above. We refer to the references cited at the beginning of the chapter for details.

A.1.3 Lebesgue integration

The Lebesgue measure can be used to define a notion of integration for suitable functions. Let us first say what "suitable" means.

Definition A.7 (Measurable function). Let $I \in \mathscr{I}$. A function $f\colon I \to [-\infty, +\infty]$ is ***measurable*** if, for every $a \in \mathbb{R}$, we have $f^{-1}([a, +\infty]) \in \mathscr{L}(I)$. •

Since it is not easy to find sets that are not measurable, it is also not so easy to define functions that are not measurable. Thus the class of measurable functions will include nearly any sort of function one is likely to encounter in practice.

Now we indicate how to integrate a certain sort of function. To do so, if S is a set and if $A \subset S$, then we define the ***characteristic function*** of A to be the function

$$\chi_A(x) = \begin{cases} 1, & x \in A, \\ 0, & x \notin A. \end{cases}$$

We now make a definition.

Definition A.8 (Simple function). Let $I \in \mathscr{I}$. A ***simple function*** is a function $f\colon I \to \mathbb{R}$ of the form

$$f = \sum_{j=1}^{k} c_j \chi_{A_j} \tag{A.1}$$

for $A_j \in \mathscr{L}(I)$ and $c_j \in \mathbb{R}$, $j \in \{1, \dots, k\}$, $k \in \mathbb{N}$. The ***integral*** of a simple function of the form (A.1) is defined by

$$\int_I f \,\mathrm{d}\lambda = \sum_{j=1}^{k} c_j \lambda(A_j).$$ •

It turns out that it is possible to approximate a measurable function taking values in $[0, +\infty]$ with an increasing sequence of simple functions. With this in mind, if $f\colon I \to [0, +\infty]$ is measurable, then we define

$$\int_I f\,\mathrm{d}\lambda = \sup\Big\{\int_I g\,\mathrm{d}\lambda \mid g \text{ is a positive simple function with } g(t) \le f(t) \text{ for } t \in I\Big\}.$$

If $f\colon I \to [-\infty, +\infty]$, then we define

$$f^+(x) = \max\{0, f(x)\}, \quad f^-(x) = -\min\{0, f(x)\}.$$

and then define

$$\int_I f\,\mathrm{d}\lambda = \int f^+\,\mathrm{d}\lambda - \int f^-\,\mathrm{d}\lambda.$$

This is the ***Lebesgue integral*** of f, and is always defined (although possibly infinite) provided that f is measurable. Let us define some language surrounding the Lebesgue integral of functions.

Definition A.9 (Integrable, essentially bounded, absolutely continuous). Let $I \in \mathscr{I}$ and let $f\colon I \to [-\infty, +\infty]$ be measurable.

(i) If at least one of $\int_I f^+\,\mathrm{d}\lambda$ or $\int_I f^-\,\mathrm{d}\lambda$ are finite, then the Lebesgue integral of f ***exists*** (and may be infinite).

(ii) If both $\int_I f^+\,\mathrm{d}\lambda$ and $\int_I f^-\,\mathrm{d}\lambda$ are infinite, then the Lebesgue integral of f ***does not exist***.

(iii) If $\int_I f^+\,\mathrm{d}\lambda < +\infty$ and $\int_I f^-\,\mathrm{d}\lambda < +\infty$, then f is ***Lebesgue integrable***, or simply ***integrable***.

(iv) If, for each compact subinterval $J \subset I$, the function $f|J$ is integrable, then f is ***locally integrable***.

(v) If there exists $M > 0$ such that $\lambda(\{t \in I \mid |f(t)| > M\}) = 0$, then f is ***essentially bounded***, and we write

$$\operatorname*{ess\,sup}_{t \in I} |f(t)| = \inf\{M \in \mathbb{R} \mid \lambda(\{t \in I \mid |f(t)| > M\}) = 0\}.$$

(vi) If there exist a locally integrable function $g\colon I \to [-\infty, +\infty]$ and some $t_0 \in I$ such that

$$f(t) = \int_{[t_0,t]} (g|[t_0,t])\,\mathrm{d}\lambda,$$

then f is ***locally absolutely continuous***. If I is bounded, then locally absolutely continuous will be used interchangeably with ***absolutely continuous***.

(vii) If there exist a locally absolutely continuous function $g\colon I \to [-\infty, +\infty]$ and some $t_0 \in I$ such that

$$f(t) = \int_{[t_0,t]} (g|[t_0,t])\,\mathrm{d}\lambda,$$

then f is ***locally absolutely differentiable***. If I is bounded, then locally absolutely differentiable will be used interchangeably with ***absolutely differentiable***. •

Remark A.10. The above concepts can readily be extended to maps $f\colon I \to \mathbb{R}^n$ for $n > 1$ by applying them to the components of the vector-valued function. We shall also see in the next section how some of the concepts can also make sense for functions on manifolds, or for manifold-valued maps. •

A.2 Vector fields with measurable time-dependence

In this section we discuss, in a formal manner, the notion of vector fields with rather general time-dependence. We also discuss bundle maps that depend generally on time. The treatment here is more thorough and complete than in the text, since it seems wise not to overburden the main body of the book with the technicalities in this section. The discussion in this section supposes the reader to be familiar with the material in Chapter 3 up to and including the definition of a vector field.

A.2.1 Carathéodory sections of vector bundles and bundle maps

For $r \in \mathbb{N} \cup \{\infty\} \cup \{\omega\}$, let M be a C^r-manifold, and let $I \in \mathscr{I}$.

It is convenient to first talk about functions. A ***Carathéodory function*** on M is a map $\varphi\colon I \times \mathsf{M} \to \mathbb{R}$ with the property that $\varphi_t\colon x \mapsto \varphi(t, x)$ is continuous for each $t \in I$ and $\varphi^x\colon t \mapsto \varphi(t, x)$ is Lebesgue measurable for each $x \in \mathsf{M}$. A Carathéodory function φ is ***locally integrally bounded*** if, for each compact subset $K \subset \mathsf{M}$, there exists a positive locally integrable function $\psi_K\colon I \to \mathbb{R}$ such that $|\varphi(t, x)| \leq \psi_K(t)$ for each $x \in K$. A Carathéodory function $\varphi\colon I \times \mathsf{M} \to \mathbb{R}$ is ***class C^r*** if φ_t is class C^r for each $t \in I$. If $r \in \mathbb{N}$, then a Carathéodory function φ is ***locally integrally of class C^r*** if it is class C^r, and if $\mathscr{L}_{X_1} \cdots \mathscr{L}_{X_r} \varphi_t$ is locally integrally bounded for all $t \in I$ and $X_1, \dots, X_r \in \Gamma^r(\mathsf{TM})$. If φ is locally integrally of class C^r for every $r \in \mathbb{N}$, this it is ***locally integrally of class C^∞***. To define what it means for a Carathéodory function $\varphi\colon I \times \mathsf{M} \to \mathbb{R}$ to be locally integrally class C^ω requires a different approach. First we require that φ_t be of class C^∞ for each $t \in I$. Let $x_0 \in \mathsf{M}$ and let $(\mathcal{U}, \phi)$ be a chart for which $x_0 \in \mathcal{U}$. In this chart, and for each $k \in \mathbb{Z}_+$, define

$$f_{\varphi,\phi,k}(t, \boldsymbol{v}) = \sum_{r=0}^{k} \frac{1}{r!} \boldsymbol{D}^r(\varphi_t \circ \phi^{-1})(\boldsymbol{x}_0) \cdot (\boldsymbol{v}, \dots, \boldsymbol{v}),$$

where $\boldsymbol{x}_0 = \phi(x_0)$ and where $\boldsymbol{v} \in \mathbb{R}^n$. If, for each x_0 and any chart $(\mathcal{U}, \phi)$ chosen as above, there exist $\rho > 0$ and a positive locally integrable function $\psi_\rho\colon I \to \mathbb{R}$ such that

1. $|f_{\varphi,\phi,k}(t, \boldsymbol{v})| \leq \psi_\rho(t)$, $k \in \mathbb{Z}_+$, provided that $\|\boldsymbol{v}\|_{\mathbb{R}^n} < \rho$, and
2. the sequence $\{f_{\varphi,\phi,k}(t, \boldsymbol{v})\}_{k \in \mathbb{Z}_+}$ converges to $\varphi_t \circ \phi^{-1}(\boldsymbol{x}_0 + \boldsymbol{v})$ for a.e. $t \in I$, provided that $\|\boldsymbol{v}\|_{\mathbb{R}^n} < \rho$,

then we say that φ is ***locally integrally of class C^ω***.

Next we talk about sections of vector bundles. Let $\pi\colon \mathsf{V} \to \mathsf{M}$ be a vector bundle. A ***Carathéodory section*** of V is a map $\xi\colon I \times \mathsf{M} \to \mathsf{V}$ with the property that $\xi(t,x) \in \mathsf{V}_x$ for each $t \in I$, and with the property that the function $\alpha \cdot \xi\colon (t,x) \mapsto \alpha(x) \cdot \xi(t,x)$ is a Carathéodory function for each $\alpha \in \Gamma^0(\mathsf{V}^*)$. A Carathéodory section ξ of V is ***locally integrally of class C^r***, $r \in \mathbb{N} \cup \{\infty\} \cup \{\omega\}$, if $\alpha \cdot \xi$ is locally integrally of class C^r for every $\alpha \in \Gamma^r(\mathsf{V}^*)$.

Our next construction concerns bundle maps. If $\pi_1 : \mathsf{V}_1 \to \mathsf{M}$ and $\pi_2 : \mathsf{V}_2 \to \mathsf{M}$ are vector bundles over the same base space, a ***Carathéodory bundle map*** over the identity from V_1 to V_2 is a map $f\colon I \times \mathsf{V}_1 \to \mathsf{V}_2$ such that

1. $\pi_2 \circ f = \pi_1$, and
2. $f \circ \xi$ is a Carathéodory section of V_2 for every Carathéodory section ξ of V_1.

A Carathéodory bundle map $f\colon \mathsf{V}_1 \to \mathsf{V}_2$ is ***locally integrally of class C^r*** if $f \circ \xi$ is a locally integrally class C^r section of V_2 for every locally integrally class C^r section ξ of V_1.

A curve $\gamma\colon I \to \mathsf{M}$ is ***locally absolutely continuous*** (resp. ***locally absolutely differentiable***) if $\varphi \circ \gamma$ is locally absolutely continuous (resp. locally absolutely differentiable) for each $\varphi \in C^r(\mathsf{M})$. Note that locally absolutely differentiable curves are, in particular, locally absolutely continuous. Let $\pi\colon \mathsf{V} \to \mathsf{M}$ be a vector bundle and let $\gamma\colon I \to \mathsf{M}$ be a locally absolutely continuous curve. A section ξ along γ is ***locally absolutely continuous*** if it is locally absolutely continuous as a curve in V. A weaker notion than that of a locally absolutely continuous section of V along γ is that of a ***locally integrable*** section of V along γ, which is a section ξ of V along γ having the property that the function $t \mapsto \alpha(\gamma(t)) \cdot \xi(t)$ is locally integrable for every $\alpha \in \Gamma^r(\mathsf{V}^*)$.

A.2.2 The time-dependent Flow Box Theorem

The classic theory of time-dependent vector fields with measurable time dependence gives the existence of integral curves for locally integrally class C^r vector fields [Sontag 1998, Appendix C]. Let us say some things about these integral curves. Suppose that $\gamma\colon I \to \mathsf{M}$ is locally absolutely continuous, that $\varphi \in C^1(\mathsf{Q})$, and that $J = [a,b]$ is a compact subinterval of I, and let $t \in]a,b[$. We then have

$$(\varphi \circ \gamma | J)(t) = \int_{[a,t]} g_{\gamma,\varphi}\,\mathrm{d}\lambda,$$

for some integrable function $g_{\gamma,\varphi}\colon J \to \mathbb{R}$. Differentiating this relation we get

$$\frac{\mathrm{d}}{\mathrm{d}t}(\varphi \circ \gamma | J)(t) = g_{\gamma,\varphi}(t), \text{ a.e. } t \in J.$$

One may verify that it is possible to write

$$\frac{\mathrm{d}}{\mathrm{d}t}(\varphi \circ \gamma | J)(t) = T_{\gamma(t)}\varphi(v_{\gamma,t}), \text{ a.e. } t \in J,$$

for some $v_{\gamma,t} \in \mathsf{T}_{\gamma(t)}\mathsf{M}$. We write $v_{\gamma,t} = \gamma'(t)$, noting that $\gamma'(t)$ is defined for almost every $t \in I$.

The following result asserts the existence and uniqueness of integral curves for vector fields with quite general time-dependence.

Theorem A.11 (Time-dependent Flow Box Theorem). *Let* $X: I \times \mathsf{M} \to \mathsf{TM}$ *be a locally integrally class* C^r *vector field,* $r \in \mathbb{N} \cup \{\infty\} \cup \{\omega\}$, *and let* $(t_0, x_0) \in \mathbb{R} \times \mathsf{M}$. *Then there exists a triple* $(\mathcal{U}, T, \Phi)$, *called a* ***flow box*** *of* X *at* (t_0, x_0), *with the following properties:*

(i) $\mathcal{U}$ *is an open subset of* M *containing* x_0;

(ii) $T > 0$ *or* $T = +\infty$;

(iii) $\Phi\colon\,]t_0 - T, t_0 + T[\, \times \mathcal{U} \to \mathsf{M}$ *is a map having the following properties:*

(a) the map $t \mapsto \Phi(t, x)$ *is locally absolutely continuous for each* $x \in \mathcal{U}$;

(b) the map $x \mapsto \Phi(t, x)$ *is of class* C^r *for each* $t \in]t_0 - T, t_0 + T[$;

(c) $\Phi(t, x) = \gamma_{t_0,x}(t)$, *where* $\gamma_{t_0,x}$ *is the integral curve of* X *at* $(t_0, x) \in \mathbb{R} \times \mathcal{U}$;

(iv) for all $t \in]t_0 - T, t_0 + T[$, $\Phi_t\colon \mathcal{U} \to \mathsf{M}$ *is a* C^r*-diffeomorphism onto its image, where* $\Phi_t(x) = \Phi(t, x)$.

Furthermore, if $(\tilde{\mathcal{U}}, \tilde{T}, \tilde{\Phi})$ *is another such triple, then* Φ *and* $\tilde{\Phi}$ *agree when restricted to* $\big(]t_0 - T, t_0 + T[\, \cap\,]t_0 - \tilde{T}, t_0 + \tilde{T}[\big) \times (\mathcal{U} \cap \tilde{\mathcal{U}})$.

If additionally $X: \mathbb{R} \times \mathsf{M} \to \mathsf{TM}$ *is of class* C^r, *then the map* Φ *has the properties:*

(iii) the map $(t, x) \mapsto \Phi(t, x)$ *is of class* C^r;

(iv) the map $t \mapsto \Phi(t, x)$ *is of class* C^{r+1} *for each* $x \in \mathcal{U}$.

For solutions to differential equations that involve an affine connection, as so often arise in this book, the following result indicates how the conclusions of Theorem A.11 can be strengthened. This result gives the stronger conclusions one can make for the trajectories *as curves on* M, given that the vector field lives on TM. One should be careful to note that the Flow Box Theorem in this case lives on TM, not on M.

Corollary A.12. *Let* $r \in \mathbb{N} \cup \{\infty\} \cup \{\omega\}$, *and let* ∇ *be a* C^r*-affine connection on* M, *let* $Y: I \times \mathsf{TM} \to \mathsf{TM}$ *be a locally integrally class* C^r*-bundle map, and let* $(t_0, v_{x_0}) \in I \times \mathsf{TM}$. *There then exist*

(i) a subinterval $J \subset I$ *containing* t_0 *in its interior, and*

(ii) a locally absolutely differentiable curve $\gamma_{(x_0,t_0)}\colon J \to \mathsf{M}$

having the properties that

(iii) $\nabla_{\gamma'(t)}\gamma'(t) = Y(t, \gamma'(t))$ *for a.e.* $t \in J$ *and*

(iv) $\gamma'(t_0) = v_{x_0}$.

Furthermore, if $\tilde{J}$ *is another such interval and if* $\tilde{\gamma}\colon \tilde{J} \to \mathsf{M}$ *is another such curve, then for* $t \in J \cap \tilde{J}$ *we have* $\gamma(t) = \tilde{\gamma}(t)$.

B

Some proofs

In this chapter we collect proofs of some of the more difficult and/or lengthy to prove results.

B.1 Proof of Theorem 4.38

Let us first show that, if γ is a minimizer, then any subarc of γ is also a minimizer. Thus let $t_1, t_2 \in [a, b]$ satisfy $a \leq t_1 < t_2 \leq b$, and define a curve $\tilde{\gamma}\colon [t_1, t_2] \to \mathsf{Q}$ by $\tilde{\gamma}(t) = \gamma(t)$ (i.e., $\tilde{\gamma}$ is the restriction of γ to $[t_1, t_2]$). With t_1 and t_2 as defined, we can also define the set of curves $C^2([t_1, t_2], \gamma(t_1), \gamma(t_2))$ analogously to our definition of $C^2([a, b], q_a, q_b)$. We also define a function $\tilde{A}_L$ on the set of curves $C^2([t_1, t_2], \gamma(t_1), \gamma(t_2))$ by

$$\tilde{A}_L(\tilde{\gamma}) = \int_{t_1}^{t_2} L(t, \tilde{\gamma}'(t))\,\mathrm{d}t.$$

We claim that if γ is a minimizer for A_L, then $\tilde{\gamma}$ is a minimizer for $\tilde{A}_L$. Indeed, if $\tilde{\gamma}$ were *not* a minimizer for $\tilde{A}_L$, then this implies the existence of a curve $\tilde{\gamma}_1 \in C^2([t_1, t_2], \gamma(t_1), \gamma(t_2))$ with the property that $\tilde{A}_L(\tilde{\gamma}_1) < \tilde{A}_L(\tilde{\gamma})$. Now define a curve $\gamma_1 \colon [a, b] \to \mathsf{Q}$ by

$$\gamma_1(t) = \begin{cases} \gamma(t), & t \in [a, t_1], \\ \tilde{\gamma}_1(t), & t \in [t_1, t_2], \\ \gamma(t), & \in [t_2, b]. \end{cases}$$

We then have

$$\begin{aligned} A_L(\gamma_1) &= \int_a^{t_1} L(t, \gamma'(t))\,\mathrm{d}t + \int_{t_1}^{t_2} L(t, \tilde{\gamma}_1'(t))\,\mathrm{d}t + \int_{t_2}^{b} L(t, \gamma'(t))\,\mathrm{d}t \\ &< \int_a^{t_1} L(t, \gamma'(t))\,\mathrm{d}t + \int_{t_1}^{t_2} L(t, \tilde{\gamma}'(t))\,\mathrm{d}t + \int_{t_2}^{b} L(t, \gamma'(t))\,\mathrm{d}t \end{aligned}$$

$$= \int_a^b L(t, \gamma'(t))\,\mathrm{d}t = A_L(\gamma).$$

This would complete our assertion that the restriction of γ to any subinterval of $[a,b]$ is itself a minimizer, except for the fact that the curve γ_1 may not be twice continuously differentiable, as it may fail to be differentiable at t_1 and t_2. However, one may show that it is possible to smooth γ_1 slightly to a curve $\bar{\gamma}_1$ so that $\bar{\gamma}_1$ is twice continuously differentiable on $[a,b]$, and so that $|A_L(\bar{\gamma}_1) - A_L(\gamma_1)| \leq \varepsilon$ for any $\varepsilon > 0$.[1] In this way, one can ensure that $A_L(\bar{\gamma}_1) < A_L(\gamma)$, contradicting the fact that γ is a minimizer.

The above argument says that γ is a minimizer along every subinterval of $[a,b]$ if it is a minimizer along the entire interval. Therefore, if $(\mathcal{U}, \phi)$ is a chart such that $\gamma(t) \in \mathcal{U}$ for some $t \in [a,b]$, and if γ is a minimizer on $[a,b]$, then it is also a minimizer for the curve restricted to an interval $[t_1, t_2]$ with the property that $\gamma(t) \in \mathcal{U}$ for $t \in [t_1, t_2]$. The outcome is that, without loss of generality, we may suppose that the curve γ lies in the domain of a chart $(\mathcal{U}, \phi)$, and we do this for the remainder of the proof.

Let $\sigma\colon J \times [a,b] \to \mathsf{Q}$ be a variation of γ and let $\sigma_s \in C^2([a,b], q_a, q_b)$ be defined by $\sigma_s(t) = \sigma(s,t)$. Consider the function f_σ defined on J by $f_\sigma(s) = A_L(\sigma_s)$. If γ is a minimizer for A_L, then $s = 0$ should be a minimum of f_σ for every variation σ of γ. We let $\boldsymbol{q}_s$ be the coordinate representative of σ_s. We have

$$\begin{aligned} 0 = \frac{\mathrm{d}}{\mathrm{d}s}\Big|_{s=0} f_\sigma(s) &= \frac{\mathrm{d}}{\mathrm{d}s}\Big|_{s=0} \int_a^b L(t, \boldsymbol{q}_s(t), \dot{\boldsymbol{q}}_s(t))\,\mathrm{d}t \\ &= \int_a^b \left(\frac{\partial L}{\partial q^i} \frac{\mathrm{d}q_s^i(t)}{\mathrm{d}s}\Big|_{s=0} + \frac{\partial L}{\partial v^i} \frac{\mathrm{d}\dot{q}_s^i(t)}{\mathrm{d}s}\Big|_{s=0} \right) \mathrm{d}t. \end{aligned}$$

Now note that

$$\frac{\mathrm{d}\dot{q}_s^i(t)}{\mathrm{d}s}\Big|_{s=0} = \frac{\mathrm{d}}{\mathrm{d}s}\Big|_{s=0} \frac{\mathrm{d}q_s^i}{\mathrm{d}t} = \frac{\mathrm{d}}{\mathrm{d}t} \frac{\mathrm{d}q_s^i(t)}{\mathrm{d}s}\Big|_{s=0}.$$

Therefore,

$$\begin{aligned} 0 = \frac{\mathrm{d}}{\mathrm{d}s}\Big|_{s=0} f_\sigma(s) &= \int_a^b \left(\frac{\partial L}{\partial q^i} \frac{\mathrm{d}q_s^i(t)}{\mathrm{d}s}\Big|_{s=0} + \frac{\partial L}{\partial v^i} \frac{\mathrm{d}}{\mathrm{d}t} \frac{\mathrm{d}q_s^i(t)}{\mathrm{d}s}\Big|_{s=0} \right) \mathrm{d}t \\ &= \int_a^b \left(\frac{\partial L}{\partial q^i} - \frac{\mathrm{d}}{\mathrm{d}t}\left(\frac{\partial L}{\partial v^i} \right) \right) \frac{\mathrm{d}q_s^i(t)}{\mathrm{d}s}\Big|_{s=0} \mathrm{d}t + \frac{\partial L}{\partial v^i} \frac{\mathrm{d}q_s^i(t)}{\mathrm{d}s}\Big|_{s=0}\Big|_{t=b}^{t=b} \\ &= \int_a^b \left(\frac{\partial L}{\partial q^i} - \frac{\mathrm{d}}{\mathrm{d}t}\left(\frac{\partial L}{\partial v^i} \right) \right) \frac{\mathrm{d}q_s^i(t)}{\mathrm{d}s}\Big|_{s=0} \mathrm{d}t, \end{aligned}$$

where in the next to last step we have used integration by parts, and in the last step we have used the fact that an infinitesimal variation vanishes at the endpoints. Since

[1] To prove this, one can choose a chart around each endpoint of $\tilde{\gamma}$ and in coordinates smooth the curve γ_1 using functions like that defined in Exercise E3.8.

$$\int_a^b \left(\frac{\partial L}{\partial q^i} - \frac{\mathrm{d}}{\mathrm{d}t}\left(\frac{\partial L}{\partial v^i}\right)\right) \frac{\mathrm{d}q_s^i(t)}{\mathrm{d}s}\Big|_{s=0} \mathrm{d}t = 0$$

for *every* variation, this means that $\frac{\mathrm{d}q_s^i(t)}{\mathrm{d}s}\Big|_{s=0}$ can be arbitrary, and so the only way for the integral to vanish is for

$$\frac{\partial L}{\partial q^i} - \frac{\mathrm{d}}{\mathrm{d}t}\left(\frac{\partial L}{\partial v^i}\right)$$

to be zero when evaluated at $(t, \boldsymbol{q}(t), \dot{\boldsymbol{q}}(t))$ for all $t \in [a, b]$.

B.2 Proof of Theorem 7.36

Let $\boldsymbol{\xi} = \{\xi_0, \xi_1, \ldots, \xi_m\}$ be indeterminates for a free Lie algebra $\mathrm{Lie}(\boldsymbol{\xi})$, which correspond to the vector fields $\mathscr{C}_\Sigma$. We can simplify many of our computations for the controllability analysis by making simplifications to a set of generators for $\mathrm{Lie}(\boldsymbol{\xi})$. We first need some notation. Let

$$\mathrm{Br}^k(\boldsymbol{\xi}) = \left\{ B \in \mathrm{Br}(\boldsymbol{\xi}) \;\Big|\; |B|_0 + \sum_{a=1}^m |B|_a = k \right\},$$

$$\mathrm{Br}_k(\boldsymbol{\xi}) = \left\{ B \in \mathrm{Br}(\boldsymbol{\xi}) \;\Big|\; |B|_0 - \sum_{a=1}^m |B|_a = k \right\}.$$

We will also need the concept of a primitive bracket.

Definition B.1 (Primitive bracket). Let $B \in \mathrm{Br}_0(\boldsymbol{\xi}) \cup \mathrm{Br}_{-1}(\boldsymbol{\xi})$ and let $B_1, B_2, B_{11}, B_{12}, B_{21}, B_{22}, \ldots,$ be the decomposition of B into its components. We shall say that B is ***primitive*** if each of its components is in $\mathrm{Br}_{-1}(\boldsymbol{\xi}) \cup \mathrm{Br}_0(\boldsymbol{\xi}) \cup \{\xi_0\}$. •

The relevant observations regarding primitive brackets are:

1. If $B \in \mathrm{Br}_{-1}(\boldsymbol{\xi})$ is primitive, then we may write $B = [B_1, B_2]$ with $B_1 \in \mathrm{Br}_{-1}(\boldsymbol{\xi})$ and $B_2 \in \mathrm{Br}_0(\boldsymbol{\xi})$ both primitive.
2. If $B \in \mathrm{Br}_0(\boldsymbol{\xi})$ is primitive, then B may have one of two forms. Either $B = [\xi_0, B_1]$ with $B_1 \in \mathrm{Br}_{-1}(\boldsymbol{\xi})$ primitive, or $B = [B_1, B_2]$ with $B_1, B_2 \in \mathrm{Br}_0(\boldsymbol{\xi})$ primitive.

Using these two rules, it is possible to construct primitive brackets of any degree. For example, the primitive brackets of degrees one through four are, up to sign

$$\begin{aligned}
&\text{Degree 1: } \{\xi_a \mid a \in \{1, \ldots, m\}\},\\
&\text{Degree 2: } \{[\xi_0, \xi_a] \mid a \in \{1, \ldots, m\}\},\\
&\text{Degree 3: } \{[\xi_a, [\xi_0, \xi_b]] \mid a, b \in \{1, \ldots, m\}\},\\
&\text{Degree 4: } \{[\xi_0, [\xi_a, [\xi_0, \xi_b]]] \mid a, b \in \{1, \ldots, m\}\} \cup\\
&\qquad\qquad \{[[\xi_0, \xi_a], [\xi_0, \xi_b]] \mid a, b \in \{1, \ldots, m\}\}.
\end{aligned}$$

From Exercise E7.2, we know that to generate $\mathrm{Lie}(\boldsymbol{\xi})$ we need only look at brackets of the form

$$[\xi_{a_k}, [\xi_{a_{k-1}}, \dots, [\xi_{a_2}, \xi_{a_1}]]], \tag{B.1}$$

where $a_i \in \{0, 1, \dots, m\}$ for $i \in \{1, \dots, k\}$. We shall see below that brackets from $\mathrm{Br}_j(\boldsymbol{\xi})$, where $j \geq 1$ or $j \leq -2$, will not be of interest to us. In particular, we shall see that, when $j \leq -2$, the brackets evaluate identically to zero. Therefore, we concentrate our attention on brackets in $\mathrm{Br}_0(\boldsymbol{\xi}) \cup \mathrm{Br}_{-1}(\boldsymbol{\xi})$ that satisfy certain requirements. We state this in the following lemma.

Lemma B.2. *Let us impose the condition on elements of* $\mathrm{Br}(\boldsymbol{\xi})$ *that we shall consider a bracket to be zero if any of its components are in* $\mathrm{Br}_{-j}(\boldsymbol{\xi})$ *for* $j \geq 2$. *Let* $B \in \mathrm{Br}_0(\boldsymbol{\xi}) \cup \mathrm{Br}_{-1}(\boldsymbol{\xi})$. *Then we may write* B *as a finite sum of primitive brackets.*

Proof. It is sufficient to prove the lemma for brackets of the form (B.1). We proceed by induction on k in (B.1). The lemma is true for $k \in \{1, 2\}$ by inspection. Now suppose the lemma true for $k \in \{1, \dots, l\}$, and let B be of the form (B.1) for $k = l + 1$. Then we have two cases. Either $B \in \mathrm{Br}_{-1}(\boldsymbol{\xi})$ or $B \in \mathrm{Br}_0(\boldsymbol{\xi})$.

We look first at the case in which $B \in \mathrm{Br}_{-1}(\boldsymbol{\xi})$. Since we are considering brackets in $\mathrm{Br}_{-2}(\boldsymbol{\xi})$ to be zero, we may write $B = [\xi_a, B']$ with $B' \in \mathrm{Br}_0(\boldsymbol{\xi})$ of the form (B.1) and $a \in \{1, \dots, m\}$. By the induction hypothesis, B is a finite sum of primitive brackets and the lemma is proved in this case, since B will then also be a finite sum of primitive brackets.

Now we look at the case in which $B \in \mathrm{Br}_0(\boldsymbol{\xi})$. There are two possibilities in this case. The first possibility is that $B = [\xi_0, B']$ with $B' \in \mathrm{Br}_{-1}(\boldsymbol{\xi})$. In this case B' is a finite sum of primitive brackets by the induction hypothesis and, therefore, B is also a finite sum of primitive brackets.

The final case is that in which $B = [\xi_{a_1}, B']$ with $B' \in \mathrm{Br}_{+1}(\boldsymbol{\xi})$ of the form (B.1). If $B' = [\xi_0, B'']$ with $B'' \in \mathrm{Br}_0(\boldsymbol{\xi})$, then, by the Jacobi identity, we have

$$B = [\xi_{a_1}, [\xi_0, B'']] = -[B'', [\xi_{a_1}, \xi_0]] - [\xi_0, [B'', \xi_{a_1}]].$$

Since $B'' \in \mathrm{Br}_0(\boldsymbol{\xi})$, by the induction hypotheses it may be written as a finite sum of primitive brackets in $\mathrm{Br}_0(\boldsymbol{\xi})$. Clearly $[\xi_{a_1}, \xi_0]$ is primitive, which proves that $[B'', [\xi_{a_1}, \xi_0]]$ is a finite sum of primitive brackets. The bracket $[B'', \xi_{a_1}]$ is in $\mathrm{Br}_{-1}(\boldsymbol{\xi})$. Therefore, by the induction hypotheses, it may be written as a finite sum of primitive brackets. Thus the term $[\xi_0, [B'', \xi_{a_1}]]$, and hence B, may be written as a finite sum of primitive brackets.

Now suppose that $B' = [\xi_{a_2}, B'']$ with $B'' \in \mathrm{Br}_{+2}(\boldsymbol{\xi})$. First look at the case in which $B'' = [\xi_0, B''']$ with $B''' \in \mathrm{Br}_{+1}(\boldsymbol{\xi})$. In this case we have

$$\begin{aligned} B = [\xi_{a_1}, [\xi_{a_2}, [\xi_0, B''']]] &= -[\xi_{a_1}, [B''', [\xi_{a_2}, \xi_0]]] - [\xi_{a_1}, [\xi_0, [B''', \xi_{a_2}]]] \\ &= [[\xi_{a_2}, \xi_0], [\xi_{a_1}, B''']] + [B''', [[\xi_{a_2}, \xi_0], \xi_{a_1}]] + \\ &\quad [[B''', \xi_{a_2}], [\xi_{a_1}, \xi_0]] + [\xi_0, [[B''', \xi_{a_2}], \xi_{a_1}]]. \end{aligned}$$

The first, third, and fourth terms can be written as finite sums of primitive brackets by the induction hypothesis, and the second term is zero by our condition that brackets in $\mathrm{Br}_{-2}(\boldsymbol{\xi})$ are taken to be zero.

If $B'' = [\xi_{a_3}, B''']$, then we keep stripping factors off of B''' until we encounter an ξ_0. When we do, we repeatedly apply the above procedure. This proves the lemma. ■

Now we shall show how it is possible to compute the brackets from $\mathrm{Br}(\boldsymbol{\xi})$ in terms of the problem data. We first present a lemma that gives the basic structure of primitive brackets. The reader will also wish to recall the ideas concerning symmetric algebras presented in Section 7.3.3. Recall from Lemma 6.34 that $\mathsf{T}_{0_q}\mathsf{TQ}$ may be canonically decomposed as $\mathsf{T}_q\mathsf{Q} \oplus \mathsf{T}_q\mathsf{Q}$. We adopt the convention that the first component in this decomposition is the tangent space to the zero section (we call this the ***horizontal*** component), and the second component is the tangent space to the fiber (we call this the ***vertical*** component).

Lemma B.3. *Suppose that $B \in \mathrm{Br}^r(\boldsymbol{\xi})$ is primitive.*

(i) If $B \in \mathrm{Br}_{-1}(\boldsymbol{\xi})$, then $\mathrm{Ev}^{\mathscr{C}_\Sigma}(B)$ is the vertical lift of a vector field on Q.

(ii) If $B \in \mathrm{Br}_0(\boldsymbol{\xi})$, then $U = \mathrm{Ev}^{\mathscr{C}_\Sigma}(B)$ has the property that, when expressed in a natural chart for TQ with coordinates $(\boldsymbol{q}, \boldsymbol{v})$, the vertical components of U are linear in the fiber coordinates $\boldsymbol{v}$ and the horizontal components are independent of $\boldsymbol{v}$. In particular, we may define a vector field on Q by $U_{\mathsf{Q}}\colon q \mapsto U(0_q) \in \mathsf{T}_q\mathsf{Q} \oplus \{0_q\} \subset \mathsf{T}_{0_q}\mathsf{TQ}$. There are two cases to consider.

(a) $B = [\xi_0, B_1]$ with $B_1 \in \mathrm{Br}_{-1}(\boldsymbol{\xi})$: Define U_1 to be the vector field on Q such that $\mathrm{Ev}^{\mathscr{C}_\Sigma}(B_1) = \mathrm{vlft}(U_1)$. Then $U(0_q) = -U_1(q)$. Let $U_2 \in \Gamma^\omega(\mathsf{TQ})$. Then $[\mathrm{vlft}(U_2), U] = \mathrm{vlft}(\langle U_1 : U_2 \rangle)$.

(b) $B = [B_1, B_2]$ with $B_1, B_2 \in \mathrm{Br}_0(\boldsymbol{\xi})$: Define $U_{1,\mathsf{Q}}, U_{2,\mathsf{Q}}$ to be the vector fields on Q corresponding to $\mathrm{Ev}^{\mathscr{C}_\Sigma}(B_1), \mathrm{Ev}^{\mathscr{C}_\Sigma}(B_2)$, respectively. Then $\mathrm{Ev}^{\mathscr{C}_\Sigma}(B)(0_q) = [U_{1,\mathsf{Q}}, U_{2,\mathsf{Q}}](q)$.

Proof. The proof is by induction on r. The result is true for $r = 1$ trivially. To prove the result for $r = 2$, we introduce some useful notation for performing the bracket calculations in coordinates. If we have two *general* vector fields

$$X_1 = X^i_{1,h}(q,v)\frac{\partial}{\partial q^i} + X^i_{1,v}(q,v)\frac{\partial}{\partial v^i}, \quad X_2 = X^i_{2,h}(q,v)\frac{\partial}{\partial q^i} + X^i_{2,v}(q,v)\frac{\partial}{\partial v^i},$$

then their Lie bracket will be represented by

$$[X_1, X_2] \sim \begin{bmatrix} \frac{\partial X^i_{2,h}}{\partial q^j} & \frac{\partial X^i_{2,h}}{\partial v^j} \\ \frac{\partial X^i_{2,v}}{\partial q^j} & \frac{\partial X^i_{2,v}}{\partial v^j} \end{bmatrix} \begin{bmatrix} X^j_{1,h} \\ X^j_{1,v} \end{bmatrix} - \begin{bmatrix} \frac{\partial X^i_{1,h}}{\partial q^j} & \frac{\partial X^i_{1,h}}{\partial v^j} \\ \frac{\partial X^i_{1,v}}{\partial q^j} & \frac{\partial X^i_{1,v}}{\partial v^j} \end{bmatrix} \begin{bmatrix} X^j_{2,h} \\ X^j_{2,v} \end{bmatrix}.$$

This is somewhat imprecise, but is convenient notationally.

If X, Y are vector fields on Q, we may compute

$$[\mathrm{vlft}(X), \mathrm{vlft}(Y)] \sim \begin{bmatrix} 0 & 0 \\ \frac{\partial Y^i}{\partial q^j} & 0 \end{bmatrix} \begin{bmatrix} 0 \\ X^j \end{bmatrix} - \begin{bmatrix} 0 & 0 \\ \frac{\partial X^i}{\partial q^j} & 0 \end{bmatrix} \begin{bmatrix} 0 \\ Y^j \end{bmatrix} = \begin{bmatrix} 0 \\ 0 \end{bmatrix}. \tag{B.2}$$

If X is a vector field on Q, we compute

$$[S, \mathrm{vlft}(X)] \sim \begin{bmatrix} 0 & 0 \\ \frac{\partial X^i}{\partial q^j} & 0 \end{bmatrix} \begin{bmatrix} v^j \\ -\Gamma^j_{kl} v^k v^l \end{bmatrix} - \begin{bmatrix} 0 & \delta^i_j \\ -\frac{\partial \Gamma^i_{kl}}{\partial q^j} v^k v^l & -(\Gamma^i_{jk} + \Gamma^i_{kj}) v^k \end{bmatrix} \begin{bmatrix} 0 \\ X^j \end{bmatrix}. \tag{B.3}$$

Inspecting (B.3) shows that $[S, \mathrm{vlft}(X)](0_q) = -X(q)$. Now let $Y \in \Gamma^\omega(\mathsf{TQ})$. We compute

$$[\mathrm{vlft}(Y), [S, \mathrm{vlft}(X)]] \sim$$
$$\begin{bmatrix} -\frac{\partial X^i}{\partial q^j} & 0 \\ \frac{\partial^2 X^i}{q^j q^k} v^k + (\frac{\partial \Gamma^i_{kl}}{\partial q^j} + \frac{\partial \Gamma^i_{lk}}{\partial q^j}) X^k v^l + (\Gamma^i_{kl} + \Gamma^i_{lk}) \frac{\partial X^k}{\partial q^j} v^l & \frac{\partial X^i}{\partial q^j} + (\Gamma^i_{kj} + \Gamma^i_{kj}) X^k \end{bmatrix} \begin{bmatrix} 0 \\ Y^j \end{bmatrix}$$
$$- \begin{bmatrix} 0 & 0 \\ \frac{\partial Y^i}{\partial q^j} & 0 \end{bmatrix} \begin{bmatrix} -X^j \\ \frac{\partial X^j}{\partial q^k} v^k + (\Gamma^j_{kl} + \Gamma^i_{lk}) X^k v^l \end{bmatrix}.$$

Reading the coefficients gives

$$[\mathrm{vlft}(Y), [S, \mathrm{vlft}(X)]] = \left(\frac{\partial Y^i}{\partial q^j} X^j + \frac{\partial X^i}{\partial q^j} Y^j + (\Gamma^i_{jk} + \Gamma^i_{kj}) X^j Y^k \right) \frac{\partial}{\partial v^i}, \tag{B.4}$$

which is the coordinate representation of $\mathrm{vlft}(\langle X : Y \rangle)$. This shows that the lemma is true for $r = 2$.

Now suppose the lemma true for $r \in \{1, \dots, l\}$ for $l \geq 2$ and let $B \in \mathrm{Br}^{l+1}(\boldsymbol{\xi})$ be primitive.

(i) Suppose that $B \in \mathrm{Br}_{-1}(\boldsymbol{\xi})$. Without loss of generality (by Property 1 of primitive brackets), we may suppose that $B = [B_1, B_2]$ with $B_1 \in \mathrm{Br}_{-1}(\boldsymbol{\xi})$ and $B_2 \in \mathrm{Br}_0(\boldsymbol{\xi})$. Then, by the induction hypotheses, we have

$$\mathrm{Ev}^{\mathscr{C}_\Sigma}(B_1) = \alpha^i(q) \frac{\partial}{\partial v^i}, \quad \mathrm{Ev}^{\mathscr{C}_\Sigma}(B_2) = \lambda^i(q) \frac{\partial}{\partial q^i} + \mu^i_j(q) v^j \frac{\partial}{\partial v^i}.$$

Now we compute

$$\mathrm{Ev}^{\mathscr{C}_\Sigma}([B_1, B_2]) \sim \begin{bmatrix} \frac{\partial \lambda^i}{\partial q^j} & 0 \\ \frac{\partial \mu^i_k}{\partial q^j} v^k & \mu^i_j \end{bmatrix} \begin{bmatrix} 0 \\ \alpha^j \end{bmatrix} - \begin{bmatrix} 0 & 0 \\ \frac{\partial \alpha^i}{\partial q^j} & 0 \end{bmatrix} \begin{bmatrix} \lambda^j \\ \mu^j_k v^k \end{bmatrix}.$$

Note that the components in the $\boldsymbol{q}$-direction are zero and the components in the $\boldsymbol{v}$-direction are only functions of $\boldsymbol{q}$. This means that this vector field is the vertical lift of a vector field on Q. This proves (i).

(ii) Suppose that $B \in \mathrm{Br}_0(\boldsymbol{\xi})$. Without loss of generality (by Property 2 of primitive brackets), we may suppose that either (ii a) $B = [\xi_0, B_1]$ with $B_1 \in \mathrm{Br}_{-1}(\boldsymbol{\xi})$ or that (ii b) $B = [B_1, B_2]$ with $B_1, B_2 \in \mathrm{Br}_0(\boldsymbol{\xi})$. Let us deal with the first case. Equation (B.3) gives $\mathrm{Ev}^{\mathscr{C}_\Sigma}_{(0_q)}(B) = -U_1(q)$, where U_1 is the vector field on Q such that $\mathrm{Ev}^{\mathscr{C}_\Sigma}(B_1) = \mathrm{vlft}(U_1)$ (such a vector field exists by (i)). For every vector field U_2 on Q, we have $[\mathrm{vlft}(U_2), [S, \mathrm{vlft}(U_1)]] = \mathrm{vlft}(\langle U_1 : U_2 \rangle)$ by (B.4). This proves (ii a).

Now suppose that we have $B_1, B_2 \in \mathrm{Br}_0(\boldsymbol{\xi})$. Then, by the induction hypotheses, we have

$$\mathrm{Ev}^{\mathscr{C}_\Sigma}(B_1) = \alpha^i(q)\frac{\partial}{\partial q^i} + \beta^i_j(q)v^j\frac{\partial}{\partial v^i}, \quad \mathrm{Ev}^{\mathscr{C}_\Sigma}(B_2) = \lambda^i(q)\frac{\partial}{\partial q^i} + \mu^i_j(q)v^j\frac{\partial}{\partial v^i}.$$

We compute

$$\mathrm{Ev}^{\mathscr{C}_\Sigma}([B_1, B_2]) \sim \begin{bmatrix} \frac{\partial \lambda^i}{\partial q^j} & 0 \\ \frac{\partial \mu^i_k}{\partial q^j}v^k & \mu^i_j \end{bmatrix} \begin{bmatrix} \alpha^j \\ \beta^j_k v^k \end{bmatrix} - \begin{bmatrix} \frac{\partial \alpha^i}{\partial q^j} & 0 \\ \frac{\partial \beta^i_k}{\partial q^j}v^k & \beta^i_j \end{bmatrix} \begin{bmatrix} \lambda^j \\ \mu^j_k v^k \end{bmatrix}.$$

The components have the order in $\boldsymbol{v}$ specified by the lemma. Also, it is clear that the vector fields on Q defined by B_1 and B_2 are

$$U_{1,\mathsf{Q}} = \alpha^i(q)\frac{\partial}{\partial q^i} \quad \text{and} \quad U_{2,\mathsf{Q}} = \lambda^i(q)\frac{\partial}{\partial q^i},$$

respectively. It is easy to see that $\mathrm{Ev}^{\mathscr{C}_\Sigma}_{(0_q)}(B) = [U_{1,\mathsf{Q}}, U_{2,\mathsf{Q}}](q)$. This completes the proof of the lemma. ■

Next we show that the primitive brackets are the *only* types of brackets we need to consider. First we look at brackets in $\mathrm{Br}_l(\boldsymbol{\xi})$ for $l \geq 1$.

Lemma B.4. *Let $l \geq 1$ be an integer and let $B \in \mathrm{Br}_l(\boldsymbol{\xi})$. Then $\mathrm{Ev}^{\mathscr{C}_\Sigma}_{(0_q)}(B) = 0$ for each $q \in \mathsf{Q}$.*

Proof. The lemma may be proved by showing that, in a natural coordinate chart for TQ, the horizontal components of $U = \mathrm{Ev}^{\mathscr{C}_\Sigma}(B)$ are polynomial in the fiber coordinates of degree l, and the vertical components of U are polynomial of degree $l+1$ in the fiber coordinates. This will follow if we can show that bracketing by ξ_a, $a \in \{1, \dots, m\}$, reduces the polynomial order of the components by one and bracketing by ξ_0 increases the polynomial order of the components by one. This is a simple calculation that follows along the same lines as the calculations done for Lemma B.3. ■

Now we look at the remaining brackets, those in $\mathrm{Br}_{-l}(\boldsymbol{\xi})$ for $l \geq 2$.

Lemma B.5. *Let $l \geq 2$ be an integer and let $B \in \mathrm{Br}^k(\boldsymbol{\xi}) \cap \mathrm{Br}_{-l}(\boldsymbol{\xi})$ for $k \geq 2$. Then $\mathrm{Ev}^{\mathscr{C}_\Sigma}(B) = 0$.*

Proof. We prove the lemma by induction on k for brackets of the form (B.1). The result makes no sense for $k = 1$ and is true for $k = 2$ by (B.2). Now suppose the lemma true for $k \in \{2, \dots, j\}$, and let $B \in \mathrm{Br}^{j+1}(\boldsymbol{\xi}) \cap \mathrm{Br}_{-l}(\boldsymbol{\xi})$ for $l \geq 2$ be of the form (B.1). Then either $B = [\xi_0, B']$ with $B' \in \mathrm{Br}_{-l-1}(\boldsymbol{\xi})$, or $B = [\xi_a, B']$ with $B' \in \mathrm{Br}_{-l+1}(\boldsymbol{\xi})$ and $a \in \{1, \dots, m\}$. In either case the result follows immediately from the induction hypotheses and (B.2). ■

The following lemma gives the proof of the theorem.

Lemma B.6. *Let* $q_0 \in \mathsf{Q}$. *Then*

$$\mathrm{Lie}^{(\infty)}(\mathscr{C}_\Sigma)_{0_{q_0}} = \mathrm{Lie}^{(\infty)}(\mathrm{Sym}^{(\infty)}(\mathcal{Y}))_{q_0} \oplus \mathrm{Sym}^{(\infty)}(\mathcal{Y})_{q_0}.$$

Proof. Write

$$\mathrm{Lie}^{(\infty)}(\mathscr{C}_\Sigma)_{0_q} = H_q(\mathscr{C}_\Sigma) \oplus V_q(\mathscr{C}_\Sigma),$$

so defining distributions $H(\mathscr{C}_\Sigma)$ and $V(\mathscr{C}_\Sigma)$ on Q. We first claim that $\mathrm{Sym}^{(\infty)}(\mathcal{Y}) \subset V(\mathscr{C}_\Sigma)$. This may be proved inductively. Define $\mathscr{S}^{(0)}(\mathscr{Y}) = \mathscr{Y}$ and inductively define

$$\mathscr{S}^{(k)}(\mathscr{Y}) = \{ \langle U_1 : U_2 \rangle \mid U_i \in \mathscr{S}^{(k_i)}(\mathscr{Y}),\ k_1 + k_2 = k - 1 \}.$$

It is trivially true that $\mathrm{vlft}(\mathscr{S}^{(1)}(\mathscr{Y})) \subset \Gamma^\omega(\mathrm{Lie}^{(\infty)}(\mathscr{C}_\Sigma))$. Now suppose that $\mathrm{vlft}(\mathscr{S}^{(k)}(\mathscr{Y})) \subset \Gamma^\omega(\mathrm{Lie}^{(\infty)}(\mathscr{C}_\Sigma))$ for $k \in \{1, \dots, l\}$ for $l \geq 1$. We see that $\mathrm{vlft}(\mathscr{S}^{(l+1)}(\mathscr{Y})) \subset \Gamma^\omega(\mathrm{Lie}^{(\infty)}(\mathscr{C}_\Sigma))$, since we may generate all elements of $\mathrm{vlft}(\mathscr{S}^{(l+1)}(\mathscr{Y}))$ by considering brackets of the form $[\mathrm{vlft}(U_1), [S, \mathrm{vlft}(U_2)]]$, where $U_i \in \mathscr{S}^{(l_i)}(\mathscr{Y})$ and $l_1 + l_2 = l - 1$. This follows from (B.4). This shows that $\mathrm{vlft}(\mathscr{S}^{(\infty)}(\mathcal{Y}) \subset \Gamma^\omega(\mathrm{Lie}^{(\infty)}(\mathscr{C}_\Sigma))$. Thus $\mathrm{Sym}^{(\infty)}(\mathcal{Y}) \subset V(\mathscr{C}_\Sigma)$, as claimed.

Next, an examination of the proof of part (ii) of Lemma B.3 shows that $H(\mathscr{Y})$ is involutive. Therefore, since the vector fields generating $H(\mathscr{Y})$ are analytic, there passes through q_0 a unique maximal integral manifold Λ_{q_0} of $H(\mathscr{Y})$ with the property that $\mathsf{T}_q(\Lambda_{q_0}) = H_q(\mathscr{Y})$ for each $q \in \Lambda_{q_0}$. In the ensuing discussion we shall suppose that Λ_{q_0} is an embedded submanifold. This can be done by working with the connected component of Λ_{q_0} containing q_0 in a sufficiently small neighborhood of q_0. If $\Lambda^T_{q_0}$ denotes the maximal integral manifold of $\mathrm{Lie}^{(\infty)}(\mathscr{C}_\Sigma)$ through 0_{q_0}, then we must have $\mathsf{T}_{0_q}(\Lambda^T_{q_0}) = \mathsf{T}_q(\Lambda_{q_0}) \oplus V_q(\mathscr{Y})$ for each $q \in \Lambda_{q_0}$. Since $\dim(\mathsf{T}_{0_q}(\Lambda^T_{q_0}))$ and $\dim(\mathsf{T}_q(\Lambda_{q_0}))$ are independent of $q \in \Lambda_{q_0}$, it follows that $\dim(V_q(\mathscr{Y}))$ must also be independent of $q \in \Lambda_{q_0}$. Thus $V(\mathscr{Y})$ is a regular subbundle of $\mathsf{T}(\Lambda_{q_0})$ containing $\mathrm{Sym}^{(\infty)}(\mathscr{Y})|\Lambda_{q_0}$. By Theorem 3.108 this means that $S|\mathsf{T}(\Lambda_{q_0})$ and $\mathrm{vlft}Y_a$, $a \in \{1, \dots, m\}$, are tangent to $\mathrm{Sym}^{(\infty)}(\mathcal{Y})|\Lambda_{q_0} \subset \mathsf{T}(\Lambda_{q_0})$. This shows that $\mathsf{T}_{0_q}(\Lambda^T_{q_0}) = \mathsf{T}_q(\Lambda_{q_0}) \oplus \mathrm{Sym}^{(\infty)}(\mathcal{Y})_{q_0}$, so giving $V(\mathscr{Y}) = \mathrm{Sym}^{(\infty)}(\mathcal{Y})$.

Finally, we see from the proof of part (ii) of Lemma B.3 that $H(\mathscr{Y})$ is the smallest involutive distribution containing $V(\mathscr{Y})$. In other words, $H_{q_0}(\mathscr{Y}) = \mathrm{Lie}^{(\infty)}(\mathrm{Sym}^{(\infty)}(\mathcal{Y}))_{q_0}$. ■

Remark B.7. Note that not all primitive brackets evaluate to a single symmetric product. Using the relation

$$[\mathrm{vlft}(Y_a), [S, \mathrm{vlft}(Y_b)]] = \mathrm{vlft}(\langle Y_a : Y_b \rangle)$$

from (7.5), one can verify, for example, that

$$[\mathrm{vlft}(Y_a), [S, [\mathrm{vlft}(Y_b), [S, \mathrm{vlft}(Y_c)]]]] = \mathrm{vlft}(\langle Y_a : \langle Y_b : Y_c \rangle \rangle). \tag{B.5}$$

Thus in this case a primitive bracket evaluates to a single symmetric product. However, using the Jacobi identity followed by the formula (B.5), one also has

$$[\mathrm{vlft}(Y_a), [[S, \mathrm{vlft}(Y_b)], [S, \mathrm{vlft}(Y_c)]]] \\ = \mathrm{vlft}\big(\langle Y_c : \langle Y_a : Y_b\rangle\rangle - \langle Y_b : \langle Y_a : Y_c\rangle\rangle\big),$$

showing that a primitive bracket can evaluate to a *sum* of symmetric products. In other words, the map sending a primitive bracket in $\mathrm{Br}_{-1}(\boldsymbol{\xi})$ to an element in $\mathrm{Sym}(\boldsymbol{\eta})$ is surjective, but not injective. •

B.3 Proof of Lemma 8.4

We will assume that $\mathrm{rank}(B) = n$, since if $\mathrm{rank}(B) < n$, then we will have proved the proposition for the induced nondegenerate form on the quotient $\mathsf{V}/\ker(B)$. One may then see that the degeneracy does not alter the result.

(i) $\Longrightarrow$ (ii) We prove this implication by induction on n. The result is vacuous when $n = 1$. For $n = 2$, we let $\mathscr{V} = \{v_1, v_2\}$ be a basis for which the matrix $[B]_{\mathscr{V}}$ has a diagonal that sums to zero. By rescaling v_1 and v_2 if necessary, we may assume that

$$[B]_{\mathscr{V}} = \begin{bmatrix} 1 & a \\ a & -1 \end{bmatrix}$$

for some $a \in \mathbb{R}$. If we define a basis $\mathscr{W} = \{w_1, w_2\}$ by

$$w_1 = \frac{-1 - a^2 + a\sqrt{1+a^2}}{2\sqrt[3]{(1+a^2)^2}} v_1 - \frac{1}{2(1+a^2)} v_2,$$
$$w_2 = (-a - \sqrt{1+a^2}) v_1 + v_2,$$

then we compute

$$[B]_{\mathscr{W}} = \begin{bmatrix} 0 & 1 \\ 1 & 0 \end{bmatrix},$$

giving this part of the lemma when $n = 2$. Now suppose this part of the lemma true for $n = k$, and let B be a maximal rank symmetric bilinear map for which there exists a basis $\mathscr{V} = \{v_1, \dots, v_{k+1}\}$ in which the sum of the diagonals of $[B]_{\mathscr{V}}$ is zero. Let us write

$$[B]_{\mathscr{V}} = \begin{bmatrix} \boldsymbol{B}_0 & \boldsymbol{a} \\ \boldsymbol{a}^T & \alpha \end{bmatrix}$$

for $\boldsymbol{a} \in \mathbb{R}^k$ and $\alpha \in \mathbb{R}$. If $\alpha = 0$, then this part of the lemma is proved, since by the induction hypotheses we may find a basis $\mathscr{W} = \{w_1, \dots, w_k, w_{k+1} = v_{k+1}\}$ such that all diagonal entries in $[B]_{\mathscr{W}}$ are zero. So we assume that $\alpha \neq 0$. The sum of the diagonals of $\boldsymbol{B}_0$ must then be $-\alpha$. Therefore, there must be an

$i_0 \in \{1, \ldots, k\}$ for which $d_{i_0} \triangleq B(e_{i_0}, e_{i_0})$ is nonzero, and whose sign is the opposite of that of α. Fix any such i_0. For $i \in \{1, \ldots, k\}$, define

$$\beta_i = -\alpha^{-1}\big(a_i + \sqrt{a_i^2 + \tfrac{\alpha^2}{k}}\big)$$

and define

$$\beta_{k+1} = -d_{i_0}^{-1}\big(a_{i_0} + \sqrt{a_{i_0}^2 - \alpha d_{i_0}}\big).$$

Now define a basis $\tilde{\mathscr{W}} = \{\tilde{w}_1, \ldots, \tilde{w}_{k+1}\}$ for V by

$$\tilde{w}_1 = v_1 + \beta_1 v_{k+1}, \ldots, \tilde{w}_k = v_k + \beta_k v_{k+1}, \; \tilde{w}_{k+1} = v_{k+1} + \beta_{k+1} v_{i_0},$$

and compute

$$\begin{aligned} B(\tilde{w}_i, \tilde{w}_i) &= B(v_i, v_i) + \tfrac{1}{k}\alpha, \qquad i \in \{1, \ldots, k\}, \\ B(\tilde{w}_{k+1}, \tilde{w}_{k+1}) &= 0. \end{aligned}$$

Now note that the sum of the diagonals of $[B]_{\tilde{\mathscr{W}}}$ is zero. Therefore, by the induction hypothesis, we may make a change of basis to $\mathscr{W} = \{w_1, \ldots, w_k, w_{k+1} = \tilde{w}_{k+1}\}$ so that

$$[B]_{\mathscr{W}} = \begin{bmatrix} \tilde{\boldsymbol{B}}_0 & \tilde{\boldsymbol{a}} \\ \tilde{\boldsymbol{a}}^T & 0 \end{bmatrix},$$

where the diagonal entries in $\tilde{\boldsymbol{B}}_0$ are all zero. This proves this part of the lemma.

(ii) $\Longrightarrow$ (iii) We proceed by induction on n. In the case in which $n = 1$ there is nothing to prove. When $n = 2$, let $\mathscr{V} = \{v_1, v_2\}$ be a basis for which $[B]_{\mathscr{V}}$ has a zero diagonal. Without loss of generality, by rescaling v_1 and v_2 if necessary, we may assume that

$$[B]_{\mathscr{V}} = \begin{bmatrix} 0 & 1 \\ 1 & 0 \end{bmatrix}.$$

If we define a basis $\mathscr{W} = \{w_1 = \frac{1}{\sqrt{2}}(v_1 + v_2), w_2 = \frac{1}{\sqrt{2}}(v_1 - v_2)\}$ for V, we see that

$$[B]_{\mathscr{W}} = \begin{bmatrix} 1 & 0 \\ 0 & -1 \end{bmatrix}.$$

Thus our assertion holds when $n = 2$. Now suppose it holds for $n = k$ and let B be a maximal rank quadratic form on a $(k+1)$-dimensional vector space V with $\mathscr{V} = \{v_1, \ldots, v_{k+1}\}$ a basis for V satisfying (ii). Then we write

$$[B]_{\mathscr{V}} = \begin{bmatrix} \boldsymbol{B}_0 & \boldsymbol{a} \\ \boldsymbol{a}^T & 0 \end{bmatrix},$$

where $\boldsymbol{B}_0 \in \mathbb{R}^{k \times k}$ is symmetric with zero diagonal and $\boldsymbol{a} \in \mathbb{R}^k$. By our induction hypothesis, we may choose a basis $\tilde{\mathscr{W}} = \{\tilde{w}_1, \ldots, \tilde{w}_k, \tilde{w}_{k+1} = v_{k+1}\}$ with the property that

$$[B]_{\tilde{\mathscr{W}}} = \begin{bmatrix} \tilde{\boldsymbol{B}}_0 & \tilde{\boldsymbol{a}} \\ \tilde{\boldsymbol{a}}^T & 0 \end{bmatrix},$$

where $\tilde{\boldsymbol{B}}_0 \in \mathbb{R}^{k\times k}$ is diagonal with all diagonal entries being $+1$ or -1, and with not all diagonal entries having the same sign, and $\tilde{\boldsymbol{a}} \in \mathbb{R}^k$. Next define a basis $\mathscr{W} = \{w_1, \dots, w_{k+1}\}$ for V by

$$w_1 = \tilde{w}_1, \dots, w_k = \tilde{w}_k, \; w_{k+1} = \tilde{w}_{k+1} - \sum_{i=1}^{k} s_i \tilde{a}_i \tilde{w}_i,$$

where

$$s_i = \begin{cases} -1, & \text{the } (i,i) \text{ element of } \tilde{B}_0 \text{ is } -1, \\ 1, & \text{the } (i,i) \text{ element of } \tilde{B}_0 \text{ is } 1. \end{cases}$$

One checks by direct calculation that

$$B(w_i, w_{k+1}) = 0, \; i \in \{1, \dots, k\}, \quad B(w_{k+1}, w_{k+1}) = -2\sum_{i=1}^{n} s_i \tilde{a}_i^2.$$

Therefore, by scaling w_{k+1}, the matrix $[B]_{\mathscr{W}}$ will be diagonal with all diagonal entries being $+1$ or -1, and not all diagonal entries will have the same sign. This is the content of (iii), and so our assertion is proven.

(iii) $\Longrightarrow$ (i) Condition (iii), with our assumption that B be maximal rank, is the assertion that we may find a basis $\mathscr{W} = \{w_1, \dots, w_n\}$ for V, where the matrix $[B]_{\mathscr{W}}$ is diagonal with all diagonal entries being either $+1$ or -1, and not all diagonal entries having the same sign. Let us treat the case in which there are more $+1$'s than -1's on the diagonal; the other case follows in exactly the same manner. By permuting the basis elements if necessary, we may then assume that

$$[B]_{\mathscr{W}} = \begin{bmatrix} 1 & 0 & \cdots & 0 & 0 & 0 & \cdots & 0 \\ 0 & -1 & \cdots & 0 & 0 & 0 & \cdots & 0 \\ \vdots & \vdots & \ddots & \vdots & \vdots & \vdots & \ddots & \vdots \\ 0 & 0 & \cdots & 1 & 0 & 0 & \cdots & 0 \\ 0 & 0 & \cdots & 0 & -1 & 0 & \cdots & 0 \\ 0 & 0 & \cdots & 0 & 0 & 1 & \cdots & 0 \\ \vdots & \vdots & \ddots & \vdots & \vdots & \vdots & \ddots & \vdots \\ 0 & 0 & \cdots & 0 & 0 & 0 & \cdots & 1 \end{bmatrix}.$$

Thus each of the (say) k -1's is paired with a $+1$ on the diagonal, with the remaining diagonal entries being $+1$. We define a basis $\mathscr{V} = \{v_1, \dots, v_n\}$ by

$$v_1 = w_1 + w_2, \; v_2 = w_1 - w_2, \dots, v_{2k-1} = w_{2k-1} + w_{2k}, \; v_{2k} = w_{2k-1} - w_{2k},$$
$$v_{2k+1} = w_{2k+1} + w_{2k}, \dots, v_n = w_n + w_{2k}.$$

One then checks by direct calculation that the diagonal entries in the matrix $[B]_{\mathscr{V}}$ are zero, which completes the proof.

B.4 Proof of Theorem 9.38

We start with a conservative bound on the controlled trajectories of the system. It will be used in the proof below. Recall that $\kappa\colon \bar{\mathbb{R}}_+ \to [0, \frac{\pi}{2}]$ is defined implicitly as the unique solution to $\kappa(x)\tan(\kappa(x)) = x$.

Lemma B.8 (Bound on trajectories). *Consider the C^ω-forced affine connection system $(\mathsf{Q}, \nabla, Y, \mathcal{D})$. Consider the solution $\gamma\colon [0,T] \to \mathsf{Q}$ to the governing equations (9.33) with initial condition $\gamma'(0) = 0_{q_0} \in \mathcal{D}$.*

Select a coordinate chart $(\mathcal{U}, \phi)$ with $q_0 \in \mathcal{U}$, let $\rho \in \mathbb{R}_+$ satisfy $B_\rho(\phi(q_0)) \subset \phi(\mathcal{U})$, and assume

$$\|Y\|_{q_0,\rho,T} T^2 < \frac{\kappa^2(\rho n^2 \|\Gamma\|_{q_0,\rho})}{n^2 \|\Gamma\|_{q_0,\rho}}. \tag{B.6}$$

Then, in the coordinate chart, $\gamma([0,T])$ is a subset of $B_\rho(\phi(q_0))$.

Proof. Let $T_0 < T$ be the smallest time at which the solution γ reaches the distance $\|\phi(\gamma(T_0)) - \phi(q_0)\| = \rho$. If the solution never reaches this distance, then $\phi(\gamma([0,T]))$ is clearly a subset of $B_\rho(\phi(q_0))$. Since $\phi(\gamma([0,T_0])) \subset B_\rho(\phi(q_0))$, for all $t \in [0,T_0]$, we have the bound $\|\frac{\mathrm{d}}{\mathrm{d}t}\phi(\gamma)(t)\| \leq y(t)$, where $y\colon [0,T_0] \to \bar{\mathbb{R}}_+$ is the solution for a.e. time to the initial value problem

$$\frac{\mathrm{d}}{\mathrm{d}t} y(t) = n^2 \|\Gamma\|_{q_0,\rho} y(t)^2 + \|Y\|_{q_0,\rho,T}, \quad y(0) = 0.$$

That is

$$y(t) = \sqrt{\frac{\|Y\|_{q_0,\rho,T}}{n^2\|\Gamma\|_{q_0,\rho}}} \tan\Big(\sqrt{\|Y\|_{q_0,\rho,T} n^2 \|\Gamma\|_{q_0,\rho}}\, t\Big).$$

Straightforward manipulations show that the condition in equation (B.6) is equivalent to $Ty(T) < \rho$. Since y is a monotone function, we also have $T_0 y(T_0) < \rho$. Note that $\|\phi(\gamma(0)) - \phi(q_0)\| = 0$ and that

$$\frac{\mathrm{d}}{\mathrm{d}t}\|\phi(\gamma(t)) - \phi(q_0)\| \leq \|\tfrac{\mathrm{d}}{\mathrm{d}t}\phi(\gamma(t))\| \leq y(t) < \frac{\rho}{T_0}$$

for all $t \in [0,T_0]$. Therefore, $\|\phi(\gamma(T_0)) - \phi(q_0)\| < \rho$ and the contradiction with the definition of T_0 is now immediate. ■

Next, let us provide a few comments on the various steps of the proof of Theorem 9.38. First, we investigate how to write the flow of a mechanical control system as the composition of more elementary flows. Two observations play a key role: the homogeneity property renders the computations tractable; the simplifying procedure can be easily repeated, giving rise to an iterative procedure. Second, we prove absolute and uniform convergence of the series expansion resulting from the first formal part of the proof. The proof of the bounds is inspired by the treatment in [Agrachev and Gamkrelidze 1978, Proposition 2.1], but it is made more complicated here by the recursive nature of the series expansion. Once the series is formally derived and it is proven to be convergent, a limiting argument leads to the final statement in equation (9.37).

Proof of Theorem 9.38.

Part I: Solution of equation (9.33) *as the composition of the certain flows.* Let $X_k, Y_k, W_k \in \Gamma^\omega(\mathbb{R}; \mathsf{TQ})$, let $k \in \mathbb{N}$, and let $v_{q,k}\colon I \to \mathsf{TQ}$ satisfy

$$\begin{aligned} v_{q,k}'(t) &= \big(S + [\mathrm{vlft}(X_k), S] + \mathrm{vlft}(Y_k) + \mathrm{vlft}(W_k)\big)(t, v_{q,k}(t)), \\ v_{q,k}(0) &= 0_{q_0}. \end{aligned} \tag{B.7}$$

At $k = 1$, if $X_1 \triangleq W_1 \triangleq 0$, and $Y_1 \triangleq Y$, then the mechanical system in equation (9.33) is equal to equation (B.7) and, therefore, $\gamma'(t) = v_{q,1}(t)$. For $k \in \mathbb{N}$, we invoke Proposition 9.6 to write

$$v_{q,k}(t) = \Phi_{0,t}^{\mathrm{vlft}(Y_k)}(v_{q,k+1}(t)), \tag{B.8}$$

for

$$\begin{aligned} v_{q,k+1}'(t) &= \big((\Phi_{0,t}^{\mathrm{vlft}(Y_k)})^*(S + [\mathrm{vlft}(X_k), S] + \mathrm{vlft}(W_k))\big)(t, v_{q,k+1}(t)), \\ v_{q,k+1}(0) &= 0_{q_0}. \end{aligned} \tag{B.9}$$

Here we compute the pull-back along the flow by means of the series in equation (9.3). This series reduces to a finite sum. From the discussion in Section 9.2.1 on the Lie bracket properties of the various vector fields, we have

$$\mathrm{ad}^{m+2}_{\mathrm{vlft}(Y_k)} S = 0, \quad \mathrm{ad}^{m+1}_{\mathrm{vlft}(Y_k)} [\mathrm{vlft}(X_k), S] = 0, \quad \mathrm{ad}^{m}_{\mathrm{vlft}(Y_k)} \mathrm{vlft}(W_k) = 0,$$

for all $m \geq 1$. With a little bookkeeping we can exploit these equalities and compute

$$\begin{aligned} (\Phi_{0,t}^{\mathrm{vlft}(Y_k)})^*&(S + [\mathrm{vlft}(X_k), S] + \mathrm{vlft}(W_k))_t \\ &= S + [\mathrm{vlft}(X_k), S]_t + \mathrm{vlft}(W_k)_t \\ &\quad + \int_0^t [\mathrm{vlft}(Y_k)_s, (S + [\mathrm{vlft}(X_k), S]_t)] \mathrm{d}s \\ &\quad + \int_0^t \int_0^{s_1} [\mathrm{vlft}(Y_k)_{s_2}, [\mathrm{vlft}(Y_k)_{s_1}, S]] \mathrm{d}s_2 \mathrm{d}s_1 \\ &= S + [\mathrm{vlft}(X_k) + \mathrm{vlft}(\underline{Y_k}), S]_t + [\mathrm{vlft}(\underline{Y_k}), [\mathrm{vlft}(X_k), S]]_t \\ &\quad + \mathrm{vlft}(W_k)_t + \int_0^t \int_0^{s_1} [\mathrm{vlft}(Y_k)_{s_2}, [\mathrm{vlft}(Y_k)_{s_1}, S]] \mathrm{d}s_2 \mathrm{d}s_1 \\ &= S + [\mathrm{vlft}(X_k) + \mathrm{vlft}(\underline{Y_k}), S]_t - \mathrm{vlft}(\langle \underline{Y_k} : X_k \rangle)_t + \mathrm{vlft}(W_k)_T \\ &\quad - \tfrac{1}{2} \mathrm{vlft}(\langle \underline{Y_k} : \underline{Y_k} \rangle)_t, \end{aligned}$$

where the $\underline{\ }$ notation is as introduced in equation (9.39). The last equality also relies on

$$\int_0^t \int_0^{s_1} [\mathrm{vlft}(Y_k)_{s_2}, [\mathrm{vlft}(Y_k)_{s_1}, S]] \mathrm{d}s_2 \mathrm{d}s_1 = -\tfrac{1}{2}\mathrm{vlft}(\langle \underline{Y_k} : \underline{Y_k} \rangle)_t,$$

which follows from an integration by parts, from the symmetry of the symmetric product, and from the calculations in Section 7.3.2. The differential equation describing the trajectory of $v_{q,k+1}$ is of the same form as equation (B.7) describing the trajectory of $v_{q,k}$, where

$$X_{k+1} = X_k + \underline{Y_k},$$
$$Y_{k+1} + W_{k+1} = -\left\langle \underline{Y_k} : X_k + \tfrac{1}{2}\underline{Y_k} \right\rangle + W_k.$$

For $k \geq 2$, the vector field X_k can be computed as $X_k = \sum_{j=1}^{k-1} \underline{Y_j}$ and, therefore,

$$Y_{k+1} + W_{k+1} = -\left\langle \underline{Y_k} : \sum_{j=1}^{k-1} \underline{Y_j} + \tfrac{1}{2}\underline{Y_k} \right\rangle + W_k. \tag{B.10}$$

Notice that, for $k \geq 2$, the quantities Y_k and W_k are not yet uniquely determined. Equation (B.10) is satisfied for all $k \geq 2$ if and only if, for all $m \geq 1$, we have

$$(Y_2 + Y_3 + \cdots + Y_{m+1}) + W_{m+1} = -\sum_{k=1}^{m} \left\langle \underline{Y_k} : \sum_{j=1}^{k-1} \underline{Y_j} + \tfrac{1}{2}\underline{Y_k} \right\rangle, \tag{B.11}$$

where we have used $W_1 = 0$. Some further manipulation leads to:

$$\begin{aligned}\sum_{k=1}^{m} \left\langle \underline{Y_k} : \sum_{j=1}^{k-1} \underline{Y_j} + \frac{1}{2}\underline{Y_k} \right\rangle &= \sum_{k=1}^{m}\sum_{j=1}^{k-1} \langle \underline{Y_k} : \underline{Y_j} \rangle + \frac{1}{2}\sum_{k=1}^{m} \langle \underline{Y_k} : \underline{Y_k} \rangle \\ &= \frac{1}{2} \sum_{j,k=1, j\neq k}^{m} \langle \underline{Y_k} : \underline{Y_j} \rangle + \frac{1}{2}\sum_{k=1}^{m} \langle \underline{Y_k} : \underline{Y_k} \rangle \\ &= \frac{1}{2} \sum_{j,k=1}^{m} \langle \underline{Y_k} : \underline{Y_j} \rangle.\end{aligned}$$

For $m \geq 1$, a selection of Y_i, $i \in \{1, \dots, m\}$, and W_{m+1} that satisfies equation (B.11) is

$$Y_i = -\frac{1}{2} \sum_{j,k=1, j+k=i}^{m} \langle \underline{Y_k} : \underline{Y_j} \rangle = -\frac{1}{2}\sum_{j=1}^{i-1} \langle \underline{Y_j} : \underline{Y_{i-j}} \rangle, \tag{B.12}$$
$$W_{m+1} = -\frac{1}{2} \sum_{j,k=1, j+k>m}^{m} \langle \underline{Y_k} : \underline{Y_j} \rangle.$$

Note that equation (B.12) is a well-defined recursive relationship, and note that the recursive definition of $\{V_k\}_{k\in\mathbb{N}}$ in equations (9.35) and (9.34) corresponds to setting $V_k = \underline{Y_k}$. The iterative procedure proves that, for any $k \geq 2$, the solution to the original mechanical system $\gamma' = v_{q,1} \colon [0, T] \to \mathsf{T}Q$ satisfies

$$\gamma'(t) = \left(\Phi_{0,t}^{\mathrm{vlft}(Y_1)} \circ \Phi_{0,t}^{\mathrm{vlft}(Y_2)} \circ \cdots \circ \Phi_{0,t}^{\mathrm{vlft}(Y_{k-1})}\right)(v_{q,k}(t)),$$

where $v_{q,k}\colon [0,T] \to \mathsf{TQ}$ is the solution to equation (B.7). The trajectory γ' is now written as the composition of k flows, and a first simplification is immediate. For all integers i, j and for all times s_1, s_2, the vector fields $\mathrm{vlft}(Y_i)_{s_1}$ and $\mathrm{vlft}(Y_j)_{s_2}$ commute. Therefore, γ is the solution to

$$\gamma'(t) = \Phi_{0,t}^{\sum_{j=1}^{k-1} \mathrm{vlft}(Y_j)}(v_{q,k}(t)). \tag{B.13}$$

A second simplification is also straightforward. The vector field in equation (B.13) is homogeneous of degree 0, i.e., it is in the form of equation (9.19) in Section 9.2.2. According to (9.20), we have, for all $t \in [0,T]$,

$$\gamma'(t) = v_{q,k}(t) + \sum_{j=1}^{k-1} \underline{Y_j}(t, \pi_{\mathsf{TQ}}(v_{q,k}(t))), \tag{B.14}$$

where the sequence of vector fields $\{Y_j\}_{j\in\mathbb{N}}$ is defined via equation (B.12), and where the curve $v_{q,k}\colon [0,T] \to \mathsf{TQ}$ is the solution to

$$v'_{q,k}(t) = \Big(S + \Big[\sum_{j=1}^{k-1} \mathrm{vlft}(\underline{Y_j}),\, S\Big] + \mathrm{vlft}(Y_k) + \mathrm{vlft}(W_k)\Big)(t, v_{q,k}(t)), \tag{B.15}$$

$$v_{q,k}(0) = 0_{q_0}.$$

Part II: Absolute and uniform convergence. Given the vector field Y, let $\Omega_1 = \{Y\}$, and, for $k \geq 2$, define recursively

$$\Omega_k = \left\{ -\tfrac{1}{2}\langle \underline{B_i} : \underline{B_{k-i}} \rangle \;\middle|\; B_i \in \Omega_i,\, B_{k-i} \in \Omega_{k-i},\, i \in \{1,\dots,k-1\} \right\}.$$

The first few sets are:

$$\Omega_1 = \{Y\}, \quad \Omega_2 = \left\{-\tfrac{1}{2}\langle \underline{Y} : \underline{Y} \rangle\right\}, \quad \Omega_3 = \left\{\tfrac{1}{4}\langle \underline{Y} : \underline{\langle \underline{Y} : \underline{Y}\rangle} \rangle\right\},$$

$$\Omega_4 = \left\{ -\tfrac{1}{8}\left\langle \underline{Y} : \underline{\left\langle \underline{Y} : \underline{\langle \underline{Y} : \underline{Y} \rangle} \right\rangle} \right\rangle, -\tfrac{1}{8}\left\langle \underline{\langle \underline{Y} : \underline{Y} \rangle} : \underline{\langle \underline{Y} : \underline{Y}\rangle} \right\rangle \right\}.$$

Next, we prove by induction that there exists a sequence $\{N_k\}_{k\in\mathbb{N}} \subset \mathbb{N}$, with the property that, for all $k \in \mathbb{N}$, the vector field Y_k is the sum of N_k vector fields belonging to Ω_k. The statement is true for $k = 1$ with $N_1 = 1$. We assume it true for all $j < k$ and prove it for k. By the induction hypothesis, we write $Y_j = \sum_{a=1}^{N_j} B_{j,a}$, where the $B_{j,a}$ are elements in Ω_j. We compute

$$\begin{aligned} Y_k &= -\frac{1}{2}\sum_{j=1}^{k-1} \langle \underline{Y_j} : \underline{Y_{k-j}} \rangle \\ &= -\frac{1}{2}\sum_{j=1}^{k-1} \Big\langle \sum_{a=1}^{N_j} \underline{B_{j,a}} : \sum_{b=1}^{N_{k-j}} \underline{B_{k-j,b}} \Big\rangle \end{aligned}$$

$$= \sum_{j=1}^{k-1}\sum_{a=1}^{N_j}\sum_{b=1}^{N_{k-j}} \underbrace{-\frac{1}{2}\left\langle \underline{B_{j,a}} : \underline{B_{k-j,b}}\right\rangle}_{\in \Omega_k}.$$

This concludes the proof by induction, where the recursive relation on N_k is

$$N_1 = 1, \quad N_k = \sum_{j=1}^{k-1} N_j N_{k-j}, \qquad k \geq 2. \tag{B.16}$$

It is known from elementary number theory, e.g., see [Andrews 1971], that the sequence $\{N_k\}_{k\in\mathbb{N}} \subset \mathbb{N}$ can be explicitly computed and bounded by

$$N_k = \frac{1}{k}\binom{2k-2}{k-1} \leq \frac{2^{2(k-1)}}{k-\frac{1}{2}}. \tag{B.17}$$

We now focus our attention on bounding from above each time-dependent vector field $B_k \in \Omega_k$. Recall the norms introduced in Section 9.1.2 computed in the coordinate chart $(\mathcal{U}, \phi)$ as specified in the theorem statement. In what follows we identify points in the manifold with their coordinate representation. We claim that there exist sequences $\{c_k\}_{k\in\mathbb{N}} \subset \mathbb{R}_+$ and $\{d_k\}_{k\in\mathbb{N}} \subset \mathbb{Z}_+$, such that

$$\|\partial^m B_k\|_{q_0,\rho',t} \leq c_k \frac{(m+d_k)!}{d_k!}\,\delta^{m+k-1}\|Y\|^k_{q_0,\rho,t} t^{2(k-1)}, \tag{B.18}$$

where we define $\delta = \max\left\{(\rho-\rho')^{-1}, \|\Gamma\|_{q_0,\rho}\right\}$, so that $\|\partial^m\Gamma\|_{q_0,\rho'} \leq m!\,\delta^{m+1}$. As discussed in Section 9.1.2, the bound in equation (B.18) is satisfied for $k=1$ and for all $m \in \mathbb{N}$, with $c_1 = 1$, $d_1 = 0$. In what follows, we provide a proof by induction on $k \geq 2$.

Any time-dependent vector field B_k, for $k \geq 2$, can be written as $B_k = -\frac{1}{2}\left\langle \underline{B_a} : \underline{B_b}\right\rangle$, for some $a, b \in \{1,\dots,k-1\}$, $a+b=k$, $B_a \in \Omega_a$, and $B_b \in \Omega_b$. Accordingly, we compute

$$\begin{aligned}
&\|\partial^m \left\langle \underline{B_a} : \underline{B_b}\right\rangle\|_{q_0,\rho',t} \\
&= \max\left\{ \operatorname{ess\,sup}\{\|\partial^m \left\langle \underline{B_a} : \underline{B_b}\right\rangle^i_s\|_{q_0,\rho'} \mid s \in [0,t]\} \mid i \in \{1,\dots,n\}\right\} \\
&\leq \max\left\{ \operatorname{ess\,sup}\left\{\left\|\partial^m\left(\frac{\partial \underline{B_a^i}}{\partial q^j}\underline{B_b^j}\right)_s\right\|_{q_0,\rho'} \,\middle|\, s\in[0,t]\right\} \,\middle|\, i \in \{1,\dots,n\}\right\} \\
&\quad + \max\left\{ \operatorname{ess\,sup}\left\{\left\|\partial^m\left(\frac{\partial \underline{B_b^i}}{\partial q^j}\underline{B_a^j}\right)_s\right\|_{q_0,\rho'} \,\middle|\, s\in[0,t]\right\} \,\middle|\, i \in \{1,\dots,n\}\right\} \\
&\quad + 2\max\left\{ \operatorname{ess\,sup}\left\{\left\|\partial^m\left(\Gamma^i_{jl}\underline{B_a^j}\,\underline{B_b^l}\right)_s\right\|_{q_0,\rho'} \,\middle|\, s\in[0,t]\right\} \,\middle|\, i \in \{1,\dots,n\}\right\}.
\end{aligned}$$

Relying on the equality

$$\frac{\mathrm{d}^m}{\mathrm{d}x^m} f(x)g(x) = \sum_{\alpha=0}^{m}\binom{m}{\alpha}\frac{\mathrm{d}^\alpha f(x)}{\mathrm{d}x^\alpha}\frac{\mathrm{d}^{m-\alpha} g(x)}{\mathrm{d}x^{m-\alpha}},$$

the first term is bounded according to

$$\begin{aligned}\left\|\partial^m\left(\frac{\partial B_a^i}{\partial q^j}\underline{B_b^j}\right)_s\right\|_{q_0,\rho'} &\leq n\sum_{\alpha=0}^{m}\frac{m!}{\alpha!(m-\alpha)!}\left\|\partial^{\alpha+1}\underline{B_a}\right\|_{q_0,\rho',s}\left\|\partial^{m-\alpha}\underline{B_b}\right\|_{q_0,\rho',s}\\ &\leq n\sum_{\alpha=0}^{m}\frac{m!}{\alpha!(m-\alpha)!}\left(c_a\frac{(\alpha+1+d_a)!}{d_a!}\,\delta^{\alpha+a}\|Y\|^a_{q_0,\rho,s}\frac{s^{2a-1}}{2a-1}\right)\\ &\quad\cdot\left(c_b\frac{(m-\alpha+d_b)!}{d_b!}\,\delta^{m-\alpha+b-1}\|Y\|^b_{q_0,\rho,s}\frac{s^{2b-1}}{2b-1}\right)\\ &\leq\frac{nc_ac_b}{(2a-1)(2b-1)}\left(\frac{m!}{d_a!d_b!}\sum_{\alpha=0}^{m}\frac{(\alpha+1+d_a)!(m-\alpha+d_b)!}{\alpha!(m-\alpha)!}\right)\\ &\quad\cdot\,\delta^{m+a+b-1}\,\|Y\|^{a+b}_{q_0,\rho,s}\,s^{2(a+b-1)}.\end{aligned}$$

The third term is bounded according to:

$$\begin{aligned}&\left\|\partial^m\left(\Gamma^i_{jl}\underline{B_a^j}\,\underline{B_b^l}\right)_s\right\|_{q_0,\rho'}\\ &\quad=n^2\sum_{\alpha=0}^{m}\sum_{\beta=0}^{\alpha}\frac{m!\,\|\partial^{m-\alpha}\Gamma\|_{q_0,\rho'}}{(m-\alpha)!\beta!(\alpha-\beta)!}\left\|\partial^\beta\underline{B_a}\right\|_{q_0,\rho',s}\left\|\partial^{\alpha-\beta}\underline{B_b}\right\|_{q_0,\rho',s}\\ &\quad\leq n^2\sum_{\alpha=0}^{m}\sum_{\beta=0}^{\alpha}\frac{m!}{(m-\alpha)!\beta!(\alpha-\beta)!}\left((m-\alpha)!\,\delta^{m-\alpha+1}\right)\\ &\qquad\cdot\left(c_a\frac{(\beta+d_a)!}{d_a!}\,\delta^{\beta+a-1}\|Y\|^a_{q_0,\rho,s}\frac{s^{2a-1}}{2a-1}\right)\\ &\qquad\cdot\left(c_b\frac{(\alpha-\beta+d_b)!}{d_b!}\,\delta^{\alpha-\beta+b-1}\|Y\|^b_{q_0,\rho,s}\frac{s^{2b-1}}{2b-1}\right)\\ &\quad\leq\frac{n^2c_ac_b}{(2a-1)(2b-1)}\left(\frac{m!}{d_a!d_b!}\sum_{\alpha=0}^{m}\sum_{\beta=0}^{\alpha}\frac{(\beta+d_a)!(\alpha-\beta+d_b)!}{\beta!(\alpha-\beta)!}\right)\\ &\qquad\cdot\,\delta^{m+a+b-1}\,\|Y\|^{a+b}_{q_0,\rho,s}\,s^{2(a+b-1)}.\end{aligned}$$

To simplify notation, let us define

$$C(l,d_1,d_2)\triangleq\sum_{a=0}^{l}\frac{(a+d_1)!(l-a+d_2)!}{a!(l-a)!}.$$

Putting all of the above together, we obtain

$$\begin{aligned}&\left\|\partial^m\left\langle\underline{B_a}:\underline{B_b}\right\rangle\right\|_{q_0,\rho',t}\\ &\leq\frac{nc_ac_b}{(2a-1)(2b-1)}\,\frac{m!}{d_a!d_b!}\,\delta^{m+a+b-1}\,\|Y\|^{a+b}_{q_0,\rho,t}\,t^{2(a+b-1)}\\ &\quad\cdot\left(C(m,d_a+1,d_b)+C(m,d_a,d_b+1)+2n\sum_{\alpha=0}^{m}C(\alpha,d_a,d_b)\right).\end{aligned}\tag{B.19}$$

Next, we introduce and prove an intermediate combinatorial relationship.

Lemma. *For $d_1, d_2 \in \mathbb{Z}_+$ and $k \in \mathbb{N}$, we have*

$$\sum_{a=0}^{k} \binom{a+d_1}{d_1} \binom{k-a+d_2}{d_2} = \binom{k+1+d_1+d_2}{k}. \tag{B.20}$$

Proof. To prove the identity we rely on the method of generating functions as described, for example, in [Andrews 1971, Section 3.4]. We claim that, for all real x with $|x| < 1$,

$$\sum_{k=0}^{+\infty} \left(\sum_{a=0}^{k} \binom{a+d_1}{d_1} \binom{k-a+d_2}{d_2} \right) x^k = \sum_{k=0}^{+\infty} \binom{k+1+d_1+d_2}{k} x^k. \tag{B.21}$$

The first step is to notice that

$$\begin{aligned}
&\sum_{k=0}^{+\infty} \Big(\sum_{a=0}^{k} \binom{a+d_1}{d_1} \binom{k-a+d_2}{d_2} \Big) x^k \\
&\qquad = \sum_{k=0}^{+\infty} \sum_{m+n=k} \binom{n+d_1}{d_1} \binom{m+d_2}{d_2} x^{n+m} \\
&\qquad = \Big(\sum_{n=0}^{+\infty} \binom{n+d_1}{d_1} x^n \Big) \Big(\sum_{m=0}^{+\infty} \binom{m+d_2}{d_2} x^m \Big).
\end{aligned}$$

Accordingly, we define

$$f_a(x) = \sum_{m=0}^{+\infty} \binom{m+a}{a} x^m,$$

and equation (B.21) is equivalent to proving that

$$f_{d_1}(x) f_{d_2}(x) = f_{d_1+d_2+1}(x). \tag{B.22}$$

In passing, we also note that the radius of convergence of f is $|x| < 1$. The second step is to study the properties of f. First we compute

$$\begin{aligned}
f_a(x) &= \sum_{m=0}^{+\infty} \frac{(m+a)!}{m!a!} x^m = \frac{1}{a!} \sum_{m=0}^{+\infty} (m+a) \cdots (m+1) x^m \\
&= \frac{1}{a!} \frac{\mathrm{d}^a}{\mathrm{d}x^a} \sum_{m=0}^{+\infty} x^{m+a} = \frac{1}{a!} \frac{\mathrm{d}^a}{\mathrm{d}x^a} \Big(x^a \sum_{m=0}^{+\infty} x^m \Big) = \frac{1}{a!} \frac{\mathrm{d}^a}{\mathrm{d}x^a} \frac{x^a}{1-x}.
\end{aligned}$$

Additionally, it is immediate to see that

$$f_0(x) = \frac{1}{1-x} \quad \Longrightarrow \quad x f_0(x) = f_0(x) - 1,$$

and consequently that

$$f_a(x) = \frac{1}{a!}\frac{\mathrm{d}^a}{\mathrm{d}x^a}\frac{1}{1-x} = \frac{1}{a!}\frac{\mathrm{d}^a}{\mathrm{d}x^a} f_0(x).$$

Finally, we prove by induction that

$$f_a(x) = f_0(x)^{a+1}. \tag{B.23}$$

For $a = 0$, the statement is obvious. We assume it true for $a \geq 1$ and compute:

$$\begin{aligned}
f_{a+1}(x) &= \frac{1}{(a+1)!}\frac{\mathrm{d}^{a+1}}{\mathrm{d}x^{a+1}}\frac{1}{1-x} = \frac{1}{(a+1)!}\frac{\mathrm{d}^a}{\mathrm{d}x^a}\left(\frac{1}{1-x}\right)^2 \\
&= \frac{1}{(a+1)!}\sum_{b=0}^{a}\binom{a}{b}\left(\frac{\mathrm{d}^b}{\mathrm{d}x^b}\frac{1}{1-x}\right)\left(\frac{\mathrm{d}^{a-b}}{\mathrm{d}x^{a-b}}\frac{1}{1-x}\right) \\
&= \frac{a!}{(a+1)!}\sum_{b=0}^{a}\left(\frac{1}{b!}\frac{\mathrm{d}^b}{\mathrm{d}x^b}\frac{1}{1-x}\right)\left(\frac{1}{(a-b)!}\frac{\mathrm{d}^{a-b}}{\mathrm{d}x^{a-b}}\frac{1}{1-x}\right) \\
&= \frac{1}{a+1}\sum_{b=0}^{a}\left(\frac{1}{1-x}\right)^{b+1}\left(\frac{1}{1-x}\right)^{a-b+1} = \left(\frac{1}{1-x}\right)^{a+2}.
\end{aligned}$$

This concludes the proof of equation (B.23), which immediately implies equation (B.22) and equation (B.21). ▼

An immediate consequence of equation (B.20) is that

$$C(l, d_1, d_2) = \frac{d_1!d_2!(l+1+d_1+d_2)!}{l!(1+d_1+d_2)!}. \tag{B.24}$$

Therefore, we compute

$$\begin{aligned}
&\frac{m!}{d_a!d_b!}\Big(C(m, d_a+1, d_b) + C(m, d_a, d_b+1) + 2n\sum_{\alpha=0}^{m} C(\alpha, d_a, d_b)\Big) \\
&= \frac{m!}{d_a!d_b!}\Big(\big((d_a+1)!d_b! + d_a!(d_b+1)!\big)\frac{(m+2+d_a+d_b)!}{m!(2+d_a+d_b)!} \\
&\quad + 2n\sum_{\alpha=0}^{m}\frac{d_a!d_b!(\alpha+1+d_a+d_b)!}{\alpha!(1+d_a+d_b)!}\Big) \\
&= \frac{(m+2+d_a+d_b)!}{(1+d_a+d_b)!} + \frac{2nm!}{(1+d_a+d_b)!}\underbrace{\sum_{\alpha=0}^{m}\frac{(\alpha+1+d_a+d_b)!}{\alpha!}}_{C(m,1+d_a+d_b,0)},
\end{aligned}$$

and, again applying equation (B.24) with $(l, d_1, d_2) = (m, 1+d_a+d_b, 0)$,

$$\frac{m!}{d_a!d_b!}\Big(C(m,d_a+1,d_b)+C(m,d_a,d_b+1)+2n\sum_{\alpha=0}^{m}C(\alpha,d_a,d_b)\Big)$$
$$=\frac{(m+2+d_a+d_b)!}{(1+d_a+d_b)!}+\frac{2nm!(1+d_a+d_b)!(m+2+d_a+d_b)!}{(1+d_a+d_b)!m!(2+d_a+d_b)!}$$
$$=\frac{(m+2+d_a+d_b)!}{(2+d_a+d_b)!}(2+2n+d_a+d_b)\,.$$

We can now return to our study of $\|\partial^m \left\langle \underline{B_a} : \underline{B_b}\right\rangle \|_{q_0,\rho',t}$, and from equation (B.19) we obtain

$$\|\partial^m \left\langle \underline{B_a} : \underline{B_b}\right\rangle \|_{q_0,\rho',t} \le \frac{nc_ac_b(2+2n+d_a+d_b)(m+2+d_a+d_b)!}{(2a-1)(2b-1)(2+d_a+d_b)!} \cdot \delta^{m+a+b-1}\|Y\|^{a+b}_{q_0,\rho,t}\,t^{2(a+b-1)}.$$

We now express everything in terms of $k=a+b$ and $B_k = -\frac{1}{2}\left\langle \underline{B_a} : \underline{B_b}\right\rangle$. We have:

$$\|\partial^m B_k\|_{q_0,\rho',t} \le \max\left\{\|\partial^m \left\langle \underline{B_a} : \underline{B_b}\right\rangle \|_{q_0,\rho',t} \;\middle|\; a+b=k\right\}.$$

Equation (B.18) is proved by defining sequences c_k and d_k such that $c_1=1$ and $d_1=0$ together with

$$d_k \le \max\left\{2+d_a+d_b \mid a+b=k\right\},$$
$$c_k \le \max\left\{\frac{nc_ac_b(2+2n+d_a+d_b)}{2(2a-1)(2b-1)} \;\middle|\; a+b=k\right\}.$$

It is immediate to see that $d_k = 2(k-1)$ satisfies the recursive requirement, so that we require c_k to satisfy $c_1=1$ together with the requirement that

$$c_k \le \max\left\{\frac{n(k+n-1)c_ac_b}{(2a-1)(2b-1)} \;\middle|\; a+b=k\right\}$$
$$= \max\left\{\frac{n(k+n-1)c_ac_{k-a}}{(2a-1)(2k-2a-1)} \;\middle|\; a\in\{1,\dots,k-1\}\right\}.$$

Note that the polynomial $p(a)=(2a-1)(2k-2a-1)$ in $a\in[1,k-1]$ assumes its minimum value of $(2k-3)$ when $a\in\{1,k-1\}$. Accordingly, a stricter requirement on c_k is

$$c_k \le \max\left\{\frac{n(k+n-1)}{2k-3}c_ac_{k-a} \;\middle|\; a\in\{1,\dots,k-1\}\right\}.$$

Since $\frac{k-1}{2k-3}\le 1$ and $\frac{n}{2k-3}\le n$ for all $k\ge 2$, a conservative selection of c_k that satisfies this requirement is provided by the sequence

$$c_1=1,\quad c_k=n(1+n)\sum_{a=1}^{k-1}c_ac_{k-a},\qquad k\ge 2.$$

Recalling the definition in equation (B.16), one can show that $c_k = (n(1+n))^{k-1}N_k$.

Finally, we summarize all the analysis in Part II and prove convergence. Evaluating for $m = 0$ the bound in equation (B.18), we have

$$\|B_k\|_{q_0,\rho',t} \leq (n(1+n))^{k-1} N_k \, \delta^{k-1} \|Y\|^k_{q_0,\rho,t} \, t^{2(k-1)},$$

and recalling the bound in equation (B.17), we compute

$$\begin{aligned} \|Y_k\|_{q_0,\rho',t} &\leq N_k \, \|B_k\|_{q_0,\rho',t} \leq (n(1+n))^{k-1} N_k^2 \, \delta^{k-1} \, \|Y\|^k_{q_0,\rho,t} \, t^{2(k-1)} \\ &\leq \frac{(2^4 n(1+n)\,\delta)^{k-1}}{(k-\frac{1}{2})^2} \, \|Y\|^k_{q_0,\rho,t} \, t^{2(k-1)}. \end{aligned}$$

An immediate consequence is that, for $(2^4 n(n+1)\delta)\,\|Y\|_{q_0,\rho,T} T^2 < 1$, the series

$$Y_\infty(t,q) = \lim_{K \to +\infty} \sum_{k=1}^{K} Y_k(t,q)$$

converges absolutely and uniformly in $t \in [0,T]$ and $q \in B^{\mathbb{C}}_{\rho'}(q_0)$.

Part III: Final limiting argument. We start by studying the behavior as $k \to +\infty$ of the right-hand side of (B.14) and of the right-hand side of the differential equation in (B.15). We shall exploit a variation of a standard result on the continuous dependence of solutions of differential equations with respect to parameter changes, see [Coppel 1965, Chapter I, Section 3]. Uniform convergence of a sequence of vector fields implies uniform convergence of the corresponding sequence of integral curves to the integral curve of the limiting vector field. In order to apply this result to the differential equation (B.15), we need to ensure that the vector field on right-hand side converges uniformly and absolutely.

Assume that the time T and input vector field Y together satisfy the bound in equation (9.36) in the theorem statement. Then Lemma B.8 guarantees that $\gamma([0,T]) \subset B_{\rho'}(q_0)$, and the analysis in Part II guarantees that the series $\sum_{k=1}^{+\infty} Y_k$ converges absolutely and uniformly on $B^{\mathbb{C}}_{\rho'}(q_0)$. Therefore, the series converges uniformly and absolutely along the curve γ. From equation (B.14), one can deduce that $\gamma(t) = \pi_{\mathsf{TQ}}(v_{q,k}(t))$, so that the series $\sum_{k=1}^{+\infty} Y_k$ converges also along $\pi_{\mathsf{TQ}} \circ v_{q,k} \colon [0,T] \to \mathsf{Q}$. Accordingly, we can take the limit as $k \to +\infty$ in equation (B.15).

Notice that, uniformly in $t \in [0,T]$ and $q \in B^{\mathbb{C}}_{\rho'}(q_0)$,

$$\lim_{k \to +\infty} Y_k(t,q) = 0, \quad \text{and} \quad \lim_{k \to +\infty} W_k(t,q) = 0,$$

and define the time-dependent vector field

$$V_\infty = \sum_{k=1}^{+\infty} V_k = \sum_{k=1}^{+\infty} Y_k.$$

Taking the limit as $k \to +\infty$ in equations (B.14) and (B.15), one obtains

$$\gamma'(t) = v_{q,\infty}(t) + V_\infty(t, \pi_{\mathsf{TQ}}(v_{q,\infty}(t))),$$

where the curve $v_{q,\infty}\colon [0,T] \to \mathsf{TQ}$ is the solution to

$$v'_{q,\infty}(t) = \big(S + [\text{vlft}(V_\infty), S]\big)(t, v_{q,\infty}(t)), \quad v_{q,\infty}(0) = 0_{q_0}. \tag{B.25}$$

According to the discussion in Section 9.2.2, the initial value problem in (B.25) can be explicitly integrated. Because $S \in \mathcal{P}_1$ and $[\text{vlft}(V_\infty), S] \in \mathcal{P}_0$, and because of the equality

$$T_{v_q}\pi_{\mathsf{TQ}} \circ [\text{vlft}(V_\infty), S](v_q) = V_\infty(q), \quad v_q \in \mathsf{TQ}.$$

the curve $v_{q,\infty}$ satisfies

$$v_{q,\infty}(t) = 0_{\zeta(t)}, \quad \text{where} \quad \zeta(t) = \Phi^{V_\infty}_{0,t}(q_0),$$

so that

$$\gamma'(t) = 0_{\zeta(t)} + V_\infty(t, \pi(0_{\zeta(t)})) = V_\infty(t, \zeta(t)).$$

The last two statements imply that $\gamma = \zeta$, and are equivalent to statement in equation (9.37) in the theorem. ■

B.5 Proof of Theorem 11.19

Part I: Non-increasing error function. We study the time derivative of $t \mapsto E_{\text{cl}}(\gamma'(t), \gamma'_{\text{ref}}(t))$ along trajectories $t \mapsto (\gamma(t), \gamma_{\text{ref}}(t)) \in \mathsf{Q} \times \mathsf{Q}$, with γ is the controlled trajectory for the closed-loop system defined by $F_{\text{PD}} + F_{\text{FF}}$. As in Lemma 11.16, we let $\dot{e}$ be the velocity error between γ and γ_{ref}; that is

$$\dot{e}(t) = \gamma'(t) - \mathcal{T}(\gamma(t), \gamma_{\text{ref}}(t)) \cdot \gamma'_{\text{ref}}(t).$$

To compute the covariant derivative of the velocity error, we will need the following result.

Lemma B.9. *Given two curves $\gamma, \eta\colon \bar{\mathbb{R}}_+ \to \mathsf{Q}$, consider the vector field $t \mapsto \mathcal{T}(\eta(t), \gamma(t)) \cdot \eta'(t)$ defined along γ. Its covariant derivative along γ satisfies*

$$\begin{aligned}\overset{\mathbb{G}}{\nabla}_{\gamma'(t)} &\mathcal{T}(\gamma(t), \eta(t)) \cdot \eta'(t) \\ &= \big(\overset{\mathbb{G}}{\nabla} \mathcal{T}(\gamma(t), \eta(t))\big) \cdot (\gamma'(t), \eta'(t)) + \frac{\mathrm{d}}{\mathrm{d}t}\left(\mathcal{T}(q, \eta(t)) \cdot \eta'(t)\right)\Big|_{q=\gamma(t)}.\end{aligned}$$

Proof. Given $w_r \in \mathsf{TQ}$, the map $t \mapsto \mathcal{T}(\gamma(t), r) \cdot w_r$ is a vector field along γ and its covariant derivative along γ is well-defined. Furthermore, given $q \in \mathsf{Q}$, the map $t \mapsto \mathcal{T}(q, \eta(t)) \cdot \eta'(t)$ is a $\mathsf{T}_q\mathsf{Q}$-valued function of time and its derivative with respect to time is well-defined. Therefore, the result follows from

$$\overset{\mathbb{G}}{\nabla}_{\gamma'(t)}\mathcal{T}(\gamma(t),\eta(t))\cdot\eta'(t) \\ = \overset{\mathbb{G}}{\nabla}_{\gamma'(t)}\left(\mathcal{T}(\gamma(t),r)\cdot w_r\right)\Big|_{w_r=\eta'(t)} + \frac{\mathrm{d}}{\mathrm{d}t}\left(\mathcal{T}(q,\eta(t))\cdot\eta'(t)\right)\Big|_{q=\gamma(t)}. \quad \blacktriangledown$$

Using the lemma and the definition of F_{PD} and F_{FF},

$$\begin{aligned}\overset{\mathbb{G}}{\nabla}_{\gamma'(t)}\dot{e}(t) &= \overset{\mathbb{G}}{\nabla}_{\gamma'(t)}\left(\gamma'(t)-\mathcal{T}(\gamma(t),\gamma_{\mathrm{ref}}(t))\cdot\gamma'_{\mathrm{ref}}(t)\right)\\ &= \mathbb{G}^\sharp(F_{\mathrm{PD}}(t,\gamma'(t))+F_{\mathrm{FF}}(t,\gamma'(t))) - \overset{\mathbb{G}}{\nabla}\mathcal{T}(\gamma(t),\gamma_{\mathrm{ref}}(t))\cdot(\gamma'(t),\gamma'_{\mathrm{ref}}(t))\\ &\quad - \frac{\mathrm{d}}{\mathrm{d}t}\left(\mathcal{T}(q,\gamma_{\mathrm{ref}}(t))\cdot\gamma'_{\mathrm{ref}}(t)\right)\Big|_{q=\gamma(t)}\\ &= -\,\mathbb{G}^\sharp\big(\mathrm{d}_1\Psi(\gamma(t),\gamma_{\mathrm{ref}}(t)) + R^\flat_{\mathrm{diss}}(\dot{e}(t))\big).\end{aligned} \tag{B.26}$$

Next, combining the results in equations (11.10) and (B.26), we compute

$$\begin{aligned}\frac{\mathrm{d}}{\mathrm{d}t}E_{\mathrm{cl}}(\gamma'(t),\gamma'_{\mathrm{ref}}(t)) &= \langle \mathrm{d}_1\Psi(\gamma(t),\gamma_{\mathrm{ref}}(t));\dot{e}(t)\rangle + \mathbb{G}\big(\dot{e}(t),\overset{\mathbb{G}}{\nabla}_{\gamma'(t)}\dot{e}(t)\big)\\ &= \langle \mathrm{d}_1\Psi(\gamma(t),\gamma_{\mathrm{ref}}(t));\dot{e}(t)\rangle - \big\langle \mathrm{d}_1\Psi(\gamma(t),\gamma_{\mathrm{ref}}(t)) + R^\flat_{\mathrm{diss}}(\dot{e}(t));\dot{e}(t)\big\rangle\\ &= -\,R_{\mathrm{diss}}(\dot{e}(t),\dot{e}(t)) \le 0.\end{aligned} \tag{B.27}$$

This proves stability with respect to the error function E_{cl}, as claimed in (ii).

Assume now there exists $t_0 \in \bar{\mathbb{R}}_+$ such that $\gamma'(t_0) = \gamma'_{\mathrm{ref}}(t_0)$. Then $\Psi(\gamma(t_0),\gamma_{\mathrm{ref}}(t_0)) = 0$ and $\dot{e}(t_0) = 0$, because $\mathcal{T}(\gamma(t_0),\gamma(t_0)) = \mathrm{id}_{\mathsf{T}_{\gamma(t_0)}\mathsf{Q}}$. In turn this implies that $E_{\mathrm{cl}}(\gamma'(t_0),\gamma'_{\mathrm{ref}}(t_0)) = 0$ and that $E_{\mathrm{cl}}(\gamma'(t),\gamma'_{\mathrm{ref}}(t)) = 0$ for all subsequent times. This proves part (i).

Part II: Modifying the error function. To construct an error function with a strictly negative-definite time derivative, we modify E_{cl} as follows. For $\varepsilon > 0$, define

$$E_{\mathrm{cl},\varepsilon}(v_q,w_r) = E_{\mathrm{cl}}(v_q,w_r) + \varepsilon\,\langle \mathrm{d}_1\Psi(q,r); v_q - \mathcal{T}(q,r)\cdot w_r\rangle\,.$$

Along $t \mapsto (\gamma'(t),\gamma'_{\mathrm{ref}}(t))$, we have

$$E_{\mathrm{cl},\varepsilon}(\gamma'(t),\gamma'_{\mathrm{ref}}(t)) = E_{\mathrm{cl}}(\gamma'(t),\gamma'_{\mathrm{ref}}(t)) + \varepsilon\,\langle \mathrm{d}_1\Psi(\gamma(t),\gamma_{\mathrm{ref}}(t));\dot{e}(t)\rangle\,.$$

Next, we compute the second time derivative of $\Psi(\gamma(t),\gamma_{\mathrm{ref}}(t))$:

$$\begin{aligned}&\frac{\mathrm{d}^2}{\mathrm{d}t^2}\Psi(\gamma(t),\gamma_{\mathrm{ref}}(t))\\ &\quad = \overset{\mathbb{G}}{\nabla}_{\gamma'(t)}\langle \mathrm{d}_1\Psi(\gamma(t),\gamma_{\mathrm{ref}}(t));\dot{e}(t)\rangle\\ &\quad = \big\langle \overset{\mathbb{G}}{\nabla}_{\gamma'(t)}\mathrm{d}_1\Psi(\gamma(t),\gamma_{\mathrm{ref}}(t));\dot{e}(t)\big\rangle + \big\langle \mathrm{d}_1\Psi(\gamma(t),\gamma_{\mathrm{ref}}(t));\overset{\mathbb{G}}{\nabla}_{\gamma'(t)}\dot{e}(t)\big\rangle,\end{aligned}$$

where

$$\begin{aligned}
&\left\langle \overset{\mathrm{G}}{\nabla}_{\gamma'(t)} \mathrm{d}_1 \Psi(\gamma(t), \gamma_{\mathrm{ref}}(t)); \dot{e}(t) \right\rangle \\
&\quad = \left\langle \overset{\mathrm{G}}{\nabla}_{\gamma'(t)} \mathrm{d}_1 \Psi(\gamma(t), r)\Big|_{r=\gamma_{\mathrm{ref}}(t)}; \dot{e}(t) \right\rangle + \left\langle \frac{\mathrm{d}}{\mathrm{d}t} \mathrm{d}_1 \Psi(q, \gamma_{\mathrm{ref}}(t))\Big|_{q=\gamma(t)}; \dot{e}(t) \right\rangle \\
&\quad = \operatorname{Hess}_1 \Psi(\gamma(t), \gamma_{\mathrm{ref}}(t)) \cdot (\gamma'(t), \dot{e}(t)) + \left\langle \frac{\mathrm{d}}{\mathrm{d}t} \mathrm{d}_1 \Psi(q, \gamma_{\mathrm{ref}}(t))\Big|_{q=\gamma(t)}; \dot{e}(t) \right\rangle,
\end{aligned}$$

and where

$$\begin{aligned}
\left\langle \frac{\mathrm{d}}{\mathrm{d}t} \mathrm{d}_1 \Psi(q, \gamma_{\mathrm{ref}}(t)); \dot{e}(t) \right\rangle &= \left\langle \mathrm{d}_1 \frac{\mathrm{d}}{\mathrm{d}t} \Psi(q, \gamma_{\mathrm{ref}}(t)); \dot{e}(t) \right\rangle \\
&= \left\langle \mathrm{d}_1 \left\langle \mathrm{d}_2 \Psi(q, \gamma_{\mathrm{ref}}(t)); \gamma'_{\mathrm{ref}}(t) \right\rangle; \dot{e}(t) \right\rangle,
\end{aligned}$$

where the first equality is justified because, in coordinates, $\frac{\partial^2 \Psi}{\partial r^j \partial q^i} = \frac{\partial^2 \Psi}{\partial q^i \partial r^j}$. We perform the following simplifications dealing with tangent vectors v_q and w_r:

$$\begin{aligned}
&\left\langle \mathrm{d}_1 \left\langle \mathrm{d}_2 \Psi(q, r); w_r \right\rangle; v_q \right\rangle = -\left\langle \mathrm{d}_1 \left\langle \mathrm{d}_1 \Psi(q, r); \mathcal{T}(q, r) \cdot w_r \right\rangle; v_q \right\rangle \\
&\quad = -\mathscr{L}_{v_q} \left\langle \mathrm{d}_1 \Psi(q, r); \mathcal{T}(q, r) \cdot w_r \right\rangle \\
&\quad = -\operatorname{Hess}_1 \Psi(q, r) \cdot (v_q, \mathcal{T}(q, r) \cdot w_r) - \left\langle \mathrm{d}_1 \Psi(q, r); \overset{\mathrm{G}}{\nabla} \mathcal{T}(q, r) \cdot (v_q, w_r) \right\rangle.
\end{aligned}$$

Therefore,

$$\begin{aligned}
&\left\langle \frac{\mathrm{d}}{\mathrm{d}t} \mathrm{d}_1 \Psi(q, \gamma_{\mathrm{ref}}(t))\Big|_{q=\gamma(t)}; \dot{e}(t) \right\rangle \\
&\qquad\qquad = -\operatorname{Hess}_1 \Psi(\gamma(t), \gamma_{\mathrm{ref}}(t)) \cdot (\dot{e}(t), \mathcal{T}(\gamma(t), \gamma_{\mathrm{ref}}(t)) \cdot \gamma_{\mathrm{ref}}(t)) \\
&\qquad\qquad\quad - \left\langle \mathrm{d}_1 \Psi(\gamma(t), \gamma_{\mathrm{ref}}(t)); \overset{\mathrm{G}}{\nabla} \mathcal{T}(\gamma(t), \gamma_{\mathrm{ref}}(t)) \cdot (\dot{e}(t), \gamma'_{\mathrm{ref}}(t)) \right\rangle.
\end{aligned}$$

In summary, we compute

$$\begin{aligned}
\frac{\mathrm{d}^2}{\mathrm{d}t^2} \Psi(\gamma(t), \gamma_{\mathrm{ref}}(t)) &= \operatorname{Hess}_1 \Psi(\gamma(t), \gamma_{\mathrm{ref}}(t)) \cdot (\gamma'(t), \dot{e}(t)) \\
&\quad - \operatorname{Hess}_1 \Psi(\gamma(t), \gamma_{\mathrm{ref}}(t)) \cdot (\dot{e}(t), \mathcal{T}(\gamma(t), \gamma_{\mathrm{ref}}(t)) \cdot \gamma_{\mathrm{ref}}(t)) \\
&\quad - \left\langle \mathrm{d}_1 \Psi(\gamma(t), \gamma_{\mathrm{ref}}(t)); \overset{\mathrm{G}}{\nabla} \mathcal{T}(\gamma(t), \gamma_{\mathrm{ref}}(t)) \cdot (\dot{e}(t), \gamma'_{\mathrm{ref}}(t)) \right\rangle \\
&\qquad\qquad + \left\langle \mathrm{d}_1 \Psi(\gamma(t), \gamma_{\mathrm{ref}}(t)); \overset{\mathrm{G}}{\nabla}_{\gamma'(t)} \dot{e}(t) \right\rangle,
\end{aligned}$$

and, therefore,

$$\begin{aligned}
\frac{\mathrm{d}^2}{\mathrm{d}t^2} \Psi(\gamma(t), \gamma_{\mathrm{ref}}(t)) &= \operatorname{Hess}_1 \Psi(\gamma(t), \gamma_{\mathrm{ref}}(t)) \cdot (\dot{e}(t), \dot{e}(t)) \\
&\quad + \left\langle \mathrm{d}_1 \Psi(\gamma(t), \gamma_{\mathrm{ref}}(t)); \overset{\mathrm{G}}{\nabla}_{\gamma'(t)} \dot{e}(t) - \overset{\mathrm{G}}{\nabla} \mathcal{T}(\gamma(t), \gamma_{\mathrm{ref}}(t)) \cdot (\dot{e}(t), \gamma'_{\mathrm{ref}}(t)) \right\rangle.
\end{aligned}$$

Substituting the value of $\overset{\mathrm{G}}{\nabla}_{\gamma'(t)} \dot{e}(t)$ from equation (B.26)

$$\frac{\mathrm{d}^2}{\mathrm{d}t^2}\Psi(\gamma(t),\gamma_{\mathrm{ref}}(t)) = \mathrm{Hess}_1\,\Psi(\gamma(t),\gamma_{\mathrm{ref}}(t))\cdot(\dot{e}(t),\dot{e}(t))$$
$$- \left\langle \mathrm{d}_1\Psi(\gamma(t),\gamma_{\mathrm{ref}}(t)); \mathbb{G}^\sharp\big(\mathrm{d}_1\Psi(\gamma(t),\gamma_{\mathrm{ref}}(t)) + R^\flat_{\mathrm{diss}}(\dot{e}(t))\big)\right\rangle$$
$$- \left\langle \mathrm{d}_1\Psi(\gamma(t),\gamma_{\mathrm{ref}}(t)); \overset{\mathbb{G}}{\nabla}\mathfrak{T}(\gamma(t),\gamma_{\mathrm{ref}}(t))\cdot(\dot{e}(t),\gamma'_{\mathrm{ref}}(t))\right\rangle$$
$$= \mathrm{Hess}_1\,\Psi(\gamma(t),\gamma_{\mathrm{ref}}(t))\cdot(\dot{e}(t),\dot{e}(t)) - \|\mathrm{d}_1\Psi(\gamma(t),\gamma_{\mathrm{ref}}(t))\|^2_{\mathbb{G}}$$
$$- \left\langle \mathrm{d}_1\Psi(\gamma(t),\gamma_{\mathrm{ref}}(t)); \mathbb{G}^\sharp(R^\flat_{\mathrm{diss}}(\dot{e}(t))) + \overset{\mathbb{G}}{\nabla}\mathfrak{T}(\gamma(t),\gamma_{\mathrm{ref}}(t))\cdot(\dot{e}(t),\gamma'_{\mathrm{ref}}(t))\right\rangle.$$

Part III: Final bounding arguments. In equation (B.27) we proved that $t \mapsto E_{\mathrm{cl}}(t)$ is a nonincreasing function. Therefore, the assumption that $E_{\mathrm{cl}}(\gamma'(t_0),\gamma'_{\mathrm{ref}}(t_0)) = L$ at $t_0 \in \bar{\mathbb{R}}_+$ implies that $\Psi(\gamma(t),\gamma_{\mathrm{ref}}(t)) \leq E_{\mathrm{cl}}(\gamma'(t),\gamma'_{\mathrm{ref}}(t)) \leq L$ for all subsequent $t \in [t_0,+\infty[$. This means that the curve $t \mapsto (\gamma(t),\gamma_{\mathrm{ref}}(t))$ remains inside the sublevel set $\Psi^{-1}(\leq L)$. Because $L < L_{\mathrm{reg}}(\Psi,\mathsf{Q})$, on $\Psi^{-1}(\leq L)$ the tensor field R_{diss} is uniformly bounded from below and the tensor fields $\mathrm{Hess}_1\Psi$, R_{diss}, and $\overset{\mathbb{G}}{\nabla}\mathfrak{T}$ are uniformly bounded from above. Accordingly, we define the strictly positive constants

$$M_1 \triangleq \inf\left\{\|R_{\mathrm{diss}}(q)\|_{\mathbb{G}}\right\} \mid (q,r)\in\Psi^{-1}(\leq L)\right\},$$
$$M_2 \triangleq \sup\left\{\|\mathrm{Hess}_1\,\Psi(q,r)\|_{\mathbb{G}}\right\} \mid (q,r)\in\Psi^{-1}(\leq L)\right\},$$
$$M_3 \triangleq \sup\left\{\|R_{\mathrm{diss}}(q)\|_{\mathbb{G}}\right\} \mid (q,r)\in\Psi^{-1}(\leq L)\right\},$$
$$M_4 \triangleq \sup\left\{\left\|\overset{\mathbb{G}}{\nabla}\mathfrak{T}(q,r)\right\|_{\mathbb{G}} \mid (q,r)\in\Psi^{-1}(\leq L)\right\}.$$

Furthermore, the assumption that $\gamma(t_0) \in \Psi^{-1}_{\gamma_{\mathrm{ref}}(t_0)}(\leq L,\gamma_{\mathrm{ref}}(t_0))$ with $L < L_{\mathrm{reg}}(\Psi,\mathsf{Q})$ implies that $\gamma(t) \in \Psi^{-1}_{\gamma_{\mathrm{ref}}(t)}(\leq L,\gamma_{\mathrm{ref}}(t))$ for all subsequent $t \in [t_0,+\infty[$. Because Ψ is uniformly quadratic, there exists $b_1 > 0$ satisfying:

$$0 < b_1\,\|\mathrm{d}_1\Psi(\gamma(t),\gamma_{\mathrm{ref}}(t))\|^2_{\mathbb{G}} \leq \Psi(\gamma(t),\gamma_{\mathrm{ref}}(t)).$$

We also define $M_5 \triangleq \sup\left\{\|\gamma'_{\mathrm{ref}}(t)\|_{\mathbb{G}} \mid t\in\bar{\mathbb{R}}_+\right\} < +\infty$.

Next, we exploit these estimates to bound from below the function $E_{\mathrm{cl},\varepsilon}$. We compute

$$E_{\mathrm{cl},\varepsilon}(\gamma'(t),\gamma'_{\mathrm{ref}}(t)) = \Psi(\gamma(t),\gamma_{\mathrm{ref}}(t)) + \frac{1}{2}\,\|\dot{e}(t)\|^2_{\mathbb{G}} + \varepsilon\,\langle \mathrm{d}_1\Psi(\gamma(t),\gamma_{\mathrm{ref}}(t));\dot{e}(t)\rangle$$
$$\geq b_1\,\|\mathrm{d}_1\Psi(\gamma(t),\gamma_{\mathrm{ref}}(t))\|^2_{\mathbb{G}} + \frac{1}{2}\,\|\dot{e}(t)\|^2_{\mathbb{G}} - \varepsilon\,\|\mathrm{d}_1\Psi(\gamma(t),\gamma_{\mathrm{ref}}(t))\|_{\mathbb{G}}\,\|\dot{e}(t)\|_{\mathbb{G}}$$
$$= \begin{bmatrix}\|\mathrm{d}_1\Psi(\gamma(t),\gamma_{\mathrm{ref}}(t))\|_{\mathbb{G}}\\ \|\dot{e}(t)\|_{\mathbb{G}}\end{bmatrix}^T \boldsymbol{P}_\varepsilon \begin{bmatrix}\|\mathrm{d}_1\Psi(\gamma(t),\gamma_{\mathrm{ref}}(t))\|_{\mathbb{G}}\\ \|\dot{e}(t)\|_{\mathbb{G}}\end{bmatrix},$$

for

$$\boldsymbol{P}_\varepsilon \triangleq \begin{bmatrix} b_1 & -\frac{\varepsilon}{2}\\ -\frac{\varepsilon}{2} & \frac{1}{2}\end{bmatrix} \in \mathbb{R}^{2\times 2}.$$

Similarly, we bound from above the time derivative of $E_{\mathrm{cl},\varepsilon}$:

$$\begin{aligned}
\frac{\mathrm{d}}{\mathrm{d}t}E_{\mathrm{cl},\varepsilon}(\gamma'(t),\gamma'_{\mathrm{ref}}(t)) &= -R_{\mathrm{diss}}(\dot{e}(t),\dot{e}(t)) \\
&\quad + \varepsilon\,\mathrm{Hess}_1\,\Psi(\gamma(t),\gamma_{\mathrm{ref}}(t))\cdot(\dot{e}(t),\dot{e}(t)) - \varepsilon\|\mathrm{d}_1\Psi(\gamma(t),\gamma_{\mathrm{ref}}(t))\|^2 \\
&\quad - \varepsilon\langle \mathrm{d}_1\Psi; \mathbb{G}^\sharp(R^\flat_{\mathrm{diss}}(\dot{e}(t))) + \overset{\mathbb{G}}{\nabla}\mathcal{T}(\gamma(t),\gamma_{\mathrm{ref}}(t))\cdot(\dot{e},\gamma'_{\mathrm{ref}})\rangle \\
&\le -(M_1-\varepsilon M_2)\,\|\dot{e}(t)\|^2_{\mathbb{G}} - \varepsilon\,\|\mathrm{d}_1\Psi(\gamma(t),\gamma_{\mathrm{ref}}(t))\|^2_{\mathbb{G}} \\
&\quad + \varepsilon\,(M_3+M_4M_5)\,\|\dot{e}(t)\|_{\mathbb{G}}\,\|\mathrm{d}_1\Psi(\gamma(t),\gamma_{\mathrm{ref}}(t))\|_{\mathbb{G}} \\
&= -\begin{bmatrix}\|\mathrm{d}_1\Psi(\gamma(t),\gamma_{\mathrm{ref}}(t))\|_{\mathbb{G}} \\ \|\dot{e}(t)\|_{\mathbb{G}}\end{bmatrix}^T \boldsymbol{Q}_\varepsilon \begin{bmatrix}\|\mathrm{d}_1\Psi(\gamma(t),\gamma_{\mathrm{ref}}(t))\|_{\mathbb{G}} \\ \|\dot{e}(t)\|_{\mathbb{G}}\end{bmatrix},
\end{aligned}$$

for

$$\boldsymbol{Q}_\varepsilon \triangleq \begin{bmatrix}\varepsilon & \frac{\varepsilon}{2}(M_3+M_4M_5) \\ \frac{\varepsilon}{2}(M_3+M_4M_5) & M_1-\varepsilon M_2\end{bmatrix} \in \mathbb{R}^{2\times 2}.$$

Now, one can show that there exists an $\overline{\varepsilon} > 0$ such that, for all $0 < \varepsilon < \overline{\varepsilon}$, the trace and the determinant of the matrices $\boldsymbol{P}_\varepsilon$ and $\boldsymbol{Q}_\varepsilon$ are positive so that the matrices $\boldsymbol{P}_\varepsilon$ and $\boldsymbol{Q}_\varepsilon$ are positive-definite. Furthermore, two applications of Exercise E2.18 show that there exist positive constants λ_1 and λ_2 such that

$$\begin{aligned}
E_{\mathrm{cl}}(\gamma'(t),\gamma'_{\mathrm{ref}}(t)) &\le \lambda_1 E_{\mathrm{cl},\varepsilon}(\gamma'(t),\gamma'_{\mathrm{ref}}(t)), \\
\frac{\mathrm{d}}{\mathrm{d}t}E_{\mathrm{cl},\varepsilon}(\gamma'(t),\gamma'_{\mathrm{ref}}(t)) &\le -\lambda_2 E_{\mathrm{cl},\varepsilon}(\gamma'(t),\gamma'_{\mathrm{ref}}(t)),
\end{aligned}$$

therefore proving that $t \mapsto E_{\mathrm{cl}}(\gamma'(t),\gamma'_{\mathrm{ref}}(t))$ converges to zero exponentially fast. This implies (iii).

B.6 Proof of Theorem 11.29

The result is a consequence of Theorem 11.19 after establishing the necessary facts. We follow the same design logic outlined in the preface to Section 11.3. From Lemma 11.27, we know that $\Psi_{\mathsf{G},r}$ is uniformly quadratic and that $L_{\mathrm{reg}}(\Psi_{\mathsf{G},r},\mathsf{G}) = L_{\mathrm{reg}}(\Psi_{\mathsf{G}},e)$. From Lemma 11.28, we know that $(\Psi_{\mathsf{G},r},\mathcal{T}_r)$ is compatible. Let us now compute the covariant derivative of $\mathcal{T}$. Let $\xi,\eta \in \mathfrak{g}$ and write

$$\begin{aligned}
\overset{\mathbb{G}_{\mathrm{I}}}{\nabla}\mathcal{T}_r(g,h)\cdot(T_eL_g(\xi),T_eL_h(\eta)) \\
&= \overset{\mathbb{G}_{\mathrm{I}}}{\nabla}_{T_eL_g(\xi)}\big(\mathcal{T}_r(g,h)\cdot(T_eL_h(\eta))\big) \\
&= \overset{\mathbb{G}_{\mathrm{I}}}{\nabla}_{T_eL_g(\xi)}\big(T_eL_g\circ \mathrm{Ad}_{g^{-1}h}(\eta)\big) \\
&= T_eL_g\Big(\overset{\mathfrak{g}}{\nabla}_\xi\,(\mathrm{Ad}_{g^{-1}h}(\eta)) + \mathscr{L}_{\xi_{\mathrm{L}}}\,\mathrm{Ad}_{g^{-1}}(\mathrm{Ad}_h(\eta))\Big) \\
&= T_eL_g\Big(\overset{\mathfrak{g}}{\nabla}_\xi\,(\mathrm{Ad}_{g^{-1}h}(\eta)) + [\mathrm{Ad}_{g^{-1}h}(\eta),\xi]\Big), \qquad \text{(B.28)}
\end{aligned}$$

where we have used the equality $\mathscr{L}_{\xi_L} \mathrm{Ad}_{g^{-1}}(\eta) = [\mathrm{Ad}_{g^{-1}}(\eta), \xi]$; see Exercise E5.19.

Next, we establish the bounds required in Theorem 11.19. First, note that the Rayleigh dissipation function is uniformly bounded from above and below because it is left-invariant and has constant norm. Second, let us show that $\overset{\mathbb{G}_{\mathbb{I}}}{\nabla}\mathcal{T}_r(g,h)$ and $\mathrm{Hess}_1 \Psi_{\mathsf{G},r}(g,h)$ are uniformly bounded from above on $\Psi_{\mathsf{G},r}^{-1}(\leq L)$. Note that $(g,h) \in \Psi_{\mathsf{G},r}^{-1}(\leq L)$ if and only if $h^{-1}g \in \Psi_{\mathsf{G}}^{-1}(\leq L)$, and recall that Ψ_{G} is proper. Smooth functions and their derivatives are uniformly bounded on a compact set. Therefore, it suffices to show that the norms of $\overset{\mathbb{G}_{\mathbb{I}}}{\nabla}\mathcal{T}_r(g,h)$ and $\mathrm{Hess}_1 \Psi_{\mathsf{G},r}(g,h)$ depend on h and g through their combined composition $h^{-1}g$. Regarding the covariant derivative of the transport map, for $\xi, \eta \in \mathfrak{g}$, we write

$$\left\|\overset{\mathbb{G}_{\mathbb{I}}}{\nabla}\mathcal{T}_r(g,h) \cdot (T_eL_g(\xi), T_eL_h(\eta))\right\|_{\mathbb{G}_{\mathbb{I}}} = \left\|\overset{\mathfrak{g}}{\nabla}_\xi \left(\mathrm{Ad}_{g^{-1}h}(\eta)\right) + [\mathrm{Ad}_{g^{-1}h}(\eta), \xi]\right\|_{\mathbb{I}}.$$

Regarding the Hessian of the tracking error function, from Lemma 11.28(i) we derive

$$\mathrm{grad}_1 \Psi_{\mathsf{G},r}(g,h) = T_eL_h(\mathrm{grad}\Psi_{\mathsf{G}}(h^{-1}g)),$$

where $\mathrm{grad}_1 \Psi_{\mathsf{G},r} = \mathbb{G}_{\mathbb{I}}^\sharp(\mathrm{d}_1\Psi_{\mathsf{G},r})$. From Definition 6.1.4 we compute, at fixed $h \in \mathsf{G}$,

$$\begin{aligned}
&\mathrm{Hess}_1 \Psi_{\mathsf{G},r}(g,h)\cdot(T_eL_g(\xi), T_eL_g(\eta)) \\
&\quad = \langle\langle T_eL_g(\xi), \overset{\mathbb{G}_{\mathbb{I}}}{\nabla}_{T_eL_g(\eta)} \mathrm{grad}_1 \Psi_{\mathsf{G},r}(g,h)\rangle\rangle \\
&\quad = \langle\langle T_eL_g(\xi), \overset{\mathbb{G}_{\mathbb{I}}}{\nabla}_{T_eL_g(\eta)} T_eL_h(\mathrm{grad}\Psi_{\mathsf{G}}(h^{-1}g))\rangle\rangle \\
&\quad = \langle\langle T_eL_{h^{-1}g}(\xi), \overset{\mathbb{G}_{\mathbb{I}}}{\nabla}_{T_eL_{h^{-1}g}(\eta)} \mathrm{grad}\Psi_{\mathsf{G}}(h^{-1}g)\rangle\rangle \\
&\quad = \mathrm{Hess}\,\Psi_{\mathsf{G}}(h^{-1}g) \cdot (T_eL_{h^{-1}g}(\xi), T_eL_{h^{-1}g}(\eta)),
\end{aligned}$$

and, therefore,

$$\|\mathrm{Hess}_1 \Psi_{\mathsf{G},r}(g,h)\|_{\mathbb{G}_{\mathbb{I}}} = \|\mathrm{Hess}\,\Psi_{\mathsf{G}}(h^{-1}g)\|_{\mathbb{G}_{\mathbb{I}}}.$$

We have now verified that our control design satisfies all assumptions of Theorem 11.19. Accordingly, we can state the corresponding stability and convergence results under appropriate feedforward and feedback controls. The statement on the closed-loop energy is straightforward because of the equality

$$\begin{aligned}
\Psi_{\mathsf{G},r}(g, \gamma_{\mathrm{ref}}(t)) &+ \tfrac{1}{2}\left\|T_eL_g\left(\xi - \mathrm{Ad}_{g^{-1}\gamma_{\mathrm{ref}}(t)}(v_{\mathrm{ref}}(t))\right)\right\|_{\mathbb{G}_{\mathbb{I}}}^2 \\
&= \Psi_{\mathsf{G}}(\gamma_{\mathrm{ref}}^{-1}(t)g) + \tfrac{1}{2}\left\|\xi - \mathrm{Ad}_{g^{-1}\gamma_{\mathrm{ref}}(t)}(v_{\mathrm{ref}}(t))\right\|_{\mathbb{I}}^2.
\end{aligned}$$

Also, note that asking for $\gamma_{\mathrm{ref}}(t_0)^{-1}\gamma(t_0)$ to belong to the sublevel set $\Psi_{\mathsf{G}}^{-1}(\leq L, e)$ is equivalent to asking for $\gamma(t_0)$ to belong to the sublevel set

$(\Psi_{\mathsf{G},r})^{-1}_{\gamma_{\mathrm{ref}}(t_0)}(\leq L, \gamma_{\mathrm{ref}}(t_0))$. Next, we claim that the control laws presented above equal the ones in Theorem 11.19 in the sense that

$$\begin{aligned} f_{\mathrm{PD}}(t,g,\xi) &= (T_eL_g)^*(F_{\mathrm{PD}}(t,T_eL_g(\xi))), \\ f_{\mathrm{FF}}(t,g,\xi) &= (T_eL_g)^*(F_{\mathrm{FF}}(t,T_eL_g(\xi))). \end{aligned}$$

Regarding the proportional control we compute

$$\begin{aligned} (T_eL_g)^*(F_{\mathrm{P}}(t,g,\xi)) &= (T_eL_g)^*(-\mathrm{d}_1\Psi_{\mathsf{G},r}(g,\gamma_{\mathrm{ref}}(t))) \\ &= -(T_eL_g)^*\Big((T_gL_{\gamma_{\mathrm{ref}}^{-1}(t)})^*(\mathrm{d}\Psi_{\mathsf{G}}(\gamma_{\mathrm{ref}}^{-1}(t)g))\Big) \\ &= -(T_eL_{\gamma_{\mathrm{ref}}^{-1}(t)g})^*(\mathrm{d}\Psi_{\mathsf{G}}(\gamma_{\mathrm{ref}}^{-1}(t)g)). \end{aligned}$$

The derivative control law is a direct consequence of the expression for the velocity error in Lemma 11.28(iii) and of the left invariance of the Rayleigh dissipation function. Regarding the feedforward control, we use Theorem 11.19 to write

$$\begin{aligned} (T_eL_g)^*(F_{\mathrm{FF}}(t,g,\xi)) = (T_eL_g)^*\mathbb{I}^\flat\Big(\overset{\mathbb{G}_{\mathbb{I}}}{\nabla}\mathcal{T}_r(g,\gamma_{\mathrm{ref}}(t))\cdot(T_eL_g(\xi),\gamma'_{\mathrm{ref}}(t)) \\ + \frac{\mathrm{d}}{\mathrm{d}t}\big(\mathcal{T}_r(g,\gamma_{\mathrm{ref}}(t))\cdot\gamma'_{\mathrm{ref}}(t)\big)\Big). \end{aligned}$$

From equation (B.28) we have

$$\begin{aligned} &\overset{\mathbb{G}_{\mathbb{I}}}{\nabla}\mathcal{T}_r(g,\gamma_{\mathrm{ref}}(t))\cdot(T_eL_g(\xi),\gamma'_{\mathrm{ref}}(t)) \\ &\qquad = T_eL_g\Big(\overset{\mathfrak{g}}{\nabla}_\xi\left(\mathrm{Ad}_{g^{-1}\gamma_{\mathrm{ref}}(t)}(v_{\mathrm{ref}}(t))\right) + [\mathrm{Ad}_{g^{-1}\gamma_{\mathrm{ref}}(t)}(v_{\mathrm{ref}}(t)),\xi]\Big). \end{aligned}$$

From the equality $\mathscr{L}_{\xi_{\mathrm{L}}}\mathrm{Ad}_g(\eta) = \mathrm{Ad}_g[\xi,\eta]$ (see Exercise E5.19), we compute

$$\begin{aligned} \frac{\mathrm{d}}{\mathrm{d}t}\left(\mathcal{T}_r(g,\gamma_{\mathrm{ref}}(t))\cdot\gamma'_{\mathrm{ref}}(t)\right) &= \frac{\mathrm{d}}{\mathrm{d}t}T_eL_g\circ\mathrm{Ad}_{g^{-1}\gamma_{\mathrm{ref}}(t)}(v_{\mathrm{ref}}(t)) \\ &= T_eL_g\circ\mathrm{Ad}_{g^{-1}}\circ\frac{\mathrm{d}}{\mathrm{d}t}\mathrm{Ad}_{\gamma_{\mathrm{ref}}(t)}(v_{\mathrm{ref}}(t)) \\ &= T_eL_g\circ\mathrm{Ad}_{g^{-1}\gamma_{\mathrm{ref}}(t)}(v'_{\mathrm{ref}}(t)). \end{aligned}$$

The final expression follows from noting that $(T_eL_g)^*\mathbb{I}^\flat = \mathbb{I}^\flat\circ T_eL_{g^{-1}}$.

B.7 Proof of Proposition 12.9

First, we claim that it is possible to choose maps $(t,q)\mapsto\eta^a(t,q),\eta^{bc}(t,q)$, $a,b,c\in\{1,\dots,m\}$, $b<c$, that are of class C^1 and for which, for $t\in\mathbb{R}$, the maps $q\mapsto\eta^a(t,q),\eta^{bc}(t,q)$, $a,b,c\in\{1,\dots,m\}$, $b<c$, are of class C^∞. These

facts are true because of Assumption 12.8(ii); see Theorem 3.114 and Exercise E3.42. Accordingly, the vector fields $W_{X,\mathrm{slow}}$ and $W_{X,\mathrm{osc}}$ satisfy the stated smoothness requirements.

Next, let us prove that equality (12.4) is satisfied. From Lemma 12.4 and from the equality in Exercise E12.3, we compute

$$\begin{aligned} Q_{\mathcal{A}_{[0,T]}}(u^a_{X,\mathrm{osc}}Y_a) = &- \sum_{a,b=1}^{m} \left\langle\!\left\langle u^a_{X,\mathrm{osc}}, u^b_{X,\mathrm{osc}} \right\rangle\!\right\rangle_{L_{\mathcal{A}}([0,T])} \langle Y_a : Y_b \rangle \\ &- 2\sum_{a,b=1}^{m} \left\langle\!\left\langle u^a_{X,\mathrm{osc}}, \mathscr{L}_{Y_a} u^b_{X,\mathrm{osc}} \right\rangle\!\right\rangle_{L_{\mathcal{A}}([0,T])} Y_b. \end{aligned} \tag{B.29}$$

From the orthonormality of $\{\varphi_i\}_{i\in\mathbb{N}}$, it is straightforward to compute

$$\left\langle\!\left\langle u^a_{X,\mathrm{osc}}, u^a_{X,\mathrm{osc}} \right\rangle\!\right\rangle_{L_{\mathcal{A}}([0,T])} = a - 1 + \sum_{i=a+1}^{m} \frac{(\eta^{ai}(t,q))^2}{4}.$$

Clearly, we have $\left\langle\!\left\langle u^a_{X,\mathrm{osc}}, u^b_{X,\mathrm{osc}} \right\rangle\!\right\rangle_{L_{\mathcal{A}}([0,T])} = \left\langle\!\left\langle u^b_{X,\mathrm{osc}}, u^a_{X,\mathrm{osc}} \right\rangle\!\right\rangle_{L_{\mathcal{A}}([0,T])}$. For $a < b$, we compute

$$\begin{aligned} \left\langle\!\left\langle u^a_{X,\mathrm{osc}}, u^b_{X,\mathrm{osc}} \right\rangle\!\right\rangle_{L_{\mathcal{A}}([0,T])} &= \sum_{i=1}^{a-1}\sum_{j=1}^{b-1} \left\langle\!\left\langle \varphi_{\mathrm{lo}(i,a)}, \varphi_{\mathrm{lo}(j,b)} \right\rangle\!\right\rangle_{L_{\mathcal{A}}([0,T])} \\ &\quad - \frac{1}{2}\sum_{i=1}^{a-1}\sum_{j=b+1}^{m} \eta^{bj} \left\langle\!\left\langle \varphi_{\mathrm{lo}(i,a)}, \varphi_{\mathrm{lo}(b,j)} \right\rangle\!\right\rangle_{L_{\mathcal{A}}([0,T])} \\ &\quad - \frac{1}{2}\sum_{i=a+1}^{m}\sum_{j=1}^{b-1} \eta^{ai} \left\langle\!\left\langle \varphi_{\mathrm{lo}(a,i)}, \varphi_{\mathrm{lo}(j,b)} \right\rangle\!\right\rangle_{L_{\mathcal{A}}([0,T])} \\ &\quad + \frac{1}{4}\sum_{i=a+1}^{m}\sum_{j=b+1}^{m} \eta^{ai}\eta^{bj} \left\langle\!\left\langle \varphi_{\mathrm{lo}(a,i)}, \varphi_{\mathrm{lo}(b,j)} \right\rangle\!\right\rangle_{L_{\mathcal{A}}([0,T])} \\ &= \sum_{i=1}^{a-1}\sum_{j=1}^{b-1} \delta_{ij}\delta_{ab} - \frac{1}{2}\sum_{i=1}^{a-1}\sum_{j=b+1}^{m} \eta^{bj}\delta_{ib}\delta_{aj} \\ &\quad - \frac{1}{2}\sum_{i=a+1}^{m}\sum_{j=1}^{b-1} \eta^{ai}\delta_{aj}\delta_{ib} + \frac{1}{4}\sum_{i=a+1}^{m}\sum_{j=b+1}^{m} \eta^{ai}\eta^{bj}\delta_{ab}\delta_{ij}. \end{aligned}$$

Because $\delta_{ab} = 0$ for $a < b$, we obtain

$$\left\langle\!\left\langle u^a_{X,\mathrm{osc}}, u^b_{X,\mathrm{osc}} \right\rangle\!\right\rangle_{L_{\mathcal{A}}([0,T])} = -\frac{1}{2}\sum_{i=1}^{a-1}\sum_{j=b+1}^{m} \eta^{bj}\delta_{ib}\delta_{aj} - \frac{1}{2}\sum_{i=a+1}^{m}\sum_{j=1}^{b-1} \eta^{ai}\delta_{aj}\delta_{ib}.$$

Because $i < a < b$, we know that $\delta_{ib} = 0$. Therefore, the first term vanishes. The only nonvanishing term corresponds to $i = b$ and $a = j$ and leads to

$$\langle\!\langle u^a_{X,\mathrm{osc}}, u^b_{X,\mathrm{osc}} \rangle\!\rangle_{L_{\mathcal{A}}([0,T])} = -\tfrac{1}{2}\eta^{ab}.$$

Next, we note that $\mathscr{L}_{Y_a} u^b_{X,\mathrm{osc}} = -\frac{1}{2}\sum_{j=b+1}^{m} \mathscr{L}_{Y_a}\eta^{bj}\varphi_{\mathrm{lo}(b,j)}$, and we compute

$$\begin{aligned}
&\langle\!\langle u^a_{X,\mathrm{osc}}, \mathscr{L}_{Y_a} u^a_{X,\mathrm{osc}} \rangle\!\rangle_{L_{\mathcal{A}}([0,T])} \\
&\quad = -\frac{1}{2}\sum_{j=a+1}^{m} \mathscr{L}_{Y_a}\eta^{aj}\Big(\sum_{i=1}^{a-1} \langle\!\langle \varphi_{\mathrm{lo}(i,a)}, \varphi_{\mathrm{lo}(a,j)} \rangle\!\rangle_{L_{\mathcal{A}}([0,T])} \\
&\qquad - \frac{1}{2}\sum_{i=a+1}^{m} \eta^{ai} \langle\!\langle \varphi_{\mathrm{lo}(a,i)}, \varphi_{\mathrm{lo}(a,j)} \rangle\!\rangle_{L_{\mathcal{A}}([0,T])} \Big) \\
&\quad = \frac{1}{4}\sum_{j=a+1}^{m} \mathscr{L}_{Y_a}\eta^{aj}\Big(\sum_{i=a+1}^{m} \eta^{ai}\delta_{ij} \Big) = \frac{1}{4}\sum_{j=a+1}^{m} \eta^{aj}\left(\mathscr{L}_{Y_a}\eta^{aj}\right).
\end{aligned}$$

For $a \neq b$, some more bookkeeping leads to

$$\begin{aligned}
\langle\!\langle u^a_{X,\mathrm{osc}}, \mathscr{L}_{Y_a} u^b_{X,\mathrm{osc}} \rangle\!\rangle_{L_{\mathcal{A}}([0,T])} &= -\frac{1}{2}\sum_{i=1}^{a-1}\sum_{j=b+1}^{m} \left(\mathscr{L}_{Y_a}\eta^{bj}\right)\delta_{ib}\delta_{aj} \\
&\quad + \frac{1}{4}\sum_{i=a+1}^{m}\sum_{j=b+1}^{m} \eta^{ai}\left(\mathscr{L}_{Y_a}\eta^{bj}\right)\delta_{ab}\delta_{ij},
\end{aligned}$$

where the second term clearly vanishes. If $a < b$, then also $i < a < b < j$ and the first term vanishes as well. If instead $a > b$, then there is one term for which $i = b$ and $a = j$. Thus

$$\langle\!\langle u^a_{X,\mathrm{osc}}, \mathscr{L}_{Y_a} u^b_{X,\mathrm{osc}} \rangle\!\rangle_{L_{\mathcal{A}}([0,T])} = \begin{cases} 0, & a < b, \\ -\frac{1}{2}\left(\mathscr{L}_{Y_a}\eta^{ba}\right), & a > b. \end{cases}$$

Finally, we substitute into (B.29) the intermediate results to obtain

$$\begin{aligned}
&Q_{\mathcal{A}_{[0,T]}}(u^a_{X,\mathrm{osc}} Y_a) \\
&\quad = -\sum_{a=1}^{m}\Big(a - 1 + \sum_{i=a+1}^{m} \frac{(\eta^{ai})^2}{4} \Big)\langle Y_a : Y_a \rangle + \frac{1}{2}\sum_{\substack{a,b=1\\ a<b}}^{m} \eta^{ab}\langle Y_a : Y_b \rangle \\
&\qquad - \frac{1}{2}\sum_{a=1}^{m}\Big(\sum_{j=a+1}^{m} \eta^{aj}(\mathscr{L}_{Y_a}\eta^{aj}) \Big) Y_a + \sum_{\substack{a,b=1\\ a>b}}^{m} (\mathscr{L}_{Y_a}\eta^{ba}) Y_b \\
&\quad = \sum_{\substack{a,b=1\\ a<b}}^{m} \eta^{ab}\langle Y_a : Y_b \rangle - \sum_{a=1}^{m}\Big(\sum_{b=1}^{m}\Big(b - 1 + \sum_{i=b+1}^{m} \frac{(\eta^{bi})^2}{4} \Big)\sigma^a_b \\
&\qquad + \sum_{b=a+1}^{m} \Big(\tfrac{1}{2}\eta^{ab}(\mathscr{L}_{Y_a}\eta^{ab}) - \mathscr{L}_{Y_b}\eta^{ab} \Big)\Big) Y_a.
\end{aligned}$$

Equality (12.4) follows from the definition of $u^a_{X,\mathrm{slow}}$.

References

Abraham, R. and Marsden, J. E. [1967] *Foundations of Mechanics*, W. A. Benjamin, Inc., New York–Amsterdam, second edition: [Abraham and Marsden 1978].

— [1978] *Foundations of Mechanics*, second edition, Addison Wesley, Reading, MA, ISBN 0-8053-0102-X.

Abraham, R., Marsden, J. E., and Ratiu, T. S. [1988] *Manifolds, Tensor Analysis, and Applications*, second edition, number 75 in Applied Mathematical Sciences, Springer-Verlag, New York–Heidelberg–Berlin, ISBN 0-387-96790-7.

Agrachev, A. A. [1990a] *Quadratic mappings in geometric control theory*, Journal of Soviet Mathematics, **51**(6), 2667–2734.

— [1990b] *The topology of quadratic mappings and Hessians of smooth mappings*, Journal of Soviet Mathematics, **49**(3), 990–1013.

— [1999] *Is it possible to recognize local controllability in a finite number of differentiations?*, in *Open Problems in Mathematical Systems and Control Theory*, pages 15–18, Communications and Control Engineering Series, Springer-Verlag, New York–Heidelberg–Berlin, ISBN 1-85233-044-9.

Agrachev, A. A. and Gamkrelidze, R. V. [1978] *The exponential representation of flows and the chronological calculus*, Mathematics of the USSR-Sbornik, **107**(4), 467–532.

— [1991] *Quadratic mappings and vector functions: Euler characteristics of level sets*, Journal of Soviet Mathematics, **55**(4), 1892–1928.

— [1993a] *Local controllability and semigroups of diffeomorphisms*, Acta Applicandae Mathematicae. An International Journal on Applying Mathematics and Mathematical Applications, **32**(1), 1–57.

— [1993b] *Local controllability for families of diffeomorphisms*, Systems & Control Letters, **20**(1), 67–76.

Agrachev, A. A. and Sachkov, Y. [2004] *Control Theory from the Geometric Viewpoint*, volume 87 of *Encyclopaedia of Mathematical Sciences*, Springer-Verlag, New York–Heidelberg–Berlin, ISBN 3-540-21019-9.

Andrews, G. E. [1971] *Number Theory*, W. B. Saunders Co., Philadelphia–London–Toronto, reprint: [Andrews 1994].

— [1994] *Number Theory*, Dover Publications, Inc., New York, ISBN 0-486-68252-8, reprint of 1971 W. B. Saunders edition.

Aneke, N. P. I., Nijmeijer, H., and de Jager, A. G. [2001] *Trajectory tracking by cascaded backstepping control for a second-order nonholonomic mechanical system*, in *Nonlinear Control in the Year 2000, Volume 1 (Paris, France)*, pages 35–47, number 258 in Lecture Notes in Control and Information Sciences, Springer-Verlag, New York–Heidelberg–Berlin, ISBN 1-85233-363-4.

Appell, P. [1900a] *Développements sur une forme nouvelle des équations de la Dynamique*, Journal de Mathématiques Pures et Appliquées. Neuvième Sér, **6**, 5–40.

— [1900b] *Sur une forme générale des équations de la Dynamique*, Journal für die Reine und Angewandte Mathematik, **121**, 310–319.

Arai, H., Tanie, K., and Shiroma, N. [1998] *Nonholonomic control of a three-DOF planar underactuated manipulator*, IEEE Transactions on Robotics and Automation, **14**(5), 681–695.

Arimoto, S. [1996] *Control Theory of Non-linear Mechanical Systems: A Passivity-Based and Circuit-Theoretic Approach*, number 49 in Oxford Engineering Science Series, Oxford University Press, Walton Street, Oxford, ISBN 0-19-856291-8.

Arimoto, S. and Miyazaki, F. [1984] *Stability and robustness of PID feedback control for robot manipulators of sensory capability*, in *Robotics Research: First International Symposium*, M. Brady and R. Paul, editors, pages 783–799, MIT Press, Cambridge, MA, ISBN 0-262-52392-2.

Arnol'd, V. I. [1966] *Sur la géométrie différentielle des groupes de Lie de dimension infinie et ses applications à l'hydrodynamique des fluides parfaits*, Université de Grenoble. Annales de l'Institut Fourier, **16**(1), 319–361.

— [1978] *Mathematical Methods of Classical Mechanics*, first edition, number 60 in Graduate Texts in Mathematics, Springer-Verlag, New York–Heidelberg–Berlin, ISBN 0-387-90314-3, second edition: [Arnol'd 1989].

— [1988] *Dynamical Systems*, volume 3 of *Encyclopedia of Mathematical Sciences*, Springer-Verlag, New York–Heidelberg–Berlin, ISBN 0-387-57241-4.

— [1989] *Mathematical Methods of Classical Mechanics*, second edition, number 60 in Graduate Texts in Mathematics, Springer-Verlag, New York–Heidelberg–Berlin, ISBN 0-387-96890-3.

— [1992] *Ordinary Differential Equations*, Springer Textbook, Springer-Verlag, New York–Heidelberg–Berlin, ISBN 3-540-54813-0, translation of the third Russian edition by R. Cooke.

Arnol'd, V. I. and Avez, A. [1989] *Ergodic Problems in Classical Mechanics*, Addison Wesley, Reading, MA, ISBN 0-201-09496-1.

Arnol'd, V. I. and Khesin, B. A. [1998] *Topological Methods in Hydrodynamics*, number 125 in Applied Mathematical Sciences, Springer-Verlag, New York–Heidelberg–Berlin, ISBN 0-387-94947-X.

Artstein, Z. [1983] *Stabilization with relaxed controls*, Nonlinear Analysis. Theory, Methods, and Applications, **7**(11), 1163–1173.

Astolfi, A. [1996] *Discontinuous control of nonholonomic systems*, Systems & Control Letters, **27**(1), 37–45.

Auckly, D. R. and Kapitanski, L. V. [2002] *On the λ-equations for matching control laws*, SIAM Journal on Control and Optimization, **41**(5), 1372–1388.

Auckly, D. R., Kapitanski, L. V., and White, W. [2000] *Control of nonlinear underactuated systems*, Communications on Pure and Applied Mathematics, **53**(3), 354–369.

Bacciotti, A. [1992] *Local Stabilizability of Nonlinear Control Systems*, number 8 in Series on Advances in Mathematics for Applied Sciences, World Scientific, Singapore–New Jersey–London–Hong Kong, ISBN 981-02-0713-1.

Baillieul, J. [1993] *Stable average motions of mechanical systems subject to periodic forcing*, in *Dynamics and Control of Mechanical Systems (Waterloo, Canada)*, M. J. Enos, editor, volume 1, pages 1–23, Fields Institute, Waterloo, Canada, ISBN 0-821-89200-2.

— [1995] *Energy methods for stability of bilinear systems with oscillatory inputs*, International Journal of Robust and Nonlinear Control, **5**(4), 285–301.

— [1998] *The geometry of controlled mechanical systems*, in *Mathematical Control Theory*, J. Baillieul and J. C. Willems, editors, pages 322–354, Springer-Verlag, New York–Heidelberg–Berlin, ISBN 0-387-98317-1.

Baillieul, J. and Lehman, B. [1996] *Open-loop control using oscillatory inputs*, in *The Control Handbook*, W. S. Levine, editor, pages 967–980, CRC Press, Boca Raton, FL, ISBN 0-8493-8570-9.

Barbashin, E. A. and Krasovskiĭ, N. N. [1952] *On global stability of motion*, Doklady Akademii Nauk SSSR, **86**(3), 453–456.

Barclay, A., Gill, P. E., and Rosen, J. B. [1998] *SQP methods and their application to numerical optimal control*, in *Variational Calculus, Optimal Control and Applications (Trassenheide, Germany)*, W. H. Schmidt, K. Heier, L. Bittner, and R. Bulirsch, editors, volume 124 of *International Series of Numerical Mathematics*, pages 207–222, Birkhäuser, Boston–Basel–Stuttgart, ISBN 3-7643-5906-4.

Başar, T., (editor) [2001] *Control Theory: Twenty-Five Seminal Papers*, John Wiley and Sons, New York, ISBN 0-780-3602-14.

Basto-Gonçalves, J. [1998] *Second-order conditions for local controllability*, Systems & Control Letters, **35**(5), 287–290.

Bates, L. M. and Śniatycki, J. Z. [1993] *Nonholonomic reduction*, Reports on Mathematical Physics, **32**(1), 444–452.

Bedrossian, N. S. and Spong, M. W. [1995] *Feedback linearization of robot manipulators and Riemannian curvature*, Journal of Robotic Systems, **12**(8), 541–552.

Bellman, R. E., Bentsman, J., and Meerkov, S. M. [1986] *Vibrational control of nonlinear systems: Vibrational stabilization*, IEEE Transactions on Automatic Control, **31**(8), 710–716.

Bentsman, J. [1987] *Vibrational control of a class of nonlinear systems by nonlinear multiplicative vibrations*, IEEE Transactions on Automatic Control, **32**(8), 711–716.

Bertsekas, D. P. [2000] *Dynamic Programming and Optimal Control*, volume 1, second edition, Athena Scientific, Nashua, NH, ISBN 1-886529-09-4.

Betts, J. T. [1998] *Survey of numerical methods for trajectory optimization*, AIAA Journal of Guidance, Control, and Dynamics, **21**(2), 193–207.

Bianchini, R. M. and Kawski, M. [2003] *Needle variations that cannot be summed*, SIAM Journal on Control and Optimization, **42**(1), 218–238.

Bianchini, R. M. and Stefani, G. [1993] *Controllability along a trajectory: A variational approach*, SIAM Journal on Control and Optimization, **31**(4), 900–927.

Bicchi, A., Marigo, A., and Piccoli, B. [2002] *On the reachability of quantized control systems*, IEEE Transactions on Automatic Control, **47**(4), 546–563.

Bierstone, E. and Milman, P. D. [1988] *Semianalytic and subanalytic sets*, Institut des Hautes Études Scientifiques. Publications Mathématiques, **67**, 5–42.

Birkhoff, G. D. [1927] *Dynamical Systems*, American Mathematical Society, Providence, RI, revised edition: [Birkhoff 1966].

— [1966] *Dynamical Systems*, American Mathematical Society, Providence, RI, revision of 1927 edition.

Blankenstein, G., Ortega, R., and van der Schaft, A. J. [2002] *The matching conditions of controlled Lagrangians and IDA-passivity based control*, International Journal of Control, **75**(9), 645–665.

Bliss, G. A. [1946] *Lectures on the Calculus of Variations*, Phoenix Science Series, University of Chicago Press, Chicago, IL.

Bloch, A. M. [2003] *Nonholonomic Mechanics and Control*, volume 24 of *Interdisciplinary Applied Mathematics*, Springer-Verlag, New York–Heidelberg–Berlin, ISBN 0-387-095535-6.

Bloch, A. M., Chang, D. E., Leonard, N. E., and Marsden, J. E. [2001] *Controlled Lagrangians and the stabilization of mechanical systems. II. Potential shaping*, IEEE Transactions on Automatic Control, **46**(10), 1556–1571.

Bloch, A. M. and Crouch, P. E. [1992] *Kinematics and dynamics of nonholonomic control systems on Riemannian manifolds*, in *Proceedings of the 32nd IEEE Conference on Decision and Control*, pages 1–5, Tucson, AZ.

— [1995] *Nonholonomic and vakonomic control systems on Riemannian manifolds*, SIAM Journal on Control and Optimization, **33**(1), 126–148.

— [1998] *Newton's law and integrability of nonholonomic systems*, SIAM Journal on Control and Optimization, **36**(6), 2020–2039.

Bloch, A. M., Krishnaprasad, P. S., Marsden, J. E., and Murray, R. M. [1996] *Nonholonomic mechanical systems with symmetry*, Archive for Rational Mechanics and Analysis, **136**(1), 21–99.

Bloch, A. M., Krishnaprasad, P. S., Marsden, J. E., and Ratiu, T. S. [1993] *Dissipation induced instabilities*, Annales de l'Institut Henri Poincaré. Analyse Non Linéaire, **11**(1), 37–90.

Bloch, A. M., Krishnaprasad, P. S., Marsden, J. E., and Sánchez de Alvarez, G. [1992] *Stabilization of rigid body dynamics by internal and external torques*, Automatica. The Journal of IFAC, **28**(4), 745–756.

Bloch, A. M., Leonard, N. E., and Marsden, J. E. [2000] *Controlled Lagrangians and the stabilization of mechanical systems. I. The first matching theorem*, IEEE Transactions on Automatic Control, **45**(12), 2253–2270.

Bloch, A. M., Reyhanoglu, M., and McClamroch, N. H. [1992] *Control and stabilization of nonholonomic dynamic systems*, IEEE Transactions on Automatic Control, **37**(11), 1746–1757.

Bode, H. W. [1945] *Network Analysis and Feedback Amplifier Design*, Van Nostrand Reinhold Co., London.

Bogoliubov, N. N. and Mitropolsky, Y. A. [1961] *Asymptotic Methods in the Theory of Non-Linear Oscillations*, Hindustan Publishing Corp., Delhi, India, translation of second Russian edition.

Bolza, O. [1961] *Lectures on the Calculus of Variations*, second edition, Chelsea, New York.

Bonnard, B. [1984] *Controllabilité de systèmes mécaniques sur les groupes de Lie*, SIAM Journal on Control and Optimization, **22**(5), 711–722.

Boothby, W. M. [1986] *An Introduction to Differentiable Manifolds and Riemannian Geometry*, second edition, Academic Press, New York, ISBN 0-121-16053-X.

Boyd, S. and Vandenberghe, L. [2004] *Convex Optimization*, Cambridge University Press, New York–Port Chester–Melbourne–Sydney, ISBN 0-521-83378-7.

Brockett, R. W. [1970] *Finite Dimensional Linear Systems*, John Wiley and Sons, New York.

— [1972] *System theory on group manifolds and coset spaces*, Journal of SIAM. Series A. Control, **10**(2), 265–284.

— [1976] *Volterra series and geometric control theory*, Automatica. The Journal of IFAC, **12**(2), 167–176.

— [1977] *Control theory and analytical mechanics*, in *The 1976 Ames Research Center (NASA) Conference on Geometric Control Theory (Moffett Field, CA)*, C. Martin and R. Hermann, editors, pages 1–48, Math Sci Press, Brookline, MA, ISBN 0-915692-721-X.

— [1978] *Feedback invariants for nonlinear systems*, in *Proceedings of the 1978 IFAC World Congress*, pages 1115–1120, Helsinki, Finland.

— [1982] *Control theory and singular Riemannian geometry*, in *New Directions in Applied Mathematics (Cleveland, OH)*, P. Hilton and G. Young, editors, pages 11–27, Springer-Verlag, New York–Heidelberg–Berlin, ISBN 0-387-90604-5.

— [1983] *Asymptotic stability and feedback stabilization*, in *Differential Geometric Control Theory (Houghton, MI)*, R. W. Brockett, R. S. Millman, and H. J. Sussmann, editors, pages 181–191, number 27 in Progress in Mathematics, Birkhäuser, Boston–Basel–Stuttgart, ISBN 3-7643-3091-0.

— [1990] *Some mathematical aspects of robotics*, in *Robotics*, R. W. Brockett, editor, pages 1–19, number 41 in Proceedings of Symposia in Applied Mathematics, American Mathematical Society, Providence, RI, ISBN 0-8218-0163-5.

— [1993] *Hybrid models for motion control systems*, in *Essays in Control: Perspectives in the Theory and its Applications*, pages 29–53, number 14 in Progress in Systems and Control Theory, Birkhäuser, Boston–Basel–Stuttgart, ISBN 0-8176-3670-6.

Brockett, R. W. and Dai, L. [1993] *Non-holonomic kinematics and the role of elliptic functions in constructive controllability*, in *Nonholonomic Motion Planning*, Z. Li and J. F. Canny, editors, pages 1–22, Kluwer Academic Publishers, Dordrecht, ISBN 0-7923-9275-2.

Bryant, R. L., Chern, S. S., Gardner, R. B., Goldschmidt, H. L., and Griffiths, P. A. [1991] *Exterior Differential Systems*, number 18 in Mathematical Sciences Research Institute Publications, Springer-Verlag, New York–Heidelberg–Berlin, ISBN 0-387-97411-3.

Bullo, F. [1998] *Nonlinear Control of Mechanical Systems: A Riemannian Geometry Approach*, Ph.D. thesis, California Institute of Technology, Pasadena, CA.

— [2000] *Stabilization of relative equilibria for underactuated systems on Riemannian manifolds*, Automatica. The Journal of IFAC, **36**(12), 1819–1834.

— [2001] *Series expansions for the evolution of mechanical control systems*, SIAM Journal on Control and Optimization, **40**(1), 166–190.

— [2002] *Averaging and vibrational control of mechanical systems*, SIAM Journal on Control and Optimization, **41**(2), 542–562.

Bullo, F., Cortés, J., Lewis, A. D., and Martínez, S. [2004] *Vector-valued quadratic forms in control theory*, in *Sixty Open Problems in Mathematical Systems and Control Theory*, V. Blondel and A. Megretski, editors, pages 315–320, Princeton University Press, Princeton, NJ, ISBN 0-691-11748-9.

Bullo, F., Leonard, N. E., and Lewis, A. D. [2000] *Controllability and motion algorithms for underactuated Lagrangian systems on Lie groups*, IEEE Transactions on Automatic Control, **45**(8), 1437–1454.

Bullo, F. and Lewis, A. D. [2002] *The linearisation of a simple mechanical control system*, Preprint.

— [2003a] *Kinematic controllability and motion planning for the snakeboard*, IEEE Transactions on Robotics and Automation, **19**(3), 494–498.

— [2003b] *Low-order controllability and kinematic reductions for affine connection control systems*, SIAM Journal on Control and Optimization, to appear.

Bullo, F. and Lynch, K. M. [2001] *Kinematic controllability and decoupled trajectory planning for underactuated mechanical systems*, IEEE Transactions on Robotics and Automation, **17**(4), 402–412.

Bullo, F. and Murray, R. M. [1999] *Tracking for fully actuated mechanical systems: A geometric framework*, Automatica. The Journal of IFAC, **35**(1), 17–34.

Bullo, F. and Žefran, M. [2002] *On mechanical systems with nonholonomic constraints and symmetries*, Systems & Control Letters, **45**(2), 133–143.

Byrnes, C. I., Isidori, A., and Willems, J. C. [1991] *Passivity, feedback equivalence, and global stabilization of minimum phase nonlinear systems*, IEEE Transactions on Automatic Control, **36**(11), 1228–1240.

Canny, J. F. [1988] *Complexity of Robot Motion Planning*, ACM Doctoral Dissertation Awards, MIT Press, Cambridge, MA, ISBN 0-262-03136-1.

Carathéodory, C. [1909] *Untersuchungen über die Grundlagen der Thermodynamik*, Mathematische Annalen, **67**, 355–386.

— [1935] *Variationsrechnung und partielle Differentialgleichungen erster Ordnung*, Teubner, Leipzig, translation: [Carathéodory 1982].

— [1982] *Calculus of Variations*, Chelsea, New York, ISBN 0-8284-0318-X, translation of *Variationsrechnung und Partielle Differentialgleichungen erster Ordnung*.

Cardin, F. and Favretti, M. [1996] *On nonholonomic and vakonomic dynamics of mechanical systems with nonintegrable constraints*, Journal of Geometry and Physics, **18**(4), 295–325.

Cattaneo, C. [1963] *Sulla struttura locale delle equazioni dinamiche di un sistema anolonomo*, Atti della Accademia Nazionale dei Lincei. Rendiconti. Classe di Scienze Fisiche, Matematiche e Naturali. Serie VIII, **34**, 396–402.

Cattaneo-Gasparini, I. [1963] *Dérivée covariante "liée" dans une V_{n-1} riemannienne à structure presque produit*, Comptes Rendus de l'Académie des Sciences. Série I. Mathématique, **256**, 2089–2091.

Čelikovsky, S. and Aranda-Bricaire, E. [1999] *Constructive nonsmooth stabilization of triangular systems*, Systems & Control Letters, **36**(1), 21–37.

Cendra, H., Marsden, J. E., and Ratiu, T. S. [2001a] *Geometric mechanics, Lagrangian reduction, and nonholonomic systems*, in *Mathematics Unlimited—2001 and Beyond*, B. Engquist and W. Schmid, editors, pages 221–273, Springer-Verlag, New York–Heidelberg–Berlin, ISBN 3-540-66913-2.

— [2001b] *Lagrangian reduction by stages*, Memoirs of the American Mathematical Society, **152**(722).

Cerven, W. T. and Bullo, F. [2003] *Constructive controllability algorithms for motion planning and optimization*, IEEE Transactions on Automatic Control, **48**(4), 575–589.

Chang, D. E., Bloch, A. M., Leonard, N. E., Marsden, J. E., and Woolsey, C. A. [2002] *The equivalence of controlled Lagrangian and controlled Hamiltonian systems*, ESAIM. Control, Optimization and Calculus of Variations, **8**, 393–422.

Chen, C.-T. [1984] *Linear System Theory and Design*, HRW Series in Electrical and Computer Engineering, Holt, Rinehart and Winston, New York–Chicago–San Francisco–Philadelphia, ISBN 0-03-060289-0.

Chen, K.-T. [1957] *Integration of paths, geometric invariants and a generalized Baker-Hausdorff formula*, Annals of Mathematics. Second Series, **65**, 163–178.

Chetaev, N. G. [1955] *Ustojchivost' dvizenija*, Gosudarstvennoe izdatelstvo tehniko-teoreticeskoj literatury, Moscow, translation: [Chetaev 1961].

— [1961] *The Stability of Motion*, Pergamon Press, Oxford–New York–Paris, translation from Russian by M. Nadler.

— [1987] *Teoreticheskaya mekhanika*, "Nauka", Moscow, translation: [Chetaev 1989].

— [1989] *Theoretical Mechanics*, Springer-Verlag, New York–Heidelberg–Berlin, ISBN 0-387-51379-5, translation from Russian by I. Aleksanova.

Chevalley, C. [1946] *Theory of Lie Groups*, number 8 in Princeton Mathematical Series, Princeton University Press, Princeton University Press, reprint: [Chevalley 1999].

— [1999] *Theory of Lie Groups*, Princeton Landmarks in Mathematics, Princeton University Press, Princeton, NJ, ISBN 0-691-08052-6, reprint of 1946 edition.

Chitour, Y. and Sussmann, H. J. [1998] *Line-integral estimates and motion planning using the continuation method*, in *Essays on Mathematical Robotics (Minneapolis,*

MN), volume 104 of *The IMA Volumes in Mathematics and its Applications*, pages 173–193, Springer-Verlag, New York–Heidelberg–Berlin, ISBN 0-387-98596-4.

Chorin, A. J. and Marsden, J. E. [1994] *A Mathematical Introduction to Fluid Mechanics*, third edition, number 4 in Texts in Applied Mathematics, Springer-Verlag, New York–Heidelberg–Berlin, ISBN 0-387-97918-2.

Chow, W.-L. [1939] *Über Systemen von linearen partiellen Differentialgleichungen erster Ordnung*, Mathematische Annalen, **117**, 98–105.

Christoffel, E. B. [1869] *Über die Tranformation der homogenen Differentialausdrücke*, Journal für die Reine und Angewandte Mathematik, **70**, 46–70.

Clairaut, A. C. [1754] *Mémoire sur l'orbite apparent du soleil autour de la Terre, an ayant égard aux perturbations produites par les actions de la Lune et des Planetes principales*, Mémoire de l'Academie des Sciences, Paris, pages 521–564.

Clarke, F. H., Ledyaev, Y. S., Sontag, E. D., and Subotin, A. I. [1997] *Asymptotic controllability implies feedback stabilization*, IEEE Transactions on Automatic Control, **42**(10), 1394–1407.

Clarke, F. H., Ledyaev, Y. S., Stern, R. J., and Wolenski, P. R. [1998] *Nonsmooth Analysis and Control Theory*, number 178 in Graduate Texts in Mathematics, Springer-Verlag, New York–Heidelberg–Berlin, ISBN 0-387-98336-8.

Coddington, E. E. and Levinson, N. [1955] *Theory of Ordinary Differential Equations*, McGraw-Hill, New York, revised edition: [Coddington and Levinson 1984].

— [1984] *Theory of Ordinary Differential Equations*, 8th edition, Robert E. Krieger Publishing Company, Huntington, NY, ISBN 0-898-74755-4.

Cohn, D. L. [1980] *Measure Theory*, Birkhäuser, Boston–Basel–Stuttgart, ISBN 0-8176-3003-1.

Coppel, W. A. [1965] *Stability and Asymptotic Behavior of Differential Equations*, D. C. Heath and Co., Boston, MA.

Coron, J.-M. [1990] *A necessary condition for feedback stabilization*, Systems & Control Letters, **14**(3), 227–232.

— [1992] *Global asymptotic stabilization for controllable systems without drift*, Mathematics of Control, Signals, and Systems, **5**(3), 295–312.

Coron, J.-M. and Praly, L. [1991] *Adding an integrator for the stabilization problem*, Systems & Control Letters, **17**(2), 89–104.

Coron, J.-M. and Rosier, L. [1994] *A relation between continuous time-varying and discontinuous feedback stabilization*, Journal of Mathematical Systems, Estimation, and Control, **4**(1), 67–84.

Cortés, J. [2002] *Geometric, Control and Numerical Aspects of Nonholonomic Systems*, number 1793 in Lecture Notes in Mathematics, Springer-Verlag, New York–Heidelberg–Berlin, ISBN 3-540-44154-9.

Cortés, J., de León, M., de Diego, D. M., and Martínez, S. [2001] *Mechanical systems subjected to generalized non-holonomic constraints*, Proceedings of the Royal Society. London. Series A. Mathematical and Physical Sciences, **457**(2007), 651–270.

— [2002] *Geometric description of vakonomic and nonholonomic dynamics. Comparison of solutions*, SIAM Journal on Control and Optimization, **41**(5), 1389–1412.

Cortés, J., Martínez, S., and Bullo, F. [2003] *On nonlinear controllability and series expansions for Lagrangian systems with dissipative forces*, IEEE Transactions on Automatic Control, **47**(8), 1396–1401.

Cortés, J., Martínez, S., Ostrowski, J. P., and Zhang, H. [2002] *Simple mechanical control systems with constraints and symmetry*, SIAM Journal on Control and Optimization, **41**(3), 851–874.

Cortés, J. and van der Schaft, A. J. [2003] *Characterization of gradient control systems*, SIAM Journal on Control and Optimization, submitted.

Craig, J. J. [1989] *Introduction to Robotics: Mechanics and Control*, second edition, Addison Wesley, Reading, MA, ISBN 0-201-09528-9.

Crampin, M. [1986] *On the concept of angular velocity*, European Journal of Physics, **7**(4), 287–293.

Crampin, M. and Pirani, F. A. E. [1986] *Applicable Differential Geometry*, number 59 in London Mathematical Society Lecture Note Series, Cambridge University Press, New York–Port Chester–Melbourne–Sydney, ISBN 0-521-23190-6.

Crouch, P. E. [1981] *Geometric structures in systems theory*, IEE Proceedings. D. Control Theory and Applications, **128**(5), 242–252.

— [1984] *Spacecraft attitude control and stabilization: Application of geometric control theory to rigid body models*, IEEE Transactions on Automatic Control, **29**(4), 321–331.

Crouch, P. E. and Byrnes, C. I. [1986] *Symmetries and local controllability*, in *Algebraic and Geometric Methods in Nonlinear Control Theory*, pages 55–75, number 29 in Mathematics and its Applications, D. Reidel Publishing Company, Dordrecht–Boston–Lancaster–Tokyo.

Crouch, P. E. and Silva Leite, F. [1991] *Geometry and the dynamic interpolation problem*, in *Proceedings of the 1991 American Control Conference*, pages 1131–1136, Boston, MA.

Curtis, M. L. [1979] *Matrix Groups*, second edition, Universitext, Springer-Verlag, New York–Heidelberg–Berlin, ISBN 0-387-96074-0.

Dalsmo, M. and van der Schaft, A. J. [1998] *On representations and integrability of mathematical structures in energy-conserving physical systems*, SIAM Journal on Control and Optimization, **37**(1), 54–91.

de León, M., Marrero, J. C., and de Diego, D. M. [1997] *Mechanical systems with nonlinear constraints*, International Journal of Theoretical Physics, **36**(4), 979–995.

DeMarco, G. [1999] *Linearization of nonholonomic systems at equilibrium points*, Rendiconti del Seminario Matematico (già "Conferenze di Fisica e di Matematica") Università e Politecnico di Torino, **57**(3), 187–198.

Do Carmo, M. P. [1992] *Riemannian Geometry*, Mathematics: Theory & Applications, Birkhäuser, Boston–Basel–Stuttgart, ISBN 0-8176-3490-8.

Donaldson, S. K. [1983] *Self-dual connections and the topology of smooth 4-manifolds*, American Mathematical Society. Bulletin. New Series, **8**(1), 81–83.

Doyle, J. C., Francis, B. A., and Tannenbaum, A. R. [1990] *Feedback Control Theory*, Macmillian, New York, ISBN 0-02-330011-6.

Dubrovin, B. A., Fomenko, A. T., and Novikov, S. P. [1990] *Modern Geometry—Methods and Applications, Part III. Introduction to Homology Theory*, number 124 in Graduate Texts in Mathematics, Springer-Verlag, New York–Heidelberg–Berlin, ISBN 0-387-97271-4.

Dugas, R. [1957] *A History of Mechanics*, Central Book Company, Inc., New York, reprint: [Dugas 1988].

— [1988] *A History of Mechanics*, Dover Publications, Inc., New York, ISBN 0-486-65632-2, reprint of the 1957 J. R. Maddox translation by Central Book Company, Inc., New York.

Duindam, V., Stramigioli, S., and Scherpen, J. M. A. [2003] *Passive compensation of nonlinear robot dynamics*, IEEE Transactions on Robotics and Automation, to appear.

Dullerud, G. E. and Paganini, F. [1999] *A Course in Robust Control Theory*, number 36 in Texts in Applied Mathematics, Springer-Verlag, New York–Heidelberg–Berlin, ISBN 0-387-98945-5.

Egeland, O. and Godhavn, J.-M. [1994] *Passivity-based adaptive attitude control of a rigid spacecraft*, IEEE Transactions on Automatic Control, **39**(4), 842–846.

Einstein, A. [1916] *Die Grundlage der allgemeinen Relativitätstheorie*, Annalen der Physik, **49**, 769–822.

Elkin, V. I. [1999] *Reduction of Nonlinear Control Systems. A Differential Geometric Approach*, number 472 in Mathematics and its Applications, Kluwer Academic Publishers, Dordrecht, ISBN 0-7923-5623-3, translation of the 1997 Russian edition by P. S. V. Naidu.

Etkin, B. [1995] *Dynamics of Flight: Stability and Control*, third edition, John Wiley and Sons, New York, ISBN 0-471-03418-5.

Evans, W. R. [1948] *Graphical analysis of control systems*, Transactions of the American Institute of Electrical Engineers, **67**, 547–551.

— [1950] *Control systems synthesis by root locus method*, Transactions of the American Institute of Electrical Engineers, **69**, 66–69.

Fantoni, I. and Lozano, R. [2001] *Non-linear Control of Underactuated Mechanical Systems*, Communications and Control Engineering Series, Springer-Verlag, New York–Heidelberg–Berlin, ISBN 1-852-33423-1.

Federer, H. [1969] *Geometric Measure Theory*, number 153 in Die Grundlehren der mathematischen Wissenschaften, Springer-Verlag, New York–Heidelberg–Berlin, reprint [Federer 1996].

— [1996] *Geometric Measure Theory*, Classics in Mathematics, Springer-Verlag, New York–Heidelberg–Berlin, ISBN 3-540-60656-4, reprint of 1969 edition.

Filippov, A. F. [1960] *Differential equations with discontinuous right-hand side*, Matematicheskiĭ Sbornik. Novaya Seriya, **51**, 99–128.

— [1988] *Differential Equations with Discontinuous Righthand Sides*, number 18 in Mathematics and its Applications (Soviet Series), Kluwer Academic Publishers, Dordrecht, ISBN 90-277-2699-X.

Fjellstad, O.-E. and Fossen, T. I. [1994] *Position and attitude tracking of AUVs: A quaternion feedback approach*, IEEE Journal of Oceanic Engineering, **19**(4),

512–518.

Fliess, M. [1981] *Fonctionnelles causales non linéaires et indéterminées non commutatives*, Bulletin de la Société Mathématique de France, **109**, 3–40.

Fliess, M., Lévine, J., Martin, P., and Rouchon, P. [1995] *Flatness and defect of non-linear systems: Introductory theory and examples*, International Journal of Control, **61**(6), 1327–1361.

Frazzoli, E., Dahleh, M. A., and Feron, E. [2003] *Maneuver-based motion planning for nonlinear systems with symmetries*, IEEE Transactions on Robotics and Automation, submitted.

Freeman, M. [1984] *Fully integrable Pfaffian systems*, Annals of Mathematics. Second Series, **119**(3), 465–510.

Frobenius, G. [1877] *Über das Pfaffsche Problem*, Journal für die Reine und Angewandte Mathematik, **82**, 230–315.

Fujimoto, K. and Sugie, T. [1998] *Canonical transformation and stabilization of generalized Hamiltonian systems*, in *Proceedings of 1998 Symposium on Nonlinear Control Systems (NOLCOS)*, pages 544–549, Enschede, The Netherlands.

— [2001a] *Canonical transformation and stabilization of generalized Hamiltonian systems*, Systems & Control Letters, **42**(3), 217–227.

— [2001b] *Stabilization of Hamiltonian systems with nonholonomic constraints based on time-varying generalized canonical transformations*, Systems & Control Letters, **44**(4), 309–319.

Gabrielian, M. S. and Krasovskiĭ, N. N. [1965] *On the problem of the stabilization of a mechanical system*, Journal of Applied Mathematics and Mechanics. Translation of the Soviet journal Prikladnaya Matematika i Mekhanika, **28**, 979–990.

Gallot, S., Hulin, D., and Lafontaine, J. [1987] *Riemannian Geometry*, Universitext, Springer-Verlag, New York–Heidelberg–Berlin, ISBN 0-387-17923-2.

Gantmacher, F. R. [1958] *Matrizenrechnung*, volume I and II of *Hochschulbcher für Mathematik*, VEB Deutscher Verlag der Wissenschaften, Berlin, translation: [Gantmacher 1959].

— [1959] *The Theory of Matrices*, volume 1 and 2, Chelsea, New York, ISBN 0-8218-1376-5 and 0-8218-2664-6, translation of German edition by K. A. Hirsch.

Gauss, C. F. [1828] *Disquisitiones generales circa superficies curva*, Commentationes Societatis Regiae Scientiarum Gottingensis recentiores, **6**, translation: [Gauss 1965].

— [1829] *Über ein neues allgemeines Grundgesatz der Mechanik*, Journal für die Reine und Angewandte Mathematik, **4**, 232–235.

— [1965] *General Investigations of Curved Surfaces*, Raven Press, Hewlett, NY, translated from Latin and German by A. Hiltebeitel and J. Morehead.

Ge, S. S., Zhendong, S., Lee, T. H., and Spong, M. W. [2001] *Feedback linearization and stabilization of second-order non-holonomic chained systems*, International Journal of Control, **74**(14), 1383–1392.

Gelfand, I. M. and Fomin, S. V. [1961] *Variatsionnoe ischislenie*, Gosudarstvennoe izdatelstvo fiziko-matematicheskoi literatury, Moscow, reprint of translation: [Gelfand and Fomin 2000].

— [2000] *Calculus of Variations*, Dover Publications, Inc., New York, ISBN 0-486-41448-5, reprint of 1963 translation from Russian by R. A. Silverman.

Giaquinta, M. and Hildebrandt, S. [1996] *Calculus of Variations*, number 310 and 311 in Grundlehren der Mathematischen Wissenschaften, Springer-Verlag, New York–Heidelberg–Berlin, ISBN 3-540-50625-X and 3-540-57961-3.

Gibbs, J. W. [1879] *On the fundamental formulae of dynamics*, American Journal of Mathematics, **2**, 49–64.

Gilbert, E. G. [1977] *Functional expansions for the response of nonlinear differential systems*, IEEE Transactions on Automatic Control, **22**(6), 909–921.

Godbillon, C. [1969] *Géométrie Différentielle et Méchanique Analytique*, Collection Méthodes. Mathématique, Hermann, Paris.

Godhavn, J.-M. and Egeland, O. [1997] *A Lyapunov approach to exponential stabilization of nonholonomic systems in power form*, IEEE Transactions on Automatic Control, **42**(7), 1028–1032.

Goldstein, H. [1980] *Classical Mechanics*, second edition, Addison Wesley, Reading, MA, ISBN 0-201-02918-9.

Goldstine, H. H. [1980] *A History of the Calculus of Variations From the 17th Through the 19th Century*, number 5 in Studies in the History of Mathematics and Physical Sciences, Springer-Verlag, New York–Heidelberg–Berlin, ISBN 0-387-90521-9.

Golubitsky, M. and Guillemin, V. [1973] *Stable Mappings and Their Singularities*, number 14 in Graduate Texts in Mathematics, Springer-Verlag, New York–Heidelberg–Berlin, ISBN 0-387-90072-1.

Golubitsky, M., Stewart, I., Buono, P.-L., and Collins, J. J. [1999] *Symmetry in locomotor central pattern generators and animal gaits*, Nature, **401**(6754), 693–695.

Goodwin, G. C., Graebe, S. F., and Salgado, M. E. [2001] *Control System Design*, Prentice-Hall, Englewood Cliffs, NJ, ISBN 0-13-958653-9.

Goodwine, B. and Burdick, J. W. [2001] *Controllability of kinematic control systems on stratified configuration spaces*, IEEE Transactions on Automatic Control, **46**(3), 358–368.

Gopal, V. and Biegler, L. T. [1998] *Large scale inequality constrained optimization and control*, IEEE Control Systems Magazine, **18**(6), 59–68.

Gràcia, X., Marín-Solano, J., and Muñoz-Lecanda, M.-C. [2003] *Some geometric aspects of variational calculus in constrained systems*, Reports on Mathematical Physics, **51**(1), 127–148.

Grauert, H. [1958] *On Levi's problem and the imbedding of real-analytic manifolds*, Annals of Mathematics. Second Series, **68**, 460–472.

Greene, R. E. and Jacobowitz, H. [1971] *Analytic isometric embeddings*, Annals of Mathematics. Second Series, **93**(1), 189–204.

Grüne, L., Sontag, E. D., and Wirth, F. R. [1999] *Asymptotic stability equals exponential stability, and ISS equals finite energy gain—if you twist your eyes*, Systems & Control Letters, **38**(2), 127–134.

Guckenheimer, J. and Holmes, P. [1990] *Nonlinear Oscillations, Dynamical Systems, and Bifurcations of Vector Fields*, number 42 in Applied Mathematical Sciences, Springer-Verlag, New York–Heidelberg–Berlin, ISBN 0-387-90819-6.

Guillemin, V. and Sternberg, S. [1990] *Symplectic Techniques in Physics*, second edition, Cambridge University Press, New York–Port Chester–Melbourne–Sydney, ISBN 0-521-38990-9.

Hahn, W. [1963] *Theory and Application of Liapunov's Direct Method*, Prentice-Hall, Englewood Cliffs, NJ, translation from German by H. H. Losenthien and S. H. Lehnigk.

— [1967] *Stability of Motion*, number 138 in Grundlehren der Mathematischen Wissenschaften, Springer-Verlag, New York–Heidelberg–Berlin, translation from German by A. P. Baartz.

Halmos, P. R. [1974a] *Measure Theory*, number 18 in Graduate Texts in Mathematics, Springer-Verlag, New York–Heidelberg–Berlin, ISBN 0-387-90088-8, reprint of 1950 edition by Litton Educational Publishing, Inc.

— [1974b] *Naïve Set Theory*, Undergraduate Texts in Mathematics, Springer-Verlag, New York–Heidelberg–Berlin, ISBN 0-387-90092-6.

— [1996] *Finite-Dimensional Vector Spaces*, second edition, Undergraduate Texts in Mathematics, Springer-Verlag, New York–Heidelberg–Berlin, ISBN 0-387-90093-4.

Halperin, D., Kavraki, L. E., and Latombe, J.-C. [1997] *Robotics*, in *Handbook of Discrete and Computational Geometry*, J. E. Goodman and J. O'Rourke, editors, chapter 41, pages 755–778, CRC Press, Boca Raton, FL, ISBN 0-8493-8524-5.

Hamberg, J. [1999] *General matching conditions in the theory of controlled Lagrangians*, in *Proceedings of the 38th IEEE Conference on Decision and Control*, pages 2519–2523, Phoenix, AZ.

Hargraves, C. R. and Paris, S. W. [1987] *Direct trajectory optimization using nonlinear programming and collocation*, AIAA Journal of Guidance, Control, and Dynamics, **10**(4), 338–342.

Hartman, P. [1964] *Ordinary Differential Equations*, John Wiley and Sons, New York, reprint of revised edition: [Hartman 2002].

— [2002] *Ordinary Differential Equations*, second edition, number 38 in Classics in Applied Mathematics, Society for Industrial and Applied Mathematics, Philadelphia, PA, ISBN 0-89871-510-5.

Hauser, J. E., Sastry, S. S., and Meyer, G. [1992] *Nonlinear control design for slightly nonminimum phase systems: Application to V/STOL aircraft*, Automatica. The Journal of IFAC, **28**(4), 665–679.

Helgason, S. [1978] *Differential Geometry, Lie Groups, and Symmetric Spaces*, number 60 in Mathematics, Academic Press, New York, ISBN 0-12-338460-5.

Hermann, R. [1960] *On the differential geometry of foliations*, Annals of Mathematics. Second Series, **72**, 445–457.

— [1962] *The differential geometry of foliations. II*, Journal of Applied Mathematics and Mechanics. Translation of the Soviet journal Prikladnaya Matematika i Mekhanika, **11**, 303–315.

— [1968] *Differential Geometry and the Calculus of Variations*, number 49 in Mathematics in Science and Engineering, Academic Press, New York.

Hermann, R. and Krener, A. J. [1977] *Nonlinear controllability and observability*, IEEE Transactions on Automatic Control, **22**(5), 728–740.

Hermes, H. [1982a] *Control systems which generate decomposable Lie algebras*, Journal of Differential Equations, **44**(2), 166–187.

— [1982b] *On local controllability*, SIAM Journal on Control and Optimization, **20**(2), 211–220.

— [1991] *Nilpotent and high-order approximations of vector field systems*, SIAM Review. A Publication of the Society for Industrial and Applied Mathematics, **33**(2), 238–264.

Hertz, H. [1894] *Die Prinzipien der Mechanik in neuem Zusammenhange dargestellt*, J. A. Barth, Leipzig, reprint of translation: [Hertz 2003].

— [2003] *The Principles of Mechanics Presented in a New Form*, Dover Publications, Inc., New York, ISBN 0-486-49557-4, translation by D. Jones and J. Walley.

Hill, G. W. [1886] *On the part of the motion of lunar perigee which is a function of the mean motions of the sun and moon*, Acta Mathematica, **8**, 1–36.

Hironaka, H. [1973] *Subanalytic sets*, in *Number Theory, Algebraic Geometry and Commutative Algebra*, pages 453–493, Kinokunya Book-Store Co., Ltd., Tokyo.

— [1975] *Triangulations of algebraic sets*, in *Algebraic geometry (Arcata, CA)*, volume 29 of *Proceedings of Symposia in Pure Mathematics*, pages 165–185, American Mathematical Society, Providence, RI.

Hirose, S. [1993] *Biologically Inspired Robots: Snake-like Locomotors and Manipulators*, Oxford University Press, Walton Street, Oxford, ISBN 0-19-856261-6.

Hirsch, M. W. [1976] *Differential Topology*, number 33 in Graduate Texts in Mathematics, Springer-Verlag, New York–Heidelberg–Berlin, ISBN 0-387-90147-5.

Hirsch, M. W. and Smale, S. [1974] *Differential Equations, Dynamical Systems, and Linear Algebra*, number 60 in Pure and Applied Mathematics, Academic Press, New York, ISBN 0-12-349550-4.

Hirschorn, R. M. and Lewis, A. D. [2001] *Geometric first-order controllability conditions for affine connection control systems*, in *Proceedings of the 40th IEEE Conference on Decision and Control*, pages 4216–4221, Orlando, FL.

— [2002] *Geometric local controllability: Second-order conditions*, Preprint.

Hörmander, L. [1973] *An Introduction to Complex Analysis in Several Variables*, second edition, North-Holland, Amsterdam–New York, ISBN 0-444-10523-9.

Horn, R. A. and Johnson, C. R. [1990] *Matrix Analysis*, Cambridge University Press, New York–Port Chester–Melbourne–Sydney, ISBN 0-521-38632-2.

Howe, R. [1983] *Very basic Lie theory*, The American Mathematical Monthly, **90**(9), 600–623.

— [1984] *Correction to "Very basic Lie theory"*, The American Mathematical Monthly, **91**(4), 247.

Hughes, P. C. and Skelton, R. E. [1980] *Controllability and observability of linear matrix-second-order systems*, Transactions of the ASME. Series E. Journal of Applied Mechanics, **47**(2), 415–420.

Hungerford, T. W. [1980] *Algebra*, number 73 in Graduate Texts in Mathematics, Springer-Verlag, New York–Heidelberg–Berlin, ISBN 0-387-90518-9.

Husemoller, D. [1994] *Fibre Bundles*, third edition, number 20 in Graduate Texts in Mathematics, Springer-Verlag, New York–Heidelberg–Berlin, ISBN 0-387-94087-1.

Iannitti, S. and Lynch, K. M. [2003] *Minimum control switch motions for the snakeboard: A case study in kinematically controllable underactuated systems*, IEEE Transactions on Robotics and Automation, to appear.

Isidori, A. [1995] *Nonlinear Control Systems*, third edition, Communications and Control Engineering Series, Springer-Verlag, New York–Heidelberg–Berlin, ISBN 3-540-19916-0.

— [1999] *Nonlinear Control Systems II*, Communications and Control Engineering Series, Springer-Verlag, New York–Heidelberg–Berlin, ISBN 1-85233-188-7.

Jacobi, C. G. J. [1884] *Vorlesungen über Dynamics*, volume Supplementband of *Gesammelte Werke*, Druck und Verlag von G. Reimer, Berlin, first printed by A. Clebsch, Leipzig, 1868, reprint: [Jacobi 1969].

— [1969] *Vorlesungen über Dynamics*, volume VIII of *Gesammelte Werke*, Chelsea, New York.

Jacobson, N. [1962] *Lie Algebras*, number 10 in Interscience Tracts in Pure and Applied Mathematics, Interscience Publishers, New York, reprint: [Jacobson 1979].

— [1979] *Lie Algebras*, Dover Publications, Inc., New York, ISBN 0-486-63832-4, reprint of 1962 Interscience edition.

Jakubczyk, B. and Respondek, W. [1980] *On linearization of control systems*, Bulletin de l'Académie Polonaise des Sciences. Série des Sciences Mathématiques, Astronomiques et Physiques, **28**(9-10), 517–522.

Jalnapurkar, S. M. and Marsden, J. E. [2000] *Stabilization of relative equilibria*, IEEE Transactions on Automatic Control, **45**(8), 1483–1491.

Jurdjevic, V. [1997] *Geometric Control Theory*, number 51 in Cambridge Studies in Advanced Mathematics, Cambridge University Press, New York–Port Chester–Melbourne–Sydney, ISBN 0-521-49502-4.

Jurdjevic, V. and Quinn, J. P. [1978] *Controllability and stability*, Journal of Differential Equations, **28**(3), 381–389.

Kailath, T. [1980] *Linear Systems*, Information and System Sciences Series, Prentice-Hall, Englewood Cliffs, NJ, ISBN 0-13-536961-4.

Kalman, R. E. [1960a] *Contributions to the theory of optimal control*, Boletín de la Sociedad Matemática Mexicana. Segunda Serie, **5**, 102–119.

— [1960b] *A new approach to linear filtering and prediction theory*, Transactions of the ASME. Series D. Journal of Basic Engineering, **82**, 35–45.

Kalman, R. E. and Bucy, R. S. [1960] *New results in linear filtering and prediction theory*, Transactions of the ASME. Series D. Journal of Basic Engineering, **83**, 95–108.

Kalman, R. E., Ho, Y.-C., and Narendra, K. S. [1963] *Controllability of linear dynamical systems*, Contributions to Differential Equations, **1**, 189–213.

Kane, T. R. [1961] *Dynamics of nonholonomic systems*, Transactions of the ASME. Series E. Journal of Applied Mechanics, **28**, 574–578.

Kane, T. R. and Levinson, D. A. [1985] *Dynamics: Theory and Applications*, McGraw-Hill Series in Mechanical Engineering, McGraw-Hill, New York.

Kane, T. R. and Wang, C. F. [1965] *On the derivation of equations of motion*, Journal of SIAM, **13**, 487–492.

Kanellakopoulos, I., Kokotovic, P. V., and Morse, A. S. [1991] *Systematic design of adaptive controllers for feedback linearizable systems*, IEEE Transactions on Automatic Control, **36**(11), 1241–1253.

— [1992] *A toolkit for nonlinear feedback design*, Systems & Control Letters, **18**(2), 83–92.

Kapitsa, P. L. [1965] *Dynamical stability of a pendulum when its point of suspension vibrates*, in *Collected Papers by P. L. Kapitsa*, volume 2, pages 714–725, Pergamon Press, Oxford–New York–Paris.

Kawski, M. [1988a] *Control variations with an increasing number of switchings*, American Mathematical Society. Bulletin. New Series, **18**(2), 149–152.

— [1988b] *Nilpotent Lie-algebras of vectorfields*, Journal für die Reine und Angewandte Mathematik, **388**, 1–17.

— [1989] *Stabilization of nonlinear systems in the plane*, Systems & Control Letters, **12**(2), 169–175.

— [1990] *The complexity of deciding controllability*, Systems & Control Letters, **15**(1), 9–14.

— [1995] *Geometric homogeneity and applications to stabilization*, in *Proceedings of 1995 Symposium on Nonlinear Control Systems (NOLCOS)*, pages 251–256, Lake Tahoe, CA.

— [1998] *Nonlinear control and combinatorics of words*, in *Geometry of Feedback and Optimal Control*, B. Jakubczyk and W. Respondek, editors, pages 305–346, number 207 in Monographs and Textbooks in Pure and Applied Mathematics, Dekker Marcel Dekker, New York, ISBN 0-824-79068-5.

Kawski, M. and Sussmann, H. J. [1997] *Noncommutative power series and formal Lie-algebraic techniques in nonlinear control theory*, in *Operators, Systems, and Linear Algebra (Kaiserslautern, Germany)*, U. Helmke, D. Pratzel-Wolters, and E. Zerz, editors, pages 111–128, B. G. Teubner, Stuttgart, ISBN 3-519-02608-2.

Kelly, S. D. and Murray, R. M. [2000] *Modelling efficient pisciform locomotion for control*, International Journal of Robust and Nonlinear Control, **10**(4), 217–241.

Khalil, H. K. [2001] *Nonlinear Systems*, third edition, Prentice-Hall, Englewood Cliffs, NJ, ISBN 0-13-067389-7.

Kharlomov, P. V. [1992] *A critique of some mathematical models of mechanical systems with differential constraints*, Journal of Applied Mathematics and Mechanics. Translation of the Soviet journal Prikladnaya Matematika i Mekhanika, **56**, 584–594.

Khatib, O. [1986] *Real-time obstacle avoidance for manipulators and mobile robots*, The International Journal of Robotics Research, **5**(1), 90–98.

Kirchhoff, G. [1869] *Über die Bewegung eines Rotationskorpers in einer Flussigkeit*, Journal für die Reine und Angewandte Mathematik, **71**, 237–273.

Kitzka, F. [1986] *An example for the application of a nonholonomic constraint of second-order in particle mechanics*, Zeitschrift für Angewandte Mathematik und Mechanik. Ingenieurwissenschaftliche Forschungsarbeiten, **66**(7), 312–314.

Klingenberg, W. [1982] *Riemannian Geometry*, number 1 in Studies in Mathematics, Walter de Gruyter, Berlin–New York, ISBN 3-11-008673-5.

Kobayashi, S. and Nomizu, K. [1963] *Foundations of Differential Geometry, Volumes I and II*, number 15 in Interscience Tracts in Pure and Applied Mathematics, Interscience Publishers, New York, ISBN 0-470-49647-9 and 0-470-49648-7.

Koditschek, D. E. [1988] *Strict global Lyapunov functions for mechanical systems*, in *Proceedings of the 1988 American Control Conference*, pages 1770–1775, Atlanta, GA.

— [1989] *The application of total energy as a Lyapunov function for mechanical control systems*, in *Dynamics and Control of Multibody Systems (Brunswick, ME)*, J. E. Marsden, P. S. Krishnaprasad, and J. C. Simo, editors, volume 97 of *Contemporary Mathematics*, pages 131–157, American Mathematical Society, Providence, RI, ISBN 0-8218-5104-7.

— [1991] *The control of natural motion in mechanical systems*, Transactions of the ASME. Series G. Journal of Dynamic Systems, Measurement, and Control, **113**(4), 547–551.

Koiller, J. [1992] *Reduction of some classical nonholonomic systems with symmetry*, Archive for Rational Mechanics and Analysis, **118**(2), 113–148.

Kolmanovsky, I. and McClamroch, N. H. [1995] *Developments in nonholonomic control problems*, IEEE Control Systems Magazine, **15**(6), 20–36.

Koon, W. S. and Marsden, J. E. [1998] *Poisson reduction for nonholonomic mechanical systems with symmetry*, Reports on Mathematical Physics, **42**(1-2), 101–134.

Kozlov, V. V. [1992] *The problem of realizing constraints in dynamics*, Journal of Applied Mathematics and Mechanics. Translation of the Soviet journal Prikladnaya Matematika i Mekhanika, **56**, 594–600.

Krantz, S. G. and Parks, H. R. [2002] *A Primer of Real Analytic Functions*, second edition, Birkhäuser Advanced Texts, Birkhäuser, Boston–Basel–Stuttgart, ISBN 0-8176-4264-1.

Krasinskiĭ, A. Y. [1988] *Stability and stabilization of equilibrium positions of nonholonomic systems*, Journal of Applied Mathematics and Mechanics. Translation of the Soviet journal Prikladnaya Matematika i Mekhanika, **52**(2), 152–158.

Krasovskiĭ, N. N. [1963] *Stability of Motion. Applications of Lyapunov's second Method to Differential Systems and Equations with Delay*, Stanford University Press, Stanford, CA.

Krener, A. J. [1974] *A generalization of Chow's theorem and the bang-bang theorem to non-linear control problems*, Journal of SIAM. Series A. Control, **12**, 43–52.

Krishnaprasad, P. S. and Tsakiris, D. P. [1995] *Oscillations,* SE(2)*-snakes and motion control*, in *Proceedings of the 34th IEEE Conference on Decision and Control*, pages 2806–11, New Orleans, LA.

— [2001] *Oscillations,* SE(2)*-snakes and motion control: A study of the roller racer*, Dynamics and Stability of Systems. An International Journal, **16**(4), 347–397.

Krstic, M., Kanellakopoulos, I., and Kokotovic, P. V. [1995] *Nonlinear and Adaptive Control Design*, John Wiley and Sons, New York, ISBN 0-471-12732-9.

Kupka, I. and Oliva, W. M. [2001] *The non-holonomic mechanics*, Journal of Differential Equations, **169**(1), 169–189.

Kurzweil, J. and Jarnik, J. [1988] *Iterated Lie brackets in limit processes in ordinary differential equations*, Resultate der Mathematik, **14**(1-2), 125–137.

Kwatny, H. G. and Blankenship, G. L. [2001] *Nonlinear Control and Analytical Mechanics: A Computational Approach*, Birkhäuser, Boston–Basel–Stuttgart, ISBN 0-8176-4147-5.

Lafferriere, G. and Sussmann, H. J. [1991] *Motion planning for controllable systems without drift*, in *Proceedings of the 1991 IEEE International Conference on Robotics and Automation*, pages 1148–1153, Sacramento, CA.

— [1993] *A differential geometric approach to motion planning*, in *Nonholonomic Motion Planning*, Z. Li and J. F. Canny, editors, pages 235–270, Kluwer Academic Publishers, Dordrecht, ISBN 0-7923-9275-2.

Lagrange, J. L. [1788] *Mécanique analytique*, Chez la Veuve Desaint, Paris, translation of revised edition: [Lagrange 1997].

— [1997] *Analytical Mechanics*, number 191 in Boston Studies in the Philosophy of Science, Kluwer Academic Publishers, Dordrecht, ISBN 0-7923-4349-2, translation of the 1811 edition in French by A. Boissonnade and V. N. Vagliente.

Lamb, H. [1895] *Hydrodynamics*, Cambridge University Press, New York–Port Chester–Melbourne–Sydney, reprint of revised edition: [Lamb 1993].

— [1993] *Hydrodynamics*, Cambridge Mathematical Library, Cambridge University Press, New York–Port Chester–Melbourne–Sydney, ISBN 0-521-45868-4, reprint of 1932 sixth edition.

Lamiraux, F. and Laumond, J.-P. [2000] *Flatness and small-time controllability of multibody mobile robots: Application to motion planning*, IEEE Transactions on Automatic Control, **45**(10), 1878–1881.

Lanczos, C. [1949] *The Variational Principles of Mechanics*, University of Toronto Press, Toronto, reprint of revised edition: [Lanczos 1986].

— [1986] *The Variational Principles of Mechanics*, Dover Publications, Inc., New York, ISBN 0-486-65067-7, reprint of fourth edition.

Lang, S. [1995] *Differential and Riemannian Manifolds*, number 160 in Graduate Texts in Mathematics, Springer-Verlag, New York–Heidelberg–Berlin, ISBN 0-387-94338-2.

Laplace, P.-S. [1799-1825] *Traité du Mécanique Céleste*, Bachelier, Paris.

LaSalle, J. P. [1968] *Stability theory for ordinary differential equations*, Journal of Differential Equations, **4**, 57–65.

LaSalle, J. P. and Lefschetz, S. [1962] *Stability by Liapunovs's Direct Method and Applications*, number 4 in Mathematics in Science and Engineering, Academic Press, New York.

Latombe, J.-C. [1991] *Robot Motion Planning*, Kluwer Academic Publishers, Dordrecht, ISBN 0-7923-9129-2.

— [1999] *Motion planning: A journey of robots, molecules, digital actors, and other artifacts*, The International Journal of Robotics Research, **18**(11), 1119–1128.

Laumond, J. P., Sekhavat, S., and Lamiraux, F. [1998] *Guidelines in nonholonomic motion planning*, in *Robot Motion Planning and Control*, J.-P. Laumond, editor, pages 1–53, number 229 in Lecture Notes in Control and Information Sciences, Springer-Verlag, New York–Heidelberg–Berlin, ISBN 3-540-76219-1.

Lee, K. K. and Arapostathis, A. [1988] *Remarks on smooth feedback stabilization of nonlinear systems*, Systems & Control Letters, **10**(1), 41–44.

Leonard, N. E. [1997] *Stability of a bottom-heavy underwater vehicle*, Automatica. The Journal of IFAC, **33**(3), 331–346.

— [1998] *Mechanics and nonlinear control: Making underwater vehicles ride and glide*, in *Proceedings of 1998 Symposium on Nonlinear Control Systems (NOLCOS)*, pages 1–6, Enschede, The Netherlands.

Leonard, N. E. and Krishnaprasad, P. S. [1995] *Motion control of drift-free, left-invariant systems on Lie groups*, IEEE Transactions on Automatic Control, **40**(9), 1539–1554.

Leonard, N. E. and Marsden, J. E. [1997] *Stability and drift of underwater vehicle dynamics: Mechanical systems with rigid motion symmetry*, Physica D. Nonlinear Phenomena, **105**, 130–162.

Lesiak, C. and Krener, A. J. [1978] *The existence and uniqueness of Volterra series for nonlinear systems*, IEEE Transactions on Automatic Control, **23**(6), 1090–1095.

Lesser, M. [1992] *A geometrical interpretation of Kane's equations*, Proceedings of the Royal Society. London. Series A. Mathematical and Physical Sciences, **436**(1896), 69–87.

Levi, M. [1998] *Geometry of Kapitsa's potentials*, Nonlinearity, **11**(5), 1365–1368.

— [1999] *Geometry and physics of averaging with applications*, Physica D. Nonlinear Phenomena, **132**(1-2), 150–164.

Levi-Civita, T. [1925] *Lezioni di Calcolo Differenziale Assoluto*, Stock, Rome, reprint of translation: [Levi-Civita 1977].

— [1977] *The Absolute Differential Calculus*, Dover Publications, Inc., New York, ISBN 0-486-63401-9, translation from Italian by M. Long.

Levi-Civita, T. and Ricci, G. [1901] *Méthodes de calcul différentiel absolu et leurs applications*, Mathematische Annalen, **54**, 125–201.

Lewis, A. D. [1995] *Aspects of Geometric Mechanics and Control of Mechanical Systems*, Ph.D. thesis, California Institute of Technology, Pasadena, CA.

— [1996] *The geometry of the Gibbs-Appell equations and Gauss's Principle of Least Constraint*, Reports on Mathematical Physics, **38**(1), 11–28.

— [1997] *Local configuration controllability for a class of mechanical systems with a single input*, in *Proceedings of the 1997 European Control Conference*, Brussels, Belgium.

— [1998] *Affine connections and distributions with applications to nonholonomic mechanics*, Reports on Mathematical Physics, **42**(1-2), 135–164.

— [1999] *When is a mechanical control system kinematic?*, in *Proceedings of the 38th IEEE Conference on Decision and Control*, pages 1162–1167, Phoenix, AZ.

— [2000a] *The category of affine connection control systems*, in *Proceedings of the 39th IEEE Conference on Decision and Control*, pages 5119–5124, Sydney, Australia.

— [2000b] *Simple mechanical control systems with constraints*, IEEE Transactions on Automatic Control, **45**(8), 1420–1436.

— [2000c] *Towards $F = ma$ in a general setting for Lagrangian mechanics*, Annales Henri Poincaré. A Journal of Theoretical and Mathematical Physics, **1**(3), 569–605.

Lewis, A. D. and Murray, R. M. [1995] *Variational principles for constrained systems: Theory and experiment*, International Journal of Non-Linear Mechanics, **30**(6), 793–815.

— [1997a] *Controllability of simple mechanical control systems*, SIAM Journal on Control and Optimization, **35**(3), 766–790.

— [1997b] *Decompositions of control systems on manifolds with an affine connection*, Systems & Control Letters, **31**(4), 199–205.

— [1999] *Controllability of simple mechanical control systems*, SIAM Review. A Publication of the Society for Industrial and Applied Mathematics, **41**(3), 555–574.

Lewis, A. D., Ostrowski, J. P., Murray, R. M., and Burdick, J. W. [1994] *Nonholonomic mechanics and locomotion: The Snakeboard example*, in *Proceedings of the 1994 IEEE International Conference on Robotics and Automation*, pages 2391–2400, San Diego, CA.

Li, P. Y. and Horowitz, R. [2001a] *Passive velocity field control (PVFC). I. Geometry and robustness*, IEEE Transactions on Automatic Control, **46**(9), 1346–1359.

— [2001b] *Passive velocity field control (PVFC). II. Application to contour following*, IEEE Transactions on Automatic Control, **46**(9), 1360–1371.

Li, Z., Montgomery, R., and Raibert, M. [1989] *Dynamics and control of a legged robot in flight phase*, in *Proceedings of the 1989 IEEE International Conference on Robotics and Automation*, pages 1816–1821, Scottsdale, AZ.

Libermann, P. and Marle, C.-M. [1987] *Symplectic Geometry and Analytical Mechanics*, number 35 in Mathematics and its Applications, D. Reidel Publishing Company, Dordrecht–Boston–Lancaster–Tokyo, ISBN 90-2772-438-5.

Lin, W., Pongvuthithum, R., and Qian, C. [2002] *Control of high-order nonholonomic systems in power chained form using discontinuous feedback*, IEEE Transactions on Automatic Control, **47**(1), 108–115.

Liu, W. [1997] *An approximation algorithm for nonholonomic systems*, SIAM Journal on Control and Optimization, **35**(4), 1328–1365.

Liu, W. and Sussmann, H. J. [1999] *Continuous dependence with respect to the input of trajectories of control-affine systems*, SIAM Journal on Control and Optimization, **37**(3), 777–803.

Lozano, R., Brogliato, B., Egeland, O., and Maschke, B. M. [2000] *Dissipative Systems Analysis and Control*, Communications and Control Engineering Series, Springer-Verlag, New York–Heidelberg–Berlin, ISBN 1-85233-285-9.

Luh, J. Y. S., Walker, M. W., and Paul, R. P. [1980] *Resolved acceleration control*, IEEE Transactions on Automatic Control, **25**(3), 468–474.

Lyapunov, A. M. [1892] *Obščaya zadača ob ustoičivosti dviženiya*, Fakul′teta i Khar′kovskogo Matematicheskogo Obshchestva, Kharkov, translation: [Lyapunov 1992].

— [1992] *The General Problem of the Stability of Motion*, Taylor & Francis, London–New York–Philadelphia–Singapore, translation from Russian by A. T. Fuller.

Lynch, K. M. [1999] *Controllability of a planar body with unilateral thrusters*, IEEE Transactions on Automatic Control, **44**(6), 1206–1211.

Lynch, K. M. and Mason, M. T. [1999] *Dynamic nonprehensile manipulation: Controllability, planning, and experiments*, The International Journal of Robotics Research, **18**(1), 64–92.

Lynch, K. M., Shiroma, N., Arai, H., and Tanie, K. [2000] *Collision-free trajectory planning for a 3-DOF robot with a passive joint*, The International Journal of Robotics Research, **19**(12), 1171–1184.

Magnus, W. [1954] *On the exponential solution of differential equations for a linear operator*, Communications on Pure and Applied Mathematics, **7**, 649–673.

Malgrange, B. [1976] *Frobenius avec singularités. I. Codimension un*, Institut des Hautes Études Scientifiques. Publications Mathématiques, **46**, 163–173.

— [1977] *Frobenius avec singularités. II. Le cas général*, Inventiones Mathematicae, **39**(1), 67–89.

Manikonda, V. and Krishnaprasad, P. S. [2002] *Controllability of a class of underactuated mechanical systems with symmetry*, Automatica. The Journal of IFAC, **38**(11), 1837–1850.

Marino, R. [1986] *On the largest feedback linearizable subsystem*, Systems & Control Letters, **6**(5), 345–351.

Marle, C.-M. [1995] *Reduction of constrained mechanical systems and stability of relative equilibria*, Communications in Mathematical Physics, **174**(2), 295–318.

Marsden, J. E. [1992] *Lectures on Mechanics*, number 174 in London Mathematical Society Lecture Note Series, Cambridge University Press, New York–Port Chester–Melbourne–Sydney, ISBN 0-521-42844-0.

Marsden, J. E. and Hoffman, M. J. [1993] *Elementary Classical Analysis*, second edition, W. H. Freeman and Company, San Francisco, CA, ISBN 0-7167-2105-8.

Marsden, J. E. and Ostrowski, J. P. [1996] *Symmetries in motion: Geometric foundations of motion control*, Nonlinear Science Today, Electronic.

Marsden, J. E. and Ratiu, T. S. [1999] *Introduction to Mechanics and Symmetry*, second edition, number 17 in Texts in Applied Mathematics, Springer-Verlag, New York–Heidelberg–Berlin, ISBN 0-387-98643-X.

Marsden, J. E. and Tromba, A. E. [2004] *Vector Calculus*, fifth edition, W. H. Freeman and Company, San Francisco, CA, ISBN 0-7167-4992-0.

Marsden, J. E. and Weinstein, A. [1974] *Reduction of symplectic manifolds with symmetry*, Reports on Mathematical Physics, **5**(1), 121–130.

Martínez, S. [2002] *Geometric Methods in Nonlinear Control Theory with Applications to Dynamic Robotic Systems*, Ph.D. thesis, Universitad Carlos III, Madrid, Spain.

Martínez, S. and Cortés, J. [2003] *Motion control algorithms for simple mechanical systems with symmetry*, Acta Applicandae Mathematicae. An International Journal on Applying Mathematics and Mathematical Applications, **76**(3), 221–264.

Martínez, S., Cortés, J., and Bullo, F. [2003a] *Analysis and design of oscillatory control systems*, IEEE Transactions on Automatic Control, **48**(7), 1164–1177.

— [2003b] *A catalog of inverse-kinematics planners for underactuated systems on matrix groups*, in *IEEE/RSJ International Conference on Intelligent Robots & Systems*, pages 625–630, Las Vegas.

Massera, J. L. [1956] *Contributions to stability theory*, Annals of Mathematics. Second Series, **64**(1), 182–206.

— [1958] *Erratum: Contributions to stability theory*, Annals of Mathematics. Second Series, **68**(1), 202.

Mathieu, É. L. [1868] *Mémoire sur le mouvement vibratoire d'une membrane de forme elliptique*, Journal de Mathématiques Pures et Appliquées. Neuvième Sér, **13**, 137–203.

Matsuda, M. [1968] *An integration theorem for completely integrable systems with singularities*, Osaka Journal of Mathematics, **5**, 279–283.

Matveev, A. S. [1999] *On the convexity of the images of quadratic mappings*, St. Petersburg Mathematical Journal, **10**(2), 343–372.

Maxwell, J. C. [1868] *On governors*, Proceedings of the Royal Society. London. Series A. Mathematical and Physical Sciences, **16**, 270–283.

Mayr, O. [1970] *The Origins of Feedback Control*, MIT Press, Cambridge, MA.

McClamroch, N. H. and Bloch, A. M. [1988] *Control of constrained Hamiltonian systems and applications to control of constrained robots*, in *Dynamical Systems Approaches to Nonlinear Problems in Systems and Circuits (Henniker, NH)*, pages 394–403, Society for Industrial and Applied Mathematics, Philadelphia, PA, ISBN 0-89871-218-1.

M'Closkey, R. T. and Murray, R. M. [1997] *Exponential stabilization of driftless nonlinear control systems using homogeneous feedback*, IEEE Transactions on Automatic Control, **42**(5), 614–628.

Meerkov, S. M. [1980] *Principle of vibrational control: Theory and applications*, IEEE Transactions on Automatic Control, **25**(4), 755–762.

Meerkov, S. M. and Tsitkin, M. Y. [1975] *The effectiveness of the method of vibrational control for the dynamic systems of the order n*, Automation and Remote Control. A translation of Avtomatika i Telemekhanika, a publication of the Academy of Sciences of the USSR, **36**(4), 525–529.

Melody, J. W., Başar, T., and Bullo, F. [2003] *On nonlinear controllability of homogeneous systems linear in the controls*, IEEE Transactions on Automatic Control, **48**(1), 139–143.

Mendelson, B. [1990] *Introduction to Topology*, Dover Books on Advanced Mathematics, Dover Publications, Inc., New York, ISBN 0-486-66352-3, reprint of 1975 edition by Allyn and Bacon.

Merkin, D. R. [1997] *Introduction to the Theory of Stability*, number 24 in Texts in Applied Mathematics, Springer-Verlag, New York–Heidelberg–Berlin, ISBN 0-387-94761-2.

Meyer, G. [1971] *Design and global analysis of spacecraft attitude control systems*, Technical Report R-361, NASA.

Meyer, K. R. [1973] *Symmetries and integrals in mechanics*, in *Dynamical Systems*, M. Peixoto, editor, pages 259–273, Academic Press, New York.

Miller, R. K. and Michel, A. N. [1980] *Asymptotic stability of systems: Results involving the system topology*, SIAM Journal on Control and Optimization, **18**(2), 181–190.

Milnor, J. W. [1956] *On manifolds homeomorphic to the 7-sphere*, Annals of Mathematics. Second Series, **64**, 399–405.

— [1959] *Differentiable structures on spheres*, American Journal of Mathematics, **81**(4), 962–972.

— [1963] *Morse Theory*, number 51 in Annals of Mathematical Studies, Princeton University Press, Princeton, NJ.

— [1965a] *Lectures on the h-Cobordism Theorem*, Princeton University Press, Princeton, NJ.

— [1965b] *Topology from the Differentiable Viewpoint*, The University Press of Virginia, Charlottesville, VA, reprint: [Milnor 1997].

— [1978] *Analytic proofs of the "Hairy Ball Theorem" and the Brouwer Fixed Point Theorem*, The American Mathematical Monthly, **85**(7), 521–524.

— [1997] *Topology from the Differentiable Viewpoint*, Princeton Landmarks in Mathematics, Princeton University Press, Princeton, NJ, ISBN 0-691-04833-9, reprint of 1965 edition.

— [2003] *Towards the Poincaré conjecture and the classification of 3-manifolds*, Notices of the American Mathematical Society, **50**(10), 1226–1233.

Milnor, J. W. and Stasheff, J. D. [1974] *Characteristic Classes*, number 76 in Annals of Mathematical Studies, Princeton University Press, Princeton, NJ, ISBN 0-691-08122-0.

Moore, G. H. [1982] *Zermelo's Axiom of Choice: Its Origins, Development, and Influence*, Springer-Verlag, New York–Heidelberg–Berlin, ISBN 0-387-90670-3.

Morin, P., Pomet, J.-B., and Samson, C. [1999] *Design of homogeneous time-varying stabilizing control laws for driftless controllable systems via oscillatory approximation of Lie brackets in closed loop*, SIAM Journal on Control and Optimization, **38**(1), 22–49.

Morin, P. and Samson, C. [2003] *Practical stabilization of driftless systems on lie groups: the transverse function approach*, IEEE Transactions on Automatic Control, **48**(9), 1496–1508.

Munkres, J. R. [1984] *Elements of Algebraic Topology*, Addison Wesley, Reading, MA, ISBN 0-201-04586-9.

Muñoz-Lecanda, M. C. and Yániz, F. J. [2002] *Dissipative control of mechanical systems: A geometric approach*, SIAM Journal on Control and Optimization, **40**(5), 1505–1516.

— [2003] *A global study of the stability of nonholonomic dissipative systems*, in *Proceedings of the Second IFAC Workshop on Lagrangian and Hamiltonian Methods for Nonlinear Control (Seville, Spain)*, A. J. van der Schaft and A. Astolfi, editors, pages 125–130, Elsevier Publishing Company, Amsterdam–London–New York.

Murray, R. M. [1994] *Nilpotent bases for a class of nonintegrable distributions with applications to trajectory generation for nonholonomic systems*, Mathematics of Control, Signals, and Systems, **7**(1), 58–75.

— [1995] *Nonlinear control of mechanical systems: A Lagrangian perspective*, in *Proceedings of 1995 Symposium on Nonlinear Control Systems (NOLCOS)*, pages 378–389, Lake Tahoe, CA.

Murray, R. M., Li, Z. X., and Sastry, S. S. [1994] *A Mathematical Introduction to Robotic Manipulation*, CRC Press, Boca Raton, FL, ISBN 0-8493-7981-4.

Murray, R. M. and Sastry, S. S. [1993] *Nonholonomic motion planning: Steering using sinusoids*, IEEE Transactions on Automatic Control, **38**(5), 700–716.

Nagano, T. [1966] *Linear differential systems with singularities and an application to transitive Lie algebras*, Journal of the Mathematical Society of Japan, **18**, 398–404.

Nash, J. [1956] *The imbedding problem for Riemannian manifolds*, Annals of Mathematics. Second Series, **63**, 20–63.

Neĭmark, J. I. and Fufaev, N. A. [1972] *Dynamics of Nonholonomic Systems*, volume 33 of *Translations of Mathematical Monographs*, American Mathematical Society, Providence, RI, ISBN 0-8218-1583-0, translation of 1967 Russian edition.

Nekhoroshev, N. N. [1977] *An exponential estimate of the time of stability of nearly-integrable Hamiltonian systems*, Russian Mathematical Surveys, **32**(6), 1–65.

Nelson, E. [1967] *Tensor Analysis*, Princeton University Press, Princeton, NJ.

Newton, I. [1687] *Philosophiae Naturalis Principia Mathematica*, S. Pepys, London, reprint of translation: [Newton 1934].

— [1934] *Principia*, volume I and II, University of California Press, Berkeley–Los Angeles–London, ISBN 0-520-00928-2 and 0-520-00929-0, from the 1729 translation by Andrew Motte.

Nijmeijer, H. and van der Schaft, A. J. [1990] *Nonlinear Dynamical Control Systems*, Springer-Verlag, New York–Heidelberg–Berlin, ISBN 0-387-97234-X.

Noakes, L., Heinzinger, G., and Paden, B. [1989] *Cubic splines on curved spaces*, IMA Journal of Mathematical Control and Information, **6**(4), 465–473.

Nyquist, H. [1932] *Regeneration theory*, Bell Laboratories Technical Journal, **11**, 126–147.

Olfati-Saber, R. [2002] *Normal forms for underactuated mechanical systems with symmetry*, IEEE Transactions on Automatic Control, **47**(2), 305–308.

Oriolo, G. and Nakamura, Y. [1991] *Control of mechanical systems with second-order nonholonomic constraints: Underactuated manipulators*, in *Proceedings of the 30th IEEE Conference on Decision and Control*, pages 2398–2403, Brighton, UK.

Orsi, R., Praly, L., and Mareels, I. M. Y. [2003] *Necessary conditions for stability and attractivity of continuous systems*, International Journal of Control, **76**(11), 1070–1077.

Ortega, J.-P. and Ratiu, T. S. [1999] *Stability of Hamiltonian relative equilibria*, Nonlinearity, **12**(3), 693–720.

— [2004] *Momentum maps and Hamiltonian reduction*, volume 222 of *Progress in Mathematics*, Birkhäuser, Boston–Basel–Stuttgart, ISBN 0-8176-4307-9.

Ortega, R., Loria, A., Nicklasson, P. J., and Sira-Ramirez, H. [1998] *Passivity-Based Control of Euler-Lagrange Systems: Mechanical, Electrical and Electromechanical Applications*, Communications and Control Engineering Series, Springer-Verlag, New York–Heidelberg–Berlin, ISBN 1-85233-016-3.

Ortega, R. and Spong, M. W. [1989] *Adaptive motion control of rigid robots: A tutorial*, Automatica. The Journal of IFAC, **25**(6), 877–888.

Ortega, R., Spong, M. W., Gómez-Estern, F., and Blankenstein, G. [2002] *Stabilization of a class of underactuated mechanical systems via interconnection and damping assignment*, IEEE Transactions on Automatic Control, **47**(8), 1218–1233.

Ortega, R., van der Schaft, A. J., Maschke, B. M., and Escobar, G. [2002] *Stabilization of port-controlled Hamiltonian systems: Energy-balancing and passivation*, Automatica. The Journal of IFAC, **38**(4), 585–596.

Ostrowski, J. P. [2000] *Steering for a class of dynamic nonholonomic systems*, IEEE Transactions on Automatic Control, **45**(8), 1492–1497.

Ostrowski, J. P. and Burdick, J. W. [1997] *Controllability tests for mechanical systems with constraints and symmetries*, Journal of Applied Mathematics and Computer Science, **7**(2), 101–127.

Paden, B. and Panja, R. [1988] *Globally asymptotically stable 'PD+' controller for robot manipulators*, International Journal of Control, **47**(6), 1697–1712.

Panier, E. R. and Tits, A. L. [1993] *On combining feasibility, descent and superlinear convergence in inequality constrained optimization*, Mathematical Programming, **59**(2), 261–276.

Papastavridis, J. G. [2002] *Analytical Mechanics: A Comprehensive Treatise on the Dynamics of Constrained Systems; for Engineers, Physicists, and Mathematicians*, Oxford University Press, Walton Street, Oxford, ISBN 0-19-512697-1.

Pappas, G. J. and Simić, S. [2002] *Consistent abstractions of affine control systems*, IEEE Transactions on Automatic Control, **47**(5), 745–756.

Pars, L. A. [1965] *A Treatise on Analytical Dynamics*, John Wiley and Sons, New York, ISBN 0-918024-07-2.

Patrick, G. W. [1992] *Relative equilibria in Hamiltonian systems: The dynamic interpretation of nonlinear stability on a reduced phase space*, Journal of Geometry and Physics, **9**(2), 111–119.

Poincaré, H. [1892-1899] *Les Méthodes Nouvelles de la Mécanique Céleste*, Gauthier-Villars, Paris, translation: [Poincaré 1993].

— [1993] *New Methods of Celestial Mechanics, Volumes 1-3*, number 13 in History of Modern Physics and Astronomy, American Institute of Physics, New York, ISBN 1-56396-114-8, 1-56396-115-6, and 1-56396-116-4, revised reprint of the 1967

English translation.

Pomet, J.-B. [1992] *Explicit design of time-varying stabilizing control laws for a class of controllable systems without drift*, Systems & Control Letters, **18**(2), 147–158.

Qian, C. and Lin, W. [2001] *Non-Lipschitz continuous stabilizers for nonlinear systems with uncontrollable unstable linearization*, Systems & Control Letters, **42**(3), 185–200.

Radford, J. [2003] *Symmetry, Reduction and Swimming in a Perfect Fluid*, Ph.D. thesis, California Institute of Technology, Pasadena, CA.

Rashevsky, P. K. [1938] *Any two points of a totally nonholonomic space may be connected by an admissible line*, Uchenye Zapiski Pedagogicheskogo Instituta imeni Libknechta, **2**, 83–94.

Rathinam, M. and Murray, R. M. [1998] *Configuration flatness of Lagrangian systems underactuated by one control*, SIAM Journal on Control and Optimization, **36**(1), 164–179.

Reif, J. H. [1979] *Complexity of the mover's problem and generalizations*, in *IEEE Symposium on Foundations of Computer Science (FOCS)*, pages 421–427, San Juan, PR.

Reyhanoglu, M., van der Schaft, A. J., McClamroch, N. H., and Kolmanovsky, I. [1999] *Dynamics and control of a class of underactuated mechanical systems*, IEEE Transactions on Automatic Control, **44**(9), 1663–1671.

Riemann, G. F. B. [1854] *Über die Hypothesen, welche der Geometrie zu Grunde liegen*, Habilitationsvortrag, Universität Göttingen, reprint in collected works: [Riemann 1990].

— [1990] *Gesammelte mathematische Werke, wissenschaftlicher Nachlass und Nachträge*, second edition, Springer-Verlag, New York–Heidelberg–Berlin, first edition in 1892 edited by H. Weber and R. Dedekind.

Rimon, E. and Koditschek, D. E. [1991] *The construction of analytic diffeomorphisms for exact robot navigation on star worlds*, Transactions of the American Mathematical Society, **327**(1), 71–116.

— [1992] *Exact robot navigation using artificial potential functions*, IEEE Transactions on Automatic Control, **8**(5), 501–518.

Rockafellar, R. T. [1970] *Convex Analysis*, Princeton Mathematical Series, Princeton University Press, Princeton, NJ, reprint: [Rockafellar 1997].

— [1997] *Convex Analysis*, Princeton Landmarks in Mathematics, Princeton University Press, Princeton, NJ, ISBN 0-691-01586-4, reprint of 1970 edition.

Rosenberg, R. M. [1977] *Analytical Dynamics of Discrete Systems*, number 4 in Mathematical Concepts and Methods in Science and Engineering, Plenum Press, New York–London, ISBN 0-306-31014-7.

Royden, H. L. [1988] *Real Analysis*, third edition, Macmillian, New York, ISBN 0-02-404151-3.

Ryan, E. P. [1994] *On Brockett's condition for smooth stabilizability and its necessity in the context of nonsmooth feedback*, SIAM Journal on Control and Optimization, **32**(6), 1597–1604.

Samson, C. [1995] *Control of chained systems application to path following and time-varying point-stabilization of mobile robots*, IEEE Transactions on Automatic Control, **40**(1), 64–77.

San Martin, L. and Crouch, P. E. [1984] *Controllability on principal fibre bundles with compact structure group*, Systems & Control Letters, **5**(1), 35–40.

Sanders, J. A. and Verhulst, F. [1985] *Averaging Methods in Nonlinear Dynamical Systems*, number 59 in Applied Mathematical Sciences, Springer-Verlag, New York–Heidelberg–Berlin, ISBN 0-387-96229-8.

Sanders, S. R., Noworolski, J. M., Liu, X. Z., and Verghese, G. C. [1991] *Generalized averaging method for power conversion circuits*, IEEE Transactions on Power Electronics, **6**(2), 251–259.

Sarychev, A. V. [2001] *Lie- and chronologico-algebraic tools for studying stability of time-varying systems*, Systems & Control Letters, **43**(1), 59–76.

Sastry, S. S. [1999] *Nonlinear Systems: Analysis, Stability and Control*, number 1 in Interdisciplinary Applied Mathematics, Springer-Verlag, New York–Heidelberg–Berlin, ISBN 0-387-98513-1.

Sattinger, D. H. and Weaver, O. L. [1986] *Lie Groups and Algebras, with Applications to Physics, Geometry and Mechanics*, number 61 in Applied Mathematical Sciences, Springer-Verlag, New York–Heidelberg–Berlin, ISBN 0-387-96240-9.

Saunders, D. J. [1989] *The Geometry of Jet Bundles*, number 142 in London Mathematical Society Lecture Note Series, Cambridge University Press, New York–Port Chester–Melbourne–Sydney, ISBN 0-521-36948-7.

Selig, J. M. [1996] *Geometrical Methods in Robotics*, Springer-Verlag, New York–Heidelberg–Berlin, ISBN 0-387-94728-0.

Serre, J.-P. [1992] *Lie Algebras and Lie Groups*, number 1500 in Lecture Notes in Mathematics, Springer-Verlag, New York–Heidelberg–Berlin, ISBN 3-540-55008-9.

Seto, D. and Baillieul, J. [1994] *Control problems in super-articulated mechanical systems*, IEEE Transactions on Automatic Control, **39**(12), 2442–2453.

Sharir, M. [1997] *Algorithmic motion planning*, in *Handbook of Discrete and Computational Geometry*, J. E. Goodman and J. O'Rourke, editors, chapter 40, pages 733–754, CRC Press, Boca Raton, FL, ISBN 0-8493-8524-5.

Shen, J. [2002] *Nonlinear Control of Multibody Systems with Symmetries via Shape Change*, Ph.D. thesis, University of Michigan, Ann Arbor, MI.

Shen, J., McClamroch, N. H., and Bloch, A. M. [2004] *Local equilibrium controllability of multibody systems controlled via shape change*, IEEE Transactions on Automatic Control, **49**(4), 506–520.

Shen, J., Sanyal, A. K., and McClamroch, N. H. [2002] *Controllability analysis of a two degree of freedom nonlinear attitude control system*, in *Proceedings of 2002 International Symposium on Mathematical Theory of Networks and Systems*, South Bend, IN.

Silva Leite, F., Camarinha, M., and Crouch, P. E. [2000] *Elastic curves as solutions of Riemannian and sub-Riemannian control problems*, Mathematics of Control, Signals, and Systems, **13**(2), 140–155.

Simo, J. C., Lewis, D. R., and Marsden, J. E. [1991] *Stability of relative equilibria I: The reduced Energy Momentum method*, Archive for Rational Mechanics and Analysis, **115**(1), 15–59.

Sipser, M. [1996] *Introduction to the Theory of Computation*, Brooks Cole, Monterey, CA, ISBN 053494728X.

Smale, S. [1970] *Topology and mechanics, I*, Inventiones Mathematicae, **10**, 305–331.

Śniatycki, J. [1998] *Nonholonomic Noether theorem and reduction of symmetries*, Reports on Mathematical Physics, **42**(1-2), 5–23.

Sontag, E. D. [1988] *Controllability is harder to decide than accessibility*, SIAM Journal on Control and Optimization, **26**(5), 1106–1118.

— [1989] *A "universal" construction of Artstein's theorem on nonlinear stabilization*, Systems & Control Letters, **13**(2), 117–123.

— [1995] *Control of systems without drift via generic loops*, IEEE Transactions on Automatic Control, **40**(7), 1210–1219.

— [1998] *Mathematical Control Theory: Deterministic Finite Dimensional Systems*, second edition, number 6 in Texts in Applied Mathematics, Springer-Verlag, New York–Heidelberg–Berlin, ISBN 0-387-98489-5.

— [1999] *Stability and stabilization: Discontinuities and the effect of disturbances*, in *Nonlinear Analysis, Differential Equations, and Control (Montreal, Canada)*, F. H. Clarke, R. J. Stern, and G. Sabidussi, editors, pages 551–598, Kluwer Academic Publishers, Dordrecht, ISBN 0-7923-5666-7.

Sontag, E. D. and Sussmann, H. J. [1986] *Time-optimal control of manipulators*, in *Proceedings of the 1986 IEEE International Conference on Robotics and Automation*, pages 1692–1697, San Francisco, CA.

— [1995] *Nonsmooth control-Lyapunov functions*, in *Proceedings of the 34th IEEE Conference on Decision and Control*, pages 2799–2805, New Orleans, LA.

Souriau, J.-M. [1970] *Structure des systèmes dynamiques*, Dunod, Paris, translation: [Souriau 1997].

— [1997] *Structure of Dynamical Systems. A Symplectic View of Physics*, number 149 in Progress in Mathematics, Birkhäuser, Boston–Basel–Stuttgart, ISBN 0-8176-3695-1, translation by C. H. Cushman-de Vries of 1970 French edition.

Spiegel, M. R. and Liu, J. M. [1998] *Schaum's Mathematical Handbook of Formulas and Tables*, second edition, McGraw-Hill, New York, ISBN 0-07-038203-4.

Spong, M. W. [1998] *Underactuated mechanical systems*, in *Control Problems in Robotics and Automation*, B. Siciliano and K. Valavanis, editors, pages 135–150, number 230 in Lecture Notes in Control and Information Sciences, Springer-Verlag, New York–Heidelberg–Berlin, ISBN 3-540-00251-0.

Spong, M. W. and Vidyasagar, M. [1989] *Robot Dynamics and Control*, John Wiley and Sons, New York, ISBN 0-471-61243-X.

Steenrod, N. [1951] *The Topology of Fibre Bundles*, number 14 in Princeton Mathematical Series, Princeton University Press, Princeton, NJ, reprint: [Steenrod 1999].

— [1999] *The Topology of Fibre Bundles*, Princeton Landmarks in Mathematics, Princeton University Press, Princeton, NJ, ISBN 0-691-00548-6, reprint of 1951 edition.

Stefan, P. [1974] *Accessible sets, orbits and foliations with singularities*, Proceedings of the London Mathematical Society. Third Series, **29**, 699–713.

Stefani, G. [1986] *On the local controllability of a scalar-input control system*, in *Theory and Applications of Nonlinear Control Systems (Stockholm, Sweden)*, C. I. Byrnes and A. Lindquist, editors, pages 167–179, North-Holland, Amsterdam–New York, ISBN 0-444-70055-2.

— [1988] *A sufficient condition for extremality*, in *Analysis and Optimization of Systems (Antibes)*, A. Bensoussan and J. L. Lions, editors, pages 270–281, number 111 in Lecture Notes in Control and Information Sciences, Springer-Verlag, New York–Heidelberg–Berlin.

Stoer, J. and Bulirsch, R. [1992] *Introduction to Numerical Analysis*, second edition, number 12 in Texts in Applied Mathematics, Springer-Verlag, New York–Heidelberg–Berlin, ISBN 0-387-97878-X.

Stramigioli, S. [2001] *Modeling and IPC Control of Interactive Mechanical Systems—A Coordinate-Free Approach*, number 266 in Lecture Notes in Control and Information Sciences, Springer-Verlag, New York–Heidelberg–Berlin, ISBN 1-85233-395-2.

Strang, G. [1980] *Linear Algebra and its Applications*, second edition, Academic Press, New York, ISBN 0-12-673660-X.

Sussmann, H. J. [1973] *Orbits of families of vector fields and integrability of distributions*, Transactions of the American Mathematical Society, **180**, 171–188.

— [1978] *A sufficient condition for local controllability*, SIAM Journal on Control and Optimization, **16**(5), 790–802.

— [1983] *Lie brackets and local controllability: A sufficient condition for scalar-input systems*, SIAM Journal on Control and Optimization, **21**(5), 686–713.

— [1986] *A product expansion of the Chen series*, in *Theory and Applications of Nonlinear Control Systems (Stockholm, Sweden)*, C. I. Byrnes and A. Lindquist, editors, pages 323–335, Elsevier Publishing Company, Amsterdam–London–New York, ISBN 0-444-70055-2.

— [1987] *A general theorem on local controllability*, SIAM Journal on Control and Optimization, **25**(1), 158–194.

— [1990a] *Real-analytic desingularization and subanalytic sets: An elementary approach*, Transactions of the American Mathematical Society, **317**(2), 417–461.

— [1990b] *Why real analyticity is important in control theory*, in *Perspectives in Control Theory (Sielpia)*, pages 315–340, number 2 in Progress in Systems and Control Theory, Birkhäuser, Boston–Basel–Stuttgart.

— [1992] *New differential geometric methods in nonholonomic path finding*, in *Systems, Models, and Feedback: Theory and Applications (Capri, Italy)*, A. Isidori and T. J. Tarn, editors, pages 365–384, number 12 in Progress in Systems and Control Theory, Birkhäuser, Boston–Basel–Stuttgart, ISBN 0-8176-3633-1.

Sussmann, H. J. and Jurdjevic, V. [1972] *Controllability of nonlinear systems*, Journal of Differential Equations, **12**, 95–116.

Swan, R. G. [1962] *Vector bundles and projective modules*, Transactions of the American Mathematical Society, **105**(2), 264–277.

Synge, J. L. [1928] *Geodesics in nonholonomic geometry*, Mathematische Annalen, **99**, 738–751.

Takegaki, M. and Arimoto, S. [1981] *A new feedback method for dynamic control of manipulators*, Transactions of the ASME. Series G. Journal of Dynamic Systems, Measurement, and Control, **103**(2), 119–125.

Terjéki, J. [1987] *Stability equivalence between damped mechanical and gradient systems*, in *Differential Equations: Qualitative Theory, Vol. I, II (Szeged, Hungary)*, B. Sz-Nagy and L. Hatvani, editors, pages 1059–1071, number 47 in Colloquia Mathematica Societatis János Bolyai, North-Holland, Amsterdam–New York.

Thomson, W. and Tait, P. G. [1867] *Treatise on Natural Philosophy*, Oxford University Press, Walton Street, Oxford.

Tilbury, D., Murray, R. M., and Sastry, S. S. [1995] *Trajectory generation for the n-trailer problem using Goursat normal form*, IEEE Transactions on Automatic Control, **40**(5), 802–819.

Troutman, J. L. [1996] *Variational Calculus and Optimal Control, Optimization with Elementary Convexity*, second edition, Undergraduate Texts in Mathematics, Springer-Verlag, New York–Heidelberg–Berlin, ISBN 0-387-94511-3.

Tyner, D. R. [2002] *Linearization of Affine Connection Control Systems*, Master's thesis, Queen's University, Kingston, ON.

Tyner, D. R. and Lewis, A. D. [2004a] *Controllability of a hovercraft model (and two general results)*, in *Proceedings of the 40th IEEE Conference on Decision and Control*, Paradise Island, Bahamas, to appear.

— [2004b] *Controllability of affine connection control systems with many or few inputs*, Preprint.

Utkin, V. I. [1992] *Sliding Modes in Control and Optimization*, Communications and Control Engineering Series, Springer-Verlag, New York–Heidelberg–Berlin, ISBN 0-387-53516-0.

van der Schaft, A. J. [1981/82] *Hamiltonian dynamics with external forces and observations*, Mathematical Systems Theory, **15**(2), 145–168.

— [1982] *Controllability and observability of affine nonlinear Hamiltonian systems*, IEEE Transactions on Automatic Control, **27**(2), 490–492.

— [1983] *Symmetries, conservation laws, and time reversibility for Hamiltonian systems with external forces*, Journal of Mathematical Physics, **24**(8), 2095–2101.

— [1985] *Controlled invariance for Hamiltonian systems*, Mathematical Systems Theory, **18**(3), 257–291.

— [1986] *Stabilization of Hamiltonian systems*, Nonlinear Analysis. Theory, Methods, and Applications, **10**(10), 1021–1035.

— [1999] *L_2-Gain and Passivity Techniques in Nonlinear Control*, second edition, number 218 in Communications and Control Engineering Series, Springer-Verlag, New York–Heidelberg–Berlin, ISBN 1-85233-073-2.

— [2000] *Port-controlled Hamiltonian systems: Towards a theory for control and design of nonlinear physical systems*, Journal of the Society of Instrument & Control

Engineers, **39**(2), 91–98.

van der Schaft, A. J. and Maschke, B. M. [1994] *On the Hamiltonian formulation of nonholonomic mechanical systems*, Reports on Mathematical Physics, **34**(2), 225–233.

van Nieuwstadt, M., Rathinam, M., and Murray, R. M. [1998] *Differential flatness and absolute equivalence of nonlinear control systems*, SIAM Journal on Control and Optimization, **36**(4), 1225–1239.

Varadarajan, V. S. [1984] *Lie Groups, Lie Algebras, and Their Representations*, number 102 in Graduate Texts in Mathematics, Springer-Verlag, New York–Heidelberg–Berlin, ISBN 0-387-90969-9.

Vela, P. A. [2003] *Averaging and Control of Nonlinear Systems (with Application to Biomimetic Locomotion)*, Ph.D. thesis, California Institute of Technology, Pasadena, CA.

Vershik, A. M. [1984] *Classical and non-classical dynamics with constraints*, in *Global Analysis Studies and Applications I: Studies and Applications*, I. G. Borisovich, Y. E. Gliklikh, Y. G. Borisovich, and I. E. Gliklikh, editors, pages 278–301, number 1108 in Lecture Notes in Mathematics, Springer-Verlag, New York–Heidelberg–Berlin, ISBN 0-387-13910-9.

von Stryk, O. and Bulirsch, R. [1992] *Direct and indirect methods for trajectory optimization*, Annals of Operations Research, **37**(1-4), 357–373.

Walker, J. A. and Schmitendorf, W. E. [1973] *A simple test for asymptotic stability in partially dissipative symmetric systems*, Transactions of the ASME. Series E. Journal of Applied Mechanics, **40**, 1120–1121.

Wan, F. Y. M. [1995] *Introduction to the Calculus of Variations and its Applications*, Chapman & Hall, New York–London, ISBN 0-534-09984-X.

Wang, D. and McClamroch, N. H. [1994] *Stability analysis of the equilibrium of a constrained mechanical system*, International Journal of Control, **60**(5), 733–746.

Wang, L. S. and Krishnaprasad, P. S. [1992] *Gyroscopic control and stabilization*, Journal of Nonlinear Science, **2**(4), 367–415.

Warner, F. W. [1971] *Foundations of Differentiable Manifolds and Lie Groups*, Scott, Foresman and Co., Glenview, IL–London, revised edition: [Warner 1989].

— [1989] *Foundations of Differentiable Manifolds and Lie Groups*, second edition, number 94 in Graduate Texts in Mathematics, Springer-Verlag, New York–Heidelberg–Berlin, ISBN 0-387-90894-3.

Weber, R. W. [1986] *Hamiltonian systems with constraints and their meaning in mechanics*, Archive for Rational Mechanics and Analysis, **91**(4), 309–335.

Wen, J. T. and Bayard, D. S. [1988] *A new class of control laws for robotic manipulators. Part I: Non-adaptive case*, International Journal of Control, **47**(5), 1361–1385.

Wen, J. T. and Kreutz-Delgado, K. [1991] *The attitude control problem*, IEEE Transactions on Automatic Control, **36**(10), 1148–1162.

Weyl, H. [1939] *The Classical Groups*, Princeton University Press, Princeton, NJ, reprint: [Weyl 1997].

— [1997] *The Classical Groups*, Princeton Landmarks in Mathematics, Princeton University Press, Princeton, NJ, ISBN 0-691-05756-7, reprint of 1939 edition.

Whitney, H. [1936] *Differentiable manifolds*, Annals of Mathematics. Second Series, **37**(3), 645–680.

Whittaker, E. T. [1904] *A Treatise on the Analytical Dynamics of Particles and Rigid Bodies*, Cambridge University Press, New York–Port Chester–Melbourne–Sydney, reprint of revised edition: [Whittaker 1988].

— [1988] *A Treatise on the Analytical Dynamics of Particles and Rigid Bodies*, Cambridge Mathematical Library, Cambridge University Press, New York–Port Chester–Melbourne–Sydney, ISBN 0-521-35883-3, reprint of 1937 fourth edition.

Willems, J. C. [1972] *Dissipative dynamical systems*, Archive for Rational Mechanics and Analysis, **45**(5), 321–351.

— [1979] *System theoretic models for the analysis of physical systems*, Ricerche di Automatica, **10**(2), 71–106.

Wilson, Jr., F. W. [1967] *The structure of the level surfaces of a Lyapunov function*, Journal of Differential Equations, **3**, 323–329.

Wonham, W. M. [1985] *Linear Multivariable Control: A Geometric Approach*, third edition, number 10 in Applications of Mathematics, Springer-Verlag, New York–Heidelberg–Berlin, ISBN 0-387-96071-6.

Yano, K. and Ishihara, S. [1973] *Tangent and Cotangent Bundles*, number 16 in Pure and Applied Mathematics, Dekker Marcel Dekker, New York, ISBN 0-8247-6071-9.

Yosida, K. [1980] *Functional Analysis*, 6th edition, number 123 in Grundlehren der Mathematischen Wissenschaften, Springer-Verlag, New York–Heidelberg–Berlin, ISBN 0-387-10210-8.

Zabczyk, J. [1989] *Some comments on stabilizability*, Applied Mathematics and Optimization. An International Journal, **19**(1), 1–9.

Zames, G. [1966] *On the input-output stability of time-varying nonlinear feedback systems. Part I: Conditions derived using concepts of loop gain, conicity, and positivity*, IEEE Transactions on Automatic Control, **11**(2), 465–476.

Zampieri, G. [2000] *Nonholonomic versus vakonomic dynamics*, Journal of Differential Equations, **163**(2), 335–347.

Žefran, M., Bullo, F., and Radford, J. [1999] *An investigation into non-smooth locomotion*, in *Proceedings of the 1999 IEEE International Conference on Robotics and Automation*, pages 2038–2043, Detroit, MI.

Zenkov, D. V. [2002] *Matching and stabilization of linear mechanical systems*, in *Proceedings of 2002 International Symposium on Mathematical Theory of Networks and Systems*, South Bend, IN.

Zenkov, D. V., Bloch, A. M., and Marsden, J. E. [1998] *The energy-momentum method for the stability of non-holonomic systems*, Dynamics and Stability of Systems. An International Journal, **13**(2), 123–165.

Zhou, K. [1996] *Robust and Optimal Control*, Prentice-Hall, Englewood Cliffs, NJ, ISBN 0-13-456567-3, with John C. Doyle and Keith Glover.

Symbol index

Subject index

Texts in Applied Mathematics

(continued from page ii)

31. *Brémaud:* Markov Chains: Gibbs Fields, Monte Carlo Simulation, and Queues.
32. *Durran:* Numerical Methods for Wave Equations in Geophysical Fluid Dynamics.
33. *Thomas:* Numerical Partial Differential Equations: Conservation Laws and Elliptic Equations.
34. *Chicone:* Ordinary Differential Equations with Applications.
35. *Kevorkian:* Partial Differential Equations: Analytical Solution Techniques, 2nd ed.
36. *Dullerud/Paganini:* A Course in Robust Control Theory: A Convex Approach.
37. *Quarteroni/Sacco/Saleri:* Numerical Mathematics.
38. *Gallier:* Geometric Methods and Applications: For Computer Science and Engineering.
39. *Atkinson/Han:* Theoretical Numerical Analysis: A Functional Analysis Framework.
40. *Brauer/Castillo-Chávez:* Mathematical Models in Population Biology and Epidemiology.
41. *Davies:* Integral Transforms and Their Applications, 3rd ed.
42. *Deuflhard/Bornemann:* Scientific Computing with Ordinary Differential Equations.
43. *Deuflhard/Hohmann:* Numerical Analysis in Modern Scientific Computing: An Introduction, 2nd ed.
44. *Knabner/Angermann:* Numerical Methods for Elliptic and Parabolic Partial Differential Equations.
45. *Larsson/Thomée:* Partial Differential Equations with Numerical Methods.
46. *Pedregal:* Introduction to Optimization.
47. *Ockendon/Ockendon:* Waves and Compressible Flow.
48. *Hinrichsen:* Mathematical Systems Theory I.
49. *Bullo/Lewis:* Geometric Control of Mechanical Systems: Modeling, Analysis, and Design for Simple Mechanical Control Systems.